Open Problems in Topology II

Open Problems in Topology II

Edited by

Elliott Pearl

ELSEVIER

Amsterdam – Boston – Heidelberg – London – New York – Oxford – Paris
San Diego – San Francisco – Singapore – Sydney – Tokyo

Elsevier
Radarweg 29, PO Box 211, 1000 AE Amsterdam, The Netherlands
The Boulevard, Langford Lane, Kidlington, Oxford OX5 1GB, UK

First edition 2007

Library of Congress Cataloging-in-Publication Data
A catalog record for this book is available from the Library of Congress

British Library Cataloguing in Publication Data
A catalogue record for this book is available from the British Library

ISBN-13: 978-0-444-52208-5

For information on all Elsevier publications
visit our website at books.elsevier.com

Transferred to Digital Printing 2010

Preface

This new book follows the 1990 volume *Open Problems in Topology* edited by J. van Mill and G.M. Reed. It builds on the success of original volume by presenting currently active research topics in topology. The contributions in this book are entirely new even though some (many?) of its problems may have been raised in other sources. As with the original volume, the intent is to provide a source of dissertation problems and to challenge the research community within and beyond topology. This volume covers a broad range of topics related to topology and examines some topics in greater depth. The problems in this volume are supposed to reflect the main trends in general topology and its applications since 1990 and I hope that they will help direct further research.

This volume was prepared with the invaluable help of several editorial advisors. They are Frédéric Mynard (General Topology), Michael Hrušák and Justin Tatch Moore (Set-theoretic Topology), Alejandro Illanes (Continuum Theory), Dikran Dikranjan (Topological Algebra), John C. Mayer (Dynamical Systems), Thomas J. Peters (Computer Science), Biagio Ricceri (Functional Analysis), and Vitalij A. Chatyrko (Dimension Theory). The advisors helped with the selection of contributors and the review of manuscripts in the respective parts of the volume.

This volume has a ninth part, *Invited Problems*, which is not organized by any particular topic. Here I invited several research groups to contribute problems representative of their interests.

The journal *Topology and its Applications* will continue to publish regular reports on the status of problems from the original book and from this new volume. Please contribute information on solutions to the respective authors and editors.

I want to thank Jan van Mill for his help in getting this project started. I thank the Fields Insititute in Toronto for their hospitality during the preparation of this volume.

Elliott Pearl, Toronto, September 2006

Contents

Part 1

General Topology

Open Problems in Topology II – Edited by E. Pearl

Selected ordered space problems

Harold Bennett and David Lutzer

1. Introduction

A *generalized ordered space* (a GO-space) is a triple $(X, \tau, <)$ where $(X, <)$ is a linearly ordered set and τ is a Hausdorff topology on X that has a base of order-convex sets. If τ is the usual open interval topology of the order $<$, then we say that $(X, \tau, <)$ is a *linearly ordered topological space* (LOTS). Besides the usual real line, probably the most familiar examples of GO-spaces are the Sorgenfrey line, the Michael line, the Alexandroff double arrow, and various spaces of ordinal numbers. In this paper, we collect together some of our favorite open problems in the theory of ordered spaces. For many of the questions, space limitations restricted us to giving only definitions and references for the question. For more detail, see [**7**]. Notably absent from our list are problems about orderability, about products of special ordered spaces, about continuous selections of various kinds, and about C_p-theory, and for that we apologize. Throughout the paper, we reserve the symbols $\mathbb{R}, \mathbb{P}$, and $\mathbb{Q}$ for the usual sets of real, irrational, and rational numbers respectively.

2. A few of our favorite things

The most important open question in GO-space theory is Maurice's problem, which Qiao and Tall showed [**16**] is closely related to several other old questions of Heath and Nyikos [**7**]. Maurice asked whether there is a ZFC example of a perfect GO-space that does not have a σ-closed-discrete dense subset. A recent paper [**9**] has shown that a ZFC example, if there is one, cannot have local density $\leq \omega_1$, and what remains is:

Question 1. *Let $\kappa > \omega_1$ be a cardinal number. Is it consistent with* ZFC *that any perfect GO-space with local density $\leq \kappa$ must have a σ-closed-discrete dense set? Equivalently, is it consistent with* ZFC *that every perfect non-Archimedean space with local density $\leq \kappa$ is metrizable? Is it consistent with* ZFC *that every perfect GO-space with local density $\leq \kappa$ and with a point-countable base is metrizable?* 1–2?

The following question is closely related to Question 1; see [**7**] for details.

Question 2 (The GO-embedding problem). *Let $\kappa > \omega_1$ be a cardinal number. Is it consistent with* ZFC *that every perfect GO-space X with local density $\leq \kappa$ embeds in some perfect LOTS? (Note: the embedding map h is not required to be monotonic and $h[X]$ is not required to be open, or closed, or dense, in the perfect LOTS.)* 3?

Let $\mathcal{M}$ be the class of all metrizable spaces. A space X is *cleavable over* $\mathcal{M}$ [**1**] if for each subset $A \subseteq X$, there is a continuous f_A from X into some member of $\mathcal{M}$ such that $f_A(x) \neq f_A(y)$ for each $x \in A$ and $y \in X \setminus A$. The property cleavable over

$\mathcal{S}$ is analogously defined, where $\mathcal{S}$ is the class of all separable metrizable spaces. It is known [**2**] that the following properties of a GO-space X are equivalent: (a) X is cleavable over $\mathcal{M}$; (b) X has a weaker metrizable topology; (c) X has a G_δ-diagonal; (d) there is a σ-discrete collection $\mathcal{C}$ of cozero subsets of X such that if $x \neq y$ are points of X, then some $C \in \mathcal{C}$ has $|C \cap \{x, y\}| = 1$.

For a GO-space X with cellularity $\leq \mathfrak{c}$, each of the following is equivalent to cleavability of X over $\mathcal{S}$: (a) X has a weaker separable metrizable topology; (b) X has a countable, point-separating cover by cozero sets; (c) X is divisible by cozero sets, i.e., for each $A \subseteq X$, there is a countable collection $\mathcal{C}_A$ of cozero subsets of X with the property that given $x \in A$ and $y \in X \setminus A$, some $C \in \mathcal{C}_A$ has $x \in C \subseteq Y \setminus \{y\}$. However, for each $\kappa > \mathfrak{c}$ there a LOTS X that is cleavable over $\mathcal{S}$ and has $c(X) = \kappa$ (Example 4.8 of [**2**]). Therefore we have:

4? **Question 3.** *Characterize GO-spaces that are cleavable over $\mathcal{S}$ without imposing restrictions on the cardinal functions of X.*

5? **Question 4.** *Characterize GO-spaces that are divisible by open sets (in which the collection $\mathcal{C}$ mentioned in (c) above consists of open sets, but not necessarily cozero-sets).*

6? **Question 5.** *Characterize GO-spaces that are cleavable over $\mathbb{R}$, i.e., in which the cleaving functions f_A can always be taken to be mappings into $\mathbb{R}$.*

Cleavability over $\mathbb{P}$ and $\mathbb{Q}$ has already been characterized in [**2**]. For compact, connected LOTS, see [**11**].

7–8? **Question 6.** *Arhangel′skiĭ has asked whether a compact Hausdorff space X that is cleavable over some LOTS (or GO-space) L must itself be a LOTS. What if X is zero-dimensional?*

Arhangel′skiĭ proved that if a compact Hausdorff space X is cleavable over $\mathbb{R}$, then X is embeddable in $\mathbb{R}$. Buzyakova [**11**] showed the same is true if $\mathbb{R}$ is replaced by the lexicographic product space $\mathbb{R} \times \{0, 1\}$.

9? **Question 7** (Buzyakova). *Is it true that a compact Hausdorff space X that is cleavable over an infinite homogeneous LOTS L must be embeddable in L?*

A topological space X is *monotonically compact* (resp., *monotonically Lindelöf*) if for each open cover $\mathcal{U}$ it is possible to choose a finite (resp., countable) open refinement $r(\mathcal{U})$ such that if $\mathcal{U}$ and $\mathcal{V}$ are any open covers of X with $\mathcal{U}$ refining $\mathcal{V}$, then $r(\mathcal{U})$ refines $r(\mathcal{V})$. It is known that any compact metrizable space is monotonically compact, and that any second countable space is monotonically Lindelöf, and that any separable GO-space is hereditarily monotonically Lindelöf [**10**].

10–13? **Question 8.** *Is every every perfect monotonically Lindelöf GO-space separable? Is every hereditarily monotonically Lindelöf GO-space separable? If there is a Souslin line, is there a compact Souslin line that is hereditarily monotonically Lindelöf and is there is a Souslin line that is not monotonically Lindelöf?*

Question 9. *If Y is a subspace of a perfect monotonically Lindelöf GO-space X, must Y be monotonically Lindelöf?* 14?

Question 10. (a) *Is the lexicographic product space $X = [0,1] \times \{0,1\}$ monotonically compact?* (b) *If X is a first-countable compact LOTS, is X monotonically Lindelöf?* (c) *If X is a first-countable monotonically compact LOTS, is X metrizable?* 15–17?

Studying properties of a space X off of the diagonal $\Delta = \{(x, x) : x \in X\}$ means studying properties of the space $X^2 \setminus \Delta$. For example one can show that a GO-space X is separable if and only $c(X^2 \setminus \Delta) = \omega$.

Question 11 ([**6**]). *Is it true that a GO-space is separable if $X^2 \setminus \Delta$ has a dense Lindelöf subspace? If X is a Souslin space, can $X^2 \setminus \Delta$ have a dense Lindelöf subspace?* 18–19?

Question 12. *In* ZFC, *is there a non-metrizable, Lindelöf LOTS X that has a countable rectangular open cover of $X^2 \setminus \Delta$ (i.e., a collection $\{U_n \times V_n : n < \omega\}$ of basic open sets in X^2 with $\bigcup\{U_n \times V_n : n < \omega\} = X^2 \setminus \Delta$)?* 20?

Under CH or $\mathfrak{b} = \omega_1$ the answer to Question 12 is affirmative [**5**].

Question 13. *Suppose X is a LOTS that is first-countable and hereditarily paracompact off of the diagonal (i.e., $X^2 \setminus \Delta$ is hereditarily paracompact). Must X have a point-countable base? Is it possible that a Souslin space can be hereditarily paracompact off of the diagonal?* 21–22?

We note that if one considers GO-spaces rather than LOTS in Question 13, then there is a consistently negative answer. Under CH, Michael [**15**] constructed an uncountable dense-in-itself subset X of the Sorgenfrey line $\mathbb{S}$ such that X^2 is Lindelöf. Because $\mathbb{S}^2$ is perfect, X^2 is perfect and Lindelöf, i.e., hereditarily Lindelöf. But X cannot have a point-countable base.

Let $\mathbb{P}_\mathbb{S}$ be the set of irrational numbers topologized as a subspace of the Sorgenfrey line. It is known [**8**] that $\mathbb{P}_\mathbb{S}$ is domain-representable, being a G_δ-subset of the Sorgenfrey line. In fact, the Sorgenfrey line is representable using a Scott domain [**12**].

Question 14. *Is the G_δ subspace $\mathbb{P}_\mathbb{S}$ of $\mathbb{S}$ also Scott-domain-representable?* 23?

Question 15 (Suggested by R. Buzyakova). *Suppose X is a GO-space that is countably compact but not compact and that every compact subset of X is a metrizable G_δ-subspace of X. Must X have a base of countable order?* 24?

Mary Ellen Rudin [**17**] proved that every compact monotonically normal space is a continuous image of a compact LOTS. Combining her result with results of Nikiel proves a generalized Hahn–Mazurkiewicz theorem, namely that a topological space X is a continuous image of a compact, connected LOTS if and only if X is compact, connected, locally connected, and monotonically normal.

Question 16. *Is there an elementary submodels proof of the generalized Hahn–Mazurkiewicz theorem above?* 25?

A space X is *weakly perfect* if for each closed subset $C \subseteq X$ there is a G_δ-subset D of X with $D \subseteq C = \mathrm{cl}_X(D)$.

26? **Question 17** ([**3**]). *Suppose Y is a subspace of a weakly perfect GO-space X. Must Y be weakly perfect?*

27? **Question 18** ([**4**]). *Suppose that X is a Lindelöf GO-space that has a small diagonal and that can be p-embedded in some LOTS. Must X be metrizable?*

28–29? **Question 19** ([**13**]). *Let $<$ be the usual ordering of $\mathbb{R}$. For which subsets $X \subseteq \mathbb{R}$ is there a tree T and linear orderings of the nodes of T so that (a) no node of T contains an order-isomorphic copy of $(X, <)$, and (b) $(X, <)$ is order-isomorphic to the branch space of T? (Both $\mathbb{R}$ and $\mathbb{P}$ are representable in this way, but $\mathbb{Q}$ is not.) Which $F_{\sigma\delta}$-subsets of $\mathbb{R}$ are order-isomorphic to the branch space of some countable tree?*

An Aronszajn line is a linearly ordered set that has cardinality ω_1, contains no order-isomorphic copy of ω_1, no copy of ω_1 with the reverse order, and no order-isomorphic copy of an uncountable set of real numbers. Such things exist in ZFC.

30–31? **Question 20** ([**14**]). *Can an Aronszajn line be Lindelöf in its open interval topology without containing a Souslin line? If an Aronszajn line has countable cellularity in its open interval topology, must the Aronszajn tree from which the line comes contain a Souslin tree?*

32? **Question 21** (Gruenhage and Zenor). *Suppose X is a LOTS with a σ-closed-discrete dense set and a continuous function $\Psi\colon (X^2 \setminus \Delta) \times X \to \mathbb{R}$ such that if $x \neq y$ are points of X, then $\Psi(x,y)(x) \neq \Psi(x,y)(y)$. (Note that this is weaker than the existence of a continuous separating family.) Must X be metrizable?*

References

[1] A. V. Arhangel′skiĭ, *The general concept of cleavability of a topological space*, Topology Appl. **44** (1992), 27–36.

[2] H. Bennett, R. Byerly, and D. Lutzer, *Cleavability in ordered spaces*, Order **18** (2001), no. 1, 1–17.

[3] H. Bennett, R. W. Heath, and D. Lutzer, *GO-spaces with σ-closed discrete dense subsets*, Proc. Amer. Math. Soc. **129** (2001), no. 3, 931–939.

[4] H. Bennett and D. Lutzer, *Diagonal conditions in ordered spaces*, Fund. Math. **153** (1997), no. 2, 99–123.

[5] H. Bennett and D. Lutzer, *Off-diagonal metrization theorems*, Topology Proc. **22** (1997), Spring, 37–58.

[6] H. Bennett and D. Lutzer, *Continuous separating families in ordered spaces and strong base conditions*, Topology Appl. **119** (2002), no. 3, 305–314.

[7] H. Bennett and D. Lutzer, *Recent developments in the topology of ordered spaces*, Recent progress in general topology, II, North-Holland, Amsterdam, 2002, pp. 83–114.

[8] H. Bennett and D. Lutzer, *Domain-representable spaces*, Fund. Math. **189** (2006), no. 3, 255–268.

[9] H. Bennett and D. Lutzer, *On a question of Maarten Maurice*, Topology Appl. **153** (2006), 1631–1638.

[10] H. Bennett, D. Lutzer, and M. Matveev, *The monotone Lindelöf property and separability in ordered spaces*, Topology Appl. **151** (2005), 180–186.

[11] R. Z. Buzyakova, *On cleavability of continua over LOTS*, Proc. Amer. Math. Soc. **132** (2004), no. 7, 2171–2181.

[12] K. Duke and D. Lutzer, *Scott-domain-representability of certain ordered spaces*, Submitted.

[13] W. Funk and D. Lutzer, *Branch space representations of lines*, Topology Appl. **151** (2005), 187–214.

[14] W. Funk and D. Lutzer, *Lexicographically ordered trees*, Topology Appl. **152** (2005), no. 3, 275–300.

[15] E. Michael, *Paracompactness and the Lindelöf property in finite and countable Cartesian products*, Compositio Math. **23** (1971), 199–214.

[16] Y.-Q. Qiao and F. D. Tall, *Perfectly normal non-metrizable non-Archimedean spaces are generalized Souslin lines*, Proc. Amer. Math. Soc. **131** (2003), no. 12, 3929–3936.

[17] M. E. Rudin, *Nikiel's conjecture*, Topology Appl. **116** (2001), no. 3, 305–331.

Open Problems in Topology II – Edited by E. Pearl

Problems on star-covering properties

Maddalena Bonanzinga and Mikhail Matveev

Introduction

In this chapter we consider some questions about properties defined in terms of stars with respect to open covers. If $\mathcal{U}$ is a cover of X, and A a subset of X, then $\mathrm{St}(A,\mathcal{U}) = \mathrm{St}^1(A,\mathcal{U}) = \bigcup\{U \in \mathcal{U} : U \cap A \neq \emptyset\}$. For $n = 1,2,\ldots$, $\mathrm{St}^{n+1}(A,\mathcal{U}) = \mathrm{St}(\mathrm{St}^n(A,\mathcal{U}),\mathcal{U})$. Even if many properties can be characterized in terms of stars (thus, paracompactness is equivalent to the requirement that every open cover has an open star refinement ([**5**], Theorem 5.1.12), and normality is equivalent to the requirement that every finite open cover has an open star refinement ([**5**], 5.1.A and 5.1.J)) we concentrate mostly on properties specifically defined by means of stars.

All spaces are assumed to be Tychonoff unless a weaker axiom of separation is indicated. If $\mathcal{A}$ is an almost disjoint family of infinite subsets of ω, then $\Psi(\mathcal{A})$ denotes the associated Ψ-space (see for example [**4**] or [**8**]). The reader is referred to [**3**] for the definitions of small uncountable cardinals such as $\mathfrak{b}$, $\mathfrak{d}$ or $\mathfrak{p}$.

Compactness-type properties

A space X is *starcompact* [**6**] if for every open cover $\mathcal{U}$ there is a finite $A \subset X$ such that $\mathrm{St}(A,\mathcal{U}) = X$. More generally, X is *n-starcompact* (where $n = 1,2,\ldots$) if for every open cover $\mathcal{U}$ there is a finite $A \subset X$ such that $\mathrm{St}^n(A,\mathcal{U}) = X$; X is *$n\frac{1}{2}$-starcompact* if for every open cover $\mathcal{U}$ there is a finite $\mathcal{O} \subset \mathcal{U}$ such that $\mathrm{St}^n(\bigcup\mathcal{O},\mathcal{U}) = X$. This terminology is from [**16**]; in [**4**], n-starcompact spaces are called strongly n-starcompact, and $n\frac{1}{2}$-starcompact spaces are called n-starcompact. A Hausdorff space is starcompact iff it is countably compact [**6**]; a Tychonoff space is $2\frac{1}{2}$-starcompact iff it is pseudocompact [**4**]. In general, none of the implications (countably compact) $\Rightarrow$ $1\frac{1}{2}$-starcompact $\Rightarrow$ 2-starcompact $\Rightarrow$ pseudocompact can be reversed (see [**4**, **16**] for examples), but in special classes of spaces the situation may be different.

Question 1 ([**20**], Problem 3.1). *Is every $1\frac{1}{2}$-starcompact Moore space compact?* 33?

Under $\mathfrak{b} = \mathfrak{c}$, the answer is affirmative [**20**].

Question 2 ([**25**]). *Does there exist a ccc pseudocompact space which is not 2-starcompact?* 34?

One can specify the previous question in this way:

Question 3. *Does there exist a pseudocompact topological group which is not 2-starcompact?* 35?

The first named author acknowledges GNSAGA-CNR, Italy for financial support.

Indeed, every pseudocompact topological group is ccc.

Lindelöf-type properties and cardinal functions

A space X is called *star-Lindelöf* if for every open cover $\mathcal{U}$ there is a countable $A \subset X$ such $\mathrm{St}(A,\mathcal{U}) = X$. More generally, the *star-Lindelöf number* of X is $\text{st-L}(X) = \min\{\tau :$ for every open cover $\mathcal{U}$ there is $A \subset X$ such that $|A| \leq \tau$ and $\mathrm{St}(A,\mathcal{U}) = X\}$. It is easily seen that for a T_1 space X, $\text{st-L}(X) \leq e(X)$, this is the reason why the star-Lindelöf number is also called *weak extent* [**9**]. In [**4**], star-Lindelöf spaces are called strongly star-Lindelöf.

A Tychonoff star-Lindelöf space can have arbitrarily large extent, but the extent of a normal star-Lindelöf space can equal at most $\mathfrak{c}$, [**17**]; alternative proofs were later given by W. Fleissner (unpublished) and in [**11**]. Consistently, even a separable normal space may have extent equal to $\mathfrak{c}$ [**12**] (obviously, a separable space is star-Lindelöf), however these questions remain open:

36–37? **Question 4** ([**12**]).

(1) *Does there exist in* ZFC *a normal star-Lindelöf space having uncountable extent? … having extent equal to* $\mathfrak{c}$*?*
(2) *Is there, in* ZFC *or consistently, a normal star-Lindelöf space having a closed discrete subspace of cardinality* $\mathfrak{c}$*?*

The next question is a natural generalization of the discussion of closed discrete subspaces.

38–39? **Question 5.** *Which Tychonoff (normal) spaces can be represented as closed subspaces of Tychonoff (normal) star-Lindelöf spaces?*

As noted by Ronnie Levy, every locally compact space is representable as a closed subspace of a Tychonoff star-Lindelöf space. This follows from the fact that $2^\kappa \setminus \{\text{point}\}$ is star-Lindelöf for every κ [**11**].

Say that X is *discretely star-Lindelöf* if every open cover $\mathcal{U}$ there is a countable, closed and discrete subspace $A \subset X$ such $\mathrm{St}(A,\mathcal{U}) = X$.

40–41? **Question 6.** *How big can be the extent of a normal discretely star-Lindelöf space? Can it be uncountable within* ZFC*?*

A Ψ-space is a consistent example [**17**].

Paracompactness-type properties

A space X is called *sr-paracompact* (G.M. Reed, [**20**]) if for every open cover $\mathcal{U}$, the cover $\{\mathrm{St}(\{x\},\mathcal{U}) : x \in X\}$ has a locally finite open refinement. More information on sr-paracompactness can be found in [**20**].

These properties were introduced by V.I. Ponomarev: X is *n-sr-paracompact* (where $n = 1, 2, \ldots$) if for every open cover $\mathcal{U}$, the cover $\{\mathrm{St}^n(\{x\},\mathcal{U})x \in X\}$ has a locally finite open refinement; X is *$(n+\frac{1}{2})$-sr-paracompact* if for every open cover $\mathcal{U}$, the cover $\{\mathrm{St}^n(U,\mathcal{U}) : U \in \mathcal{U}\}$ has a locally finite open refinement.

Question 7 (V.I. Ponomarev). *For which n can one construct an n-sr-paracompact Tychonoff space which is not $(n-\frac{1}{2})$-sr-paracompact? (Here n is either $2, 3, \ldots$ of the form $m + \frac{1}{2}$ where $m = 1, 2, \ldots$).* 42?

Property (a)

A space X has *property (a)* (or is an (a)-space) if for every open cover $\mathcal{U}$ and for every dense subspace $D \subset X$, there is a closed in X and discrete $E \subset D$ such that $\mathrm{St}(E, \mathcal{U}) = X$ [**15**].

For countably compact spaces, "closed and discrete" means "finite"; a space X is called acc (which is abbreviation for *absolutely countably compact*) if for every open cover $\mathcal{U}$ and for every dense subspace $D \subset X$, there is a finite $E \subset D$ such that $\mathrm{St}(E, \mathcal{U}) = X$ [**13**]. Replacing "finite" with "countable" provides the definition of *absolute star-Lindelöfness* [**1**].

In many ways property (a) is similar to normality [**10, 15, 18, 21**] even if there are exceptions, thus every Tychonoff space can be embedded into a Tychonoff (a)-space as a closed subspace [**14**] while normality is a closed-hereditary property. Regular closed subspaces of (a)-spaces are discussed in [**22**]. A normal countably compact space need not be acc [**19**]. Every monotonically normal space has property (a) [**21**].

Here is one parallel with normality: if X is a countably paracompact (a)-space, and Y a compact metrizable space, then $X \times Y$ has property (a) as well [**10**]. A space X is called *(a)-Dowker* if X is has property (a) while $X \times (\omega + 1)$ does not [**15**].

Question 8 ([**10**]). *Do (a)-Dowker spaces exist?* 43?

Another parallel with normality is the restriction on cardinalities of closed discrete subspaces. The classical Jones lemma says that if D is a dense subspace, and H a closed discrete subspace in a space, then $2^{|H|} \leq 2^{|D|}$. The situation with normality replaced by property (a) is discussed in [**7, 8, 15, 23, 24**]. A separable (a)-space cannot contain a closed discrete subspace of cardinality $\mathfrak{c}$ [**15**].

Question 9 ([**7**]). *Is it consistent that there is an (a)-space X having a closed discrete subspace of cardinality κ^+ so that $\kappa = d(X)$ and $2^\kappa < 2^{\kappa^+}$? In particular, is it consistent with $2^\omega < 2^{\omega_1}$ that a separable (a)-space may contain an uncountable closed discrete subspace?* 44–45?

Question 10 ([**7**]). *Is it consistent that there is an (a)-space X having a closed discrete subspace of cardinality κ^+ so that $\kappa = c(X)\chi(X)$ and $2^\kappa < 2^{\kappa^+}$? In particular, is it consistent with $2^\omega < 2^{\omega_1}$ that a first countable ccc (a)-space may contain an uncountable closed discrete subspace?* 46–47?

An interesting special case of the problem under consideration is property (a) for Ψ-spaces. It is easily seen that if $\mathcal{A}$ is a *maximal* almost disjoint family, then $\Psi(\mathcal{A})$ is not an (a)-space. If $|\mathcal{A}| < \mathfrak{p}$ then $\Psi(\mathcal{A})$ is an (a)-space, and there is a family $\mathcal{A}$ of cardinality $\mathfrak{p}$ such that $\Psi(\mathcal{A})$ is an (a)-space [**24**]. The question whether $\mathfrak{p}$ can be substituted by $\mathfrak{d}$ remains open [**8**]. A characterization of families $\mathcal{A}$ for which $\Psi(\mathcal{A})$ is an (a)-space is given in [**24**]; it is shown in [**7**] that if there is an uncountable $\mathcal{A}$ for which $\Psi(\mathcal{A})$ is an (a)-space, then there is a dominating family of cardinality $\mathfrak{c}$ in ${}^{\omega_1}\omega$. The question is open whether the converse is true.

48? **Question 11** ([**8**]). *Must $\Psi(\mathcal{A})$ be an (a)-space if it is normal or countably paracompact?*

Here is another instance of the general problem:

49? **Question 12** (Attributed to P. Szeptycki in [**8**]). *Is there a characterization of those subspaces X of $\mathbb{R}$ for which the subspaces of the Niemytzki plane obtained by removing the points outside of X in the bottom line are (a)-spaces?*

The analogy between normality and property (a) suggests also these questions:

50–51? **Question 13.** *Characterize Tychonoff spaces X for which the product $X \times \beta X$ is an (a)-space (or an acc space).*

52? **Question 14.** *How big can be the extent of a star-Lindelöf (a)-space?*

The first question is motivated by Tamano's theorem ([**5**], 5.1.38), the second one by the results on the extent of normal star-Lindelöf spaces mentioned above.

Naturally, Question 13 has this modification: characterize Tychonoff spaces X for which the product $X \times cX$ is an (a)-space (or an acc space) for every compactification cX. It is easily seen that $\omega_1 \times (\omega_1 + 1)$ is not acc [**13**]. Some positive results can be found in [**26**]; thus, the product of an acc space and a compact sequential space is acc.

Acknowledgment. The authors express gratitude to Ronnie Levy, Samuel Gomes Da Silva and Paul Szeptycki for useful discussions.

References

[1] M. Bonanzinga, *Star-Lindelöf and absolutely star-Lindelöf spaces*, Questions Answers Gen. Topology **16** (1998), no. 2, 79–104.

[2] M. Bonanzinga and M. V. Matveev, *Centered-Lindelöfness versus star-Lindelöfness*, Comment. Math. Univ. Carolin. **41** (2000), no. 1, 111–122.

[3] E. K. van Douwen, *The integers and topology*, Handbook of set-theoretic topology, North-Holland, Amsterdam, 1984, pp. 111–167.

[4] E. K. van Douwen, G. M. Reed, A. W. Roscoe, and I. J. Tree, *Star covering properties*, Topology Appl. **39** (1991), no. 1, 71–103.

[5] R. Engelking, *General topology*, Heldermann Verlag, Berlin, 1989.

[6] W. M. Fleischman, *A new extension of countable compactness*, Fund. Math. **67** (1970), 1–9.

[7] S. Gomes Da Silva, *Property (a) and dominating families*, Comment. Math. Univ. Carolin. **46** (2005), no. 4, 667–684.

[8] S. Gomes Da Silva, *On the presence of countable paracompactness, normality and property (a) in spaces from almost disjoint families*, Questions Answers Gen. Topology, To appear.

[9] R. E. Hodel, *Combinatorial set theory and cardinal function inequalities*, Proc. Amer. Math. Soc. **111** (1991), no. 2, 567–575.

[10] W. Just, M. V. Matveev, and P. J. Szeptycki, *Some results on property (a)*, Topology Appl. **100** (2000), no. 1, 67–83.

[11] G. Kozma, *On removing one point from a compact space*, Houston J. Math. **30** (2004), no. 4, 1115–1126.

[12] R. Levy and M. Matveev, *Weak extent in normal spaces*, Comment. Math. Univ. Carolin. **46** (2005), no. 3, 497–501.

[13] M. V. Matveev, *Absolutely countably compact spaces*, Topology Appl. **58** (1994), no. 1, 81–92.

[14] M. V. Matveev, *Embeddings into (a)-spaces and acc spaces*, Topology Appl. **80** (1997), no. 1–2, 169–175.

[15] M. V. Matveev, *Some questions on property (a)*, Questions Answers Gen. Topology **15** (1997), no. 2, 103–111.

[16] M. V. Matveev, *A survey on star covering properties*, Preprint 330, Topology Atlas, 1998.

[17] M. V. Matveev, *How weak is weak extent?*, Topology Appl. **119** (2002), no. 2, 229–232.

[18] M. V. Matveev, O. I. Pavlov, and J. K. Tartir, *On relatively normal spaces, relatively regular spaces, and on relative property* (a), Topology Appl. **93** (1999), no. 2, 121–129.

[19] O. Pavlov, *A normal countably compact not absolutely countably compact space*, Proc. Amer. Math. Soc. **129** (2001), no. 9, 2771–2775.

[20] G. M. Reed, *Set-theoretic problems in Moore spaces*, Open problems in topology, North-Holland, Amsterdam, 1990, pp. 163–181.

[21] M. E. Rudin, I. S. Stares, and J. E. Vaughan, *From countable compactness to absolute countable compactness*, Proc. Amer. Math. Soc. **125** (1997), no. 3, 927–934.

[22] W.-X. Shi, Y.-K. Song, and Y.-Z. Gao, *Spaces embeddable as regular closed subsets into acc spaces and (a)-spaces*, Topology Appl. **150** (2005), 19–31.

[23] P. J. Szeptycki, *Soft almost disjoint families*, Proc. Amer. Math. Soc. **130** (2002), no. 12, 3713–3717.

[24] P. J. Szeptycki and J. E. Vaughan, *Almost disjoint families and property (a)*, Fund. Math. **158** (1998), no. 3, 229–240.

[25] I. J. Tree, *Constructing regular 2-starcompact spaces that are not strongly 2-star-Lindelöf*, Topology Appl. **47** (1992), no. 2, 129–132.

[26] J. E. Vaughan, *On the product of a compact space with an absolutely countably compact space*, Papers on general topology and applications (Amsterdam, 1994), Ann. New York Acad. Sci., vol. 788, New York Acad. Sci., 1996, pp. 203–208.

Open Problems in Topology II – Edited by E. Pearl

Function space topologies

Dimitris N. Georgiou, Stavros D. Iliadis and Frédéric Mynard

1. Splitting and admissible topologies

Let Y and Z be two fixed topological spaces. By $C(Y,Z)$ we denote the set of all continuous maps from Y to Z. If t is a topology on the set $C(Y,Z)$, then the corresponding topological space is denoted by $C_t(Y,Z)$.

Let X be a space. To each map $g\colon X\times Y\to Z$ which is continuous in $y\in Y$ for each fixed $x\in X$, we associate the map $g^*\colon X\to C(Y,Z)$ defined as follows: for every $x\in X$, $g^*(x)$ is the map from Y to Z such that $g^*(x)(y)=g(x,y)$, $y\in Y$. Obviously, for a given map $h\colon X\to C(Y,Z)$, the map $h^\diamond\colon X\times Y\to Z$ defined by $h^\diamond(x,y)=h(x)(y)$, $(x,y)\in X\times Y$, satisfies $(h^\diamond)^*=h$ and is continuous in y for each fixed $x\in X$. Thus, the above association (defined in [**19**]) between the mappings from $X\times Y$ to Z that are continuous in y for each fixed $x\in X$, and the mappings from X to $C(Y,Z)$ is one-to-one.

In 1946, R. Arens [**2**] introduced the notion of an admissible topology: a topology t on $C(Y,Z)$ is called *admissible*[1] if the map $e\colon C_t(Y,Z)\times Y\to Z$, called the evaluation map, defined by $e(f,y)=f(y)$, is continuous.

In 1951, R. Arens and J. Dugundji [**1**] introduced the notion of a splitting topology:[2] a topology t on $C(Y,Z)$ is called *splitting*indexsplitting topology if for every space X, the continuity of a map $g\colon X\times Y\to Z$ implies the continuity of the map $g^*\colon X\to C_t(Y,Z)$. They also proved that a topology t on $C(Y,Z)$ is admissible if and only if for every space X, the continuity of a map $h\colon X\to C_t(Y,Z)$ implies that of the map $h^\diamond\colon X\times Y\to Z$. (For more on the notions of splitting and admissible topologies see also the books [**16, 17, 36**].) If in the above definitions it is assumed that the space X belongs to a fixed class $\mathcal{A}$ of topological spaces, then the topology t is called *$\mathcal{A}$-splitting* or *$\mathcal{A}$-admissible*, respectively (see [**25**]). We call two classes of spaces $\mathcal{A}_1$ and $\mathcal{A}_2$ *equivalent*, in symbols $\mathcal{A}_1\sim\mathcal{A}_2$, if a topology t on $C(Y,Z)$ is $\mathcal{A}_1$-splitting if and only if t is $\mathcal{A}_2$-splitting and t is $\mathcal{A}_1$-admissible if and only if t is $\mathcal{A}_2$-admissible (see [**25**]).

Each topology t on a set X defines a *topological convergence class*, denoted here by $\mathcal{C}(t)$, consisting of all pairs $(\mathcal{F},s)$ where $\mathcal{F}$ is a filter on X converging topologically to $s\in X$. A filter $\mathcal{F}$ on $C(Y,Z)$ *converges continuously*[3] to a function f if $e(\mathcal{F}\times\mathcal{G})$ (where $e\colon C(Y,Z)\times Y\to Z$ is the evaluation map) converges to $f(x)$ in Z whenever $\mathcal{G}$ is a filter convergent to x in X. By $\mathcal{C}^*$ we denote the

[1] Such a topology is called *jointly continuous* or *conjoining* by some authors.

[2] In [**1**] such a topology is called *proper*.

[3] Continuous convergence can alternatively be described in terms of convergence of nets in $C(X,Z)$, as given by O. Frink (see [**20, 32**]): a net $S=\{f_\lambda:\lambda\in\Lambda\}$ in $C(Y,Z)$ converges continuously to $f\in C(Y,Z)$ if and only if for every $y\in Y$ and for every open neighborhood W of $f(y)$ in Z there exists an element $\lambda_0\in\Lambda$ and an open neighborhood V of y in Y such that for every $\lambda\ge\lambda_0$, we have $f_\lambda(V)\subseteq W$. Notice that this is the description used in [**1**].

class of all pairs $(\mathcal{F}, f)$ such that $\mathcal{F}$ is a filter on $C(Y, Z)$ converging continuously to $f \in C(Y, Z)$. In general, the class $\mathcal{C}^*$ is not a topological convergence class (e.g., [**1**]). The *topological modification of the continuous convergence*,[4] denoted here by $t(\mathcal{C}^*)$, is obtained in a standard way, that is, a subset U of $C(Y, Z)$ is open in $t(\mathcal{C}^*)$ if $U \in \mathcal{F}$ whenever $(\mathcal{F}, f) \in \mathcal{C}^*$. See [**3, 4**] for details on continuous convergence.

It is well known (e.g., [**1**]) that a topology t on $C(Y, Z)$ is splitting if and only if $\mathcal{C}^* \subseteq \mathcal{C}(t)$. Also, a topology t on $C(Y, Z)$ is admissible if and only if $\mathcal{C}(t) \subseteq \mathcal{C}^*$. Using the above characterizations, one can prove the following results (1–4) concerning topologies on $C(Y, Z)$:

(1) Each splitting topology is contained in each admissible topology [**1**].
(2) The topology $t(\mathcal{C}^*)$ is the greatest splitting topology.
(3) The intersection of all admissible topologies coincides with the greatest splitting topology (see [**18, 42**]).
(4) A topology t is simultaneously splitting and admissible if and only if $\mathcal{C}^*$ is a topological convergence class, if and only if $\mathcal{C}(t) = \mathcal{C}^*$ (see [**1**]).
(5) There exist Y and Z such that the greatest splitting topology is not admissible and, therefore, there does not always exists a simultaneously splitting and admissible topology. For example, such spaces are $Y = \mathbb{Q}$ and $Z = \mathbb{R}$, where $\mathbb{Q}$ is the set of rational numbers and $\mathbb{R}$ is the set of real numbers with the usual topology (see [**2, 19**]).
(6) If Z is a T_i-space, $i = 0, 1, 2$, then $C_{t(\mathcal{C}^*)}(Y, Z)$ is a T_i-space (see [**16**]).

Also, it is known that (see [**25**]):

(7) For every class $\mathcal{A}$ of spaces, there exists the greatest $\mathcal{A}$-splitting topology which is denoted by $t(\mathcal{A})$.
(8) If t is an $\mathcal{A}$-splitting and $\mathcal{A}$-admissible topology and $C_t(Y, Z) \in \mathcal{A}$, then $t = t(\mathcal{A})$.
(9) There exists a space X such that $\mathcal{A} \sim \mathcal{T}$, where $\mathcal{T}$ is the class of all spaces and $\mathcal{A}$ the singleton $\{X\}$. Therefore, $t(\mathcal{C}^*) = t(\{X\})$.
(10) Let $\mathcal{A}$ be the class of spaces consisting of the Sierpiński space and all completely regular spaces with only one non-isolated point. Then $\mathcal{A} \sim \mathcal{T}$. Therefore, $t(\mathcal{C}^*) = t(\mathcal{A})$.

53? **Problem 1** (See [**1**]). *Characterize the greatest splitting topology* $t(\mathcal{C}^*)$ *on* $C(Y, Z)$ *directly in terms of the topological structures of* Y *and* Z.

54–55? **Problem 2.** *Is the space* $C_{t(\mathcal{C}^*)}(Y, Z)$ *regular (respectively, Tychonoff) in the case where* Z *is regular (respectively, Tychonoff)?*

56? **Problem 3** (See [**25**]). *Characterize, in terms of the topological structures of* Y *and* Z, *spaces* X *such that the singleton* $\{X\}$ *is equivalent to a well-known (or given) class of spaces.*

In particular,

[4] In terms of nets, a subset U of $C(Y, Z)$ is open in $t(\mathcal{C}^*)$ if and only if for every pair $(\{f_\lambda : \lambda \in \Lambda\}, f) \in \mathcal{C}^*$, where $f \in U$, there exists $\lambda_0 \in \Lambda$ such that $f_\lambda \in U$, for every $\lambda \geq \lambda_0$.

Problem 4. *Characterize, in terms of the topological structures of Y and Z, spaces X such that $\{X\} \sim T$.* 57?

Problem 5. *Characterize classes $\mathcal{A}$ of spaces such that $C_{t(\mathcal{A})}(Y, Z) \in \mathcal{A}$.* 58?

Problem 6. *Characterize classes $\mathcal{A}$ of spaces such that $t(\mathcal{A})$ is $\mathcal{A}$-admissible and $C_{t(\mathcal{A})}(Y, Z) \in \mathcal{A}$.* 59?

See [**22**, **25**] for more information and problems concerning $\mathcal{A}$-splitting and $\mathcal{A}$-admissible topologies.

2. The greatest splitting, compact-open, and Isbell topologies

In this paper, by a *compact* space we mean a space such that each open cover has a finite subcover. Also, by a *locally compact* space we mean a space such that each point of it has an open neighborhood with compact closure. In these definitions we do not assume any separation axiom.

The *compact-open topology* on $C(Y, Z)$, denoted here by t_{co}, was defined by R.H. Fox in 1945 (see [**19**]): a subbasis for t_{co} is the family of all sets of the form

$$(K, U) = \{f \in C(Y, Z) : f(K) \subseteq U\},$$

where K is a compact subset of Y and U is an open subset of Z. It is well known that:

(1) The compact-open topology is always splitting (see [**1**, **19**]).
(2) If Y is a regular locally compact space (and, therefore, Y is a T_1-space), then the topology t_{co} is admissible (see [**1**, **2**, **19**]). In this case (as observed in [**1**]), the compact-open topology coincides with the greatest splitting topology.
(3) If Z is regular (respectively, Tychonoff), then the space $C_{t_{\text{co}}}(Y, Z)$ is regular (respectively, Tychonoff) (see [**2**, **16**]).

In 1972, D. Scott defined a topology on a partially ordered set L which is known as the Scott topology (see, for example, [**27**]). If L is the set $\mathcal{O}(Y)$ of all open sets of the space Y partially ordered by inclusion, then the Scott topology coincides with a topology defined in 1970 by B.J. Day and G.M. Kelly (see [**10**]): a subset $\mathbb{H}$ of $\mathcal{O}(Y)$ is an element of this topology (that is, the Scott topology) if and only if: (α) the conditions $U \in \mathbb{H}$, $V \in \mathcal{O}(Y)$, and $U \subseteq V$ imply $V \in \mathbb{H}$, and (β) for every collection of open sets of Y, whose union belongs to $\mathbb{H}$, there are finitely many elements of this collection whose union also belongs to $\mathbb{H}$. In [**14**] the elements of the Scott topology are called *compact families.*

The *Isbell topology* on $C(Y, Z)$, denoted here by t_{Is}, was defined by J.R. Isbell in 1975 (see [**27**, **31**, **36**]): a subbasis for t_{Is} is the family of all sets of the form

$$(\mathbb{H}, U) = \{f \in C(Y, Z) : f^{-1}(U) \in \mathbb{H}\},$$

where $\mathbb{H}$ is an element of the Scott topology on $\mathcal{O}(Y)$ and U is an open subset of Z (see [**33**]).

Problem 7 (See [**34**] for the case of regular spaces). *Is the space $C_{t_{Is}}(Y, Z)$ regular (respectively, Tychonoff) when Z is a regular (respectively, Tychonoff) space?* 60–61?

A subset B of a space X is called *bounded* or *relatively compact* (see, for example, [**34**]) if every open cover of X contains a finite subcover of B. A space X is called *corecompact* if for every open neighborhood U of a point $x \in X$ there exists an open neighborhood $V \subseteq U$ of x such that V is bounded in the space U. (These spaces were introduced by various authors under different names: quasi-locally compact spaces in [**45**], spaces with property C in [**10**], semi-locally bounded spaces in [**31**], CL-spaces in [**28**], and corecompact spaces in [**29**]). It is known that:

(1) The compact-open topology is contained in the Isbell topology (see, for example, [**36**]), that is, $t_{\mathrm{co}} \leq t_{\mathrm{Is}}$.
(2) The Isbell topology is always splitting (see, for example, [**34, 36, 43**]), that is, $t_{\mathrm{Is}} \leq t(\mathcal{C}^*)$.
(3) Let $\mathbf{S} = \{0, 1\}$ with the topology $\{\emptyset, \{0,1\}, \{0\}\}$ be the *Sierpiński space.* The set $C(Y, \mathbf{S})$ can be identified with $\mathcal{O}(Y)$ (via the indicator functions of open sets). Then, the Isbell topology on $C(Y, \mathbf{S})$ is the Scott topology on $\mathcal{O}(Y)$, and $t_{\mathrm{Is}} = t(\mathcal{C}^*)$ for $Z = \mathbf{S}$ (see [**10, 41, 43**]). If $C(Y, \mathbf{S})$ is identified with the set $\mathcal{C}(Y)$ of closed subsets of Y, then the Isbell topology becomes the upper Kuratowski topology (see [**14**]).
(4) The space Y is corecompact if and only if the Isbell topology on $C(Y, Z)$ is admissible, if $Z = \mathbf{S}$ (see [**29, 34, 43**]), or if Z is any topological space containing $\mathbf{S}$ as a subspace (see [**43**]).

The notion of a *consonant* space and some of its properties were introduced in [**14**]. The following equivalent conditions can serve as a definition of consonant spaces:

(a) Y is consonant;
(b) The compact-open topology coincides with the Isbell topology on $C(Y, \mathbf{S})$;
(c) The compact-open topology coincides with the Isbell topology on $C(Y, Z)$ for every space Z.

A space Y is called *Z-consonant*[5] if the compact-open topology coincides with the Isbell topology on $C(Y, Z)$. Notice that consonance coincides with $\mathbf{S}$-consonance.

62–63? **Problem 8.** *For what spaces Z does Z-consonance imply consonance? In particular, is it the case if Z is locally finite and contains* $\mathbf{S}$*? What if Z is an Alexandroff space (that is, a space each point of which has a minimal open neighborhood)?*

A space Y is called *Z-concordant* if the Isbell topology coincides with the greatest splitting $t(\mathcal{C}^*)$ topology on $C(Y, Z)$. A space Y is called *concordant* if it is Z-concordant for every space Z. Of course, every space is $\mathbf{S}$-concordant.

64–67? **Problem 9.** *For what spaces Z is the condition of Z-concordance trivial (satisifed by every topological space)? In particular, is it the case when Z is a finite space?*

[5]This name was used for a different notion in [**39**], namely for what we call Z-harmonic in the present paper.

A locally finite space? An Alexandroff space? A space containing the Sierpiński space?

Problem 10. *Is there a space Z such that Z-concordance implies concordance?* 68?

Note that if Y is corecompact (and Z is an arbitrary space), then the Isbell topology on $C(Y, Z)$ is admissible (e.g., [**34, 43**]) and, therefore, coincides with the greatest splitting topology. Hence, corecompact spaces are concordant. A space Y is called *Z-harmonic* if it is both Z-consonant and Z-concordant. A space Y is called *harmonic* if it is Z-harmonic for every space Z. Note that regular locally compact spaces are harmonic (e.g., [**1**]). Consonance is a well-studied notion for which most questions have been solved. For instance it is known that:

(1) A topological space Y is consonant if and only if every compact family is compactly generated. (A compact family $\mathbb{H}$ of open subsets of Y is called compactly generated (see [**14**]) if there exists a family $\mathcal{K}$ of compact subsets of Y such that $\mathbb{H}$ coincides with the set of all open subsets of Y containing an element of $\mathcal{K}$.)
(2) Every Čech complete (that is, a Tychonoff space which is a G_δ-subset of one of its compactifications), every locally Čech complete, every regular k_ω-space (that is, a space X for which there is a countable family $\{K_i : i \in \omega\}$ of compact subsets such that $C \subseteq X$ is closed if and only if $C \cap K_i$ is closed in K_i for every $i \in \omega$), and every regular locally k_ω-space are consonant. In particular, every complete metrizable space is consonant (see [**14, 40**]).
(3) There exist non-consonant hereditarily Baire separable metrizable spaces, F_σ-discrete metrizable spaces, regular σ-compact analytic first countable spaces (see [**6**]). The Sorgenfrey line and the set of rationals are not consonant (see [**5, 8**]).
(4) The consonant spaces are exactly $\mathcal{Q}$-covering images of regular locally compact spaces, where $f\colon X \to Y$ is called $\mathcal{Q}$-covering (see [**7**]) if for every compact family $\mathbb{H}_Y$ on Y, there exists a compact family $\mathbb{H}_X$ on X such that $U \in \mathbb{H}_Y$ whenever U is open in Y and $f^{-1}(U) \in \mathbb{H}_X$.

However, the following problem remains:

Problem 11 (See [**6**]). *Is there in* ZFC *a metrizable consonant topological space which is not completely metrizable?* (There are consistent such examples; see [**6**].) 69?

In contrast, very little is known on Z-consonance and Z-concordance for a general space Z. However, we know:

(1) If Y is a completely regular and $\mathbb{R}$-harmonic space, then Y is consonant (see [**13, 39**]). But the converse is false:
(2) The space $\mathbb{N}^\omega$ is consonant (in particular $\mathbb{N}$-consonant) but not $\mathbb{N}$-concordant, hence not $\mathbb{N}$-harmonic [**18**]. Similarly, $\mathbb{R}^\omega$ is consonant (in particular $\mathbb{R}$-consonant), but not $\mathbb{R}$-concordant, hence not $\mathbb{R}$-harmonic [**21**].

(3) Let P be a disjoint sum of countably many spaces Q_i. Then P^ω is not P-harmonic (see [**21**]). If the spaces Q_i are moreover Čech complete, P^ω is not even P-concordant (see [**21**]).

The (too) general problem is of course to characterize Z-consonance, Z-concordance and Z-harmonicity of a space Y in terms of the topologies of Y and Z. In particular:

70? **Problem 12.** *Find sufficient conditions for* $\mathbb{R}$*-harmonicity,* $\mathbb{R}$*-consonance, and* $\mathbb{R}$*-concordance.*

71–73? **Problem 13.** *Find completely regular* $\mathbb{R}$*-harmonic spaces that are not locally compact (or not even of point-countable type or not even a q-space (in the sense of E. Michael; see [**37**])).*

74? **Problem 14.** *Under what conditions on spaces Y and Z, does Z-harmonicity of Y imply that Y is consonant?*

75? **Problem 15.** *Is there a completely regular space that is* $\mathbb{R}$*-concordant but not consonant?*

76–78? **Problem 16.** *For a given space Z, under what condition on a space Y does:*

(1) *Z-consonance of Y imply that Y is consonant?*
(2) *Z-concordance of Y imply that Y is concordant?*
(3) *Z-harmonicity of Y imply that Y is harmonic?*

79–81? **Problem 17.** *Find a class of maps $\mathcal{Q}$ such that X is* $\mathbb{R}$*-harmonic (*$\mathbb{R}$*-consonant,* $\mathbb{R}$*-concordant) if and only if there exists a map $f\colon Z \to X$ of $\mathcal{Q}$ with Z regular locally compact.*

[**11, 12, 15, 24–26, 38, 39, 44**] are other papers related to these questions.

References

[1] R. Arens and J. Dugundji, *Topologies for function spaces*, Pacific J. Math. **1** (1951), 5–31.
[2] R. F. Arens, *A topology for spaces of transformations*, Ann. of Math. (2) **47** (1946), 480–495.
[3] R. Beattie and H. P. Butzmann, *Convergence structures and applications to functional analysis*, Kluwer Academic Publishers, Dordrecht, 2002.
[4] E. Binz, *Continuous convergence on $C(X)$*, Springer, Berlin, 1975.
[5] A. Bouziad, *Borel measures in consonant spaces*, Topology Appl. **70** (1996), no. 2–3, 125–132.
[6] A. Bouziad, *Consonance and topological completeness in analytic spaces*, Proc. Amer. Math. Soc. **127** (1999), no. 12, 3733–3737.
[7] A. Bouziad, *H-trivial spaces*, Topology Appl. **132** (2003), no. 2, 159–174.
[8] C. Costantini and S. Watson, *On the dissonance of some metrizable spaces*, Topology Appl. **84** (1998), 259–268.
[9] B. Day, *A reflection theorem for closed categories*, J. Pure Appl. Algebra **2** (1972), no. 1, 1–11.
[10] B. Day and G. M. Kelly, *On topologically quotient maps preserved by pullbacks or products*, Proc. Cambridge Philos. Soc. **67** (1970), 553–558.
[11] G. Di Maio, L. Holá, D. Holý, and R. A. McCoy, *Topologies on the space of continuous functions*, Topology Appl. **86** (1998), no. 2, 105–122.

[12] G. Di Maio, E. Meccariello, and S. Naimpally, *Hyper-continuous convergence in function spaces*, Questions Answers Gen. Topology **22** (2004), no. 2, 157–162.

[13] S. Dolecki, G. H. Greco, and A. Lechicki, *Sur la topologie de la convergence supérieure de Kuratowski*, C. R. Acad. Sci. Paris Sér. I Math. **312** (1991), no. 12, 923–926.

[14] S. Dolecki, G. H. Greco, and A. Lechicki, *When do the upper Kuratowski topology (homeomorphically, Scott topology) and the co-compact topology coincide?*, Trans. Amer. Math. Soc. **347** (1995), no. 8, 2869–2884.

[15] S. Dolecki and F. Mynard, *Hyperconvergences*, Appl. Gen. Topol. **4** (2003), no. 2, 391–419.

[16] J. Dugundji, *Topology*, Allyn and Bacon Inc., Boston, 1978.

[17] R. Engelking, *General topology*, Heldermann Verlag, Berlin, 1989.

[18] M. Escardó, J. Lawson, and A. Simpson, *Comparing Cartesian closed categories of (core) compactly generated spaces*, Topology Appl. **143** (2004), 105–145.

[19] R. H. Fox, *On topologies for function spaces*, Bull. Amer. Math. Soc. **51** (1945), 429–432.

[20] O. Frink, Jr., *Topology in lattices*, Trans. Amer. Math. Soc. **51** (1942), 569–582.

[21] D. N. Georgiou and S. D. Iliadis, *On the compact-open and finest splitting topologies*, 2005, Preprint. To appear in Topology Appl.

[22] D. N. Georgiou and S. D. Iliadis, *Some problems concerning splitting and admissible topologies*, Questions Answers Gen. Topology **23** (2005), no. 2, 101–105.

[23] D. N. Georgiou and S. D. Iliadis, *On finest splitting and admissible topologies for some function spaces*, 2006, Submitted.

[24] D. N. Georgiou, S. D. Iliadis, and B. K. Papadopoulos, *On dual topologies*, Topology Appl. **140** (2004), no. 1, 57–68.

[25] D. N. Georgiou, S. D. Iliadis, and B. K. Papadopulos, *Topologies of function spaces*, Zap. Nauchn. Sem. S.-Peterburg. Otdel. Mat. Inst. Steklov. (POMI) **208** (1993), 82–97, Translation: J. Math. Sci. (New York) 81 (1996) 2506–2514.

[26] D. N. Georgiou and B. K. Papadopoulos, *A note on the finest splitting topology*, Questions Answers Gen. Topology **15** (1997), no. 2, 137–144.

[27] G. Gierz, K. H. Hofmann, K. Keimel, J. D. Lawson, M. W. Mislove, and D. S. Scott, *Continuous lattices and domains*, Cambridge University Press, Cambridge, 2003.

[28] K. H. Hofmann, *Continuous lattices, topology and topological algebra*, Topology Proc. **2** (1977), no. 1, 179–212 (1978).

[29] K. H. Hofmann and J. D. Lawson, *The spectral theory of distributive continuous lattices*, Trans. Amer. Math. Soc. **246** (1978), 285–310.

[30] S. D. Iliadis and B. K. Papadopoulos, *The continuous convergence on function spaces*, Panamer. Math. J. **4** (1994), no. 3, 33–42.

[31] J. R. Isbell, *Function spaces and adjoints*, Math. Scand. **36** (1975), no. 2, 317–339.

[32] C. Kuratowski, *Sur la notion de limite topologique d'ensembles*, Ann. Soc. Polon. Math. **21** (1948), 219–225 (1949).

[33] P. Th. Lambrinos, *The bounded-open topology on function spaces*, Manuscripta Math. **36** (1981), no. 1, 47–66.

[34] P. Th. Lambrinos and B. Papadopoulos, *The (strong) Isbell topology and (weakly) continuous lattices*, Continuous lattices and their applications (Bremen, 1982), Lecture Notes in Pure and Appl. Math., vol. 101, Marcel Dekker Inc., New York, 1985, pp. 191–211.

[35] J. D. Lawson and M. W. Mislove, *Problems in domain theory and topology*, Open problems in topology, North-Holland, Amsterdam, 1990, pp. 349–372.

[36] R. A. McCoy and I. Ntantu, *Topological properties of spaces of continuous functions*, Lecture Notes in Math., vol. 1315, Springer, Berlin, 1988.

[37] E. Michael, *A quintuple quotient quest*, General Topology and Appl. **2** (1972), 91–138.

[38] F. Mynard, *Coreflectively modified continuous duality applied to classical product theorems*, Appl. Gen. Topol. **2** (2001), no. 2, 119–154.

[39] F. Mynard, *First-countability sequentiality and tightness of the upper Kuratowski convergence*, Rocky Mountain J. Math. **33** (2003), no. 3, 1011–1038.

[40] T. Nogura and D. Shakhmatov, *When does the Fell topology on a hyperspace of closed sets coincide with the meet of the upper Kuratowski and the lower Vietoris topologies?*, Topology Appl. **70** (1996), no. 2–3, 213–243.

[41] B. K. Papadopoulos, *Proper topologies on the set S^Y*, Glas. Mat. Ser. III **23(43)** (1988), no. 1, 143–146.

[42] H. Render, *Nonstandard topology on function spaces with applications to hyperspaces*, Trans. Amer. Math. Soc. **336** (1993), no. 1, 101–119.

[43] F. Schwarz and S. Weck, *Scott topology, Isbell topology, and continuous convergence*, Continuous lattices and their applications (Bremen, 1982), Lecture Notes in Pure and Appl. Math., vol. 101, Marcel Dekker Inc., New York, 1985, pp. 251–271.

[44] D. Scott, *Continuous lattices*, Toposes, algebraic geometry and logic (Halifax, N.S., 1971), Lecture Notes in Math., vol. 274, Springer, Berlin, 1972, pp. 97–136.

[45] A. J. Ward, *Problem*, Proceedings of the International Symposium on Topology and its Applications (Herceg-Novi, 1968) (D. R. Kurepa, ed.), Savez Drustava Mat. Fiz. i Astronom., Beograd, 1969, p. 352.

[46] P. Wilker, *Adjoint product and hom functors in general topology*, Pacific J. Math. **34** (1970), 269–283.

Open Problems in Topology II – Edited by E. Pearl

Spaces and mappings: special networks

Chuan Liu and Yoshio Tanaka

Introduction

It has been over 40 years after Arhangel'skiĭ [**1**] published the seminal paper "Mappings and Spaces". The problems in the paper stimulated many developments in General Topology. Some of these problems have been solved, and some are still open. In this chapter, we shall survey some results on various generalization of metric spaces and their quotient spaces in terms of k-networks, weak bases, and weak topologies. We pose some related problems that are interesting to the authors.

A collection $\mathcal{P}$ of subsets of a space X is a *network* (or *net*) for X if, whenever $x \in U$ with U open, there is $P \in \mathcal{P}$ with $x \in P \subset U$. A space is a *σ-space* if it has a σ-locally finite network, and a *cosmic space* if it has a countable network. k-networks are particularly useful special networks. Recall that $\mathcal{P}$ is a *k-network* if, whenever $K \subset U$ with K compact and U open, there is a finite $\mathcal{P}' \subset \mathcal{P}$ with $K \subset \bigcup \mathcal{P}' \subset U$. A base is a k-network. A space is an *$\aleph$-space* if it has a σ-locally-finite k-network, and an *$\aleph_0$-space* if it has a countable k-network.

k-networks have played an important role in the study of various kinds of quotient spaces of metric spaces and of generalized metric spaces and their metrizability. For related surveys, see [**29**, **31**, **72**, **75**, **77**], etc. Readers may refer to [**9**, **10**, **19**] for unstated definitions and terminology in this paper.

All spaces are regular T_1-spaces, and maps are continuous surjections.

Mappings

A collection $\mathcal{P}$ of subsets of a space X is *point-countable* (resp. *compact-countable*; *star-countable*) if each point (resp. compact subset; member in $\mathcal{P}$) meets only countably many elements of $\mathcal{P}$. "*Point-finite*" or "*compact-finite*" are defined similarly. A family $\mathcal{P} = \{A_\alpha : \alpha \in I\}$ is *closure-preserving* (abbr. CP) if for any $J \subset I$, $\bigcup\{\overline{A_\alpha} : \alpha \in J\} = \overline{\bigcup\{A_\alpha : \alpha \in J\}}$. $\mathcal{P}$ is *hereditarily closure-preserving* (abbr. HCP) if for any $J \subset I$, $\{B_\alpha : B_\alpha \subset A_\alpha, \alpha \in J\}$ is closure-preserving.

A cover $\mathcal{P}$ of X is a *cs-network* (resp. *cs*-network*) if, whenever L is a convergent sequence with the limit point x such that $L \cup \{x\} \subset U$ with U open in X, then for some $P \in \mathcal{P}$, $x \in P \subset U$, and L is eventually (resp. frequently) in P. ("Frequently" means P contains a subsequence of L.)

A space X is *determined by a cover* $\mathcal{P}$ [**20**], if X has the *weak topology* with respect to $\mathcal{P}$; that is, $G \subset X$ is open in X if $G \cap P$ is open in P for each $P \in \mathcal{P}$. Here, we can replace "open" by "closed". For basic properties of weak topologies, see [**9**, **72**, **79**]. A space is *sequential* (resp. a *k-space*) if it is determined by compact metric subsets (resp. compact subsets). As is well-known, sequential spaces (resp. k-spaces) are characterized as quotient images of (locally compact) metric spaces (resp. locally compact spaces). A k-space X is sequential if each point of X is a G_δ-set [**48**], or X has a point-countable k-network [**20**]. A space is *Fréchet* (or *Fréchet–Urysohn*) if whenever $A \subset X$ with $x \in \overline{A}$, there is a sequence in A converging to the point x. Fréchet spaces are sequential.

Let $f: X \to Y$ be a map. Then f is an *s-map* (resp. *compact map*; *Lindelöf*; countable-to-one) if each fiber $f^{-1}(y)$ is separable (resp. compact; Lindelöf; countable). Also, f is a *compact-covering map* if each compact subset of Y is an image of some compact subset of X.

In 1966, Michael [**47**] characterized quotient (compact-covering) images of separable metric spaces as k-and-$\aleph_0$-spaces. In [**1**], Arhangel′skiĭ posed an important problem: "How does one characterize, in intrinsic terms, quotient s-images of metric spaces?" Hoshina [**23**], Gruenhage, Michael and Tanaka [**20**] gave characterizations for this problem, and other topologists gave many characterizations for various kinds of quotient images of metric spaces. In 1987, Tanaka [**71**] obtained a concise characterization by means of cs^*-networks: A space X is a quotient s-image of a metric space iff X is a sequential space with a point-countable cs^*-network. On the other hand, Michael and Nagami [**50**] posed a classical problem related to quotient s-images of metric spaces: "Is a quotient s-image of a metric space a quotient, compact-covering s-image of a metric space?" Michael [**49**] showed that a space X is a quotient, compact-covering s-image of a metric space if X is a k-space with a point-countable closed k-network. Lin and Liu [**33**] showed that the same result also holds if X is a sequential space with a point-countable cs-network. Assuming the existence of the σ'-set (in [**60**]), Chen [**6**] constructed a quotient s-image of a metric space that is not a quotient, compact-covering s-image of a metric space. He also constructed a *Hausdorff* counterexample in [**5**].

82? **Problem 1.** *Let X be a quotient s-image of a metric space. If X is Fréchet, is X a compact-covering s-image of a metric space?*

Let us recall canonical quotient spaces, the *sequential fan* S_ω and the *Arens' space* S_2. For an infinite cardinal α, S_α is the space obtained from the topological sum of α many convergent sequences by identifying all limit points. While, S_2 is the space obtained from the topological sum of $\{L_n : n < \omega\}$, L_n are the convergent sequence $\{1/n : n \in \mathbb{N}\} \cup \{0\}$, by identifying each $1/n \in L_0$ with $0 \in L_n$ $(n \geq 1)$.

The space S_{ω_1} is a Fréchet space which has a point-countable k-network, but does not have any point-countable cs^*-network (in view of [**70**]). A quotient s-image of a metric space has a point-countable k-network, and contains no copy of S_{ω_1} in view of [**20**], also it has a point-countable cs^*-network, but the authors don't even know whether the answer of the following problem is positive among sequential spaces. If the answer is positive, then so is the answer to [**20**, Question 10.2].

83? **Problem 2.** *Let X be a Fréchet space with a point-countable k-network. If X contains no closed copy of S_{ω_1}, does X have a point-countable cs^*-network?*

A collection $\mathcal{B} = \bigcup\{\mathcal{B}_x(n) : x \in X, n \in \mathbb{N}\}$ of subsets of X is an $\aleph_0$-*weak base* if (a) for each $n, x \in X$, $\mathcal{B}_x(n)$ is closed under finite intersections with $x \in \bigcap \mathcal{B}_x(n)$, and (b) $U \subset X$ is open iff for each $x \in U$, $n \in \mathbb{N}$, there is $B_x(n) \in \mathcal{B}_x(n)$ with $x \in B_x(n) \subset U$. A space X is *$\aleph_0$-weakly first-countable* [**67**] (or *weakly quasi-first-countable* [**64**]) if the $\mathcal{B}_x(n)$ is countable for each $x \in X, n \in \mathbb{N}$. If $\mathcal{B}_x(n) = \mathcal{B}_x(1)$ for each $x \in X, n \in \mathbb{N}$, the collection $\mathcal{B}$ is a *weak base* as defined by Arhangel′skiĭ [**1**], and X is *weakly first-countable* (or, *g-first-countable* [**65**]) if $\mathcal{B}_x(1)$ is countable for each $x \in X$. A space X is *symmetric* (or *symmetrizable*) [**1**], if there is a real-valued function d defined on $X \times X$ such that (a) $d(x,y) = d(y,x) \geq 0$, here $d(x,y) = 0$ iff $x = y$, and (b) $U \subset X$ is open iff for each $x \in U$, $\{y \in X : d(x,y) < 1/n\} \subset U$ for some $n \in \mathbb{N}$. A first-countable space or a symmetric space is weakly first-countable, and a weakly first-countable space is $\aleph_0$-weakly first-countable, hence sequential [**64**]. A space X is *g-metrizable* (resp. *g-second countable*) [**65**] if it has a σ-locally finite weak base (resp. a countable weak base). The space S_2 is g-second countable, but not Fréchet. The space S_ω is a Fréchet space with a countable $\aleph_0$-weak base, but it is not g-first-countable. A g-metrizable space is symmetrizable, and a Fréchet g-metrizable space is metrizable [**65**].

Problem 3. *Let X be a Fréchet space with a σ-locally finite $\aleph_0$-weak base. Is X a closed countable-to-one image of a metric space?* 84?

Lašnev [**26**] gave the first characterization for closed images of metric spaces. Foged [**12**] characterized *Lašnev spaces* (= closed images of metric spaces) as Fréchet spaces with a σ-HCP k-network. A space is a Fréchet $\aleph$-space iff it is a closed s-image of a metric spaces [**17, 27**]. Liu [**38**] characterized an $\aleph_0$-weakly first-countable Lašnev space X as a closed, σ-compact image (i.e., each fiber is σ-compact) of a metric space. The above problem is equivalent to a question whether every closed, σ-compact image of a metric space is a closed countable-to-one image of a metric space.

A space X is *stratifiable* (or an *M_3-space*) if for each open subset U of X, there is a sequence $\{S(U,n) : n \in \mathbb{N}\}$ of open subsets such that (a) $U = \bigcup\{S(U,n) : n \in \mathbb{N}\} = \bigcup\{\overline{S(U,n)} : n \in \mathbb{N}\}$, and (b) if $U \subset V$, then $S(U,n) \subset S(V,n)$ for all $n \in \mathbb{N}$.

If in the above definition the sets $S(U,n)$ are required to be closed instead of open, then X is *semi-stratifiable* [**7**]. X is *k-semistratifiable* [**45**] if whenever $C \subset U$ with C compact and U open in X, $C \subset S(U,n)$ for some $n \in \mathbb{N}$. An M_3-space, or a space with a σ-CP k-network is k-semistratifiable. A k-semistratifiable space is a σ-space [**16**], and a σ-space is semi-stratifiable. A classical problem whether a space with a σ-CP base (= M_1-space) is equivalent to a space with a σ-cushioned pair-base (= M_3-space) remains open. A space with a σ-CP network (= σ-space) need not be equivalent to a space with a σ-cushioned pair-network (= semi-stratifiable space); see [**19**]. A k-semistratifiable space is equivalent to a space with a σ-cushioned pair-k-network [**14, 15**].

Problem 4. *Is a Fréchet M_3-space (hence, M_1-space) with a point-countable k-network a Lašnev space?* 85?

A Lašnev space is a Fréchet M_3-space with a point-countable k-network. A positive answer to the above problem implies an affirmative answer to Junnila and

Yun's question [**25**]: "Is every Fréchet M_3-space with a point-countable closed k-network an $\aleph$-space?"

In view of Foged's characterization for Lašnev spaces, can we generalize this to g-metrizable domains? Every closed image of an $\aleph$-space, in particular, g-metrizable space, has a σ-HCP k-network [**71**]. But, for the converse, the authors don't even know whether a k-and-$\aleph_0$-space is a closed image of a g-second countable space.

86? **Problem 5.** *Is a k-space with a σ-HCP k-network characterized as a closed image of a g-metrizable space?*

k-networks and weak bases

A Fréchet k-semistratifiable space is an M_3-space with a σ-CP k-network [**14**, **15**].

87? **Problem 6** ([**14**]). *Does a k-semistratifiable space have a σ-CP k-network?*

A paracompact space with a σ-locally countable k-network is an $\aleph$-space. But, a g-metrizable (hence, k-semistratifiable) space need not be normal [**13**].

88? **Problem 7** ([**31**]). *Is a k-and-k-semistratifiable space with a σ-locally countable k-network an $\aleph$-space?*

A space X is a g-metrizable space iff X is a weakly first-countable $\aleph$-space [**11**] iff X is a weakly first-countable space with a σ-HCP k-network [**28**, **73**] iff X has a σ-HCP weak base [**37**] iff X has a σ-compact-finite weak base by closed subsets [**35**] iff X is a k-space with a σ-HCP k-network and contains no closed copy of S_ω [**39**].

89–90? **Problem 8.**

(1) *Let X be a k-space with a σ-locally countable k-network. If X contains no closed copy of S_ω, does X have a σ-locally countable weak base?*
(2) ([**31**]) *Does a k-and-$\aleph$-space have a σ-CP weak base?*

For (1) in the above problem, it is positive if the k-network is σ-locally finite, and X has a σ-locally countable base if X contains no closed copy of S_ω and no S_2, in particular, X is first-countable. A weakly first-countable space (hence, it contains no copy of S_ω) with a σ-compact-finite k-network need not have a point-countable weak base [**34**].

91? **Problem 9.** *Let X be a weakly first-countable space. Is any weak base for X a k-network?*

Liu [**37**] showed that not every weak base is a k-network. But, in most cases, a weak base is a k-network. Among spaces whose every compact subset is Fréchet (e.g., spaces whose points are G_δ-sets), any weak base is a k-network [**75**]. The authors don't even know whether a weak base is a k-network among sequential spaces.

92? **Problem 10.** *Let Y be an open and closed image of a (compact) weakly first-countable space. Is Y weakly first-countable?*

A compact symmetrizable space is metrizable [**1**], but a compact weakly first-countable space need not be first-countable under CH [**24**].

It was asked in [**74**] whether weak first-countability (or symmetrizability) is preserved by open and closed maps. If every point of the domain is a G_δ-set, weak first-countability and symmetrizability are preserved by open and closed maps [**74**]. A perfect image of a first-countable symmetrizable space need not be weakly first-countable. Liu and Lin [**41**] showed that an open (resp. open compact) image of a countable g-second countable space (resp. symmetrizable space) need not be weakly first-countable. A similar counterexample is also obtained by Sakai [**61**].

Problem 11. *Is a space with a σ-compact-finite weak base g-metrizable?* 93?

A space with a σ-compact-finite weak base by closed subsets is g-metrizable. Under CH, a separable space with a σ-compact-finite weak base is g-metrizable [**36**], but the authors don't know whether this is true in ZFC.

A space is *meta-Lindelöf* if every open cover has a point-countable open refinement.

Problem 12. *Let X be a space with a σ-compact-finite weak base (generally, a k-space with a σ-compact finite k-network). Is X meta-Lindelöf, or a σ-space?* 94–95?

The above problem for the parenthetic part was posed in [**43**]. For this problem, if X is a quotient Lindelöf image of a locally ω_1-compact[1] space (in particular, if X is locally ω_1-compact), then X is a paracompact σ-space. If the character $\chi(X)$ of X is less than or equal to ω_1 (e.g., X is locally separable under CH), then X is hereditarily meta-Lindelöf. A k-space with a σ-HCP k-network, in particular, a g-metrizable space is hereditarily meta-Lindelöf [**58**]. Lin [**32**] improved this result by showing that a k-and-k-semistratifiable space is hereditarily meta-Lindelöf.

Problem 13. *Let X be a space with σ-compact-finite weak base. Is X k-semistratifiable, or every point of X a G_δ-set?* 96?

Spaces determined or dominated by certain covers

For a cover $\mathcal{P}$ of a space X, let us call $\mathcal{P}$ a *determining cover* [**81**] if X is determined by $\mathcal{P}$. An open cover is a determining cover. A space with a determining cover by sequential spaces (resp. k-spaces) is a sequential space (resp. k-space). For a closed cover $\mathcal{F}$ of a space X, X is *dominated* by $\mathcal{F}$ [**46**] if $\mathcal{F}$ is a CP cover, and any $\mathcal{P} \subset \mathcal{F}$ is a determining cover of the union of $\mathcal{P}$. Let us call the closed cover $\mathcal{F}$ a *dominating cover* [**81**]. A HCP closed cover, or an increasing determining closed cover $\{F_n : n \in \mathbb{N}\}$ is a dominating cover. A CW-complex has a dominating cover by compact metric spaces. A space with a dominating cover by paracompact spaces (resp. normal spaces) is paracompact (resp. normal) [**46**, **52**], and similar presevations also hold for M_3-spaces [**3**], σ-spaces [**68**], etc. For summaries on determining or dominating covers in terms of their preservations by

[1] A space X is ω_1-*compact* if any uncountable subset has an accumulation point in X.

maps, subsets, or products, see [**81**]. A space with a point-countable determining cover by k-and-$\aleph$-spaces, generally, a quotient Lindelöf image of a k-and-$\aleph$-space has a point-countable k-network [**71**], and a space with a dominating cover by $\aleph$-spaces has a σ-CP k-network [**77**]. A space with a point-finite determining cover by metric spaces, generally, a quotient compact image of a metric space is symmetric.

We shall recall a classical problem of whether closed subsets (or points) in a symmetric space X are G_δ-sets, which was raised in 1966 by E. Michael. (For some questions on symmetric spaces, see [**4**], etc.). Michael's problem is positive (actually, X is hereditarily Lindelöf) if X is ω_1-compact [**55**], and is also positive (points in X are G_δ-sets) if $\chi(X) \leq \omega_1$ [**66**]. The problem below was posed around 1980 by Y. Tanaka. The answer to (a) and (c) of this problem are negative under X being *Hausdorff* [**2**]. In 1978, Davis, Gruenhage and Nyikos [**8**] answered Michael's problem negatively. They gave a symmetric space X which has a point-finite determining closed cover by metric spaces (thus, X is a quotient compact image of a metric space), but X is not countably metacompact, thus X has a closed set which is not a G_δ-set, and it is not submetacompact ($= \theta$-refinable) (also, see [**19**]), and they gave also a *Hausdorff* symmetric space which has a point-finite determining closed cover by metric spaces, but it has a point which is not a G_δ-set. A separable space with a point-finite determining cover by compact metric spaces need not be normal, or meta-Lindelöf [**20**]. Also, every first-countable space with a point-finite determining closed and open cover by metric spaces need not be normal. Let X be a space with a point-countable determining cover by cosmic spaces (e.g., X is a quotient s-image of a locally separable metric space), or a space with a point-countable determining closed cover by σ-spaces. Then, points of X are G_δ-sets if $\chi(X) \leq \omega_1$, and X is a σ-space if X is ω_1-compact, or submetacompact with $\chi(X) \leq \omega_1$.

97–100? **Problem 14.** *Let X be a quotient compact image of a locally compact metric space (equivalently, a space with a point-finite determining cover by compact metric subsets).*

(a) *Is each closed subset of X a G_δ-set?*
(b) *Is each point of X a G_δ-set?*
(c) *Is X a subparacompact space (or, σ-space)?*

Let X be a k-space. When X has a star-countable k-network, X has a dominating cover by $\aleph_0$-spaces [**59**], so X is a paracompact σ-space. When X has a σ-HCP k-network, generally, X is a k-semistratifiable space, X is meta-Lindelöf. When X is a topological group with a point-countable k-network, then X is metrizable if the sequential order of X is countable [**63**]. When X is a topological group with a point-countable k-network (or a point-countable determining cover) by cosmic spaces, X is the topological sum of cosmic spaces [**42**]. A space X has a σ-compact-finite k-network if X has a star-countable or σ-HCP k-network, or X has a dominating cover by spaces with a σ-compact-finite k-network [**43**] (or [**77**]).

Problem 15. *Let G be a topological group which is a k-space with a σ-compact-finite k-network, or a space with a point-countable determining cover by metric spaces. Is G paracompact (or, meta-Lindelöf)?* 101–102?

A space X is an *A-space* [**51**], if, whenever $\{A_n : n \in \mathbb{N}\}$ is a decreasing sequence with $x \in \overline{A_n \setminus \{x\}}$ for all $n \in \mathbb{N}$, there exist $B_n \subset A_n$ such that $\bigcup\{\overline{B_n} : n \in \mathbb{N}\}$ is not closed in X, and X is an *inner-closed A-space* if the B_ns are closed in X. A weakly first-countable space is an A-space, and a countably bi-quasi-k-spaces (of [**48**]) is an inner-closed A-space [**51**]. In terms of the spaces S_ω and S_2, let us review some results stated in [**77**] mainly. Let X be a sequential space. Then X is an A-space iff it contains no (closed) copy of S_ω. When points of X are G_δ-sets, or X has a point-countable k-network, X is Fréchet iff it contains no (closed) copy of S_2. Let X be a k-space with a point-countable k-network. Then X has a point-countable base iff X contains no closed copy of S_ω and no S_2, equivalently, X is an inner-closed A-space. When X is symmetric, X has a point-countable base (equivalently, X is developable [**22**]) iff X is Fréchet. A k-space X is metrizable if X is an M-space with a point-countable k-network, or an inner-closed A-space having a point-countable k-network by separable subsets or a σ-compact-finite k-network. For a space X having a dominating cover by metric spaces or a point-countable determining cover by locally separable metric spaces, X is metrizable if it is an inner-closed A-space. Let G be a topological group (thus, G contains a closed copy of S_ω iff it contains a closed copy of S_2). Then G is metrizable if it is a weakly first-countable space [**56**], or a k-and-A-space (or a Fréchet space) with a point-countable k-network. For a topological group G having a dominating or point-countable determining closed cover $\mathcal{P}$ by bi-sequential spaces (of [**48**]) (e.g., first-countable spaces), G is metrizable if G is a Fréchet space or an A-space. If in the above statement the cover $\mathcal{P}$ is a point-countable determining cover, the points are G_δ-sets, and G is an A-space then G is again metrizable. G is also metrizable if the cover $\mathcal{P}$ is a point-finite determining cover [**57**]. For the following problem, it is positive if $\mathcal{P}$ is point-finite or closed in (a), and so is if all elements of $\mathcal{P}$ are metric, or cosmic in (b).

Problem 16. *Let X be a space with a point-countable determining cover $\mathcal{P}$. If* 103–104?
the following (a) or (b) holds, is X metrizable?

(a) *X is paracompact first-countable, and the elements of $\mathcal{P}$ are metric.*
(b) *X is a topological group which is an A-space, and the elements of $\mathcal{P}$ are first-countable spaces* ([**57**]).

Products of k-spaces having certain k-networks

k-networks are countably productive [**20**]. Weak bases need not be productive, but they are (countably) productive if the product spaces are sequential. Symmetric spaces (resp. sequential spaces) are countably productive if the product spaces are k-spaces [**56**] (resp. [**81**]). A pair (X, Y) of spaces satisfies the *Tanaka condition* [**29**, **31**] if one of the following holds: (a) both X and Y are first-countable, (b) either X or Y is locally compact, and (c) both X and Y are locally k_ω-spaces.

Here, a space X is a *k_ω-space* if it has a countable determining cover by compact subsets (equivalently, X is a quotient image of a locally compact Lindelöf space). Every space with a countable determining closed cover by locally compact subsets is locally k_ω [**76**]. For a pair (X, Y) satisfying the Tanaka condition, $X \times Y$ is a k-space, assuming X is a k-space and Y is locally compact, and vice versa in (b). Tanaka [**69**] proved that for k-and-$\aleph$-spaces X, Y, the product $X \times Y$ is a k-and-$\aleph$-spaces iff (X, Y) satisfies the Tanaka condition. Gruenhage [**18**] proved that a set-theoretic axiom $\mathfrak{b} = \omega_1$ weaker than CH is equivalent to the statement that for Lašnev spaces X, Y, the product $X \times Y$ is a k-space iff (X, Y) satisfies the Tanaka condition. Liu and Tanaka [**44**] improved Gruenhage's result by showing the result is valid for k-spaces with a compact-countable k-network. Lin and Liu [**33**] proved that for sequential spaces X, Y with a point-countable cs-network (e.g., X, Y are k-and-$\aleph$-spaces), $X \times Y$ is a sequential space iff (X, Y) satisfies the Tanaka condition, and they gave a counterexample that under $\mathfrak{b} > \omega_1$, there exist quotient finite-to-one images X, Y of locally compact metric spaces with $X \times Y$ sequential, but (X, Y) does not satisfy the Tanaka condition. Under CH, Shibakov [**62**] also obtained a counterexample for spaces with a point-countable closed k-network. Tanaka [**76**] gave analogous characterizations for $X \times Y$ to be a k-space if X, Y have point-countable k-networks such that σ-compact closed subsets are $\aleph_0$-spaces (e.g., spaces with a compact-countable k-network, or spaces with a point-countable cs-network) in terms of Tanaka condition, and X^2 is a k-space iff X is first-countable or locally k_ω.

105–106? **Problem 17.**

(1) *Let X be a quotient s-image of a metric space, in particular let X be a space with a point-countable determining cover by (compact) metric subsets. If X^2 is a k-space, is X first-countable or locally k_ω?*

(2) *Let X be a k-space with a point-countable k-network. What is a necessary and sufficient condition for X^2 to be a k-space?*

A bi-k-space [**48**] is a generalization of first-countable spaces and locally compact spaces. For sequential spaces X, Y which are closed images of paracompact bi-k-spaces, Gruenhage's result remains valid, but replace "first-countable" by "bi-k" in the Tanaka condition, and X^2 is a k-space iff X is bi-k or locally k_ω [**78**]. (Some questions on products of k-spaces were posed in [**78**]). For a sequential space X and a bi-k-space Y, $X \times Y$ is a k-space iff X is a Tanaka space [2] (in [**54**]), or Y is locally countably compact ([**77**] cf. [**53**]). For products of weak topologies, see [**80, 81**].

For a space X with a point-countable determining cover by cosmic spaces, $\chi(X) \leq 2^c$, $c = 2^\omega$. But, for each $\alpha \geq \omega$, there is a symmetric space X with a point-finite determining closed cover by metric spaces such that $\chi(X) > \alpha$, and $\chi(X) > c$ when we replace "metric spaces" by "compact metric spaces". Let X

[2] A space X is a *Tanaka space* if for a decreasing sequence $\{A_n : n \in \mathbb{N}\}$ with $x \in \overline{A_n \setminus \{x\}}$ for all $n \in \mathbb{N}$, there exist $x_n \in A_n$ such that the sequence $\{x_n : n \in \mathbb{N}\}$ converges to some point in X.

be a space with a point-countable k-network. When X^2 is a k-space, X is first-countable or locally σ-compact (in view of [**76**]), thus $\chi(X) \leq c$. When X^ω is a k-space, X is first-countable [**30**, **44**]. Now, let X be a symmetric space having property $(*)$: any separable closed subset is a space whose points are G_δ-sets. A symmetric space has $(*)$ under CH, or if it is meta-Lindelöf or collectionwise Hausdorff. When X^2 is a k-space, $\chi(X) \leq 2^c$. When X^ω is a k-space, X is first-countable. The authors don't know whether a symmetric space Y which contains no (closed) copy of the space S_2 is first-countable (here, Y is first-countable when Y has $(*)$). If this is positive, then the above results hold without $(*)$.

Problem 18. 107–108?

(1) *Let X^2 be a symmetric space. Is $\chi(X) \leq c$ (or $\chi(X) \leq 2^c$)?*
(2) *Let X^ω be a symmetric space. Is X first-countable?*

A space X has *countable tightness* if whenever $x \in \overline{A}$, there is a countable subset $C \subset A$ with $x \in \overline{C}$. A space has countable tightness iff it has a determining cover by countable subsets [**48**]. A sequential space or a hereditarily separable space has countable tightness. For spaces X, Y having countable tightness, if $X \times Y$ is a k-space, then $X \times Y$ has countable tightness, and the converse holds when X, Y have a dominating cover by locally compact spaces. While, for a closed map $f\colon X \to Y$ with X strongly collectionwise Hausdorff, let Y^2 have countable tightness, then each boundary $\partial f^{-1}(y)$ is c-compact (ω_1-compact if Y is sequential) [**21**]. Every product of spaces with a countable determining cover by locally separable metric subsets has countable tightness. But a product of a space with a point-finite determining cover (or a dominating cover) by ω_1 many compact metric subsets may not have countable tightness. Liu and Lin [**40**] proved that the axiom $\mathfrak{b} = \omega_1$ is equivalent to the assertion that for k-spaces X, Y with a point-countable k-network by cosmic subsets (e.g., X, Y have a point-countable determining cover by locally separable metric subsets), $X \times Y$ has countable tightness iff one of X, Y is first-countable, or both X, Y are locally cosmic, and X^2 has countable tightness iff X is locally cosmic. They also showed that the axiom is equivalent to the assertion that for spaces X, Y with a dominating cover by metric subsets, if $X \times Y$ has a countable tightness, then one of X, Y is first-countable, or both X, Y have a countable dominating cover by metric subsets (equivalently, X, Y are $\aleph$-spaces). The converse of this assertion holds if the answer to (1) in the following problem is positive.

Problem 19. 109–110?

(1) *Let X be a space with a countable determining closed cover by metric subsets. Does X^2 have countable tightness?*
(2) *Let X be a k-space with a point-countable k-network, in particular, let X be a (Fréchet) space with a point-countable (countable) determining closed cover by metric subsets. What is a necessary and sufficient condition for X^2 to have countable tightness?*

References

[1] A. V. Arhangel'skiĭ, *Mappings and spaces*, Uspekhi Mat. Nauk **21** (1966), no. 4(130), 133–184, Translation: Russian Math. Surveys 21 (1966) no. 4, 115–162.

[2] D. A. Bonnett, *A symmetrizable space that is not perfect*, Proc. Amer. Math. Soc. **34** (1972), 560–564.

[3] C. R. Borges, *On stratifiable spaces*, Pacific J. Math. **17** (1966), 1–16.

[4] D. K. Burke and S. W. Davis, *Questions on symmetrizable spaces*, Questions Answers Gen. Topology **1** (1983), no. 2, 113–118.

[5] H. Chen, *Weak neighborhoods and Michael–Nagami's question*, Houston J. Math. **25** (1999), no. 2, 297–309.

[6] H. Chen, *Compact-covering maps and k-networks*, Proc. Amer. Math. Soc. **131** (2003), no. 8, 2623–2632.

[7] G. D. Creede, *Concerning semi-stratifiable spaces*, Pacific J. Math. **32** (1970), 47–54.

[8] S. W. Davis, G. Gruenhage, and P. J. Nyikos, *G_δ sets in symmetrizable and related spaces*, General Topology and Appl. **9** (1978), no. 3, 253–261.

[9] J. Dugundji, *Topology*, Allyn and Bacon Inc., Boston, 1966.

[10] R. Engelking, *General topology*, Heldermann Verlag, Berlin, 1989.

[11] L. Foged, *On g-metrizability*, Pacific J. Math. **98** (1982), no. 2, 327–332.

[12] L. Foged, *A characterization of closed images of metric spaces*, Proc. Amer. Math. Soc. **95** (1985), no. 3, 487–490.

[13] L. Foged, *Normality in k-and-ℵ spaces*, Topology Appl. **22** (1986), no. 3, 223–240.

[14] G. Gao, *On k-networks and bases*, J. Suzhou University **2** (1986), 107–111.

[15] Z. M. Gao, *A result on k-semistratifiable spaces*, Questions Answers Gen. Topology **3** (1986), no. 2, 137–143.

[16] Z. M. Gao, *The closed images of metric spaces and Fréchet ℵ-spaces*, Questions Answers Gen. Topology **5** (1987), no. 2, 281–291.

[17] Z. M. Gao and Y. Hattori, *A characterization of closed s-images of metric spaces*, Questions Answers Gen. Topology **4** (1987), no. 2, 147–151.

[18] G. Gruenhage, *k-spaces and products of closed images of metric spaces*, Proc. Amer. Math. Soc. **80** (1980), no. 3, 478–482.

[19] G. Gruenhage, *Generalized metric spaces*, Handbook of set-theoretic topology, North-Holland, Amsterdam, 1984, pp. 423–501.

[20] G. Gruenhage, E. Michael, and Y. Tanaka, *Spaces determined by point-countable covers*, Pacific J. Math. **113** (1984), no. 2, 303–332.

[21] G. Gruenhage and Y. Tanaka, *Products of k-spaces and spaces of countable tightness*, Trans. Amer. Math. Soc. **273** (1982), no. 1, 299–308.

[22] R. W. Heath, *On spaces with point-countable bases*, Bull. Acad. Polon. Sci. Sér. Sci. Math. Astronom. Phys. **13** (1965), 393–395.

[23] T. Hoshina, *On the quotient s-images of metric spaces*, Sci. Rep. Tokyo Kyoiku Daigaku Sect. A **10** (1970), 265–268 (1970).

[24] N. N. Jakovlev, *On the theory of o-metrizable spaces*, Dokl. Akad. Nauk SSSR **229** (1976), no. 6, 1330–1331, Translation: Soviet Math. Dokl. 17 (1976) 1217–1219.

[25] H. Junnila and Z. Q. Yun, *ℵ-spaces and spaces with σ-hereditarily closure-preserving k-network*, Topology Appl. **44** (1992), 209–215.

[26] N. Lašnev, *Closed images of metric spaces*, Dokl. Akad. Nauk SSSR **170** (1966), 505–507, Translation: Soviet Math. Dokl. 7 (1966) 1219–1221.

[27] S. Lin, *Mapping theorems on ℵ-spaces*, Topology Appl. **30** (1988), no. 2, 159–164.

[28] S. Lin, *On g-metrizable spaces*, Chinese Ann. Math. Ser. A **13** (1992), no. 3, 403–409.

[29] S. Lin, *Generalized metric spaces and mappings*, Chinese Science Press, Beijing, 1995.

[30] S. Lin, *A note on the Arens' space and sequential fan*, Topology Appl. **81** (1997), no. 3, 185–196.

[31] S. Lin, *Point-countable coverings and sequence-covering maps*, Chinese Science Press, Beijing, 2002.

[32] S. Lin, *Covering properties of k-semistratifiable spaces*, Topology Proc. **29** (2005), no. 1, 199–206.

[33] S. Lin and C. Liu, *On spaces with point-countable cs-networks*, Topology Appl. **74** (1996), 51–60.

[34] S. Lin and Y. Tanaka, *Point-countable k-networks, closed maps, and related results*, Topology Appl. **59** (1994), no. 1, 79–86.

[35] C. Liu, *Spaces with a σ-compact finite k-network*, Questions Answers Gen. Topology **10** (1992), no. 1, 81–87.

[36] C. Liu, *Notes on g-metrizable spaces*, Topology Proc. **29** (2005), no. 1, 207–215.

[37] C. Liu, *On weak bases*, Topology Appl. **150** (2005), 91–99.

[38] C. Liu, *Notes on closed maps*, Houston J. Math., To appear.

[39] C. Liu and M. M. Dai, *g-metrizability and* S_ω, Topology Appl. **60** (1994), no. 2, 185–189.

[40] C. Liu and S. Lin, *On the countable tightness of product spaces*, Acta Math. Sin. (Engl. Ser.) **21** (2005), no. 4, 929–936.

[41] C. Liu and S. Lin, *On countable-to-one maps*, Topology Appl. (2006), Article in press.

[42] C. Liu, M. Sakai, and Y. Tanaka, *Topological groups with a certain point-countable cover*, Topology Appl. **119** (2002), no. 2, 209–217.

[43] C. Liu and Y. Tanaka, *Spaces having σ-compact-finite k-networks, and related matters*, Topology Proc. **21** (1996), 173–200.

[44] C. Liu and Y. Tanaka, *Star-countable k-networks, compact-countable k-networks, and related results*, Houston J. Math. **24** (1998), no. 4, 655–670.

[45] D. Lutzer, *Semimetrizable and stratifiable spaces*, General Topology and Appl. **1** (1971), no. 1, 43–48.

[46] E. Michael, *Continuous selections. I*, Ann. of Math. (2) **63** (1956), 361–382.

[47] E. Michael, *$\aleph_0$-spaces*, J. Math. Mech. **15** (1966), 983–1002.

[48] E. Michael, *A quintuple quotient quest*, General Topology and Appl. **2** (1972), 91–138.

[49] E. Michael, *$\aleph_0'$-spaces and a function space theorem of R. Pol*, Indiana Univ. Math. J. **26** (1977), no. 2, 299–306.

[50] E. Michael and K. Nagami, *Compact-covering images of metric spaces*, Proc. Amer. Math. Soc. **37** (1973), 260–266.

[51] E. Michael, R. C. Olson, and F. Siwiec, *A-spaces and countably bi-quotient maps*, Dissertationes Math. **133** (1976), 43 pp.

[52] K. Morita, *On spaces having the weak topology with respect to closed coverings. II*, Proc. Japan Acad. **30** (1954), 711–717.

[53] F. Mynard, *Strongly sequential spaces*, Comment. Math. Univ. Carolin. **41** (2000), no. 1, 143–153.

[54] F. Mynard, *More on strongly sequential spaces*, Comment. Math. Univ. Carolin. **43** (2002), no. 3, 525–530.

[55] S. Nedev, *Symmetrizable spaces and final compactness*, Dokl. Akad. Nauk SSSR **175** (1967), 532–534, Translation: Soviet Math. Dokl. 8 (1967) 890–892.

[56] S. Ĭ. Nedev, *o-metrizable spaces*, Trudy Moskov. Mat. Obshch. **24** (1971), 201–236, Translation: Trans. Moscow Math. Soc. 24 (1971), 213–247.

[57] T. Nogura, D. Shakhmatov, and Y. Tanaka, *Metrizability of topological groups having weak topologies with respect to good covers*, Topology Appl. **54** (1993), 203–212.

[58] L.-X. Peng, *The hereditarily Metalindelöf property of k-spaces with σ-HCP closed k-networks*, Topology Proc. **26** (2001/02), no. 1, 307–310.

[59] M. Sakai, *On spaces with a star-countable k-network*, Houston J. Math. **23** (1997), no. 1, 45–56.

[60] M. Sakai, *A special subset of the real line and regularity of weak topologies*, Topology Proc. **23** (1998), Spring, 281–287.

[61] M. Sakai, *Counterexamples on generalized metric spaces*, Sci. Math. Jpn. **64** (2006), no. 1, 73–76.

[62] A. Shibakov, *Sequentiality of products of spaces with point-countable k-networks*, Topology Proc. **20** (1995), 251–270.

[63] A. Shibakov, *Metrizability of sequential topological groups with point-countable k-networks*, Proc. Amer. Math. Soc. **126** (1998), no. 3, 943–947.

[64] R. Sirois-Dumais, *Quasi- and weakly-quasi-first-countable spaces*, Topology Appl. **11** (1980), no. 2, 223–230.

[65] F. Siwiec, *On defining a space by a weak base*, Pacific J. Math. **52** (1974), 233–245.

[66] R. M. Stephenson, Jr., *Symmetrizable, F-, and weakly first countable spaces*, Canad. J. Math. **29** (1977), no. 3, 480–488.

[67] S. A. Svetlichnyĭ, *Intersection of topologies and metrizability in topological groups*, Vestnik Moskov. Univ. Ser. I Mat. Mekh. (1989), no. 4, 79–81, Translation: Moscow Univ. Math. Bull. 44 (1989) no. 4, 78–80.

[68] Y. Tanaka, *On local properties of topological spaces*, Sci. Rep. Tokyo Kyoiku Daigaku Sect. A **11** (1972), 106–116.

[69] Y. Tanaka, *A characterization for the products of k- and $\aleph_0$-spaces and related results*, Proc. Amer. Math. Soc. **59** (1976), no. 1, 149–155.

[70] Y. Tanaka, *Closed maps on metric spaces*, Topology Appl. **11** (1980), no. 1, 87–92.

[71] Y. Tanaka, *Point-countable covers and k-networks*, Topology Proc. **12** (1987), no. 2, 327–349.

[72] Y. Tanaka, *Metrization II*, Topics in general topology, North-Holland, Amsterdam, 1989, pp. 275–314.

[73] Y. Tanaka, *σ-hereditarily closure-preserving k-networks and g-metrizability*, Proc. Amer. Math. Soc. **112** (1991), no. 1, 283–290.

[74] Y. Tanaka, *Closed maps and symmetric spaces*, Questions Answers Gen. Topology **11** (1993), no. 2, 215–233.

[75] Y. Tanaka, *Theory of k-networks*, Questions Answers Gen. Topology **12** (1994), no. 2, 139–164.

[76] Y. Tanaka, *Products of k-spaces having point-countable k-networks*, Topology Proc. **22** (1997), Spring, 305–329.

[77] Y. Tanaka, *Theory of k-networks. II*, Questions Answers Gen. Topology **19** (2001), no. 1, 27–46.

[78] Y. Tanaka, *Products of k-spaces, and questions*, Comment. Math. Univ. Carolin. **44** (2003), no. 2, 335–345.

[79] Y. Tanaka, *Quotient spaces and decompositions*, Encyclopedia of general topology (K. P. Hart, J. Nagata, and J. E. Vaughan, eds.), Elsevier Science, 2004, pp. 43–46.

[80] Y. Tanaka, *Products of weak topologies*, Topology Proc. **29** (2005), no. 1, 361–376.

[81] Y. Tanaka, *Determining covers and dominating covers*, Questions Answers Gen. Topology **24** (2006), no. 2.

Open Problems in Topology II – Edited by E. Pearl

Extension problems of real-valued continuous functions

Haruto Ohta and Kaori Yamazaki

1. Introduction

By a space we mean a completely regular T_1-space. A subset A of a space X is said to be C-*embedded* in X if every real-valued continuous function on A extends continuously over X, and A is said to be C^*-*embedded* in X if every bounded real-valued continuous function on A extends continuously over X. The aim of this paper is to collect some open questions concerning C-, C^*-embeddings and extension properties which can be described by extensions of real-valued continuous functions.

Let $\mathbb{N}$, $\mathbb{Q}$, $\mathbb{R}$ and $\mathbb{I}$ denote the sets of natural numbers, rationals, reals, and the closed unit interval, respectively, with the usual topologies. Let ω be the first infinite cardinal. For undefined terms on generalized metric spaces, see [**9**]. General terminology and notation will be used as in [**6**].

2. C-embedding versus C^*-embedding

This section overlaps partly with the survey [**28**]; here, we update information about status of the questions and add some new questions. It is not difficult to construct an example of a closed set which is C^*-embedded but not C-embedded (for example, see [**26**, Construction 2.3]). It is, however, interesting to ask if C^*-embedding implies C-embedding under certain circumstances.

Question 1. *Is every C^*-embedded subset of a first countable space C-embedded?* 111?

Note that every C^*-embedded subset of a first countable space is closed. Kulesza–Levy–Nyikos [**18**] proved that if $\mathfrak{b} = \mathfrak{s} = \mathfrak{c}$, then there exists a maximal almost disjoint family $\mathcal{R}$ of infinite subsets of $\mathbb{N}$ such that every countable set of nonisolated points of the space $\mathbb{N} \cup \mathcal{R}$ is C^*-embedded. Since every set of nonisolated points of $\mathbb{N} \cup \mathcal{R}$ is discrete and $\mathbb{N} \cup \mathcal{R}$ is pseudocompact, those countable sets are not C-embedded. Thus, Question 1 has a negative answer under $\mathfrak{b} = \mathfrak{s} = \mathfrak{c}$, but it remains open whether there exists a counterexample in ZFC. Kulesza–Levy–Nyikos [**18**] also proved that, assuming the Product Measure Extension Axiom (PMEA), there exists no infinite discrete C^*-embedded subset of a pseudocompact space of character less than $\mathfrak{c}$. Since every C^*- but not C-embedded subset contains an infinite discrete C^*-embedded subset, this implies that no pseudocompact space, in particular, no space $\mathbb{N} \cup \mathcal{R}$, can be a counterexample to Question 1 under PMEA (see [**17**] for related results). On the other hand, by Tietze's extension

---

Research of the second author is partially supported by the Ministry of Education, Science, Sports and Culture, Grant-in-Aid for Young Scientists (B), No. 16740028.

theorem, no normal space can also be a counterexample to Question 1. Thus, it is natural to ask if typical examples of first countable, non-normal spaces contain C^*-embedded subsets which are not C-embedded. In [**28**] the first author proved that every C^*-embedded subset of the Niemytzki plane NP is C-embedded and asked the following questions, which have been unanswered as yet.

112? **Question 2.** *Is every C^*-embedded subset of the square S^2 of the Sorgenfrey line C-embedded in S^2?*

113? **Question 3.** *Is every C^*-embedded subset of the product $\mathbb{R}_{\mathbb{Q}} \times \mathbb{N}^\omega$ of the Michael line with the space of irrationals C-embedded in $\mathbb{R}_{\mathbb{Q}} \times \mathbb{N}^\omega$?*

It was proved in [**28**] that if a space X contains a pair of disjoint closed sets, one of which is countable discrete, which cannot be separated by disjoint open sets, then the absolute $E(X)$ of X contains a closed C^*-embedded subset which is not C-embedded. Hence, the absolutes $E(\mathrm{NP})$, $E(S^2)$ and $E(\mathbb{R}_{\mathbb{Q}} \times \mathbb{N}^\omega)$ of the Niemytzki plane, the Sorgenfrey plane and Michael's product space, respectively, contain C^*- but not C-embedded closed sets.

Another interesting case of the relationship between C^*- and C-embeddings is a closed rectangle in a product. Indeed, the next question asked in [**28**] and the one after next are still open.

114? **Question 4.** *Let A be a C-embedded closed subset of a space X and Y a space such that $A \times Y$ is C^*-embedded in $X \times Y$. Then, is $A \times Y$ C-embedded in $X \times Y$?*

115? **Question 5.** *Let A (resp. B) be a C-embedded closed subset of a space X (resp. Y) such that $A \times B$ is C^*-embedded in $X \times Y$. Then, is $A \times B$ C-embedded in $X \times Y$?*

It is known that the answer to Question 4 (resp. 5) is positive in each of the following cases (1)–(4) (resp. (5) and (6)):

(1) Y is the product of a σ-locally compact (i.e., the union of countably many locally compact closed subspaces), paracompact space with a metric space ([**28**, Corollary 4.10]). In particular, Y is either σ-locally compact, paracompact ([**43**, Theorem 1.1]) or a metric space ([**11**, Theorem 1.1]).
(2) $Y \approx Y^2$ and Y contains an infinite compact set ([**16**, Theorem 2.1]).
(3) X is a normal P-space and Y is a paracompact Σ-space ([**43**, Theorem 1.2]).
(4) X is a normal weak $P(\omega)$-space and Y is K-analytic ([**48**, Theorem 4.2]).
(5) Y is locally compact and paracompact (combine [**21**, Theorem 4] with [**43**, Theorem 1.1]).
(6) Y is a metric space (combine [**11**, Theorem 1.1] with [**38**, Theorem 4]).

Question 4 is a special case of Question 5, and the authors do not know if the answer to Question 5 is positive in each of the cases (1)–(4) above. As for another candidate for positive cases, the following question naturally arises; the case where Y is a paracompact M-space was asked by Gutev and the first author in [**11**, Problem 1].

Question 6. *Is an answer to Question 4 positive if either the space X or Y is assumed to be a paracompact M-space (or equivalently, a paracompact p-space)?* 116?

Recall from [**13**] that a subset A of a space X is said to be U^ω-*embedded* in X if for every real-valued continuous function g on A there exists a real-valued continuous function f on X such that $g(x) \le f(x)$ for each $x \in A$ (see [**11**, Lemma 2.5]). It is known that A is C-embedded in X if and only if A is C^*- and U^ω-embedded in X, i.e., $C = C^* + U^\omega$. Gutev and the first author [**11**] proved that if A is a C-embedded subset of a space X and Y is a metric space, then $A \times Y$ is C^*-embedded in $X \times Y$ if and only if $A \times Y$ is U^ω-embedded in $X \times Y$. Neither the 'if' part nor the 'only if' part of this result is known to be true for any generalized metric spaces Y. In particular, the following question is open (see [**11**, Problems 1 and 2] for related questions).

Question 7. *Let A be a C-embedded subset of a space X and Y a stratifiable space.* 117–118?

(i) *Is $A \times Y$ C^*-embedded in $X \times Y$ provided that $A \times Y$ is U^ω-embedded in $X \times Y$?*
(ii) *Is $A \times Y$ U^ω-embedded in $X \times Y$ provided that $A \times Y$ is C^*-embedded in $X \times Y$?*

Next, we turn to questions on infinite products. For an infinite cardinal γ, let us consider the following condition $p(\gamma)$.

$p(\gamma)$: For every collection of pairs (X_α, A_α), $\alpha < \gamma$, of a space X_α and its closed subset A_α with $|A_\alpha| > 1$, if the product $A = \prod_{\alpha<\gamma} A_\alpha$ is C^*-embedded in the product $X = \prod_{\alpha<\gamma} X_\alpha$, then A is C-embedded in X.

Question 8. *Is $p(\omega)$ true?* 119?

Note that:

(i) A positive answer to Question 5 answers Question 8 positively, and
(ii) if $p(\omega)$ is true, then $p(\gamma)$ is true for all infinite cardinals γ.

Before showing these, let us agree on a notation. For spaces Y_α, $\alpha < \gamma$, and $M \subseteq \gamma$, we write $Y^{(M)} = \prod_{\alpha \in M} Y_\alpha$. The latter fact (ii) is obvious, since every uncountable product $\prod_{\alpha<\gamma} Y_\alpha$ can be considered as the countable product $\prod_{n<\omega} Y^{(I_n)}$ by dividing γ into countably many nonempty sets I_n, $n < \omega$. To show the former fact (i), let X_i, $i < \omega$, be spaces and $A_i \subseteq X_i$ with $|A_i| > 1$ for each $i < \omega$, and assume that the product $A = \prod_{i<\omega} A_i$ is C^*-embedded in $X = \prod_{i<\omega} X_i$. Divide ω into two infinite sets I and J, and take an infinite compact set $K \subseteq A^{(J)}$. Then, $A^{(I)} \times K$ is C^*-embedded in $X^{(I)} \times K$, because $A^{(I)} \times K$ is C^*-embedded in $A = A^{(I)} \times A^{(J)}$ by [**38**, Theorem 3] and A is C^*-embedded in X. Hence, it follows from [**24**, Lemma 2.8] that $A^{(I)}$ is C-embedded in $X^{(I)}$. Similarly, $A^{(J)}$ is C-embedded in $X^{(J)}$. Consequently, if Question 5 has a positive answer, then $A = A^{(I)} \times A^{(J)}$ is C-embedded in $X = X^{(I)} \times X^{(J)}$, and hence, we have $p(\omega)$.

The authors do not know if the converse of (i) is also true; however, they conjecture that the answer to Question 5 is negative but that to Question 8 is positive.

As was remarked in [**11**], we often find an interesting relationship between C^*- and U^ω-embedding which is parallel to that between normality and countable paracompactness. For example, the result stated before Question 7 above is parallel to the theorem of Morita–Rudin–Starbird (see [**35**]) asserting that, for a normal countably paracompact space X and a metric space Y, the normality of $X \times Y$ is equivalent to the countable paracompactness of $X \times Y$. Thus, the questions we discuss in this section can be considered as relativizations of questions on the relationship between normality and countable paracompactness. In this aspect, the following questions are inspired by theorems of Nagami [**25**, Corollary 1.6] and Zenor [**50**, Theorem A].

120? **Question 9.** *For each $i < \omega$, let X_i be a space and A_i a closed subset of X_i with $|A_i| > 1$. Assume that the product $A = \prod_{i<\omega} A_i$ is U^ω-embedded in the product $X = \prod_{i<\omega} X_i$ and $\prod_{i\le n} A_i$ is C-embedded in $\prod_{i\le n} X_i$ for each $n < \omega$. Is, then, A C-embedded in X?*

For an infinite cardinal γ, a subset A is said to be P^γ-*embedded* in X if for every normal cover $\mathcal{U}$ of A with $|\mathcal{U}| \le \gamma$, there exists a normal cover $\mathcal{V}$ of X such that $\{V \cap A : V \in \mathcal{V}\}$ refines $\mathcal{U}$. It is known that A is P^ω-embedded in X if and only if A is C-embedded in X, i.e., $P^\omega = C$. For more information about P^γ-embedding, see [**1**, **14**].

121? **Question 10.** *For each $i < \omega$, let X_i be a space and A_i a closed subset of X_i with $|A_i| > 1$. Let γ be an infinite cardinal. Assume that the product $A = \prod_{i<\omega} A_i$ is C^*-embedded in the product $X = \prod_{i<\omega} X_i$ and $\prod_{i\le n} A_i$ is P^γ-embedded in $\prod_{i\le n} X_i$ for each $n < \omega$. Is, then, A P^γ-embedded in X?*

A positive answer to Question 8 answers Question 10 positively. To show this, we need a definition from [**22**, **27**]. A collection $\mathcal{U}$ of subsets of X is *uniformly locally finite* in X if there exists a normal cover $\mathcal{V}$ of X such that each member of $\mathcal{V}$ intersects at most finitely many members of $\mathcal{U}$. Now, it suffices to show that for every collection of pairs (X_i, A_i), $i < \omega$, of a space X_i and its closed subset A_i with $|A_i| > 1$, if $A = \prod_{i<\omega} A_i$ is C-embedded in $X = \prod_{i<\omega} X_i$ and $A^{(n+1)} = \prod_{i\le n} A_i$ is P^γ-embedded in $X^{(n+1)} = \prod_{i\le n} X_i$ for each $n < \omega$, then A is P^γ-embedded in X. Let $\mathcal{U}$ be a normal cover of A with $|\mathcal{U}| \le \gamma$. We may assume that $\mathcal{U}$ is a uniformly locally finite cozero-set cover of A, since every normal cover has a uniformly locally finite cozero-set refinement (see [**22**, Proposition 4.2]). For each $U \in \mathcal{U}$, there exists a cozero-set $G(U)$ in X such that $G(U) \cap A = U$. Pick up $a_i \in A_i$ for each $i < \omega$. For every $n < \omega$ and every $U \in \mathcal{U}$, define $V_n(U) = \{(x_0, x_1, \ldots, x_n) \in A^{(n+1)} : (x_0, x_1, \ldots, x_n, a_{n+1}, a_{n+2}, \ldots) \in U\}$. Since $\{V_n(U) : U \in \mathcal{U}\}$ is a uniformly locally finite cozero-set cover of $A^{(n+1)}$ and $A^{(n+1)}$ is P^γ-embedded in $X^{(n+1)}$, it follows from [**46**, Theorem 1.3] that there exists a locally finite cozero-set cover $\{H_n(U) : U \in \mathcal{U}\}$ of $X^{(n+1)}$ such that $H_n(U) \cap A^{(n+1)} = V_n(U)$ for each $U \in \mathcal{U}$.

Set $\mathcal{H} = \{p_n^{-1}(H_n(U)) \cap G(U) : U \in \mathcal{U}, n < \omega\}$, where each $p_n : X \to X^{(n+1)}$ is the projection. Then, $\bigcup \mathcal{H}$ is a cozero-set in X containing A, because $\mathcal{H}$ is σ-locally finite. Since A is C-embedded in X, by [**8**, Theorem 1.18] there exists a cozero-set G in X such that $A \cap G = \emptyset$ and $\bigcup \mathcal{H} \cup G = X$. Finally, putting $\mathcal{H}_1 = \mathcal{H} \cup \{G\}$, we obtain a σ-locally finite cozero-set cover (and hence, a normal cover) $\mathcal{H}_1$ of X such that $\{H \cap A : H \in \mathcal{H}_1\}$ refines $\mathcal{U}$. Hence, A is P^γ-embedded in X.

3. π-embedding and $\pi_{\mathcal{Z}}$-embeddings

In this section, let A denote a subset of a space X and γ an infinite cardinal. We are concerned with extension properties which can be described by extensions of real-valued continuous functions via product spaces. More precisely, for a class $\mathcal{Z}$ of spaces, we say that A is *$\pi_{\mathcal{Z}}$-embedded* in X if $A \times Y$ is C^*-embedded in $X \times Y$ for every $Y \in \mathcal{Z}$ (see [**32**]). When $\mathcal{Z}$ is the class of all spaces, $\pi_{\mathcal{Z}}$-embedded subsets are called *π-embedded* omitting $\mathcal{Z}$. Particularly interesting cases are when $\mathcal{Z}$ is one of the following classes:

$$\begin{aligned} \mathcal{C}_\gamma &= \text{the class of all compact spaces } Y \text{ with } w(Y) \le \gamma, \\ \mathcal{M} &= \text{the class of all metric spaces}, \\ \mathcal{P} &= \text{the class of all paracompact } M\text{-spaces}. \end{aligned}$$

Morita–Hoshina [**24**] and Przymusiński [**30**] proved the following theorem which shows that $P^\gamma = \pi_{\mathcal{C}_\gamma} = \pi_{\{\mathbb{I}^\gamma\}}$ and $C = \pi_{\mathcal{C}_\omega} = \pi_{\{\mathbb{I}^\omega\}}$.

Theorem 1 (Morita–Hoshina and Przymusiński). *Let A be a subset of a space X and γ an infinite cardinal. Then, the following are equivalent:*

(1) *A is P^γ-embedded in X,*
(2) *$A \times Y$ is C^*-embedded in $X \times Y$ for every $Y \in \mathcal{C}_\gamma$,*
(3) *$A \times \mathbb{I}^\gamma$ is C^*-embedded in $X \times \mathbb{I}^\gamma$.*

A question that naturally arises after Theorem 1 is what extension property can be expressed as $\pi_{\mathcal{Z}}$-embedding for some class $\mathcal{Z}$. This is the motivation of the next two questions. Following [**5**], we write $Y \in AE(X, A)$ if any map $f : A \to Y$ extends continuously over X, and by AR we mean an absolute retract for the class of metrizable spaces. Morita [**20**] and Przymusiński [**30**] proved that A is P^γ-embedded in X if and only if $Y \in AE(X, A)$ for all complete ARs Y with $w(Y) \le \gamma$. By contrast, Sennott [**36**] defined A to be *M^γ-embedded* in X if $Y \in AE(X, A)$ for all ARs Y with $w(Y) \le \gamma$. It is known ([**37**, **41**]) that every $\pi_{\mathcal{M}}$- and P^γ-embedded subset is M^γ-embedded, but the converse is not true as we state below.

Question 11. *For every infinite cardinal γ, does there exist a class $\mathcal{Z}_\gamma$ of spaces* 122?
such that for every pair (X, A) of a space X and its closed subspace A, A is M^γ-embedded in X if and only if A is $\pi_{\mathcal{Z}_\gamma}$-embedded in X?

Dydak [**4**] defined A to be *P^γ(locally finite)-embedded* (resp. *P^γ(point-finite)-embedded*) in X if every locally finite (resp. point-finite) partition of unity α on A with $|\alpha| \le \gamma$ extends to a locally finite (resp. point-finite) partition of unity on X.

The second author [**47**] showed that for every γ, there exists a class $\mathcal{Z}_\gamma$ of spaces such that P^γ(locally finite) $= \pi_{\mathcal{Z}_\gamma}$, where the case of $\gamma = \omega$ was essentially due to Przymusiński [**33**]. Thus, every π-embedded subset is P^γ(locally finite)-embedded for every γ. The authors proved in [**29**] that A is P^γ(point-finite)-embedded in X if and only if $Y \in AE(X, A)$ for all σ-complete ARs Y with $w(Y) \leq \gamma$. Hence, P^γ(point-finite)-embedding is located between M^γ- and P^γ-embeddings. Now, it will be natural to ask the following question.

123? **Question 12.** *For every infinite cardinal γ, does there exist a class $\mathcal{Z}_\gamma$ of spaces such that for every pair (X, A) of a space X and its closed subspace A, A is P^γ(point-finite)-embedded in X if and only if A is $\pi_{\mathcal{Z}_\gamma}$-embedded in X?*

Sennott [**37**] showed that Theorem 1 above remains true if the class $\mathcal{C}_\gamma$ is replaced by the class $\mathcal{L}_\gamma$ of locally compact, paracompact spaces Y with $w(Y) \leq \gamma$. Since P^γ-embedding is defined in terms of normal covers of A and X only, Theorem 1 can be considered as an internal characterization of $\pi_{\mathcal{C}_\gamma}$-embedded subsets and $\pi_{\mathcal{L}_\gamma}$-embedded subsets. Gutev and the first author [**11**] gave an internal characterization of $\pi_{\mathcal{M}}$-embedded subsets, and the second author [**48**] gave an internal characterization of $\pi_{\mathcal{CM}}$-embedded subsets, where $\mathcal{CM}$ is the class of complete metric spaces. However, no internal characterization of $\pi_{\mathcal{P}}$-embedded subsets is known, in particular, the following question asked by Waśko [**41**, Problem 1.10] is unanswered.

124? **Question 13** (Waśko). *Characterize those spaces whose every closed set is $\pi_{\mathcal{P}}$-embedded.*

A subset A is said to be *P-embedded* in X if A is P^γ-embedded in X for every γ. M-, P(locally finite)- and P(point-finite)-embeddings are defined similarly. It is known that closed subsets of spaces in enough wide classes are P(locally finite)- or $\pi_{\mathcal{P}}$-embedded. For example, every closed set in a collectionwise normal, countably paracompact space is P(locally finite)-embedded (see [**34**, Theorem 2]), and every closed set in a paracompact, perfectly normal space (or more generally, a paracompact P-space in the sense of Morita [**19**]) X is $\pi_{\mathcal{P}}$-embedded since $X \times Y$ is normal for every paracompact M-space Y (see [**19**, Theorem 6.5]). By contrast, little seems to be known about π-embedding. Starbird [**38**] proved that every compact subset of a space is π-embedded, and Michael proved that every closed set in a metric space is π-embedded (see [**38**, Theorem 4]), while a closed set in the product of a compact space with a metric space need not be π-embedded (see the example (5) below). The positive results by Starbird and Michael have been generalized to two directions. Morita [**21**] generalized Starbird's result by proving that every locally compact, paracompact, P-embedded subset A of a space X is π-embedded in X. He also proved that the set $\mathbb{Q}$ is not π-embedded in the Michael line $\mathbb{R}_{\mathbb{Q}}$, which shows that local compactness of A cannot be weakened to σ-local compactness in his result. The following questions, however, remain open.

125? **Question 14** (Hoshina [**15**]). *If X is the image of a locally compact, paracompact space under a closed continuous map, then is every closed set of X π-embedded?*

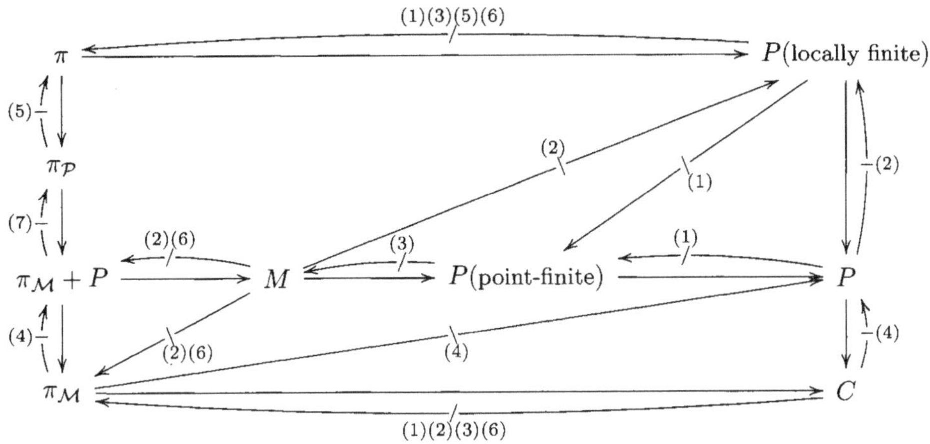

FIGURE 1. Extension properties between π- and C-embeddings.

Question 15. *Is every closed set of a σ-locally compact, paracompact space π-embedded?* 126?

In [**3**, Example 1.1] van Douwen constructed an example of a locally compact, collectionwise normal, submetrizable space Λ such that the product $\Lambda \times \mathbb{N}^\omega$ is not normal. It is worth noting that the space Λ contains a closed set which is not M-embedded (and hence, not π-embedded), because if all closed subsets of Λ were M-embedded, then Λ is perfectly normal by [**36**, Corollary 5], and consequently, $\Lambda \times \mathbb{N}^\omega$ must be normal. As another direction, Fujii [**7**] proved that every closed set of a stratifiable space is π-embedded, and Stares [**39**] generalized this to closed sets of spaces called decreasing (G) spaces. The notion of a decreasing (G) space goes back to [**2**] and defined explicitly in [**40**].

Question 16 (Fujii). *Is every closed set of a paracompact σ-space π-embedded?* 127?

Fujii [**7**] also studied the problem what generalized ordered space has the property that every closed set is π-embedded, and proved that a locally compact, generalized ordered space is the case. Gruenhage, Hattori and the first author [**10**] determined π-embedded subsets of generalized ordered spaces, and showed that a separable (and hence, perfectly normal), Čech-complete, paracompact, generalized ordered space can have a non π-embedded subset. See also [**12**, **23**, **31**] for related topics.

The diagram in Figure 1 illustrates the relationship among the extension properties discussed in this section, where $A \to B$ means that every A-embedded subset is B-embedded, and $A \not\leftarrow_{(i)} B$ means that the example (i) listed below shows that not all B-embedded subsets are A-embedded. For the proof of each implication, see [**4**, **15**, **41**].

(1) The set $\mathbb{Q}$ of rationals in the Michael line $\mathbb{R}_\mathbb{Q}$ is P(locally finite)-embedded in $\mathbb{R}_\mathbb{Q}$ because $\mathbb{R}_\mathbb{Q}$ is paracompact, while $\mathbb{Q}$ is neither $\pi_\mathcal{M}$-embedded

([**21**]) nor P^ω(point-finite)-embedded ([**49**, Corollary 3.3]). An example of a paracompact space having a non-P^ω(point-finite)-embedded closed subset was first given by Dydak [**4**, Example 12.14].

(2) Przymusiński and Wage [**34**, Example 3] constructed an example of a collectionwise normal space Z with a closed set K which is not P^ω(locally finite)-embedded. The second author proved in [**47**, Corollary 2.5] that every $\pi_{\mathcal{M}_\omega}$-embedded set, where $\mathcal{M}_\omega$ is the class of all separable metric spaces, is P^ω(locally finite)-embedded (see Przymusiński [**33**] for a closed set of a normal space). Hence, the set K is not $\pi_{\mathcal{M}_\omega}$-embedded in Z. The authors [**29**] proved that all closed subsets of the space Z are M-embedded.

(3) The Bernstein set A in the space $\mathbb{R}_A$ is P(point-finite)-embedded but not M^ω-embedded ([**29**, Example 3.4]). Similarly to the Michael line, A is P(locally finite)-embedded in $\mathbb{R}_A$.

(4) Every closed subset of a perfectly normal space X is $\pi_{\mathcal{M}}$-embedded, since $X \times M$ is normal for every metric space M. Hence, a perfectly normal, non-collectionwise normal space contains a closed set which is $\pi_{\mathcal{M}}$-embedded but not P-embedded (for example, see [**6**, Problem 5.5.3]).

(5) Waśko [**41**, Example 1.7] proved that the product space $\beta\mathbb{N}\times\mathbb{N}^\omega$ contains a closed set X which is not π-embedded. Since the product of $\beta\mathbb{N}\times\mathbb{N}^\omega$ with a paracompact M-spaces is paracompact, the set X is $\pi_{\mathcal{P}}$-embedded. For a simpler example, see [**10**, Example 1].

(6) Waśko [**41**, Example 2.5] constructed an example of a Lindelöf space having a zero-set Y which is not $\pi_{\mathcal{M}}$-embedded. The set Y is M-embedded, since P-embedded zero-sets are M-embedded (see [**36**, Corollary 1]).

(7) Waśko [**42**] constructed an example of a $\pi_{\mathcal{M}}$- and P-embedded subset which is not $\pi_{\mathcal{P}}$-embedded.

Concerning the diagram in Figure 1 there remain the following two questions, which were first asked by the second author [**45**, Problems 3.6.5 and 3.6.6].

128? **Question 17.** *Is every $\pi_{\mathcal{M}}$- and P-embedded subset P(locally finite)-embedded?*

129? **Question 18.** *Is every $\pi_{\mathcal{P}}$-embedded subset P(locally finite)-embedded?*

4. Miscellaneous questions

In this section, we give two questions which particularly interest the authors. The first one was asked by the second author [**44, 45**] and was asked again in [**29**].

130? **Question 19.** *For every uncountable cardinal γ, if A is P^γ- and P^ω(locally finite)-embedded in X, then is A P^γ(locally finite)-embedded in X?*

Przymusiński–Wage [**34**] called a normal space X *(functionally) Katětov* if for every closed subset A of X every locally finite open (cozero-set) cover of A can be extended to a locally finite open cover of X, and called X *countably (functionally) Katětov* if it satisfies the same condition with coverings assumed to be countable. They proved that a space X is functionally Katětov if and only if every closed set in X is P(locally finite)-embedded in X, and X is countably functionally Katětov if and only if every closed set in X is P^ω(locally finite)-embedded in X (see [**44**, **45**, **47**] for more about this topic in general context). Thus, a positive answer to Question 19 implies that every collectionwise normal, countably functionally Katětov space is functionally Katětov. A question whether every collectionwise normal, countably Katětov space is Katětov is also open (see [**34**, Question 3]). The authors [**29**] proved that a similar question to Question 19 for P^γ(point-finite)-embedding has a positive answer.

Question 20 (Dydak [**4**]). *For every infinite cardinal γ, if A is P^γ(point-finite)-* 131?
embedded in X, then is $A \times \mathbb{I}$ P^γ(point-finite)-embedded in $X \times \mathbb{I}$?

The second author [**44**] proved that the answer to a similar question to Question 20 for P^γ(locally finite)-embedding is positive. The authors [**29**] proved that Question 20 is equivalent to the following question: For every infinite γ, does every P^γ-embedded subset A of a space X satisfying condition (b) satisfy (a) below?

(a) For every subset B of X, which is the inverse image of an analytic set in a metric space M under a continuous map $f\colon X \to M$, with $B \cap A = \emptyset$, there exists a cozero-set U in X such that $B \subseteq U$ and $U \cap A = \emptyset$.

(b) For every subset B of X, which is the intersection of countably many cozero-sets in X, with $B \cap A = \emptyset$, there exists a cozero-set U in X such that $B \subseteq U$ and $U \cap A = \emptyset$.

They showed in [**29**] that assuming the continuum hypothesis (CH) there exists a P-embedded subset satisfying (b) but not (a). Thus, the answer to Question 20 is negative under CH, but it is still open whether there exists an example in ZFC.

References

[1] R. A. Alò and H. L. Shapiro, *Normal topological spaces*, Cambridge University Press, New York, 1974.

[2] P. J. Collins and A. W. Roscoe, *Criteria for metrisability*, Proc. Amer. Math. Soc. **90** (1984), no. 4, 631–640.

[3] E. K. van Douwen, *A technique for constructing honest locally compact submetrizable examples*, Topology Appl. **47** (1992), no. 3, 179–201.

[4] J. Dydak, *Extension theory: the interface between set-theoretic and algebraic topology*, Topology Appl. **74** (1996), 225–258.

[5] J. Dydak, *Cohomological dimension theory*, Handbook of geometric topology (R. J. Daverman and R. B. Sher, eds.), North-Holland, Amsterdam, 2002, pp. 423–470.

[6] R. Engelking, *General topology*, Heldermann Verlag, Berlin, 1989.

[7] S. Fujii, *π-embedding and Dugundji extension theorem*, Master's thesis, University of Tsukuba, 1984.

[8] L. Gillman and M. Jerison, *Rings of continuous functions*, D. Van Nostrand Co., Princeton, NJ, 1960.

[9] G. Gruenhage, *Generalized metric spaces*, Handbook of set-theoretic topology, North-Holland, Amsterdam, 1984, pp. 423–501.

[10] G. Gruenhage, Y. Hattori, and H. Ohta, *Dugundji extenders and retracts on generalized ordered spaces*, Fund. Math. **158** (1998), no. 2, 147–164.

[11] V. Gutev and H. Ohta, *Does C^*-embedding imply C-embedding in the realm of products with a non-discrete metric factor?*, Fund. Math. **163** (2000), no. 3, 241–265.

[12] Y. Hattori, *π-embeddings and Dugundji extension theorems for generalized ordered spaces*, Topology Appl. **84** (1998), 43–54.

[13] T. Hoshina, *Spaces with a property related to uniformly local finiteness*, Tsukuba J. Math. **6** (1982), no. 1, 51–62.

[14] T. Hoshina, *Extensions of mappings. II*, Topics in general topology, North-Holland, Amsterdam, 1989, pp. 41–80.

[15] T. Hoshina, *Extensions of mappings*, Recent progress in general topology (Prague, 1991), North-Holland, Amsterdam, 1992, pp. 405–416.

[16] T. Hoshina and K. Yamazaki, *Extensions of functions on product spaces*, Topology Appl. **82** (1998), 195–204.

[17] J. Kulesza and R. Levy, *Separation in Ψ-spaces*, Topology Appl. **42** (1991), no. 2, 101–107.

[18] J. Kulesza, R. Levy, and P. J. Nyikos, *Extending discrete-valued functions*, Trans. Amer. Math. Soc. **324** (1991), no. 1, 293–302.

[19] K. Morita, *Products of normal spaces with metric spaces*, Math. Ann. **154** (1964), 365–382.

[20] K. Morita, *On generalizations of Borsuk's homotopy extension theorem*, Fund. Math. **88** (1975), no. 1, 1–6.

[21] K. Morita, *On the dimension of the product of topological spaces*, Tsukuba J. Math. **1** (1977), 1–6.

[22] K. Morita, *Dimension of general topological spaces*, Surveys in general topology (G. M. Reed, ed.), Academic Press, New York, 1980, pp. 297–336.

[23] K. Morita, *Absolutely P- and M-embeddable spaces*, Questions Answers Gen. Topology **1** (1983), no. 1, 3–10.

[24] K. Morita and T. Hoshina, *P-embedding and product spaces*, Fund. Math. **93** (1976), no. 1, 71–80.

[25] K. Nagami, *Countable paracompactness of inverse limits and products*, Fund. Math. **73** (1972), no. 3, 261–270.

[26] N. Noble, *Countably compact and pseudo-compact products*, Czechoslovak Math. J. **19(94)** (1969), 390–397.

[27] H. Ohta, *Topologically complete spaces and perfect maps*, Tsukuba J. Math. **1** (1977), 77–89.

[28] H. Ohta, *Extension properties and the Niemytzki plane*, Appl. Gen. Topol. **1** (2000), no. 1, 45–60.

[29] H. Ohta and K. Yamazaki, *Extension of point-finite partitions of unity*, Fund. Math. **191** (2006), no. 3, 187–199.

[30] T. C. Przymusiński, *Collectionwise normality and extensions of continuous functions*, Fund. Math. **98** (1978), no. 1, 75–81.

[31] T. C. Przymusiński, *Product spaces*, Surveys in general topology, Academic Press, New York, 1980, pp. 399–429.

[32] T. C. Przymusiński, *Extending functions from products with a metric factor and absolutes*, Pacific J. Math. **101** (1982), no. 2, 463–475.

[33] T. C. Przymusiński, *A solution to a problem of E. Michael*, Pacific J. Math. **114** (1984), no. 1, 235–242.

[34] T. C. Przymusiński and M. L. Wage, *Collectionwise normality and extensions of locally finite coverings*, Fund. Math. **109** (1980), no. 3, 175–187.

[35] M. E. Rudin and M. Starbird, *Products with a metric factor*, General Topology and Appl. **5** (1975), no. 3, 235–248.

[36] L. I. Sennott, *On extending continuous functions into a metrizable AE*, General Topology and Appl. **8** (1978), no. 3, 219–228.

[37] L. I. Sennott, *Some remarks on M-embedding*, Topology Proc. **3** (1978), no. 2, 507–520 (1979).

[38] M. Starbird, *Extending maps from products*, Studies in topology (Charlotte, NC, 1974), Academic Press, New York, 1975, pp. 559–564.

[39] I. S. Stares, *Extension of functions and generalized metric spaces*, D.Phil. thesis, Oxford University, 1994.

[40] I. S. Stares, *Concerning the Dugundji extension property*, Topology Appl. **63** (1995), no. 2, 165–172.

[41] A. Waśko, *Extensions of functions defined on product spaces*, Fund. Math. **124** (1984), no. 1, 27–39.

[42] A. Waśko, *Extensions of functions from products with compact or metric factors*, Fund. Math. **125** (1985), no. 1, 81–88.

[43] K. Yamazaki, *C^*-embedding and C-embedding on product spaces*, Tsukuba J. Math. **21** (1997), no. 2, 515–527.

[44] K. Yamazaki, *Extensions of partitions of unity*, Topology Proc. **23** (1998), Spring, 289–313.

[45] K. Yamazaki, *Extensions of mappings on product spaces*, Ph.D. thesis, University of Tsukuba, 2000.

[46] K. Yamazaki, *Controlling extensions of functions and C-embedding*, Topology Proc. **26** (2001/02), no. 1, 323–341.

[47] K. Yamazaki, *P(locally-finite)-embedding and rectangular normality of product spaces*, Topology Appl. **122** (2002), no. 1–2, 453–466.

[48] K. Yamazaki, *Products of weak P-spaces and K-analytic spaces*, 2002, Preprint.

[49] K. Yamazaki, *Extending point-finite partitions of unity*, Houston J. Math. **29** (2003), no. 2, 353–359.

[50] P. Zenor, *Countable paracompactness in product spaces*, Proc. Amer. Math. Soc. **30** (1971), 199–201.

Open Problems in Topology II – Edited by E. Pearl

$L\Sigma(\kappa)$-spaces

Oleg Okunev

All spaces are assumed to be Tychonoff (= completely regular Hausdorff). We use terminology and notation as in [**2**].

A *multivalued mapping* from a set X to a set Y is a mapping that assigns to each point of X a subset of Y (not necessarily non-empty). For a multivalued mapping $p\colon X \to Y$ and a set A in X the *image of A under p* is $p(A) = \bigcup\{p(x) : x \in A\}$. A multivalued mapping $p\colon X \to Y$ is *compact-valued* if all images of points are compact, and is *upper semicontinuous* if for every open set V in Y, the set $\{x \in X : p(x) \subset V\}$ is open. It is easy to verify that a composition of compact-valued upper semicontinuous mappings is compact-valued upper semicontinuous.

It is not difficult to prove also that compact-valued upper semicontinuous mappings are exactly compositions of continuous mappings, inverses of closed embeddings and inverses of perfect mappings (see, e.g., Proposition 1.1 in [**3**]; thus the phrase "Y is an image of X under a compact-valued upper semicontinuous mapping" may be viewed as an abbreviation for "Y is a continuous image of a closed subset of a perfect preimage of X", or "Y is a continuous image of a closed subset of the product of X with a compact space" (see [**3**] for details). In particular, Lindelöf Σ-spaces are exactly images under compact-valued upper semicontinuous mappings of second-countable spaces [**1**].

The main theorem in [**5**] says that t-equivalent spaces (that is, spaces whose spaces of continuous functions with the topology of pointwise convergence are homeomorphic) are related via finite-valued upper semicontinuous mappings. Hence we arrive at the idea of looking at the classes of spaces that may be obtained from second-countable spaces by applying compact-valued upper semicontinuous mappings with additional restrictions on the images of points. Anyway, the classes of all spaces that may be obtained as images of second-countable spaces under finite-valued or metrizable-compact-valued upper semicontinuous mappings appear quite natural. Thus, the following definition [**3**]:

Let κ be a cardinal (finite or infinite). A space X is an *$L\Sigma(\leq\kappa)$-space* if there is a second-countable space M and a compact-valued upper semicontinuous mapping $p\colon M \to X$ such that $p(M) = X$ and $w(p(x)) \leq \kappa$ for every $x \in X$ ($w(\cdot)$ denotes weight). A space X is an *$L\Sigma(\kappa)$-space* if it is an $L\Sigma(\leq\kappa)$-space and is not an $L\Sigma(\leq\lambda)$-space for any $\lambda < \kappa$.

A space X is an *$L\Sigma(<\kappa)$-space* if there is a second-countable space M and a compact-valued upper semicontinuous mapping $p\colon M \to X$ such that $p(M) = X$ and $w(p(x)) < \kappa$ for every $x \in X$.

Of course, for finite κ the definition says really that the images of points under p have at most κ points. The class $L\Sigma(<\omega)$ is the class of all images of second-countable spaces under finite-valued upper semicontinuous mappings. The definitions also admit natural reformulations in terms of networks modulo compact covers in the spirit of the seminal article of K. Nagami [**4**]; see [**3**].

Obviously $L\Sigma(\leq 1)$-spaces are exactly the spaces with a countable network. Surprisingly, the class of $L\Sigma(2)$-spaces is already rich enough to include the Double Arrow space, the one-point compactification $A(\mathfrak{c})$ of the discrete space of cardinality $\mathfrak{c}$, and all one-point compactifiations of all Ψ-like spaces [**3**]. It is easy to see that the classes $L\Sigma(\leq\kappa)$ are invariant with respect to closed subspaces, continuous images, countable unions, and, for infinite κ, countable products. For $\kappa \geq \mathfrak{c}$, the class $L\Sigma(\leq\kappa)$ coincides with the class of all Lindelöf Σ-spaces of cardinality $\leq \kappa$.

Some classes similar to $L\Sigma(\leq\omega)$ were studied by M. Tkačenko [**6**], and by V. Tkachuk, who proved, in particular, that all Eberlein compacta of cardinality $\leq \mathfrak{c}$ are $L\Sigma(\leq\omega)$-spaces [**7**].

A few natural questions about the classes $L\Sigma(\omega)$ and $L\Sigma(n)$ are resolved in [**3**]; however, many more remain open and appear quite interesting.

It is shown in [**3**] that for every $n \in \omega$, the space $A(\omega_1)^n$ is an $L\Sigma(n{+}1)$-space, but $A(\omega_2)^2$ is not an $L\Sigma(3)$-space (thus, it is an $L\Sigma(4)$-space if $\omega_2 \leq \mathfrak{c}$). However, for many individual spaces it remains unclear what $L\Sigma$-classes they belong to.

132? **Problem 1.** *Let X be the Double Arrow space. Is it true that $X \times X$ is an $L\Sigma(4)$-space?*

133? **Problem 2.** *Let X be a linearly ordered $L\Sigma(n)$-space. Must $X \times X$ be an $L\Sigma(n^2)$-space?*

134? **Problem 3.** *Let $\mathcal{A}$ be an almost disjoint family of subsets of ω of cardinality ω_1, and let X be the one-point compactification of the Ψ-like space corresponding to $\mathcal{A}$. Can $X \times X$ be a $L\Sigma(3)$-space? Can $X \times X$ be a $L\Sigma(4)$-space?*

It is easy to see that the free topological group of an $L\Sigma(n)$-space, $n \geq 2$, is an $L\Sigma(<\omega)$-space without a countable network.

135? **Problem 4.** *Let G be a topological $L\Sigma(\leq n)$-group for some $n \in \omega$. Must G have a countable network?*

136? **Problem 5.** *Suppose $C_p(X)$ is an $L\Sigma(\leq n)$-space for some $n \in \omega$. Must X have a countable network?*

As mentioned above, V. Tkachuk proved in [**7**] that every Eberlein compact space of cardinality $\leq \mathfrak{c}$ is an $L\Sigma(\leq\omega)$-space; on the other hand, not all Corson compacta are $L\Sigma(\leq\omega)$ [**3**].

137? **Problem 6.** *Are all Rosenthal compact spaces $L\Sigma(\leq\omega)$?*

138? **Problem 7.** *Assume* $\mathsf{MA}(\omega_1)$. *Is it true that every scattered compact space of cardinality ω_1 and height n, $n \in \omega$, belongs to $L\Sigma(\leq n{+}1)$?*

A positive answer to Problem 7 for $n = 3$ is proved in [**3**].

All examples of $L\Sigma(n)$-spaces that the author knows are finite unions of subspaces that admit continuous bijections onto second-countable spaces.

139? **Problem 8.** *Let $n \in \omega$. Is it true that every $L\Sigma(\leq n)$-space is a union of $\leq n$ subspaces that admit continuous bijections onto second-countable spaces?*

Problem 9. *Let $n \in \omega$. Is it true that every $L\Sigma(\leq n)$-space is a union of finitely many subspaces that admit continuous bijections onto second-countable spaces?* 140?

Problem 10. *Is it true that every $L\Sigma(<\omega)$-space is a union of countably many subspaces that admit continuous bijections onto second-countable spaces?* 141?

It is proved in [**3**] that $L\Sigma(\leq\omega)$-spaces have no uncountable free sequences; in particular, every compact $L\Sigma(\leq\omega)$-space has countable tightness. On the other hand, [**3**] contains an example that shows that σ-compact $L\Sigma(<\omega)$-spaces may have arbitrary tightness.

Problem 11. *Let X be an $L\Sigma(n)$-space for some $n \in \omega$. Can X have uncountable tightness?* 142?

It is immediate from the main theorem in [**5**] that the class $L\Sigma(<\omega)$ is preserved by the t-equivalence relation; the reason is that this class is invariant with respect to images under finite-valued upper semicontinuous mappings.

Problem 12. *Let X and Y be t-equivalent spaces and suppose X is an $L\Sigma(\leq\omega)$-space. Must Y be a an $L\Sigma(\leq\omega)$-space?* 143?

Problem 13. *Let X be an $L\Sigma(\leq\omega)$-space and let $p\colon X \to Y$ be a finite-valued upper semicontinuous mapping such that $p(X) = Y$. Must Y be an $L\Sigma(\leq\omega)$-space?* 144?

Problem 14. *Let X be an $L\Sigma(\leq\omega)$-space and let $p\colon X \to Y$ be an upper semicontinuous mapping such that $p(X) = Y$ and $p(x)$ is compact metrizable for every $x \in X$. Must Y be an $L\Sigma(\leq\omega)$-space?* 145?

An interesting particular case of Problem 13 was pointed out by C. Paniagua Ramírez:

Problem 15. *Let X be an $L\Sigma(\leq\omega)$-space. Must the Alexandroff double of X be an $L\Sigma(\leq\omega)$-space?* 146?

A negative answer to Problem 15 would yield negative answers to Problems 12–14.

Problem 16. *Is the class of compact spaces in $L\Sigma(<\omega)$ absolute?* 147?

It is proved in [**3**] that the closely related class of $KL\Sigma(<\omega)$-spaces is absolute; a space X is called a *$KL\Sigma(<\omega)$-space* if it is the image under a finite-valued upper semicontinuous mapping of a metrizable *compact* space. The classes of compact $L\Sigma(<\omega)$-spaces and $KL\Sigma(<\omega)$-spaces are different: every $KL\Sigma(\leq\omega)$-space is Fréchet. This is essentially proved in [**7**], while the one-point compactification of a Mrówka space is a compact space in $L\Sigma(2)$ without the Fréchet property.

M. Tkačenko asked in [**6**] (in a different terminology) whether every compact $L\Sigma(\leq\omega)$-space has a dense metrizable subspace. It is shown in [**3**] that every $KL\Sigma(<\omega)$-space has a dense metrizable subspace, and, on the other hand, that the negation of $\mathsf{MA}(\omega_1)$ implies the existence of a $KL\Sigma(\leq\omega)$-space without a dense metrizable subspace. Thus, the following is essentially what remains of the original question of M. Tkačenko:

148? **Problem 17.** *Does there exist in* ZFC *a compact $L\Sigma(\leq\omega)$-space without a dense metrizable subspace?*

149? **Problem 18.** *Does there exist in* ZFC *a $KL\Sigma(\leq\omega)$-space without a dense metrizable subspace?*

References

[1] A. V. Arhangel′skiĭ, *The structure and classification of topological spaces and cardinal invariants*, Uspekhi Mat. Nauk **33** (1978), no. 6(204), 29–84, Translation: Russian Math. Surveys 33 (1978) 33–95.

[2] R. Engelking, *General topology*, Heldermann Verlag, Berlin, 1989.

[3] W. Kubiś, O. Okunev, and P. Szeptycki, *On some classes of Lindelöf Σ-spaces*, Topology Appl. **153** (2006), no. 14, 2574–2590.

[4] K. Nagami, *Σ-spaces*, Fund. Math. **65** (1969), 169–192.

[5] O. Okunev, *Tightness of compact spaces is preserved by the t-equivalence relation*, Comment. Math. Univ. Carolin. **43** (2002), no. 2, 335–342.

[6] M. G. Tkachenko, *P-approximable compact spaces*, Comment. Math. Univ. Carolin. **32** (1991), no. 3, 583–595.

[7] V. V. Tkachuk, *A glance at compact spaces which map "nicely" onto the metrizable ones*, Topology Proc. **19** (1994), 321–334.

Open Problems in Topology II – Edited by E. Pearl

Problems on (ir)resolvability

Oleg Pavlov

Introduction. Resolvability hierarchy

A topological space is called *τ-resolvable* (*resolvable* if $\tau = 2$) if it contains τ disjoint dense subsets. Clearly, X cannot be τ-resolvable if $\tau > \Delta(X)$, where the *dispersion character* $\Delta(X)$ is the minimal cardinality of a nonempty open subset of X. If X is $\Delta(X)$-resolvable, then it is called *maximally resolvable*. Although all known *natural* examples of topological spaces are maximally resolvable, quite a few methods are known for constructing spaces that are *irresolvable*, that is, not resolvable (or have even stronger irresolvability properties as defined below). The main problem is to endow these spaces with interesting additional properties. The converse line of research is to describe classes of spaces that only contain resolvable (ω-resolvable, maximally resolvable) spaces, or spaces resolvable in some other sense.

Given a topological property $\mathcal{P}$, a space (X, τ) is called *maximal* $\mathcal{P}$ if (X, τ) has $\mathcal{P}$ but (X, τ') does not have it for any stronger topology τ'. A space that is maximal with respect to being dense in itself is simply called *maximal*. Maximality implies irresolvability, and there are a few properties in between. For example, X is an *MI-space* if X is dense in itself and every dense subset of X is open. X is an *SI-space* (or *strongly irresolvable*, or *hereditarily irresolvable*) if X is dense in itself and does not contain resolvable subsets. X is *open-hereditarily irresolvable* if every nonempty open subset is irresolvable[1].

We focus on resolvability in three classes of spaces: connected, with Baire property, and homogeneous, as well as on structure resolvability. All considered spaces contain more than one point!

Resolvability of connected spaces

A typical method of constructing an irresolvable (MI-, maximal) space X involves refining the topology of another space Y. In fact, it is often possible to refine the topology of Y in such a way that X is an MI-space and for a natural condensation $f\colon X \to Y$ and every regular open set $U \subseteq X$, the image $f(U)$ is open in Y. In particular, if Y is connected, then so is X. This method is not useful for producing a regular X since f, as defined above, is a homeomorphism

[1] It is required by the definition that an MI-space is dense in itself. It turns out that X is an MI-space if and only if X is dense in itself and every subset of X is an intersection of an open and a closed set. Spaces that have the latter property (not necessarily dense in themselves) are called *submaximal*, see Bourbaki. There is a tendency of late to replace the term "MI-space" with "submaximal" (that is, to require that a submaximal space is dense in itself). We will use more traditional term "MI-space", however, as it is both well-established and goes back to Hewitt. Also, some authors call an open-hereditarily irresolvable space an SI-space.

if X is regular. Indeed, the following problem is one of the most central in the resolvability theory:

150? **Question 1.** *Is there an infinite regular connected irresolvable space?*

Comfort and García-Ferreira posed a similar question for Tychonoff spaces, in [**10**, Question 8.13]. Yaschenko noted that every infinite Tychonoff locally connected space is $\mathfrak{c}$-resolvable; Costantini proved in [**14**] that every infinite regular locally connected space is ω-resolvable. For relevant Hausdorff examples, see [**2**, **19**, **30**, **40**, **52**].

We established in [**41**] that if a regular space X is of countable extent and $\Delta(X) \geq \omega_2$, then X is ω-resolvable. In particular, an infinite connected Lindelöf space is ω-resolvable if the negation of CH is assumed.

151? **Question 2.** *Is there an infinite connected Lindelöf irresolvable space?*

152? **Question 3.** *Is there an infinite regular connected MI-space?*

The latter problem is due to Arhangel′skiĭ and Collins [**5**]. They showed that for small spaces the answer is negative.

El′kin proved in [**19**] that for every infinite cardinal κ there exists a maximal connected Hausdorff space of dispersion character κ. It was shown in [**24**, **52**] that there is a connected topology on the real line that is maximal connected. However, the following old problems (see the above mentioned papers) remain open:

153? **Question 4.** *Is there an infinite regular connected space that is maximal connected?*

154? **Question 5.** *Is there a regular connected topology on the real line that is finer than the usual topology and also maximal connected?*

155? **Question 6.** *Is there an infinite maximal regular connected space?*

156? **Question 7.** *Is there a maximal regular connected topology on the real line that is finer than the usual topology?*

Baire property and a problem of Katětov

Katětov asked in [**29**] whether there exists a Hausdorff dense in itself space X such that every real-valued continuous function on X is continuous at some point. Malykhin noted in [**35**] that this problem has the affirmative solution if and only if there exists a Hausdorff Baire irresolvable space. Kunen, Szymanski and Tall established in [**32**, **33**] that the existence of such a Baire space (and also of a zero-dimensional open-hereditarily irresolvable Baire space) is equiconsistent with the existence of a measurable cardinal. Since every irresolvable space contains a nonempty open SI-subspace, this means that the existence of either a Hausdorff or zero-dimensional SI Baire space is equiconsistent with the existence of a measurable cardinal. It is easy to see that the topology of any Hausdorff open-hereditarily irresolvable Baire space can be refined to the topology of a Hausdorff Baire space that is maximal. On the other hand, every regular nodec (let alone submaximal) Baire space is scattered according to [**5**, Theorem 7.7].

While Hausdorff countably compact irresolvable (not Baire) spaces exist in ZFC, see [**37**, **41**], the existence of regular irresolvable feebly compact (every locally finite family of open sets is finite) or Tychonoff irresolvable pseudocompact spaces have not been established even consistently.

Question 8. *Is there an infinite regular feebly compact irresolvable space?* 157?

Question 9. *Is there an infinite Tychonoff pseudocompact irresolvable space?* 158?

The latter question was posed by Comfort and García-Ferreira in [**11**].

Question 10. *Is there an infinite regular connected feebly compact irresolvable* 159?
space?

Question 11. *Is there an infinite Tychonoff connected pseudocompact irresolvable* 160?
space?

Question 12 (Gruenhage). *Is there a regular open-hereditarily resolvable Baire* 161?
topology on the set ω_1*?*

Bolstein proved in [**8**] (also see [**22**, **53**]) that there exists a real-valued function (that can even be chosen to have a countable range) on X that is discontinuous at every point if and only if X is *almost resolvable*, that is, if and only if X is a countable union of subsets with empty interiors. Malykhin attempted in [**35**] to prove that every dense in itself space is almost resolvable (and hence, to answer Katětov's question in the negative) if a Kurepa family exists on every regular cardinal; see discussion in [**5**]. We mimicked Malykhin's method by using Ulam matrices and showed that every dense in itself ccc space is a union of countably many subsets with empty interior under CH, and every ccc space of cardinality ω_1—in ZFC.

Question 13. *Is there a combinatorial principle that implies that every dense in* 162?
itself space is almost resolvable?

Groups and homogeneous spaces

There are many interesting results and questions about resolvability in the class of all topological groups. It is known that if a dense in itself countable topological group or abelian topological group is not resolvable, then it contains a countable open Boolean subgroup, and, further, there is a P-point in ω^*. Hence, consistently, every dense in itself countable topological group and abelian topological group is ω-resolvable. Since there is a countable topological group that is a maximal space if $\mathfrak{p} = \mathfrak{c}$ (a brilliant example of Malykhin [**36**], also see [**39**]), the existence of either a topological group that is a maximal space or an irresolvable topological group is consistent and independent from the axioms of ZFC.

The next question is due to Malykhin.

163? **Question 14.** *Is there an irresolvable (SI-, MI-) topological group of uncountable dispersion character?*

Such a group must be nonabelian.

164? **Question 15.** *Is there an irresolvable topological group in* ZFC*?*

Protasov [**44**] constructed a countable regular homogeneous maximal space, and also Hausdorff homogeneous maximal spaces of arbitrary dispersion character.

165? **Question 16** (Protasov [**44**]). *Is there a regular left-group of an uncountable dispersion character that is a maximal space?*

A space is maximal if and only if it is both an MI-space and extremally disconnected. Every MI-space (in particular, every maximal space) is nodec (every nowhere dense subset is discrete), see [**25**]. The following two old problems are still open (see [**3–5**, **46**]):

166? **Question 17.** *Is there a dense in itself extremally disconnected topological group in* ZFC*?*

Note that, consistently, every countable discrete subset of an extremally disconnected topological group is closed [**59**].

167? **Question 18.** *Is there a dense in itself nodec topological group in* ZFC*?*

Under MA, every countable abelian group admits a nondiscrete nodec topology according to [**55**] and [**47**, Chapter 2].

El′kin proved in [**17**] that a space X is maximal if and only if for every $x \in X$, the set of all pierced open neighborhoods $\{U \setminus \{x\} : U \in \tau(x)\}$ of x forms a base of an ultrafilter on X. Hence, if a maximal space has a uncountable pseudocharacter at some point, then this pseudocharacter is an Ulam-measurable cardinal, and the cardinality of X is Ulam-measurable as well.

168–169? **Question 19.** *Is there a maximal topological space of a uncountable pseudocharacter? A homogeneous one?*

Comfort and van Mill [**13**] called a topological group G *absolutely resolvable* if it contains two disjoint subsets that are dense in every group topology on G; G is *strongly resolvable* if it is resolvable in every nondiscrete group topology. Clearly, every absolutely resolvable topological group is strongly resolvable. The authors of [**13**] proved that an abelian topological group is strongly resolvable whenever it does not contain a countable Boolean subgroup (see [**58**] for a generalization) and posed the following problem:

170? **Question 20.** *Characterize algebraically the absolutely resolvable groups.*

Compactness, products, miscellaneous questions

Many compactness-type properties imply maximal resolvability of a space. For example, every Hausdorff k-space is maximally resolvable [**48**]. The following question is due to Malykhin.

Question 21. *Is every Lindelöf space of an uncountable dispersion character resolvable?* 171?

The answer is positive if the dispersion character is at least ω_2 according to [**41**]. Also, it was recently established in [**28**] that every hereditarily Lindelöf space of uncountable dispersion character is maximally resolvable.

It is not hard to construct a Hausdorff countably compact irresolvable space, see [**37**, **41**]. On the other hand, every regular countably compact space is ω_1-resolvable, see [**49**]. The following question is due to Comfort:

Question 22. *Is every Tychonoff countably compact space 2^ω-resolvable? Maximally resolvable?* 172–173?

This question is open in the class of all regular spaces as well.

Bešlagić and Levy [**6**] established that the existence of dense in themselves (Tychonoff, zero-dimensional) infinite spaces X and Y such that $X \times Y$ is irresolvable is equiconsistent with the existence of a measurable cardinal. On the other hand, Malykhin [**34**] observed that $X \times Y$ is resolvable whenever X and Y are dense in themselves isodyne[2] spaces of equal cardinality. He constructed a countable T_1 space X such that $X \times X$ is not 4-resolvable[3].

Question 23. *Is there a Hausdorff (regular) isodyne dense in itself space X such that $X \times X$ is not ω-resolvable? Is there such a countable space?* 174–175?

Question 24. *Is there a T_1 (Hausdorff, regular) isodyne dense in itself space X such that $X \times X$ is not 3-resolvable? Is there such a countable space?* 176–177?

There are no metamathematical reasons to expect that X is κ^+-resolvable if X is κ-resolvable. There are no metamathematical reasons not to expect that X is λ-resolvable if λ is a limit cardinal and X is κ-resolvable for every $\kappa < \lambda$. Indeed, the statement "X is λ-resolvable if λ is a limit cardinal and X is κ-resolvable for every $\kappa < \lambda$" has been proven for all limit λ that are regular or of countable cofinality, see [**7**, **26**, **27**].

Question 25 ([**27**]). *Is X λ-resolvable if λ is a singular cardinal of uncountable cofinality and X is κ-resolvable for every $\kappa < \lambda$? Is it true for $\lambda = \aleph_{\omega_1}$?* 178–179?

Structure resolvability

One of the ways to generalize resolvability concepts is to allow disjoint dense sets to have nonempty, albeit small, intersections[4]. This leads to completely new

---

[2]A space X is called *isodyne* if $\Delta(X) = |X|$. It is easy to see that every dense in itself countable T_1 space is isodyne.

[3]Malykhin used a characterization of k-resolvable, not $k+1$-resolvable spaces in terms of ultrafilters. There are similar useful characterizations, also due El'kin [**17**, **18**], for irresolvable and SI-spaces. Namely, a dense in itself space X is irresolvable iff its topology contains a base for an ultrafilter on X; X is an SI-space iff every free ultrafilter of open sets forms a base for an ultrafilter on X. Schröder [**50**] removed the requirement of being dense in itself from some of these characterizations.

[4]In most considered cases, the intersections must be elements of certain ideals of small sets. See [**15**] for an introductory study of resolvability modulo an ideal.

effects, such as a space X admitting greater than $\Delta(X)$ many subsets that are disjoint modulo small sets. To this end, Malykhin [**38**] called a space X *extraresolvable* if there is a family $\mathcal{D}$ of dense subsets of X such that (i) $|\mathcal{D}| > \Delta(X)$, and (ii) the intersection of every two elements of $\mathcal{D}$ is nowhere dense in X. Comfort and García-Ferreira called a space X *strongly extraresolvable* if condition (ii) in the definition of extraresolvability is replaced with a stronger one: $|D_0 \cap D_1| \leq \mathrm{nwd}(X)$, for every distinct $D_0, D_1 \in \mathcal{D}$. Here $\mathrm{nwd}(X)$, the *nowhere density number* of X is the smallest cardinality of a subset of X that is not nowhere dense in X. The following two questions were posed in [**11, 23**]

180–181? **Question 26.** *Is there an extraresolvable space that is not maximally resolvable? A strongly extraresolvable space?*

The authors of [**12**] gave a consistent positive solution to the first part of the question.

182? **Question 27.** *Is there a compact first countable extraresolvable space in* ZFC*?*

The latter question is due to Alas. Under CH, a dense in itself space is extraresolvable if it is countably tight and $\mathrm{nwd}(X)$ is uncountable, see [**23**]. In particular, any Suslin line is extraresolvable in a model of CH. On the other hand, compact metric spaces and compact topological groups are not extraresolvable.

A powerful method for constructing examples with numerous applications was introduced in [**27**] (also, see [**12**]). In particular, for every cardinal λ there exists a Tychonoff λ-resolvable space that is not λ^+-extraresolvable (hence, not λ^+-resolvable).

Literature

The systematic study of resolvability began with works of Hewitt [**25**] and Katětov [**29**]. Many fundamental results were discovered by El′kin and Malykhin in the 1970s. Also, substantial advances in understanding maximal Hausdorff connected spaces were made in [**19, 24, 52**]. [**9**] is a comprehensive survey of that time of maximal properties; see [**31**] for further references. More recently, important results regarding Baire spaces were obtained in [**32, 33, 51**], and, regarding groups, in several papers of Protasov ([**42–45, 47**]) and Zelenyuk ([**54–58**]); also see [**13**] and the survey [**46**]. The article [**5**] jump-started the study of submaximal and MI-spaces. More references on structure resolvability and applications of independent families (which became a ubiquitous tool for constructing spaces with resolvability type properties) can be found in [**12**]. Several old difficult resolvability problems were solved in [**27, 28**].

So far, ZFC regular maximal spaces have been constructed by van Douwen [**16**] (a countable one), El′kin in [**20**] and [**21**] (of arbitrary dispersion character), Protasov [**45**] (a homogenous one), and by the authors of [**1**]. Any construction with new properties will certainly be of interest.

Acknowledgment. The author is grateful to colleagues for useful suggestions, especially to Alan Dow, who helped enormously, Wistar Comfort, Ronnie Levy, Igor Protasov, Angel Tamariz-Mascarúa, and others.

References

[1] O. T. Alas, I. V. Protasov, M. G. Tkachenko, V. V. Tkachuk, R. G. Wilson, and I. V. Yaschenko, *Almost all submaximal groups are paracompact and σ-discrete*, Fund. Math. **156** (1998), no. 3, 241–260.

[2] D. R. Anderson, *On connected irresolvable Hausdorff spaces*, Proc. Amer. Math. Soc. **16** (1965), 463–466.

[3] A. V. Arhangel′skiĭ, *An extremal disconnected bicompactum of weight $\mathfrak{c}$ is inhomogeneous*, Dokl. Akad. Nauk SSSR **175** (1967), 751–754, Translation: Soviet Math. Dokl. 8 (1967) 897–900.

[4] A. V. Arhangel′skiĭ, *Groupes topologiques extrémalement discontinus*, C. R. Acad. Sci. Paris Sér. A-B **265** (1967), A822–A825.

[5] A. V. Arhangel′skiĭ and P. J. Collins, *On submaximal spaces*, Topology Appl. **64** (1995), no. 3, 219–241.

[6] A. Bešlagić and R. Levy, *Irresolvable products*, Papers on general topology and applications (Gorham, ME, 1995), Ann. New York Acad. Sci., vol. 806, New York Acad. Sci., 1996, pp. 42–48.

[7] K. P. S. Bhaskara Rao, *On ℵ-resolvabiltiy*, 1994, Preprint.

[8] R. Bolstein, *Sets of points of discontinuity*, Proc. Amer. Math. Soc. **38** (1973), 193–197.

[9] D. E. Cameron, *A survey of maximal topological spaces*, Topology Proc. **2** (1977), no. 1, 11–60.

[10] W. W. Comfort and S. García-Ferreira, *Resolvability: a selective survey and some new results*, Topology Appl. **74** (1996), 149–167.

[11] W. W. Comfort and S. García-Ferreira, *Strongly extraresolvable groups and spaces*, Topology Proc. **23** (1998), Summer, 45–74.

[12] W. W. Comfort and W. Hu, *Resolvability properties via independent families*, 2006, To appear.

[13] W. W. Comfort and J. van Mill, *Groups with only resolvable group topologies*, Proc. Amer. Math. Soc. **120** (1994), no. 3, 687–696.

[14] C. Costantini, *On the resolvability of locally connected spaces*, Proc. Amer. Math. Soc. **133** (2005), no. 6, 1861–1864.

[15] J. Dontchev, M. Ganster, and D. Rose, *Ideal resolvability*, Topology Appl. **93** (1999), no. 1, 1–16.

[16] E. K. van Douwen, *Applications of maximal topologies*, Topology Appl. **51** (1993), no. 2, 125–139.

[17] A. G. El′kin, *Ultrafilters and irresolvable spaces*, Vestnik Moskov. Univ. Ser. I Mat. Mekh. **24** (1969), no. 5, 51–56, Translation: Moscow Univ. Math. Bull. 24 (1969) no. 5–6, 37–40.

[18] A. G. El′kin, *k-resolvable spaces that are not maximally resolvable*, Dokl. Akad. Nauk SSSR **195** (1970), 274–277, Translation: Soviet Math. Dokl. 11 (1970) 1452–1456.

[19] A. G. El′kin, *Maximal connected Hausdorff spaces*, Mat. Zametki **26** (1979), no. 6, 939–948, Translation: Math. Notes 26 (1979) no. 6, 974–978.

[20] A. G. El′kin, *Regular maximal spaces*, Mat. Zametki **27** (1980), no. 2, 301–305, Translation: Math. Notes 27 (1980) no. 2, 150–151.

[21] A. G. El′kin, *Some topologies on an infinite set*, Uspekhi Mat. Nauk **35** (1980), no. 3(213), 179–183, Translation: Russian Math. Surveys 35 (1980) no. 3, 225–230.

[22] J. Foran and P. Liebnitz, *A characterization of almost resolvable spaces*, Rend. Circ. Mat. Palermo (2) **40** (1991), no. 1, 136–141.

[23] S. García-Ferreira, V. I. Malykhin, and A. H. Tomita, *Extraresolvable spaces*, Topology Appl. **101** (2000), no. 3, 257–271.

[24] J. A. Guthrie, H. E. Stone, and M. L. Wage, *Maximal connected expansions of the reals*, Proc. Amer. Math. Soc. **69** (1978), no. 1, 159–165.

[25] E. Hewitt, *A problem of set-theoretic topology*, Duke Math. J. **10** (1943), 309–333.

[26] A. Illanes, *Finite and ω-resolvability*, Proc. Amer. Math. Soc. **124** (1996), no. 4, 1243–1246.

[27] I. Juhász, L. Soukup, and Z. Szentmiklóssy, *𝒟-forced spaces: a new approach to resolvability*, Topology Appl. **153** (2006), no. 11, 1800–1824.

[28] I. Juhász, L. Soukup, and Z. Szentmiklóssy, *Resolvability of spaces having small spread or extent*, Topology Appl. (2006), Article in press.

[29] M. Katětov, *On topological spaces containing no disjoint dense sets*, Rec. Math. [Mat. Sb.] **21(63)** (1947), 3–12 (Russian).

[30] G. J. Kennedy and S. D. McCartan, *Maximal connected expansions*, Proc. Galway Topology Colloquium (1997), Topology Atlas, Toronto, 1998, p. 5 pp.

[31] G. J. Kennedy and S. D. McCartan, *Singular sets and maximal topologies*, Proc. Amer. Math. Soc. **127** (1999), no. 11, 3375–3382.

[32] K. Kunen, A. Szymański, and F. Tall, *Baire irresolvable spaces and ideal theory*, Ann. Math. Sil. (1986), no. 14, 98–107.

[33] K. Kunen and F. Tall, *On the consistency of the non-existence of Baire irresolvable spaces*, 1998, Preprint.

[34] V. I. Malykhin, *Products of ultrafilters and indecomposable spaces*, Mat. Sb. **90(132)** (1973), 106–116, Translation: Sb. Math. 19 (1974) 105–115.

[35] V. I. Malykhin, *The decomposability of a product of two spaces, and a certain problem of Katětov*, Dokl. Akad. Nauk SSSR **222** (1975), no. 5, 1033–1036, Translation: Soviet Math. Dokl. 16 (1975) 725–729.

[36] V. I. Malykhin, *Extremally disconnected and nearly extremally disconnected groups*, Dokl. Akad. Nauk SSSR **220** (1975), 27–30, Translation: Soviet Math. Dokl. 16 (1975) 21–25.

[37] V. I. Malykhin, *Resolvability of A-, CA- and PCA-sets in compacta*, Topology Appl. **80** (1997), no. 1–2, 161–167.

[38] V. I. Malykhin, *Irresolvability is not descriptively good*, 1998, Preprint.

[39] V. I. Malykhin, *Irresolvable countable spaces of weight less than $\mathfrak{c}$*, Comment. Math. Univ. Carolin. **40** (1999), no. 1, 181–185.

[40] K. Padmavally, *An example of a connected irresolvable Hausdorff space*, Duke Math. J. **20** (1953), 513–520.

[41] O. Pavlov, *On resolvability of topological spaces*, Topology Appl. **126** (2002), no. 1–2, 37–47.

[42] I. V. Protasov, *Partitions of direct products of groups*, Ukraïn. Mat. Zh. **49** (1997), no. 10, 1385–1395.

[43] I. V. Protasov, *Indecomposable topologies on groups*, Ukraïn. Mat. Zh. **50** (1998), no. 12, 1646–1655.

[44] I. V. Protasov, *Maximal topologies on groups*, Sibirsk. Mat. Zh. **39** (1998), no. 6, 1368–1381, Translation: Sib. Math. J. 39 (1998) no. 6, 184–1194.

[45] I. V. Protasov, *Irresolvable left-topological groups*, Ukraïn. Mat. Zh. **52** (2000), no. 6, 758–765.

[46] I. V. Protasov, *Algebra in the Stone–Čech compactification: application to topologies on group*, 2006, Preprint.

[47] I. V. Protasov and E. G. Zelenyuk, *Topologies on groups determined by sequences*, VNTL Publishers, Lviv, 1999.

[48] E. G. Pytkeev, *On maximally resolvable spaces*, Tr. Mat. Inst. Steklova **154** (1983), 209–213, Translation: Proc. Steklov Inst. Math. 154 (1985) 225–230.

[49] E. G. Pytkeev, *Resolvability of countably compact regular spaces*, Proc. Steklov Inst. Math. **Suppl. 2** (2002), S152–S154.

[50] J. Schröder, *On sub-, pseudo- and quasimaximal spaces*, Comment. Math. Univ. Carolin. **39** (1998), no. 1, 197–206.

[51] S. Shelah, *Baire irresolvable spaces and lifting for a layered ideal*, Topology Appl. **33** (1989), no. 3, 217–221.

[52] P. Simon, *An example of maximal connected Hausdorff space*, Fund. Math. **100** (1978), no. 2, 157–163.

[53] A. Tamariz-Mascarúa and H. Villegas-Rodríguez, *Spaces of continuous functions, box products and almost-ω-resolvable spaces*, Comment. Math. Univ. Carolin. **43** (2002), no. 4, 687–705.

[54] E. G. Zelenyuk, *Topological groups with finite semigroups of ultrafilters*, Mat. Stud. **6** (1996), 41–52.

[55] E. G. Zelenyuk, *Topological groups in which each nowhere dense subset is closed*, Mat. Zametki **64** (1998), no. 2, 207–211, Translation: Math. Notes 64 (1998) no. 2, 177–180.

[56] E. G. Zelenyuk, *Decomposability of topological groups*, Ukraïn. Mat. Zh. **51** (1999), no. 1, 41–47.

[57] E. G. Zelenyuk, *Partitions of groups into absolutely dense subsets*, Mat. Zametki **67** (2000), no. 5, 706–711, Translation: Math. Notes 67 (2000) no. 5, 599–602.

[58] E. G.. Zelenyuk, *On partitions of groups into dense subsets*, Topology Appl. **126** (2002), no. 1–2, 327–339.

[59] Y. Zelenyuk, *On extremally disconnected topological groups*, Topology Appl. **153** (2006), no. 14, 2382–2385.

Open Problems in Topology II – Edited by E. Pearl

Topological games and Ramsey theory

Marion Scheepers

Introduction

As of now (2005) it has been thirty years since Telgársky's paper on spaces defined by topological games [**74**], nearly seventy years since Rothberger's introduction of the Rothberger property in [**49**], seventy years since the explicit introduction of the Banach–Mazur game [**41**], seventy-five years since Ramsey's theorem [**47**], and eighty years since Hurewicz's paper on the Menger basis property [**34**] and more than eight-five years since Borel's introduction of strong measure zero sets [**9**].

Here I describe a number of questions about topological games inspired by these sources. This problem survey is neither comprehensive, nor exhaustive: In particular, only problems related to the classical selection principle $\mathsf{S}_1(\cdot,\cdot)$ are featured. I have completely neglected surveying problems from the important selection principles $\mathsf{S}_{\mathrm{fin}}(\cdot,\cdot)$, selective screenability and several others. Moreover, I did not venture into the equally important areas of relativized selection principles, star selection principles, or selection principles in algebraic structures with a compatible topology. Much more could have been written and asked about the Banach–Mazur game and several important related games, and much more could have been reported and asked about set-picking games. And virtually nothing has been explicitly said or asked about the behavior of topological games under various topological constructions. Clearly this is an important topic, and there is much about it in the literature. In the interest of space I have selected a few problems and attempted to sketch a context for these which would be suggestive enough to lead the reader's imagination to the vast unexplored, and likely yet unimagined areas of game theory, selection principles and Ramsey theory in mathematics.

Notation. $\overline{A}$ denotes the closure of A. $|A|$ denotes the cardinality of A. For a positive integer n, $[A]^n$ denotes the set $\{S \subseteq A : |S| = n\}$. The symbol $[A]^{<\aleph_0}$ denotes the collection of finite subsets of A. The symbol $A \subset B$ means A is a subset of B, but is not equal to B; $A \subseteq B$ allows for the possibility that A and B are equal. For a property P and a space X, the symbol $X \models P$ denotes that the space X has the property P. For a set A the symbol ${}^{<\omega}A$ denotes the set of finite sequences with terms from A.

Acknowlegement. A paper which surveys a number of open problems from an area of mathematics obviously will draw heavily on the literature and experience of other workers in these areas. Readers familiar with Fred Galvin, Gary Gruenhage or Rasti Telgársky, will recognize in this paper my large intellectual debt to these three mathematicians who have greatly contributed to the enjoyment of the mathematics I describe here.

1. Basic concepts

Consider the following: Two players, named ONE and TWO, will play a game. Two sets, A and B are given. One of the rules of the game is: ONE's moves in the game will be to choose, when it is ONE's turn, an element of A; TWO's moves in the game will be to choose, whenever it is TWO's turn, an element of B. Some restrictions may apply as to what choices of elements of A or B are legal for either player. Such restrictions must be specified as part of the rules of the game. The assumption is that no restrictions apply, unless they are explicitly stated among the rules of the game. When there are restrictions, the rule is that the first player to not obey the restrictions loses, and the game stops when such a "bad" move is made by a player. Another rule of the game is: The game will have infinitely many innings, one inning per positive integer. In the n-th inning ONE first makes a move by choosing an $O_n \in A$, and TWO then responds by choosing a $T_n \in B$. In this way the players construct a sequence

$$O_1, T_1, O_2, T_2, \ldots, O_n, T_n, \ldots$$

Such a sequence is said to be a *play of the game.*

A rule is needed to define who wins, and who loses a play of the game in which the players followed all stated restrictions. One way to give such a rule is to specify a set Σ of sequences where the odd-numbered entries of the sequence are elements of A, and the even numbered entries are elements of B, and to then declare one of the players as the winner if the play of the game is a member of Σ.

The most important concept related to the notion of an infinite game is the notion of a *strategy.* A *perfect information* strategy for player ONE is a function F whose domain is the set of finite sequences of elements of B, including the empty sequence $\langle\rangle$, and whose values are members of A. A play of the game is said to be an F-play if:

(1) $O_1 = F(\langle\rangle)$ and
(2) For each n, $O_{n+1} = F(T_1, \ldots, T_n)$.

Analogously, a perfect information strategy for player TWO is a function G whose domain is the set of finite nonempty sequences of elements of A. A play of the game is said to be a G-play if for each n, $T_n = G(O_1, \ldots, O_n)$.

A perfect information strategy uses the entire history of the opponent's moves to compute the move the player should make next. Since perfect information strategies are by far the most commonly considered ones in the literature, we shall adopt the convention that "strategy" means "perfect information strategy". For other types of strategies we will use explicit terminology to indicate the type of strategy considered.

A strategy F for ONE is a winning strategy if ONE wins each F-play in which both players followed all rules. A strategy G for TWO is a winning strategy if TWO wins each G-play in which both players followed all rules. A game is said to be determined if one of the players has a winning strategy. Else, the game is said to be undetermined.

Even though each play of a game results in a win for one or the other of the players, this does not mean the game is determined. To prove determinacy of a game one must show the existence of a winning strategy for one of the players. Thus the statement that a player does not have a winning strategy is formally weaker than the statement that the opponent does have a winning strategy. In the literature a number of very good theorems have been missed by authors who did not make this distinction.

Two games, say P and P', are *equivalent* if:

- ONE has a winning strategy in P if, and only if, ONE has a winning strategy in P', and
- TWO has a winning strategy in P if, and only if, TWO has a winning strategy in P'.

And two games, P and P', are *dual* if:

- ONE has a winning strategy in P if, and only if, TWO has a winning strategy in P', and
- TWO has a winning strategy in P if, and only if, ONE has a winning strategy in P'.

For an infinite set S let $\mathcal{A}$ and $\mathcal{B}$ be collections whose members are families of subsets of S. The symbol $\mathsf{S}_1(\mathcal{A},\mathcal{B})$ denotes the statement:

> For each sequence $(A_n : n < \infty)$ there is a sequence $(b_n : n < \infty)$ such that for each n, $b_n \in A_n$, and the set $\{b_n : n \in \infty\} \in \mathcal{B}$.

$\mathsf{S}_1(\mathcal{A},\mathcal{B})$ is an example of a selection principle.

In the game $\mathsf{G}_1(\mathcal{A},\mathcal{B})$ two players, ONE and TWO, play an inning per positive integer. In the n-th inning ONE first chooses an $O_n \in \mathcal{A}$, and then TWO responds with a $T_n \in O_n$. A play $O_1, T_1, \ldots, O_n, T_n, \ldots$ is won by TWO if $\{T_n : n < \infty\} \in \mathcal{B}$; else, ONE wins. If ONE has no winning strategy in $\mathsf{G}_1(\mathcal{A},\mathcal{B})$ then $\mathsf{S}_1(\mathcal{A},\mathcal{B})$ holds. The converse implication is not always true. When it is true, this gives a powerful tool to analyse the selection principle.

2. The point-picking ganes

Several mathematicians have introduced examples of "point-picking" games. In these games a topological space X and a subset H of X are given. ONE picks in the n-th inning a nonempty open subset O_n of X such that $H \subset O_n$, and TWO picks an element $x_n \in O_n$. ONE is declared the winner if TWO's chosen set of points has an appropriate property.

2.1. Berner–Juhász-style point-picking games. We assume for these games that all spaces are T_3, and have no isolated points. In [**6**] Berner and Juhász consider such games where $H = \emptyset$, and the four properties "dense", "dense in itself" (i.e., has no isolated points), "somewhere dense" and "not discrete" (i.e., clusters at some point). We discuss here the game for exactly one of these properties: "dense". In the notation of [**6**] the game $\mathsf{G}^D_\omega(X)$ is the game where ONE wins if TWO's chosen set of points $\{x_n : n < \infty\}$ is dense in X, and TWO wins otherwise. Define the following family, $\mathcal{D} := \{D \subset X : D \text{ is dense}\}$.

Using the techniques of [**66**] one can prove that $\mathsf{G}^D_\omega(X)$ and $\mathsf{G}_1(\mathcal{D},\mathcal{D})$ are dual games. The π-weight of a space X, denoted $\pi(X)$, is the minimal cardinality of a family $\mathcal{U}$ of nonempty open subsets of X such that for each nonempty open set $V \subset X$ there is a $U \in \mathcal{U}$ with $U \subseteq V$. Here is what is known for player TWO in the dual game (Theorem 2.1 of [**6**]):

Theorem 1 (Berner–Juhasz). *Let X be a T_3-space with no isolated points. TWO has a winning strategy in $\mathsf{G}_1(\mathcal{D},\mathcal{D})$ if, and only if, $\pi(X) = \aleph_0$.*

The hereditary density of X, denoted $\delta(X)$, is the least infinite cardinal κ such that each dense subset of X contains a dense subset of cardinality at most κ. Since ONE has an obvious winning strategy in the game $\mathsf{G}_1(\mathcal{D},\mathcal{D})$ if $\delta(X) > \aleph_0$, the only interesting situation regarding winning strategies for ONE is when $\pi(X) > \aleph_0 = \delta(X)$:

183? **Problem 1** (Berner–Juhasz). *Is there in* ZFC *an example of a T_3 topological space X for which the game $\mathsf{G}_1(\mathcal{D},\mathcal{D})$ is undetermined?*

Here is a summary of attacks on this still open problem.

2.1.1. *Attack 1: Axiom $\diamondsuit$ and HFD spaces.* Consider the Tychonoff product of ω_1 copies of the two-point discrete space $2 = \{0,1\}$. According to [**35**] a subspace X of 2^{ω_1} is an *HFD* if there is for each countable subset S of X a countable ordinal $\gamma(S)$ such that for any finite function σ from $\omega_1 \setminus \gamma(S)$ to $\{0,1\}$, there is some element f of S with $\sigma \subset f$.

In [**6**] an HFD space in which neither player has a winning strategy is constructed using the axiom $\diamondsuit$. It was shown in [**66**] that every HFD satisfies $\mathsf{S}_1(\mathcal{D},\mathcal{D})$. And in [**6**] also, using CH, an HFD space is constructed for which ONE has a winning strategy in $\mathsf{G}_1(\mathcal{D},\mathcal{D})$. Thus, for HFDs the property $\mathsf{S}_1(\mathcal{D},\mathcal{D})$ is not equivalent to ONE not having a winning strategy in $\mathsf{G}_1(\mathcal{D},\mathcal{D})$.

2.1.2. *Attack 2: $\mathfrak{p} = \mathfrak{c}$ and irresolvable spaces.* A second interesting attack on this problem came from Dow and Gruenhage, [**14**]: A space is *irresolvable* if no two dense subsets of it are disjoint. A little bit is known about irresolvable spaces in this connection. For a family $\mathcal{A}$ of sets, the symbol $\mathsf{Split}(\mathcal{A},\mathcal{B})$ denotes the statement that for each $A \in \mathcal{A}$ there are elements B_1 and B_2 of $\mathcal{B}$ such that $B_1 \cap B_2 = \emptyset$, and $B_1 \cup B_2 \subset A$.

Theorem 2 ([**66**]). *Let X be a T_3-space. Each of the following implies that X satisfies* $\mathsf{Split}(\mathcal{D},\mathcal{D})$.

(1) *TWO has a winning strategy in $\mathsf{G}_1(\mathcal{D},\mathcal{D})$ on X.*
(2) *ONE has no winning strategy in $\mathsf{G}_1(\mathcal{D},\mathcal{D})$ on X^2.*

Theorem 2 immediately implies that for an irresolvable T_3-space X, TWO has no winning strategy in $\mathsf{G}_1(\mathcal{D},\mathcal{D})$ on X, and ONE has a winning strategy in $\mathsf{G}_1(\mathcal{D},\mathcal{D})$ on X^2. Dow and Gruenhage prove

Theorem 3 (Dow–Gruenhage). *If there is a countable T_3-space which is irresolvable, and for which ONE has no winning strategy in $\mathsf{G}_1(\mathcal{D},\mathcal{D})$, then there is a semi-selective filter which is a countable intersection of ultrafilters on $\mathbb{N}$.*

They prove, on the one hand, that it is consistent that there are no such semi-selective filters, and thus that for every countable T_3-irresolvable space X, ONE has a winning strategy in $\mathsf{G}_1(\mathcal{D},\mathcal{D})$ on X. And on the other hand they show that $\mathfrak{p} = \mathfrak{c}$, i.e., $\mathsf{MA}(\sigma\text{-centered})$, implies the existence of a countable irresolvable T_3-space for which ONE has no winning strategy in $\mathsf{G}_1(\mathcal{D},\mathcal{D})$.

For irresolvable spaces the following problem is open:

Problem 2. *Let X be a countable irresolvable T_3-space with $\delta(X) = \aleph_0$. Does it satisfy $\mathsf{S}_1(\mathcal{D},\mathcal{D})$ if, and only if, ONE has no winning strategy in $\mathsf{G}_1(\mathcal{D},\mathcal{D})$?* 184?

And for an irresolvable T_3-space X with $\delta(X) = \aleph_0$, and with no isolated points, we have $\pi(X) \geq \mathfrak{r}$, the *reaping number*, see for example Proposition 34 of [**66**]. By the example in [**73**] we also know that there are such irresolvable spaces X for which $\pi(X) \leq \mathfrak{i}$, the minimal cardinality for a maximal independent family on $\mathbb{N}$. Let the *irresolvability number* be the greatest cardinal number κ such that for each irresolvable T_3 topology on $\mathbb{N}$ with no isolated points, the π-weight is at least κ. Thus:

$$\mathfrak{r} \leq \text{irresolvability number} \leq \mathfrak{i}.$$

Problem 3. *Is the irresolvability number equal to $\mathfrak{r}$?* 185?

2.1.3. *Attack 3:* cov($\mathcal{M}$) *and function spaces.* Let $\mathcal{M}$ denote the ideal of first category subsets of the real line. By the Baire category theorem, the real line is not covered by countably many first category subsets. Thus cov($\mathcal{M}$), the least cardinality of a family of first category subsets of the real line covering the real line, is uncountable.

Theorem 4 (Juhász). *For any cardinal $\kappa \geq \aleph_0$, the following are equivalent:*

(1) *In each T_3-space X with $\delta(X) = \aleph_0$ and $\pi(X) \leq \kappa$, ONE has no winning strategy in $\mathsf{G}_1(\mathcal{D},\mathcal{D})$.*
(2) $\kappa < \mathrm{cov}(\mathcal{M})$.

Thus by Theorem 1 and Theorem 4: If $\aleph_1 < \mathrm{cov}(\mathcal{M})$ then there are examples of T_3-spaces for which neither player has a winning strategy in the game $\mathsf{G}_1(\mathcal{D},\mathcal{D})$—take a T_3-topology on $\mathbb{N}$ such that $\pi(\mathbb{N}) = \aleph_1$. The inequality $\aleph_1 < \mathrm{cov}(\mathcal{M})$ is known to be independent of ZFC. There are well-studied examples of spaces which can be used to give such examples, and which display a connection of these games with Ramsey theory (to be discussed later). The study of these examples also give a new and illuminating proof of Theorem 4. Namely, consider $C_p(X)$ for a separable metric space X. An open cover $\mathcal{U}$ for X is said to be an ω-cover if X is not in $\mathcal{U}$, but for each finite subset F of X there is a $U \in \mathcal{U}$ such that $F \subseteq U$. Define Ω to be the collection of Ω-covers of X.

Theorem 5 ([**66**], Theorem 13). *For a separable metric space X the following are equivalent:*

(1) *X has property $\mathsf{S}_1(\Omega,\Omega)$.*
(2) *ONE has no winning strategy in the game $\mathsf{G}_1(\Omega,\Omega)$ on X.*
(3) *$C_p(X)$ has property $\mathsf{S}_1(\mathcal{D},\mathcal{D})$.*
(4) *On $C_p(X)$, ONE has no winning strategy in the game $\mathsf{G}_1(\mathcal{D},\mathcal{D})$.*

The hypothesis that X is a separable metric space cannot be weakened to X being a Tychonoff space. A Tychonoff space X of R. Pol [**46**] satisfies $\mathsf{S}_1(\Omega,\Omega)$, see Proposition 15 of [**66**], but $C_p(X)$ does not satisfy $\mathsf{S}_1(\mathcal{D},\mathcal{D})$.

In conclusion to the survey of Problem 1, note that by Theorems 5 and 6 of [**66**] we may restrict attention to countable spaces X, and thus we may assume that we are considering T_3 topologies on the set of positive integers, $\mathbb{N}$.

2.2. Gruenhage-style point-picking games. In a series of papers [**27–29**] Gruenhage considers for $H \neq \emptyset$ the following winning conditions for player ONE: TWO's set of points "converges to H", or "clusters at H". We will discuss here only the case when H is a one-element set. We first discuss the game where TWO's set of chosen points converge to H; the game which requires that TWO's chosen points cluster at H is discussed after that.

2.2.1. *When TWO's set of points converge.* In Gruenhage's notation $\mathsf{G}(\{x\},X)$ denotes the game where ONE wins if TWO's chosen set of points $\{x_n : n \in \infty\}$ converges to x. To describe this game in terms of the game $\mathsf{G}_1(\mathcal{A},\mathcal{B})$, fix an $x \in X$ and define the following families

$$\Omega_x := \{D \subset X : x \in \overline{D} \setminus D\}$$
$$\Gamma_x := \{D \subset X : \text{for each neighborhood } U \text{ of } x,\ D \setminus U \text{ is finite}\}$$

Thus, Ω_x is the collection of subsets of X clustering at x. Using the techniques of [**66**] one can prove that $\mathsf{G}(\{x\},X)$ and $\mathsf{G}_1(\Omega_x,\Gamma_x)$ are dual games. This game, though in the same family of ideas, is quite different from the Berner–Juhasz style point-picking game: There are several known ZFC examples of spaces X for which the game $\mathsf{G}_1(\Omega_x,\Gamma_x)$ is undetermined at some $x \in X$. Some of these are surveyed in Gruenhage's survey paper [**30**]. And the existence of winning strategies for TWO in the game $\mathsf{G}_1(\Omega_x,\Gamma_x)$ have been shown to have deep connections with several important concepts. We first briefly survey the game from the point of view of player ONE, and then discuss it from the point of view of player TWO.

The *tightness* of X at x, denoted by $t(X,x)$, is the least cardinal κ such that each subset A of X with $x \in \overline{A}$ contains a subset B with $|B| \leq \kappa$, and $x \in \overline{B}$. X has *countable tightness at* x if $t(X,x) = \aleph_0$. The *tightness of* X is $t(X) = \sup\{t(X,x) : x \in X\}$. X is a countably tight space if $t(X) = \aleph_0$.

The minimal cardinality of a neighborhood base for a point x of a space X is denoted $\chi(X,x)$. A space is discrete if for all $x \in X$ we have $\chi(X,x) = 1$. And a space is first countable if for each $x \in X$ we have $\chi(X,x) \leq \aleph_0$.

Theorem 6. *For an infinite cardinal number κ the following are equivalent:*

(1) $\kappa < \mathfrak{p}$.
(2) *For each T_1-space X with $t(X) = \aleph_0$, for each point $x \in X$ such that $\chi(X,x) = \kappa$, ONE has no winning strategy in the game $\mathsf{G}_1(\Omega_x,\Gamma_x)$.*

According to the following theorem it suffices to study the existence of winning strategies of ONE in $\mathsf{G}_1(\Omega_x, \Gamma_x)$ with respect to topologies on $\mathbb{N}$, the set of positive integers.

Theorem 7 (Sharma). *Let X be a space of countable tightness. Then at some $x \in X$ ONE has a winning strategy in $\mathsf{G}_1(\Omega_x, \Gamma_x)$ if, and only if, there is a countable subspace Y of X such that $x \in Y$ and ONE has a winning strategy in $\mathsf{G}_1(\Omega_x, \Gamma_x)$ played in Y.*

Sharma further proved in [**72**] the following satisfying result that studying winning strategies of ONE amounts to studying the selection principle.

Theorem 8 (Sharma). *For a point $x \in X$, the following are equivalent:*

(1) *ONE has no winning strategy in the game $\mathsf{G}_1(\Omega_x, \Gamma_x)$.*
(2) *X has property $\mathsf{S}_1(\Omega_x, \Gamma_x)$.*

Fix a topological space X and a non-isolated point $x \in X$. A collection $\mathcal{P}$ of *nonempty* subsets of X is said to be a π-network at x if there is for each open set $U \subset X$ with $x \in U$, a $P \in \mathcal{P}$ with $P \subseteq U$ and $x \notin P$. And $\mathcal{P}$ is said to *converge* to x if $\mathcal{P}$ is infinite and for each open neighborhood U of x the set $\{P \in \mathcal{P} : P \not\subseteq U\}$ is finite. We shall use the following notation:

$$\Pi_x := \{\mathcal{F} \subset [X]^{<\aleph_0} : \mathcal{F} \text{ is a } \pi\text{-network at } x\},$$
$$\mathfrak{C}_x := \{\mathcal{F} \in \Pi_x : \mathcal{F} \text{ converges to } x\}.$$

According to Reznichenko and Sipacheva X is *Fréchet–Urysohn for finite sets* at x if each element of Π_x contains a subset which is an element of $\mathfrak{C}_x$ [**48**] and according to Dow and Steprañs X is said to be *groupwise Fréchet* [**15**]. By results of [**32, 48**], X has this property at x if, and only if, it satisfies the selection principle $\mathsf{S}_1(\Pi_x, \mathfrak{C}_x)$.

To translate this to our current arena, define for a space X the Pixley–Roy space over X, denoted $PR(X)$: The underlying set for $PR(X)$ is $[X]^{<\aleph_0}$. The topology on $PR(X)$ is defined by declaring the following neighborhood bases for elements of $PR(X)$. Let $F \in PR(X)$ as well as an open set $U \supset F$ be given. Then $[F, U] := \{G \in PR(X) : F \subseteq G \subseteq U\}$ is a neighborhood of F. For each $x \in X$ we have the following correspondences between:

(1) $\Omega_{\{x\}}$ for $PR(X)$, and Π_x:
  (a) $(\mathcal{F} \in \Pi_x) \Leftrightarrow (\{\{x\} \cup F : F \in \mathcal{F}\} \in \Omega_{\{x\}})$.
  (b) $(A \in \Omega_{\{x\}}) \Leftrightarrow (\{F \setminus \{x\} : F \in A\} \in \Pi_x)$.
(2) $\Gamma_{\{x\}}$ for $PR(X)$, and $\mathfrak{C}_x$:
  (a) $(\mathcal{F} \in \mathfrak{C}_x) \Leftrightarrow (\{\{x\} \cup F : F \in \mathcal{F}\} \in \Gamma_{\{x\}})$.
  (b) $(A \in \Gamma_{\{x\}}) \Leftrightarrow (\{F \setminus \{x\} : F \in A\} \in \mathfrak{C}_x)$.

Then evidently X has property $\mathsf{S}_1(\Pi_x, \mathfrak{C}_x)$ if, and only if, $PR(X)$ has property $\mathsf{S}_1(\Omega_{\{x\}}, \Gamma_{\{x\}})$. And then by Sharma's theorem we have that this is equivalent to ONE not having a winning strategy in the game $\mathsf{G}_1(\Omega_{\{x\}}, \Gamma_{\{x\}})$ on $PR(X)$. Compare this remark with Theorem 17 of [**32**]. One of the fundamental open questions regarding this example is

186? **Problem 4** (Gruenhage–Szeptycki). *Is there a* ZFC *example of a countable but not first countable space X such that ONE has no winning strategy in* $\mathsf{G}_1(\Omega_{\{x\}}, \Gamma_{\{x\}})$ *at each $x \in X$?*

As explained in [**32**], a positive solution to this problem would also give a positive solution in ZFC to an old question of Malykhin: Is there a countable topological group which is Fréchet–Urysohn, but not metrizable?

The following is another important example in relating Gruenhage's game to other areas of the field of selection principles. An open cover of a topological space is said to be a *γ-cover* if it is infinite, and each infinite subset of it covers the space. Define Γ to be the collection of open γ-covers of X.

The following theorem contains contributions from a number of people. The constant function with value zero is denoted $\mathbf{0}$.

Theorem 9. *Let X be a Tychonoff space. The following are equivalent:*

(1) *$C_p(X)$ satisfies* $\mathsf{S}_1(\Omega_{\mathbf{0}}, \Gamma_{\mathbf{0}})$.
(2) *On $C_p(X)$ ONE has no winning strategy in the game* $\mathsf{G}_1(\Omega_{\mathbf{0}}, \Gamma_{\mathbf{0}})$.
(3) *X has property* $\mathsf{S}_1(\Omega, \Gamma)$.
(4) *On X ONE has no winning strategy in* $\mathsf{G}_1(\Omega, \Gamma)$.

The equivalence of 1 and 3 is from [**26**]. The equivalence of 1 and 2 can be derived from Sharma's theorem 8. And the equivalence of 3 and 4 can be derived from Theorem 3.4 of [**40**].

Now we discuss player TWO. A space is said to be *metalindelöf* if every open cover has an open point-countable refinement covering the space. Assume that we are working with a space X which is locally compact, non-compact and Hausdorff. Then X has a one-point compactification X^*. Let the symbol ∞ denote the additional point such that $X^* = X \cup \{\infty\}$. Neighborhoods of ∞ are of the form $(X \setminus C) \cup \{\infty\}$ where $C \subset X$ is compact. The following beautiful result is from [**28**]:

Theorem 10 (Gruenhage). *For a non-compact, locally compact Hausdorff space X with $t(X) = \aleph_0$ the following are equivalent:*

(1) *On X^* TWO has a winning strategy in* $\mathsf{G}_1(\Omega_\infty, \Gamma_\infty)$.
(2) *X is metalindelöf.*

It is not known to what extent $t(X) = \aleph_0$ is necessary:

187? **Problem 5** (Gruenhage). *Can there be a locally compact Hausdorff space which is not metalindelöf, and yet TWO has a winning strategy in the game* $\mathsf{G}_1(\Omega_\infty, \Gamma_\infty)$?

In [**30**, Question 3.7], Gruenhage suggests as candidates for solving Problem 5 open, non-metalindelöf subsets of $\beta(\omega) \setminus \omega$.

The following result also illustrates the importance of the concepts being surveyed here. A compact space X is *Corson compact* if there is an infinite cardinal number κ such that X is homeomorphic to a subspace of $\{f \in \mathbb{R}^\kappa : \{\alpha < \kappa : f(\alpha) \neq 0\}$ is countable$\}$. For X a set, let X_Δ denote the set $X^2 \setminus \{(x,x) : x \in X\}$. If X is compact, then X_Δ is locally compact.

Theorem 11 (Gruenhage). *For a compact space X the following are equivalent:*

(1) *X is Corson compact.*
(2) *On X^*_Δ TWO has a winning strategy in* $\mathsf{G}_1(\Omega_\infty, \Gamma_\infty)$.
(3) *Every subspace of X^2 is metalindelöf.*

A result of [**26**] together with a result of Galvin gives:

Theorem 12 (Gerlits–Nagy, Galvin). *For a first countable $T_{3\frac{1}{2}}$ space X the following are equivalent:*

(1) *TWO has a winning strategy in* $\mathsf{G}_1(\Omega_0, \Gamma_0)$ *on* $C_p(X)$.
(2) *X is countable.*

The product theory for spaces in which the game $\mathsf{G}_1(\Omega_x, \Gamma_x)$ is undetermined is not well understood. Several ZFC examples of spaces X and Y are known for which ONE does not have a winning strategy in the game $\mathsf{G}_1(\Omega_x, \Gamma_x)$ at any x in either of X or Y, but ONE has a winning strategy in $\mathsf{G}_1(\Omega_{(x,y)}, \Gamma_{(x,y)})$ at some point of $X \times Y$. As to finite powers: Nogura proved in [**42**] the following nice result.

Theorem 13 (Nogura). *If for each n ONE has no winning strategy in* $\mathsf{G}_1(\Omega_x, \Gamma_x)$ *at any $x \in X^n$, then ONE has no winning strategy in* $\mathsf{G}_1(\Omega_x, \Gamma_x)$ *at any $x \in X^\omega$.*

But the following problem is apparently unsolved:

Problem 6 (Gruenhage). *Is there in* ZFC *for each $n > 1$ a space X such that in* 188?
X^n ONE has no winning strategy in $\mathsf{G}_1(\Omega_x, \Gamma_x)$ *at any $x \in X^n$, but there is an $x \in X^{n+1}$ such that ONE has a winning strategy in* $\mathsf{G}_1(\Omega_x, \Gamma_x)$*?*

A ZFC example is known for $n = 1$.

2.2.2. *When TWO's set of points cluster.* In p. 345 of [**27**] Gruenhage mentions the following modification of the game $\mathsf{G}(\{x\}, X)$: ONE and TWO play as before, but the winning condition is different: ONE wins if the set of points $\{x_n : n < \infty\}$ chosen by TWO *clusters* at x. This means that each neighborhood of x contains an $x_n \neq x$. We may assume that x is not an isolated point in X. Furthermore, we may assume that TWO never chooses an $x_n = x$. Let us denote this game by $\mathsf{G}_{\text{cl}}(\{x\}, X)$.

Techniques of [**66**] show that $\mathsf{G}_{\text{cl}}(\{x\}, X)$ and $\mathsf{G}_1(\Omega_x, \Omega_x)$ are dual games. M. Sakai introduced the notion of *countable strong fan tightness* in [**50**]: A space X has countable strong fan tightness at $x \in X$ if $\mathsf{S}_1(\Omega_x, \Omega_x)$ holds. Arguments similar to Sharma's show that it suffices to study existence of winning strategies for player ONE in the game $\mathsf{G}_1(\Omega_x, \Omega_x)$ with respect to topologies on $\mathbb{N}$. But countable strong fan tightness is not equivalent to ONE not having a winning strategy in the corresponding game (see pp. 250–251 of [**61**]):

Theorem 14. *There is a T_1 topology τ on $\mathbb{N}$ such that* $(\mathbb{N}, \tau)$ *has* $\mathsf{S}_1(\Omega_1, \Omega_1)$*, and yet ONE has a winning strategy in* $\mathsf{G}_1(\Omega_1, \Omega_1)$.

For nice topological spaces the equivalence is restored. For example:

Theorem 15. *For $T_{3\frac{1}{2}}$-space X the following are equivalent:*

(1) *$C_p(X)$ has property $\mathsf{S}_1(\Omega_o, \Omega_o)$.*
(2) *ONE has no winning strategy in the game $\mathsf{G}_1(\Omega_o, \Omega_o)$.*
(3) *X has property $\mathsf{S}_1(\Omega, \Omega)$.*
(4) *ONE has no winning strategy in the game $\mathsf{G}_1(\Omega, \Omega)$ played on X.*

The equivalence of 1 and 3 is from [**50**], and the equivalence with the other statements is obtained in [**61**].

Theorem 16. *For an infinite cardinal number κ, the following are equivalent:*
(1) $\kappa < \mathrm{cov}(\mathcal{M})$.
(2) *For each T_1-space X with $t(X) = \aleph_0$ and for each $x \in X$ such that $\chi(X, x) = \kappa$, ONE has no winning strategy in $\mathsf{G}_1(\Omega_x, \Omega_x)$.*

In p. 345 of [**27**] Gruenhage points out that TWO has a winning strategy in $\mathsf{G}_1(\Omega_x, \Gamma_x)$ if, and only if, TWO has a winning strategy in $\mathsf{G}_1(\Omega_x, \Omega_x)$, but that the situation for player ONE is different. There are several examples illustrating this. For example, work of Gerlits and Nagy, of Sakai and others can be used as follows to illustrate this: Consider $C_p(X)$ for sets X of real numbers. By Theorem 9, $C_p(X) \models \mathsf{S}_1(\Omega_0, \Gamma_0)$ if, and only if, $X \models \mathsf{S}_1(\Omega, \Gamma)$. But by Theorem 15 $C_p(X) \models \mathsf{S}_1(\Omega_0, \Omega_0)$ if and only if $X \models \mathsf{S}_1(\Omega, \Omega)$. And in [**37**] we use the Continuum Hypothesis to construct a Lusin set X such that $X \models \mathsf{S}_1(\Omega, \Omega) + \neg\mathsf{S}_1(\Omega, \Gamma)$.

Recall that a space is a *Fréchet* space if for each nonempty subset A, a point x is in the closure of A if, and only if, there is a sequence in A converging to x. By a result of Gerlits and Nagy, $C_p(X)$ being Fréchet is equivalent to $C_p(X) \models \mathsf{S}_1(\Omega_0, \Gamma_0)$. In general $\mathsf{S}_1(\Omega_x, \Gamma_x)$ at each $x \in X$ implies that X is Fréchet, but the converse is not true. One might inquire if $\mathsf{S}_1(\Omega_x, \Omega_x)$ at each x, plus being a Fréchet space, implies that $\mathsf{S}_1(\Omega_x, \Gamma_x)$ holds at each x. Hrušák proved in [**33**] that it is consistent relative to the consistency of ZFC that there are counterexamples:

A family $\mathcal{J}$ of subsets of ω is a *free ideal* if $\omega = \bigcup \mathcal{J}$, $\mathcal{J}$ is closed under finite unions, subsets of elements of $\mathcal{J}$ are elements of $\mathcal{J}$, and ω is not a member of $\mathcal{J}$. The free ideal $\mathcal{J}$ is said to be *+-Ramsey* if for every family $T \subset {}^{<\omega}\omega$ which has the property that for each $\sigma \in T$ the set $\{n : \sigma^\frown(n) \in T\}$ is not a member of $\mathcal{J}$, there is a function $f\colon \omega \to \omega$ such that $\{f(n) : (f(0), \ldots, f(n)) \in T\}$ is not a member of $\mathcal{J}$.

Given any free ideal $\mathcal{J}$ on ω, endow the set $\omega \cup \{\infty\}$ with a topology as follows: Declare each element of ω to be an isolated point, and declare neighborhoods of ∞ to be all sets of the form $\{\infty\} \cup J$ where $\omega \setminus J$ is in $\mathcal{J}$. The symbol $X(\mathcal{J})$ denotes this space.

Almost disjoint families are often used as sources for free ideals on ω. A family $\mathcal{A}$ of infinite subsets of ω is *almost disjoint* if the intersection of any two distinct elements of $\mathcal{A}$ is finite. Call an infinite almost disjoint family $\mathcal{A}$ a *covering* almost disjoint family of $\omega = \bigcup \mathcal{A}$. Subsets of unions of finite subfamilies of a covering almost disjoint family $\mathcal{A}$ form a free ideal on ω. Let $\mathcal{J}(\mathcal{A})$ denote this free ideal.

Every almost disjoint family is contained in a maximal one. If $\mathcal{B}$ is a maximal almost disjoint family then $\omega \setminus \bigcup \mathcal{B}$ is finite, we may assume this is empty, and thus that $\mathcal{B}$ is a covering almost disjoint family. Hrušák showed that if there is a maximal almost disjoint family $\mathcal{B}$ such that $\mathcal{J}(\mathcal{B})$ is +-Ramsey, then there exists a corresponding almost disjoint family $\mathcal{A}$ such that $X(\mathcal{J}(\mathcal{A}))$ is a Fréchet space and ONE has no winning strategy in $\mathsf{G}_1(\Omega_\infty, \Omega_\infty)$, but ONE has a winning strategy in $\mathsf{G}_1(\Omega_\infty, \Gamma_\infty)$.

It is consistent relative to the consistency of ZFC that there is a maximal almost disjoint family $\mathcal{B}$ on ω such that $\mathcal{J}(\mathcal{B})$ is +-Ramsey. It is an open problem whether such a maximal almost disjoint family can be found in ZFC.

Problem 7 (Hrušák). *Is there in* ZFC *a maximal almost disjoint family $\mathcal{B}$ on ω such that $\mathcal{J}(\mathcal{B})$ is +-Ramsey?* 189?

In [**30**] Gruenhage finds in ZFC examples of almost disjoint families $\mathcal{A}$ such that the spaces $X(\mathcal{J}(\mathcal{A}))$ are Fréchet and ONE has no winning strategy in $\mathsf{G}_1(\Omega_\infty, \Omega_\infty)$, but ONE has a winning strategy in $\mathsf{G}_1(\Omega_\infty, \Gamma_\infty)$.

3. Modifications of the point-picking games

There are several very interesting modifications of point-picking games. Instead of having TWO pick a point from ONE's open set, TWO may choose certain types of subsets of ONE's open sets. We now briefly survey two of these modifications and selected open problems.

3.1. TWO chooses nonempty compact sets.

3.1.1. *When TWO's selected sets converge.* Let X be a locally compact, non-compact Hausdorff space with one-point compactification $X^* = X \cup \{\infty\}$. Consider the generalization of $\mathsf{G}(\{\infty\}, X^*)$ where for each n, in inning n ONE chooses an open set O_n containing ∞, and TWO chooses a nonempty compact subset T_n of X such that $T_n \subset O_n$. ONE wins a play $O_1, T_1, \ldots, O_n, T_n, \ldots$ if for each open set O containing ∞, all but finitely many T_n are subsets of O. Gruenhage introduced this game in the following guise in [**29**], denoted there $G^*(X)$: Players ONE and TWO play an inning per positive integer. In the n-th inning ONE first chooses a compact subset O_n of X, and then TWO responds by choosing a compact subset T_n of X such that $O_n \cap T_n = \emptyset$. A play $(O_1, T_1, \ldots, O_n, T_n, \ldots)$ is won by ONE if $\{T_n : n < \infty\}$ is a locally finite family of compact subsets of X. Else, TWO wins.

To discuss the beautiful duality theory that is emerging for this example, we introduce the following concepts and notation: Let $\mathcal{R}$ be a collection of nonempty subsets of a set S. A family $\mathcal{F}$ of nonempty subsets of S is said to be *$\mathcal{R}$-avoiding* if there is for each $R \in \mathcal{R}$ an $F \in \mathcal{F}$ such that $R \cap F = \emptyset$. If S is a non-compact topological space and $\mathcal{R}$ is the collection of nonempty compact subsets of S, then a $\mathcal{R}$-avoiding family is also said to be *compact-avoiding*.

$$\Omega^*_\infty := \{\mathcal{F} : \mathcal{F} \text{ is compact-avoiding family of nonempty compact subsets of } X\}.$$
$$\Gamma^*_\infty := \{\mathcal{F} \in \Omega^*_\infty : \mathcal{F} \text{ is locally finite}\}.$$

A family of nonempty subsets of a space is said to be *discrete* if each element of the space has a neighborhood which has nonempty intersection with at most one member of the family. It is easy to see that elements of Γ^*_∞ are families of compact subsets of X converging to ∞. It can also be shown that for locally compact non-compact Hausdorff spaces, $\mathsf{S}_1(\Omega^*_\infty, \Gamma^*_\infty)$ implies: For each sequence $(A_n : n < \infty)$ of elements of Ω^*_∞ there is a sequence $(C_n : n < \infty)$ such that for each n we have $C_n \in A_n$, and $\{C_n : n < \infty\}$ is a discrete family converging to ∞. Indeed, one may even show there is a sequence of such C_n together with corresponding neighborhoods U_n of C_n (in X) such that $\{U_n : n < \infty\}$ is a discrete family of sets converging to ∞. Thus, by Theorem 2.3 of [**31**], for a locally compact Hausdorff space $\mathsf{S}_1(\Omega^*_\infty, \Gamma^*_\infty)$ on X^* is exactly the *moving off property* on X. In [**2**] we proved $G^*(X)$ and $\mathsf{G}_1(\Omega^*_\infty, \Gamma^*_\infty)$ are dual games. Thus Theorem 5 of [**29**] translates to

Theorem 17 (Gruenhage). *For a non-compact, locally compact T_2-space X, the following are equivalent:*

(1) *TWO has a winning strategy in $\mathsf{G}_1(\Omega^*_\infty, \Gamma^*_\infty)$.*
(2) *X is paracompact.*

As to the situation for player ONE we have the following: For a topological space X the symbol $C_k(X)$ denotes the space of continuous real-valued functions with the following topology: For $f \in C(X)$, for $K \subset X$ compact, and for $\epsilon > 0$, $[f, K, \epsilon] = \{g \in C(X) : (\forall x \in K)(|f(x) - g(x)| < \epsilon)\}$. The sets of the form $[f, K, \epsilon]$ form a basis for a topology on $C(X)$, and $C_k(X)$ denotes this space endowed with the compact-open topology. The symbol $C_o(X)$ denotes the following subspace of $C_k(X)$:

$$C_o(X) := \{f \in C(X) : (\forall \epsilon > 0)(\{x \in X : |f(x)| \geq \epsilon\} \text{ is compact})\}.$$

Theorem 18. *For a locally compact space X the following are equivalent:*

(1) *$C_o(X)$ satisfies $\mathsf{S}_1(\Omega_\mathbf{o}, \Gamma_\mathbf{o})$.*
(2) *ONE has no winning strategy in the game $\mathsf{G}_1(\Omega_\mathbf{o}, \Gamma_\mathbf{o})$ on $C_o(X)$.*
(3) *ONE has no winning strategy in the game $\mathsf{G}_1(\Omega^*_\infty, \Gamma^*_\infty)$ on X.*
(4) *X satisfies $\mathsf{S}_1(\Omega^*_\infty, \Gamma^*_\infty)$.*

The equivalence of 1 and 2 follows from Sharma's Theorem, Theorem 8. The equivalence of 3 and 4 appeared in [**31**] and independently in [**2**]. The equivalence of 1 and 4 appeared implicitly in [**43**]. We will later discuss rather surprising equivalences of the statements in Theorem 18 with completeness properties of the space $C_k(X)$. And now we have the following open problem stated in slightly different form [**31**]:

190? **Problem 8** (Gruenhage–Ma). *Is there a* ZFC *example of a locally compact T_4-space in which the game $\mathsf{G}_1(\Omega^*_\infty, \Gamma^*_\infty)$ is undetermined?*

3.1.2. *When TWO's selected sets cluster.* For the above example we have the following from [**52**] for player ONE:

Theorem 19. *For non-compact, locally compact T_2-spaces X the following are equivalent:*

(1) *$C_o(X)$ satisfies $\mathsf{S}_1(\Omega_\mathbf{0}, \Omega_\mathbf{0})$.*
(2) *ONE has no winning strategy in the game $\mathsf{G}_1(\Omega_\mathbf{0}, \Omega_\mathbf{0})$ on $C_o(X)$.*

(3) *ONE has no winning strategy in the game* $\mathsf{G}_1(\Omega^*_\infty, \Omega^*_\infty)$ *on* X.

(4) X *satisfies* $\mathsf{S}_1(\Omega^*_\infty, \Omega^*_\infty)$.

Problem 9. *Characterize those non-compact, locally compact Hausdorff spaces* 191?
for which ONE has no winning strategy in the game $\mathsf{G}_1(\Omega^*_\infty, \Omega^*_\infty)$.

Experience suggests that this is an important class of spaces.

3.2. TWO selects nonempty open sets. In [**12**] Daniels, Kunen and Zhou consider the following game $\mathsf{G}(X)$: In the n-th inning ONE chooses a nonempty open set $O_n \subset X$, and TWO responds with a nonempty open set $T_n \subset O_n$. They play an inning per positive integer. ONE wins a play $O_1, T_1, \ldots, O_n, T_n, \ldots$ if $\bigcup_{n<\infty} T_n$ is dense in X. Else, TWO wins. The following families of open sets are naturally associated with this game.

$$\mathfrak{D} := \{\mathcal{U} : \bigcup \mathcal{U} \text{ is dense and each set in } \mathcal{U} \text{ is open but not dense}\}$$

$$\mathfrak{D}_\Omega := \{\mathcal{U} \in \mathfrak{D} : \text{for each finite family of nonempty open sets } \mathcal{F} \text{ there is a } U \in \mathcal{U} \text{ such that for each } V \in \mathcal{F},\ U \cap V \neq \emptyset \text{ and } U \text{ not dense}\}$$

Using Theorem 4.1 of [**12**] and Lemma 13 of [**70**] one sees that $G(X)$ and $\mathsf{G}_1(\mathfrak{D}, \mathfrak{D})$ are equivalent games. The *cellularity* of X, denoted $c(X)$, is the minimal cardinal κ such that each pairwise disjoint family of open subsets of X has cardinality at most κ. Each element of $\mathfrak{D}$ has a countable subset in $\mathfrak{D}$ if, and only if, X has countable cellularity. Thus $\mathsf{S}_1(\mathfrak{D}, \mathfrak{D})$ implies $c(X) = \aleph_0$. By Theorem 14 of [**70**]:

Theorem 20. *A topological space satisfies* $\mathsf{S}_1(\mathfrak{D}, \mathfrak{D})$ *if, and only if, ONE has no winning strategy in the game* $\mathsf{G}_1(\mathfrak{D}, \mathfrak{D})$.

It was shown in [**70**] that if each finite power of a space has $\mathsf{S}_1(\mathfrak{D}, \mathfrak{D})$, then the space has $\mathsf{S}_1(\mathfrak{D}_\Omega, \mathfrak{D}_\Omega)$ (in fact, ONE has no winning strategy in the game $\mathsf{G}_1(\mathfrak{D}_\Omega, \mathfrak{D}_\Omega)$). The converse is not true: W. Just (unpublished) constructed from $\diamondsuit$ a Souslin line L which has property $\mathsf{S}_1(\mathfrak{D}_\Omega, \mathfrak{D}_\Omega)$. Since $\mathsf{S}_1(\mathfrak{D}_\Omega, \mathfrak{D}_\Omega)$ implies $\mathsf{S}_1(\mathfrak{D}, \mathfrak{D})$, and since $\mathsf{S}_1(\mathfrak{D}, \mathfrak{D})$ implies $c(X) = \aleph_0$, L^2 does not have property $\mathsf{S}_1(\mathfrak{D}, \mathfrak{D})$.

Problem 10. *Is there a space with property* $\mathsf{S}_1(\mathfrak{D}, \mathfrak{D})$, *but not* $\mathsf{S}_1(\mathfrak{D}_\Omega, \mathfrak{D}_\Omega)$? 192?

The existence of topological spaces solving Problem 10 is not provable in ZFC, [**51**].

Theorem 21. *Martin's Axiom implies: If a space has property* $\mathsf{S}_1(\mathfrak{D}, \mathfrak{D})$, *then all of its finite powers have property* $\mathsf{S}_1(\mathfrak{D}_\Omega, \mathfrak{D}_\Omega)$.

Problem 11. *Is there a space with* $\mathsf{S}_1(\mathfrak{D}_\Omega, \mathfrak{D}_\Omega)$, *and yet ONE has a winning* 193?
strategy in the game $\mathsf{G}_1(\mathfrak{D}_\Omega, \mathfrak{D}_\Omega)$?

Problem 12. *Is it consistent that there are Souslin lines, but no Souslin line has* 194?
property $\mathsf{S}_1(\mathfrak{D}_\Omega, \mathfrak{D}_\Omega)$?

The following result from [**70**] concludes this brief survey of $\mathsf{G}_1(\mathfrak{D}, \mathfrak{D})$:

Theorem 22. *For an infinite cardinal number κ the following are equivalent:*

(1) $\kappa < \text{cov}(\mathcal{M})$.

(2) *For each T_1-space X with $c(X) = \aleph_0$ and $\pi(X) = \kappa$, ONE has no winning strategy in the game $\mathsf{G}_1(\mathfrak{D}, \mathfrak{D})$.*

4. The set-picking games

Several mathematicians have introduced examples of "set-picking" games. In these games a topological space X is given. ONE picks in the n-th inning a point $x_n \in X$ and TWO picks a subset T_n of X such that $x_n \in T_n$. ONE is declared the winner if TWO's family of sets $\{T_n : n < \infty\}$ has an appropriate property.

4.1. Galvin–Telgársky style set-picking games. We now assume that all spaces are Lindelöf, and have no isolated points. In the early 1970s Galvin, and independently Telgàrsky, introduced the following game, known as the *point-open game*, for a topological space X: Players ONE and TWO play an inning per positive integer. In the n-th inning ONE first chooses a point, x_n from X. TWO responds by choosing an open set U_n with $x_n \in U_n$. The play $x_1, U_1, x_2, U_2, \ldots$ is won by ONE if $\{U_n : n \in \infty\}$ is a cover of X. Else, TWO wins. Some of Galvin's results were published in [**23**], and Telgàrsky's results were subsumed in [**74**]. Define $\mathcal{O}$ to be the collection of open covers of the space X.

Theorem 23 (Galvin). *For any topological space the point-open game and the game $\mathsf{G}_1(\mathcal{O}, \mathcal{O})$ are dual to each other.*

Interestingly, Rothberger introduced the selection principle $\mathsf{S}_1(\mathcal{O}, \mathcal{O})$ forty years earlier in [**49**]. And in [**45**] Pawlikowski proves:

Theorem 24 (Pawlikowski). *A topological space has property $\mathsf{S}_1(\mathcal{O}, \mathcal{O})$ if, and only if, ONE has no winning strategy in the game $\mathsf{G}_1(\mathcal{O}, \mathcal{O})$.*

In [**23**] Galvin, and in Theorem 6.3 of [**74**] Telgàrsky proves: If X is a Lindelöf space in which each element is an intersection of countably many open sets, then TWO has a winning strategy in $\mathsf{G}_1(\mathcal{O}, \mathcal{O})$ if, and only if, X is countable.

4.2. Tkachuk style set-picking games. In [**77**] Tkachuk introduced two more set-picking games, denoted by Ω and Θ. The players ONE and TWO play an inning per positive integer.

4.2.1. *The game* Θ. In the game Θ ONE chooses in inning n an $x_n \in X$, and TWO responds with an open set U_n such that $x_n \in U_n$. A play $x_1, U_1, x_2, U_2, \ldots, x_n, U_n, \ldots$ is won by ONE if $\{U_n : n \in \infty\}$ has a dense union in X. Else, TWO wins.

In Theorem 3.3 of [**77**] Tkachuk proves: The games Θ and $\mathsf{G}_1(\mathcal{O}, \mathfrak{D})$ are dual to each other. The corresponding selection principle $\mathsf{S}_1(\mathcal{O}, \mathfrak{D})$ was considered several years earlier by Daniels in [**10**]. She considered it for Pixley–Roy hyperspaces over sets of real numbers and called this property *weakly* C$''$. For Pixley–Roy spaces the following things are known:

Theorem 25. *For a set X of real numbers the following are equivalent:*

(1) $PR(X) \models \mathsf{S}_1(\mathcal{O}, \mathfrak{D})$.
(2) $PR(X) \models$ *ONE has no winning strategy in* $\mathsf{G}_1(\mathcal{O}, \mathfrak{D})$.
(3) $PR(X) \models \mathsf{S}_1(\mathfrak{D}, \mathfrak{D})$.
(4) $PR(X) \models$ *ONE has no winning strategy in* $\mathsf{G}_1(\mathfrak{D}, \mathfrak{D})$.
(5) $PR(X) \models \mathsf{S}_1(\mathfrak{D}_\Omega, \mathfrak{D}_\Omega)$.
(6) $PR(X) \models$ *ONE has no winning strategy in* $\mathsf{G}_1(\mathfrak{D}_\Omega, \mathfrak{D}_\Omega)$.
(7) $X \models \mathsf{S}_1(\Omega, \Omega)$.
(8) *ONE has no winning strategy in* $\mathsf{G}_1(\Omega, \Omega)$.
(9) *For each n,* $X^n \models \mathsf{S}_1(\mathcal{O}, \mathcal{O})$.
(10) *For each n ONE has no winning strategy in* $\mathsf{G}_1(\mathcal{O}, \mathcal{O})$ *on* X^n.

The equivalence of 1 and 9 was given in [**10**]; the equivalence of 7 and 9 is from [**50**]. The equivalence of 9 and 10 is from Pawlikowski's Theorem. The equivalence of 7 and 8 is from [**61**]. The other equivalences are from [**70**]. This establishes a connection between this set-picking game, the point-open game, and the earlier discussed modification of the point-picking games.

In [**65**] this game and selection principle were connected with Borel's classical notion of strong measure zero. According to Borel a metric space X has strong measure zero if there is for each sequence $(\epsilon_n : n \in \infty)$ of positive reals a partition $X = \bigcup_{n \in \infty} X_n$ such that each X_n has diameter less than ϵ_n. Borel conjectured that only countable sets of real numbers have this property. Borel's Conjecture was shown to be independent of ZFC. In [**65**] each dense subset X of the unit interval was associated with a corresponding compact subspace $T(X)$ of the Alexandroff double of the unit interval, $[0, 1]$.

Theorem 26. *For a dense subset X of* $[0, 1]$ *the following are equivalent:*

(1) *X has Borel strong measure zero.*
(2) $T(X)$ *has the property* $\mathsf{S}_1(\mathcal{O}, \mathfrak{D})$.
(3) *ONE has no winning strategy in* $\mathsf{G}_1(\mathcal{O}, \mathfrak{D})$.

It was also shown in [**65**] that TWO has a winning strategy in $\mathsf{G}_1(\mathcal{O}, \mathfrak{D})$ on $T(X)$ if, and only if, X is countable.

4.2.2. *The game* Ω. In this game ONE chooses in inning n an $x_n \in X$, and TWO responds with an open set U_n such that $x_n \in \overline{U}_n$. A play $x_1, U_1, x_2, U_2, \ldots, x_n, U_n, \ldots$ is won by ONE if $\{U_n : n \in \infty\}$ has a dense union in X. Else, TWO wins. Define

$$\mathcal{K} := \{\mathcal{U} \in \mathfrak{D} : X = \bigcup\{\overline{U} : U \in \mathcal{U}\}\},$$

$$\mathcal{K}_\Omega := \{\mathcal{U} \in \mathcal{K} : \text{for each finite } F \subset X \text{ there is a } U \in \mathcal{U} \text{ with } F \subset \overline{U}\}.$$

In Theorem 3.3 of [**77**] Tkachuk proves that the games Ω and $\mathsf{G}_1(\mathcal{K}, \mathfrak{D})$ are dual to each other. The game $\mathsf{G}_1(\mathcal{K}, \mathfrak{D})$ is further analysed in [**70**], where the following two results are proven: A space is *weakly Lindelöf* if each open cover contains a countable subset with dense union. Call a space *weakly $\mathcal{K}$-Lindelöf* if each element of $\mathcal{K}$ has a countable subset with dense union.

Theorem 27. *For an infinite cardinal number* κ *the following are equivalent:*

(1) $\kappa < \text{cov}(\mathcal{M})$.
(2) *For each T_3-space which is weakly $\mathcal{K}$-Lindelöff and has π-weight κ, ONE has no winning strategy in the game* $\mathsf{G}_1(\mathcal{K}, \mathfrak{D})$.
(3) *For each T_3-space which is weakly Lindelöf and has π-weight κ, ONE has no winning strategy in the game* $\mathsf{G}_1(\mathcal{O}, \mathfrak{D})$.

Theorem 28. *For any $T_{3\frac{1}{2}}$-space with cellularity at least* $\text{unif}(\mathcal{SMZ})$, *ONE has a winning strategy in the game* $\mathsf{G}_1(\mathcal{K}, \mathfrak{D})$.

Theorem 28 suggests defining the cardinal number $\mathfrak{j}$ by:

> $\mathfrak{j}$ is the least cardinal number κ such that for any $T_{3\frac{1}{2}}$-space with cellularity at least κ, ONE has a winning strategy in the game $\mathsf{G}_1(\mathcal{K}, \mathfrak{D})$.

Theorems 27 and 28 show $\text{cov}(\mathcal{M}) \leq \mathfrak{j} \leq \text{unif}(\mathcal{SMZ})$. It is known to be consistent that these lower- and upper-bounds on $\mathfrak{j}$ are not equal. Thus:

195? **Problem 13.** *Is $\mathfrak{j}$ a specific one of the cardinals* $\text{cov}(\mathcal{M})$ *or* $\text{unif}(\mathcal{SMZ})$*?*

The selection principle $\mathsf{S}_1(\mathcal{K}, \mathcal{K})$ and corresponding game $\mathsf{G}_1(\mathcal{K}, \mathcal{K})$ are considered in [**68**]. We say that a space X is *$\mathcal{K}$-Lindelöf* if each element of $\mathcal{K}$ has a countable subset which is in $\mathcal{K}$. Also, recall that a set of real numbers is a *Lusin set* if it is uncountable, but its intersection with each nowhere dense set is countable. The following is proved in [**68**]:

Theorem 29. *For an uncountable set X of real numbers the following are equivalent:*

(1) $\mathsf{S}_1(\mathcal{K}, \mathcal{K})$ *holds.*
(2) *ONE has no winning strategy in* $\mathsf{G}_1(\mathcal{K}, \mathcal{K})$.
(3) *X is $\mathcal{K}$-Lindelöf.*
(4) *X is a Lusin set.*

Call a space *$\mathcal{K}_\Omega$-Lindelöf* if each element of $\mathcal{K}_\Omega$ has a countable subset which is in $\mathcal{K}_\Omega$. Each set of reals which is $\mathcal{K}_\Omega$-Lindelöf is $\mathcal{K}$-Lindelöf, and thus a Lusin set. W. Just proved that the converse is false in a strong sense: If there is a Lusin set at all, then there is one which is not $\mathcal{K}_\Omega$-Lindelöf. Thus, the Continuum Hypothesis implies there is a Lusin set which does not satisfy $\mathsf{S}_1(\mathcal{K}_\Omega, \mathcal{K}_\Omega)$. But in [**68**] it is shown that $\diamondsuit$ implies that there is a Lusin set which satisfies $\mathsf{S}_1(\mathcal{K}_\Omega, \mathcal{K}_\Omega)$. The ease with which Lusin sets without $\mathsf{S}_1(\mathcal{K}_\Omega, \mathcal{K}_\Omega)$ are obtained suggests:

196? **Problem 14.** *Is it consistent that there are Lusin sets, but none are $\mathcal{K}_\Omega$-Lindelöf?*

5. Further restrictions on ONE

In [**78**] Tsaban initiated an investigation of the following scenario: Instead of being given a family $\mathcal{A}$ and a family $\mathcal{B}$, one is given a sequence $(\mathcal{A}_n : n < \infty)$ of families, as well as the family $\mathcal{B}$. The game $\mathsf{G}_1(\mathcal{A}_n : n < \infty, \mathcal{B})$ is played as before, except that in the n-th inning ONE must choose a set $O_n \in \mathcal{A}_n$, and TWO responds with $T_n \in O_n$. A play is won by TWO if $\{T_n : n < \infty\}$ is a member of $\mathcal{B}$; else, ONE wins. The corresponding selection principle $\mathsf{S}_1(\mathcal{A}_n : n < \infty, \mathcal{B})$

states that for each sequence $(\mathcal{O}_n : n < \infty)$ such that for each n we have $\mathcal{O}_n \in \mathcal{A}_n$, there is a sequence $(T_n : n < \infty)$ such that for each n we have $T_n \in \mathcal{O}_n$, and $\{T_n : n < \infty\} \in \mathcal{B}$.

This innovation poses several new challenges. Here is a sample game-theoretic problem: For a fixed integer n, an open cover $\mathcal{U}$ of a space X is said to be an n-cover if there is for each n-element subset F of X a set $U \in \mathcal{U}$ such that $F \subseteq U$, and X is not a member of $\mathcal{U}$. Define $\Omega[n]$ is the collection of open n-covers of X.

It is evident that for each n we have $\Omega \subseteq \Omega[n+1] \subseteq \Omega[n]$. According to Galvin and Miller [**24**] a set of real numbers is a *strong γ-set* if there is an increasing sequence $(k_n : n < \infty)$ of positive integers such that $\mathsf{S}_1(\Omega[k_n] : n < \infty, \Gamma)$ holds. It is not hard to see that $\mathsf{S}_1(\Omega[k_n] : n < \infty, \Gamma)$ holds if, and only if, $\mathsf{S}_1(\Omega[n] : n < \infty, \Gamma)$ holds [**78**].

Problem 15 (Tsaban). *Does $\mathsf{S}_1(\Omega[n] : n < \infty, \Gamma)$ imply that ONE has no winning strategy in the game $\mathsf{G}_1(\Omega[n] : n < \infty, \Gamma)$?* 197?

6. Banach–Mazur games

The famous *Banach–Mazur game* made its debut as Problem 43 in The Scottish Book [**41**]. This game is an example of the general class of descending chain games, and is defined as follows on topological space X: In the first inning ONE first chooses a nonempty open subset O_1 of X; then TWO responds with a nonempty open set $T_1 \subseteq O_1$. In the $(n+1)$-th inning ONE chooses a nonempty open set $O_{n+1} \subseteq T_n$, and TWO responds with a nonempty open set $T_{n+1} \subseteq O_{n+1}$. A play

$$O_1, T_1, O_2, T_2, \ldots, O_n, T_n, \ldots$$

is won by TWO if $\bigcap_{n<\infty} T_n \neq \emptyset$; otherwise, ONE wins. This game will be denoted $\mathrm{BM}(X)$.

Recall that a space is a *Baire space* if the intersection of any countable family of dense open subses of X is dense. The following is a well-known fact about $\mathrm{BM}(X)$:

Theorem 30. *Let X be a topological space.*

(1) *ONE has a winning strategy in* $\mathrm{BM}(X)$ *if, and only if, X has a nonempty open subset which is of the first category in itself.*
(2) *If TWO has a winning strategy in* $\mathrm{BM}(X)$*, then all powers of X are, even in the box topology, are Baire spaces.*
(3) *If TWO has a winning strategy in* $\mathrm{BM}(X)$*, then for any Baire space Y, $X \times Y$ is a Baire space.*

Part (1) is due to Banach and was given the current general form by Oxtoby. Parts (2) and (3) are from [**80**].

In Problem 67 of [**41**] Banach proposes another game in the same spirit, which was generalized to its present form by Galvin in the 1970s: Galvin's generalization is as follows: Let S be an infinite set and let J be a free ideal on S. ONE starts the game by choosing a set $O_1 \subset S$ with $O_1 \notin J$. TWO responds by choosing a set $T_1 \subseteq O_1$ with $T_1 \notin J$. In general, with $O_1, T_1, \ldots, O_n, T_n$ chosen, ONE chooses in inning $n+1$ a set $O_{n+1} \subseteq T_n$ with $O_{n+1} \notin J$, and then TWO responds with $T_{n+1} \subseteq O_{n+1}$ and $T_{n+1} \notin J$, and so on. A play

$$O_1, T_1, O_2, T_2, \ldots, O_n, T_n, \ldots$$

is won by TWO if $\bigcap_{n<\infty} T_n \neq \emptyset$; otherwise, ONE wins. This is the *Banach–Galvin game*, denoted $\mathrm{BG}(J)$.

And we also introduce a third game, the *Galvin–Ulam game*: Let S be an infinite set and let $\kappa \geq 2$ be a cardinal number. The game $\mathrm{GU}(S, \kappa)$ proceeds as follows: In the first inning, ONE partitions S into at most κ disjoint pieces. Then TWO responds by choosing one of these pieces, say T_1, and partitions T_1 into at most κ disjoint pieces. This completes inning 1. Then ONE opens the second inning by choosing one of the pieces of TWO's partition of T_1, say O_1, and ONE partitions O_1 into at most κ disjoint pieces. Then TWO responds by choosing one of these pieces, say T_2, and partitions T_2 into at most κ disjoint pieces, and so on. The players play an inning per positive integer in this way, producing a sequence $T_1, O_1, \ldots, T_n, O_n, \ldots$. Player ONE wins if $\bigcap_{n<\infty} T_n \neq \emptyset$, and else TWO wins. This game is denoted $\mathrm{GU}(S, \kappa)$. Ulam's original game, described in [**79**], was $\mathrm{GU}(S, 2)$, and he asked if TWO has a winning strategy in $\mathrm{GU}(\omega_1, 2)$.

These three games are related as follows:

Theorem 31 (Baumgartner). *If there is a free ideal J on an infinite set S such that TWO has a winning strategy in the game* $\mathrm{BG}(J)$*, then ONE has a winning strategy in the game* $\mathrm{GU}(S, \aleph_0)$.

Theorem 32 (Galvin). *Let λ be an infinite cardinal number. If there is a set S of cardinality κ such that ONE has a winning strategy in the game* $\mathrm{GU}(S, \lambda)$*, then: For every topological space X with $\pi(X) \leq \lambda$, if X^κ is, in the box topology, a Baire space, then TWO has a winning strategy in the game* $\mathrm{BM}(X)$.

Taking Theorems 31 and 32 together we obtain

Corollary 33. *If there is a free ideal J on an infinite set S such that TWO has a winning strategy in the game* $\mathrm{BG}(J)$*, then for any space X with $\pi(X) = \aleph_0$ the following are equivalent:*

(1) *TWO has a winning strategy in* $\mathrm{BM}(X)$.
(2) *$X^{|S|}$ is, in the box topology, a Baire space.*

The equivalence of 1 and 2 in this corollary gives a very elegant characterization of countable weight spaces for which TWO has a winning strategy in the Banach–Mazur game. This brings us to the following conjecture of Galvin:

198? **Conjecture 1** (Galvin). *There is for each infinite cardinal number λ a cardinal number κ such that the following are equivalent for each space of π-weight λ:*

(1) *TWO has a winning strategy in* $\mathrm{BM}(X)$.
(2) *X^κ is a Baire space in the box topology.*

The status of this conjecture is as follows:

Theorem 34 (Gray, Solovay). *If "ZFC + there is an infinite set S such that ONE has a winning strategy in the game* $\mathrm{GU}(S,\aleph_0)$*" is consistent, then "ZFC + there is an uncountable measurable cardinal" is consistent.*

Theorem 35 (Magidor). *If "ZFC + there is an uncountable measurable cardinal" is consistent, then "ZFC + there is an infinite set S such that ONE has a winning strategy in the game* $\mathrm{GU}(S,\aleph_0)$*" is consistent.*

Thus, the hypothesis to Baumgartner's theorem is consistent relative to the consistency of ZFC plus the existence of a measurable cardinal. In a personal communication (1995) Magidor explained to me a proof of the fact that if "ZFC + there is a proper class of measurable cardinals" is consistent, then "ZFC + Galvin's Conjecture" is consistent. But it may be that Galvin's conjecture is simply a theorem of ZFC: Currently there is no evidence to the contrary.

The theory of the Banach–Mazur game is also intimately related to point-picking and set-picking games. One of the avenues between the Banach–Mazur game and these games is via the spaces $C_k(X)$ and $C_0(X)$ associated with X. Namely, there are the following two theorems:

Theorem 36 (Ma). *Let X be a non-compact locally compact T_3-space. The following are equivalent:*

(1) *TWO has a winning strategy in* $\mathrm{BM}(C_k(X))$.
(2) *TWO has a winning strategy in* $\mathsf{G}_1(\Omega^*_\infty,\Gamma^*_\infty)$.
(3) *X is paracompact.*

It follows from Ma's theorem and Galvin's Conjecture that for every locally compact T_2-space X, if a sufficiently large box-topology power of $C_k(X)$ is a Baire space, then X is paracompact. The consistency results on Galvin's Conjecture suggest that the successor of the π-weight of $C_k(X)$ would be a sufficiently large power. Thus:

Problem 16. *Is it true for each locally compact T_2-space that if the box-topology power of $\pi(C_k(X))^+$ copies of $C_k(X)$ is Baire, then X is paracompact?* 199?

Theorem 37 (Gruenhage–Ma). *Let X be a non-compact locally compact T_3-space. The following are equivalent:*

(1) *ONE has no winning strategy in* $\mathrm{BM}(C_k(X))$.
(2) *ONE has no winning strategy in* $\mathsf{G}_1(\Omega^*_\infty,\Gamma^*_\infty)$.
(3) *X satisfies* $\mathsf{S}_1(\Omega^*_\infty,\Gamma^*_\infty)$.
(4) *$C_o(X)$ satisfies* $\mathsf{S}_1(\Omega_o,\Gamma_o)$.
(5) *ONE has no winning strategy in the game* $\mathsf{G}_1(\Omega_o,\Gamma_o)$ *on* $C_o(X)$.

In Theorem 37 the equivalence of 1, 2 and 3 is due to Gruenhage and Ma, while 4 is due to Nyikos, and the equivalence of 4 and 5 follows from Sharma's Theorem. Not all the implications in Theorems 36 and 37 require that X be locally compact.

Problem 17 (Gruenhage). *For $T_{3\frac{1}{2}}$-spaces, is it true that:* 200–201?

(1) *ONE has no winning strategy in* $\mathrm{BM}(C_k(X))$ *if, and only if, ONE has no winning strategy in* $\mathsf{G}_1(\Omega^*_\infty, \Gamma^*_\infty)$*?*
(2) *TWO has a winning strategy in* $\mathrm{BM}(C_k(X))$ *if, and only if, TWO has a winning strategy in* $\mathsf{G}_1(\Omega^*_\infty, \Gamma^*_\infty)$*?*

Thus, the following is an alternative formulation of Problem 8

202? **Problem 18** (Gruenhage–Ma). *Are there locally compact T_4-spaces X for which* $\mathrm{BM}(C_k(X))$ *is undetermined?*

7. The length of games

The game $\mathsf{G}_1(\mathcal{A}, \mathcal{B})$ as described before has an inning per positive integer; that is, it is of length ω. In those examples where ONE does not have a winning strategy, one could ask if TWO would have a winning strategy if the game were allowed to run for more innings. Let α be an infinite ordinal number. Then $\mathsf{G}_1^\alpha(\mathcal{A}, \mathcal{B})$ is the following modification of $\mathsf{G}_1(\mathcal{A}, \mathcal{B})$: Players ONE and TWO play an inning per ordinal $\gamma < \alpha$. In the γ-th inning ONE first chooses $O_\gamma \in \mathcal{A}$ and then TWO responds with $T_\gamma \in O_\gamma$. A play

$$O_0, T_0, \ldots, O_\gamma, T_\gamma, \ldots$$

is won by TWO if $\{T_\gamma : \gamma < \alpha\}$ is in $\mathcal{B}$; else, ONE wins.

For a space X for which the selection hypothesis $\mathsf{S}_1(\mathcal{A}, \mathcal{B})$ holds, define:

(1) $\mathrm{tp}_{\mathsf{S}_1(\mathcal{A},\mathcal{B})}(X) = \min\{\alpha : \text{TWO has a winning strategy in } \mathsf{G}_1^\alpha(\mathcal{A}, \mathcal{B})\}$.
(2) $\mathrm{SP}_{\mathsf{S}_1(\mathcal{A},\mathcal{B})}(X) = \{\mathrm{tp}_{\mathsf{S}_1(\mathcal{A},\mathcal{B})}(Y) : Y \subseteq X\}$.

Then $\mathrm{tp}_{\mathsf{S}_1(\mathcal{A},\mathcal{B})}(X)$ is the $\mathsf{S}_1(\mathcal{A}, \mathcal{B})$-*type* of X and $\mathrm{SP}_{\mathsf{S}_1(\mathcal{A},\mathcal{B})}(X)$ is the $\mathsf{S}_1(\mathcal{A}, \mathcal{B})$-*spectrum* of X. In [**11**] Daniels and Gruenhage initiate a study of $\mathrm{SP}_{\mathsf{S}_1(\mathcal{O},\mathcal{O})}(\mathbb{R})$, the $\mathsf{S}_1(\mathcal{O}, \mathcal{O})$-spectrum of the real line. In [**3**] Baldwin proves:

Theorem 38 (Baldwin, CH). $\mathrm{SP}_{\mathsf{S}_1(\mathcal{O},\mathcal{O})}(\mathbb{R}) \supseteq \omega \cup \{\alpha \leq \omega_1 : \alpha \textit{ a limit ordinal}\}$.

On the other hand, the Borel Conjecture implies that $\mathrm{SP}_{\mathsf{S}_1(\mathcal{O},\mathcal{O})}(\mathbb{R}) = \omega + 1 \cup \{\omega_1\}$. Also $\mathsf{MA}(\sigma\text{-centered}) + \neg\mathsf{CH}$ implies that $\mathrm{SP}_{\mathsf{S}_1(\mathcal{O},\mathcal{O})}(\mathbb{R}) = \omega + 1 \cup \{\omega_1\}$, see Theorem 7 of [**11**]. In [**71**] I conjectured

203? **Conjecture 2.** *In* ZFC, $\mathrm{SP}_{\mathsf{S}_1(\mathcal{O},\mathcal{O})}(\mathbb{R}) \subseteq \omega \cup \{\alpha \leq \omega_1 : \alpha \textit{ a limit ordinal}\}$.

Thus: Borel's Conjecture implies that $\mathrm{SP}_{\mathsf{S}_1(\mathcal{O},\mathcal{O})}(\mathbb{R})$ is as small as possible, and by Conjecture 2 CH implies that $\mathrm{SP}_{\mathsf{S}_1(\mathcal{O},\mathcal{O})}(\mathbb{R})$ is as large as possible. It would be very interesting if there were axiomatic circumstances under which $\mathrm{SP}_{\mathsf{S}_1(\mathcal{O},\mathcal{O})}(\mathbb{R})$ is neither of these possibilities.

204? **Problem 19.** *Is the following consistent relative to the consistency of* ZFC*?*

$$\omega + 1 \cup \{\omega_1\} \subset \mathrm{SP}_{\mathsf{S}_1(\mathcal{O},\mathcal{O})}(\mathbb{R}) \subset \omega \cup \{\alpha \leq \omega_1 : \alpha \textit{ a limit ordinal}\}.$$

There are, in ZFC, topological spaces X for which $tp_{\mathsf{S}_1(\mathcal{O},\mathcal{O})}(X)$ is an infinite successor ordinal.

Theorem 39. *For each ordinal $\alpha \leq \omega \cdot 2$ there is a topological space X_α such that $tp_{\mathsf{S}_1(\mathcal{O},\mathcal{O})}(X_\alpha) = \alpha$.*

Conjecture 3. *There exists a space X for which* $\mathrm{SP}_{\mathsf{S}_1(\mathcal{O},\mathcal{O})}(X) = \omega_1 + 1$. 205?

In [**6**] Berner and Juhasz initiated the investigation of $\mathrm{tp}_{\mathsf{S}_1(\mathcal{D},\mathcal{D})}(X)$ for topological spaces X. In [**67**] I prove

Theorem 40. *For a set X of real numbers we have the following:*

$$\mathrm{tp}_{\mathsf{S}_1(\mathcal{D},\mathcal{D})}(C_p(X)) = \mathrm{tp}_{\mathsf{S}_1(\Omega,\Omega)}(X) = \mathrm{tp}_{\mathsf{S}_1(\Omega_{\mathbf{0}},\Omega_{\mathbf{0}})}(C_p(X)).$$

The advantage of this theorem is that it allows for an instant transfer of information regarding game-length properties among the three different games $\mathsf{G}_1^\alpha(\mathcal{D},\mathcal{D})$, $\mathsf{G}_1^\alpha(\Omega,\Omega)$ and $\mathsf{G}_1^\alpha(\Omega_x,\Omega_x)$.

Then it is shown in [**67**] that the Continuum Hypothesis implies $\{1,\omega,\omega^2,\omega_1\} \subseteq \mathrm{SP}_{\mathsf{S}_1(\Omega,\Omega)}(\mathbb{R})$. Borel's Conjecture implies that $\mathrm{SP}_{\mathsf{S}_1(\Omega,\Omega)}(\mathbb{R}) = \{1,\omega,\omega_1\}$.

Conjecture 4. *The Continuum Hypothesis implies* 206?

$$\mathrm{SP}_{\mathsf{S}_1(\Omega,\Omega)}(\mathbb{R}) = \{\omega^\alpha : 0 \leq \alpha < \omega_1\} \cup \{\omega_1\}.$$

One might think that Theorem 38 and some relationship between $\mathrm{tp}_{\mathsf{S}_1(\mathcal{O},\mathcal{O})}(X)$ and $\mathrm{tp}_{\mathsf{S}_1(\Omega,\Omega)}(X)$ might be used to give a proof of Conjecture 4. But there is a catch: It can happen that $\mathrm{tp}_{\mathsf{S}_1(\mathcal{O},\mathcal{O})}(X) = \omega \cdot 2 < \omega_1 = \mathrm{tp}_{\mathsf{S}_1(\Omega,\Omega)}(X)$. In [**81**] Zapletal proves that it is consistent with CH that there is for each ordinal $\alpha < \omega_1$ a subset X_α of the real line, equipped with a refinement of the usual topology, such that $\mathrm{tp}_{\mathsf{S}_1(\mathcal{D},\mathcal{D})} = \omega^\alpha$. This does not quite prove Conjecture 4.

Besides a few isolated results that I did not mention here, this is just about the extent to which $\mathrm{SP}_{\mathsf{S}_1(\mathcal{A},\mathcal{B})}(X)$ and $\mathrm{tp}_{\mathsf{S}_1(\mathcal{A},\mathcal{B})}$ has been studied. My motive behind introducing the several examples of $\mathsf{S}_1(\mathcal{A},\mathcal{B})$ before is to give the reader some suggestion of how much is still to be investigated.

8. Memory restrictions on the winning player

Consider a two-player game of length ω in which one of the players has a winning strategy. As was described earlier, a winning strategy uses as information all preceding moves of the opponent to determine the next move of a player. These are also called *perfect information* winning strategies. But what if a player has a limited memory capacity?

This natural question has been considered often enough that some standard terminology has evolved for certain types of memory limited strategies. For example, a strategy of a player which uses only the most recent move of the opponent is said to be a *tactic*. Fix a positive integer k. A strategy which uses the most recent $\leq k$ moves of the opponent as only information is said to be a k-tactic.

Thus, if we are talking about player ONE, a strategy F would be a k-tactic for ONE if it is a function with domain the set of finite sequences of length at most k of legal moves of player TWO, including the empty sequence, and with range the set of legal moves of ONE: Thus ONE would use F as follows to play the game: In inning 1, ONE plays $F(\langle\rangle) = O_1$; With T_1 TWO's response in inning 1, ONE plays $F(T_1) = O_2$ in inning 2, and so on. In general, in the $(j+1)$-th inning ONE plays $O_{j+1} = F(T_1,\ldots,T_j)$ for $j \leq k$, and for later innings ONE plays $O_{i+k+1} = F(O_{i+1}, \ldots, O_{i+k})$. The notion of a k-tactic for TWO is defined similarly. A k-tactic F for a player is a *winning k-tactic* if each play of the game during which the player used F, is won by that player. In this terminology, a tactic is a 1-tactic.

Fleissner and Kunen [**22**] observed that in the Banach–Mazur game on a space, ONE has a winning (perfect information) strategy if, and only if, ONE has a winning 1-tactic. G. Debs [**13**] discovered an example of a topological space in which TWO has a winning perfect information strategy in the Banach–Mazur game, but not a winning 1-tactic. In all known examples of this phenomenon, TWO actually has a winning 2-tactic. Telgársky conjectured [**76**]:

207? **Conjecture 5** (Telgársky)**.** *There is for each k a topological space X_k such that TWO has a winning $(k+1)$-tactic in the Banach–Mazur game, but not a winning $(k+1)$-tactic.*

One of the consequences of Telgársky's conjecture is that there should exist a topological space in which TWO has a winning perfect information strategy in the Banach–Mazur game, but there is no k for which TWO has a winning k-tactic. No such examples are known.

208? **Problem 20.** *Is there a topological space X such that TWO has a winning perfect information strategy in* $\mathrm{BM}(X)$*, but there is no k such that TWO has a winning k-tactic in* $\mathrm{BM}(X)$*?*

Progress on either of these problems would of course be progress on the following:

209? **Problem 21.** *Is there a topological space X such that TWO has a winning perfect information strategy in* $\mathrm{BM}(X)$*, but does not have a winning 2-tactic?*

A strategy for a player is said to be a *Markov strategy* if it takes as input only the most recent move of the opponent and the number of the inning in progress. Thus, F is a Markov strategy for ONE if $O_1 = F(\langle\rangle, 1)$, $O_2 = F(T_1, 2)$, ..., $O_{n+1} = F(T_n, n+1)$, and so on. And F is a Markov strategy for TWO if for each n, $T_n = F(O_n, n)$. A Markov k-tactic is a strategy which uses the inning number as well as at most the k most recent moves of the opponent. The following is known for the Banach–Mazur game:

Theorem 41. *Let X be a topological space. If TWO has a winning Markov k-tactic in* $\mathrm{BM}(X)$*, then TWO has a winning k-tactic in* $\mathrm{BM}(X)$.

The case $k = 1$ is due to Galvin and Telgársky [**25**], and the case when $k > 1$ is due to Bartoszyński, Just and Scheepers [**4**].

Debs' example is of the kind where from a given topology τ on an appropriate space X, and from a free ideal J on X one defines τ_J to be the set $\{U \setminus M : U \in \tau \text{ and } M \in J\}$. Under appropriate conditions τ_J is a topology. Debs' example is of the form (X, τ_J). This motivated considering the following situation: Let J be a free ideal on a set S, and let $\langle J\rangle$ be its σ-completion. Then S has a T_1 topology for which J is exactly the ideal of nowhere dense sets, and $\langle J\rangle$ is the σ-ideal of first category (= meager) sets. Consider the game $MG(J)$ (the *monotonic game* on J), an example of a *meager-nowhere dense game* in which in the n-th inning ONE first chooses a meager set O_n, and then TWO responds with a nowhere dense set T_n. ONE must further obey the rule that for each n, $O_n \subset O_{n+1}$ but $O_{n+1} \neq O_n$. A play $(O_1, T_1, \ldots, O_n, T_n, \ldots)$ is won by TWO if $\bigcup_{n<\infty} O_n \subseteq \bigcup_{n<\infty} T_n$; else ONE wins. A standard diagonalization argument shows that TWO has a winning strategy in $MG(J)$. The question is if TWO has a winning strategy relying on less than perfect information. It can be shown that TWO has a winning 1-tactic if, and only if, $J = \langle J \rangle$. And there are examples where TWO does not have a winning 1-tactic, but does have a winning 2-tactic. There are also examples where TWO does not have a winning 2-tactic in $MG(J)$, while it is independent of ZFC whether TWO has a winning k-tactic for some $k > 2$. In all examples where it could be shown that if TWO has a winning k-tactic for some k, then indeed TWO has a winning 3-tactic. Thus:

Conjecture 6 (3-tactic conjecture). *For every free ideal J, if TWO has a winning* 210?
k-tactic in $MG(J)$, then TWO has a winning 3-tactic in $MG(J)$.

It is not even known if the 3-tactic conjecture is consistent.

The countable-finite game on a set S is the game $MG(J)$ where $J = [S]^{<\aleph_0}$. For this example Koszmider proved in [**38**] that when $|S| < \aleph_\omega$ then TWO has a winning 2-tactic in $MG(J)$, and he proved that it is consistent that TWO has a winning 2-tactic in $MG(J)$ for all sets S. The following conjecture might be simply a ZFC theorem:

Conjecture 7 (Countable-finite conjecture). *For every infinite set S, TWO has* 211?
a winning 2-tactic in the countable-finite game on S.

A strategy F for player TWO is a *coding strategy* if TWO uses it as follows to play: $T_1 = F(O_1, \langle\rangle)$, and for all n, $T_{n+1} = F(O_{n+1}, T_n)$. Thus TWO remembers the most recent move by ONE and by TWO when deciding how to play next.

Theorem 42 (Galvin–Telgársky). *If TWO has a winning strategy in* $\mathrm{BM}(X)$, *then TWO has a winning coding strategy in* $\mathrm{BM}(X)$.

If in $MG(J)$ we relax the requirements on ONE, and only require that ONE plays in each inning so that $O_n \subseteq O_{n+1}$, we obtain the game $\mathrm{WMG}(J)$ (the *weakly monotonic game*), where still TWO wins if $\bigcup_{n<\infty} O_n \subseteq \bigcup_{n<\infty} T_n$. The Generalized Continuum Hypothesis (GCH) implies that for every free ideal on an infinite set S, TWO has a winning coding strategy in $\mathrm{WMG}(J)$. There is evidence to suggest that the additional hypothesis GCH is not needed. Thus:

Conjecture 8 (Coding Strategy Conjecture). *For each free ideal J on an infinite* 212?
set S, TWO has a winning coding strategy in $\mathrm{WMG}(J)$.

Quite a bit more information about meager-nowhere dense games can be obtained from [**4**, **53**–**58**, **60**].

Winning strategies of limited memory are of intrinsic interest, but often have very nice applications. To illustrate this, consider the following two beautiful results of Gruenhage: A space is said to be *metacompact* if every open cover has an open point-finite refinement covering the space. And a space is *σ-metacompact* if every open cover has an open refinement which covers the space and is a union of countably many point-finite families.

Theorem 43 (Gruenhage). *For a locally compact, non-compact space X the following are equivalent:*

(1) *TWO has a winning* 1*-tactic in* $\mathsf{G}_1(\Omega_\infty, \Gamma_\infty)$ *on* X^*.
(2) X *is metacompact.*

Theorem 44 (Gruenhage). *For a locally compact space X the following are equivalent:*

(1) *TWO has a winning Markov* 1*-tactic in* $\mathsf{G}_1(\Omega_\infty, \Gamma_\infty)$ *on* X^*.
(2) X *is σ-metacompact.*

A compact space X is *Eberlein compact* if there is an infinite cardinal number κ such that X is homeomorphic to a subspace of $\{f \in \mathbb{R}^\kappa : (\forall \epsilon > 0)(\{\alpha < \kappa : |f(\alpha)| > \epsilon\} \text{ is finite})\}$.

Corollary 45 (Gruenhage). *Let X be a compact space. The following are equivalent;*

(1) *TWO has a winning Markov strategy in* $\mathsf{G}_1(\Omega^*_\infty, \Gamma^*_\infty)$ *on* X^*_Δ.
(2) X *is Eberlein compact.*

9. Ramsey theory

In [**47**] F.P. Ramsey published the now famous Ramsey Theorem: For any infinite set S, for any positive integers m and n, and for any function $f\colon [S]^n \to \{1, \dots, m\}$, there is an infinite set $B \subseteq A$ and an $i \in \{1, \dots, m\}$ such that f is constant, of value i, on $[B]^n$.

This result and the inventiveness of several mathematicians has lead to the field now known as Ramsey Theory. Ramsey Theory shows up in many contexts in Mathematics, and it is also intimately related to selection principles and their corresponding games. Though several instances of selection principles have been characterized by Ramseyan theorems, much is still to be discovered about this connection. We briefly survey what is known about Ramsey theory in the context of the selection principle $\mathsf{S}_1(\mathcal{A}, \mathcal{B})$ and mention a few specific problems. In all cases the game $\mathsf{G}_1(\mathcal{A}, \mathcal{B})$ plays a central role in characterizing a selection principle by a Ramseyan theorem.

9.1. The partition relation $\mathcal{A} \to (\mathcal{B})^n_k$. Let an infinite set S and families $\mathcal{A}$ and $\mathcal{B}$ be given as before. For positive integers m and n, the symbol $\mathcal{A} \to (\mathcal{B})^n_k$ denotes the statement that for each $A \in \mathcal{A}$, and for each function $f\colon [A]^n \to \{1, \dots, k\}$, there is an $i \in \{1, \dots, k\}$ and a set $B \in \mathcal{B}$ such that $B \subseteq A$, and f is constant of value i on $[B]^n$.

The simplest possible case for this partition relation, namely for positive integer k, $\mathcal{A} \to (\mathcal{A})^1_k$, is also known as the pigeonhole principle. Now not all elements of $\mathcal{A}$ may have the property that when partitioned into two pieces, at least one of the pieces is again in $\mathcal{A}$. We use a special symbol to denote this subcollection of $\mathcal{A}$, namely $\mathcal{A}_\Omega = \{A \in \mathcal{A} : \text{for any partition } A = A_1 \cup A_2,\ A_1 \in \mathcal{A} \text{ or } A_2 \in \mathcal{A}\}$.

For many of the $\mathcal{A}$ we have considered here, $\mathcal{A} = \mathcal{A}_\Omega$, but also for many others considered here, $\mathcal{A} \neq \mathcal{A}_\Omega$. For example, when $\mathcal{A} \in \{\Gamma_x, \Omega_x, \Gamma, \Omega, \mathfrak{D}_\Omega, \mathcal{K}_\Omega\}$, then $\mathcal{A} = \mathcal{A}_\Omega$. A subcollection of $\mathcal{A}$ for which the Ramseyan partition relation is trivial is the collection $\mathcal{A}_\Gamma = \{A \in \mathcal{A} : \text{for any infinite subset } B \subseteq A \text{ we have } B \in \mathcal{A}\}$. For $\mathcal{A} \in \{\Gamma_x, \Gamma\}$, we have $\mathcal{A} = \mathcal{A}_\Gamma$. By Ramsey's Theorem we have for any family $\mathcal{A}$ for which $\mathcal{A}_\Gamma$ is nonempty, that for all m and n the partition relation $\mathcal{A}_\Gamma \to (\mathcal{A}_\Gamma)^n_k$ holds. Already the partition relation $\mathcal{A}_\Omega \to (\mathcal{A}_\Gamma)^2_2$ gives significant new information. It is namely equivalent with the statement that each element of $\mathcal{A}_\Omega$ has a subset which is an element of $\mathcal{A}_\Gamma$. For $\mathcal{A} = \Omega_x$ for the point x in the topological space X, $\mathcal{A}_\Gamma = \Gamma_x$, and so the partition relation characterizes being Fréchet–Urysohn at the point x. Partition relations of the form $\mathcal{A}_\Omega \to (\mathcal{B})^n_k$ for $\mathcal{B} \subset \mathcal{A} \setminus \mathcal{A}_\Gamma$ are much harder to come by, and give much more information.

Here are some examples where equivalence between the selection principle and the corresponding Ramseyan partition relation have been established. Here, X is a topological space.

Selection Principle	**Ramseyan partition relation**	**Source**
$X \models \mathsf{S}_1(\mathcal{O}, \mathcal{O})$	$(\forall k)(\Omega \to (\mathcal{O})^2_k)$	[**69**]
$X \models \mathsf{S}_1(\Omega, \Omega)$	$(\forall n)(\forall k)(\Omega \to (\Omega)^n_k)$	[**37, 59**]
$X \models \mathsf{S}_1(\Omega, \Omega^{gp})$	$(\forall n)(\forall k)(\Omega \to (\Omega^{gp})^n_k)$	[**39**]
$X \models \mathsf{S}_1(\Lambda, \mathcal{O}^{gp})$	$(\forall k)(\Omega \to (\mathcal{O}^{gp})^2_k)$	[**39**]
$X \models \mathsf{S}_1(\Lambda, \mathcal{O}^{wgp})$	$(\forall k)(\Omega \to (\mathcal{O}^{wgp})^2_k)$	[**1**]
$X \models \mathsf{S}_1(\mathfrak{D}, \mathfrak{D})$	$(\forall n)(\forall k)(\mathfrak{D}_\Omega \to (\mathfrak{D}_\Omega)^n_k)$	[**51**]
$C_p(X) \models \mathsf{S}_1(\Omega_{\mathbf{0}}, \Omega_{\mathbf{0}})$	$(\forall n)(\forall k)(\Omega_{\mathbf{0}} \to (\Omega_{\mathbf{0}})^n_k)$	[**61**]
$C_p(X) \models \mathsf{S}_1(\Omega_{\mathbf{0}}, \Omega_{\mathbf{0}}^{gp})$	$(\forall n)(\forall k)(\Omega_{\mathbf{0}} \to (\Omega_{\mathbf{0}}^{gp})^n_k)$	[**39**]

Very little is known about the partition relation $\mathcal{K}_\Omega \to (\mathcal{K})^2_n$. In [**68**] it is shown for uncountable sets of real numbers X:

(1) If X is $\mathcal{K}_\Omega$-Lindelöf then for each n, $\mathcal{K}_\Omega \to (\mathcal{K})^2_n$.
(2) $\mathcal{K}_\Omega \to (\mathcal{K})^2_2$ implies that X is a Lusin set.

Problem 22. *Could there be a Lusin set which does not satisfy the partition relation* $\mathcal{K}_\Omega \to (\mathcal{K})^2_2$*?* 213?

Problem 23. *Could there be a Lusin set which satisfies the partition relation* $\mathcal{K}_\Omega \to (\mathcal{K})^2_2$*, but which is not* $\mathcal{K}_\Omega$*-Lindelöf?* 214?

9.2. Polarized partition relations. Let k be a positive integer and let $\mathcal{A}_1, \mathcal{A}_2, \mathcal{B}_{1,1}, \mathcal{B}_{2,1}, \ldots, \mathcal{B}_{1,k}$ and $\mathcal{B}_{2,k}$ be collections of subsets of an infinite set S. Then the the symbol $\binom{\mathcal{A}_1}{\mathcal{A}_2} \to \begin{pmatrix} \mathcal{B}_{1,1} & \ldots & \mathcal{B}_{1,k} \\ \mathcal{B}_{2,1} & \ldots & \mathcal{B}_{2,k} \end{pmatrix}^{1,1}$ denotes the *polarized partition relation*, defined as follows: For each $A_1 \in \mathcal{A}_1$, for each $A_2 \in \mathcal{A}_2$ and for each $f\colon A_1 \times A_2 \to \{1, \ldots, k\}$ there are an $i \in \{1, \ldots, k\}$ and sets $B_1 \subseteq A_1$ and $B_2 \subseteq A_2$ such that $B_1 \in \mathcal{B}_{1,i}$, $B_2 \in \mathcal{B}_{2,i}$, and $\{f(x,y) : (x,y) \in B_1 \times B_2\} = \{i\}$.

The polarized partition relation was introduced in Section 9 of [**21**] and thoroughly studied in [**18**] for ordinals. Another class of partition relation known as the *square-bracket* partition relation was introduced in Section 18 of [**19**]. These three classes of partition relations were considered mostly in connection with cardinal or ordinal numbers, although the ordinary partition relation and the square bracket relation have also been extensively studied in the theory of ultrafilters (see for example [**5**, **7**, **8**]) and the theory of linear order types (see for example [**20**] and [**17**]). We discuss here a hybrid of the polarized and of the square bracket relation (hinted at on p. 59 of [**16**]) in connection with certain open covers of topological spaces.

This partition relation is defined as follows: Let $\mathcal{A}_1, \mathcal{A}_2, \mathcal{B}_1$ and $\mathcal{B}_2$ be collections of subsets of the set S and let k and ℓ be positive integers. Then the symbol $\binom{\mathcal{A}_1}{\mathcal{A}_2} \rightarrow \begin{bmatrix}\mathcal{B}_1\\ \mathcal{B}_2\end{bmatrix}^{1,1}_{k/<\ell}$ denotes the statement that for each $A_1 \in \mathcal{A}_1$, for each $A_2 \in \mathcal{A}_2$ and for each $f\colon A_1 \times A_2 \rightarrow \{1,\dots,k\}$ there are sets $B_1 \subseteq A_1$, $B_2 \subseteq A_2$, and $J \subseteq \{1,\dots,k\}$ such that $B_1 \in \mathcal{B}_1$, $B_2 \in \mathcal{B}_2$, $|J| \leq \ell$ and $\{f(x,y) : (x,y) \in B_1 \times B_2\} \subseteq J$.

Theorem 46. *If X has property $\mathsf{S}_1(\Omega,\Omega)$, then $\binom{\Omega}{\Omega} \rightarrow \begin{bmatrix}\Omega\\ \Omega\end{bmatrix}^{1,1}_{k/<3}$ holds.*

215? **Problem 24.** *Does the polarized partition relation characterize $\mathsf{S}_1(\Omega,\Omega)$?*

References

[1] L. Babinkostova, Lj. D. R. Kočinac, and M. Scheepers, *Combinatorics of open covers. VIII*, Topology Appl. **140** (2004), no. 1, 15–32.

[2] L. Babinkostova, Lj. D. R. Kočinac, and M. Scheepers, *Notes on selection principles in topology. I. Paracompactness*, J. Korean Math. Soc. **42** (2005), no. 4, 709–721.

[3] S. Baldwin, *Possible point-open types of subsets of the reals*, Topology Appl. **38** (1991), no. 3, 219–223.

[4] T. Bartoszyński, W. Just, and M. Scheepers, *Covering games and the Banach–Mazur game: K-tactics*, Canad. J. Math. **45** (1993), no. 5, 897–929.

[5] J. E. Baumgartner and A. D. Taylor, *Partition theorems and ultrafilters*, Trans. Amer. Math. Soc. **241** (1978), 283–309.

[6] A. J. Berner and I. Juhász, *Point-picking games and HFDs*, Models and sets (Aachen, 1983), Lecture Notes in Math., vol. 1103, Springer, Berlin, 1984, pp. 53–66.

[7] A. Blass, *Ultrafilter mappings and their Dedekind cuts*, Trans. Amer. Math. Soc. **188** (1974), 327–340.

[8] A. Blass, *Amalgamation of nonstandard models of arithmetic*, J. Symbolic Logic **42** (1977), no. 3, 372–386.

[9] E. Borel, *Sur la classification des ensembles de mesure nulle*, Bull. Soc. Math. France **47** (1919), 97–125.

[10] P. Daniels, *Pixley–Roy spaces over subsets of the reals*, Topology Appl. **29** (1988), no. 1, 93–106.

[11] P. Daniels and G. Gruenhage, *The point-open type of subsets of the reals*, Topology Appl. **37** (1990), no. 1, 53–64.

[12] P. Daniels, K. Kunen, and H. X. Zhou, *On the open-open game*, Fund. Math. **145** (1994), no. 3, 205–220.

[13] G. Debs, *Stratégies gagnantes dans certains jeux topologiques*, Fund. Math. **126** (1985), no. 1, 93–105.

[14] A. Dow and G. Gruenhage, *A point-picking game and semi-selective filters*, Topology Proc. **14** (1989), no. 2, 221–238.

[15] A. Dow and J. Steprāns, *Countable Fréchet α_1-spaces may be first countable*, Arch. Math. Logic **32** (1992), no. 1, 33–50.

[16] P. Erdős, A. Hajnal, A. Máté, and R. Rado, *Combinatorial set theory: partition relations for cardinals*, North-Holland, Amsterdam, 1984.

[17] P. Erdős, A. Hajnal, and E. C. Milner, *Partition relations for η_α and for $\aleph_\alpha$-saturated models*, Theory of sets and topology (in honour of Felix Hausdorff, 1868–1942), VEB Deutsch. Verlag Wissensch., Berlin, 1972, pp. 95–108.

[18] P. Erdős, A. Hajnal, and E. C. Milner, *On set systems having paradoxical covering properties*, Acta Math. Acad. Sci. Hungar. **31** (1978), no. 1–2, 89–124.

[19] P. Erdős, A. Hajnal, and R. Rado, *Partition relations for cardinal numbers*, Acta Math. Acad. Sci. Hungar. **16** (1965), 93–196.

[20] P. Erdős, E. C. Milner, and R. Rado, *Partition relations for η_α-sets*, J. London Math. Soc. **3** (1971), 193–204.

[21] P. Erdős and R. Rado, *A partition calculus in set theory*, Bull. Amer. Math. Soc. **62** (1956), 427–489.

[22] W. G. Fleissner and K. Kunen, *Barely Baire spaces*, Fund. Math. **101** (1978), no. 3, 229–240.

[23] F. Galvin, *Indeterminacy of point-open games*, Bull. Acad. Polon. Sci. Sér. Sci. Math. Astronom. Phys. **26** (1978), no. 5, 445–449.

[24] F. Galvin and A. W. Miller, *γ-sets and other singular sets of real numbers*, Topology Appl. **17** (1984), no. 2, 145–155.

[25] F. Galvin and R. Telgársky, *Stationary strategies in topological games*, Topology Appl. **22** (1986), no. 1, 51–69.

[26] J. Gerlits and Zs. Nagy, *Some properties of $C(X)$. I*, Topology Appl. **14** (1982), no. 2, 151–161.

[27] G. Gruenhage, *Infinite games and generalizations of first-countable spaces*, General Topology and Appl. **6** (1976), no. 3, 339–352.

[28] G. Gruenhage, *Covering properties on $X^2 \setminus \Delta$, W-sets, and compact subsets of Σ-products*, Topology Appl. **17** (1984), no. 3, 287–304.

[29] G. Gruenhage, *Games, covering properties and Eberlein compacts*, Topology Appl. **23** (1986), no. 3, 291–297.

[30] G. Gruenhage, *The story of a topological game,*, To appear.

[31] G. Gruenhage and D. K. Ma, *Baireness of $C_k(X)$ for locally compact X*, Topology Appl. **80** (1997), no. 1–2, 131–139.

[32] G. Gruenhage and P. J. Szeptycki, *Fréchet–Urysohn for finite sets*, Topology Appl. **151** (2005), 238–259.

[33] M. Hrušák, *Selectivity of almost disjoint families*, Acta Univ. Carolin. Math. Phys. **41** (2000), no. 2, 13–21.

[34] W. Hurewicz, *Über eine Verallgemeinerung des Borelschen Theorems*, Math. Z. **24** (1925), 401–425.

[35] I. Juhasz, *Consistency results in topology*, Handbook of mathematical logic (J. Barwise, ed.), North-Holland, Amsterdam, 1977.

[36] I. Juhász, *On point-picking games*, Topology Proc. **10** (1985), no. 1, 103–110.

[37] W. Just, A. W. Miller, M. Scheepers, and P. J. Szeptycki, *The combinatorics of open covers. II*, Topology Appl. **73** (1996), no. 3, 241–266.

[38] P. Koszmider, *On coherent families of finite-to-one functions*, J. Symbolic Logic **58** (1993), no. 1, 128–138.

[39] Lj. D. R. Kočinac and M. Scheepers, *Combinatorics of open covers. VII. Groupability*, Fund. Math. **179** (2003), no. 2, 131–155.

[40] D. K. Ma, *The Cantor tree, the γ-property, and Baire function spaces*, Proc. Amer. Math. Soc. **119** (1993), no. 3, 903–913.

[41] D. Mauldin (ed.), *The Scottish Book*, Birkhäuser, Boston, 1981.

[42] T. Nogura, *The product of $\langle\alpha_i\rangle$-spaces*, Topology Appl. **21** (1985), no. 3, 251–259.

[43] P. J. Nyikos, *Moving-off collections and spaces of continuous functions*, Preprint.

[44] J. C. Oxtoby, *The Banach–Mazur game and Banach category theorem*, Contributions to the theory of games, vol. 3, Princeton University Press, Princeton, NJ, 1957, pp. 159–163.

[45] J. Pawlikowski, *Undetermined sets of point-open games*, Fund. Math. **144** (1994), no. 3, 279–285.

[46] R. Pol, *A function space $C(X)$ which is weakly Lindelöf but not weakly compactly generated*, Studia Math. **64** (1979), no. 3, 279–285.

[47] F. P. Ramsey, *On a problem of formal logic*, Proc. London Math. Soc. **30** (1930), 264–286.

[48] E. A. Reznichenko and O. V. Sipacheva, *Properties of Fréchet–Uryson type in topological spaces, groups and locally convex spaces*, Vestnik Moskov. Univ. Ser. I Mat. Mekh. (1999), no. 3, 32–38, Translation: Moscow Univ. Math. Bull. 54 (1999) no. 3, 33–38.

[49] F. Rothberger, *Eine Verschärfung der Eigenschaft C*, Fund. Math. **30** (1938), 50–55.

[50] M. Sakai, *Property C'' and function spaces*, Proc. Amer. Math. Soc. **104** (1988), no. 3, 917–919.

[51] M. Scheepers, *Cellularity and a selection principle*, In preparation.

[52] M. Scheepers, *Selection principles and Baire spaces*, In preparation.

[53] M. Scheepers, *Concerning n-tactics in the countable-finite game*, J. Symbolic Logic **56** (1991), no. 3, 786–794.

[54] M. Scheepers, *Meager-nowhere dense games. II. Coding strategies*, Proc. Amer. Math. Soc. **112** (1991), no. 4, 1107–1115.

[55] M. Scheepers, *Meager-nowhere dense games. I. n-tactics*, Rocky Mountain J. Math. **22** (1992), no. 3, 1011–1055.

[56] M. Scheepers, *Meager-nowhere dense games. III. Remainder strategies*, Topology Proc. **17** (1992), 215–231.

[57] M. Scheepers, *Meager nowhere-dense games. IV. n-tactics (continued)*, J. Symbolic Logic **59** (1994), no. 2, 603–605.

[58] M. Scheepers, *Meager-nowhere dense games. V. Coding strategies again*, Quaest. Math. **17** (1994), no. 4, 419–435.

[59] M. Scheepers, *Combinatorics of open covers. I. Ramsey theory*, Topology Appl. **69** (1996), no. 1, 31–62.

[60] M. Scheepers, *Meager-nowhere dense games. VI. Markov k-tactics*, Illinois J. Math. **40** (1996), no. 2, 182–193.

[61] M. Scheepers, *Combinatorics of open covers. III. Games, $C_p(X)$*, Fund. Math. **152** (1997), no. 3, 231–254.

[62] M. Scheepers, *Large open covers and a polarized partition relation*, Quaest. Math. **20** (1997), no. 4, 707–714.

[63] M. Scheepers, *Open covers and the square bracket partition relation*, Proc. Amer. Math. Soc. **125** (1997), no. 9, 2719–2724.

[64] M. Scheepers, *Rothberger's property and partition relations*, J. Symbolic Logic **62** (1997), no. 3, 976–980.

[65] M. Scheepers, *Combinatorics of open covers. IV. Subspaces of the Alexandroff double of the unit interval*, Topology Appl. **83** (1998), no. 1, 63–75.

[66] M. Scheepers, *Combinatorics of open covers. VI. Selectors for sequences of dense sets*, Quaest. Math. **22** (1999), no. 1, 109–130.

[67] M. Scheepers, *The length of some diagonalization games*, Arch. Math. Logic **38** (1999), no. 2, 103–122.

[68] M. Scheepers, *Lusin sets*, Proc. Amer. Math. Soc. **127** (1999), no. 1, 251–257.

[69] M. Scheepers, *Open covers and partition relations*, Proc. Amer. Math. Soc. **127** (1999), no. 2, 577–581.

[70] M. Scheepers, *Combinatorics of open covers. V. Pixley–Roy spaces of sets of reals, and ω-covers*, Topology Appl. **102** (2000), no. 1, 13–31.

[71] M. Scheepers, *Topological games*, Encyclopedia of general topology (K. P. Hart, J. Nagata, and J. E. Vaughan, eds.), Elsevier Science, 2004, pp. 439–442.

[72] P. L. Sharma, *Some characterizations of W-spaces and w-spaces*, General Topology and Appl. **9** (1978), no. 3, 289–293.

[73] P. J. Szeptycki, *Comment on: "Dense subsets of product spaces" [Questions Answers Gen. Topology 11 (1993) no. 2, 135–138] by Z. Piotrowski*, Questions Answers Gen. Topology **13** (1995), no. 2, 221–222.

[74] R. Telgársky, *Spaces defined by topological games*, Fund. Math. **88** (1975), no. 3, 193–223.

[75] R. Telgársky, *Spaces defined by topological games. II*, Fund. Math. **116** (1983), no. 3, 189–207.

[76] R. Telgársky, *Topological games: on the 50th anniversary of the Banach–Mazur game*, Rocky Mountain J. Math. **17** (1987), no. 2, 227–276.

[77] V. V. Tkachuk, *Some new versions of an old game*, Comment. Math. Univ. Carolin. **36** (1995), no. 1, 177–196.

[78] B. Tsaban, *Strong γ-sets and other singular spaces*, Topology Appl. **153** (2005), no. 4, 620–639.

[79] S. Ulam, *Combinatorial analysis in infinite sets and some physical theories*, SIAM Rev. **6** (1964), 343–355.

[80] H. E. White, Jr., *Topological spaces that are α-favorable for a player with perfect information*, Proc. Amer. Math. Soc. **50** (1975), 477–482.

[81] J. Zapletal, *Transfinite open-point games*, Topology Appl. **111** (2001), no. 3, 289–297.

Open Problems in Topology II – Edited by E. Pearl

Selection principles and special sets of reals

Boaz Tsaban

1. Introduction

The field of *Selection Principles in Mathematics* started with Scheepers' identification and classification of common prototypes for selection hypotheses appearing in classical and modern works. For surveys of the field see [**30**, **50**, **63**].

The main four prototypes in the field are defined as follows. Fix a topological space X, and let $\mathscr{A}$ and $\mathscr{B}$ each be a collection of covers of X. Following are properties which X may or may not have [**45**].

$\binom{\mathscr{A}}{\mathscr{B}}$: Every member of $\mathscr{A}$ has a subset which is a member of $\mathscr{B}$.

$\mathsf{S}_1(\mathscr{A},\mathscr{B})$: For each sequence $\{\mathcal{U}_n\}_{n\in\mathbb{N}}$ of members of $\mathscr{A}$, there exist members $U_n \in \mathcal{U}_n$, $n \in \mathbb{N}$, such that $\{U_n : n \in \mathbb{N}\} \in \mathscr{B}$.

$\mathsf{S}_{\text{fin}}(\mathscr{A},\mathscr{B})$: For each sequence $\{\mathcal{U}_n\}_{n\in\mathbb{N}}$ of members of $\mathscr{A}$, there exist finite subsets $\mathcal{F}_n \subseteq \mathcal{U}_n$, $n \in \mathbb{N}$, such that $\bigcup_{n\in\mathbb{N}} \mathcal{F}_n \in \mathscr{B}$.

$\mathsf{U}_{\text{fin}}(\mathscr{A},\mathscr{B})$: For each sequence $\{\mathcal{U}_n\}_{n\in\mathbb{N}}$ of members of $\mathscr{A}$ which do not contain a finite subcover, there exist finite subsets $\mathcal{F}_n \subseteq \mathcal{U}_n$, $n \in \mathbb{N}$, such that $\{\bigcup \mathcal{F}_n : n \in \mathbb{N}\} \in \mathscr{B}$.

When $\mathscr{A}$ and $\mathscr{B}$ vary through topologically significant collections, we obtain properties studied in various contexts by many authors. We give some examples.

Fix a topological space X, and let $\mathcal{O}$ denote the collection of all open covers of X. In the case of metric spaces, $\mathsf{S}_{\text{fin}}(\mathcal{O},\mathcal{O})$ is the property shown by Hurewicz [**27**] to be equivalent to Menger's basis property [**36**], and $\mathsf{S}_1(\mathcal{O},\mathcal{O})$ is Rothberger's property traditionally known as C'' [**42**], a property related to Borel's strong measure zero [**14**].

Considering special types of covers we obtain additional properties. Henceforth, by *cover of* X we mean a nontrivial one, i.e., such that X itself is not a member of the cover. An open cover $\mathcal{U}$ of X is an ω*-cover* if for each finite $F \subseteq X$, there is $U \in \mathcal{U}$ such that $F \subseteq U$. $\mathcal{U}$ is a γ*-cover* of X if it is infinite and for each $x \in X$, x is a member of all but finitely many members of $\mathcal{U}$. Let Ω and Γ denote the collections of all ω-covers and γ-covers of X, respectively. Then $\mathsf{U}_{\text{fin}}(\mathcal{O},\Gamma)$ is the Hurewicz property [**28**], and $\mathsf{S}_1(\Omega,\Gamma)$ is the Gerlits–Nagy γ-property, introduced in the context of function spaces [**22**]. Additional properties of these types were studied by Arhangel′skiĭ, Sakai, and others. Some of the properties are relatively new.

The field of selection principles studies the interrelations between all properties defined by the above selection prototypes as well as similar ones, and properties which do not fall into this category but that can be related to properties which do.

Supported by the Koshland Fellowship.

In its broadest sense, the field (and even just its problems) cannot be surveyed in a single book chapter. We will restrict attention to its part dealing with sets of real numbers.[1] Even there, we omit several important topics. Two of them—topological Ramsey theory and topological game theory—are discussed in Scheepers' chapter.

While all problems we mention are about sets of real numbers, some of them deal with sets of reals not defined by selection principles, and belong to the more classical era of the field. Naturally, we usually mention problems we are more familiar with.

The references we give are usually an accessible account of the problem or related problems, but not necessarily the original source (which is usually cited in the given reference). In fact, most of the problems have been around much before being documented in a publication. Thus, most of the problems posed here should be considered folklore.

The current chapter is a comprehensively revised and updated version of our earlier survey [**56**].

2. The Scheepers Diagram problem

Each of the properties mentioned in Section 1, where $\mathscr{A}$, $\mathscr{B}$ range over $\mathcal{O}$, Λ, Ω, Γ, is either void or equivalent to one in the Scheepers Diagram in Figure 1 (where an arrow denotes implication) [**29**, **45**].

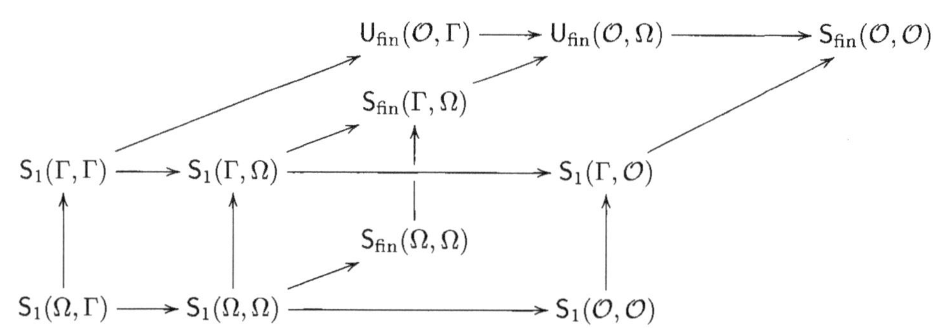

FIGURE 1. The Scheepers Diagram

Almost all implications which do not appear in Figure 1, and are not compositions of existing implications, are not provable: Assuming CH, there are sets of reals witnessing that [**29**]. Only the following two implications remain unsettled.

216–217? **Problem 2.1** ([**29**]).

(1) *Is* $\mathsf{U}_{\mathrm{fin}}(\mathcal{O},\Omega) = \mathsf{S}_{\mathrm{fin}}(\Gamma,\Omega)$*?*
(2) *And if not, does* $\mathsf{U}_{\mathrm{fin}}(\mathcal{O},\Gamma)$ *imply* $\mathsf{S}_{\mathrm{fin}}(\Gamma,\Omega)$*?*

---

[1] This includes separable zero-dimensional metric spaces, since such spaces are homeomorphic to subsets of the irrational numbers.

By a *Borel cover* of X we mean a cover of X consisting of Borel subsets of X. Let $\mathcal{B}$, $\mathcal{B}_\Omega$, $\mathcal{B}_\Gamma$ denote the collections of *countable Borel* covers, ω-covers, and γ-covers of X, respectively. Since we are dealing with sets of reals, we may assume that all *open* covers we consider are countable [**58**]. It follows that when $\mathscr{A}$, $\mathscr{B}$ range over $\mathcal{B}$, $\mathcal{B}_\Omega$, $\mathcal{B}_\Gamma$ we get properties stronger than the corresponding ones when $\mathscr{A}$, $\mathscr{B}$ range over $\mathcal{O}$, Ω, Γ. In the Borel case, more equivalences are known and the diagram in Figure 2 is complete [**51**].

$$
\begin{array}{ccccc}
\mathsf{S}_1(\mathcal{B}_\Gamma,\mathcal{B}_\Gamma) & \longrightarrow & \mathsf{S}_1(\mathcal{B}_\Gamma,\mathcal{B}_\Omega) & \longrightarrow & \mathsf{S}_{\rm fin}(\mathcal{B},\mathcal{B}) \\
\uparrow & & \uparrow & & \uparrow \\
 & & \mathsf{S}_{\rm fin}(\mathcal{B}_\Omega,\mathcal{B}_\Omega) & & \\
 & & \uparrow & & \\
\mathsf{S}_1(\mathcal{B}_\Omega,\mathcal{B}_\Gamma) & \longrightarrow & \mathsf{S}_1(\mathcal{B}_\Omega,\mathcal{B}_\Omega) & \longrightarrow & \mathsf{S}_1(\mathcal{B},\mathcal{B})
\end{array}
$$

FIGURE 2. The Scheepers Diagram in the Borel case

In particular, the answer to the Borel counterpart of Problem 2.1 is positive.

Problem 2.1 can be reformulated in terms of topological properties of sets generating Borel non-σ-compact groups [**69**]. This is related to the following problem.

Problem 2.2 ([**69**]). *Can a Borel non-σ-compact subgroup of a Polish group be* 218? *generated by a subspace satisfying* $\mathsf{U}_{\rm fin}(\mathcal{O},\Gamma)$*?*

A set $X \subseteq \mathbb{R}$ satisfies $\mathsf{U}_{\rm fin}(\mathcal{O},\Gamma)$ if, and only if, for each G_δ $G \subseteq \mathbb{R}$ containing X, there is a σ-compact $F \subseteq \mathbb{R}$ with $X \subset F \subset G$ [**29**].

Problem 2.3 ([**6**]). *Assume that $X \subseteq B \subseteq \mathbb{R}$, B is Borel, and X satisfies* 219–220? $\mathsf{U}_{\rm fin}(\mathcal{O},\Gamma)$. *Must there be a σ-compact F with $X \subset F \subset B$? What if B is $F_{\sigma\delta}$?*

A positive answer for the first part of Problem 2.3 implies a negative answer for Problem 2.2. A positive answer for its second part implies a positive answer for Problem 2.1(2).

3. Examples without special set-theoretic hypotheses

3.1. Dichotomic examples. Let $\mathcal{J}$ be a property of sets of reals. Sometimes there is a set-theoretic hypothesis P independent of ZFC, that can be used to construct an $X \in \mathcal{J}$, and such that its negation $\neg P$ also implies the existence of some $Y \in \mathcal{J}$ (possibly on trivial grounds). In this case, the *existence* of an $X \in \mathcal{J}$ is a theorem of ZFC.

The hypotheses used in the dichotomic arguments are often related to combinatorial cardinal characteristics of the continuum. See [**13**] for a survey of these. The *critical cardinality* of a nontrivial family $\mathcal{J}$ of sets of reals is

$$\mathrm{non}(\mathcal{J}) = \min\{|X| : X \subseteq \mathbb{R}, X \notin \mathcal{J}\}.$$

Figure 3 indicates the critical cardinalities of the properties in the Scheepers diagram 1 (we use $\mathcal{M}$ for the ideal of meager sets of reals). The critical cardinalities in the Borel case are equal to those in the open case.

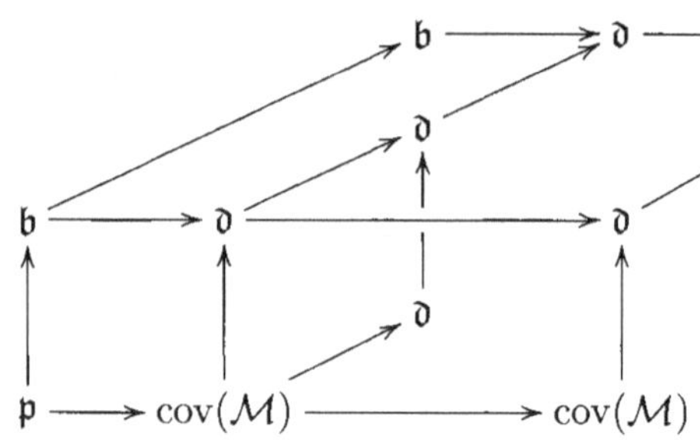

FIGURE 3. Critical cardinalities in the Scheepers Diagram

Dichotomic arguments imply the existence (in ZFC) of a set of reals X satisfying $\mathsf{S}_1(\Gamma,\Gamma)$ such that $|X| = \mathfrak{t}$ [**46**], and a set of reals satisfying $\mathsf{S}_{\mathrm{fin}}(\Omega,\Omega)$ such that $|X| = \mathrm{cf}(\mathfrak{d})$ [**65**]. Now, $\mathrm{non}(\mathsf{S}_1(\Gamma,\Gamma)) = \mathfrak{b}$, $\mathrm{non}(\mathsf{S}_{\mathrm{fin}}(\Omega,\Omega)) = \mathfrak{d}$, and it is consistent that $\mathfrak{b} > \mathfrak{t}$ and $\mathfrak{d} > \mathrm{cf}(\mathfrak{d})$. Thus, these existence results are not satisfactory.

221? **Problem 3.1** ([**12**]). *Does there exist (in* ZFC*) a set of reals X satisfying* $\mathsf{S}_1(\Gamma,\Gamma)$ *such that* $|X| = \mathfrak{b}$*?*

222? **Problem 3.2** ([**65**]). *Does there exist (in* ZFC*) a set of reals satisfying* $\mathsf{S}_{\mathrm{fin}}(\Omega,\Omega)$ *such that* $|X| = \mathfrak{d}$*?*

3.2. Direct constructions. Constructions which do not appeal to a dichotomy are philosophically much more pleasing.

There is a direct construction of a set of reals H satisfying $\mathsf{U}_{\mathrm{fin}}(\mathcal{O},\Gamma)$ such that $|H| = \mathfrak{b}$ (and such that H does not contain a perfect set) [**10**]. All finite powers of this set H satisfy $\mathsf{U}_{\mathrm{fin}}(\mathcal{O},\Gamma)$ [**12**]. In fact, H can be chosen as a subgroup or even a subfield of $\mathbb{R}$ [**62**, **65**]. There is also a direct construction of a set of reals M satisfying $\mathsf{S}_{\mathrm{fin}}(\mathcal{O},\mathcal{O})$ but not $\mathsf{U}_{\mathrm{fin}}(\mathcal{O},\Gamma)$, such that $|M| = \mathfrak{d}$ [**65**].

223? **Problem 3.3** ([**65**]). *Is there a direct (non-dichotomic) construction of a set of reals M satisfying* $\mathsf{S}_{\mathrm{fin}}(\Omega,\Omega)$ *but not* $\mathsf{U}_{\mathrm{fin}}(\mathcal{O},\Gamma)$*?*

3.3. The Borel case. Let $\mathcal{J}$ be a property of sets of reals. *Borel's Conjecture for $\mathcal{J}$* is the statement "All elements of $\mathcal{J}$ are countable". For all but three of the properties in the Borel case, Borel's Conjecture is consistent.

224? **Problem 3.4** ([**40**]). *Is Borel's Conjecture for* $\mathsf{S}_{\mathrm{fin}}(\mathcal{B},\mathcal{B})$ *consistent?*

This is the same as asking whether it is consistent that each uncountable set of reals can be mapped onto a dominating subset of $\mathbb{N}^{\mathbb{N}}$ by a Borel function [**51**]. Problem 3.4 is also open for $\mathsf{S}_1(\mathcal{B}_\Gamma,\mathcal{B}_\Omega)$.

Problem 3.5 (Magidor). *Is Borel's Conjecture for* $\mathsf{S}_{\mathrm{fin}}(\mathcal{B},\mathcal{B})$ *equivalent to Borel's* 225?
Conjecture for $\mathsf{S}_1(\mathcal{B}_\Gamma,\mathcal{B}_\Omega)$*?*

Problem 3.6 ([**51**]). *Is Borel's Conjecture for* $\mathsf{S}_{\mathrm{fin}}(\mathcal{B}_\Omega,\mathcal{B}_\Omega)$ *consistent?* 226?

4. Examples from CH or MA

Consider the Borel case (Figure 2). For each set of reals X, we can put $\bullet$ in each place in the diagram where the property is satisfied, and $\circ$ elsewhere. There are 14 settings consistent with the arrows in the diagram, and they are all listed in Figure 4.

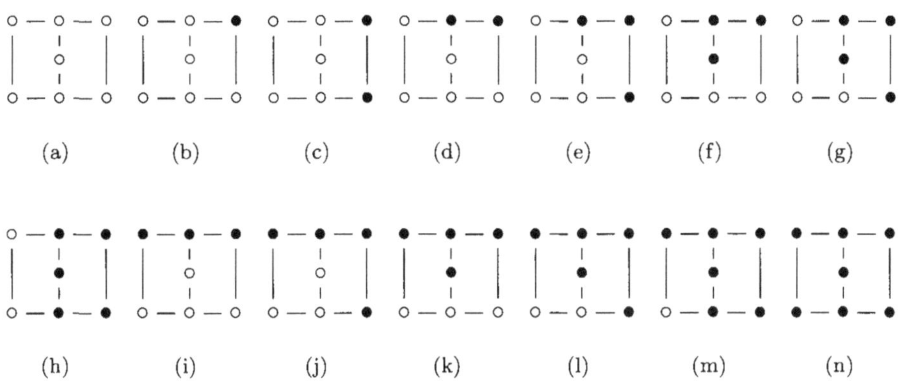

FIGURE 4. The consistent settings

Setting (a) is realized by $\mathbb{R}\setminus\mathbb{Q}$, i.e., $\mathbb{N}^{\mathbb{N}}$.

Assume CH. Settings (c), (h), and (i) were realized in [**29**], Setting (k) was realized in [**62**], and Setting (n) was realized in [**15, 39, 41**]. To realize Setting (b), take a set L as in Setting (c) and a set S as in Setting (i), and take $X = L\cup S$. As $\mathsf{S}_{\mathrm{fin}}(\mathcal{B},\mathcal{B})$ is additive, X satisfies this property. But since $\mathsf{S}_1(\mathcal{B}_\Gamma,\mathcal{B}_\Omega)$ and $\mathsf{S}_1(\mathcal{B},\mathcal{B})$ are hereditary for subsets [**12**], X does not satisfy any of these. It seems that using forcing-theoretic arguments similar to those of [**15**], we can realize Settings (f) and (m).

Problem 4.1. *Does* CH *imply a realization of the settings (d), (e), (g), (j), and* 227–231?
(l)?

All constructions mentioned above can be carried out using MA. Except perhaps Setting (n).

Problem 4.2 ([**39**]). *Does* MA *imply the existence of an uncountable set of reals* 232?
satisfying $\mathsf{S}_1(\mathcal{B}_\Omega,\mathcal{B}_\Gamma)$*?*

5. The δ-property

For a sequence $\{X_n\}_{n\in\mathbb{N}}$ of subsets of X, define $\liminf X_n = \bigcup_m \bigcap_{n\geq m} X_n$. For a family $\mathcal{F}$ of subsets of X, $L(\mathcal{F})$ denotes its closure under the operation lim inf. X has the *δ-property* [**22**] if for each OPEN ω-cover $\mathcal{U}$ of X, $X \in L(\mathcal{U})$.

Clearly, $\binom{\Omega}{\Gamma}$ implies the δ-property. $\mathsf{S}_1(\Omega,\Gamma) = \binom{\Omega}{\Gamma}$ [**22**].

233? **Problem 5.1** ([**22**]). *Is the δ-property equivalent to $\binom{\Omega}{\Gamma}$?*

Miller points out that, as a union of an increasing sequence of sets with the δ-property has again the δ-property, a negative answer to the following problem implies a negative answer to Problem 5.1.

234? **Problem 5.2** ([**39**]). *Does every union of an increasing sequence $\{X_n\}_{n\in\mathbb{N}}$ of sets satisfying $\binom{\Omega}{\Gamma}$ satisfy $\binom{\Omega}{\Gamma}$?*

The answer is positive in the Borel case [**62**].

6. Preservation of properties

6.1. Heredity. A property of sets of reals is *hereditary* if for each set of reals X satisfying the property, all subsets of X satisfy that property. None of the selection hypotheses involving open covers is provably hereditary [**12**]. However, the property $\mathsf{S}_1(\mathcal{B},\mathcal{B})$ as well as all properties of the form $\Pi(\mathcal{B}_\Gamma,\mathcal{B})$ are hereditary [**12**] (but $\mathsf{S}_1(\mathcal{B}_\Omega,\mathcal{B}_\Gamma)$ is not [**39**]).

235–236? **Problem 6.1** ([**12, 39**]). *Is $\mathsf{S}_1(\mathcal{B}_\Omega,\mathcal{B}_\Omega)$ or $\mathsf{S}_{\rm fin}(\mathcal{B}_\Omega,\mathcal{B}_\Omega)$ hereditary?*

All properties in the Scheepers Diagram in Figure 1, except for the following two, are known to be hereditary for F_σ subsets‘[**66**].

237–238? **Problem 6.2** ([**66**]). *Are $\mathsf{S}_{\rm fin}(\Gamma,\Omega)$ and $\mathsf{S}_1(\Gamma,\Omega)$ hereditary for F_σ subsets?*

The Borel versions of all properties are hereditary for arbitrary Borel subsets [**51**].

6.2. Finite powers. $\mathsf{S}_1(\Omega,\Gamma)$, $\mathsf{S}_1(\Omega,\Omega)$, and $\mathsf{S}_{\rm fin}(\Omega,\Omega)$ are the only properties in the open case which are preserved under taking finite powers [**29**].

239–240? **Problem 6.3** ([**51**]). *Are any of the classes $\mathsf{S}_1(\mathcal{B}_\Omega,\mathcal{B}_\Omega)$ or $\mathsf{S}_{\rm fin}(\mathcal{B}_\Omega,\mathcal{B}_\Omega)$ preserved by finite powers?*

Assume that X satisfies $\mathsf{S}_1(\mathcal{B}_\Omega,\mathcal{B}_\Omega)$ and $Y \subseteq X$. If $\mathsf{S}_1(\mathcal{B}_\Omega,\mathcal{B}_\Omega)$ is preserved by finite powers, then X^k satisfies $\mathsf{S}_1(\mathcal{B}_\Omega,\mathcal{B}_\Omega)$, and in particular $\mathsf{S}_1(\mathcal{B},\mathcal{B})$, for all k. As $\mathsf{S}_1(\mathcal{B},\mathcal{B})$ is hereditary, Y^k satisfies $\mathsf{S}_1(\mathcal{B},\mathcal{B})$ for all k. It follows that Y satisfies $\mathsf{S}_1(\mathcal{B}_\Omega,\mathcal{B}_\Omega)$ [**51**]. Similar assertions for $\mathsf{S}_{\rm fin}(\mathcal{B}_\Omega,\mathcal{B}_\Omega)$ and $\mathsf{S}_{\rm fin}(\mathcal{B},\mathcal{B})$ also hold [**51**]. Thus, a positive answer to Problem 6.3 implies a positive answer to Problem 6.1.

241? **Problem 6.4** ([**51**]). *Is $\mathsf{S}_1(\mathcal{B}_\Omega,\mathcal{B}_\Gamma)$ preserved by finite powers?*

The corresponding problems for the other classes are settled in the negative [**51**].

6.3. Products. Some positive results are available for products of sets. E.g., if $X, Y \subseteq \mathbb{R}$ have strong measure zero and X also satisfies $\mathsf{U}_{\mathrm{fin}}(\mathcal{O}, \Gamma)$, then $X \times Y$ has strong measure zero [**47**].

Problem 6.5 ([**47**]). *Assume that* $X, Y \subseteq \mathbb{R}$ *satisfy* $\mathsf{S}_1(\mathcal{O}, \mathcal{O})$, *and* X *also satisfies* $\mathsf{U}_{\mathrm{fin}}(\mathcal{O}, \Gamma)$. *Does it follow that* $X \times Y$ *satisfies* $\mathsf{S}_1(\mathcal{O}, \mathcal{O})$? 242?

It is not even known whether a positive answer follows when X satisfies $\mathsf{S}_1(\Omega, \Gamma)$.

The following problem withstood considerable attacks by several mathematicians. The property in it is equivalent to the Gerlits–Nagy $(*)$ property, and is also equivalent to $\mathsf{S}_1(\Omega, \mathcal{O}^{\gamma\text{-gp}})$ [**32**].

Problem 6.6. *Is* $\mathsf{U}_{\mathrm{fin}}(\mathcal{O}, \Gamma) \cap \mathsf{S}_1(\mathcal{O}, \mathcal{O})$ *preserved by finite products?* 243?

A positive answer here implies a positive answer to Problem 6.5. It is not even known whether $\mathsf{U}_{\mathrm{fin}}(\mathcal{O}, \Gamma) \cap \mathsf{S}_1(\mathcal{O}, \mathcal{O})$ is preserved by finite *powers*.

None of the properties in Figure 1 is provably preserved by finite products [**11**, **48**, **51**, **58**]. Borel's conjecture implies a consistently positive answer for $\mathsf{S}_1(\mathcal{O}, \mathcal{O})$ and below it.

Problem 6.7 (Scheepers). *Are any of the* $\mathsf{S}_{\mathrm{fin}}$ *or* $\mathsf{U}_{\mathrm{fin}}$ *type properties in the Scheepers Diagram in Figure 1 consistently preserved by finite products?* 244?

A natural place to check Problem 6.7 for $\mathsf{S}_{\mathrm{fin}}(\mathcal{O}, \mathcal{O})$ is Miller's model (in which, by the way, $\mathsf{U}_{\mathrm{fin}}(\mathcal{O}, \Omega) = \mathsf{S}_{\mathrm{fin}}(\mathcal{O}, \mathcal{O})$ [**66**, **70**]).

Assume that Y has Hausdorff dimension zero. The assumption that X satisfies $\mathsf{S}_1(\Omega, \Gamma)$ does not imply that $X \times Y$ has Hausdorff dimension zero. However, if X satisfies $\mathsf{S}_1(\{\mathcal{O}_n\}_{n\in\mathbb{N}}, \Gamma)$,[2] then $X \times Y$ has Hausdorff dimension zero [**68**].

Problem 6.8 ([**68**]). *Assume that* $|X| < \mathfrak{p}$. *Is it true that for each* Y *with Hausdorff dimension zero,* $X \times Y$ *has Hausdorff dimension zero?* 245?

Problem 6.9 (Krawczyk). *Is it consistent (relative to* ZFC*) that there are uncountable sets of reals satisfying* $\mathsf{S}_1(\Omega, \Gamma)$, *but for each such set* X *and each set* Y *with Hausdorff dimension zero,* $X \times Y$ *has Hausdorff dimension zero?* 246?

6.4. Unions. The question of which of the properties in Figure 1 is provably preserved under taking finite or countable unions (i.e., is *additive* or *σ-additive*) is completely settled. Some of the classes which are not provably additive are *consistently* additive [**61**].

Problem 6.10 ([**61**]). *Is* $\mathsf{S}_{\mathrm{fin}}(\Omega, \Omega)$ *consistently additive?* 247?

The problem is also open for $\mathsf{S}_1(\Gamma, \Omega)$ and $\mathsf{S}_{\mathrm{fin}}(\Gamma, \Omega)$.

Problem 6.11 ([**61**]). *Is* $\mathsf{S}_{\mathrm{fin}}(\mathcal{B}_\Omega, \mathcal{B}_\Omega)$ *consistently additive?* 248?

---

[2] $\mathsf{S}_1(\{\mathcal{O}_n\}_{n\in\mathbb{N}}, \Gamma)$ is the *strong γ-property* [**21**, **60**]: For each sequence $\{\mathcal{U}_n\}_{n\in\mathbb{N}}$ where for each n, $\mathcal{U}_n$ is an open n-cover of X (i.e., each $F \subseteq X$ with $|F| \leq n$ is contained in some member of $\mathcal{U}_n$), there are $U_n \in \mathcal{U}_n$, $n \in \mathbb{N}$, such that $\{\mathcal{U}_n : n \in \mathbb{N}\}$ is a γ-cover of X.

In some cases, there remains the task to determine the *exact* additivity number. The *additivity number* of a nontrivial family $\mathcal{J}$ of sets of reals is

$$\mathrm{add}(\mathcal{J}) = \min\{|\mathcal{F}| : \mathcal{F} \subseteq \mathcal{I} \text{ and } \bigcup \mathcal{F} \notin \mathcal{J}\}.$$

$\max\{\mathfrak{b}, \mathfrak{g}\} \leq \mathrm{add}(\mathsf{S}_{\mathrm{fin}}(\mathcal{O},\mathcal{O})) \leq \mathrm{cf}(\mathfrak{d})$, $\mathfrak{h} \leq \mathrm{add}(\mathsf{S}_1(\Gamma,\Gamma)) \leq \mathfrak{b}$, and $\mathrm{add}(\mathcal{N}) \leq \mathrm{add}(\mathsf{S}_1(\mathcal{O},\mathcal{O}))$ [**61**].

249? **Problem 6.12** ([**61**]). *Is* $\mathrm{add}(\mathsf{S}_{\mathrm{fin}}(\mathcal{O},\mathcal{O})) = \max\{\mathfrak{b}, \mathfrak{g}\}$*?*

250? **Problem 6.13** ([**49**]). *Is* $\mathrm{add}(\mathsf{S}_1(\Gamma,\Gamma)) = \mathfrak{b}$*?*

Problem 6.13 is related to Problem 9.1 below. The answer for the Borel version of Problem 6.13 is positive.

251? **Problem 6.14** ([**9**]). *Is it consistent that* $\mathrm{add}(\mathcal{N}) < \mathrm{add}(\mathsf{S}_1(\mathcal{O},\mathcal{O}))$*?*

Another type of problem is exemplified by the following problem. It is easy to see that if X satisfies $\mathsf{S}_1(\Omega,\Gamma)$ and D is countable, then $X \cup D$ satisfies $\mathsf{S}_1(\Omega,\Gamma)$.

252–253? **Problem 6.15** (Miller, Tsaban). *Assume that X satisfies* $\mathsf{S}_1(\Omega,\Gamma)$ *and* $|D| < \mathfrak{p}$. *Does $X \cup D$ satisfy* $\mathsf{S}_1(\Omega,\Gamma)$*? Is it true under* MA *when* $|D| = \aleph_1$*?*

Recently, F. Jordan proved that for each D, the following are equivalent:

(1) $X \cup D$ satisfies $\mathsf{S}_1(\Omega,\Gamma)$ for each X satisfying $\mathsf{S}_1(\Omega,\Gamma)$;
(2) $X \times D$ satisfies $\mathsf{S}_1(\Omega,\Gamma)$ for each X satisfying $\mathsf{S}_1(\Omega,\Gamma)$.

7. Modern types of covers

7.1. τ-covers. Recall that by "cover of X" we mean one not containing X as an element. $\mathcal{U}$ is a *large* cover of X if each $x \in X$ is covered by infinitely many members of $\mathcal{U}$. It is a *τ-cover* of X if, in addition, for each $x, y \in X$, either $\{U \in \mathcal{U} : x \in U, y \notin U\}$ is finite, or else $\{U \in \mathcal{U} : y \in U, x \notin U\}$ is finite [**54**]. Let T denote the collection of open τ-covers of X. Then $\Gamma \subseteq \mathrm{T} \subseteq \Omega$.

The most important problem concerning τ-covers is the following.

254? **Problem 7.1** ([**57**]). *Is* $\binom{\Omega}{\Gamma} = \binom{\Omega}{\mathrm{T}}$*?*

This problem is related to many problems posed in [**39, 54, 55, 58, 60, 68**], etc. The best known result in this direction is that $\binom{\Omega}{\mathrm{T}}$ implies $\mathsf{S}_{\mathrm{fin}}(\Gamma,\mathrm{T})$ [**55**].

To state a modest form of Problem 7.1, note that if $\binom{\Omega}{\mathrm{T}}$ implies $\mathsf{S}_{\mathrm{fin}}(\mathrm{T},\Omega)$, then $\binom{\Omega}{\mathrm{T}} = \mathsf{S}_{\mathrm{fin}}(\Omega,\mathrm{T})$.

255? **Problem 7.2** ([**55**]). *Is* $\binom{\Omega}{\mathrm{T}} = \mathsf{S}_{\mathrm{fin}}(\Omega,\mathrm{T})$*?*

256? **Problem 7.3** (Scheepers). *Does* $\mathsf{S}_1(\Omega,\mathrm{T})$ *imply* $\mathsf{U}_{\mathrm{fin}}(\mathcal{O},\Gamma)$*?*

There are many more problems of this type, and they are summarized in [**37**].

Not much is known about the preservation of the new properties under set-theoretic operations. Miller [**39**] proved that assuming CH, there exists a set of reals X satisfying $\mathsf{S}_1(\mathcal{B}_\Omega,\mathcal{B}_\Gamma)$ and a subset Y of X such that Y does not satisfy $\binom{\Omega}{\mathrm{T}}$.

Together with the remarks preceding Problem 6.1, we have that the only classes (in addition to those in Problem 6.1) for which the heredity problem is not settled are the following ones.

Problem 7.4 ([**12**]). *Are any of the properties* $\mathsf{S}_1(\mathcal{B}_\mathrm{T},\mathcal{B}_\Gamma)$, $\mathsf{S}_1(\mathcal{B}_\mathrm{T},\mathcal{B}_\mathrm{T})$, $\mathsf{S}_1(\mathcal{B}_\mathrm{T},\mathcal{B}_\Omega)$, 257–262?
$\mathsf{S}_1(\mathcal{B}_\mathrm{T},\mathcal{B})$, $\mathsf{S}_{\mathrm{fin}}(\mathcal{B}_\mathrm{T},\mathcal{B}_\mathrm{T})$, *or* $\mathsf{S}_{\mathrm{fin}}(\mathcal{B}_\mathrm{T},\mathcal{B}_\Omega)$, *hereditary?*

Here are the open problems regarding unions.

Problem 7.5 ([**61**]). *Are any of the properties* $\mathsf{S}_1(\mathrm{T},\mathrm{T})$, $\mathsf{S}_{\mathrm{fin}}(\mathrm{T},\mathrm{T})$, $\mathsf{S}_1(\Gamma,\mathrm{T})$, 263–272?
$\mathsf{S}_{\mathrm{fin}}(\Gamma,\mathrm{T})$, *and* $\mathsf{U}_{\mathrm{fin}}(\mathcal{O},\mathrm{T})$ *(or any of their Borel versions) additive?*

It is consistent that $\mathsf{U}_{\mathrm{fin}}(\mathcal{O},\Gamma)=\mathsf{U}_{\mathrm{fin}}(\mathcal{O},\mathrm{T})$, and therefore $\mathsf{U}_{\mathrm{fin}}(\mathcal{O},\mathrm{T})$ is consistently σ-additive [**71**]. $\mathsf{S}_1(\mathrm{T},\mathrm{T})$ is preserved under taking finite unions if, and only if, $\mathsf{S}_1(\mathrm{T},\mathrm{T})=\mathsf{S}_1(\mathrm{T},\Gamma)$ [**37**].

Here are the open problems regarding powers.

Problem 7.6. *Are any of the following properties preserved under taking finite powers?* 273–279?

(1) $\mathsf{S}_1(\Omega,\mathrm{T})$, *or* $\mathsf{S}_{\mathrm{fin}}(\Omega,\mathrm{T})$,
(2) $\mathsf{S}_1(\mathrm{T},\Gamma)$, $\mathsf{S}_1(\mathrm{T},\mathrm{T})$, $\mathsf{S}_1(\mathrm{T},\Omega)$, $\mathsf{S}_{\mathrm{fin}}(\mathrm{T},\mathrm{T})$, *or* $\mathsf{S}_{\mathrm{fin}}(\mathrm{T},\Omega)$,

Most of these problems are related to Problem 7.1.

A solution to any of the problems involving τ-covers must use new ideas, since this family of covers is not as amenable as the classical ones. In [**55**] it is shown that if we use an amenable variant of τ-covers (called τ^*-covers, see below), then most of the corresponding problems can be solved.

7.2. τ^*-covers. $Y\subseteq[\mathbb{N}]^{\aleph_0}$ is *linearly refinable* if for each $y\in Y$ there exists an infinite subset $\hat{y}\subseteq y$ such that the family $\hat{Y}=\{\hat{y}:y\in Y\}$ is linearly (quasi)ordered by $\subseteq^*$. A cover $\mathcal{U}=\{U_n:n\in\mathbb{N}\}$ of X is a *τ^*-cover* of X if it is large, and the family of all sets $\{n:x\in U_n\}$, $x\in X$, is linearly refinable. T^* is the collection of all countable open τ^*-covers of X.

Every analytic space satisfies $\binom{\mathrm{T}}{\Gamma}$ [**54**].

Problem 7.7 ([**55**]). *Does* $\{0,1\}^{\mathbb{N}}$ *satisfy* $\binom{\mathrm{T}^*}{\Gamma}$*?* 280?

7.3. Groupable covers. Groupability notions for covers appear naturally in the studies of selection principles [**2, 31, 32, 59**].

A cover $\mathcal{U}$ of X is *multifinite* if there exists a partition of $\mathcal{U}$ into infinitely many finite covers of X.

Let ξ be γ, τ, or ω. A cover $\mathcal{U}$ of X is *ξ-groupable* if it is multifinite, or there is a partition of $\mathcal{U}$ into finite sets, $\mathcal{U}=\bigcup_{n\in\mathbb{N}}\mathcal{F}_n$, such that $\{\bigcup\mathcal{F}_n\}_{n\in\mathbb{N}}$ is a ξ-cover of X. Denote the collection of ξ-groupable open covers of X by $\mathcal{O}^{\xi\text{-gp}}$. Then $\mathcal{O}^{\gamma\text{-gp}}\subseteq\mathcal{O}^{\tau\text{-gp}}\subseteq\mathcal{O}^{\omega\text{-gp}}$.

$\mathsf{S}_{\mathrm{fin}}(\mathcal{O},\mathcal{O})=\mathsf{S}_{\mathrm{fin}}(\Omega,\mathcal{O})$ [**45**], $\mathsf{U}_{\mathrm{fin}}(\mathcal{O},\Gamma)=\mathsf{S}_{\mathrm{fin}}(\Omega,\mathcal{O}^{\gamma\text{-gp}})$ [**32**], and $\mathsf{U}_{\mathrm{fin}}(\mathcal{O},\Omega)=\mathsf{S}_{\mathrm{fin}}(\Omega,\mathcal{O}^{\omega\text{-gp}})$ [**2**]. A positive answer to the following problem is consistent [**71**].

Problem 7.8. *Is* $\mathsf{U}_{\mathrm{fin}}(\mathcal{O},\mathrm{T})=\mathsf{S}_{\mathrm{fin}}(\Omega,\mathcal{O}^{\tau\text{-gp}})$*?* 281?

$\mathsf{S}_1(\Omega, \mathcal{O}^{\omega\text{-gp}})$ is strictly stronger than $\mathsf{S}_1(\mathcal{O}, \mathcal{O})$ [**60**]. $\mathsf{S}_1(\Omega, \mathcal{O}^{\omega\text{-gp}}) = \mathsf{U}_{\text{fin}}(\mathcal{O}, \Omega) \cap \mathsf{S}_1(\mathcal{O}, \mathcal{O})$ [**60**], so the following problem can also be stated in classical terms.

282? **Problem 7.9** ([**2**]). *Is* $\mathsf{S}_1(\Omega, \Omega) = \mathsf{S}_1(\Omega, \mathcal{O}^{\omega\text{-gp}})$*?*

$\mathsf{U}_{\text{fin}}(\mathcal{O}, \Gamma) = \binom{\Lambda}{\mathcal{O}^{\Upsilon\text{-gp}}}$ [**59**]. Zdomskyy proved that a positive answer to the following problem follows from NCF.

283? **Problem 7.10.** *Is* $\mathsf{U}_{\text{fin}}(\mathcal{O}, \Omega) = \binom{\Lambda}{\mathcal{O}^{\omega\text{-gp}}}$*?*

8. Splittability

Assume that $\mathscr{A}$ and $\mathscr{B}$ are collections of covers of a space X. The following property was introduced in [**45**], in connection to Ramsey Theory.

$\mathsf{Split}(\mathscr{A}, \mathscr{B})$: Every cover $\mathcal{U} \in \mathscr{A}$ can be split into two disjoint subcovers $\mathcal{V}$ and $\mathcal{W}$, each containing an elements of $\mathscr{B}$.

If we consider this prototype with $\mathscr{A}, \mathscr{B} \in \{\Lambda, \Omega, \mathrm{T}, \Gamma\}$, we obtain 16 properties, each of which being either trivial or equivalent to one in Figure 5. In this diagram, the dotted implications are open. The implication (1) in this diagram holds if, and only if, its implication (2) holds, and if (1) (and (2)) holds, then (3) holds, either.

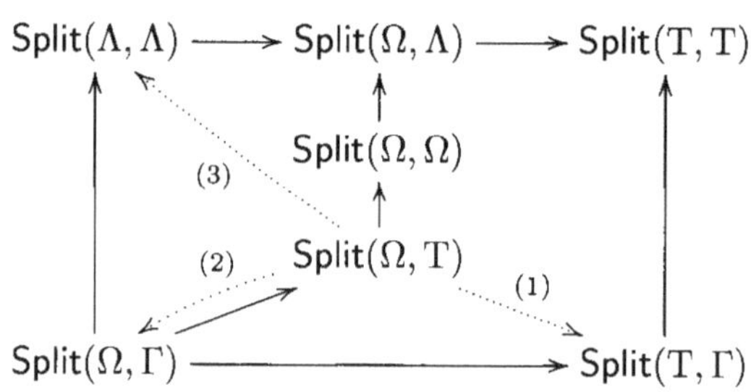

FIGURE 5. The splitting properties

284–285? **Problem 8.1** ([**58**]).

(1) *Does* $\mathsf{Split}(\Omega, \mathrm{T})$ *imply* $\mathsf{Split}(\mathrm{T}, \Gamma)$*?*
(2) *And if not, then does* $\mathsf{Split}(\Omega, \mathrm{T})$ *imply* $\mathsf{Split}(\Lambda, \Lambda)$*?*

The product of a σ-compact X with Y satisfying $\mathsf{U}_{\text{fin}}(\mathcal{O}, \mathscr{B})$ ($\mathscr{B} \in \{\mathcal{O}, \Omega, \Gamma\}$) satisfies $\mathsf{U}_{\text{fin}}(\mathcal{O}, \mathscr{B})$ [**61, 66**].

286? **Problem 8.2** (Zdomskyy). *Assume that X is compact and Y satisfies* $\mathsf{Split}(\Lambda, \Lambda)$. *Does $X \times Y$ satisfy* $\mathsf{Split}(\Lambda, \Lambda)$*?*

Problem 8.2 is also open for the other splitting properties.

287? **Problem 8.3** ([**61**]). *Improve the lower bound or the upper bound in the inequality* $\aleph_1 \leq \operatorname{add}(\mathsf{Split}(\Omega, \Lambda)) \leq \mathfrak{c}$.

Problem 8.4 ([**61**]). *Can the lower bound* $\mathfrak{u}$ *on* $\mathrm{add}(\mathsf{Split}(\mathrm{T},\mathrm{T}))$ *be improved?* 288?

All problems below are settled for the properties which do not appear in them.

Problem 8.5 ([**58**]). *Is* $\mathsf{Split}(\Lambda,\Lambda)$ *additive?* 289?

$\mathsf{Split}(\Lambda,\Lambda)$ is consistently additive [**61**, **70**].

Problem 8.6 ([**58**]). *Are any of the properties* $\mathsf{Split}(\mathcal{B}_\Omega,\mathcal{B}_\Lambda)$, $\mathsf{Split}(\mathcal{B}_\Omega,\mathcal{B}_\Omega)$, 290–293? $\mathsf{Split}(\mathcal{B}_\mathrm{T},\mathcal{B}_\mathrm{T})$, *and* $\mathsf{Split}(\mathcal{B}_\mathrm{T},\mathcal{B}_\Gamma)$ *hereditary?*

Problem 8.7 ([**58**]). *Are any of the properties* $\mathsf{Split}(\Omega,\Omega)$, $\mathsf{Split}(\Omega,\mathrm{T})$, *or* $\mathsf{Split}(\mathrm{T},$ 294–296? $\mathrm{T})$ *preserved under taking finite powers?*

9. Function spaces and local-global principles

Let X be a topological space, and $x \in X$. A subset A of X *converges* to x, $x = \lim A$, if A is infinite, $x \notin A$, and for each neighborhood U of x, $A \setminus U$ is finite. Consider the following collections: $\Omega_x = \{A \subseteq X : x \in \mathrm{cl}(A) \setminus A\}$ and $\Gamma_x = \{A \subseteq X : |A| = \aleph_0 \text{ and } x = \lim A\}$. $\Gamma_x \subseteq \Omega_x$. The following implications hold, and none further [**4**].

$$\begin{array}{ccccccc}
\mathsf{S}_1(\Gamma_x,\Gamma_x) & \longrightarrow & \mathsf{S}_{\mathrm{fin}}(\Gamma_x,\Gamma_x) & \longrightarrow & \mathsf{S}_1(\Gamma_x,\Omega_x) & \longrightarrow & \mathsf{S}_{\mathrm{fin}}(\Gamma_x,\Omega_x) \\
\uparrow & & \uparrow & & \uparrow & & \uparrow \\
\mathsf{S}_1(\Omega_x,\Gamma_x) & \longrightarrow & \mathsf{S}_{\mathrm{fin}}(\Omega_x,\Gamma_x) & \longrightarrow & \mathsf{S}_1(\Omega_x,\Omega_x) & \longrightarrow & \mathsf{S}_{\mathrm{fin}}(\Omega_x,\Omega_x)
\end{array}$$

In the current section, when we write $\Pi(\mathscr{A}_x,\mathscr{B}_x)$ without specifying x, we mean $(\forall x)\Pi(\mathscr{A}_x,\mathscr{B}_x)$. $\mathsf{S}_{\mathrm{fin}}(\Omega_x,\Omega_x)$ is Arhangel′skiĭ's *countable fan tightness*, and $\mathsf{S}_1(\Omega_x, \Omega_x)$ is Sakai's *countable strong fan tightness*. $\mathsf{S}_1(\Gamma_x,\Gamma_x)$ and $\mathsf{S}_{\mathrm{fin}}(\Gamma_x,\Gamma_x)$ are Arhangel′skiĭ's properties α_2 and α_4, respectively.

In the remainder of this section, X will always denote a *subset of* $\mathbb{R}\setminus\mathbb{Q}$. The set of all real-valued functions on X, denoted $\mathbb{R}^X$, is equipped with the Tychonoff product topology. $C_p(X)$ is the subspace of $\mathbb{R}^X$ consisting of the continuous real-valued functions on X. The topology of $C_p(X)$ is known as the *topology of pointwise convergence*. The constant zero element of $C_p(X)$ is denoted $\mathbf{0}$.

For some of the pairs $(\mathscr{A},\mathscr{B}) \in \{\Omega,\Gamma\}^2$ and $\Pi \in \{\mathsf{S}_1,\mathsf{S}_{\mathrm{fin}}\}$, it is known that $C_p(X)$ satisfies $\Pi(\mathscr{A}_\mathbf{0},\mathscr{B}_\mathbf{0})$ if, and only if, X satisfies $\Pi(\mathscr{A},\mathscr{B})$ (see [**50**] for a summary).

Fremlin's s_1 for X and Bukovský's wQN for X are equivalent to $\mathsf{S}_1(\Gamma_\mathbf{0},\Gamma_\mathbf{0})$ for $C_p(X)$ [**20**, **49**]. In a manner similar to the observation made in Section 3 of [**49**], a positive solution to Problem 6.13 should imply a positive solution to the following problem.

Problem 9.1 ([**19**]). *Assume that* $\kappa < \mathfrak{b}$, *and for each* $\alpha < \lambda$, $C_p(X_\alpha)$ *satisfies* 297? $\mathsf{S}_1(\Gamma_\mathbf{0},\Gamma_\mathbf{0})$. *Does* $C_p(\bigcup_{\alpha<\kappa} X_\alpha)$ *satisfy* $\mathsf{S}_1(\Gamma_\mathbf{0},\Gamma_\mathbf{0})$*?*

If X satisfies $\mathsf{S}_1(\Gamma,\Gamma)$, then $C_p(X)$ satisfies $\mathsf{S}_1(\Gamma_\mathbf{0},\Gamma_\mathbf{0})$ [**49**].

Problem 9.2 ([**49**]). *Is* $\mathsf{S}_1(\Gamma_\mathbf{0},\Gamma_\mathbf{0})$ *for* $C_p(X)$ *equivalent to* $\mathsf{S}_1(\Gamma,\Gamma)$ *for* X*?* 298?

If the answer is positive, then Problems 6.13 and 9.1 coincide. There are several partial solutions to Problem 9.2: First, if $C_p(X)$ is *hereditarily* $\mathsf{S}_1(\Gamma_0,\Gamma_0)$, then X satisfies $\mathsf{S}_1(\Gamma,\Gamma)$ [**24**]. Second, $\mathsf{S}_1(\Gamma_0,\Gamma_0)$ for $C_p(X)$ is equivalent to $\bar{\mathsf{S}}_1(\Gamma,\Gamma)$ for X, where $\bar{\mathsf{S}}_1$ is like S_1 with the following additional restriction on the given γ-covers $\mathcal{U}_n$: For each n, the family of closures of the elements of $\mathcal{U}_{n+1}$ refines $\mathcal{U}_n$ [**17**]. Finally, $\mathsf{S}_1(\Gamma_0,\Gamma_0)$ for $C_p(X)$ is equivalent to $\mathsf{S}_1(C_\Gamma,C_\Gamma)$ for X, where C_Γ is the collection of *clopen* γ-covers of X [**43**]. This reduces Problem 9.2 to the question whether $\mathsf{S}_1(\Gamma,\Gamma)=\mathsf{S}_1(C_\Gamma,C_\Gamma)$.

The following also seems to be open.

299? **Problem 9.3** (Scheepers). *Is* $\mathsf{S}_1(\Gamma_0,\Omega_0)$ *for* $C_p(X)$ *equivalent to* $\mathsf{S}_1(\Gamma,\Omega)$ *for* X*?*

$\mathsf{S}_1(\Gamma_0,\Omega_0)$ for $C_p(X)$ is equivalent to $\mathsf{S}_1(C_\Gamma,C_\Omega)$ for X, where C_Ω is the collection of clopen ω-covers of X [**43**], so we really want to know whether $\mathsf{S}_1(\Gamma,\Omega)=\mathsf{S}_1(C_\Gamma,C_\Omega)$.

A topological space Y is *κ-Fréchet* if it satisfies $\binom{O(\Omega_x)}{\Gamma_x}$, where $O(\Omega_x)$ is the family of elements of Ω_x which are open.

300? **Problem 9.4** ([**44**]). *What is the minimal cardinality of a set* $X\subseteq\mathbb{R}$ *such that* $C_p(X)$ *does not satisfy* $\binom{O(\Omega_x)}{\Gamma_x}$*?*

The answer is at least $\mathfrak{b}$ [**44**].

There are many additional important questions about these and related kinds of local-global principles. Some of them are surveyed in [**23**].

10. Topological groups

Let $\mathcal{O}_{\rm nbd}$ denote the covers of G of the form $\{g\cdot U : g\in G\}$, where U is a neighborhood of the unit element of G. Okunev has introduced the property $\mathsf{S}_{\rm fin}(\mathcal{O}_{\rm nbd},\mathcal{O})$, traditionally called *o-boundedness* or *Menger-boundedness.* Let $\Omega_{\rm nbd}$ denote the covers of G of the form $\{F{\cdot}U : F\in[G]^{<\aleph_0}\}$, where U is a neighborhood of the unit element of G, such that for each $F\in[G]^{<\aleph_0}$, $F\cdot U\neq G$. Kočinac has introduced $\mathsf{S}_1(\Omega_{\rm nbd},\Omega)$, $\mathsf{S}_1(\Omega_{\rm nbd},\Gamma)$, and $\mathsf{S}_1(\mathcal{O}_{\rm nbd},\mathcal{O}_{\rm nbd})$, traditionally called *Scheepers-boundedness*, *Hurewicz-boundedness*, and *Rothberger-boundedness.*

The relations among these boundedness properties and their topological counterparts were studied in many papers, see [**1, 3, 5, 25, 26, 34, 35, 62, 69**], and references therein. In particular, the following diagram of implications is complete.

$$\begin{array}{ccccc}
\mathsf{S}_1(\Omega_{\rm nbd},\Gamma) & \longrightarrow & \mathsf{S}_1(\Omega_{\rm nbd},\Omega) & \longrightarrow & \mathsf{S}_{\rm fin}(\mathcal{O}_{\rm nbd},\mathcal{O}) \\
 & & & & \uparrow \\
 & & & & \mathsf{S}_1(\mathcal{O}_{\rm nbd},\mathcal{O}_{\rm nbd})
\end{array}$$

$\mathsf{S}_{\rm fin}(\mathcal{O}_{\rm nbd},\mathcal{O})$ is not provably preserved under cartesian products [**5, 33, 62**].

301? **Problem 10.1** (Tkačenko). *Are there, in* ZFC, *groups* G,H *satisfying* $\mathsf{S}_{\rm fin}(\mathcal{O}_{\rm nbd},\mathcal{O})$ *such that* $G\times H$ *does not satisfy* $\mathsf{S}_{\rm fin}(\mathcal{O}_{\rm nbd},\mathcal{O})$*?*

A topological group G satisfies $\mathsf{S}_1(\Omega_{\rm nbd}, \Omega)$ if, and only if, all finite powers of G satisfy $\mathsf{S}_{\rm fin}(\mathcal{O}_{\rm nbd}, \mathcal{O})$ [**3**]. Thus, the case where $G = H$ in Problem 10.1 is related to the following problem.

Problem 10.2 ([**35**]). *Is $\mathsf{S}_1(\Omega_{\rm nbd}, \Omega) = \mathsf{S}_{\rm fin}(\mathcal{O}_{\rm nbd}, \mathcal{O})$ for separable metrizable* 302–303?
groups? Specifically:

(1) *Does* CH *imply the existence of a separable metrizable group G satisfying $\mathsf{S}_{\rm fin}(\mathcal{O}_{\rm nbd}, \mathcal{O})$ but not $\mathsf{S}_1(\Omega_{\rm nbd}, \Omega)$?*
(2) *Is it consistent that $\mathsf{S}_{\rm fin}(\mathcal{O}_{\rm nbd}, \mathcal{O}) = \mathsf{S}_1(\Omega_{\rm nbd}, \Omega)$ for separable metrizable groups?*

If G is *analytic* and does not satisfy $\mathsf{S}_1(\Omega_{\rm nbd}, \Gamma)$, then G^2 does not satisfy $\mathsf{S}_{\rm fin}(\mathcal{O}_{\rm nbd}, \mathcal{O})$. Thus, for analytic groups, $\mathsf{S}_1(\Omega_{\rm nbd}, \Gamma) = \mathsf{S}_1(\Omega_{\rm nbd}, \Omega)$ [**7**]. Moreover, for analytic *abelian* groups, $\mathsf{S}_{\rm fin}(\mathcal{O}_{\rm nbd}, \mathcal{O}) = \mathsf{S}_1(\Omega_{\rm nbd}, \Gamma)$ [**7**]. For general analytic groups this is open.

Problem 10.3 ([7]). *Is there an analytic group satisfying $\mathsf{S}_{\rm fin}(\mathcal{O}_{\rm nbd}, \mathcal{O})$ but not* 304?
$\mathsf{S}_1(\Omega_{\rm nbd}, \Gamma)$?

It seems that $\mathbb{Z}^{\mathbb{N}}$ for boundedness properties of topological groups is like $\mathbb{R}$ for topological and measure-theoretic notions of smallness [**35**]. Thus, unless otherwise indicated, all of the problems in the remainder of this section are concerning subgroups of $\mathbb{Z}^{\mathbb{N}}$.

Say that $G \leq \mathbb{Z}^{\mathbb{N}}$ is *bounded* if $\{|g| : g \in G\}$ is bounded (with respect to $\leq^*$), where $|g|$ denotes the absolute value of g. For subgroups of $\mathbb{Z}^{\mathbb{N}}$:

(1) G satisfies $\mathsf{S}_1(\Omega_{\rm nbd}, \Gamma)$ if, and only if, G is bounded [**1**].
(2) G satisfies $\mathsf{S}_1(\mathcal{O}_{\rm nbd}, \mathcal{O}_{\rm nbd})$ if, and only if, G has strong measure zero [**3**].

Problem 10.4 ([**35**]). *Is it consistent that there is $G \leq \mathbb{Z}^{\mathbb{N}}$ such that G has strong* 305?
measure zero, is unbounded, and does not satisfy $\mathsf{S}_{\rm fin}(\mathcal{O}, \mathcal{O})$?

Problem 10.5 ([**35**]). *Is it consistent that there is $G \leq \mathbb{Z}^{\mathbb{N}}$ such that G has* 306?
strong measure zero and satisfies $\mathsf{S}_{\rm fin}(\mathcal{O}, \mathcal{O})$, but is unbounded and does not satisfy $\mathsf{S}_1(\mathcal{O}, \mathcal{O})$?

Some open problems involve only the standard covering properties. The following problem is related to Problem 3.2.

Problem 10.6 ([**62**]). *Is there (in* ZFC*) a group $G \leq \mathbb{Z}^{\mathbb{N}}$ of cardinality $\mathfrak{d}$ satisfying* 307?
$\mathsf{S}_{\rm fin}(\mathcal{O}, \mathcal{O})$?

Problem 10.7 ([**62**]). *Does* CH *imply the existence of a a group $G \leq \mathbb{Z}^{\mathbb{N}}$ of* 308–309?
cardinality $\mathfrak{c}$ satisfying $\mathsf{S}_1(\mathcal{B}_\Omega, \mathcal{B}_\Gamma)$, or at least $\mathsf{S}_1(\Omega, \Gamma)$?

Some approximations to Problem 10.7 are given in [**62**]: CH implies the existence of groups satisfying $\mathsf{S}_1(\mathcal{B}_\Omega, \mathcal{B}_\Omega)$ and of groups satisfying $\mathsf{S}_1(\mathcal{B}_\Gamma, \mathcal{B}_\Gamma)$ in all finite powers. The answer for Problem 10.7 is positive if it is for 5.2. It is also positive for the property (δ). To get a complete positive answer, it suffices to construct a set $X \subseteq \mathbb{Z}^{\mathbb{N}}$ such that all finite powers of X satisfy $\mathsf{S}_1(\mathcal{B}_\Omega, \mathcal{B}_\Gamma)$. Thus, it suffices to have a positive answer for Problem 6.4.

Finally, recall Problem 2.2, and see the other problems in [**69**].

11. Cardinal characteristics of the continuum

We mention here several problems in the field which are connected to selection principles.

The main open problem in the field is the *Minimal Tower Problem*. This problem has motivated the study of τ-covers.

310? **Problem 11.1** ([**18**]). *Is it consistent that* $\mathfrak{p} < \mathfrak{t}$*?*

Shelah is currently working on a possible positive solution to this problem.

The study of τ^*-covers, a variant of τ-covers, led to the following problem. A family $\mathcal{F} \subseteq [\mathbb{N}]^{\aleph_0}$ is *linearly refinable* if for each $A \in \mathcal{F}$ there exists an infinite subset $\hat{A} \subseteq A$ such that the family $\hat{\mathcal{F}} = \{\hat{A} : A \in \mathcal{F}\}$ is linearly (quasi)ordered by $\subseteq^*$. $\mathfrak{p}^*$ is the minimal size of a centered family in $[\mathbb{N}]^{\aleph_0}$ which is not linearly refinable.

$\mathfrak{p} = \min\{\mathfrak{p}^*, \mathfrak{t}\}$, and $\mathfrak{p}^* \leq \mathfrak{d}$ [**55**].

311? **Problem 11.2** ([**52, 55**]). *Is* $\mathfrak{p} = \mathfrak{p}^*$*?*

A family $\mathcal{A} \subseteq ([\mathbb{N}]^{\aleph_0})^{\mathbb{N}}$ is a τ*-family* if for each n, $\{A(n) : A \in \mathcal{A}\}$ is linearly ordered by $\subseteq^*$. A family $\mathcal{A} \subseteq ([\mathbb{N}]^{\aleph_0})^{\mathbb{N}}$ is *o-diagonalizable* if there is $g \in \mathbb{N}^{\mathbb{N}}$ such that $(\forall A \in \mathcal{A})(\exists n)\ g(n) \in A(n)$. Let $\mathfrak{od}$ denote the minimal cardinality of a τ-family which is not o-diagonalizable. $\mathrm{non}(\mathsf{S}_1(\mathrm{T}, \mathcal{O})) = \mathfrak{od}$ [**37**]. $\mathfrak{od}$ is the "tower version" of $\mathrm{cov}(\mathcal{M})$: If we replace "linearly ordered by $\subseteq^*$" by "centered" in the definition of $\mathfrak{od}$, then we obtain $\mathrm{cov}(\mathcal{M})$. Thus, $\mathrm{cov}(\mathcal{M}) \leq \mathfrak{od}$. If $\mathrm{cov}(\mathcal{M}) = \aleph_1$, then $\mathfrak{od} = \aleph_1$ either [**37**].

312? **Problem 11.3** ([**37**]). *Is it consistent that* $\mathrm{cov}(\mathcal{M}) < \mathfrak{od}$*?*

Another variant of the minimal tower problem is the following. For a cardinal number $\kappa > 1$ (finite or infinite), define $\mathfrak{p}_\kappa$ to be the minimal cardinality of a centered subset of $[\mathbb{N}]^{\aleph_0}$ which cannot be partitioned into less than κ sets each having a pseudo-intersection.

It is easy to see that $\mathfrak{p} = \mathfrak{p}_2 = \mathfrak{p}_3 = \cdots = \mathfrak{p}_{\aleph_0}$, and $\mathfrak{p}_\mathfrak{t} = \mathfrak{t}$. It turns out that $\mathfrak{p} = \mathfrak{p}_{\aleph_1}$ [**53**]. We get a hierarchy of cardinals between $\mathfrak{p}$ and $\mathfrak{t}$:

$$\mathfrak{p} = \mathfrak{p}_{\aleph_1} \leq \mathfrak{p}_{\aleph_2} \leq \cdots \leq \mathfrak{p}_\mathfrak{t} = \mathfrak{t}.$$

313? **Problem 11.4.** *Is* $\mathfrak{p}_\mathfrak{p} = \mathfrak{t}$*?*

Finally, consider the following Ramsey-theoretic cardinal: For a subset Y of $\mathbb{N}^{\mathbb{N}}$ and $g \in \mathbb{N}^{\mathbb{N}}$, we say that g *avoids middles* in Y if:

(1) for each $f \in Y$, $g \not\leq^* f$;
(2) for all $f, h \in Y$ at least one of the sets $\{n : f(n) < g(n) \leq h(n)\}$ and $\{n : h(n) < g(n) \leq f(n)\}$ is finite.

$\mathrm{add}(\mathfrak{X}, \mathfrak{D})$ is the minimal cardinality κ of a dominating $Y \subseteq \mathbb{N}^{\mathbb{N}}$ such that for each partition of Y into κ many pieces, there is a piece such that no g avoids middles in that piece. This cardinal is studied in [**52**].

314? **Problem 11.5** ([**52**]). *Is* $\mathrm{cov}(\mathcal{M}) \leq \mathrm{add}(\mathfrak{X}, \mathfrak{D})$*?*

12. Additional problems and other special sets of reals

If a set of reals X satisfies $\mathsf{S}_{\rm fin}(\mathcal{O},\mathcal{O})$, then for each continuous image Y of X in $\mathbb{N}^{\mathbb{N}}$, Y is not dominating, that is, the set $G=\{g\in\mathbb{N}^{\mathbb{N}}:(\exists f\in Y)\ g\leq^* f\}$ is not equal to $\mathbb{N}^{\mathbb{N}}$ [**27**]. In fact, G satisfies $\mathsf{S}_{\rm fin}(\mathcal{O},\mathcal{O})$ [**67**].

If X satisfies $\mathsf{U}_{\rm fin}(\mathcal{O},\Omega)$, then for each continuous image Y of X in $\mathbb{N}^{\mathbb{N}}$, $\{g\in\mathbb{N}^{\mathbb{N}}:(\exists k)(\exists f_1,\dots,f_k\in Y)\ g\leq^*\max\{f_1,\dots,f_k\}\}$ is not comeager [**66**].

Problem 12.1 ([**66**]). *Assume that X satisfies $\mathsf{U}_{\rm fin}(\mathcal{O},\Omega)$. Does it follow that* 315?
$G=\{g\in\mathbb{N}^{\mathbb{N}}:(\exists k)(\exists f_1,\dots,f_k\in Y)\ g\leq^*\max\{f_1,\dots,f_k\}\}$ satisfies $\mathsf{U}_{\rm fin}(\mathcal{O},\Omega)$?

We now give a short selection of problems on special sets of reals. See [**38**] or the cited references for the definitions.

$X\subseteq\mathbb{R}$ is a *ν-set* if for each $Y\subseteq X$ which is nowhere dense in X, Y is countable (i.e., X is Luzin relative to itself). Every continuous image of a ν-set has the property assumed in the following problem.

Problem 12.2 ([**16**]). *Assume that $X\subseteq\mathbb{R}$, and for each $Y\subseteq X$, Y is concen-* 316?
trated on a countable subset of Y. Does it follow that X is a continuous image of a ν-set?

Problem 12.3 ([**8**]). *Is it consistent that* $\mathrm{cov}(\mathcal{M})=\aleph_1<\mathfrak{c}=\aleph_{\omega_1}$, *and there is a* 317?
$\mathfrak{c}$-Luzin set (i.e., L with $|L|=\mathfrak{c}$ and $|L\cap M|<\mathfrak{c}$ for all meager $M\subseteq\mathbb{R}$)?

Problem 12.4 ([**8**]). *Assume that every strong measure zero set of reals is meager-* 318?
additive. Does Borel's Conjecture follow?

The assumption in the last problem implies that $\mathrm{cov}(\mathcal{M})=\mathrm{non}(SMZ)<\mathrm{cf}(\mathcal{M})$.

If $X,Y\subseteq\{0,1\}^{\mathbb{N}}$ are meager-additive, then $X\times Y$ is a meager-additive subset of $\{0,1\}^{\mathbb{N}}\times\{0,1\}^{\mathbb{N}}$. The same is true for null-additive subsets of $\{0,1\}^{\mathbb{N}}$. For the real line this is open.

Problem 12.5 ([**64**]). *Assume that $X,Y\subseteq\mathbb{R}$ are meager- (respectively, null-)* 319–320?
additive. Does it follow that $X\times Y$ is meager- (respectively, null-) additive?

Weiss proved that every meager-additive subset of the Cantor space, when viewed as a subset of $\mathbb{R}$ (where each $f\in\{0,1\}^{\mathbb{N}}$ is identified with $\sum_n f(n)/2^n$), is meager-additive (with respect to the usual addition in $\mathbb{R}$); and similarly for null-additive.

Problem 12.6 ([**64**]). *Assume that $X\subseteq\mathbb{R}$ is meager- (respectively, null-) addi-* 321–322?
tive, and $X\subseteq[0,1]$. Does it follow that X is meager- (respectively, null-) additive when viewed as a subset of $\{0,1\}^{\mathbb{N}}$?

For a set H, define $H_x=\{y:(x,y)\in H\}$.

Problem 12.7 (Bartoszyński). *Assume that $X\subseteq\mathbb{N}^{\mathbb{N}}$ is nonmeager and $Y\subseteq\mathbb{N}^{\mathbb{N}}$* 323?
is dominating. Is there a Borel set $H\subseteq\mathbb{N}^{\mathbb{N}}\times\mathbb{N}^{\mathbb{N}}$ such that every meager set is contained in H_x for some $x\in X\cup Y$?

Acknowledgments. We thank Ljubisa Kočinac, Nadav Samet, Marion Scheepers, and Lyubomyr Zdomskyy for their useful comments on this chapter.

References

[1] L. Babinkostova, *Metrizable groups and strict o-boundedness*, 2005, Preprint. To appear in Mat. Vesnik.

[2] L. Babinkostova, Lj. D. R. Kočinac, and M. Scheepers, *Combinatorics of open covers. VIII*, Topology Appl. **140** (2004), no. 1, 15–32.

[3] L. Babinkostova and M. Kočinac, Lj. D. R. Scheepers, *Combinatorics of open covers (XI): Menger- and Rothberger-bounded groups*, 2004, Preprint. To appear in Topology Appl.

[4] L. Babinkostova, M. Scheepers, and B. Tsaban, *Local-global principles in topology (I)*, In preparation.

[5] T. Banakh, P. Nickolas, and M. Sanchis, *Filter games and pathological subgroups of a countable product of lines*, J. Austral. Math. Soc., To appear.

[6] T. Banakh and L. Zdomskyy, *Separation properties between the σ-compactness and Hurewicz property*, 2006, Preprint.

[7] T. Bankah and L. Zdomskyy, *Selection principles and infinite games on multicovered spaces and their applications*, Book in progress.

[8] T. Bartoszyński and H. Judah, *Strong measure zero sets*, Set theory of the reals (Ramat Gan, 1991), Bar-Ilan University, Ramat Gan, 1993, pp. 13–62.

[9] T. Bartoszyński and H. Judah, *On the cofinality of the smallest covering of the real line by meager sets. II*, Proc. Amer. Math. Soc. **123** (1995), no. 6, 1879–1885.

[10] T. Bartoszynski and S. Shelah, *Continuous images of sets of reals*, Topology Appl. **116** (2001), no. 2, 243–253.

[11] T. Bartoszynski, S. Shelah, and B. Tsaban, *Additivity properties of topological diagonalizations*, J. Symbolic Logic **68** (2003), no. 4, 1254–1260.

[12] T. Bartoszyński and B. Tsaban, *Hereditary topological diagonalizations and the Menger–Hurewicz conjectures*, Proc. Amer. Math. Soc. **134** (2006), no. 2, 605–615.

[13] A. Blass, *Combinatorial cardinal characteristics of the continuum*, Handbook of set theory (M. Foreman, A. Kanamori, and M. Magidor, eds.), Springer, Berlin,, To appear.

[14] E. Borel, *Sur la classification des ensembles de mesure nulle*, Bull. Soc. Math. France **47** (1919), 97–125.

[15] J. Brendle, *Generic constructions of small sets of reals*, Topology Appl. **71** (1996), no. 2, 125–147.

[16] J. B. Brown and G. V. Cox, *Continuous images of Lusin sets*, Proc. Amer. Math. Soc. **89** (1983), no. 1, 52–54.

[17] L. Bukovský and J. Haleš, *QN-space, wQN-space, and coverings properties*, Preprint.

[18] E. K. van Douwen, *The integers and topology*, Handbook of set-theoretic topology, North-Holland, Amsterdam, 1984, pp. 111–167.

[19] D. H. Fremlin, *Sequential convergence in $C_p(X)$*, Comment. Math. Univ. Carolin. **35** (1994), no. 2, 371–382.

[20] D. H. Fremlin, *SSP and wQN*, 2003, Preprint.

[21] F. Galvin and A. W. Miller, *γ-sets and other singular sets of real numbers*, Topology Appl. **17** (1984), no. 2, 145–155.

[22] J. Gerlits and Zs. Nagy, *Some properties of $C(X)$. I*, Topology Appl. **14** (1982), no. 2, 151–161.

[23] G. Gruenhage, *Products of Fréchet spaces*, Topology Proc. **30** (2006), no. 2.

[24] J. Haleš, *On Scheepers' conjecture*, Acta Univ. Carolin. Math. Phys. **46** (2005), no. 2, 27–31.

[25] C. Hernández, *Topological groups close to being σ-compact*, Topology Appl. **102** (2000), no. 1, 101–111.

[26] C. Hernández, D. Robbie, and M. G. Tkachenko, *Some properties of o-bounded and strictly o-bounded groups*, Appl. Gen. Topol. **1** (2000), no. 1, 29–43.

[27] W. Hurewicz, *Über eine Verallgemeinerung des Borelschen Theorems*, Math. Z. **24** (1925), 401–425.

[28] W. Hurewicz, *Über Folgen stetiger Funktionen*, Fund. Math. **9** (1927), 193–204.

[29] W. Just, A. W. Miller, M. Scheepers, and P. J. Szeptycki, *The combinatorics of open covers. II*, Topology Appl. **73** (1996), no. 3, 241–266.

[30] Lj. D. R. Kočinac, *Selected results on selection principles*, Proceedings of the 3rd Seminar on Geometry & Topology, Azarb. Univ. Tarbiat Moallem, Tabriz, 2004, pp. 71–104.

[31] Lj. D. R. Kočinac and M. Scheepers, *Function spaces and a property of Reznichenko*, Topology Appl. **123** (2002), no. 1, 135–143.

[32] Lj. D. R. Kočinac and M. Scheepers, *Combinatorics of open covers. VII. Groupability*, Fund. Math. **179** (2003), no. 2, 131–155.

[33] A. Krawczyk and H. Michalewski, *Linear metric spaces close to being σ-compact*, Technical report 46, Institute of Mathematics, Warsaw University, 2001, `http://www.mimuw.edu.pl/english/research/reports/tr-imat/46/products.ps`.

[34] A. Krawczyk and H. Michalewski, *An example of a topological group*, Topology Appl. **127** (2003), no. 3, 325–330.

[35] M. Machura and B. Tsaban, *The combinatorics of the Baire group*, Submitted. `arXiv: math.GN/0508146`

[36] K. Menger, *Einige überdeckungssätze der Punktmengenlehre*, Sitzungsberichte der Wiener Akademie **133** (1924), 421–444.

[37] H. Mildenberger, S. Shelah, and B. Tsaban, *The combinatorics of τ-covers*, Topology Appl. (2006), Article in press.

[38] A. W. Miller, *Special subsets of the real line*, Handbook of set-theoretic topology, North-Holland, Amsterdam, 1984, pp. 201–233.

[39] A. W. Miller, *A nonhereditary Borel-cover γ-set*, Real Anal. Exchange **29** (2003/04), no. 2, 601–606.

[40] J. Pawlikowski and I. Recław, *Parametrized Cichoń's diagram and small sets*, Fund. Math. **147** (1995), no. 2, 135–155.

[41] I. Recław, *Every Lusin set is undetermined in the point-open game*, Fund. Math. **144** (1994), no. 1, 43–54.

[42] F. Rothberger, *Sur les familles indénombrables de suites de nombres naturels et les problèmes concernant la propriété C*, Proc. Cambridge Philos. Soc. **37** (1941), 109–126.

[43] M. Sakai, *The sequence selection properties of $C_p(X)$*, Topology Appl. (2006), Article in press.

[44] M. Sakai, *Special subsets of reals characterizing local properties of function spaces*, Quad. Mat., To appear.

[45] M. Scheepers, *Combinatorics of open covers. I. Ramsey theory*, Topology Appl. **69** (1996), no. 1, 31–62.

[46] M. Scheepers, *$C_p(X)$ and Arhangel′skiĭ's α_i-spaces*, Topology Appl. **89** (1998), no. 3, 265–275.

[47] M. Scheepers, *Finite powers of strong measure zero sets*, J. Symbolic Logic **64** (1999), no. 3, 1295–1306.

[48] M. Scheepers, *The length of some diagonalization games*, Arch. Math. Logic **38** (1999), no. 2, 103–122.

[49] M. Scheepers, *Sequential convergence in $C_p(X)$ and a covering property*, East-West J. Math. **1** (1999), no. 2, 207–214.

[50] M. Scheepers, *Selection principles and covering properties in topology*, Note Mat. **22** (2003), no. 2, 3–41.

[51] M. Scheepers and B. Tsaban, *The combinatorics of Borel covers*, Topology Appl. **121** (2002), no. 3, 357–382.

[52] S. Shelah and B. Tsaban, *Critical cardinalities and additivity properties of combinatorial notions of smallness*, J. Appl. Anal. **9** (2003), no. 2, 149–162.

[53] P. Simon and B. Tsaban, *The minimal cardinality where the Pytkeev property fails*, Proc. Amer. Math. Soc., To appear.

[54] B. Tsaban, *A topological interpretation of* $\mathfrak{t}$, Real Anal. Exchange **24** (1998), no. 1, 391–404.

[55] B. Tsaban, *Selection principles and the minimal tower problem*, Note Mat. **22** (2003), no. 2, 53–81.

[56] B. Tsaban, *Selection principles in mathematics: A milestone of open problems*, Note Mat. **22** (2003), no. 2, 179–208.

[57] B. Tsaban (ed.), *SPM bulletin*, vol. 1, 2003. `arXiv:math.GN/0301011`

[58] B. Tsaban, *The combinatorics of splittability*, Ann. Pure Appl. Logic **129** (2004), 107–130.

[59] B. Tsaban, *The Hurewicz covering property and slaloms in the Baire space*, Fund. Math. **181** (2004), no. 3, 273–280.

[60] B. Tsaban, *Strong* γ*-sets and other singular spaces*, Topology Appl. **153** (2005), no. 4, 620–639.

[61] B. Tsaban, *Additivity numbers of covering properties*, 2006, Preprint. To appear in Quad. Mat. `arXiv:arXiv:math.GN/0604451`

[62] B. Tsaban, *o-bounded groups and other topological groups with strong combinatorial properties*, Proc. Amer. Math. Soc. **134** (2006), no. 3, 881–891.

[63] B. Tsaban, *Some new directions in infinite-combinatorial topology*, Set theory (Centre de Recerca Matemàtica Barcelona, 2003–2004) (J. Bagaria and S. Todorčević, eds.), Birkhäuser, Boston, 2006, pp. 225—256.

[64] B. Tsaban and T. Weiss, *Products of special sets of real numbers*, Real Anal. Exchange **30** (2004/05), no. 2, 819–835.

[65] B. Tsaban and L. Zdomskyy, *Scales, fields, and a problem of Hurewicz*, 2005, Preprint. Submitted for publication. `arXiv:math.GN/0507043`

[66] B. Tsaban and L. Zdomskyy, *Combinatorial images of sets of reals and semifilter trichotomy*, 2006, Preprint. Submitted for publication. `arXiv:math.GN/0604536`

[67] B. Tsaban and L. Zdomskyy, *Menger's covering property and groupwise density*, J. Symbolic Logic **71** (2006), 1053–1056.

[68] T. Weiss and B. Tsaban, *Topological diagonalizations and Hausdorff dimension*, Note Mat. **22** (2003), no. 2, 83–92.

[69] L. Zdomskyy, *Can a Borel group be generated by a Hurewicz subspace?*, 2005, Preprint. To appear in Mat. Stud.

[70] L. Zdomskyy, *A semifilter approach to selection principles*, Comment. Math. Univ. Carolin. **46** (2005), no. 3, 525–539.

[71] L. Zdomskyy, *A semifilter approach to selection principles II:* τ^**-covers*, Comment. Math. Univ. Carolin. **47** (2006), no. 3, 539–547.

Part 2

Set-theoretic Topology

Open Problems in Topology II – Edited by E. Pearl

Introduction: Twenty problems in set-theoretic topology

Michael Hrušák and Justin Tatch Moore

Every healthy mathematical discipline needs a short and concise list of its central problems to maintain its focus. These problems are presumably hard to solve and indicative of the major directions in the field. Ideally, the problems themselves form these directions. In the case of set-theoretic topology, such problems have always been there. However, over the course of the years these problems may have shifted out of focus.

When we were asked to edit the set-theoretic topology section of this book, we decided to do things a little differently. While the first book has been widely successful, merely mimicking the old format did not seem quite the approach we wanted. Rather than compiling a list of completely new problems, we have revisited old ones.

We have assembled a list of 20 major problems in the area, which grew out of discussions we have had at various conferences in 2004 and 2005. While the list is biased by our personal preferences, we have tried to correct this somewhat by discussing it with a number of premier researchers in the area of set-theoretic topology and incorporating their suggestions.

The subsequent articles in this section are strongly inspired by the list, even though they do not exactly follow it; a short article written by a specialist in the area dedicated to one of these *classical problems* containing motivation, historical background and references as well as a list of related questions and test problems.

Problem 1 (Efimov). *Does every infinite compact space either contain a convergent sequence or a copy of $\beta\mathbb{N}$?*

Problem 2 (Arhangel′skiĭ–Franklin). *Is there a finite bound on the sequential order of compact sequential spaces?*

Problem 3 (Fremlin). *Does every perfectly normal compact space admit an at most* 2*-to-*1 *map onto a metric space?*

Problem 4 (Gruenhage). *Does the class of uncountable first countable spaces have a basis consisting of a set of reals with the separable metric, the discrete, and the Sorgenfrey topologies?*

Problem 5 (van Douwen). *Do all compact homogeneous spaces have cellularity at most $\mathfrak{c}$?*

The first listed author gratefully acknowledges support received from a CONACyT grant 46337-F and a PAPIIT grant IN106705. The second author gratefully acknowledges support received from an NSF grant DMS–0401893.

Problem 6 (W. Rudin). *Does every infinite compact homogeneous space contain a convergent sequence?*

Problem 7 (Howes–Miščenko). *Is there a linearly Lindelöf, non-Lindelöf normal space?*

Problem 8 (M.E. Rudin). *Is there a normal space of cardinality $\aleph_1$ whose product with $[0,1]$ is not normal?*

Problem 9 (Malykhin). *Is every separable Fréchet group metrizable?*

Problem 10 (Arhangel′skiĭ). *Is there a non-discrete extremally disconnected topological group?*

Problem 11 (Dow). *Is every extremally disconnected image of $\mathbb{N}^*$ separable?*

Problem 12 (Szymański). *Can ω^* and ω_1^* be homeomorphic?*

Problem 13 (Scarborough–Stone). *Is every product of sequentially compact spaces countably compact?*

Problem 14 (van Douwen). *Is every Lindelöf space a D-space?*

Problem 15 (Ceder). *Are all stratifiable spaces M_1?*

Problem 16 (Nyikos). *Is there a separable, countably compact, first countable space which is not compact?*

Problem 17 (Hušek). *Is every compact space with a small diagonal metrizable?*

Problem 18 (van Mill–Wattel). *Is every space which admits a continuous weak selection weakly orderable?*

Problem 19 (Erdős–Shelah). *Is there a completely separable MAD family?*

Problem 20 (Michael). *Is there a Lindelöf space which has a non-Lindelöf product with the space of irrationals?*

We do not include any definitions nor bibliography on most of these problems as they are to be treated in the subsequent articles by specialists and/or are well known and had been dealt with in the previous *Open problems in topology* book. The only exceptions to this rule are Problem 10 and Problem 17.

The notion of small diagonal was introduced, and Problem 17 was formulated by M. Hušek in [**2**]. We do not feel the need to elaborate more on this problem as G. Gruenhage in a recent and easily available paper [**1**] not only proves new relevant results, but also surveys old results and provides test questions as well as extensive bibliography on the subject.

We will, however, say a few words about Problem 10. Recall that a topological space is *extremally disconnected* if the closures of any two disjoint open sets are disjoint. The first example of a non-discrete (countable) extremally disconnected topological group was constructed using CH by Sirota in 1969 [**5**]. In 1972, A. Louveau [**3**] showed that the existence of a selective ultrafilter on ω is sufficient for Sirota's example. Recently, Y. Zelenyuk [**6**] showed that the existence of a countable non-discrete extremally disconnected topological group implies the existence of a P-point in ω^*, hence, *it is undecidable in* ZFC *whether every countable extremally disconnected topological group is discrete.* Malykhin in [**4**] showed that each (non-discrete) extremely disconnected topological group has an open Boolean subgroup, According to P. Simon, it is also an open question, whether *there is (even consistently) a non-discrete extremally disconnected topological group of size* $\aleph_1$.

Finally, we would like to thank to all of the *interviewed*, most of all Alan Dow and Gary Gruenhage, for helping us compose the list, to all the contributors for their articles and timely submissions, and to Elliott Pearl for putting all of this together.

Enjoy!

Michael Hrušák and Justin Tatch Moore

References

[1] G. Gruenhage, *Spaces having a small diagonal*, Topology Appl. **122** (2002), no. 1–2, 183–200.

[2] M. Hušek, *Topological spaces without κ-accessible diagonal*, Comment. Math. Univ. Carolin. **18** (1977), no. 4, 777–788.

[3] A. Louveau, *Sur un article de S. Sirota*, Bull. Sci. Math. (2) **96** (1972), 3–7.

[4] V. I. Malykhin, *Extremally disconnected and nearly extremally disconnected groups*, Dokl. Akad. Nauk SSSR **220** (1975), 27–30, Translation: Soviet Math. Dokl. 16 (1975) 21–25.

[5] S. M. Sirota, *A product of topological groups, and extremal disconnectedness*, Mat. Sb. **79(121)** (1969), 179–192, Translation: Sb. Math. 8 (1969) 169–180.

[6] Y. Zelenyuk, *On extremally disconnected topological groups*, Topology Appl. **153** (2006), no. 14, 2382–2385.

Open Problems in Topology II – Edited by E. Pearl

Thin-tall spaces and cardinal sequences

Joan Bagaria

1. Introduction

Recall that the *α-Cantor–Bendixson derivative* of a topological space X, denoted by X^α, is defined inductively on the ordinals α as follows: $X^0 = X$, $X^{\alpha+1}$ is the set of accumulation points of X^α, and for limit α, $X^\alpha = \bigcap\{X^\beta : \beta < \alpha\}$.

The *height* of X, denoted by $\mathrm{ht}(X)$, is the least α such that $X^\alpha = X^{\alpha+1}$.

Let $I_\alpha(X) = X^\alpha \setminus X^{\alpha+1}$ be the αth *level* of X. Thus, $I_\alpha(X)$ is the set of isolated points of the subspace $X^\alpha = X \setminus \bigcup\{I_\beta(X) : \beta < \alpha\}$.

The *cardinal sequence* of X is $CS(X) = \langle |I_\beta(X)| : \beta < \mathrm{ht}(X)\rangle$.

X is *scattered—emph* if and only if $X^\alpha = \emptyset$ for some ordinal α. Thus, if X is scattered, then $\mathrm{ht}(X)$ is the first ordinal α such that $I_\alpha(X)$ is empty.

Note that the height of a space X is the same as the height of the scattered subspace $Y = \bigcup_{\alpha<\mathrm{ht}(X)} I_\alpha(X)$. Moreover, $I_\alpha(Y) = I_\alpha(X)$, for all $\alpha < \mathrm{ht}(X) = \mathrm{ht}(Y)$. Thus, to study the possible heights and cardinal sequences of arbitrary topological spaces, one may restrict the study, without loss of generality, to the class of scattered spaces.

Notice that if X is Hausdorff, then each level $I_\alpha(X)$ is infinite, except, maybe, for the last one. If, moreover, X is compact, then $\mathrm{ht}(X)$ is a successor ordinal and the last level must be finite.

The *width* of a scattered space X, denoted by $\mathrm{wd}(X)$, is the supremum of the cardinalities of the levels of X. Thus, $\mathrm{wd}(X)$ is always a cardinal. X is *thin-tall* if $\mathrm{wd}(X) < \mathrm{ht}(X)$.

In this note we are concerned with a series of problems regarding the existence of Hausdorff scattered spaces of a given height and with a given cardinal sequence. Notice that if X is scattered, Hausdorff, locally-compact, but not compact, then by taking the one-point compactification we obtain a compact space with the same cardinal sequence, except for the last level, which contains only one point. Thus, instead of looking at compact spaces, it makes more sense to concentrate on locally-compact scattered Hausdorff spaces, henceforth LCS spaces, for then we can avoid having to mention all the time the last finite level.

1.1. Limitations. There are strong limitations on the height and the cardinal sequence a space can have. The main one is the following:

Lemma 1.1. *Suppose κ is an infinite cardinal. If X is a regular scattered space such that $|I_\alpha(X)| \le \kappa$ for some $\alpha < \mathrm{ht}(X)$, then $|X^\alpha| \le 2^\kappa$.*

Partially supported by the research projects BFM2002-03236 and MTM 2005-01025 of the Spanish *Ministerio de Educación y Ciencia*, and 2002SGR 00126 and 2005SGR-00738 of the *Generalitat de Catalunya*.

PROOF. Fix $\alpha < \mathrm{ht}(X)$ with $|I_\alpha(X)| \leq \kappa$. For each $x \in I_\beta(X)$, $\beta > \alpha$, there is a closed neighborhood $\mathcal{U}$ of x such that $\mathcal{U} \cap I_\beta(X) = \{x\}$. Thus, x is the unique point in $I_\beta(X)$ that belongs to the closure of $\mathcal{U} \cap I_\alpha(X)$. This shows that every point in X^α is uniquely determined by some subset of $I_\alpha(X)$. Hence, $|X^\alpha| \leq 2^\kappa$. □

So, for instance, if the Continuum Hypothesis (CH) holds, then there cannot be any regular scattered space (hence no LCS space) X with at least one countable level α and $|X^\alpha| > \aleph_1$.

It has been recently shown by I. Juhász, S. Shelah, L. Soukup, and Z. Szentmiklóssy [**7**] that for scattered spaces that are regular or zero-dimensional, all cardinal sequences are possible, subject only to the restriction imposed by Lemma 1.1 above.

In the case of LCS spaces, a further limitation is given by the following Lemma, which makes the study of the cardinal sequences of those spaces much more difficult, and therefore interesting.

Lemma 1.2. *If X is an LCS space with $CS(X) = \langle \kappa_\alpha : \alpha < \eta \rangle$, then:*

(1) *$\kappa_{\alpha+1} \leq \kappa_\alpha^{\aleph_0}$, for all $\alpha + 1 < \eta$.*

(2) *If $\delta < \eta$ is a limit ordinal of cofinality λ and C is a sequence of order-type λ converging to δ, then $\kappa_\delta \leq \prod\{\kappa_\alpha : \alpha \in C\}$.*

PROOF. (1) Given a point $x \in I_{\alpha+1}(X)$, let $\mathcal{U}$ be a compact neighborhood of x with $I_{\alpha+1}(X) \cap \mathcal{U} = \{x\}$. Fix any countable infinite sequence S contained in $\mathcal{U} \cap I_\alpha(X)$. Then, by compactness of $\mathcal{U}$, S converges to x. This shows that every point in $I_{\alpha+1}(X)$ is uniquely determined by a countable sequence of points in $I_\alpha(X)$. Hence, $\kappa_{\alpha+1} \leq \kappa_\alpha^{\aleph_0}$.

(2) Given $x \in I_\delta(X)$, let $\mathcal{U}$ be a compact neighborhood of x with $I_{\alpha+1}(X) \cap \mathcal{U} = \{x\}$. For each $\alpha \in C$, pick a point $x_\alpha \in \mathcal{U} \cap I_\alpha(X)$. Then, by compactness of $\mathcal{U}$, the sequence $\{x_\alpha : \alpha \in C\}$ converges to x. Thus, every point in $I_\delta(X)$ is uniquely determined by an element of the product $\prod\{I_\alpha(X) : \alpha \in C\}$. Hence, $\kappa_\delta \leq \prod\{\kappa_\alpha : \alpha \in C\}$. □

As a corollary we have that if X is an LCS space with cardinal sequence $\langle \kappa_\alpha : \alpha < \eta \rangle$, then for every $\alpha < \beta < \omega_1$, $\kappa_\beta \leq \kappa_\alpha^{\aleph_0}$.

2. Spaces of countable height

The problem of building LCS spaces with a given cardinal sequence was completely solved by LaGrange [**13**] for spaces of countable height. The limitations imposed by Lemma 1.2 above characterize all the possible sequences. Namely, if $\theta = \langle \kappa_\alpha : \alpha < \eta \rangle$ is any sequence of infinite cardinals, with η countable, then there exists an LCS space with cardinal sequence θ if and only if for every $\alpha < \beta < \eta$, $\kappa_\beta \leq \kappa_\alpha^{\aleph_0}$.

It is easy to see that any uncountable cardinal κ endowed with the order topology is an example of an LCS space whose cardinal sequence has length κ and all levels are of cardinality κ.

Furthermore, J.C. Martínez [**19**] has recently shown that for every sequence $\theta = \langle \kappa_\alpha : \alpha < \eta \rangle$ of infinite cardinals such that all cardinals in the sequence are greater or equal than the cardinality of η, it is always possible to force an LCS space with cardinal sequence θ.

Thus, the interesting problem is to build LCS spaces of uncountable height in which at least one of the levels has cardinality less than the cardinality of the height of the space. In particular, thin-tall LCS spaces.

3. Spaces of height $< \omega_2$

In 1968, R. Telgársky asked if there exists a thin-tall LCS space. After some positive results using extra set-theoretic hypotheses, the problem was solved by M. Rajagopalan [**20**] by building a space of height ω_1 and countable width. A simpler proof was provided by I. Juhász and W. Weiss [**9**], who also showed that for every uncountable ordinal $\alpha < \omega_2$ there is an LCS space of height α with all levels countable.

Further, I. Juhász and W. Weiss have obtained a complete characterization of the cardinal sequences of LCS spaces of height ω_1, namely, they are all possible, with the only restriction imposed by Lemma 1.2.

Theorem 3.1 (I. Juhász and W. Weiss [**10**]). *Let $\theta = \langle \kappa_\alpha : \alpha < \omega_1 \rangle$. Then there is an LCS space X with $CS(X) = \theta$ iff for every $\alpha < \beta < \omega_1$, $\kappa_\beta \leq \kappa_\alpha^{\aleph_0}$.*

Since the condition that for every $\alpha < \beta < \omega_1$, $\kappa_\beta \leq \kappa_\alpha^{\aleph_0}$ can be easily forced, we have as a corollary that for height ω_1 an LCS space with *any* cardinal sequence can always be forced, a result that was first proved by J.C. Martínez in [**15**]. So, the question is what happens at heights $> \omega_1$.

Let us also mention that A. Dow and P. Simon [**5**] show that there exist $2^{\aleph_1}$-many, the largest possible number, LCS spaces of width ω and height ω_1.

In view of Lemmas 1.1 and 1.2, the following is the best possible absolute result, i.e., not depending on cardinal arithmetic, about the existence of LCS spaces of height $< \omega_2$ with arbitrary cardinal sequences:

Theorem 3.2 (J. Bagaria [**1**] and J.C. Martínez [**16**], independently). *Let $\eta < \omega_2$, and let $\theta = \langle \kappa_\alpha : \alpha < \eta \rangle$ be a sequence of infinite cardinals such that for every $\alpha < \eta$, $\kappa_\alpha \leq \omega_1$. Then, there exists an LCS space with cardinal sequence θ.*

It follows from [**11**] (see also [**2**, Theorem 3.2]), that Theorem 3.2 above cannot be extended to sequences of length $\omega_1 + 1$. Indeed, it is consistent with ZFC that there is no LCS space of height $\omega_1 + 1$ whose levels have all cardinality $\leq \omega_1$, except for the last one, which has cardinality ω_2. This explains the condition imposed in the following Theorem on the levels of uncountable cofinality. Juhász and Weiss have conjectured that this is the best possible result (in ZFC) on cardinal sequences for LCS spaces of height $< \omega_2$.

Theorem 3.3 (I. Juhász and W. Weiss [**10**]). *Let $\eta < \omega_2$, and let $\theta = \langle \kappa_\alpha : \alpha < \eta \rangle$ be a sequence of cardinals such that for each $\alpha < \beta < \eta$, $\kappa_\beta \leq \kappa_\alpha^\omega$, and $\kappa_\alpha \leq \omega_1$ whenever $cf(\alpha) = \omega_1$. Then there is an LCS space X with $CS(X) = \theta$.*

A complete characterization of cardinal sequences for LCS spaces of height $\eta < \omega_2$ has been obtained under GCH by I. Juhász, L. Soukup, and W. Weiss [**8**]. See the chapter of L. Soukup [**24**] in this volume.

4. Spaces of height ω_2

The existence of an LCS space of height ω_2 and countable width requires the negation of CH (see Lemma 1.1), but it does not follow from it. Indeed, W. Just [**11**] showed that by adding ω_2 Cohen reals over a model of CH there is no such space in the forcing extension. Further, in [**2**] it is shown that the non-existence of such a space is consistent with Martin's Axiom (MA) and the continuum large. More surprisingly, I. Juhász, S. Shelah, L. Soukup, and Z. Szentmiklóssy [**7**] have recently proved that after adding Cohen reals to a model of CH every LCS space has at most ω_1-many countable levels. Therefore, there is no hope of extending the results stated in the previous section (Theorems 3.1 and 3.3) to LCS spaces of height $\geq \omega_2$.

The problem of the consistency of the existence of an LCS space of height ω_2 and countable width was solved in 1987 by J. Baumgartner and S. Shelah [**2**] in a groundbreaking work. They show that such a space can always be forced (see Section 7 for more details).

Theorem 4.1 (J. Baumgartner and S. Shelah [**2**]). *There is a σ-closed$*$ccc forcing notion that forces an LCS space of height ω_2 and countable width.*

A slight modification of the Baumgartner–Shelah construction (see [**1**]) yields that for every sequence $\theta = \langle \kappa_\alpha : \alpha < \omega_2 \rangle$ such that $\kappa_\alpha \in \{\omega, \omega_1\}$, all $\alpha < \omega_2$, one can force the existence of an LCS space with cardinal sequence θ. And L. Soukup (unpublished) has shown that for every sequence θ of length ω_2 such that each cardinal in the sequence is ω, ω_1, or ω_2, it is consistent that there is an LCS space with cardinal sequence θ.

However, not much more is known about other possible cardinal sequences.

324? **Problem 4.2.** *What are the possible cardinal sequences of LCS spaces of height ω_2?*

Ideally, one would like to have a complete characterization (in ZFC) of all the possible sequences. Notice, however, that even for spaces of height strictly between ω_1 and ω_2 the problem is not yet completely solved (see Theorem 3.3 and our remarks above).

The most interesting open problem about spaces of height ω_2 is the following:

325? **Problem 4.3.** *Does there exist (in* ZFC*) an LCS space of width ω_1 and height ω_2?*

Such a space exists in L ([**12**]), and it can always be forced while making the continuum large ([**2**], see [**1**]). Hence, the existence of such a space is consistent with both CH and not-CH. However, the answer to the problem is not known even under GCH. The best result so far in this direction is the recent difficult construction by I. Juhász, S. Shelah, L. Soukup, and Z. Szentmiklóssy [**6**], of an LCS space of height ω_2 with only ω_1 isolated points, i.e., with the first level of cardinality ω_1. This solved an old problem of Juhász. Other versions of the problem, still open, are the following (see [**6**]):

Problem 4.4. *Does there exist (in* ZFC*) an LCS space with only ω_1 isolated points and of height α, for every $\alpha < \omega_3$?* 326?

Problem 4.5. *Does there exist (in* ZFC*) an LCS space with only ω_2 isolated points and of height ω_3?* 327?

5. Spaces of height $> \omega_2$

The best result to date on the consistency of the existence of LCS spaces of countable width and height greater than ω_2 is the following:

Theorem 5.1 (J.C. Martínez [**18**]). *There is a σ-closed $*$ ccc forcing notion that forces that for every $\alpha < \omega_3$ there is an LCS space of height α and width ω.*

The proof is again based on the Baumgartner–Shelah forcing construction, but it needs quite a bit of extra work to ensure that the forcing does not collapse cardinals. For this, Martínez uses the notion of *tree of intervals*, introduced in [**17**], which allows one to work with different Δ-functions (see Section 7 below for the definition) at ordinals of cofinality ω_2 in a coherent way. In a recent work, L. Soukup [**25**] has given another proof, also based on the Baumgartner–Shelah construction. Instead of using trees of intervals he proves a rather general *lifting theorem* which allows one to lift any reasonable forcing construction of an LCS space of height a regular cardinal κ and countable width to one of a space of height any $\alpha < \kappa^+$ and countable width.

The following is a long-standing open problem:

Problem 5.2. *Is the existence of an LCS space of width ω and height ω_3 consistent?* 328?

This is a wide open question. In fact, it is not even known if there is any bound on the possible heights of an LCS space of countable width. Lemma 1.1 puts the bound on 2^ω, but of course 2^ω can consistently take arbitrarily high values.

As for spaces of uncountable width, a variation of the Baumgartner forcing, this time with infinite forcing conditions, was used by J.C. Martínez [**14**] to force, for any given regular cardinal κ, an LCS space of height κ^+ with all levels of cardinality κ. Moreover, P. Koepke and J.C. Martínez [**12**] showed that such spaces also exist in the constructible universe L by building them using simplified morasses. Further, J.C. Martínez [**17**] shows that for every infinite regular cardinal κ, one can always force, while preserving cardinals, an LCS space of width κ and height any ordinal $\alpha < \kappa^{++}$.

But this does not yet solve the following:

Problem 5.3. *Is the existence of an LCS space of width ω_1 and height ω_3 consistent?* 329?

Since, apparently, there is nothing equivalent to a Δ-function (see Section 7 below) for ω_3, trying to lift the construction of Baumgartner–Shelah [**2**] one cardinal up will not work.

6. A basic construction

Many of the known constructions of LCS spaces are based on the following Definition and Lemma, which are essentially due to J. Baumgartner [**2**].

Definition 6.1. Given $\theta = \langle \kappa_\alpha : \alpha < \lambda \rangle$, where each κ_α is an infinite cardinal, we say that a poset $(T, \leq)$ is a *$LCS(\theta)$-structure* if

(1) $T = \bigcup\{T_\alpha : \alpha < \lambda\}$, where $T_\alpha = \{\alpha\} \times \kappa_\alpha$;
(2) for every pair of distinct elements s, t of T there exists a finite subset of T, denoted by $i\{s,t\}$, such that:
  (a) if $u \in i\{s,t\}$, then $u \leq s, t$,
  (b) if $u \leq s, t$, then there exists $v \in i\{s,t\}$ such that $u \leq v$;
(3) if $s \in T_\alpha$, $t \in T_\beta$ and $s < t$, then $\alpha < \beta$;
(4) for every $\alpha < \beta < \lambda$, if $t \in T_\beta$, then the set $\{s \in T_\alpha : s < t\}$ is infinite.

Lemma 6.2. *If there exists a $LCS(\theta)$-structure, then there exists an LCS space X with $CS(X) = \theta$.*

Proof. Suppose $(T, \leq)$ is a $LCS(\theta)$-structure. For each $t \in T$, let $C(t) = \{s \in T : s \leq t\}$. Let

$$B = \{C(t) \setminus (C(s_1) \cup \cdots \cup C(s_n)) : n < \omega,\ s_1, \ldots, s_n \in T,\ s_1, \ldots, s_n < t\}.$$

B is a clopen base for a Hausdorff topology $\mathcal{T}$ on T. Let $X = (T, \mathcal{T})$. For each $t \in T$, $C(t)$ is a compact neighborhood of t. Hence, X is locally-compact. Further, if Z is a non-empty subspace of X, we can always find $t \in Z$ with $C(t) \cap Z = \{t\}$. i.e., t is an isolated point of Z. Hence, X is scattered. Finally, by (4), for each $\alpha < \lambda$, $X^\alpha \setminus X^{\alpha+1} = T_\alpha$ is the α-level of X. So, X has height λ and $CS(X) = \theta$. □

Notice that $X = (T, \mathcal{T})$ is zero-dimensional, right-separated, non-compact, and if it has uncountable height, it is not even Lindelöf. For X to be ccc, κ_0 must be countable. By taking the one-point compactification of X, we obtain a scattered Boolean space with cardinal sequence θ, plus a top level with only one element.

It is an interesting open question whether every LCS space can be obtained from an LCS structure, i.e., whether for every sequence θ of infinite cardinals, the existence of an LCS space with cardinal sequence θ implies the existence of a $LCS(\theta)$ structure.

7. Forcing an LCS space

The main advantage of working with LCS structures is that they are amenable to the forcing technique. Indeed, there is a natural notion of forcing ([**2**]) that produces a $LCS(\theta)$-structure, provided, of course, that the cardinals involved are not collapsed.

Definition 7.1. Fix a sequence $\theta = \langle \kappa_\alpha : \alpha < \lambda \rangle$ of infinite cardinals, and let $T = \bigcup\{T_\alpha : \alpha < \lambda\}$, where $T_\alpha = \{\alpha\} \times \kappa_\alpha$, all $\alpha < \lambda$. Let $\mathbb{P} = \mathbb{P}_\theta$ be the set of all $p = (x_p, \leq_p, i_p)$ such that:

(0) $x_p \in [T]^{<\omega}$.

(1) $\leq_p$ is a partial ordering on x_p satisfying:

  (a) if $s \in T_\alpha$, $t \in T_\beta$ and $s <_p t$, then $\alpha < \beta$.

(2) $i_p \colon [x_p]^2 \to [x_p]^{<\omega}$ is such that for every $s, t \in x_p$ with $s \neq t$, the following hold:

  (a) if $v \in i_p\{s,t\}$, then $v \leq_p s, t$;

  (b) if $u \leq_p s, t$, then there is $v \in i_p\{s,t\}$ such that $u \leq_p v$.

The ordering is given by:

$$p \leq q \text{ iff } x_p \supseteq x_q,\ \leq_p \cap (x_q \times x_q) = \leq_q \text{ and } i_p \restriction [x_q]^2 = i_q$$

As long as forcing with $\mathbb{P}$ does not collapse any of the cardinals κ_α, nor λ, a $LCS(\theta)$-structure will exist in any $\mathbb{P}$-generic extension. To see this, let $p = (x_p, \leq_p, i_p) \in \mathbb{P}$. Pick $t \in T \setminus x_p$, and define $q = (x_q, \leq_q, i_q) \in \mathbb{P}$ as follows: $x_q = x_p \cup \{t\}$, $\leq_q = \leq_p$, $i_q\{u,v\} = i_p\{u,v\}$ if $u, v \in x_p$, and $i_q\{t,u\} = \emptyset$, for all $u \in x_p$. It is clear that q is a condition stronger than p with $t \in x_q$. This shows that for every $t \in T$, the set $D_t = \{p \in \mathbb{P} : t \in x_p\}$ is dense in $\mathbb{P}$. So, if $G \subseteq \mathbb{P}$ is generic for all the D_t, $t \in \mathbb{P}$, the relation $\leq = \bigcup\{\leq_p : p \in G\}$ and the function $i = \bigcup\{i_p : p \in G\}$ witness that $(T, \leq)$ satisfies (1), (2) and (3) of Definition 6.1. We claim that $(T, \leq)$ also satisfies (4). For suppose $\alpha < \beta < \lambda$ and $t \in T_\beta$. Given $p \in \mathbb{P}$ with $t \in x_p$, choose $s \in T_\alpha \setminus x_p$. Now define $q \in \mathbb{P}$ as follows: $x_q = x_p \cup \{s\}$, $\leq_q = \leq_p \cup \{(s,u) : u \in x_p,\ t \leq_p u\}$, $i_q\{u,v\} = i_p\{u,v\}$ if $u, v \in x_p$, $i_q\{s,u\} = \{s\}$ if $t \leq_p u$, and $i_q\{s,u\} = \emptyset$ otherwise. It can be easily checked that q is a condition stronger than p which forces $s \leq t$. A genericity argument now shows that (4) of Definition 6.1 holds for $(T, \leq)$.

When forcing with $\mathbb{P}_\theta$ to obtain an $LCS(\theta)$-structure, the main problem is to show that one does not collapse any of the cardinals involved in the sequence θ. Unfortunately, if θ has length $> \omega_1$, then ω_1 is always collapsed: Let S be a countable subset of T_{ω_1}. For each $\alpha < \omega_1$, the set

$$D_\alpha = \{p \in \mathbb{P}_\theta : \exists s, t \in S \cap x_p\ i_p\{s,t\} = \{(\alpha, 0)\}\}$$

is dense. Hence, if $\mathbb{P}_\theta$ does not collapse ω_1, in any generic extension there are $s, t \in S$ such that for some (actually, uncountably-many) $\alpha \neq \beta$, $i\{s,t\} = \{(\alpha,0)\} = \{(\beta,0)\}$, which is impossible. A similar argument shows that if the length of θ is greater than some uncountable cardinal κ, then $\mathbb{P}_\theta$ forces that all cardinals $\leq \kappa$ are countable. Thus, in order to force an $LCS(\theta)$-structure of uncountable height, one usually needs to do something to avoid collapsing cardinals. If the height is $< \omega_2$, this doesn't seem to be much of a problem, since such structures either exist in ZFC (Theorem 3.2), or very likely they can be forced by modifying the cardinal arithmetic accordingly (see Theorems 3.1 and 3.3, although the construction given by Juhász–Weiss does not produce LCS structures directly). But for heights $\geq \omega_2$ the only solution is to restrict the i-function, namely, for every p and every distinct

$s, t \in x_p$, the elements of $i_p\{s,t\}$ must be chosen within a countable subset of T. Needless to say, the choice has to be done very carefully to ensure that condition (4) of Definition 6.1 is satisfied by the generic LCS structure.

A solution to this problem for T of height ω_2 was provided by J. Baumgartner and S. Shelah [**2**]. They introduce the notion of a Δ-function on pairs of ordinals $< \omega_2$, and show that the existence of such a function can be forced over a model of CH while preserving cardinals. More recently, using S. Todorčević's ρ function, B. Veličković has shown that the existence of a slightly stronger form of a Δ-function is actually a consequence of $\Box_{\omega_1}$ (see [**3**]).

Definition 7.2. A function $f\colon [\omega_2]^2 \to [\omega_2]^{\leq\omega}$ is called a (strong) Δ-*function* if $f\{\alpha,\beta\} \subseteq \min\{\alpha,\beta\}$, for all $\alpha, \beta < \omega_2$, and for any uncountable set D of finite subsets of ω_2 there exists an uncountable $E \subseteq D$ such that for every distinct $a, b \in E$, for all $\alpha \in a \setminus b$, for all $\beta \in b \setminus a$, and for all $\tau \in a \cap b$,

(1) if $\alpha, \beta > \tau$, then $\tau \in f\{\alpha,\beta\}$;
(2) if $\beta > \tau$, then $f\{\alpha,\tau\} \subseteq f\{\alpha,\beta\}$;
(3) if $\alpha > \tau$, then $f\{\beta,\tau\} \subseteq f\{\alpha,\beta\}$.

By modifying item (2) in the definition of the poset $\mathbb{P}$ (see Definition 7.1) to require that for every $s, t \in x_p$, $i_p\{s,t\} \subseteq f(\{s,t\})$, where f is a Δ-function, Baumgartner–Shelah show that the poset that forces a LCS structure of height ω_2 with all levels countable is ccc, and therefore it does not collapse cardinals. This yields a proof of Theorem 4.1.

For T of height greater than ω_2, but smaller than ω_3, Δ-functions can still be used ([**18**]), but there are also other approaches ([**25**]). Δ-functions will very likely still play an important rôle in the complete solution to Problem 4.2, and perhaps also in Problems 5.2 and 5.3, although this seems less likely. It is generally felt that a positive solution to Problems 5.2 and 5.3 would probably require new methods.

A possible approach to Problem 4.3 would be to use the combinatorial techniques introduced in [**6**]. Another approach would be to try to lift the proof of Theorem 3.1 given in [**1**], in which LCS spaces of height $< \omega_2$ and width $\leq \omega_1$ are built by generically expanding LCS structures. These expansions are obtained by forcing over countable models of a fragment of ZFC, and so the whole construction is carried out in ZFC. However, to build a space of height ω_2 and width ω_1, one would now have to expand structures of size ω_1 by means of generic extensions of models of a fragment of ZFC also of size ω_1. But to do this in ZFC one needs to work with countable conditions, which introduces some new technical difficulties.

8. Final remarks

8.1. Stone duality. A Boolean algebra is *superatomic* iff every homomorphic image is atomic. Equivalently, iff its Stone space is scattered. If B is a superatomic Boolean algebra, the height and the cardinal sequence of B may be defined as, respectively, the height and the cardinal sequence of its Stone space. However, given a Boolean algebra B, one can also define, by induction on α, the

α-Cantor–Bendixson ideals J^α of B as follows: $J^0 = \{0\}$. $J^{\alpha+1}$ is the ideal generated by J^α together with all the $b \in B$ such that b/J^α is an atom in B/J^α. If α is a limit ordinal, J^α is the union of all J^β, $\beta < \alpha$. B is superatomic iff $B = J^\alpha$, for some α. The height of B is the least α such that $J^\alpha = J^{\alpha+1}$. The α-th level of B is the set of atoms of B/J^α, which we denote by $J_\alpha(B)$. The cardinal sequence of B is $CS(B) = \langle |J_\alpha(B)| : \alpha < \mathrm{ht}(B)\rangle$.

Many important results for the theory of LCS spaces and cardinal sequences are part of the literature on superatomic Boolean algebras. However, all these results can be easily translated to the language of topological scattered spaces, via Stone duality. Conversely, all results on LCS spaces have a direct translation to the language of superatomic Boolean algebras.

The class of Boolean algebras arising, via Stone duality, from the LCS spaces obtained from LCS structures coincides with the classes studied in [**4**].

For more information about superatomic Boolean algebras we refer the reader to J. Roitman's survey article [**21**].

8.2. PCF structures. There is a particularly important class of LCS structures, the so-called PCF structures. They play an essential rôle in the proof of Shelah's celebrated theorem on cardinal arithmetic. Namely: if $\aleph_\omega$ is a strong limit, then $2^{\aleph_\omega} < \aleph_{\omega_4}$. A PCF structure is an LCS structure with some extra requirements ([**22**]). The non-existence of a PCF structure of size ω_3 (compare with Problem 5.2 above) would improve Shelah's bound on $2^{\aleph_\omega}$ to $\aleph_{\omega_3}$.

References

[1] J. Bagaria, *Locally-generic Boolean algebras and cardinal sequences*, Algebra Universalis **47** (2002), no. 3, 283–302.

[2] J. E. Baumgartner and S. Shelah, *Remarks on superatomic Boolean algebras*, Ann. Pure Appl. Logic **33** (1987), no. 2, 109–129.

[3] M. Bekkali, *Topics in set theory*, Lecture Notes in Math., vol. 1476, Springer, Berlin, 1991.

[4] R. Bonnet and M. Rubin, *On well-generated Boolean algebras*, Ann. Pure Appl. Logic **105** (2000), 1–50.

[5] A. Dow and P. Simon, *Thin-tall Boolean algebras and their automorphism groups*, Algebra Universalis **29** (1992), no. 2, 211–226.

[6] I. Juhász, S. Shelah, L. Soukup, and Z. Szentmiklóssy, *A tall space with a small bottom*, Proc. Amer. Math. Soc. **131** (2003), no. 6, 1907–1916.

[7] I. Juhász, S. Shelah, L. Soukup, and Z. Szentmiklóssy, *Cardinal sequences and Cohen real extensions*, Fund. Math. **181** (2004), no. 1, 75–88.

[8] I. Juhász, L. Soukup, and W. Weiss, *Cardinal sequences of length $< \omega_2$ under GCH*, Fund. Math. **189** (2006), no. 1, 35–52.

[9] I. Juhász and W. Weiss, *On thin-tall scattered spaces*, Colloq. Math. **40** (1978), no. 1, 63–68.

[10] I. Juhász and W. Weiss, *Cardinal sequences*, Ann. Pure Appl. Logic (2006), Article in press.

[11] W. Just, *Two consistency results concerning thin-tall Boolean algebras*, Algebra Universalis **20** (1985), no. 2, 135–142.

[12] P. Koepke and J. C. Martínez, *Superatomic Boolean algebras constructed from morasses*, J. Symbolic Logic **60** (1995), no. 3, 940–951.

[13] R. LaGrange, *Concerning the cardinal sequence of a Boolean algebra*, Algebra Universalis **7** (1977), no. 3, 307–312.

[14] J. C. Martínez, *A consistency result on thin-tall superatomic Boolean algebras*, Proc. Amer. Math. Soc. **115** (1992), no. 2, 473–477.

[15] J. C. Martínez, *On uncountable cardinal sequences for superatomic Boolean algebras*, Arch. Math. Logic **34** (1995), no. 4, 257–261.

[16] J. C. Martínez, *On cardinal sequences of scattered spaces*, Topology Appl. **90** (1998), 187–196.

[17] J. C. Martínez, *A forcing construction of thin-tall Boolean algebras*, Fund. Math. **159** (1999), no. 2, 99–113.

[18] J. C. Martínez, *A consistency result on thin-very tall Boolean algebras*, Israel J. Math. **123** (2001), 273–284.

[19] J. C. Martínez, *A consistency result on cardinal sequences of scattered Boolean spaces*, Math. Log. Q. **51** (2005), no. 6, 586–590.

[20] M. Rajagopalan, *A chain compact space which is not strongly scattered*, Israel J. Math. **23** (1976), no. 2, 117–125.

[21] J. Roitman, *Superatomic Boolean algebras*, Handbook of Boolean algebras, Vol. 3, North-Holland, Amsterdam, 1989, pp. 719–740.

[22] J. Ruyle, *Cardinal sequences of PCF structures*, Doctoral dissertation, University of California, Riverside, 1998.

[23] S. Shelah, *Cardinal arithmetic*, The Clarendon Press Oxford University Press, New York, 1994.

[24] L. Soukup, *Cardinal sequences and universal spaces*, Open problems in topology II (E. Pearl, ed.), Elsevier Science.

[25] L. Soukup, *A lifting theorem on forcing LCS spaces,*, To appear in the Hajnal–T. Sós Conference Volume.

Open Problems in Topology II – Edited by E. Pearl

Sequential order

Alan Dow

There are four very natural countable convergence conditions that are very well known. A space is *first countable* if every point has a countable local base. A space is *Fréchet–Urysohn* if each point x is in the closure of a set exactly when there is a sequence $\{a_n : n \in \omega\}$ from the set converging to the point, denoted $a_n \to x$. A space has *countable tightness* if a point is in the closure of a set exactly when there is a countable subset of the given set which also has the point in the closure. The fourth condition is the sequential property. The definition of this property already sets it apart from the previous three because it can not be stated just in terms of a fixed point and which sets it is in the closure of. A subset A of a space X is *sequentially closed* if each sequence from A which converges in X will converge to a point of A. A space is *sequential* if each sequentially closed subset is closed. In a sequential space, the closure of a set A can be computed by iterating the operation of adding limit points of converging sequences (one might say the Fréchet-Urysohn operation). This gives rise to the notion of the sequential order of a space.

Definition ([**1**]). For a subset A of a space X, define for each ordinal α, $A^{(\alpha)}$ recursively as follows. Set $A^{(0)}$ to be A; for limit α, let $A^{(\alpha)} = \bigcup_{\beta<\alpha} A^{(\beta)}$; and

$$A^{(\alpha+1)} = \{x \in X : \text{there exists } \{a_n : n \in \omega\} \subset A^{(\alpha)} \text{ such that } a_n \to x\}.$$

It is immediate that for each A and each space X, the set $A^{(\omega_1)}$ is sequentially closed and $A^{(\alpha)} \subset A^{(\omega_1)}$ for all α. We define the *sequential order* of A in X, denoted $\mathrm{so}(A, X)$, to be the minimum α such that $A^{(\alpha)}$ is sequentially closed. The sequential order of a (sequential) space X is the supremum of $\mathrm{so}(A, X)$ for all $A \subset X$.

The question that is the basis of this article is the following.

Question 1. *Is there a compact sequential space with sequential order greater than two?* 330?

It was shown by Baškirov [**2**] to follow from CH that there is a compact space of sequential order ω_1. In the author's view, this investigation is made more interesting when one recalls Balogh's result that the Proper Forcing Axiom, PFA, implies that each compact space of countable tightness is sequential.

Question 2. *Does* PFA *imply that there is a finite bound to the sequential order of compact sequential spaces?* 331?

It is shown in [**3**] that Martin's Axiom implies there are compact sequential spaces of sequential order four. The next question is likely more accessible than the others.

The author acknowledges support provided by NSF grant DMS-0103985.

332? **Question 3.** *Is it consistent with the failure of* CH *to have compact sequential spaces with sequential order ω_1?*

333? **Question 4.** *Is it consistent with $\mathfrak{b} = \mathfrak{c} > \omega_1$ or $\mathfrak{s} = \mathfrak{c} > \omega_1$ to have compact sequential spaces with infinite sequential order (or sequential order larger than four)?*

The requirement that the spaces be compact is quite essential in the above questions. It is not difficult to construct spaces with a given countable sequential order.

Proposition ([**1**]). *Suppose that $\{\alpha_n : n \in \omega\} \subset \omega_1$ and for each n, there is a countable sequential space with sequential order at least α_n. Then there is a countable sequential space with sequential order greater than each α_n.*

Although this result is easy to prove, it is instructive to recall that the topology could not be first countable if we are expecting sequential order larger than two. We may assume that the sequence $\{\alpha_n : n \in \omega\}$ is monotone increasing and for each n, fix a countable space X_n and an $A_n \subset X_n$ such that $\text{so}(A_n, X_n) \geq \alpha_n$. For each n, let x_n be a point of X_n in $A^{(\alpha_n)} \setminus A^{(\beta_n)}$ where the β_ns are chosen so that the sequence $\{\beta_n + 1 : n \in \omega\}$ has the same supremum as the sequence of α_ns. Our desired space X will consist of a new point x together with the topological sum of the (pairwise disjoint) X_ns. A subset A of $\bigcup_{n\in\omega} X_n$ will have x as a limit only if x_n is in the closure of $A \cap X_n$ for infinitely many n. The space X is sequential (and completely regular if each X_n is completely regular) and no sequence disjoint from $\{x_n : n \in \omega\}$ will converge to x.

It is really quite remarkable that the largest sequential order that has been found in the class of compact sequential spaces is two. Any example of a compact sequential space that is not Fréchet–Urysohn is such an example. The best known example is to simply take the one point compactification of any of the well-known Ψ-spaces. There does appear to be a natural connection to the study of maximal almost disjoint (*mad*) families. If ω is a subset of a compact sequential space X then there are mad families on ω consisting of sequences that converge in X. Therefore if $\text{so}(\omega, X) > 1$, then clopen neighborhoods of the points which are $\omega^{(2)} \setminus \omega^{(1)}$ will themselves contain mad families of sequences from $\omega^{(1)}$ while not splitting any elements of the original mad family on ω. This complex interaction between these families appears to be at the root of the problems.

Each of the examples that has been constructed to witness sequential order has been scattered. Moreover, if A is the set of isolated points in a compact scattered space X, then these examples have had the property that each of the scattering levels of X correspond naturally to the iterations in the sequential order hierarchy, i.e., the sets $\{A^{(\alpha+1)} \setminus A^{(\alpha)} : \alpha < \text{so}(A, X)\}$ are also the scattering levels of X. Note, by contrast, that the space ω_1 (the space of countable ordinals) has scattering height ω_1 but sequential order one.

334? **Question 5.** *If there is a compact sequential space with sequential order α, is there such a space which is scattered? Can it be arranged so that in addition, for* *each integer $n < \alpha$, $A^{(n+1)} \setminus A^{(n)}$ is the set of points on the $n+1$-st scattering level of X?*

It may be interesting to ask each of the above questions in the context of countably compact spaces.

Question 6. *For which $\alpha \leq \omega_1$ is there a sequentially compact sequential regular space with sequential order α?* 335?

References

[1] A. V. Arhangel'skiĭ and S. P. Franklin, *Ordinal invariants for topological spaces*, Michigan Math. J. **15** (1968), 313–320, addendum, ibid. 15 (1968) 506.

[2] A. I. Baškirov, *The classification of quotient maps and sequential bicompacta*, Dokl. Akad. Nauk SSSR **217** (1974), 745–748, Translation: Soviet Math. Dokl. 15 (1974) 1104–1109.

[3] A. Dow, *Sequential order under MA*, Topology Appl. **146/147** (2005), 501–510.

Open Problems in Topology II – Edited by E. Pearl

On D-spaces

Todd Eisworth

Introduction

So what makes a mathematical problem interesting? This question always provokes spirited debate among mathematicians no matter where it is posed. Of course there is no good answer to it (as English critic and author William Hazlitt says, "Whatever interests, is interesting."), but there are certainly some mathematical questions that arouse the curiosity of almost anyone who comes in contact with them, questions that tempt with the simplicity of of their formulation, tantalize with promises of an elegant solution if only one can look at the problem in just the right way, and taunt with the number of excellent mathematicians who have examined the question in the past and failed to solve it. The theory of D-spaces is replete with such questions, and in this short note we will examine a few of them.

What *is* a D-space? A topological space[1] X is a *D-space* if for every *neighborhood assignment* $\{N(x) : x \in X\}$ (that is, $N(x)$ is an open neighborhood of x for each $x \in X$) there is a closed discrete subset D of X such that $X = \bigcup\{N(x) : x \in D\}$. The concept goes back to work of van Douwen [**7**]. One of the first things said about D-spaces in the cited paper is the following:

> "Up to now no satisfactory example of a space which is not a D-space is known, where by satisfactory example we mean an example having a covering property at least as strong as metacompactness or subparacompactness."

Over twenty years later, the situation is much the same. We still lack a basic understanding of the relationship between covering properties and the state of being a D-space. In fact, as Fleissner and Stanley noted in 2001 [**8**]:

> "Besides the trivial observation that a compact T_1-space is a D-space, there are no proofs known that a covering property implies D-space."

This state of affairs is the main topic of the following short note.

Questions about D-spaces

The first subject we address is our lack of understanding about the relationship between covering properties and the state of being a D-space. As Fleissner and Stanley say, we simply lack theorems that say that such-and-such covering property implies that a space is a D-space. Moreover, we lack the techniques that would

[1]We assume that all spaces under consideration are at least regular.

The author acknowledges support from NSF grant DMS-0506063.

allow us to build counterexamples to such alleged theorems. To wit, all of the following problems (essentially from [**7**]) remain open:

336? **Question 1.** *Is every (hereditarily) Lindelöf space a D-space?*

337? **Question 2.** *Is every paracompact space a D-space?*

338? **Question 3.** *Does there exist a subparacompact or metacompact space which is not a D-space?*

In view of the last question, we recall that a space X is *subparacompact* if every open covering of X can be refined by a σ-discrete closed covering.

Arhangel'skiĭ [**2**] has recently addressed the relationship between covering properties and D-spaces. He adds the following questions to van Douwen's list:

339? **Question 4** ([**2**, Problem 1.18]). *Is every countably metacompact weakly θ-refinable (Tychonoff) space a D-space?*

Recall that a space X is *weakly θ-refinable* if for each open covering $\mathcal{C}$ of X, there exists a sequence $\langle \mathcal{C}_n : n < \omega \rangle$ of open coverings of X, each refining $\mathcal{C}$, such that for every $x \in X$ there is a $k < \omega$ with $\mathcal{C}_k$ point-finite at x.

A σ-metrizable space is weakly θ-refinable, so this suggests the related question:

340? **Question 5** ([**2**, Problem 1.21]). *Is every countably metacompact σ-metrizable space a D-space?*

Finally, let us recall that a space X is *screenable* if every open covering of X has an open σ-disjoint refinement. Arhangel'skiĭ and Buzyakova [**3**] establish that every space with a point-countable base is in fact a D-space; in particular, every space with a σ-disjoint base is a D-space, and so the following question is natural:

341? **Question 6** ([**2**, Problem 1.22]). *Is every screenable (Tychonoff) space a D-space?*

We next turn to a problem of Buzyakova concerning cardinal invariants and their relation to D-spaces. We start with the observations that every compact space is trivially a D-space, and that a countably compact D-space is compact. These facts follow immediately from the easy fact that $l(X) = e(X)$ for a D-space (where $l(X)$, the *Lindelöf number of* X, is the smallest infinite cardinal τ such that every open covering of X contains a subcovering of cardinality $\leq \tau$, and $e(X)$, the *extent of* X, is defined to be the supremum of cardinalities of closed discrete subsets of X).

The converse is not true; Buzyakova notes in [**5**] that if we take the product $X = D(\omega_1) \times \omega_1$ (where $D(\omega_1)$ is a discrete space of cardinality ω_1), then $l(X) = e(X) = \aleph_1$, but X is not a D-space. To see this last fact, note that X contains a closed copy of ω_1 and it is straightforward to see that this precludes X being a D-space. However, she notes that the following question may be of interest:

342? **Question 7** ([**5**, Question 3.6]). *Suppose that $l(Y) = e(Y)$ for every subspace Y of X. Is X then a D-space?*

It is worth mentioning Buzyakova's main result from [**5**]: if X is compact, then $C_p(X)$ (the space of continuous real-valued functions on X with the topology of pointwise convergence) is hereditarily a D-space. Her result is quite strong; for example, it allows one to immediately deduce the following two well-known theorems:

Theorem (Baturov [4]). *If X is compact, then $l(Y) = e(Y)$ for every subspace Y of $C_p(X)$.*

Theorem (Grothendieck [**9**]). *If X is compact and Y is a countably compact subspace of $C_p(X)$, then Y is compact.*

Of course, both theorems follow immediately from Buzyakova's result using simple properties of D-spaces, and perhaps this helps to make the case that D-spaces are a class worthy of more research.

Arhangel′skiĭ also puts forward the following question regarding D-spaces and spaces of the form $C_p(X)$:

Question 8 ([**2**, Question 1.23]). *Suppose that X is a Tychonoff space with $C_p(X)$* 343?
Lindelöf. Is $C_p(X)$ then a D-space?

Many other questions on D-spaces have recently appeared in the literature and we do not have space to consider them all. We refer the reader instead to [**1**–**3**, **5**] for more comprehensive coverage, and limit ourselves to the following intriguing questions of Arhangel′skiĭ concerning unions of D-spaces.

Question 9. *Suppose that a (Hausdorff, regular, Tychonoff) T_1-space is the union* 344?
of two subspaces which are both D-spaces. Is then X a D-space as well?

Question 10. *Suppose X is countably compact, and $X = \bigcup_{n<\omega} X_n$ where each* 345?
X_n is a D-space. Is X compact?

Arhangel′skiĭ conjectures that a positive answer to the first is highly unlikely, but he and Buzyakova [**3**] have shown that if a regular T_1 space X is the union of a finite collection of metrizable subspaces, then X is a D-space. Regarding the second question, Gary Gruenhage [**10**] has shown that a positive answer results if we require that X is a finite union of D-spaces.

Stickyness

The fact that Gruenhage's article [**10**] ends the previous section is serendipitous, for we want to examine his techniques in more detail in this section. In [**10**], Gruenhage develops a general framework based on earlier work of Fleissner and Stanley [**8**] and implicit in Buzyakova's work that seems to handle the most important theorems of the form "a space X of such-and-such a type must be a D-space". He uses these methods to solve many open problems asked by Arhangel′skiĭ, Buzyakova, and others. We briefly outline his techniques, as they should be helpful to anyone tackling problems in this area.

Let X be a space. A binary relation R on X is *nearly good* if $x \in x\overline{A}$ implies that $x\,R\,y$ for some $y \in A$. If N is a neighborhood assignment on X, and X' and

D are subsets of X, then we say that D is *N-sticky mod R on X'* if whenever x is in X' and $x\,R\,y$ for some $y \in D$, then in fact $x \in N(D)$, i.e., $N(D)$ swallows all R-predecessors of each $y \in D$ that lie in X'. If $X' = X$, then we say[2] that D is *N-sticky mod R*.

The simplest example illustrating this definition is to let N be a neighborhood assignment on X, and define

$$(*) \qquad x\,R\,y \iff y \in N(x).$$

This particular R is nearly good, and a set D is N-sticky mod R if $x \in N(D)$ whenever $N(x) \cap D \neq \emptyset$.[3]

Theorem (Gruenhage). *Let N be a neighborhood assignment for X. Suppose as well that R is a nearly good relation on X such that every nonempty closed subset F of X contains a nonempty closed discrete subset D that is N-sticky mod R on F. Then there is a closed discrete D^* in X with $X = N(D^*)$.*

This theorem has some strength. Consider, for example, the case where X is left-separated and N is a neighborhood assignment on X, without loss of generality with $N(x) \subseteq [x, \infty)$ (the interval is defined using the order that left-separates X). Let R be as in $(*)$. Given a nonempty closed subset F of X, let x be the least element of F. Then the closed discrete set $D = \{x\}$ is N-sticky mod R on F, and so from the preceding theorem we conclude that X is a D-space.

We get more powerful results using the next theorem, which is also taken from [**10**]. The statement of the following theorem makes reference to N-close sets, where N is a neighborhood assignment on X. We say that a subset Z of X is *N-close* if $Z \subseteq N(x)$ for every $x \in Z$.

Theorem (Gruenhage). *Let N be a neighborhood assignment on X, and suppose there is a nearly good relation R on X such that for any $y \in X$, we can express the set $R^{-1}(y) \setminus N(y)$ as a countable union of N-close sets. Then there is a closed discrete D such that $X = N(D)$.*

We give one more easy example from [**10**] illustrating how powerful this result is. The key is that one can vary the relation R in order to get results in different situations.

Recall that a space X satisfies *open* (G) if each point $x \in X$ has a countable neighborhood base $\mathcal{B}_x$ such that whenever $x \in \overline{A}$ and $N(x)$ is a neighborhood of x, then there is an $a \in A$ and $B \in \mathcal{B}_a$ for which $x \in B \subseteq N(x)$.

Theorem. *Any space satisfying open (G) is a D-space.*

Proof. Given a neighborhood assignment N for such a space X, one defines

$$x\,R\,y \iff \text{there exists a } B \in \mathcal{B}_y \text{ such that } x \in B \subseteq N(x).$$

---

[2]Gruenhage deals with a generalization of this situation, where R is a relation from X to $[X]^{<\omega}$; this generalization allows him to capture more examples, but we shall deal only with the simpler version.

[3]This is what Fleissner and Stanley referred to as "N-sticky" in [**8**].

This choice of R is nearly good because X satisfies open (G). Furthermore, for each $B \in \mathcal{B}_y$, we can let $C(B)$ be the set of all $x \in B$ for which $B \subseteq N(x)$. The set $C(B)$ is N-close, and $R^{-1}(y) = \bigcup\{C(B) : B \in \mathcal{B}_y\}$. From the theorem cited earlier, we conclude that X is a D-space. □

It is clear that any space with a point-countable base satisfies open (G)[4] and so such spaces are D-spaces, a fact first shown by Arhangel'skiĭ and Buzyakova [**3**].

Gruenhage's paper contains a wealth of other results; for example, he shows that all Corson compacta are hereditarily D-spaces, and that $C_p(X)$ is hereditarily a D-space whenever X is a Lindelöf Σ-space. We refer the reader to [**10**] for the details.

General remarks

The current state of knowledge about D-spaces is full of asymmetries. We are rich with theorems that state that certain types of spaces are D-spaces, but we are lacking theorems of the form "If X is a D-space, then ...". We have many results that state that spaces with certain types of bases are D-spaces, but there are no substantial theorems saying that spaces satisfying certain covering properties are D-spaces. We have fairly general techniques for proving that something is a D-space, but we are sorely in need of more techniques for building spaces that are not D-spaces. Correcting these asymmetries should provide the next generation of general topologists with ample work.

Finally, I thought I would drop the authorial "we" for a moment, just to say that I, too, pulled out a pencil and scrap paper when I first heard the question of whether a regular Lindelöf space must be a D-space. I was sure I could see how the proof would go, and then later that night I reversed my opinion and thought I could see how a counterexample might work. Good problems are like this—they are interesting because they interest!

References

[1] A. V. Arhangel'skiĭ, *D-spaces and finite unions*, Proc. Amer. Math. Soc. **132** (2004), no. 7, 2163–2170.

[2] A. V. Arhangel'skiĭ, *D-spaces and covering properties*, Topology Appl. **146/147** (2005), 437–449.

[3] A. V. Arhangel'skiĭ and R. Z. Buzyakova, *Addition theorems and D-spaces*, Comment. Math. Univ. Carolin. **43** (2002), no. 4, 653–663.

[4] D. P. Baturov, *Subspaces of function spaces*, Vestnik Moskov. Univ. Ser. I Mat. Mekh. (1987), no. 4, 66–69, Translation: Moscow Univ. Math. Bull. 42 (1987) no. 4, 75–78.

[5] R. Z. Buzyakova, *Hereditary D-property of function spaces over compacta*, Proc. Amer. Math. Soc. **132** (2004), no. 11, 3433–3439.

[6] P. J. Collins, G. M. Reed, and A. W. Roscoe, *The point-countable base problem*, Open problems in topology, North-Holland, Amsterdam, 1990, pp. 237–250.

[7] E. K. van Douwen and W. F. Pfeffer, *Some properties of the Sorgenfrey line and related spaces*, Pacific J. Math. **81** (1979), no. 2, 371–377.

[8] W. G. Fleissner and A. Stanley, *D-spaces*, Topology Appl. **114** (2001), no. 3, 261–271.

[4]The converse is an open question, see [**6**].

[9] A. Grothendieck, *Critères de compacité dans les espaces fonctionnels généraux*, Amer. J. Math. **74** (1952), 168–186.

[10] G. Gruenhage, *A note on D-spaces*, Topology Appl. **153** (2006), no. 13, 90–96.

Open Problems in Topology II – Edited by E. Pearl

The fourth head of $\beta\mathbb{N}$

Ilijas Farah

Start from $\mathbb{N}$, the space of natural numbers with the discrete topology, and consider its Čech–Stone compactification, $\beta\mathbb{N}$. This is the compactification of $\mathbb{N}$ such that every $f\colon \mathbb{N} \to [0,1]$ has a unique continuous extension $\tilde{f}\colon \beta\mathbb{N} \to [0,1]$. For the rest of this note all maps are continuous. In his introduction to $\beta\mathbb{N}$ ([**28**]), Jan van Mill called it a three-headed monster. The first head shows under the Continuum Hypothesis, CH, and it is 'smiling and friendly' since CH easily resolves problems about $\beta\mathbb{N}$ (more precisely, as easily as solutions to problems about $\beta\mathbb{N}$ get). The second head is the 'ugly head of independence' as Paul Erdős used to call it (the head which, in van Mill's own words, 'constantly tries to confuse you'). The smallest, third, head is the ZFC-head of $\beta\mathbb{N}$. It provides those few facts about $\beta\mathbb{N}$ that can be resolved without applying additional set-theoretic axioms. Ever since Shelah's groundbreaking results discussed below we are witnessing the emergence of the fourth head of $\beta\mathbb{N}$: A coherent theory of $\beta\mathbb{N}$ deduced from forcing axioms (or Ramseyan axioms) with strong rigidity phenomena for $\beta\mathbb{N}$ and similar Čech–Stone compactifications.

The reader is assumed to have only basic familiarity with the topology of $\beta\mathbb{N}$ and axiomatic set theory (see e.g., [**11**]).

1. Trivial continuous maps

Let us start with a concrete problem, naturally stated as a problem about the Čech–Stone remainder $\mathbb{N}^* = \beta\mathbb{N} \setminus \mathbb{N}$. A map $f\colon \mathbb{N}^* \to \mathbb{N}^*$ is *trivial* if there is a function $h\colon \mathbb{N} \to \beta\mathbb{N}$ such that $f = \tilde{h} \restriction \mathbb{N}^*$. Assuming CH, it is very easy to construct nontrivial maps and even nontrivial autohomeomorphisms of $\mathbb{N}^*$. In [**29**], Shelah constructed a model of ZFC in which all autohomeomorphisms of $\mathbb{N}^*$ are trivial (of course, assuming there is a model of ZFC). In other words, he showed that a nontrivial autohomeomorphism of $\mathbb{N}^*$ cannot be constructed without using some additional set-theoretic axioms.

Question 1. *Is it possible to construct a nontrivial map $f\colon \mathbb{N}^* \to \mathbb{N}^*$ without using additional set-theoretic axioms?* 346?

Another way of stating this question (and similarly all the questions stated below) is: Assuming there is a model of ZFC, is there a model of ZFC in which there are no nontrivial maps $f\colon \mathbb{N}^* \to \mathbb{N}^*$? To an untrained eye this question may appear to be ad hoc, but please read on.

The existence of a nontrivial $f\colon \mathbb{N}^* \to \mathbb{N}^*$ implies the existence of both a nontrivial surjection $f\colon \mathbb{N}^* \to \mathbb{N}^*$ and a nontrivial injection $f\colon \mathbb{N}^* \to \mathbb{N}^*$ (dualize

Partially supported by NSERC.

examples of [**13**, §3.2]). However, a nontrivial $f\colon \mathbb{N}^* \to \mathbb{N}^*$ that is both a surjection and an injection cannot be constructed without using additional set-theoretic axioms, by Shelah's result.

347? **Question 2** (Dow). *Is it possible to construct a nonseparable extremally disconnected image of* $\mathbb{N}^*$ *without using additional set-theoretic axioms?*

If Question 1 has a negative answer, so does Question 2. This is because the assumption that all maps $f\colon \mathbb{N}^* \to \mathbb{N}^*$ are trivial implies every extremally disconnected image of $\mathbb{N}^*$ is separable (see [**13**, Proposition 4.11.7]). Under the Proper Forcing Axiom, PFA,[1] all extremally disconnected images of $\mathbb{N}^*$ have countable cellularity (see e.g., [**34**]). M. Bell has constructed a nonseparable zero-dimensional image of $\mathbb{N}^*$ with countable cellularity ([**2**]). An $f\colon (\mathbb{N}^*)^\kappa \to (\mathbb{N}^*)^\lambda$ is *trivial* if it is of the form $\tilde{h}$ for some $h\colon \mathbb{N}^\kappa \to \mathbb{N}^\lambda$. Note that $\tilde{h}$ need not exist for an arbitrary h; see e.g., [**16**]. The triviality of all $f\colon \mathbb{N}^* \to \mathbb{N}^*$ is equivalent to its self-strengthening asserting that every map $f\colon (\mathbb{N}^*)^\kappa \to (\mathbb{N}^*)^\lambda$ is trivial, where κ and λ are any two cardinals, finite or infinite (see [**16**]). This is a consequence of a phenomenon conjectured by van Douwen ([**4**]) and proved in [**16**]: If $f\colon (\mathbb{N}^*)^\kappa \to K$ for a compact space K, then the domain can be decomposed into finitely many clopen sets so that f depends on at most one coordinate on each one of the pieces. This remains true if $\mathbb{N}^*$ is replaced with any $\beta\mathbb{N}$-*space*: a space with the property that the closure of any infinite discrete subspace is homeomorphic to $\beta\mathbb{N}$.

An appealing variation on Question 2 is the following (a copy of $\mathbb{N}^*$ in a compact space X is *nontrivial* if it is nowhere dense and not of the form $\overline{D} \setminus D$ for a countable discrete $D \subseteq X$).

348? **Question 3** (van Douwen). *Is it possible to construct a nontrivial copy of* $\mathbb{N}^*$ *inside* $\mathbb{N}^*$ *without using additional set-theoretic axioms?*

This is closely related to Question 1. If every $f\colon \mathbb{N}^* \to \mathbb{N}^*$ is trivial then every copy of $\mathbb{N}^*$ inside $\mathbb{N}^*$ is trivial. Conversely, if all copies of $\mathbb{N}^*$ inside $\mathbb{N}^*$ are trivial and all autohomeomorphisms of $\mathbb{N}^*$ are trivial, then all injections $f\colon \mathbb{N}^* \to \mathbb{N}^*$ are trivial, and therefore all maps $f\colon \mathbb{N}^* \to \mathbb{N}^*$ are trivial. Under CH nontrivial copies of $\mathbb{N}^*$ exist in abundance. A natural way of assuring that a copy X of $\mathbb{N}^*$ is nontrivial is to make it into a P-set (i.e., a set such that every G_δ superset of X includes an open neighbourhood of X). Todorčević's Open Coloring Axiom, OCA, implies that no copy of $\mathbb{N}^*$ is a P-set ([**21**] for consistency, [**25**] from OCA; see also [**13**, Corollary 3.5.5]). As pointed out by A. Dow, it is not easy to find a nontrivial copy of $\mathbb{N}^*$ anywhere.

K.P. Hart noted that every $f\colon \mathbb{N}^* \to 2^\kappa$ is trivial, because every continuous $f\colon \mathbb{N}^* \to \{0,1\}$ extends to $\beta\mathbb{N}$. Therefore every copy of $\mathbb{N}^*$ in any 2^κ is trivial. He has also suggested a line of attack to Question 2: construct continuous $f\colon \mathbb{N}^* \to 2^{\mathfrak{c}}$ such that $f[\mathbb{N}^*]$ is extremally disconnected and the extension of f to $\beta\mathbb{N}$ sends $\mathbb{N}$ into $f[\mathbb{N}^*]$.

[1] The exact statements PFA, OCA and MA can be found in [**13**, **20**, **34**], or any up-to-date text on combinatorial set theory. Todorčević's OCA is different from its namesake introduced in [**1**]. Note that PFA implies both OCA and MA.

2. A partial result

Assuming MA and OCA the following was proved in [**13**], building on the work of Shelah–Steprāns, Todorčević, Just and Veličković (a map $f\colon \alpha^* \to \gamma^*$ is *trivial* if $f = \tilde{h} \restriction \alpha^*$ for some $h\colon \alpha \to \beta\gamma$).

Theorem 1 (OCA + MA). *For any two locally compact countable spaces α and γ and $f\colon \alpha^* \to \gamma^*$ there is a clopen partition $\alpha^* = U \dot\cup V$ such that $f \restriction U$ is trivial and $f[V]$ is nowhere dense.*

This easily implies that under OCA+ MA $(\alpha^*)^\kappa$ maps onto $(\gamma^*)^\lambda$ if and only if this is witnessed by a trivial map. Also, $(\alpha^*)^\kappa$ and $(\gamma^*)^\lambda$ are homeomorphic if and only if this is witnessed by a trivial map. It is not difficult to characterize when such a trivial map exists; see [**13**, Theorem 4.5.1] for one-dimensional versions and [**16**, Theorem 4.6] for (a bit more difficult) higher-dimensional versions. Therefore Theorem 1 and analogous results reduce the highly complex problem of the existence of continuous maps between large topological spaces to a simple problem of countable combinatorics. The interest in our questions largely derives from such reductions of complexity. Just ([**22**]) first proved the consistency of the statement '$(\mathbb{N}^*)^d$ does not map onto $(\mathbb{N}^*)^{d+1}$ for any $d \in \mathbb{N}$.' He used a rather weak consequence of Theorem 1 proved in [**24**]. As an additional motivation for the program discussed here, the reader is invited to compare the complex proof of [**22**] with the straightforward calculation of the same result from a consequence of Theorem 1 given in [**13**, Theorem 4.6.1]; see also [**16**, **17**].

Parovičenko's theorem implies that under CH any two remainders α^* and γ^* of locally compact countable spaces are homeomorphic (as long as they are both nonempty. However, powers $(\alpha^*)^\kappa$ and $(\alpha^*)^\gamma$ are homeomorphic if and only if $\kappa = \gamma$ ([**4**]).

Question 4. *Is it possible to construct a nontrivial map between Čech–Stone* 349?
remainders of locally compact countable spaces without using additional axioms of set theory?

Again, the conclusion is equivalent to its self-strengthening asserting all maps between powers of such spaces are trivial. Admittedly, the restriction to the class of countable locally compact spaces (also known as 'countable ordinals') is ad hoc. An analogue of Theorem 1 holds for a slightly wider class of spaces; see [**13**, §4.10]. Pending an answer to Question 4, I will refrain from fantasizing about the widest class of spaces for which analogous rigidity results can be proved (but see [**13**, §§4.10–4.11]). It is nevertheless worth mentioning that PFA implies all autohomeomorphisms of D^* are trivial for every discrete space D ([**36**]).

3. Rigidity phenomena for quotients $\mathcal{P}(\mathbb{N})/I$

Via the Stone duality, all of the above discussion could be recast in terms of Boolean algebras. The space $\mathbb{N}^*$ is the Stone space of the Boolean algebra $\mathcal{P}(\mathbb{N})/\operatorname{Fin}$, where Fin is the ideal of finite subsets of $\mathbb{N}$. Hence Question 1 asks whether for every homomorphism $\Phi\colon \mathcal{P}(\mathbb{N})/\operatorname{Fin} \to \mathcal{P}(\mathbb{N})/\operatorname{Fin}$ there exists a sequence $\{\mathcal{U}_n\}$ of ultrafilters on $\mathbb{N}$ such that $\Phi([A]_{\mathrm{Fin}}) = [\{n : A \in \mathcal{U}_n\}]_{\mathrm{Fin}}$. Such a homomorphism is said to have an *additive lifting*. It is not difficult to see that each homomorphism $\Phi\colon \mathcal{P}(\mathbb{N})/\operatorname{Fin} \to \mathcal{P}(\mathbb{N})/\mathcal{I}$ for a countably generated ideal $\mathcal{I}$ has an additive lifting if and only if each homomorphism $\Phi\colon \mathcal{P}(\mathbb{N}) \to \mathcal{P}(\mathbb{N})/\mathcal{I}$ for a countably generated ideal $\mathcal{I}$ has an additive lifting (see [**8**, Theorem 3.3]). The algebraic reformulation suggests asking for which ideals $\mathcal{I}$ on $\mathbb{N}$ the following assertion is true:

($C_{\mathcal{I}}$) Every homomorphism $\Phi\colon \mathcal{P}(\mathbb{N}) \to \mathcal{P}(\mathbb{N})/\mathcal{I}$ has an additive lifting.

The topological dual of this statement, namely the assertion that every continuous map g from a closed subset F of $\mathbb{N}^*$ into $\beta\mathbb{N}$ is trivial, is still meaningful for very simple subspaces F. For example, if F is an intersection (or a closure of the union) of countably many clopen sets then the analogue of Theorem 1 holds: g is a direct sum of a trivial map and a map with a nowhere dense range ([**13**, Theorem 3.9.2]). Again CH trivializes the question. I have conjectured that PFA implies ($C_{\mathcal{I}}$) for every analytic ideal $\mathcal{I}$. (Consider $\mathcal{P}(\mathbb{N})$ with the Cantor-set topology; a set is *analytic* if it is a continuous image of the irrationals.) I reluctantly refrain from discoursing on this subject any further. The survey [**19**] of this conjecture is still up-to-date.

It seems plausible that all questions stated here have a positive answer, and more precisely that it follows from PFA.

4. Further results

At present it is unclear what the limitations of the rigidity phenomena described above are. It is, however, clear that they go a bit beyond the outlined framework. A small number of rigidity results for $\mathbb{N}^*$ were recently proved by combining lifting results with other techniques. For example, assuming OCA and MA, Dow and Hart have proved that a Čech–Stone remainder of a locally compact, σ-compact space X is a continuous image of $\mathbb{N}^*$ if and only if X is homeomorphic to a sum of $\mathbb{N}$ with a compact space ([**9**]). From the same assumptions they also deduced that the Stone space of the Lebesgue measure algebra is not a continuous image of $\mathbb{N}^*$ ([**10**]). Both statements contradict the conclusion of Parovičenko's theorem. I proved ([**14**, §8]) that OCA implies $\operatorname{Exp}(\mathbb{N}^*)$ is not a continuous image of $\mathbb{N}^*$, thus confirming a conjecture of M. Bell. Dow ([**7**]) proved that PFA implies every two-to-one image of $\mathbb{N}^*$ is *trivial*: whenever $f\colon \mathbb{N}^* \to X$ is such that each fibre has exactly two points then X is homeomorphic to $\mathbb{N}^*$ and moreover f has to be a trivial map in the sense of Question 1. Needless to say, CH implies the existence of nontrivial two-to-one maps. In Dow's result it is important that each fibre has exactly two points; van Douwen ([**5**]) has constructed a nontrivial $f\colon \mathbb{N}^* \to X$ such that each fibre has at most two points. It is not known whether it is possible that a two-to-one image of $\mathbb{N}^*$ is not homeomorphic to $\mathbb{N}^*$; curiously enough, a negative answer follows from CH ([**12**]).

Question 5. *Is it possible to construct a nontrivial n-to-one map on $\mathbb{N}^*$ for some $n \in \mathbb{N}$ without using additional axioms of set theory?* 350?

This subject would not be complete without the insight provided by Murray Bell.

Question 6 (M. Bell). *Is it possible to construct an extremally disconnected image of a zero set in $\mathbb{N}^*$ that is not an image of $\mathbb{N}^*$ without using additional set-theoretic axioms?* 351?

Some partial answers to this question were obtained in [**8**]. The *rectangle algebra* is the algebra of subsets of $\mathbb{N}^{\mathbb{N}}$ generated by rectangles, $\prod_{i=1}^{\infty} A_i$.

Question 7 (M. Bell). *Is it possible to show that the Stone space of the rectangle algebra is a continuous image of $\mathbb{N}^*$ without using additional set-theoretic axioms?* 352?

I would conjecture that in both cases there is either a (properly defined) 'trivial' map witnessing the connection or appropriate axioms imply there is no surjection.

5. Conclusion

The main purpose of this note was to draw the attention of topologists and set theorists to an emerging canonical theory of spaces and Boolean algebras closely related to $\mathbb{N}^*$ and $\mathcal{P}(\mathbb{N})/\mathrm{Fin}$ respectively. Under CH, two such spaces that could possibly be related (via a homeomorphism or a surjection) are indeed related. A variant of this sweeping claim is a consequence of Woodin's Σ^2_1-absoluteness theorem ([**15**, **37**]); see [**13**, §2.1] and also §5.1 below.

This note is about the other extreme situation. For spaces X and Y define the notion of 'trivial' map $f\colon X \to Y$. This notion should be simple so that deciding the existence of a trivial homeomorphism/surjection between X and Y is reasonably easy, and that the statement 'there is a trivial isomorphism (or surjection) between X and Y' is absolute between sufficiently closed models of ZFC. It should also be well-chosen so that forcing (or Ramseyan) axioms imply every isomorphism (or surjection) between X and Y has to be trivial. This 'ideal' scenario is rather flexible. For example, in the case when $X = Y$ it serves to completely describe the group of all autohomeomorphisms of X: take the case when $X = \mathbb{N}^*$ ([**29**]), from where all this has started.

In most situations it is sufficient to prove that sufficiently strong forcing (or Ramseyan) axioms imply that if there is a homeomorphism (or surjection) $f\colon X \to Y$ then a trivial homeomorphism (or surjection) exists. In many concrete cases this is a theorem; see [**13**, §2.1]. The existence of a trivial connecting map is typically a $\mathbf{\Sigma}^1_2$ statement, and therefore absolute by Shoenfield's Absoluteness Theorem. Having a general lifting theorem greatly simplifies the question whether two spaces are homeomorphic or otherwise related. Compare e.g., [**23**], where a weak lifting theorem was supplemented by a technical tour de force argument and the proof of the same result in [**13**, Corollary 3.4.4, Proposition 1.13.13] where a strong lifting theorem was supplemented by straightforward computations. Isolating the notion of 'trivial' connecting map is not necessary to prove rigidity results. Take, for example, the Dow–Hart result on the Stone space of the Lebesgue measure algebra ([**10**]).

The problems of determining the relation between lifting statements such as those considered above and other set-theoretic statements are difficult and well-studied, but this is an another story (see e.g., [**31–33**]).

5.1. Metamathematics. The phenomenon that CH resolves so many questions about $\mathbb{N}^*$ has a metamathematical explanation or two. Model-theoretically (see [**3**] for model-theoretic background), $\mathcal{P}(\mathbb{N})/\operatorname{Fin}$ is a countably saturated Boolean algebra, hence CH implies it is a saturated model of the (complete) theory of atomless Boolean algebras. Clopen algebras of other Parovičenko spaces are also countably saturated, and this allows one to apply back-and-forth methods to relate these and similar algebras.

Another explanation is of a different nature. Instead of giving a technical device for constructing maps, it implies that maps that can be constructed in some models of set theory can also be constructed using CH. Hence CH is an optimal assumption for finding such maps. Let X and Y be spaces whose basic open sets can be coded by real numbers; for example, Čech–Stone compactifications of countable, locally compact spaces, as well as their finite powers, are of this form. A continuous map $f\colon X \to Y$ can be coded by a set of pairs of basic open subsets of X and Y, and therefore by a set of real numbers, C_f. Statements like 'f is onto' or 'f is a homeomorphism' are *projective in* C_f: they can be expressed using quantification over the real numbers only. Thus 'α^* and γ^* are homeomorphic' (for countable ordinals α and γ) is equivalent to a statement of the form $(\exists C \subseteq \mathbb{R})\, \phi(C)$ for a statement ϕ projective in C. A statement of this syntactical form is called a Σ^2_1-*statement*. Using a large cardinal assumption, Woodin proved ([**37**], see [**15**] or [**27**]) that if a Σ^2_1 statement can be forced then it holds in every forcing extension that satisfies CH.

Consequences of OCA + MA fit together forming a coherent picture of $\mathbb{N}^*$ and related spaces, with their rigidity properties maximized (see [**13**, §2.1] for an overview). A satisfactory metamathematical explanation of this phenomenon is yet to be found, but the current state of our understanding suggests that the ability to make gaps in quotient algebras indestructible is of central importance. The gap-freezing technique was developed in [**34**, §8] (cf. [**1**]) and first employed in this context in [**30**]. See [**6**] for analysis of gaps in $[\kappa]^\omega/\operatorname{Fin}$ or [**18**] for gaps in quotients of the form $\mathcal{P}(\mathbb{N})/\mathcal{I}$; for an approach compatible with CH see [**35**].

Notably, OCA and MA both hold in Woodin's canonical model for the negation of CH ([**38**]; see [**26**]). A discussion of the wider metamathematical context is beyond the scope of this article and it appears elsewhere ([**19**], [**13**, §2.1]).

Acknowledgement. I would like to thank A. Dow, K.P. Hart, J. van Mill, J.T. Moore and J. Steprāns for interest and remarks.

References

[1] U. Abraham, M. Rubin, and S. Shelah, *On the consistency of some partition theorems for continuous colorings, and the structure of* $\aleph_1$*-dense real order types*, Ann. Pure Appl. Logic **29** (1985), no. 2, 123–206.

[2] M. Bell, *Two Boolean algebras with extreme cellular and compactness properties*, Canad. J. Math. **35** (1983), no. 5, 824–838.

[3] C. C. Chang and H. J. Keisler, *Model theory*, North-Holland, Amsterdam, 1990.

[4] E. K. van Douwen, *Prime mappings, number of factors and binary operations*, Dissertationes Math. **199** (1981), 35 pp.

[5] E. K. van Douwen, *Applications of maximal topologies*, Topology Appl. **51** (1993), no. 2, 125–139.

[6] A. Dow, *Extending real-valued functions in* $\beta\kappa$, Fund. Math. **152** (1997), no. 1, 21–41.

[7] A. Dow, *Two to one images of* $\beta\mathbb{N}\setminus\mathbb{N}$ *under PFA*, Israel J. Math., To appear.

[8] A. Dow and I. Farah, *Is* $\mathcal{P}(\omega)$ *a subalgebra?*, Fund. Math. **183** (2004), no. 2, 91–108.

[9] A. Dow and K. P. Hart, ω^* *has (almost) no continuous images*, Israel J. Math. **109** (1999), 29–39.

[10] A. Dow and K. P. Hart, *The measure algebra does not always embed*, Fund. Math. **163** (2000), no. 2, 163–176.

[11] A. Dow and K. P. Hart, *The Čech–Stone compactifications of* $\mathbb{N}$ *and of* $\mathbb{R}$, Encyclopedia of general topology, Elsevier Science, 2004, pp. 213–217.

[12] A. Dow and G. Techanie, *Two-to-one continuous images of* $\mathbb{N}^*$, Fund. Math. **186** (2005), no. 2, 177–192.

[13] I. Farah, *Analytic quotients: theory of liftings for quotients over analytic ideals on the integers*, Mem. Amer. Math. Soc. **148** (2000), no. 702, 177 pp.

[14] I. Farah, *Liftings of homomorphisms between quotient structures and Ulam stability*, Logic Colloquium '98 (Prague), Assoc. Symbol. Logic, Urbana, IL, 2000, pp. 173–196.

[15] I. Farah, *A proof of the* Σ^2_1*-absoluteness theorem*, 2001, Notes.

[16] I. Farah, *Dimension phenomena associated with* $\beta\mathbb{N}$*-spaces*, Topology Appl. **125** (2002), no. 2, 279–297.

[17] I. Farah, *Powers of* $\mathbb{N}^*$, Proc. Amer. Math. Soc. **130** (2002), no. 4, 1243–1246.

[18] I. Farah, *Luzin gaps*, Trans. Amer. Math. Soc. **356** (2004), no. 6, 2197–2239.

[19] I. Farah, *Rigidity conjectures*, Logic Colloquium 2000, Assoc. Symbol. Logic, Urbana, IL, 2005, pp. 252–271.

[20] K. P. Hart and F. D. Tall, *Consistency results in topology, I: Quotable principles*, Encyclopedia of general topology (K. P. Hart, J. Nagata, and J. E. Vaughan, eds.), Elsevier Science, 2004, pp. 419–422.

[21] W. Just, *Nowhere dense P-subsets of* ω, Proc. Amer. Math. Soc. **106** (1989), no. 4, 1145–1146.

[22] W. Just, *The space* $(\omega^*)^{n+1}$ *is not always a continuous image of* $(\omega^*)^n$, Fund. Math. **132** (1989), no. 1, 59–72.

[23] W. Just, *Repercussions of a problem of Erdős and Ulam on density ideals*, Canad. J. Math. **42** (1990), no. 5, 902–914.

[24] W. Just, *A modification of Shelah's oracle-c.c. with applications*, Trans. Amer. Math. Soc. **329** (1992), no. 1, 325–356.

[25] W. Just, *A weak version of AT from OCA*, Set theory of the continuum (Berkeley, 1989), Springer, New York, 1992, pp. 281–291.

[26] P. Larson, *Showing OCA in* $\mathbb{P}_{\max}$*-style extensions*, Kobe J. Math. **18** (2001), no. 2, 115–126.

[27] P. Larson, *The stationary tower*, American Mathematical Society, Providence, RI, 2004.

[28] J. van Mill, *An introduction to* $\beta\omega$, Handbook of set-theoretic topology, North-Holland, Amsterdam, 1984, pp. 503–567.

[29] S. Shelah, *Proper forcing*, Lecture Notes in Math., vol. 940, Springer, Berlin, 1982.

[30] S. Shelah and J. Steprāns, *PFA implies all automorphisms are trivial*, Proc. Amer. Math. Soc. **104** (1988), no. 4, 1220–1225.

[31] S. Shelah and J. Steprāns, *Nontrivial homeomorphisms of* $\beta\mathbb{N} \setminus \mathbb{N}$ *without the continuum hypothesis*, Fund. Math. **132** (1989), no. 2, 135–141.

[32] S. Shelah and J. Steprāns, *Martin's axiom is consistent with the existence of nowhere trivial automorphisms*, Proc. Amer. Math. Soc. **130** (2002), no. 7, 2097–2106.

[33] J. Steprāns, *The autohomeomorphism group of the Čech–Stone compactification of the integers*, Trans. Amer. Math. Soc. **355** (2003), no. 10, 4223–4240.

[34] S. Todorčević, *Partition problems in topology*, American Mathematical Society, Providence, RI, 1989.

[35] S. Todorčević, *A dichotomy for P-ideals of countable sets*, Fund. Math. **166** (2000), no. 3, 251–267.

[36] B. Veličković, *OCA and automorphisms of* $\mathcal{P}(\omega)/\mathrm{fin}$, Topology Appl. **49** (1993), no. 1, 1–13.

[37] W. H. Woodin, σ_1^2*-absoluteness*, Handwritten note of May 1985.

[38] W. H. Woodin, *The continuum hypothesis. II*, Notices Amer. Math. Soc. **48** (2001), no. 7, 681–690.

Open Problems in Topology II – Edited by E. Pearl

Are stratifiable spaces M_1?

Gary Gruenhage

The Bing–Nagata–Smirnov characterization of metrizable spaces as the regular spaces having a σ-locally finite (or σ-discrete) base was one of the seminal results of the early 1950s in general topology. Then the late 1950s saw E. Michael's characterizations of paracompactness in regular spaces via σ-locally finite, σ-closure-preserving, and other related types of refinements. Clearly motivated by these now classical results, in 1961 Michael's student J. Ceder [**2**] introduced the following class of spaces as a natural generalization of metrizable spaces:

Definition. A regular space X is an M_1*-space* if it has σ-closure-preserving base.[1]

M_1-spaces are paracompact by one of Michael's theorems, and it is easy to see that closed sets are G_δ, so they are also perfectly normal. An important subclass of M_1-spaces is the class of closed images of metrizable spaces [**29**].

However, Ceder could not show that M_1-spaces are hereditary, even for closed subspaces. To see the problem, note that the trace of a closure-preserving collection on a closed subset need not be closure preserving (there are easy examples in the plane illustrating this). Nor could he show that they are preserved by nice mappings such as closed or even perfect mappings. Thus he also considered two formally larger classes, which he called M_2-spaces and M_3-spaces, respectively. These classes had more technical definitions, but otherwise, they had essentially the same topological properties and they had the advantage of being preserved by arbitrary subspaces as well as closed mappings.

Definition. A collection $\mathcal{B}$ is a *quasi-base* for X if whenever $x \in U$, U open, there is $B \in \mathcal{B}$ with $x \in \mathrm{int}(B) \subset B \subset U$. A regular space X is an M_2*-space* if it admits a σ-closure-preserving quasi-base $\mathcal{B}$ (which may be taken to consist of closed sets).

Note that M_2-spaces are hereditary, since the trace of a closure-preserving collection of *closed* sets on a subspace is closure-preserving in the subspace.

Recall that B is a regular closed set if $B = \mathrm{cl}(\mathrm{int}(B))$. If $\mathcal{B}$ is a σ-closure-preserving quasi-base of regular closed sets, it is easy to check that the interiors form a σ-closure-preserving base. So if M_2 is really more general than M_1, it comes from allowing members of the quasi-base to have nonempty *outliers* $B \setminus \mathrm{cl}(\mathrm{int}(B))$. Note that such outliers can help make a collection closure-preserving; e.g., a collection $D_0, D_1, \ldots$ of disks in the plane converging to a point p is not closure-preserving, but $\{D_n \cup \{p\} : n \in \omega\}$ is closure-preserving.

The M_3-spaces were defined by Ceder as the regular spaces having a σ-cushioned pair-base, though the following characterization of Borges [**1**], who showed that

[1]Recall that a collection $\mathcal{U}$ is *closure-preserving* if $\mathrm{cl}(\bigcup\mathcal{U}') = \bigcup\{\mathrm{cl}(U) : U \in \mathcal{U}'\}$ for any subcollection $\mathcal{U}'$ of $\mathcal{U}$, and is σ*-closure-preserving* if it is a countable union of closure-preserving collections.

M_3-spaces have many other good properties (e.g., they satisfy the Dugundji Extension Theorem), and renamed them *stratifiable spaces*, provides a more elegant definition:

Definition. A T_1-space X is an M_3-*space* (or *stratifiable*) iff one can assign to each closed set H a decreasing sequence $U_n(H)$, $n \in \omega$, of open sets satisfying:

(1) $H = \bigcap_{n\in\omega} U_n(H) = \bigcap_{n\in\omega} \mathrm{cl}(U_n(H))$;
(2) $H \subset K \Rightarrow U_n(H) \subset U_n(K)$.

Since the first condition characterizes perfect normality, stratifiable spaces can be thought of as the class of monotonically perfectly normal spaces. They are also exactly the monotonically normal σ-spaces (σ-*spaces* are spaces having a σ-discrete network).

Ceder didn't know if any of these classes were in fact different. In the mid-1970s, the author [**6**] and Junnila [**15**] independently proved that stratifiable and M_2-spaces are the same. But to this day, it is not known if stratifiable and M_1-spaces are the same.

353? **Problem 1.** *Are stratifiable (equivalently, M_2-) spaces M_1?*

Since stratifiable spaces have turned out to be one of the most useful and important classes of generalized metrizable spaces, an answer to the problem would be of great interest, and if positive, would render many papers on the subject obsolete.

1. Equivalent questions

As mentioned in the introduction, Ceder was led to define M_2- and M_3-spaces because he could not show that M_1-spaces were preserved by some basic topological operations. In fact, certain preservation statements are equivalent to Problem 1:

Theorem 1.1. *The following statements are equivalent:*

(1) *Stratifiable spaces are M_1;*
(2) *Every (closed) subspace of an M_1-space is M_1;*
(3) *Perfect (closed) images of M_1-spaces are M_1.*

The above equivalences follow immediately from the the fact that stratifiable spaces are preserved by subspaces and closed images, along with the following very pretty result of Heath and Junnila [**10**]:

Theorem 1.2. *Every stratifiable space X is a closed subspace of an M_1-space Z such that $Z \setminus X$ consists of isolated points and there is a perfect retraction $r\colon Z \to X$.*

Here are some other equivalences:

Theorem 1.3. *The following statements are equivalent:*

(1) *Stratifiable spaces are M_1;*

(2) *Every point of every stratifiable space has a (σ-)closure-preserving local base;*

(3) *Every closed subset of every stratifiable space has a (σ-)closure-preserving outer base.*

Here, an *outer base* for a subset H of X is a collection $\mathcal{U}$ of open supersets of H such that every open superset of H contains a member of $\mathcal{U}$. That these statements are equivalent follows fairly easily from the fact that stratifiable spaces are paracompact σ-spaces, preservation under closed mappings, and using the following recent and important result of Mizokami [**19**]:

Theorem 1.4. *Every closed subset of an M_1-space has a closure-preserving outer base.*

For some time the class $\mathcal{P}$ of M_1-spaces in which every closed subset has a closure-preserving outer base was studied; by Theorem 1.4, every M_1-space is in $\mathcal{P}$.

2. Related classes and partial results

One of the most important early partial results on Problem 1 was the following result of Itō [**13**]:

Theorem 2.1. *The following are equivalent for a stratifiable space X:*

(1) *Every closed subset of X has a closure-preserving outer base (and hence X is M_1);*

(2) *Every point of X has a closure-preserving local base.*

Using the fact the stratifiable spaces are paracompact σ-spaces, it is easy to see that if every closed subset of a stratifiable space X has a closure-preserving outer base, then X is M_1 (by Theorem 1.4, the converse also holds). So Itō's result says it suffices that every point have a closure-preserving local base. E.g., first-countable stratifiable spaces are M_1.

Mizokami, Shimane, and Kitamura [**21**], extending a result of the first two of these authors [**20**], have improved the first-countable result to sequential spaces and more:

Theorem 2.2. *A space X is M_1 if it is stratifiable and has the following property:*

(δ) *Whenever U is dense open in X and $x \in X \setminus U$, there is a closure-preserving collection $\mathcal{F}$ of closed subsets of X that is a network at x, such that $\mathrm{cl}(F \cap U) = F$ for every $F \in \mathcal{F}$.*

Note that this result extends Itō's, for if $\mathcal{B}$ is a closure-preserving local base at x, and U is dense open, then $\mathcal{F} = \{\mathrm{cl}(B) : B \in \mathcal{B}\}$ witnesses property (δ). It is easy to observe that every Fréchet space satisfies (δ); less obvious is that sequential stratifiable spaces satisfy (δ) [**20**]. More generally, a stratifiable space satisfies (δ) (see [**21**]) if it has the following property, which has been called *weak approximation by points (WAP)* [**27**]:

(WAP) If A is not closed, there exists $B \subset A$ such that $\mathrm{cl}(B) \setminus A$ is exactly one point.

There are classes of spaces formally stronger than M_1 for which it is as yet undetermined whether every M_3-(or sometimes even every M_1-) space belongs to the class. The most pertinent of these classes, so it seems at present, is the class of μ-spaces, introduced by Nagami [**22**] for dimension-theoretic reasons.

Definition. A space X is *F_σ-metrizable* if it is a countable union of closed metrizable subspaces, and X is a μ-space if it is homeomorphic to a subspace of a countable product of paracompact F_σ-metrizable spaces.

I showed [**7**] that stratifiable F_σ-metrizable spaces are M_1. The following two results extend this:

Theorem 2.3.
(1) *Stratifiable μ-spaces are M_1;*
(2) *A stratifiable space is M_1 if it is a countable union of closed M_1 subspaces.*

The second result follows from Mizokami's theorem in [**17**] that a stratifiable space which is a countable union of closed subspaces in the class $\mathcal{P}$ is M_1, together with his more recent result mentioned earlier that every M_1-space is in $\mathcal{P}$.

Theorem 2.3(1) is due to Mizokami [**17**]; Junnila and Mizokami [**14**] subsequently showed that a couple of other subclasses of stratifiable spaces that had been studied in the literature are μ-spaces. Tamano [**30**] obtained the following useful internal characterization of μ-spaces:

Theorem 2.4. *The following are equivalent:*
(1) *X is a stratifiable μ-space;*
(2) *X has a base $\mathcal{B} = \bigcup_{n\in\omega} \mathcal{B}_n$, where each $\mathcal{B}_n$ is* mosaical, *i.e., there is a σ-discrete cover $\mathcal{F}_n$ of X such that $F \cap B \neq \emptyset \iff F \subset B$ for every $F \in \mathcal{F}_n$ and $B \in \mathcal{B}_n$.*

There are spaces having a countable network which are not μ-spaces [**32, 33**]; but we don't know the answer to:

354? **Problem 2.** *Is every stratifiable space a μ-space?*

Since the class of μ-spaces is hereditary, by the Heath–Junnila theorem it is equivalent to ask if every M_1-space is a μ-space. Obviously a positive answer to Problem 2 settles Problem1. An important partial result is that spaces having a σ-closure-preserving clopen base, which are called *M_0-spaces*, are μ-spaces [**12, 16**]. The class of M_0-spaces turns out to coincide with the class of stratifiable μ-spaces X with $\dim X = 0$; also, every stratifiable μ-space is a perfect image of an M_0-space [**17**].

Consider the following string of containments, where $\mathcal{M}_i$ denotes the class of M_i-spaces, $\mathcal{S}$ ($\mathcal{S}\mu$) is the class of stratifiable (stratifiable μ-) spaces, and $\mathcal{PM}_0$ ($\mathcal{CM}_0$) is the class of perfect (closed) images of M_0-spaces:

$$\mathcal{M}_0 \subset \mathcal{S}\mu \subset \mathcal{PM}_0 \subset \mathcal{CM}_0 \subset \mathcal{M}_1 \subset \mathcal{S}.$$

It is not known if any of these containments other than the leftmost are strict. Indeed, parts of this line could collapse, maybe all the way from $\mathcal{S}$ to $\mathcal{S}\mu$. But it could also happen, e.g., that $\mathcal{S} = \mathcal{M}_1 \neq \mathcal{CM}_0$. That every space in $\mathcal{CM}_0$ is M_1, in fact hereditarily M_1, follows from the observation that M_0-spaces are hereditary, and the following result of Itō [**11**]:

Theorem 2.5. *If every closed subset of an M_1-space X is M_1, then every closed image of X is M_1.*

The class $\mathcal{PM}_0$ would equal $\mathcal{S}\mu$ if the following old question of Nagami [**22**] had a positive answer:

Problem 3. *Are μ-spaces preserved by perfect mappings?* 355?

This seems to be open even for closed mappings. A partial result is that the closed image of a stratifiable F_σ-metrizable space is a μ-space [**14**].

Another interesting subclass of stratifiable spaces was introduced by Oka [**25**]:

Definition. A stratifiable space X is in the class $\mathcal{EM}_3$ if there is a σ-closure-preserving collection $\mathcal{E}$ satisfying:

Whenever $x \in U$, U open, there is $\mathcal{F} \subset \mathcal{E}$ such that $\bigcup \mathcal{F}$ is closed, and $x \in X \setminus \bigcup \mathcal{F} \subset U$.

Oka's motivation for defining $\mathcal{EM}_3$ was dimension-theoretic; he proved the following:

Theorem 2.6.

(1) $\dim X = \operatorname{Ind} X$ *for every* $X \in \mathcal{EM}_3$*;*
(2) $\mathcal{EM}_3$ *is the class of perfect (or closed) images of (strongly)* 0*-dimensional stratifiable spaces;*
(3) $\mathcal{EM}_3$ *is hereditary, countably productive, and preserved by closed maps.*

It follows that the class $\mathcal{EM}_3$ fits between $\mathcal{CM}_0$ and $\mathcal{S}$; but it is not known if it is equal to either one or both, nor is its relation to $\mathcal{M}_1$ known. If $\mathcal{EM}_3 = \mathcal{M}_1$, then it would follow from the Heath–Junnila theorem that $\mathcal{S} = \mathcal{M}_1$. Also note that $\mathcal{S} = \mathcal{EM}_3$ iff every stratifiable space is the closed (or perfect) image of a (srongly) 0-dimensional stratifiable space. The following dimension-theoretic questions are also open:

Problem 4. 356–359?

(1) *Let X be strongly* 0*-dimensional. If $X \in \mathcal{M}_1$, must $X \in \mathcal{M}_0$? What if $X \in \mathcal{PM}_0$?*
(2) *Is every M_1-space the perfect (or closed) image of a (strongly)* 0*-dimensional M_1-space?*
(3) *For an M_1-space X, is it true that* $\operatorname{Ind}(X) \leq n$ *iff X has a σ-closure-preserving base $\mathcal{B}$ such that for every $B \in \mathcal{B}$,* $\operatorname{Ind}(\partial B) \leq n-1$*?*
(4) *Does* $\dim X = \operatorname{Ind} X$ *for all stratifiable X? What if X is separable?*

If the answer to (2) is positive, then by the Heath–Junnila result, every stratifiable space is also the closed image of a 0-dimensional M_1-space, so it would follow that $\mathcal{S} = \mathcal{EM}_3$. (3) is known to hold for stratifiable μ-spaces [**17**]. (4) is known to consistently fail (e.g., under CH) for (non-stratifiable) spaces having a countable network ([**3**]; see also [**4**]).

3. Function spaces and a possible counterexample

Gartside and Reznichencko [**5**] investigated stratifiability of function spaces, and in particular, proved that the space $C_k(X)$ of all real-valued continuous functions on X with the compact-open topology is stratifiable whenever X is Polish (complete separable metric). They show this first when X is the space $\mathbb{P}$ of irrationals, and then use the fact that any Polish space Y is the continuous image of $\mathbb{P}$, and hence $C_k(Y)$ embeds in $C_k(\mathbb{P})$. It is quite interesting that their proof of stratifiability of $C_k(\mathbb{P})$ gives no clue as to its M_1-ness, and no one has yet been able to determine if $C_k(\mathbb{P})$ is M_1 or not.

360? **Problem 5** ([**5**]). *Is $C_k(\mathbb{P})$ an M_1-space?*

It is also not known if $C_k(\mathbb{P})$ is a μ-space or in $\mathcal{EM}_3$. A negative answer to Problem 5 of course solves Problem 1 in the negative. There are unpublished results of Balogh and Gruenhage, Gartside, Nyikos, and Tamano showing that collections built in some ways from standard basic open sets won't work; e.g., no collection of sets consisting of finite unions of standard basic open sets of $C_k(\mathbb{P})$ can form either a σ-closure-preserving or σ-mosaical base. In the positive direction, the author and Tamano [**9**] have shown that $C_k(X)$ is a μ-space whenever X is σ-compact Polish.

Possibly, one could show that $C_k(\mathbb{P})$ is M_1 by showing it has property (δ). However, while it is known that $C_k(\mathbb{P})$ is not sequential [**26**], the following is open even for σ-compact Polish spaces:

361? **Problem 6.** *If X is Polish, does $C_k(X)$ have the WAP property?*

Gartside and Reznichenko asked if a converse of their result is true:

362? **Problem 7.** *If X is separable metrizable, and $C_k(X)$ is stratifiable, must X be Polish?*

This is still unsettled, but Nyikos [**23**] has shown that $C_k(X)$ is not stratifiable for any separable metric X which contains a 0-dimensional closed subspace with no uncountable compact sets (e.g., a closed subspace homeomorphic to the rationals); a corollary is that the answer to Problem 7 is positive for coanalytic subsets of Polish spaces.

4. Some final remarks

We close with a few more remarks about Problem 1 and some suggestions for further reading. A brief survey with proofs of basic results on stratifiable and M_1-spaces is included in [**8**]. Much more extensive and highly recommended is Tamano's survey [**31**], which includes among other things proofs of Itō's theorems as well as most of the results we mentioned on μ-spaces and the class $\mathcal{EM}_3$. Also discussed there are some classes of M_1-spaces that fall between stratifiable μ-spaces and hereditarily M_1-spaces that are defined in terms of special bases. More recent surveys, which like this one do not include proofs, have been written by Mizokami [**18**] and Nyikos (see Classic Problem IV in [**24**]).

Anyone hoping to prove that stratifiable implies M_1 should become familiar with techniques in [**21**] and/or [**20**], many of which were also used in the important paper [**19**]. Large parts of these arguments involve fattening up closure-preserving collections of closed sets to collections which have certain combinatorial and regularity properties (with the goal of building closure-preserving collections of regular closed sets). It is difficult to characterize these techniques briefly, so we only mention some tools that are common to not only these arguments but many that preceded these. Monotone normality is heavily exploited. Any stratifiable space has a weaker metrizable topology; constructing weaker metrizable topologies having certain close relations to the given topology is frequently useful. Another important tool is the following key lemma in Itō's proof of Theorem 2.1: given a closure-preserving collection $\mathcal{B}$ of closed sets, there is a σ-discrete set D such that, for every $B \in \mathcal{B}$, $D \cap B$ is dense in B. Also, building networks with special properties can be useful; oft-used here is the result in [**28**] that any closure-preserving collection $\mathcal{B}$ of closed sets in a stratifiable space is mosaical (see Theorem 2.4(2) for the meaning of mosaical).

For stratifiable spaces, separability and Lindelöfness, as well as the hereditary versions, are equivalent, and these are in turn equivalent to having a countable network. So Problem 1 would seem to split naturally into two cases, the countable network case and the σ-discrete but uncountable network case. However, there seems to be no evidence that these cases will turn out any differently or that the countable network case is any easier. Indeed, $C_k(\mathbb{P})$, which presently the only specific space known to be stratifiable that is not known to be M_1, has a countable network.

Finally, we remark that it seems doubtful that the answer will turn out to be independent of ZFC; the only known consistency result in the area is due to N. Zhong [**34**], who showed that that stratifiable spaces of cardinality less than $\mathfrak{b}$ are M_1.

References

[1] C. R. Borges, *On stratifiable spaces*, Pacific J. Math. **17** (1966), 1–16.

[2] J. G. Ceder, *Some generalizations of metric spaces*, Pacific J. Math. **11** (1961), 105–125.

[3] G. Delistathis and S. Watson, *A regular space with a countable network and different dimensions*, Trans. Amer. Math. Soc. **352** (2000), no. 9, 4095–4111.

[4] A. Dow and K. P. Hart, *Cosmic dimensions*, 2005, Preprint. `arXiv:math.GN/0509099`

[5] P. Gartside and E. A. Reznichenko, *Near metric properties of function spaces*, Fund. Math. **164** (2000), no. 2, 97–114.

[6] G. Gruenhage, *Stratifiable spaces are M_2*, Topology Proc. **1** (1977), 221–226.

[7] G. Gruenhage, *On the $M_3 \Rightarrow M_1$ question*, Topology Proc. **5** (1980), 77–104 (1981).

[8] G. Gruenhage, *Generalized metric spaces*, Handbook of set-theoretic topology, North-Holland, Amsterdam, 1984, pp. 423–501.

[9] G. Gruenhage and K. Tamano, *If X is σ-compact Polish, then $C_k(X)$ has a σ-closure-preserving base*, Topology Appl. **151** (2005), 99–106.

[10] R. W. Heath and H. J. K. Junnila, *Stratifiable spaces as subspaces and continuous images of M_1-spaces*, Proc. Amer. Math. Soc. **83** (1981), no. 1, 146–148.

[11] M. Itō, *The closed image of a hereditary M_1-space is M_1*, Pacific J. Math. **113** (1984), no. 1, 85–91.

[12] M. Itō, *M_0-spaces are μ-spaces*, Tsukuba J. Math. **8** (1984), no. 1, 77–80.

[13] M. Itō, *M_3-spaces whose every point has a closure preserving outer base are M_1*, Topology Appl. **19** (1985), no. 1, 65–69.

[14] H. Junnila and T. Mizokami, *Characterizations of stratifiable μ-spaces*, Topology Appl. **21** (1985), no. 1, 51–58.

[15] H. J. K. Junnila, *Neighbornets*, Pacific J. Math. **76** (1978), no. 1, 83–108.

[16] H. J. K. Junnila, *A characterization of M_0-spaces*, Proc. Amer. Math. Soc. **91** (1984), no. 3, 481–484.

[17] T. Mizokami, *On M-structures*, Topology Appl. **17** (1984), no. 1, 63–89.

[18] T. Mizokami, *One development of the M_3 vs. M_1 problem*, Topology Atlas Inv. Contrib. **8** (2003), 1–8. arXiv:math.GN/0312457

[19] T. Mizokami, *On closed subsets of M_1-spaces*, Topology Appl. **141** (2004), 197–206.

[20] T. Mizokami and N. Shimane, *On the M_3 versus M_1 problem*, Topology Appl. **105** (2000), no. 1, 1–13.

[21] T. Mizokami, N. Shimane, and Y. Kitamura, *A characterization of a certain subclass of M_1-spaces*, JP J. Geom. Topol. **1** (2001), no. 1, 37–51.

[22] K. Nagami, *Normality of products*, Actes du Congrès International des Mathématiciens (Nice, 1970), Tome 2, Gauthier-Villars, Paris, 1971, pp. 33–37.

[23] P. J. Nyikos, *On some non-Archimedean spaces of Alexandroff and Urysohn*, Topology Appl. **91** (1999), no. 1, 1–23.

[24] P. J. Nyikos, *Classic problems—25 years later. I*, Topology Proc. **26** (2001/02), no. 1, 345–356.

[25] S. Oka, *Dimension of stratifiable spaces*, Trans. Amer. Math. Soc. **275** (1983), no. 1, 231–243.

[26] R. Pol, *Normality in function spaces*, Fund. Math. **84** (1974), no. 2, 145–155.

[27] P. Simon, *On accumulation points*, Cahiers Topologie Géom. Différentielle Catég. **35** (1994), no. 4, 321–327.

[28] F. Siwiec and J. Nagata, *A note on nets and metrization*, Proc. Japan Acad. **44** (1968), 623–627.

[29] F. G. Slaughter, Jr., *The closed image of a metrizable space is M_1*, Proc. Amer. Math. Soc. **37** (1973), 309–314.

[30] K. Tamano, *On characterizations of stratifiable μ-spaces*, Math. Japon. **30** (1985), no. 5, 743–752.

[31] K. Tamano, *Generalized metric spaces. II*, Topics in general topology, North-Holland, Amsterdam, 1989, pp. 367–409.

[32] K. Tamano, *A cosmic space which is not a μ-space*, Topology Appl. **115** (2001), no. 3, 259–263.

[33] K. Tamano and S. Todorčević, *Cosmic spaces which are not μ-spaces among function spaces with the topology of pointwise convergence*, Topology Appl. **146/147** (2005), 611–616.

[34] N. Zhong, *Small M_3-space is M_1*, Questions Answers Gen. Topology **12** (1994), no. 1, 113–115.

Open Problems in Topology II – Edited by E. Pearl

Perfect compacta and basis problems in topology

Gary Gruenhage and Justin Tatch Moore

An interesting example of a compact Hausdorff space that is often presented in beginning courses in topology is the unit square $[0,1] \times [0,1]$ with the lexicographic order topology. The closed subspace consisting of the top and bottom edges is perfectly normal. This subspace is often called the Alexandroff double arrow space. It is also sometimes called the "split interval", since it can be obtained by splitting each point x of the unit interval into two points x_0, x_1, and defining an order by declaring $x_0 < x_1$ and using the induced order of the interval otherwise. The top edge of the double arrow space minus the last point is homeomorphic to the Sorgenfrey line, as is the bottom edge minus the first point. Hence it has no countable base, so being compact, is nonmetrizable. There is an obvious two-to-one continuous map onto the interval.

There are many other examples of nonmetrizable perfectly normal compacta, if extra set-theoretic hypotheses are assumed. The most well-known is the Souslin line (compactified by adding a first and last point). Filippov [**6**] showed that the space obtained by "resolving" each point of a Luzin subset of the sphere S^2 into a circle by a certain mapping is a perfectly normal locally connected nonmetrizable compactum (see also Example 3.3.5 in [**38**]). Moreover a number of authors have obtained interesting examples under CH (or sometimes something stronger); see, e.g., Filippov and Lifanov [**17**], Fedorchuk [**5**], and Burke and Davis [**3**].

At some point, researchers began to wonder if there is a sense in which minor variants of the double arrow space are the only ZFC examples of perfectly normal nonmetrizable compacta. A first guess was made by David Fremlin, who asked if it is consistent that every perfectly normal compact space is the continuous image of the product of the double arrow space with the unit interval. But this was too strong: Watson and Weiss [**39**] constructed a counterexample (which looked like the double arrow space with a countable set of isolated points added in a certain way). Finally, the following question, also due to Fremlin, became the central one:

Question 1 ([**9**]). *Is it consistent that every perfect compactum admits a continuous and at-most-two-to-one map onto a metric space?* 363?

We call a space which does admit an at-most-two-to-one continuous map onto a metric space *premetric of order* 2.

Gruenhage noticed a close connection with what is now being called the "basis problem" for uncountable first-countable spaces:

Question 2. *Is it consistent that every uncountable first-countable regular space contains either an uncountable discrete subspace, or a fixed uncountable subspace of the real line or of the Sorgenfrey line?* 364?

The first author acknowledges support from NSF grant DMS-0405216 and the second author from NSF grant DMS-0401893.

In other words, might there be a three-element basis for uncountable first-countable regular spaces? One might be tempted to remove the requirement of first-countability in this question, but this is not possible by Moore's ZFC L-space [**21**]. It's clear that if there is any three element basis, it must be the three mentioned in Question 2. The connection to Fremlin's problem is this: a positive answer to the basis problem for first-countable spaces implies a positive answer to Fremlin's conjecture, and Fremlin's conjecture is equivalent, under PFA, to the basis conjecture for subspaces of perfectly normal compacta [**12**].

As is suggested by some previous partial results, it is possible that PFA or Martin's Maximum MM could imply positive answers to these questions. Fremlin [**7**] showed that under MM, any perfectly normal compactum admits a map to a metric space M whose fibers have cardinality two or less on a comeager subset of M. Gruenhage [**11**] showed that even without first-countability, PFA implies a positive answer to the basis problem in the class of cometrizable spaces[1] (later, Todorčević [**30**] proved that this follows from OCA, a consequence of PFA).

It turns out that there is an axiom, namely Woodin's Axiom $(*)$ [**40**], which is a provably optimal set theoretic hypothesis in the sense that if either of these questions can be shown to have a positive answer in some suitably robust model,[2] then $(*)$ implies a positive answer. It is important to note that questions for which $(*)$ is optimal in this sense are ones which are of a certain logical form and which reduce to spaces of size and weight not greater than $\aleph_1$. This includes not only these two questions, but most of the ones that follow. Thus we have decided to state them in the form "Does $(*)$ imply ...", though this is of course usually not the way they originally appeared. In practice, $(*)$ can be rather difficult to apply directly; our formulation can be taken to be an essentially equivalent way of asking if the statements can be proved consistent. See the last section for further discussion of $(*)$.

1. Perfect compacta

Predating Fremlin's problem are two other basic questions about perfectly normal compacta:

365? **Question 3** $(*)$. *If $X \times Y$ is perfect and compact, then is either X or Y metrizable?*

366? **Question 4** $(*)$. *Is every locally connected perfect compactum metrizable?*

The first question is due to Przymusinski [**23**] and the variant of the second which asks if $\mathsf{MA}_{\aleph_1}$ gives a positive answer has been attributed to Rudin (see [**22**]). If $(*)$ implies a positive answer to either the basis problem or to Fremlin's problem, then both of these questions also have positive answers [**10, 12**].

---

[1] A space (X, τ) is *cometrizable* if there is a weaker metric topology τ' such that every point has a τ-neighborhood base consisting of τ'-closed sets.

[2] In particular, if PFA or MM implies a positive answer or if such an answer can be forced.

A consistent positive answer to Question 4 would imply a consistent positive answer to the following question, which appears in [**18**][3] (see also Problem 6.12 in [**24**]). But in this case we don't know any consistency results, positive or negative:

Question 5. *If a compact convex subset of a locally convex topological vector space is perfectly normal, must it be metrizable?* 367?

It is perhaps worth noting that Helly's space of nondecreasing functions from $[0,1]$ into $[0,1]$ with the pointwise topology is compact, convex, separable, and first-countable but not metrizable.

Concerning Przymusinski's question, suppose that there are disjoint uncountable $A_0, A_1 \subseteq [0,1]$ such that there is no monotonic injection of an uncountable subset of A_0 into A_1. Abraham and Shelah have shown in [**2**] that such pairs of subsets of $[0,1]$ can exist in a model of $\mathsf{MA}_{\aleph_1}$. On the other hand, Todorčević proved in [**28**] that if X_0 and X_1 are obtained as in the split interval construction, but with only the points of A_0 and A_1 split, then $X_0 \times X_1$ is perfectly normal. Hence $\mathsf{MA}_{\aleph_1}$ is not sufficient for a positive answer to Przymusinski's question.

Since no uncountable subspace of the Sorgenfrey line is embeddable in a locally connected perfect compactum [**10**], a positive answer to the following would give a positive answer to Question 4:

Question 6 (∗). *Does every nonmetrizable perfect compactum contain a copy of an uncountable subspace of the Sorgenfrey line?* 368?

The difference between maps with metric fibers and with $\leq$ 2-point fibers in this context is unclear:

Question 7 (∗). *Does every perfect compactum admit a map into a metric space with metric fibers?* 369?

Question 8 (∗). *If K is a perfect compactum which maps into a metric space with metric fibers, must K admit an at-most-two-to-one map into a metric space?* 370?

A compact Souslin line K is a perfectly normal compactum which does not map onto a metric space with metric fibers [**27**]. Filippov's CH example mentioned in the introduction admits an obvious map onto a compact metric space with metric fibers, but is not premetric of order two.

A weaker form of Question 7 can be stated as follows. Suppose that $K \subseteq [0,1]^{\omega_1}$ is a perfect compactum. For $f \neq g \in K$, let $\Delta(f,g)$ denote the least α such that $f(\alpha) \neq g(\alpha)$, and define

$$T(K) = \{f \restriction \alpha : f \in K \text{ and } \exists g \in K(\alpha < \Delta(f,g) < \omega_1)\}.$$

Question 9 (∗). *If K is a nonmetrizable perfect compactum, can $T(K)$ contain an Aronszajn subtree?* 371?

---

[3]The author of this question is not clear; it seems to have already been known to MacGibbon in [**18**], but this was the earliest reference we could locate.

This question appears in [4] along with a number of related questions. See also the article from North Bay in this volume [**15**].

372? **Question 10** (*). *If X is a perfect compactum and $Y \subseteq X^2$ is scattered, must Y have Cantor–Bendixson rank less than ω_1? What if Y is assumed to be locally compact?*

Assuming CH, Gruenhage has constructed an example of a perfect compactum X whose square is a hereditarily normal, hereditarily separable space [**13**]. In fact, X is premetric of order 2 and X^2 contains a locally compact, locally countable S-space. It is possible to show, however, that Question 10 has a positive answer for compacta which are premetric of order 2 ((*) is required for this deduction).

It is also not known if Fremlin's problem can be reduced to the 0-dimensional case, which motivates the following two questions, the latter suggested by Todorčević.

373? **Question 11.** *Is it consistent*[4] *that every perfect compactum is the continuous image of a 0-dimensional perfect compactum?*

374? **Question 12** (*). *Does every nonmetrizable perfect compactum contains a closed subspace with uncountably many clopen sets?*

2. Uncountable spaces

Call a space X *functionally countable* if every continuous real-valued function defined on X has countable range.

375? **Question 13** (*). *Is every first-countable hereditarily functionally countable space countable?*

376? **Question 14** (*). *Does every uncountable functionally countable subspace of a countably tight compact space have an uncountable discrete subspace?*

Obviously any uncountable hereditarily functionally countable space has countable spread, and a first-countable example is a counterexample to the basis conjecture. Any uncountable left-separated subspace of a Souslin line is a consistent example of such a space. Currently the only known ZFC example of an uncountable functionally countable space with no uncountable discrete subspace is Moore's L-space, which is hereditarily functionally countable. Assuming $\mathsf{MA}_{\aleph_1}$, it is known that there are no first-countable L-spaces [**25**] and that any compactification of an L-space maps continuously onto $[0,1]^{\omega_1}$ [**8**, 44A] (see [**34**, p. 68]). Under (*), any functionally countable first-countable space of countable spread must be both hereditarily Lindelöf and hereditarily separable, and any uncountable one would also be a counterexample to the basis conjecture.

[4](*) may not necessarily be an optimal hypothesis for giving a positive solution to this problem, since we cannot assume without loss of generality that the space has weight $\aleph_1$. It still seems likely, however, that a forcing axiom is an appropriate hypothesis to yield a positive solution.

Question 15. *Is it consistent that every uncountable first-countable space of countable spread either contains an uncountable subspace of the Sorgenfrey line or has a countable network?* 377?

If a positive answer to this question is consistent with $\mathsf{MA}_{\aleph_1}$, then this would also give a positive answer to the basis question, since $\mathsf{MA}_{\aleph_1}$ implies that any uncountable space with a countable network contains a uncountable separable metrizable subspace [**11**]. As with the basis conjecture, under PFA [**11**] (or even OCA [**30**]), this question has a positive answer in the class of cometrizable spaces, even without the first-countable assumption.

Question 15 is related to some other questions concerning when spaces have a countable network. Recall that a subset Y of a space X is *weakly separated* if one can assign to each $y \in Y$ a neighborhood U_y of y such that $y \neq z$ implies $y \notin U_z$ or $z \notin U_y$. Note that if X has a countable network, then X does not contain an uncountable weakly separated subspace. The converse of this was asked by Tkachenko [**26**]:

Question 16. *Is it consistent that a space with no uncountable weakly separated subspace must have a countable network?* 378?

Unlike Question 15, this is open even in the non-first-countable case. Todorčević discusses this question in [**30**] and states without proof that under PFA, if no finite power of a space X has an uncountable weakly separated subspace, then X has a countable network. Juhasz, Soukup, and Szentmiklóssy [**14**] obtained the same result under $\mathsf{MA}_{\aleph_1}$ for spaces of size and weight $\leq \aleph_1$. Note that it follows that under PFA (under $\mathsf{MA}_{\aleph_1}$ for spaces of size and weight $\leq \aleph_1$), Question 15 and Question 2 are equivalent.

The following also remain unsolved:

Question 17. (a) *Is it consistent that X has a countable network if X^2 has no uncountable discrete subspace?* (b) *What if X^ω is hereditarily separable and hereditarily Lindelöf?* 379–380?

Question 17(b) is an old question of Arhangel′skiĭ [**1**]. Todorčević [**30**] has shown that there are cometrizable counterexamples to these questions, as well as Question 16, as long as $\mathfrak{b} \neq \omega_2$. These questions are also open in the first-countable case, and in that case, a positive answer to Question 15 with PFA implies a positive answer to these as well.

3. Approaches, axiomatics, further reading

It should be emphasized that analysis of these problems would benefit greatly from a combinatorial reformulation or approximation, particularly one which is Ramsey theoretic in nature. If there are positive solutions, Todorčević's method of building forcings with models as side conditions will likely provide the basic framework. The standard source is [**30**]; further reading can also be found in [**31**, **32**]. The methods of [**20**] can be considered as a continuation of this theme.

In [**33**], Todorčević has given positive answers to Fremlin's question and the basis problem in the rather broad class of spaces that can be represented as relatively compact subsets of the class $\mathcal{B}_1(X)$ of all Baire class 1 functions on some Polish space X endowed with the topology of pointwise convergence. Compact subsets of such $\mathcal{B}_1(X)$ are sometimes called Rosenthal compacta since one interpretation of the famous Rosenthal ℓ_1-theorem says that the double dual ball of a separable Banach space containing no copy of ℓ_1 equipped with the weak* topology is one example of such a compactum. The class also contains the split interval, the one point compactification of a discrete set of size at most $2^{\aleph_0}$, and is closed under the operations of taking countable products and closed subspaces. Todorčević proves that if K is a Rosenthal compactum with no uncountable discrete subspaces, then K is perfect and premetric of order at most 2; moreover, if K is not metrizable, then it contains a full copy of the split interval.

Unlike the broader class of regular spaces, questions about Rosenthal compacta can typically be settled in the framework of ZFC. The analysis in [**33**], however, has a strong set-theoretic theme and a number of the arguments presented there may give some insight into how to approach some of the problems in this article. The reader may also find [**36**, **37**] informative in a similar manner.

While a complete understanding of Woodin's axiom $(*)$ is probably not necessary for an analysis of these problems, it is worth making a few more remarks about it. Axiom $(*)$ is the assertion that $L(\mathcal{P}(\omega_1))$, is a generic extension of $L(\mathbb{R})$ by the $\mathcal{P}_{\max}$ forcing. Many questions in this article can be cast in the language of $H(\aleph_1^+)$, the collection of sets of hereditary cardinality at most $\aleph_1$, since it is often possible to assume without loss of generality that the weight and possibly the cardinality of the space is at most $\aleph_1$. Furthermore, the assertions in the questions typically are Π_2 in their complexity; they have a pair $\forall X \exists Y$ of unbounded quantifiers followed by bounded quantification.[5] The $\mathcal{P}_{\max}$ forcing has the effect of making $H(\aleph_1^+)$ satisfy all Π_2 sentences which are *Ω-consistent.* Being Ω-consistent is a natural strengthening of "has a well founded model"; a precise definition can be found in [**40**]. For our purposes it is sufficient to say that if a statement can always be forced over any ground model with sufficient large cardinals, then it is Ω-consistent. All the forcing axioms and nearly all consistency results in set-theoretic topology fit this description. Large cardinals are needed for the analysis of $\mathcal{P}_{\max}$ but these can often be avoided in applications if one wishes to obtain consistency results instead.

Another interesting property of the $\mathcal{P}_{\max}$ extension is its minimality. If G is $\mathcal{P}_{\max}$-generic over $L(\mathbb{R})$ and X is any new element of $H(\aleph_1^+)$, then $L(\mathbb{R})[X] = L(\mathbb{R})[G]$. Since a C-sequence on ω_1 can never be in $L(\mathbb{R})$ under appropriate large cardinal hypotheses, the $\mathcal{P}_{\max}$ extension is always of the form $L(\mathbb{R})[C]$ where

[5] X usually takes the form of a space, Y usually takes the form of either a substructure (e.g., an uncountable discrete subspace) or a connecting map (e.g., an embedding from an canonical space into X). The bounded quantification is usually made over the base and/or set of points in X.

C is some C-sequence on ω_1.[6] In this context, $L(\mathbb{R})$ is a model in which the Axiom of Choice fails and which satisfies strong Ramsey-theoretic statements (e.g., ω_1 is measurable and in particular Ramsey's theorem holds for ω_1). This gives an *a posteriori* explanation as to the role of Todorčević's method of minimal walks [**29**] in building counterexamples such as Moore's L-space [**21**]. This method involves an analysis of a number of two place functions which are recursively defined on C-sequences. It is likely that this method will be useful in constructing counterexamples related to the above questions. The reader is referred to [**35**] for further information.

It also seems plausible that a hypothesis such as the following may be useful in constructing an informative counterexample to some of these questions:

$\mho$: There are continuous $f_\alpha\colon \alpha \to \omega$ $(\alpha < \omega_1)$ such that if $E \subseteq \omega_1$ is closed and unbounded, then there is a δ in E such that $f_\delta(E \cap \delta) = \omega$.

The object postulated by this axiom can naturally be used to strengthen the combinatorial objects constructed using the method of minimal walks. Since quantification is only over the closed unbounded filter, this axiom cannot be negated by ccc forcing and hence is consistent with $\mathsf{MA}_{\aleph_1}$. It is even immune to Axiom A forcings and many forcings built using models as side conditions (see, e.g., [**30**]). It therefore cannot be used to construct, e.g., an S-space. It has been used to construct a counterexample to Shelah's basis conjecture for the uncountable linear orders [**19**]. Whether $\mho$ can be used to construct a counterexample can, in general, be used as a litmus test for whether the more involved methods presented in [**20**] are needed to eliminate counterexamples (as opposed to the more user-friendly techniques of [**30**]). This axiom was also useful in constructing an L-space which later was the prototype for the ZFC construction in [**21**].

References

[1] A. V. Arhangel'skiĭ, *The structure and classification of topological spaces and cardinal invariants*, Uspekhi Mat. Nauk **33** (1978), no. 6(204), 29–84, Translation: Russian Math. Surveys 33 (1978) 33–95.

[2] U. Avraham and S. Shelah, *Martin's axiom does not imply that every two $\aleph_1$-dense sets of reals are isomorphic*, Israel J. Math. **38** (1981), no. 1–2, 161–176.

[3] D. K. Burke and S. W. Davis, *Compactifications of symmetrizable spaces*, Proc. Amer. Math. Soc. **81** (1981), no. 4, 647–651.

[4] D. Daniel, J. Nikiel, L. B. Treybig, H. M. Tuncali, and E. D. Tymchatyn, *On perfectly normal compacta*, Questions Answers Gen. Topology **23** (2005), no. 1, 1–14.

[5] V. V. Fedorchuk, *Perfectly normal compact space without intermediate dimensions*, Bull. Acad. Polon. Sci. Sér. Sci. Math. Astronom. Phys. **23** (1975), no. 9, 975–979.

[6] V. V. Filippov, *Perfectly normal bicompacta*, Dokl. Akad. Nauk SSSR **189** (1969), 736–739, Translation: Soviet Math. Dokl. 10 (1969) 1503–1507.

[7] D. H. Fremlin, *Notes on Martin's Maximum*, Unpublished notes.

[8] D. H. Fremlin, *Consequences of Martin's axiom*, Cambridge University Press, Cambridge, 1984.

[9] D. H. Fremlin, *Problem list*, 2005.

[6] A *C-sequence* (on ω_1) is a sequence C_α $(\alpha < \omega_1)$ such that C_α is a cofinal subset of α and if $\gamma < \alpha$, then $C_\alpha \cap \gamma$ is finite.

[10] G. Gruenhage, *On the existence of metrizable or Sorgenfrey subspaces*, General topology and its relations to modern analysis and algebra, VI (Prague, 1986), Heldermann Verlag, Berlin, 1988, pp. 223–230.

[11] G. Gruenhage, *Cosmicity of cometrizable spaces*, Trans. Amer. Math. Soc. **313** (1989), no. 1, 301–315.

[12] G. Gruenhage, *Perfectly normal compacta, cosmic spaces, and some partition problems*, Open problems in topology, North-Holland, Amsterdam, 1990, pp. 85–95.

[13] G. Gruenhage and P. J. Nyikos, *Normality in X^2 for compact X*, Trans. Amer. Math. Soc. **340** (1993), no. 2, 563–586.

[14] I. Juhász, L. Soukup, and Z. Szentmiklóssy, *Forcing countable networks for spaces satisfying $R(X^\omega) = \omega$*, Comment. Math. Univ. Carolin. **37** (1996), no. 1, 159–170.

[15] A. Karasev, H. Tuncali, and V. Valov, *Topology in North Bay: some problems in continuum theory, dimension theory and selections*, Open problems in topology II (E. Pearl, ed.), Elsevier Science.

[16] P. Larson, *Forcing over models of determinacy*, Handbook of set theory (M. Foreman, A. Kanamori, and M. Magidor, eds.), Springer, Berlin,, To appear.

[17] I. K. Lifanov and V. V. Filippov, *Two examples for the theory of the dimension of bicompacta*, Dokl. Akad. Nauk SSSR **192** (1970), 26–29, Translation: Soviet Math. Dokl. 11 (1970) 573–576.

[18] B. MacGibbon, *A criterion for the metrizability of a compact convex set in terms of the set of extreme points*, J. Funct. Anal. **11** (1972), 385–392.

[19] J. T. Moore, *Persistent counterexamples to basis conjectures*, Notes of August 2004.

[20] J. T. Moore, *A five element basis for the uncountable linear orders*, Ann. of Math. (2) **163** (2006), no. 2, 669–688.

[21] J. T. Moore, *A solution to the L space problem*, J. Amer. Math. Soc. **19** (2006), no. 3, 717–736.

[22] P. J. Nyikos, *Problem K6*, Topology Proc. **7** (1982), 385.

[23] T. C. Przymusiński, *Products of normal spaces*, Handbook of set-theoretic topology, North-Holland, Amsterdam, 1984, pp. 781–826.

[24] D. Shakhmatov, *Compact spaces and their generalizations*, Recent progress in general topology (Prague, 1991), North-Holland, Amsterdam, 1992, pp. 571–640.

[25] Z. Szentmiklóssy, *S-spaces and L-spaces under Martin's axiom*, Topology, Vol. II (Budapest, 1978), Colloq. Math. Soc. János Bolyai, vol. 23, North-Holland, Amsterdam, 1980, pp. 1139–1145.

[26] M. G. Tkachenko, *Chains and cardinals*, Dokl. Akad. Nauk SSSR **239** (1978), no. 3, 546–549, Translation: Soviet Math. Dokl. 19 (1978) 382–385.

[27] V. V. Tkachuk, *A glance at compact spaces which map "nicely" onto the metrizable ones*, Topology Proc. **19** (1994), 321–334.

[28] S. Todorčević, *Remarks on cellularity in products*, Compositio Math. **57** (1986), no. 3, 357–372.

[29] S. Todorčević, *Partitioning pairs of countable ordinals*, Acta Math. **159** (1987), no. 3–4, 261–294.

[30] S. Todorčević, *Partition problems in topology*, American Mathematical Society, Providence, RI, 1989.

[31] S. Todorčević, *A classification of transitive relations on ω_1*, Proc. London Math. Soc. **73** (1996), no. 3, 501–533.

[32] S. Todorčević, *Countable chain condition in partition calculus*, Discrete Math. **188** (1998), 205–223.

[33] S. Todorčević, *Compact subsets of the first Baire class*, J. Amer. Math. Soc. **12** (1999), no. 4, 1179–1212.

[34] S. Todorčević, *Chain-condition methods in topology*, Topology Appl. **101** (2000), no. 1, 45–82.

[35] S. Todorčević, *Coherent sequences*, Handbook of set theory (M. Foreman, A. Kanamori, and M. Magidor, eds.), Springer, Berlin,, To appear.

[36] S. Todorčević and C. Uzcátegui, *Analytic topologies over countable sets*, Topology Appl. **111** (2001), no. 3, 299–326.

[37] S. Todorčević and C. Uzcátegui, *Analytic k-spaces*, Topology Appl. **146/147** (2005), 511–526.

[38] S. Watson, *The construction of topological spaces: planks and resolutions*, Recent progress in general topology (Prague, 1991), North-Holland, Amsterdam, 1992, pp. 673–757.

[39] S. Watson and W. Weiss, *A topology on the union of the double arrow space and the integers*, Topology Appl. **28** (1988), no. 2, 177–179.

[40] W. H. Woodin, *The axiom of determinacy, forcing axioms, and the nonstationary ideal*, Walter de Gruyter & Co., Berlin, 1999.

Open Problems in Topology II – Edited by E. Pearl

Selection problems for hyperspaces

Valentin Gutev and Tsugunori Nogura

For a T_1-space X, let $\mathcal{F}(X)$ be the set of all non-empty closed subsets of X. Usually, we endow $\mathcal{F}(X)$ with the *Vietoris topology* τ_V, and call it the *Vietoris hyperspace* of X. Let us recall that τ_V is generated by all collections of the form

$$\langle \mathcal{V} \rangle = \Big\{ S \in \mathcal{F}(X) : S \subset \bigcup \mathcal{V} \text{ and } S \cap V \neq \varnothing, \text{ whenever } V \in \mathcal{V} \Big\},$$

where $\mathcal{V}$ runs over the finite families of open subsets of X.

Suppose that $\Phi\colon Y \to \mathcal{F}(X)$ is a map, usually called a *set-valued mapping*, or a *multi-map*, and, sometimes, a *multifunction*. Once $\mathcal{F}(X)$ has been topologized, Φ becomes a function between topological spaces, and it makes sense to talk about its continuity and other topological properties. In 1951 Ernest Michael [**29**] raised the following general question:

Question 1 (Michael, [**29**, Question 6.1]). *When is it possible to find a continuous $f\colon Y \to X$, such that $f(y) \in \Phi(y)$ for all $y \in Y$ (i.e., a selection for Φ)?*

As he wrote in his paper [**29**], a sufficient condition for this to be possible is that both the following hold: Φ is continuous, and there exists a "selection" from $\mathcal{F}(X)$ to X. The problem is thus reduced to two simpler ones; the second of which is concerned only with the space X, and has nothing to do with the space Y or the function Φ. This second problem is now known as the Selection Problem for Hyperspaces, and here we will mainly discuss different aspects of it.

In the sequel, all spaces are assumed to be infinite and at least Hausdorff, while any subset $\mathcal{D} \subset \mathcal{F}(X)$ will carry the relative Vietoris topology τ_V as a subspace of the hyperspace $\mathcal{F}(X)$. A map $f\colon \mathcal{D} \to X$ is a *selection* for $\mathcal{D}$ if $f(S) \in S$ for every $S \in \mathcal{D}$. A selection $f\colon \mathcal{D} \to X$ is *continuous* if it is continuous with respect to the relative Vietoris topology τ_V on $\mathcal{D}$. Sometimes, for reasons of convenience, we also say that f is a *Vietoris continuous* selection for $\mathcal{D}$ to stress the attention that f is continuous with respect to the topology τ_V.

For a subset $\mathcal{D} \subset \mathcal{F}(X)$, we will use $Sel[\mathcal{D}]$ to denote the set of all Vietoris continuous selections for $\mathcal{D}$. Also, we will use the following special subsets of $\mathcal{F}(X)$, where $n \geq 1$.

$$\begin{aligned} \mathcal{F}_n(X) &= \{S \in \mathcal{F}(X) : |S| \leq n\}, \\ [X]^n &= \{S \subset X : |S| = n\}, \\ \mathcal{C}(X) &= \{S \in \mathcal{F}(X) : S \text{ is compact}\}. \end{aligned}$$

Note that we may identify X with the set $[X]^1 = \mathcal{F}_1(X)$, and, in fact, X is homeomorphic to the space $(\mathcal{F}_1(X), \tau_V)$. The latter means that the Vietoris topology is *admissible*, see [**29**].

1. Weak selections and orderability

A space X is *orderable* (or *linearly orderable*) if the topology of X coincides with the open interval topology on X generated by a linear ordering on X. A space X is *suborderable* (or *generalized ordered*) if it can be embedded into an orderable space. Following [**31**], we say that a space X is *weakly orderable* if there exists a coarser orderable topology on X. In all these cases, we call the corresponding linear order on X *compatible* for the topology of X, or, merely, a compatible order for X. Finally, let us recall that a selection $f\colon \mathcal{F}_2(X) \to X$ is usually called a *weak selection* for X.

381? **Question 2** (van Mill and Wattel, [**31**]). *Is a space X weakly orderable provided it has a continuous weak selection $f : \mathcal{F}_2(X) \to X$?*

The motivation for this question goes back to Eilenberg [**9**] and Michael [**29**]. Namely, every weak selection $f\colon \mathcal{F}_2(X) \to X$ generates a natural order-like relation $\preceq_f$ on X [**29**, Definition 7.1] by letting for $x, y \in X$ that $x \preceq_f y$ iff $f(\{x,y\}) = x$. For simplicity, we write that $x \prec_f y$ if $x \preceq_f y$ and $x \neq y$. Unfortunately, in general, this relation is not transitive, and, in fact, the clopen subsets of X are an indication for the *non-transitivity* of $\preceq_f$, [**21**, Proposition 2.2]. Here are some particular classes of spaces for which Question 2 was resolved in positive.

Connected spaces. In 1941 Eilenberg [**9**, Theorem I] proved that a connected space X is weakly orderable iff the subset $P(X) = \{(x,y) \in X\times X : x \neq y\}$ of its square $X \times X$, obtained by deleting the diagonal, is not connected. In fact, he showed [**9**, (3.1)] that if X is connected, then $P(X)$ has exactly two connected components A and B such that $\Lambda(A) = B$, where $\Lambda\colon P(X) \to P(X)$ is defined by $\Lambda(x,y) = (y,x)$, $(x,y) \in X \times X$. Hence, a connected weakly orderable space has precisely two compatible orders, which are inverse of each other. Relying on these results, Michael [**29**, Lemma 7.1] proved that a connected space X is weakly orderable iff $Sel[\mathcal{F}_2(X)] \neq \varnothing$ by showing that, in this case, the relation $\preceq_f$ is a compatible linear order on X for every $f \in Sel[\mathcal{F}_2(X)]$. Thus, in particular, a connected weakly orderable space X has precisely two continuous weak selections (i.e., $|Sel[\mathcal{F}_2(X)]| = 2$).

Locally connected spaces. Eilenberg [**9**, (8.1) and (8.2)] proved that a connected and locally connected space X is orderable iff $P(X)$ is not connected. According to results of Michael [**29**, Lemmas 7.2 and 7.4], this implies that a connected and locally connected space X is orderable iff $Sel[\mathcal{F}_2(X)] \neq \varnothing$. In fact, Michael [**29**, Lemma 7.4] proved that a locally connected space X, with $Sel[\mathcal{F}_2(X)] \neq \varnothing$, must be weakly orderable. Finally, Nogura and Shakhmatov [**34**, Theorem 4 and Remark 16] proved that a locally connected space X is orderable iff $Sel[\mathcal{F}_2(X)] \neq \varnothing$. They also demonstrated [**34**, Example 8] that there exists a separable completely metrizable connected space X, which is not orderable, but $Sel[\mathcal{F}(X)] \neq \varnothing$ (hence, $Sel[\mathcal{F}_2(X)] \neq \varnothing$ as well).

Compact spaces. Michael's characterization of weakly orderable spaces [**29**] implies that a connected compact space X is orderable iff $Sel[\mathcal{F}_2(X)] \neq \varnothing$. The question if connectedness is essential for compact spaces was raised by van Douwen [**8**], and resolved by van Mill and Wattel [**31**, Theorem 1.1] by proving that a compact space X is orderable iff $Sel[\mathcal{F}_2(X)] \neq \varnothing$.

Pseudocompact spaces. E.K. van Douwen [**8**, Theorem 1.2] proved that a countably compact space X, with $Sel[\mathcal{F}_2(X)] \neq \varnothing$, must be sequentially compact, hence, for these spaces, countable compactness is equivalent to sequential compactness. On the other hand, van Mill and Wattel [**32**, Theorems 2.2 and 3.1] proved that a Tychonoff space X is suborderable iff there exists an $f \in Sel[\mathcal{F}_2(X)]$ such that, for every $p \in \beta X \setminus X$, f can be extended to a continuous weak selection for $X \cup \{p\}$ (i.e., $f = g \restriction \mathcal{F}_2(X)$ for some $g \in Sel[\mathcal{F}_2(X \cup \{p\})]$), where βX is the Čech–Stone compactification of X. Weak selections $f \in Sel[\mathcal{F}_2(X)]$ with this property were called *locally uniform* [**32**], and were characterized in an inner manner as well (see [**32**, Theorem 2.2]). Finally, let us mention that, by [**36**, Proposition 3.2], if βX is orderable, then X must be a pseudocompact normal space (hence, countably compact as well), while, by Glicksberg's theorem [**16**, Theorem 3.1], $\beta(X \times X) = \beta X \times \beta X$ for every Tychonoff space X for which $X \times X$ is pseudocompact. On this base, it was obtained in [**4**, Theorem 1.16] (using direct arguments) and in [**33**, Theorem 2.1] (relying on Glicksberg's theorem mentioned above) that βX is orderable iff $X \times X$ is pseudocompact and $Sel[\mathcal{F}_2(X)] \neq \varnothing$. Finally, solving [**4**, Problems 5.1 and 5.2], García-Ferreira and Sanchis [**15**, Theorem 2.3] proved that $X \times X$ is pseudocompact if X is itself pseudocompact and $Sel[\mathcal{F}_2(X)] \neq \varnothing$, hence βX is orderable (and X suborderable) iff X is a pseudocompact space, with $Sel[\mathcal{F}_2(X)] \neq \varnothing$. Thus, by [**4**, Proposition 1.19], every pseudocompact space, with $Sel[\mathcal{F}_2(X)] \neq \varnothing$, must be sequentially compact.

Countable spaces. It was proved in [**14**, Theorem 3.1] that a countable space X is weakly orderable iff it has a continuous weak selection, while, by [**14**, Example 3.6], there exists a non-regular countable space X, with $Sel[\mathcal{F}(X)] \neq \varnothing$.

Going back to Question 2, it should be mentioned that local compactness and local connectedness are equivalent for connected weakly orderable spaces [**4**, Proposition 1.18] (see, also [**9**]). From this point of view, we have the following particular question for locally compact spaces.

Question 3 ([**24**]). *Let X be a locally compact space, with $Sel[\mathcal{F}_2(X)] \neq \varnothing$. Then,* 382?
is X weakly orderable?

Related to Question 3, it was proved in [**4**, Theorem 4.6] (under the Diamond Principle) that there exists a monotonically normal, locally compact and locally countable space X which is not suborderable, but $Sel[\mathcal{F}(X)] \neq \varnothing$ (hence, $Sel[\mathcal{F}_2(X)] \neq \varnothing$ as well).

E. Michael [**29**, Lemma 7.5.1] proved that $Sel[\mathcal{C}(X)] \neq \varnothing$ for every weakly orderable space X. Hence, if the answer to Question 2 is in positive, then $Sel[\mathcal{F}_2(X)] \neq \varnothing$ should imply that $Sel[\mathcal{C}(X)] \neq \varnothing$ (and, in particular, that

$Sel[\mathcal{F}_n(X)] \neq \varnothing$ for every $n > 2$). This was the reason for the following question.

383? **Question 4** ([**24**]). *Does there exist a space X such that $Sel[\mathcal{F}_2(X)] \neq \varnothing$, but $Sel[\mathcal{F}_n(X)] = \varnothing$ for some $n > 2$?*

Clearly, a positive solution to Question 4 will imply a negative solution to Question 2, however Question 4 is open even when $n = 3$. Related to this, it was obtained in [**13**, Corollary 4.1] that $Sel[\mathcal{F}_3(X)] \neq \varnothing$ provided $Sel[\mathcal{F}_2(X)] \neq \varnothing$ and $Sel\left[[X]^3\right] \neq \varnothing$. One of the main obstacles in this particular case is that a selection $f \in Sel[\mathcal{F}_2(X)]$ may generate a triple of distinct points $x, y, z \in X$ such that

$$\cdots \prec_f z \prec_f x \prec_f y \prec_f z \prec_f \cdots$$

where $\prec_f$ is the order-like relation corresponding to f. This may explain the hypothesis "$Sel\left[[X]^3\right] \neq \varnothing$" in the mentioned result of [**13**]. Further, since every weak selection for X is continuous on the singletons of X (see, for instance, [**24**, Proposition 1.4]), we always have that $Sel[\mathcal{F}_2(X)] \neq \varnothing$ provided $Sel\left[[X]^2\right] \neq \varnothing$. Motivated by this, the following question was asked in [**13**].

384? **Question 5** ([**13**]). *Let X be a space such $Sel[\mathcal{F}_n(X)] \neq \varnothing$ and $Sel\left[[X]^{n+1}\right] \neq \varnothing$ for some $n \geq 3$. Then, is it true that $Sel[\mathcal{F}_{n+1}(X)] \neq \varnothing$?*

On the other hand, using methods of Graph Theory and Flows in Networks, it was obtained in [**27**, Theorem 4.1] that $Sel[\mathcal{F}_4(X)] \neq \varnothing$ provided $Sel[\mathcal{F}_3(X)] \neq \varnothing$. Hence, the following question is also of a particular interest.

385? **Question 6.** *Let X be a space such that $Sel[\mathcal{F}_{2n+1}(X)] \neq \varnothing$ for some $n \geq 2$. Then, is it true that $Sel[\mathcal{F}_{2n+2}(X)] \neq \varnothing$?*

A somewhat different approach to Question 2 was applied in [**21**]. Namely, if X is weakly orderable and $\mathcal{T}$ is the topology on X, then there should exist an ordered topology $\mathcal{T}_*$ on X, with $\mathcal{T}_* \subset \mathcal{T}$, and, clearly, X has a continuous weak selection with respect to $\mathcal{T}_*$ (when X is endowed with $\mathcal{T}_*$, and $\mathcal{F}_2(X)$ with the Vietoris topology generated by $\mathcal{T}_*$). On the other hand, by [**21**, Corollary 3.2], if $\mathcal{T}$ is the topology on a space X and $f \in Sel[\mathcal{F}_2(X)]$, then f is continuous with respect to any topology $\tilde{\mathcal{T}}$, which is finer than $\mathcal{T}$. Finally, going back to Eilenberg [**9**, (7.1)], given an ordered set $(X, \preceq)$, among all topologies on X which lead to a weakly orderable space with respect $\preceq$, there is a coarsest one. Motivated by this, the following question was asked in [**21**].

386? **Question 7** ([**21**]). *Let X be a space, $f \in Sel[\mathcal{F}_2(X)]$, and let $\mathcal{T}$ be the topology on X. Does there exist a topology $\mathcal{T}_* \subset \mathcal{T}$ on X such that f is continuous with respect to $\mathcal{T}_*$, and $\mathcal{T}_*$ is the coarsest topology on X with this property?*

For a selection $f \in Sel[\mathcal{F}_2(X)]$, the relation $\preceq_f$ generates a natural "f-open interval" topology $\mathcal{T}_f$ on X [**21**], which was called a *selection* topology. The selection topology $\mathcal{T}_f$ is always a regular topology on X [**26**, Corollary 2.3], and always $\mathcal{T}_f \subset \mathcal{T}$ [**21**, Theorem 3.5], where $\mathcal{T}$ is the original topology on X. Thus, the selection topology $\mathcal{T}_f$ is coarser than any topology on X with respect to which f is continuous, hence it was a possible candidate for a solution to Question 7. However, there exists a space X and $f \in Sel[\mathcal{F}_2(X)]$ such that f is not continuous with respect to $\mathcal{T}_f$, [**21**, Example 3.6] and [**26**, Theorem 4.1].

Clearly, if X is a compact space, with $Sel[\mathcal{F}_2(X)] \neq \varnothing$, and $\mathcal{T}$ is the topology on X, then $\mathcal{T}_f = \mathcal{T}$ for every $f \in Sel[\mathcal{F}_2(X)]$. According to Eilenberg [**9**] and Michael [**29**], the same is true if X is connected and locally connected, and $Sel[\mathcal{F}_2(X)] \neq \varnothing$. However, this fails for locally connected spaces [**26**, Example 3.6]; for connected spaces [**26**, Example 3.7]; also for sequentially compact spaces [**26**, Example 3.12], which was behind the motivation for the following question.

Question 8 ([**26**]). *Does there exist a space X, with $Sel[\mathcal{F}_2(X)] \neq \varnothing$, which* 387?
is neither compact nor connected and locally connected, but $\mathcal{T}_f = \mathcal{T}$ for every $f \in Sel[\mathcal{F}_2(X)]$, where $\mathcal{T}$ is the topology on X?

2. Topological well-ordering and selections

There is a principal difference between continuous selections for $\mathcal{F}_2(X)$ and continuous selections for $\mathcal{F}(X)$. According to van Mill and Wattel [**31**, Theorem 1.1], for a compact space X, we have that $Sel[\mathcal{F}(X)] \neq \varnothing$ iff $Sel[\mathcal{F}_2(X)] \neq \varnothing$. However, this is not true even for spaces which are both connected and locally connected. Indeed, it was shown in [**10**, Proposition 5.1] that $Sel[\mathcal{F}(\mathbb{R})] = \varnothing$, while $Sel[\mathcal{C}(\mathbb{R})] \neq \varnothing$ because $\mathbb{R}$ is an ordered space [**29**, Lemma 7.5.1]. In fact, Michael [**29**, Lemma 7.3] proved that if X is a connected space and $f \in Sel[\mathcal{F}(X)]$, then X is not only weakly orderable with respect to the linear order $\preceq_f$ on X, but $f(F)$ is the $\preceq_f$-first element of F for every closed set $F \in \mathcal{F}(X)$. Such selections f were called *monotone* in [**21**], and were characterized topologically with the property that if $G, F \in \mathcal{F}(X)$ and $f(F) \in G \subset F$, then $f(G) = f(F)$. On the other hand, an ordered space Z with the property that every non-empty closed subset of Z has a first element is called *topologically well-ordered* [**10**]. Clearly, every compact ordered space is topologically well-ordered, while by the mentioned Michael's result, if X is connected and $f \in Sel[\mathcal{F}(X)] \neq \varnothing$, then X is topologically well-ordered with respect to the open interval topology generated by $\preceq_f$. In fact, X must have a stronger property because, by [**29**, Lemma 7.3], every non-empty closed (with respect to the original topology on X) subset of X must have a $\preceq_f$-first element. Indeed, it was demonstrated in [**21**, Example 5.3] that there is a connected separable metrizable space X and a linear order $\preceq$ on X such that $Sel[\mathcal{F}(X)] = \varnothing$, but the ordered topology on X generated by $\preceq$ is coarser than the original topology of X, and X equipped with this topology is a topologically well-ordered space. This led to the following concept in [**21**]: A space X is *Sorgenfrey well-orderable* if there exists a linear order $\preceq$ on X such that

(i) the ordered topology $\mathcal{T}_{\preceq}$ is coarser that the original topology $\mathcal{T}$ of X,
(ii) X is topologically well-ordered with respect to $\mathcal{T}_{\preceq}$,

(iii) if $x \in X$ and $x \prec \sup_{\preceq} X$, then for every $\mathcal{T}$-neighbourhood V of x there exists a point $y \in X$ such that $x \prec y$ and $\{z \in X : x \preceq z \prec y\} \subset V$.

388? **Question 9.** *Let X be a space, with $Sel[\mathcal{F}(X)] \neq \varnothing$. Then, is X Sorgenfrey well-orderable?*

The following may be useful considering this question:

- By [**21**, Theorem 5.1] (see, also, [**29**, Lemma 7.5.1]), $Sel[\mathcal{F}(X)] \neq \varnothing$ provided X is Sorgenfrey well-orderable;
- By [**31**, Theorem 1.1], a compact space X is orderable (hence, Sorgenfrey well-orderable) iff $Sel[\mathcal{F}(X)] \neq \varnothing$;
- By [**21**, Theorem 5.1] and [**29**, Lemmas 7.3 and 7.5.1], a connected space X is Sorgenfrey well-orderable iff $Sel[\mathcal{F}(X)] \neq \varnothing$;
- By [**29**, Lemmas 7.4 and 7.5.1] and [**34**, Theorem 4], a locally connected space X is topologically well-orderable (hence, Sorgenfrey well-orderable) iff $Sel[\mathcal{F}(X)] \neq \varnothing$.

Motivated by the pseudocompact case (recall, every pseudocompact space X, with $Sel[\mathcal{F}_2(X)] \neq \varnothing$, is weakly orderable and sequentially compact), the following particular question is also of a certain interest.

389? **Question 10.** *Let X be a sequentially compact space, with $Sel[\mathcal{F}(X)] \neq \varnothing$. Then, is X Sorgenfrey well-orderable?*

The topologically well-ordered spaces are a natural generalization of ordinal spaces, hence the selection problem for them is naturally related to that one of ordinal spaces. Recently, topologically well-ordered spaces and ordinal spaces were successfully characterized also in terms of continuous selections for the *Fell hyperspace topology*; for these and related results, also some further open problems, we refer the interested reader to [**1–3**, **11**, **12**, **18**, **23**].

3. Selections and disconnectedness-like properties

One of the main problems that remains to be dealt with is about selections for the Vietoris hyperspace on spaces that may have many clopen sets. The cardinality of $Sel[\mathcal{F}(X)]$ may provide some information for this, but mainly when it is finite. Suppose that X is a space, with $Sel[\mathcal{F}(X)] \neq \varnothing$. The following hold:

- If X is connected, then $|Sel[\mathcal{F}(X)]| \leq 2$, [**29**, Lemmas 7.2 and 7.3];
- The set $Sel[\mathcal{F}(X)]$ is finite if and only if X has finitely many connected components, [**35**, Theorem 1];
- If X is infinite and connected, then $|Sel[\mathcal{F}(X)]| = 2$ if and only if X is compact, [**34**, Theorem 1].

For some other relations between the cardinality of $Sel[\mathcal{F}(X)]$ and X, we refer the interested reader to [**14**, **34**, **35**].

On the other hand, all known selection constructions are based on some extreme principle, and our knowledge about particular members of $Sel[\mathcal{F}(X)]$ is at present mainly reduced to this. Here are some results about such "extreme"-like selections for the Vietoris hyperspace on spaces X, with $Sel[\mathcal{F}(X)] \neq \varnothing$:

- If X is zero-dimensional, then the set $\{f(X) : f \in Sel[\mathcal{F}(X)]\}$ is dense in X, [**22**, Theorem 1.3];
- If the set $\{f(X) : f \in Sel[\mathcal{F}(X)]\}$ is dense in X, then X is totally disconnected, [**22**, Theorem 1.5];
- If for every point $x \in X$ there exists an $f_x \in Sel[\mathcal{F}(X)]$, with $f_x^{-1}(x) = \{S \in \mathcal{F}(X) : x \in S\}$, then X is zero-dimensional, see the proof of [**22**, Theorem 1.4];
- If X is first countable and zero-dimensional, then for every $x \in X$ there exists an $f_x \in Sel[\mathcal{F}(X)]$, with $f_x^{-1}(x) = \{S \in \mathcal{F}(X) : x \in S\}$, [**22**, Theorem 1.4];
- If X is separable, then it is first countable and zero-dimensional iff for every $x \in X$ there exists an $f_x \in Sel[\mathcal{F}(X)]$, with $f_x^{-1}(x) = \{S \in \mathcal{F}(X) : x \in S\}$, [**14**, Theorem 3.5] and [**22**, Theorem 1.4];
- If X is countable, then it is metrizable and scattered iff for every $x \in X$ there exists an $f_x \in Sel[\mathcal{F}(X)]$, with $f_x^{-1}(x) = \{S \in \mathcal{F}(X) : x \in S\}$, [**14**, Theorem 3.5].

Recall that a space X is *totally disconnected* if each singleton of X is the intersection of clopen subsets of X; also that X is *zero-dimensional* if it has a base of clopen sets (i.e., if $\operatorname{ind}(X) \le 0$, where $\operatorname{ind}(X)$ is the *small inductive dimension* of X). Here are a few actual questions motivated by these results.

Question 11 ([**22**]). *Does there exist a space X which is not zero-dimensional,* 390?
but the set $\{f(X) : f \in Sel[\mathcal{F}(X)]\}$ is dense in X?

Question 12 ([**24**]). *Let X be a totally disconnected space, with $Sel[\mathcal{F}(X)] \neq \varnothing$.* 391?
Is the set $\{f(X) : f \in Sel[\mathcal{F}(X)]\}$ dense in X?

Related to these problems, recall that a collection $\mathcal{P}$ of open subsets of a space X is a *π-base* for X if every non-empty open subset $U \subset X$ contains some non-empty $V \in \mathcal{P}$. Recently, it was proved in [**19**, Theorem 2.1] that if X is a space, with $Sel[\mathcal{F}(X)] \neq \varnothing$, then the set $\{f(X) : f \in Sel[\mathcal{F}(X)]\}$ is dense in X iff X has a clopen π-base. In particular, if X is a space, with a clopen π-base and $Sel[\mathcal{F}(X)] \neq \varnothing$, then it must be totally disconnected, [**19**, Corollary 2.3]. This gives rise to the following alternative reading of Question 12.

Question 13. *Let X be a totally disconnected space, with $Sel[\mathcal{F}(X)] \neq \varnothing$. Then, is it true that X has a clopen π-base?*

The above questions are especially interesting concerning the Selection Problem for the Vietoris hyperspace on homogeneous spaces X, with $Sel[\mathcal{F}(X)] \neq \varnothing$. Namely, for such spaces X, we have that $X = \{f(X) : f \in Sel[\mathcal{F}(X)]\}$. This was used in [**19**, Corollary 3.2] to show that if X is a homogeneous metrizable space, with $Sel[\mathcal{F}(X)] \neq \varnothing$, then X must be zero-dimensional; and in [**19**, Corollary 3.3] to show that if, in addition, X is also separable, then $Sel[\mathcal{F}(X)] \neq \varnothing$ iff X is a discrete space, or a discrete sum of copies of the Cantor set, or the irrational line. Hence, a separable homogeneous metrizable space X, with $Sel[\mathcal{F}(X)] \neq \varnothing$, must be strongly zero-dimensional and orderable. Here, a space X is *strongly zero-dimensional* if its covering dimension $\dim(X)$ is at most zero.

392? **Question 14.** *Does there exist a homogeneous metrizable space X such that $Sel[\mathcal{F}(X)] \neq \varnothing$, but X is not orderable?*

A zero-dimensional orderable metrizable space X must be also strongly zero-dimensional, while, by a result van Mill, Pelant and Pol [**30**], a metrizable space X must be completely metrizable provided $Sel[\mathcal{F}(X)] \neq \varnothing$. Finally, according to [**6**, **10**], $Sel[\mathcal{F}(X)] \neq \varnothing$ for every completely metrizable strongly zero-dimensional space. Thus, we have also the following natural question.

393? **Question 15** ([**24**]). *Does there exist a zero-dimensional completely metrizable space X such that* $\dim(X) \neq 0$ *and $Sel[\mathcal{F}(X)] \neq \varnothing$?*

Concerning metrizable spaces, the interested reader is also referred to [**5**, **7**, **17**, **20**] for some further questions about the Selection Problem on metric-generated hyperspace topologies.

Going back to Question 14, another aspect is to look for properties that follow from orderability, and to see which of them may depend on continuous selections. Clearly, $\operatorname{ind}(X) \leq 1$ for every orderable space X, which leads us to the following natural question (see, [**24**]).

394? **Question 16.** *Let X be a space, with $Sel[\mathcal{F}(X)] \neq \varnothing$. Then, is it true that* $\operatorname{ind}(X) \leq 1$*?*

Recently, it was established in [**28**] that, for every $n < \omega$, there exists a *Polish space* X_n (i.e., separable and completely metrizable) such that $\dim(X_n) = n$ and $Sel[\mathcal{C}(X_n)] \neq \varnothing$; also that there exists a strongly infinite-dimensional Polish space X, with $Sel[\mathcal{C}(X)] \neq \varnothing$. In fact, all these results are based on natural examples of totally disconnected spaces of arbitrary covering dimension. On the other hand, it was obtained in [**28**], that $Sel[\mathcal{F}_n(X)] \neq \varnothing$ for every second countable totally disconnected space X, but we don't know if this result can be extended to selections for $\mathcal{C}(X)$.

395? **Question 17.** *Let X be a second countable totally disconnected space. Then, is it true that $Sel[\mathcal{C}(X)] \neq \varnothing$?*

Another natural class of homogeneous spaces is given by topological groups. Below we summarize some of the results in this direction:

- A pseudocompact topological group G, with a continuous weak selection, is either finite, or topologically homeomorphic to the Cantor set, [**4**, Corollary 1.27] (for an alternative proof in the compact case, see [**25**, Corollary 5.6]);
- If G is a locally pseudocompact topological group, with a continuous weak selection, then it is locally compact, metrizable, and orderable, [**4**, Theorem 1.25];

- A topological group G, with $Sel[\mathcal{F}(G)] \neq \varnothing$, is zero-dimensional and metrizable iff there exists a selection $f \in Sel[\mathcal{F}(G)]$ such that $f^{-1}(p) = \{S \in \mathcal{F}(G) : p \in S\}$ for some (any) $p \in G$, [**25**, Corollary 5.5];
- A locally compact topological group G is totally disconnected and orderable iff $Sel[\mathcal{F}(G)] \neq \varnothing$, [**19**, Corollary 3.4].

Motivated by these results, we have also the following question.

Question 18. *Characterize those topological groups G which admit continuous selections for $\mathcal{F}(G)$. In particular, is it true that a topological group G is zero-dimensional provided $Sel[\mathcal{F}(G)] \neq \emptyset$?* 396?

References

[1] G. Artico and U. Marconi, *Selections and topologically well-ordered spaces*, Topology Appl. **115** (2001), no. 3, 299–303.

[2] G. Artico, U. Marconi, R. Moresco, and J. Pelant, *Selectors and scattered spaces*, Topology Appl. **111** (2001), no. 1–2, 35–48.

[3] G. Artico, U. Marconi, and J. Pelant, *Zero-selectors and GO spaces*, Topology Appl. **152** (2005), no. 3, 243–257.

[4] G. Artico, U. Marconi, J. Pelant, L. Rotter, and M. G. Tkachenko, *Selections and suborderability*, Fund. Math. **175** (2002), no. 1, 1–33.

[5] D. Bertacchi and C. Costantini, *Existence of selections and disconnectedness properties for the hyperspace of an ultrametric space*, Topology Appl. **88** (1998), no. 3, 179–197.

[6] M. M. Choban, *Many-valued mappings and Borel sets. I*, 1970, Translation: Trans. Moscow Math. Soc. 22 (1970) 258–280, pp. 229–250.

[7] C. Costantini and V. Gutev, *Recognizing special metrics by topological properties of the "metric"-proximal hyperspace*, Tsukuba J. Math. **26** (2002), no. 1, 145–169.

[8] E. K. van Douwen, *Mappings from hyperspaces and convergent sequences*, Topology Appl. **34** (1990), no. 1, 35–45.

[9] S. Eilenberg, *Ordered topological spaces*, Amer. J. Math. **63** (1941), 39–45.

[10] R. Engelking, R. W. Heath, and E. Michael, *Topological well-ordering and continuous selections*, Invent. Math. **6** (1968), 150–158.

[11] S. Fujii, *Characterizations of ordinal spaces via continuous selections*, Topology Appl. **122** (2002), no. 1–2, 143–150.

[12] S. Fujii and T. Nogura, *Characterizations of compact ordinal spaces via continuous selections*, Topology Appl. **91** (1999), no. 1, 65–69.

[13] S. García-Ferreira, V. Gutev, and T. Nogura, *Extensions of 2-point selections*, New Zealand J. Math. (2006), Accepted for publication.

[14] S. García-Ferreira, V. Gutev, T. Nogura, M. Sanchis, ana A. Tomita, *Extreme selections for hyperspaces of topological spaces*, Topology Appl. **122** (2002), no. 1–2, 157–181.

[15] S. García-Ferreira and M. Sanchis, *Weak selections and pseudocompactness*, Proc. Amer. Math. Soc. **132** (2004), no. 6, 1823–1825.

[16] I. Glicksberg, *Stone–Čech compactifications of products*, Trans. Amer. Math. Soc. **90** (1959), 369–382.

[17] V. Gutev, *Selections and hyperspace topologies via special metrics*, Topology Appl. **70** (1996), no. 2–3, 147–153.

[18] V. Gutev, *Fell continuous selections and topologically well-orderable spaces II*, Proc. Ninth Prague Topological Symposium (2001), Topology Atlas, Toronto, 2002, pp. 157–163.

[19] V. Gutev, *Approaching points by continuous selections*, J. Math. Soc. Japan (2006), Accepted for publication.

[20] V. Gutev and T. Nogura, *Selections for Vietoris-like hyperspace topologies*, Proc. London Math. Soc. **80** (2000), no. 1, 235–256.

[21] V. Gutev and T. Nogura, *Selections and order-like relations*, Appl. Gen. Topol. **2** (2001), no. 2, 205–218.

[22] V. Gutev and T. Nogura, *Vietoris continuous selections and disconnectedness-like properties*, Proc. Amer. Math. Soc. **129** (2001), no. 9, 2809–2815.

[23] V. Gutev and T. Nogura, *Fell continuous selections and topologically well-orderable spaces*, Mathematika **51** (2004), no. 1–2, 163–169.

[24] V. Gutev and T. Nogura, *Some problems on selections for hyperspace topologies*, Appl. Gen. Topol. **5** (2004), no. 1, 71–78.

[25] V. Gutev and T. Nogura, *Selection pointwise-maximal spaces*, Topology Appl. **146/147** (2005), 397–408.

[26] V. Gutev and T. Nogura, *A topology generated by selections*, Topology Appl. **153** (2005), no. 5–6, 900–911.

[27] V. Gutev and T. Nogura, *Weak selections and flows in networks*, 2005, Submitted for publication.

[28] V. Gutev and T. Nogura, *Selections and totally disconnected spaces*, 2006, In preparation.

[29] E. Michael, *Topologies on spaces of subsets*, Trans. Amer. Math. Soc. **71** (1951), 152–182.

[30] J. van Mill, J. Pelant, and R. Pol, *Selections that characterize topological completeness*, Fund. Math. **149** (1996), no. 2, 127–141.

[31] J. van Mill and E. Wattel, *Selections and orderability*, Proc. Amer. Math. Soc. **83** (1981), no. 3, 601–605.

[32] J. van Mill and E. Wattel, *Orderability from selections: another solution to the orderability problem*, Fund. Math. **121** (1984), no. 3, 219–229.

[33] K. Miyazaki, *Continuous selections on almost compact spaces*, Sci. Math. Jpn. **53** (2001), no. 3, 489–494.

[34] T. Nogura and D. Shakhmatov, *Characterizations of intervals via continuous selections*, Rend. Circ. Mat. Palermo (2) **46** (1997), no. 2, 317–328.

[35] T. Nogura and D. Shakhmatov, *Spaces which have finitely many continuous selections*, Boll. Un. Mat. Ital. A (7) **11** (1997), no. 3, 723–729.

[36] M. Venkataraman, M. Rajagopalan, and T. Soundararajan, *Orderable topological spaces*, General Topology and Appl. **2** (1972), 1–10.

Open Problems in Topology II – Edited by E. Pearl

Efimov's problem

Klaas Pieter Hart

Introduction

In their memoir [**1**, p. 54] Alexandroff and Urysohn asked "*existe-il un espace compact (bicompact) ne contenant aucun point* ($\varkappa$)*?*" and went on to remark "*La resolution affirmative de ce problème nous donnerait une exemple des espaces compacts (bicompacts) d'une nature toute differente de celle des espaces connus jusqu'à présent*". The 'compact' of that memoir is our countably compact, 'bicompact' is '*compact Hausdorff*' and a *$\varkappa$-point* is one that is the limit of a nontrivial convergent sequence. A look through the examples in [**1**] will reveal a few familiar classics: the ordinal space ω_1 and the corresponding Long line, the double circumference, the Tychonoff plank (in disguise), the lexicographically ordered square, and the Double Arrow space. The geometric nature of the constructions made the introduction of nontrivial convergent sequences practically unavoidable and it turns out that the remark was quite correct as we will see below.

The question was answered by Tychonoff [**16**] and Čech [**3**] using the very same space, though their presentations were quite different. Tychonoff took for every $x \in (0,1)$ its binary expansion $0.a_1(x)a_2(x)\ldots a_n(x)\ldots$ (favouring the one that ends in zeros), thus creating a countable set $\{a_n : n \in \mathbb{N}\}$ of points in the *Tychonoff cube* $[0,1]^{(0,1)}$, whose closure is the required space. Čech developed what we now call the *Čech–Stone compactification*, denoted βX, of completely regular spaces and showed that $\beta\mathbb{N}$, where $\mathbb{N}$ is the discrete space of the natural numbers, has no convergent sequences.

A natural question is whether one has to go to such great lengths to construct a compact Hausdorff space without convergent sequences. This then is Efimov's problem, raised in [**8**].

Efimov's problem. *Does every infinite compact Hausdorff space contain either a nontrivial convergent sequence or else a copy of* $\beta\mathbb{N}$*?* 397?

It should be noted that Efimov raised his problem not in the context sketched above but as part of a program to determine when Čech–Stone compactifications of discrete spaces were embeddable in certain compact Hausdorff spaces.

For the rest of this note all convergent sequences will be assumed to be nontrivial, so that adjective will not be used. Basic information about $\beta\mathbb{N}$ may be found in [**7**].

1. Attacking the problem

Efimov's problem may of course be cast in the form of an implication: If a compact Hausdorff space does not contain any convergent sequences must it then contain a copy of $\beta\mathbb{N}$?

Let us consider an infinite compact Hausdorff space X that does not contain any convergent sequences. As any infinite Hausdorff space, it does contain an infinite relatively discrete subspace; we take a countably infinite subset of that subspace and identify it with $\mathbb{N}$. The sequence $\langle n\rangle_n$ does not converge, so we can take two distinct accumulation points, x_0 and x_1, of $\mathbb{N}$. Take neighbourhoods U_0 and U_1 of x_0 and x_1 respectively with disjoint closures and put $A_0 = U_0 \cap \mathbb{N}$ and $A_1 = U_1 \cap \mathbb{N}$. Thus we find that in an infinite compact Hausdorff space without convergent sequences every countably infinite discrete subset has two infinite subsets with disjoint closures. To get a copy of $\beta\mathbb{N}$ one should construct an infinite discrete subset with the property that any two disjoint subsets have disjoint closures. To appreciate how difficult this may be we continue our construction.

We have our two disjoint subsets of $\mathbb{N}$ with disjoint closures. We iterate the procedure above and determine, recursively, a family $\{A_s : s \in {}^{<\omega}2\}$, where ${}^{<\omega}2$ is the binary tree of finite sequences of zeros and ones, that satisfies

- if $s \subseteq t$ then $A_t \subseteq A_s$, and
- $\operatorname{cl} A_{s*0} \cap \operatorname{cl} A_{s*1} = \emptyset$.

Using this family one defines, for every point x in the Cantor set ${}^{\omega}2$, a closed set $F_x = \bigcap_n \operatorname{cl} A_{x\restriction n}$. By construction the (nonempty) closed sets F_x are disjoint and we see that the cardinality of X must be at least $\mathfrak{c}$. In fact, with some care one can arrange matters so that

- $F = \bigcup_x F_x$ is closed, and
- mapping the points of F_x to x gives a continuous map from F onto ${}^{\omega}2$.

Using the Tietze–Urysohn theorem one can employ this map to obtain a continuous surjection from X onto the unit interval $\mathbb{I}$ or even the Hilbert cube ${}^{\omega}\mathbb{I}$. As we will see below what is needed is a continuous map onto the Tychonoff cube ${}^{\mathfrak{c}}\mathbb{I}$; however, naïvely, the Hilbert cube is best possible. Though the construction above can be continued for (at least) ω_1 many steps to show that $\mathbb{N}$ has at least $2^{\aleph_1}$ many accumulation points, the examples below show that it will not necessarily yield a map onto the next cube ${}^{\omega_1}\mathbb{I}$.

To get a copy of $\beta\mathbb{N}$ inside X more is needed, as Efimov himself established in [**8**] when he characterized the spaces that do contain such a copy. On the one hand the space $\beta\mathbb{N}$ admits a continuous map onto the *Cantor cube* ${}^{\mathfrak{c}}2$ and thence onto the Tychonoff cube ${}^{\mathfrak{c}}\mathbb{I}$; the Tietze–Urysohn theorem may then be applied to produce a continuous map from the ambient space onto this cube. On the other hand assume that X maps onto ${}^{\mathfrak{c}}\mathbb{I}$. Since the cube contains a copy of $\beta\mathbb{N}$, a standard argument produces a closed subset F of X and an irreducible map from F onto $\beta\mathbb{N}$. Because $\beta\mathbb{N}$ is *extremally disconnected* this map is a homeomorphism.

It follows that the following statements about a compact Hausdorff X are equivalent:

(1) X contains a copy of $\beta\mathbb{N}$,
(2) X maps onto ${}^{\mathfrak{c}}\mathbb{I}$,
(3) some closed subset of X maps onto ${}^{\mathfrak{c}}2$, and
(4) there is a *dyadic system* $\{\langle F_{\alpha,0}, F_{\alpha,1}\rangle : \alpha < \mathfrak{c}\}$ of closed sets in X.

The dyadic system satisfies, by definition,

- $F_{\alpha,0} \cap F_{\alpha,1} = \emptyset$ for all α, and
- $\bigcap_{\alpha \in \operatorname{dom} p} F_{\alpha,p(\alpha)} \neq \emptyset$, whenever p is a finite partial function from $\mathfrak{c}$ to 2.

To deduce (4) from (3) simply set $F_{\alpha,i} = f^{\leftarrow}(\pi_\alpha^{\leftarrow}(i))$, where f is the map onto ${}^{\mathfrak{c}}2$ and π_α is the projection onto the αth coordinate. Conversely, this same formula implicitly defines a continuous map from $\bigcap_{\alpha<\mathfrak{c}}(F_{\alpha,0} \cup F_{\alpha,1})$ onto ${}^{\mathfrak{c}}2$.

In [**15**] Shapirovskiĭ added another condition to this list: there is a closed set F such that $\pi\chi(x, F) \geq \mathfrak{c}$ for all points of F. Here $\pi\chi(x, F)$ is the *π-character* of x (in F): the minimum cardinality of a family $\mathcal{U}$ of nonempty open sets such that every neighbourhood of x contains an elements of $\mathcal{U}$ (the elements of $\mathcal{U}$ need not be neighbourhoods of x).

2. Counterexamples

There are several consistent counterexamples to Efimov's problem. This of course precludes an unqualified positive answer and leaves us with two possibilities: a real, ZFC, counterexample or the consistency of a positive answer.

Here is a list of the better-known counterexamples.

(1) For every natural number n there is a compact Hausdorff space X_n with the property that every infinite closed subset has covering dimension n. As both the convergent sequence and $\beta\mathbb{N}$ are zero-dimensional neither can be a subspace of X_n. This example was constructed by Fedorčuk in [**10**] using the Continuum Hypothesis (CH).

(2) Another example, this time with the aid of $\diamondsuit$, was constructed by Fedorčuk in [**11**]. The space is a compact S-space of size $2^{\mathfrak{c}}$ without convergent sequences. As $\beta\mathbb{N}$ is not hereditarily separable it cannot be embedded into this space.

(3) Yet another counterexample was constructed by Fedorčuk in [**12**] using a principle he called the Partition Hypothesis. In present day terms this is the conjunction of $\mathfrak{s} = \aleph_1$ and $2^{\aleph_0} = 2^{\aleph_1}$. Here $\mathfrak{s}$ is the *splitting number*, the minimum cardinality of a *splitting family*, that is, a family $\mathcal{S}$ of subsets of $\mathbb{N}$ such that for every infinite subset A of $\mathbb{N}$ there is $S \in \mathcal{S}$ such that both $A \cap S$ and $A \setminus S$ are infinite. Fedorčuk's principle holds in the Cohen model. The title of [**12**] makes it completely clear why this is a counterexample to Efimov's question: no convergent sequences and the space is simply too small to contain $\beta\mathbb{N}$.

(4) In [**6**] Dow weakened Fedorčuk's hypothesis substantially, at the cost of a more elaborate construction, to the conjunction of $\operatorname{cf}([\mathfrak{s}]^{\aleph_0}, \subset) = \mathfrak{s}$ and $2^{\mathfrak{s}} < 2^{\mathfrak{c}}$. The former equality says that there are $\mathfrak{s}$ many countable subsets of $\mathfrak{s}$ so that each countable subset of $\mathfrak{s}$ is contained in one of them.

All four counterexamples arise as limits of suitable inverse systems, where at each stage some or all convergent sequences are dealt with. In the first two constructions the CH allows one to do some (clever) bookkeeping so that every potential convergent sequence in the limit is dealt with at some intermediate stage. In the third and fourth example the inverse system is ω_1 long but at every stage all convergent sequences in the space constructed that far are dealt with; the cofinality assumption on $\mathfrak{s}$ enables one to do this by splitting just $\mathfrak{s}$ objects. The final space then has at most $2^{\mathfrak{s}}$ points, so that the power assumption prevents $\beta\mathbb{N}$ from being a subspace.

A simpler version of the third space was given by van Douwen and Fleissner in [**4**] using $2^{\aleph_0} = 2^{\aleph_1}$ plus a version of $\mathfrak{s} = \aleph_1$ for the Cantor set ${}^{\omega}2$: there should be a family $\{U_\alpha : \alpha < \omega_1\}$ of open sets such that for every convergent sequence s there is an α for which $s \cap U_\alpha$ and $s \setminus \operatorname{cl} U_\alpha$ are infinite. This example is indeed simpler than the others: after copying the sets U_α to each cube ${}^{\beta}2$ (where $\omega \le \beta < \omega_1$) one can simply write down a formula for the example, as a subspace of the Cantor cube ${}^{\omega_1}2$. Indeed, choose, for each $\beta \ge \omega$, a homeomorphism $h_\beta\colon {}^{\omega}2 \to {}^{\beta}2$ and, for all α, put $U_{\beta,\alpha} = h_\beta[U_\alpha]$. Furthermore let $b\colon \omega_1 \times \omega_1 \to \omega_1$ be a bijection with the property that $b(\alpha, \beta) = \gamma$ implies $\beta \le \gamma$. Now the space X is the subspace of ${}^{\omega_1}2$ consisting of those points x that satisfy

$$x\bigl(b(\alpha,\beta)\bigr) = 0 \text{ implies } x \restriction \beta \in \operatorname{cl} U_{\beta,\alpha} \text{ and}$$
$$x\bigl(b(\alpha,\beta)\bigr) = 1 \text{ implies } x \restriction \beta \in \operatorname{cl} V_{\beta,\alpha}$$

where $V_{\beta,\alpha} = {}^{\beta}2 \setminus \operatorname{cl} U_{\beta,\alpha}$.

3. Is there still a problem?

The condition $\operatorname{cf}\bigl([\mathfrak{s}]^{\aleph_0}, \subset\bigr) = \mathfrak{s}$ in used in Dow's example is quite weak; indeed, if it were false an inner model with a measurable cardinal would have to exist. This is explained in [**13**]: if there is any cardinal κ of uncountable cofinality for which $\operatorname{cf}\bigl([\kappa]^{\aleph_0}, \subset\bigr) > \kappa$ then the *Covering Lemma* fails badly: not just for L but for any inner model that satisfies the Generalized Continuum Hypothesis.

One might therefore be tempted to conclude that Efimov's problem is all but solved, especially in the absence of large cardinals. However, that completely disregards the necessary inequality $2^{\mathfrak{s}} < 2^{\mathfrak{c}}$; without it the guarantee that the example does not contain $\beta\mathbb{N}$ is gone. We are thus lead to consider situations where $2^{\mathfrak{s}} = 2^{\mathfrak{c}}$, or even $\mathfrak{s} = \mathfrak{c}$. The best-known of these is of course when Martin's Axiom (MA) holds and, indeed, it is not (yet) known what the effect of MA + ¬CH (or even PFA) is on Efimov's problem.

398? **Question 1.** *Does* MA + ¬CH *(or* PFA*) imply that a compact Hausdorff space without convergent sequences contains a copy of $\beta\mathbb{N}$?*

As noted above, this is equivalent to asking whether such a space admits a continuous map onto ${}^{\mathfrak{c}}\mathbb{I}$. Also, as shown below, under MA every countable and discrete subset of a compact Hausdorff space without convergent sequences has $2^{\mathfrak{c}}$ accumulation points. As such a space does admit a continuous map onto ${}^{\omega}\mathbb{I}$ and considering the adage "MA makes cardinals below $\mathfrak{c}$ countable", it may be worthwhile to investigate the following weaker question first.

Question 2. *Does* MA + ¬CH *(or* PFA*) imply that a compact Hausdorff space* 399?
without convergent sequences maps onto ${}^{\omega_1}\mathbb{I}$ *or even onto each cube* ${}^{\kappa}\mathbb{I}$ *for* $\kappa < \mathfrak{c}$*?*

In case Question 1 has a positive answer it becomes of interest how much of MA is actually needed. The equalities $\mathfrak{s} = \mathfrak{c}$ and $\mathfrak{t} = \mathfrak{c}$ seem to suggest themselves as possible candidates; the former by the rôle of $2^{\mathfrak{s}} < 2^{\mathfrak{c}}$ in the examples of Fedorčuk and Dow and the latter by the fact, shown below, that a countable discrete set in a compact Hausdorff space without convergent sequences has at least $2^{\mathfrak{t}}$ accumulation points. The meaning of $\mathfrak{t}$ will be explained below.

In the CH-type constructions mentioned in Section 2, where one deals with one convergent sequence at a time, the preferred thing to do is to blow up the limit to a larger set, every point of which will be an accumulation point of the sequence. The most frugal thing to do would be to split the limit into just two points. This is called a simple extension and an inverse limit construction where at each step one performs a simple extension never leads to a space that can be mapped onto ${}^{\omega_1}\mathbb{I}$ *unless* the initial space in the system already does so. In Boolean algebraic form this result is due to Koppelberg [**14**]; in [**6**] one finds a topological proof and the following question. The definition of 'simple extension' in [**6**] does not mention the two-point limitation but it is used in the proof.

Question 3. *Is it consistent that every such simple inverse limit contains a con-* 400?
vergent sequence? Does PFA *imply this?*

4. Larger cardinals

In [**8**] Efimov considered the general problem of characterizing when a space contains a copy of $\beta\kappa$, where κ is any infinite cardinal with the discrete topology. The characterization of embeddability of $\beta\mathbb{N}$ given by Efimov as discussed in Section 1 remains valid in the general situation, as does Shapirovskiĭ's characterization of being able to map a space onto a Tychonoff cube of a given weight.

Many generalizations of Efimov's problem suggest themselves but they will never be as succinct as the original question. Given an uncountable cardinal κ an audacious question would be: *Does every compact Hausdorff space contain either the Alexandroff (one-point) compactification* $\alpha\kappa$ *of* κ *or a copy of* $\beta\kappa$*?*

This would also be a foolish question: an arbitrary compact Hausdorff space need not contain a relatively discrete subset of cardinality κ. A better question would therefore be

Question 4. *Does every large enough compact Hausdorff space contain either the* 401?
Alexandroff compactification $\alpha\kappa$ *of* κ *or else a copy of* $\beta\kappa$*?*

This of course begs the question what 'large enough' should mean. Therefore one should first investigate for what class of spaces Question 4 actually makes sense. The answer will have to involve some kind of *structural* description of 'large enough' because for every cardinal κ the ordinal space $\kappa + 1$ contains neither $\alpha\omega_1$ nor $\beta\omega_1$, so that size alone does not seem to matter.

Efimov's question is a structural question in disguise: if a compact Hausdorff space does not contain a convergent sequence then can one find a dyadic system of cardinality $\mathfrak{c}$? One may disregard the structural part and concentrate on the cardinality part only to get a weaker version of Efimov's question:

402? **Question 5.** *If in a compact Hausdorff space every countable and discrete set has more than one accumulation point must there be such a set with* $2^{\mathfrak{c}}$ *accumulation points?*

The naïve construction from Section 1 shows that one always gets at least $2^{\aleph_1}$ accumulation points and, naïve though it may be, it does show that the answer to this question is positive under MA: one gets a family $\{F_s : s \in {}^{<\mathfrak{c}}2\}$ of closed sets indexed by the complete complete binary tree of height $\mathfrak{c}$ and such that always $F_{s*0} \cap F_{s*1} = \emptyset$; in this way one obtains a pairwise disjoint family $\{F_x : x \in {}^{\mathfrak{c}}2\}$ of nonempty closed sets, all contained in the derived set of the initial countable and discrete set. To be precise, the construction can be continued all the way up to the cardinal $\mathfrak{t}$, which is, by definition, the minimum cardinal κ for which there is a sequence $\langle A_\alpha : \alpha < \kappa\rangle$ of infinite subsets of $\mathbb{N}$ such that $A_\alpha \subset^* A_\beta$ whenever $\beta < \alpha$ but for which no infinite set A exists with $A \subset^* A_\alpha$ for all α. This shows, as promised above, that the discrete set has at least $2^{\mathfrak{t}}$ accumulation points.

One cannot simply copy Question 5 to larger cardinals: if α is a compact ordinal space and D a (discrete) subset of uncountable size κ then D has κ many accumulation points. However, we can build the partial result on Question 5 into its translation. The strongest version that we get is the following.

403? **Question 6.** *Let κ be an infinite cardinal. For what compact Hausdorff spaces X is the following implication valid? If $|D^d| > \kappa$ for all discrete subsets of size κ then $|D^d| \geq 2^{2^\kappa}$ for some discrete subset of size κ.*

Here D^d denotes the set of accumulation points of D. Various (weaker) versions of this question can be obtained by inserting $|D^d| \geq \lambda$ in the antecedent and $|D^d| \geq \mu$ into its consequent for cardinals λ and μ that satisfy $\kappa < \lambda < \mu \leq 2^{2^\kappa}$.

A related notion was defined by Arhangel'skiĭ in [**2**]: denote by $g(X)$ the supremum of cardinalities of closures of discrete subsets of the space X; Arhangel'skiĭ asked whether $g(X) = |X|$ for compact Hausdorff spaces. The following question combines this with a sup = max problem.

404? **Question 7.** *When does a compact Hausdorff space X have a discrete subset D such that $|\mathrm{cl}\, D| = |X|$?*

Efimov [**9**] showed that this is true for dyadic spaces (provided every inaccessible cardinal is strongly inaccessible); in [**5**] Dow shows that relatively small (cardinality at most $\aleph_\omega$) compact Hausdorff spaces of countable tightness do have such discrete subsets and also gives some consistent examples of compact Hausdorff spaces X of cardinality $\aleph_2$ with $g(X) \leq \aleph_1$.

Acknowledgement. Thanks to Alan Dow for his help in improving some of the questions raised in this note.

References

[1] P. S. Alexandroff and P. S. Urysohn, *Mémoire sur les espaces topologiques compacts*, Verh. Kon. Akad. Wetensch. Amsterdam **14** (1929), no. 1, 96 pp.

[2] A. V. Arhangel'skiĭ, *An extremal disconnected bicompactum of weight $\mathfrak{c}$ is inhomogeneous*, Dokl. Akad. Nauk SSSR **175** (1967), 751–754, Translation: Soviet Math. Dokl. 8 (1967) 897–900.

[3] E. Čech, *On bicompact spaces*, Ann. of Math. (2) **38** (1937), no. 4, 823–844.

[4] E. K. van Douwen and W. G. Fleissner, *Definable forcing axiom: an alternative to Martin's axiom*, Topology Appl. **35** (1990), no. 2–3, 277–289.

[5] A. Dow, *Closures of discrete sets in compact spaces*, Studia Sci. Math. Hungar. **42** (2005), no. 2, 227–234.

[6] A. Dow, *Efimov spaces and the splitting number*, Topology Proc. **29** (2005), no. 1, 105–113.

[7] A. Dow and K. P. Hart, *The Čech–Stone compactifications of $\mathbb{N}$ and of $\mathbb{R}$*, Encyclopedia of general topology, Elsevier Science, 2004, pp. 213–217.

[8] B. Efimov, *The imbedding of the Stone–Čech compactifications of discrete spaces into bicompacta*, Dokl. Akad. Nauk SSSR **189** (1969), 244–246, Translation: Soviet Math. Dokl. 10 (1969) 1391–1394.

[9] B. Efimov, *Subspaces of dyadic bicompacta*, Dokl. Akad. Nauk SSSR **185** (1969), 987–990, Translation: Soviet Math. Dokl. 10 (1969) 453–456.

[10] V. V. Fedorchuk, *A bicompactum whose infinite closed subsets are all n-dimensional*, Mat. Sb. **96(138)** (1975), 41–62, Translation: Sb. Math. 25 (1976) 37–57.

[11] V. V. Fedorchuk, *Completely closed mappings, and the compatibility of certain general topology theorems with the axioms of set theory*, Mat. Sb. **99(141)** (1976), no. 1, 3–33, Translation: Sb. Math. 28 (1977) 1–26.

[12] V. V. Fedorchuk, *A compact space having the cardinality of the continuum with no convergent sequences*, Math. Proc. Cambridge Philos. Soc. **81** (1977), no. 2, 177–181.

[13] W. Just, A. R. D. Mathias, K. Prikry, and P. Simon, *On the existence of large p-ideals*, J. Symbolic Logic **55** (1990), no. 2, 457–465.

[14] S. Koppelberg, *Minimally generated Boolean algebras*, Order **5** (1989), no. 4, 393–406.

[15] B. È. Šapirovskiĭ, *Mappings onto Tikhonov cubes*, Uspekhi Mat. Nauk **35** (1980), 122–130, Translation: Russian Math. Surveys 35 (1980) 145–156.

[16] A. Tychonoff, *Über einen Funktionenraum*, Math. Ann. **111** (1935), 762–766.

Open Problems in Topology II – Edited by E. Pearl

Completely separable MAD families

Michael Hrušák and Petr Simon

An infinite family $\mathcal{A} \subseteq [\omega]^\omega$ is *almost disjoint* if any two of its distinct elements have finite intersection. A family $\mathcal{A}$ is *maximal almost disjoint (MAD)* if it is almost disjoint and for every $X \in [\omega]^\omega$ there is an $A \in \mathcal{A}$ such that $A \cap X$ is infinite.

There are almost disjoint (hence also MAD) families of cardinality $\mathfrak{c}$ and many MAD families with special combinatorial and/or topological properties can be constructed using set-theoretic assumptions like CH, MA or $\mathfrak{b} = \mathfrak{c}$. However, special MAD families are notoriously difficult to construct in ZFC alone. The reason being the lack of a device ensuring that a recursive construction of a MAD family would not prematurely terminate, an object that would serve a similar purpose as independent linked families do for the construction of special ultrafilters (see [**15**]). The notion of a completely separable MAD family is a candidate for such a device and, moreover, is an interesting notion in its own right.

MAD families provide a powerful tool for topological constructions. Not only for the study of $\beta\mathbb{N} \setminus \mathbb{N}$, the Čech–Stone remainder of the discrete countable space ([**1**, **2**]) but also, typically via the corresponding Ψ-space [**8**], the study of convergence properties of topological spaces [**6**, **21**].

1. The main problem

The notion of completely separable MAD family was introduced by S.H. Hechler [**9**] in 1971:

Definition. An infinite MAD family $\mathcal{A}$ on ω is *completely separable* if for every subset $M \subseteq \omega$ either there is an $A \in \mathcal{A}$ with $A \subseteq M$ or there is a finite subfamily $\mathcal{B} \subseteq \mathcal{A}$ with $M \subseteq \bigcup \mathcal{B}$.

A year later, P. Erdős and S. Shelah asked the central problem of this article:

Problem 1. *Does there exist a completely separable MAD family in* ZFC*?* 405?

A MAD family $\mathcal{A}$ on ω is of *true cardinality* $\mathfrak{c}$ if for every subset $M \subseteq \omega$ the set $\{A \in \mathcal{A} : |M \cap A| = \omega\}$ is either finite or of size $\mathfrak{c}$. It is easily seen that every completely separable MAD family is of true cardinality $\mathfrak{c}$. On the other hand, the existence of a MAD family of true cardinality $\mathfrak{c}$ readily implies the existence of a completely separable MAD family.

An almost disjoint family $\mathcal{A}$ is *nowhere MAD* if for every $X \subseteq \omega$ either $X \subseteq^* \bigcup \mathcal{B}$ for some finite $\mathcal{B} \subseteq \mathcal{A}$ or there is a $B \in [X]^\omega$ almost disjoint from all elements of $\mathcal{A}$. Given a cardinal number κ, a MAD family $\mathcal{A}$ is *κ-partitionable* if $\mathcal{A}$ can be

The first listed author gratefully acknowledges support received from a CONACyT grant 46337-F and a PAPIIT grant IN106705 partitioned into κ subfamilies $\{\mathcal{A}_\xi : \xi < \kappa\}$ such that $\mathcal{A} \setminus \mathcal{A}_\xi$ is nowhere MAD for every $\xi < \kappa$.

Note that a MAD family $\mathcal{A}$ is $\mathfrak{c}$-partitionable if and only if it is of true cardinality $\mathfrak{c}$. This motivated A. Dow to ask how close can we get to constructing a $\mathfrak{c}$-partitionable MAD family:

406? **Problem 2.** *For which κ is there a κ-partitionable MAD family?*

In [**21**] it is shown that 2-partitionable families exist in ZFC. This was later extended by E. van Douwen to show that ω-partitionable MAD familes exist. In fact, one can, in ZFC, construct a $\mathfrak{t}$-partitionable MAD family (personal communication by A. Dow), but it is not known whether there is a $\mathfrak{b}$-partitionable MAD family.

Problem 1 has a close connection to the disjoint refinement property. Given two familes $\mathcal{M}, \mathcal{A} \subseteq [\omega]^\omega$, we say that the family $\mathcal{A}$ *refines* the family $\mathcal{M}$, if for each $M \in \mathcal{M}$ there is an $A \in \mathcal{A}$ such that $A \subseteq M$. We say that a family $\mathcal{M} \subseteq [\omega]^\omega$ has an *almost disjoint refinement* if there is an almost disjoint family $\mathcal{A} \subseteq [\omega]^\omega$ which refines $\mathcal{M}$. Clearly, not every $\mathcal{M} \subseteq [\omega]^\omega$ has an almost disjoint refinement, $\mathcal{M} = [\omega]^\omega$ being an example. It is known that every uniform ultrafilter on ω has an almost disjoint refinement [**3**]; to present other examples, we need a definition. Given $\mathcal{R} \subseteq [\omega]^\omega$, let

$$\mathcal{I}^+(\mathcal{R}) = \{M \subseteq \omega : |\{R \in \mathcal{R} : |M \cap R| = \omega\}| \geq \omega\}.$$

If $\mathcal{R}$ is an infinite partition of ω into infinite sets, then $\mathcal{I}^+(\mathcal{R})$ has an almost disjoint refinement [**3**].

Hence, a MAD family $\mathcal{A}$ is completely separable if an only if $\mathcal{A}$ is an almost disjoint refinement of $\mathcal{I}^+(\mathcal{A})$. However, the following problem is open, too:

407? **Problem 3.** *Let $\mathcal{A}$ be an infinite MAD family on ω. Does there exist an almost disjoint refinement of $\mathcal{I}^+(\mathcal{A})$?*

Note that this is the strongest possible formulation of an almost disjoint refinement property: Given an arbitrary family $\mathcal{M} \subseteq [\omega]^\omega$, which has an almost disjoint refinement $\mathcal{B}$, it is easy to find a MAD family $\mathcal{A}$ with $\mathcal{M} \subseteq \mathcal{I}^+(\mathcal{A})$; simply replace each $B \in \mathcal{B}$ by infinitely many disjoint subsets of it and extend to a maximal almost disjoint family.

If Problem 3 has an positive answer, then so does Problem 1, in a very strong sense:

Theorem ([**2**]). *The following statements are equivalent:*

1. *For every MAD family $\mathcal{A}$ on ω, $\mathcal{I}^+(\mathcal{A})$ has an almost disjoint refinement.*
2. *There is a set $\mathcal{S} \subseteq [\omega]^\omega$ satisfying* (a) *each infinite $M \subseteq \omega$ contains a member of $\mathcal{S}$ and* (b) *every infinite MAD family $\mathcal{A} \subseteq \mathcal{S}$ is completely separable.*

In particular, if $\mathcal{I}^+(\mathcal{A})$ has an almost disjoint refinement for every MAD family $\mathcal{A}$, then there is a completely separable MAD family. It is another open problem, whether this implication can be reversed.

There are consistent affirmative answers to Problem 3. Each of the following assumptions on cardinal invariants

$$\mathfrak{a} = 2^{\omega}, \quad \mathfrak{b} = \mathfrak{d}, \quad \mathfrak{d} \leq \mathfrak{a}, \quad \mathfrak{s} = \omega_1$$

implies that Problem 3 has a positive answer [**2, 23**].

Also, if one relaxes maximality, then one can find an infinite AD family $\mathcal{A}$ which is completely separable in the sense that every set belonging to $\mathcal{I}^+(\mathcal{A})$ contains an element of $\mathcal{A}$ [**2**].

In conversation, A. Dow remarked that most applications of completely separable MAD familes require that they can be recursively constructed, rather then that they just exist, which seems to be similar to Problem 3:

Problem 4. *Suppose that $\mathcal{I}^+(\mathcal{A})$ has an almost disjoint refinement for every MAD family $\mathcal{A}$. Can every nowhere MAD family be extended to a completely separable MAD family?* 408?

2. Topological connection

An equivalent formulation of Problem 3 has been asked also in a purely topological language. The space $\beta\mathbb{N} \setminus \mathbb{N}$ is not extremally disconnected, so it must contain a point which belongs to the intersection of closures of two disjoint open sets. In 1967, R.S. Pierce asked in [**19**], whether it is possible to show, without using the Continuum Hypothesis, that there are *3-points* in $\beta\mathbb{N} \setminus \mathbb{N}$, i.e points which lie simultaneously in the closure of three pairwise disjoint open sets. N. Hindman [**11**] then showed that there are not only 3-points, but even $\mathfrak{c}$-points in $\beta\mathbb{N} \setminus \mathbb{N}$ in 1969, and finally B. Balcar and P. Vojtáš [**3**] proved that every point in $\beta\mathbb{N} \setminus \mathbb{N}$ is a $\mathfrak{c}$-point in 1980.

Meanwhile, S.H. Hechler started to consider nowhere dense sets instead of points and showed that under MA, if S is any nowhere dense subset of $\beta\mathbb{N} \setminus \mathbb{N}$, then there exists a family of $\mathfrak{c}$ pairwise disjoint open sets each of which contains S in its closure [**10**]. Call such a set a *$\mathfrak{c}$-set*. It is easy to show that the following is nothing but a topological reformulation of Problem 3:

Problem 5. *Is every nowhere dense subset of $\beta\mathbb{N} \setminus \mathbb{N}$ a $\mathfrak{c}$-set?*

The topological language allows to formulate a seemingly easier problem, also open till now:

Problem 6. *Is every nowhere dense subset of $\beta\mathbb{N} \setminus \mathbb{N}$ a 2-set?* 409?

Still, this is not the end of the story. A.I. Veksler introduced the following order on the family of all nowhere dense subsets of a topological space: If C, D are nowhere dense in X, let $C < D$ mean that $C \subseteq D$ and C is nowhere dense in D. He studied this order in a series of paper; we quote here [**25**] as a sample. A theorem from [**22**] says that the next problem is again Problem 3 in disguise:

Problem 7. *Is it true that a family of all nowhere dense subsets of $\beta\mathbb{N} \setminus \mathbb{N}$, when ordered by $<$, has no maximal elements?*

While studying the sequential order of compact spaces A. Dow [**6**] introduced the notion of a totally MAD family:

Definition. Given $\mathcal{A}$ and $\mathcal{B}$ infinite families of subsets of ω, say that $\mathcal{A}$ is *totally bounded* with respect to $\mathcal{B}$, if for each infinite $\mathcal{B}' \subseteq \mathcal{B}$ and each $h \in \mathcal{B}'^{\omega}$, there is an $A \in \mathcal{A}$ such that $A \cap \bigcup\{B \setminus h[B] : B \in \mathcal{B}'\}$ is infinite. A MAD family $\mathcal{A}$ is *totally MAD* if for each infinite $\mathcal{B} \subseteq \mathcal{A}$ no subset of cardinality less than $\mathfrak{c}$ is totally bounded with respect to $\mathcal{B}$.

Dow showed that a totally MAD family exists assuming $\mathfrak{b} = \mathfrak{c}$, noted that every totally MAD family has a refinement which is a completely separable MAD family and asked:

410? **Problem 8.** *Is there a totally MAD family in* ZFC*? Does* $\mathfrak{b} = \omega_1$ *imply there is a totally MAD family?*

A positive answer to Dow's second problem, implies a positive answer to the following weak form of Problem 1.

411? **Problem 9.** *Is there a completely separable MAD family assuming* $\mathfrak{c} \leq \omega_2$*?*

3. MAD families in forcing extensions

While (as mentioned in the introduction) MAD families with strong combinatorial properties are hard to come by in ZFC, there is also a definite lack of negative (i.e., consistency) results. In this section we present some of the open test problems for understanding the behavior of MAD families in forcing extensions. The first of these problems is due to J. Steprāns [**24**]:

412? **Problem 10.** *Is there a Cohen-indestructible MAD family in* ZFC*?*

K. Kunen [**16**] showed that the answer is positive under CH. J. Steprāns showed that the answer is also positive in any model obtained by adding $\aleph_1$-many Cohen reals. Each of $\mathfrak{b} = \mathfrak{c}$, $\mathfrak{a} < \text{cov(meagre)}$ and $\diamondsuit(\mathfrak{d})$ ([**13**, **14**, **17**]) is also sufficient for the positive answer. The problem has the following combinatorial translation:

Theorem ([**13**, **17**]). *The following statements are equivalent for a MAD family* $\mathcal{A}$.

(1) $\mathcal{A}$ *is Cohen-indestructible.*
(2) *For every* $f : \mathbb{Q} \to \omega$ *there is an* $A \in \mathcal{A}$ *such that* $f^{-1}[A]$ *is somewhere dense.*

Surprisingly, it is not even known whether there is (in ZFC) a MAD family which survives some forcing extension adding new reals (equivalently, a single Sacks real extension):

413? **Problem 11.** *Is there a Sacks-indestructible MAD family in* ZFC*?*

A flawed proof of this appeared in [**13**]. This and other flaws of the paper were rectified in [**5**]. The following old problem (sometimes attributed to J. Roitman) can be formulated as a problem about cardinal invariants of the continuum:

Problem 12. *Does* $\mathfrak{d} = \omega_1$ *imply* $\mathfrak{a} = \omega_1$*?* 414?

Consult [**12**, **18**] for some partial positive results. Recently S. Shelah [**20**] using a novel technique of iteration along templates showed that $\mathfrak{d} < \mathfrak{a}$ is relatively consistent with ZFC. J. Brendle [**4**] presented an axiomatic treatment of Shelah's technique and showed that it can not be used to solve Problem 12.

We would like to thank Alan Dow for commenting on a preliminary version of the paper and for providing interesting questions.

References

[1] B. Balcar, J. Pelant, and P. Simon, *The space of ultrafilters on* $\mathbb{N}$ *covered by nowhere dense sets*, Fund. Math. **110** (1980), no. 1, 11–24.

[2] B. Balcar and P. Simon, *Disjoint refinement*, Handbook of Boolean algebras, Vol. 2, North-Holland, Amsterdam, 1989, pp. 333–388.

[3] B. Balcar and P. Vojtáš, *Almost disjoint refinement of families of subsets of* $\mathbb{N}$, Proc. Amer. Math. Soc. **79** (1980), no. 3, 465–470.

[4] J. Brendle, *Mad families and iteration theory*, Logic and algebra, Contemp. Math., vol. 302, American Mathematical Society, Providence, RI, 2002, pp. 1–31.

[5] J. Brendle and S. Yatabe, *Forcing indestructibility of MAD families*, Ann. Pure Appl. Logic **132** (2005), no. 2–3, 271–312.

[6] A. Dow, *Sequential order under MA*, Topology Appl. **146/147** (2005), 501–510.

[7] P. Erdős and S. Shelah, *Separability properties of almost-disjoint families of sets*, Israel J. Math. **12** (1972), 207–214.

[8] L. Gillman and M. Jerison, *Rings of continuous functions*, D. Van Nostrand Co., Princeton, NJ, 1960.

[9] S. H. Hechler, *Classifying almost-disjoint families with applications to* $\beta N \setminus N$, Israel J. Math. **10** (1971), 413–432.

[10] S. H. Hechler, *Generalizations of almost disjointness, c-sets, and the Baire number of* $\beta N \setminus N$, General Topology and Appl. **8** (1978), no. 1, 93–110.

[11] N. Hindman, *On the existence of c-points in* $\beta N \setminus N$, Proc. Amer. Math. Soc. **21** (1969), 277–280.

[12] M. Hrušák, *Another* $\diamondsuit$*-like principle*, Fund. Math. **167** (2001), no. 3, 277–289.

[13] M. Hrušák, *MAD families and the rationals*, Comment. Math. Univ. Carolin. **42** (2001), no. 2, 345–352.

[14] M. Hrušák and S. García Ferreira, *Ordering MAD families a la Katětov*, J. Symbolic Logic **68** (2003), no. 4, 1337–1353.

[15] K. Kunen, *Ultrafilters and independent sets*, Trans. Amer. Math. Soc. **172** (1972), 299–306.

[16] K. Kunen, *Set theory*, North-Holland, Amsterdam, 1980.

[17] M. S. Kurilić, *Cohen-stable families of subsets of integers*, J. Symbolic Logic **66** (2001), no. 1, 257–270.

[18] J. T. Moore, M. Hrušák, and M. Džamonja, *Parametrized* $\diamondsuit$ *principles*, Trans. Amer. Math. Soc. **356** (2004), no. 6, 2281–2306.

[19] R. S. Pierce, *Modules over commutative regular rings*, American Mathematical Society, Providence, RI, 1967.

[20] S. Shelah, *Two cardinal invariants of the continuum* ($\mathfrak{d} < \mathfrak{a}$) *and FS linearly ordered iterated forcing*, Acta Math. **192** (2004), no. 2, 187–223.

[21] P. Simon, *A compact Fréchet space whose square is not Fréchet*, Comment. Math. Univ. Carolin. **21** (1980), no. 4, 749–753.

[22] P. Simon, *A note on nowhere dense sets in* ω^*, Comment. Math. Univ. Carolin. **31** (1990), no. 1, 145–147.

[23] P. Simon, *A note on almost disjoint refinement*, Acta Univ. Carolin. Math. Phys. **37** (1996), no. 2, 89–99.

[24] J. Steprāns, *Combinatorial consequences of adding Cohen reals*, Set theory of the reals (Ramat Gan, 1991), Bar-Ilan University, Ramat Gan, 1993, pp. 583–617.

[25] A. I. Veksler, *Maximal nowhere dense sets and their applications to problems of existence of remote points and of weak P-points*, Math. Nachr. **150** (1991), 263–275.

Open Problems in Topology II – Edited by E. Pearl

Good, splendid and Jakovlev

Istvan Juhász and William A.R. Weiss

1. The size of good and splendid spaces

It is a trivial observation that a compact space that is locally countable is actually countable, while the (ordered) space ω_1 of countable ordinals shows that this fails if compact is replaced with countably compact. This prompted E. van Douwen to ask to following two questions in the mid 1970s.

Problem 1. *Is there a countably compact and locally countable T_3 space of size continuum? Are there such spaces of arbitrarily large cardinality?* 415–416?

The reason why he asked the first question was that he could construct such a space under the assumption $\mathfrak{b} = \mathfrak{c}$.

In [7] partial answers to these questions were given and the following terminology was introduced.

Definition. A countably compact and locally countable T_3 space is called *good*. A good space is *splendid* if countable subsets have countable (or equivalently, compact) closures. (Of course, ω_1 is splendid.) $G(\kappa)$ (resp. $S(\kappa)$) is the statement that there is a good (resp. splendid) space of size κ.

Good spaces are locally compact and first countable, that is where T_3 is needed.

Below we summarize what is known about the sizes of good and splendid spaces, most of it from [7] or [8].

Proposition 1.1.

(1) *For $\kappa > \omega$, $G(\kappa)$ implies $\mathrm{cf}([\kappa]^\omega, \subset) = \kappa$, hence $\mathrm{cf}(\kappa) > \omega$ and if $\kappa \geq \mathfrak{c}$ then even $\kappa^\omega = \kappa$.*

(2) *For all $n < \omega$ we have $S(\omega_n)$.*

(3) *If $\mathfrak{b} = \mathfrak{c}$ then for all $n < \omega$ we have $G(\mathfrak{c}^{+n})$.*

(4) *If for any κ with $\mathrm{cf}(\kappa) = \omega < \kappa$ we have both $\mathrm{cf}([\kappa]^\omega, \subset) = \kappa^+$ and $\square_\kappa$, then $S(\lambda)$ holds whenever $cf(\lambda) > \omega$.*

(5) *If $\kappa > \aleph_\omega$ and $G(\kappa)$ holds then $\kappa \geq \mathrm{cf}([\aleph_\omega]^\omega, \subset)$.*

(6) *The Chang conjecture variant $(\aleph_{\omega+1}, \aleph_\omega) \twoheadrightarrow (\aleph_1, \aleph_0)$ implies that $S(\kappa)$ fails for all $\kappa \geq \aleph_\omega$.*

(7) *The existence of a supercompact cardinal implies the consistency of $\mathsf{MA} + \aleph_{\omega+1} < \mathfrak{c} + \neg G(\aleph_{\omega+1})$.*

(8) *If $\mathbb{P}$ is the forcing that iteratively adds ω_1 dominating reals to any ground model V then, in $V^{\mathbb{P}}$, the statement $G(\kappa)$ holds for all κ such that $\kappa^\omega = \kappa$.*

The first author was supported by OTKA grant no. 37758.

It has recently been observed by P. Nyikos that the first part of (1) immediately follows from the following (folklore?) fact that is probably due to Sierpiński: For any κ, $\mathrm{cf}([\kappa]^\omega, \subset) = \min\{|\mathcal{C}| : \mathcal{C} \text{ is } \omega\text{-hitting in } [\kappa]^\omega\}$, where $\mathcal{C}$ is ω-hitting means that for any $A \in [\kappa]^\omega$ there is $C \in \mathcal{C}$ with $|A \cap C| = \omega$. Indeed, if we have a good space X with underlying set κ then the family $\mathcal{C}$ of all compact open sets in X is clearly ω-hitting and of size κ. Note that (5) obviously follows from (1).

(4) was proved in [**7**] assuming $\mathsf{V} = \mathsf{L}$ and Nyikos noticed that the proof given there goes through under the weaker assumption of (4). The latter holds, e.g., if the covering lemma over the core model is valid, hence large cardinals are necessary to get a $\kappa > \aleph_\omega$ with $\mathrm{cf}(\kappa) > \omega$ such that $S(\kappa)$ fails.

Concerning (6) and (7) we note that the consistency of the Chang conjecture variant $(\aleph_{\omega+1}, \aleph_\omega) \twoheadrightarrow (\aleph_1, \aleph_0)$ had been established from a 2-huge cardinal that is much stronger than a supercompact.

(5) raises the following question.

417? **Problem 2.** *Is there a good space of cardinality* $\mathrm{cf}([\aleph_\omega]^\omega, \subset)$*?*

2. Connections to other problems

It is very natural to raise the following question.

418? **Problem 3.** *Is there a* ZFC *example of a good space that is not splendid?*

Of course, such a space has a separable closed subspace that can not be compact, hence it is an example asked for by Nyikos, see [**9**] in this volume.

The good spaces of size $\mathfrak{c}$ constructed by van Douwen from $\mathfrak{b} = \mathfrak{c}$ are separable, hence non-splendid. The published version (see [**4**, Example 13.1]) has an extra property: $\mathfrak{b} = \mathfrak{c}$ implies that for any ultrafilter $u \in \omega^*$ there is a separable good space X_u of size $\mathfrak{c}$ which is not u-compact. In [**7**] this was (partially) strengthened as follows: If, in addition to $\mathfrak{b} = \mathfrak{c}$, one also has $2^{\mathfrak{c}} < \mathfrak{c}^{+\omega}$ then there is a single good space X that is not u-compact for any $u \in \omega^*$. It follows then that neither $\prod\{X_u : u \in \omega^*\}$ nor $X^{2^{\mathfrak{c}}}$ is countably compact, hence these spaces provide (consistent) *good* counterexamples to the Scarborough–Stone problem, see [**10**] in this volume. One can also show that the Vietoris hyperspace $H(X)$ of the latter space X is not countably compact, either.

If X is a splendid space of size κ (w.l.o.g. its underlying set is κ) then the collection $\mathcal{C}$ of all compact (hence countable) and open subsets of X is easily seen to form a Kurepa family, that is the trace of $\mathcal{C}$ on any countable subset of κ is countable. Also, $\mathcal{C}$ is clearly cofinal in $[\kappa]^{\leqslant\omega}$. Conversely, as was shown in [**5**], the existence of a cofinal Kurepa family in $[\kappa]^{\leqslant\omega}$ that is also closed under finite intersections and is well-founded by inclusion implies $S(\kappa)$. The above family $\mathcal{C}$ is clearly closed under finite intersections but, contrary to the claim made in [**5**], it is not clear that well-foundedness of it may also be assumed. Thus we are lead to the following question.

419? **Problem 4.** *Is $S(\kappa)$ equivalent to the existence of a cofinal Kurepa family over κ?*

The following interesting result was (essentially) proved in [**3**]. If $G(\kappa)$ holds and there is a κ-sized maximal almost disjoint family of subsets of ω then there is also a separable, crowded, sequentially compact T_2 space of cardinality κ. (Compare this to the well-known fact that any crowded and countably compact T_3 space has size at least $\mathfrak{c}$.) This observation prompted the authors of [**3**] to ask if the smallest size of such a T_2 space is actually equal to $\mathfrak{a}$. In view of part (7) of Proposition 1.1, if $\mathfrak{a} = \aleph_{\omega+1} < \mathrm{cf}([\aleph_\omega]^\omega, \subset)$ (that may consistently occur) then we know at least as much that the method they used to get such a space of size $\mathfrak{a}$ can not be applied anymore.

Finally we mention the very recent article [**1**] concerning some problems that seem to be very closely related to those about the possible sizes of good spaces. Our definitions slightly deviate from those given in [**1**] because we think that ours are more natural (and more general).

Definition. An uncountable T_3 space X is called a *Jakovlev* space iff $X = \bigcup^*_{n<\omega} L_n$ (here * denotes disjoint union) such that

- every $x \in L_n$ has a countable neighbourhood U with $U \setminus \{x\} \subset \bigcup_{i<n} L_i$;
- every infinite $A \subset L_n$ has a limit point.

$J(\kappa)$ stands for the statement that there is a Jakovlev space of cardinality κ.

So Jakovlev spaces are locally countable and partially countably compact. It is easy to see that any Jakovlev space is locally compact and hence first countable, and that the one-point compactification of a Jakovlev space is weakly first countable (the family $\{\bigcup_{i\geq n} L_i : n < \omega\}$ forms a weak base at the point-at-infinity $*$), while every neighbourhood of $*$ is co-countable, hence the compactification is not first countable at $*$. (The assumption of uncountability of a Jakovlev space was made because of this.) Thus Jakovlev spaces (of size $> \mathfrak{c}$) answer a couple of very old questions of Arhangel'skiĭ from [**2**], asking if weakly first countable but not first countable compact T_2 spaces (of size $> \mathfrak{c}$) exist. The first example of such a space was constructed from CH by Jakovlev in [**6**], that explains the terminology.

Let us now give a summary of the results from [**1**].

Proposition 2.1.

(1) *$J(\kappa)$ implies $\kappa \geq \mathfrak{b}$.*
(2) *$\mathfrak{b} = \mathfrak{c} \Rightarrow J(\mathfrak{c}^+) \Rightarrow J(\mathfrak{c})$.*
(3) *If $\mathbb{C}$ is the forcing that adds ω_1 Cohen reals to any ground model V then $V^{\mathbb{C}} \models J(\omega_1)$.*
(4) *If $\mathbb{P}$ is the forcing that iteratively adds ω_1 dominating reals to any ground model V then, in $V^{\mathbb{P}}$, the statement $J(\kappa)$ holds for all κ such that $\kappa^\omega = \kappa$. If V also satisfies* GCH *then $J(\kappa)$ holds in $V^{\mathbb{P}}$ for all cardinals $\kappa > \omega$.*

Thus we see that there are both analogies and differences about the problems concerning the possible sizes of good and Jakovlev spaces, respectively. The biggest difference at present is that while good, even splendid, spaces exist in ZFC, the same question concerning Jakovlev spaces remains open. We close our paper by formulating this problem and a few related others.

420–423? **Problem 5.** *Is there a Jakovlev space in* ZFC*? Are $J(\mathfrak{c})$ or $J(\mathfrak{c}^+)$ provable in* ZFC*? Does $J(\mathfrak{c})$ imply $J(\mathfrak{c}^+)$? Do Jakovlev spaces of arbitrarily large size exist in* ZFC*?*

References

[1] U. Abraham, I. Gorelic, and I. Juhász, *On Jakovlev spaces*, Israel J. Math. **152** (2006), 205–219.

[2] A. V. Arhangel′skiĭ, *Mappings and spaces*, Uspekhi Mat. Nauk **21** (1966), no. 4(130), 133–184, Translation: Russian Math. Surveys 21 (1966) no. 4, 115–162.

[3] A. Bešlagić, E. K. van Douwen, J. W. L. Merrill, and W. S. Watson, *The cardinality of countably compact Hausdorff spaces*, Topology Appl. **27** (1987), no. 1, 1–10.

[4] E. K. van Douwen, *The integers and topology*, Handbook of set-theoretic topology, North-Holland, Amsterdam, 1984, pp. 111–167.

[5] A. Dow, *Set theory in topology*, Recent progress in general topology (Prague, 1991), North-Holland, Amsterdam, 1992, pp. 167–197.

[6] N. N. Jakovlev, *On the theory of o-metrizable spaces*, Dokl. Akad. Nauk SSSR **229** (1976), no. 6, 1330–1331, Translation: Soviet Math. Dokl. 17 (1976) 1217–1219.

[7] I. Juhász, Zs. Nagy, and W. Weiss, *On countably compact, locally countable spaces*, Period. Math. Hungar. **10** (1979), no. 2–3, 193–206.

[8] I. Juhász, S. Shelah, and L. Soukup, *More on countably compact, locally countable spaces*, Israel J. Math. **62** (1988), no. 3, 302–310.

[9] P. J. Nyikos, *First countable, countably compact, noncompact spaces*, Open problems in topology II (E. Pearl, ed.), Elsevier Science.

[10] J. Vaughan, *The Scarborough–Stone problem*, Open problems in topology II (E. Pearl, ed.), Elsevier Science.

Open Problems in Topology II – Edited by E. Pearl

Homogeneous compacta

Jan van Mill

0. Introduction

All spaces under discussion are Tychonoff.

A space X is *homogeneous* if for all $x, y \in X$ there is a homeomorphism $f\colon X \to X$ such that $f(x) = y$. So, loosely speaking, all points in X are topologically equivalent.

Many of the familiar classical objects in topology are homogeneous: manifolds without boundary, the Cantor set, the rational numbers, the irrational numbers, the universal Menger continua (Bestvina [**7**]), the Hilbert cube and connected manifolds modeled on it (Keller [**29**]), the pseudo-arc (Bing [**8**]), the circle of pseudo-arcs (Bing and Jones [**9**]), and topological groups. Besides topological groups and large products of the spaces we just mentioned, all of the examples of homogeneous spaces that immediately come to mind are *metrizable*. This is not by accident. Homogeneity is best understood in the presence of metrizability and plays a significant role there. See for example Daverman [**12**], Bessaga and Pełczyński [**6**] and Toruńczyk [**45**, **46**] for evidence of this in the study of the topology of both finite- and infinite-dimensional manifolds.

Outside the class of metrizable spaces, homogeneity is not a well understood notion and our knowledge is very limited. There are various examples of exotic homogeneous compacta; a few recent ones can be found in Kunen [**32**], de la Vega and Kunen [**49**], and van Mill [**36**]. And there are many theorems that imply that certain spaces are inhomogeneous, quite often based on delicate cardinality considerations. See de la Vega [**48**], Ridderbos [**39**], Juhász, Nyikos and Szentmiklóssy [**28**], Arhangel′skiĭ [**5**], and Arhangel′skiĭ, Ridderbos and van Mill [**37**], for some recent results. But there does not seem to be any unifying theme emerging yet.

In this note we will discuss some longstanding open problems about the structure of homogeneous compacta. The fact that these questions are all stated in very simple topological terms fully demonstrates that there is still a lot of work to be done in this area.

1. Rudin's problem

In [**40**], Walter Rudin proved that under the Continuum Hypothesis (abbreviated CH) the Čech–Stone remainder $\mathbb{N}^* = \beta\mathbb{N} \setminus \mathbb{N}$ of the natural numbers $\mathbb{N}$ with the discrete topology is not homogeneous. He proved the existence of two types of points in $\mathbb{N}^*$ with evident different topological behavior: the P-points and the non-P-points. In [**21**], Frolík established the inhomogeneity of $\mathbb{N}^*$ in ZFC; moreover, he showed that $\mathbb{N}^*$ decomposes into $2^{\mathfrak{c}}$ equivalence classes under autohomeomorphisms. See e.g., Comfort [**10**], Wimmers [**50**], and Kunen [**30**] for additional information on the inhomogeneity of $\mathbb{N}^*$.

In [**41**], Rudin returned to $\beta\mathbb{N}$, and asked whether the inhomogeneity of $\mathbb{N}^*$ was a consequence of the fact that it contains no convergent sequences. This problem has been open for almost half of a century now, and is known as *Rudin's Problem.*

424? **Problem 1.** *Does every infinite homogeneous compact space contain a nontrivial convergent sequence?*

See Rudin's own very informative discussion of the problem in [**42**, pp. 157–159].

By Kuz′minov [**33**] (see also Uspenskij [**47**]), every compact group is *dyadic*, i.e., a continuous image of a Cantor cube. This easily implies that every infinite compact group does contain a nontrivial convergent sequence, hence Rudin's Problem has an affirmative answer for groups.

Another (but, related) approach to this is given not by the dyadic approach but by the fact that a compact group of weight τ contains a copy of 2^τ. It was proved independently by Efimov [**16**], Gerlits [**22**] and Hagler [**23**] that if X is a dyadic space of weight τ, then X contains a copy of 2^τ if and only if X cannot be expressed as the union of countably many closed subspaces of weight less than τ. Now let us assume that X is a homogeneous dyadic space of weight $\tau \geq \omega$ (for example, a compact group of weight τ). Suppose that X can be written as $X = \bigcup_{n<\omega} X_n$, where each X_n is closed and has weight less than τ. The Baire Category Theorem implies that some X_n has nonempty interior, and then it easily follows by homogeneity and compactness that X has weight less than τ, a contradiction. So X contains a copy of 2^τ, and hence a nontrivial convergent sequence (these considerations are well-known of course). So Rudin's Problem not only has an affirmative answer for groups, but even for the much broader class of dyadic spaces. For a nice modern proof that every compact group of weight τ contains a copy of 2^τ, see Shakhmatov [**44**]. Observe that not all homogeneous dyadic spaces are groups: every Tychonoff cube $\mathbb{I}^\tau$ for $\tau \geq \omega$ is homogeneous (being a product of Hilbert cubes) and clearly dyadic, and has the fixed-point property by Brouwer's Theorem so cannot be a topological group. There are even zero-dimensional examples of such spaces, see Pašenkov [**38**].

Interestingly, the class of homogeneous dyadic spaces is about the only general class of homogeneous compacta for which Problem 1 was answered. Even for separable spaces it is unknown, or for spaces with countable π-weight. The point is, as I said in the introduction, that our knowledge of nonmetrizable homogeneous compacta is very limited. Examples of Fedorčuk [**19**, **20**] show that even in small compact spaces, the existence of nontrivial convergent sequences is a delicate matter.

2. Van Douwen's problem

The existence of Haar measure on a compact group clearly implies that it has countable cellularity. This prompts the question whether there are homogeneous compacta of uncountable cellularity. It was answered by Maurice [**35**], who constructed homogeneous ordered compacta of cellularity continuuum. Different examples can now be found by using the following highly nontrivial result of Dow and Pearl [**15**]: if X is zero-dimensional and first countable, then X^ω is homogeneous. Indeed, let X be the Alexandroff double of the Cantor set. Then X is zero-dimensional, compact, first countable, and has cellularity $\mathfrak{c}$. Hence by the Dow–Pearl Theorem, its countable infinite product X^ω is a homogeneous compact space with cellularity $\mathfrak{c}$.

The proof that these spaces are homogeneous is very strongly based on the fact that they are first countable. By Arhangel'skiĭ's Theorem from [**1**], first countable compacta have cardinality at most $\mathfrak{c}$. Large homogeneous compacta can be found for example by forming large products of homogeneous first countable compacta and compact groups. Then one increases many cardinal functions, among them trivially cardinality, but cellularity remains bounded by $\mathfrak{c}$ (Engelking [**17**, Theorem 2.3.17]). The following problem, known as *van Douwen's Problem*, is therefore natural.

Problem 2. *Is there a compact homogeneous space with cellularity greater than* $\mathfrak{c}$*?* 425?

This problem has been open already for more than 30 years. I remember that I got a letter from van Douwen when I was a PhD-student in which he mentioned it (unfortunately, I cannot find the letter anymore). He wrote me that one can form a Souslin circle which is homogeneous by collapsing the endpoints of a certain Souslin continuum to a single point. The square of this space has cellularity ω_1. He then asked about cellularity $\mathfrak{c}$ in ZFC, and bigger.

Let us remark that no bound is known for the cellularity of homogeneous compacta. It would already be a fantastic achievement if it could be shown that every homogeneous compactum has cellularity at most, say, $\beth_{\omega_1}$.

3. Arhangel'skiĭ's problem

Many good questions about homogeneity can be found in Arhangel'skiĭ [**3**]. I focus here on Conjecture 1.17 only since I find it of particular interest.

Problem 3. *Is every homogeneous compact space of countable tightness first countable?* 426?

It was shown recently by de la Vega [**48**] that every homogeneous compact space of countable tightness has cardinality at most $\mathfrak{c}$. This interesting result may suggest that the answer to Problem 3 is yes.

In fact, de la Vega showed that if X is a homogeneous compactum, then $|X| \le 2^{t(X)}$. Since by a deep result of Šapirovskiĭ [**43**] for every compact space X we have $\pi_\chi(X) \le t(X)$, it is natural to ask whether tightness can be replaced by π-character in this result. This is an interesting question due to de la Vega.

Ismail [**26**, 1.13] and Hart and Kunen [**24**, 2.5.1(2)] pointed out that if X is a homogeneous compactum, then $|X| = 2^{\chi(X)}$. This is a consequence of the classical Čech–Pospišil Theorem, see Juhász [**27**, 3.16], and Arhangel'skiĭ's Theorem from [**1**]. So if X is a homogeneous compactum of cardinality at most $\mathfrak{c}$, then X is first countable under CH. Hence as de la Vega observed in [**48**], the answer to Problem 3 is yes under CH. Malyhin [**34**] constructed consistent examples of Fréchet–Urysohn compact spaces that are not first countable at any point. This result makes one wonder whether the problem could be independent. This is not impossible since homogeneity is not free from set theory: there is by van Mill [**36**] (see also Hart and Ridderbos [**25**]) a compact space X in ZFC such that X is homogeneous under $\mathsf{MA} + \neg\mathsf{CH}$, but not under CH.

4. Continuous images of homogeneous compacta

Since every compact metrizable space is a continuous image of the homogeneous Cantor set, the following question is quite natural. I have no idea who asked it.

427? **Problem 4.** *Is every compact space a continuous image of a compact homogeneous space?*

It is clear that 'yes' to Problem 4 implies 'yes' to Problem 2.

It is sometimes possible to improve the homogeneity properties of a space by considering products of it. For example, if $X = \omega + 1$ then it is not homogeneous, while X^ω is, being homeomorphic to the Cantor set. Another example is $X = \mathbb{I}$. Then X^ω is the Hilbert cube which is homogeneous by Keller's Theorem from [**29**]. In fact, as we saw above, by the Dow–Pearl Theorem from [**15**], this trick works for every zero-dimensional, first countable space. Sometimes one has to take large products to finally arrive at a homogeneous space. Consider for example the space $X = 2^\kappa \oplus \{0,1\}$, the topological sum of 2^κ and a two-point space, where κ is an infinite cardinal. Then X^μ for every $\mu < \kappa$ is not homogeneous, while X^κ is (this is an observation of Ridderbos). Kunen [**31**] has shown that in this way it is impossible to improve the homogeneity properties of any infinite compact F-space. See also van Douwen [**13**], Farah [**18**] and Arhangel′skiĭ [**4**] for related results. A characterization of those spaces X for which some power X^κ is homogeneous is unknown. In general, this question seems beyond reach. Whether it makes sense to formulate it for certain special classes of (compact) spaces is unclear to me.

It is not true that every compact space is a retract of a compact homogeneous space. Motorov has shown that the familiar $\sin\frac{1}{x}$-continuum in the plane is not a retract of any homogeneous compact space (Arhangel′skiĭ [**2**]). It is an interesting problem whether there is a compact *zero-dimensional* space that is not a retract of a homogeneous compact space.

5. Remarks

Problems 1–4 are stated for compact spaces. This is essential. For Problems 1, 2 and 4 consider a large discrete space, and for Problem 3 a Σ-product in 2^{ω_1} (Engelking [**17**, 3.10.D]). Even if we add familiar compactness properties such as countable compactness or pseudocompactness, then there are appropriate examples. For Problem 3, simply observe that a Σ-product in 2^{ω_1} is countably compact (even ω-bounded). In Comfort and van Mill [**11**] it was shown that every compact space is a retract of a homogeneous, countably compact space. This implies that compactness is essential in Problems 2 and 4. Van Douwen [**14**] proved that under MA there is a countably compact topological group without nontrivial convergent sequences, hence compactness is essential in Problem 1 as well (we touch a delicate problem here since it is not known whether there is a countably compact topological group without nontrivial convergent sequences in ZFC).

I am indebted to Wis Comfort, Klaas Pieter Hart, Ken Kunen and Guit-Jan Ridderbos for some helpful comments.

References

[1] A. V. Arhangel′skiĭ, *The power of bicompacta with first axiom of countability*, Dokl. Akad. Nauk SSSR **187** (1969), 967–970, Translation: Soviet Math. Dokl. 10 (1969) 951–955.

[2] A. V. Arhangel′skiĭ, *Cell structures and homogeneity*, Mat. Zametki **37** (1985), no. 4, 580–586, Translation: Math. Notes 37 (1985) no. 4, 321–324.

[3] A. V. Arhangel′skiĭ, *Topological homogeneity. Topological groups and their continuous images*, Uspekhi Mat. Nauk **42** (1987), no. 2(254), 69–105, Translation: Russian Math. Surveys 42 (1987) no. 2, 83–131.

[4] A. V. Arhangel′skiĭ, *A weak algebraic structure on topological spaces and cardinal invariants*, Topology Proc. **28** (2004), no. 1, 1–18.

[5] A. V. Arhangel′skiĭ, *Homogeneity of powers of spaces and the character*, Proc. Amer. Math. Soc. **133** (2005), no. 7, 2165–2172.

[6] C. Bessaga and A. Pełczyński, *Selected topics in infinite-dimensional topology*, PWN, Warsaw, 1975.

[7] M. Bestvina, *Characterizing k-dimensional universal Menger compacta*, Mem. Amer. Math. Soc. **71** (1988), no. 380, 110 pp.

[8] R H Bing, *A homogeneous indecomposable plane continuum*, Duke Math. J. **15** (1948), 729–742.

[9] R H Bing and F. B. Jones, *Another homogeneous plane continuum*, Trans. Amer. Math. Soc. **90** (1959), 171–192.

[10] W. W. Comfort, *Ultrafilters: some old and some new results*, Bull. Amer. Math. Soc. **83** (1977), no. 4, 417–455.

[11] W. W. Comfort and J. van Mill, *On the product of homogeneous spaces*, Topology Appl. **21** (1985), no. 3, 297–308.

[12] R. J. Daverman, *Decompositions of manifolds*, Academic Press, Orlando, FL, 1986.

[13] E. K. van Douwen, *Nonhomogeneity of products of preimages and π-weight*, Proc. Amer. Math. Soc. **69** (1978), no. 1, 183–192.

[14] E. K. van Douwen, *The product of two countably compact topological groups*, Trans. Amer. Math. Soc. **262** (1980), no. 2, 417–427.

[15] A. Dow and E. Pearl, *Homogeneity in powers of zero-dimensional first-countable spaces*, Proc. Amer. Math. Soc. **125** (1997), no. 8, 2503–2510.

[16] B. A. Efimov, *Mappings and imbeddings of dyadic spaces*, Mat. Sb. **103(145)** (1977), no. 1, 52–68, Translation: Sb. Math. 32 (1977) 45–57.

[17] R. Engelking, *General topology*, Heldermann Verlag, Berlin, 1989.

[18] I. Farah, *Dimension phenomena associated with $\beta\mathbb{N}$-spaces*, Topology Appl. **125** (2002), no. 2, 279–297.

[19] V. V. Fedorchuk, *The cardinality of hereditarily separable bicompacta*, Dokl. Akad. Nauk SSSR **222** (1975), no. 2, 302–305, Translation: Soviet Math. Dokl. 16 (1975) 651–655.

[20] V. V. Fedorchuk, *A compact space having the cardinality of the continuum with no convergent sequences*, Math. Proc. Cambridge Philos. Soc. **81** (1977), no. 2, 177–181.

[21] Z. Frolík, *Sums of ultrafilters*, Bull. Amer. Math. Soc. **73** (1967), 87–91.

[22] J. Gerlits, *On subspaces of dyadic compacta*, Studia Sci. Math. Hungar. **11** (1976), no. 1–2, 115–120.

[23] J. Hagler, *On the structure of S and $C(S)$ for S dyadic*, Trans. Amer. Math. Soc. **214** (1975), 415–428.

[24] J. E. Hart and K. Kunen, *Bohr compactifications of discrete structures*, Fund. Math. **160** (1999), no. 2, 101–151.

[25] K. P. Hart and G. J. Ridderbos, *A note on an example by van Mill*, Topology Appl. **150** (2005), 207–211.

[26] M. Ismail, *Cardinal functions of homogeneous spaces and topological groups*, Math. Japon. **26** (1981), no. 6, 635–646.

[27] I. Juhász, *Cardinal functions in topology—ten years later*, Mathematisch Centrum, Amsterdam, 1980.

[28] I. Juhász, P. J. Nyikos, and Z. Szentmiklóssy, *Cardinal restrictions on some homogeneous compacta*, Proc. Amer. Math. Soc. **133** (2005), no. 9, 2741–2750.

[29] O.-H. Keller, *Die Homoiomorphie der kompakten konvexen Mengen im Hilbertschen Raum*, Math. Ann. **105** (1931), no. 1, 748–758.

[30] K. Kunen, *Weak P-points in* $\mathbb{N}^*$, Topology, Vol. II (Budapest, 1978), Colloq. Math. Soc. János Bolyai, vol. 23, North-Holland, Amsterdam, 1980, pp. 741–749.

[31] K. Kunen, *Large homogeneous compact spaces*, Open problems in topology, North-Holland, Amsterdam, 1990, pp. 261–270.

[32] K. Kunen, *Compact L-spaces and right topological groups*, Topology Proc. **24** (1999), Summer, 295–327 (2001).

[33] V. Kuz′minov, *Alexandrov's hypothesis in the theory of topological groups*, Dokl. Akad. Nauk SSSR **125** (1959), 727–729.

[34] V. I. Malykhin, *A Fréchet–Uryson compact set without points of countable character*, Mat. Zametki **41** (1987), no. 3, 365–376, Translation: Math. Notes 41 (1987) no. 3, 210–216.

[35] M. A. Maurice, *Compact ordered spaces*, Mathematisch Centrum, Amsterdam, 1964.

[36] J. van Mill, *On the character and π-weight of homogeneous compacta*, Israel J. Math. **133** (2003), 321–338.

[37] J. van Mill, *On the cardinality of power homogeneous compacta*, Topology Appl. **146/147** (2005), 421–428.

[38] V. V. Pašenkov, *Extensions of bicompacta*, Dokl. Akad. Nauk SSSR **214** (1974), 44–47, Translation: Soviet Math. Dokl. 15 (1974) 43–47.

[39] G. J. Ridderbos, *On the cardinality of power homogeneous Hausdorff spaces*, 2005, Preprint.

[40] W. Rudin, *Homogeneity problems in the theory of Čech compactifications*, Duke Math. J. **23** (1956), 409–419.

[41] W. Rudin, *Averages of continuous functions on compact spaces*, Duke Math. J. **25** (1958), 197–204.

[42] W. Rudin, *The way I remember it*, American Mathematical Society, Providence, RI, 1997.

[43] B. Šapirovskiĭ, *On the tightness, π-weight and related notions*, Uč. Zap. Latv. Univ. **257** (1976), 88–89.

[44] D. Shakhmatov, *A direct proof that every infinite compact group G contains* $\{0,1\}^{w(G)}$, Papers on general topology and applications (Flushing, NY, 1992), Ann. New York Acad. Sci., vol. 728, New York Acad. Sci., 1994, pp. 276–283.

[45] H. Toruńczyk, *On CE-images of the Hilbert cube and characterization of Q-manifolds*, Fund. Math. **106** (1980), no. 1, 31–40.

[46] H. Toruńczyk, *Characterizing Hilbert space topology*, Fund. Math. **111** (1981), no. 3, 247–262.

[47] V. V. Uspenskij, *Why compact groups are dyadic*, General topology and its relations to modern analysis and algebra, VI (Prague, 1986), Heldermann Verlag, Berlin, 1988, pp. 601–610.

[48] R. de la Vega, *A new bound on the cardinality of homogeneous compacta*, Topology Appl. **153** (2006), no. 12, 2118–2123.

[49] R. de la Vega and K. Kunen, *A compact homogeneous S-space*, Topology Appl. **136** (2004), 123–127.

[50] E. L. Wimmers, *The Shelah P-point independence theorem*, Israel J. Math. **43** (1982), no. 1, 28–48.

Open Problems in Topology II – Edited by E. Pearl

Compact spaces with hereditarily normal squares

Justin Tatch Moore

In 1948, Katětov proved the following metrization theorem.

Theorem 1 ([**3**]). *If X is a compact space*[1] *and every subspace of X^3 is normal, then X is metrizable.*

This is an immediate consequence of the following two results which are of independent interest.

Theorem 2 ([**3**]). *If $X \times Y$ is hereditarily normal, then either X is perfectly normal or else every countable subspace of Y is closed and discrete.*

Theorem 3 ([**7**]). *If X is a compact space and the diagonal is a G_δ subset of X^2, then X is metrizable.*

Katětov then asked whether the dimension in his theorem could be lowered to 2. In [**1**] Gruenhage and Nyikos present two examples which show that consistently this is not possible.

Theorem 4 ([**1**]). *If there is a Q-set then there is a separable compact space X such that X^2 contains an uncountable discrete subspace and yet has every subspace normal.*

Theorem 5 ([**1**]). *If the Continuum Hypothesis is true, then there is a non-metrizable compact space X such that every subspace of X^2 is separable and normal.*

The first construction is due to Nyikos and is optimal in the sense that the existence of such a space implies the existence of a Q-set [**1**]. The second construction is due to Gruenhage and does not obviously require the full strength of the Continuum Hypothesis.

In [**4**], Larson and Todorčević proved that it is consistent that Katětov's problem has a positive answer.

Theorem 6 ([**4**]). *It is relatively consistent with* ZFC *that if X is a compact space and X^2 is hereditarily normal, then X is metrizable.*

The solution they give represents a set-theoretic breakthrough. The purpose of this section is to suggest how one might obtain a positive solution to Katětov's problem via an analysis which is almost purely topological. The broader goal is to obtain a better understanding of hereditary and perfect normality in compact topological spaces.

I will begin by giving a list of questions which have so far have not received much attention. I was made aware of most if not all of them by Todorčević.

[1] In this article, all spaces are assumed to be regular.

The research presented in this paper was supported by NSF grant DMS–0401893.

428? **Question 1.** *If X is compact and X^2 is hereditarily normal, must X be separable?*

Recall that a space X is *premetric of degree* ≤ 2 iff there is a continuous map f from X into a metric space such that the preimage of any point contains at most two elements. Both Gruenhage's and Nyikos's examples in [**1**] are premetric of degree ≤ 2.

429? **Question 2** (See [**8**]). *If X is compact and X^2 is hereditarily normal, must X be premetric of degree ≤ 2?*

430? **Question 3.** *If there is a compact non-metrizable X which is premetric of degree ≤ 2 such that X^2 is hereditarily normal, must there exist either a Q-set or a Luzin set?*

In each case, a positive answer to the question is a consequence of a positive answer to Katětov's problem and hence is consistent by [**4**]. The hope is that it is possible to prove positive answers to these questions in ZFC.

Notice that a counterexample to Question 1 is necessarily a compact L-space. While a Suslin line comes to mind as a candidate for an example, M.E. Rudin has shown that this is not possible: if L is a compact Suslin line, then L^2 is not hereditarily normal [**6**]. Interestingly, however, $2^{\aleph_0} < 2^{\aleph_1}$ implies that a counterexample to Question 1 must have a square which does not satisfy the countable chain condition. This is a consequence of the following results of Shapirovskii and Todorčević.

Theorem 7 (See [**11**]). *The regular open algebra of any hereditarily normal ccc space has size at most continuum.*

Theorem 8 ([**10**]). *If X is compact and X^2 does not contain an uncountable discrete subspace, then X is separable.*

Observe that a positive answer to Question 2 would give a positive answer to Question 1 since every premetric compactum of degree ≤ 2 is separable.

Question 3 is motivated Theorem 10 below which shows that Gruenhage's construction requires the existence of a Luzin set. Observe that it is relatively easy to obtain a model of set theory in which there are no Q-sets or Luzin sets; this is true after adding $\aleph_2$ random reals to any model, for instance. Hence a positive solution of the above questions would yield a different solution to Katětov's problem.

We will now revisit Gruenhage's example mentioned above. The construction is closely based around a well known construction of Kunen.

Theorem 9 ([**2**]). *If the Continuum Hypothesis is true, then there is a strengthening of the topology on $\mathbb{R}$ to a topology which is locally countable, locally compact, and such that the difference between the closure of a set in this and the usual topology is countable. In particular such a space is hereditarily separable but not Lindelöf.*

M. Wage observed that the construction could be carried out on an arbitrary uncountable set of reals instead of just $\mathbb{R}$ assuming the Continuum Hypothesis.

Such spaces have come to be known as *Kunen lines.* Gruenhage's construction is connected in the sense that his X^2 contains a subspace Z which maps 2–1 onto a Kunen line where the underlying set of reals is a Luzin set.

In order to state Theorem 10 concisely, I will first introduce some notation.

Suppose that X and Y are topological spaces and $f: X \to Y$ is continuous. Define Δ_f to be all pairs (x_0, x_1) in X^2 such that $f(x_0) = f(x_1)$. The function $f_*: \Delta_f \to Y$ is defined by $f_*(x, y) = f(x) = f(y)$.

If f is the identity function, then Δ_f is the diagonal and the subscript is suppressed, giving the standard notation.

Theorem 10. *Suppose that X is a compact non-metrizable space such that X^2 is hereditarily normal; X is premetric of degree ≤ 2; and the quotient of $\Delta_f \setminus \Delta$ by f_* is a Kunen line. Then there is a Luzin set.*

It is not clear whether Gruenhage's construction can be carried out from the existence of a Luzin set. Todorčević has shown that an analogue of Wage's construction can be carried out if $\mathfrak{b} = \aleph_1$, an assumption which follows from the existence of a Luzin set.

Theorem 11 ($\mathfrak{b} = \aleph_1$, **[9]**). *If X is a set of reals of size $\aleph_1$, then there is a refinement of the metric topology which is locally compact, locally countable, perfectly normal and hereditarily separable in all of its finite powers.*

Carrying out Gruenhage's construction assuming only the existence of a Luzin set seems to be a considerably more subtle matter—see my note **[5]** for some limited progress. I conjecture that this is possible.

PROOF OF THEOREM 10. Let X be given as in the statement of the theorem and $f: X \to K$ witness that X is premetric of degree ≤ 2. If U is an open subset of X and $\{x_0, x_1\}$ is a pair of points in X then we say that U *splits* $\{x_0, x_1\}$ if both U and $X \setminus \overline{U}$ contain an element of $\{x_0, x_1\}$. Since X is non-metric and compact, it is possible to recursively select points z_ξ in K and open sets U_ξ in X such that U_ξ splits $f^{-1}(z_\xi)$ but does not split $f^{-1}(z_\eta)$ if $\xi < \eta < \omega_1$. Let $Z = \{z_\xi : \xi < \omega_1\}$ and let V_n $(n < \omega)$ enumerate a base for the topology on K. By removing points from Z if necessary, we may assume that it has no countable neighborhoods.

Observe that if $f(x) = z_\xi$ then one of the collections

$$\{f^{-1}(V_n) \cap U_\xi : x \in f^{-1}(V_n)\}, \quad \{f^{-1}(V_n) \setminus \overline{U_\xi} : x \in f^{-1}(V_n)\}$$

intersects to the singleton $\{x\}$ and hence forms a local base for x. Also observe that since $f^{-1}(V_n)$ does not split any pair of the form $f^{-1}(z)$ for $z \in K$, sets of the form $f^{-1}(V_n) \cap U_\xi$ and $f^{-1}(V_n) \setminus \overline{U_\xi}$ can split $f^{-1}(z_\eta)$ only when $\eta \leq \xi$.

Suppose that Z is not a Luzin set in $\mathrm{cl}_K(Z)$. It suffices to show that X^2 is not hereditarily normal. To this end, let $E \subseteq K$ be a closed set such that $E \cap Z$ is relatively nowhere dense and uncountable. Define the following sets

$$G = \{(x_0, x_1) \in \Delta_f : x_0 \neq x_1 \text{ and } f_*(x_0, x_1) \in E \cap Z\}$$
$$H = \{(x, x) \in X^2 : f(x) \notin E\}.$$

Clearly $\overline{G} \cap H = G \cap \overline{H} = \emptyset$. It is sufficient to show that if $W \subseteq X^2$ is open and contains H then $\overline{W} \cap G$ is nonempty.

By shrinking W if necessary, we may assume that is a union of sets of the form

$$((f^{-1}(V_n) \cap U_\xi) \times (f^{-1}(V_n) \cap U_\xi)) \cup ((f^{-1}(V_n) \setminus \overline{U_\xi}) \times (f^{-1}(V_n) \setminus \overline{U_\xi}))$$

for $n < \omega$ and $\xi < \omega_1$ such that $V_n \cap E = \emptyset$. Since X^2 is hereditarily normal, it follows from [**3**] that X is perfect and therefore that W is a countable union of such sets. Let δ be an upper bound for all $\xi < \omega_1$ required in this union. If $\delta < \xi < \omega_1$ and (x_0, x_1) is in $\Delta_f \setminus \Delta$ with $f_*(x_0, x_1) = z_\xi$, then (x_0, x_1) is in W provided that z_ξ is not in E. Put $D = \{z_\xi : \xi \leq \delta\}$.

By our assumption on $\Delta_f \setminus \Delta$, the closure of $Z \setminus (E \cup D)$ in the metric topology and in the quotient topology induced by f_* differ by a countable set D'. Since E is nowhere dense, D is countable, and Z has no countable neighborhoods, Z is contained in the metric closure of $Z \setminus (E \cup D)$.

I will now show that if (x_0, x_1) is in $\Delta_f \setminus \Delta$ with $f_*(x_0, x_1)$ in $Z \setminus (D \cup D')$, then either (x_0, x_1) or (x_0, x_1) is in the closure of W. This finishes the proof since there is a (x_0, x_1) such that $f_*(x_0, x_1)$ is in $Z \setminus (D \cup D')$ and both (x_0, x_1) and (x_1, x_0) are in G. To this end, suppose that (x_0, x_1) are given as above and let $z = f_*(x_0, x_1)$. Since z is not in D', z is a limit point of $Z \setminus (E \cup D)$ in the quotient topology since it is in the metric topology. This means that there is an element of $f_*^{-1}(z)$ which is in the closure of the f_*-preimage of $Z \setminus (E \cup D)$. Since this preimage is contained in W, either (x_0, x_1) or (x_1, x_0) is in the closure of W as desired. □

References

[1] G. Gruenhage and P. J. Nyikos, *Normality in X^2 for compact X*, Trans. Amer. Math. Soc. **340** (1993), no. 2, 563–586.

[2] I. Juhász, K. Kunen, and M. E. Rudin, *Two more hereditarily separable non-Lindelöf spaces*, Canad. J. Math. **28** (1976), no. 5, 998–1005.

[3] M. Katětov, *Complete normality of Cartesian products*, Fund. Math. **35** (1948), 271–274.

[4] P. Larson and S. Todorčević, *Katětov's problem*, Trans. Amer. Math. Soc. **354** (2002), no. 5, 1783–1791.

[5] J. T. Moore, *Splitting Todorčević's set mapping from $\mathfrak{b} = \omega_1$*, 2004, Unpublished note.

[6] M. E. Rudin, *Hereditary normality and Souslin lines*, General Topology and Appl. **10** (1979), no. 1, 103–105.

[7] V. E. Šneĭder, *Continuous images of Suslin and Borel sets. Metrization theorems*, Dokl. Akad. Nauk SSSR **50** (1945), 77–79.

[8] S. Todorčević, *Chain-condition methods in topology. II.*

[9] S. Todorčević, *Partition problems in topology*, American Mathematical Society, Providence, RI, 1989.

[10] S. Todorčević, *Free sequences*, Topology Appl. **35** (1990), no. 2–3, 235–238.

[11] S. Todorčević, *Chain-condition methods in topology*, Topology Appl. **101** (2000), no. 1, 45–82.

Open Problems in Topology II – Edited by E. Pearl

The metrization problem for Fréchet groups

Justin Tatch Moore and Stevo Todorčević

1. Introduction

Let us begin this paper by recalling the following classical metrization theorem of Birkhoff and Kakutani.

Theorem 1. *Every first countable group is metrizable.*

In this article, we will be interested in the extent to which the assumption of first countability in this theorem can be weakened. Recall that a Hausdorff[1] topological space X is *Fréchet* if whenever x is a limit point of $A \subseteq X$, there is a sequence a_n $(n < \omega)$ of elements of A which converges to x. This is a natural weakening of first countability which has been extensively studied in the literature. It turns out that this property by itself is not sufficient to ensure the metrizability of a topological group.

Example 1. *The direct sum of ω_1 copies of the circle group $(\mathbb{R}/\mathbb{Z}, +)$ is a σ-compact Fréchet group which is not first countable.*

Such an example, however is easily ruled out by requiring that the group be separable. Hence we arrive at the following problem posed by Malykhin in 1978.

Problem 1. *Is every separable Fréchet group metrizable?* 431?

The requirement of separability in Malykhin's problem can be replaced by a more restrictive notion without changing the problem as the following proposition shows.

Proposition. *If every countable Fréchet group is metrizable, then so is every separable Fréchet group.*

We have already noted that some countability requirement is necessary in the formulation of Malykhin's problem. Later Todorčević provided an example with an additional striking property.

Example 2 ([**13**]). *There are two σ-compact Fréchet groups whose product is not countably tight.*

This highlights an auxiliary consideration, the productivity of the Fréchet property in groups, which will also be part of our focus.

It was known from the beginning that Malykhin's problem is really a consistency question. At the time he posed Problem 1, Malykhin was aware of the following consistent counterexample.

[1]All spaces in this article are assumed to be Hausdorff.

The first author acknowledges the support of NSF grant DMS-0401893

Example 3 ($\mathfrak{p} > \omega_1$). *If G is a separable metrizable group with at least two elements, then G^{ω_1} is a separable Fréchet group which is not metrizable.*

On the other hand, Shibakov later showed that CH can also be used to generate counterexamples with additional properties.

Example 4 ([**9**], CH). *There are separable Fréchet groups whose product is not Frećhet.*

It is also not clear what role the group structure plays in Malykhin's problem. Recall that $[\omega]^{<\omega}$, the finite subsets of ω, is a group when equipped with the operation $\triangle$ of symmetric difference.

432? **Question 1.** *Is there a topology which makes $([\omega]^{<\omega}, \triangle)$ a nonmetrizable Fréchet group?*

433? **Question 2.** *Suppose that there is a separable nonmetrizable Fréchet group. Is there a topology which makes $([\omega]^{<\omega}, \triangle)$ a nonmetrizable Fréchet group?*

These questions really ask about the existence of certain filters, known as FUF filters in the literature, on $[\omega]^{<\omega}$ (see [**8**]).

Since there is already a recent survey [**8**] by Shakhmatov on convergence in settings where there is additional algebraic structure, we will focus on a scenario for proving a positive answer to Malykhin's problem and refer the reader to that article for further information.

2. The role of gaps

Suppose that G is a countable topological group on ω with 0 serving as the identity. A central object in the study of Malykhin's problem is the ideal $\mathcal{I}_G$ of subsets of ω which do not accumulate to 0. It is easily verified that G is Fréchet iff $\mathcal{I}_G^{\perp\perp} = \mathcal{I}_G$.[2]

Recall that two families $\mathcal{A}$ and $\mathcal{B}$ of subsets of ω form a *gap* if they are orthogonal (i.e. $\mathcal{B} \subseteq \mathcal{A}^\perp$) and yet there is no $C \subseteq \omega$ such that every element of $\mathcal{A}$ is almost contained in C and every element of $\mathcal{B}$ is almost disjoint from C. If $\mathcal{A}^{\perp\perp}$ is countably generated, then the gap is said to be *Rothberger*. Hence Malykhin's problem is equivalent to asking whether there is a countable Fréchet group G such that $\mathcal{I}_G$ does not form a Rothberger gap with $\mathcal{I}_G^\perp$.

Todorčević's Open Coloring Axiom is an assertion which has a strong influence on the structure of gaps in $\mathcal{P}(\omega)$:

> If X is a separable metric space and $G \subseteq [X]^2$ is an open graph on X, then either G is countably chromatic or else has an uncountable complete subgraph.

This axiom was defined in [**12**], where its influence on gaps is presented. Moreover, if the underlying set of reals is analytic, then OCA is provable in ZFC for

[2] If $\mathcal{A}$ is a family of subsets of ω, we let $\mathcal{A}^\perp$ denote the collection of all subsets of ω which have finite intersection with every element of $\mathcal{A}$.

all open graphs on X. This can either be deduced[3] from the consistency proof of OCA or proved directly as in [**3**]. Not surprisingly, it is possible to carry out a parallel analysis of gaps in $\mathcal{P}(\omega)$ in which $\mathcal{A}$ is analytic. This led to the proof of the following effective solution to Malykhin's problem. Recall that a countable topological space X is *analytic* if its topology is an analytic subset of $\mathcal{P}(X)$ when identified with 2^X.

Theorem 2 ([**16**]). *Suppose that G is a separable group with a countable dense analytic subspace. Then G is metrizable.*

Countable metrizable spaces are always analytic but in general this is a considerably larger class which includes a number of important test spaces. For example the countable sequential fan and Arens space are examples of analytic topologies. In fact a countable space is analytic if and only if it can be embedded into $C_p(X)$ for some Polish space X.

While Malykhin's problem can only have a consistent positive solution and Theorem 2 is a ZFC theorem, the analysis of the combinatorial difficulties seems likely to be similar. The reader is referred to [**15**, **16**] for applications of OCA which are closely related to the subject matter.

3. Other convergence properties

There are two other weakenings of first countability which are important in the present context.

Definition. A topological space is said to have the *weak diagonal sequence property*[4] if whenever S_i $(i < \omega)$ is a collection of sequences which converges to a given point x, there is a sequence S_∞ which converges to x such that $S_i \cap S_\infty$ is nonempty for infinitely many $i < \omega$. If S_∞ can always be selected so as to intersect every S_i, then the space is said to have the *diagonal sequence property.*[5]

In the general setting of topological spaces, they are unrelated to the Fréchet property. Nyikos demonstrated that this is not the case is the more restrictive setting of topological groups.

Theorem 3 ([**6**]). *Fréchet groups have the weak diagonal sequence property.*

Whether Fréchet groups have the stronger of these properties, however, is unclear and may be closely related to Malykhin's problem.

Question 3 ([**7**]). *Is it consistent that every countable Fréchet group has the* 434?
diagonal sequence property?

---

[3]This is a consequence of the following observations: (1) the partial order for forcing an instance of OCA (see [**12**, §8]) can be modified so that the resulting homogeneous set is moreover relatively closed, (2) analytic sets have the perfect set property yielding a perfect homogeneous set in the extension, and (3) by Shoenfeld's absoluteness theorem [**10**], the homogeneous set exists in the ground model.

[4]This property is often referred to as α_4.

[5]This property is often referred to as α_2.

435? **Question 4** ([**7**]). *Is it consistent that every countable Fréchet group with the diagonal sequence property is metrizable?*

Both of these questions could also be also asked without the assumption of countability. Let us note the following reformulation of a result of Szlenk (see [**11**] and [**14**, p. 65], or [**16**, p. 516]) that there is a positive solution to the effective version of Question 4.

Theorem 4. *Every analytic Fréchet space with the diagonal sequence property is first countable.*

An important question becomes whether (and how much) Theorem 3 can be strengthened. The ultimate target is to demonstrate that, consistently, separable Fréchet groups are *bi-sequential*: whenever $\mathcal{U}$ is a convergent ultrafilter, there is a sequence U_n $(n < \omega)$ of elements of $\mathcal{U}$ which converges to the same point as $\mathcal{U}$. This property is easily shown to be productive and strengthens both the Fréchet and diagonal sequence properties. Furthermore, in the class of topological groups, this condition is as strong as metrizability, as the following result of Arhangel′skiĭ and Malykhin [**1**] demonstrates.

Theorem 5. *Bi-sequential groups are first countable and therefore metrizable.*

PROOF. Clearly we may assume that G has no isolated points and therefore that the nowhere dense subsets of G extend the cofinite filter. Suppose that g is in G and let $\mathcal{U}$ be an ultrafilter converging to g which is disjoint from the collection of nowhere dense subsets of G. Applying the bi-sequentiality of G, let U_n $(n < \omega)$ be a sequence of elements of $\mathcal{U}$ which converge to g. Let V_n be the interior of the closure of U_n and set $W_n = V_n * V_n^{-1}$. It follows then that $\{W_n : n < \omega\}$ forms a countable neighborhood base at 0 and hence G is first countable. By the Birkhoff–Kakutani theorem, every first countable group is metrizable, finishing the proof. □

The point is, however, that it may be more natural to verify that the group at hand is bi-sequential and indeed this is the approach taken in [**16**].

We will now recall a set-theoretic definition. Now suppose that X is a countable set. A collection $\mathcal{H}$ of subsets of X is a *co-ideal* (on X) if

(1) $\mathcal{H}$ is closed under supersets relative to X and
(2) If Z is in $\mathcal{H}$ and $Z = \bigcup_{i<n} Z_i$, then there is an $i < n$ such that Z_i is in Z.

Definition. A co-ideal $\mathcal{H}$ on X is said to be *selective* if it satisfies the following two additional conditions:

p^+**:** Whenever Z_n $(n < \omega)$ is a decreasing sequence of elements of $\mathcal{H}$, then there is a Z_∞ in $\mathcal{H}$ such that $Z_\infty \setminus Z_n$ is finite for all $n < \omega$.
q^+**:** Whenever Z is in $\mathcal{H}$ and $\phi : Z \to \omega$ is finite-to-one, there is a $Z_* \subseteq Z$ in $\mathcal{H}$ such that $\phi \restriction Z_*$ is one-to-one.

If G is a topological group, define

$$\mathcal{H}_G = \{X \subseteq G : X \text{ accumulates at } 0\}.$$

It is easily verified that $\mathcal{H}_G$ is always a co-ideal, even if G is not a group and 0 is replaced by an arbitrary limit point in the space. The following proposition shows that in the context which is of interest to us, this co-ideal is selective.

Proposition ([**16**]). *If G is a countable Fréchet group, then $\mathcal{H}_G$ is a selective co-ideal.*

4. In search of a test model

Examples 1 and 3 suggest the need for a model which allows for the failure of certain consequences of PFA close to $\mathsf{MA}_{\aleph_1}(\sigma\text{-centered})$ while maintaining other consequences, and OCA in particular. Models with these general properties were considered by Larson and Todorčević in [**4, 5**] and were obtained by forcing with a Souslin tree over a model of a strong fragment of PFA. Note, however, that since $\mathfrak{p} > \omega_1$ in their ground model and since examples solving Malykhin's problem are preserved in forcing extensions which do not add reals, such models can not yield a solution to Malykhin's problem. The problem is essentially that one needs the conjunction of OCA and together with a failure of $\mathfrak{p} > \omega_1$ which is more pertinent to the problem at hand. Michael Hrušák has suggested the following problems in relation to this.

Question 5. *Is* OCA *consistent with the assertion that every ω-splitting family in* 436?
$[\omega]^\omega$ contains an ω-splitting subfamily of size $\aleph_1$?

Question 6. *Assume* OCA *and that every ω-splitting family in $[\omega]^\omega$ contains an* 437?
ω-splitting subfamily of size $\aleph_1$. Is every separable Fréchet group metrizable?

References

[1] A. V. Arhangel′skiĭ and V. I. Malykhin, *Metrizability of topological groups*, Vestnik Moskov. Univ. Ser. I Mat. Mekh. (1996), no. 3, 13–16, Translation: Moscow Univ. Math. Bull. 51 (1996) no. 3, 9–11.

[2] A. Dow, *Set theory in topology*, Recent progress in general topology (Prague, 1991), North-Holland, Amsterdam, 1992, pp. 167–197.

[3] Q. Feng, *Homogeneity for open partitions of pairs of reals*, Trans. Amer. Math. Soc. **339** (1993), no. 2, 659–684.

[4] P. Larson and S. Todorčević, *Chain conditions in maximal models*, Fund. Math. **168** (2001), no. 1, 77–104.

[5] P. Larson and S. Todorčević, *Katětov's problem*, Trans. Amer. Math. Soc. **354** (2002), no. 5, 1783–1791.

[6] P. J. Nyikos, *Metrizability and the Fréchet–Urysohn property in topological groups*, Proc. Amer. Math. Soc. **83** (1981), no. 4, 793–801.

[7] D. Shakhmatov, *α_i-properties in Fréchet–Urysohn topological groups*, Topology Proc. **15** (1990), 143–183.

[8] D. Shakhmatov, *Convergence in the presence of algebraic structure*, Recent progress in general topology, II, North-Holland, Amsterdam, 2002, pp. 463–484.

[9] A. Shibakov, *Countable Fréchet topological groups under CH*, Topology Appl. **91** (1999), no. 2, 119–139.

[10] J. R. Shoenfield, *The problem of predicativity*, Essays on the foundations of mathematics, Hebrew University, Magnes Press, Jerusalem, 1961, pp. 132–139.

[11] W. Szlenk, *The non-existence of a separable reflexive Banach space universal for all separable reflexive Banach spaces*, Studia Math. **30** (1968), 53–61.

[12] S. Todorčević, *Partition problems in topology*, American Mathematical Society, Providence, RI, 1989.

[13] S. Todorčević, *Some applications of S and L combinatorics*, The work of Mary Ellen Rudin (Madison, WI, 1991), Ann. New York Acad. Sci., vol. 705, New York Acad. Sci., 1993, pp. 130–167.

[14] S. Todorčević, *Analytic gaps*, Fund. Math. **150** (1996), no. 1, 55–66.

[15] S. Todorčević, *A proof of Nogura's conjecture*, Proc. Amer. Math. Soc. **131** (2003), no. 12, 3919–3923.

[16] S. Todorčević and C. Uzcátegui, *Analytic k-spaces*, Topology Appl. **146/147** (2005), 511–526.

Open Problems in Topology II – Edited by E. Pearl

Čech–Stone remainders of discrete spaces

Peter J. Nyikos

1. Introduction

The study of Čech–Stone remainders has long been a major theme in set-theoretic topology. A whole book [**13**] was published that primarily dealt with the remainder $\omega^* = \beta\omega \setminus \omega$ of the countable discrete space ω, and discussion of this remainder takes up a sizable chunk of a book that was published back in 1960 [**8**]. It is remarkable that one of the most basic questions about it is still unsolved:

Problem 1. *Is it consistent that ω^* is homeomorphic to ω_1^*?* 438?

Here ω_1^* refers to the Čech–Stone remainder of a discrete space of cardinality ω_1. What makes this problem all the more remarkable is that if we put any other pair of distinct infinite cardinals for $\{\omega, \omega_1\}$, even if one of this pair is one of the members of the new pair, the answer is negative. Moreover, this has been known since the late 1970s. Since most of the research that established this and other nontrivial facts detailed later was done by Polish and Czech mathematicians [**2**, **6**, **7**], I decided to break with the usual American custom and use the expression "Čech–Stone" in place of "Stone–Čech."

An interesting alternative formulation of Problem 1 in ZFC is:

Problem 1′. *Is it consistent that the Boolean algebras $\mathcal{P}(\omega)/\mathrm{fin}$ and $\mathcal{P}(\omega_1)/[\omega_1]^{<\omega}$ are isomorphic?*

In the absence of the Axiom of Choice (AC) the two problems are not equivalent: what passes for the Čech–Stone remainders could be empty, while the quotient algebras are both uncountable. It would be interesting if the Boolean Algebra version had a positive answer in ZF while the answer to both versions is negative in ZFC. While our primary interest is what happens in ZFC, I will be making remarks about what to watch out for if AC is not assumed. The theory would have a varying flavor depending on what weakenings of AC are assumed. Three natural weakenings are: (1) the Boolean Prime Ideal Theorem, which assures that every discrete space has a Čech–Stone compactification; (2) the existence of right inverses to the two quotient maps; and (3) the axiom of dependent choices DC, which implies the Countable AC (= AC for countable collections of sets).

Three other weakenings of AC dovetail well with Problem 1′. Let $\mathsf{wAC}(\kappa, \lambda)$ stand for the axiom that there is a choice function for collections of $\leq \kappa$ sets, each of cardinality $\leq \lambda$). Then $\mathsf{wAC}(2^{\omega_1}, \omega_1)$ implies (2) above; $\mathsf{wAC}(2^{\omega}, \omega)$ implies that the quotient map from $\mathcal{P}(\omega)$ to $\mathcal{P}(\omega)/\mathrm{fin}$ has a right inverse; and we will see some proofs which go through if $\{\kappa, \lambda\} \subset \{\omega, \omega_1\}$. An interesting but farfetched scenario is that of Problem 1′ having a Yes answer in the absence of (2), yet for the quotient map from $\mathcal{P}(\omega)$ to $\mathcal{P}(\omega)/\mathrm{fin}$ to have a right inverse.

The following natural variation on Problem 1 is also unsolved:

439? **Problem 2.** *Is it consistent for $U(\kappa)$ and $U(\lambda)$ to be homeomorphic for different κ, λ?*

Here $U(\kappa)$ stands for the set of uniform ultrafilters on κ—those of which every member is of cardinality κ. It is very easy to show that there is no homeomorphism if $\mathrm{cf}(\kappa) \neq \mathrm{cf}(\lambda)$ (see Theorem 2.1) but even the case $\kappa = \omega$, $\lambda = \aleph_\omega$ has resisted all attempts at a solution. The Boolean algebra version of Problem 2 is left as an exercise for the reader; facts from Section 2 make this exercise trivial.

The conventional wisdom is that Problems 1 and 2 have negative answers, so I could easily have worded Problem 1, "Is it a theorem of ZFC that ω^* is not homeomorphic to ω_1^*?" and used a similar wording for Problem 2. However, I am recommending that we treat the claim that ω^* and ω_1^* *are* homeomorphic as an axiom, the way Rothberger boldly treated $\mathfrak{p} > \omega_1$ even though all the evidence then available (including Godel's proof of consistency of CH) suggested it was false. Recall also how Bing, unaware of Rothberger's research, published an example of a nonmetrizable separable normal Moore space on the assumption that there exists a Q-set, known to be contradicted by the natural-seeming axiom $2^\omega < 2^{\omega_1}$; and how Mary Ellen Rudin published the first example of a Dowker space assuming the existence of a Souslin tree, also not known at the time to be consistent. Accordingly I formulate:

Axiom Ω: ω^* is homeomorphic to ω_1^*.

Nowadays people are reluctant to assume axioms so boldly, except perhaps in the case of large cardinal axioms, but there are certain advantages to this approach. It encourages researchers to publish consequences of the axiom in the optimistic hope that some day the axiom may turn out to be consistent, so that if the breakthrough does happen, we will have a whole body of different statements known to be simultaneously consistent. On the other hand, if the axiom should turn out to be false, the proof that this is so will probably be a proof by contradiction, building upon consequences of the axiom that are already known.

Clearly, Axiom Ω implies $2^\omega = 2^{\omega_1}$: the weights of ω^* and ω_1^* are 2^ω and 2^{ω_1} respectively. Two other easy consequences of Axiom Ω are that there is a complete ω_1-tower (Theorem 2.2) and, in contrast, that there is a Q-set. The contrast is heightened in both directions in Section 3, by Theorems 3.1 and 3.3 respectively. A great many natural questions about the implications of Axiom Ω remain unanswered; a short list is given in Section 5.

We assume AC except where explicitly stated otherwise. Lower-case Gothic letters designate small uncountable cardinals [**5**, **12**].

2. Some basics

We recall some basic facts about the Čech–Stone compactifications of discrete spaces. The underlying set for βD, where D is a discrete space, is the set of all ultrafilters on D. A base for the topology is the collection of all sets of the form $[A] = \{p \in \beta D : A \in p\}$. This makes βD into a compact space, which is 0-dimensional because each $[A]$ is clopen. The discrete space D is identified with the set of principal ultrafilters and is the dense set of isolated points of βD. The *Čech–Stone remainder* $\beta D \setminus D$ is designated D^*, and $[A] \setminus A$ ($= [A] \setminus D$) is designated A^*. The sets of the form A^* thus form a base for the relative topology on D^*.

The collection $CO(X)$ of clopen sets of any topological space X is a Boolean algebra under the usual operations of $\cup$, $\cap$, and complementation. The following facts are well known, easy to prove, and useful; for instance, (b) clearly implies the equivalence of Problems 1 and 1′.

(a) The unary operation $[\cdot]$ is a Boolean algebra isomorphism from $\mathcal{P}(D)$ to $CO(\beta D)$.

(b) The unary operation * is a Boolean algebra homomorphism from $\mathcal{P}(D)$ onto $CO(D^*)$, whose kernel is $[D]^{<\omega}$.

(c) For each infinite cardinal $\kappa \leq |D|$, the set of ultrafilters whose smallest members are of cardinality $< \kappa$ is a dense open subspace of D^*. Therefore, $U(D)$ is a nowhere dense closed subspace of βD.

(d) If we define $\widehat{A}$ as $\{[A] \cap U(D) : A \subset D\}$ then $\widehat{\ }$ is a Boolean algebra homomorphism from $\mathcal{P}(D)$ onto $CO(U(D))$, whose kernel is $[D]^{<|D|}$.

(e) If X and Y are compact 0-dimensional spaces and $\phi \colon X \to Y$ is continuous, and $\phi^{\leftarrow} \colon CO(Y) \to CO(X)$ is defined by $\phi^{\leftarrow}(K) = \phi^{-1}(K)$ then $\phi^{\leftarrow}$ is a Boolean algebra homomorphism. Moreover, $\phi^{\leftarrow}$ is injective iff ϕ is surjective, and vice versa.

(f) A continuous bijection between compact Hausdorff spaces is a homeomorphism. Consequently, $\phi^{\leftarrow}$ is an isomorphism iff ϕ is a homeomorphism.

A good understanding of ω^* and ω_1^* calls for skill in shuttling back and forth between $\mathcal{P}(D)$ and $CO(D^*)$, using the * operation to "go upstairs" from D to D^* and implicitly using various choice functions to "go downstairs" by labeling clopen subsets of D^* as A^*, etc. This is where the various weakenings of AC come in. [In their absence, we translate topological language on ω^* by using the natural correspondence between open sets and ideals, and between clopen sets and elements (or principal ideals).]

A similar shuttle works for $\mathcal{P}(D)$ and $U(D)$. We see it operating in the proof of the following theorem [**4**].

Theorem 2.1. *If* $\mathrm{cf}(\kappa) \neq \mathrm{cf}(\lambda)$, *then* $U(\kappa)$ *and* $U(\lambda)$ *are not homeomorphic.*

Proof. Let $\mathrm{cf}(\kappa) < \mathrm{cf}(\lambda)$. Partition λ into $\mathrm{cf}(\kappa)$ sets A_α of cardinality λ. The (disjoint) clopen sets $\widehat{A_\alpha}$ upstairs have dense union in $U(\lambda)$ because every member of $[\lambda]^\lambda$ meets some A_α in a set of size λ. On the other hand, no family of $\mathrm{cf}(\kappa)$ disjoint clopen subsets $\widehat{B_\alpha}$ of $U(\kappa)$ is dense in $U(\kappa)$: there is a κ-element subset of κ which meets B_α in a set of cardinality $|\alpha|$ and is disjoint from all earlier B_β. □

In analyzing ω_1^* the dense open subspace $S(\omega_1) = \omega_1^* \setminus U(\omega_1)$, known as the *space of subuniform ultrafilters*, plays an important role. It is the union of the ascending ω_1-chain $\{\alpha^* : \alpha \text{ is a countable limit ordinal}\}$ of clopen sets. In other words, it is what I call an ω_1-oval: A union of a chain of clopen sets in a Čech–Stone remainder of a discrete space is an *oval* and is a *κ-oval* if the chain has cofinality κ.

In particular, the ω-ovals are the cozero sets. The small uncountable cardinal $\mathfrak{t}$ can be characterized as the least κ such there is a dense κ-oval in ω^*. With this in mind it is easy to see:

Theorem 2.2. Axiom Ω *implies* $\mathfrak{t} = \omega_1$.

The Boolean algebra version is that there is an ideal generated by an ω_1-chain in $\mathcal{P}(\omega)/\text{fin}$ that meets every nonzero ideal.

Another shuttle goes between $\mathcal{P}(D)$ (or $CO(D^*)$) and $\mathcal{P}(\omega \times \omega)$. I call it the *RH transfer* in honor of Rothberger and Hechler, who made good use of it.

Let $\mathcal{A} = \{A_n : n \in \omega\}$ be a family of subsets of ω such that $A_n^\# = A_n \backslash \bigcup_{i=0}^{n-1} A_i$ is infinite for all n. An *RH transfer* of $\mathcal{A}$ to $\omega \times \omega$ is a bijection $\psi\colon \omega \to \omega \times \omega$ which distributes the elements of $\omega \setminus \bigcup_{n=0}^{\infty} A_n$ into the bottom row $\omega \times \{0\}$, and sends $A_n^\#$ into the n+1-st column $\{n\} \times \omega$.

In an RH transfer, subsets of ω that are almost disjoint from all the A_n are characterized by their images being dominated by the graph of a function. The transfer and the definition of the function can all be defined in ZF, taking advantage of the listing of $\mathcal{A}$ and the well-ordering on ω. It is when we combine the transfers with moves downstairs that some form of AC is required. The following simple theorem [**9**] only requires $\mathsf{wAC}(\omega, \omega)$ in a move downstairs followed by composing one RH transfer with the inverse of another, followed by a move upstairs.

Theorem 2.3. *Any two ω-ovals in ω^* are homeomorphic; moreover, there is a permutation of ω whose extension to $\beta\omega$ is a homeomorphism taking one to the other.*

3. Some consequences of Axiom Ω

There is some confusion about whether "Q-set" is understood to include "uncountable," so I have suggested extending the usual list of Gothic-letter small cardinals to include $\mathfrak{q}$. The trouble is, there are two natural and useful rivals for what $\mathfrak{q}$ could designate. So I recommend using subscripts, as follows: $\mathfrak{q}_0$ = the least cardinal κ for which there is a set of reals of size κ that is not a Q-set. $\mathfrak{q}_1$ = the least cardinal κ for which no set of reals of size κ is a Q-set.

Theorem 3.1. Axiom Ω *implies* $\mathfrak{q}_1 > \omega_1$.

This theorem is a corollary of a much stronger theorem mentioned (but not proved) in [**10**]. Call a family $\mathcal{A}$ of ω_1-many denumerable subsets of ω a *strong Q-sequence* if every 2-coloring of the members of $\mathcal{A}$ is uniformizable. This means that if $f_A \to \{0, 1\}$ is given for each $A \in \mathcal{A}$, there is a function $f\colon \omega \to \{0, 1\}$ such that $f(a) = f_A(a)$ for all but finitely many a in each $A \in \mathcal{A}$. Clearly any strong Q-sequence is an AD family, by which I mean a collection of denumerable sets such that the intersection of any two is finite. We will see below how a Q-set of cardinality ω_1 is intimately associated with the special case where each f_A is constant.

Unlike $\mathfrak{q}_1 > \omega_1$, the existence of a strong Q-sequence does not follow from MA_{ω_1} and indeed is incompatible with it [**10**]. In contrast:

Theorem 3.2. Axiom Ω *implies there is a strong Q-sequence.*

PROOF. For each countable limit ordinal α let $A_\alpha = [\alpha, \alpha + \omega)$. Obviously, every 2-coloring (indeed every coloring!) of the individual A_α is uniformizable. Upstairs in ω_1^*, uniformizability of every 2-coloring translates into the following: for each choice of clopen $C_\alpha \subset A_\alpha^*$ there is a clopen K such that $K \cap A_\alpha = C_\alpha$ for all α. [Just let C_α be the remainder of the support of f_{A_α}, etc.] Assuming Axiom Ω we shuttle over to ω^* with a homeomorphism ψ. The images of the A_α move downstairs to an AD family on ω which is easily seen to be a strong Q-sequence by a translation like that above. □

Among the many statements equivalent to $\mathfrak{q}_1 > \omega_1$ is the existence of a separable nonmetrizable normal Moore space, as well as the existence of the special case where the Moore space is locally compact and its set of nonisolated points is a closed discrete space; see [**11**] and its references in Section II. This special case has a nice alternative characterization as a normal uncountable Ψ-like space: A *Ψ-like space* is a locally compact, locally countable space X in which ω is a dense set of isolated points and $X \setminus \omega$ is closed discrete.

We can associate an AD family $\mathcal{A}$ on ω with the nonisolated points of a Ψ-like space X, with each $A \in \mathcal{A}$ associated with a point p_A such that $A \cup \{p_A\}$ is a compact open neighborhood of p_A. Normality of X then translates to uniformizability of every 2-coloring of $\mathcal{A}$ in which each f_A is constant. In this way, Theorem 3.1 is made to follow from Theorem 3.2.

Rothberger showed that $\mathfrak{q}_0 \leq \mathfrak{b}$. Since $\mathfrak{b} \leq \mathfrak{d}$, the following theorem shows that Axiom Ω implies $\mathfrak{q}_0 = \omega_1$.

Theorem 3.3. Axiom Ω *implies* $\mathfrak{d} = \omega_1$.

The proof of this theorem in [**4**] starts with the assumption that κ^* is homeomorphic to ω^* (where κ is regular uncountable). It makes a downstairs move that implicitly uses $\mathsf{wAC}(\omega_1, \omega_1)$ and then explicitly constructs a κ-scale in $({}^\omega\omega, <^*)$, i.e., a cofinal family of order type κ under the order $<^*$ of strict eventual domination. Recall that $\mathfrak{d}$ is the least (uncountable) cardinality of a cofinal family in this order. Thus Theorem 3.3 is established.

The case of general κ is then made in [**4**] to lead to a contradiction in every case except $\kappa = \omega_1$, as part of a sequence of proofs that culminates in the theorem mentioned in the paragraph following Problem 1.

There is a topological route to Theorem 3.3 via the following theorem:

Theorem 3.4. *Let κ be a regular cardinal. The following are equivalent.*

(1) *There is a κ-scale.*
(2) *The exterior of some (hence every) ω-oval in ω^* is a κ-oval.*

(3) *There is an ω-oval E and a family $\mathcal{C}$ of κ disjoint clopen sets in ω^* such that every clopen set containing E also contains all but $< \kappa$ members of $\mathcal{C}$, but also every subfamily of $< \kappa$ members of $\mathcal{C}$ is missed by some clopen set containing E.*

This theorem needs only ZF for the forward implications but the reverse implications both seem to require moves downstairs utilizing $\mathsf{wAC}(\kappa, \omega)$. The topological proof of Theorem 3.3 is finished by finding a pair E, $\mathcal{C}$ in ω_1^* that answers to the description in (3) of Theorem 3.4, with $\kappa = \omega_1$. The following pair needs nothing beyond ZF: $E = \bigcup_{n=0}^{\infty} A_n^*$ where $A_n = \{\alpha + n : \alpha \in \Lambda \cup \{0\}\}$ (where Λ stands for the set of countable limit ordinals) and $\mathcal{C} = \{C_\alpha : \alpha \in \Lambda\}$ where $C_\alpha = [\omega \cdot \alpha, \omega \cdot (\alpha+1))^*$. In particular, if K is a clopen set that meets uncountably many C_α, and $K = B^*$, then B meets uncountably may intervals $[\omega \cdot \alpha, \omega \cdot (\alpha+1))$ and so it meets some A_n in an uncountable set. Therefore, $K \cap E \neq \emptyset$, and any clopen set that contains E consequently must contain all but countably many C_α. Since we only moved downstairs with K, this much is true in ZF as long as one substitutes, if necessary, members of $\mathcal{P}(\omega_1)/[\omega_1]^{<\omega}$ for the clopen sets.

In contrast to this explicit example, the modification of (2) which puts ω_1^* in place of ω^* is actually equivalent to $\mathfrak{d} = \omega_1$ as will be explained in the next section. This may be behind the fact that I have been unable to show (3) implies (2) without going downstairs.

Theorem 3.5. Axiom Ω *implies that there is a dense ω_1-oval in ω^* whose complement does not contain any P-points.*

There do exist models where this consequence of Axiom Ω holds. In fact, in [**13**, 7.15] a proof attributed to Mary Ellen Rudin shows how the set Z of ultrafilters on ω which do not contain any sets of density 0 is nowhere dense in ω and is a P-set without P-points of ω^*. Under CH, the complement of Z is an ω_1-oval.

The key to Theorem 3.5 is that no uniform ultrafilter on ω_1 is a P-point of ω_1^*: take a sequence of partitions P_n of ω_1 into 2^n uncountable pieces such that the common refinement of P_n is the partition into singletons. Thus, in ω_1, the subspace $U(\omega_1)$ fits the description of the complement.

440? **Problem 3.** *Is there a model of $\mathfrak{d} = \omega_1$ in which there is a strong Q-sequence and also a dense oval as described in Theorem 3.5?*

4. Implications for ω_1^*

There is a variation on RH transfer for ω_1 that helps with the analysis of ω-ovals in ω_1^*. Let $\mathcal{A} = \{A_n : n \in \omega\}$ be a family of subsets of ω_1 such that $A_n^\# = A_n \setminus \bigcup_{i=0}^{n-1} A_i$ is uncountable for all n. An *RH-like transfer* of $\mathcal{A}$ to $\omega_1 \times \omega_1$ is a bijection $\psi : \omega_1 \to \omega \times \omega_1$ which does one of two things, depending on whether $A_\infty = \omega_1 \setminus \bigcup_{n=0}^{\infty} A_n$ is countable or uncountable. If it is countable, ψ distributes the elements of $\omega_1 \setminus \bigcup_{n=0}^{\infty} A_n$ into the bottom row $\omega \times \{0\}$, and sends $A_n^\#$ into the

$n+1$-st column $\{n\} \times \omega_1$. If A_∞ is uncountable, ψ sends it onto $\{0\} \times \omega_1$ and $A_n^\#$ onto $\{n+1\} \times \omega_1$.

This transfer is good for analyzing ω-ovals in ω_1^* that meet $U(\omega_1)$ in a noncompact subset. The first case represents ovals whose closure contains all of $U(\omega_1)$. If the exterior of such an oval is also an oval, then there is an almost-ascending sequence $\{A_\alpha : \alpha \in \omega_1\}$ of countable subsets of $\omega \times \omega_1$ such that every set that is almost disjoint from the columns of $\omega \times \omega_1$ is a subset of some A_α. This implies $\mathfrak{d} = \omega_1$, as a look at the traces of the A_α on $\omega \times \omega$ shows. The converse is also easy for those used to the arguments in **[5]** involving β and $\mathfrak{d}$. In the second case, where A_∞ is uncountable, one looks at sets of the form $A_\alpha \cup A_\infty$ to arrive at the same conclusions. Also, ω-ovals which do not meet $U(\omega_1)$ can be encapsuled in the remainder of a countable set; then, if $\mathfrak{d} = \omega_1$ we get the conclusion that *every* ω-oval in ω_1^* has an ω_1-oval exterior. Of course, Axiom Ω gives the same conclusion even more easily, thanks to Theorems 2.3 and 3.3. The former theorem also shows (1) implies (5) in the following theorem, and together with the modified RH transfer in this section it easily implies (4) is equivalent to (5). The other implications, all of which are very easy, are shown in **[4]**.

Theorem 4.1. *The following are equivalent.*

(1) Axiom Ω
(2) *Any two nonempty clopen subsets of* ω_1^* *are homeomorphic.*
(3) *There is an autohomeomorphism of* ω_1^* *that does not take* $U(\omega_1)$ *to itself.*
(4) *There is an autohomeomorphism of* ω_1^* *such that* $U(\omega_1)$ *is disjoint from its image.*
(5) *For any two* ω*-ovals in* ω_1^**, there is an autohomeomorphism of* ω_1^* *taking one to the other.*

We call an autohomeomorphism of ω^* or ω_1^* *nontrivial* if it cannot be induced by a 1-1 function from ω (resp. ω_1) to itself.

Corollary 1. Axiom Ω *implies that* ω_1^* *has nontrivial autohomeomorphisms.*

Problem 4. *Does* Axiom Ω *imply that* ω^* *has nontrivial autohomeomorphisms?* 441?

One might think that there are bijections from ω_1 to itself whose effect on ω_1^* cannot be mimicked by functions from ω to itself acting on ω^*, but appearances may be deceiving. The search for bijections without mimics is especially challenging in models where there are strong Q-sequences.

5. Some more open problems about Axiom Ω

Of the endless questions one might ask about the implications of Axiom Ω, the following seem especially natural to me:

Problem 5. *Does* Axiom Ω *place any restrictions on* 2^ω *besides the usual one (it cannot have countable cofinality) and the denial of* CH*?* 442?

Problem 6. *Does* Axiom Ω *imply that there are (or are not?!) P-points in* ω^**?* 443?

Theorem 3.5 shows that every P-point of ω_1^* is in $S(\omega_1)$.

444–447? **Problem 7.** *Does* Axiom Ω *have any implications for the small uncountable cardinals* $\mathfrak{a}$, $\mathfrak{i}$, $\mathfrak{r}$, $\mathfrak{u}$*?*

448–449? **Problem 8.** *Does* Axiom Ω *negate* ♣*?* †*?*

Axiom † ("*stick*") states that there is a family $\mathcal{A}$ of ω_1 countable subsets of ω_1 such that every *uncountable* subset of ω_1 contains some member of $\mathcal{A}$, while ♣ adds the condition that $\mathcal{A}$ is a ladder system.

Note that Axiom Ω + † implies $\mathfrak{r} = \omega_1$, since $\mathfrak{r}$ is the least cardinality of a π-base for a free ultrafilter on ω. [A family of sets witnessing † is a π-base for every uniform ultrafilter on ω_1.]

450? **Problem 9.** *Does* Axiom Ω *imply that there is a family of more than* ω_1 *disjoint clopen sets in* ω_1^**, each of which meets* $U(\omega_1)$*?*

If the answer to this problem is Yes, then so is the answer to Problem 8. In contrast, if Axiom Ω implies all disjoint clopen families of cardinality $\mathfrak{c}$ have (all but $< \mathfrak{c}$) members missing $U(\omega_1)$, we must look elsewhere than the density example for a mimic of $U(\omega_1)$ in ω^*, because that nowhere dense P-set Z is met by a family of $\mathfrak{c}$-many disjoint clopen subsets of ω^*. To see this, partition ω into two subsets, such that in both of them the ratio of numbers $< n$ to n gets arbitrarily close to both 0 and 1. Repeat this process countably many times, and diagonalize to get $\mathfrak{c}$-many almost disjoint subsets of ω in which this same phenomenon happens. Each one is in some member of Z.

451? **Problem 10.** *Does* Axiom Ω *have any implications for the cardinals in Cichoń's diagram [***12***] that are not below* $\mathfrak{d}$*?*

452? **Problem 11.** *Does* Axiom Ω *imply that there are no* ω_2*-ovals in* ω^**?*

453? **Problem 12.** *Does* Axiom Ω *imply that there is an* ω_1*-oval in* ω^* *whose exterior is also an* ω_1*-oval?*

In [**3**] it is shown (in effect) that $\mathfrak{t} = \omega_1$ is equivalent to there being a pair of disjoint ω_1-ovals in ω^* whose union is dense in ω^*, but there are models of $\mathfrak{t} = \omega_1$ where neither can be the exterior of the other. I am unaware of any such models where $\mathfrak{d} = \omega_1$, however.

If ♣ holds, the subspace $S(\omega_1)$ of ω_1^* can be split into two disjoint ω_1-ovals, each of which has all of $U(\omega_1)$ in its closure, making each one the exterior of the other. On the other hand, this is impossible if what is called $(*)$ in [**1**] holds. But in this latter case there may be ways of constructing disjoint ω_1-ovals inside A^* for some countable A, each of which is the exterior in A^* of the other; this is easy to do under CH, which is compatible with $(*)$. Then the union of one oval with the complement of A^* in ω_1^* is an ω_1-oval whose exterior in ω_1^* is the other oval.

6. Notes on Problem 2

Comparatively little research has been done on the implications of a Yes answer to Problem 2. Unlike with Problem 1, there is no end of pairs κ, λ that are candidates for an affirmative answer. Here we content ourselves with a few observations about the pair $\omega, \aleph_\omega$. Most of what we will say carries over to any other pair that is not eliminated by Theorem 2.1.

First, an argument similar to the construction of a Bernstein set shows that if $\mathcal{K}$ is a collection of κ sets of cardinality κ, then there is a pair of disjoint sets which meets each one in a set of cardinality κ. So if $U(\aleph_\omega)$ is homeomorphic to $U(\omega)$ $(= \omega^*)$, the reaping number $\mathfrak{r}$ is $\geq \aleph_{\omega+1}$.

The other observations have to do with the variety of dense ovals in $U(\aleph_\omega)$, summarized in the following theorem. In any model where it is homeomorphic to $U(\omega)$ $(= \omega^*)$, we get the same variety in ω^*, in marked contrast to the little we know about ovals in ω^* if Axiom Ω holds (see Problem 11).

Theorem 6.1. *$U(\aleph_\omega)$ has dense κ-ovals for all κ such that $\omega < \kappa < \aleph_\omega$ and also $\kappa = \mathfrak{b}$ and also for all regular κ between $\aleph_\omega$ and* $\min\{\mathrm{cf}[\aleph_\omega]^\omega, \aleph_{\omega_1}\}$.

For $\kappa = \omega_n$ $(n > 0)$, use $\aleph_\omega \times \omega_n$ and let $C_\alpha = \aleph_\omega \times \alpha$; a cofinality argument shows that every subset of $\aleph_\omega \times \omega_n$ of cardinality $\aleph_\omega$ meets some C_α in a set of cardinality $\aleph_\omega$. The remainders of the C_α in $U(\aleph_\omega)$ union up to a dense κ-oval.

For regular κ from $\aleph_\omega$ to the minimum of $\mathrm{cf}[\aleph_\omega]^\omega$ and $\aleph_{\omega_1}$, use the powerful result of pcf theory that there are subsequences of $\{\aleph_n : n \in \omega\}$ where the product has a scale of the desired length. Let C_α be the part of $\aleph_\omega \times \omega$ below the graph of f_α.

For $\kappa = \mathfrak{b}$ use $A_\omega \times \omega$ and functions f_α $\alpha < \mathfrak{b}$ that are constant on $[\omega_n, \omega_{n+1})$ and nondecreasing, and are well-ordered by the order of eventual domination. Here too, let C_α be the part of $\aleph_\omega \times \omega$ below the graph of f_α.

This last argument works for any κ for which there is a $<^*$-unbounded $<^*$-well-ordered family of increasing functions of cofinality κ in ω^ω.

References

[1] U. Abraham and S. Todorčević, *Partition properties of ω_1 compatible with CH*, Fund. Math. **152** (1997), no. 2, 165–181.

[2] B. Balcar and R. Frankiewicz, *To distinguish topologically the spaces m^*. II*, Bull. Acad. Polon. Sci. Sér. Sci. Math. Astronom. Phys. **26** (1978), no. 6, 521–523.

[3] A. Błaszczyk and A. Szymański, *Hausdorff's gaps versus normality*, Bull. Acad. Polon. Sci. Sér. Sci. Math. **30** (1982), no. 7–8, 371–378.

[4] W. W. Comfort, *Compactifications: recent results from several countries*, Topology Proc. **2** (1977), no. 1, 61–87 (1978).

[5] E. K. van Douwen, *The integers and topology*, Handbook of set-theoretic topology, North-Holland, Amsterdam, 1984, pp. 111–167.

[6] R. Frankiewicz, *To distinguish topologically the space m^**, Bull. Acad. Polon. Sci. Sér. Sci. Math. Astronom. Phys. **25** (1977), no. 9, 891–893.

[7] R. Frankiewicz, *Assertion Q distinguishes topologically ω^* and $\mathfrak{m}^*$ when $\mathfrak{m}$ regular and* $\mathfrak{m} > \omega$, Colloq. Math. **38** (1978), no. 2, 175–177.

[8] L. Gillman and M. Jerison, *Rings of continuous functions*, D. Van Nostrand Co., Princeton, NJ, 1960.

[9] S. H. Hechler, *A dozen small uncountable cardinals*, TOPO 72—general topology and its applications, Lecture Notes in Math., vol. 378, Springer, Berlin, 1974, pp. 207–218.

[10] J. Steprāns, *Strong Q-sequences and variations on Martin's axiom*, Canad. J. Math. **37** (1985), no. 4, 730–746.

[11] F. D. Tall, *Normality versus collectionwise normality*, Handbook of set-theoretic topology, North-Holland, Amsterdam, 1984, pp. 685–732.

[12] J. E. Vaughan, *Small uncountable cardinals and topology*, Open problems in topology, North-Holland, Amsterdam, 1990, pp. 195–218.

[13] R. C. Walker, *The Stone–Čech compactification*, Springer, New York, 1974.

Open Problems in Topology II – Edited by E. Pearl

First countable, countably compact, noncompact spaces

Peter J. Nyikos

The main problem in this article is one of my all-time favorites. To drum up interest in it, I announced at the 1986 Prague Topological Symposium (Toposym) that I was offering a US$500 prize for a solution during the following ten years [**9**]. There was essentially no progress on the problem all during those ten years, and so at the 1996 Toposym I raised the award to US$1000 during the following ten years. Those ten years have almost passed with no progress on the problem at all to the best of my knowledge, and I am hereby removing all time limits on the $1000 award and am contemplating raising it. Here is the problem that is the focus of all this largesse:

Problem 1. *Does* ZFC *imply the existence of a separable, first countable, countably compact, noncompact Hausdorff* (T_2) *space?* 454?

A mild put-down of general topology one hears from time to time is that there are too many adjectives in a typical problem or theorem. For me, however, one of the charms of general topology is that there are so many theorems and problems one can understand with no more than a typical undergraduate textbook in general topology as a resource. The adjectives used here definitely fall under that heading; the concepts are like second nature to many of us, and I have little mental pictures that I associate to each one to help keep arguments straight.

I will soon cut down on the number of adjectives in the alternative wording below, but the ones in the original wording are all implicitly there. The usual topology on ω_1 satisfies everything except separability. The Novak–Teresaka space described in Vaughan's article [**15**] satisfies everything except first countability. If one refines the cofinite topology on ω_1 by making initial segments open, then the resulting space satisfies everything except T_2 and is T_1. The remaining two properties are obviously necessary also to have an open problem. Also, the question mentions ZFC because there is a multitude of consistent examples of spaces as in Problem 1; see Sections 1 and 2. In fact, Problem 1 is one of a small but growing number of topological problems for which a negative answer is known to entail $(2^\omega =)\mathfrak{c} \geq \aleph_3$, yet $\mathfrak{c} = \aleph_3$ has not been ruled out. See Section 4.

For reasons having nothing to do with the aforementioned put-down, it is convenient to introduce the following concepts:

A space X is *ω-bounded* if every countable subset has compact closure, and *strongly ω-bounded* if every σ-compact subset has compact closure.

Also, after this paragraph, *space* will mean T_2-space. In fact, it could almost as easily mean T_3-space because of the well-known fact that T_2 implies T_3 for first countable, countably compact spaces [**1**, **2**]. (However, this doesn't work with "locally countable" in place of "first countable," as shown by an example of mine, mentioned in Vaughan's article [**15**] and done in detail in [**14**].) I could also have gone quite far in the opposite direction: an easy proof by contrapositive shows that every first countable space in which convergent sequences have unique limits is T_2. With this convention, we give the negative version of Problem 1 thus:

Problem 1′. *Is it consistent that every first countable, countably compact space is ω-bounded?*

Unlike with Problem 1, we have no consistency results either way for the following problem:

455? **Problem 2.** *Is there a first countable, ω-bounded space that is not strongly ω-bounded?*

The following ZFC example shows that first countability cannot be dropped from this problem.

Example. *Let p be a weak P-point in $\omega^* = \beta\omega \setminus \omega$ that is not a P-point. Then $\omega^* \setminus \{p\}$ is locally compact, ω-bounded, and not strongly ω-bounded.*

1. Consistent good examples for Problem 1

The best-known consistent examples for Problem 1 are the separable uncountable good spaces that are also discussed in the chapters by Juhász and Weiss [**7**] and by Vaughan [**15**].

A space is *good* if it is countably compact, locally countable, and T_3. A space is *splendid* if it is good and ω-bounded.

Clearly, a good space is noncompact iff it is uncountable, and so a good space is splendid iff every countable subset has countable closure. It follows that a ZFC example of good space that is not splendid would solve Problem 1. We do not know whether the added generality in Problem 1 is necessary—in other words, part (a) of the following problem is unsolved; so is part (b):

456–457? **Problem 3.** *Is it consistent that there is a countably compact, first countable space that is not ω-bounded and yet* (a) *every good space is splendid, or* (b) *every locally compact, countably compact, first countable space is ω-bounded?*

Clearly, every good space is locally compact. In [**11**] I gave a general construction of separable good spaces that are not splendid, referring to such spaces as *Ostaszewski–van Douwen spaces*. This is the Ostaszewski construction by induction explained in detail in Vaughan's article [**15**], with one modification stemming from the fact that we do not care whether the spaces are r-compact for some ultrafilter r.

At the αth step ($\alpha < \mathfrak{c}$) we have a locally compact, locally countable noncompact space $(\alpha, \mathcal{T}_\alpha)$ with ω as a dense set of isolated points. Having earlier listed all sequences s on $\mathfrak{c}$ in a $\mathfrak{c}$-sequence, we take the first sequence s_β without a cluster point; if there is none, then $(\alpha, \mathcal{T}_\alpha)$ is a good space. Otherwise, the crucial question is whether $(\alpha, \mathcal{T}_\alpha)$ is a wD space:

A space X is *wD* if for every infinite closed discrete subspace D there is an infinite $E \subset D$ for which there is a discrete family of open sets U_e such that $U_e \cap E = \{e\}$ for all $e \in E$.

If $(\alpha, \mathcal{T}_\alpha)$ is a wD space, then we select an infinite E as above, and choose U_e to be a compact (hence countable) open neighborhood of e, cutting down E if necessary so that when we make $\bigcup_{e\in E} U_e \cup \{\alpha\}$ the one-point compactification of $\bigcup_{e\in E} U_e$ and add α to $(\alpha, \mathcal{T}_\alpha)$, the resulting space $(\alpha+1, \mathcal{T}_{\alpha+1})$ remains noncompact. Once $\alpha \geq \omega_1$ this reason for cutting down E becomes obsolete (although we may have other reasons for cutting it down, see below) because $(\alpha, \mathcal{T}_\alpha)$ is automatically noncompact, being uncountable and locally countable. If we can define $(\alpha, \mathcal{T}_\alpha)$ for all $\alpha < \mathfrak{c}$ then we have a good space at the end, because every sequence has been given a cluster point. In the article by Vaughan [**15**], various models are given where the construction ends in a good space, either by continuing all the way to $\mathfrak{c}$ or stopping earlier with all sequences having cluster points. In particular, in any model where $\mathfrak{b} = \mathfrak{c}$, the construction can (if so desired) continue to stage $\mathfrak{c}$ [**3**, Theorem 13.4]. In some models we have no choice, such as models where $\mathfrak{p} = \mathfrak{c}$; see [**4**] where it is shown that every T_3 separable, countably compact space of Lindelöf number $< \mathfrak{p}$ is compact. In others we do have a choice (such as models where $\mathfrak{p} = \omega_1$, see below).

If at some point $(\alpha, \mathcal{T}_\alpha)$ is not a wD space, and we do not yet have a good space, then we have to scrap it and try again. It cannot be extended to a space for Problem 1 because of the following fact:

Theorem 1.1 ([**11**, Theorem 1.3]). *Every subspace of a first countable, countably compact space is a wD space.*

Moreover, in a certain sense, we *have* to modify various choices of U_e and/or the set to which E is cut down and/or the order in which s_β is listed: as explained in [**11**], every separable good noncompact space admits a construction such as we have gone through just now.

This is not to say that there might not be other ways of constructing the same space. In [**10**] there is a construction which begins with a *splendid* space of cardinality $\mathfrak{c}$ and repeatedly tears chunks from it, attaching the chunks to a copy of ω which will be dense in the intermediate spaces. As explained in [**11**], however, the same obstacle of an intermediate non-wD space might be encountered before we have a countably compact space on our hands. There is the added incovenience that in some models obtained using enormously large cardinals, all *splendid* spaces are of cardinality less than $\mathfrak{c}$; see Section 3.

A different alternative construction is in models of $\mathfrak{p} = \omega_1$, which is equivalent [**3**] to $\mathfrak{t} = \omega_1$, i.e., there is a decreasing mod-finite ω_1-tower on ω. This is a family $\{A_\alpha : \alpha \in \omega_1\}$ of infinite subsets of ω such that $A_\alpha \subset^* A_\beta$ whenever $\alpha > \beta$, and such that no infinite subset $\subset^* A_\alpha$ for all $\alpha < \omega_1$. Given such a family, one constructs a Franklin–Rajagopalan (FR) space as explained in Vaughan's article [**15**], a countably compact space with ω as a dense set of isolated points and set of non-isolated points homeomorphic to ω_1. Like all FR-spaces it is locally compact, and so it is a good space. As usual, it is defined "all in one go" in [**15**] but it could also be constructed by the Ostaszewski technique and provides us with an example where the construction ends before stage $\mathfrak{c}$ in any model of $\mathfrak{t} = \omega_1 < \mathfrak{c}$.

Yet another approach to constructing good spaces is in the last section of [**11**]: construct good spaces in ground models or intermediate models with a view to them being preserved in forcing extensions. Unfortunately, I know of no progress here since [**11**] was written; the results sketched there have not even been published yet.

2. Other consistent constructions for Problem 1

In this section we summarize constructions of separable, countably compact, first countable, noncompact spaces that are not good. None to date has given us examples in any model where there are good examples for Problem 1.

Example ([**11**, p. 139]). *Modify the Ostaszewski construction to begin with $2^\omega \times \omega$, to serve as a dense subspace for the rest of the inductive construction. At stage α we could take advantage of the fact that the union of the U_e-analogues is homeomorphic to $2^\omega \times \omega$ and to compactify it by identifying it with an open subspace of a copy $\mathbb{C}_\alpha$ of the Cantor set 2^ω, the rest of $\mathbb{C}_\alpha$ being disjoint from the space we have constructed thus far. The special case where there is only point in the rest of $\mathbb{C}_\alpha$ is especially close to the Ostaszewski construction, allowing for $[\omega, \alpha)$ to be the rest of the space at stage α.*

Example ([**8**, Example 3.11]). *Begin with an open ball in $\mathbb{R}^n$ $(n \geq 2)$ and recursively add copies of $[0, 1)$ in a way that makes the spaces we build into n-manifolds with the original open ball as a dense subspace.*

CH and a few forcing models are enough to give us countably compact noncompact manifolds (hence locally compact in addition to being first countable), but this seems too restrictive a method of constructing spaces for Problem 1. The following problem from [**11**], for example, is still unsolved.

458? **Problem 4.** *Is it consistent for there to be a countably compact manifold of weight $> \omega_1$?*

Lacking a Yes answer, we are stymied in all models of $\mathfrak{p} > \omega_1$, while $\mathfrak{p} = \omega_1$ is already enough to give us a good FR-space (see above). Section 6 of [**11**] details the main hurdles to any solution of Problem 4. The key problem is that wD is no longer good enough to continue the construction if it is not yet countably compact at stage α; one needs for there to be a subsequence of s_α contained in a closed copy of the closed ball minus a single point, and there are ZFC examples of weight ω_1 where there is no such subsequence.

Example ([**11**, Section 5]). *A countably compact, first countable linearly ordered space Y is attached to ω so that ω is a dense set of isolated points. If $\mathfrak{t} > \omega_1$ then Y is densely linearly ordered and nowhere locally compact.*

Making the whole space countably compact relies on the existence of numerous tight $(\omega_1, \mathfrak{c}^*)$-gaps and $(\mathfrak{c}, \omega_1^*)$-gaps. Some progress has been made in this direction—see the solution to Problem 10 of [**11**]—but the models involved have good solutions to Problem 1 in them, and so Problem 3 remains open.

3. Arbitrarily large first countable, locally compact, countably compact spaces

In this section we turn to some related problems which may involve solutions to Problem 1. The following problem was featured in [**9**]:

Problem 5. *Can there be an upper bound on the cardinalities of locally compact, first countable, countably compact spaces?* 459?

Without "locally compact" this would have an easy answer: take any regular cardinal and remove all its points of first countability. Problem 5 was originally motivated by Arhangel′skiĭ's famous solution to Alexandroff's old problem of whether there is an upper bound on the cardinalities of first countable, compact spaces. Arhangel′skiĭ showed that $\mathfrak{c}$ is the upper bound. Both Problem 5 and Problem 6 below are generalizations of Alexandroff's problem. Problem 5 is also a generalization of the second part of Problem 1 in Juhász and Weiss's article [**7**], which asks the same question (in negated form) of good spaces.

Problem 6. *Can there be an upper bound on the cardinalities of locally compact, first countable, ω-bounded spaces?* 460?

As far as we know, it may be consistent that $\mathfrak{c}$ is the upper bound in Problem 5 or Problem 6 as well; compare the first part of Problem 1 in Juhász's article. In any case, a positive solution even to Problem 6 would require the use of some very large cardinals.

Back when I first started thinking about Problem 5, I had not yet heard of the joint work of Juhász, Nagy and Weiss [**5**] which produced arbitrarily large splendid spaces, which are more than enough for a consistent No answer to Problem 5. We now know that their construction works under e.g., Covering(V, K); for details see the chapter by Juhász and Weiss [**7**]. Thus it is consistent with $\mathfrak{c}$ being anything reasonable and requires large cardinals for its negation.

On the other hand, we also know [**6**, **15**] that the Chang Conjecture variant $(\aleph_{\omega+1}, \aleph_\omega) \to (\aleph_1, \aleph_0)$ destroys all splendid spaces of size $\geq \aleph_\omega$. This variant, called the CCV below, has been shown consistent assuming a 2-huge cardinal. This use of the CCV has not been extended to a solution of Problem 6. The best we have so far is:

Theorem 3.1 ([**6**]). *If the* CCV *holds, then every locally compact, locally hereditarily Lindelöf, ω-bounded space is of Lindelöf degree $< \aleph_\omega$ and hence of cardinality $< \max\{\aleph_\omega, c^+\}$*

For convenience I will temporarily adopt the following expressions: A space is *amenable* if it is locally compact, locally hereditarily Lindelöf and countably compact and *fine* if it is locally compact, locally hereditarily Lindelöf and ω-bounded.

Clearly, every splendid space is fine and every good space is amenable. A corollary of Theorem 3.1 is that the CCV implies every amenable space of Lindelöf degree $> \aleph_\omega$ contains a separable, countably compact, noncompact subspace. This suggests that a ZFC construction of an amenable space of Lindelöf degree $\geq \aleph_{\omega+1}$ (in particular, a good space of cardinality $\geq \aleph_{\omega+1}$) would solve Problem 1. But it is conceivable that there may be one construction that works assuming the CCV, and another that works if the CCV fails, and which produces an ω-bounded space in some models. This may not be the end of the story, however. Every amenable space that is not ω-bounded contains a separable, noncompact closed subspace of cardinality $\leq \mathfrak{c}$. So if forcing is enough to destroy all such spaces, it seems plausible that a poset of modest size would be enough to do the trick. But the CCV is not destroyed by forcing by a set of cardinality lower than the first uncountable measurable cardinal. So it may not be a major step from the ZFC construction of an amenable space to an affirmative solution to Problem 1, or at least to a proof that large cardinals are needed to get a negative solution.

However, there is a far more sensational possibility: the hypothetical ZFC construction may actually be of a fine (perhaps even splendid) space, thereby showing that the CCV is inconsistent and hence so is the existence of 2-huge cardinals. Such a discovery would set off a major flurry of activity in large cardinal theory, as set theorists search for a natural lower bound for cardinals that are in jeopardy, so to speak.

On the other hand, the time may be ripe for lowering the large cardinal needed for the nonexistence of arbitrarily large fine (or at least splendid) spaces. A great deal has happened since [**6**] was published, including the discovery of Woodin cardinals and the equiconsistency of the Axiom of Determinacy (AD) with that of infinitely many Woodin cardinals. Recall that the consistency of AD was once thought to call for cardinals far larger than even 2-huge cardinals, and now it is known to call for something less than even a supercompact cardinal. If the consistency of nonexistence of arbitrarily large amenable spaces could be lowered this much, it would make their set-theoretic independence secure in the opinion of most set theorists.

A special case of *fine*, implicit in our next theorem [**12**], might give impetus to this quest. It uses the concept of a *Kurepa family*—a family $\mathcal{K}$ of denumerable sets which is uncluttered in the following way: for each countable $A \subset \bigcup \mathcal{K}$ the family $\mathcal{K} \restriction A = \{A \cap K : K \in \mathcal{K}\}$ is countable. A Kurepa family is called cofinal if it is $\subset$-cofinal in $[\bigcup \mathcal{K}]^{\omega}$.

Theorem 3.2. *Let κ be an infinite cardinal. The following are equivalent.*

(a) *There is a cofinal Kurepa family of cardinality κ.*

(b) *There is a locally metrizable, ω-bounded, 0-dimensional space of weight κ.*

The proof uses Stone Duality and the fact that a compact 0-dimensional space is metrizable iff it has at most countably many clopen sets.

461? **Problem 7.** *Can 0-dimensionality be dropped from Theorem 3.2?*

In Juhasz and Weiss's article [**7**] the problem is posed whether (a) is equivalent to there being a splendid space of cardinality κ. Note that the spaces described in (b) are intermediate between splendid and fine spaces. The simple structure of cofinal Kurepa families suggests that a lowering of consistency strength as above may be within reach. The following problem may be especially tractable:

Problem 8. *Is there a cofinal Kurepa family on* $\mathbb{R}$*?* 462?

4. Towards negative answers

In Section 1 we saw that there are good spaces that are not ω-bounded if either $\mathfrak{t} = \omega_1$ or $\mathfrak{b} = \mathfrak{c}$. This, together with the Pigeonhole Principle applied to the well known fact [**3**] that $\mathfrak{t} \leq \mathfrak{b}$, implies that there are good spaces if either $\mathfrak{c} = \omega_1$ or $\mathfrak{c} = \omega_2$. This means that a negative answer to Problem 1 cannot be obtained by iterated forcing with countable supports: because this very popular and sophisticated method of producing models of set theory makes $\mathfrak{c} \leq \omega_2$. The main alternative, finite support forcing, can lead to models of $\mathfrak{c} \geq \omega_3$, but only if a tail of the iteration consists of ccc posets. One drawback is that ccc posets are a rather restrictive class. The technique of mixed supports does allow for some non-ccc posets while still leading to models of $\mathfrak{c} \geq \omega_3$, but we still have no good mixed-support candidates for negative answers to Problem 1. Another drawback of ccc forcing with finite supports is that Cohen reals are added at each limit stage, and these produce ω_1-towers, all of which need to be destroyed if we want a counterexample to Problem 1. They *can* be destroyed without making $\mathfrak{b} = \mathfrak{c}$ or even $\mathfrak{b} = \mathfrak{d}$: Section 7 of [**11**] gives one simple way, but with some choices of ground model the final model still has good spaces that are not ω-bounded, and we still do not know whether all choices give good spaces.

On the other hand, models of $\mathfrak{c} = \omega_2$ have not been eliminated as candidates for negative answers to Problem 4, nor to the following problem:

Problem 9. *Is there a scattered, countably compact T_3 space that can be continuously mapped onto* $[0,1]$*?* 463?

While $\mathfrak{b} = \mathfrak{c}$ is enough to construct good examples for this problem [**3**, Theorem 13.4], no example at all has been constructed just assuming $\mathfrak{t} = \omega_1$ and we do not know whether $\mathfrak{c} = \omega_2$ is enough to construct one.

For countably compact Tychonoff spaces, admitting a continuous function onto $[0,1]$ is equivalent to having a non-scattered Stone–Čech compactification, and this is of interest in the geometry of Banach spaces of continuous functions [**13**]. Any example for Problem 9 contains a separable subspace that is also an example (just take the closure of any countable subspace whose image is the set of rational points in $[0,1]$). Since no compact scattered space can be mapped onto $[0,1]$, no example is ω-bounded.

All the (consistent) examples for Problem 9 thus far constructed are first countable, in fact good, and I conjecture that Problem 9 is reducible to the locally countable, hence good case, making it a special case of Problem 1. A minimality argument involving the Cantor-Bendixson derivatives of a scattered space shows that if there is an example for Problem 9, it has a subspace Y with the same properties in which each point has a neighborhood which does not admit of a continuous map onto $[0,1]$. Such a neighborhood meets at most countably many fibers $\pi^{-1}(r)$, since otherwise it would have a copy of the Cantor set in its image, and therefore it could be mapped onto $[0,1]$. If Y has a point all of whose neighborhoods are uncountable [in other words, if Y is not a good space] then every neighborhood of that point must meet some fiber $\pi^{-1}(r)$ in an uncountable subset; moreover, $\pi^{-1}(r)$ is of Lindelöf degree $\mathfrak{c}$ because there are $\mathfrak{c}$ disjoint crowded countable subsets of $[0,1]$ with r in their closure.

In contrast, it is very easy to construct a separable, locally countable, scattered, countably compact (T_2) space that can be mapped continuously onto $[0,1]$, using the technique of [**14**] which adds points to compactify countable discrete subspaces and uses the resulting copies of $\omega+1$ to define a weak base for the topology. This highlights the importance of first countability in Problem 1 and of the T_3 separation axiom in Problem 9. It appears that the natural techniques for producing separable, locally countable, countably compact spaces either tie up the whole space into one compact (and countable) package, or else they tie up countable subsets as loosely as possible. The intermediate situation, where a fairly tight but noncompact tying-up is required, is where the challenging problems lie.

References

[1] P. S. Alexandroff and P. S. Urysohn, *Mémoire sur les espaces topologiques compacts*, Verh. Kon. Akad. Wetensch. Amsterdam **14** (1929), no. 1, 96 pp.

[2] C. E. Aull, *A note on countably paracompact spaces and metrization*, Proc. Amer. Math. Soc. **16** (1965), 1316–1317.

[3] E. K. van Douwen, *The integers and topology*, Handbook of set-theoretic topology, North-Holland, Amsterdam, 1984, pp. 111–167.

[4] S. H. Hechler, *On some weakly compact spaces and their products*, General Topology and Appl. **5** (1975), 83–93.

[5] I. Juhász, Zs. Nagy, and W. Weiss, *On countably compact, locally countable spaces*, Period. Math. Hungar. **10** (1979), no. 2–3, 193–206.

[6] I. Juhász, S. Shelah, and L. Soukup, *More on countably compact, locally countable spaces*, Israel J. Math. **62** (1988), no. 3, 302–310.

[7] I. Juhász and W. Weiss, *Good, splendid and Jakovlev*, Open problems in topology II (E. Pearl, ed.), Elsevier Science.

[8] P. J. Nyikos, *The theory of nonmetrizable manifolds*, Handbook of set-theoretic topology, North-Holland, Amsterdam, 1984, pp. 633–684.

[9] P. J. Nyikos, *Progress on countably compact spaces*, General topology and its relations to modern analysis and algebra, VI (Prague, 1986), Heldermann Verlag, Berlin, 1988, pp. 379–410.

[10] P. J. Nyikos, *On first countable, countably compact spaces. II. Remainders in a van Douwen construction and P-ideals*, Topology Appl. **35** (1990), no. 2–3, 185–196.

[11] P. J. Nyikos, *On first countable, countably compact spaces. III. The problem of obtaining separable noncompact examples*, Open problems in topology, North-Holland, Amsterdam, 1990, pp. 127–161.

[12] P. J. Nyikos, *Generalized Kurepa and MAD families and topology*, 2005, Preprint.

[13] P. J. Nyikos and J. J. Schäffer, *Flat spaces of continuous functions*, Studia Math. **42** (1972), 221–229.

[14] P. J. Nyikos and J. E. Vaughan, *The Scarborough–Stone problem for Hausdorff spaces*, Topology Appl. **44** (1992), 309–316.

[15] J. Vaughan, *The Scarborough–Stone problem*, Open problems in topology II (E. Pearl, ed.), Elsevier Science.

Open Problems in Topology II – Edited by E. Pearl

Linearly Lindelöf problems

Elliott Pearl

Introduction

The regularity of cardinals plays a fine role in characterizations of compactness properties of topological spaces. While a Hausdorff space is compact if and only every infinite subset has a complete accumulation point, Lindelöf spaces cannot be characterized similiarly. [A point x is a *complete accumulation point* of a set A if $|A \cap U| = |A|$ for every neighborhood U of x. A space is *Lindelöf* if every open cover has a countable subcover.]

This role was demonstrated by P.S. Alexandroff and P.S. Urysohn who proved in their fundamental memoir [**1**, p. 17] that for a regular cardinal κ, the following properties are equivalent in a Hausdorff space: every subset of size κ has a complete accumulation point; every decreasing sequence of length κ of nonempty closed sets has nonempty intersection; every open cover of size κ has a subcover of strictly smaller size. It follows that if every uncountable subset of a space contains a complete accumulation point then the space is Lindelöf, and every Lindelöf space is *finally compact in the sense of accumulation points* (each uncountable subset of regular cardinality has a complete accumulation point). They noted that $\aleph_\omega$ with the order topology is Lindelöf but does not contain any complete accumulation points (in itself).

In 1962, A. Miščenko [**18**] proved that these properties are different by describing a space which is finally compact in the sense of accumulation points but not Lindelöf.

Meanwhile, in 1967 W.B. Sconyers considered properties of ordered open covers. A space is *linearly Lindelöf* if every increasing open cover $\{U_\alpha : \alpha \in \kappa\}$ has a countable subcover (by increasing, we mean that $\alpha < \beta < \kappa$ implies $U_\alpha \subseteq U_\beta$). According to Alexandroff and Urysohn's results, a space is finally compact in the sense of accumulation points if and only if it is linearly Lindelöf and it is the latter name that is now used popularly.

The main problem.

Question 1. *Is a normal, linearly Lindelöf space Lindelöf?* 464?

This is the famous open problem. A counterexample must be a Dowker space, as noted by Miščenko (see below). This problem, and its equivalent formulations, were discussed in Norman Howes' paper [**10**]. Mary Ellen Rudin has mentioned this problem in print several times.

Properties. A space is linearly Lindelöf iff every open cover contains a subcover of countable cofinality. A space is linearly Lindelöf iff whenever an open cover of cardinality κ has no subcover of cardinality $< \kappa$ then the cofinality of κ must be countable.

A linearly Lindelöf space is Lindelöf if and only it is countably metacompact. Every linearly Lindelöf space with a point-countable base is Lindelöf. Every linearly Lindelöf space with network weight strictly less than $\aleph_\omega$ is Lindelöf. In a linearly Lindelöf space, every closed discrete subspace is countable. Every meta-Lindelöf, linearly Lindelöf space is Lindelöf. Every locally metrizable, linearly Lindelöf space is hereditarily Lindelöf (hence separable metrizable too).

A.N. Karpov [**11**] proved that a countable product of Čech-complete, linearly Lindelöf spaces is linearly Lindelöf.

M. Matveev [**17**] proved that if a Tychonoff space Y is countably 1-paracompact in every Tychonoff space X that contains Y as a closed subspace, then Y is linearly Lindelöf. A subspace Y of X is *countably* 1*-paracompact in* X if for every countable open cover $\mathcal{U}$ of X there is an open cover $\mathcal{V}$ of X which refines $\mathcal{U}$ and is locally finite at the points of Y.

G. Gruenhage [**9**, Theorem 4.2] proved that if a space has countable extent and is the union of finitely many D-spaces, then the space is linearly Lindelöf.

P. Lipparini [**16**, Theorem 1.3] proved that if a product of spaces is linearly Lindelöf then all but at most countably many factors are compact.

Examples. A.S. Miščenko's [**18**] example is $\bigcup_{k=1}^{\infty}(\prod_{i=1}^{k}(\omega_i+1)\times\prod_{i=k+1}^{\infty}\omega_i)$ as a subspace of the product $\prod_{i=1}^{\infty}(\omega_i+1)$, where each ordinal ω_i has the order topology. It is Tychonoff, linearly Lindelöf but not Lindelöf. Miščenko also showed that every linearly Lindelöf space with a particular countable shrinking property (equivalent to countable metacompactness) is Lindelöf.

Another example of a Tychonoff, non-Lindelöf, linearly Lindelöf space was discovered independently by G. Gruenhage and R. Buzyakova. It is even a topological group. Here is a description. Let $D=\{0,1\}$ be the discrete two-point space. Consider the Tychonoff product $D^{\aleph_\omega}$ For each $x\in D^{\aleph_\omega}$, let A_x denote the set of all $\alpha\in\aleph_\omega$ such that the corresponding α-coordinate of x is 1. Now take the subspace of $D^{\aleph_\omega}$ consisting of all points x such that the cardinality of A_x is $<\aleph_\omega$.

Local compactness

In the proceedings of the Colloquium on Topology (Keszthely, 1972), L. Babai and A. Máté asked: *Is there a locally compact space which is not Lindelöf and is such that any of its subsets of a cardinality not cofinal to ω has a complete accumulation point?* They noted that without assuming that the space is locally compact the answer is yes. In [**3**–**5**], A.V. Arhangel'skiĭ and R. Buzyakova also asked if every locally compact, linearly Lindelöf space is Lindelöf.

ZFC examples. K. Kunen [**14**, **15**] constructed two Hausdorff, locally compact, linearly Lindelöf spaces which are not Lindelöf. Actually, Kunen proved two versions of the following theorem.

Theorem. *There is a compact Hausdorff space X with a point p such that the character of p in X is uncountable and equal to the weight of X, and for all regular uncountable κ, no κ-sequence of points distinct from p converges to p.*

Given such a compact Hausdorff space X and a point p, $X \setminus \{p\}$ is locally compact and linearly Lindelöf but not Lindelöf. In the first version, X is an inverse limit of the spaces $\beta\beth_n$ and the point p is a thread of weak $P_{\beth_n}$-points in $\beta\beth_n$. Here X has weight $\beth_\omega$. In the second version, X is an inverse limit of spaces of weight $< \aleph_\omega$ and the point p is a thread of weak $P_{\aleph_n}$-points. Here X has weight $\aleph_\omega$. The second version is obtained from the first by applying the elementary submodel method.

A conditional theorem. P. Nyikos [**20**] proved that it is consistent (relative to the existence of large cardinals) that every locally compact, linearly Lindelöf, normal space is Lindelöf. This result is a consequence of the theory of antidiamond principles developed by T. Eisworth and P. Nyikos [**8**]. In particular, the following axiom is consistent if it is consistent that there is a supercompact cardinal. It is a consequence of PFA but is also compatible with CH.

Axiom $\mathcal{P}$. *For every P-ideal $\mathcal{I}$ on an uncountable set X, either there is an uncountable subset A of X such that every countable subset of A is in $\mathcal{I}$, or X is the union of countably many sets $\{B_n : n \in \omega\}$ such that $B_n \cap I$ is finite for all n and for all $I \in \mathcal{I}$.*

A collection $\mathcal{I}$ of countable subsets of a set X is a *P-ideal* if it is downward closed with respect to $\subset$, closed under finite unions, and has the property that if $\{I_n : n \in \omega\} \in [\mathcal{I}]^\omega$ then there exists $J \in \mathcal{I}$ such that $I_n \subset^* J$ for all n.

In the context of locally compact Hausdorff spaces, this axiom can be applied to the ideal of countable subsets of X with compact closure.

First Trichotomy Theorem (Axiom $\mathcal{P}$). *Let X be a Hausdorff, locally compact space. Then at least one of the following is true: X is the countable union of ω-bounded subspaces, X has an uncountable closed discrete subspace, or X has a countable subset with non-Lindelöf closure.*

With the additional assumption that $\mathfrak{c} < \aleph_\omega$ this trichotomy theorem can be applied to show that every normal, locally compact, linearly Lindelöf space is Lindelöf. Note that in every known model of $\mathcal{P}$, $\mathfrak{c}$ is either $\aleph_1$ or $\aleph_2$.

Sequentially linearly Lindelöf spaces

M. Kojman and V. Lubitch [**12**] introduced the stronger notion of a sequentially linearly Lindelöf space. A space is *sequentially linearly Lindelöf* if it satisfies the following property for all uncountable regular cardinals $\kappa \leq w(X)$: for every $A \subset X$ of cardinality κ there exists a $B \subset A$ of cardinality κ that converges to a point $x \in X$. A set $B \subset X$ is said to *converge* to point $x \in X$ if $|B \setminus U| < |B|$ for all open sets U containing x.

Kojman and Lubitch proved that the existence of a good $(\aleph_\omega, \aleph_{\omega+1})$-scale implies the existence of a sequentially linearly Lindelöf topology on $\aleph_{\omega+1}$ which is not Lindelöf. This is done by extracting a subspace from Miščenko's space. This method is analogous to the method of Kojman and Shelah [**13**] of using an $(\aleph_\omega, \aleph_{\omega+1})$-scale (which always exists in ZFC) to extract a small Dowker space from Rudin's ZFC example of a Dowker space. Good $(\aleph_\omega, \aleph_{\omega+1})$-scales exist in many models of ZFC. In particular, since good $(\aleph_\omega, \aleph_{\omega+1})$-scales exist in L and are preserved by ccc and by $\aleph_1$-closed forcing, adding $\aleph_{\omega+1}$ Cohen reals to a model of $\mathsf{V} = \mathsf{L}$ followed by $\aleph_{\omega+2}$ subsets to $\aleph_1$ yields a model with $2^{\aleph_\omega} > \aleph_{\omega+1} = \mathfrak{c}$ and a sequential linearly Lindelöf topology on $\aleph_{\omega+1}$ that is not Lindelöf. Such a topology on $\aleph_{\omega+1}$, together with $\aleph_{\omega+1} = \mathfrak{c}$, can be used to produce a realcompact and linearly Lindelöf topology on $\aleph_{\omega+1}$ that is not Lindelöf.

465? **Question 2** (Kojman and Lubitch). *Can one prove in* ZFC *alone the existence of a sequentially linearly Lindelöf space that is not Lindelöf?*

Discretely Lindelöf spaces

In a survey of relative topological properties [**2**], Arhangel'skiĭ introduced this property (among others) related to Lindelöfness. X is *discretely Lindelöf* if the closure of every discrete subspace of X is a Lindelöf space. This property was originally called *strongly discretely Lindelöf.* Note that the analogous property of discrete compactness would coincide with compactness.

Arhangel'skiĭ proved the following lemma from which it follows that every discretely Lindelöf space is linearly Lindelöf: if X is discretely Lindelöf then every open cover whose cardinality does not have countable cofinality has a subcover of strictly smaller cardinality.

466–468? **Question 3** ([**2**, Problem 14]). *Is every regular (or Tychonoff, or normal), discretely Lindelöf space Lindelöf?*

In [**3**, Corollary 3.5], Arhangel'skiĭ and Buzyakova showed that every Tychonoff, countably tight, discretely Lindelöf space is Lindelöf. More generally [**3**, Theorem 3.6], every Tychonoff, discretely Lindelöf space of tightness less than $\aleph_\omega$ is Lindelöf.

469? **Question 4** (Arhangel'skiĭ). *Is every locally compact, discretely Lindelöf space Lindelöf?*

Estimating cardinality

In [**5**, Theorem 3.1], Arhangel'skiĭ and Buzyakova generalize Arhangel'skiĭ famous theorem to prove that the cardinality of every first countable, linearly Lindelöf space is at most $\mathfrak{c}$. In a later article, they prove a more general result ([**4**, Theorem 2.1]): if X is a linearly Lindelöf, sequential space, then $|X| \leq \mathfrak{c}$ iff $\psi(X) \leq \mathfrak{c}$. Here $\psi(X)$ is the pseudocharacter of X.

Z.T. Balogh proved a stronger result too. Call a space $[\kappa, \lambda]$-*linearly Lindelöf* if every cover by open sets increasing in well-order type $\leq \lambda$ has a subcover of cardinality $\leq \kappa$ ($\kappa < \lambda$ are infinite cardinals).

Theorem ([**6**, Theorem 3.2]). *Let κ be an infinite cardinal. Suppose that X is a $[\kappa, 2^\kappa]$-linearly Lindelöf (Tychonoff) space such that $t(X) \leq \kappa$, $\psi(X) \leq 2^\kappa$ and $|\overline{S}| \leq 2^\kappa$ for every $S \in [X]^{\leq\kappa}$. Then $|X| \leq 2^\kappa$.*

Balogh applies this theorem to prove [**6**, Corollary 3.7] that if X is a $[\aleph_0, \mathfrak{c}]$-linearly Lindelöf, sequential space with $\psi(X) \leq \mathfrak{c}$, then $|X| \leq \mathfrak{c}$.

Question 5 (Arhangel'skiĭ). *Is it true in* ZFC *that every first countable, ω_1-Lindelöf, Hausdorff space has cardinality at most $\mathfrak{c}$?* 470?

A space is ω_1-*Lindelöf* iff every open cover of cardinality $\aleph_1$ has a countable subcover. Under CH, ω_1-Lindelöf Tychonoff spaces of countable tightness are Lindelöf. Buzyakova [7, Corollary 3.5] proved that every first countable, ω_1-Lindelöf, Hausdorff space has Lindelöf degree at most $\mathfrak{c}$ (and therefore such spaces have size at most $2^{\mathfrak{c}}$).

Question 6 ([**7**, Problem 3.7]). *If X is a separable, ω_1-Lindelöf, Hausdorff space of countable tightness and countable pseudocharacter is it true that $|X| \leq \mathfrak{c}$?* 471?

Buzyakova [7, Theorem 3.10] proved that if X is a realcompact, ω_1-Lindelöf space of countable tightness and countable pseudocharacter then $|X| \leq \mathfrak{c}$.

The Hušek number

In [**3**], Arhangel'skiĭ and Buzyakova define the *Hušek number* of X, $\mathrm{Hus}(X)$, as the supremum of $\mathrm{Hus}(x, X)$, where x runs over X; $\mathrm{Hus}(x, X)$ is the smallest infinite cardinal κ such that for every subset $A \subset X$ such that $|A|$ is a regular cardinal not less than κ and A does not contain x, there is an open neighborhood U of x such that $|A \setminus U| = |A|$. Equivalently, $\mathrm{Hus}(x, X)$ is the smallest infinite cardinal κ such that if A, an infinite subset of X of regular cardinality, converges to x, then the cardinality of A is less than κ. The Hušek number is interesting because if X is a compact Hausdorff space and $x \in X$ then $\mathrm{Hus}(x, X) \leq \aleph_1$ iff $X \setminus \{x\}$ is linearly Lindelöf, and if $\mathrm{Hus}(x, X) = \aleph_0$ then x is an isolated point of X. Kunen's examples of locally compact, linearly Lindelöf, non-Lindelöf spaces are based on the construction of a compact Hausdorff space X that has a point x such that $\mathrm{Hus}(x, X) \leq \aleph_1$ and X is not necessarily first countable at x.

Question 7 ([**3**, Question 4]). *Let X be a compact Hausdorff space such that* $\mathrm{Hus}(X) \leq \aleph_1$. *Is it then true that the cardinality of X is not greater than $\mathfrak{c}$?* Yes, if CH. 472?

Question 8 ([**3**, Question 5]). *Is it true in* ZFC *that every compact Hausdorff space X such that* $\mathrm{Hus}(X) \leq \aleph_1$ *is sequential?* Yes, if MA. 473?

The following problem was considered by the authors but did not appear in their article [**3**].

Question 9 (Arhangel'skiĭ and Buzyakova). *If a compactum has a lineary Lindelöf co-diagonal is it then metrizable?* 474?

A consistent realcompact example

Question 10. *Is there a* ZFC *example of a realcompact, linearly Lindelöf space that is not Lindelöf?* 475?

The Kojman–Lubitch construction is a consistent example.

Arhangel′skiĭ and Buzyakova [**3**, Example 15] had contructed a consistent example (under $\mathfrak{c} = 2^{\aleph_\omega}$) of a hereditarily realcompact, linearly Lindelöf, Tychonoff topology on $[0,1]$ that is not Lindelöf. Let X denote the Buzyakova–Gruenhage example. The idea of the proof is to construct a subspace H of $X \times [0,1]$ such that the projection on the first coordinate maps H onto X while the projection of H on the second coordinate is one-to-one (and even onto). The points of H are chosen by a transfinite recursion of length $\mathfrak{c}$ and the construction requires $\mathfrak{c} = 2^{\aleph_\omega}$.

A consistent first countable example

Assuming $\mathsf{MA} + \aleph_\omega < \mathfrak{c}$, O. Pavlov [**21**] constructed a finer-than-usual topology on a subset of the Cantor set that is first countable, linearly Lindelöf but not Lindelöf. This gives a negative answer to [**4**, Question 3] and [**3**, Question 9].

476–477? **Question 11** ([**21**, Question 4]). *Does the existence of a first countable, linearly Lindelöf, not Lindelöf space follow from $\aleph_\omega < \mathfrak{c}$? From $2^{\aleph_\omega} = \mathfrak{c}$?*

478? **Question 12** ([**21**, Question 5]). *Does the existence of linearly Lindelöf, not Lindelöf space of countable pseudocharacter follow from $\aleph_\omega < \mathfrak{c}$?*

479? **Question 13** ([**21**, Question 6]). *Is there a pseudocompact, first countable, linearly Lindelöf space that is not Lindelöf (equivalently, not compact)?*

Lindelöf problems

Which additional conditions force a linearly Lindelöf space to be Lindelöf?

480? **Question 14** ([**4**, Question 1], [**4**, Question 3]). *Is it true in* ZFC *that every locally compact, first countable, linearly Lindelöf space is Lindelöf?*

481? **Question 15** ([**4**, Question 4]). *Is there a nonmetrizable, linearly Lindelöf space with a base of countable order? Equivalently, is every linearly Lindelöf space with a base of countable order Lindelöf?*

482? **Question 16** (Arhangel′skiĭ and Buzyakova). *Is every realcompact, locally compact, linearly Lindelöf space Lindelöf?*

Note that Kunen's example is pseudocompact.

483? **Question 17** (J.T. Moore). *Are there two Lindelöf spaces whose product is linearly Lindelöf but not Lindelöf?*

Moore [**19**] gave examples if $\mathfrak{c} > \aleph_\omega$. This is in fact an equivalent of $\mathfrak{c} > \aleph_\omega$ if one of the factors is required to be a separable metric space.

484? **Question 18** (J.T. Moore). *Is there a linearly Lindelöf, non-Lindelöf space of size $\aleph_\omega$?*

485? **Question 19** (J.T. Moore). *Is there a normal, linearly Lindelöf, non-Lindelöf space of weight $\aleph_\omega$?*

486? **Question 20** (Arhangel′skiĭ). *Is the product of any two (or countably many) linearly Lindelöf p-spaces Lindelöf?*

Acknowledgments

I thank R. Buzyakova, K.P. Hart, M. Kojman, K. Kunen, J. Moore, P. Nyikos and O. Pavlov for help in preparing this survey.

References

[1] P. S. Alexandroff and P. S. Urysohn, *Mémoire sur les espaces topologiques compacts*, Verh. Kon. Akad. Wetensch. Amsterdam **14** (1929), no. 1, 96 pp.

[2] A. V. Arhangel'skiĭ, *A generic theorem in the theory of cardinal invariants of topological spaces*, Comment. Math. Univ. Carolin. **36** (1995), no. 2, 303–325.

[3] A. V. Arhangel'skiĭ and R. Z. Buzyakova, *Convergence in compacta and linear Lindelöfness*, Comment. Math. Univ. Carolin. **39** (1998), no. 1, 159–166.

[4] A. V. Arhangel'skiĭ and R. Z. Buzyakova, *On some properties of linearly Lindelöf spaces*, Topology Proc. **23** (1998), Summer, 1–11.

[5] A. V. Arhangel'skiĭ and R. Z. Buzyakova, *On linearly Lindelöf and strongly discretely Lindelöf spaces*, Proc. Amer. Math. Soc. **127** (1999), no. 8, 2449–2458.

[6] Z. T. Balogh, *On density and the number of G_δ-points in somewhat Lindelöf spaces*, Topology Proc. **27** (2003), no. 1, 9–14.

[7] R. Z. Buzyakova, *Cardinalities of some Lindelöf and ω_1-Lindelöf T_1/T_2-spaces*, Topology Appl. **143** (2004), 209–216.

[8] T. Eisworth and P. J. Nyikos, *Antidiamond principles and topological applications*, 2002, Preprint.

[9] G. Gruenhage, *A note on D-spaces*, 2005, Preprint. `arXiv:math.GN/0503275`

[10] N. R. Howes, *Ordered coverings and their relationships to some unsolved problems in topology*, Proc. Wash. State Univ. conference on general topology, University of Wasington, Pullman, WA, 1970, pp. 60–68.

[11] A. N. Karpov, *Countable products of Čech-complete linearly Lindelöf and initially compact spaces*, Vestnik Moskov. Univ. Ser. I Mat. Mekh. (2000), no. 5, 7–9, Translation: Moscow Univ. Math. Bull. 55 (2000) no. 5, 5–7.

[12] M. Kojman and V. Lubitch, *Sequentially linearly Lindelöf spaces*, Topology Appl. **128** (2003), no. 2–3, 135–144.

[13] M. Kojman and S. Shelah, *A ZFC Dowker space in $\aleph_{\omega+1}$: an application of PCF theory to topology*, Proc. Amer. Math. Soc. **126** (1998), no. 8, 2459–2465.

[14] K. Kunen, *Locally compact linearly Lindelöf spaces*, Comment. Math. Univ. Carolin. **43** (2002), no. 1, 155–158.

[15] K. Kunen, *Small locally compact linearly Lindelöf spaces*, 2005, pp. 193–198. `arXiv:math.GN/0402066`

[16] P. Lipparini, *Compact factors in finally compact products of topological spaces*, 2006, pp. 1365–1382.

[17] M. V. Matveev, *Answering a question on relative countable paracompactness*, Questions Answers Gen. Topology **18** (2000), no. 2, 305–307. `arXiv:math.GN/0006196`

[18] A. Miščenko, *On finally compact spaces*, Dokl. Akad. Nauk SSSR **145** (1962), 1224–1227, Translation: Soviet Math. Dokl. 3 (1962) 1199–1202.

[19] J. T. Moore, *Some of the combinatorics related to Michael's problem*, Proc. Amer. Math. Soc. **127** (1999), no. 8, 2459–2467.

[20] P. J. Nyikos, *Normal, locally compact, linearly Lindelöf spaces may be Lindelöf*, 2002, Preprint.

[21] O. Pavlov, *A first countable linearly Lindelöf not Lindelöf topological space*, Houston J. Math., To appear.

Open Problems in Topology II – Edited by E. Pearl

Small Dowker spaces

Paul J. Szeptycki

A normal space whose product with the closed unit interval $\mathbb{I}$ is not normal is called a *Dowker space*. The spaces are named for Hugh Dowker who proved a number of characterizations of the class of spaces for which $X \times \mathbb{I}$ is normal. One of his motivations for studying this class of spaces was the study of the so-called insertion property: X has the insertion property whenever $f, g\colon X \to \mathbb{R}$ such that $f < g$, g lower semi-continuous and f upper semi-continuous, there is a continuous $h\colon X \to \mathbb{R}$ such that $f < h < g$. Dowker proved the following characterization ([**17**]).

Theorem. *For a normal space, the following are equivalent*

(1) X *is countably paracompact.*
(2) $X \times Y$ *is normal for all infinite, compact metric spaces* Y.
(3) $X \times Y$ *is normal for some infinite compact metric space* Y.
(4) X *has the insertion property.*

The homotopy extension property also fueled interest in normality of products $X \times \mathbb{I}$ (later Starbird [**46**] and Morita [**34**] showed that normality of X was sufficient to for the homotopy extension property).

Dowker was the first to raise the question when he asked (in a footnote to [**17**]): "It would be interesting to have an example of a normal Hausdorff space that is not countably paracompact." Indeed. The connection between normality and countable paracompactness is quite subtle and constructing a normal not countably paracompact space has turned out to be a difficult and deep problem.

The first few examples of Dowker spaces were all constructed by M.E. Rudin. In [**39**], assuming the existence of Suslin tree, she constructed a locally countable realcompact Dowker space of size $\aleph_1$. At the time Suslin's hypothesis was still open, but, nonetheless, this was the first construction of any kind of Dowker space. Moreover, it established a blueprint for the construction of many Dowker spaces to come. The construction is also fairly flexible and can be modified to obtain locally compact, first countable examples assuming the existence of a Suslin tree.

Rudin's next example was the first, and for many years only, ZFC Dowker space [**40**]. An easy to describe subspace of the box product $\Box\{\omega_n + 1 : 0 < n < \omega\}$, it is strongly collectionwise normal ([**23**]) and orthocompact ([**24**]), but it has very few other nice properties. For example, the cardinality of the space is ${\aleph_\omega}^{\aleph_0}$ and all other natural local and global cardinal functions on this space are equally large.

Although this example answered Dowker's original question, instead of closing the book on the Dowker space problem, it raised what has come to be known as the 'small Dowker space problem.' A small Dowker space means any Dowker space with some small local or global cardinal invariant. Here small is usually interpreted

The author acknowledges support from NSERC grant 238944.

as countable, ω_1 or less than or equal continuum. Each of the following problems was raised in one of Rudin's early papers and they are still open in ZFC:

487? **Problem 1.** *Is there a Dowker space of size* $\aleph_1$*?*

488? **Problem 2.** *Is there a first countable Dowker space?*

489? **Problem 3.** *Is there a separable Dowker space?*

490? **Problem 4.** *Is there a locally compact Dowker space?*

Since most of the consistent examples of Dowker spaces of size ω_1 are locally countable it is natural to ask whether such spaces exist in ZFC:

491? **Problem 5.** *Is there a locally countable Dowker space?*

Rudin's ZFC example and the questions it raised, led to a flurry of activity fueled by the set-theoretic developments of the 1970s and 1980s. During that time many very special consistent examples were constructed under a variety of different assumptions. In addition, a few other ZFC examples were obtained by modifying Rudin's example ([**16**, **19**, **25**, **29**, **30**]). No truly new ZFC example of any kind of Dowker space, small or large, was constructed until the celebrated example of Balogh published more than 20 years after Rudin's example [**3**]. Balogh's example is of size continuum and is the first ZFC example of a small Dowker space. While the example itself is, in the words of Jerry Vaughan "a milestone in set-theoretic topology" [**50**], more remarkable is the main technique invented for the construction. Balogh developed an elementary submodel technique based on ideas of M.E. Rudin. This technique is quite flexible (at least for Balogh) who previously used it to construct Q-set spaces, and later to solve Morita's conjectures [**4**, **6**]. He also employed it to construct a few other more specialized small Dowker spaces in ZFC: a screenable example [**5**] and a hereditarily collectionwise normal, hereditarily meta-Lindelöf, hereditarily realcompact example [**8**]. Perhaps the best reference for somebody wishing to learn the technique is the posthumous [**7**]. While all the examples are of size $2^{\aleph_0}$, none is first countable, locally compact, locally countable or separable. So although Balogh's examples are small Dowker spaces, they do not settle any of the main Questions 1–5 in ZFC.

Except for a few consistent examples constructed quite recently (that I will mention below) that is pretty much the short history of the small Dowker space problem to date. Returning to the main Questions 1–5, it is remarkable how few independence results have been obtained. There are a handful of independence results showing that certain constructions cannot yield ZFC examples, but there are very few quotable results. For example, PFA implies there is no hereditarily separable Dowker space (since all such examples are hereditarily Lindelof assuming PFA [**49**]). Slightly more technical, but also important is Balogh's result (see [**2**]) that under $\mathsf{MA} + \neg\mathsf{CH}$ there is no locally countable locally compact submetrizable Dowker space of size $\aleph_1$ (a space is *submetrizable* if it has a weaker metric topology). Note that an example announced in [**28**] assuming $\diamondsuit$ has all these properties. However, none of the modifications of Rudin's example, nor Balogh's examples (see the discussion in [**37**]) are submetrizable.

Problem 6. *Is there a submetrizable Dowker space?* 492?

Illustrating how weak our independence results seem, none rules out a single ZFC example giving a positive answer to Problems 1–5 simultaneously:

Problem 7. *Is there a* ZFC *example of a locally compact, locally countable (hence* 493? *first countable) separable Dowker space of size* $\aleph_1$*?*

This gives some support to conjecture positive answers to the main questions. Further evidence is given by an example of Chris Good [**22**]: assuming that no inner model contains a measurable cardinal, there is a locally compact, locally countable (hence first countable) zero-dimensional, collectionwise normal, σ-discrete Dowker space. Therefore, the consistency of no first countable Dowker space, or no locally compact Dowker space would require at least a measurable cardinal. Of course, Good's example is large in cardinality and sheds no light on the existence of a Dowker spaces of size $\aleph_1$, or a separable example.

Part of the difficulty in showing the consistency of no Dowker space of size $\aleph_1$ is that we have no natural set-theoretic formulation of the problem (compare with the S-space and L-space problems). This shortcoming was touched upon by the following question of Rudin [**42**]:

Problem 8. *Find a purely set-theoretic translation of the assertion: "There is a* 494? *Dowker space of size* $\aleph_1$*."*

Rudin also asked whether the existence of a Dowker space of size $\aleph_1$ implies the existence of a Souslin tree [**42**]. We now know the answer to this question is 'no' (there are models of CH without Suslin trees and CH implies the existence of some very nice Dowker spaces of size $\aleph_1$, e.g., [**28**]). However, it is possible to revive Rudin's question if we strengthen the hypotheses: (local compactness may be a particularly important assumption, see the discussion on locally compact Dowker spaces below). A positive answer to the following version of Rudin's question is possible (but would be surprising):

Problem 9. *Does the existence of (some kind of) locally compact (maybe add* 495? *locally countable, realcompact, σ-discrete) Dowker space of size* $\aleph_1$ *imply the existence of a Suslin tree?*

While there are many sufficient conditions established for the existence of certain small Dowker spaces, establishing any kind of necessary conditions would seem to require some new ideas.

It is also not known if either MA + ¬CH or PFA say something about Dowker spaces of size $\aleph_1$:

Problem 10. *Does* MA + ¬CH *or* PFA *imply that there are no Dowker spaces of* 496? *size* ω_1*?*

Barring a ZFC example of size $\aleph_1$, it is natural to ask what is the minimum cardinality of a Dowker space. For the purposes of brevity, let us denote this cardinal by $\mathfrak{D}$. Rudin's example is of cardinality ${\aleph_\omega}^\omega$ and Balogh's example is of size continuum, and it is consistent that $2^{\aleph_0} = \aleph_\omega{}^{\aleph_0}$. Kojman and Shelah constructed a subspace of Rudin's ZFC example using a scale in $\prod_{n\in\omega}\omega_n$ of cardinality $\aleph_{\omega+1}$ (such scales exist in ZFC [**45**]). Later Kojman and Lubitch constructed a similar example with even nicer properties [**29**]. These examples gives the current best absolute bound for $\mathfrak{D}$.

497? **Problem 11** (Kojman and Shelah; Watson). *Is it consistent that every Dowker space is of cardinality $\geq \aleph_{\omega+1}$?*

Balogh's example of size continuum shows that $\mathfrak{D}$ may be considered a cardinal invariant of the continuum. But is it a natural cardinal invariant? Are there any provable relations to the well-known cardinal invariants? This has been touched on before: Todorčević described an example of a Dowker space constructed from a Lusin set [**49**] and asked whether $\mathfrak{b} = \omega_1$ sufficed for the construction (it was shown in [**47**] that no). By a construction of Rudin, it is consistent with MA_{ω_3} that there is a Dowker space of cardinality ω_2 (this construction was described in [**40**] but later retracted since it was not known at the time whether the assumptions used were consistent—however, they are). So, if there is some provable inequality between a known cardinal invariant and $\mathfrak{D}$, the latter would need be the smaller. E.g., we can ask:

498? **Problem 12.** *Is there a Dowker space of cardinality $\leq \mathfrak{p}$?*

A curious fact is that of all the Dowker spaces constructed from CH (e.g., [**10, 12, 28, 47**]), none of them is locally compact. And of all the Dowker spaces constructed from some variant of $\diamondsuit$ (e.g, [**11, 13, 15, 21, 22, 27, 39, 43, 51**]) all are either locally compact or can easily be made so. This led Balogh and Nyikos to ask

499? **Problem 13.** *Does* CH *imply the existence of a locally compact Dowker space?*

Nyikos recently found a general method for constructing hereditarily normal, locally compact Dowker spaces. The construction appears in [**36**] where he discusses the locally compact Dowker space problem.In particular, assuming $\diamondsuit$ Nyikos constructs a locally compact, hereditarily normal, hereditarily separable, locally countable, hereditarily strongly collectionwise Hausdorff Dowker space of cardinality $\aleph_1$. The example can in addition be made so that its one-point compactification is Fréchet–Urysohn and satisfies the convergence property α_1 of Arhangel'skiĭ. In [**18**] it is shown consistent with CH that there are no such examples. Thus CH is not sufficient to construct some types of locally compact Dowker spaces. In particular it is consistent with CH that there are no locally compact, ω_1-compact Dowker spaces with the property that in the one point compactification, the point at infinity is a α_1-point.

Finally, we list a number of other small Dowker space problems that seem completely open. Most of these have been mentioned several times in the literature, the most important being:

500? **Problem 14.** *Is there a Dowker space with a σ-disjoint base?*

An equivalent question is whether a normal space with a σ-disjoint base is paracompact. This question is completely open. There could be a ZFC example but in Rudin's survey article from the first *Open Problems in Topology* she conjectures a theorem.

Closely related is the following problem due to Mike Reed:

Problem 15 (Reed). *If a normal space is a union of a countable family of open metrizable subspaces, must it be metrizable?* 501?

A counterexample would need be a Dowker space with a σ-disjoint base. Reed has a regular example in ZFC and can prove, consistently, that there are no counterexamples of size $< \mathfrak{c}$ [**38**].

A class of spaces still not well understood are para-Lindelof spaces. The following problem arose from Caryn Navy's examples of normal para-Lindelöf non-paracompact spaces constructed in [**35**]. All of them are countably paracompact raising the question:

Problem 16 (Tall; Watson). *Is there a para-Lindelof Dowker space?* 502?

The following question of Kemoto finds motivation from the fact that normal subspaces of finite products of ordinals are countably paracompact and whether countably paracompact subspaces of finite products of ω_1 are normal seems to be a difficult problem [**9**]:

Problem 17 (Kemoto). *Can ω_1^ω contain a Dowker subspace?* 503?

The result of Gardner–Pfeffer that if there are no real-valued measurable cardinals, then every normal, countably paracompact weakly theta-refinable space is realcompact [**20**] raises a natural Dowker space question: can countable paracompactness be eliminated from this result. A negative answer to the following question would show that countable paracompactness is needed.

Problem 18 (Nyikos). *Is Balogh's screenable Dowker space realcompact?* 504?

Another motivation for this question comes Robert Blair's question whether all perfectly normal spaces of cardinality less than the first measurable are real compact [**14**]. There are consistent counterexamples to Blair's question (e.g., Ostaszewski's space and the more recent [**26**]) but none in ZFC. Since Balogh's method for constructing ZFC Dowker spaces is also tailor-made to construct perfectly normal spaces, it would be interesting to know how to build in the failure of realcompactness using his method.

An old question of Michael and Arhangel′skiĭ asks whether points are G_δ sets in a regular symmetrizable space Given a symmetrizable Dowker space, adding a point at infinity in the natural way would provide a counterexample to the Michael–Arhangel′skiĭ problem.

Problem 19 (Davis). *Is there a symmetrizable Dowker space?* 505?

Mařík's theorem states that in a normal, countably paracompact space every Baire measure on X admits an extension to a closed-regular Borel measure [**33**].

So a space has come to be called a Mařík space if every Baire measure can be so extended. There are consistent examples to the following problems constructed in [**1**] but it is open whether there are ZFC examples.

506? **Problem 20.** *Is there a Mařík Dowker space?*

507? **Problem 21.** *Is there a normal space such that every Baire measure can be extended to a Borel measure but it is not Mařík? (It necessarily would be Dowker.)*

References

[1] J. M. Aldaz, *Borel extensions of Baire measures*, Fund. Math. **154** (1997), no. 3, 275–293.

[2] Z. T. Balogh, *Locally nice spaces under Martin's axiom*, Comment. Math. Univ. Carolin. **24** (1983), no. 1, 63–87.

[3] Z. T. Balogh, *A small Dowker space in ZFC*, Proc. Amer. Math. Soc. **124** (1996), no. 8, 2555–2560.

[4] Z. T. Balogh, *Nonshrinking open covers and K. Morita's third conjecture*, Topology Appl. **84** (1998), 185–198.

[5] Z. T. Balogh, *A normal screenable nonparacompact space in ZFC*, Proc. Amer. Math. Soc. **126** (1998), no. 6, 1835–1844.

[6] Z. T. Balogh, *Nonshrinking open covers and K. Morita's duality conjectures*, Topology Appl. **115** (2001), no. 3, 333–341.

[7] Z. T. Balogh, *A natural Dowker space*, Topology Proc. **27** (2003), no. 1, 1–7.

[8] Z. T. Balogh, *Dowker spaces*, Encyclopedia of general topology (K. P. Hart, J. Nagata, and J. E. Vaughan, eds.), Elsevier Science, 2004, pp. 354–356.

[9] Z. T. Balogh, T. Eisworth, G. Gruenhage, O. Pavlov, and P. Szeptycki, *Uniformization and anti-uniformization properties of ladder systems*, Fund. Math. **181** (2004), no. 3, 189–213.

[10] M. Bell, *On the combinatorial principle* $P(\mathfrak{c})$, Fund. Math. **114** (1981), no. 2, 149–157.

[11] A. Bešlagić, *A Dowker product*, Trans. Amer. Math. Soc. **292** (1985), no. 2, 519–530.

[12] A. Bešlagić, *Another Dowker product*, Topology Appl. **36** (1990), no. 3, 253–264.

[13] A. Bešlagić, *Yet another Dowker product*, Topology Appl. **56** (1994), no. 1, 35–43.

[14] R. L. Blair, *Spaces in which special sets are z-embedded*, Canad. J. Math. **28** (1976), no. 4, 673–690.

[15] P. de Caux, *A collectionwise normal weakly θ-refinable Dowker space which is neither irreducible nor realcompact*, Topology Proc. **1** (1976), 67–77.

[16] A. Dow and J. van Mill, *An extremally disconnected Dowker space*, Proc. Amer. Math. Soc. **86** (1982), no. 4, 669–672.

[17] C. H. Dowker, *On countably paracompact spaces*, Canad. J. Math. **3** (1951), 219–224.

[18] T. Eisworth and P. J. Nyikos, *Antidiamond principles and topological applications*, 2002, Preprint.

[19] W. G. Fleissner, T. LaBerge, and A. Stanley, *Killing normality with a Cohen real*, Topology Appl. **72** (1996), no. 2, 173–181.

[20] R. J. Gardner and W. F. Pfeffer, *Borel measures*, Handbook of set-theoretic topology, North-Holland, Amsterdam, 1984, pp. 961–1043.

[21] C. Good, *Dowker spaces, anti-Dowker spaces, products and manifolds*, Topology Proc. **20** (1995), 123–143.

[22] C. Good, *Large cardinals and small Dowker spaces*, Proc. Amer. Math. Soc. **123** (1995), no. 1, 263–272.

[23] K. P. Hart, *Strong collectionwise normality and M. E. Rudin's Dowker space*, Proc. Amer. Math. Soc. **83** (1981), no. 4, 802–806.

[24] K. P. Hart, *More on M. E. Rudin's Dowker space*, Proc. Amer. Math. Soc. **86** (1982), no. 3, 508–510.

[25] K. P. Hart, H. Junnila, and J. van Mill, *A Dowker group*, Comment. Math. Univ. Carolin. **26** (1985), no. 4, 799–810.

[26] F. Hernández-Hernández and T. Ishiu, *A perfectly normal nonrealcompact space consistent with* $\mathsf{MA}_{\aleph_1}$, Topology Appl. **143** (2004), 175–188.

[27] I. Juhasz, *Consistency results in topology*, Handbook of mathematical logic (J. Barwise, ed.), North-Holland, Amsterdam, 1977.

[28] I. Juhász, K. Kunen, and M. E. Rudin, *Two more hereditarily separable non-Lindelöf spaces*, Canad. J. Math. **28** (1976), no. 5, 998–1005.

[29] M. Kojman and V. Lubitch, *Sequentially linearly Lindelöf spaces*, Topology Appl. **128** (2003), no. 2-3, 135–144.

[30] M. Kojman and S. Shelah, *A ZFC Dowker space in* $\aleph_{\omega+1}$*: an application of PCF theory to topology*, Proc. Amer. Math. Soc. **126** (1998), no. 8, 2459–2465.

[31] T. Laberge, *ACRIN space, Ostaszewski spaces, and preservation of normality*, Ph.D. thesis, University of Kansas, 1992.

[32] T. LaBerge, *An ACRIN Dowker space*, Topology Proc. **18** (1993), 173–178.

[33] J. Mařík, *The Baire and Borel measure*, Czechoslovak Math. J. **7(82)** (1957), 248–253.

[34] K. Morita, *On generalizations of Borsuk's homotopy extension theorem*, Fund. Math. **88** (1975), no. 1, 1–6.

[35] C. Navy, *Paralindelöf versus paracompact*, Thesis, University of Wisconsin, 1981.

[36] P. J. Nyikos, *Hereditarily normal, locally compact Dowker spaces*, Topology Proc. **24** (1999), Spring, 261–276.

[37] P. J. Nyikos, *Applications of some strong set-theoretic axioms to locally compact* T_5 *and hereditarily scwH spaces*, Fund. Math. **176** (2003), no. 1, 25–45.

[38] G. M. Reed, *Some old favorites*, Contributed problems (Zoltán Balogh Memorial Topology Conference), 2002, `http://notch.mathstat.muohio.edu/balogh/`, pp. 19–20.

[39] M. E. Rudin, *Countable paracompactness and Souslin's problem*, Canad. J. Math. **7** (1955), 543–547.

[40] M. E. Rudin, *A normal space* X *for which* $X \times I$ *is not normal*, Fund. Math. **73** (1972), no. 2, 179–186.

[41] M. E. Rudin, *A separable Dowker space*, Symposia Mathematica, Vol. XVI (Roma, 1973), Academic Press, London, 1974, pp. 125–132.

[42] M. E. Rudin, *Souslin trees and Dowker spaces*, Topics in topology (Keszthely, 1972), Colloq. Math. Soc. János Bolyai, vol. 8, North-Holland, Amsterdam, 1974, pp. 557–562.

[43] M. E. Rudin, *A normal screenable nonparacompact space*, Topology Appl. **15** (1983), no. 3, 313–322.

[44] M. E. Rudin, *Dowker spaces*, Handbook of set-theoretic topology, North-Holland, Amsterdam, 1984, pp. 761–780.

[45] S. Shelah, *Cardinal arithmetic*, The Clarendon Press Oxford University Press, New York, 1994.

[46] M. Starbird, *The Borsuk homotopy extension theorem without the binormality condition*, Fund. Math. **87** (1975), no. 3, 207–211.

[47] P. J. Szeptycki, *A Dowker space from a Lusin set*, Topology Appl. **58** (1994), no. 2, 173–179.

[48] P. J. Szeptycki and W. Weiss, *Dowker spaces*, The work of Mary Ellen Rudin (Madison, WI, 1991), Ann. New York Acad. Sci., vol. 705, New York Acad. Sci., 1993, pp. 119–129.

[49] S. Todorčević, *Partition problems in topology*, American Mathematical Society, Providence, RI, 1989.

[50] J. Vaughan, *Review of Z. T. Balogh,* A small Dowker space in ZFC*, Proc. Amer. Math. Soc. 124 (1996) 2555–2560*, Math. Reviews (1996), 96j:54026.

[51] W. Weiss, *Small Dowker spaces*, Pacific J. Math. **94** (1981), no. 2, 485–492.

Open Problems in Topology II – Edited by E. Pearl

Reflection of topological properties to $\aleph_1$

Franklin D. Tall

Large cardinals exhibit *reflection* phenomena: roughly speaking, if some universal property holds below κ, it holds everywhere; alternatively, if there is an object with some property, there is one of size less than κ. A standard set-theoretic technique is to collapse a large cardinal κ to be *small*, e.g., $\aleph_1, \aleph_2$, or $2^{\aleph_0}$, and to see if some particular instance of reflection holds. The prototypical example of this is the proof in [**6**] that if a supercompact cardinal is Lévy-collapsed to ω_2, then for every regular cardinal $\lambda \geq \omega_2$ and every stationary set S of ω-cofinal ordinals in λ, there is an $\alpha < \lambda$ of cofinality ω_1 such that $S \cap \alpha$ is stationary in α. One naturally wonders whether such phenomena exist in topology. The simplest such interesting question—due to Fleissner [**13**]—is

Problem 1. *Is every first countable $\aleph_1$-collectionwise Hausdorff space collectionwise Hausdorff?* 508?

There are easy counterexamples if one omits "first countable." More interesting, but apparently harder to deal with is Hamburger's

Problem 2. *If X is a first countable space with every subspace of size $\aleph_1$ metrizable, is X metrizable?* 509?

However, Shelah has proven (among many other results in [**29**]):

Proposition 1. *There is a first countable nonmetrizable space of size $\aleph_2$ such that every subspace of size $\leq \aleph_1$ is metrizable if and only if there is a first countable Hausdorff space which is $\aleph_1$-collectionwise Hausdorff but is not $\aleph_2$-collectionwise Hausdorff.*

"First countable" is the topological analogue of "ω-cofinal". In proofs involving topological reflection, it is useful because it ensures that the *reflection* of a space is a subspace of the original space (Defining the reflection of a space requires getting further into large cardinals than we wish to do here. For a first countable space X sitting on some $\lambda \geq \kappa$, κ supercompact, say $X = \langle \lambda, \mathcal{T} \rangle$, it would just be the subspace $\langle \alpha, \mathcal{T} | \alpha \rangle$, for some $\alpha < \kappa$. See [**11**]).

The rationale for thinking at least some reflection questions are consistent is an intuition that we have so little comprehension of uncountable cardinals, that from our perspective, they all look like $\aleph_1$. Axioms expressing this intuition can be found in [**17**, **31**].

A third old chestnut is

Problem 3. *Is $2^{\aleph_0} = \aleph_2$ consistent with every normal Moore space being metrizable?* 510?

Research supported by NSERC grant A-7354.

It is not immediately obvious that the third problem is of the same genre as the first two, but it is. The standard—so far unsuccessful—technique for attacking these is to collapse a supercompact to ω_2; for the first two, via Lévy-collapse; for the third via the Mitchell-collapse [**1**, **26**] which combines the Lévy collapse and adding Cohen reals. Since after adding $\aleph_2$ Cohen reals, normal Moore spaces of size $\aleph_1$ are metrizable [**12**], a positive solution to Problem 3 as well as to Problem 2 could be obtained if one could positively answer

511? **Problem 4.** *Does countably closed forcing preserve nonmetrizability for first countable spaces?*

This would go via the standard "reflection plus preservation" argument, as in [**11**], Lévy- or Mitchell-collapsing a supercompact cardinal to ω_2.

Countably closed forcing *does* preserve nonmetrizability for spaces of size $\aleph_1$ [**23**]; this leads to the following result.

Proposition 2 ([**23**]). *Suppose there is a nonmetrizable X which can be made metrizable by countably closed forcing. Then every subspace of X of size* $\aleph_1$ *is metrizable.*

Unfortunately, while looking at [**29**] while preparing this note, I realized that Shelah had in his Sections 1.3 and 1.4 solved Problem 4 under a mild assumption. First of all, he constructs an example:

Proposition 3 ([**29**]). *Suppose there is a strong limit cardinal* λ *of uncountable cofinality such that* $2^\lambda = \lambda^+$. *Then there is a first countable Hausdorff space of size* λ^+ *which is not collectionwise Hausdorff, but is* $\aleph_1$*-collectionwise Hausdorff, and indeed has every subspace of size* $\aleph_1$ *metrizable.*

Shelah does not mention the last fact, but since the space is the union of a closed discrete subspace with a set of isolated points, the metrizability follows. (Incidentally, Shelah's definition of "collectionwise Hausdorff" in the paper is actually what is usually called "hereditarily collectionwise Hausdorff," but on a quick perusal, does not seem to affect the results he states.) He then remarks that collapsing λ^+ to $\aleph_1$ with countable conditions will make the space collectionwise Hausdorff, which—by the argument above—will make it metrizable.

Shelah's example makes it highly unlikely that Lévy-collapsing a supercompact to ω_2 via countable conditions will solve Problems 1 and 2, since one would somehow have to make use of GCH failing at cardinals of uncountable cofinality. Probably there are ZFC-counterexamples. With regard to Problem 3, someone should follow up on Watson's idea that Fleissner's proof [**15**] that CH implies that the existence of a normal nonmetrizable Moore space could possibly be modified to use only that $2^{\aleph_0} \leq \aleph_2$.

There are relevant partial results on preservation via countably closed forcing; for example, Shelah [**27**] proves by a variation of the proof in [**6**] alluded to earlier,

Lemma. *Countably closed forcing preserves noncollectionwise Hausdorffness for first countable spaces with local density* $\leq \aleph_1$.

This yields

Proposition 4 ([**27**]). *If it is consistent there is a supercompact cardinal, it is consistent that every first countable $\aleph_1$-collectionwise Hausdorff space of local density $\leq \aleph_1$ is collectionwise Hausdorff.*

There are other presentations and generalizations of this result, see [**10, 16**]. Studying the various proofs may give some insight into how to construct a counterexample, which will necessarily not be *locally* small.

Another partial result in which one throws in extra conditions so that noncollectionwise Hausdorffness is preserved is in [**23**], improving [**25**]:

Definition. A space is *stationarily θ-collectionwise Hausdorff* if for each closed discrete $\{x_\alpha : \alpha < \theta\}$, and each stationary $S \subseteq \theta$, there is a stationary $T \subseteq S$, such that $\{x_\alpha : \alpha \in T\}$ is separated.

Proposition 5 ([**23**]). *Let θ be a regular cardinal. If X is not stationarily θ-collectionwise Hausdorff, then X is not θ-collectionwise Hausdorff in any countably closed extension.*

This yields, for example

Proposition 6 ([**25**]). *If it's consistent there is a supercompact cardinal, it's consistent that every first countable $\aleph_1$-collectionwise Hausdorff space is weakly collectionwise Hausdorff.*

(Recall that a space is *weakly collectionwise Hausdorff* if every closed discrete subspace includes a separated (by disjoint open sets) subspace of the same cardinality.)

Actually, it is not certain that Junqueira's result is an improvement over that of Laberge and Landver. I do not know offhand of an example of say a stationarily $\aleph_1$-collectionwise Hausdorff space that is not $\aleph_1$-collectionwise Hausdorff. Speaking of examples, the standard example—easier than Shelah's—which shows that the reflection problems we have mentioned cannot have positive answers in ZFC is due to Fleissner [**14**]. $E(\omega_2)$ is the axiom which asserts that *there is a stationary $E \subseteq \{\alpha \in \omega_2 : \mathrm{cf}(\alpha) = \omega\}$, such that for no limit $\delta < \omega_2$ is $E \cap \delta$ stationary in δ.* The failure of $E(\omega_2)$ is equiconsistent with the existence of a Mahlo cardinal [**21**].

Example. *Given such an E, for each $\alpha \in E$, choose $s_\alpha \colon \omega \to \alpha$ to be a strictly increasing sequence converging to α. Let $D = \{s_\alpha|m : \alpha \in E, m \in \omega\}$. Let $X = E \cup D$. The points of D are taken to be isolated; the nth neighbourhood of $\alpha \in E$ is $N(\alpha, n) = \{\alpha\} \cup \{s_\alpha|m : m > n\}$. X is a locally compact, locally countable Moore space, which is $\aleph_1$-collectionwise Hausdorff but not (stationarily) $\aleph_2$-collectionwise Hausdorff. Every subspace of X of size $\leq \aleph_1$ is metrizable.*

See [**9, 25, 29**] for other examples of interest. Let us also note that connections between Problem 1 and problems concerning sequential fans are established in [**25**].

Balogh [**4, 5**] proposes a number of problems of the form, "if X is countably tight (or first countable) and every subspace of size $\leq \aleph_1$ has some covering or base property, does X?", and obtains some interesting partial results using Fleissner's Axiom R [**16**].

One of these, raised by several authors (see e.g., [**8**]) is particularly interesting:

512? **Problem 5.** *Is it consistent that if every subspace of size* $\leq \aleph_1$ *of a first countable space has a point-countable base, then the whole space does?*

As usual, a non-reflecting stationary set of ω-cofinal ordinals in ω_2 is a counterexample.

Aside from the standard approach [**6**, **27**] of collapsing a supercompact cardinal and proving a preservation lemma, an alternate approach is to Foreman–Laver-collapse a huge cardinal [**18**]. The flavour of this is to obtain an object of size $\aleph_2$ being the union of $\aleph_1$ nice subsets, if objects of size $\aleph_1$ are nice. A prototypical example is

Proposition 7 ([**31**]). *Assuming the consistency of a huge cardinal, there is a model in which every first countable* $\aleph_1$*-collectionwise Hausdorff space has the property that closed discrete subspaces of size* $\aleph_2$ *are the union of* $\aleph_1$ *subspaces, each of which is separated by disjoint open sets.*

The same idea yields a more obviously reflective result:

Proposition 8 ([**29**, **31**]). *Assuming the consistency of a huge cardinal, there is a model in which every first countable weakly* $\aleph_1$*-collectionwise Hausdorff space is weakly* $\aleph_2$*-collectionwise Hausdorff.*

In another direction, we have

Proposition 9 ([**31**]). *Assuming the consistency of a huge cardinal, there is a model in which every* $\aleph_1$*-collectionwise normal Moore space of size* $\aleph_2$ *is metrizable.*

It is not clear whether the limitation to $\aleph_2$ can be consistently removed, say by assuming some sufficiently large cardinal. See [**31**]. Proposition 3 tends to make one think not. Also, Shelah [**29**] has shown there is a first countable Hausdorff space which is $(2^{\aleph_0})^+$-weakly collectionwise Hausdorff, but is not weakly collectionwise Hausdorff. The difficulty first occurs at $\aleph_\omega$, assuming GCH, but, in any event, occurs at some singular cardinals of cofinality ω.

Assuming the existence of a huge cardinal κ with $j(\kappa)$ supercompact (see [**31**] for an explanation of "j" and applications of this stronger hypothesis), one can get both the power of the Foreman–Laver collapse (of a huge cardinal to ω_1) and the power of the Lévy-collapse of a supercompact cardinal to ω_2.

Another major problem which possibly involves reflection is Arhangel′skiĭ's:

513? **Problem 6.** *Is it consistent that every Lindelöf* T_2 *(*T_3*?) space with points* G_δ *has cardinality* $\leq 2^{\aleph_0}$*?*

By assuming the Continuum Hypothesis, which is reasonable in this context, we again have a question involving $\aleph_1$.

514? **Problem 7.** *Is it true (or consistent via countably closed forcing) that Lindelöf* T_2 *spaces with points* G_δ *cannot be made non-Lindelöf by countably closed forcing?*

If so, then in the "true" case and probably in the consistent one, Arhangel′skiĭ's problem can be settled affirmatively by Lévy-collapsing a supercompact cardinal to ω_2 [**32**].

There is a combinatorial version of Problem 7 which does not mention forcing [**32**].

Definition. A *covering tree* for a Lindelöf space X is a collection of open sets $\{U_f : f \in {}^{<\omega_1}\omega\}$, such that for each $\alpha \in \omega_1$ and each $f \in {}^{\alpha}\omega$, $\{U_{f\cup\{\langle\alpha,n\rangle\}} : n < \omega\}$ covers X. For $f \in {}^{\leq\omega_1}\omega$, the f-branch $\mathcal{B}_f$ is defined to be $\{U_{f|\beta} : \beta \in \operatorname{dom} f\}$.

Problem 7′. *Is it true that for each Lindelöf T_2 space X with points G_δ, that for each covering tree T of X, $\mathcal{B}_f$ is a cover for some $f \in {}^{<\omega_1}\omega$?*

A related problem also due to Arhangel′skiĭ is

Problem 8. *Does every Lindelöf first countable T_1 space have cardinality $\leq 2^{\aleph_0}$?* 515?

Relevant partial results are:

Proposition 10 ([**3**]). *There is no Lindelöf space with points G_δ of size greater than or equal to the first measurable cardinal.*

Proposition 11. *It's consistent (via countably closed forcing [**19, 22, 28**] or $\mathsf{V} = \mathsf{L}$ [**30, 33**]) that there is a Lindelöf 0-dimensional space with points G_δ of size $\aleph_2 = (2^{\aleph_0})^+$.*

Note that the claim by Morgan referred to in [**32**] that in L one could get such spaces of size $\aleph_n$, for every n, has been withdrawn.

Proposition 12 ([**28, 32**]). *By Lévy-collapsing a supercompact cardinal to ω_2 and then adding $\kappa \geq \aleph_3$ Cohen subsets of ω_1, one obtains a model in which there are no Lindelöf spaces with points G_δ of size $\geq \aleph_2$ and $< \kappa$.*

Another reflection problem involving Lindelöf spaces is due to Hajnal and Juhász [**20**]:

Problem 9. *Does every Lindelöf space have a Lindelöf subspace of size $\aleph_1$?* 516?

By adding topological hypotheses, one can get positive results [**7**]. There is a consistent 0-dimensional counterexample [**24**]. There are only very weak positive consistency results, assuming first countability (!) [**7**]. I bet there is a ZFC counterexample.

Presently available methods for dealing with the problems discussed have more or less reached a dead end. Perhaps positive consistency results can be obtained by a clever use of countably closed elementary submodels in order to prove forcing–preservation results, exploiting first countability or points G_δ. On the other hand, such dead ends are often a signal that there are counterexamples to be obtained by someone taking a fresh look at the problems.

References

[1] U. Abraham, *Aronszajn trees on* $\aleph_2$ *and* $\aleph_3$, Ann. Pure Appl. Logic **24** (1983), no. 3, 213–230.

[2] A. V. Arhangel'skiĭ, *The power of bicompacta with first axiom of countability*, Dokl. Akad. Nauk SSSR **187** (1969), 967–970, Translation: Soviet Math. Dokl. 10 (1969) 951–955.

[3] A. V. Arhangel'skiĭ and V. I. Ponomarev, *Fundamentals of general topology*, D. Reidel Publishing Co., Dordrecht, 1984.

[4] Z. T. Balogh, *Locally nice spaces and axiom R*, Topology Appl. **125** (2002), no. 2, 335–341.

[5] Z. T. Balogh, *Reflecting point-countable families*, Proc. Amer. Math. Soc. **131** (2003), no. 4, 1289–1296.

[6] J. E. Baumgartner, *A new class of order types*, Ann. Math. Logic **9** (1976), no. 3, 187–222.

[7] J. E. Baumgartner and F. D. Tall, *Reflecting Lindelöfness*, Topology Appl. **122** (2002), no. 1–2, 35–49.

[8] P. J. Collins, G. M. Reed, and A. W. Roscoe, *The point-countable base problem*, Open problems in topology, North-Holland, Amsterdam, 1990, pp. 237–250.

[9] A. Dow, *An empty class of nonmetric spaces*, Proc. Amer. Math. Soc. **104** (1988), no. 3, 999–1001.

[10] A. Dow, *An introduction to applications of elementary submodels to topology*, Topology Proc. **13** (1988), no. 1, 17–72.

[11] A. Dow, F. D. Tall, and W. Weiss, *New proofs of the consistency of the normal Moore space conjecture. I*, Topology Appl. **37** (1990), no. 1, 33–51.

[12] A. Dow, F. D. Tall, and W. Weiss, *New proofs of the consistency of the normal Moore space conjecture. II*, Topology Appl. **37** (1990), no. 2, 115–129.

[13] W. G. Fleissner, *On* λ *collection Hausdorff spaces*, Topology Proc. **2** (1977), no. 2, 445–456.

[14] W. G. Fleissner, *Separation properties in Moore spaces*, Fund. Math. **98** (1978), no. 3, 279–286.

[15] W. G. Fleissner, *If all normal Moore spaces are metrizable, then there is an inner model with a measurable cardinal*, Trans. Amer. Math. Soc. **273** (1982), no. 1, 365–373.

[16] W. G. Fleissner, *Left separated spaces with point-countable bases*, Trans. Amer. Math. Soc. **294** (1986), no. 2, 665–677.

[17] M. Foreman, *Potent axioms*, Trans. Amer. Math. Soc. **294** (1986), no. 1, 1–28.

[18] M. Foreman and R. Laver, *Some downwards transfer properties for* $\aleph_2$, Adv. Math. **67** (1988), no. 2, 230–238.

[19] I. Gorelic, *The Baire category and forcing large Lindelöf spaces with points* G_δ, Proc. Amer. Math. Soc. **118** (1993), no. 2, 603–607.

[20] A. Hajnal and I. Juhász, *Remarks on the cardinality of compact spaces and their Lindelöf subspaces*, Proc. Amer. Math. Soc. **59** (1976), no. 1, 146–148.

[21] L. Harrington and S. Shelah, *Some exact equiconsistency results in set theory*, Notre Dame J. Formal Logic **26** (1985), no. 2, 178–188.

[22] I. Juhász, *Cardinal functions. II*, Handbook of set-theoretic topology, North-Holland, Amsterdam, 1984, pp. 63–109.

[23] L. R. Junqueira, *Upwards preservation by elementary submodels*, Topology Proc. **25** (2000), Spring, 225–249.

[24] P. Koszmider and F. D. Tall, *A Lindelöf space with no Lindelöf subspace of size* $\aleph_1$, Proc. Amer. Math. Soc. **130** (2002), no. 9, 2777–2787.

[25] T. LaBerge and A. Landver, *Reflection and weakly collectionwise Hausdorff spaces*, Proc. Amer. Math. Soc. **122** (1994), no. 1, 291–302.

[26] W. Mitchell, *Aronszajn trees and the independence of the transfer property*, Ann. Math. Logic **5** (1972), 21–46.

[27] S. Shelah, *Remarks on* λ*-collectionwise Hausdorff spaces*, Topology Proc. **2** (1977), no. 2, 583–592 (1978).

[28] S. Shelah, *On some problems in general topology*, Set theory (Boise, ID, 1992–1994), Contemp. Math., vol. 192, American Mathematical Society, Providence, RI, 1996, pp. 91–101.

[29] S. Shelah, *Remarks on $\aleph_1$-CWH not CWH first countable spaces*, Set theory (Boise, ID, 1992–1994), Contemp. Math., vol. 192, American Mathematical Society, Providence, RI, 1996, pp. 103–145.

[30] S. Shelah and L. Stanley, *S-forcing. I. A "black-box" theorem for morasses, with applications to super-Souslin trees*, Israel J. Math. **43** (1982), no. 3, 185–224.

[31] F. D. Tall, *Topological applications of generic huge embeddings*, Trans. Amer. Math. Soc. **341** (1994), no. 1, 45–68.

[32] F. D. Tall, *On the cardinality of Lindelöf spaces with points G_δ*, Topology Appl. **63** (1995), no. 1, 21–38.

[33] D. J. Velleman, *Morasses, diamond, and forcing*, Ann. Math. Logic **23** (1982), no. 2–3, 199–281 (1983).

Open Problems in Topology II – Edited by E. Pearl

The Scarborough–Stone problem

Jerry E. Vaughan

1. Introduction

In their paper from 1966, C.T. Scarborough and A.H. Stone posed the following problem:

Problem 1.1 (Scarborough–Stone [**37**]). *Must every product of sequentially compact spaces be countably compact?* 517?

This problem has been restated many times (e.g., [**44**, **46**]), yet major portions of it remain open. For instance

Problem 1.2. *In* ZFC, *is there a family of sequentially compact Tychonoff spaces whose product is not countably compact?* 518?

Problem 1.2 is stated differently from Problem 1.1 because under the assumptions of various set-theoretic conditions, there are families of sequentially compact Tychonoff spaces (even T_6) whose product is not countably compact. In this article, we discuss the current state of knowledge about the Scarborough–Stone problem.

Terms not defined here can be found in the references [**3**, **15**, **44**]). By a T_5-space we mean a hereditarily normal space. By a T_6-space we mean a perfectly normal space. We assume all spaces are Hausdorff.

The following definitions will be central to this article. A space is *countably compact* provided every countable open cover of X has a finite subcover (equivalently, if every sequence has a cluster point). A space is *sequentially compact* provided every sequence has a convergent subsequence. For $r \in \omega^*$, $x \in X$ and $s\colon \omega \to X$, we say that x is the *r-limit* of s provided for every neighborhood U of x, $\{n \in \omega : s(n) \in U\} \in r$ (in a Hausdorff space the r-limit of a sequence is unique). A space is *r-compact* if every sequence has an r-limit. A space is *ω-bounded* provided every countable set has compact closure or, equivalently, if the space is r-compact for every $r \in \omega^*$.

While countable compactness is not productive (see Section 2 below), both r-compactness and ω-boundedness are productive properties.

Theorem 1.3 (Bernstein [4]). *Every product of r-compact spaces is r-compact, hence countably compact.*

Corollary 1.4. *Every product of ω-bounded spaces is ω-bounded, hence countably compact.*

Also, if we place a restriction on the number of factors in the product, we have the following Theorem which gives additional motivation for Problem 1.2.

Theorem 1.5. (a) *A product of countably many sequentially compact spaces is sequentially compact.* (b) (Scarborough–Stone) *The product of at most ω_1 sequentially compact spaces is countably compact.*

For $A, B \in [\omega]^\omega$ we write $A \subset^* B$ provided $A \setminus B$ is finite (the mod-finite order). A *tower* on ω is an ordered family of infinite subsets of ω, $\{A_\alpha : \alpha < \kappa\}$, such that for every $\alpha < \beta$, $A_\beta \subset^* A_\alpha$, and there does not exist an infinite $A \subset \omega$ such that $A \subset^* A_\alpha$ for all $\alpha < \kappa$.

We finish the introduction by recalling a few more definitions.

An ultrafilter r is called a *T-point* provided it contains a tower. The cardinal $\mathfrak{t}$ is defined to be the smallest cardinal κ such that there exists a tower indexed by κ (see [**8**]). In any space a closed nowhere dense set H is called a *T_κ-set* provided H is the intersection of a decreasing chain of κ many clopen neighborhoods of H. We will also say that H is a *T-set* provided H is a T_κ-set for some $\kappa \geq \omega_1$. In this paper, we only consider these notions in the space ω^*.

We also need the mod finite order on ${}^\omega\omega$: We say $f <^* g$ provided $\{n \in \omega : g(n) < f(n)\}$ is finite. The cardinal $\mathfrak{b}$ is defined to be the smallest cardinal κ such that there exists an unbounded family of size κ in ${}^\omega\omega$ (see [**8**]).

The proof of Theorem 1.5 also shows that (a) the product of fewer than $\mathfrak{t}$ sequentially compact spaces is sequentially compact, and (b) the product of no more than $\mathfrak{t}$ sequentially compact spaces is countably compact. See [**8**, 13.1].

2. How to construct a product that is not countably compact

We now describe a technique for constructing a family of spaces whose product is not countably compact.

A family of pairs $\{(X_\alpha, s_\alpha) : \alpha < \kappa\}$, where X_α is a space, and s_α is a sequence in X_α is called an *SS-family* provided for every $r \in \omega^*$ there exists $\alpha < \kappa$ such that s_α has no r-limit in X_α.

Lemma 2.1. *A product $\Pi_{\alpha<\kappa} X_\alpha$ is not countably compact if and only if for all $\alpha < \kappa$ there exists a sequence $s_\alpha \colon \omega \to X_\alpha$ such that $\{(X_\alpha, s_\alpha) : \alpha < \kappa\}$ is an SS-family.*

It follows that the Scarborough–Stone problem reduces to the problem of constructing an SS-family of sequentially compact spaces.

We illustrate the use of an SS-family with a well known example.

Example 2.2 (Novak–Terasaka [**25**, **40**]). *There exist two countably compact Tychonoff spaces X_0, X_1 whose product is not countably compact.*

The spaces X_0 and X_1 are constructed so that that both are subsets of $\beta\omega$, and $X_0 \cap X_1 = \omega$. Let s denote the identity function on ω. Then s has the property that for every $r \in \beta\omega$, the r-limit of s exists (and obviously equals r) and is in X_0 or X_1 but not both. It follows that the family $\{(X_0, s), (X_1, s)\}$ is an SS-family. By Lemma 2.1, $X_0 \times X_1$ is not countably compact.

We know only two ways to construct (consistent) SS-families of sequentially compact spaces: the Ostaszewski construction (§ 3) and the Franklin–Rajagopalan construction (§ 4). Each construction uses a set-theoretic property that is not a theorem of ZFC; so neither construction can produce a counterexample to Problem 1.1 in ZFC. Each construction works in some models where the other does not, but combining the two constructions will not produce a counterexample in ZFC either (see §6).

3. The Ostaszewski construction

Fix $r \in \omega^*$. We discuss the Ostaszewski construction [**33**, **43**] of a sequentially compact space which is not r-compact. The construction produces zero-dimensional Hausdorff spaces, hence Tychonoff spaces. The underlying set in the Ostaszewski construction is $\mathfrak{c}$, and the construction proceeds by induction on $\mathfrak{c}$. At each step in the induction one has a topology $\mathcal{T}_\alpha$ on an ordinal $\alpha < \mathfrak{c}$ such that $\mathcal{T}_\alpha$ is a locally compact, locally countable, Hausdorff space (hence first countable and zero-dimensional). In the space $X_\alpha = (\alpha, \mathcal{T}_\alpha)$ there will be a next sequence s with no cluster point (else we are done). Ostazwewski's construction defines a topology $\mathcal{T}_{\alpha+1}$ on $\alpha + 1$ so that some subsequence of s converges to the new point α. To define neighborhoods of α, we consider the range of s (which is a closed discrete subset of X_α), and expand each $s(n)$ to an open set U_n. The expanded family $\{U_n : n \in \omega\}$ will be used to define a local base at α in $X_{\alpha+1} = (\alpha+1, \mathcal{T}_{\alpha+1})$ and must be chosen to make sure $X_{\alpha+1}$ is a locally compact, locally countable, Hausdorff space (hence zero-dimensional and Hausdorff). The way to insure $\mathcal{T}_\alpha$ is as desired is to choose the family $\{U_n : n \in \omega\}$ to be a discrete (locally finite) family in $\mathcal{T}_\alpha$. Choosing $\{U_n : n \in \omega\}$ to be discrete is difficult in general. This can be achieved, however, if the following weak form of normality holds in X_α.

Definition (R.L. Moore [**23**]). A space X has *property D* provided for every closed discrete countable set $Y \subset X$, there exists a discrete family of open sets $\{U_y : y \in Y\}$ such that $y \in U_y$ for all $y \in Y$.

Thus we need that each X_α has property D. Let $\{U_n^{s(n)} : n \in \omega\}$ be a decreasing local base at $s(n)$. Then for each $x \in X_\alpha$, there exists $f_x \in {}^\omega\omega$ and a neighborhood V of x such that V intersects at most one of the sets $U_{f_x}^{s(n)}(n)$ for $y \in H$. Several people observed that if there is a function f that dominates $\{f_x : x \in H\}$ in ${}^\omega\omega$ then $\{U_{f(n)}^{s(n)} : n \in \omega\}$ is a locally finite expansion of $\{s(n) : n \in \omega\}$ ([**8**, **19**, **22**]). In particular, a dominating f can be found if $|X| < \mathfrak{b}$. Van Douwen expressed this as follows:

Lemma 3.1 (van Douwen [**8**, 12.2]). *If X is a first countable space and $|X| < \mathfrak{b}$ then X has property D. Moreover there exists a first countable zero-dimensional Hausdorff space such that $|X| = \mathfrak{b}$ and X does not have property D.*

If we assume $\diamondsuit$, which is much stronger than $\mathfrak{b} = \mathfrak{c}$, we obtain the following result using the Ostaszewski construction.

Theorem 3.2 (Vaughan [**43**]). *Assume $\diamondsuit$. There exists a family $\{X_r : r \in \omega^*\}$ such that each X_r is a sequentially compact T_6-space that is not r-compact. Thus $\Pi_{r\in\omega^*} X_r$ is not countably compact.*

Achieving zero-dimensional Hausdorff spaces at each step in the Ostaszewski construction seems to require an extra set-theoretic hypothesis. This is because at step α, the space $X_\alpha = (\alpha, \mathcal{T}_\alpha)$ has cardinality less than $\mathfrak{c}$, but to apply van Douwen's lemma, we need that the space has cardinality less than $\mathfrak{b}$. Thus the assumption $\mathfrak{b} = \mathfrak{c}$ seems natural. Nevertheless, we discuss below some models of $\mathfrak{b} < \mathfrak{c}$ where the Ostaszewski construction works.

Juhász, Nagy and Weiss [**18**] called a space X *good* provided it is regular, countably compact and locally countable; hence also first countable, locally compact, and zero-dimensional [**18**]. As we discussed above, assuming $\mathfrak{b} = \mathfrak{c}$, van Douwen constructed for each $r \in \omega^*$ a good space which is not r-compact; thus the family of all these spaces provided a counterexample to Scarborough–Stone (Lemma 2.1). Moreover Juhász, Nagy and Weiss constructed a single good space that is not r-compact for any $r \in \omega^*$ (they assume Martin's Axiom and $2^{\mathfrak{c}} < \aleph_\omega$, but only use MA to get $\mathfrak{b} = \mathfrak{c}$). Also K. Kunen and A. Berner (unpublished, see [**49**]), came up with a similar single space assuming the first two steps of GCH. All these constructions are modifications of the Ostaszewski construction, and $\mathfrak{b} = \mathfrak{c}$ holds in these models.

Models of $\mathfrak{b} < \mathfrak{c}$ are known where Ostaszewski's construction yields a family of sequentially compact spaces whose product is not countably compact. In 1988, K. Kunen showed us that in models obtained from models of CH by adding at least ω_1 Cohen reals, the Ostaszewski construction can be used to produce a counterexample to the Scarborough–Stone problem (unpublished). In 2002 Alan Dow noticed that Kunen's method worked in models obtained from CH by adding random reals (also unpublished). It would be interesting to ask about other models, but it is not clear this would solve the remaining portions of the Scarborough–Stone problem.

Another approach by Juhász, Shelah and Soukup [**19**] shows that if one iteratively adds ω_1 dominating reals to any model, then at each step of the induction, a dominating function can be found for the family $\{f_x : x \in X_\alpha\}$ in the next model, which can be used to get the discrete expansion described above. Thus in this model the Ostaszewski construction produces a counterexample to the Scarborough–Stone problem, and if we assume $\neg$CH in the ground model, we have $\mathfrak{b} = \omega_1 < \mathfrak{c}$ in this model.

In [**31**] Nyikos weakened property D by requiring that the expanded family $\{U_i : i \in \omega\}$ be a pairwise disjoint collection instead of a discrete collection. This weaker property is called *ω-collectionwise Hausdorff*. An analogue of Lemma 3.1 can be proved in ZFC with no restriction on the cardinality of X. This means that we can continue the induction maintaining the ω-collectionwise Hausdorff property at each step. Hence the analogous Ostaszewski construction can be continued through $\mathfrak{c}$ and this can be done in ZFC and yields the next result. We mention, however, that we pay a price for using the weaker property because the constructed spaces are Hausdorff, but not regular.

Theorem 3.3 (ZFC, Nyikos and Vaughan [**31**]). *There exists a family of $2^{\mathfrak{c}}$ sequentially compact Hausdorff spaces whose product is not countably compact.*

A space is called a *Urysohn space* if any pair of points can be separated by neighborhoods with disjoint closures. The Scarborough–Stone problem is only half solved in the class of Urysohn spaces. The Hausdorff spaces in [**31**] are not regular, but we do not know if they can be made to be Urysohn in ZFC (see [**47**]).

4. Franklin–Rajagopalan spaces

The second way to construct a family of sequentially compact spaces whose product is not countably compact uses Franklin–Rajagopalan spaces [**11**] (FR-spaces). Let $\mathcal{T} = \{T_\alpha : \alpha < \kappa\}$ be a tower on ω. The underlying set for the FR-space built from $\mathcal{T}$ is $\omega \cup \kappa$, (where we consider ω and κ to be disjoint sets). The topology is defined by declaring ω to be a set of isolated points, and basic neighborhoods of $\alpha \in \kappa$, are defined by $N(\alpha, \beta, F) = (\beta, \alpha] \cup T_\beta \setminus (T_\alpha \cup F)$ where $\beta < \alpha$ and F is a finite subset of ω. Clearly the subspace topology on κ is its usual order topology (see [**31**]). Given a tower $\mathcal{T}$, let $X(\mathcal{T})$ denote the FR-space constructed from $\mathcal{T}$.

There are two main ideas in using FR-spaces to solve the Scarborough–Stone problem. (1) A sequentially compact FR-space $X(\mathcal{T})$ is not r-compact if and only if there exists an ultrafilter u such that $u \supset \mathcal{T}$, and u is below r in the Rudin–Keisler order on ultrafilters [**31**, Theorem 1.2]. (2) There are certain models of set theory in which ω^* can be covered by a family of T-sets, in fact by fewer than $\mathfrak{c}$ many T-sets (hence every $r \in \omega^*$ contains a tower). These are models of Hechler [**16**], and models obtained from adding ω_1 Cohen reals to a model of $\mathfrak{b} > \omega_1$ (for example models of MA+¬CH [**1**]). We do not know if the Ostaszewski construction works in these models, but we have the following

Theorem 4.1 (Nyikos and Vaughan [**31**]). *If a family of T-sets covers ω^*, then the product of the corresponding family of sequentially compact FR-spaces is not countably compact.*

In models where ω^* can be covered by fewer than $\mathfrak{c}$ many T-sets, we have the somewhat surprising result that there exists a product of fewer than $\mathfrak{c}$ sequentially compact spaces whose product is not countably compact.

5. Martin's axiom and proper forcing

In this section we consider the positive solutions to the Scarborough–Stone problem in the classes of T_6 and T_5-spaces.

Theorem 5.1 (W. Weiss [**48**]). *Assume* MA + ¬CH *(or its consequence $\mathfrak{p} > \omega_1$). Every countably compact T_6-space is compact. Hence every product of countably compact T_6-spaces is countably compact.*

Theorem 5.1 and Theorem 3.2 provide a complete solution for T_6-spaces:

Theorem 5.2. *Whether or not every product of sequentially compact T_6-spaces is countably compact is consistent with and independent of* ZFC.

The Scarborough–Stone problem for T_5-spaces was solved using the Proper Forcing Axiom (PFA), which is stronger than MA + ¬CH.

Theorem 5.3 (Nyikos, Soukup, Veličović [**30**, Cor. 1.4]). *Assume* PFA. *In a countably compact T_5-space every countable set has compact (Fréchet–Urysohn) closure. Thus every product of countably compact T_5-spaces is ω-bounded, hence countably compact.*

Theorem 5.3 and Theorem 3.2 provide a complete solution for T_5-spaces:

Theorem 5.4. *Whether or not every product of sequentially compact T_5-spaces is countably compact is consistent with and independent of* ZFC.

6. Concluding comments

Theorem 5.1 and Theorem 5.2 tell us that under certain set-theoretic assumptions countable compactness implies a productive compactness property (and therefore the Scarborough–Stone problem has a positive answer). In the case of T_6-spaces, the productive property is compactness, and in the case of T_5-spaces the productive property is ω-boundedness. Note that only countable compactness, not sequential compactness, is involved.

We also note that as we weaken the separation from T_6 to T_5 in Theorems 5.1 and 5.3, we had to use much stronger set-theory, going from $\mathfrak{p} > \omega_1$ to PFA. Nyikos remarked to us that we can weaken $\mathfrak{p} > \omega_1$ in Theorem 5.1 by thinking of the statement "every countably compact T_6-space is compact" as a set-theoretic property. This property is weaker than $\mathfrak{p} > \omega_1$ because T. Eisworth has shown that it is consistent with CH [**10**].

In order to give a positive (consistent) solution to the Scarborough–Stone problem in the class of Tychonoff spaces it will be necessary to use some of the strength in sequential compactness because in the class of Tychonoff spaces (unlike the T_5 and T_6 cases) it is not consistent that countable compactness by itself implies a productive compactness property, as the Novak–Terasaka example shows (Example 2.2).

Let us note that it is consistent that no mixture of spaces produced by the Ostaszewski construction and the Franklin–Rajagopalan construction will yield a normal counterexample to the Scarborough–Stone problem. To see this, suppose we have a family $\mathcal{F}$ of sequentially compact spaces, each of which is either an FR-space or has countable tightness (such as the Ostaszewski spaces constructed using $\mathfrak{b} = \mathfrak{c}$, a consequence of PFA). Then assuming PFA, the product of the spaces in $\mathcal{F}$ is countably compact. This follows from some known results. Nyikos proved that under PFA, every countably compact, countably tight, normal space is ω-bounded [**2**, **28**]. Kunen showed that assuming MA + $\neg$CH + $\diamondsuit(\mathfrak{c}, \omega_1\text{-limits})$ (a consequence of PFA [**3**, Theorem 7.13]) that there exists a single $u \in \omega^*$ such that every sequentially compact FR-space is u-compact (unpublished, see [**31**]). Thus the family $\mathcal{F}$ consists entirely of ω-bounded spaces and u-compact spaces. Hence the product of the spaces in $\mathcal{F}$ is u-compact, and therefore countably compact.

Obviously there are many variations on the Scarborough–Stone problem. We state one more:

Problem 6.1. *Must every product of sequentially compact topological groups be countably compact?* 519?

This problem was raised in [**46**, Problem 347], but has not received much attention. There are, however, a number of consistent examples of not countably compact products of countably compact groups (e.g., see [**7**, **14**, **42**]).

We have included in our References some citations related to the Scarborough–Stone problem which we did not have space to discuss.

References

[1] B. Balcar, R. Frankiewicz, and C. Mills, *More on nowhere dense closed P-sets*, Bull. Acad. Polon. Sci. Sér. Sci. Math. **28** (1980), no. 5–6, 295–299 (1981).

[2] Z. T. Balogh, A. Dow, D. H. Fremlin, and P. J. Nyikos, *Countable tightness and proper forcing*, Bull. Amer. Math. Soc. (N.S.) **19** (1988), no. 1, 295–298.

[3] J. E. Baumgartner, *Applications of the proper forcing axiom*, Handbook of set-theoretic topology, North-Holland, Amsterdam, 1984, pp. 913–959.

[4] A. R. Bernstein, *A new kind of compactness for topological spaces*, Fund. Math. **66** (1970), 185–193.

[5] D. Booth, *A Bolean view of sequential compactness*, Fund. Math. **85** (1974), no. 2, 99–102.

[6] K. M. Devi, P. R. Meyer, and M. Rajagopalan, *When does countable compactness imply sequential compactness?*, General Topology and Appl. **6** (1976), no. 3, 279–289.

[7] E. K. van Douwen, *The product of two countably compact topological groups*, Trans. Amer. Math. Soc. **262** (1980), no. 2, 417–427.

[8] E. K. van Douwen, *The integers and topology*, Handbook of set-theoretic topology, North-Holland, Amsterdam, 1984, pp. 111–167.

[9] A. Dow, I. Juhász, L. Soukup, and Z. Szentmiklóssy, *More on sequentially compact implying pseudoradial*, Topology Appl. **73** (1996), no. 2, 191–195.

[10] T. Eisworth, *CH and first countable, countably compact spaces*, Topology Appl. **109** (2001), no. 1, 55–73.

[11] S. P. Franklin and M. Rajagopalan, *Some examples in topology*, Trans. Amer. Math. Soc. **155** (1971), 305–314.

[12] Z. Frolík, *The topological product of countably compact spaces*, Czechoslovak Math. J. **10(85)** (1960), 329–338.

[13] J. Ginsburg and V. Saks, *Some applications of ultrafilters in topology*, Pacific J. Math. **57** (1975), no. 2, 403–418.

[14] K. P. Hart and J. van Mill, *A countably compact topological group H such that $H \times H$ is not countably compact*, Trans. Amer. Math. Soc. **323** (1991), no. 2, 811–821.

[15] K. P. Hart, J. Nagata, and J. E. Vaughan (eds.), *Encyclopedia of general topology*, Elsevier Science, Amsterdam, 2004.

[16] S. H. Hechler, *On the existence of certain cofinal subsets of ${}^{\omega}\omega$*, Axiomatic set theory (Los Angeles, CA, 1967), American Mathematical Society, Providence, RI, 1974, pp. 155–173.

[17] S. H. Hechler, *On some weakly compact spaces and their products*, General Topology and Appl. **5** (1975), 83–93.

[18] I. Juhász, Zs. Nagy, and W. Weiss, *On countably compact, locally countable spaces*, Period. Math. Hungar. **10** (1979), no. 2–3, 193–206.

[19] I. Juhász, S. Shelah, and L. Soukup, *More on countably compact, locally countable spaces*, Israel J. Math. **62** (1988), no. 3, 302–310.

[20] I. Juhász and W. Weiss, *Good, splendid and Jakovlev*, Open problems in topology II (E. Pearl, ed.), Elsevier Science.

[21] K. Kunen, *Set theory*, North-Holland, Amsterdam, 1980.

[22] K. Kunen, 1988, Conversation, Madison, WI.

[23] R. L. Moore, *Foundations of point set theory*, American Mathematical Society, Providence, RI, 1962.
[24] N. Noble, *Countably compact and pseudo-compact products*, Czechoslovak Math. J. **19(94)** (1969), 390–397.
[25] J. Novák, *On the Cartesian product of two compact spaces*, Fund. Math. **40** (1953), 106–112.
[26] P. J. Nyikos, *First countable, countably compact, noncompact spaces*, Open problems in topology II (E. Pearl, ed.), Elsevier Science.
[27] P. J. Nyikos, *Notes on the Scarborough–Stone problem*, 1988, Interim Report, Prague Topological Symposium 3.
[28] P. J. Nyikos, *Progress on countably compact spaces*, General topology and its relations to modern analysis and algebra, VI (Prague, 1986), Heldermann Verlag, Berlin, 1988, pp. 379–410.
[29] P. J. Nyikos, B. Shapirovskiĭ, Z. Szentmiklóssy, and B. Veličković, *Complete normality and countable compactness*, Topology Proc. **17** (1992), 395–403.
[30] P. J. Nyikos, L. Soukup, and B. Veličković, *Hereditary normality of* $\gamma\mathbb{N}$*-spaces*, Topology Appl. **65** (1995), no. 1, 9–19.
[31] P. J. Nyikos and J. E. Vaughan, *Sequentially compact, Franklin–Rajagopalan spaces*, Proc. Amer. Math. Soc. **101** (1987), no. 1, 149–155.
[32] P. J. Nyikos and J. E. Vaughan, *The Scarborough–Stone problem for Hausdorff spaces*, Topology Appl. **44** (1992), 309–316.
[33] A. J. Ostaszewski, *On countably compact, perfectly normal spaces*, J. London Math. Soc. **14** (1976), no. 3, 505–516.
[34] M. Rajagopalan, *Some outstanding problems in topology and the* V*-process*, Categorical topology (Mannheim, 1975), Springer, Berlin, 1976, pp. 501–517.
[35] M. Rajagopalan and R. G. Woods, *Products of sequentially compact spaces and the* V*-process*, Trans. Amer. Math. Soc. **232** (1977), 245–253.
[36] V. Saks, *Products of countably compact spaces*, Topology Proc. **4** (1979), no. 2, 553–575 (1980).
[37] C. T. Scarborough and A. H. Stone, *Products of nearly compact spaces*, Trans. Amer. Math. Soc. **124** (1966), 131–147.
[38] A. Szymański and M. Turzański, βN *and sequential compactness*, Colloq. Math. **35** (1976), no. 2, 205–208.
[39] F. D. Tall, *Weakly collectionwise Hausdorff spaces*, Topology Proc. **1** (1977), 295–304.
[40] H. Terasaka, *On Cartesian product of compact spaces*, Osaka Math. J. **4** (1952), 11–15.
[41] S. Todorčević, *Partition problems in topology*, American Mathematical Society, Providence, RI, 1989.
[42] A. H. Tomita, *Square of countably compact groups without non-trivial convergent sequences*, Topology Appl. **153** (2005), no. 1, 107–122.
[43] J. E. Vaughan, *Products of perfectly normal, sequentially compact spaces*, J. London Math. Soc. **14** (1976), no. 3, 517–520.
[44] J. E. Vaughan, *Countably compact and sequentially compact spaces*, Handbook of set-theoretic topology, North-Holland, Amsterdam, 1984, pp. 569–602.
[45] J. E. Vaughan, *A countably compact, first countable, nonnormal* T_2*-space which is almost compact*, Glas. Mat. Ser. III **23(43)** (1988), no. 1, 147–152.
[46] J. E. Vaughan, *Small uncountable cardinals and topology*, Open problems in topology, North-Holland, Amsterdam, 1990, pp. 195–218.
[47] J. E. Vaughan, *Two step iteration of almost disjoint families*, J. Symbolic Logic **69** (2004), no. 1, 81–90.
[48] W. Weiss, *Countably compact spaces and Martin's axiom*, Canad. J. Math. **30** (1978), no. 2, 243–249.
[49] W. Weiss, *Review of J. E. Vaughan,* Products of perfectly normal, sequentially compact spaces, *J. London Math. Soc. 14 (1976) 517–520*, Math. Reviews **57** (1979), # 4098.

Part 3

Continuum Theory

Open Problems in Topology II – Edited by E. Pearl

Questions in and out of context

David P. Bellamy

These are problems which have seemed to me important, interesting, or both for years. Some of them are just isolated questions which may not lead anywhere; some are missing links in different developments. Sometimes, I am aware of how they fit into the general scheme of topology; sometimes I am not. Some I express as conjectures; some just as questions. If a question is not my own, I have given the name of the person who first posed it if I know who it is.

Conjecture 1. *Let X be a tree-like continuum and let $f\colon X \to X$ be a continuous* 520?
map. Then there exists an indecomposable continuum $W \subseteq X$ such that $f(W) \cap W \neq \emptyset$.

I think that this is the last remaining hope of getting something akin to a fixed point theorem for all tree-like continua.

Conjecture 2 (Norman Passmore). *Let P be a pseudo-arc and let $C(P)$ denote* 521?
its hyperspace of subcontinua. Let $X \subseteq C(P)$ be a pseudo-arc. Then X is a subset of a Whitney level for some Whitney map μ.

This was posed by Norman Passmore, in his Ph.D. dissertation [**5**] in 1976, and is still open.

Conjecture 3 (Howard Cook). *Let $A \subseteq \mathbb{R}^n$ be a compact set with the property* 522?
that every component of A is either a point or a pseudo-arc. Then there exists a pseudo-arc P with $A \subseteq P \subset \mathbb{R}^n$.

Howard Cook proved the $\mathbb{R}^2$ case of this in his dissertation [**3**]. I have discussed this with Howard Cook in conversation and he thinks the higher-dimensional case would be a good dissertation problem, also. He suspects that it is probably easier for $n > 2$ than for $n = 2$.

Conjecture 4. *Let $A \subseteq \mathbb{R}^n$ and suppose A is compact and every component of* 523?
A is a hereditarily indecomposable continuum. Then there exists a hereditarily indecomposable continuum M with $A \subseteq M \subset \mathbb{R}^{n+1}$.

I think this is easier than Conjecture 3; I can do it if A has only finitely many components. This is, of course, just a natural generalization of Conjecture 3.

Conjecture 5. *Every finite-dimensional hereditarily indecomposable continuum* 524?
can be embedded into a finite product of pseudo-arcs.

It is true that every hereditarily indecomposable continuum can be embedded into a product of pseudo-arcs; a countable product if the continuum is metric. Another variation follows. An affirmative answer to this would be a satisfying analog to what is known about embedding compact finite-dimensional metric spaces into Euclidean spaces. Conjecture 6 is, of course, strictly stronger than Conjecture 5. Conjecture 7 is another special case.

525? **Conjecture 6.** *Every n-dimensional hereditarily indecomposable continuum can be embedded into a product of $2n + 1$ pseudo-arcs.*

526? **Conjecture 7.** *Every planar hereditarily indecomposable continuum embeds into a product of two pseudo-arcs.*

527? **Conjecture 8.** *Let X be a chainable continuum. Then there is a base of connected open neighborhoods of the diagonal in $X \times X$; that is, if Δ denotes the diagonal of $X \times X$ and U is open in $X \times X$ with $\Delta \subseteq U$ then there is a connected open $V \subseteq X \times X$ such that $\Delta \subseteq V \subseteq U$.*

This is true for the arc, all Knaster continua, and the pseudo-arc for what appear to be completely different reasons. It is not true for solenoids, which are, of course, circularly chainable but not chainable.

528? **Problem 9.** *Let $\mathbb{H}^*$ denote the Stone–Čech remainder of $[0, \infty)$. Let M be the orbit of any point of $\mathbb{H}^*$ under the action of its homeomorphism group. Is M connected?*

This construction will be revisited in Problem 26. Many other spaces besides $[0, \infty)$ could, of course, be used here.

A continuum X is *invertible* iff given any nonempty open set $U \subseteq X$ there is a homeomorphism $h\colon X \to X$ such that $h(X \setminus U) \subseteq U$.

529? **Problem 10** (Sam Nadler). *Is the pseudo-arc the only invertible chainable continuum?*

It is not the only invertible tree-like continuum.

530? **Problem 11.** *Does every homogeneous tree-like continuum have the fixed-point property?*

The only known homogeneous tree-like continuum is a pseudo-arc.

531? **Problem 12.** *Is there a dendroid M in the plane such that the accessible points of M are precisely the endpoints of M, and the set of endpoints is totally disconnected?*

There is a dendroid K in the plane with the property that a point is accessible if and only if it is an endpoint, and in which the set of endpoints is connected [**1**]. There is also a continuum in the plane with a totally disconnected set of accessible points.

A topological space W is *widely connected* if and only if every nondegenerate connected subset of W is dense in W (P.M. Swingle, [**7**, **8**]).

The only known examples of widely connected sets are as follows:

Let X be a metric indecomposable continuum. Let $\mathcal{C}$ be the set of composants of X. Let $\mathcal{S}$ be the collection of closed subsets A of X such that $X \setminus A$ is not connected. Let $\varphi\colon \mathcal{C} \to \mathcal{S}$ be a bijection. The set $\{C \cap \varphi(C) : C \in \mathcal{C}\}$ is a collection of pairwise disjoint nonempty sets. Let W be a choice set on this collection. W is widely connected.

Conjecture 13. *Let W be a widely connected separable metric space. Then W can be densely embedded into some metric indecomposable continuum.* 532?

Problem 14. *Is the Stone–Čech compactification of a widely connected space necessarily an indecomposable continuum?* 533?

Of course, assume complete regularity or normality.

Problem 15. *Are there widely connected spaces of arbitrarily large cardinality?* 534?

This is motivated by Michel Smith's work. He proved that there are indecomposable continua with arbitrarily large numbers of composants [**6**].

An *absolute suspension* is a continuum X such that given any $p, q \in X$ with $p \neq q$, X has the structure of a suspension with p, q as vertices. Precisely, given any $p, q \in X$ with $p \neq q$, there exists a space Y and a homeomorphism $H\colon X \to \Sigma Y$, the suspension of Y, with $H(p), H(q)$ being the vertices of ΣY.

Conjecture 16 (de Groot, [**4**]). *Every finite-dimensional absolute suspension is a sphere.* 535?

Conjecture 17 (Janusz Łysko). *The suspension of an absolute suspension is itself an absolute suspension.* 536?

The next four problems involve homogeneous arcwise connected continua.

Problem 18 (R.L. Wilson). *Is there a uniquely arcwise connected homogeneous compact Hausdorff continuum?* 537?

Such an example cannot be metric [**2**].

Problem 19 (Lewis Lum). *Let X be a metric arcwise connected homogeneous continuum which is not a simple closed curve. Must X contain simple closed curves of arbitrarily small diameter?* 538?

Conjecture 20. *Let X be a homogeneous arcwise connected continuum which is not a simple closed curve. Let U be an open set in X and let M be an arc component of U. Then M is cyclicly connected.* 539?

Problem 21. *Let X be a homogeneous arcwise connected continuum which is not a simple closed curve. Is every arcwise connected open subset of X also cyclicly connected?* 540?

Let X be a continuum, and let T be the well-known set function. The statement *T is continuous for X* means that $T\colon 2^X \to 2^X$ is continuous in the Vietoris topology.

Conjecture 22. *If T is continuous restricted to the hyperspace of subcontinua of X, then T is continuous for X.* 541?

Conjecture 23 (Nadler–Bellamy). *Let X be a homogeneous one-dimensional continuum. Then T is continuous for X.* 542?

Two points x and y of a continuum X are in the same *tendril class* iff there is a nowhere dense continuum $W \subseteq X$ with $p, q \in W$.

543? **Problem 24.** *Is there a continuum X with a proper dense open tendril class?*

544? **Problem 25.** *If every tendril class of a continuum X is dense, and X has more than one tendril class, is X necessarily indecomposable?*

A space X is *thin* iff for all $A, B \subseteq X$, if A is homeomorphic to B, there exists a homeomorphism $H: X \to X$ such that $H(A) = B$.

The only known examples of thin spaces are: finite discrete spaces, finite indiscrete spaces, and products of one of each of these.

545? **Problem 26** (P.H. Doyle). *Does there exist an infinite thin space?*

More specifically, is there a complete metric space X and a point $p \in \beta X \setminus X$ such that the orbit of p under the homeomorphism group of either βX or of $\beta X \setminus X$ is an infinite thin space? Based upon what one can readily prove about thin spaces, this is as simple as an example could possibly be.

References

[1] D. P. Bellamy, *An interesting plane dendroid*, Fund. Math. **110** (1980), no. 3, 191–208.

[2] D. P. Bellamy and L. Lum, *The cyclic connectivity of homogeneous arcwise connected continua*, Trans. Amer. Math. Soc. **266** (1981), no. 2, 389–396.

[3] H. Cook, *On the most general plane closed point set through which it is possible to pass a pseudo-arc*, Ph.D. thesis, University of Texas at Austin, 1962.

[4] J. de Groot, *On the topological characterization of manifolds*, General topology and its relations to modern analysis and algebra, III (Prague, 1971), Academia, Prague, 1972, pp. 155–158.

[5] N. S. Passmore, *Embedding products into hyperspaces*, Ph.D. thesis, University of Delaware, 1976.

[6] M. Smith, *Generating large indecomposable continua*, Pacific J. Math. **62** (1976), no. 2, 587–593.

[7] P. M. Swingle, *The existence of widely connected and biconnected semigroups*, Proc. Amer. Math. Soc. **11** (1960), 243–248.

[8] P. M. Swingle, *Widely connected and biconnected semigroups*, Proc. Amer. Math. Soc. **11** (1960), 249–254.

Open Problems in Topology II – Edited by E. Pearl

An update on the elusive fixed-point property

Charles L. Hagopian

R.H. Bing's expository article [**11**], "The elusive fixed point property," which appeared in the American Mathematical Monthly in 1969, has been an invaluable guide to generations of mathematicians. It consists of twelve questions and a variety of related theorems and examples. A space has the *fixed-point property* if each map of the space into itself has a fixed point. A *continuum* is a nondegenerate compact connected metric space. Bing was interested in the fundamental problem of determining which continua have the fixed-point property. We review Bing's questions, some results that followed the publication of [**11**], and some related unsolved problems.

Bing's Question 1. *Is there a 2-dimensional polyhedron with the fixed-point property which has even Euler characteristic?* 546?

Question 1 is still open. It was motivated by W. Lopez's example [**68**] of an 8-dimensional polyhedron with even Euler characteristic that has the fixed-point property.

After stating O.H. Hamilton's theorem [**51**] that each arc-like continuum has the fixed-point property, Bing asked the following question.

Bing's Question 2. *Does each tree-like continuum have the fixed-point property?*

Bing referred to Question 2 as one of the most interesting unsolved problems in geometric topology. The first results related to Question 2 following the publication of [**11**] were positive. In 1970, H. Cook [**21**] proved every λ-dendroid (hereditarily decomposable hereditarily unicoherent continuum) is tree-like. R. Mańka [**70**] in 1976 generalized theorems of K. Borsuk [**14**] and Hamilton [**50**] [**11**, Th. 9] by proving that every λ-dendroid has the fixed-point property.

In 1979, D.P. Bellamy [**8**] answered Question 2 in the negative. Modifying a 6-adic solenoid, Bellamy [**8**] constructed a nonplanar tree-like continuum that admits a fixed-point-free map. Bellamy [**8**] used this example and an inverse limit technique of J.B. Fugate and L. Mohler [**29**] to construct a second tree-like continuum that admits a fixed-point-free homeomorphism. The following problems remain unsolved.

Problem 1. *Can Bellamy's second example be embedded in the plane?* 547?

Problem 2. *Does every triod-like continuum have the fixed-point property?* 548?

Bing believed that special classes of triod-like continua provide a good starting point for an investigation. In a conversation with the author, he asked the following four questions.

Suppose M is an inverse limit space of triods.

(a) Does M have the fixed-point property if each bonding map is the identity on the set of end points of the triod? 549?

550? (b) Does M have the fixed-point property if each bonding map is the identity on the end points and center point of the triod?

551? (c) Does M have the fixed-point property if each bonding map is the identity on one arm of the triod?

(d) Does M have the fixed-point property if each bonding map is the identity on two arms of the triod?

Questions (a), (b), and (c) are still open. In 1984, M.M. Marsh [**77**] answered (d) in the affirmative. A more general theorem for inverse limits of n-ods in [**78**] gives a partial answer to (c). Marsh [**80**] in 1989 further generalized these results for inverse limits of n-ods and fans.

Bellamy's second example is an arc continuum (each of its proper subcontinua is an arc).

552? **Problem 3.** *Does every triod-like arc continuum have the fixed-point property?*

553? **Problem 4.** *Do there exist two commuting maps of a triod onto itself that do not have a coincidence point?*

If there exist two maps with these properties, then there exists an inverse limit space of triods with a fixed-point-free map induced by a commuting non-square (parallelogram) diagram, thus answering Problem 2 in the negative. E. Dyer [**22**] in 1956 stated Problem 4 for a tree instead of a triod. According to a result established in 1981 by Fugate and T.B. McLean [**28**, Th. 1.11], every map induced on an inverse limit of trees by commuting squares has a fixed point.

Problems 2 and 3 remain unsolved when the word "triod-like" is replaced by "n-od like." One invariant arc component of Bellamy's first tree-like continuum contains a fan. It is not known if Bellamy's first example is fan-like.

554? **Problem 5.** *Does every fan-like continuum have the fixed-point property?*

In 1980, L.G. Oversteegen and J.T. Rogers [**95**] defined a tree-like continuum as an inverse limit of planar curves with a fixed-point-free map that is almost induced by commuting squares. Their geometrically explicit construction was given to motivate the study of planar embeddings. Oversteegen and Rogers [**96**] in 1982 defined another example as an inverse limit of trees with a fixed-point-free map induced by commuting non-square parallelograms.

During the period from 1992 to 2000, P. Minc constructed a variety of related tree-like continua. He defined a tree-like continuum that admits fixed-point-free maps with arbitrarily small trajectories [**86**], a tree-like continuum that admits a periodic-point-free homeomorphism [**87**], and a weakly chainable (continuous image of a chainable continuum) tree-like continuum [**88**] and a hereditarily indecomposable tree-like continuum [**89**] that admit fixed-point-free maps.

555? **Problem 6** (Minc). *Does every weakly chainable tree-like arc continuum have the fixed-point property?*

556? **Problem 7** (Minc). *Does there exist a tree-like arc continuum that admits a periodic-point-free homeomorphism?*

In 1993, L. Fearnley and D.G. Wright [**26**] gave a geometric description of Bellamy's continuum.

After Bing's publication [**11**], there were many applications of the following theorem of W. Holsztynski [**52**].

Theorem 1. *Every inverse limit space of ANRs with universal bonding maps has the fixed-point property.*

In 1981, C.A. Eberhart and Fugate [**24**] used Theorem 1 to establish the fixed-point property for inverse limit spaces defined with weakly confluent bonding maps between the same tree. They [**24**] extended this result to weakly arc preserving bonding maps between different trees and Marsh [**79**] further extended it to allow for certain types of arc folding by the bonding maps. Finally, in 1998, Eberhart and Fugate [**23**] established necessary and sufficient conditions for the bonding maps between trees to be universal. It is often overlooked that Theorem 1 had previously been established for compact polyhedra by J. Mioduszewski and M. Rochowski [**90**, p. 69], which is sufficient for these applications.

A. Granas [**32**] in 1968 proved that every approximative absolute retract continuum has the fixed-point property. In 2004, J.J. Charatonik and J.R. Prajs [**19**, Th. 3.3] showed each absolute retract for the class of tree-like continua is an approximative absolute retract. They [**19**, Th. 3.8] also showed that every inverse limit space of trees with confluent bonding maps is an absolute retract for tree-like continua. This gives a proof (which does not involve Theorem 1) of Eberhart and Fugate's theorem [**24**, Ths. 5,7] that every inverse limit space of trees with confluent bonding maps has the fixed-point property.

Problem 8 (J.J. Charatonik, W.J. Charatonik, and Prajs). *Is every absolute retract for hereditarily unicoherent continua a tree-like continuum?* 557?

Problem 9 (J.J. Charatonik, W.J. Charatonik, and Prajs). *Does every absolute retract for hereditarily unicoherent continua have the fixed-point property?* 558?

Every tree-like continuum that is an absolute retract for hereditarily unicoherent continua is an absolute retract for tree-like continua. Hence an affirmative answer to Problem 8 would also answer Problem 9 in the affirmative [**19**, Th. 3.3].

In 1981, Fugate and McLean [**28**] generalized a theorem of P.A. Smith [**105**] by proving every periodic homeomorphism of a tree-like continuum has a nonvoid connected fixed-point set. They [**28**] also proved every hereditarily indecomposable tree-like continuum has the fixed-point property for pointwise periodic homeomorphisms.

Problem 10 (Fugate and McLean). *Does every tree-like continuum have the fixed-point property for pointwise periodic homeomorphisms?* 559?

Problem 11 (Bellamy). *If M is a homogeneous tree-like continuum, must M have the fixed-point property?* 560?

In 1976, the author [**38**] used E.G. Effros's theorem [**25**] and Hamilton's argument [**51**] to show every homogeneous almost chainable continuum has the fixed-point property. I.W. Lewis [**67**] in 1981 improved this result by showing all such continua are arc-like. Nevertheless, the method in [**38**] provides a natural approach to Problem 11. In 1982, Oversteegen and E.D. Tymchatyn [**97**] proved the answer to Problem 11 is yes if M is planar. They accomplished this by showing every homogeneous nonseparating plane continuum has zero span [**66**].

Bing's third question is often referred to as the plane fixed-point problem. It has been called the most interesting outstanding problem in plane topology [**11**].

561? **Bing's Question 3.** *Does the intersection of each decreasing sequence of disks have the fixed-point property?*

A plane continuum is the intersection of a decreasing sequence of disks if and only if it does not separate the plane. Hence it is called a nonseparating plane continuum. An affirmative answer to Question 3 would provide a beautiful generalization to the 2-dimensional version of Brouwer's fixed-point theorem [**4**, **59**].

Generalizing a theorem of Hamilton [**50**] [**11**, Th. 10], in 1967–1970, H. Bell [**5**], K. Sieklucki [**104**], and S.D. Iliadis [**55**] proved the following theorem.

Theorem 2. *For every fixed-point-free map f of a nonseparating plane continuum M into M there exists an indecomposable continuum I in the boundary of M such that $f(I) = I$.*

Theorem 2 follows from the existence of an outchannel in the complement of M. Bell developed the theory of variation to simplify and possibly extend his proof. Recently, J.C. Mayer, Oversteegen, and Tymchatyn [**84**] used the theory of prime ends to prove Bell's basic variation results. They [**84**] established Bell's theorem that a nonseparating plane continuum that admits a fixed-point-free map has exactly one outchannel and its variation must be -1.

If a continuum can be embedded in the plane in a way that excludes the existence of an outchannel, then it cannot be a counterexample to Question 3 [**15**]. A. Lelek [**66**] showed that a counterexample must also have positive span. The natural embedding of W.T. Ingram's [**57**] atriodic tree-like plane continuum with positive span does not allow for an outchannel. In 1982, Mayer [**83**] defined an atriodic tree-like plane continuum with positive span that might allow for the existence of an outchannel.

562? **Problem 12** (Mayer). *Does Mayer's continuum have the fixed-point property?*

In 1978, Iliadis [**54**] used a construction that eliminates the possibility of an outchannel in his proof of the following theorem.

Theorem 3. *If K is a nonseparating plane continuum, then there exists a nonseparating plane continuum Q containing K such that every component of $Q \setminus K$ is a half-open arc and Q has the fixed-point property.*

A *cut disk* is a nonseparating plane continuum obtained by deleting one dense canal from a disk. Imagine starting on the boundary of a disk in the plane and digging a canal (a simply connected open set) that gets narrower as it goes into the disk. If we stop digging at any finite length, the resulting continuum will be a disk, and by Brouwer's theorem has the fixed-point property. To be dense, the canal must be infinitely long and get arbitrarily close to all remaining points in the disk. The accessible boundary of a cut disk is a one-to-one continuous image of the real line. Every cut disk is an indecomposable continuum.

Problem 13 (Bell). *Does every cut disk have the fixed-point property?* 563?

Borsuk [**12**] proved that every locally connected nonseparating plane continuum has the fixed-point property by showing that all such continua are retracts of a disk. Borsuk and J. Stallings [**108**] were following this approach to Question 3 when they asked if every nonseparating plane continuum is an almost continuous retract of a disk.

A function f of a space Y into a space Z is *almost continuous* if for each open set G in $Y \times Z$ that contains f there exists a map g of Y into Z such that G contains g.

Since every almost continuous retract of a disk has the fixed-point property [**108**], an affirmative answer to Borsuk and Stallings' question would have answered Question 3 in the affirmative. In 1984, V. Akis [**1**] proved a plane continuum formed by a disk and a ray spiraling to its boundary is not an almost continuous retract of a disk. Akis and D. Curtis [**3**] in 1987 did the same for a spiral to a triod, thus showing the answer to Borsuk and Stallings' question is no, even with the added assumption the nonseparating plane continuum is tree-like. However, this approach is not completely ruled out by these results. Using a theorem of Akis [**1**] and H. Rosen [**101**], recently B.D. Garrett [**30**] showed every nonseparating plane continuum is an almost continuous retract of a continuum that is an almost continuous retract of a disk. Although channels are not mentioned in Garrett's proof, he used an arc-attaching technique similar to Iliadis's construction for Theorem 3 and his intermediate continuum does not allow for an outchannel. Garrett's theorem and an affirmative solution to the following problem would provide an affirmative answer to Question 3.

Problem 14 (Garrett). *If Y is a continuum that is an almost continuous retract of a disk and a continuum Z is an almost continuous retract of Y, must Z have the fixed-point property?* 564?

In 1971, the author [**33**] used Theorem 2 to prove that every arcwise connected nonseparating plane continuum has the fixed-point property. Shortly thereafter, the author [**34**] and J. Kransinkiewicz [**62**] established the fixed-point property for all δ-connected nonseparating plane continua. B. Knaster and S. Mazurkiewicz [**60**] defined a continuum to be *δ-connected* if each pair of its points belong to a hereditarily decomposable subcontinuum.

C. Kuratowski [**64**] defined a continuum M to be of *type* λ if M is irreducible and every indecomposable continuum in M is a continuum of condensation. If a continuum M is of type λ, then M admits a unique monotone upper semi-continuous decomposition to an arc with the property that each element of the decomposition has void interior relative to M [**65**, Th. 3, p. 216]. The elements of this decomposition are called *tranches*. A continuum M is *λ-connected* if for each pair p and q of its points there is a type λ subcontinuum of M irreducible between p and q.

Every λ-connected plane continuum is δ-connected [**39**, Th. 2]. Hence every λ-connected nonseparating plane has the fixed-point property.

Suppose M is a continuum of type λ and each tranche of M has the fixed-point property. In 1971, G.R. Gordh [**31**, Th. 3B.1] proved every monotone map of M onto M has a fixed point. To answer a question of Gordh, the author [**46**] in 2003 defined a nonplanar continuum of type λ that admits a fixed-point-free map and has the condition that each of its tranches has the fixed-point property. Recently, planar continua with these properties have been constructed by the author and Mańka [**48**] and V. Martinez-de-la-Vega [**82**].

565? **Problem 15.** *Must a plane continuum of type λ have the fixed-point property if none of its tranches separates the plane and each tranche has the fixed-point property?*

566? **Problem 16.** *Must a plane continuum of type λ have the fixed-point property if each of its tranches has the fixed-point property and its decomposition is continuous?*

In 1976, H. Bell [**6**, **7**] generalized a theorem of Cartwright and Littlewood [**18**] [**11**, Th. 11] by proving every homeomorphism of a nonseparating plane continuum onto itself that can be extended to a homeomorphism of the plane has a fixed point. Bell announced in 1984 that the Cartwright and Littlewood Theorem extends to the class of all holomorphic maps of the plane (see also Akis [**2**]). R. Fokkink, Mayer, Oversteegen, and Tymchatyn [**27**] recently introduced the class of oriented maps of the plane and showed that among perfect maps of the plane the oriented maps are exactly the compositions of open maps and monotone maps. Each invariant nonseparating plane continuum under a positively-oriented map of the plane must contain a fixed point [**27**].

Every tree-like plane continuum is a nonseparating plane continuum. Hence an affirmative solution to Problem 1 would answer Question 3 in the affirmative. A partial solution to either Problem 2 or 3 derived by adding the assumption that the continuum is planar would be a breakthrough.

In 1990, Minc [**85**] generalized the results of [**33**] and [**51**] by proving that every weakly chainably connected nonseparating plane continuum has the fixed-point property. A continuum is *weakly chainably connected* if each pair of its points is contained in a weakly chainable subcontinuum. Noting this theorem fails for tree-like continua in general [**88**] and there exists a hereditarily indecomposable tree-like continuum without the fixed-point property [**89**], Minc asked the following question.

567? **Problem 17.** *Does every hereditarily indecomposable nonseparating plane continuum have the fixed-point property?*

Every tree-like continuum and every nonseparating plane continuum is disk-like. Bellamy's continuum [**8**] was the first known example of a disk-like continuum without the fixed-point property. R. Bennett [**9**] in 1966 proved every locally connected disk-like continuum is planar and therefore has the fixed-point property [**12**]. Since every arcwise connected disk-like plane continuum has the fixed-point property [**33**], it is natural to ask the following question [**37**].

Problem 18. *Does every arcwise connected disk-like continuum have the fixed-point property?* 568?

S.B. Nadler [**93**, Th. 3.2] in 1980 used Theorem 1 to prove that every inverse limit space of disks with weakly confluent bonding maps has the fixed-point property.

Dyer [**22**] in 1956 proved every product of arc-like continua has the fixed-point property. Suppose X and Y are δ-connected continua and $X \times Y$ is disk-like. In 1975, the author [**35**] proved X and Y are arc-like. Hence $X \times Y$ has the fixed-point property.

Problem 19. *Does every disk-like product of two continua have the fixed-point property?* 569?

Marsh and R. Escobedo independently proved each product of two zero span continua has the fixed-point property. In 2004, Marsh [**81**] extended this result to all products of zero span continua.

Let $C(X)$ be the hyperspace of compact connected subsets of a continuum X. In 1962, J. Segal [**103**] proved $C(X)$ has the fixed-point property if X is arc-like. In 1972, Rogers [**98**] showed if X is the circle with a spiral, then $C(X)$ is homeomorphic to the cone over X. This gave the first example of a hyperspace $C(X)$ without the fixed-point property. Shortly thereafter, Nadler and Rogers [**94**] proved if X is a disk with a spiral, then $C(X)$ does not have the fixed-point property, thus giving an example of a continuum X with the fixed-point property whose hyperspace $C(X)$ admits a fixed-point-free map. Rogers [**98**] also proved if X is a hereditarily indecomposable continuum, then both $C(X)$ and the cone over X have the fixed-point property. Krasinkiewicz [**63**] and Rogers [**99, 100**] in 1974 proved $C(X)$ has the fixed-point property if X is circle-like. In 1983, Marsh [**76**] applied his concept of s-connectedness to show the cone over a circle-like continuum must have the fixed-point property. J. Bustamente, R. Escobedo, and F. Macias-Romero [**17**] recently used s-connectedness to show that $C(X)$ has the fixed-point property if X has zero span.

The author [**36**] in 1975 proved the following theorem.

Theorem 4. *Suppose X is a δ-connected continuum and $C(X)$ can be ϵ-mapped (for each $\epsilon > 0$) into the plane. Then X is either arc-like or circle-like.*

It follows that $C(X)$ is disk-like [**102**] and has the fixed-point property [**63, 99, 100, 103**].

Problem 20. *Can the assumption that X is δ-connected be removed from Theorem 4?* 570?

A continuum is *uniquely arcwise connected* if it is arcwise connected and does not contain a simple closed curve. The $\sin\frac{1}{x}$ circle (Warsaw circle) is the simplest example of a uniquely arcwise connected plane continuum that separates the plane. In [**11**], Bing gave a dog-chases-rabbit argument to show the $\sin\frac{1}{x}$ circle has the fixed-point property. Then Bing asked his fourth question which had previously been raised by G.S. Young [**110**].

Bing's Question 4. *Does each uniquely arcwise connected plane continuum have the fixed-point property?*

In 1979, the author [**40**] used the dog-chases-rabbit principle to answer Question 4 in the affirmative. The proof involved a continuous image of a ray defined by Borsuk [**14**] and a nested sequence of polygonal disks constructed by Sieklucki [**104**].

Bing [**11**, Th. 14, Th. 15] defined a 1-dimensional uniquely arcwise connected nonplanar continuum X with the fixed-point property and a disk D such that $D \cap X$ is an arc and $D \cup X$ does not have the fixed-point property [**47**, **73**].

Bing's Question 5. *Does the product of Bing's continuum X and an interval have the fixed-point property?*

In 1970, W.L. Young [**111**] answered Question 5 in the affirmative. Bing's continuum X is similar to an earlier example of G.S. Young [**110**]. Other examples of uniquely arcwise connected continua without the fixed-point property have been given by the author and Mańka [**47**], Holsztynski [**53**], Mańka [**73**], Mohler and Oversteegen [**92**], and M. Sobolewski [**106**]. Bing's sixth question was also raised by G.S. Young [**110**].

Bing's Question 6. *Must every homeomorphism of a uniquely arcwise connected continuum into itself have a fixed point?*

In 1976, Mohler [**91**] used the Markov–Katutani theorem (measure theory) to answered Question 6 in the affirmative.

In 1986, the author [**41**] proved the following theorem.

Theorem 5. *Every deformation of a uniquely arcwise connected continuum into itself has a fixed point.*

571? **Problem 21.** *Does every disk-like continuum have the fixed-point property for deformations?*

572? **Problem 22.** *Does every hereditarily decomposable continuum that does not contain a simple closed curve have the fixed-point property for deformations?*

A map of a continuum that sends each arc-component into itself is called an *arc-component preserving map*. Note that every deformation of a continuum into itself is an arc-component preserving map. The author established the fixed-point property for deformations of tree-like continua [**45**], nonseparating plane continua [**42**, Th. 4.1], indecomposable plane continua [**42**, Th. 4.12], and plane continua that do not contain a simple closed curve [**43**, Th. 15] by proving that all continua in these classes have the fixed-point property for arc-component preserving maps. G.S. Young's example [**110**] shows that deformations cannot be replaced by arc-component preserving maps in Theorem 5 or Problem 22.

A set is *simply connected* if it is arcwise connected and its fundamental group is trivial. In 1996, the author [**44**] proved the following general theorem.

Theorem 6. *Suppose M is a plane continuum, $\mathcal{D}$ is a decomposition of M, and each element of $\mathcal{D}$ is simply connected. Then every map of M that sends each element of $\mathcal{D}$ into itself has a fixed point.*

It follows from Theorem 6 that every simply connected plane continuum has the fixed-point property. In fact, an arcwise connected plane continuum has a trivial fundamental group if and only if it has the fixed-point property. This result answers a question of Mańka [**71**].

Since Bing's continuum X in Question 5 is not embeddable in the plane, he asked the following question.

Bing's Question 7. *If C is a plane continuum with the fixed-point property and D is a disk such that $C \cap D$ is an arc, must $C \cup D$ have the fixed-point property?*

Recently, the author and Prajs [**49**] defined a continuum Ω that answered Question 7 in the negative.

Lopez [**68**] proved there is a polyhedoron P with the fixed-point property and a disk D such that $P \cap D$ is an arc and $P \cup D$ does not have the fixed-point property. To show that $P \cup D$ does not have the fixed-point property, Lopez used the following theorem of F. Wecken [**109**] [**11**, Th. 16].

Theorem 7. *Every connected polyhedron without local separating points and Euler characteristic 0 admits a fixed-point-free map.*

Since the dimension of Lopez's example is 17, Bing asked for something simpler.

Bing's Question 8. *What is the lowest dimension for such a polyhedron P as* 573?
promised by Lopez's theorem?

Question 8 remains open. In a conversation, Bing conjectured the answer is 2. In 1990, Mańka [**72**] defined two rational arcwise connected continua X and Y with the fixed-point property such that $X \cap Y$ is contractible and $X \cup Y$ admits a fixed-point-free map.

Problem 23 (Mańka). *Suppose X and Y are 1-dimensional plane continua with* 574?
the fixed-point property and $X \cap Y$ is arcwise connected. Must $X \cup Y$ have the fixed-point property?

Bing's Question 9. *If a compact 1-dimensional continuum has the fixed-point property, does its cartesian product with an arc?*

Bing's Question 10. *If a plane continuum has the fixed-point property, does its* 575?
cartesian product with an arc?

M. Sobolewski [**107**] recently answered Question 9 with a counterexample. Question 10 remains open. Since Sobolewski's continuum contains a solenoid and sequences of Cook continua [**20**], it is clearly not embeddable in the plane. The product of the plane continuum Ω defined in [**49**] and an arc has the fixed-point property. R. Knill's continuum [**61**] [**11**, Th. 23], the can-with-a-skirt, provides a negative answer when the dimension is changed to 2 in Question 9.

In 2002, Mańka [**74**] proved the product of a λ-dendroid and an arc must have the fixed-point property.

576? **Problem 24.** *If M is a tree-like continuum with the fixed-point property, must $M \times [0,1]$ have the fixed-point property?*

577? **Problem 25.** *If M is a uniquely arcwise connected continuum with the fixed-point property, must $M \times [0,1]$ have the fixed-point property?*

578? **Problem 26.** *Must the product of a uniquely arcwise connected plane continuum and an arc have the fixed-point property?*

579? **Problem 27.** *If a plane continuum is simply connected, must its product with an arc have the fixed-point property?*

Theorem 7 and Lopez's example [**68**] shows there is a polyhedron with the fixed-point property such that its cartesian product with an arc admits a fixed-point-free map. This motivated Bing's eleventh question.

Bing's Question 11. *If P and Q are polyhedra without local separating points but with the fixed-point property, must $P \times Q$ have the fixed-point property?*

In 1971, G.E. Bredon [**16**] gave a negative answer to Question 11. Unlike earlier examples, Bredon's polyhedral factors satisfy the Shi condition. Every polyhedron of the same homotopy type as a Bredon factor has the fixed-point property.

Bing [**11**] reviewed S. Kinoshita's continuum [**58**], the can-with-a-roll-of-toilet-paper. It is contractible, does not have the fixed-point-property, and its cone does not have the fixed-point property. Since Kinoshita's fixed-point-free map is not one-to-one, Bing asked the following question.

Bing's Question 12. *Must every homeomorphism of a contractible continuum onto itself leave some point fixed?*

Borsuk [**13**] in 1935 defined a 3-dimensional acyclic continuous curve that admits a fixed-point-free homeomorphism. Bing [**10**] [**11**, Th. 20], in 1967 defined a 2-dimensional continuum with the same properties. Bing and Borsuk's continua are not contractible. Knill's cone-with-a-skirt [**61**] [**11**, Th. 21] is a contractible 2-dimensional continuum without the fixed-point property. However, Knill's fixed-point-free map is not a homeomorphism. In 1972, J.M. Łysko [**69**] gave a negative answer to Question 12. Łysko's continuum which contained two rolls of thickened toilet paper has dimension 3.

Problem 28 (Łysko and Mańka). *Must every homeomorphism of a 2-dimensional contractible continuum have a fixed point?* 580?

At the 2006 Spring Topology and Dynamical Systems Conference, Bellamy asked the following question.

Problem 29. *Must the cone over an arc continuum have the fixed-point property?* 581?

The cone-with-a-skirt is actually the cone over a spiral to a circle in the plane. By filling in the bounded complementary domain of the circle, one gets a non-separating plane continuum with the fixed-point property whose cone admits a fixed-point-free map [**61**] [**11**, Th. 22]. Knill [**61**] in 1967 mentioned the classical problem of determining whether the cone over each tree-like continuum has the fixed-point property.

A. Illanes [**56**] recently solved this problem by showing the cone over a plane continuum X consisting of a spiral to a triod admits a fixed-point-free map. To answer a question of Rogers [**99**, p. 234], Illanes also showed the hyperspace $C(X)$ does not have the fixed-point property. Illanes's example is closely related to the following problem.

Problem 30 (Prajs). *Must the cone over a dendroid have the fixed-point property?* 582?

By attaching to Illanes's plane continuum X a disjoint open arc that joins the center point of the triod in X with the starting point of the spiral in X, we can define a uniquely arcwise connected continuum Y in Euclidean 3-space with the fixed-point property. Since the cone over Y can be retracted onto the cone over X, the cone over Y does not have the fixed-point property.

Problem 31. *Must the cone over a uniquely arcwise connected plane continuum have the fixed-point property?* 583?

Problem 32. *Must the cone over a simply connected plane continuum have the fixed-point property?* 584?

The first version of this paper was presented at the 2006 Spring Topology and Dynamical Systems Conference in Greensboro. Recently, the author received a preprint of a survey article on Bing's twelve questions by Roman Mańka [**75**].

References

[1] V. N. Akis, *Fixed point theorems and almost continuity*, Fund. Math. **121** (1984), no. 2, 133–142.

[2] V. N. Akis, *On the plane fixed point problem*, Topology Proc. **24** (1999), Spring, 15–31.

[3] V. N. Akis and D. W. Curtis, *A characterisation of the absolute quasiretracts in a class of acyclic continua*, Bull. Austral. Math. Soc. **36** (1987), no. 2, 227–237.

[4] W. L. Ayres, *Some generalizations of Scherer fixed-point theorem*, Fund. Math. **16** (1930), 332–336.

[5] H. Bell, *On fixed point properties of plane continua*, Trans. Amer. Math. Soc. **128** (1967), 539–548.

[6] H. Bell, *A fixed point theorem for plane homeomorphisms*, Bull. Amer. Math. Soc. **82** (1976), no. 5, 778–780.

[7] H. Bell, *A fixed point theorem for plane homeomorphisms*, Fund. Math. **100** (1978), no. 2, 119–128.

[8] D. P. Bellamy, *A tree-like continuum without the fixed-point property*, Houston J. Math. **6** (1980), no. 1, 1–13.

[9] R. Bennett, *Locally connected 2-cell and 2-sphere-like continua*, Proc. Amer. Math. Soc. **17** (1966), 674–681.

[10] R. H. Bing, *Challenging conjectures*, Amer. Math. Monthly **74** (1967), no. 1, part II, 56–64.

[11] R. H. Bing, *The elusive fixed point property*, Amer. Math. Monthly **76** (1969), 119–132.

[12] K. Borsuk, *Einige Satze uber stetige Streckenbilder*, Fund. Math. **18** (1932), 198–213.

[13] K. Borsuk, *Sur un continu acyclique qui se lasisse transformer topologiquement en lui-meme sans points invariants*, Fund. Math. **24** (1935), 51–58.

[14] K. Borsuk, *A theorem on fixed points*, Bull. Acad. Polon. Sci. Cl. III **2** (1954), 17–20.

[15] B. Brechner and J. C. Mayer, *The prime end structure of indecomposable continua and the fixed point property*, General topology and modern analysis, Academic Press, New York, 1981, pp. 151–168.

[16] G. E. Bredon, *Some examples for the fixed point property*, Pacific J. Math. **38** (1971), 571–575.

[17] J. Bustamante, R. Escobedo, and F. Macías-Romero, *A fixed point theorem for Whitney blocks*, Topology Appl. **125** (2002), no. 2, 315–321.

[18] M. L. Cartwright and J. E. Littlewood, *Some fixed point theorems*, Ann. of Math. (2) **54** (1951), 1–37.

[19] J. J. Charatonik and J. R. Prajs, *AANR spaces and absolute retracts for tree-like continua*, Czechoslovak Math. J. **55(130)** (2005), no. 4, 877–891.

[20] H. Cook, *Continua which admit only the identity mapping onto non-degenerate subcontinua*, Fund. Math. **60** (1967), 241–249.

[21] H. Cook, *Tree-likeness of dendroids and λ-dendroids*, Fund. Math. **68** (1970), 19–22.

[22] E. Dyer, *A fixed point theorem*, Proc. Amer. Math. Soc. **7** (1956), 662–672.

[23] C. Eberhart and J. B. Fugate, *Universal maps on trees*, Trans. Amer. Math. Soc. **350** (1998), no. 10, 4235–4251.

[24] C. A. Eberhart and J. B. Fugate, *Weakly confluent maps on trees*, General topology and modern analysis, Academic Press, New York, 1981, pp. 209–215.

[25] E. G. Effros, *Transformation groups and C^*-algebras*, Ann. of Math. (2) **81** (1965), 38–55.

[26] L. Fearnley and D. G. Wright, *Geometric realization of a Bellamy continuum*, Bull. London Math. Soc. **25** (1993), no. 2, 177–183.

[27] R. Fokkink, J. C. Mayer, L. G. Oversteegen, and E. D. Tymchatyn, *Fixed points for positively oriented mapings of the plane*, 2005, Preprint.

[28] J. B. Fugate and T. B. McLean, *Compact groups of homeomorphisms on tree-like continua*, Trans. Amer. Math. Soc. **267** (1981), no. 2, 609–620.

[29] J. B. Fugate and L. Mohler, *A note on fixed points in tree-like continua*, Topology Proc. **2** (1977), no. 2, 457–460 (1978).

[30] B. D. Garrett, *Almost continuous retracts of almost continuous retracts and the fixed point property for certain non-separating plane continua*, Topology Proc. **28** (2004), no. 1, 133–142.

[31] G. R. Gordh, *Monotone decompositions of irreducible Hausdorff continua*, Ph.D. dissertation, University of California, Riverside, 1971.

[32] A. Granas, *Fixed point theorems for the approximative ANRs*, Bull. Acad. Polon. Sci. Sér. Sci. Math. Astronom. Phys. **16** (1967), 15–19.

[33] C. L. Hagopian, *A fixed point theorem for plane continua*, Bull. Amer. Math. Soc. **77** (1971), 351–354.

[34] C. L. Hagopian, *Another fixed point theorem for plane continua*, Proc. Amer. Math. Soc. **31** (1972), 627–628.

[35] C. L. Hagopian, *Disk-like products of λ connected continua. I*, Proc. Amer. Math. Soc. **51** (1975), 448–452.

[36] C. L. Hagopian, *A fixed point theorem for hyperspaces of λ connected continua*, Proc. Amer. Math. Soc. **53** (1975), no. 1, 231–234.

[37] C. L. Hagopian, *Fixed point problems for disk-like continua*, Amer. Math. Monthly **83** (1976), no. 6, 471–473.

[38] C. L. Hagopian, *The fixed-point property for almost chainable homogeneous continua*, Illinois J. Math. **20** (1976), no. 4, 650–652.

[39] C. L. Hagopian, *Mapping theorems for plane continua*, Topology Proc. **3** (1978), no. 1, 117–122 (1979).

[40] C. L. Hagopian, *Uniquely arcwise connected plane continua have the fixed-point property*, Trans. Amer. Math. Soc. **248** (1979), no. 1, 85–104.

[41] C. L. Hagopian, *The fixed-point property for deformations of uniquely arcwise connected continua*, Topology Appl. **24** (1986), 207–212.

[42] C. L. Hagopian, *Fixed points of arc-component-preserving maps*, Trans. Amer. Math. Soc. **306** (1988), no. 1, 411–420.

[43] C. L. Hagopian, *Fixed points of plane continua*, Rocky Mountain J. Math. **23** (1993), no. 1, 119–186.

[44] C. L. Hagopian, *The fixed-point property for simply connected plane continua*, Trans. Amer. Math. Soc. **348** (1996), no. 11, 4525–4548.

[45] C. L. Hagopian, *The fixed-point property for deformations of tree-like continua*, Fund. Math. **155** (1998), no. 2, 161–176.

[46] C. L. Hagopian, *Irreducible continua without the fixed-point property*, Bull. Polish Acad. Sci. Math. **51** (2003), no. 2, 121–127.

[47] C. L. Hagopian and R. Mańka, *Simple spirals on double Warsaw circles*, Topology Appl. **128** (2003), no. 2–3, 93–101.

[48] C. L. Hagopian and R. Mańka, *Rational irreducible plane continua without the fixed-point property*, Proc. Amer. Math. Soc. **133** (2005), no. 2, 617–625.

[49] C. L. Hagopian and J. R. Prajs, *A fixed-point anomaly in the plane*, Fund. Math. **186** (2005), no. 3, 233–249.

[50] O. H. Hamilton, *Fixed points under transformations of continua which are not connected im kleinen*, Trans. Amer. Math. Soc. **44** (1938), no. 1, 18–24.

[51] O. H. Hamilton, *A fixed point theorem for pseudo-arcs and certain other metric continua*, Proc. Amer. Math. Soc. **2** (1951), 173–174.

[52] W. Holsztyński, *Universal mappings and fixed point theorems*, Bull. Acad. Polon. Sci. Sér. Sci. Math. Astronom. Phys. **15** (1967), 433–438.

[53] W. Holsztyński, *Fixed points of arcwise connected spaces*, Fund. Math. **64** (1969), 289–312.

[54] S. D. Iliadis, *On the fixed point property of plane continua*, Preprint. Also see *Abstracts of the International Congress of Mathematicians*, Helsinki, Finland, 1978.

[55] S. D. Iliadis, *Positions of continua on the plane and fixed points*, Vestnik Moskov. Univ. Ser. I Mat. Mekh. **25** (1970), no. 4, 66–70.

[56] A. Illanes, *A tree-like continuum whose cone admits a fixed-point-free map*, Houston J. Math., To appear.

[57] W. T. Ingram, *An atriodic tree-like continuum with positive span*, Fund. Math. **77** (1972), no. 2, 99–107.

[58] S. Kinoshita, *On some contractible continua without fixed point property*, Fund. Math. **40** (1953), 96–98.

[59] V. Klee and S. Wagon, *Old and new unsolved problems in plane geometry and number theory*, Mathematical Assocication of America, Washington, DC, 1991.

[60] B. Knaster and S. Mazurkiewicz, *Sur un problème concernant les transformations continues*, Fund. Math. **21** (1933), 85–90.

[61] R. J. Knill, *Cones, products and fixed points*, Fund. Math. **60** (1967), 35–46.

[62] J. Krasinkiewicz, *Concerning the boundaries of plane continua and the fixed point property*, Bull. Acad. Polon. Sci. Sér. Sci. Math. Astronom. Phys. **21** (1973), 427–431.

[63] J. Krasinkiewicz, *On the hyperspaces of snake-like and circle-like continua*, Fund. Math. **83** (1974), no. 2, 155–164.

[64] C. Kuratowski, *Thèorie des continus irreducibles entre deux points II*, Fund. Math. **10** (1927), 225–276.

[65] K. Kuratowski, *Topology. Volume II*, Academic Press, New York, 1968.

[66] A. Lelek, *Disjoint mappings and the span of spaces*, Fund. Math. **55** (1964), 199–214.

[67] W. Lewis, *Almost chainable homogeneous continua are chainable*, Houston J. Math. **7** (1981), no. 3, 373–377.

[68] W. Lopez, *An example in the fixed point theory of polyhedra*, Bull. Amer. Math. Soc. **73** (1967), 922–924.

[69] J. M. Łysko, *An example of a contractible continuum without fixed point property for homeomorphisms*, Bull. Acad. Polon. Sci. Sér. Sci. Math. Astronom. Phys. **20** (1972), 663–666.

[70] R. Mańka, *Association and fixed points*, Fund. Math. **91** (1976), no. 2, 105–121.

[71] R. Mańka, *Some problems concerning fixed points*, Bull. Acad. Polon. Sci. Sér. Sci. Math. Astronom. Phys. **26** (1978), no. 5, 429–435.

[72] R. Mańka, *On the additivity of the fixed point property for 1-dimensional continua*, Fund. Math. **136** (1990), no. 1, 27–36.

[73] R. Mańka, *On spirals and fixed point property*, Fund. Math. **144** (1994), no. 1, 1–9.

[74] R. Mańka, *Cylinders over λ-dendroids have the fixed point property*, Fund. Math. **173** (2002), no. 2, 101–112.

[75] R. Mańka, *The topological fixed point property*, 2006, Preprint.

[76] M. M. Marsh, *s-connected spaces and the fixed point property*, Topology Proc. **8** (1983), no. 1, 85–97.

[77] M. M. Marsh, *A fixed point theorem for inverse limits of fans*, Proc. Amer. Math. Soc. **91** (1984), no. 1, 139–142.

[78] M. M. Marsh, *A fixed-point theorem for inverse limits of simple n-ods*, Topology Appl. **24** (1986), 213–216.

[79] M. M. Marsh, *u-mappings on trees*, Pacific J. Math. **127** (1987), no. 2, 373–387.

[80] M. M. Marsh, *ϵ-mappings onto a tree and the fixed point property*, Topology Proc. **14** (1989), no. 2, 265–277.

[81] M. M. Marsh, *Products of span zero continua and the fixed point property*, Proc. Amer. Math. Soc. **132** (2004), no. 6, 1849–1853.

[82] V. Martínez-de-la-Vega, 2004, In a seminar at California State University, Sacramento.

[83] J. C. Mayer, *Embeddings and prime end structure of chainable continua*, Houston J. Math. **8** (1982), no. 2, 221–253.

[84] J. C. Mayer, L. G. Oversteegen, and E. D. Tymchatyn, *Variation and uniqueness of outchannels*, 2003, Preprint.

[85] P. Minc, *A fixed point theorem for weakly chainable plane continua*, Trans. Amer. Math. Soc. **317** (1990), no. 1, 303–312.

[86] P. Minc, *A tree-like continuum admitting fixed point free maps with arbitrarily small trajectories*, Topology Appl. **46** (1992), no. 2, 99–106.

[87] P. Minc, *A periodic point free homeomorphism of a tree-like continuum*, Trans. Amer. Math. Soc. **348** (1996), no. 4, 1487–1519.

[88] P. Minc, *A weakly chainable tree-like continuum without the fixed point property*, Trans. Amer. Math. Soc. **351** (1999), no. 3, 1109–1121.

[89] P. Minc, *A hereditarily indecomposable tree-like continuum without the fixed point property*, Trans. Amer. Math. Soc. **352** (2000), no. 2, 643–654.

[90] J. Mioduszewski and M. Rochowski, *Remarks on fixed point theorem for inverse limit spaces*, Colloq. Math. **9** (1962), 67–71.

[91] L. Mohler, *The fixed point property for homeomorphisms of 1-arcwise connected continua*, Proc. Amer. Math. Soc. **52** (1975), 451–456.

[92] L. Mohler and L. G. Oversteegen, *Open and monotone fixed point free maps on uniquely arcwise connected continua*, Proc. Amer. Math. Soc. **95** (1985), no. 3, 476–482.

[93] S. B. Nadler, Jr., *Universal mappings and weakly confluent mappings*, Fund. Math. **110** (1980), no. 3, 221–235.

[94] S. B. Nadler, Jr. and J. T. Rogers, Jr., *A note on hyperspaces and the fixed point property*, Colloq. Math. **25** (1972), 255–257.

[95] L. G. Oversteegen and J. T. Rogers, Jr., *An inverse limit description of an atriodic tree-like continuum and an induced map without a fixed point*, Houston J. Math. **6** (1980), no. 4, 549–564.

[96] L. G. Oversteegen and J. T. Rogers, Jr., *Fixed-point-free maps on tree-like continua*, Topology Appl. **13** (1982), no. 1, 85–95.

[97] L. G. Oversteegen and E. D. Tymchatyn, *Plane strips and the span of continua. I*, Houston J. Math. **8** (1982), no. 1, 129–142.

[98] J. T. Rogers, Jr., *The cone = hyperspace property*, Canad. J. Math. **24** (1972), 279–285.

[99] J. T. Rogers, Jr., *Hyperspaces of arc-like and circle-like continua*, Topology conference (Blacksburg, Va., 1973), Springer, Berlin, 1974, pp. 231–235.

[100] J. T. Rogers, Jr., *Whitney continua in the hyperspace $C(X)$*, Pacific J. Math. **58** (1975), no. 2, 569–584.

[101] H. Rosen, *Nonseparating almost continuous retracts of I^n*, Proc. Amer. Math. Soc. **91** (1984), no. 1, 118–122.

[102] J. Segal, *Hyperspaces of the inverse limit space*, Proc. Amer. Math. Soc. **10** (1959), 706–709.

[103] J. Segal, *A fixed point theorem for the hyperspace of a snake-like continuum*, Fund. Math. **50** (1961), 237–248.

[104] K. Sieklucki, *On a class of plane acyclic continua with the fixed point property*, Fund. Math. **63** (1968), 257–278.

[105] P. A. Smith, *Fixed-point theorems for periodic transformations*, Amer. J. Math. **63** (1941), 1–8.

[106] M. Sobolewski, *A uniquely arcwise connected continuum without the fixed point property*, Bull. Polish Acad. Sci. Math. **34** (1986), no. 5–6, 307–313.

[107] M. Sobolewski, *A curve with the fixed point property whose cylinder admits a fixed point free map*, Houston J. Math. **31** (2005), no. 1, 239–253.

[108] J. Stallings, *Fixed point theorems for connectivity maps*, Fund. Math. **47** (1959), 249–263.

[109] F. Wecken, *Fixpunktklassen. III. Mindestzahlen von Fixpunkten*, Math. Ann. **118** (1942), 544–577.

[110] G. S. Young, *Fixed-point theorems for arcwise connected continua*, Proc. Amer. Math. Soc. **11** (1960), 880–884.

[111] W. L. Young, *A product space with the fixed point property*, Proc. Amer. Math. Soc. **25** (1970), 313–317.

Open Problems in Topology II – Edited by E. Pearl

Hyperspaces of continua

Alejandro Illanes

Introduction

A *continuum* is a compact, connected metric space with more than one point. Throughout this paper X will denote a continuum, the *hyperspaces* of X are defined as: $2^X = \{A \subset X : A \text{ is closed and nonempty}\}$, $C(X) = \{A \in 2^X : A \text{ is connected}\}$, $F_n(X) = \{A \in 2^X : A \text{ has at most } n \text{ points}\}$, $C_n(X) = \{A \in 2^X : A \text{ has at most } n \text{ components}\}$.

The hyperspace 2^X is equipped with the Hausdorff metric H. The space $F_n(X)$ is called *n-symmetric product* and $C_n(X)$ is called *n-fold hyperspace.*

Hyperspaces of continua have been widely studied. The book [**19**] surveys most of what was known on this topic up to 1999. There are many open problems on hyperspaces. The interested reader can find many of these problems in [**19**, **29**]. At the end of [**19**], there are comments about the problems posed in [**29**].

In this paper we focus on problems which have been posed since 1999.

The concepts not defined here are taken as in [**30**].

Whitney and diameter maps

A *Whitney map* for 2^X (resp., $C(X)$) is a continuous function $\mu\colon 2^X \to [0,\infty)$ (resp., $\mu\colon C(X) \to [0,\infty)$) such that:

(a) $\mu(\{p\}) = 0$, for each $p \in X$,

(b) if $A, B \in 2^X$ (resp., $C(X)$) and $A \subset B \neq B$, when $\mu(A) < \mu(B)$.

Whitney maps are important tools in the study of hyperspaces, they were introduced by H. Whitney (in [**37**]) in a context different of hyperspaces. As can be seen in [**29**, 0.50.1, 0.50.2 and 0.50.3] and [**19**, 13.5, 13.7 and 13.8], every continuum X admits Whitney maps for 2^X. A general problem in this area is to determine for which continua there is a really simple way of defining a Whitney map for $C(X)$. For instance, it is easy to see that the diameter map $\operatorname{diam}\colon C([0,1]) \to [0,1]$ is a Whitney map.

Problem 1. *Is the arc the only continuum X for which there exists a metric such* 585?
that $\operatorname{diam}\colon C(X) \to [0,\infty)$ *is a Whitney map?*

It is easy to check that if X is the $\sin(\frac{1}{x})$-continuum (the closure in the Euclidean plane $\mathbb{R}^2$ of the graph of the function $\sin(\frac{1}{x})$, defined on the interval $(0,1]$) and if $\pi_1, \pi_2\colon \mathbb{R}^2 \to \mathbb{R}$ are the projections on the first and second coordinate, respectively, then the function $\mu\colon C(X) \to [0,3]$ given by $\mu(A) = \operatorname{diam}(\pi_1(A)) + \operatorname{diam}(\pi_2(A))$ is a Whitney map. This motivates the following definition: a continuum Y is said to admit an *n-determined Whitney map* provided that there exist continuous functions $f_1, \dots, f_n\colon Y \times Y \to [0,\infty)$ such that the map $\mu\colon C(Y) \to [0,\infty)$ given by $\mu(A) = \operatorname{diam}(f_1(A)) + \cdots + \operatorname{diam}(f_n(A))$ is a Whitney map. From the observation above, the $\sin(\frac{1}{x})$-continuum admits a 2-determined Whitney map and it can be seen that a simple closed curve S is not n-determined for any positive integer n. This motivates the following problem.

586? **Problem 2.** *Characterize those continua which admit a 2-determined Whitney map. Characterize those continua which admit an n-determined Whitney map, for some positive integer n.*

In [**29**, Remark 14.67, p. 471], S.B. Nadler, Jr., observed that if S is a unit circle in $\mathbb{R}^2$, then the diameter map from 2^S onto the interval $[0, 2]$ is not an open map (if $p, q, r \in S$ are the vertices of an equilateral triangle, then diam attains a local minimum at $\{p, q, r\}$). Answering a question by S.B. Nadler, Jr., in [**29**, Question 14.68], W.J. Charatonik and A. Samulewicz proved that if X is the suspension over a discrete compact set, then X admits a metric for which the diameter map from 2^X onto $[0, \operatorname{diam}(X)]$ is open. In particular, a closed simple curve S admits a metric for which the diameter map is open. The following questions remain open.

587? **Problem 3** ([**6**, Problem 5.10]). *Does the suspension of any compact metric space (continuum) admit an open diameter mapping? In particular, does the n-dimensional sphere ($n > 1$) admit such a mapping?*

588? **Problem 4** ([**6**, Problem 5.19]). *Does every dendrite (local dendrite, graph, locally connected continuum) admit an open diameter mapping?*

Cones, products and hyperspaces

Using a number of results found by several authors, in [**22**] and [**18**], A. Illanes and M. de J. López gave a complete description of those continua X for which there exists a finite dimensional continuum Z such that $C(X)$ is homeomorphic to the cone over the cone of Z. With respect to products, A. Illanes has shown [**11**] that a continuum X has the properties that $C(X)$ is finite dimensional and it is homeomorphic to the product of two nondegenerate continua if and only if X is an arc or a simple closed curve.

It is natural to ask (assuming that $\dim C_n(X) < \infty$ and $n > 2$) for which continua X, is $C_n(X)$ homeomorphic to a cone or to the product of two nondegenerate continua? In [**26**], some partial answers to this question are given. There are only a few examples on this topic, they are described below.

(a) $C_2([0, 1])$ is homeomorphic to $[0, 1]^4$ (R. Shori, [**14**, Lemma 2.2]).
(b) $C_2(S^1)$ (where S^1 is a simple closed curve) is homeomorphic to the cone over a solid torus ([**17**]).
(c) $C_n(X)$ is homeomorphic to a cone when X is the cone over a 0-dimensional compact space ([**25**]).

The following questions remain open.

589? **Problem 5.** *Is $C_n(S^1)$ homeomorphic to a cone for some $n > 2$ (for all $n > 2$)? It would be interesting to know the answer to this problem for $n = 3$.*

Problem 6. *Is there a finite graph X different from simple m-ods, $[0,1]$ or S^1 such that $C_n(X)$ is homeomorphic to a cone or to the product of two nondegenerate continua, for some $n > 1$?* 590?

Problem 7 ([**26**, Question 3.8]). *Does there exist a hereditarily decomposable continuum X that is different from simple m-ods, $[0,1]$ or S^1 such that $C_n(X)$ is homeomorphic to the cone over a finite-dimensional continuum for some integer $n > 1$?* 591?

Problem 8 ([**26**, Question 3.7]). *Does there exist an indecomposable continuum X such that $C_n(X)$ is homeomorphic to the cone over a finite-dimensional continuum for some integer $n > 1$?* 592?

Problem 9 ([**26**, Question 4.12]). *Does there exist a continumm X that is not an arc, for which there is an integer $n > 1$ such that $C_n(X)$ is homeomorphic to the product of two finite-dimensional continua?* 593?

In this area, it would be interesting to characterize those continua X for which $F_n(X)$ is homeomorphic to the cone over some continuum Z and it would be also interesting to characterize those continua X for which $F_n(X)$ is homeomorphic to the product of two nondegenerate continua. There are only a few results on this direction. E. Castañeda showed in [**4**] that: (a) if X is a finite graph, then $F_2(X)$ is the cone over a continuum Y if and only if X is a simple m-od or an arc and, (b) if X is a finite graph, then $F_2(X)$ is a product of two nondegenerate continua if and only if X is an arc. On the other hand, it is known (see [**1**, Theorem 6]) that $F_3([0,1])$ is homeomorphic to $[0,1]^3$. In [**3**, Lemma 1], E. Castañeda proved that If X is a simple m-od, then $F_2(X)$ is homeomorphic to the cone over a continuum Z, this result was extended in [**25**] for every $F_n(X)$. The following problems are open.

Problem 10 ([**4**, Question 3.15]). *Is $[0,1]$ the only finite graph such that $F_3(X)$ is a product of two nondegenerate continua?* 594?

Problem 11. *Do there exist a finite graph X and an integer $n \geq 4$ such that $F_n(X)$ is a product of two nondegenerate continua?* 595?

Problem 12. *Are the simple m-ods and the arc the only finite graphs for which $F_n(X)$ ($n \geq 3$) is the cone over a continuum Y? This problem is interesting even for $n = 3$.* 596?

Problem 13 ([**25**, Question 3.5]). *Does there exist a continuum X which is not the cone over a compactum such that $F_n(X)$ is homeomorphic to the cone over a finite-dimensional continuum for some integer $n \geq 2$?* 597?

Means

A *mean* is a continuous function $m\colon F_2(X) \to X$ such that $m(\{x\}) = x$ for each $x \in X$. The main problem in this area is to charaterize those continua X which admit a mean. Many authors have studied this problem and there are a number of open problems on this area. In this section we only include some of my favorite questions, all of them appeared in [**20**], where the following results were obtained.

(a) Each dendrite admits a monotone mean, while the harmonic fan admits no monotone mean.
(b) Each n-cell, as well as the dyadic solenoid, admits a mean that is monotone and open, simultaneously.
(c) Each simple n-od, as well as the Cantor fan admits no open mean.
(d) The harmonic fan admits a confluent mean.

The interested reader can find more information and problems about means in [**19**, Ch. XII, Section 76] and [**5**].

598? **Problem 14** ([**20**, Question 2.3]). *Suppose that X is a dendroid and it admits a monotone mean, does it follow that X is a dendrite?*

599? **Problem 15** ([**20**, Question 3.8]). *Does each tree admit an open mean?*

600? **Problem 16** ([**20**, Question 3.9]). *Does there exist a dendrite X such that X is not a tree and X admits and open mean?*

601? **Problem 17** ([**20**, Question 4.2]). *Does there exist a continuum X such that X admits a mean but X does not admit a confluent mean?*

Fixed point property

B. Knaster asked in *The Scottish book*, in 1952, whether $C(X)$ must have the fixed point property when X has the fixed point property. In [**32**] S.B. Nadler, Jr. and J.T. Rogers, Jr. showed that if Y is the union of a disk D with a ray surrounding the boundary of D, then Y the fixed point property but $C(Y)$ and 2^Y do not have the fixed point property. So, J.T. Rogers asked if $C(X)$ has the fixed point when X is a tree-like continuum [**28**, Problem 446, p. 307]. Recently, the author has answered this question in [**9**] by showing that if Z is the union of a simple triod T with a ray surrounding it, then $C(Z)$ does not have the fixed point property. It is not known if the statements (a) $C(X)$ has the fixed point property and, (b) 2^X has the fixed point property; are equivalent [**29**, Question 7.12, p. 299].

602? **Problem 18.** *Let Z be as in the previous paragraph. Does 2^Z have the fixed point property?*

With respect to symmetric products, in the paper in which they were introduced [**1**], K. Borsuk and S. Ulam asked whether $F_n(X)$ has the fixed point property when X has the fixed point property. This question was solved by J. Oledzki in 1988 [**33**], who gave an example of a 2-dimensional continuum X with the fixed point property such that $F_2(X)$ does not have the fixed point property. With this in mind, S.B. Nadler, Jr. has offered the following list of new questions (among others) in [**31**].

603? **Problem 19** ([**31**, 8.12, p. 119]). *Does $F_n(X)$ have the fixed point property when X is a circle-like continuum with the fixed point property?*

Problem 20 ([**31**, 8.13, p. 119]). *Does $F_n(X)$, $n \geq 2$, have the fixed point property when X is hereditarily indecomposable continuum with the fixed point property?* 604?

The answer to Problem 20 is not known even when X is the pseudo-arc and $n \geq 3$.

Problem 21 ([**31**, 8.14, p. 120]). *Is there a 1-dimensional continuum X with the fixed point property such that $F_n(X)$ does not have the fixed point property for some n?* 605?

Problem 22 ([**31**, 8.15, p. 120]). *If X is a tree-like continuum with the fixed point property then does $F_n(X)$ have the fixed point property?* 606?

Problem 23 ([**31**, 8.16, p. 120]). *Is there a continuum X such that $F_n(X)$ has the fixed point property for some $n > 1$ and, yet, $F_m(X)$ does not have the fixed point property for some m?* 607?

Problem 24 ([**31**, 8.17, p. 121]). *Is there a continuum X such that X and $F_2(X)$ have the fixed point property but $F_3(X)$ does not have the fixed point property?* 608?

Problem 25 ([**31**, 8.23, p. 122]). *If X is a continuum such that $C(X)$ has the fixed point property, then does $C_n(X)$ have the fixed point property for each n?* 609?

Problem 26 ([**31**, 8.24, p. 123]). *If X is a continuum such that $F_n(X)$ has the fixed point property for all n, then does $C_n(X)$ have the fixed point property for all n?* 610?

Let Z be as in the first paragraph at the beginning of this section (a simple triod with a ray surrounding it). Since it has been proved [**9**] that $C(Z)$ does not have the fixed point property, it would be interesting to know if $F_n(Z)$ has the fixed point property for each n. If this is true, then Z would provide a negative answer to Problem 26. If this is false, then Problem 22 would be solved in the negative.

Problem 27. *Let Z be as in the previous paragraph, does $F_n(Z)$ have the fixed point property for all n?* 611?

Problem 28 ([**31**, 8.26, p. 123]). *If X is a continuum such that $F_n(X)$ has the fixed point property for all n, then does 2^X have the fixed point property?* 612?

Problem 29 ([**31**, p. 77]). *Does $F_n(X)$ have the fixed point property when X is arc-like and $n \geq 3$?* 613?

It is known that if X is arc-like, then $F_2(X)$ has the fixed point property ([**19**, 22.25, p. 199]).

Mappings between hyperspaces

In [**29**, 22.25, p. 199], S.B. Nadler, Jr., discussed the problem of when there exists a continuous map from one of the continua $C(X)$, 2^X or X onto another. On this topic, I. Krzemińska and J.R. Prajs [**21**] have shown that there exists a uniformly path connected continuum X such that X is not a continuous image of $C(X)$. Recently, A. Illanes [**12**], has constructed a continuum X such that $C(X)$ is not a continuous image of X. The following questions remain open.

614? **Problem 30** ([**21**, 22.25, p. 199]). *If $f\colon X \to Y$ is a continuous surjection between continua, does there exist a continuous surjection $g\colon C(X) \to C(Y)$?*

615? **Problem 31** ([**21**, Question 3, p. 61]). *Given a continuum X, does there exist a continuous surjection $f\colon C(X) \to C(C(X))$?*

Unicoherence of $F_2(S^1)$

Answering a question by A. Illanes and A. García-Máynez, E. Castañeda showed (in [**2**]) that if S_1 and S_2 are the circles in the plane, centered at the origin, with radius 1 and 2, respectively and, R is a topological copy of the real line such that one end surrounds asymptotically the circle S_1 and the other end surrounds S_2, then $X = S_1 \cup S_2 \cup R$ is unicoherent but $F_2(X)$ is not unicoherent. It is known that if X is a locally connected unicoherent continuum, then $F_2(X)$ is unicoherent [**10**]. A discussion on this topic can be found in [**7**]. The following questions are open.

616? **Problem 32** ([**2**, Problem 1, p. 66]). *Does there exist an indecomposable continuum X such that $F_2(X)$ is not unicoherent?*

617? **Problem 33** ([**2**, Problem 2, p. 66]). *Does there exist a hereditarily unicoherent continuum X such that $F_2(X)$ is not unicoherent?*

618? **Problem 34** (J.J. Charatonik, [**2**, Problem 3, p. 66]). *Does there exist a hereditarily unicoherent, hereditarily decomposable continuum X such that $F_2(X)$ is not unicoherent?*

Locating cells in hyperspaces

In [**23**], S. López made a very detailed study of those continua X for which the element X in $C(X)$ has a neighborhood in $C(X)$ which is a 2-cell. The following question remains open.

619? **Problem 35** ([**23**, Question 10, p. 189]). *Suppose that there is a neighborhood $\mathcal{D}$ of X in $C(X)$ such that $\mathcal{D}$ is embeddable in the plane. Does X have a neighborhood in $C(X)$ which is a 2-cell?*

Locating m-cells in hyperspaces has been an important tool in the study of hyperspaces. An *m-od* in a continuum X is a subcontinuum B for which there exists a subcontinuum $A \subset B$ such that $B - A$ contains at least m components. When $C_1, \dots, C_m$ are components of $B - A$, taking an order arc α_i from A to $A \cup \mathrm{cl}_X(C_i)$, for each $i \in \{1, \dots, m\}$ (that is, $\alpha_i\colon [0,1] \to C(X)$ is a continuous function such that $\alpha_i(0) = A$, $\alpha_i(1) = A \cup \mathrm{cl}_X(C_i)$ and $\alpha_i(s) \subsetneq \alpha_i(t)$ if $s < t$, for the existence of order arcs see [**19**, Theorem 15.3]) and defining $\varphi\colon [0,1]^m \to C(X)$ by $\varphi((t_1, \dots, t_m)) = \alpha_1(t_1) \cup \dots \cup \alpha_m(t_m)$, it is easy to see that φ is an embedding. Thus, if there exists an m-od in X, then there exists an m-cell in $C(X)$. The converse of this implication is also true (see [**19**, Theorem 70.1]). Therefore, we have a complete intrinsic charaterization of those continua X for which there exists an m-cell in $C(X)$. It would be interesting to have a similar characterization for the hyperspaces $C_n(X)$. With the idea described above it can be proved that if $B_1, \ldots, B_n$ are pairwise disjoint subcontinua of X such that each B_i is an r_i-od, then there exists a $(r_1+\cdots+r_n)$-cell in $C_n(X)$. The problem here is to determine if this is the only way to obtain cells in $C_n(X)$. Thus we have the following problem.

Problem 36. *Suppose that X is a continuum such that there exists an m-cell in $C_n(X)$, then does there exist pairwise disjoint subcontinua $B_1, \ldots, B_n$ of X such that each B_i is an r_i-od and $m = r_1 + \cdots + r_n$?* 620?

Unique hyperspaces

The continuum X is said to have *unique hyperspace* $C(X)$ $(2^X, C_n(X)$ and $F_n(X)$, respectively) provided that if Y is a continuum and $C(X)$ $(2^X, C_n(X)$ and $F_n(X)$, respectively) is homeomorphic to $C(Y)$ $(2^Y, C_n(Y)$ and $F_n(Y)$, respectively), then X is homeomorphic to Y. A discussion on what is known about unique hyperspaces can be found in [**13**]. The following questions are open.

Problem 37 ([**13**, p. 77]). *Do hereditarily indecomposable continua X have unique hyperspace $F_2(X)$?* 621?

Problem 38 ([**8**, p. 93]). *Let X and Y be dendrites whose respective sets of end-points are closed. Suppose that $C_2(X)$ is homeomorphic to $C_2(Y)$, then does it follow that X is homeomorphic to Y?* 622?

The respective question for $C_n(X)$ instead of $C_2(X)$ with $n \neq 2$ has been solved in the affirmative in [**8**, Theorem 5.24].

Problem 39. *Let X be a dendrite and let Y be a continuum such that $F_n(X)$ is homeomorphic to $F_n(Y)$, for some $n \geq 3$. Does it follow that Y is also a dendrite?* 623?

The respective question for $n = 2$ was answered in the positive in [**13**].

Problem 40. *Let X and Y be a dendrites. Suppose that the respective sets of ordinary points (that is, sets of non-ramification points) are open and $F_n(X)$ is homeomorphic to $F_n(Y)$, for some $n \geq 3$. Does it follow that X is homeomorphic to Y?* 624?

The respective question for $n = 2$ was answered in the positive in [**13**].

Problem 41. *Let X and Y be metric compactifications of the ray $[0, 1)$. Suppose that $F_3(X)$ and $F_3(Y)$ are homeomorphic. Does it follow that X and Y are homeomorphic?* 625?

The respective question for $n \neq 3$ has been recently solved in the affirmative by J.M. Martínez-Montejano.

Problem 42. *Let X be a metric compactification of the ray $[0, 1)$ and let Y be a continuum. Suppose that $F_n(X)$ is homeomorphic to $F_n(Y)$ for some $n > 1$. Does it follow that Y is a metric compactification of the ray $[0, 1)$?* 626?

1. Miscellaneous problems

In [**3**], E. Castañeda, showed that if X is a locally connected continuum, then $F_2(X)$ can be embedded in $\mathbb{R}^3$ if and only if X can be embedded in the figure-8 curve (the continuum obtained by joining two simple closed curves by a point). The following problem is open.

627? **Problem 43.** *Can $F_2(X)$ be embedded in $\mathbb{R}^4$ for each finite graph X?*

A topological property $\mathcal{P}$ is said to be *sequential decreasing Whitney property* provided that if μ is a Whitney map for $C(X)$, $\{t_n\}_{n=1}^{\infty}$ is a sequence in the interval $(t, \mu(X))$ such that $\lim t_n = t$ and each Whitney level $\mu^{-1}(t_n)$ has property $\mathcal{P}$, then $\mu^{-1}(t)$ has property $\mathcal{P}$. Sequential decreasing Whitney properties were introduced and studied by F. Orozco-Zitli in [**34**], where he posed the following problem.

628? **Problem 44** ([**34**, Question 1, p. 305]). *Is the property of being a hereditarily arcwise connected continuum a sequential decreasing Whitney property?*

Let $A_s(X) = \{A \in C(X) : A \text{ is an arc}\} \cup F_1(X)$. It is known that if X is a dendrite, then $A_s(X)$ is homeomorphic to $F_2(X)$, by associating each arc with its set of end-points. On the other hand, in [**15**], it has been proved that if X is a dendroid with only one ramification point and $F_2(X)$ is homeomorphic to $A_s(X)$, then X is a dendrite. So the following question arises naturally.

629? **Problem 45** (A. Soto, [**15**, Question 1, p. 308]). *If X is a dendroid such that $F_2(X)$ and $A_s(X)$ are homeomorphic, must X be a dendrite?*

Given a continuous function between continua $f\colon X \to Y$ the *induced mapping* $2^f\colon 2^X \to 2^Y$ is defined by $2^f(A) = f(A)$ (the image of A under f). A wide discussion on what has been done about induced mappings can be found in [**19**, Ch. XII, Section 77].

630? **Problem 46.** *Suppose that X is hereditarily indecomposable and $F\colon 2^X \to 2^X$ is a homeomorphism, is it true that there exist a homeomorphism $f\colon X \to X$ such that $F = 2^f$?*

The continuum X is said to be *zero-dimensional closed set aposyndetic* provided that for each zero-dimensional closed subset A of X and for each $p \in X - A$, there exists a subcontinuum M of X such that $p \in \operatorname{int}(M)$ and $M \cap A = \emptyset$. Answering a question by J. Goodykoontz, Jr., recently, J.M. Martínez-Montejano has shown that the hyperspaces 2^X and $C_n(X)$ (for all n) are zero-dimensional closed set aposyndetic and he offered the following question.

631? **Problem 47** ([**24**, Question 3.1]). *Let $n \geq 3$. Is $F_n(X)$ zero-dimensional closed set aposyndetic?*

632? **Problem 48** ([**16**, Problem 1, p. 180]). *Are there integers $1 < n < m$ and continua X and Y such that* $\dim C(X)$ *is finite and $C_n(X)$ is homeomorphic to $C_m(Y)$?*

Some partial answers to this question are given in [**16**].

Problem 49 ([**16**, Problem 2, p. 180]). *Do there exist two non-homeomorphic* 633?
continua X and Y such that $C_n(X)$ is homeomorphic to $C_n(Y)$, $\dim C_n(X)$ *is finite and $n > 1$?*

Given $p \in X$, in [**36**], the following map was considered: $\varphi_p \colon X \to F_2(X)$, given by $\varphi_p(x) = \{p, x\}$. In the same paper, P. Pellicer-Covarrubias proved that [**36**, Lemma 5.3] if X is contractible, then φ_p is a deformation retraction in $F_2(X)$. This motivates the following problem.

Problem 50 ([**36**, p. 291]). *Suppose that φ_p is a deformation retraction, does this* 634?
imply that X is contractible?

Given $p \in X$, let $C(p, X) = \{A \in C(X) : p \in A\}$ and let $\mathcal{K}(X) = \{C(p, X) \in C(C(X)) : p \in X\}$. Spaces of the form $\mathcal{K}(X)$ were studied by P. Pellicer-Covarrubias in [**35**], where she posed the following problems.

Problem 51 ([**35**, p. 284]). *Let T be a simple triod. Does there exists a continuum* 635?
X such that X is not homeomorphic to T and $\mathcal{K}(X)$ is homeomorphic to $\mathcal{K}(T)$? If so, must X be indecomposable?

Problem 52 ([**35**, p. 284]). *Let G be a finite graph. Does there exists a continuum* 636?
X such that X is not homeomorphic to G and $\mathcal{K}(X)$ is homeomorphic to $\mathcal{K}(G)$? If so, must X be indecomposable?

We finish this list of problems including an interesting problem of Continuum Theory (not of hyperspaces) which has not been posed in the literature. For the particular case when X is the pseudo-arc, this question was solved in [**27**].

Problem 53. *Is it true that for each continuum X there exists an uncountable* 637?
family of pairwise non-homeomorphic metric compactifications of the ray $[0, 1)$ with remainder X.

References

[1] K. Borsuk and S. Ulam, *On symmetric products of topological spaces*, Bull. Amer. Math. Soc. **37** (1931), 875–882.

[2] E. Castañeda, *A unicoherent continuum whose second symmetric product is not unicoherent*, Topology Proc. **23** (1998), Spring, 61–67.

[3] E. Castañeda, *Embedding symmetric products in Euclidean spaces*, Continuum theory (Denton, TX, 1999), Lecture Notes in Pure and Appl. Math., vol. 230, Marcel Dekker Inc., New York, 2002, pp. 67–79.

[4] E. Castañeda, *Symmetric products as cones and products*, Topology Proc. **28** (2004), no. 1, 55–67.

[5] J. J. Charatonik, *Some problems concerning means on topological spaces*, Topology, measures, and fractals (Warnemünde, 1991), Akademie-Verlag, Berlin, 1992, pp. 166–177.

[6] W. J. Charatonik and A. Samulewicz, *On size mappings*, Rocky Mountain J. Math. **32** (2002), no. 1, 45–69.

[7] A. García-Máynez and A. Illanes, *A survey on unicoherence and related properties*, An. Inst. Mat. Univ. Nac. Autónoma México **29** (1989), 17–67 (1990).

[8] D. Herrera-Carrasco, *Hiperespacios de dendritas*, Doctoral dissertation, Facultad de Ciencias, Universidad Nacional Autónoma de México, November, 2005.

[9] A. Illanes, *A tree-like continuum whose hyperspace of subcontinua has a fixed-point-free map*, Preprint.

[10] A. Illanes, *Multicoherence of symmetric products*, An. Inst. Mat. Univ. Nac. Autónoma México **25** (1985), 11–24.

[11] A. Illanes, *Hyperspaces which are products*, Topology Appl. **79** (1997), no. 3, 229–247.

[12] A. Illanes, *A continuum whose hyperspace of subcontinua is not g-contractible*, Proc. Amer. Math. Soc. **130** (2002), no. 7, 2179–2182.

[13] A. Illanes, *Dendrites with unique hyperspace* $F_2(X)$, JP J. Geom. Topol. **2** (2002), no. 1, 75–96.

[14] A. Illanes, *The hyperspace* $C_2(X)$ *for a finite graph* X *is unique*, Glas. Mat. Ser. III **37(57)** (2002), no. 2, 347–363.

[15] A. Illanes, *Hyperspaces of arcs and two-point sets in dendroids*, Topology Appl. **117** (2002), no. 3, 307–317.

[16] A. Illanes, *Comparing n-fold and m-fold hyperspaces*, Topology Appl. **133** (2003), no. 3, 179–198.

[17] A. Illanes, *A model for the hyperspace* $C_2(S^1)$, Questions Answers Gen. Topology **22** (2004), no. 2, 117–130.

[18] A. Illanes and M. de J. López, *Hyperspaces homeomorphic to cones. II*, Topology Appl. **126** (2002), no. 3, 377–391.

[19] A. Illanes and S. B. Nadler, Jr., *Hyperspaces*, Marcel Dekker Inc., New York, 1999.

[20] A. Illanes and L. C. Simón, *Means with special properties*, Houston J. Math. **29** (2003), no. 2, 313–324.

[21] I. Krzemińska and J. R. Prajs, *On continua whose hyperspace of subcontinua is* σ*-locally connected*, Topology Appl. **96** (1999), no. 1, 53–61.

[22] M. de J. López, *Hyperspaces homeomorphic to cones*, Topology Appl. **126** (2002), no. 3, 361–375.

[23] S. López, *Hyperspaces locally 2-cell at the top*, Continuum theory (Denton, TX, 1999), Lecture Notes in Pure and Appl. Math., vol. 230, Marcel Dekker Inc., New York, 2002, pp. 173–190.

[24] J. M. M. Martínez-Montejano, *Zero-dimensional closed set aposyndesis and hyperspaces*, Houston J. Math. **32** (2006), no. 4, 1101–1105.

[25] S. Macías, *Fans whose hyperspaces are cones*, Topology Proc. **27** (2003), no. 1, 217–222.

[26] S. Macías and S. B. Nadler, Jr., *n-fold hyperspaces, cones and products*, Topology Proc. **26** (2001/02), no. 1, 255–270.

[27] V. Martínez-de-la-Vega, *An uncountable family of metric compactifications of the ray with remainder pseudo-arc*, Topology Appl. **135** (2004), 207–213.

[28] J. van Mill and G. M. Reed (eds.), *Open problems in topology*, North-Holland, Amsterdam, 1990.

[29] S. B. Nadler, Jr., *Hyperspaces of sets*, Marcel Dekker Inc., New York, 1978.

[30] S. B. Nadler, Jr., *Continuum theory*, Marcel Dekker Inc., New York, 1992.

[31] S. B. Nadler, Jr., *The fixed point property for continua*, Textos, vol. 30, Sociedad Matemática Mexicana, 2005.

[32] S. B. Nadler, Jr. and J. T. Rogers, Jr., *A note on hyperspaces and the fixed point property*, Colloq. Math. **25** (1972), 255–257.

[33] J. Oledzki, *On symmetric products*, Fund. Math. **131** (1988), no. 3, 185–190.

[34] F. Orozco-Zitli, *Sequential decreasing Whitney properties*, Continuum theory (Denton, TX, 1999), Lecture Notes in Pure and Appl. Math., vol. 230, Marcel Dekker Inc., New York, 2002, pp. 297–306.

[35] P. Pellicer-Covarrubias, *The hyperspaces* $C(p, X)$, Topology Proc. **27** (2003), no. 1, 259–285.

[36] P. Pellicer-Covarrubias, *Retractions in hyperspaces*, Topology Appl. **135** (2004), 277–291.

[37] H. Whitney, *Regular families of curves, I*, Proc. Natl. Acad. Sci. **18** (1932), 275–278.

Open Problems in Topology II – Edited by E. Pearl

Inverse limits and dynamical systems

W. T. Ingram

1. Introduction

Throughout this article, we use the term *continuum* to mean a compact, connected subset of a metric space; by a *mapping* we mean a continuous function. A continuum is *decomposable* provided it is the union of two of its proper subcontinua; a continuum is *indecomposble* if it is not decomposable. A continuum is *hereditarily decomposable* if each of its subcontinua is decomposable.

If $X_1, X_2, X_3, \ldots$ is a sequence of metric spaces and $f_1, f_2, f_3, \ldots$ is a sequence of mappings such that $f_i\colon X_{i+1} \to X_i$ for $i = 1, 2, 3, \ldots$, by the *inverse limit* of the inverse limit sequence $\{X_i, f_i\}$ is meant the subset of the product space $\prod_{i>0} X_i$ that contains the point $(x_1, x_2, x_3, \ldots)$ if and only if $f_i(x_{i+1}) = x_i$ for each positive integer i. The inverse limit of the inverse limit sequence $\{X_i, f_i\}$ is denoted by $\varprojlim\{X_i, f_i\}$. For convenience of notation, we will use boldface characters to denote sequences. Thus, if $s_1, s_2, s_3, \ldots$ is a sequence, we denote this sequence by $\mathbf{s}$. By this convention, the point $(x_1, x_2, x_3, \ldots)$ of an inverse limit space will also be denoted by $\mathbf{x}$, the sequence of factor spaces by $\mathbf{X}$ and the sequence of bonding maps by $\mathbf{f}$. For brevity, we will denote the inverse limit space by $\varprojlim \mathbf{f}$.

A problem set is invariably personal and reflects the interests of the compiler of the set. So it is with this collection of problems. Because of recent developments in the use of inverse limits in certain kinds of models in economics, in Section 7 we include some problems arising from this although we have not personally contributed anything to these applications. Instead we rely on some who have made contributions for problems that reflect the current state of this research.

2. Characterization of chainability

Although it is not the original definition of chainability we take as our definition that a continuum is *chainable* to be that the continuum is homeomorphic to an inverse limit on intervals; a continuum is *tree-like* provided it is homeomorphic to an inverse limit on trees. A continuum is *unicoherent* provided it is true that if it is the union of two subcontinua H and K then $H \cap K$ is connected; a continuum is *hereditarily unicoherent* provided every subcontinuum of it is unicoherent. A continuum M is a *triod* provided there is a subcontinuum H of M so that $M \setminus H$ has at least three components; a continuum is *atriodic* provided it contains no triod. It is immediate that chainable continua are tree-like. It is well known that chainable continua are atriodic and tree-like continua are hereditarily unicoherent.

Several characterizations of chainability of a continuum exist. These include (1) (the original definition) for each $\varepsilon > 0$ there is a finite collection of open sets $C_1, C_2, \ldots, C_n$ covering M such that $\operatorname{diam}(C_i) < \varepsilon$ for $1 \le i \le n$ and $C_i \cap C_j \ne \emptyset$ if and only if $|i - j| \le 1$ and (2) for each positive number ε there is a map f_ε of the continuum to $[0, 1]$ such that if t is in $[0, 1]$ then the diameter of $f_\varepsilon^{-1}(f_\varepsilon(t))$ is less than ε. Notably missing is a characterization involving a list of internal topological properties of the continuum. For example, in case the continuum is hereditarily decomposable, RH Bing [**3**, Theorem 11] proved that the continuum is chainable if and only if it is atriodic and hereditarily unicoherent. This characterization for hereditarily decomposable continua is satisfying in that it is given in terms of internal topological properties of the continuum.

638? **Problem 1.** *Characterize chainability of a continuum in terms of internal topological properties of the continuum.*

J.B. Fugate [**10**] extended Bing's result from the class of hereditarily decomposable continua to the class of those continua having the property that every indecomposable subcontinuum is chainable. Thus, Problem 1 may be solved by characterizing chainability of indecomposable continua. Case and Chamberlin [**6**] gave a characterization of tree-like continua as those one-dimensional continua for which every mapping to a one-dimensional polyhedron is inessential (i.e., homotopic to a constant map). J. Krasinkiewicz later proved that a one-dimensional continuum is tree-like if and only if every mapping of it to a figure-8 (the union of two circles with a one-point intersection) is inessential [**26**]. Although these characterizations of tree-likeness do not involve "internal" topological properties, it would still be of significant interest to characterize chainability among tree-like continua. Since tree-like continua are hereditarily unicoherent, atriodicity is a natural candidate for one of the properties on a list of characterizing properties. That atriodicity alone is not sufficient was shown in [**14**].

One significant attempt at characterizing chainability involves the notion of the span of a continuum. If M is a continuum, the *span* of M is the least upper bound of $\{\varepsilon \geq 0 :$ there is a subcontinuum C of $M \times M$ such that $p_1(C) = p_2(C)$ and $\operatorname{dist}(x, y) \geq \varepsilon$ for all (x, y) in $C\}$ (p_1 and p_2 denote the projections of $M \times M$ into its factors). The following problem on span remains open even though it was featured [**8**] in the first volume of *Open problems in topology.*

639? **Problem 2.** *If the span of a continuum is 0, is M chainable?*

A. Lelek intoduced span in [**28**] and proved that chainable continua have span 0. Although span 0 is a topological property, in some real sense it is not "internal". Consequently, if one were to settle Problem 2 in the affirmative, the nature of the definition of span would, in this author's opinion, leave work to be done on Problem 1. That said, Problem 2 is significant in its own right and not only because it has become an "old" problem. For instance, a positive solution would tell us that we know all of the homogeneous plane continua [**34**].

3. Plane embedding

In thinking about Problem 1 and in light of the Case–Chamberlin theorem [**6**] characterizing tree-likeness, the author began a quest to settle the question whether atriodic tree-like continua are chainable. That investigation led to an example of an atriodic tree-like continuum that is not chainable [**14**]. Span turned out to be just the tool needed to show that the example obtained is not chainable.

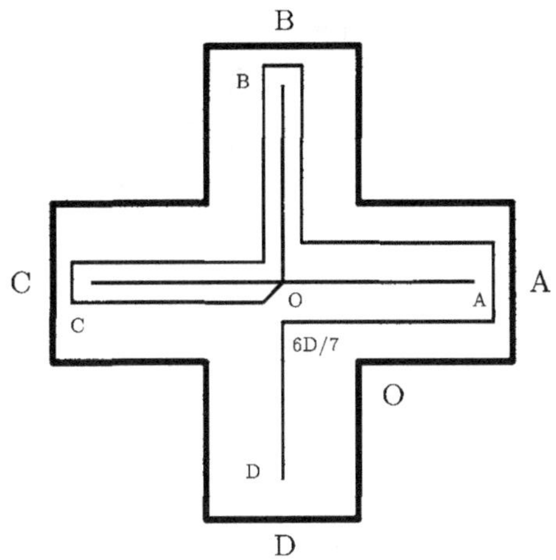

FIGURE 1. Bing's example of a nonplanar tree-like continuum

However, span was not the first tool the author tried to use. In fact, two other properties of chainable continua first came to mind: planarity and the fixed point property. Bing showed that chainable continua can be embedded in the plane [**3**] and O.H. Hamilton showed that chainable continua have the fixed point property [**12**]. The author chose to try to employ Bing's result and construct an atriodic tree-like continuum that cannot be embedded in the plane. This leads to our next problem.

Problem 3. *Characterize those tree-like continua that can be embedded in the plane.* 640?

The reader interested in this problem should be familiar with Brian Raines' work [**35**] on local planarity of inverse limits of graphs.

Of course, tree-like continua that cannot be embedded in the plane are well known. Arguably the simplest of these may be one given by Bing [**3**]. This example consists of a ray with remainder a simple triod together with an arc that intersects the union of the ray and the triod only at the junction point of the triod. A map of the 4-od that produces in its inverse limit a simple triod and a ray having the simple triod as a remainder is shown in Figure 1. Bing's example is obtained by attaching an arc to this inverse limit at the point $(O, O, O, \ldots)$ and otherwise misses the inverse limit.

Finding an atriodic example presents somewhat more of a challenge. Although other non-planar atriodic tree-like continua are known, an example may be constructed in the following way. Let M be the atriodic tree-like continuum with positive symmetric span that the author constructed in [**14**] and let C be the product of M with a Cantor set. A construction of Laidacker [**27**] produces an atriodic tree-like continuum M' that contains C. Dušan Repovš, Arkadij B. Skopenkov, and Evgenij V. Ščepin prove in [**36**] that the plane does not contain uncountably

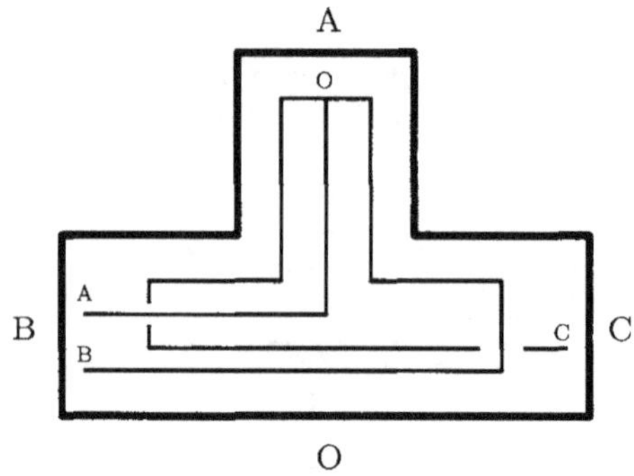

FIGURE 2. A first attempt at a nonplanar atriodic tree-like continuum

many mutually exclusive tree-like continua with positive symmetric span so the continuum M' is non-planar.

In order to tackle Problem 3, it would be helpful to have some examples of planar continua that "look" like they might be non-planar as well as some simpler examples of non-planar atriodic tree-like continua to study. The remainder of this section is devoted to some examples and possible examples.

The author's first attempts to construct an atriodic tree-like continuum that cannot be embedded in the plane failed (as have many subsequent attempts). We briefly describe an early attempt. The picture in Figure 2 is a schematic drawing of a mapping f of a simple triod $T = [OA] \cup [OB] \cup [OC]$ onto itself. The action of the function is to take the first half of $[OA]$ onto $[AO]$ with $f(O) = A$ and the second half of $[OA]$ onto $[OB]$ with $f(A) = B$; f takes the first third of $[OB]$ onto $[AO]$, the next sixth onto the first half of $[OC]$, the next sixth folds back to O and the final third is taken onto $[OB]$; f takes the first third of $[OC]$ onto $[AO]$, the next third half way out $[OB]$ and back, and the final third onto $[OC]$. The resulting inverse limit is atriodic, but it is a chainable continuum because $f \circ f$ factors through $[0,1]$ (i.e., there are maps $g\colon T \to [0,1]$ and $h\colon [0,1] \to T$ so that $f = h \circ g$). Though the schematic of f cannot be drawn in the plane, $\varprojlim \mathbf{f}$ being chainable is planar.

An alternative to Bing's non-planar tree-like continuum mentioned above is the following. Take a continuum consisting of two mutually exclusive rays each having the same simple triod as remainder but the rays wind around the triod in opposite directions. The resulting tree-like continuum is non-planar. This observation suggests the following way possibly to construct a non-planar atriodic tree-like continuum. The continuum is an inverse limit on a simple 5-od, $[OA] \cup [OB] \cup [OC] \cup [OD] \cup [OE]$. Restricted to the triod $[OA] \cup [OB] \cup [OC]$ our 5-od map is just the triod map that the author used in [**14**] to obtain an atriodic tree-like continuum that is not chainable. We use the other two arms of the 5-od to obtain rays that wind in "opposite" directions onto that example. We provide a schematic diagram of the map in Figure 3. The author does not know if the resulting inverse limit is non-planar.

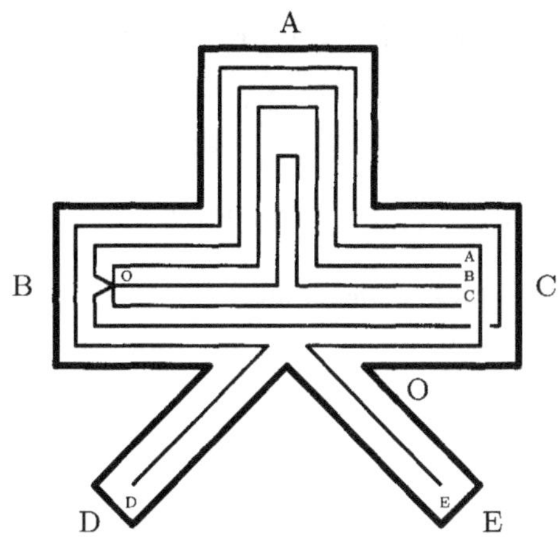

FIGURE 3. A nonplanar example?

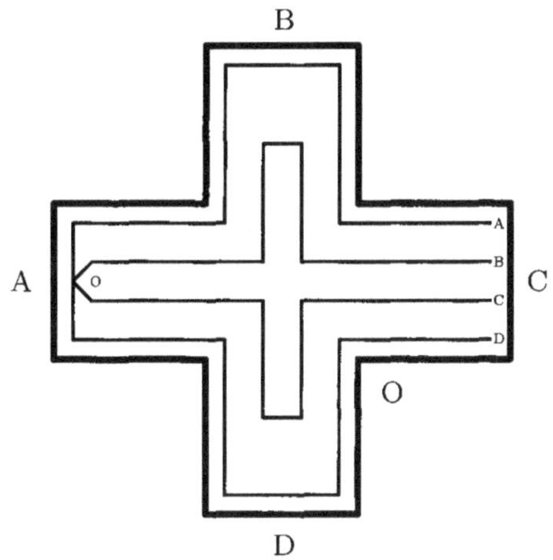

FIGURE 4. A twisted example

There is a somewhat simpler possibility that results from an inverse limit on 4-ods. The author does not know if the resulting inverse limit space is non-planar. The bonding map f (shown in a schematic in Figure 4) has the interesting feature that, although it can be "drawn in the plane", f^2 cannot be "drawn in the plane". This appears to be caused by a twist of the arms of the 4-od imposed by the bonding map. Unfortunately, as our second example shows (see Figure 2), not being able to "draw" a schematic of the bonding map in the plane does not guarantee that the inverse limit is non-planar.

4. Inverse limits on $[0,1]$

Considerable interaction between dynamacists and continuum theorists has occurred in the past fifteen or twenty years. Inverse limits appeal to dynamacists in part because they allow one to transform the study of a dynamical system consisting of a space and a mapping of that space into itself into the study of a (perhaps more complicated) space, the inverse limit, and a homeomorphism, the shift, of that space into itself. Considerations in dynamics have led to extensive investigations of parameterized families of maps. Many of these are maps of $[0,1]$ into itself and include the *logistic* family and the *tent* family. Interest in these families also rekindled the author's interest in inverse limits on $[0,1]$ using a constant sequence of bonding maps in which that bonding map is chosen from one of those two families or from one of several other families of piecewise linear maps including the families f_t for $0 \leq t \leq 1$, g_t for $0 \leq t \leq 1$, f_{ab} (also denoted g_{bc} by the author and others) where both parameters come from $[0,1]$, and the family of permutation maps. In this article we provide definitions only for the tent family (in the next paragraph) and the permutation maps (in the next section). For definitions of the families not discussed further in this article and information on some of the inverse limits generated by these families see [**15**–**17**, **21**]. With one exception these are families of unimodal maps, a class of maps of special interest in dynamics. Permutation maps are Markov maps, a class also of interest in dynamics. A map is *monotone* provided its point inverses are connected; a map $f\colon [0,1] \to [0,1]$ is *unimodal* provided f is not monotone and there is a point c, $0 < c < 1$, such that $f \restriction [0,c]$ and $f \restriction [c,1]$ are monotone. A map $f\colon [0,1] \to [0,1]$ is *Markov* provided there is a finite subset $\{x_1 = 0, x_2, \ldots, x_n = 1\}$ with $x_i < x_{i+1}$ and $f \restriction [x_i, x_{i+1}]$ is monotone for $1 \leq i < n$.

Tent maps are unimodal maps of $[0,1]$ constructed as follows. Choose a number s from $[0,1]$ and let $f_s\colon [0,1] \to [0,1]$ be the piecewise linear map that passes through $(0,0)$, $(1/2, s)$, and $(1,0)$. (Specifically, f_s is given by $f_s(x) = 2sx$ for $0 \leq x \leq 1/2$ and $f_s(x) = (2-2x)s$ for $1/2 \leq x \leq 1$.)

One problem involving the tent family has sparked considerable interest and has given rise to a large number of partial results.

641? **Problem 4.** *If f_s and f_t are tent maps with $\varprojlim \mathbf{f_s}$ and $\varprojlim \mathbf{f_t}$ homeomorphic, is $s = t$?*

This has been settled in a number of cases including Lois Kailhofer's proof for maps that have periodic critical points [**24**]. Štimac has announced a positive solution if the maps have preperiodic critical points.

The collection of inverse limits arising from the tent family is rich in its variety. Barge, Brucks, and Diamond have shown that there are uncountably many parameter values at which the inverse limit is so complicated that it contains a copy of every continuum arising as an inverse limit space from a tent family core (see the next paragraph) [**2**]. In spite of the presence of complicated topology at some parameter values, progress has been made on Problem 4 when the orbit of the critical point is infinite. B. Raines has begun a systematic study of these inverse limits and has made some significant progress for certain parameter values. The author acknowledges private correspondence with Professor Raines that provided some of the problems in this section as well as some of the information on the literature related to these problems.

If f_s is a tent map $\varprojlim \mathbf{f_s}$ is the closure of a topological ray. Except for $s = 1$ the inverse limit is a decomposable continuum and if R is the ray that is dense in the inverse limit, $\overline{R} - R$ is a proper subcontinuum that results from the inverse limit on $[f_s(s), s]$ using the restriction of f_s to that interval as the bonding map. We refer to $\varprojlim \mathbf{f_s} \restriction [f_s(s), s]$ as the *core* of $\varprojlim \mathbf{f_s}$ and the map $f_s \restriction [f_s(s), s]$ as a *tent core*. Sometimes the tent core is rescaled to be the map of $[0, 1]$ onto itself given by $f_s(x) = sx + 2 - s$ for $0 \le x \le 1 - 1/s$ and $f_s(x) = s - sx$ for $1 - 1/s \le x \le 1$. Since the critical point is different depending on one's perspective, it is simply denoted by c in the remainder of this section.

Raines' approach to the case that the orbit of the critical point c is infinite has been to look at the ω-limit set of c, $\omega(c) = \bigcap_{n=1}^{\infty} \overline{\{f^k(c) : k \ge n\}}$. When the orbit of c is infinite, $\omega(c) = [0, 1]$ or $\omega(c)$ is totally disconnected. If the orbit is infinite and $\omega(c)$ is totally disconnected, $\omega(c)$ may be a countable set, a Cantor set, or the union of a countable set and a Cantor set. It is in the case that $\omega(c) = [0, 1]$ that the Barge, Brucks and Diamond phenomenon of [**2**] occurs (i.e., there are parameter values at which the inverse limit of the tent map contains a copy of every continuum that arises as an inverse limit space from a tent family core).

Problem 5 (Raines). *Suppose f is a tent core with critical point c such that $\omega(c) = [0, 1]$. If C is a composant of $\varprojlim \mathbf{f}$, does C contain a copy of every continuum that arises as an inverse limit space of a tent family core?* 642?

Problem 6 (Raines). *Suppose f is a unimodal map with critical point c. Give necessary and sufficient conditions on c so that $\varprojlim \mathbf{f}$ contains a copy of every continuum that arises as an inverse limit space in a tent family core.* 643?

In case f is a tent core with critical point c and $\omega(c)$ is countable or the union of a countable set and a Cantor set, it is known that the inverse limit is an indecomposable arc continuum without end points (by an *arc continuum* we mean a continuum such that every proper subcontinuum is an arc). Good, Knight, and Raines have shown [**11**] that there are uncountably many members of the tent family cores with $\omega(c)$ countable that have topologically different inverse limits.

In case f is a tent core with critical point c and $\omega(c)$ is a Cantor set, the inverse limit is indecomposable but it may have end points. If it has end points the set of end points is uncountable [**5**]. The subcontinua of $\varprojlim \mathbf{f}$ can be quite complicated as demonstrated in [**4**]. This gives rise to the next problem.

Problem 7 (Raines). *Let f be a tent core with critical point c and $\omega(c)$ a Cantor set. Classify all possible subcontinua of $\varprojlim \mathbf{f}$.* 644?

We close this section with one final problem. If n is a positive integer and σ is a permutation on the set $\{1, 2, \ldots, n\}$, define a map $f_\sigma \colon [0, 1] \to [0, 1]$ in the following way: (1) for $1 \le i \le n$, let $a_i = (i-1)/(n-1)$, (2) let $f_\sigma(a_i) = a_{\sigma(i)}$, and

(3) extend f_σ linearly to all of $[0, 1]$. We call a map so constructed a *permutation map*. These maps are all Markov maps and many interesting continua result as the inverse limit space based on a permutation map. In [**18**] the author began a study of the inverse limits spaces that result from using a permutation map in an inverse limit. By brute force, all continua arising from permutations based on 3, 4, or 5 elements were determined.

645? **Problem 8.** *Classify all continua arising from permutation maps.*

5. The property of Kelley

A continuum M with metric d is said to have the *Property of Kelley* provided if $\varepsilon > 0$ there is a positive number δ such that if p and q are points of M and $d(p, q) < \delta$ and H is a subcontinuum of M containing p then there is a subcontinuum K of M containing q such that $\mathcal{H}(H, K) < \varepsilon$ ($\mathcal{H}$ denotes the Hausdorff distance on the hyperspace of subcontinua $C(M)$). This property that we now call the Property of Kelley was introduced by J. Kelley in his study of hyperspaces, but it is a nice continuum approximation property in its own right. The author considered the property in [**19, 20, 23**]. While presenting the results that appeared in [**19, 20**] in seminar, the author was asked the following question by W.J. Charatonik.

646? **Problem 9** (Charatonik). *Is there a characterization of the Property of Kelley in terms of the inverse limit representation of the continuum?*

The author briefly tried to distill a sufficient condition from the proofs in the papers in [**19, 20**] but never found a satisfying theorem. Nonetheless, it would be of interest to be able to determine the presence of the Property of Kelley based on some easily checked conditions on the bonding maps in an inverse limit representation of the continuum. Private communication with W.J. Charatonik indicates that he and a student have obtained some sufficient conditions on an inverse limit sequence to guarantee that the inverse limit have the Property of Kelley.

Permutation maps were defined in Section 4. In [**18**] it was shown that if f is a permutation map based on a permutation on 3, 4, or 5 elements, then $\varprojlim \mathbf{f}$ has the Property of Kelley. This leads us to ask the following question.

647? **Problem 10.** *Do all permutation maps produce continua with the Property of Kelley?*

6. Inverse limits with upper semi-continuous bonding functions

W.S. Mahavier introduced inverse limits with upper semi-continuous bonding functions in [**29**] but as inverse limits on closed subsets of $[0, 1] \times [0, 1]$. In that article he showed that inverse limits on closed subsets of $[0, 1] \times [0, 1]$ are inverse limits on $[0, 1]$ using upper semi-continuous closed set valued functions as bonding functions. In a subsequent paper [**22**], Mahavier and the author extended the definition to the setting of inverse limits on compact Hausdorff spaces using upper semi-continuous closed set valued bonding functions. If Y is a compact Hausdorff space, 2^Y denotes the collection of all closed subsets of Y. If X and Y are compact Hausdorff spaces, a function $f\colon X \to 2^Y$ is called *upper semi-continuous* at the point x of X provided if O is an open set in Y that contains $f(x)$ then there is an open set U in X that contains x and $f(t)$ is a subset of O for every t in U. If $X_1, X_2, X_3, \ldots$ is a sequence of compact Hausdorff spaces and $f_1, f_2, f_3, \ldots$ is a sequence of upper semi-continuous functions such that $f_i\colon X_{i+1} \to 2^{X_i}$ for each i, by the inverse limit of the inverse sequence $\{X_i, f_i\}$ is meant the subset of $\prod_{i>0} X_i$ that contains the point $\mathbf{x} = (x_1, x_2, x_3, \ldots)$ if and only if $x_i \in f(x_{i+1})$. The reader will note that in case the functions are single valued, this definition reduces to the usual definition of an inverse limit. Beyond the collection of chainable continua that occur with single valued bonding functions, many interesting examples of continua result from inverse limits on $[0,1]$ with upper semi-continuous bonding functions that cannot occur with single valued functions. Among these are the Hilbert cube, the Cantor fan, a 2-cell with a sticker, and the Hurewicz continuum H that has the property that if M is a metric continuum there is a subcontinuum K of H and a mapping of K onto M. The example that produces a 2-cell with an attached arc leads to the following problem.

Problem 11. *Is there an upper semi-continuous function $f\colon [0,1] \to 2^{[0,1]}$ such that $\varprojlim \mathbf{f}$ is a 2-cell?* 648?

Admittedly, this problem is rather more specific than most in this article, but perhaps it can serve as a starting point for an interesting investigation of these new and different inverse limits.

We end this section with a problem inspired by considerations from Section 7. Some models in economics are not well-defined either forward in time or backward in time [**7**], [**37**]. Some models consist of the union of two mappings that have no point in common. Perhaps an investigation of these new inverse limits using these models would be helpful to economists as well as a way to begin work on our next problem.

Problem 12. *Suppose $f\colon [0,1] \to 2^{[0,1]}$ is an upper semi-continuous function that is the union of two mappings of $[0,1]$. What can be said about $\varprojlim \mathbf{f}$?* 649?

7. Applications of inverse limits in Economics

An exciting recent development in inverse limits is the development of models in economics in which the state of the model at time t is related to its state at time $t+1$ by some non-invertible mapping f. A solution to the model is an infinite sequence $x_1, x_2, x_3, \ldots$ such that $f(x_{t+1}) = x_t$ for $t = 1, 2, 3, \ldots$. So the set of solutions is the inverse limit on the state space using the map f as a bonding map. These models have arisen in cash-in-advance models [**25**] and overlapping generations models [**30, 31**] studied by various economists. These models generally fall into a category of models described by economists as having "backward dynamics" or as models involving "backward maps". Economists are interested in the inverse limit because it contains as its points all future states predicted by the model. The author acknowledges private correspondence with

Judy Kennedy and Brian Raines used in the development of this section and appreciates the contribution of problems by both of them. The problems that they contributed are labeled below with their names. Of course, any errors or misstatement of problems are solely the responsibility of the author.

If $f\colon X \to X$ and $g\colon X \to X$ are maps of a topological space X, we say that f and g are *topologically conjugate* provided there is a homeomorphism $h\colon X \to X$ such that $f \circ h = h \circ g$. If f and g are topologically conjugate, a homeomorphism h such that $f \circ h = h \circ g$ is called a conjugacy.

650? **Problem 13** (Kennedy–Stockman). *Suppose $f\colon [0,1] \to [0,1]$ and $g\colon [0,1] \to [0,1]$ are topologically conjugate. How does one* construct *a homeomorphism h so that $f \circ h = h \circ g$?*

This problem deserves attention independent of the interest by economists. For economists the *existence* of a conjugacy is not sufficient information for carrying out some of the computations they need such as the computation of measures and then integrals for utility functions. Specific questions related to this problem and asked by Kennedy and Stockman include:

(1) Can the conjugacy be constructed by means of a sequence of approximations?
(2) If f and g are piecewise differentiable, must the conjugacy be piecewise differentiable?

The next problem is related to Problem 5 above.

651? **Problem 14** (Kennedy–Stockman). *Do continua that contain copies of every inverse limit that arises in a tent family core occur as inverse limits in the cash-in-advance model [**25**] or the overlapping generations model [**33**]?*

Some models in economics are based on relations instead of functions so neither forward nor backward dynamics is well defined. In particular the Christiano–Harrison model [**7**] and a Stockman model [**37**] fit this scenario. Perhaps inverse limits with upper semi-continuous bonding functions (see Section 6) could be employed in an analysis of these models. Consequently, we reiterate Problem 12.

In considering models in economics, measure theory will likely play an important role for several reasons one of which we have already mentioned. For instance, when economists consider models involving backward dynamics, they would like to be able to "rank" the inverse limit spaces in some meaningful way, perhaps by using "natural" invariant measures. For a survey of literature on such measures see [**13**]. When comparing two inverse limit spaces but with a precise meaning of "better" to be determined, Kennedy and Stockman ask the following.

652? **Problem 15** (Kennedy–Stockman). *Suppose policy A and policy B in economics lead to different inverse limit spaces. Determine which of the inverse limits is "better".*

With a precise meaning of "complex" to be determined, they also ask.

Problem 16 (Kennedy–Stockman). *For an economics model, what is the measure of the set of initial conditions that lead to "complex" dynamics?* 653?

Problem 17 (Kennedy–Stockman). *In a model from economics, if an equilibrium point (i.e., point in the inverse limit) is chosen at random, what is the probability that it is "complex"?* 654?

One search for appropriate measures on the inverse limit space centers on somehow making use of measures already developed. Kennedy and Stockman have recently succeeded in "lifting" given measures for interval maps to measures on the corresponding inverse limit spaces although they remark that such measures on the inverse limit space apparently are already known, see [**9**]. For an introduction to measures for interval maps see [**1**, Sections 6.4–6.6]. See also [**13**]. Kennedy and Stockman ask if there exist other useful measures one might consider, particularly in non-chaotic situations.

Recall that if $f\colon X \to X$ is a mapping of a metric space and A is a closed subset of X such that $f[A] = A$, then we call A an *invariant* set. If A is a closed invariant subset of X, then the *basin of attraction* of A is $\{x \in X : \omega(x) \subset A\}$. A subset B of X is *nowhere dense* in X provided $\overline{B}$ does not contain a nonempty open set. A subset M of X is said to be *residual* in X provided $X - M$ is the union of countably many nowhere dense subsets. A closed invariant subset of X is called a *topological attractor* [**33**] for f provided the basin of attraction for A contains a residual subset of X and if A' is another closed invariant subset of X then the common part of the basin of attraction of A' and the basin of attraction of A is the union of at most countably many nowhere dense sets. For more information of topological attractors and metric attractors (defined below), see [**33**].

One possible tool for analyzing an inverse limit arising in a model from economics lies in the shift homeomorphism. There are two shifts and they are inverses of each other. Specifically, below we are referring to the shift $\sigma\colon \varprojlim \mathbf{f} \to \varprojlim \mathbf{f}$ given by $\sigma(\mathbf{x}) = (x_2, x_3, x_4, \ldots)$. Raines asks the following.

Problem 18 (Raines). *Let f be a map of the interval. Find necessary and sufficient conditions for $\varprojlim \mathbf{f}$ to admit a proper subset that is a topological attractor for the shift homeomorphism.* 655?

Problem 19 (Raines). *Let f be a unimodal map of the interval. Classify all of the topological attractors for the shift homeomorphism on $\varprojlim \mathbf{f}$.* 656?

Not all models from economics involve one-dimensional spaces. This prompts the following problem.

Problem 20 (Raines). *Let f be a map of $[0,1] \times [0,1]$. Identify topological attractors in $\varprojlim \mathbf{f}$ under the shift homeormorphism.* 657?

If X is a metric space with a measure μ, $f\colon X \to X$ is a mapping and A is a closed invariant subset of X, then A is called a *metric attractor* for f provided the basin of attraction for A has positive measure and and if A' is another closed invariant subset of X then the common part of the basin of attraction of A' and the basin of attraction of A has measure zero.

658–659? **Problem 21** (Raines). *In the previous two problems, change the phrase "topological attractor" to "metric attractor".*

References

[1] K. T. Alligood, T. D. Sauer, and J. A. Yorke, *Chaos*, Springer, New York, 1997.

[2] M. Barge, K. Brucks, and B. Diamond, *Self-similarity in inverse limit spaces of the tent family*, Proc. Amer. Math. Soc. **124** (1996), no. 11, 3563–3570.

[3] R H Bing, *Snake-like continua*, Duke Math. J. **18** (1951), 653–663.

[4] K. Brucks and H. Bruin, *Subcontinua of inverse limit spaces of unimodal maps*, Fund. Math. **160** (1999), no. 3, 219–246.

[5] H. Bruin, *Planar embeddings of inverse limit spaces of unimodal maps*, Topology Appl. **96** (1999), no. 3, 191–208.

[6] J. H. Case and R. E. Chamberlin, *Characterizations of tree-like continua*, Pacific J. Math. **10** (1960), 73–84.

[7] L. Christiano and S. Harrison, *Chaos, sunspots and automatic stabilizers*, J. Monetary Economics **44** (1999), 3–31.

[8] H. Cook, W. T. Ingram, and A. Lelek, *Eleven annotated problems about continua*, Open problems in topology, North-Holland, Amsterdam, 1990, pp. 295–302.

[9] I. P. Cornfeld, S. V. Fomin, and Ya. G. Sinai, *Ergodic theory*, Springer, New York, 1982.

[10] J. B. Fugate, *Decomposable chainable continua*, Trans. Amer. Math. Soc. **123** (1966), 460–468.

[11] C. Good, R. Knight, and B. Raines, *Non-hyperbolic one-dimensional invariant sets with uncountably infinite collections of inhomogeneities*, Preprint.

[12] O. H. Hamilton, *A fixed point theorem for pseudo-arcs and certain other metric continua*, Proc. Amer. Math. Soc. **2** (1951), 173–174.

[13] B. R. Hunt, J. A. Kennedy, T.-Y. Li, and H. E. Nusse, *SLYRB measures: natural invariant measures for chaotic systems*, Phys. D **170** (2002), no. 1, 50–71.

[14] W. T. Ingram, *An atriodic tree-like continuum with positive span*, Fund. Math. **77** (1972), no. 2, 99–107.

[15] W. T. Ingram, *Inverse limits*, Sociedad Matemática Mexicana, México, 2000.

[16] W. T. Ingram, *Inverse limits on* $[0,1]$ *using piecewise linear unimodal bonding maps*, Proc. Amer. Math. Soc. **128** (2000), no. 1, 279–286.

[17] W. T. Ingram, *Inverse limits on circles using weakly confluent bonding maps*, Topology Proc. **25** (2000), Spring, 201–211.

[18] W. T. Ingram, *Invariant sets and inverse limits*, Topology Appl. **126** (2002), no. 3, 393–408.

[19] W. T. Ingram, *Inverse limits and a property of J. L. Kelley. I*, Bol. Soc. Mat. Mexicana (3) **8** (2002), no. 1, 83–91.

[20] W. T. Ingram, *Inverse limits and a property of J. L. Kelley. II*, Bol. Soc. Mat. Mexicana (3) **9** (2003), no. 1, 135–150.

[21] W. T. Ingram and W. S. Mahavier, *Interesting dynamics and inverse limits in a family of one-dimensional maps*, Amer. Math. Monthly **111** (2004), no. 3, 198–215.

[22] W. T. Ingram and W. S. Mahavier, *Inverse limits of upper semi-continuous set valued functions*, Houston J. Math. **32** (2006), no. 1, 119–130.

[23] W. T. Ingram and D. D. Sherling, *Two continua having a property of J. L. Kelley*, Canad. Math. Bull. **34** (1991), no. 3, 351–356.

[24] L. Kailhofer, *A classification of inverse limit spaces of tent maps with periodic critical points*, Fund. Math. **177** (2003), no. 2, 95–120.

[25] J. Kennedy, D. R. Stockman, and J. A. Yorke, *Inverse limits and an implicitly defined difference equation from economics*, Topology Appl. , To appear.

[26] J. Krasinkiewicz, *On one-point union of two circles*, Houston J. Math. **2** (1976), no. 1, 91–95.

[27] M. Laidacker, *Imbedding compacta into continua*, Topology Proc. **1** (1977), 91–105.

[28] A. Lelek, *Disjoint mappings and the span of spaces*, Fund. Math. **55** (1964), 199–214.

[29] W. S. Mahavier, *Inverse limits with subsets of* $[0,1] \times [0,1]$, Topology Appl. **141** (2004), 225–231.

[30] A. Medio, *The problem of backward dynamics in economics models*, 2000, Preprint.

[31] A. Medio, *Invariant probability distributions in economic models: a general result*, Macroeconomic Dynamics **8** (2004), 162–187.

[32] A. Medio and B. Raines, *Backward dynamics in economics. The inverse limit approach*, J. Economic Dynamics and Control (2006), Article in press.

[33] A. Medio and B. Raines, *Inverse limit spaces arising from problems in economics*, Topology Appl. (2006), Article in press.

[34] L. G. Oversteegen and E. D. Tymchatyn, *Plane strips and the span of continua. I*, Houston J. Math. **8** (1982), no. 1, 129–142.

[35] B. Raines, *Local planarity in one-dimensional continua*, Preprint.

[36] D. Repovš, A. B. Skopenkov, and E. V. Ščepin, *On uncountable collections of continua and their span*, Colloq. Math. **69** (1995), no. 2, 289–296.

[37] D. R. Stockman, *Balanced-budget rules: cycles and complex dynamics*, Preprint.

Open Problems in Topology II – Edited by E. Pearl

Indecomposable continua

Wayne Lewis

Except for discussion near the end of this article, we shall consider a *continuum* to be a compact connected metric space. In discussion near the end of this article we shall consider a *non-metric continuum* to be a compact connected nonmetrizable Hausdorff space.

A continuum is *indecomposable* if it is not the union of two proper subcontinua. It is *hereditarily indecomposable* if each of its subcontinua is indecomposable.

Hereditary equivalence

The best studied hereditarily indecomposable continuum is the *pseudo-arc*. Moise [**62**] showed that the pseudo-arc is *hereditarily equivalent*, i.e., homeomorphic to each of its nondegenerate subcontinua, and gave it its name because it shares this property with the arc.

Question 1. *Is the pseudo-arc the only nondegenerate continuum other than the arc which is hereditarily equivalent?* 660?

It follows from results by Henderson [**26**] and Cook [**18**] that any nondegenerate hereditarily equivalent continuum other than the arc must be hereditarily indecomposable and tree-like.

Mohler and Oversteegen [**61**] have constructed examples of non-metric decomposable hereditarily equivalent continua, including one which is not a Hausdorff arc. Smith [**89**] has constructed a non-metric hereditarily indecomposable hereditarily equivalent continuum, obtained as an inverse limit of ω_1 copies of the pseudo-arc. Oversteegen and Tymchatyn [**65**] have shown that any planar hereditarily equivalent continuum must be close to being chainable, i.e., must be weakly chainable and have symmetric span zero.

Homogeneity

Bing [**9**] has characterized the pseudo-arc as a nondegenerate hereditarily indecomposable chainable continuum, i.e., any such continuum must be homeomorphic to the continuum constructed by Moise, and to a continuum constructed earlier by Knaster [**32**] to show that hereditarily indecomposable continua exist. Bing [**7**] and Moise [**63**] have independently shown that the pseudo-arc is homogeneous and Bing [**10**] has shown that it is the only nondegenerate homogeneous chainable continuum. This latter result has been generalized by this author [**43**] to show that the pseudo-arc is the only nondegenerate homogeneous almost chainable ontinuum.

The first characterization of the pseudo-arc has been used extensively and it would be useful to know if it can be generalized.

Question 2. *Is the pseudo-arc the only nondegenerate hereditarily indecomposable weakly chainable continuum?* 661?

A weakly chainable continuum can be described in terms of a defining sequence of open covers. Fearnley [**22**] and Lelek [**38**] have shown that a continuum is *weakly chainable* if and only if it is the continuous image of a chainable continuum, and hence of the pseudo-arc. Thus the above question can be rephrased as "If X is a nondegenerate hereditarily indecomposable continuum which is the continuous image of the pseudo-arc, is X itself a pseudo-arc?"

This question is of interest not just in terms of the images of the pseudo-arc or a possible generalization of a known characterization of the pseudo-arc. It is central to the classification of homogeneous plane continua and is a special case of a family of questions of interest.

662? **Question 3.** *Is the pseudo-arc the only nondegenerate homogeneous nonseparating plane continuum?*

Hagopian [**23**] has shown that every nonseparating homogeneous plane continuum is hereditarily indecomposable. Oversteegen and Tymchatyn [**66**] have shown that every such continuum is weakly chainable. More on the status of the classification of homogeneous plane continua or homogeneous one-dimensional continua can be found in survey articles by this author [**51**] and by Rogers [**79, 80**].

Rogers [**77**] has shown that every homogeneous hereditarily indecomposable continuum is tree-like. Krupski and Prajs [**36**] have shown that every homogeneous tree-like continuum is hereditarily indecomposable. Both of these results are independent of whether the continuum is planar.

663? **Question 4.** *Is the pseudo-arc the only nondegenerate homogeneous tree-like continuum?*

While it is known from the result of Oversteegen and Tymchatyn that any such continuum which is planar must be weakly chainable, such is not yet known to be the case for the possibly nonplanar case.

664? **Question 5.** *Is every homogeneous tree-like continuum weakly chainable?*

A continuous surjection $f\colon X \to Y$ between continua is *confluent* if, for each subcontinuum H of Y and each component C of $f^{-1}(H)$, $f(C) = H$, The class of confluent maps includes the classes of open maps and of monotone maps. Cook [**16**] has shown that a continuum Y is hereditarily indecomposable if and only if every continuous surjection from a continuum onto Y is confluent. Thus, any hereditarily indecomposable continuum which is weakly chainable is the confluent image of the pseudo-arc. McLean [**57**] has shown that the confluent image of a tree-like continuum is tree-like. A positive answer to the following question would show that the pseudo-arc is the only nondegenerate hereditarily indecomposable weakly chainable continuum, providing positive answers to Questions 2 and 3.

665? **Question 6.** *Is the confluent image of a chainable continuum chainable?*

Bing [**9**] and Rosenholtz [**81**], respectively, have shown that monotone maps and open maps preserve chainability. Confluent maps also preserve indecomposability, hereditary indecomposability and atriodicity. One of the most general forms of this family of questions is due to Mohler [**60**].

Question 7. *Is every weakly chainable atriodic tree-like continuum chainable?* 666?

Minc [**59**] has provided a partial answer by showing that any atriodic weakly chainable continuum which is an inverse limit of trees with simplicial bonding maps is chainable.

As indicated above, Rogers [**77**] has shown that every homogeneous hereditarily indecomposable continuum is tree-like. However, there are homogeneous indecomposable continua which are not tree-like, e.g., solenoids or solenoids of pseudo-arcs. However, all such known nondegenerate continua are one-dimensional, leading Rogers to ask the following two questions. (Any product of nondegenerate continua is aposyndetic and hence decomposable. Thus there can be no indecomposable homogeneous analog of the Hilbert cube.)

Question 8 ([**79**]). *Is each nondegenerate homogeneous indecomposable continuum one-dimensional?* 667?

Question 9 ([**78**]). *Is each homogeneous indecomposable cell-like continuum tree-like?* 668?

Every known nondegenerate homogeneous indecomposable continuum (whether hereditarily indecomposable or not) is circle-like. Also, every known homogeneous plane continuum (whether indecomposable or not, whether separating the plane or not) is circle-like. The family of homogeneous circle-like continua has been completely classified [**25**, **46**, **74**] as the circle, solenoids, circle of pseudo-arcs and solenoids of pseudo-arcs, with the circle of pseudo-arcs and for each solenoid the solenoid of pseudo-arcs being unique.

Question 10. *Is every nondegenerate homogeneous indecomposable continuum circle-like?* 669?

Question 11. *Is every nondegenerate homogeneous plane continuum circle-like?* 670?

Burgess [**14**] has shown that every continuum which is both circle-like and tree-like is chainable. Recalling Bing's result [**10**] that the only nondegenerate homogeneous chainable continuum is the pseudo-arc, positive answers to each of Question 9 and 10 would imply that the only nondegenerate indecomposable homogeneous cell-like continuum is the pseudo-arc while a positive answer to Question 10 implies a positive answer to Question 11. A positive answer to Question 11 implies that the classification of homogeneous plane continua is complete with the four known examples of the point, simple closed curve, pseudo-arc and circle of pseudo-arcs.

A space X is said to be *homogeneous with respect to the class $\mathcal{M}$ of maps* if for every $x, y \in X$ there exists a continuous surjection $f\colon X \twoheadrightarrow X$ of X onto itself with $f(x) = y$ and $f \in \mathcal{M}$. (Usual homogeneity is thus homogeneity with respect to homeomorphisms.) J. Charatonik and Maćkowiak [**15**] have shown that the result of Bing can be strengthened to show that the pseudo-arc is the only nondegenerate chainable continuum which is homogeneous with respect to confluent maps. Several examples are known of continua which are not homogeneous but which are homogeneous with respect to various classes of maps. Prajs [**71**] has shown that the disc is homogeneous with respect to open maps, while Seaquist [**82**] has shown that it is not homogeneous with respect to monotone maps. Prajs [**72**] and Seaquist [**83**] have independently shown that the Sierpiński universal plane curve is homogeneous with respect to open monotone maps. J. Charatonik has asked the following question, with special interest in possible generalized homogeneity of the pseudo-circle.

671? **Question 12.** *Does there exist an hereditarily indecomposable continuum which is not homogeneous but which is homogeneous with respect to continuous functions (and hence with respect to confluent maps)?*

ϵ-premaps

Determining whether a nondegenerate weakly chainable hereditarily indecomposable continuum must be chainable, and hence a pseudo-arc, is equivalent to determining if such a continuum can be mapped onto a chainable continuum with arbitrarily small point inverses. The pseudo-arc has a strong version of a converse property. If $f\colon P \to X$ is a continuous surjection from the pseudo-arc P onto the nondegenerate continuum X and $\epsilon > 0$, there exists a homeomorphism $h\colon P \to P$ such that $\operatorname{diam}(f \circ h)^{-1}(x) < \epsilon$ for every $x \in X$, i.e., $f \circ h$ is an ϵ-map of P onto X. For such a continuous surjection f we shall call the composition $f \circ h$ an *ϵ-premap* corresponding to f. It can be shown that any nondegenerate continuum sharing this property with the pseudo-arc P that every continuous surjection onto a nondegenerate continuum has a corresponding ϵ-premap must be chainable and indecomposable.

672? **Question 13.** *Is the pseudo-arc the only nondegenerate continuum with the property that for every $\epsilon > 0$ every continuous surjection onto a nondegenerate continuum has a corresponding ϵ-premap?*

Fixed points

Indecomposable continua play an important role in the study of the fixed-point property. We expect this topic to be more thoroughly covered in a separate article in this volume and include only the following questions. The first two are due to Łysko [**54**].

673? **Question 14.** *Does there exist a continuum X with the fixed point property such that $X \times P$ (P = pseudo-arc) does not have the fixed point property?*

There are known examples [**11**] of continua with the fixed point property whose product with the unit interval $[0,1]$ does not have the fixed point property. However, these do not translate to such examples for products with the pseudo-arc and the structure of an hereditarily indecomposable continuum places restrictions on maps of products of it with other continua. The above question is a special case of the following. The case of the pseudo-arc, or a nondegenerate hereditarily indecomposable continuum with the fixed point property, seems especially challenging.

Question 15. *If X is a nondegenerate continuum with the fixed point property,* 674?
does there always exist a nondegenerate continuum Y with the fixed point property such that $X \times Y$ does not have the fixed point property?

A continuum X has the *complete invariance property* if every nonempty closed subset of X is the complete fixed point set of some self-map of X. Martin and Nadler [**55**] have shown that every two-point set is a fixed point set for some continuous self-map of the pseudo-arc. Cornette [**20**] has shown that every subcontinuum of the pseudo-arc is a retract of the pseudo-arc. Toledo [**92**] has shown that every subcontinuum of the pseudo-arc is the fixed point set of a periodic homeomorphism of the pseudo-arc. This author [**42**] has shown that there are proper subsets of the pseudo-arc with nonempty interior which are the fixed point sets of homeomorphisms.

The following question due to Martin and Nadler [**55**] is also of interest in the special case of self-homeomorphisms of the pseudo-arc.

Question 16. *Does the pseudo-arc have the complete invariance property?* 675?

Maps of products

Bellamy and Łysko [**6**] have shown that every homeomorphism of $P \times P$, where P is the pseudo-arc, is a composition of a product of homeomorphisms on the individual factors with a permutation of the factors. Bellamy and Kennedy [**5**] have extended this to a product of an arbitrary number of copies of the pseudo-arc. The arguments in both cases use specific properties of the pseudo-arc. However, the considerations motivating this investigation stem from the structure of hereditarily indecomposable continua in general.

Question 17. *If $X = \prod_{\alpha \in A} X_\alpha$ is a product of hereditarily indecomposable con-* 676?
tinua, is every homeomorphism of X a composition of a product of homeomorphism on the individual factor spaces with a permutation of the factors?

The following question about the structure of products of the pseudo-arc is due to Bellamy.

Question 18. *Does every nondegenerate subcontinuum of P^n, the product of* 677?
finitely many copies of the pseudo-arc, contain a pseudo-arc?

A continuum X is *pseudo-contractible* if there exists a continuum Y, points a and b in Y, a point $x_0 \in X$ and a continuous function $f\colon X \times Y \to X$ such that $f(x,a) = x$ for each $x \in X$ and $f(x,b) = x_0$ for each $x \in X$. There exist examples of continua which are pseudo-contractible but not contractible, e.g., the spiral around a disk. Sobolewski [**91**] has shown that no nondegenerate chainable continuum other than the arc is pseudo-contractible. In particular, the pseudo-arc and the Knaster-type indecomposable continua are not pseudo-contractible.

678? **Question 19.** *Does there exist a nondegenerate (hereditarily) indecomposable continuum which is pseudo-contractible?*

Concerning maps of products of pseudo-arcs, Łysko [**53**] has also asked the following question.

679? **Question 20.** *Assume that P is the pseudo-arc and that $r\colon P \times P \to \Delta = \{(x,y) \in P \times P | x = y\}$ is a continuous retraction. Must r be of the form $r(x,y) = (x,x)$ for all (x,y) or $r(x,y) = (y,y)$ for all (x,y)?*

Homeomorphism groups

This author [**48**] has used properties of homeomorphisms of $P \times M$, where P is a pseudo-arc and M a continuum, to show that $H(P)$, the topological group of self-homeomorphisms of the pseudo-arc, does not contain a nondegenerate continuum. The following is a variation of questions asked by Brechner [**12**], Krasinkiewicz [**35**] and this author [**45**].

680? **Question 21.** *Is $H(P)$, the topological group of all self-homeomorphisms of the pseudo-arc P, totally disconnected?*

681? **Question 22.** *Does $C(P)$, the space of all continuous functions from the pseudo-arc into itself, contain any nondegenerate connected sets other than collections of constant maps?*

The above two questions are also of interest for nondegenerate hereditarily indecomposable continua in general, not just for the pseudo-arc.

The Menger universal curve M has quite different local structure from an hereditarily indecomposable continuum. For it, the complexity of this local structure has allowed Brechner [**12**] to show that $H(M)$, the topological group of all self-homeomorphisms of M, is totally disconnected and Oversteegen and Tymchatyn [**67**] to show that $H(M)$ is one-dimensional.

682? **Question 23.** *What is the dimension of $H(P)$, the topological group of all self-homeomorphisms of the pseudo-arc P?*

There is one aspect in which $H(P)$ differs from $H(M)$. Using essential maps onto simple closed curves in M, Brechner [**12**] has shown that given any two distinct self-homeomorphisms f and g of the Menger curve M there exists $\epsilon > 0$ and a separation of $H(M)$ into sets A and B, with $f \in A$, $g \in B$, and $\operatorname{dist}(A,B) > \epsilon$, where distance is measured by the sup metric. For the pseudo-arc P [**49**], given any homeomorphism $h\colon P \to P$ and any $\epsilon > 0$, there exist self-homeomorphisms $h_1, h_2, \ldots, h_n$ of the pseudo-arc such that $h = h_n \circ \cdots \circ h_2 \circ h_1$ and $\operatorname{dist}(h_i, \mathrm{id}_P) < \epsilon$ for each $1 \le i \le n$. No method of classifying "essential" maps or homeomorphism has been identified which is inherent to the structure of hereditarily indecomposable continua, though the composant structure of such continua and their subcontinua imposes strong constraints on continuous families of maps or homeomorphisms.

It would also be of interest to know more about the algebraic structure of the topological group $H(P)$ of all self-homeomorphisms of the pseudo-arc. This author [**50**] has shown that every inverse limit of finite solvable groups acts effectively on the pseudo-arc. Thus, for every positive integer n, there exists a period n homeomorphism of the pseudo-arc [**47**]. Though the pseudo-arc is chainable, in the construction of periodic homeomorphisms it is convenient to view it as an inverse limit of n-ods, with the period n homeomorphisms being realized as the restrictions of period n rotations of the plane. The pseudo-arc also admits [**44**] effective p-adic Cantor group actions. The smallest group not known to act effectively on the pseudo-arc is A_5, the alternating group on 5 symbols, a simple group of order 60.

Question 24. *Does every compact zero-dimensional topological group act effectively on the pseudo-arc?* 683?

The following question is due to Brechner [**13**]. It is also of interest without the assumption that the homeomorphism is periodic, or with an assumption of nth roots for $n \geq 2$.

Question 25. *Does each periodic homeomorphism h of the pseudo-arc have a square root, i.e., a homeomorphism g such that $g \circ g = h$?* 684?

While there are limited results on families of homeomorphisms of the pseudo-arc, it has more often been possible to determine if there exists a homeomorphism of the pseudo-arc with specific properties. If none of the properties involves extendability to a homeomorphism of the plane or other Euclidean space and if the chainability of the pseudo-arc and the relations between composants of the pseudo-arc or of its subcontinua are not inconsistent with the desired set of properties, a homeomorphism with the desired properties can usually be shown to exist.

While many of the questions posed so far are specifically phrased in terms of the pseudo-arc, the corresponding versions for such continua as the pseudo-circle, pseudo-solenoids or other continua all of whose nondegenerate proper subcontinua are pseudo-arcs are also of interest.

Q-like continua

A continuum X is *Q-like* for the polyhedron Q if, for each $\epsilon > 0$, there exists a continuous surjection $f \colon X \to Q$ such that $\operatorname{diam}(f^{-1}(q)) < \epsilon$ for each $q \in Q$. We have been using the term *chainable*, which is equivalent to *arc-like*. It is known that the pseudo-arc P is Q-like for every nondegenerate connected polyhedron Q. It is also true, but not so often recognized, than any nondegenerate chainable continuum which is either indecomposable or 2-indecomposable is Q-like for every nondegenerate connected polyhedron Q. (A continuum is *2-indecomposable* if it is the union of two proper subcontinua each of which is indecomposable.)

Ingram [**27**] has constructed an atriodic simple triod-like continuum which is not chainable (arc-like). He [**28**] has constructed a family of $\mathfrak{c} = 2^{\aleph_0}$ distinct such continua such that the members of the family can be embedded disjointly in the plane. It is a classic result of Moore [**64**] that there does not exist an uncountable family of disjoint triods in the plane. Ingram's family also has the property that any continuum can be continuously mapped onto at most countably many members of the family.

Ingram has constructed such atriodic simple triod-like nonchainable continua with the property that every nondegenerate proper subcontinuum is an arc, as well as ones such that every nondegenerate proper subcontinuum is a pseudo-arc [**29, 30**].

Minc [**58**] has constructed an atriodic simple 4-od-like continuum which is not simple triod-like. His example also has the property that every nondegenerate proper subcontinuum is an arc and is obtained from an inverse limit of simple 4-ods with simplicial bonding maps and the same single step bonding map each time.

685? **Question 26.** *Does there, for every $n \geq 2$, exist an atriodic simple $(n+1)$-od-like continuum which is not simple n-od-like? such an example which is planar? such an example with the property that every nondegenerate proper subcontinuum is an arc? a pseudo-arc?*

One can place a partial order on the family of nondegenerate connected topological graphs where $G_1 \leq G_2$ if there is a continuous surjection $f\colon G_2 \to G_1$ with each nondegenerate $f^{-1}(g)$, $g \in G_1$, being a connected subgraph of G_2. Under this partial order, if $G_1 \leq G_2$, then every G_1-like continuum is G_2-like. The arc and the simple closed curve are the minimals elements in this partial order.

686? **Question 27.** *If G is a nondegenerate connected topological graph, does there exist an atriodic G-like continuum which is not H-like for any graph $H < G$?*

Hyperspaces

Kelley [**31**] has shown that both indecomposability and hereditary indecomposability can be characterized in terms of the hyperspace $C(X)$ of a nondegenerate continuum X. The nondegenerate continuum X is indecomposable if and only if $C(X) \setminus \{X\}$ is not arcwise connected. The nondegenerate continuum X is hereditarily indecomposable if and only if $C(X)$ is uniquely arcwise connected.

Eberhart and Nadler [**21**] have shown that for every nondegenerate hereditarily indecomposable continuum X the hyperspace $C(X)$ is either two-dimensional or infinite-dimensional. This author [**41**] has shown that every nondegenerate hereditarily indecomposable continuum is the open, monotone image of a one-dimensional hereditarily indecomposable continuum. Thus, there exist one-dimensional hereditarily indecomposable continua with infinite-dimensional hyperspaces. Levin and Sternfeld [**40**] have shown that every continuum of dimension two or greater has infinite-dimensional hyperspace and Levin [**39**] has shown that every two-dimensional continuum contains a one-dimensional subcontinuum with infinite-dimensional hyperspace.

Question 28. *What is a characterization, either algebraic or topological, of the one-dimensional hereditarily indecomposable continua with infinite-dimensional hyperspaces?* 687?

The dimension-raising maps constructed by the author involve "collapsing holes" to raise dimension. There must be many such "holes." Rogers [**75**] has shown that if X is a one-dimensional hereditarily indecomposable continuum with finitely generated first Čech cohomology then $C(X)$ is two-dimensional. Tymchatyn [**93**] has shown that if X is a nondegenerate hereditarily indecomposable plane continuum then $C(X)$ can be embedded in $\mathbb{R}^3$.

Krasinkiewicz [**33**] has shown that if X is a nondegenerate hereditarily indecomposable continuum then $C(X)$ doe not contain any subcontinuum homeomorphic to $Y \times [0,1]$ for a nondegenerate continuum Y. He has asked the following question.

Question 29. *If X is an hereditarily indecomposable continuum, can $C(X)$ ever contain the product of two nondegenerate continua?* 688?

Hereditarily indecomposable continua are characterized by their hyperspaces of subcontinua in the sense that X and Y are homeomorphic if and only if $C(X)$ and $C(Y)$ are homeomorphic. The composant structure of an hereditarily indecomposable continuum and of its subcontinua imposes a branching which occurs everywhere in the hyperspace. Ball, Hagler and Sternfeld [**1**] have shown that, while distinct hereditarily indecomposable continua have distinct hyperspaces of subcontinua, there is a natural ultrametric which can be placed on the hyperspace $C(X)$ of an hereditarily indecomposable continuum which yields a topology finer than that normally placed on $C(X)$ by the Hausdorff metric and with the property that $C(X)$ and $C(Y)$ are homeomorphic and in fact isometric for any nondegenerate hereditarily indecomposable continua X and Y.

Dimensions greater than one

Hereditarily indecomposable continua of dimension greater than one have complex structure. For example, it is known from results of Mason, Walsh and Wilson [**56**] that no such continuum can be P-like for any polyhedron P, i.e., if such a continuum is expressed as an inverse limit of polyhedra, the factor spaces in the inverse sequence must get increasingly complex.

The following questions are due to Krasinkiewicz [**34**].

Question 30. *Suppose X is an hereditarily indecomposable continuum such that* $\dim(X) = n \geq 2$. *Does there exist an essential map from X onto the n-dimensional sphere S^n? (By a result of Krasinkiewicz, if* $\dim(X) > n$, *then there does exist such an essential map onto S^n.)* 689?

Question 31. *If X is an hereditarily indecomposable continuum and A is a subcontinuum of X, is it true that* $\mathrm{Sh}(A) \leq \mathrm{Sh}(X)$? *(*$\mathrm{Sh}(A)$ *denotes the shape of A.)* 690?

691? **Question 32.** *If X is an hereditarily indecomposable continuum and A is a subcontinuum of X, is it true that A is a shape retract of X? Under what conditions is A a(n open) retract of X?*

692? **Question 33.** *If X is an hereditarily indecomposable continuum with* $\dim(X) \geq 2$ *does there exist a continuous surjection from X onto a nontrivial solenoid? onto a nontrivial pseudo-solenoid?*

The pseudo-arc has a very rich collection of self-homeomorphisms. There are other hereditarily indecomposable continua with very few homeomorphisms. Cook [**17**] has constructed one-dimensional hereditarily indecomposable continua with very few self-maps and no nonidentity self-homeomorphisms. One example has the property that any map between subcontinua is either a constant or a retract onto subcontinua, while another has the stronger property that any map between subcontinua is either a constant or the identity map of a subcontinuum onto itself.

Pol [**69**] has constructed, for every positive integer n, hereditarily indecomposable continua of arbitrary positive dimension, whose groups of self-homeomorphisms are cyclic groups of order n. She [**70**] has recently constructed, for every positive integer n, an hereditarily indecomposable one-dimensional continuum X_n with exactly n continuous self-surjections each of which is a homeomorphism and such that X_n admits an atomic map onto the pseudo-arc. (A mapping $f\colon X \to Y$ between continua is *atomic* if, for each subcontinuum K of X such that $f(K)$ is nondegenerate, $f^{-1}(f(K)) = K$.)

693? **Question 34.** *Which finite groups are the complete homeomorphism groups of hereditarily indecomposable continua? For each such group, does there exist an hereditarily indecomposable continuum with that group as its homeomorphism group such that every continuous self-surjection of the continuum is a homeomorphism? such a continuum of arbitrary positive dimension?*

Renska [**73**] has constructed for every $m = 2, 3, \ldots, \infty$ an m-dimensional hereditarily indecomposable Cantor manifold Y_m whose only self-homeomorphism is the identity. Pol [**68**] has constructed for every such m an m-dimensional hereditarily indecomposable continuum whose only continuous self-surjection is the identity. For each such m she has constructed $\mathfrak{c} = 2^{\aleph_0}$ such continua which are pairwise incomparable by continuous maps.

694? **Question 35.** *Does there for each* $m = 2, 3, \ldots, \infty$ *exist an m-dimensional hereditarily indecomposable Cantor manifold whose only continuous self-surjection is the identity?*

Non-metric continua

If x is a point of the nondegenerate continuum X, the *composant C_x of X corresponding to the point x* is the union of all proper subcontinua of X containing the point x. If X is a nondegenerate decomposable continuum, then X always has either exactly one composant, in the case when X is not irreducible between any pair of points, or exactly three distinct composants, in the case when X is irreducible between some pair of points, with any two of the composants intersecting. This is the case whether X is metrizable or not.

For indecomposable continua, the case is quite different. A nondegenerate indecomposable metric continuum always has $\mathfrak{c} = 2^{\aleph_0}$ distinct composants, which form a partition of the continuum with each composant being a dense first category set in the continuum.

Bellamy [**3**] has constructed a non-metric indecomposable continuum with exactly two composants and, by identifying a point in each composant [**4**], a non-metric indecomposable continuum with only one composant. Smith [**87**] has constructed a non-metric hereditarily indecomposable continuum with exactly two composants. Bellamy's identification to produce a single composant produces a decomposable subcontinuum and so cannot be used to obtain an hereditarily indecomposable continuum.

Question 36. *Does there exist a non-metric hereditarily indecomposable continuum with only one composant?* 695?

Smith [**85**] has also shown that for every infinite cardinal α there is an indecomposable continuum with exactly 2^α composants.

If $\mathbb{H} = [0,1)$ and $\mathbb{H}^*$ is the Stone–Čech remainder $\mathbb{H}^* = \beta(\mathbb{H}) \setminus \mathbb{H}$, then $\mathbb{H}^*$ is an indecomposable continuum. However, in set theory determined by ZFC alone it is not possible to determine the number of composants of $\mathbb{H}^*$. There are consistency results by Rudin, Blass, Banakh, Mildenberger and Shelah showing that the number of composants of H^* can be 1, 2 or $2^\mathfrak{c}$. Banakh and Blass [**2**] have shown that the number of composants of $\mathbb{H}^*$ must be either finite or $2^\mathfrak{c}$, but it is not known that if the number of composants is finite it must be one or two.

Question 37. *If X is an indecomposable non-metric continuum with only finitely many composants, must X have at most two composants?* 696?

Question 38. *If X is an indecomposable non-metric continuum with infinitely many composants, must the number of composants of X be 2^α for some infinite cardinal α?* 697?

By considering inverse limits of pseudo-arcs indexed by ω_1, Smith [**89**] has constructed a non-metric hereditarily indecomposable homogeneous hereditarily equivalent continuum. Thus this continuum shares many of the properties of the metric pseudo-arc. He has also constructed an inverse limit of pseudo-arcs indexed by ω_1 which is a non-metric hereditarily indecomposable continuum which is neither homogeneous nor hereditarily equivalent. The first example has $\mathfrak{c} = 2^{\aleph_0}$ composants, while the second example has only two composants.

The following questions are due to Smith.

Question 39. *Are there non-metric indecomposable hereditarily equivalent continua other than the inverse limit of ω_1 pseudo-arcs constructed by Smith?* 698?

699? **Question 40.** *Are there non-metric homogeneous chainable continua other than the inverse limit of ω_1 pseudo-arcs constructed by Smith? In particular, is there an inverse limit on a large set of chainable continua which is homogeneous?*

700? **Question 41.** *How many topologically distinct continua obtainable as inverse limits of pseudo-arcs indexed by ω_1 are there?*

Bing [**8**] has shown that every metric continuum of dimension greater than one contains an hereditarily indecomposable continuum. Smith [**84**, **88**, **90**] has a number of results and examples for products of non-metric continua and for non-metric continua of various dimensions showing that the analogous situation is not true, or that there may exist indecomposable continua but not hereditarily indecomposable continua. This is an area deserving much further investigation.

Conclusion

Kuratowski [**37**] wrote in 1973 about the theory of indecomposable continua:

> "It is one of the most developed and, to my mind, most beautiful branches of topology. It has attracted the attention of such distinguished mathematicians as P.S. Alexandrov, RH Bing, D. van Dantzig, G.W. Henderson, S. Mazurkiewicz, E.E. Moise, R.L. Moore, and, among the younger generation, C. Hagopian, J. Krasinkiewicz, Rogers and many others.
>
> "This is not surprising: by means of indecomposable continua one has succeeded in solving many earlier problems and in opening up new, extraordinarily rich topics.
>
> "In particular, great interest has been aroused (and this is even more noteworthy) by *hereditarily* indecomposable continua, i.e. those whose every subcontinuum is indecomposable (B. Knaster gave the first example of an hereditarily indecomposable continuum in 1922 in "Fund. Math." 3)."

Except that individuals are not as young as they once were, this is just as true more than 30 years later as it was when Kuratowski wrote it and indecomposable continua continue to richly reward anyone willing to investigate them.

References

[1] R. N. Ball, J. N. Hagler, and Y. Sternfeld, *The structure of atoms (hereditarily indecomposable continua)*, Fund. Math. **156** (1998), no. 3, 261–278.

[2] T. Banakh and A. Blass, *The number of near-coherence classes of ultrafilters is either finite or $2^{\mathfrak{c}}$*, 2005, Preprint. To appear in the proceedings of the set theory year at CRM (Barcelona).

[3] D. P. Bellamy, *A non-metric indecomposable continuum*, Duke Math. J. **38** (1971), 15–20.

[4] D. P. Bellamy, *Indecomposable continua with one and two composants*, Fund. Math. **101** (1978), no. 2, 129–134.

[5] D. P. Bellamy and J. A. Kennedy, *Factorwise rigidity of products of pseudo-arcs*, Topology Appl. **24** (1986), 197–205.

[6] D. P. Bellamy and J. M. Łysko, *Factorwise rigidity of the product of two pseudo-arcs*, Topology Proc. **8** (1983), no. 1, 21–27.

[7] R H Bing, *A homogeneous indecomposable plane continuum*, Duke Math. J. **15** (1948), 729–742.

[8] R H Bing, *Higher-dimensional hereditarily indecomposable continua*, Trans. Amer. Math. Soc. **71** (1951), 267–273.

[9] R H Bing, *Snake-like continua*, Duke Math. J. **18** (1951), 653–663.

[10] R H Bing, *Each homogeneous nondegenerate chainable continuum is a pseudo-arc*, Proc. Amer. Math. Soc. **10** (1959), 345–346.

[11] R H Bing, *The elusive fixed point property*, Amer. Math. Monthly **76** (1969), 119–132.

[12] B. L. Brechner, *On the dimensions of certain spaces of homeomorphisms*, Trans. Amer. Math. Soc. **121** (1966), 516–548.

[13] B. L. Brechner, *Problem 8 in Continuum Theory Problems*, Topology Proc. **8** (1983), 361–394.

[14] C. E. Burgess, *Chainable continua and indecomposability*, Pacific J. Math. **9** (1959), 653–659.

[15] J. J. Charatonik and T. Maćkowiak, *Confluent and related mappings on arc-like continua—an application to homogeneity*, Topology Appl. **23** (1986), no. 1, 29–39.

[16] H. Cook, *Weakly confluent mappings and dimension*, 1977 Spring Topology Conference Abstracts, Baton Rouge (1977) 31.

[17] H. Cook, *Continua which admit only the identity mapping onto non-degenerate subcontinua*, Fund. Math. **60** (1967), 241–249.

[18] H. Cook, *Tree-likeness of hereditarily equivalent continua*, Fund. Math. **68** (1970), 203–205.

[19] H. Cook, W. T. Ingram, and A. Lelek, *A list of problems known as Houston problem book*, Continua (Cincinnati, OH, 1994), Lecture Notes in Pure and Appl. Math., vol. 170, Marcel Dekker Inc., New York, 1995, pp. 365–398.

[20] J. L. Cornette, *Retracts of the pseudo-arc*, Colloq. Math. **19** (1968), 235–239.

[21] C. Eberhart and S. B. Nadler, Jr., *The dimension of certain hyperspaces*, Bull. Acad. Polon. Sci. Sér. Sci. Math. Astronom. Phys. **19** (1971), 1027–1034.

[22] L. Fearnley, *Characterizations of the continuous images of the pseudo-arc*, Trans. Amer. Math. Soc. **111** (1964), 380–399.

[23] C. L. Hagopian, *Indecomposable homogeneous plane continua are hereditarily indecomposable*, Trans. Amer. Math. Soc. **224** (1976), no. 2, 339–350 (1977).

[24] C. L. Hagopian, *Atriodic homogeneous continua*, Pacific J. Math. **113** (1984), no. 2, 333–347.

[25] C. L. Hagopian and J. T. Rogers, Jr., *A classification of homogeneous, circle-like continua*, Houston J. Math. **3** (1977), no. 4, 471–474.

[26] G. W. Henderson, *Proof that every compact decomposable continuum which is topologically equivalent to each of its nondegenerate subcontinua is an arc*, Ann. of Math. (2) **72** (1960), 421–428.

[27] W. T. Ingram, *An atriodic tree-like continuum with positive span*, Fund. Math. **77** (1972), no. 2, 99–107.

[28] W. T. Ingram, *An uncountable collection of mutually exclusive planar atriodic tree-like continua with positive span*, Fund. Math. **85** (1974), no. 1, 73–78.

[29] W. T. Ingram, *Hereditarily indecomposable tree-like continua*, Fund. Math. **103** (1979), no. 1, 61–64.

[30] W. T. Ingram, *Hereditarily indecomposable tree-like continua. II*, Fund. Math. **111** (1981), no. 2, 95–106.

[31] J. L. Kelley, *Hyperspaces of a continuum*, Trans. Amer. Math. Soc. **52** (1942), 22–36.

[32] B. Knaster, *Un continu dont tout sous-continu est indécomposable*, Fund. Math. **3** (1922), 247–286.

[33] J. Krasinkiewicz, *On the hyperspaces of hereditarily indecomposable continua*, Fund. Math. **84** (1974), no. 3, 175–186.

[34] J. Krasinkiewicz, *Mapping properties of hereditarily indecomposable continua*, Houston J. Math. **8** (1982), no. 4, 507–516.

[35] J. Krasinkiewicz, *Problem 154 in A list of problems known as Houston problem book*, Continua (Cincinnati, OH, 1994), Lecture Notes in Pure and Appl. Math., vol. 170, Marcel Dekker Inc., New York, 1995, pp. 365–398.

[36] P. Krupski and J. R. Prajs, *Outlet points and homogeneous continua*, Trans. Amer. Math. Soc. **318** (1990), no. 1, 123–141.

[37] K. Kuratowski, *A half century of Polish mathematics*, Pergamon Press, Elmsford, NY, 1980.

[38] A. Lelek, *On weakly chainable continua*, Fund. Math. **51** (1962), 271–282.

[39] M. Levin, *Certain finite-dimensional maps and their application to hyperspaces*, Israel J. Math. **105** (1998), 257–262.

[40] M. Levin and Y. Sternfeld, *The space of subcontinua of a 2-dimensional continuum is infinite-dimensional*, Proc. Amer. Math. Soc. **125** (1997), no. 9, 2771–2775.

[41] W. Lewis, *Monotone maps of hereditarily indecomposable continua*, Proc. Amer. Math. Soc. **75** (1979), no. 2, 361–364.

[42] W. Lewis, *Stable homeomorphisms of the pseudo-arc*, Canad. J. Math. **31** (1979), no. 2, 363–374.

[43] W. Lewis, *Almost chainable homogeneous continua are chainable*, Houston J. Math. **7** (1981), no. 3, 373–377.

[44] W. Lewis, *Homeomorphism groups and homogeneous continua*, Topology Proc. **6** (1981), no. 2, 335–344 (1982).

[45] W. Lewis, *Continuum theory problems*, Topology Proc. **8** (1983), no. 2, 361–394.

[46] W. Lewis, *Homogeneous circlelike continua*, Proc. Amer. Math. Soc. **89** (1983), no. 1, 163–168.

[47] W. Lewis, *Periodic homeomorphisms of chainable continua*, Fund. Math. **117** (1983), no. 1, 81–84.

[48] W. Lewis, *Pseudo-arcs and connectedness in homeomorphism groups*, Proc. Amer. Math. Soc. **87** (1983), no. 4, 745–748.

[49] W. Lewis, *Most maps of the pseudo-arc are homeomorphisms*, Proc. Amer. Math. Soc. **91** (1984), no. 1, 147–154.

[50] W. Lewis, *Compact group actions on chainable continua*, Houston J. Math. **11** (1985), no. 2, 225–236.

[51] W. Lewis, *The classification of homogeneous continua*, Soochow J. Math. **18** (1992), no. 1, 85–121.

[52] W. Lewis, *The pseudo-arc*, Bol. Soc. Mat. Mexicana (3) **5** (1999), no. 1, 25–77.

[53] J. M. Łysko, *Maps of products of continua*, Preprint.

[54] J. M. Łysko, *Problem 42 in Continuum Theory Problems*, Topology Proc. **8** (1983), 361–394.

[55] J. R. Martin and S. B. Nadler, Jr., *Examples and questions in the theory of fixed-point sets*, Canad. J. Math. **31** (1979), no. 5, 1017–1032.

[56] A. Mason, J. J. Walsh, and D. C. Wilson, *Inverse limits which are not hereditarily indecomposable*, Proc. Amer. Math. Soc. **83** (1981), no. 2, 403–407.

[57] T. B. McLean, *Confluent images of tree-like curves are tree-like*, Duke Math. J. **39** (1972), 465–473.

[58] P. Minc, *An atriodic simple-4-od-like continuum which is not simple-triod-like*, Trans. Amer. Math. Soc. **338** (1993), no. 2, 537–552.

[59] P. Minc, *On weakly chainable inverse limits with simplicial bonding maps*, Proc. Amer. Math. Soc. **119** (1993), no. 1, 281–289.

[60] L. Mohler, *Problem 171 in A list of problems known as Houston problem book*, Continua (Cincinnati, OH, 1994), Lecture Notes in Pure and Appl. Math., vol. 170, Marcel Dekker Inc., New York, 1995, pp. 365–398.

[61] L. Mohler and L. G. Oversteegen, *On hereditarily decomposable hereditarily equivalent nonmetric continua*, Fund. Math. **136** (1990), no. 1, 1–12.

[62] E. E. Moise, *An indecomposable plane continuum which is homeomorphic to each of its nondegenerate subcontinua*, Trans. Amer. Math. Soc. **63** (1948), 581–594.

[63] E. E. Moise, *A note on the pseudo-arc*, Trans. Amer. Math. Soc. **67** (1949), 57–58.

[64] R. L. Moore, *Concerning triodic continua in the plane*, Fund. Math. **13** (1928), 261–263.

[65] L. G. Oversteegen and E. D. Tymchatyn, *Plane strips and the span of continua. I*, Houston J. Math. **8** (1982), no. 1, 129–142.

[66] L. G. Oversteegen and E. D. Tymchatyn, *On span and weakly chainable continua*, Fund. Math. **122** (1984), no. 2, 159–174.

[67] L. G. Oversteegen and E. D. Tymchatyn, *On the dimension of certain totally disconnected spaces*, Proc. Amer. Math. Soc. **122** (1994), no. 3, 885–891.

[68] E. Pol, *The non-existence of common models for some classes of higher-dimensional hereditarily indecomposable continua*, Preprint.

[69] E. Pol, *Hereditarily indecomposable continua with exactly n autohomeomorphisms*, Colloq. Math. **94** (2002), no. 2, 225–234.

[70] E. Pol, *Hereditarily indecomposable continua with finitely many continuous surjections*, Bol. Soc. Mat. Mexicana (3) **11** (2005), no. 1, 139–147.

[71] J. R. Prajs, *On open homogeneity of closed balls*, 1992, Preprint.

[72] J. R. Prajs, *Plane continua homogeneous with respect to light open mappings*, 1993, Preprint.

[73] M. Reńska, *Rigid hereditarily indecomposable continua*, Topology Appl. **126** (2002), no. 1–2, 145–152.

[74] J. T. Rogers, Jr., *Solenoids of pseudo-arcs*, Houston J. Math. **3** (1977), no. 4, 531–537.

[75] J. T. Rogers, Jr., *Weakly confluent mappings and finitely-generated cohomology*, Proc. Amer. Math. Soc. **78** (1980), no. 3, 436–438.

[76] J. T. Rogers, Jr., *Homogeneous, separating plane continua are decomposable*, Michigan Math. J. **28** (1981), no. 3, 317–322.

[77] J. T. Rogers, Jr., *Homogeneous hereditarily indecomposable continua are tree-like*, Houston J. Math. **8** (1982), no. 3, 421–428.

[78] J. T. Rogers, Jr., *Cell-like decompositions of homogeneous continua*, Proc. Amer. Math. Soc. **87** (1983), no. 2, 375–377.

[79] J. T. Rogers, Jr., *Homogeneous continua*, Topology Proc. **8** (1983), no. 1, 213–233.

[80] J. T. Rogers, Jr., *Classifying homogeneous continua*, Topology Appl. **44** (1992), 341–352.

[81] I. Rosenholtz, *Open maps of chainable continua*, Proc. Amer. Math. Soc. **42** (1974), 258–264.

[82] C. R. Seaquist, *Monotone homogeneity and planar manifolds with boundary*, Topology Appl. **81** (1997), no. 1, 47–53.

[83] C. R. Seaquist, *Monotone open homogeneity of Sierpiński curve*, Topology Appl. **93** (1999), no. 2, 91–112.

[84] M. Smith, *On non-metric indecomposable and hereditarily indecomposable subcontinua of products of long Hausdorff arcs*, Preprint.

[85] M. Smith, *Generating large indecomposable continua*, Pacific J. Math. **62** (1976), no. 2, 587–593.

[86] M. Smith, *Large indecomposable continua with only one composant*, Pacific J. Math. **86** (1980), no. 2, 593–600.

[87] M. Smith, *A hereditarily indecomposable Hausdorff continuum with exactly two composants*, Topology Proc. **9** (1984), no. 1, 123–143.

[88] M. Smith, *Hausdorff hypercubes which do not contain arcless continua*, Proc. Amer. Math. Soc. **95** (1985), no. 1, 109–114.

[89] M. Smith, *On nonmetric pseudo-arcs*, Topology Proc. **10** (1985), no. 2, 385–397.

[90] M. Smith, *Continua arbitrary products of which do not contain nondegenerate hereditarily indecomposable continua*, Topology Proc. **13** (1988), no. 1, 137–160.

[91] M. Sobolewski, *Pseudoarc is not pseudocontractible*, 1991, Preprint.

[92] J. A. Toledo, *Inducible periodic homeomorphisms of tree-like continua*, Trans. Amer. Math. Soc. **282** (1984), no. 1, 77–108.

[93] E. D. Tymchatyn, *Hyperspaces of hereditarily indecomposable plane continua*, Proc. Amer. Math. Soc. **56** (1976), 300–302.

Open Problems in Topology II – Edited by E. Pearl

Open problems on dendroids

Veronica Martínez-de-la-Vega and Jorge M. Martínez-Montejano

Dedicated to the memory of our friend, colleague and teacher, Professor Janusz J. Charatonik.

1. Introduction

A *continuum* is a compact, connected, metric space. A *dendroid* is an arcwise connected and hereditarily unicoherent continuum. A *dendrite* is defined as a locally connected dendroid. Dendroids were defined by B. Knaster in the Topology Seminar in Wrocław in the late 1950s. One of the most assiduous participants of this Seminar was Prof. J.J. Charatonik, who wrote his Doctoral Dissertation and over fifty papers on dendroids. Many of his doctoral students, such as S.T. Czuba, T. Maćkowiak, P. Krupski and J. Prajs, also have done many contributions to this field. Even though dendroids are one-dimensional and most of them can be geometrically realized, there are many properties and intrinsic characterizations of them which are still unknown. The purpose of this paper is to give a survey of some results and open problems in this interesting area of Continuum Theory.

2. The problem

We begin with a problem that many authors consider as one of the most importance in the study of dendroids. Before posing the problem it is interesting to note that in the early 1960s, B. Knaster saw dendroids as those continua that for which for every $\varepsilon > 0$ there exists a tree T and an ε-retraction $r\colon X \to T$ (an ε-retraction is a retraction such that $\operatorname{diam}(r^{-1}(t)) < \varepsilon$ for every $t \in T$). The contemporary definition of dendroids (the one given above) was formulated in a more convenient way.

Problem 1. *Let X be a dendroid and $\varepsilon > 0$. Are there a subtree T of X and an ε-retraction from X onto T?* 701?

Let us note that Fugate in [**26, 27**] has some partial positive answers to this problem and that a positive answer to it would give an affirmative answer to a variety of other problems about dendroids (see, e.g., Problem 27).

3. Mappings on dendrites

Characterize dendrites among dendroids is one of the eldest problems in the study of dendroids, over 60 characterizations of dendrites can be found in [**10**]. Also in [**10**] a survey of some open problems is made.

Problem 2 ([**10**, Problem 2.14]). *Characterize all dendrites X having the property that every open image of X is homeomorphic to X.* 702?

Problem 3 ([**10**, Question 3.6]). *Does every monotonely homogeneous dendrite (Definition 3.8) contain a homeomorphic copy of the dendrite L_0?* 703?

A topological space X is said to be *chaotic (strongly chaotic)* if for any two distinct points p and q of X there exists open neighborhoods U of p and V of q such that no open subset of U is homeomorphic to any open subset (subset) of V; and *rigid (strongly rigid)* if the only homeomorphism of X onto (into) X is the identity map.

704–705? **Problem 4** ([**10**, Problem 4.5]). *Give any structural characterization of (strongly) chaotic and of (strongly) rigid dendrites.*

Consider the following conditions (ω_0) and (ω).

(ω_0) For every compact space Y, for every light open mapping $f\colon Y \to f(Y)$ with $X \subset f(Y)$ and for every point $y_0 \in f^{-1}(X) \subset Y$ there exists a homeomorphic copy X' of X in Y with $y_0 \in X'$ such that the restriction $f|_{X'}\colon X' \to f(X') = X$ is a homeomorphism.

(ω) For every compact space Y, for every light open mapping $f\colon Y \to f(Y)$ with $X \subset f(Y)$ there exists a homeomorphic copy X' of X in Y such that the restriction $f|_{X'}\colon X' \to f(X') = X$ is a homeomorphism.

Now consider conditions $(\omega_0(\mathcal{M}))$ and $(\omega_0(\mathcal{M}))$ regarding a continuum X and a class $\mathcal{M}$ of light mappings. Which can be defined replacing the phrase "light open mapping $f\colon Y \to f(Y)$" by "light mapping $f\colon Y \to f(Y)$ in $\mathcal{M}$."

Note that if $\mathcal{O}$ stands for the class of open mappings, then $(\omega_0(\mathcal{O}))$ and $(\omega(\mathcal{O}))$ coincide with conditions (ω_0) and (ω).

One can also consider conditions (γ_0) and (γ) obtained from (ω_0) and (ω), respectively, by replacing the phrase "every compact space Y" with "every continuum Y", and define condition (δ) X is a dendrite.

In [**18**, Statement 1] it is shown why conditions (δ), (ω_0), (ω), (γ_0) and (γ) are equivalent. So the following problem is posed.

706? **Problem 5** ([**18**, Problem 2]). *Does the equivalence in [**18**, Statement 1] remain true if we replace openness of the light mapping f by a less restrictive condition? In other words, for what (larger) classes $\mathcal{M}$ of light mappings are conditions (δ), (ω_0), (ω), (γ_0) and (γ) equivalent?*

Also by [**18**, Observation 4] if the class $\mathcal{M}$ contains the class $\mathcal{O}$ of open mappings, and if, for a continuum X, implication $(\delta) \Rightarrow (\omega_0(\mathcal{M}))$ holds, then all the conditions (δ), (ω_0), (ω), $(\omega_0(\mathcal{M}))$ and $(\omega(\mathcal{M}))$ are equivalent. Thus Problem 5 reduces to:

707? **Problem 6** ([**18**, Problem 5]). *For what classes $\mathcal{M}$ of mappings containing the class $\mathcal{O}$ does each dendrite X satisfy condition $(\omega_0(\mathcal{M}))$ (i.e., the implication $(\delta) \Rightarrow (\omega_0(\mathcal{M}))$ holds)?*

In [**7**] the following result of K. Omiljanowski is proved.

Theorem 3.1. *Let a dendrite Y be such that all ramification points of Y are of order 3 and the set $R(Y)$ of all ramification points of Y is discrete. If a dendrite X can be mapped onto Y under a monotone mapping, then X contains a homeomorphic copy of Y.*

In [**2**] J.J. Charatonik proved the following theorem.

Theorem 3.2. *Let D be a dendrite. For every compact space X and for every light open surjective mapping $f: X \to Y$ with $D \subset Y$ there is a homeomorphic copy D' of D in X such that the restriction $f|_{D'}: D' \to f(D')$ is a homeomorphism.*

The inverse implication of Theorem 3.2 was proved in [**11**, Corollary 10] and [**18**, Theorem 16]. It is interesting to ask if Theorem 3.1 can be reversed. So the natural problem is.

Problem 7 ([**2**, Problem 1.3]). *Characterize all dendrites Y having the property* 708?
that if a dendrite X can be mapped onto Y under a monotone mapping, then X contains a homeomorphic copy of Y.

Using the notation and all the classes of mappings defined in [**2**], the following Corollary is proved.

Corollary 3.3 ([**2**, Corollary 3.2]). *Let a continuum Y satisfy the conditions of Theorem 3.1. If a dendrite X can be mapped onto Y under a mapping that belongs to one of the classes of mappings $\mathcal{OM}$, $\mathcal{C}$, Loc$\mathcal{C}$, $\mathcal{QM}$, $\mathcal{WM}$, then X contains a homeomorphic copy of Y.*

Problem 8 ([**2**, Question 3.3]). *Let a continuum Y satisfy conditions of The-* 709?
orem 3.1 and X a dendrite that can be mapped onto Y under a semi-confluent mapping. Must then X contain a homeomorphic copy of Y?

Notice that Theorem 3.1 cannot be extended to continua X that contain simple closed curves, not even if X is locally connected or X is a linear graph [**53**, Ch. X, §3, Ex., p. 189]). Using all the above ideas one of the natural questions is:

Problem 9 ([**2**, Question 3.7]). *Let a continuum Y satisfy conditions of Theo-* 710?
rem 3.1 and let a continuum X be such that if X can be mapped onto Y under a monotone mapping, then X contains a homeomorphic copy of Y. Must then X be a dendrite? If not under what additional assumptions does the implication hold?

Another important and interesting question related to Theorem 3.1 is if the implication in the result can be reversed. A partial answer is given.

Theorem 3.4 ([**2**, Theorem 4.1]). *Let a dendrite Y have the property that for each dendrite X if X can be mapped onto Y under a monotone mapping, then X contains a homeomorphic copy of Y. Then either Y is an arc or all the ramification points of Y are of order 3.*

And so the following question arises.

Problem 10 ([**2**, Question 4.2]). *Let a dendrite Y have the same property as in* 711?
Theorem 3.4. Must then Y either be an arc or have the set $R(Y)$ of ramification points discrete?

Given a space X and a map $f: X \to X$. A point x of X is said to be *fixed* if $f(x) = x$, *periodic (of period n)* provided that there is $n \in \mathbb{N}$ such that $f^n(x) = x$

(and $f^k(x) \neq x$ for $k < n$), *recurrent*, provided that for each open set U containing x there is $n \in \mathbb{N}$ such that $f^n(x) \in U$, *eventually periodic of period* n provided that there exists $m \in \mathbb{N} \cup \{0\}$ such that $f^m(x)$ is a periodic point of f of period n, *eventually periodic* provided that there is $n \in \mathbb{N}$ such that x is an eventually periodic point of period $n \in \mathbb{N}$ for f and a *non-wandering* point of f provided that for any open set U containing x there exists $y \in U$ and $n \in \mathbb{N}$ such that $f^n(y) \in U$. For a mapping $f: X \to X$ the sets of fixed, periodic, recurrent, eventually periodic and non-wandering points of f will be denoted by $F(f)$, $P(f)$, $R(f)$, $EP(f)$ and $\Omega(f)$, respectively.

Also a space X is said to have the *periodic-recurrent property* (PR-property) provided that for every mapping $f: X \to X$ the equality $\mathrm{cl}(P(f)) = \mathrm{cl}(R(f))$ holds, the *non-wandering-periodic property* (ΩP-property) provided that for every mapping $f: X \to X$ the equality $\Omega(f) = P(f)$ holds and the *non-wandering-eventually periodic property* (ΩEP-property) provided that for every mapping $f: X \to X$ the inclusion $\Omega(f) \subset \mathrm{cl}_X(EP(f))$ is satisfied.

In [**16**, Proposition 2.11] it is proved that a dendrite X has the PR-property if and only if X does not contain any copy of the Gehman dendrite; and the following problem is posed.

712? **Problem 11** ([**16**, Problem 2.14]). *Give an internal (i.e., structural) characterization of dendrites with ΩP-property.*

In [**16**, Corollary 3.6] it is proved that if X is a tree and f a mapping $f: X \to X$ such that $\Omega(f)$ is finite, then $\mathrm{card}(P(f)) = \mathrm{card}(\Omega(f))$. It is not known if the assumption of the finiteness of the set $\Omega(f)$ is or is not essential. So, the following question is posed.

713? **Problem 12** ([**16**, Question 3.7]). *Do there exist a tree X and a mapping $f: X \to X$ such that $\Omega(f)$ is infinite while $P(f)$ is finite?*

In [**16**, Theorem 4.6] it is proved that if X is a dendrite such that for each mapping $f: X \to X$ the equality $\mathrm{card}(P(f)) = \mathrm{card}(\Omega(f))$ holds, then X is a tree. So, it is asked if the converse is true.

714? **Problem 13** ([**16**, Question 4.7]). *Is it true that for each tree X the assertion that for each mapping $f: X \to X$ the equality* $\mathrm{card}(P(f)) = \mathrm{card}(\Omega(f))$ *holds?*

Given a dendroid X, we define $E(X)$ as the set of endpoints, $O(X)$ the set of ordinary points and $R(X)$ the set of ramification points of X. In [**7**, Proposition 4.7] J.J. Charatonik proved the following.

Proposition 3.5. *Let x and y be any two points of the standard universal dendrite $X = D_m$ for $m \in \{3, 4, \dots, \omega\}$. Then there is a homeomorphism $h: X \to X$ such that $h(x) = y$ if and only if one of the following conditions is satisfied: $x, y \in E(X)$; $x, y \in O(X)$; $x, y \in R(X)$.*

To this respect the following problem remains unsolved.

715? **Problem 14** ([**7**, Question 4.9]). *What dendrites X have the property that for each two points x and y of X there exists a homeomorphism $h: X \to X$ with*

$h(x) = y$ if and only if both these points are either end points, or ordinary points or ramification points of X?

It is known that:

Theorem 3.6 ([**7**, Corollary 5.5]). *Let D_m be the standard universal dendrite of order $m \in \{3, 4, \dots, \omega\}$. Then each monotone surjection of D_m onto itself is a near homeomorphism if and only if $m = 3$.*

A map $f\colon X \to Y$ is a *near homeomorphism* provided that for $\varepsilon > 0$ there exists a homeomorphism $h\colon X \to Y$ such that $\sup\{d(f(x), h(x) : x \in X\} < \varepsilon$. The following problem is still open.

Problem 15 ([**7**, Problem 5.1]). *What dendrites X have the property that each monotone mapping of X onto itself is a near homeomorphism?* 716?

Let $\mathcal{M}$ be a class of mappings. Two continua X and Y are said to be *equivalent with respect to $\mathcal{M}$* if there are a mapping in $\mathcal{M}$ from X onto Y and a mapping in $\mathcal{M}$ from Y onto X. A class $\mathcal{M}$ of mappings is said to be *neat* if all homeomorphisms are in $\mathcal{M}$ and the composition of any two mappings in $\mathcal{M}$ is also in $\mathcal{M}$. Therefore, if a neat class $\mathcal{M}$ of mappings is given, then a family of continua is decomposed into disjoint equivalence classes in the sense that two continua belong to the same class provided that they are equivalent with respect to $\mathcal{M}$. A continuum is said to be *isolated with respect to $\mathcal{M}$* provided the above mentioned class to which X belongs consists of X only.

In [**7**, Theorems 6.7 and 6.14] it is shown that universal dendrites are not isolated with respect to monotone mappings. The following problem remain open.

Problem 16 ([**7**, Problem 6.1]). *Find all dendrites which are isolated with respect to monotone mappings.* 717?

About the previous problem the authors have the following conjecture.

Conjecture 3.7. *A dendrite X is isolated with respect to monotone mappings if and only if $|R(X)| < \infty$.* 718?

Definition 3.8. Let $\mathcal{M}$ be a class of mappings. A continuum X is *homogeneous with respect to $\mathcal{M}$* provided that for every two points p and q of X there is a surjective mapping $f\colon X \to X$ such that $f(p) = q$ and $f \in \mathcal{M}$.

In [**7**, Theorem 7.1] it is proved that any standard universal dendrite D_m of order $m \in \{3, 4, \dots, \omega\}$ is homogeneous with respect to monotone mappings. After this it is observed that each m-od is an example of a dendrite which is not homogeneous with respect to confluent, and therefore to monotone, mappings. Then the following problem is posed.

Problem 17 ([**7**, Question 7.2]). *What dendrites are homogeneous with respect to monotone mappings?* 719?

4. Maps onto dendroids

In [**32**, Theorem 2], J. Heath and V. Nall proved the following.

Theorem 4.1. *There does not exist a (exactly) 2-to-1 map from a hereditarily decomposable continuum onto a dendroid.*

They asked the following.

720? **Problem 18** ([**32**, p. 288]). *Is there an indecomposable continuum I that admits a map onto a dendroid X such that the inverse of each point in the range contains at most two points?*

A negative answer to Problem 18 would have strengthened Theorem 4.1. P. Minc has shown that this is the case in some special situations and he says that it would be interesting to partially answer Problem 18 for the case of chainable continua. He proved the next two theorems.

Theorem 4.2 ([**44**, Corollary 3.12]). *Let f be a map of an indecomposable continuum Y into a plane dendroid P. Then there is a point $p \in P$ such that $f^{-1}(p)$ is uncountable.*

Theorem 4.3 ([**44**, Corollary 3.9 and Remark 3.10]). *Let K be either any Knaster type continuum or any solenoid. Suppose that f is a map of K onto an arbitrary dendroid X. Then there is a point $x \in X$ such that $f^{-1}(x)$ consists of at least three points.*

Also, P. Minc pointed out the following: Theorem 4.3 shows that some indecomposable continua do not admit 2-or-1-to-1 maps on dendroids. On the other hand, it is easy to construct such maps from many standard examples of chainable hereditarily decomposable continua. So he posed the next problem.

721? **Problem 19** ([**45**, p. 289]). *Is it true that a chainable continuum is hereditarily decomposable if and only if it admits a 2-or-1-to-1 map onto a dendroid?*

Let us note that in [**45**, Theorem 1.1] he proved that every chainable continuum can be mapped into a dendroid in such a way that all point-inverses consist of at most three points.

5. Contractibility

The symbols $\mathrm{L}_{\sup}$, $\mathrm{L}_{\inf}$ and L_t mean the upper limit, the lower limit and the topological limit. A dendroid X is said to be: (a) *smooth*, (b) *semi-smooth*, (c) *weakly smooth*, if there exists a point $p \in X$ such that for every $a \in X$ and each convergent sequence $\{a_n\}_{n\in\mathbb{N}} \subset X$, with $a_n \to a$ we have: (a) $\mathrm{L}_t\, pa_n = pa$, (b) $\mathrm{L}_{\sup}\, pa_n$ is an arc, (c) $\mathrm{L}_{\inf}\, pa_n = pb$ for some $b \in X$. A dendroid X is said to be *pointwise smooth* if for each $x \in X$ there exists a point $p(x) \in X$ such that for each convergent sequence x_n convergent to a point a, the sequence of arcs $p(x)a_n$ is convergent and $\mathrm{L}_t\, pa_n = pa$. The next theorem is well known.

Theorem 5.1 ([**5**, **6**]). *Every contractible one-dimensional continuum is a dendroid.*

It is known that the inverse implication is not true.

The main problem related to contractibility of dendroids is the following.

Problem 20 ([**8**, p. 28]). *Find a structural characterization of contractible dendroids.* 722?

Given a dendroid X, a mapping $H\colon X \times [0,1] \to X$ such that $H(x,0) = x$ for each point $x \in X$ is called a *deformation*. A nonempty proper subset A of a dendroid X is said to be *homotopically fixed* provided that for every deformation $H\colon X \times [0,1] \to X$ we have that $H(A \times [0,1]) = A$. A nonempty subset A of a dendroid X is said to be *homotopically steady* provided that for every deformation $H\colon X \times [0,1] \to X$ we have that $A \subset H(X \times \{1\})$. Denote by $\mathcal{D}(X)$ the family of all deformations on X. Define, $\mathcal{K}(X)$, the *kernel of steadiness* of X by $\mathcal{K}(X) = \bigcap\{H(X \times \{1\}) : H \in \mathcal{D}(X)\}$.

A nonempty proper subcontinuum A of a dendroid X is called an R^i*-continuum* (where $i \in \{1,2,3\}$) if there exist an open set U containing A and two sequences $\{C_n^1 : n \in \mathbb{N}\}$ and $\{C_n^2 : n \in \mathbb{N}\}$ of components in U such that

$$A = \begin{cases} \operatorname{L_{sup}} C_n^1 \cap \operatorname{L_{sup}} C_n^2 & \text{for } i = 1, \\ \operatorname{L_t} C_n^1 \cap \operatorname{L_t} C_n^2 & \text{for } i = 2, \\ \operatorname{L_{inf}} C_n^1 & \text{for } i = 3. \end{cases}$$

It is well known that if a dendroid X contains a homotopically fixed subset, then X is not contractible [**14**, Proposition 1]]) and that each R^i-continuum of a dendroid X (where $i \in \{1,2,3\}$) is a homotopically fixed subset of X [**21**, Theorem 3].

J.J. Charatonik and A. Illanes proved [**15**, Theorem 4.3] that each contractible space has empty kernel of steadiness and asked:

Problem 21 ([**15**, Question 4.5]). *Does every non-contractible dendroid have nonempty kernel of steadiness?* 723?

Also, it is shown that [**15**, Example 4.7] there is a plane dendroid X and a subcontinuum A of X such that A is an R^i-continuum in X for each $i = 1,2,3$, so it is homotopically fixed, while not homotopically steady. So the following questions arise.

Problem 22 ([**15**, Questions 4.9]). (a) *Does the existence of a homotopically fixed subset in a dendroid imply the existence of a homotopically steady subset?* 724–725?
(b) *What are the interrelations between R^i-continua and homotopically steady subsets of dendroids? More precisely, let an R^i-continuum A (for some $i \in \{1,2,3\}$) be contained in a dendroid X. Must A contain a nonempty homotopically steady subset of X?*

The following question (asked by W.J. Charatonik) is related to Problem 21.

Problem 23 ([**15**, Question 4.19]). *Given a dendroid X with a nondegenerate kernel $\mathcal{K}(X)$ of steadiness, is the dendroid $X/\mathcal{K}(X)$ always contractible?* 726?

A point p of a dendroid X is called a *Q-point* of X provided that there exists a sequence of points p_n of X converging to p such that $\mathrm{L}_{\sup} pp_n \neq \{p\}$ and if for each $n \in \mathbb{N}$ the arc $p_n q_n$ is irreducible between p_n and the continuum $\mathrm{L}_{\sup} pp_n$, then the sequence of points q_n converges also to p. The following problem is open.

727? **Problem 24** ([**8**, p. 30]). *Is it true that if a dendroid has a Q-point, then it is non-contractible?*

By [**23**, Corollary 3.10], it is known that if a dendroid is hereditarily contractible, then it is pointwise smooth. So, the following questions arise naturally.

728? **Problem 25** ([**23**, Question 3.11]). *Does pointwise smoothness of dendroids imply their hereditary contractibility?*

729? **Problem 26** ([**14**, Question 13]). *Find an intrinsic characterization of hereditarily contractible dendroids.*

6. Hyperspaces

Given a continuum X, the *hyperspace* 2^X of X is defined by $2^X = \{A \subset X : A \text{ is nonempty and closed}\}$. We consider 2^X with the Hausdorff metric H. Other hyperspaces considered here are $C(X) = \{A \in 2^X : A \text{ is connected}\}$, and, for each $n \in \mathbb{N}$, $F_n(X) = \{A \in 2^X : A \text{ has at most } n \text{ elements}\}$.

In [**48**, Theorem 6.18], S.B. Nadler, Jr. proved that $C(X)$, 2^X and $F_n(X)$ have the fixed point property when X is either a smooth dendroid or a fan. Right below the proof, S.B. Nadler, Jr. asks the following.

730–732? **Problem 27.** *Do $C(X)$, 2^X and $F_n(X)$ have the fixed point property when X is a dendroid?*

We note that a positive answer would follow from an affirmative answer to Problem 1.

7. Property of Kelley

A continuum X is said to have the *property of Kelley at a point* $x \in X$ provided that for each sequence of points x_n converging to x and for each continuum K in X containing the point x there is a sequence of continua K_n in X with $x_n \in K_n$ for each $n \in \mathbb{N}$ and converging to K. A continuum X is said to have the property of Kelley if it has the property at each point $x \in X$.

Given a dendroid X, and $x \in X$, we define the *Jones function* $T(x) = \{y \in X : \text{if there exists } A \in C(X) \text{ such that } y \in \mathrm{int}_X(A) \text{ then } x \in A\}$.

A *λ-dendroid* is a hereditarily unicoherent and hereditarily decomposable continuum.

S.T. Czuba proved the following implications.

Theorem 7.1 ([**25**, Lemma 2 and Corollary 5]). *Let X be a dendroid. Then:*

$$X \text{ has the property of Kelley} \Longrightarrow \forall x, y \in X \text{ if } xy \cap T(x) \neq \{x\}, \text{ then } y \in T(x)$$
$$\Longrightarrow X \text{ is smooth}$$
$$\Longrightarrow X \text{ is locally connected at some point.}$$

In Theorem 7.1, the assumption that X is a dendroid is essential: a λ-dendroid obtained as a compactification of the Cantor fan minus its vertex such that the remainder is an arc has the property of Kelley and is non-smooth (it is not locally connected at any point). Concerning to this J.J. Charatonik asked the following.

Problem 28 ([**8**, Question 5.20]). *For what continua X does the property of Kelley imply local connectedness of X at some point?* 733?

Fans having the property of Kelley have been characterized in [**3**, **9**]. But there are no known characterizations of dendroids having the property of Kelley. So we have the following.

Problem 29. *Characterize dendroids having the property of Kelley.* 734?

8. Retractions

In [**12**, Theorem 3.1 and Theorem 3.3], J.J. Charatonik et al proved the following.

Theorem 8.1. *Let X be a one-dimensional continuum. If there is a retraction from $C(X)$ (2^X) onto X, then X is a uniformly arcwise connected dendroid.*

And, in [**28**, Theorem 2.9], J.T. Goodykoontz, Jr. showed the following.

Theorem 8.2. *Every smooth fan X is a deformation retract of 2^X.*

Also, there are known examples of a non-smooth fan X such that there is no retraction from 2^X onto X [**1**, Example 3.7] and of a non-planable smooth dendroid for which there is no retraction from 2^X onto X [**12**, Example 5.52]. So, in [**12**], J.J. Charatonik et al asked the following.

Problem 30. *For what smooth dendroids X does there exist a deformation retraction from 2^X onto X?* 735?

Let X be a continuum. A retraction $r\colon 2^X \to X$ is said to be *associative* provided that $r(A \cup B) = r(\{r(A)\} \cup B)$ for every $A, B \in 2^X$.

Let X be a hereditarily unicoherent continuum. A retraction $r\colon 2^X \to X$ is said to be *internal* provided that $r(A) \in I(A)$ for each $A \in 2^X$, where $I(A)$ denotes the continuum irreducible with respect to containing A.

It is known [**12**, Theorem 3.21] that the Mohler–Nikiel universal smooth dendroid admits an associative retraction and, as we mentioned above, that there is a smooth dendroid which admits no retraction from 2^X onto X. So, the following problem arises.

736? **Problem 31** ([**12**, Problem 5.57]). *Characterize smooth dendroids X admitting a retraction from 2^X onto X.*

Also, since the Mohler–Nikiel universal smooth dendroid have the property of Kelley, the following problem arises.

737? **Problem 32** (Problem 3.23, [**12**]). *Let X be a dendroid with the property of Kelley. Does there exist a retraction $r\colon 2^X \to X$?*

9. Means

Given a Hausdorff space X, a *mean* μ on X is defined as a map $\mu\colon X \times X \to X$ such that for each $x, y \in X$ we have that $\mu(x,x) = x$ and $\mu(x,y) = \mu(y,x)$.

The natural question that comes with the definition is: which spaces, especially metric continua admit a mean? This question has been around for more than half of a century and has been answered for a very small class of spaces. So, the main problem about means is the following.

738–739? **Problem 33** ([**12**, Problems 5.28 and 5.50]). *Characterize metric continua (in particular dendroids) that admit a mean.*

For a continuum X the existence of a mean $\mu\colon X \times X \to X$ is equivalent to the existence of a retraction $r\colon F_2(X) \to X$, where the two concepts are related to each other by the equality $\mu(x,y) = r(\{x,y\})$. In this respect, the existence of a retraction $r\colon 2^X \to X$ implies the existence of a mean but it is not known if the inverse implication is true. So we have the following problem.

740? **Problem 34** (Question 5.44, [**12**]). *Does there exist a dendroid X which admits a mean and for which there is no retraction from 2^X onto X?*

Related to the previous problem and Theorem 8.1, we have the next problems.

741? **Problem 35** ([**12**, Question 5.48]). *Let X be a dendroid admitting a mean. Must X be uniformly arcwise connected?*

742? **Problem 36** ([**12**, Question 5.49], compare to Problem 32). *Let X be a dendroid with the property of Kelley. Does there exist a mean $\mu\colon X \times X \to X$?*

A mean $\mu\colon X \times X \to X$ is said to be *associative* provided that $\mu(x, \mu(y,z)) = \mu(\mu(x,y), z)$ for every $x, y, z \in X$.

It is known that

Theorem 9.1 ([**12**, Theorem 5.3]). *Let X be a locally connected continuum. Then the following conditions are equivalent:*

(1) *X is an absolute retract;*
(2) *there is a retraction $r\colon 2^X \to X$.*

Moreover, if X is one-dimensional, then each of them is equivalent to any of the following:

(3) X is a dendrite;

(4) there exists an associative retraction $r\colon 2^X \to X$;
(5) there exists an associative mean $\mu\colon X \times X \to X$;
(6) there exists a mean $\mu\colon X \times X \to X$.

About Theorem 9.1, J.J. Charatonik et al asked the following.

Problem 37 ([**12**, Question 5.38]). *Assume that X is locally connected. Does (6) imply (5)? Does (5) imply (1)?* 743–744?

It is known that there is a smooth dendroid admitting no mean [**12**, Example 5.52] and that the Mohler–Nikiel universal smooth dendroid admits an associative mean [**12**, Theorem 3.21 and Proposition 5.16]. So the following problem arises.

Problem 38 ([**12**, Problem 5.56]). *Characterize smooth dendroids admitting a mean.* 745?

A mean μ on a dendroid X is said to be internal if for each $x, y \in X$, $\mu(x, y) \in xy$. M. Bell and S. Watson give an example of a contractible and selectible fan which admits a mean while it does not admit neither an associative mean nor an internal mean [**1**, Example 4.8]. So they asked the following.

Problem 39 ([**1**, Problem 4.3]). *Does a selectible dendroid have a mean? Does a contractible dendroid have a mean?* 746–747?

10. Selections

A continuous selection for a family $\mathcal{H} \subset 2^X$ is a map $s\colon \mathcal{H} \to X$ such that $s(A) \in A$ for each $A \in \mathcal{H}$. A continuum X is said to be *selectible* provided that it admits a continuous selection for $C(X)$.

In [**49**], S.B. Nadler, Jr. and L.E. Ward, Jr. proved the following.

Theorem 10.1.

(1) *Every selectible continuum is a dendroid;*
(2) *A locally connected continuum is selectible if and only if it is a dendrite;*
(3) *Each selectible dendroid is a continuous image of the Cantor fan, hence it is uniformly arcwise connected.*

A selection $s\colon \mathcal{H} \to X$, where $\mathcal{H} \subset 2^X$, is said to be rigid provided that if $A, B \in \mathcal{H}$ and $s(B) \in A \subset B$, then $s(A) = s(B)$.

In [**52**], L.E. Ward, Jr. showed the following.

Theorem 10.2. *A continuum X is a smooth dendroid if and only if there exists a rigid selection for $C(X)$.*

On the other hand, there is an uniformly arcwise connected dendroid which is not selectible and there is a non-smooth dendroid admitting non-rigid selections for its hyperspace of subcontinua [**8**, Figure 17 and Figure 18 respectively].

In [**47**], S.B. Nadler, Jr. posed the following problem (which is still open).

Problem 40. *Give an internal characterization of selectible dendroids (of selectible fans).* 748–749?

In [**41**], T. Maćkowiak gave an example of a contractible and non-selectible dendroid, J.J. Charatonik asked for an example with these and additional properties.

750–752? **Problem 41** ([**8**, Question 8.7]). *Is there a contractible and non-selectible dendroid which is* (a) *planable,* (b) *hereditarily contractible,* (c) *a fan?*

An *open selection* is a selection that also is an open map. In [**42**], it is shown that a smooth fan X admits an open selection if and only if X is locally connected. Regarding this topic, the following problems are still unsolved.

753? **Problem 42** ([**42**, Problem 1]). *If X is a finite tree, then does X admit an open selection?*

754? **Problem 43** ([**42**, Problem 2]). *Can a non-locally connected dendroid admit an open selection?*

Let D be a dendrite and $\Sigma(D)$ the space of selections of D. Trying to give new tools to solve Problem 40, J.E. McParland proved that for each dendrite D, the space $\Sigma(D)$ (a) is not compact [**43**, Theorem 3.9], (b) is nowhere dense in $D^{C(D)}$ [**43**, Theorem 3.10], (c) is not dense in $D^{C(D)}$ [**43**, Theorem 3.11] and (d) is not an arc [**43**, Theorem 4.3]. For us it is natural to present the following problem.

755? **Problem 44.** *Give a wider variety of properties with which $\Sigma(D)$ is endowed. In particular, is $\Sigma(D)$ homeomorphic to ℓ_2?*

11. Smooth dendroids

S.T. Czuba showed some relations among the different types of smoothness (see definitions of Section 5) and proved:

Theorem 11.1 ([**20**, Theorem 1]). *A fan is pointwise smooth if and only if it is smooth.*

Theorem 11.2 ([**23**, Theorem 4.6]). *If a dendroid is pointwise smooth and weakly smooth, then it is also semi-smooth.*

Corollary 11.3 ([**23**, Corollary 4.9]). *If a dendroid X is pointwise smooth and semi-smooth, then it is also weakly smooth.*

Consider the following definitions. Let $\mathcal{T}$ be a property and $\mathcal{A}$ a class of continua. $\mathcal{T}$ is *finite (countable) in the class $\mathcal{A}$* if there is a finite (countable) set $\mathcal{F} \subset \mathcal{A}$ such that a member X of $\mathcal{A}$ has property $\mathcal{T}$ if and only if X contains a homeomorphic copy of some member of $\mathcal{F}$. A class $\mathcal{A}$ has a *common model M* under continuous mapping if there is a continuum M belonging to $\mathcal{A}$ with the property that every member of $\mathcal{A}$ is a continuous image of M. A class $\mathcal{A}$ has a *universal element U*, if there is a a continuum U belonging to $\mathcal{A}$ with property that every member $\mathcal{A}$ can be homeomorphically embedded into U.

Now consider the following classes of continua: (a) dendroids, (b) fans, (c) smooth dendroids, (d) smooth fans, (e) semi-smooth dendroids, (f) semi-smooth fans, (g) weakly smooth dendroids, (h) weakly smooth fans, (i) pointwise smooth dendroids, (j) uniformly arcwise connected dendroids, (k) uniformly arcwise connected fans.

Some of the questions which remain unanswered are.

Problem 45. *Does there exist a common model for the classes (a), (b), (e)* 756?
and (f)?

In [**13**, Theorem 11] and [**36**] it is shown that classes (c), (d), (j), (k) have a common model.

Problem 46. *Does there exist a universal element for the classes (b), (e), (f),* 757?
(g), (h), (i), (j) and (k)?

A universal element is known for classes (c) [**30**, **46**] and (d) [**13**, Theorem 10]. Class (a) does not have a universal element, see [**35**].

12. Planability

Considering planability of dendroids, in 1959, B. Knaster posed the following question, which is still unsolved.

Problem 47. *Characterize dendroids that can be embedded in the plane.* 758?

In [**39**] T. Maćkowiak showed that there is no universal element in the class of plane smooth dendroids and in [**31**] L. Habiniak proved that there is no plane dendroid containing all plane smooth dendroids. Using the same definitions of Section 11 the following is still an open problem.

Problem 48 ([**13**, p. 307]). *Is the property of non-embeddability in the plane finite* 759?
in the classes (e), (f), (g), (h), (i)?

It is known that the property of non-embeddability in the plane is not finite for classes (a), (c) and (j) (see [**38**]) and for classes (b), (d) (see [**17**]).

Now, we move to different concepts. Lelek proved.

Theorem 12.1 ([**37**, Theorem p. 307]). *If the set $E(X)$ of all end points of a dendroid X is not a $G_{\delta\sigma\delta}$-set then X is non-planable.*

However, the following problem is still open.

Problem 49 ([**19**, [Problem 1]). *Does there exist a dendroid X such that the set* 760?
$E(X)$ is not a $G_{\delta\sigma\delta}$-set?

Also, T. Maćkowiak asked:

Problem 50 ([**39**]). *Is planability of dendroids (fans) an invariant property with* 761–762?
respect to open mappings?

A negative answer to Problem 50 for finite graphs is known in [**53**, p. 189]. Also considering continuous images of dendroids J.J. Charatonik proved in [**4**] that a monotone image of a planable λ-dendroid (dendroid, fan) is a planable λ-dendroid (dendroid, fan).

13. Shore sets

A subset A of a dendroid X is said to be a *shore set* provided that for each $\varepsilon > 0$, there exists a subcontinuum B of X such that $B \cap A = \emptyset$ and $H(B, X) < \varepsilon$.

Answering a question of I. Puga-Espinosa, A. Illanes proved in [**33**] the following.

Theorem 13.1. *If X is a dendroid and $A_1, A_2, \ldots, A_m$ are pairwise disjoint shore subcontinua of X, then $A_1 \cup A_2 \cup \cdots \cup A_m$ is a shore set.*

He also gave an example [**33**, Example 5] which shows that it is necessary to require pairwise disjointness in the previous theorem. The following natural problem is still open.

763? **Problem 51** ([**33**, Question 6]). *Is the union of two disjoint closed shore subsets of a dendroid X also a shore set?*

References

[1] M. Bell and S. Watson, *Not all dendroids have means*, Houston J. Math. **22** (1996), no. 1, 39–50.

[2] J. Charatonik, J, *Dendrites and montone mappings*, Math. Pannon. **15** (2004), no. 1, 115–125.

[3] J. Charatonik, J and W. J. Charatonik, *The property of Kelley for fans*, Bull. Polish Acad. Sci. Math. **36** (1988), no. 3–4, 169–173 (1989).

[4] J. J. Charatonik, *A theorem on monotone mappings of planable λ-dendroids*, Bull. Acad. Polon. Sci. Sér. Sci. Math. Astronom. Phys. **24** (1976), no. 3, 171–172.

[5] J. J. Charatonik, *Problems and remarks on contractibility of curves*, General topology and its relations to modern analysis and algebra, IV (Prague, 1976), Part B, Association of Czechoslovak Mathematicians and Physicists, Prague, 1977, pp. 72–76.

[6] J. J. Charatonik, *Contractibility of curves*, Matematiche (Catania) **46** (1991), no. 2, 559–592 (1993).

[7] J. J. Charatonik, *Monotone mappings of universal dendrites*, Topology Appl. **38** (1991), no. 2, 163–187.

[8] J. J. Charatonik, *On acyclic curves. A survey of results and problems*, Bol. Soc. Mat. Mexicana (3) **1** (1995), no. 1, 1–39.

[9] J. J. Charatonik and W. J. Charatonik, *Fans with the property of Kelley*, Topology Appl. **29** (1988), no. 1, 73–78.

[10] J. J. Charatonik and W. J. Charatonik, *Dendrites*, XXX National Congress of the Mexican Mathematical Society (Spanish) (Aguascalientes, 1997), Sociedad Matemática Mexicana, México, 1998, pp. 227–253.

[11] J. J. Charatonik, W. J. Charatonik, and P. Krupski, *Dendrites and light open mappings*, Proc. Amer. Math. Soc. **128** (2000), no. 6, 1839–1843.

[12] J. J. Charatonik, W. J. Charatonik, K. Omiljanowski, and J. R. Prajs, *Hyperspace retractions for curves*, Dissertationes Math. **370** (1997), 34 pp.

[13] J. J. Charatonik and C. Eberhart, *On smooth dendroids*, Fund. Math. **67** (1970), 297–322.

[14] J. J. Charatonik and Z. Grabowski, *Homotopically fixed arcs and the contractibility of dendroids*, Fund. Math. **100** (1978), no. 3, 229–237.

[15] J. J. Charatonik and A. Illanes, *Homotopy properties of curves*, Comment. Math. Univ. Carolin. **39** (1998), no. 3, 573–580.

[16] J. J. Charatonik and A. Illanes, *Mappings on dendrites*, Topology Appl. **144** (2004), 109–132.

[17] J. J. Charatonik, L. T. Januszkiewicz, and T. Maćkowiak, *An uncountable collection of nonplanable smooth dendroids*, Bull. Acad. Polon. Sci. Sér. Sci. Math. Astronom. Phys. **25** (1977), no. 2, 147–149.

[18] J. J. Charatonik and P. Krupski, *Dendrites and light mappings*, Proc. Amer. Math. Soc. **132** (2004), no. 4, 1211–1217.

[19] J. J. Charatonik and Z. Rudy, *Remarks on non-planable dendroids*, Colloq. Math. **37** (1977), no. 2, 205–216.

[20] S. T. Czuba, *The notion of pointwise smooth dendroids*, Uspekhi Mat. Nauk **34** (1979), no. 6(210), 215–217, Translation: Russian Math. Surveys 34 (1979) no. 6, 169–171.

[21] S. T. Czuba, *R-continua and contractibility of dendroids*, Bull. Acad. Polon. Sci. Sér. Sci. Math. **27** (1979), no. 3–4, 299–302.

[22] S. T. Czuba, *R^i-continua and contractibility*, Proceedings of the International Conference on Geometric Topology (Warsaw, 1978), PWN, Warsaw, 1980, pp. 75–79.

[23] S. T. Czuba, *On pointwise smooth dendroids*, Fund. Math. **114** (1981), no. 3, 197–207.

[24] S. T. Czuba, *On dendroids for which smoothness, pointwise smoothness and hereditary contractibility are equivalent*, Comment. Math. Prace Mat. **25** (1985), no. 1, 27–30.

[25] S. T. Czuba, *On dendroids with Kelley's property*, Proc. Amer. Math. Soc. **102** (1988), no. 3, 728–730.

[26] J. B. Fugate, *Retracting fans onto finite fans*, Fund. Math. **71** (1971), no. 2, 113–125.

[27] J. B. Fugate, *Small retractions of smooth dendroids onto trees*, Fund. Math. **71** (1971), no. 3, 255–262.

[28] J. T. Goodykoontz, Jr., *Some retractions and deformation retractions on 2^X and $C(X)$*, Topology Appl. **21** (1985), no. 2, 121–133.

[29] B. G. Graham, *On contractible fans*, Fund. Math. **111** (1981), no. 1, 77–93.

[30] J. Grispolakis and E. D. Tymchatyn, *A universal smooth dendroid*, Bull. Acad. Polon. Sci. Sér. Sci. Math. Astronom. Phys. **26** (1978), no. 12, 991–998.

[31] L. J. Habiniak, *There is no plane dendroid containing all plane smooth dendroids*, Bull. Acad. Polon. Sci. Sér. Sci. Math. **30** (1982), no. 9–10, 465–470 (1983).

[32] J. Heath and V. C. Nall, *Tree-like continua and 2-to-1 maps*, Proc. Amer. Math. Soc. **132** (2004), no. 1, 283–289.

[33] A. Illanes, *Finite unions of shore sets*, Rend. Circ. Mat. Palermo (2) **50** (2001), no. 3, 483–498.

[34] A. Illanes, *Hyperspaces of arcs and two-point sets in dendroids*, Topology Appl. **117** (2002), no. 3, 307–317.

[35] J. Krasinkiewicz and P. Minc, *Nonexistence of universal continua for certain classes of curves*, Bull. Acad. Polon. Sci. Sér. Sci. Math. Astronom. Phys. **24** (1976), no. 9, 733–741.

[36] W. Kuperberg, *Uniformly pathwise connected continua*, Studies in topology (Charlotte, NC, 1974), Academic Press, New York, 1975, pp. 315–324.

[37] A. Lelek, *On plane dendroids and their end points in the classical sense*, Fund. Math. **49** (1960), 301–319.

[38] T. Maćkowiak, *A certain collection of non-planar fans*, Bull. Acad. Polon. Sci. Sér. Sci. Math. Astronom. Phys. **25** (1977), no. 6, 543–548.

[39] T. Maćkowiak, *Planable and smooth dendroids*, General topology and its relations to modern analysis and algebra, IV (Prague, 1976), Part B, Association of Czechoslovak Mathematicians and Physicists, Prague, 1977, pp. 260–267.

[40] T. Maćkowiak, *Continuous selections for $c(x)$*, Bull. Polish Acad. Sci. Math. **26** (1978), 547–551.

[41] T. Maćkowiak, *Contractible and nonselectible dendroids*, Bull. Polish Acad. Sci. Math. **33** (1985), no. 5–6, 321–324.

[42] V. Martínez-de-la-Vega, *Open selections on smooth fans*, Topology Appl. **153** (2006), no. 8, 1214–1235.

[43] J. E. McParland, *The selection space of a dendroid. I*, Continuum theory (Denton, TX, 1999), Lecture Notes in Pure and Appl. Math., vol. 230, Marcel Dekker Inc., New York, 2002, pp. 245–269.

[44] P. Minc, *Bottlenecks in dendroids*, Topology Appl. **129** (2003), no. 2, 187–209.

[45] P. Minc, *Mapping chainable continua onto dendroids*, Topology Appl. **138** (2004), 287–298.

[46] L. Mohler and J. Nikiel, *A universal smooth dendroid answering a question of J. Krasinkiewicz*, Houston J. Math. **14** (1988), no. 4, 535–541.

[47] S. B. Nadler, Jr., *Hyperspaces of sets*, Marcel Dekker Inc., New York, 1978.

[48] S. B. Nadler, Jr., *The fixed point property for continua*, Textos, vol. 30, Sociedad Matemática Mexicana, 2005.

[49] S. B. Nadler, Jr. and L. E. Ward, Jr., *Concerning continuous selections*, Proc. Amer. Math. Soc. **25** (1970), 369–374.

[50] S. B. Nadler, Jr. and L. E. Ward, Jr., *Concerning exactly* $(n, 1)$ *images of continua*, Proc. Amer. Math. Soc. **87** (1983), no. 2, 351–354.

[51] L. G. Oversteegen, *Noncontractibility of continua*, Bull. Acad. Polon. Sci. Sér. Sci. Math. Astronom. Phys. **26** (1978), no. 9–10, 837–840.

[52] L. E. Ward, Jr., *Rigid selections and smooth dendroids*, Bull. Acad. Polon. Sci. Sér. Sci. Math. Astronom. Phys. **19** (1971), 1041–1044.

[53] G. T. Whyburn, *Analytic topology*, American Mathematical Society, Providence, RI, 1942.

Open Problems in Topology II – Edited by E. Pearl

$\frac{1}{2}$-Homogeneous continua

Sam B. Nadler, Jr.

1. Introduction

A space is *homogeneous* provided that for any two of its points, there is a homeomorphism of the space onto itself taking one of the points to the other point. Intuitively, spaces that are homogeneous look the same everywhere.

Homogeneity is a classical topic in continuum theory. For information about homogeneous continua, I refer the reader to the article by Janusz Prajs in this book. We will discuss $\frac{1}{2}$-homogeneity, a notion that is closely related to homogeneity.

We give the formal definition of $\frac{1}{2}$-homogeneity in a moment. First, we note a visible property of the closed unit n-dimensional ball B^n in Euclidean n-space: For any two points in the sphere S^{n-1} $(= \partial B^n)$ or in $B^n \setminus S^{n-1}$, there is a homeomorphism of B^n onto B^n taking one of the points to the other, but there is no homeomorphism of B^n onto B^n taking a point of S^{n-1} to a point of $B^n \setminus S^{n-1}$ (a formal proof of the last fact uses Invariance of Domain [**6**, p. 95, VI9]. The abstract formulation of this property of B^n is the definition of $\frac{1}{2}$-homogeneity, which we give next.

Let $\mathcal{H}(X)$ denote the group of homeomorphisms of a space X onto itself. An *orbit* of X is the action of $\mathcal{H}(X)$ at a point x of X, meaning $\{h(x) : h \in \mathcal{H}(X)\}$ for a given point $x \in X$. We say X is $\frac{1}{2}$*-homogeneous* provided that X has exactly two orbits. More generally, for a positive integer n, X is said to be $\frac{1}{n}$*-homogeneous* provided that X has exactly n orbits. Thus, the $\frac{1}{1}$-homogeneous spaces are the homogeneous spaces.

We give some simple examples of $\frac{1}{2}$-homogeneous continua: A figure eight (the join of two simple closed curves at a point); a θ-curve; the Hawaiian earring (a null sequence of simple closed curves joined at a point); the Sierpiński universal curve [**10**]; the compactification of $\mathbb{R}^1$ with two disjoint circles as remainder for which the ray $[0, \infty)$ continually winds in a clockwise (or counterclockwise) direction as it approaches one circle and the other ray $(-\infty, 0]$ does the same as it approaches the other circle.

Regarding the last example, we note that if the rays approach the circles changing direction after each complete revolution (only), then the compactification is $\frac{1}{3}$-homogeneous. We note another situation in which two related constructions give different results: The suspension over any nonlocally connected homogeneous continuum is $\frac{1}{2}$-homogeneous, but the cone over such a continuum is not $\frac{1}{2}$-homogeneous when it is finite dimensional (see Theorem 4.8).

Until recently, there were only two papers about $\frac{1}{2}$-homogeneity, [**10**, **18**]. In the past two years, four more papers have been written. The four recent papers fit into three categories: $\frac{1}{2}$-homogeneous continua with cut points ([**16**, **17**]), $\frac{1}{2}$-homogeneous cones [**14**], and $\frac{1}{2}$-homogeneous hyperspaces [**15**]. We survey the main results and discuss open problems in each category separately. We provide detailed references for all results and problems that do not originate here.

2. Notation and terminology

A *compactum* is a nonempty compact metric space. A *continuum* is a connected compactum. We assume that the reader is somewhat familiar with continuum theory. Most notation and terminology that we use is standard and can be found in [**11**, **13**, **20**]. However, we note the following items (other notation and terminology that is not standard is presented as it comes up):

A *cut point* (separating point) of a connected space is a point whose removal disconnects the space.

The *remainder of a compactification* Y of a space Z is $Y \setminus Z$. Let X be a compactification of $\mathbb{R}^1$, and let R denote the open, dense copy of $\mathbb{R}^1$ in X. For any point $r \in R$, the closure in X of a component of $R \setminus \{r\}$ is called an *end of the compactification* X. (Thus, up to homeomorphism, there are at most two ends.)

The symbol $\mathrm{ord}_p(X)$ denotes the *order of the space* X *at* p; $\mathrm{ord}_p(X) \leq \omega$ means that p has arbitrarily small open neighborhoods whose boundaries are finite [**11**, p. 274].

The symbols AR and ANR stand for absolute retract and absolute neighborhood retract, respectively.

A continuum Y is *n-homogeneous* (n a positive integer) provided that for any two n-element subsets A and B of Y, there is a homeomorphism h of Y onto Y such that $h(A) = B$ [**19**]. A continuum Y is *n-homogeneous at a point* $p \in Y$ (n a positive integer) provided that for any two n-element subsets A and B of Y such that $p \in A \cap B$, there is a homeomorphism h of Y onto Y such that $h(A) = B$ and $h(p) = p$ (this notion originates in [**16**]).

A *finite graph* is a 1-dimensional compact connected polyhedron.

A *bouquet of continua* Y is a continuum X with a cut point c such that the closure of each component of $X \setminus \{c\}$ is homeomorphic to Y. The *Hawaiian earring* is the unique locally connected bouquet of infinitely many simple closed curves.

3. $\frac{1}{2}$-Homogeneous continua with cut points

We denote the subspace of all cut points of a continuum X by $\mathrm{Cut}(X)$.

Recall that every continuum has noncut points [**13**, p. 89, 6.6]; thus, when a $\frac{1}{2}$-homogeneous continuum X has cut points, the two orbits of X must be $\mathrm{Cut}(X)$ and its complement (the set of all noncut points of X).

In [**16**] the general stucture of $\frac{1}{2}$-homogeneous continua with cut points and the structure of their two orbits was described in detail. In addition, it was determined how the two orbits are situated in X. Nevertheless, as we will see, there are still open questions about the structure of such continua.

We state the results from [**16**] in the four theorems that follow. The first theorem lays the foundation for the next three theorems. We note that the first theorem shows (implicitly) that if a $\frac{1}{2}$-homogeneous continuum has a cut point, then it has either uncountably many cut points or only one cut point.

Theorem 3.1 ([**16**, 6.1]). *Let X be a $\frac{1}{2}$-homogeneous continuum with at least one cut point.*

(1) *If $|\mathrm{Cut}(X)| > 1$, then $\mathrm{Cut}(X)$ is homeomorphic to $\mathbb{R}^1$, $\mathrm{Cut}(X)$ is both open and dense in X, the orbit of all noncut points of X is the union of two disjoint, homeomorphic and homogeneous continua (possibly single points, in which case X is an arc, and the ends of the compactification X are mutually homeomorphic.*

(2) *If $\mathrm{Cut}(X) = \{c\}$, then the closures of the components of $X \setminus \{c\}$ are mutually homeomorphic and are (each) 2-homogeneous at c; furthermore, if $\mathrm{ord}_c(X) \leq \omega$, then X is a locally connected bouquet of simple closed curves (thus, X is the Hawaiian earring when $X \setminus \{c\}$ has infinitely many components).*

In connection with part (2) of Theorem 3.1, we note that the closures of the components of $X \setminus \{c\}$ need not be homogeneous: Attach two disjoint copies of a pinched 2-sphere together at the pinched points [**16**, 7.4]; the continuum obtained from the attachment is easily seen to be $\frac{1}{2}$-homogeneous (see Theorem 3.4).

The next theorem isolates the properties in Theorem 3.1 that are relevant to the structure of $\mathrm{Cut}(X)$ for any $\frac{1}{2}$-homogeneous continuum.

Theorem 3.2 ([**16**, 6.2]). *If X is a $\frac{1}{2}$-homogeneous continuum, then either $\mathrm{Cut}(X)$ is homeomorphic to $\mathbb{R}^1$ and $\mathrm{Cut}(X)$ is both open and dense in X or $\mathrm{Cut}(X)$ consists of at most one point; furthermore, if $\mathrm{Cut}(X)$ consists of a single point c, then $\mathrm{ord}_c(X) \geq 4$ and $\mathrm{ord}_c(X)$ is even if $\mathrm{ord}_c(X)$ is an integer.*

The following theorem characterizes all $\frac{1}{2}$-homogeneous continua with more than one cut point:

Theorem 3.3 ([**16**, 6.4]). *Let X be a continuum with more than one cut point. Then X is $\frac{1}{2}$-homogeneous if and only if X is an arc or X is a compactification of $\mathbb{R}^1$ whose remainder is the union of two disjoint, nondegenerate, homeomorphic continua and the ends of X are mutually homeomorphic and $\frac{1}{3}$-homogeneous.*

Our final theorem from [**16**] is a partial characterization of $\frac{1}{2}$-homogeneous continua with more than one cut point:

Theorem 3.4 ([**16**, 6.5]). *Let X be a continuum with only one cut point c. Assume that the components of $X \setminus \{c\}$ form a null sequence. Then X is $\frac{1}{2}$-homogeneous if and only if the closures of the components of $X \setminus \{c\}$ are mutually homeomorphic and are (each) 2-homogeneous at c.*

In the following example, we show that the assumption in Theorem 3.4 that the components of $X \setminus \{c\}$ form a null sequence is required and is restrictive.

Example 3.5 ([**16**, 7.2 and 7.3]). Let $Z = \{0, 1, \frac{1}{2}, \ldots, \frac{1}{n}, \ldots\}$, let C be the Cantor set, let S^1 be the unit circle, and fix a point $p \in S^1$. The quotient space $X = (Z \times S^1)/(Z \times \{p\})$ has only one cut point $c = Z \times \{p\}$ and the closures of the components of $X \setminus \{c\}$ are mutually homeomorphic and are (each) 2-homogeneous at c; however, X is not $\frac{1}{2}$-homogeneous since X is locally connected at some noncut points but not at others. On the other hand, $Y = (C \times S^1 / (C \times \{p\})$ is a $\frac{1}{2}$-homogeneous continuum with only one cut point $c = C \times \{p_0\}$ and, yet, the components of $Y \setminus \{c\}$ do not form a null sequence. The continuum Y also shows that the assumption that $\mathrm{ord}_c(X) \leq \omega$ in part (2) of Theorem 3.1 is required.

Now, we come to some questions about Theorem 3.3 and Theorem 3.4.

The characterization in Theorem 3.3 would be enhanced if we had a solution to the following problem (the two ways of stating the problem are equivalent by [**16**, 4.7]):

764? **Problem 3.6** ([**16**, Section 7]). Find intrinsic conditions that characterize all $\frac{1}{3}$-homogeneous compactifications of $[0, \infty)$. In other words, When is the remainder of a compactification of $[0, \infty)$ an orbit of the compactification?

It may be that any inherent characterization of $\frac{1}{3}$-homogeneous compactifications of $[0, \infty)$ would be too technical to be useful. In fact, we do not know the answer to Problem 3.6 when the remainder of the compactification is a simple closed curve; the problem is pinpointed in the following question:

765? **Problem 3.7** ([**16**, 7.1]). Consider a compactification of the ray $[0, \infty)$ with the circle S^1 as remainder such that every point of S^1 is a limit of points in the ray at which the ray reverses direction for at least one full revolution about S^1. Can such a compactification be $\frac{1}{3}$-homogeneous?

We ask about extending Theorem 3.4:

766? **Problem 3.8** ([**16**, Section 7]). Characterize (inherently) all $\frac{1}{2}$-homogeneous continua with only one cut point.

Regarding Theorem 3.4 as well as part (2) of Theorem 3.1, we would like a solution to the following problem:

767? **Problem 3.9** ([**16**, Section 7]). Characterize (inherently) the continua that are 2-homogeneous at a point.

We discuss three theorems from [**17**] that characterize particular continua in terms of $\frac{1}{2}$-homogeneity. The first two theorems characterize the arc.

Theorem 3.10 ([**17**, 3.6]). *The arc is only $\frac{1}{2}$-homogeneous semilocally connected continuum with more than one cut point.*

Theorem 3.11 ([**17**, 4.6]). *The arc is only $\frac{1}{2}$-homogeneous hereditarily decomposable continuum whose nondegenerate proper subcontinua are arc-like.*

The assumption of being hereditarily decomposable in Theorem 3.11 is required: The arc of pseudo-arcs is a $\frac{1}{2}$-homogeneous arc-like continuum [**17**, 4.8].

768? **Problem 3.12** ([**17**, 4.9]). Is there a $\frac{1}{2}$-homogeneous indecomposable arc-like continuum?

By Theorem 3.11, there is no $\frac{1}{2}$-homogeneous hereditarily decomposable circle-like continuum. The arc of pseudo-arcs with the end tranches identified to a point is an example of a $\frac{1}{2}$-homogeneous decomposable circle-like continuum [**17**, 4.8]. These observations lead to the following question:

Problem 3.13 ([**17**, 4.10]). Is there a $\frac{1}{2}$-homogeneous indecomposable circle-like continuum? 769?

The question of determining all $\frac{1}{2}$-homogeneous arc-like continua or circle-like continua is implicit from Theorem 3.11, Problem 3.12 and Problem 3.13.

Our next theorem characterizes the Hawaiian earring.

Theorem 3.14 ([**17**, 3.12]). *Let X be a $\frac{1}{2}$-homogeneous hereditarily locally connected continuum with a cut point that is not a finite graph. Then X is the Hawaiian earring.*

I do not know if having a cut point is required for Theorem 3.14:

Problem 3.15. Is the Hawaiian earring the only $\frac{1}{2}$-homogeneous hereditarily locally connected continuum that is not a finite graph? 770?

Let us note a lemma that follows easily from Theorem 3.1 and Theorem 3.10:

Lemma 3.16. *A $\frac{1}{2}$-homogeneous finite graph with at least one cut point is either an arc or a bouquet of finitely many simple closed curves.*

The following variation on Theorem 3.14 is an immediate consequence of Lemma 3.16 and Theorem 3.14:

Theorem 3.17. *Let X be a hereditarily locally connected continuum with a cut point. Then X is $\frac{1}{2}$-homogeneous if and only if X is an arc or a bouquet of simple closed curves (that is, a finite bouquet or the Hawaiian earring).*

Lemma 3.16 raises a question about finite graphs. Patkowska [**18**, p. 25, Theorem 1] claims "Moreover, we find a full classification of all $\frac{1}{2}$-homogeneous polyhedra by means of homogeneous multigraphs." However, the meaning of the claim (and its verification) does not seem to be in [**18**], even for the case of finite graphs. Lemma 3.16 takes care of the case of finite graphs with a cut point; however, we do not know about cyclic finite graphs:

Problem 3.18. What are all the $\frac{1}{2}$-homogeneous finite graphs that have no cut point? 771?

4. $\frac{1}{2}$-Homogeneous cones

We denote the cone over a compactum X by Cone(X) and its vertex by v_X. The question of when the cone over a continuum is $\frac{1}{2}$-homogeneous was investigated for the first time in [**15**]. So far, there are no other papers about this topic.

The main results in [**15**] fall into three categories: 1-dimensional continua, an ANR theorem for finite-dimensional compacta, and continua with conditions weaker than being atriodic in some nonempty open set. We discuss many of the results and open problems from [**15**].

Note that Cone(B^n) and Cone(S^n) are $(n+1)$-cells and, hence, are $\frac{1}{2}$-homogeneous. One of the main results from [**15**] is that B^1 and S^1 are the only 1-dimensional continua whose cones are $\frac{1}{2}$-homogeneous:

Theorem 4.1 ([**15**, 6.1]). *Let X be a 1-dimensional continuum. Then* Cone(X) *is $\frac{1}{2}$-homogeneous if and only if X is an arc or a simple closed curve.*

The natural analogue for all finite-dimensional continua of Theorem 4.1 is false:

Example 4.2 ([**15**, 1.1]). For each integer $n \geq 4$, let $X = \text{Cone}(S^{n-1}/A)$, where A is an arc in the $(n-1)$-sphere S^{n-1} such that the fundamental group of $S^{n-1} \setminus A$ is nontrivial. Then X is an n-dimensional AR that is not a manifold and, yet, Cone(X) is $\frac{1}{2}$-homogeneous since Cone(X) an $(n+1)$-cell [**2**, p. 26, 4.4].

Theorem 4.1 and Example 4.2 lead us to a question for dimensions 2 and 3 as well as a question for any finite dimension:

772? **Problem 4.3** ([**15**, 1.2]). If X is a continuum, even a Peano continuum, of dimension $n = 2$ or 3 such that Cone(X) is $\frac{1}{2}$-homogeneous, must X be an n-cell or an n-sphere?

773? **Problem 4.4** ([**15**, 1.3]). If the cone over a finite-dimensional continuum, even a Peano continuum, is $\frac{1}{2}$-homogeneous, then is the cone an n-cell?

We note several corollaries to Theorem 4.1. (It is not obvious why Corollary 4.5 is a consequence of Theorem 4.1; to see why it is uses some technical lemmas that we do not include here.)

Corollary 4.5 ([**15**, 6.2]). *Let X be a nondegenerate continuum that contains only finitely many simple closed curves. Then* Cone(X) *is $\frac{1}{2}$-homogeneous if and only if X is an arc or a simple closed curve.*

Corollary 4.6 ([**15**, 6.3]). *If X is a nondegenerate tree-like continuum, then* Cone(X) *is $\frac{1}{2}$-homogeneous if and only if X is an arc.*

Corollary 4.7 ([**15**, 6.7]). *The only circle-like continuum whose cone is $\frac{1}{2}$-homogeneous is a simple closed curve.*

Next, we turn our attention to the class of finite-dimensional compacta whose cones are $\frac{1}{2}$-homogeneous. We begin with an ANR theorem and a corollary for all finite-dimensional compacta; we show that the theorem and the corollary do not extend to infinite-dimensional compacta.

Theorem 4.8 ([**15**, 3.5]). *Let X be a finite-dimensional compactum. If* Cone(X) *is $\frac{1}{2}$-homogeneous, then X is an ANR.*

Corollary 4.9 ([**15**, 3.7]). *If X is a finite-dimensional compactum such that* Cone(X) *is $\frac{1}{2}$-homogeneous, then* Cone(X) *is an AR.*

Theorem 4.8 and Corollary 4.9 do not extend to infinite dimensions even when X is locally connected. We will give an example that is based on the following result (which we state slightly differently than in [**15**]):

Theorem 4.10 ([**15**, 3.8]). *If Y is a homogeneous compactum and Q is the Hilbert cube, then* $\text{Cone}(Y \times Q)$ *is either homogeneous or $\frac{1}{2}$-homogeneous.*

Example 4.11 ([**15**, 3.10]). Let $X = M \times Q$, where M is the 1-dimensional Menger universal curve and Q is the Hilbert cube. Then X is locally connected and $\text{Cone}(X)$ is $\frac{1}{2}$-homogeneous (by Theorem 4.10), but X and $\text{Cone}(X)$ are not ANRs. Furthermore, Theorem 4.8 and Corollary 4.9 fail badly in infinite dimensions in that X (hence, $\text{Cone}(X)$) need not even be locally connected: Let $X = Y \times Q$, where Y is a nonlocally connected homogeneous continuum Y (e.g., the dyadic solenoid or the pseudo-arc [**3**]); then $\text{Cone}(X)$ is $\frac{1}{2}$-homogeneous (by Theorem 4.10), but X and $\text{Cone}(X)$ are not locally connected.

We note another corollary to Theorem 4.8 and a problem concerning the corollary.

Corollary 4.12 ([**15**, 3.6]). *If X is a finite-dimensional contractible continuum such that* $\text{Cone}(X)$ *is $\frac{1}{2}$-homogeneous, then X is an AR.*

Unlike Theorem 4.8 and Corollary 4.9, we do not know if Corollary 4.12 extends to infinite dimensions:

Problem 4.13 ([**15**, 3.11]). If X is a contractible continuum such that $\text{Cone}(X)$ 774–775
is $\frac{1}{2}$-homogeneous, then is X an AR? What about with the additional assumption that X is locally connected?

Finally, we discuss results for continua that satisfy conditions that are weaker than being atriodic. For the first result, we note names for two special continua: (1) the *hairy point* is the union of a null sequence of countably infinitely many arcs all emanating from the same point and otherwise disjoint from one another; (2) the *null comb* is the continuum homeomorphic to the union of the line segments in the plane from $(0,0)$ to $(1,0)$ and from $(\frac{1}{n},0)$ to $(\frac{1}{n},\frac{1}{n})$ for each $n = 1, 2, \ldots$.

Theorem 4.14 ([**15**, 6.4]). *Let X be a nondegenerate continuum with a nonempty open set U such that U does not contain a hairy point or a null comb. Then* $\text{Cone}(X)$ *is $\frac{1}{2}$-homogeneous if and only if X is an arc or a simple closed curve.*

Corollary 4.15 ([**15**, 6.5]). *Let X be a nondegenerate continuum that contains no simple triod in some nonempty open set U. Then* $\text{Cone}(X)$ *is $\frac{1}{2}$-homogeneous if and only if X is an arc or a simple closed curve.*

Corollary 4.16 ([**15**, 6.6]). *Let X be a nondegenerate continuum with a nonempty open set U such that every nondegenerate subcontinuum of U is arc-like. Then* $\text{Cone}(X)$ *is $\frac{1}{2}$-homogeneous if and only if X is an arc or a simple closed curve.*

We complete this section by stating two problems that are natural from what we have discussed (the problems are not explicitly stated elsewhere).

776? **Problem 4.17.** Find more classes of continua whose cones are $\frac{1}{2}$-homogeneous.

777? **Problem 4.18.** Find classes of continua whose suspensions are $\frac{1}{2}$-homogeneous.

5. $\frac{1}{2}$-Homogeneous hyperspaces

For a continuum X with metric d, the *hyperspace* $C(X)$ is the space of all subcontinua of X with the Hausdorff metric ([**9**] or [**12**]).

It has been known for some time when $C(X)$ is homogeneous, namely, if and only if X is a Peano continuum in which every arc is nowhere dense or, equivalently, $C(X)$ is the Hilbert cube. (This was proved in [**12**, p. 564, 17.2] using [**5**, p. 22, 4.1].) The next logical step from the point of view of homogeneity-type properties is to inquire into when $C(X)$ is $\frac{1}{2}$-homogeneous. This was investigated for the first time in [**14**]. One of our principal tools used in [**14**] is the theory of layers (or tranches) of irreducible hereditarily decomposable continua [**11**, pp. 190–219].

Two simple continua for which $C(X)$ is $\frac{1}{2}$-homogeneous are the arc and the simple closed curve; in both cases, $C(X)$ is a 2-cell [**9**, pp. 33–35]. The two main results in [**14**], which we state next, suggest that there are very few continua X for which $C(X)$ is $\frac{1}{2}$-homogeneous and, in fact, that the arc and the simple closed curve may be the only ones.

Theorem 5.1 ([**14**, 3.1]). *If X is a locally connected continuum, then $C(X)$ is $\frac{1}{2}$-homogeneous if and only if X is an arc or a simple closed curve.*

Theorem 5.2 ([**14**, 5.1]). *Let X be a nondegenerate atriodic continuum. Then $C(X)$ is $\frac{1}{2}$-homogeneous if and only if X is an arc or a simple closed curve.*

Corollary 5.3 ([**14**, 5.2]). *Let X be a continuum such that each nondegenerate proper subcontinuum of X is arc-like. Then $C(X)$ is $\frac{1}{2}$-homogeneous if and only if X is an arc or a simple closed curve.*

Corollary 5.3 shows that when X is arc-like (circle-like, atriodic tree-like), then $C(X)$ is $\frac{1}{2}$-homogeneous if and only if X is an arc (a simple closed curve, an arc, respectively) [**14**, 5.3, 5.4, 5.6].

Corollary 5.4 ([**14**, 5.7]). *Let X be a continuum such that* $\dim C(X) = 2$. *Then $C(X)$ is $\frac{1}{2}$-homogeneous if and only if X is an arc or a simple closed curve.*

778–779? **Problem 5.5** ([**15**, Section 1]). If X is a continuum such that $C(X)$ is $\frac{1}{2}$-homogeneous, then is X an arc or a simple closed curve? What about when $C(X)$ is finite dimensional?

We note two results that give information about the second part of Problem 5.5 (two other such results are [**14**, 3.9 and 3.12]).

Theorem 5.6 ([**14**, 3.10]). *Let X be a decomposable continuum such that* $\dim C(X) < \infty$. *If $C(X)$ is $\frac{1}{2}$-homogeneous, then X is hereditarily decomposable.*

Theorem 5.7 ([**14**, 3.11]). *Let X be a nonlocally connected continuum such that* $\dim C(X) < \infty$ *and $C(X)$ is $\frac{1}{2}$-homogeneous. Then every nondegenerate proper subcontinuum of X is decomposable.*

A number of questions were asked at the end of [**14**]. The purpose of some of the questions was to indicate directions that might lead to solutions or partial solutions to Problem 5.5. We summarize a few such questions:

Problem 5.8 ([**14**]). Let X be a continuum such that $C(X)$ is $\frac{1}{2}$-homogeneous. Is $\dim C(X) < \infty$ and, in fact, is $\dim C(X) = 2$ [**14**, 6.2]? Is X decomposable [**14**, 6.4]? Is $\dim X = 1$ and, in fact, must X be hereditarily decomposable [**14**, 6.5]? Must X be hereditarily decomposable when $\dim C(X) < \infty$ [**14**, 6.5]? 780–783?

(Note: In [**14**], the third part of 6.5 says, "Is $\dim C(X) < \infty$?", which was already asked in [**14**, 6.2]; the way the third part of [**14**, 6.5] is stated in Problem 5.8 is what was meant.)

Regarding Problem 5.8, X can not be hereditarily indecomposable [**14**, 3.3]. However, we do not know if X can contain a nondegenerate hereditarily indecomposable continuum—if it does not, then $\dim X = 1$ [**4**, p. 270, Theorem 5].

In investigating when $C(X)$ is $\frac{1}{2}$-homogeneous, it is important to have conditions under which various elements of $C(X)$ belong or do not belong to the manifold interior of a 2-cell. Acosta showed that if X is an atriodic continuum, then no singleton, $\{x\}$, belongs to the manifold interior of any 2-cell in $C(X)$ (weakened form of [**1**, p. 40, Theorem 3]). Acosta's result was important for the proof of Theorem 5.2; in view of that, the following question was asked in [**14**]:

Problem 5.9 ([**14**, 6.6]). What conditions on continua X (other than being atriodic) or on points $p \in X$ are necessary and/or sufficient for $\{p\}$ not to belong to the manifold interior of a 2-cell in $C(X)$? 784?

A point p of a finite graph X is as in Problem 5.9 if and only if $\mathrm{ord}_p(X) \leq 2$; however, for the point $p = (0,0)$ in the null comb X (defined preceding Theorem 4.14), $\mathrm{ord}_p(X) = 1$ and, yet, $\{p\}$ belongs to an n-cell in $C(X)$ for every n by [**9**, p. 40, 6.4]. As noted in [**14**], Problem 5.9 for n-cells in $C(X)$ is open as well.

We also note the following question about 2-cells in $C(X)$:

Problem 5.10 ([**14**, 6.7]). Is there a continuum X such that $\dim C(X) < \infty$ and, for every $x \in X$, $\{x\}$ is a point of the manifold interior of a 2-cell in $C(X)$? 785?

There is no reason to restrict the study of $\frac{1}{2}$-homogeneous hyperspaces to the hyperspace $C(X)$. Several special hyperspaces other than $C(X)$ are of general interest—the hyperspace 2^X of all nonempty compact subsets of a continuum X (with the Hausdorff metric), the n-fold hyperspace $C_n(X)$ of all elements of 2^X with at most n components, and the n-fold symmetric product $F_n(X)$ of all elements of 2^X with at most n points. Our final question concerns these hyperspaces.

Problem 5.11 ([**14**, 6.8]). For what continua X are the hyperspaces 2^X, $C_n(X)$ or $F_n(X)$ $\frac{1}{2}$-homogeneous ($n > 1$ for the case of $F_n(X)$)? What about $\frac{1}{m}$-homogeneity for any integer $m > 1$? 786–787?

The case of $\frac{1}{2}$-homogeneity for $C_2(X)$ seems especially interesting: $C_2([0,1])$ is $\frac{1}{2}$-homogeneous since $C_2([0,1])$ is a 4-cell ([**7**, p. 349, Lemma 2.2], due to

R.M. Schori); however, $C_2(S^1)$ is not $\frac{1}{2}$-homogeneous since $C_2(S^1)$ is the cone over a solid torus [**8**]. This (naively) suggests that $C_2(X)$ may only be $\frac{1}{2}$-homogeneous when X is an arc. In fact, this is true when X is locally connected (a proof is in the comments following Question 6.8 of [**14**]).

References

[1] G. Acosta, *Continua with unique hyperspace*, Continuum theory (Denton, TX, 1999), Lecture Notes in Pure and Appl. Math., vol. 230, Marcel Dekker Inc., New York, 2002, pp. 33–49.

[2] F. D. Ancel and S. B. Nadler, Jr., *Cones that are cells, and an application to hyperspaces*, Topology Appl. **98** (1999), 19–33.

[3] R H Bing, *A homogeneous indecomposable plane continuum*, Duke Math. J. **15** (1948), 729–742.

[4] R H Bing, *Higher-dimensional hereditarily indecomposable continua*, Trans. Amer. Math. Soc. **71** (1951), 267–273.

[5] D. W. Curtis and R. M. Schori, *Hyperspaces of Peano continua are Hilbert cubes*, Fund. Math. **101** (1978), no. 1, 19–38.

[6] W. Hurewicz and H. Wallman, *Dimension Theory*, Princeton University Press, Princeton, NJ, 1941.

[7] A. Illanes, *The hyperspace $C_2(X)$ for a finite graph X is unique*, Glas. Mat. Ser. III **37(57)** (2002), no. 2, 347–363.

[8] A. Illanes, *A model for the hyperspace $C_2(S^1)$*, Questions Answers Gen. Topology **22** (2004), no. 2, 117–130.

[9] A. Illanes and S. B. Nadler, Jr., *Hyperspaces*, Marcel Dekker Inc., New York, 1999.

[10] J. Krasinkiewicz, *On homeomorphisms of the Sierpiński curve*, Prace Mat. **12** (1969), 255–257.

[11] K. Kuratowski, *Topology. Volume II*, Academic Press, New York, 1968.

[12] S. B. Nadler, Jr., *Hyperspaces of sets*, Marcel Dekker Inc., New York, 1978.

[13] S. B. Nadler, Jr., *Continuum theory*, Marcel Dekker Inc., New York, 1992.

[14] S. B. Nadler, Jr. and P. Pellicer-Covarrubias, *Hyperspaces with exactly two orbits*, Glasnik Mat. **41** (2006), 141–157.

[15] S. B. Nadler, Jr. and P. Pellicer-Covarrubias, *Cones that are $\frac{1}{2}$-homogeneous*, Houston J. Math., To appear.

[16] S. B. Nadler, Jr., P. Pellicer-Covarrubias, and I. Puga, *$\frac{1}{2}$-homogeneous continua with cut points*, Topology Appl., To appear.

[17] V. Neumann-Lara, P. Pellicer-Covarrubias, and I. Puga, *On $\frac{1}{2}$-homogeneous continua*, Topology Appl. **153** (2006), no. 14, 2518–2527.

[18] H. Patkowska, *On $\frac{1}{2}$-homogeneous ANR-spaces*, Fund. Math. **132** (1989), no. 1, 25–58.

[19] G. S. Ungar, *On all kinds of homogeneous spaces*, Trans. Amer. Math. Soc. **212** (1975), 393–400.

[20] G. T. Whyburn, *Analytic topology*, American Mathematical Society, Providence, RI, 1942.

Open Problems in Topology II – Edited by E. Pearl

Thirty open problems in the theory of homogeneous continua

Janusz R. Prajs

Broadly understood symmetry is an archetypical quality abundant both in nature and human creativity. In particular, its presence in mathematics is overwhelming. Functions, formulas and spaces with special symmetric properties, as a rule, tend to be more important and have more applications than others. In geometry, symmetry manifests through invariance with respect to certain isometric transformations. Since the concept of an isometry is not topological, one can ask what topological properties could possibly represent symmetry in this broad meaning. Which topological spaces would have *strong* symmetric properties? We propose the following answer: the richer the group of self-homeomorphisms of a topological space, the more *symmetric* the space. This answer, which naturally corresponds to geometric symmetry, leads us to the concept of *topological homogeneity* introduced by Sierpiński [**25**, p. 16]. A topological space X is *homogeneous* provided for each $x, y \in X$ there exists a homeomorphism $h\colon X \to X$ such that $h(x) = y$. This definition identifies a fundamental class of spaces with rich groups of self-homeomorphisms.

The systematic study of homogeneous spaces began with the question of Knaster and Kuratowski [**6**] whether the simple close curve is the only nondegenerate, homogeneous plane continuum. Since then, classifying homogeneous continua became a classic topic, which now is an important area in *continuum theory*. The restriction to the study of homogeneous continua is reasonable indeed. First, the class of *all* homogeneous spaces is so vast, that one cannot expect many important results about that class as whole. Thus the class of compact, metrizable spaces, which have the most common applications, is a natural choice. As it was shown by Michael Mislove and James Rogers [**13**, **14**], each compact, metrizable homogeneous space is a product of a finite set or the Cantor set, and a *homogeneous continuum*, that is, a homogeneous compact, connected, metric space. This makes investigating homogeneous continua particularly important.

Let us notice that the class of homogeneous continua is a natural generalization of the two following important classes of spaces, both in the focus of classic, mainstream study in topology and mathematics: (1) closed, connected manifolds, and (2) compact, connected topological groups (including Lie groups). The significance of these classes provides further motivation for the study of homogeneous continua.

The author was supported in part by the NSF grant DMS-0405374. The author was supported in part by assigned time from the Chair of the Department of Mathematics and Statistics, Dr. Doraiswamy Ramachandran, at California State University, Sacramento

Despite excellent motivation and persistent effort since early 1920s, progress in understanding homogeneous continua and finding new examples has been slow, though rewarded with occasional unexpected turns and spectacular breakthroughs. The purpose of this article is to contribute to this effort by offering questions and possible directions for future research. We do not focus our attention, however, on the most classic questions such as the ones about homogeneous plane continua, (hereditarily) indecomposable homogeneous continua, or hereditarily decomposable homogeneous continua. Excellent references to these problems can be found in [**9**, **12**, **23**].

The problems collected in this paper are divided into two parts. In Section 2 we present miscellaneous problems, some of which already have been published, and some are new. In the author's view, these questions may have potential to become a part of the mainstream study of homogeneous continua in the future. The remaining part of the paper is devoted to a new line of study of homogeneous continua, initiated in [**20**, **21**] and based on the duality of *filament* and *ample* subcontinua. As it is shown in [**22**], this new research is related to the past applications of *aposyndesis* to homogeneous continua. The idea is to identify major archetypical classes of homogeneous continua related to the structure of their filament subcontinua, and investigate properties of these classes. After presenting definitions and summary of basic facts in Section 3, we propose and discuss questions related to this new approach in Section 4.

The author acknowledges collaboration with Keith Whittington on filament and ample subcontinua of homogeneous continua. In fact, a substantial part of the questions from Section 4 emerged from this collaboration.

1. Preliminaries

A *continuum* is a compact, connected, nonempty metric space. Continua of dimension 1 are called *curves*. If X is a continuum, $C(X)$ will denote the hyperspaces consisting of all subcontinua of X under the Hausdorff metric. The definition of a *homogeneous space* is given in the introduction. A space X is 2-*homogeneous* if for every $x_1, x_2, y_1, y_2 \in X$ with $x_1 \neq x_2$ and $y_1 \neq y_2$ there exists a homeomorphism $h\colon X \to X$ such that $h(\{x_1, x_2\}) = \{y_1, y_2\}$.

Though we do not explicitly use the Effros Theorem in this paper, it is a fundamental tool applied in the proofs of many cited results, and can be very helpful when attacking problems involving homogeneous continua. Therefore we recall it here. If X is a homogeneous continuum, then for every positive ε, there is a number δ, called an *Effros number* for ε, such that for each pair of points with $d(x, y) < \delta$, there is some homeomorphism $f\colon X \to X$ that carries x to y and such that $d(z, f(z)) < \varepsilon$ for each $z \in X$. This is called the *Effros Theorem*. It follows from the more general statement that for each $x \in X$, the evaluation map, $g \mapsto gx$, from the homeomorphism group onto X is open. The latter follows from [**5**, Theorem 2]. (See also [**26**, Theorem 3.1].)

2. Fourteen miscellaneous problems

By Mazurkiewicz's theorem [**11**] the simple closed curve is the only locally connected, nondegenerate homogeneous continuum in the plane. An analogous result in 3-space is yet to be found. Note that, by Anderson's result [**1**], 1-dimensional locally connected continua are precisely the simple closed curve and Menger curve. In the first question the *Pontryagin sphere* appears. The Pontryagin sphere has several equivalent definitions. For instance, let S be the Sierpiński universal plane curve, also known as *Sierpiński's carpet*, in its standard geometric construction in the unit square $[0,1] \times [0,1]$. The quotient space obtained from S by identifying each pair of points a, b such that a and b are in the boundary of the same complementary domain of S in the plane, and a and b have at least one coordinate the same, is a *Pontryagin sphere*. Another way to define the Pontryagin sphere is to take two Pontryagin disks defined in [**15**, pp. 608–609] and glue them together along their combinatorial boundary. It is known that the Pontryagin sphere is homogeneous.

Question 1. *If X is a homogeneous, locally connected, 2-dimensional continuum in 3-space, is X either a 2-manifold, or a Pontryagin sphere?* 788?

Question 2. *Is every nondegenerate, simply connected homogeneous continuum in 3-space homeomorphic to 2-sphere $\mathbb{S}^2$?* 789?

The four next questions refer to the important class of path-connected homogeneous continua, and they are essential in the non-locally connected case. Krystyna Kuperberg asked [**7**, Problem 2, p. 630] whether each path-connected homogeneous continuum is locally connected. This question was answered in the negative in [**17**]. The following related question seems to provide a similar type of challenge.

Question 3. *If X is a simply connected homogeneous continuum, is X locally connected?* 790?

The following question was explored in the past by David Bellamy, who obtained a strong partial result [**2**].

Question 4. *If X is a path-connected homogeneous continuum, is X uniformly path-connected?* 791?

Since all uniformly path-connected continua are weakly chainable, a positive answer to the previous question would imply one to the next question.

Question 5. *If X is a path-connected homogeneous continuum, is X weakly chainable?* 792?

The path-connected example $\mathbb{P}$ from [**17**] has a natural projection onto the Menger curve such that $\mathbb{P}$ has a unique path lifting property with respect to this projection. It is not known whether each homogeneous path-connected continuum admits such a map. We ask the following.

793? **Question 6.** *Let X be a homogeneous path-connected (1-dimensional) continuum. Does X admit an open surjective map $f\colon X \to Y$ onto a locally connected continuum Y such that X has the unique path lifting property with respect to f?*

The next three questions are related to each other. They ask about the existence of certain inverse limit representations for some homogeneous continua. To formulate the first of these problems, which originally appeared in [**16**], we need some definitions. A surjective map $f\colon X \to Y$ is called *confluent* if for every continuum K in Y and every $p \in f^{-1}(K)$ there exists a continuum $C \subset X$ such that $p \in C$ and $f(C) = K$. A continuum X is *confluently graph-like* provided for every $\varepsilon > 0$ there is a confluent map of X to a graph with point inverses having diameters less than ε. A continuum is called *confluently graph-representable* if it can be represented as the inverse limit of graphs with confluent bonding maps. By one of the main results of [**16**] the property "confluently graph-like" is equivalent to "confluently graph-representable" for continua.

794? **Question 7.** *If X is a homogeneous curve that contains an arc, is X confluently graph-like?*

795? **Question 8** (J.H. Case [**4**]). *If X is a homogeneous curve that contains an arc, can X be represented as inverse limit of either simple closed curves or topological Menger curves with covering bonding maps?*

796? **Question 9.** *If X is a homogeneous continuum such that each point of X has a neighborhood whose components are n-manifolds (Menger manifolds, Hilbert cube manifolds), is X the inverse limit of n-manifolds (Menger manifolds, Hilbert cube manifolds) with covering bonding maps?*

Known examples suggest that the three following questions may admit positive answers. A counterexample would provide an even more spectacular result.

797? **Question 10.** *Does every nondegenerate (1-dimensional) homogeneous continuum have a nondegenerate weakly chainable subcontinuum?*

798? **Question 11.** *Does every nondegenerate homogeneous continuum contain either an arc or a nondegenerate (hereditarily) indecomposable subcontinuum?*

799? **Question 12.** *Does every homogeneous curve contain either an arc or a proper, nondegenerate terminal subcontinuum?*

The remaining two problems in this section are new. The next one, interesting by its own right, appears in connection to the study of *filament sets*, and is related to Problems 22 and 23 from Section 4.

800? **Question 13.** *If a homogeneous continuum X is a finite (equivalently, countable) union of its indecomposable subcontinua, is X indecomposable?*

We say the group of self-homeomorphisms $H(X)$ of a space X *respects* a partition $\mathcal{G}$ of X if $h(G) \in \mathcal{G}$ for every $h \in H(X)$ and $G \in \mathcal{G}$. Given a subcontinuum K of a space X, let $\mathcal{H}_K = \{h(K) : h \in H(X)\}$. For every $x, y \in X$ we write

$x \sim_K y$ provided that $x = y$ or there are continua $K_1, \ldots, K_n \in \mathcal{H}_K$ such that $K_1 \cup \cdots \cup K_n$ is connected and $x, y \in K_1 \cup \cdots \cup K_n$. Note that $\sim_K$ is an equivalence. The equivalence classes of $\sim_K$ are called *K-components*. The space X is *K-connected* if X is the only K-component in X. It is an immediate observation the partition into K-components is respected by self-homeomorphisms of X. If K_1, K_2 are two subcontinua of a continuum X, we write $K_1 \simeq K_2$ provided the K_1-components and K_2-components are identical. Note that $\simeq$ is an equivalence in $C(X)$. We have the trivial structure of $\{p\}$-components generated by singletons $\{p\}$, which is the trivial decomposition into singletons, and which we usually ignore. For every homogeneous space X we assign the cardinality $\kappa(X)$ of the collection of the equivalence classes of $\simeq$ represented by nondegenerate subcontinua of X. Thus $\kappa(X) = 0$ when X is a singleton. It can easily be observed that $\kappa(\mathbb{S}^1) = 1$ for the unit circle $\mathbb{S}^1$. As an exercise please note that $\kappa(X) \geq 2$ if X is indecomposable, and $\kappa(X) \geq 3$ if X is the circle of pseudo-arcs. (In fact $\kappa(X) = 3$ in the latter case.) The following proposition can easily be shown.

Proposition 2.1. *If X is a nondegenerate 2-homogeneous continuum, then $\kappa(X) = 1$. In particular, if X is either a manifold, the Menger curve, or the Hilbert cube, then $\kappa(X) = 1$.*

Using the Homeomorphism Extension Theorem [**10**], one can show that for the Menger curve $\mathbb{M}$ we have $\kappa(\mathbb{M} \times \mathbb{M}) = 1$, even though $\mathbb{M} \times \mathbb{M}$ is not 2-homogeneous [**8**].

Question 14. *Let X be a homogeneous continuum with $\kappa(X) = 1$. Must X be* 801–802?
path-connected? Must X be locally-connected?

3. Filament sets: definitions and basic properties

In this section we provide basic concepts and facts involved in a new line of study of homogeneous continua, initiated in [**20**, **21**] and based on the duality of *filament* and *ample* subcontinua. We begin with the following definitions of certain subsets of a continuum X, which are crucial in the remaining part of the paper. With an exception of (iv), they were introduced in [**20**].

(i) A subcontinuum F of X is called *filament* if there exists a neighborhood N of F such that the component of N containing F has empty interior.
(ii) A set $Y \subset X$ is called *filament* if every subcontinuum of Y is filament in X.
(iii) A set $Z \subset X$ is called *co-filament* if $X \setminus Z$ is a filament set in X.
(iv) A subcontinuum G of X is called *almost filament* if G is the limit, in the sense of the Hausdorff distance, of filament continua in X.
(v) A subcontinuum A of X is called *ample* if every neighborhood N of A contains a continuum B such that $A \subset \operatorname{int}(B) \subset B \subset N$.

The three following propositions summarize the most fundamental properties [**20**] of the introduced concepts. Part (b) of Proposition 3.1 was originally proved in [**27**].

Proposition 3.1. *Let X be a homogeneous continuum.*

(a) *A subcontinuum K of X is ample if and only if K is not filament.*
(b) *The set of all pairs $(x, y) \in X \times X$ such that only ample subcontinua of X can contain both x and y is a dense G_δ subset of $X \times X$.*
(c) *Each ample subcontinuum of X contains a minimal ample subcontinuum.*
(d) *Each closed filament set in X has a filament neighborhood in X.*
(e) *The collection of filament subcontinua of X is an open, connected subset of $C(X)$.*
(f) *The subspace of $C(X)$ of ample subcontinua of X is a compact absolute retract.*

Proposition 3.2. *For every continuum X the following conditions are equivalent:*

(a) *X is indecomposable.*
(b) *X is the only ample subcontinuum of X.*
(c) *Every nonempty subset of X is co-filament.*

Proposition 3.3. *For every homogeneous continuum X the following conditions are equivalent:*

(a) *X is locally connected.*
(b) *Every subcontinuum of X is ample.*
(c) *X is the only closed, co-filament subset of X.*

Next we define *filament composants* in continua, which generalize the concept of composants of indecomposable continua (see Proposition 3.4 below).

(vi) Given a point $p \in X$, the union of the filament continua in X containing p is called the *filament composant* of X determined by p, and denoted by $\mathrm{Fcs}(p)$.

Please note that the filament composants of a decomposable continuum are different from the (usual) composants. We recall the following fundamental properties of filament composants (see [**20**, Proposition 1.8]).

Proposition 3.4. *Let X be a continuum and $p \in X$. If $\mathrm{Fcs}(p)$ is nonempty, it is a countable union of filament continua, each containing p. Thus each filament composant is a first-category F_σ subset of X. If X is indecomposable, the composants and filament composants of X are identical.*

Employing the concept of a filament continuum, we define some classes of continua.

(vii) A continuum X is *filament additive* provided for each two filament subcontinua F_1 and F_2 with nonempty intersection, the union $F_1 \cup F_2$ is filament.
(viii) A continuum X is called *filament connected* if for each two points $p, q \in X$ there are filament continua $F_1, \dots, F_n$ in X such that $p, q \in F_1 \cup \cdots \cup F_n$ and the union $F_1 \cup \cdots \cup F_n$ is connected.

Most known homogeneous curves are filament additive. The first non-filament additive homogeneous curve was defined in [**17**]. In higher dimensions, each product of at least two homogeneous, non-locally connected continua is non-filament additive [**21**]. Filament additive continua and filament connected continua are disjoint classes of spaces.

(ix) A continuum X is called *filamentable* if either X is a singleton, or X has a filament subcontinuum whose complement is filament.

The diagram in Figure 1 below represents a classification scheme of homogeneous continua introduced in [**19**]. It is based on the concept of a *co-filament continuum*, that is, a co-filament, compact, connected set. Homogeneous continua form a spectrum having at its ends Class I with the richest collection of co-filament subcontinua, and Class IV with the smallest one. The following conditions define the corresponding classes:

(I) *Every subcontinuum is co-filament*;

(II) *Contains non-co-filament subcontinua, and also subcontinua that simultaneously are co-filament and filament*;

(III) *All co-filament subcontinua are ample and some of them are proper*; and

(IV) *The whole space is the only co-filament subcontinuum.*

Homogeneous Continua			
Filamentable		Nonfilamentable	
Class I	Class II	Class III	Class IV
Ia: singleton, indecomposable, filamentable, aposyndetic	IIa: filamentable, decomposable, aposyndetic	IIIa: nonfilamentable, with proper co-filament subcontinua, aposyndetic	IVa: locally connected, nondegenerate, nonfilamentable, aposyndetic
Ib: indecomposable, nondegnerarate, filamentable, nonaposyndetic	IIb: filamentable, decomposable, nonaposyndetic	IIIb: nonfilamentable, with proper co-filament subcontinua, nonaposyndetic	IVb: no proper co-filament subcontinua, nonfilamentable, nonaposyndetic

FIGURE 1. Classification of homogeneous continua

The eight classes indicated in the diagram (Figure 1) are mutually disjoint and each of them is nonempty. If a continuum belongs to a class labeled with b, its aposyndetic decomposition quotient space is in the corresponding class labeled with a. By a recent result of Rogers [**24**], and by Anderson's characterization of locally connected homogeneous curves [**1**], all members of Class IVb have their aposyndetic quotient spaces homeomorphic to either the circle $\mathbb{S}^1$, or the Menger curve $\mathbb{M}$. Below, we list at least one example of spaces belonging to each class. Note that the selected examples have dimension less than or equal to 1.

(Ia) A singleton.
(Ib) The pseudo-arc and solenoids.
(IIa) The Case continuum.
(IIb) The continuous curve of pseudo-arcs with the Case continuum as the quotient space.
(IIIa) The path-connected continuum $\mathbb{P}$ from [**17**].
(IIIb) The continuous curve of pseudo-arcs with $\mathbb{P}$ as the quotient space.
(IVa) The circle $\mathbb{S}^1$ and Menger curve $\mathbb{M}$.
(IVb) The continuous curves of pseudo-arcs with $\mathbb{S}^1$ and $\mathbb{M}$ as the quotient spaces.

Finally, note that the extreme classes, Classes I and IV, are exclusively composed of filament additive continua. Thus the property "filament connected" can only occur in Classes II and III. Each of the Classes IIa, IIb, IIIa and IIIb has both filament additive and non-filament additive members.

4. Filament sets: sixteen questions

In the previous section we presented the most fundamental concepts and facts related to the new line research, in the area of homogeneous continua, introduced in [**20**, **21**] and continued in [**18**, **19**, **22**]. In this section we collect problems that are related to this new research. We begin with a question posed in [**21**].

803? **Question 15.** *Is every homogeneous continuum either filament additive or filament connected?*

This intriguing problem has a positive solution in Classes I, II and IV. Obviously, in the filament additive part of Class III this question is also answered in the affirmative. It is interesting that Class III is the only one of the four, where some other problems remain unsolved. For instance, a possible counterexample to a classic question by Józef Krasinkiewicz and Piotr Minc whether a nondegenerate, hereditarily decomposable, homogeneous continuum must be a simple closed curve, would have to be in the non-filament additive part of Class III [**19**, **21**]. It is not accidental that the path-connected continuum from [**17**] is again in that part of Class III because every non-locally connected, path-connected homogeneous continuum is in there [**19**]. Class III and, in particular, its non-filament additive part remain mystery areas, which deserve special attention in the future.

The four following problems seem to be essential to understand the filament structure of homogeneous continua.

804? **Question 16.** *If K is a subcontinuum of an almost filament continuum L in a homogeneous continuum X, is K almost filament?*

805? **Question 17.** *If X is a homogeneous continuum with dense filament composants, is X almost filament?*

806? **Question 18.** *If X is a homogeneous continuum and $x \in X$, is the filament composant* $\mathrm{Fcs}(x)$ *a first-category subset of the closure* $\mathrm{cl}(\mathrm{Fcs}(x))$*?*

Question 19. *If X is a homogeneous, non-locally connected continuum, does there exist a nondegenerate subcontinuum K of X such that for every filament subcontinuum F of X intersecting K the union $K \cup F$ is filament?* 807?

The two next questions are about products. If K is a subcontinuum of the product $X \times Y$ of continua X and Y, and at least one of the two projections of K is a filament subcontinuum of the corresponding space, then K is filament in $X \times Y$ [**21**]. The converse is not necessarily true. Indeed, David Bellamy and Janusz Łysko observed in [**3**] that if X is a non-circle solenoid, then the diagonal of the product $X \times X$ is filament even though both projections are ample in the corresponding spaces. In view of these facts the following question is of interest.

Question 20. *Let X and Y be homogeneous continua, $\pi_X\colon X \times Y \to X$ and $\pi_Y\colon X \times Y \to Y$ the projections, and F a filament subcontinuum of the product $X \times Y$. Is either $\pi_X(F)$ almost filament in X, or $\pi_Y(F)$ almost filament in Y?* 808?

A number of rules are known that indicate where the product of given two homogeneous continua X and Y may belong, in the diagram from the previous section. For example (IVa) $\times$ (IVa) $\subset$ (IVa), which is well known, and it means that if $X, Y \in$ Class IVa, then $X \times Y \in$ Class IVa. In [**19**] it is shown that (IVa) $\times$ (IVb) $\subset$ (IIIa). It is also observed that if X is filamentable, then so is $X \times Y$. Thus, for instance, (IIa) $\times$ (IIIb) $\subset$ (IIa) and (IIb) $\times$ (IVb) $\subset$ (IIa), etc. The following is an interesting open question in this area.

Question 21. *If X and Y are nonfilamentable homogeneous continua, is the product $X \times Y$ nonfilamentable?* 809?

The next four questions are interconnected. It can be observed that in a homogeneous continuum a minimal ample subcontinuum with nonempty interior would have to be indecomposable. The existence of such subcontinuum would imply that the space is the finite union of indecomposable subcontinua. Thus the four following questions are also related to Question 13.

Question 22. *If X is a homogeneous continuum such that a minimal ample subcontinuum of X has nonempty interior, must X be indecomposable?* 810?

It can be shown that if a homogeneous continuum has a finite co-filament subset, then each minimal ample subcontinuum has nonempty interior. Therefore, the next question would be answered in the affirmative if the previous one was.

Question 23. *If X is a homogeneous continuum having a finite co-filament set C, must X be indecomposable? What if C has at most two elements?* 811?

In the next two questions we focus on some converse directions to the ones of Questions 22 and 23, respectively.

Question 24. *If Y is an indecomposable subcontinuum, with nonempty interior, of a homogeneous continuum X, must Y be a minimal ample subcontinuum of X?* 812?

813? **Question 25.** *Let X be a homogeneous continuum such that every closed co-filament set in X is infinite. Does every minimal ample subcontinuum of X have empty interior in X?*

In a filament additive continuum all minimal ample subcontinua are indecomposable [**21**]. It is not known whether the converse is true.

814? **Question 26.** *Let X be a homogeneous continuum such that every minimal ample subcontinuum of X is indecomposable. Must X be filament additive?*

In [**20**] it has been proved that for a homogeneous continuum X the collection $\mathcal{A}(X)$ of ample subcontinua of X, as a subspace of $C(X)$, is an AR. It is interesting to ask the following.

815? **Question 27.** *Is the collection $\mathcal{A}(X)$ of ample subcontinua of a homogeneous continuum X a deformation retract of $C(X)$?*

The only known examples of homogeneous continua having the collection of minimal ample subcontinua closed belong to Class IV. Therefore, the following question naturally appears.

816–818? **Question 28.** *Let X be a homogeneous continuum, $\mathcal{A}_0(X)$ be the collection of minimal ample subcontinua of X, and assume $\mathcal{A}_0(X)$ is a closed subset of $C(X)$. Is $\mathcal{A}_0(X)$ a partition of X? Is $\mathcal{A}_0(X)$ the Jones aposyndetic decomposition of X? Does X belong to Class IV?*

Our knowledge about the important class of homogeneous path-connected continua, especially in the non-locally connected case, is still very limited. It may be worth to explore the direction of the following question.

819? **Question 29.** *Let X be a homogeneous continuum having an ample, locally connected (path-connected) subcontinuum. Is X path-connected?*

The last question of the paper is related to Question 9 from Section 2. It employs the concept of *micro-local connectedness*. A continuum X is *micro-locally connected at p* provided there exists an open neighborhood U of p such that the component of U containing p is locally connected at p. The micro-local connectedness at p implies that X is micro-locally connected (everywhere) whenever X is homogeneous. Note that X from Question 9 is micro-locally connected. In the following question, the micro-local connectivity of the space implies the *filament local product structure* [**18**], i.e., points have neighborhoods homeomorphic to the product $K \times \mathcal{C}$, where K is a continuum and $\mathcal{C}$ is the Cantor set. Moreover, the micro-local connectivity of the space also implies that K can be locally connected. In Question 9, additionally, K can be an n-cell (a Menger continuum, the Hilbert cube). For instance, solenoids and the Case continuum are spaces for which the hypotheses of Questions 9 and 30 hold.

820? **Question 30.** *If X is a micro-locally connected homogeneous continuum, is X the inverse limit of locally connected continua with covering bonding maps?*

References

[1] R. D. Anderson, *One-dimensional continuous curves and a homogeneity theorem*, Ann. of Math. (2) **68** (1958), 1–16.

[2] D. P. Bellamy, *Short paths in homogeneous continua*, Topology Appl. **26** (1987), no. 3, 287–291.

[3] D. P. Bellamy and J. M. Łysko, *Connected open sets in products of indecomposable continua*, 2005, Preprint.

[4] J. H. Case, *Another* 1*-dimensional homogeneous continuum which contains an arc*, Pacific J. Math. **11** (1961), 455–469.

[5] E. G. Effros, *Transformation groups and* C^**-algebras*, Ann. of Math. (2) **81** (1965), 38–55.

[6] B. Knaster and C. Kuratowski, *Probléme 2*, Fund. Math. **1** (1920), 223.

[7] K. Kuperberg, *A locally connected microhomogeneous nonhomogeneous continuum*, Bull. Acad. Polon. Sci. Sér. Sci. Math. **28** (1980), no. 11–12, 627–630 (1981).

[8] K. Kuperberg, W. Kuperberg, and W. R. R. Transue, *On the* 2*-homogeneity of Cartesian products*, Fund. Math. **110** (1980), no. 2, 131–134.

[9] W. Lewis, *Continuum theory problems*, Topology Proc. **8** (1983), no. 2, 361–394.

[10] J. C. Mayer, L. G. Oversteegen, and E. D. Tymchatyn, *The Menger curve. Characterization and extension of homeomorphisms of non-locally-separating closed subsets*, Dissertationes Math. **252** (1986), 45 pp.

[11] S. Mazurkiewicz, *Sur les continus homogènes*, Fund. Math. **5** (1924), 137–146.

[12] J. van Mill and G. M. Reed (eds.), *Open problems in topology*, North-Holland, Amsterdam, 1990.

[13] M. W. Mislove and J. T. Rogers, Jr., *Local product structures on homogeneous continua*, Topology Appl. **31** (1989), no. 3, 259–267.

[14] M. W. Mislove and J. T. Rogers, Jr., *Addendum: "Local product structures on homogeneous continua"*, Topology Appl. **34** (1990), no. 2, 209.

[15] W. J. R. Mitchell, D. Repovš, and E. V. Ščepin, *On* 1*-cycles and the finite dimensionality of homology* 4*-manifolds*, Topology **31** (1992), no. 3, 605–623.

[16] L. G. Oversteegen and J. R. Prajs, *On confluently graph-like compacta*, Fund. Math. **178** (2003), no. 2, 109–127.

[17] J. R. Prajs, *A homogeneous arcwise connected non-locally-connected curve*, Amer. J. Math. **124** (2002), no. 4, 649–675.

[18] J. R. Prajs, *Filament local product structures in homogeneous continua*, 2005, Preprint.

[19] J. R. Prajs, *Co-filament subcontinua in homogeneous spaces*, 2006, Preprint.

[20] J. R. Prajs and K. Whittington, *Filament sets and homogeneous continua*, 2005, Preprint.

[21] J. R. Prajs and K. Whittington, *Filament additive homogeneous continua*, 2006, Preprint. To appear in Indiana Univ. Math. J.

[22] J. R. Prajs and K. Whittington, *Filament sets, aposyndesis, and the decomposition theorem of Jones*, Trans. Amer. Math. Soc. (2006), Article in press.

[23] J. T. Rogers, Jr., *Classifying homogeneous continua*, Topology Appl. **44** (1992), 341–352.

[24] J. T. Rogers, Jr., *Higher dimensional aposyndetic decompositions*, Proc. Amer. Math. Soc. **131** (2003), no. 10, 3285–3288.

[25] W. Sierpiński, *Sur une propriété topologique des ensembles dénombrables denses en soi*, Fund. Math. **1** (1920), 11–28.

[26] G. S. Ungar, *On all kinds of homogeneous spaces*, Trans. Amer. Math. Soc. **212** (1975), 393–400.

[27] K. Whittington, *Open connected sets in homogeneous spaces*, Houston J. Math. **27** (2001), no. 3, 523–531.

Part 4

Topological Algebra

Open Problems in Topology II – Edited by E. Pearl

Problems about the uniform structures of topological groups

Ahmed Bouziad and Jean-Pierre Troallic

1. Introduction

In the introduction of their fundamental paper entitled "Pseudocompactness and uniform continuity in topological groups" published in 1966, W.W. Comfort and K.A. Ross asserted, without proving it, that if every real-valued left uniformly continuous function on a topological group G is right uniformly continuous, then the left and right uniform structures on G coincide. Let us recall that the family of all $\{(x,y) \in G \times G : x^{-1}y \in V\}$, with V a neighborhood of the identity element e in G, is a basis of the left uniform structure $\mathcal{L}_G$ on G; a basis of the right uniform structure $\mathcal{R}_G$ on G is obtained by replacing "$x^{-1}y$" by "xy^{-1}". Actually, forty years later, and despite many mathematicians' efforts, it still isn't known whether this property is true or false. The aim of this paper is to take stock of this problem, to present a few new ideas in order to study it, and to raise certain questions connected with the subject.

The well-known class of all (Hausdorff) balanced topological groups is denoted by [SIN]. A topological group G is said to be *balanced* (or a *[SIN]-group*) if $\mathcal{L}_G = \mathcal{R}_G$, or, equivalently, if every neighborhood of the identity element contains a neighborhood which is invariant under all inner automorphisms of G. (Cf. for instance [**37**].) Following Protasov [**35**], we will say that G is *functionally balanced* (or an *[FSIN]-group*) if every bounded real-valued left uniformly continuous function on G is right uniformly continuous, and we will say that G is *strongly functionally balanced* if every real-valued left uniformly continuous function on G is right uniformly continuous (Itzkowitz [**23**]). The symbol [FSIN] (respectively [SFSIN]) stands for the class of all functionally (respectively strongly functionally) balanced topological groups. It is plain that [SIN] is a subclass of [SFSIN] and that [SFSIN] is a subclass of [FSIN].

All the questions below are motivated by the following "Itzkowitz's Problem", which was first raised by Itzkowitz in [**21**]:

Question 1. *Is [SFSIN] = [SIN]?* 821?

In fact, we will especially consider the following bounded version of the problem (and from now on, the phrase "the main problem" will denote this case):

Question 2. *Is [FSIN] = [SIN]?* 822?

Let us recall that between 1988 and 1992, Itzkowitz [**21**], Milnes [**33**] and Protasov [**35**] provided independent proofs that every locally compact functionally balanced topological group is balanced; in 1997, and in another direction, Megrelishvili, Nickolas and Pestov [**32**] proved that every locally connected functionally balanced topological group is balanced. Improvements of these results were presented by Itzkowitz [**22**] in his survey on the subject published in 1998. We will, of course, highlight the progress made after this date.

2. The two versions of Itzkowitz's problem

Considering a bounded version *and* an unbounded version of the problem was not immediately deemed useful; for example, in the locally compact case in [**21**], or in the locally connected case in [**32**], it was stated that any G in [SFSIN] is balanced, although the proof works with [FSIN] instead of [SFSIN]. Nevertheless, it remains possible that the two problems are only one.

823? **Question 3.** *Is [SFSIN] = [FSIN]?*

Clearly, a negative answer to Question 3 would imply a very strong negative answer to Question 2.

In 1991, Protasov [**35**] gave a positive answer to Question 2 for the class of almost metrizable groups (a class which contains both that of locally compact groups and of metrizable groups) by using the following very interesting characterization of [FSIN]: let $\mathcal{V}_G(e)$ denote the neighborhood system of the identity element e of a given topological group G; then G is a member of [FSIN] if and only if G satisfies the Protasov and Saryev criterion [**36**], that is to say, if and only if for all $A \subset G$ and $V \in \mathcal{V}_G(e)$, there is $U \in \mathcal{V}_G(e)$ such that $UA \subset AV$. The part played by this criterion in the problem was made completely clear in [**4**, **6**] when observing, after an immediate re-writing, that it means the equality of the proximities on G induced by the left and right uniformities.

Another way to formulate Protasov and Saryev's criterion consists in saying that for each $A, B \subset G$ and $V \in \mathcal{V}_G(e)$ such that $AV \subset B$, there is $U \in \mathcal{V}_G(e)$ such that $UA \subset B$. This formulation leads us to propose the following criterion for G to be strongly functionally balanced. It is easily derivable from the work of Leader [**31**]. To state it, the following terminology is needed. A sequence $(A_n)_{n\in\mathbb{N}}$ of subsets of G is said to be *strongly left increasing* (respectively *strongly right increasing*) if there is $V \in \mathcal{V}_G(e)$ such that $A_nV \subset A_{n+1}$ (respectively $VA_n \subset A_{n+1}$) for each $n \in \mathbb{N}$.

Proposition. *The following conditions are equivalent for any topological group G.*

(1) *G is strongly functionally balanced.*
(2) *Every sequence of subsets of G which is strongly left increasing is strongly right increasing.*

A topological group is said to be non-Archimedean if there exists a base for the neighborhood system of the identity element which consists of open subgroups. It was discovered by Hernández [**13**] that every G in [SFSIN] which is non-Archimedean and $\aleph_0$-bounded is balanced. This result was recently extended in [**40**] to every non-Archimedean group which is strongly functionally generated by the set of all its subspaces of countable o-tightness. For all we know, these are the only instances where *unbounded* uniformly continuous functions were really involved. Let us take the opportunity to state the criterion for balancedness that Hernández established beforehand: A topological group G is balanced (i.e., $\bigcap_{x\in G} xVx^{-1} \in \mathcal{V}_G(e)$ for all $V \in \mathcal{V}_G(e)$) if and only if $\bigcap_{a\in A} aVa^{-1} \in \mathcal{V}_G(e)$ for every left uniformly discrete subset A of G and every $V \in \mathcal{V}_G(e)$. This very important criterion was already implicitly used in [**32**], and explicitly formulated in [**22**]. Let us recall that a subset A of G is said to be *left uniformly discrete* if there is $V \in \mathcal{V}_G(e)$ such that aV and bV are disjoint whenever $a, b \in A$ and $a \neq b$.

From now on, we will only consider the bounded version of Itzkowitz's Problem, it being understood that most of the problems raised below obviously admit an unbounded version.

3. Some remarks about [FSIN] and [SIN]

Let G be a topological group. If (and only if) for any precompact uniform space Y, every left uniformly continuous mapping of G into Y is right uniformly continuous, then $G \in$ [FSIN]. If Y runs through the larger class of all bounded uniform spaces, a characterization of [SIN] is obtained. Before specifying this point, let us recall some definitions and properties.

A uniform space Y is said to be *bounded* if all real-valued uniformly continuous functions on Y are bounded. Another concept in its right place here is that of injective uniform space: Y is said to be *injective* if whenever A is a subspace of a uniform space X, any uniformly continuous mapping of A into Y has a uniformly continuous extension to X [**18**]. It can be shown that if Y is injective, then Y is bounded. The most familiar example of an injective uniform space is that of the unit interval $[0, 1]$; this fact, proved by Katětov [**27**, **28**], is here of great significance since the belonging of the topological group G to [FSIN] means that the left and right uniformities on G induce the same proximity on G. Another standard injective uniform space is the metric Hedgehog $H(A)$ over a set A, that is the set of all (a, x) ($a \in A$, $0 \le x \le 1$), $A \times \{0\}$ being reduced to a point, with the metric $d((a, x), (a, y)) = |x - y|$ and $d((a, x), (b, y)) = x + y$ if $a \neq b$.

A family $(A_i)_{i\in I}$ of subsets of G is said to be *left uniformly discrete* if there is $V \in \mathcal{V}_G(e)$ such that A_iV and A_jV are disjoint whenever $i, j \in I$ and $i \neq j$. As already said in Section 2, a subset A of G is *left uniformly discrete* if the family $(\{a\})_{a\in A}$ is left uniformly discrete. The subset A of G is said to be *right thin* (in G) if $\bigcap_{a\in A} aVa^{-1}$ is a neighborhood of the identity element e for every $V \in \mathcal{V}_G(e)$. Right uniform discreteness and left thinness are defined similarly. Finally, the subset A of G is said to be *lower uniformly discrete* if there is $V \in \mathcal{V}_G(e)$ such that VaV and VbV are disjoint whenever $a, b \in A$ and $a \neq b$.

The following is a key lemma; knowing whether it can be extended to any left uniformly discrete subset A of G is equivalent to Question 2 since, as said in Section 2, if every left uniformly discrete subset of G is right thin, then G is balanced [**22**].

Lemma ([**5**]). *Every lower uniformly discrete subset of a functionally balanced group G is right thin in G.*

Proposition. *Let G be a topological group. Then the following are equivalent:*

(1) *G is balanced.*
(2) *For any bounded uniform space Y, every left uniformly continuous mapping $f: G \to Y$ is right uniformly continuous.*
(3) *For any injective uniform space Y, every left uniformly continuous mapping $f: G \to Y$ is right uniformly continuous.*
(4) *For any set A, every left uniformly continuous mapping $f: G \to H(A)$ is right uniformly continuous.*
(5) *Any left uniformly discrete family of subsets of G is right uniformly discrete.*

PROOF. Obviously, (1) implies (2). Any injective uniform space being bounded, (2) implies (3). Since $H(A)$ is injective, (3) implies (4). The implication (4) $\Rightarrow$ (5) holds for any two uniformities on a given set X (in place of the left and right uniformities on G); see [**8**] or [**30**]. Finally, let us suppose that (5) holds; then any two subsets of G which are left proximal are right proximal (so that $G \in$ [FSIN]), and every left uniformly discrete subset of G is lower uniformly discrete; therefore, by the above key lemma, (1) is satisfied. (Note that the equivalence between (1), (2) and (3) is also a consequence of the well-known fact that every uniform space can be embedded in an injective uniform space [**18**].) □

In view of the previous proposition, Question 2 could be stated as follows:

Question 4. *Let G be a functionally balanced group. Is any left uniformly discrete family of subsets of G right uniformly discrete?*

Let us say that a topological group G is injective (respectively bounded) if the uniform space $(G, \mathcal{L}_G)$ (or, equivalently, $(G, \mathcal{R}_G)$) is injective (respectively bounded). Since any injective uniform space is proximally fine [**17**], the answer to Question 2 is positive for all injective topological groups.

Proposition. *Every injective topological group which is functionally balanced is balanced.*

824? **Question 5.** *Is it true, more generally, that every bounded member of [FSIN] belongs to [SIN]?*

It should be pointed out that the answer to the following question is not clear.

825–826? **Question 6.** *Is every injective (respectively bounded) topological group functionally balanced?*

4. The class [ASIN]

A natural approach to the main problem is to consider any class [C] of topological groups which contains the class [SIN] as closely as possible, and try to prove that [C] also includes the class [FSIN]. The *dual* problem consists in examining whether the inclusion [C] $\cap$ [FSIN] $\subset$ [SIN] holds, the class [C] now being as wide as possible. To illustrate that idea, let us consider the class [ASIN] of all *almost balanced* topological groups, that is the class of all topological groups G for which the identity element e has at least a right thin neighborhood in G. Clearly, the class [ASIN] is much larger than [SIN]. The following is established in [**5**].

Theorem. *[ASIN]* $\cap$ *[FSIN]* $=$ *[SIN].*

In view of that property, Question 2 becomes: Is [FSIN] $\subset$ [ASIN]? Moreover, for any class of topological groups contained in [ASIN], a positive answer to the main problem holds; it applies, for instance, to every topological group which is locally precompact or which contains an open subgroup belonging to [SIN]. Here is another interesting class contained in [ASIN]:

Proposition. *Let G be a topological group. Suppose that the identity element of G has a neighborhood V such that every bounded real-valued continuous function on G is left uniformly continuous, when restricted to V. Then G is a member of [ASIN].*

It is proved by Itzkowitz and Tkachuk in [**26**] that every uniformly functionally complete topological group G is balanced; of course, this follows from the previous proposition. Recall that G is said to be *uniformly functionally complete* [**26**], or with property U [**29**], if every real-valued (or, equivalently [**7**], every *bounded* real-valued) continuous function on G is left uniformly continuous.

5. A few other questions

In this section we collect some concrete open questions related to the main problem.

Question 7. *Let G be a functionally balanced group. Is every left uniformly discrete subset of G right uniformly discrete?* 827?

Question 8. *Let G be a functionally balanced group. Is every left precompact subset of G right precompact?* 828?

Question 9. *Let G be a functionally balanced group. Let us suppose that every left precompact subset of G is right precompact. Is G balanced then?* 829?

The main statement of a recent paper by Itzkowitz [**24**] consists in saying that if $G \in$ [FSIN] and if every left uniformly discrete subset of G is right uniformly discrete, then $G \in$ [SIN]. This is to be compared with the implication (5) $\Rightarrow$ (1) in the first proposition of Section 3 above. We must admit that one point in Itzkowitz's argumentation eluded us.

It is well known that any balanced topological group is isomorphic (topologically and algebraically) to a subgroup of a product of balanced metrizable topological groups [**9**]. (See [**37**] for details.) This suggests the two following questions:

Question 10. *Is every member of [FSIN] isomorphic to a subgroup of a product of metrizable topological groups?* 830?

The dual question is:

831? **Question 11.** *Let G be a member of [FSIN] which is isomorphic to a subgroup of a product of metrizable topological groups. Is G balanced?*

A topological group G is said to be $\aleph_0$-*bounded* if for every neighborhood V of the identity element, there is a countable subset A of G such that $G = AV$. It is well known that G is $\aleph_0$-bounded if and only if it is isomorphic to a subgroup of a product of metrizable separable topological groups [**10**].

832? **Question 12.** *Let G be a member of [FSIN] which is $\aleph_0$-bounded. Is G balanced?*

In fact, the question arises even in the following simple case.

833? **Question 13.** *Is every countable member of [FSIN] balanced?*

Let us remark that a positive answer to Question 13 will imply a positive answer to Question 4 for $\aleph_0$-bounded groups.

6. A representative case

The fact that a metrizable topological group which is functionally balanced is balanced belongs to the theory of uniform spaces. The same remark holds, more generally, for every topological group such that the neighborhood system of the identity element has a linearly ordered base, and this follows from the proximally fineness (proved in [**1**]) of every uniform space $(X,\mathcal{U})$ such that $\mathcal{U}$ has a linearly ordered base. The above positive answer to Question 2 for the class of all injective topological groups rested on the same sort of argument. The following combinatorial lemma, established in [**1**], is essential for the approach of these results.

Lemma. *Let $(X,\mathcal{U})$ be a uniform space and $W \in \mathcal{U}$ a symmetric entourage. Let us suppose that $(x_\alpha, y_\alpha) \notin W^3$ for all $\alpha \in \Gamma$, where Γ is a set of regular cardinal, and $x_\alpha, y_\alpha \in X$. Then, there is a set $\Gamma' \subset \Gamma$ with the same cardinal as Γ such that $(x_\alpha, y_\beta) \notin W$ for all $\alpha, \beta \in \Gamma'$.*

Surprisingly enough, in the context of topological groups, and in order to tackle our main problem, that lemma may be used under very general topological conditions. Let us illustrate this by the proposition below. Let us say that a subset Y of a topological group G is *relatively o-radial in G* if for every $y \in Y$ and every family $(O_i)_{i\in I}$ of open subsets of G such that $y \in \mathrm{cl}(\bigcup_{i\in I} O_i \cap Y) \setminus \bigcup_{i\in I} \mathrm{cl}(O_i)$, there is a set $J \subset I$ of regular cardinality such that for every neighborhood V of y in G, we have $|\{j \in J : O_j \cap V = \emptyset\}| < |J|$. Obviously, we will say that G is *o-radial* if it is relatively o-radial in itself. Radial spaces are defined in [**15**] (cf. also [**2**]); every radial topological group is o-radial. If Y is relatively o-Malykhin in G (as defined in [**6**]), that is for instance the case if Y is left (or right) precompact [**6**], then Y is relatively o-radial in G. All locally precompact topological groups, all q-groups (as defined in [**12**]), are o-Malykhin (i.e., relatively o-Malykhin in themselves), and therefore o-radial.

Proposition. *Every functionally balanced o-radial topological group is balanced.*

PROOF. Suppose that G is not balanced. Then there is a symmetric open $W \in \mathcal{V}_G(e)$ such that for every $U \in \mathcal{V}_G(e)$ one can find $g_U, h_U \in G$ with $g_U^{-1}h_U \in U$ and $g_U h_U^{-1} \notin W^6$. Clearly, $e \in \mathrm{cl}(\bigcup_{U\in\mathcal{V}_G(e)} g_U^{-1}Wh_U)$. Since G is o-radial, there is $\Gamma \subset \mathcal{V}_G(e)$ such that the cardinal of Γ is regular and $e \in \mathrm{cl}(\bigcup_{U\in\Gamma'} g_U^{-1}Wh_U)$ for every $\Gamma' \subset \Gamma$ with the same cardinal as Γ. By the previous lemma, there is a set $\Gamma'' \subset \Gamma$ with the same cardinal as Γ such that $g_U h_V^{-1} \notin W^2$ for all $U, V \in \Gamma''$. Let us put $A = \{g_U : U \in \Gamma''\}$ and $B = W\{h_V : V \in \Gamma''\}$; then A and B are left, but not right, proximal which contradicts the functional balance of G. □

In fact, if G is an o-radial topological group, then the uniform space $(G, \mathcal{L}_G)$ is proximally fine, and that remains true if G is more generally strongly functionally generated (in the sense of [**3**]) by its relatively o-radial subspaces.

References

[1] E. M. Alfsen and O. Njåstad, *Totality of uniform structures with linearly ordered base*, Fund. Math. **52** (1963), 253–256.

[2] A. V. Arhangel'skiĭ, *Some properties of radial spaces*, Mat. Zametki **27** (1980), no. 1, 95–104, Translation: Math. Notes 27 (1980) no. 1, 50–54.

[3] A. V. Arhangel'skiĭ, *Topological function spaces*, Kluwer Academic Publishers, Dordrecht, 1992.

[4] A. Bouziad and J. P. Troallic, *Functional equicontinuity and uniformities in topological groups*, Topology Appl. **144** (2004), 95–107.

[5] A. Bouziad and J.-P. Troallic, *Nonseparability and uniformities in topological groups*, Topology Proc. **28** (2004), no. 2, 343–359.

[6] A. Bouziad and J. P. Troallic, *Left and right uniform structures on topological groups*, Topology Appl. **153** (2006), no. 13, 2351–2361.

[7] W. W. Comfort and K. A. Ross, *Pseudocompactness and uniform continuity in topological groups*, Pacific J. Math. **16** (1966), 483–496.

[8] Z. Frolík, *Basic refinements of the category of uniform spaces*, TOPO 72—general topology and its applications, Lecture Notes in Math., vol. 378, Springer, Berlin, 1974, pp. 140–158.

[9] M. I. Graev, *Theory of topological groups. I. Norms and metrics on groups. Complete groups. Free topological groups*, Uspekhi Mat. Nauk **5** (1950), no. 2(36), 3–56, Translation: Free topological groups, Amer. Math. Soc. Transl. 35 (1951).

[10] I. I. Guran, *Topological groups similar to Lindelöf groups*, 1981, Translation: Soviet Math. Dokl. 23 (1981) 173–175, pp. 1305–1307.

[11] G. Hansel and J.-P. Troallic, *Equicontinuity, uniform structures and countability in locally compact groups*, Semigroup Forum **42** (1991), no. 2, 167–173.

[12] G. Hansel and J.-P. Troallic, *Sequential criteria for the equality of uniform structures in q-groups*, Topology Appl. **57** (1994), no. 1, 47–52.

[13] S. Hernández, *Topological characterization of equivalent uniformities in topological groups*, Topology Proc. **25** (2000), Spring, 181–188.

[14] S. Hernández, *Uniformly continuous mappings defined by isometries of spaces of bounded uniformly continuous functions*, Houston J. Math. **29** (2003), no. 1, 149–155.

[15] H. Herrlich, *Quotienten geordneter Räume und Folgenkonvergenz*, Fund. Math. **61** (1967), 79–81.

[16] S. Hingano, *On uniformities and uniformly continuous functions on factor-spaces of topological groups*, Topology Proc. **29** (2005), no. 2, 521–526.

[17] M. Hušek, *Factorization of mappings (products of proximally fine spaces)*, Seminar Uniform Spaces (Prague, 1973–1974), Mat. Ústav Československé Akad. Věd, Prague, 1975, pp. 173–190.

[18] J. R. Isbell, *Uniform spaces*, American Mathematical Society, Providence, RI, 1964.

[19] G. Itzkowitz, *Continuous measures, Baire category, and uniform continuity in topological groups*, Pacific J. Math. **54** (1974), no. 2, 115–125.

[20] G. Itzkowitz, *Uniform structure in topological groups*, Proc. Amer. Math. Soc. **57** (1976), no. 2, 363–366.

[21] G. Itzkowitz, *Uniformities and uniform continuity on topological groups*, General topology and applications (Staten Island, NY, 1989), Lecture Notes in Pure and Appl. Math., vol. 134, Marcel Dekker Inc., New York, 1991, pp. 155–178.

[22] G. Itzkowitz, *On balanced topological groups*, Topology Proc. **23** (1998), Spring, 219–233.

[23] G. Itzkowitz, *Projective limits and balanced topological groups*, Topology Appl. **110** (2001), no. 2, 163–183.

[24] G. Itzkowitz, *Functional balance, discrete balance, and balance in topological groups*, Topology Proc. **28** (2004), no. 2, 569–577.

[25] G. Itzkowitz, S. Rothman, H. Strassberg, and T. S. Wu, *Characterization of equivalent uniformities in topological groups*, Topology Appl. **47** (1992), no. 1, 9–34.

[26] G. Itzkowitz and V. V Tkachuk, *Uniformly functionally complete and uniformly* $\mathbb{R}$*-factorizable groups*, Preprint.

[27] M. Katětov, *On real-valued functions in topological spaces*, Fund. Math. **38** (1951), 85–91.

[28] M. Katětov, *Correction to "On real-valued functions in topological spaces" (Fund. Math. 38 (1951) 85–91)*, Fund. Math. **40** (1953), 203–205.

[29] J. M. Kister, *Uniform continuity and compactness in topological groups*, Proc. Amer. Math. Soc. **13** (1962), 37–40.

[30] M. Kosina and P. Pták, *Intrinsic characterization of distal spaces*, Seminar Uniform Spaces (Prague, 1973–1974), Mat. Ústav Československé Akad. Véd, Prague, 1975, pp. 217–231.

[31] S. Leader, *Spectral structures and uniform continuity*, Fund. Math. **60** (1967), 105–115.

[32] M. Megrelishvili, P. Nickolas, and V. Pestov, *Uniformities and uniformly continuous functions on locally connected groups*, Bull. Austral. Math. Soc. **56** (1997), no. 2, 279–283.

[33] P. Milnes, *Uniformity and uniformly continuous functions for locally compact groups*, Proc. Amer. Math. Soc. **109** (1990), no. 2, 567–570.

[34] W. H. Previts and T. S. Wu, *Notes on balanced groups*, Bull. Inst. Math. Acad. Sinica **29** (2001), no. 2, 105–123.

[35] I. V. Protasov, *Functionally balanced groups*, Mat. Zametki **49** (1991), no. 6, 87–91, Translation: Math. Notes 49 (1991) no. 6, 614–616.

[36] I. V. Protasov and A. Saryev, *The semigroup of closed subsets of a topological group*, Izv. Akad. Nauk Turkmen. SSR Ser. Fiz.-Tekhn. Khim. Geol. Nauk (1988), no. 3, 21–25.

[37] W. Roelcke and S. Dierolf, *Uniform structures in topological groups and their quotients*, McGraw-Hill, New York, 1981.

[38] J. P. Troallic, *Sequential criteria for equicontinuity and uniformities on topological groups*, Topology Appl. **68** (1996), no. 1, 83–95.

[39] J. P. Troallic, *Equicontinuity, uniform continuity and sequences in topological groups*, Topology Appl. **93** (1999), no. 3, 179–190.

[40] J. P. Troallic, *Equicontinuity and balanced topological groups*, Topology Appl. **135** (2004), 63–71.

Open Problems in Topology II – Edited by E. Pearl

On some special classes of continuous maps

Maria Manuel Clementino and Dirk Hofmann

1. Special morphisms of Top

Triquotient maps were introduced by E. Michael in [**24**] as those continuous maps $f\colon X \to Y$ for which there exists a map $(\_)^\sharp\colon \mathcal{O}X \to \mathcal{O}Y$ such that, for every U, V in the lattice $\mathcal{O}X$ of open subsets of X:

(T1) $U^\sharp \subseteq f(U)$,
(T2) $X^\sharp = Y$,
(T3) $U \subseteq V \Rightarrow U^\sharp \subseteq V^\sharp$,
(T4) $(\forall y \in U^\sharp)\ (\forall \Sigma \subseteq \mathcal{O}X \text{ directed})\ f^{-1}(y) \cap U \subseteq \bigcup \Sigma \Rightarrow (\exists S \in \Sigma)\ y \in S^\sharp$.

It is easy to check that, if $f\colon X \to Y$ is an open surjection, then the direct image $f(\_)\colon \mathcal{O}X \to \mathcal{O}Y$ satisfies (T1)–(T4). If $f\colon X \to Y$ is a retraction, so that there exists a continuous map $s\colon Y \to X$ with $f \circ s = 1_Y$, then $(\_)^\sharp := s^{-1}(\_)$ satisfies (T1)–(T4). Moreover, if $f\colon X \to Y$ is a proper surjection (by proper map we mean a closed map with compact fibres: see [**2**]), then $U^\sharp := Y \setminus f(X \setminus U)$ fulfills (T1)–(T4). That is, open surjections, retractions and proper surjections are triquotient maps. But there are triquotient maps which are neither of these maps (cf. [**3**, **15**] for examples). However, we do not know whether these three classes of maps describe completely triquotient maps, in the sense we describe now:

Question 1. *Is it true that any triquotient map can be factored through (possibly infinitely many) open surjections, proper surjections and retractions?* 834?

T. Plewe in [**26**] related triquotient maps to Topological Grothendieck Descent Theory (see [**17**]). We recall that a continuous map $f\colon X \to Y$ is an *effective descent* map if its pullback functor $f^*\colon \mathbf{Top}/Y \to \mathbf{Top}/X$, that assigns to each $g\colon W \to Y$ its pullback along f, is monadic. If f^* is premonadic, then f is a *descent map* (see [**18**, **19**]). Descent maps are exactly universal quotient maps [**10**], or pullback-stable quotient maps, that is quotient maps whose pullback along any map is still a quotient. We point out here that this class of maps was introduced independently by B. Day and M. Kelly [**10**], by E. Michael [**23**], under the name *biquotient maps*, and by O. Hájek [**11**], as *limit lifting maps*. Effective descent maps turned out to be very difficult to describe topologically. The only characterisation that is known is due to J. Reiterman and W. Tholen [**27**] and uses heavily ultrafilter convergence. (We will concentrate on ultrafilter convergence later in this work.)

Problem 2. *Find a characterisation of topological effective descent maps in terms of the topologies or the Kuratowski closures.* 835?

The authors acknowledge partial financial assistance by Centro de Matemática da Universidade de Coimbra/FCT and Unidade de Investigação e Desenvolvimento Matemática e Aplicações da Universidade de Aveiro/FCT.

One possible approach to this problem could be via the existence of a map between the topologies that resembles the map $(\_)^\sharp$ introduced by E. Michael to define triquotient maps. Indeed, it is interesting to notice that one can characterise several of these classes of morphisms using a map $\mathcal{O}X \to \mathcal{O}Y$ as follows: a continuous map $f\colon X \to Y$ is:

(1) a universal quotient map (= biquotient = limit lifting = descent) if and only if there exists a map $(\_)^\sharp\colon \mathcal{O}X \to \mathcal{O}Y$ satisfying (T1)–(T3) and
  (U4) $(\forall y \in Y)\ (\forall \Sigma \subseteq \mathcal{O}X \text{ directed})\ f^{-1}(y) \subseteq \bigcup \Sigma \Rightarrow (\exists S \in \Sigma)\ y \in S^\sharp$.
(2) a proper surjection if and only if there exists a map $(\_)^\sharp\colon \mathcal{O}X \to \mathcal{O}Y$ satisfying (T1)–(T3) and, for every $U \in \mathcal{O}X$
  (P4) $(\forall y \in U^\sharp)\ (\forall \Sigma \subseteq \mathcal{O}X \text{ directed})\ f^{-1}(y) \cap U \subseteq \bigcup \Sigma \Rightarrow (\exists S \in \Sigma)\ y \in S^\sharp$ and $f^{-1}(y) \subseteq S$.
(3) an open surjection if and only if there exists a map $(\_)^\sharp\colon \mathcal{O}X \to \mathcal{O}Y$ satisfying (T1)–(T3) and, for every $U \in \mathcal{O}X$
  (O4) $(\forall y \in f(U))\ (\forall \Sigma \subseteq \mathcal{O}X \text{ directed})\ f^{-1}(y) \cap U \subseteq \bigcup \Sigma \Rightarrow (\exists S \in \Sigma)\ y \in S^\sharp$.

836? **Problem 3.** *Describe effective descent maps via the existence of a map $(\_)^\sharp$ similar to those described above.*

A possible approach to Problem 3 might be making use of the following result, that we could prove only for maps between finite topological spaces. It is based on the existence of a map between the lattices of *locally closed subsets* (i.e., the subsets which are an intersection of an open and a closed subset), which we will denote by $\mathcal{LC}(\_)$.

Theorem 1. *If X and Y are finite spaces, a continuous map $f\colon X \to Y$ is an effective descent map if and only if, for every pullback $g\colon W \to Z$ of f, there exists a map $(\_)^\sharp\colon \mathcal{LC}(W) \to \mathcal{LC}(Z)$ such that, for every $A, B \in \mathcal{LC}(W)$,*

(1) $A^\sharp \subseteq g(A)$;
(2) $W^\sharp = Z$;
(3) $A \subseteq B \Rightarrow A^\sharp \subseteq B^\sharp$;
(4) $(\forall z \in Z)\ g^{-1}(z) \subseteq A \Rightarrow z \in A^\sharp$.

We believe that the work of G. Richter [**28**] may be inspiring to attack Problem 3.

In the pioneer work [**15**], G. Janelidze and M. Sobral describe several classes of maps using convergence, whenever X and Y are finite spaces.

Theorem 2 ([**15**]). *If X and Y are finite spaces, a continuous map $f\colon X \to Y$ is:*

(a) *a universal quotient if and only if, for every $y_0, y_1 \in Y$ with $y_1 \to y_0$, there exist $x_0, x_1 \in X$ such that $x_1 \to x_0$, $f(x_0) = y_0$ and $f(x_1) = y_1$;*
(b) *an effective descent map if and only if, for every $y_0, y_1, y_2 \in Y$ with $y_2 \to y_1 \to y_0$, there exist $x_0, x_1, x_2 \in X$ such that $x_2 \to x_1 \to x_0$ and $f(x_i) = y_i$, for $i = 0, 1, 2$;*

(c) *a triquotient map if and only if, for every* $y_0, \ldots, y_n \in Y$ *with* $y_n \to \cdots \to y_0$*, there exist* $x_0, \ldots, x_n \in X$ *such that* $x_n \to \cdots \to x_0$ *and* $f(x_i) = y_i$*, for* $i = 0, \ldots, n$.

(d) *a proper map if and only if, for every* $x_1 \in X$ *and* $y_0 \in Y$ *with* $f(x_1) \to y_0$*, there exists* $x_0 \in X$ *such that* $x_1 \to x_0$ *and* $f(x_0) = y_0$*;*

(e) *an open map if and only if, for every* $x_0 \in X$ *and* $y_1 \in Y$ *with* $y_1 \to f(x_0)$*, there exists* $x_1 \in X$ *such that* $x_1 \to x_0$ *and* $f(x_1) = y_1$*;*

(f) *a perfect map if and only if, for every* $x_1 \in X$ *and* $y_0 \in Y$ *with* $f(x_1) \to y_0$*, there exists a unique* $x_0 \in X$ *such that* $x_1 \to x_0$ *and* $f(x_0) = y_0$*;*

(g) *a local homeomorphism (or étale map) if and only if, for every* $x_0 \in X$ *and* $y_1 \in Y$ *with* $y_1 \to f(x_0)$*, there exists a unique* $x_1 \in X$ *such that* $x_1 \to x_0$ *and* $f(x_1) = y_1$.

We recall that a continuous map $f\colon X \to Y$ is *perfect* if it is proper and Hausdorff (i.e., if $f(x) = f(x')$ and $x \neq x'$, there exists $U, V \in \mathcal{O}X$ with $x \in U$, $x' \in V$ and $U \cap V = \emptyset$), and that it is a *local homeomorphism*, or an *étale map*, if it is open and, for each $x \in X$, there exists $U \in \mathcal{O}X$ such that $x \in U$ and $f_{\restriction U}\colon U \to f(U)$ is a homeomorphism.

This work led us to investigate the extension of these characterisations to maps between (infinite) topological spaces. The right setting to use convergence turned out to be the ultrafilter convergence. As a side result we also obtained, together with W. Tholen, a useful characterisation of exponentiable maps via convergence (see [**9**]) we will mention in Section 2.

The results corresponding to (a), (d), (e), (f) were either known or easy to obtain; in fact, the characterisation of universal quotient maps using convergence is the basis for the definition of limit lifting maps by Hájek, and the descriptions of open, proper and perfect maps are straightforward (see [**4**]). Statement (b) corresponds, in the infinite case, to the Reiterman–Tholen characterisation of effective descent maps. Indeed, although this is not completely evident in the original formulation [**27**], the notions and techniques introduced in [**4**] clarify completely the analogy between these characterisations. In the latter paper, we also generalized the Janelidze–Sobral–Clementino characterisation of triquotient maps (c) to the infinite case, as we will explain later.

After that, only a characterisation of local homeomorphism, using ultrafilter convergence, remained unknown to us. To explain the problem we first state the characterisation of proper, perfect and open maps.

Proposition 3.

(1) *A continuous map* $f\colon X \to Y$ *is proper (perfect) if and only if, for each ultrafilter* $\mathfrak{x}$ *in* X *with* $f[\mathfrak{x}] \to y$ *in* Y*, there exists a (unique)* x *in* X *such that* $\mathfrak{x} \to x$ *and* $f(x) = y$*:*

$$\begin{array}{ccc} X & \quad & \mathfrak{x} \dashrightarrow x \\ {\scriptstyle f}\downarrow & & \downarrow \qquad \vdots \\ Y & & f[\mathfrak{x}] \longrightarrow y \end{array}$$

(2) *A continuous map $f\colon X \to Y$ is open if and only if, for each $x \in X$ and each ultrafilter $\mathfrak{y}$ in Y with $\mathfrak{y} \to f(x)$ in Y, there exists an ultrafilter $\mathfrak{x}$ in X such that $\mathfrak{x} \to x$ and $f(\mathfrak{x}) = \mathfrak{y}$:*

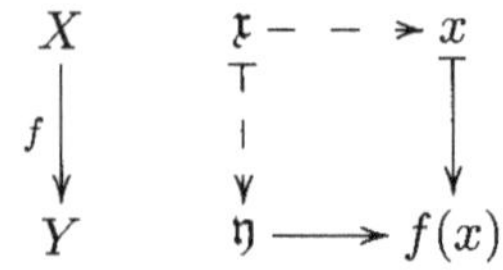

Analysing the characterisations of proper and perfect map, we conjectured that, from the characterisation of open map, one could obtain a characterisation of local homeomorphism imposing the unicity of the lifting of the convergence $\mathfrak{y} \to f(x)$.

Indeed, the parallelism between the two situations becomes evident once we observe that, if we denote by $\delta_f\colon X \to X \times_Y X$, $x \mapsto (x,x)$, the continuous map induced by the pullback property of the (pullback) diagram below

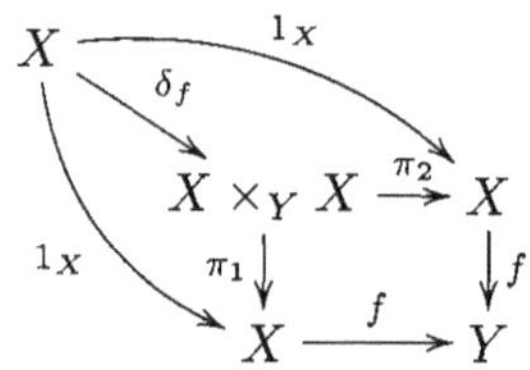

then: a continuous map $f\colon X \to Y$ is perfect if and only if f and δ_f are proper maps; a continuous map $f\colon X \to Y$ is a local homeomorphism if and only if f and δ_f are open maps.

Eventually we have shown, together with G. Janelidze [**8**], that our conjecture was wrong. Calling a continuous map $f\colon X \to Y$ having the unicity of the lifting of $\mathfrak{y} \to f(x)$ described above a *discrete fibration* (using the parallelism with categorical discrete fibrations), one has:

Proposition 4 ([**8**]). *Every local homeomorphism is a discrete fibration (and the converse is false).*

We could prove that the two notions coincide under some conditions on the domain of the map. For that, given a cardinal number λ, we call a topological space X a *λ-space* if the character of X is at most λ and each subset of X with cardinality less than λ is closed.

Theorem 5 ([**8**]). *If X is a λ-space, for some cardinal λ, then, for continuous maps with domain X, local homeomorphisms and discrete fibrations coincide.*

Among λ-spaces one has the indiscrete spaces (= 0-spaces), the Alexandrov spaces (= 1-spaces) and the first countable T_1-spaces (= $\aleph_0$-spaces).

837? **Problem 4.** *Characterise those topological spaces X such that, for a continuous map $f\colon X \to Y$, f is a local homeomorphism if and only if it is a discrete fibration.*

In order to formulate more results and problems in this context we need to consider *iterations* of the ultrafilter convergence. The right way of doing this is making it a functorial process. There are two natural choices in this direction. We may use the ultrafilter functor $U\colon \mathbf{Rel} \to \mathbf{Rel}$, which assigns to each set X its set of ultrafilters UX, and to each relation $r\colon X \nrightarrow Y$ the corresponding relation $Ur\colon UX \nrightarrow UY$ (see for instance [**1**]). For maps $f\colon X \to Y$, $Uf\colon UX \to UY$ is the usual map; for simplicity we write $Uf(\mathfrak{x}) = f[\mathfrak{x}]$. The most valuable functor in this study is the functor $\mathrm{Conv}\colon \mathbf{Top} \to \mathbf{URS}$, where **URS** is the category of ultrarelational spaces and convergence preserving maps (see [**4**]), which assigns to each topological space X the space $\mathrm{Conv}(X)$; here $\mathrm{Conv}(X)$ is the set consisting of pairs $(\mathfrak{x}, x)$, where x is a point and $\mathfrak{x}$ is an ultrafilter converging to x in X, equipped with a convergence structure as follows: first we consider the map $p\colon \mathrm{Conv}(X) \to X$ with $p(\mathfrak{x}, x) = x$; an ultrafilter $\mathfrak{X}$ converges to $(\mathfrak{x}, x)$ in $\mathrm{Conv}(X)$ if $p[\mathfrak{X}] = \mathfrak{x}$. Each continuous map $f\colon X \to Y$ induces a map $\mathrm{Conv}(f)\colon \mathrm{Conv}(X) \to \mathrm{Conv}(Y)$ with $(\mathfrak{x}, x) \mapsto (f[\mathfrak{x}], f(x))$, which preserves the convergence structure (see [**4**] for details).

It is clear that we can consider instead $\mathrm{Conv}\colon \mathbf{URS} \to \mathbf{URS}$. Furthermore, the map $p\colon \mathrm{Conv}(X) \to X$ preserves the structure, so that it defines a natural transformation $p\colon \mathrm{Conv} \to 1_{\mathbf{URS}}$.

This functor Conv is an excellent tool to describe our classes of maps via their lifting of convergence. For that we need to consider (possibly transfinite) iterations of the functor $\mathrm{Conv}\colon \mathbf{URS} \to \mathbf{URS}$, as described in [**4**].

For an ordinal number α we call a continuous map $f\colon X \to Y$ between ultrarelational (or topological) spaces *α-surjective* if, for every $\beta < \alpha$, $\mathrm{Conv}^\beta(f)\colon \mathrm{Conv}^\beta(X) \to \mathrm{Conv}^\beta(Y)$ is surjective; $f\colon X \to Y$ is *Ω-surjective* if $\mathrm{Conv}^\alpha(f)$ is surjective for every ordinal α.

Theorem 6 ([**4**]). *For a continuous map $f\colon X \to Y$ between topological spaces,*

(1) *f is 1-surjective if and only if it is surjective;*
(2) *f is 2-surjective if and only if it is a universal quotient map (if and only if it is a descent map);*
(3) *f is 3-surjective if and only if it is an effective descent map;*
(Ω) *f is Ω-surjective if and only if it is a triquotient map.*

Similarly to Problem 2, we may formulate the following

Problem 5. *Study the properties of the classes of 4-surjective, ..., n-surjective, ω-surjective maps, and possible characterisations of these classes using the topologies (or even the sequential closures).* 838?

Concerning assertion (Ω) above, it is shown in [**4**] that, for a continuous map $f\colon X \to Y$, f is a Ω-surjection if and only if f is a λ_Y-surjection, where λ_Y is the successor of the cardinal of Y. This covers the result already known for a continuous map between finite spaces: f is a triquotient (hence Ω-surjective) if and only if it is ω-surjective (see [**3, 15**]).

839? **Problem 6.** *Characterise those topological spaces Y such that, for a continuous map $f: X \to Y$, f is Ω-surjective if and only if it is ω-surjective.*

Using the functor U: **Rel** $\to$ **Rel** instead of Conv: **URS** $\to$ **URS**, one can also iterate U and formulate the notions of U-α-surjective map, for any continuous map $f: X \to Y$ between topological spaces. (We will keep the name α-surjective for Conv-α-surjective maps.)

It is easy to check that every 3-surjective map is U-3-surjective, i.e., every effective descent map is U-3-surjective.

840? **Question 7.** *Is every U-3-surjective map an effective descent map?*

841? **Question 8.** *If the answer to the previous question is negative, is the class of effective descent maps the least pullback-stable class containing the U-3-surjective maps?*

Furthermore, the functor Conv may be also useful to characterise local homeomorphisms as special discrete fibrations. For instance, one may ask the following:

842? **Question 9.** *Is every continuous map f such that both f and* Conv(f) *are discrete fibrations a local homeomorphism?*

There is another problem in Topological Descent Theory, described in the sequel, that justifies the study of local homeomorphisms, or étale maps, using convergence.

The notion of an effective (global-)descent map can be generalised by considering instead of all morphisms with codomain Y a well-behaved subclass E(Y) of morphisms. One important example is obtained by taking E the class of all étale maps, so that E(Y) is the category of étale bundles over the space Y. A continuous map $f: X \to Y$ is called an *effective étale-descent* map if the pullback functor f^* : E(Y) $\to$ E(X) is monadic. For finite spaces X and Y, the problem of characterising effective étale-descent maps was solved by G. Janelidze and M. Sobral:

Theorem 7 ([**16**]). *The morphism $f: X \to Y$ in* **FinTop** *is an effective étale-descent map if and only if the functor φ: $\mathsf{Z}(\mathrm{Eq}(f)) \to Y$ is an equivalence of categories.*

Here Eq(f) is the equivalence relation on X induced by f, and $\mathsf{Z}(\mathrm{Eq}(f))$ is the category having as objects the points of X; a morphism from x to x' is an equivalence class of *zigzags* in X (see Figure 1).

Now φ: $\mathsf{Z}(\mathrm{Eq}(f)) \to Y$ is an equivalence of categories if and only if

(1) $f: X \to f(X)$ is a quotient map,
(2) $\mathsf{Z}(\mathrm{Eq}(f))$ is a preorder and
(3) $f: X \to Y$ is essentially surjective on objects (i.e., for every $y \in Y$ there exists $x \in X$ such that $f(x) \to y \to f(x)$).

The obvious question is now how to transport this result into the context of all topological spaces.

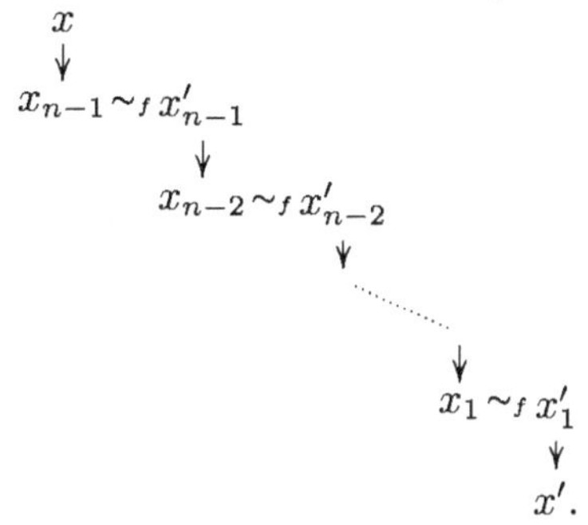

FIGURE 1. An equivalence class of zigzags

Question 10. *Characterise effective étale-descent maps $f\colon X \to Y$ between arbitrary topological spaces.* 843?

A possible solution to the problem above requires most likely translations of *point-convergence* notions and arguments to *(ultra)filter-convergence* ones. Both notions, of local homeomorphism and quotient map, should be considered in this problem via ultrafilter convergence. We have already mentioned the study of local homeomorphisms using convergence developed in [**8**]; possible descriptions of quotients, and their relations to zigzags, are studied in [**13**].

2. Corresponding morphisms in related categories

We study now the *same* problems in categories related to **Top**. Here *same* will be used with two meanings: either we consider characteristic categorical properties of the morphisms, or we deal with topological categories whose objects and morphisms have a description similar to the (ultrafilter) convergence description of **Top**. (We will not focus in this latter subject since it would take us too far from our purpose here. But we refer the reader to [**5**].)

We start by considering some important supercategories (improvements) of **Top**. The interest in these categories has its roots in the fact that many categorically defined constructions either cannot be carried out in **Top** or destroy properties of spaces or maps. In order to perform these constructions, topologists move (temporarily) outside **Top** into larger but better behaved environments such as the category **PsTop** of pseudotopological spaces and continuous (i.e., convergence preserving) maps. Recall that a *pseudotopology* on a set X may be described as a convergence relation $\mathfrak{x} \to x$ between ultrafilters $\mathfrak{x}$ on X and points $x \in X$ such that the principal ultrafilter $\dot{x}$ converges to x. The category **PsTop** contains **Top** as a full and reflective subcategory; in fact, it is the quasitopos hull of **Top** (see [**12**] for details). Being in particular a quasitopos, **PsTop** is locally cartesian closed and therefore the class of effective descent morphisms coincides with the class of quotient maps.

A pseudotopology on a set X is called a *pretopology* if it is closed under intersections in the sense that $\bigcap_{\mathfrak{y}\to x}\mathfrak{y}\subseteq\mathfrak{x}$ implies $\mathfrak{x}\to x$, for each ultrafilter $\mathfrak{x}\in UX$ and each $x\in X$. Hence convergence to a point x is completely determined by the neighbourhood filter $\bigcap_{\mathfrak{y}\to x}\mathfrak{y}$. Together with continuous maps pretopological spaces form the category **PrTop**. In [**12**] is is shown that **PrTop** is the extensional topological hull of **Top**, that is, the smallest extensional topological category containing **Top** *nicely*. However, in contrary to **PsTop**, the category **PrTop** is not cartesian closed. Exponentiable pretopological spaces are characterised in [**22**] as those spaces where each point has a smallest neighbourhood. The map version of this result is established in [**29**]: it states that a continuous map $f\colon X\to Y$ between pretopological spaces is exponentiable if and only if each $x\in X$ has a neighbourhood V such that, for each ultrafilter $\mathfrak{x}$ in X, if $V\in\mathfrak{x}$ and $f[\mathfrak{x}]\to f(x)$ in Y, then $\mathfrak{x}\to x$ in X. Whereas exponentiable objects and morphisms are fully understood in **PrTop**, effective descent maps have not been described yet.

844? **Question 11.** *Characterise effective descent maps $f\colon X\to Y$ between pretopological spaces.*

The study of these classes of maps is also interesting in metric-like structures. Together with metric spaces we also consider premetric spaces. By a *premetric space* we mean a set X together with a map $a\colon X\times X\to[0,\infty]$ such that $a(x,x)=0$ and $a(x,z)\le a(x,y)+a(y,z)$, for any $x,y,z\in X$; that is, a premetric is a, possibly infinite, reflexive and not necessarily non-symmetric distance. We consider now the categories **Met**, of metric spaces and non-expansive maps, and **PMet**, of premetric spaces and non-expansive maps.

Exponentiable and effective descent maps between metric, and more generally premetric, spaces are characterised in [**7**] and [**6**] respectively. We list here the results which might serve, together with the corresponding results for topological spaces, as a guideline for the study of these classes of maps in approach spaces as outlined below.

Theorem 8. *A non-expansive map $f\colon(X,a)\to(Y,b)$ between premetric spaces is exponentiable in* **PMet** *if and only if, for each $x_0,x_2\in X$, $y_1\in Y$ and $u_0,u_1\in\mathbb{R}$ such that $u_0\ge b(f(x_0),y_1)$, $u_1\ge b(y_1,f(x_2))$ and $u_0+u_1=\max\{a(x_0,x_2),b(f(x_0),y_1)+b(y_1,f(x_2))\}<\infty$,*

$$(\forall\varepsilon>0)\,(\exists x_1\in f^{-1}(y_1))\ a(x_0,x_1)<u_0+\varepsilon\ \text{and}\ a(x_1,x_2)<u_1+\varepsilon.$$

Theorem 9. *A non-expansive map $f\colon(X,a)\to(Y,b)$ between metric spaces is exponentiable in* **Met** *if and only if it is exponentiable in* **PMet** *and has bounded fibres.*

Theorem 10. *A morphism $f\colon(X,a)\to(Y,b)$ in* **PMet** *(***Met***) is an effective descent map if and only if*

$$(\forall y_0,y_1,y_2\in Y)\ b(y_2,y_1)+b(y_1,y_0)=\inf\{a(x_2,x_1)+a(x_1,x_0):x_i\in f^{-1}(y_i)\}.$$

Approach spaces were introduced by R. Lowen [**20**] as a natural generalisation of both topological and metric spaces. They can be defined in many different ways; however, the most convenient presentation for our purpose uses ultrafilter convergence (see also [**5**]): an *approach space* (X, a) is a pair consisting of a set X and a *numerified* convergence structure $a\colon UX \times X \to [0, \infty]$ such that $0 \geq a(\dot{x}, x)$ and $Ua(\mathfrak{X}, \mathfrak{x}) + a(\mathfrak{x}, x) \geq a(m_X(\mathfrak{X}), x)$. A map $f\colon X \to Y$ between approach spaces (X, a) and (Y, b) is called *non-expansive* if $a(\mathfrak{x}, x) \geq b(Uf(\mathfrak{x}), f(x))$, for all $\mathfrak{x} \in UX$ and $x \in X$. We denote by **App** the category of approach spaces and non-expansive maps. So far, in **App** little is known about exponentiable objects and morphisms and nothing about effective descent morphisms, though one may conjecture that a combination of the known results in **Top** and **Met** will provide characterisations of these classes of objects and maps in **App**.

Question 12. *Characterise exponentiable objects and maps in* **App**. 845?

Question 13. *Characterise effective descent maps in* **App**. 846?

Exponentiable objects in approach theory are studied in [**14**, **21**], and the following sufficient condition is obtained.

Theorem 11 ([**14**]). *An approach space (X, a) is exponentiable provided that, for each $\mathfrak{X} \in U^2X$, $x \in X$ with $a(m_X(\mathfrak{X}), x) < \infty$, each $\gamma_0, \gamma_1 \in [0, \infty)$ with $\gamma_1 + \gamma_0 = a(m_X(\mathfrak{X}), x)$, and each $\varepsilon > 0$, there exists an ultrafilter $\mathfrak{x}$ such that $Ua(\mathfrak{X}, \mathfrak{x}) \leq \gamma_1 + \varepsilon$ and $a(\mathfrak{x}, x) \leq \gamma_0 + \varepsilon$.*

We conjecture that the condition above is also necessary for (X, a) to be exponentiable. A first step towards a solution to Question 13 is to define the functor Conv in the context of approach theory. This, by the way, would also open the door to carry the notion of triquotient map to **App**.

We turn now our attention to the category **Unif** of uniform spaces and uniformly continuous maps. The question regarding exponentiable maps was settled by S. Niefield [**25**].

In the result below $(X, \mathcal{U})$ and $(Y, \mathcal{V})$ are uniform spaces, with $\mathcal{U}$ and $\mathcal{V}$ the sets of entourages of the uniformities. For $f\colon X \to Y$, $A \subseteq X \times X$ and $y, y' \in Y$, let $A_{y,y'} := A \cap (f^{-1}(y) \times f^{-1}(y'))$.

Theorem 12. *A morphism $f\colon X \to Y$ in* **Unif** *is exponentiable if and only if there exists $U_0 \in \mathcal{U}$ satisfying*

(1) *for all $U \in \mathcal{U}$, $V_U = \{(y, y') \in Y \times Y : U_{0y,y'} = U_{yy'}\} \in \mathcal{V}$,*
(2) *there exists $V_0 \in \mathcal{V}$ such that the projection $\pi_1\colon U_{0yy'} \to f^{-1}(y)$ is a surjection whenever $(y, y') \in V_0$.*

In particular we have that a uniform space is exponentiable if and only its uniformity has a smallest entourage. However, nothing is known about effective descent maps in **Unif**.

Question 14. *Characterise effective descent maps $f\colon X \to Y$ in* **Unif**. 847?

References

[1] M. Barr, *Relational algebras*, Reports of the Midwest Category Seminar, IV, Lecture Notes in Math., vol. 137, Springer, Berlin, 1970, pp. 39–55.

[2] N. Bourbaki, *Topologie générale*, Hermann & Cie, Paris, 1961.

[3] M. M. Clementino, *On finite triquotient maps*, J. Pure Appl. Algebra **168** (2002), no. 2–3, 387–389.

[4] M. M. Clementino and D. Hofmann, *Triquotient maps via ultrafilter convergence*, Proc. Amer. Math. Soc. **130** (2002), no. 11, 3423–3431.

[5] M. M. Clementino and D. Hofmann, *Topological features of lax algebras*, Appl. Categ. Structures **11** (2003), no. 3, 267–286.

[6] M. M. Clementino and D. Hofmann, *Effective descent morphisms in categories of lax algebras*, Appl. Categ. Structures **12** (2004), no. 5–6, 413–425.

[7] M. M. Clementino and D. Hofmann, *Exponentiation in* $\mathbb{V}$*-categories*, Topology Appl. **153** (2006), no. 16, 3113–3128.

[8] M. M. Clementino, D. Hofmann, and G. Janelidze, *Local homeomorphisms via ultrafilter convergence*, Proc. Amer. Math. Soc. **133** (2005), no. 3, 917–922.

[9] M. M. Clementino, D. Hofmann, and W. Tholen, *The convergence approach to exponentiable maps*, Port. Math. (N.S.) **60** (2003), no. 2, 139–160.

[10] B. Day and G. M. Kelly, *On topologically quotient maps preserved by pullbacks or products*, Proc. Cambridge Philos. Soc. **67** (1970), 553–558.

[11] O. Hájek, *Notes on quotient maps*, Comment. Math. Univ. Carolin. **7** (1966), 319–323.

[12] H. Herrlich, E. Lowen-Colebunders, and F. Schwarz, *Improving Top: PrTop and PsTop*, Category theory at work (Bremen, 1990), Heldermann Verlag, Berlin, 1991, pp. 21–34.

[13] D. Hofmann, *An algebraic description of regular epimorphisms in topology*, J. Pure Appl. Algebra **199** (2005), 71–86.

[14] D. Hofmann, *Exponentiation for unitary structures*, Topology Appl. **153** (2006), no. 16, 3180–3202.

[15] G. Janelidze and M. Sobral, *Finite preorders and topological descent. I*, J. Pure Appl. Algebra **175** (2002), 187–205.

[16] G. Janelidze and M. Sobral, *Finite preorders and topological descent. II. Étale descent*, J. Pure Appl. Algebra **174** (2002), no. 3, 303–309.

[17] G. Janelidze, M. Sobral, and W. Tholen, *Beyond Barr exactness: effective descent morphisms*, Categorical foundations, Cambridge University Press, 2004, pp. 359–405.

[18] G. Janelidze and W. Tholen, *How algebraic is the change-of-base functor?*, Category theory (Como, 1990), Lecture Notes in Math., vol. 1488, Springer, Berlin, 1991, pp. 174–186.

[19] G. Janelidze and W. Tholen, *Facets of descent. I*, Appl. Categ. Structures **2** (1994), no. 3, 245–281.

[20] R. Lowen, *Approach spaces*, The Clarendon Press Oxford University Press, New York, 1997.

[21] R. Lowen and M. Sioen, *On the multitude of monoidal closed structures on* **UAP**, Topology Appl. **137** (2004), 215–223.

[22] E. Lowen-Colebunders and G. Sonck, *Exponential objects and Cartesian closedness in the construct Prtop*, Appl. Categ. Structures **1** (1993), no. 4, 345–360.

[23] E. Michael, *Bi-quotient maps and Cartesian products of quotient maps*, Ann. Inst. Fourier (Grenoble) **18** (1968), no. 2, 287–302.

[24] E. Michael, *Complete spaces and tri-quotient maps*, Illinois J. Math. **21** (1977), 716–733.

[25] S. B. Niefield, *Cartesianness: topological spaces, uniform spaces, and affine schemes*, J. Pure Appl. Algebra **23** (1982), no. 2, 147–167.

[26] T. Plewe, *Localic triquotient maps are effective descent maps*, Math. Proc. Cambridge Philos. Soc. **122** (1997), no. 1, 17–43.

[27] J. Reiterman and W. Tholen, *Effective descent maps of topological spaces*, Topology Appl. **57** (1994), no. 1, 53–69.

[28] G. Richter, *Exponentiable maps and triquotients in* **Top**, J. Pure Appl. Algebra **168** (2002), no. 1, 99–105.

[29] G. Richter, *A characterization of exponentiable maps in PrTop*, Appl. Categ. Structures **11** (2003), no. 3, 261–265.

Open Problems in Topology II – Edited by E. Pearl

Dense subgroups of compact groups

W. W. Comfort

1. Introduction

The symbol $\mathbf{G}$ here denotes the class of all infinite groups, and $\mathbf{TG}$ denotes the class of all infinite topological groups which satisfy the T_0 separation property. Each element of $\mathbf{TG}$ is, then, a Tychonoff space (i.e., a completely regular, Hausdorff space) [**50**, 8.4]. We say that $G = (G, \mathcal{T}) \in \mathbf{TG}$ is *totally bounded* (some authors prefer the expression *pre-compact*) if for every $U \in \mathcal{T} \setminus \{\emptyset\}$ there is a finite set $F \subseteq G$ such that $G = FU$. Our point of departure is the following portion of *Weil's Theorem* [**77**]: Every totally bounded $G \in \mathbf{TG}$ embeds as a dense topological subgroup of a compact group $\overline{G}$; this is unique in the sense that for every compact group $\widetilde{G}$ containing G densely there is a homeomorphism-and-isomorphism $\psi\colon \overline{G} \twoheadrightarrow \widetilde{G}$ fixing G pointwise.

As usual, a space is *ω-bounded* if each of its countable subsets has compact closure. For $G = (G, \mathcal{T}) \in \mathbf{TG}$ we write $G \in \mathbf{C}$ [resp., $G \in \boldsymbol{\Omega}$; $G \in \mathbf{CC}$; $G \in \mathbf{P}$; $G \in \mathbf{TB}$] if $(G, \mathcal{T})$ is compact [resp., ω-bounded; countably compact; pseudocompact; totally bounded]. And for $\mathbf{X} \in \{\mathbf{C}, \boldsymbol{\Omega}, \mathbf{CC}, \mathbf{P}, \mathbf{TB}\}$ and $G \in \mathbf{G}$ we write $G \in \mathbf{X}'$ if G admits a group topology $\mathcal{T}$ such that $(G, \mathcal{T}) \in \mathbf{X}$. The class-theoretic inclusions

$$\mathbf{C} \subseteq \boldsymbol{\Omega} \subseteq \mathbf{CC} \subseteq \mathbf{P} \subseteq \mathbf{TB} \subseteq \mathbf{TG} \quad \text{and}$$
$$\mathbf{C}' \subseteq \boldsymbol{\Omega}' \subseteq \mathbf{CC}' \subseteq \mathbf{P}' \subseteq \mathbf{TB}' \subseteq \mathbf{TG}' = \mathbf{G}$$

are easily established. (For $\mathbf{P} \subseteq \mathbf{TB}$, see [**31**]. For $\mathbf{TG}' = \mathbf{G}$, impose on an arbitrary $G \in \mathbf{G}$ the discrete topology.)

We deal here principally with (dense) subgroups of groups $G \in \mathbf{C}$, that is, with $G \in \mathbf{TB}$. Given $G \in \mathbf{G}$ we write $\mathfrak{tb}(G) := \{\mathcal{T} : (G, \mathcal{T}) \in \mathbf{TB}\}$. It is good to remember that $\mathfrak{tb}(G) = \emptyset$ and $|\mathfrak{tb}(G)| = 1$ are possible (for different G); see 2.3(a) and 5.8(a)(2) below.

The symbol $\mathbf{A}$ is used as a prefix to indicate an Abelian hypothesis. Thus, for emphasis and clarity: The expression $G \in \mathbf{AG}$ may be read "G is an infinite Abelian group", and $G \in \mathbf{ACC}'$ may be read "G is an infinite Abelian group which admits a countably compact Hausdorff group topology."

For $G, H \in \mathbf{G}$, we write $G =_{\text{alg}} H$ to indicate that G and H are algebraically isomorphic. For (Tychonoff) spaces X and Y, we write $X =_{\text{top}} Y$ to indicate that the spaces X and Y are homeomorphic. The relation $G =_{\text{alg}} H$ promises nothing whatever about the underlying topologies (if any) on G and H; similarly, the relation $X =_{\text{top}} Y$ is blind to ambient algebraic considerations (if any). We say that $G, H \in \mathbf{TG}$ are topologically isomorphic, and we write $G \cong H$, if some bijection between G and H establishes simultaneously both $G =_{\text{alg}} H$ and $G =_{\text{top}} H$.

We distinguish between *Problems* and *Questions*. As used here, a *Problem* is open-ended in flavor, painted with a broad brush; different worthwhile contributions ("solutions") might lead in different directions. In constrast, a *Question* here is relatively limited in scope, stated in narrow terms; the language suggests that a "Yes" or "No" answer is desired—although, as we know from experience, that response may vary upon passage from one axiom system to another.

With thanks and appreciation I acknowledge helpful comments received on preliminary versions of this paper from: Dikran Dikranjan, Frank Gould, Kenneth Kunen, Gábor Lukács, Jan van Mill, Dieter Remus, and Javier Trigos-Arrieta.

2. Groups with topologies of pre-assigned type

848–852? **Problem 2.1.** *Let* $\mathbf{X} \in \{\mathbf{C}, \mathbf{\Omega}, \mathbf{CC}, \mathbf{P}, \mathbf{TB}\}$. *Characterize algebraically the groups in* $\mathbf{X}'$.

853–857? **Problem 2.2.** *Let* $\mathbf{X} \in \{\mathbf{AC}, \mathbf{A\Omega}, \mathbf{ACC}, \mathbf{AP}, \mathbf{ATB}\}$. *Characterize algebraically the groups in* $\mathbf{X}'$.

Discussion 2.3. (a) That $\mathbf{ATB}' = \mathbf{AG}$ is easily seen (as in Theorem 3.1(a) below, for example, using the fact that $\mathrm{Hom}(G, \mathbb{T})$ separates points of G whenever $G \in \mathbf{AG}$). That the inclusion $\mathbf{TB}' \subsetneq \mathbf{G}$ is proper restates the familiar fact that there are groups $G \in \mathbf{G}$ whose points are not distinguished by homomorphisms into compact (Hausdorff) groups; in our notation, these are G such that $\mathfrak{tb}(G) = \emptyset$. For example: according to von Neumann and Wigner [**59**], [**50**, 22.22(h)], every homomorphism h from the (discrete) special linear group $G := SL(2, \mathbb{C})$ to a compact group satisfies $|h[G]| = 1$.

(b) It is a consequence of the Čech–Pospisil Theorem [**9**] (see also [**43**, Problem 3.12.11], [**51**, 28.58]) that every $G \in \mathbf{C}$ satisfies $|G| = 2^{w(G)}$. Thus in order that $G \in \mathbf{G}$ satisfy $G \in \mathbf{C}'$ it is necessary that $|G|$ have the form $|G| = 2^{\kappa}$.

(c) The algebraic classification of the groups in $\mathbf{AC}'$ is complete. The full story is given in [**50**, §25].

(d) It is well known that every $G \in \mathbf{P}'$ satisfies $|G| \geq \mathfrak{c}$. See [**28**] or [**13**, 6.13] for an explicit proof, and see [**8**], [**42**, 1.3] for earlier, more general results.

(e) The fact that every pseudocompact space satisfies the conclusion of the Baire Category Theorem has two consequences relating to Problems 2.1 and 2.2. (1) If $G \in \mathbf{P}'$, the cardinal number $|G| = \kappa$ cannot be a strong limit cardinal with $\mathrm{cf}(\kappa) = \omega$ [**42**]. (2) every torsion group in $\mathbf{AP}'$ is of bounded order [**29**, 7.4].

(f) No complete characterization of the groups in $\mathbf{P}'$ (nor even in $\mathbf{AP}'$) yet exists, but the case of the torsion groups in $\mathbf{AP}'$ is well understood ([**26**, **40**, **41**]): A torsion group $G \in \mathbf{AG}$ of bounded order is in $\mathbf{P}'$ iff for each of its p-primary constituents $G(p)$ each infinite cardinal number of the form $\kappa := |p^k \cdot G(p)|$ satisfies $\bigoplus_\kappa \mathbb{Z}(p) \in \mathbf{P}'$. (Thus for example, as noted in [**26**], if p is prime and κ is a strong limit cardinal of countable cofinality, then $\bigoplus_{2^\kappa} \mathbb{Z}(p^2) \oplus \bigoplus_\kappa \mathbb{Z}(p) \in \mathbf{AP}'$ while $\bigoplus_{2^\kappa} \mathbb{Z}(p) \oplus \bigoplus_\kappa \mathbb{Z}(p^2) \notin \mathbf{P}'$.)

(g) An infinite closed subgroup H of $G \in \mathbf{X} \in \{\mathbf{C}, \mathbf{\Omega}, \mathbf{CC}, \mathbf{TB}\}$ satisfies $H \in \mathbf{X}$, but the comparable assertion for $\mathbf{X} = \mathbf{P}$ is false [**32**]. Indeed every $H \in \mathbf{TB}$ embeds as a closed topological subgroup of a group $G \in \mathbf{P}$ ([**20**, **72**, **75**]).

(h) Examples are easily found in ZFC showing that the inclusions $\mathbf{AC} \subseteq \mathbf{A\Omega} \subseteq \mathbf{ACC} \subseteq \mathbf{AP} \subseteq \mathbf{ATB}$ are proper. (See also in this connection 3.3(h) below.) As is indicated in [**15**, 3.10], the inclusions $\mathbf{AC}' \subseteq \mathbf{A\Omega}'$ and $\mathbf{ACC}' \subseteq \mathbf{AP}' \subseteq \mathbf{ATB}'$ are proper in ZFC, but the examples cited there from the literature to show $\mathbf{A\Omega}' \neq \mathbf{ACC}'$ rest on either CH [**70**, **71**] or MA [**73**].

Question 2.4. *Is there in* ZFC *a group* $G \in \mathbf{ACC}' \setminus \mathbf{A\Omega}'$? 858?

3. Topologies induced by groups of characters

For $G \in \mathbf{G}$, we use notation as follows.

- $\mathcal{H}(G) := \mathrm{Hom}(G, \mathbb{T})$, the set of homomorphisms from G to the circle group $\mathbb{T}$.
- $\mathcal{S}(G)$ is the set of point-separating subgroups of $\mathcal{H}(G)$.
- When $A \in \mathcal{S}(G)$, $\mathcal{T}_A$ is the smallest topology on G with respect to which each element of A is continuous.
- When $(G, \mathcal{T}) \in \mathbf{TG}$, $\widehat{(G, \mathcal{T})}$ is the set of $\mathcal{T}$-continuous funtions in $\mathcal{H}(G)$.

These symbols are well-defined for arbitrary groups G, but (in view of the privileged status of the group $\mathbb{T}$) typically they are useful only when G is Abelian. The fact that the groups $A \in \mathcal{S}(G)$ are required to separate points ensures that the topology $\mathcal{T}_A$ satisfies the T_0 separation property required throughout this article; indeed, the evaluation map $e_A \colon G \to \mathbb{T}^A$ (given by $e_A(x)_h = h(x)$ for $x \in G$, $h \in A$) is an injective homomorphism, and $\mathcal{T}_A$ is the topology inherited by G (identified in this context with $e_A[G]$) from $\mathbb{T}^A$. When $\mathcal{H}(G)$ is given the (compact) topology inherited from $\mathbb{T}^G$, a subgroup $A \subseteq \mathcal{H}(G)$ satisfies $A \in \mathcal{S}(G)$ iff A is dense in $\mathcal{H}(G)$ (cf. [**30**, 1.9]).

The point of departure for our next problem is this theorem.

Theorem 3.1 ([**30**]). *Let* $G \in \mathbf{AG}$. *Then*

(a) $A \in \mathcal{S}(G) \Rightarrow (G, \mathcal{T}_A) \in \mathbf{ATB}$;
(b) $(G, \mathcal{T}) \in \mathbf{ATB} \Rightarrow \exists\, A \in \mathcal{S}(G)$ *such that* $\mathcal{T} = \mathcal{T}_A$;
(c) $A \in \mathcal{S}(G) \Rightarrow \widehat{(G, \mathcal{T}_A)} = A$; *and*
(d) $A \in \mathcal{S}(G) \Rightarrow w(G, \mathcal{T}_A) = |A|$.

It follows from Theorem 3.1(c) that the map $A \mapsto \mathcal{T}_A$ from $\mathcal{S}(G)$ to $\mathfrak{tb}(G)$ is an order-preserving bijection between posets, so $\mathfrak{tb}(G)$ is large. That theme is noted and developed at length in [**4**, **5**, **25**, **27**, **30**, **34**, **63**], where the following results (among many others) are given for such G: (a) From $|\mathcal{H}(G)| = 2^{|G|}$ ([**53**], [**44**, 47.5], [**50**, 24.47]) and $|\mathcal{S}(G)| = 2^{2^{|G|}}$ ([**54**], [**5**, 4.3]) follows $|\mathfrak{tb}(G)| = 2^{2^{|G|}}$; (b) from any set of $2^{2^{|G|}}$-many elements $\mathcal{T}_A \in \mathfrak{tb}(G)$, some $2^{2^{|G|}}$-many of the spaces $(G, \mathcal{T}_A)$ are pairwise nonhomeomorphic; (c) each of the two posets $(\mathfrak{tb}(G), \subseteq)$ and $(\mathcal{P}(\mathcal{P}(|G|)), \subseteq)$ embeds into the other, so any question relating to the existence of a chain or anti-chain or well-ordered set in $\mathfrak{tb}(G)$ of prescribed cardinality is independent of the algebraic structure of G and is equivalent to the corresponding strictly set-theoretic question in the poset $(\mathcal{P}(\mathcal{P}(|G|)), \subseteq)$.

859–863? **Problem 3.2.** *Fix $G \in \mathbf{AG}$ and fix $\mathbf{X} \in \{\mathbf{AC}, \mathbf{A\Omega}, \mathbf{ACC}, \mathbf{AP}, \mathbf{ATB}\}$. For which $A \in \mathcal{S}(G)$ is $(G, \mathcal{T}_A) \in \mathbf{X}$?*

Discussion 3.3. (a) From 3.1(a), the answer for $\mathbf{X} = \mathbf{ATB}$ is "all such A".

(b) As mentioned earlier, every group in $\mathbf{C}'$ (whether or not Abelian) satisfies $|G| = 2^{w(G)}$. Thus for many $G \in \mathbf{AG}$ (for example, those with cardinality not of the form 2^κ) the answer for $\mathbf{X} = \mathbf{AC}$ is "no such A".

(c) In parallel with (b), 2.3(e) above indicates that for many $G \in \mathbf{AG}$ there is no $A \in \mathcal{S}(G)$ such that $(G, \mathcal{T}_A) \in \mathbf{P}$.

(d) Peripherally or directly, several authors address these questions: Given $G \in \mathbf{AG}$, (1) find $A \in \mathcal{S}(G)$ such that $(G, \mathcal{T}_A) \in \mathbf{AP}$; or, (2) for which $h \in \mathcal{H}(G)$ is there $A \in \mathcal{S}(G)$ such that $h \in A$ and $(G, \mathcal{T}_A) \in \mathbf{AP}$? With no pretense toward completeness I mention in this connection [**20**, 4.1], [**29**, 5.11, 6.5], [**45**], [**49**, §3, §4], [**17**, 3.6, 3.10], [**57**].

(e) For each of the five classes $\mathbf{X}$ considered in Problem 3.2, the continuous homomorphic image of each $G \in \mathbf{X}$ is itself in $\mathbf{X}$. Thus when $B \subseteq A \in \mathcal{S}(G)$ and $B \in \mathcal{S}(G)$, from $(G, \mathcal{T}_A) \in \mathbf{X}$ follows $(G, \mathcal{T}_B) \in \mathbf{X}$. Since a compact (Hausdorff) topology is minimal among Hausdorff topologies, it is immediate that if $(G, \mathcal{T}_A) \in \mathbf{AC}$ and $A \supseteq B \in \mathcal{S}(G)$, then $A = B$.

(f) That remark leads naturally to less trivial considerations. We say as usual that a group $(G, \mathcal{T}) \in \mathbf{TG}$ is *minimal*, and we write $(G, \mathcal{T}) \in \mathbf{M}$, if no ($T_0$) topological group topology on G is strictly coarser than $\mathcal{T}$. The difficult question, whether (in our notation) $\mathbf{AM} \subseteq \mathbf{ATB}$ holds, occupied The Bulgarian School for nearly 15 years, finding finally its positive solution in 1984 [**61**]. (An earlier example had shown that the relation $\mathbf{M} \subseteq \mathbf{TB}$ is false.) The relevance of the class $\mathbf{M}$ to Problem 3.2 is the obvious fact that $(G, \mathcal{T}_A) \in \mathbf{AM}$ iff A is minimal in $\mathcal{S}(G)$. We noted in (e) that $\mathbf{C} \subseteq \mathbf{M}$, so $\mathbf{AC}' \subseteq \mathbf{AM}'$, but many $G \in \mathbf{AG} = \mathbf{ATB}'$ are not in $\mathbf{AM}'$: the groups $\mathbb{Q}^n$ $(n < \omega)$, $\mathbb{Z}(p^\infty)$ are examples. For a careful study, with proofs and historical commentary and a comprehensive bibliography, of the groups which are/are not in the classes $\mathbf{M}$, $\mathbf{AM}$, $\mathbf{M}'$, $\mathbf{AM}'$, including a proof of the theorem $\mathbf{AM} \subseteq \mathbf{ATB}$, the reader should consult [**38**]. See also [**36**] for background on the principal remaining outstanding problem in this area: Which reduced, torsion-free $G \in \mathbf{AG}$ are in $\mathbf{AM}'$?

(g) Let D be a dense subgroup of $E \in \mathbf{ATG}$. Since each $h \in \widehat{D}$ is uniformly continuous and $\mathbb{T}$ is complete, each such h extends (uniquely) to $\overline{h} \in \widehat{E}$; we have, then, $\widehat{D} =_{\text{alg}} \widehat{E}$. Now let $\{G_i : i \in I\} \subseteq \mathbf{ATB}$ and set $G := \prod_{i \in I} G_i$. From the uniqueness aspect of Weil's theorem we have $\overline{G} = \prod_{i \in I} \overline{G_i}$; further, the relation $\widehat{\overline{G}} = \bigoplus_{i \in I} \widehat{\overline{G_i}}$ is well known. (Here $\bigoplus_{i \in I} \widehat{\overline{G_i}}$ is given the discrete topology. See [**50**, 23.21] for a comprehensive treatment.) It follows from the Tychonoff Product Theorem that if each $G_i \in \mathbf{X} \in \{\mathbf{C}, \mathbf{\Omega}, \mathbf{TB}\}$ then also $G \in \mathbf{X}$; it is shown in [**30**] that, similarly, if each $G_i \in \mathbf{P}$ then also $G \in \mathbf{P}$.

The relevance of the foregoing paragraph to Problem 3.2 is this: If $\mathbf{X} \in \{\mathbf{C}, \mathbf{\Omega}, \mathbf{P}, \mathbf{TB}\}$ and $\{G_i : i \in I\} \subseteq \mathbf{AG}$, and if $A_i \in \mathcal{S}(G_i)$ makes $(G_i, \mathcal{T}_{A_i}) \in \mathbf{X}$, then $A := \bigoplus_{i \in I} A_i$ makes also $(G, \mathcal{T}_A) \in \mathbf{X}$. This indicates a global coherence or stability in passing among the groups $G \in \mathbf{AG}$ while seeking $A \in \mathcal{S}(G)$ for which $(G, \mathcal{T}_A) \in \mathbf{X}$. [Remark. In contrast to the classes $\mathbf{C}$, $\mathbf{\Omega}$, $\mathbf{P}$ and $\mathbf{TB}$, the issue of "the closure under the formation of products" in the class $\mathbf{CC}$ is much more complex and subtle. See [**39**] for an extended discussion and many relevant problems; see also Question 5.1 below.]

(h) The algebraic structure of a group $A \in \mathcal{S}(G)$ is not of itself sufficient to determine whether or not $(G, \mathcal{T}_A) \in \mathbf{X}$. For examples to that effect, begin with $K := \{0,1\}^{\mathfrak{c}}$ in its usual compact topology and define $G := \bigoplus_{\mathfrak{c}} \{0,1\}$. (0) Let G_0 be the Σ-product in K; (1) Let D be a countable, dense subgroup of K and let $G_1 \in \mathbf{CC}$ satisfy $|G_1| \leq \mathfrak{c}$ and $D \subseteq G_1 \subseteq K$. [Such a group G_1 may be defined by choosing for each countable set $C \subseteq D$ an accumulation point p_C of C in K, forming the subgroup of K generated by D and all points p_C, and iterating the process over the countable ordinals; see [**32**] for details, and see [**65**] for the modified argument furnishing such G_1 which is p-compact (in the sense of 5(**A**) below) for pre-assigned $p \in \omega^*$.] (2) Let G_2 be a proper, dense pseudocompact subgroup of G_0. [Such a group is given in [**29**, 7.3], [**21**].] (3) Write $K = K_0 \times K_1$ with $K_i \cong K$, let H and E be the natural copies of G_0 and D in K_0 and K_1, respectively, and set $G_3 := H \times E \subseteq K$. Then each G_i is dense in K, with $|G_i| \leq \mathfrak{c}$ by construction. For $0 \leq i \leq 2$ we have $|G_i| \geq \mathfrak{c}$ by 2.3(d), and also $|G_3| = |G_0| \cdot |D| \geq \mathfrak{c}$. Thus each G_i is a Boolean group with $|G_i| = \mathfrak{c}$, so $G_i =_{\text{alg}} G$ for $0 \leq i \leq 3$ [**50**, A.25]. Let $\mathcal{T}_i$ be the topology on G_i inherited from K, and (using Theorem 3.1(b) above) let $A_i \in \mathcal{S}(G)$ satisfy $(G, \mathcal{T}_{A_i}) \cong (G_i, \mathcal{T}_i)$. Again, since A_i is a Boolean group with $|A_i| = w(G, \mathcal{T}_{A_i}) = \mathfrak{c}$, we have $A_i =_{\text{alg}} \oplus_{\mathfrak{c}} \{0,1\}$ for $0 \leq i \leq 3$. Then we have

$(G, \mathcal{T}_{A_0}) \cong G_0 \in \mathbf{\Omega} \setminus \mathbf{C}$. (Proof. $|G_0| = \mathfrak{c} < 2^{\mathfrak{c}} = |K|$.)

$(G, \mathcal{T}_{A_1}) \cong G_1 \in \mathbf{CC} \setminus \mathbf{\Omega}$. (Proof. D is dense in G_1 with $|D| = \omega$, so $G_1 \in \mathbf{\Omega}$ gives the contradiction $G_1 = K$.)

$(G, \mathcal{T}_{A_2}) \cong G_2 \in \mathbf{P} \setminus \mathbf{CC}$. (Proof. The general result given in [**12**], [**29**, 3.3] shows that no proper, G_δ-dense subset of G_0 is countably compact.)

$(G, \mathcal{T}_{A_3}) \cong G_3 \in \mathbf{TB} \setminus \mathbf{P}$. (Proof. If $G_3 \in \mathbf{P}$ then also its continuous image $E = \pi_1[G_3] \in \mathbf{P}$, which with $|E| = \omega < \mathfrak{c}$ contradicts 2.3(d).)

Without specific reference to any of the classes $\mathbf{X}$ here considered, two questions now arise naturally. (See also Problem 5.6 below for a related query.)

Problem 3.4. *Fix $G \in \mathbf{AG}$. For which $A, B \in \mathcal{S}(G)$ does the relation $(G, \mathcal{T}_A) \cong$* 864?
$(G, \mathcal{T}_B)$ hold?

Problem 3.5. *Fix $G \in \mathbf{AG}$. For which $A, B \in \mathcal{S}(G)$ does the relation $(G, \mathcal{T}_A) =_{\text{top}}$* 865?
$(G, \mathcal{T}_B)$ hold?

Discussion 3.6. (a) A simple example, easily generalized, will suffice to indicate a distinction between Problems 3.4 and 3.5. Choose $A, B \in \mathcal{S}(\mathbb{Z})$ such that $|A| = |B| = \omega$ and $A \neq_{\text{alg}} B$. The spaces $(\mathbb{Z}, \mathcal{T}_A)$, $(\mathbb{Z}, \mathcal{T}_B)$ are then countable and metrizable without isolated points, hence according to a familiar theorem of Sierpiński [**66**] (see also [**43**, Exercise 6.2.A(d)]) are homeomorphic. So here $(\mathbb{Z}, \mathcal{T}_A) =_{\text{top}} (\mathbb{Z}, \mathcal{T}_B)$ and $(\mathbb{Z}, \mathcal{T}_A) =_{\text{alg}} (\mathbb{Z}, \mathcal{T}_B)$, but $(\mathbb{Z}, \mathcal{T}_A) \cong (\mathbb{Z}, \mathcal{T}_B)$ fails since from Theorem 3.1(c) we have $\widehat{(\mathbb{Z}, \mathcal{T}_A)} = A \neq_{\text{alg}} B = \widehat{(\mathbb{Z}, \mathcal{T}_B)}$.

(b) It is evident from Theorem 3.1(d) that if $(G, \mathcal{T}_A) =_{\text{top}} (G, \mathcal{T}_B)$ then $|A| = |B|$. The converse fails, however, even in the case $G = \mathbb{Z}$: It is shown in [**62**] that there is a set $\{A_\eta : \eta < \mathfrak{c}\} \subseteq \mathcal{S}(\mathbb{Z})$, with each $A_\eta =_{\text{alg}} \bigoplus_{\mathfrak{c}} \mathbb{Z}$, such that $(\mathbb{Z}, \mathcal{T}_{A_\eta}) \neq_{\text{top}} (\mathbb{Z}, \mathcal{T}_{A_{\eta'}})$ for $\eta < \eta' < \mathfrak{c}$; one may arrange further that all, or none, of the spaces $(\mathbb{Z}, \mathcal{T}_\eta)$ contain a nontrivial convergent sequence. Similar results in a more general context are given in [**24**].

(c) In groups of the form $\mathcal{H}(G) \in \mathbf{AC}$ with $G \in \mathbf{AG}$, Dikranjan [**37**] has introduced a strong density concept, called $\mathfrak{g}$-density, which is enjoyed by certain $A \in \mathcal{S}(G)$. We omit the formal definition here, but we note that $A \in \mathcal{S}(G)$ is $\mathfrak{g}$-dense in $\mathcal{H}(G)$ iff no nontrivial sequence converges in $(G, \mathcal{T}_A)$ [**37**, 4.22]. Thus for certain pairs $A, B \in \mathcal{S}(G)$, $\mathfrak{g}$-density successfully responds to Problem 3.5: If $(G, \mathcal{T}_A) =_{\text{top}} (G, \mathcal{T}_B)$, then A is $\mathfrak{g}$-dense in $\mathcal{H}(G)$ iff B is $\mathfrak{g}$-dense in $\mathcal{H}(G)$.

For additional theorems on $\mathfrak{g}$-dense and $\mathfrak{g}$-closed subgroups of the groups $\mathcal{H}(G) \in \mathbf{AC}$, see [**3**, **56**].

4. Extremal phenomena

It is obvious that a group $(G, \mathcal{T}) \in \mathbf{C}$ admits neither a proper, dense subgroup in $\mathbf{C}$ nor a strictly larger group topology $\mathcal{U}$ such that $(G, \mathcal{U}) \in \mathbf{C}$. A brief additional argument ([**33**, 3.1], [**29**, 3.2]) shows that if $(G, \mathcal{T}) \in \mathbf{P}$ is metrizable (equivalently: satisfies $w(G, \mathcal{T}) = \omega$), then (1) $(G, \mathcal{T}) \in \mathbf{C}$ and (2) $(G, \mathcal{T})$ admits neither a proper, dense subgroup in $\mathbf{P}$ nor a strictly larger group topology $\mathcal{U}$ such that $(G, \mathcal{U}) \in \mathbf{P}$. Indeed, for $(G, \mathcal{T}) \in \mathbf{AP}$ these conditions are equivalent [**21**]: (a) $(G, \mathcal{T})$ admits a proper, dense subgroup in $\mathbf{P}$; (b) there is a topology $\mathcal{U}$ on G, strictly refining $\mathcal{T}$, such that $(G, \mathcal{U}) \in \mathbf{P}$; (c) $w(G, \mathcal{T}) > \omega$. Some questions then arise naturally.

866? **Question 4.1.** *Do the three conditions just listed from [**21**] remain equivalent for $(G, \mathcal{T}) \in \mathbf{P}$ when G is not assumed to be Abelian?*

When $G \in \mathbf{AG}$ is given the topology $\mathcal{T}_A$ with $A = \mathcal{H}(G)$, every subgroup of G is closed; so, $(G, \mathcal{T}_A) \in \mathbf{ATG}$ has no proper dense subgroup. As to the existence of strict refinements, it is clear for every $G \in \mathbf{TB}'$ that $\mathfrak{tb}(G)$ has a largest (= biggest) topology. (When $G \in \mathbf{AG} \subseteq \mathbf{TB}'$ this is $\mathcal{T} = \mathcal{T}_A$ with $A = \mathcal{H}(G)$.) The class $\mathbf{X} = \mathbf{TB}$, then, is included in this next Problem largely to preserve the symmetry developed in this essay; in this immediate context it deserves minimal attention. As to Problem 4.3, we leave it to the reader to determine which pairs $\mathbf{X}, \mathbf{Y}$ give rise to questions of interest and which to questions which are silly or without content.

867–868? **Problem 4.2.** *Fix $\mathbf{X} \in \{\mathbf{\Omega}, \mathbf{CC}, \mathbf{TB}\}$, and let $(G, \mathcal{T}) \in \mathbf{X}$. Find pleasing necessary and/or sufficient conditions that* (a) *$(G, \mathcal{T})$ has a proper, dense subgroup in $\mathbf{X}$; and/or* (b) *there is a topology $\mathcal{U}$ on G, strictly refining $\mathcal{T}$, such that $(G, \mathcal{U}) \in \mathbf{X}$.*

Or even, more generally:

Problem 4.3 ([**15**]). *Fix* $\mathbf{X}, \mathbf{Y} \in \{\mathbf{C}, \mathbf{\Omega}, \mathbf{CC}, \mathbf{P}, \mathbf{TB}\}$, *and let* $(G, \mathcal{T}) \in \mathbf{X}$. *Find* 869–870?
pleasing necessary and/or sufficient conditions that (a) $(G, \mathcal{T})$ *has a proper, dense subgroup in* $\mathbf{Y}$*; and/or* (b) *there is a topology* $\mathcal{U}$ *on* G*, strictly refining* $\mathcal{T}$*, such that* $(G, \mathcal{U}) \in \mathbf{Y}$.

Remark 4.4. We noted already in 3.3(h) that according to [**12**], [**29**, 3.3] the Σ-product in a group of the form G^κ with $G \in \mathbf{C}$ and $\kappa \geq \omega$ admits no proper dense subgroup in **CC**.

5. Miscellaneous questions

(A). Recall first a definition of A.R. Bernstein [**6**]: For an ultrafilter $p \in \omega^* := \beta(\omega) \setminus \omega$, a Tychonoff space X is *p-compact* if for every (continuous) $f : \omega \to X$ the Stone–Čech extension $\overline{f} : \beta(\omega) \to \beta(X)$ satisfies $\overline{f}(p) \in X$. (See [**13, 14, 65**] for references to related tools introduced by Frolík, by Katětov, and by V. Saks.) It is known [**11, 47, 65**] that for a set $\{X_i : i \in I\}$ of Tychonoff spaces, every product of the form $\prod_{i \in I} (X_i)^{\kappa_i}$ is countably compact iff there is $p \in \omega^*$ such that each X_i is p-compact. It is then immediate, as in [**13**, 8.9], that the class **CC** is closed under the formation of (arbitrary) products if and only if there is $p \in \omega^*$ such that each $G \in \mathbf{CC}$ is p-compact. Thus we are led to a bizarre question.

Question 5.1 ([**14**, 1.A.1]). *Is it consistent with the axioms of* ZFC *that there is* 871?
$p \in \omega^*$ *such that every countably compact group is p-compact?*

Discussion 5.2. The chapter [**39**] makes reference to work of E. van Douwen, J. van Mill, K.P. Hart, A. Tomita and others giving the existence of models of ZFC in which some product of finitely many elements from the class **ACC** fails to be in **ACC**. See also [**46**] for a similar conclusion based on the existence of a selective ultrafilter $p \in \omega^*$. It is evident from **(A)** above that in any of these models of ZFC no ultrafilter p as in Question 5.1 can exist.

(B). It was noted in 3.3(g) that when D is a dense subgroup of $G \in \mathbf{ATG}$, the map $\phi : \widehat{G} \twoheadrightarrow \widehat{D}$ given by $\phi(h) = h|D$ establishes the equality $\widehat{G} =_{\text{alg}} \widehat{D}$; clearly ϕ is continuous when $\widehat{D}$ and $\widehat{G}$ are given their respective compact-open topologies. Following [**22, 23**], we say that $G \in \mathbf{ATG}$ is *determined* if for each dense subgroup D of G the map $\phi : \widehat{G} \twoheadrightarrow \widehat{D}$ is a homeomorphism. The principal theorem in this area, given in [**10**] and [**1**] independently, is this: Every metrizable $G \in \mathbf{ATG}$ is determined. (The exact generalization of that theorem to the (possibly) non-Abelian context is given in [**55**]: For every dense subgroup D of metrizable $G \in \mathbf{TG}$ and for every compact Lie group K, one has $\mathrm{Hom}(D, K) \cong \mathrm{Hom}(G, K)$ when those groups are given the compact-open topology.) The authors of [**22, 23**] have noted the existence of many nonmetrizable determined $G \in \mathbf{ATG}$ (some of them compact), and they raised this question (see also [**18**, §6]).

Question 5.3. *Is the product of finitely many determined groups in* **ATG** *nec-* 872–873?
essarily determined? If $G \in \mathbf{ATG}$ *is determined, must* $G \times G$ *be determined?*

Discussion 5.4. Concerning Question 5.3, it is known [**74**] that the product of two determined groups in **ATG**, of which one is discrete, is again determined.

Now let non($\mathcal{N}$) be the least cardinal κ such that some set $X \subseteq \mathbb{T}$ with $|X| = \kappa$ has positive outer (Haar) measure. It is known [**22, 23**] that no $G \in \mathbf{AC}$ with $w(G) \geq \text{non}(\mathcal{N})$ is determined, so if $\text{non}(\mathcal{N}) = \aleph_1$ (in particular, if CH holds) then $G \in \mathbf{AC}$ is determined iff $w(G) = \omega$ (i.e., iff G is metrizable). The authors of [**22, 23**] asked Question 5.5, a sharpened version of [**18**, 6.1].

874–876? **Question 5.5.** *Is there in* ZFC *a cardinal* κ *such that* $G \in \mathbf{AC}$ *is determined iff* $w(G) < \kappa$*? Is* $\kappa = \text{non}(\mathcal{N})$*? Is* $\kappa = \aleph_1$*?*

(C). The remark in 3.6(a) shows for $G, H \in \mathbf{ATG}$ that the conditions $G =_{\text{alg}} H$, $G =_{\text{top}} H$ do not together ensure that G and H are topologically isomorphic, even when G and H carry the topologies induced by $\widehat{G}$ and $\widehat{H}$, respectively. The following problem then arises naturally.

877–878? **Problem 5.6.** (a) *Find interesting necessary and/or sufficient conditions on* $G, H \in \mathbf{TG}$ *to ensure that if* $G =_{\text{alg}} H$ *and* $G =_{\text{top}} H$ *then necessarily* $G \cong H$. (b) *Find interesting necessary and/or sufficient conditions on* $G, H \in \mathbf{ATG}$ *to ensure that if* $G =_{\text{alg}} H$ *and* $G =_{\text{top}} H$ *then necessarily* $G \cong H$.

Problem 5.6 relates to pairs from **TG**. A similar problem focuses on a fixed $G \in \mathbf{TG}$, as follows.

879–880? **Problem 5.7.** (a) *For which* $G \in \mathbf{TG}$ *do the conditions* $H \in \mathbf{TG}$, $G =_{\text{alg}} H$ *and* $G =_{\text{top}} H$ *guarantee that* $G \cong H$*?* (b) *For which* $G \in \mathbf{ATG}$ *do the conditions* $H \in \mathbf{ATG}$, $G =_{\text{alg}} H$ *and* $G =_{\text{top}} H$ *guarantee that* $G \cong H$*?*

Discussion 5.8. (a) There are many theorems in the literature showing that certain $G \in \mathbf{G}$ admit a topology with certain pre-assigned properties, further that any two such topologies $\mathcal{T}_0$, $\mathcal{T}_1$ satisfy $(G, \mathcal{T}_0) =_{\text{top}} (G, \mathcal{T}_1)$, or $(G, \mathcal{T}_0) \cong (G, \mathcal{T}_1)$, or even $\mathcal{T}_0 = \mathcal{T}_1$. [Remark. In this last case, every automorphism of G is a $\mathcal{T}_0$-homeomorphism.] We cite six results of this and similar flavor; clearly, these relate closely to the issues raised in Problems 5.6 and 5.7. (1) Van der Waerden [**76**] gave examples of groups $(G, \mathcal{T}) \in \mathbf{C}$ such that $\mathfrak{tb}(G) = \{\mathcal{T}\}$. (2) Groups as in (1) are necessarily metrizable [**48, 58**], but there exist $G \in \mathbf{G}$ of arbitrary cardinality $\geq \mathfrak{c}$ with $|\mathfrak{tb}(G)| = 1$: [**27**, 3.17] shows that for every family $\{G_i : i \in I\} \subseteq \mathbf{C}$ of algebraically simple, non-Abelian Lie groups, the only topology $\mathcal{T}$ on the group $H := \bigoplus_{i \in I} G_i$ for which $(H, \mathcal{T}) \in \mathbf{TB}$ is the topology inherited from the (usual compact) topology on $\prod_{i \in I} G_i$. See also [**64**] for additional relevant references. (3) [Hulanicki, Orsatti] On an *Orsatti group*—i.e., a group G of the form $G =_{\text{alg}} \prod_{p \in \mathbb{P}} (\mathbb{Z}_p^{k_p} \times F_p)$ with $\mathbb{Z}_p$ the p-adic integers and with finite $F_p \in \mathbf{AG}$ such that $p \cdot F_p = \{0\}$—the obvious natural topology $\mathcal{T}$ making $(G, \mathcal{T}) \in \mathbf{C}$ is the only topology making $G \in \mathbf{C}$; conversely, every $G \in \mathbf{AG}$ with a unique topology $\mathcal{T}$ making $(G, \mathcal{T}) \in \mathbf{C}$ is an Orsatti group. (4) [Stewart] If $G \in \mathbf{G}$ admits a connected topology $\mathcal{T}$ such that $(G, \mathcal{T}) \in \mathbf{C}$ and the center of G is totally disconnected, then $\mathcal{T}$ is the only topology making $(G, \mathcal{T}) \in \mathbf{C}$. (5) [Scheinberg] If $G, H \in \mathbf{ATG}$ are connected and (locally) compact and $G =_{\rm top} H$, then $G \cong H$. (6) Every totally disconnected $G \in \mathbf{C}$ with $w(G) = \kappa$ satisfies $G =_{\rm top} \{0,1\}^\kappa$; if further there is $p \in \mathbb{P}$ such that $p \cdot G = \{0\}$, then $G \cong (\mathbb{Z}(p))^\kappa$.

(b) We note *en passant* that in a model of ZFC with distinct cardinals κ_i ($i = 0, 1$) such that $2^{\kappa_0} = 2^{\kappa_1}$, the group $G = \bigoplus_{2^{\kappa_i}} \{0,1\} =_{\rm alg} \{0,1\}^{\kappa_i}$ admits totally disconnected (compact) topologies $\mathcal{T}_i$ such that $w(G, \mathcal{T}_i) = \kappa_i$, hence $(G, \mathcal{T}_0) \neq_{\rm top} (G, \mathcal{T}_1)$.

(c) For a detailed discussion of the results cited in (a)(3)–(a)(6), with references to works of Stewart, Hulanicki, Scheinberg, Orsatti and others, see [**19**, **35**, **38**, **50**, **52**, **64**].

(d) Suppose for some $G \in \mathbf{G}$ and for one of the classes $\mathbf{X} \in \{\mathbf{C}, \mathbf{\Omega}, \mathbf{CC}, \mathbf{P}\}$ that (1) $G \in \mathbf{X}'$ and (2) every two topologies $\mathcal{T}_0, \mathcal{T}_1$ making $(G, \mathcal{T}_i) \in \mathbf{X}$ satisfy $(G, \mathcal{T}_0) \cong (G, \mathcal{T}_1)$. Then $G = (G, \mathcal{T}_0)$ is an example of the sort sought in Problem 5.7. [Proof. From $(G, \mathcal{T}_0) \in \mathbf{X}$ and $G =_{\rm top} H \in \mathbf{TG}$ follows $H \in \mathbf{X}$, and any (hypothesized) isomorphism $\phi \colon H \twoheadrightarrow G$ induces on G a topology $\mathcal{T}_1$ such that ϕ is a homeomorphism and $(G, \mathcal{T}_1) \in \mathbf{X}$. Then $(G, \mathcal{T}_0) \cong (G, \mathcal{T}_1) \cong H$.]

(D). A subgroup G of $K \in \mathbf{TG}$ is said to be *essentially dense* [resp., *totally dense*] in K if $|G \cap N| > 1$ [resp., $G \cap N$ is dense in N] for every closed, normal, nontrivial subgroup N of G. Given $K \in \mathbf{TG}$, the *essential density* $ed(K)$ [resp., the *total density* $td(K)$] of K is the cardinal number

$$ed(K) := \min\{|G| : G \text{ is an essentially dense subgroup of } K\},$$
$$td(K) := \min\{|G| : G \text{ is a totally dense subgroup of } K\}.$$

In contrast with the properties studied heretofore in this article—compactness, pseudocompactness, and so forth—the properties of essential and total density are not intrinsic to a group $G \in \mathbf{TG}$: They must be investigated relative to an enveloping group $K \in \mathbf{TG}$. It is known [**2**, **60**, **67**] for G dense in $K \in \mathbf{TG}$ that $G \in \mathbf{M}$ iff $K \in \mathbf{M}$ and G is essentially dense in K. Hence, since $\mathbf{C} \subseteq \mathbf{M}$, for $G \in \mathbf{AG}$ these properties are equivalent: (1) $(G, \mathcal{T}_A) \in \mathbf{AM}$; (2) $(G, \mathcal{T}_A)$ is essentially dense in $\overline{(G, \mathcal{T}_A)}$; (3) A is minimal in $\mathcal{S}(G)$. We are drawn to the companion problem for total density.

Problem 5.9. *For which $A \in \mathcal{S}(G)$ is $(G, \mathcal{T}_A)$ totally dense in $\overline{(G, \mathcal{T}_A)}$?* 881?

It is shown in [**16**] that there are $G \in \mathbf{ATB}$ such that $ed(G) < td(G)$, and that consistently such $G \in \mathbf{AP}$ exist. The authors of [**16**] leave several related questions unanswered, however, including these.

Question 5.10 ([**16**]). (a) *Is there, in* ZFC *or in augmented axiom systems, a group $G \in \mathbf{ACC}$ such that $ed(G) < td(G)$?* (b) *Is there in* ZFC *a group $G \in \mathbf{AP}$ such that $ed(G) < td(G)$?* 882–883?

Discussion 5.11. Every $G \in \mathbf{AC}$ satisfies $ed(G) = td(G)$ [**69**]. The paper [**7**] provides much additional useful background for 5.9 and 5.10.

References

[1] L. Aussenhofer, *Contributions to the duality theory of Abelian topological groups and to the theory of nuclear groups*, Dissertationes Math. **384** (1998), 113 pp.

[2] B. Banaschewski, *Minimal topological algebras*, Math. Ann. **211** (1974), 107–114.

[3] G. Barbieri, D. Dikranjan, C. Milan, and H. Weber, *Convergent sequences in precompact group topologies*, Appl. Gen. Topol. **6** (2005), no. 2, 149–169.

[4] A. Berarducci, D. Dikranjan, M. Forti, and S. Watson, *Cardinal invariants and independence results in the poset of precompact group topologies*, J. Pure Appl. Algebra **126** (1998), 19–49.

[5] S. Berhanu, W. W. Comfort, and J. D. Reid, *Counting subgroups and topological group topologies*, Pacific J. Math. **116** (1985), no. 2, 217–241.

[6] A. R. Bernstein, *A new kind of compactness for topological spaces*, Fund. Math. **66** (1970), 185–193.

[7] E. Boschi and D. Dikranjan, *Essential subgroups of topological groups*, Comm. Algebra **28** (2000), no. 10, 4941–4970.

[8] F. S. Cater, P. Erdős, and F. Galvin, *On the density of λ-box products*, General Topology and Appl. **9** (1978), no. 3, 307–312.

[9] E. Čech and B. Pospísil, *Sur les espaces compacts. II. Sur les charactères des points dans les espaces* I, Publ. Fac. Sci. Univ. Masaryk **258** (1938), 14 pp.

[10] M. J. Chasco, *Pontryagin duality for metrizable groups*, Arch. Math. (Basel) **70** (1998), no. 1, 22–28.

[11] W. W. Comfort, *Review of J. Ginsburg and V. Saks,* Some applications of ultrafilters in topology, *Pacific J. Math. 57 (1975) 403–418*, Math. Reviews **52** (1976), 227–228, #1633.

[12] W. W. Comfort, *Some questions and answers in topological groups*, Memorias del Seminario Especial de Topologia (J. Bracho and C. Prieto, eds.), vol. 5, Instituto de Matemáticas de la Universidad Nacional Autónoma de México, 1983, pp. 131–149.

[13] W. W. Comfort, *Topological groups*, Handbook of set-theoretic topology, North-Holland, Amsterdam, 1984, pp. 1143–1263.

[14] W. W. Comfort, *Problems on topological groups and other homogeneous spaces*, Open problems in topology, North-Holland, Amsterdam, 1990, pp. 313–347.

[15] W. W. Comfort, *Tampering with pseudocompact groups*, Topology Proc. **28** (2004), no. 2, 401–424.

[16] W. W. Comfort and D. Dikranjan, *Essential density and total density in topological groups*, J. Group Theory **5** (2002), no. 3, 325–350.

[17] W. W. Comfort and J. Galindo, *Extremal pseudocompact topological groups*, J. Pure Appl. Algebra **197** (2005), 59–81.

[18] W. W. Comfort, S. Hernández, D. Remus, and F. J. Trigos-Arrieta, *Some open questions on topological groups*, Nuclear groups and Lie groups (Madrid, 1999), Heldermann Verlag, Lemgo, 2001, pp. 57–76.

[19] W. W. Comfort, K. H. Hofmann, and D. Remus, *Topological groups and semigroups*, Recent progress in general topology (Prague, 1991), North-Holland, Amsterdam, 1992, pp. 57–144.

[20] W. W. Comfort and J. van Mill, *On the existence of free topological groups*, Topology Appl. **29** (1988), no. 3, 245–265.

[21] W. W. Comfort and J. van Mill, *Extremal pseudocompact Abelian groups are metrizable*, Proc. Amer. Math. Soc., To appear.

[22] W. W. Comfort, S. U. Raczkowski, and F. J. Trigos-Arrieta, *Concerning the dual group of a dense subgroup*, Proc. Ninth Prague Topological Symposium (2001), Topology Atlas, Toronto, 2002, pp. 22–35.

[23] W. W. Comfort, S. U. Raczkowski, and F. J. Trigos-Arrieta, *The dual group of a dense subgroup*, Czechoslovak Math. J. **54(129)** (2004), no. 2, 509–533.

[24] W. W. Comfort, S. U. Raczkowski, and F. J. Trigos-Arrieta, *Making group topologies with, and without, convergent sequences*, Appl. Gen. Topol. **7** (2007), 109–124.

[25] W. W. Comfort and D. Remus, *Long chains of Hausdorff topological group topologies*, J. Pure Appl. Algebra **70** (1991), no. 1–2, 53–72.

[26] W. W. Comfort and D. Remus, *Abelian torsion groups with a pseudocompact group topology*, Forum Math. **6** (1994), no. 3, 323–337.

[27] W. W. Comfort and D. Remus, *Long chains of topological group topologies—a continuation*, Topology Appl. **75** (1997), no. 1, 51–79.

[28] W. W. Comfort and L. C. Robertson, *Cardinality constraints for pseudocompact and for totally dense subgroups of compact topological groups*, Pacific J. Math. **119** (1985), no. 2, 265–285.

[29] W. W. Comfort and L. C. Robertson, *Extremal phenomena in certain classes of totally bounded groups*, Dissertationes Math. **272** (1988), 48 pp.

[30] W. W. Comfort and K. A. Ross, *Topologies induced by groups of characters*, Fund. Math. **55** (1964), 283–291.

[31] W. W. Comfort and K. A. Ross, *Pseudocompactness and uniform continuity in topological groups*, Pacific J. Math. **16** (1966), 483–496.

[32] W. W. Comfort and V. Saks, *Countably compact groups and finest totally bounded topologies*, Pacific J. Math. **49** (1973), 33–44.

[33] W. W. Comfort and T. Soundararajan, *Pseudocompact group topologies and totally dense subgroups*, Pacific J. Math. **100** (1982), no. 1, 61–84.

[34] D. Dikranjan, *Chains of pseudocompact group topologies*, J. Pure Appl. Algebra **124** (1998), 65–100.

[35] D. Dikranjan, *Compactness and connectedness in topological groups*, Topology Appl. **84** (1998), 227–252.

[36] D. Dikranjan, *Recent advances in minimal topological groups*, Topology Appl. **85** (1998), 53–91.

[37] D. Dikranjan, *Closure operators in topological groups related to von Neumann's kernel*, Topology Appl. **153** (2006), no. 11, 1930–1955.

[38] D. Dikranjan, I. R. Prodanov, and L. N. Stoyanov, *Topological groups*, Marcel Dekker Inc., New York, 1990.

[39] D. Dikranjan and D. Shakhmatov, *Selected topics from the structure theory of topological groups*, Open problems in topology II (E. Pearl, ed.), Elsevier Science.

[40] D. Dikranjan and D. Shakhmatov, *Pseudocompact topologizations of groups*, Zb. Rad. (1990), no. 4, 83–93.

[41] D. Dikranjan and D. Shakhmatov, *Algebraic structure of pseudocompact groups*, Mem. Amer. Math. Soc. **133** (1998), no. 633, 83 pp.

[42] E. K. van Douwen, *The weight of a pseudocompact (homogeneous) space whose cardinality has countable cofinality*, Proc. Amer. Math. Soc. **80** (1980), no. 4, 678–682.

[43] R. Engelking, *General topology*, Heldermann Verlag, Berlin, 1989.

[44] L. Fuchs, *Infinite Abelian groups, Vol. I*, Academic Press, New York, 1970.

[45] J. Galindo, *Dense pseudocompact subgroups and finer pseudocompact group topologies*, Sci. Math. Jpn. **55** (2002), no. 3, 627–640.

[46] S. García-Ferreira, A. H. Tomita, and S. Watson, *Countably compact groups from a selective ultrafilter*, Proc. Amer. Math. Soc. **133** (2005), no. 3, 937–943.

[47] J. Ginsburg and V. Saks, *Some applications of ultrafilters in topology*, Pacific J. Math. **57** (1975), no. 2, 403–418.

[48] J. E. Hart and K. Kunen, *Bohr compactifications of non-abelian groups*, Topology Proc. **26** (2001/02), no. 2, 593–626.

[49] S. Hernández and S. Macario, *Dual properties in totally bounded abelian groups*, Arch. Math. (Basel) **80** (2003), no. 3, 271–283.

[50] E. Hewitt and K. A. Ross, *Abstract harmonic analysis. Vol. I*, Springer, Berlin, 1963.

[51] E. Hewitt and K. A. Ross, *Abstract harmonic analysis. Vol. II*, Springer, Berlin, 1970.

[52] K. H. Hofmann and S. A. Morris, *The structure of compact groups*, Walter de Gruyter & Co., Berlin, 1998.

[53] S. Kakutani, *On cardinal numbers related with a compact Abelian group*, Proc. Imp. Acad. Tokyo **19** (1943), 366–372.

[54] L. Ja. Kulikov, *On the theory of Abelian groups of arbitrary cardinality*, Rec. Math. [Mat. Sb.] **9** (1941), 165–185 (Russian).

[55] G. Lukács, *On homomorphism spaces of metrizable groups*, J. Pure Appl. Algebra **182** (2003), no. 2–3, 263–267.

[56] G. Lukács, *Precompact abelian groups and topological annihilators*, 2005, Preprint.

[57] G. Lukács, *Notes on duality theories of abelian groups*, 2006, Work in progress. `arXiv: math.GN/0605149`

[58] W. Moran, *On almost periodic compactifications of locally compact groups*, J. London Math. Soc. **3** (1971), 507–512.

[59] J. von Neumann and E. P. Wigner, *Minimally almost periodic groups*, Ann. of Math. (2) **41** (1940), 746–750.

[60] I. R. Prodanov, *Precompact minimal group topologies and p-adic numbers*, Annuaire Univ. Sofia Fac. Math. **66** (1971/72), 249–266 (1974).

[61] I. R. Prodanov and L. N. Stojanov, *Every minimal abelian group is precompact*, C. R. Acad. Bulgare Sci. **37** (1984), no. 1, 23–26.

[62] S. U. Raczkowski, *Totally bounded topological group topologies on the integers*, Topology Appl. **121** (2002), no. 1–2, 63–74.

[63] D. Remus, *Die Anzahl von T_2-präkompakten Gruppentopologien auf unendlichen abelschen Gruppen*, Rev. Roumaine Math. Pures Appl. **31** (1986), no. 9, 803–806.

[64] D. Remus, *The rôle of W. Wistar Comfort in the theory of topological groups*, Topology Appl. **97** (1999), no. 1–2, 31–49.

[65] V. Saks, *Ultrafilter invariants in topological spaces*, Trans. Amer. Math. Soc. **241** (1978), 79–97.

[66] W. Sierpiński, *Sur une propriété topologique des ensembles dénombrables denses en soi*, Fund. Math. **1** (1920), 11–28.

[67] R. M. Stephenson, Jr., *Minimal topological groups*, Math. Ann. **192** (1971), 193–195.

[68] T. E. Stewart, *Uniqueness of the topology in certain compact groups*, Trans. Amer. Math. Soc. **97** (1960), 487–494.

[69] L. N. Stoyanov, *A property of precompact minimal abelian groups*, Annuaire Univ. Sofia Fac. Math. Méc. **70** (1975/76), 253–260 (1981).

[70] M. G. Tkachenko, *On pseudocompact topological groups*, Interim Report of the Prague Topological Symposium, 2, Math. Institute Czech. Acad. Sci., Prague, 1987, p. 18 pp.

[71] M. G. Tkachenko, *Countably compact and pseudocompact topologies on free abelian groups*, Izv. Vyssh. Uchebn. Zaved. Mat. (1990), no. 5, 68–75, Translation: Soviet Math. (Iz. VUZ) 34 (1990) no. 5, 79–86.

[72] M. G. Tkachenko, *On dimension of locally pseudocompact groups and their quotients*, Comment. Math. Univ. Carolin. **31** (1990), no. 1, 159–166.

[73] M. G. Tkachenko and I. Yaschenko, *Independent group topologies on abelian groups*, Topology Appl. **122** (2002), no. 1–2, 425–451.

[74] F. J. Trigos-Arrieta, *Determined Abelian topological groups*, 2006, Work in progress.

[75] M. I. Ursul, *Embedding of locally precompact groups into locally pseudocompact groups*, Izv. Akad. Nauk Moldav. SSR Ser. Fiz.-Tekhn. Mat. Nauk (1989), no. 3, 54–56.

[76] B. L. van der Waerden, *Stetigkeitssätze für halbeinfache Liesche Gruppen*, Math. Z. **36** (1933), 780–786.

[77] A. Weil, *Sur les espaces à structure uniforme et sur la topologie générale*, Hermann & Cie, Paris, 1937.

Open Problems in Topology II – Edited by E. Pearl

Selected topics from the structure theory of topological groups

Dikran Dikranjan and Dmitri Shakhmatov

This article contains open problems and questions covering the following topics: the dimension theory of topological groups, pseudocompact and countably compact group topologies on Abelian groups, with or without nontrivial convergent sequences, categorically compact groups, sequentially complete groups, the Markov–Zariski topology, the Bohr topology, and transversal group topologies. *All topological groups considered in this chapter are assumed to be Hausdorff.*

1. Dimension theory of topological groups

We highlight here our favourite problems from the dimension theory of topological groups.

Problem 1 ([**1**]). *Is* $\operatorname{ind} G = \operatorname{Ind} G = \dim G$ *for a topological group* G *with a countable network?* 884?

The classical result of Pasynkov says that $\operatorname{ind} G = \operatorname{Ind} G = \dim G$ for a (locally) compact group G [**50**].

Question 2 ([**52**]). *Is* $\operatorname{ind} G = \operatorname{Ind} G = \dim G$ *for a* σ*-compact group* G*?* 885?

This is a delicate question since there exists an example of a precompact topological group G such that G is a Lindelöf Σ-space, $\dim G = 1$ but $\operatorname{ind} G = \operatorname{Ind} G = \infty$ [**52**, **53**]. Even the following particular case of Question 2 seems to be open.

Question 3 (M.J. Chasco). *If a topological group* G *is a* k_ω*-space, must* $\operatorname{ind} G = \operatorname{Ind} G = \dim G$*?* 886?

Recall that X is a k_ω*-space* provided that there exists a countable family $\{K_n : n \in \omega\}$ of compact subspaces of X such that a subset U of X is open in X if and only if $U \cap K_n$ is open in K_n for every $n \in \omega$.

Question 4 ([**53**]). *Is* $\operatorname{ind} G = \operatorname{Ind} G$ *for a Lindelöf group* G*?* 887?

The answer to Question 4 is positive if G is a Lindelöf Σ-space (in particular, a σ-compact space), so only the inequality $\operatorname{ind} G \leq \dim G$ must be proved in order to answer Questions 2 or 3 positively.

The first named author was partially supported by the project MIUR 2005 "Anelli commutativi e loro moduli: teoria moltiplicativa degli ideali, metodi omologici e topologici." The second named author acknowledges partial financial support from the Grant-in-Aid for Scientific Research no. 155400823 by the Japan Society for the Promotion of Science.

888? **Problem 5** (Old problem). *If H is a subgroup of a topological group G, is then* $\dim H \leq \dim G$*?*

The answer is positive if H is $\mathbb{R}$-factorizable [**60**] (in particular, precompact [**54**]).

889? **Question 6** ([**55**]). *Suppose that X is a separable metric space with* $\dim X \leq n$. *Is there a separable metric group G that contains X as a closed subspace such that* $\dim G \leq 2n+1$*?*[1]

Without the requirement that X is closed in G the answer is positive due to the Nöebeling–Pontryagin theorem: X is a subspace of the topological group $\mathbb{R}^{2n+1}$. The separability in the above question is essential: There exists a metric space X of weight ω_1 such that $\dim X = 1$ and X cannot be embedded into any finite-dimensional topological group [**42**].

The next question is the natural group analogue of the classical result about the existence of the universal space of a given weight and covering dimension.

890? **Question 7** ([**55**]). *Let τ be an infinite cardinal and n be a natural number. Is there an (Abelian) topological group $H_{\tau,n}$ with* $\dim H_{\tau,n} \leq n$ *and* $w(H_{\tau,n}) \leq \tau$ *such that every (Abelian) topological group G satisfying* $\dim G \leq n$ *and* $w(G) \leq \tau$ *is topologically and algebraically isomorphic to a subgroup of $H_{\tau,n}$?*

The special case of the above question when $\tau = \omega$ is due to Arhangel′skiĭ [**1**].

Transfinite inductive dimensions have many peculiar properties in topological groups [**56**]. For example, (i) if G is a locally compact group having small transfinite inductive dimension trind(G), then G must be finite-dimensional, and (ii) if G is a separable metric group having large transfinite inductive dimension trInd(G), then G must be finite-dimensional as well. It is not clear if (ii) holds for trind(G) instead of trInd(G):

891? **Problem 8** ([**56**]). *For which ordinals α does there exist a separable metric group G_α whose small transfinite inductive dimension* trind(G_α) *equals α?*

The reader is referred to [**41, 55, 56**] for additional open problems in the dimension theory of topological groups.

2. Pseudocompact and countably compact group topologies on Abelian groups

We denote by $\mathcal{C}$ the class of Abelian groups that admit a countably compact group topology, and use $\mathcal{P}$ to denote the class of Abelian groups that admit a pseudocompact group topology.

The next two problems are the most fundamental problems in this area:

[1] We were kindly informed by T. Banakh that Question 6 has been recently answered in the negative in [**2**]: There exists a 1-dimensional separable metric space (namely, the hedgehog with countably many spines) which cannot be embedded into any finite-dimensional topological group as a closed subspace.

Problem 9 ([**19**]). *Describe the algebraic structure of members of the class $\mathcal{P}$.* 892?

Problem 10. *Describe the algebraic structure of members of the class $\mathcal{C}$.* 893?

Despite a substantial progress on Problem 9 for particular classes of groups achieved in [**7**, **8**, **16–18**], the general case is still very far from the final solution. (We refer the reader to [**4**] for further reading on this topic.)

Let G be an Abelian group. As usual $r(G)$ denotes the *free rank* of G. For every natural number $n \geq 1$ define $G[n] = \{g \in G : ng = 0\}$ and $nG = \{ng : g \in G\}$. Every group G from the class $\mathcal{C}$ satisfies the following two conditions [**16**, **18**, **26**]:

PS: Either $r(G) \geq \mathfrak{c}$ or $G = G[n]$ for some $n \in \omega \setminus \{0\}$.

CC: For every pair of integers $n \geq 1$ and $m \geq 1$ the group $mG[n]$ is either finite or has size at least $\mathfrak{c}$.

It is totally unclear if these are the only necessary conditions on a group from the class $\mathcal{C}$:

Question 11. *Is it true that an Abelian group G belongs to $\mathcal{C}$ if and only if G satisfies both* **PS** *and* **CC**? 894?

Question 12 ([**22**]). *Is it true in* ZFC *that an Abelian group G of size at most $2^{\mathfrak{c}}$ belongs to $\mathcal{C}$ if and only if G satisfies both* **PS** *and* **CC**? 895?

Question 12 has a positive *consistent* answer [**22**].

Assuming MA, there exist countably compact Abelian groups G, H such that $G \times H$ is not countably compact [**31**]. Therefore, our next question could be viewed as a *weaker form* of productivity of countable compactness in topological groups that still has a chance for a positive answer in ZFC.

Question 13 ([**19**]). *If G and H belong to $\mathcal{C}$, must then their product $G \times H$ also belong to $\mathcal{C}$?* 896?

In fact, one can consider a much bolder hypothesis:

Question 14 ([**19**]). *Is $\mathcal{C}$ closed under* arbitrary *products? That is, if G_i belongs to $\mathcal{C}$ for each $i \in I$, does then $\prod_{i \in I} G_i$ belong to $\mathcal{C}$?* 897?

The next question provides a slightly less bold conjecture:

Question 15 ([**19**]). (i) *Is there a cardinal τ having the following property: A product $\prod_{i \in I} G_i$ belongs to $\mathcal{C}$ provided that $\prod_{j \in J} G_j$ belongs to $\mathcal{C}$ whenever $J \subseteq I$ and $|J| \leq \tau$?* 898–899?

(ii) *Does the statement in item* (i) *hold true when $\tau = \mathfrak{c}$ or $\tau = 2^{\mathfrak{c}}$?*

Of course Question 14 simply asks if the statement in item (i) of Question 15 holds true when $\tau = 1$. It might be worth noting that Question 15 is motivated by a theorem of Ginsburg and Saks [**35**]: A product $\prod_{i \in I} X_i$ of topological spaces X_i is countably compact provided that $\prod_{j \in J} G_j$ is countably compact whenever $J \subseteq I$ and $|J| \leq 2^{\mathfrak{c}}$.

A partial positive answer to Question 14 has been given in [**20**]: It is consistent with ZFC that, for every family $\{G_i : i \in I\}$ of groups with $2^{|I|} \leq 2^{\mathfrak{c}}$ such that

G_i belongs to $\mathcal{C}$ and $|G_i| \leq 2^{\mathfrak{c}}$ for each $i \in I$, the product $\prod_{i\in I} G_i$ also belongs to $\mathcal{C}$. A similar result for smaller products and smaller groups has been proved in [**26**, Theorem 5.6] under the assumption of MA. In particular, if the groups G and H in Question 13 are additionally assumed to be of size at most $2^{\mathfrak{c}}$, then the positive answer to this restricted version of Question 13 is consistent with ZFC [**20**].

Recall that an Abelian group G is *algebraically compact* provided that one can find an Abelian group H such that $G \times H$ admits a compact group topology. Algebraically compact groups form a relatively narrow subclass of Abelian groups (for example, the group $\mathbb{Z}$ of integers is not algebraically compact). On the other hand, every Abelian group G is *algebraically pseudocompact*; that is, one can find an Abelian group H such that $G \times H \in \mathcal{P}$ [**18**, Theorem 8.15]. It is unclear if this result can be strengthened to show that every Abelian group is *algebraically countably compact*:

900? **Question 16** ([**22**]). *Given an Abelian group G, can one always find an Abelian group H such that $G \times H \in \mathcal{C}$?*

Recall that an Abelian group G is *divisible* provided that for every $g \in G$ and each positive integer n one can find $h \in G$ such that $nh = g$. An Abelian group is *reduced* if it does not have non-zero divisible subgroups. Every Abelian group G admits a unique representation $G = D(G) \times R(G)$ into the maximal divisible subgroup $D(G)$ of G (the so-called *divisible part of* G) and the reduced subgroup $R(G) \cong G/D(G)$ of G (the so-called *reduced part of* G). It is well-known that an Abelian group G admits a compact group topology if and only if both its divisible part $D(G)$ and its reduced part $R(G)$ admit a compact group topology. However, there exist groups G and H that belong to $\mathcal{P}$ but neither $D(G)$ nor $R(H)$ belong to $\mathcal{P}$ [**18**, Theorem 8.1(ii)]. This was *strengthened* in [**22**, **26**] as follows: It is consistent with ZFC that there exist groups G' and H' from the class $\mathcal{C}$ such that neither $D(G')$ nor $R(H')$ belong to $\mathcal{P}$. These results leave open the following:

901–902? **Problem 17** ([**19**]). *In* ZFC, *give an example of groups G and H from the class $\mathcal{C}$ such that:*

(i) *$D(G)$ does not belong to $\mathcal{C}$ (or even $\mathcal{P}$),*
(ii) *$R(H)$ does not belong to $\mathcal{C}$ (or even $\mathcal{P}$).*

Even the following question is also open:

903–904? **Question 18** ([**19**]). *Let G be a group in $\mathcal{C}$.*

(i) *Is it true that either $D(G)$ or $R(G)$ belongs to $\mathcal{C}$?*
(ii) *Must either $D(G)$ or $R(G)$ belong to $\mathcal{P}$?*

We note that item (ii) of the last question is a strengthening of Question 9.8 from [**18**]. Even consistent results related to the last question are currently unavailable.

An Abelian group G is *torsion* provided that $G = \bigcup\{G[n] : n \in \omega, n \geq 1\}$, and is *torsion-free* provided that $\bigcup\{G[n] : n \in \omega, n \geq 1\} = \{0\}$.

Question 19 ([**22**]). *Is there a torsion Abelian group that is in $\mathcal{P}$ but not in $\mathcal{C}$?* 905?

Question 20 ([**22**]). *Is there a torsion-free Abelian group that is in $\mathcal{P}$ but not in $\mathcal{C}$?* 906?

It is consistent with ZFC that a group Questions 19 and 20 must have size strictly bigger than $2^{\mathfrak{c}}$ [**22**].

Problem 21 ([**22**]). (i) *Describe in* ZFC *the algebraic structure of separable countably compact Abelian groups.* 907–908?

(ii) *Is it true in* ZFC *that an Abelian group G admits a separable countably compact group topology if and only if $|G| \leq 2^{\mathfrak{c}}$ and G satisfies both* **PS** *and* **CC**?

A consistent positive solution to Problem 21(ii) is given in [**22**].

3. Properties determined by convergent sequences

It is well-known that infinite compact groups have (lots of) nontrivial convergent sequences. There exists an example (in ZFC) of a pseudocompact Abelian group without nontrivial convergent sequences [**58**]. While there are plenty of consistent examples of countably compact groups without nontrivial convergent sequences [**10**, **22**, **26**, **31**, **39**, **46**, **59**, **63**], the following remains a major open problem in this area:

Problem 22. *Does there exist, in* ZFC, *a countably compact group without nontrivial convergent sequences?* 909?

Recall that a (Hausdorff) topological group G is *minimal* if G does not admit a strictly weaker (Hausdorff) group topology. Even though a countably compact, minimal Abelian group need not be compact, it can be shown that it must contain a nontrivial convergent sequence. More generally, one can show that an infinite, countably compact, minimal nilpotent group has a nontrivial convergent sequence. Whether "nilpotent" can be dropped remains unclear.

Problem 23. *Must an infinite, countably compact, minimal group contain a nontrivial convergent sequence?* 910?

The next question may be considered as a countably compact (or pseudocompact) *heir* of the fact that compact groups have nontrivial convergent sequences that still has a chance of a positive answer in ZFC.

Question 24 ([**22**]). *Let G be an infinite group admitting a countably compact (or a pseudocompact) group topology. Does G have a countably compact (respectively, pseudocompact) group topology that contains a nontrivial convergent sequence?* 911?

The next question goes in the opposite direction:

Question 25 ([**22**]). (i) *Does every group G admitting a pseudocompact group topology have also a pseudocompact group topology without nontrivial convergent sequences (without infinite compact subsets)?* 912–913?

(ii) *Does every group G admitting a countably compact group topology have also a countably compact group topology without nontrivial convergent sequences (without infinite compact subsets)?*

Question 25(ii) has a consistent positive answer in the special case when $|G| \leq 2^{\mathfrak{c}}$ [**22**]. The part "without nontrivial convergent sequences" of item (ii) of our next question has appeared in [**10**].

914–916? **Question 26.** (i) *When does a compact Abelian group G admit a proper dense subgroup H without nontrivial convergent sequences? without infinite compact subsets?*

(ii) *When does a compact Abelian group G admit a proper dense pseudocompact subgroup H without nontrivial convergent sequences? without infinite compact subsets?*

(iii) *When does a compact Abelian group G admit a proper dense countably compact subgroup H without nontrivial convergent sequences? without infinite compact subsets?*

In GCH, a precompact group H such that $w(H) < w(H)^\omega$ has a nontrivial convergent sequence [**47**]. Thus $w(G) = w(G)^\omega$ is a necessary condition for the group G to have a subgroup H as in Question 26. This condition alone is not sufficient: If $K = \prod_{n\in\omega} \mathbb{Z}_{2^n}$ and τ is an infinite cardinal, then every dense subgroup H of $G = K^\tau$ has a nontrivial convergent sequence [**10**] (here $\mathbb{Z}_m$ denotes the cyclic group $\mathbb{Z}/m\mathbb{Z}$). Many partial results towards solution of Question 26 are given in [**10, 33**].

917–918? **Question 27.** (i) *If a compact Abelian group has a proper dense pseudocompact subgroup without nontrivial convergent sequences, does it also have a proper dense pseudocompact subgroup without infinite compact subsets?*

(ii) *If a compact Abelian group has a proper dense countably compact subgroup without nontrivial convergent sequences, does it also have a proper dense countably compact subgroup without infinite compact subsets?*

Now we relax item (ii) to get the following:

919? **Question 28.** *Is the existence of a countably compact Abelian group without nontrivial convergent sequences equivalent to the existence of a countably compact Abelian group without infinite compact subsets?*

In connection with the last four questions we should note that, under MA, an infinite compact space of size at most $\mathfrak{c}$ contains a nontrivial convergent sequence.

A topological group G is called *sequentially complete* [**24, 25**] if G is sequentially closed in every (Hausdorff) group that contains G as a topological subgroup. Obviously, every topological group without nontrivial convergent sequences is sequentially complete. Moreover, sequential completeness is preserved under taking arbitrary direct products and closed subgroups [**24**].

Denote by $\mathcal{S}$ the class of closed subgroups of the products of countably compact Abelian groups. Since countably compact groups are sequentially complete and precompact, every group from the class $\mathcal{S}$ is sequentially complete and precompact.

Question 29 ([**25**]). (i) *Does every precompact sequentially complete Abelian group G belong to $\mathcal{S}$?* 920–921?

(ii) *What is the answer to* (i) *if one additionally assumes that $|G| \leq \mathfrak{c}$?*

Every precompact Abelian group is both a quotient group and a continuous isomorphic image of some sequentially complete precompact Abelian group [**25**, Theorem B]. This motivates the following:

Question 30 ([**25**]). *Is every precompact Abelian group G:* 922–924?

(i) *a quotient of a group from $\mathcal{S}$?*
(ii) *a continuous homomoprhic image of group from $\mathcal{S}$?*
(iii) *a continuous isomorphic image of group from $\mathcal{S}$?*

Item (iii) of Question 30 has a positive answer when $|G| \leq \mathfrak{c}$ [**25**, Theorem A], and more generally, if $|G|$ is a non-measurable cardinal [**61**].

4. Categorically compact groups

A topological group G is *categorically compact* (briefly, *c-compact*) if for each topological group H the projection $G \times H \to H$ sends closed subgroups of $G \times H$ to closed subgroups of H [**29**]. Obviously, compact groups are c-compact. To establish the converse is the main open problem in this area:

Problem 31. (i) *Are c-compact groups compact?* 925–926?

(ii) *Are nondiscrete c-compact groups compact?*

Item (i) has appeared in [**29**]. Two related weaker versions are also open:

Question 32. *Is every (nondiscrete) c-compact group minimal?* 927?

Question 33. *Does every nondiscrete c-compact group have a nontrivial convergent sequence?* 928?

A positive answer to Problem 31 in the Abelian case makes recourse to the deep theorem of precompactness of Prodanov and Stoyanov [**15**]. Similar to (usual) compactness, taking products, closed subgroups and continuous homomorphic images preserves c-compactness [**29**] (a proof of the productivity of c-compactness was obtained independently also in [**3**] in a much more general setting). Therefore, a positive answer to Question 32 would imply that every closed subgroup H of a c-compact group is *totally minimal*, i.e., all quotient groups of H are minimal. At present we only know that separable c-compact groups are totally minimal (and complete) [**29**].

Lukacs [**45**] resolved Problem 31 positively for maximally almost periodic groups. Moreover, he showed that it suffices to solve this problem only for second countable groups (analogously, the case of locally compact SIN-groups, is reduced to that of countable discrete groups [**45**]). (Recall that a *SIN-group* is a topological group for which the left and right uniformities coincide.) According to [**45**], in Question 33 it suffices to consider only the nondiscrete c-compact groups that have no nontrivial continuous homomorphisms into compact groups.

Connected locally compact c-compact groups are compact [**29**]. Hence the connected locally compact group $SL_2(\mathbb{R})$ is not categorically compact, although it is separable and totally minimal [**15**]. Nothing is known about c-compactness of disconnected locally compact groups. In fact, even the discrete case is wide open:

929? **Question 34** ([**29**]). *Is every discrete c-compact group finite (finitely generated, of finite exponent, countable)?*

One can prove that a countable discrete group G is c-compact if and only if every subgroup of G is totally minimal [**29**]. Therefore, the negative answer to this question is equivalent to the existence of an infinite group G such that no subgroup or quotient group of G admits a nondiscrete Hausdorff group topology (this is a stronger version of the famous Markov problem on the existence of a countably infinite group without nondiscrete Hausdorff group topologies).

A group G is *h-complete* if all continuous homomorphic images of G are complete, and G is *hereditarily h-complete* if every closed subgroup of G is h-complete. c-compact groups are hereditarily h-complete, and the inverse implication holds for SIN-groups (in particular, Abelian groups) [**29**].

Both c-compactness and h-completeness are stable under products, and h-completeness also has the the so-called *three space property*: If K is a closed normal subgroup of a topological group G such that both K and the quotient group G/K are h-complete, then G is h-complete. This leaves open:

930? **Question 35** ([**29**, Question 4.3]). *If K is a closed normal subgroup of a topological group G such that both K and the quotient group G/K are c-compact, must G be c-compact as well?*

Nilpotent (in particular, Abelian) h-complete groups are compact, while solvable c-compact groups are compact [**29**]. This motivates the following:

931? **Question 36** ([**29**, Question 3.13]). *Are solvable h-complete groups c-compact?*

5. The Markov–Zariski topology

According to Markov [**48**], a subset S of a group G is called:

(a) *elementary algebraic* if there exist an integer $n > 0$, $a_1, \ldots, a_n \in G$ and $\varepsilon_1, \ldots, \varepsilon_n \in \{-1, 1\}$ such that

$$S = \{x \in G : x^{\varepsilon_1} a_1 x^{\varepsilon_2} a_2 \ldots a_{n-1} x^{\varepsilon_n} = a_n\},$$

(b) *algebraic* if S is an intersection of finite unions of elementary algebraic subsets,

(c) *unconditionally closed* if S is closed in *every* Hausdorff group topology of G,

(d) *potentially dense* if G admits a Hausdorff group topology $\mathcal{T}$ on G such that S is dense in $(G, \mathcal{T})$.

The family of algebraic subsets of a group G coincides with the family of closed subsets of a T_1 topology $\mathfrak{Z}_G$ on G, called the *Zariski topology*. The family of unconditionally closed subsets of G coincides with the family of closed subsets of a

T_1 topology $\mathfrak{M}_G$ on G, namely the infimum (taken in the lattice of all topologies on G) of all Hausdorff group topologies on G. We call $\mathfrak{M}_G$ the *Markov topology* of G. Analogously, let $\mathfrak{P}_G$ be the infimum of all precompact Hausdorff group topologies on G (if G admits no such topologies let $\mathfrak{P}_G$ denote the discrete topology of G). It seems natural to call $\mathfrak{P}_G$ the *precompact Markov topology* of G. Note that $(G, \mathfrak{Z}_G)$, $(G, \mathfrak{M}_G)$ and $(G, \mathfrak{P}_G)$ are quasi-topological groups, i.e., the inversion and shifts are continuous.

Clearly, $\mathfrak{Z}_G \subseteq \mathfrak{M}_G \subseteq \mathfrak{P}_G$. If G is Abelian, then $\mathfrak{Z}_G = \mathfrak{M}_G = \mathfrak{P}_G$ [**21**]. Markov has attributed the equality $\mathfrak{Z}_G = \mathfrak{M}_G$ in the Abelian case to Perel'man but the proof has never appeared in print. Another proof was recently announced by Sipacheva [**41**]. In the particular case when G is almost torsion-free[2] the equality $\mathfrak{Z}_G = \mathfrak{M}_G$ was earlier proved in [**62**].

Problem 37 ([**48**]). *Does $\mathfrak{Z}_G = \mathfrak{M}_G$ hold true for an arbitrary group G?* 932?

The answer is positive when G is countable [**48**]. A consistent counterexample to this question was announced quite recently by Sipacheva [**57**] (see also [**41**]).

Let $\mathcal{M}$ denote the class of groups G with $\mathfrak{Z}_G = \mathfrak{M}_G$.

Question 38. *For which infinite cardinals τ does the permutation group $S(\tau)$ of a set of size τ belong to $\mathcal{M}$?* 933?

The answer is positive for $\tau = \omega$ [**23**].

Question 39. *For which uncountable cardinals τ does the free group of rank τ belong to $\mathcal{M}$?* 934?

Question 40. (i) *Is $\mathcal{M}$ closed under taking subgroups? In particular, do all subgroups of $S(\omega)$ belong to $\mathcal{M}$?* 935–936?

(ii) *Is $\mathcal{M}$ closed under taking (finite) direct products?*

Question 41. (i) *What is the minimal cardinality of a group G such that $\mathfrak{Z}_G \neq \mathfrak{M}_G$?* 937–939?

(ii) *Is $\mathfrak{c}$ such a cardinality in* ZFC*?*

(iii) *Is ω_1 such a cardinality in* ZFC*?*

Let G be a group and $\mathcal{T}$ be any Hausdorff group topology on G. Then $\mathfrak{M}_G \subseteq \mathcal{T}$, and therefore the $\mathcal{T}$-closure of a subset of G must be contained in its $\mathfrak{M}_G$-closure. In other words, the $\mathfrak{M}_G$-closure of a given set $S \subseteq G$ is the *biggest* subset of G that one could possibly hope to attain by taking the closure of S in *any* Hausdorff group topology on G. This naturally leads to a question whether the $\mathfrak{M}_G$-closure of S can actually be *realized* by taking the closure of S in *some* Hausdorff group topology on G.

Question 42. *Let G be a group of size at most $2^{\mathfrak{c}}$ and $\mathcal{E}$ a countable family of subsets of G. Can one find a Hausdorff group topology $\mathcal{T}_{\mathcal{E}}$ on G such that the $\mathcal{T}_{\mathcal{E}}$-closure of every $E \in \mathcal{E}$ coincides with its $\mathfrak{M}_G$-closure?* 940?

---

[2] An Abelian group G is *almost torsion-free* if $G[n]$ is finite for every $n > 1$.

For an Abelian group G the answer is positive, and in fact the topology $\mathcal{T}_{\mathcal{E}}$ in this case can be chosen to be precompact [**21**].

The counterpart of Question 42 for $\mathfrak{Z}_G$ instead of $\mathfrak{M}_G$ has a negative (consistent) answer. Indeed, let G be an infinite group such that $\mathfrak{Z}_G \neq \mathfrak{M}_G$ and $\mathfrak{M}_G$ is discrete. (The recent example of Sipacheva [**57**] would do.) Let e be the unit element of G. Then $G \setminus \{e\}$ cannot be $\mathfrak{Z}_G$-closed. Indeed, if it were, then $\{e\}$ would be $\mathfrak{Z}_G$-open, and so by homogeneity of $\mathfrak{Z}_G$, the topology $\mathfrak{Z}_G$ would be discrete, implying $\mathfrak{Z}_G = \mathfrak{M}_G$, a contradiction. So the $\mathfrak{Z}_G$-closure of $G \setminus \{e\}$ must coincide with G. On the other hand, the only Hausdorff group topology $\mathcal{T}$ on G is the discrete topology, and so the $\mathcal{T}$-closure of $G \setminus \{e\}$ is $G \setminus \{e\}$.

Let us consider now the counterpart of Question 42 for $\mathfrak{P}_G$ instead of $\mathfrak{M}_G$.

941? **Question 43.** *Let G be a group of size at most $2^{\mathfrak{c}}$ having at least one precompact Hausdorff group topology, and let $\mathcal{E}$ be a countable family of subsets of G. Can one find a precompact Hausdorff group topology $\mathcal{T}_{\mathcal{E}}$ on G such that the $\mathcal{T}_{\mathcal{E}}$-closure of every $E \in \mathcal{E}$ coincides with its $\mathfrak{P}_G$-closure?*

Again, for an Abelian group G the answer is positive [**21**].

Thereafter, we consider only Abelian groups and refer to the three equivalent topologies $\mathfrak{Z}_G = \mathfrak{M}_G = \mathfrak{P}_G$ as the *Markov–Zariski topology*, denoting it by $\mathfrak{T}_G$. For an infinite Abelian group G, $\mathfrak{T}_G$ is neither Hausdorff, nor a group topology on G, but still has various nice properties, e.g., the space $(G, \mathfrak{T}_G)$ is hereditarily compact, hereditarily separable and Fréchet–Urysohn, moreover it has only finitely many connected components, and each component is an irreducible space [**21**].

942? **Problem 44.** *Let G be an Abelian group with $|G| \leq 2^{\mathfrak{c}}$ and $\mathcal{E}$ a family of subsets of G with $|\mathcal{E}| < 2^{|G|}$. Does there exist a precompact Hausdorff group topology $\mathcal{T}_{\mathcal{E}}$ on G such that the $\mathcal{T}_{\mathcal{E}}$-closure of each $E \in \mathcal{E}$ coincides with its $\mathfrak{T}_G$-closure?*

As was mentioned before, the answer is positive for countable families $\mathcal{E}$ [**21**]. Moreover, it was shown that if $|G| \leq \mathfrak{c}$, one can choose the approximating topology $\mathcal{T}_{\mathcal{E}}$ to be even metric.

The inequality $|\mathcal{E}| < 2^{|G|}$ in the above problem is essential. Indeed, let G be an infinite Abelian group. If one takes as $\mathcal{E}$ the family of *all* subsets of G, then the existence of a Hausdorff group topology $\mathcal{T}_{\mathcal{E}}$ on G such that the $\mathcal{T}_{\mathcal{E}}$-closure of each $E \in \mathcal{E}$ coincides with its $\mathfrak{T}_G$-closure would obviously imply that $\mathcal{T}_{\mathcal{E}} = \mathfrak{T}_G$. Thus $\mathfrak{T}_G$ would become Hausdorff, a contradiction.

The restriction on the cardinality of the group G in Questions 42, 43 and Problem 44 is obviously necessary since the closure of a countable set in a Hausdorff topology cannot exceed $2^{\mathfrak{c}}$.

The problem of characterization of the potentially dense subsets S of a group G goes back to Markov [**48**] who proved that every infinite subset of $\mathbb{Z}$ is potentially dense. This was extended in [**62**] to Abelian groups G of size $\leq \mathfrak{c}$ that are either of prime exponent or almost torsion-free. Tkachenko and Yaschenko asked in [**62**] whether the restriction $|G| \leq \mathfrak{c}$ can be relaxed to $|G| \leq 2^{\mathfrak{c}}$. To clarify better the problem, let us drop all additional restrictions on the Abelian group G. Obviously, if S is potentially dense in G, then $|G| \leq 2^{2^{|S|}}$ and S must be $\mathfrak{T}_G$-dense in G. It is not clear if these two conditions are not only necessary but also sufficient for potential density.

Question 45. *Let G be an Abelian group and let S be an infinite subset of G such that $|G| \leq 2^{2^{|S|}}$ and S is $\mathfrak{T}_G$-dense in G.* 943–944?

(i) *Is S potentially dense in G?*

(ii) *Does there exist a Hausdorff precompact group topology $\mathcal{T}$ such that S is $\mathcal{T}$-dense in G?*

A positive answer to both items of this question in the case of a countable set S has been given in [**21**] thereby providing a positive answer to the above mentioned question from [**62**].

6. Bohr topologies of Abelian groups

Let G be an Abelian group. Following E. van Douwen [**32**], we denote by $G^\#$ the group G equipped with the Bohr topology, i.e., the initial topology with respect to the family of all homomorphisms of G into the circle group $\mathbb{T}$. It is a well known fact, due to Glicksberg (see also [**34**] in this volume), that $G^\#$ has no infinite compact subsets (in particular, no nontrivial convergent sequences). Therefore, $G^\#$ is always sequentially complete. For future reference, we mention two fundamental properties of the Bohr topology for arbitrary Abelian groups G, H:

(i) the Bohr topology of $G \times H$ coincides with the product topology of $G^\# \times H^\#$;

(ii) if H is a subgroup of G, then H is closed in $G^\#$ and its topology as a topological subgroup of $G^\#$ coincides with that of $H^\#$.

E. van Douwen [**49**] posed the following challenging problem (see also [**34**]): If G and H are Abelian groups of the same size, must $G^\#$ and $H^\#$ be homeomorphic? A negative solution was obtained in [**43**] and independently, in [**30**]: $(\mathbb{V}_p^\omega)^\#$ and $(\mathbb{V}_q^\omega)^\#$ are not homeomorphic for different primes p and q. (For every positive integer m and a cardinal κ, $\mathbb{V}_m^\kappa$ denotes the direct sum of κ many copies of the group $\mathbb{Z}_m$.) Motivated by this, let us call a pair G, H of infinite Abelian groups:

(1) *Bohr-homeomorphic* if $G^\#$ and $H^\#$ are homeomorphic,

(2) *weakly Bohr-homeomorphic* if $G^\#$ can be homeomorphically embedded into $H^\#$ and $H^\#$ can also be homeomorphically embedded into $G^\#$.

Obviously, Bohr-homeomorphic groups are weakly Bohr-homeomorphic, and the status of the converse implication is totally unclear (see Question 49(ii)). As we shall see in the sequel, weak Bohr-homeomorphism provides a more flexible tool for studying the Bohr topology than the more *rigid* notion of Bohr-homeomorphism, e.g, $(\mathbb{V}_p^\omega)^\#$ and $(\mathbb{V}_q^\omega)^\#$ are not even weakly Bohr-homeomorphic for different primes p and q.

If $G^\#$ homeomorphically embeds into $H^\#$ and H is a bounded torsion group, then G must also be a bounded torsion group [**37**]. In particular, boundedness is invariant under weak Bohr-homeomorphisms, i.e., if G is a bounded Abelian group and the pair G, H are weakly Bohr-homeomorphic, than H must be bounded. Therefore, when studying weak Bohr-homeomorphisms (and thus Bohr-homeomorphisms), without any loss of generality whatsoever, one can consider completely separately bounded torsion Abelian groups and non-bounded Abelian groups.

We start first with the class of bounded torsion Abelian groups. According to Prüfer's theorem, every infinite bounded group has the form $\prod_{i=1}^{n} \mathbb{V}_{m_i}^{\kappa_i}$ for certain integers $m_i > 0$ and cardinals κ_i. For this reason, and in view of items (i) and (ii), the study of the Bohr topology of the bounded Abelian groups can be focused on the groups $\mathbb{V}_m^\kappa$.

For bounded Abelian groups G, H the following two algebraic conditions play a prominent role.

(3) $|mG| = |mH|$ whenever $m \in \mathbb{N}$ and $|mG| \cdot |mH| \geq \omega$.

(4) $\mathrm{eo}(G) = \mathrm{eo}(H)$ and $r_p(G) = r_p(H)$ for all p with $r_p(G) + r_p(H) \geq \omega$, where $\mathrm{eo}(G)$ is the *essential order* of G [**9**, **37**], i.e., *the smallest positive integer* m with mG finite.

Since a pair G, H satisfies (3) iff each one of these groups has a finite-index subgroup that is isomorphic to a subgroup of the other [**9**], we call such pairs of bounded Abelian groups G and H *weakly isomorphic* [**9**]. By (ii), weakly isomorphic bounded Abelian groups are weakly Bohr-homeomorphic. According to [**9**], weakly Bohr-homeomorphic bounded Abelian groups satisfy (4), i.e.,

$$\text{weakly isomorphic} \Rightarrow \text{weakly Bohr-homeomorphic} \Rightarrow (4).$$

Let us discuss the opposite implications. For countable Abelian groups G, H the second part of (4) becomes vacuous, while $eo(G) = eo(H)$ yields that G, H are weakly isomorphic. Analogously, one can see that (4) for groups of square-free essential order implies weak isomorphism and Bohr-homeomorphism. Hence all four properties (1)–(4) coincide for bounded Abelian groups that have square-free essential order [**9**, **37**]. Therefore, the invariant $\mathrm{eo}(G)$ alone allows for a complete classification (up to Bohr-homeomorphism) of all bounded Abelian groups of this class.

The situation changes completely even for the simplest *uncountable* bounded Abelian groups of essential order 4. Indeed, $G = \mathbb{V}_4^{\omega_1}$ and $H = \mathbb{V}_2^{\omega_1} \times \mathbb{V}_4^{\omega}$ are not weakly isomorphic, because $\omega_1 = |2G| > |2H| = \omega$. However, we do not know whether these groups are weakly Bohr-homeomorphic:

945? **Question 46.** *Can* $(\mathbb{V}_4^{\omega_1})^{\#}$ *be homeomorphically embedded into* $(\mathbb{V}_2^{\omega_1} \times \mathbb{V}_4^{\omega})^{\#}$*?*

Here is the question in the most general form:

946? **Question 47.** *Given a cardinal* $\kappa \geq \omega$ *and an integer* $s > 1$*, are* $\mathbb{V}_{p^s}^{\kappa}$ *and* $\mathbb{V}_p^{\kappa} \times \mathbb{V}_{p^s}^{\omega}$ *weakly Bohr-homeomorphic? Can this depend on* p*?*

If the answer to Question 47 is positive for all p, then bounded Abelian groups G and H would be weakly Bohr-homeomorphic if and only if (4) holds.

The next question is an equivalent form of the strongest negative answer to Question 47.

Question 48. *Assume that p is a prime number, $k > 1$ is an integer, κ and λ are infinite cardinals such that $(\mathbb{V}_{p^k}^{\kappa})^{\#}$ can be homeomorphically embedded into $(\mathbb{V}_{p^{k-1}}^{\kappa} \times \mathbb{V}_{p^k}^{\lambda})^{\#}$. Must then inequality $\lambda \geq \kappa$ hold?* 947?

Note that a positive answer to Question 48 answers negatively Questions 46 and 47.

The countable groups $\mathbb{V}_4^{\omega}$ and $\mathbb{V}_2^{\omega} \times \mathbb{V}_4^{\omega}$ are obviously weakly isomorphic, hence weakly Bohr-homeomorphic (see the discussion above).

Question 49. (i) ([**43**]) *Are $\mathbb{V}_4^{\omega}$ and $\mathbb{V}_2^{\omega} \times \mathbb{V}_4^{\omega}$ Bohr-homeomorphic?* 948–949?
(ii) *Are weakly Bohr-homeomorphic bounded groups* always *Bohr-homeomorphic?*

Question 50. *Suppose that G and H are bounded Abelian groups such that $G^{\#}$ homeomorphically embeds into $H^{\#}$. Does there exist a subgroup G' of G of finite index that algebraically embeds into H?* 950?

Note that a positive answer to this question would imply, in particular, that weak Bohr-homeomorphism coincides with weak isomorphism. Hence a positive answer to this question would imply a positive answer to Question 48.

Now we leave the *bounded world* and turn to the class of non-bounded groups. According to Hart and Kunen [**40**], two Abelian groups G and H are *almost isomorphic* if G and H have isomorphic finite index subgroups. This definition is motivated by the fact that almost isomorphic Abelian groups are always Bohr-homeomorphic [**40**]. The converse implication fails. Indeed, $\mathbb{Q}$ and $\mathbb{Q}/\mathbb{Z} \times \mathbb{Z}$ are Bohr-homeomorphic [**6**], and yet these groups are not almost isomorphic. It is nevertheless unclear if the reverse implication holds for bounded groups.

Question 51 ([**43**]). *Are Bohr-homeomorphic bounded Abelian groups almost isomorphic?* 951?

The question on whether the pairs $\mathbb{Z}, \mathbb{Z}^2$ and $\mathbb{Z}, \mathbb{Q}$ are Bohr-homeomorphic is raised in [**5**, **34**]. Let us consider here the version for weak Bohr-homeomorphisms:

Question 52. (i) *Are $\mathbb{Z}$ and $\mathbb{Q}$ weakly Bohr-homeomorphic?* 952–953?
(ii) *Are $\mathbb{Z}$ and $\mathbb{Q}/\mathbb{Z}$ (weakly) Bohr-homeomorphic?*

A positive answer to item (i) of Question 52 would yield that all torsion-free Abelian groups of a fixed finite free rank are weakly Bohr-homeomorphic. If both items have a positive answer, then the weak Bohr-homeomorphism class of $\mathbb{Z}^{\#}$ would comprise the class of all Abelian groups G of finite rank[3] such that either G is non-torsion or G contains a copy of the group $\mathbb{Q}/\mathbb{Z}$. (In particular, all finite powers of $\mathbb{Z}$, $\mathbb{Q}$ and $\mathbb{Q}/\mathbb{Z}$ along with their finite products would become weakly Bohr-homeomorphic.)

Many nice properties of $\mathbb{Z}^{\#}$ can be found in [**44**]. For a fast growing sequence a_n in $\mathbb{Z}^{\#}$ the range is a closed discrete set of $\mathbb{Z}^{\#}$ (see [**34**] for further properties of the lacunary sets in $\mathbb{Z}^{\#}$), whereas for a polynomial function $n \mapsto a_n = P(n)$ the range has no isolated points [**44**, Theorem 5.4]. Moreover, the range $P(\mathbb{Z})$ is closed

[3]I.e., there exists $n \in \omega$ such that $r_0(G) \leq n$ and $|G[p]| \leq p^n$ for every prime p.

when $P(x) = x^k$ is a monomial. For quadratic polynomials $P(x) = ax^2 + bx + c$ ($a, b, c \in \mathbb{Z}$, $a \neq 0$) the situation is already more complicated: the range $P(\mathbb{Z})$ is closed iff there is at most one prime that divides a, but does not divide b [**44**, Theorem 5.6]. This leaves open the general question.

954? **Problem 53.** *Characterize the polynomials $P(x) \in \mathbb{Z}[x]$ such that $P(\mathbb{Z})$ is closed in $\mathbb{Z}^\#$.*

Answering a question of van Douwen, Gladdines [**38**] found a closed countable *subset* of $(\mathbb{V}_2^\omega)^\#$ that is not a retract of $(\mathbb{V}_2^\omega)^\#$, while Givens [**36**] proved that every infinite $G^\#$ contains a closed countable subset that is not a retract of $G^\#$. However, the question remains open in the case of *subgroups*:

955? **Question 54** (Question 81, [**49**]). *If H is a countable subgroup of an Abelian group G, must $H^\#$ be a retract of $G^\#$?*

An affirmative answer to this question of E. van Douwen was obtained in [**6**] in the case when H is finitely generated (see also [**13**] for other partial results and open problems). The general case is still open.

We refer the reader to [**12**, **14**] for further information about Bohr topology.

7. Miscellanea

Two nondiscrete topologies τ_1 and τ_2 on a set X are called *transversal* if $\tau_1 \cup \tau_2$ generates the discrete topology on X. A precompact group topology on a group does not admit a transversal group topology, and under certain natural conditions the converse is also true [**27**].

956? **Question 55** ([**28**]). *Characterize locally compact groups that admit a transversal group topology.*

This question is resolved for locally compact Abelian groups [**27**] and for connected locally compact groups [**28**].

There exists a locally Abelian group G and a compact normal subgroup K of G such that G does not admit a transversal group topology while G/K does have a transversal group topology [**27**, Example 5.4]. The inverse implication remains unclear:

957? **Question 56** ([**28**]). *If G is a topological group that admits a transversal group topology and K is a compact normal subgroup of G, does also G/K admit a transversal group topology?*

The answer is positive when $G = K \times H$ for some subgroup H of G [**27**], or when G is a locally compact Abelian group (argue as in the proof of the implication (d) $\Rightarrow$ (c) of [**27**, Corollary 6.7]).

8–959? **Question 57** ([**28**]). (i) *Is it true that no minimal group topology admits a transversal group topology?*

(ii) *Does the topology of the unitary group of an infinite-dimensional Hilbert space admit a transversal group topology?*

The answer to item (i) is positive in the Abelian case.

The quasi-components (respectively, the connected components) of the Abelian pseudocompact groups are precisely all (connected) precompact groups [**11**]. The non-Abelian case remains unclear:

Problem 58 ([**11**]). *Describe the connected components and the quasi-components of pseudocompact groups.* 960?

Given a group G, let $\mathcal{H}(G)$ denote the family of all Hausdorff group topologies on G, and $\mathcal{P}(G)$ the family of all precompact Hausdorff group topologies on G. Note that $\mathcal{H}(G)$ and $\mathcal{P}(G)$ are partially ordered sets with respect to set-theoretic inclusion of topologies.

Question 59. *Suppose that G and H are infinite Abelian groups. Must the groups G and H be (algebraically) isomorphic* 961–962?

(i) *if the posets $\mathcal{H}(G)$ and $\mathcal{H}(H)$ are isomorphic?*

(ii) *if the posets $\mathcal{P}(G)$ and $\mathcal{P}(H)$ are isomorphic?*

A relevant information (and the origin of this question) may be found in [**51**].

Acknowledgement. We thankfully acknowledge helpful comments on a preliminary version of this paper offered to us by M.J. Chasco, W. Comfort, K. Kunen, M. Megrelishvili, V. Pestov, E. Reznichenko and V. Uspenskij.

References

[1] A. V. Arhangel′skiĭ, *Topological homogeneity. Topological groups and their continuous images*, Uspekhi Mat. Nauk **42** (1987), no. 2(254), 69–105, Translation: Russian Math. Surveys 42 (1987) no. 2, 83–131.

[2] T. Banakh and L. Zdomskyy, *Closed locally path-connected subspaces of finite-dimensional topological groups are locally compact*, 2006, Preprint.

[3] M. M. Clementino and W. Tholen, *Tychonoff's theorem in a category*, Proc. Amer. Math. Soc. **124** (1996), no. 11, 3311–3314.

[4] W. W. Comfort, *Dense subgroups of compact groups*, Open problems in topology II (E. Pearl, ed.), Elsevier Science.

[5] W. W. Comfort, S. Hernández, D. Remus, and F. J. Trigos-Arrieta, *Some open questions on topological groups*, Nuclear groups and Lie groups (Madrid, 1999), Heldermann Verlag, Lemgo, 2001, pp. 57–76.

[6] W. W. Comfort, S. Hernández, and F. J. Trigos-Arrieta, *Cross sections and homeomorphism classes of abelian groups equipped with the Bohr topology*, Topology Appl. **115** (2001), no. 2, 215–233.

[7] W. W. Comfort and D. Remus, *Imposing pseudocompact group topologies on abelian groups*, Fund. Math. **142** (1993), no. 3, 221–240.

[8] W. W. Comfort and D. Remus, *Abelian torsion groups with a pseudocompact group topology*, Forum Math. **6** (1994), no. 3, 323–337.

[9] L. De Leo and D. Dikranjan, *The staightening theorem for bounded codomains*, Preprint.

[10] J. J. Dijkstra and J. van Mill, *Groups without convergent sequences*, Topology Appl. **74** (1996), 275–282.

[11] D. Dikranjan, *Compactness and connectedness in topological groups*, Topology Appl. **84** (1998), 227–252.

[12] D Dikranjan, *Continuous maps in the Bohr topology*, Appl. Gen. Topol. **2** (2001), no. 2, 237–270.

[13] D. Dikranjan, *A class of abelian groups defined by continuous cross sections in the Bohr topology*, Rocky Mountain J. Math. **32** (2002), no. 4, 1331–1355.

[14] D. Dikranjan, *Van Douwen's problems related to the Bohr topology*, Proc. Ninth Prague Topological Symposium (2001), Topology Atlas, Toronto, 2002, pp. 37–50.

[15] D. Dikranjan, I. R. Prodanov, and L. N. Stoyanov, *Topological groups: characters, dualities and minimal group topologies*, Marcel Dekker Inc., New York, 1990.

[16] D. Dikranjan and D. Shakhmatov, *Pseudocompact topologizations of groups*, Zb. Rad. (1990), no. 4, 83–93.

[17] D. Dikranjan and D. Shakhmatov, *Pseudocompact topologies on groups*, Topology Proc. **17** (1992), 335–342.

[18] D. Dikranjan and D. Shakhmatov, *Algebraic structure of pseudocompact groups*, Mem. Amer. Math. Soc. **133** (1998), no. 633, 83 pp.

[19] D. Dikranjan and D. Shakhmatov, *Compact-like group topologies on Abelian groups*, Contributed problems (Zoltán Balogh Memorial Topology Conference), 2002, `http://notch.mathstat.muohio.edu/balogh/`, pp. 6–9.

[20] D. Dikranjan and D. Shakhmatov, *Categorical aspects of the class of Abelian groups admitting a countably compact group topology*, 2003, Preprint.

[21] D. Dikranjan and D. Shakhmatov, *On the Markov–Zariski topology of an abelian group*, 2003, Preprint.

[22] D. Dikranjan and D. Shakhmatov, *Forcing hereditarily separable compact-like group topologies on abelian groups*, Topology Appl. **151** (2005), 2–54.

[23] D. Dikranjan and D. Shakhmatov, *Markov and Zariski topologies on the permutation group of integers*, 2006, Preprint.

[24] D. Dikranjan and M. G. Tkachenko, *Sequential completeness of quotient groups*, Bull. Austral. Math. Soc. **61** (2000), no. 1, 129–150.

[25] D. Dikranjan and M. G. Tkachenko, *Varieties generated by countably compact abelian groups*, Proc. Amer. Math. Soc. **130** (2002), no. 8, 2487–2496.

[26] D. Dikranjan and M. G. Tkachenko, *Algebraic structure of small countably compact abelian groups*, Forum Math. **15** (2003), no. 6, 811–837.

[27] D. Dikranjan, M. G. Tkachenko, and I. Yaschenko, *On transversal group topologies*, Topology Appl. **153** (2005), no. 5–6, 786–817.

[28] D. Dikranjan, M. G. Tkachenko, and I. Yaschenko, *Transversal group topologies on non-abelian groups*, Topology Appl. **153** (2006), no. 17, 3338–3354.

[29] D. Dikranjan and V. V. Uspenskij, *Categorically compact topological groups*, J. Pure Appl. Algebra **126** (1998), 149–168.

[30] D. Dikranjan and S. Watson, *A solution to van Douwen's problem on Bohr topologies*, J. Pure Appl. Algebra **163** (2001), no. 2, 147–158.

[31] E. K. van Douwen, *The product of two countably compact topological groups*, Trans. Amer. Math. Soc. **262** (1980), no. 2, 417–427.

[32] E. K. van Douwen, *The maximal totally bounded group topology on G and the biggest minimal G-space, for abelian groups G*, Topology Appl. **34** (1990), no. 1, 69–91.

[33] J. Galindo and S. García-Ferreira, *Compact groups containing dense pseudocompact subgroups without non-trivial convergent sequences*, Preprint.

[34] J. Galindo, S. Hernández, and T. S. Wu, *Recent results and open questions relating Chu duality and Bohr compactifications of locally compact groups*, Open problems in topology II (E. Pearl, ed.), Elsevier Science.

[35] J. Ginsburg and V. Saks, *Some applications of ultrafilters in topology*, Pacific J. Math. **57** (1975), no. 2, 403–418.

[36] B. N. Givens, *Closed subsets which are not retracts in the Bohr topology*, Topology Appl. **129** (2003), no. 1, 11–14.

[37] B. N. Givens and K. Kunen, *Chromatic numbers and Bohr topologies*, Topology Appl. **131** (2003), no. 2, 189–202.

[38] H. Gladdines, *Countable closed sets that are not a retract of $G^{\#}$*, Topology Appl. **67** (1995), no. 2, 81–84.

[39] A. Hajnal and I. Juhász, *A separable normal topological group need not be Lindelöf*, General Topology and Appl. **6** (1976), no. 2, 199–205.

[40] J. E. Hart and K. Kunen, *Bohr compactifications of discrete structures*, Fund. Math. **160** (1999), no. 2, 101–151.

[41] K. Kozlov, E. Reznichenko, and O. Sipacheva, *Moscow questions on topological algebra*, Open problems in topology II (E. Pearl, ed.), Elsevier Science.

[42] J. Kulesza, *Spaces which do not embed in topological groups of the same dimension*, Topology Appl. **50** (1993), no. 2, 139–145.

[43] K. Kunen, *Bohr topologies and partition theorems for vector spaces*, Topology Appl. **90** (1998), 97–107.

[44] K. Kunen and W. Rudin, *Lacunarity and the Bohr topology*, Math. Proc. Cambridge Philos. Soc. **126** (1999), no. 1, 117–137.

[45] G. Lukács, *Hereditarily h-complete groups*, Topology Appl. **145** (2004), 177–189.

[46] V. I. Malykhin, *An example of a topological group*, Topological spaces and their mappings, Latvijskij Gosudarstvennyj Universitet, Riga, 1981, pp. 120–123.

[47] V. I. Malykhin and L. B. Shapiro, *Pseudocompact groups without convergent sequences*, Mat. Zametki **37** (1985), no. 1, 103–109, Translation: Math. Notes 37 (1985) no. 1, 59–62.

[48] A. A. Markov, *On unconditionally closed sets*, C. R. (Doklady) Acad. Sci. URSS (N.S.) **44** (1944), 180–181.

[49] J. van Mill, *Open problems in van Douwen's papers*, Collected papers. Vol. I, II, North-Holland, Amsterdam, 1994.

[50] B. A. Pasynkov, *The coincidence of various definitions of dimensionality for locally bicompact groups*, Dokl. Akad. Nauk SSSR **132** (1960), 1035–1037, Translation: Soviet Math. Dokl. 1 (1960) 720–722.

[51] D. Shakhmatov, *Character and pseudocharacter in minimal topological groups*, Mat. Zametki **38** (1985), no. 6, 908–914, Translation: Math. Notes 39 (1986) 465–470.

[52] D. Shakhmatov, *A problem of coincidence of dimensions in topological groups*, Topology Appl. **33** (1989), no. 1, 105–113.

[53] D. Shakhmatov, *Zerodimensionality of free topological groups and topological groups with noncoinciding dimensions*, Bull. Polish Acad. Sci. Math. **37** (1989), no. 7–12, 497–506.

[54] D. Shakhmatov, *Imbeddings into topological groups preserving dimensions*, Topology Appl. **36** (1990), no. 2, 181–204.

[55] D. Shakhmatov, *A survey of current researches and open problems in the dimension theory of topological groups*, Questions Answers Gen. Topology **8** (1990), no. 1, 101–128.

[56] D. Shakhmatov, *A note on transfinite inductive dimensions in topological groups*, Topology Appl. **54** (1993), 213–220.

[57] O. V. Sipacheva, *Consistent solution of Markov's problem about algebraic sets*, 2006, Preprint. `arXiv:math.GR/0605558`

[58] S. M. Sirota, *A product of topological groups, and extremal disconnectedness*, Mat. Sb. **79(121)** (1969), 179–192, Translation: Sb. Math. 8 (1969) 169–180.

[59] M. G. Tkachenko, *Countably compact and pseudocompact topologies on free abelian groups*, Izv. Vyssh. Uchebn. Zaved. Mat. (1990), no. 5, 68–75, Translation: Soviet Math. (Iz. VUZ) 34 (1990) no. 5, 79–86.

[60] M. G. Tkachenko, *Subgroups, quotient groups and products of $\mathbb{R}$-factorizable groups*, Tôpology Proc. **16** (1991), 201–231.

[61] M. G. Tkachenko, *Provisional solution to a Comfort–van Mill problem*, Topology Appl. **125** (2002), no. 1, 171–182.

[62] M. G. Tkachenko and I. Yaschenko, *Independent group topologies on abelian groups*, Topology Appl. **122** (2002), no. 1–2, 425–451.

[63] A. H. Tomita, *Two countably compact topological groups: one of size $\aleph_\omega$ and the other of weight $\aleph_\omega$ without non-trivial convergent sequences*, Proc. Amer. Math. Soc. **131** (2003), no. 8, 2617–2622.

Open Problems in Topology II – Edited by E. Pearl

Recent results and open questions relating Chu duality and Bohr compactifications of locally compact groups

Jorge Galindo, Salvador Hernández and Ta-Sun Wu

1. Introduction

In this paper we collect some problems that have appeared in the context of harmonic analysis on locally compact groups but can be understood, and perhaps solved, adopting topological methods. Naturally, this will also produce some genuine topological questions that can be handled using methods of harmonic analysis. We start with a simple example that illustrates quite well the interplay between the two subjects. Consider the group $\mathbb{Z}$ of integers and let us agree to say that a sequence $(n_k) \in \mathbb{Z}$ converges to n_0 when the sequence (t^{n_k}) converges to t^{n_0} for all $t \in \mathbb{T} = \{t \in \mathbb{C} : |t| = 1\}$. Are there convergent sequences under this definition?

It may appear that finding some convergent sequence should not be difficult. Suppose however that $\{n_k\}$ is a sequence which goes to 0. Then, by hypothesis, the sequence of functions $\{t^{n_k}\}$ converges pointwise to 1 on $\mathbb{T}$. Or, equivalently, the sequence of functions $\{e^{i2\pi n_k x}\}$ converges pointwise to 1 on the interval $[0, 1]$. Applying Lebesgue's Dominated Convergence Theorem, it follows that the sequence $\{0\} = \{\int_0^1 e^{i2\pi n_k x} dx\}$ converges to $\int_0^1 dx = 1$, which is a contradiction.

Quite surprisingly we have seen that the definition of convergence given above on $\mathbb{Z}$ produces no nontrivial convergent sequences. This convergence actually stems from the initial topology generated by the functions $n \mapsto t^n$ of $\mathbb{Z}$ into $\mathbb{T}$. It is called the *Bohr topology* of $\mathbb{Z}$ (denoted $\mathbb{Z}^\sharp$) and is the largest precompact (and, therefore, nondiscrete) group topology that can be defined on the integers. Even though this topology has been widely studied recently, we are still far from understanding it well in general.

There are other *more topological* approaches to show the absence of nontrivial convergent sequences in $\mathbb{Z}^\sharp$. The one we shall focus on in this paper is based on a careful study of the mappings $n \mapsto t^n$ of $\mathbb{Z}$ into $\mathbb{T}$. When a sequence of integers $\{m_j\}$ is lacunary, i.e., $\frac{m_{j+1}}{m_j} > q > 1$, the subset $A = \{m_j : j \in \mathbb{N}\}$ lives in $\mathbb{Z}^\sharp$ as an *interpolation subset*: that is to say, every real-valued bounded function f (regardless of its continuity) defined on A can be extended to a continuous function $\overline{f}$ of $\mathbb{Z}^\sharp$ into $\mathbb{R}$ (alternatively, we can say that the subset $\{m_j : j \in \mathbb{N}\}$ is C^*-*embedded* in $\mathbb{Z}^\sharp$). It is easily verified that a convergent sequence cannot be an interpolation set

The two first named authors acknowledge partial financial support by the Spanish Ministry of Science (including FEDER funds), grant MTM2004-07665-C02-01; and Fundació Caixa Castelló (Bancaja), grant P1 1B2005-22.

and, since every sequence contains many lacunary subsequences (and, therefore, interpolation sets), it follows that there are no convergent sequences in $\mathbb{Z}^\sharp$.

This property actually extends to all abelian groups. If G is an abelian group, let us denote by $G^\sharp$, the group G equipped with its maximal precompact group topology and by bG the completion of $G^\sharp$. In [**11**] van Douwen initiated a detailed analysis of the topological properties of $G^\sharp$ and, in doing so, he disclosed to general topologists a collection of questions that had by then been in consideration in harmonic analysis for at least 30 years. He in particular proved the following theorem that we take as our starting point.

Theorem 1.1 ([**11**]). *If G is an abelian group, every $A \subset G$ contains a subset D with $|D| = |A|$ that is relatively discrete and C^*-embedded in bG.*

2. Basic definitions

2.1. On Chu duality. Chu duality, called unitary duality by Chu [**5**], is based on giving a certain topological and algebraic structure to the set of finite dimensional representations of a topological group G. Denote to that end by G_n^x the set of all continuous n-dimensional unitary representations of G. It follows from a result of Goto [**20**] that the set G_n^x, equipped with the compact-open topology, is a locally compact space. The space $G^x = \bigsqcup_{n<\omega} G_n^x$ (as a topological sum) is called the *Chu dual* of G [**5**].

The algebraic structure of G^x is given by two standard operations: the direct sum and the tensor product of representations, that are induced by the corresponding operations between finite dimensional operators.

- $(\pi \oplus \pi')(x) = \pi(x) \oplus \pi'(x)$, for all $\pi, \pi' \in G^x$ and $x \in G$.
- $(\pi \otimes \pi')(x) = \pi(x) \otimes \pi'(x)$, for all $\pi, \pi' \in G^x$ and $x \in G$.

There is also a concept of equivalence for representations that is often useful: two representations $\pi_1, \pi_2 \in G_n^x$ are said to be *(unitarily) equivalent*, in symbols $\pi_1 \sim \pi_2$, when there is a unitary matrix U such that $\pi_1(x) = U^{-1}\pi_2(x)U$ for all $x \in G$. This clearly defines an equivalence relation in G^x.

The main feature of Chu duality is the construction of a *bidual* of G from the Chu dual G^x. Denoting by $\mathcal{U} = \bigsqcup_{n<\omega} \mathcal{U}(n)$, the topological sum of the spaces $\mathcal{U}(n)$ of $n \times n$ unitary matrices (topologized as subsets of $\mathbb{C}^{n^2}$), this bidual consists of the so-called continuous *quasi-representations*, i.e., mappings $Q : G^x \to \mathcal{U}$ satisfying:

- $Q[G_n^x] \subset \mathcal{U}(n)$.
- $Q(\pi \oplus \pi') = Q(\pi) \oplus Q(\pi')$, for all $\pi, \pi' \in G^x$.
- $Q(\pi \otimes \pi') = Q(\pi) \otimes Q(\pi')$, for all $\pi, \pi' \in G^x$.
- $Q(U^{-1}\pi U) = U^{-1}Q(\pi)U$, $\pi \in G_n^x$, $U \in \mathcal{U}(n)$.

See [**5**] or [**34**] or [**35**] for details.

The set of all *continuous* quasi-representations of G equipped with the compact-open topology is a topological group with pointwise multiplication as composition law, called the *Chu quasi-dual group* of G and denoted by G^{xx}. The evaluation map $\epsilon \colon G \to G^{xx}$ establishes a group homomorphism between G and G^{xx} that gives a measure of how strongly finite dimensional representations determine the structure of G. This homomorphism is one-to-one if and only if continuous finite dimensional representations separate points of G. Groups with that property are usually called maximally almost periodic, or MAP for short, and constitute the natural scope of Chu duality. The map ϵ is always continuous on compacta (an application of Ascoli's theorem) and hence ϵ is continuous for every locally compact group. When $\epsilon\colon G \to G^{xx}$ is in addition open and surjective (i.e., it is an isomorphism of topological groups) G is said to *satisfy Chu duality* or to be *Chu reflexive* (or simply *Chu*). Using this terminology, one has [**5**] that LCA groups and compact groups satisfy Chu duality (Chu duality actually reduces to the dualities of Pontryagin and Tannaka–Kreĭn respectively for such groups). There is a duality theory for nonabelian groups which is based on infinite-dimensional representations (a recent account of duality theory of locally compact groups is given in [**13**]). We shall not touch on this duality here.

2.2. The Bohr compactification. The Bohr compactification of a topological group G, can be defined as a pair (bG, b) where bG is a compact Hausdorff group and b is a continuous homomorphism from G onto a dense subgroup of bG such that every continuous homomorphism $h\colon G \to K$ into a compact group K extends to a continuous homomorphism $h^b\colon bG \to K$, making the lower triangle in the following diagram commutative:

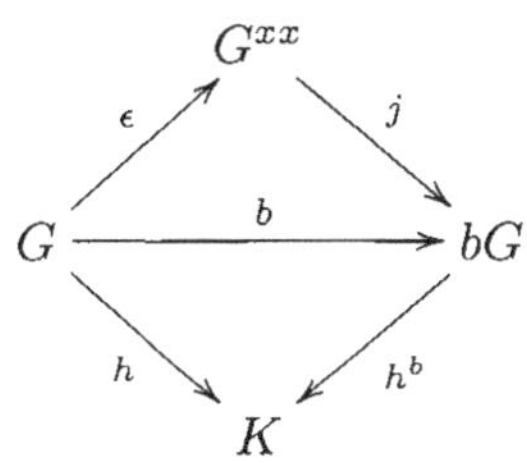

The upper triangle of this diagram gives the relation between Bohr compactifications and Chu duality. Chu [**5**] proved that the group of all quasi-representations of G (*continuous or not*), equipped with the topology of pointwise convergence on G^x provides a realization of bG. As G^{xx} consists of *continuous* quasi-representations, the inclusion homomorphism $j\colon G^{xx} \to bG$ that appears in the above diagram is clearly continuous and one-to-one.

The topology that b induces on G, will be referred to as the *Bohr topology*. Since $b = j \circ \epsilon$, the map b will be one-to-one exactly when ϵ is, in other words, the Bohr topology will be Hausdorff precisely when G is MAP. Since compact groups (and, in particular, bG) are completely determined by their finite-dimensional representations (this is Tannaka–Kreĭn duality), the Bohr topology of a group G may also be defined as the one that G inherits from its embedding in the product $\mathcal{U}^{G^x}$. We refer to [**34**, V, §14] or to [**35**] for a careful examination of bG and its properties.

3. Abelian groups

In the case of Abelian groups, the notions introduced above become essentially simpler. This is due to the fact that, for Abelian groups, all irreducible representations are one dimensional; that is, homomorphisms into the *torus*, $\mathbb{T}$, the group of all complex numbers of modulus one. These one dimensional representations are called *characters* and are the building blocks of the duality theory of Abelian groups (see [**55**]).

Let (G, τ) be an arbitrary topological abelian group. A *character* on (G, τ) is a continuous homomorphism χ from G to the torus $\mathbb{T}$. The set $\widehat{G}$ of all characters, equipped with the compact open topology, is a topological group with pointwise multiplication as the composition law, which is called the *dual group* of (G, τ). There is a natural evaluation homomorphism $\epsilon\colon G \to \widehat{\widehat{G}}$ of G into its bidual group. We say that a topological abelian group (G, τ) satisfies *Pontryagin–van Kampen duality* if the evaluation map ϵ is a topological isomorphism onto. The Pontryagin–van Kampen theorems establish that every LCA group satisfies P–vK duality.

In [**11**] van Douwen proved, among other things, the remarkable Theorem 1.1. Except for the standing abelian hypothesis, his proofs of results concerning $\sharp$-groups made no use whatsoever of specific algebraic properties. This probably led him to ask whether two groups G_1 and G_2 with the same cardinality should have $G_1^\sharp$ and $G_2^\sharp$ homeomorphic. Some years later Kunen [**37**] and, independently, Dikranjan and Watson [**10**], gave examples of *torsion* groups with the same cardinality having nonhomeomorphic $\sharp$-spaces. Still, much remains unknown. Actually, it is not yet known what happens with some utterly elementary groups:

963–964? **Question 1.** *Are the spaces $\mathbb{Z}^\sharp$ and $(\mathbb{Z} \times \mathbb{Z})^\sharp$ are homeomorphic? What about the spaces $\mathbb{Q}^\sharp$ and $\mathbb{Z}^\sharp$?*

One consequence of Theorem 1.1 is that $\sharp$-groups cannot contain infinite compact subsets. Indeed, the closure $\mathrm{cl}_{bG}\, D$ of a discrete and C^*-embedded subset D of bG is homeomorphic to βD, the Stone–Čech compactification of the *discrete* space D. If A is a compact subset of $G^\sharp$ and $D \subset A$ is as in Theorem 1.1, we obviously have $\mathrm{cl}_{bG}\, D \subset A$. But $|\mathrm{cl}_{bG}\, D| = |\beta D| = 2^{2^{|D|}} = 2^{2^{|A|}}$. This is a particular case of a general fact true for any LCA group, a pivotal result indeed about the Bohr topology of LCA groups.

Theorem 3.1 (Glicksberg, 1962 [**19**]). *Let G be an LCA group. If $A \subset G$ is compact in bG, then A is compact in G.*

Theorem 3.1 in its full generality can actually be deduced from Theorem 1.1 and, conversely, Theorem 1.1 follows from Theorem 3.1, by way of Rosenthal's ℓ^1 theorem. These relations will be explored in Section 4. Further properties concerning the Bohr topology of a LCA group can be found in [**6, 7, 17, 30**].

4. Nonabelian groups

Here we will focus on determining to what extent the results concerning duality theory and Bohr topology of abelian groups can be extended to the noncommutative context. The first contributions to these program have been given by Chu [**5**], Heyer [**34, 35**], Landstad [**39**], Moran [**43**], Poguntke [**46–48**], and Roeder [**53**]. Recent contributions to the subject can be found in [**8, 18, 25–27, 32, 51, 52, 62**]. Nevertheless, we do not know how Chu reflexive groups are placed within the class of locally compact groups and this is one of the major difficulties for understanding Chu duality. Therefore, the main question here is:

Question 2. *Characterize the (necessarily MAP) locally compact groups that sat-* 965?
isfy Chu duality.

Obviously Question 2 leaves open a good number of other questions about Chu duality. Firstly, we give a brief account on the subject. A topological space X is called *hemicompact* if there is a countable family of compact subsets $\{K_n : n \in \omega\}$ of G such that every compact subset L of G is contained in some K_n.

Proposition 4.1. *Let G be a locally compact MAP group.*

(1) *If G is discrete (resp. metrizable), then G_n^x is compact (resp. hemicompact).*
(2) *Conversely, if G is compact then each equivalence class defined by $\sim$ is open. Therefore, the quotient space $G^x/\sim$ is discrete.*
(3) *If G is second countable then G_n^x and G^{xx} are second countable. As a consequence G^x is metrizable. In this case G is Chu-reflexive if and only if the evaluation map ϵ is onto.*
(4) *G^{xx} need not be locally compact, even for countable G, [**32, 52**].*

Now, we recall a notion due to Takahashi [**58**] in order to obtain a representation of the Chu quasi-dual for some classes of groups. For each locally compact group G, Takahashi has constructed a locally compact group G_T called *Takahashi quasi dual* such that G_T is maximally almost periodic, and G'_T is compact. The category of locally compact groups with these two properties is denoted by TAK. If $n > 1$ and $D \in \mathrm{Hom}_c(G, U(n))$ then the sets $t_n(D; U) = \{D \otimes \chi : \chi \in U\}$, U any neighborhood of the identity in the group G_1^x, form a fundamental system of neighbourhoods of D for a topology in $\mathrm{Hom}_c(G, U(n))$. We denote by G_n^t the set $\mathrm{Hom}_c(G, U(n))$ equipped with this topology and the symbol G^t denotes the topological sum of the spaces G_n^t, for $n \in \mathbb{N}$. A *unitary mapping* on G^t is a continuous mapping $p \colon G^t \to \mathcal{U}$ conserving the main operations between unitary representations (see [**46**] for details). The set of all unitary mappings on G^t equipped with the compact-open topology is a topological group, with pointwise multiplication as the composition law, which is usually called the *Takahashi quasi-dual group* of G and is denoted by G_T. It is easily verified that $G^{xx} \subset G_T \subset bG$. The Takahashi duality theorem establishes that $G \cong G_T$ if $G \in$ TAK. A detailed discussion and extension of this theory has been given by Poguntke in [**46**].

Concerning Chu duality, the first difficulty is to identify the quasidual group G^{xx} of a locally compact group G. Some extreme situations, totally alien to the abelian case may actually appear, as for instance that $G^{xx} = bG$ or, what is the same (see [**18**, **32**]), that $\widehat{G}_n$ is discrete for every n. Next follows some examples that illustrate the different situations that may arise.

Example 4.2 (Moran [**43**]). *Let $\{p_i\}$ be an infinite sequence of distinct prime numbers ($p_i > 2$), and let F_i be the projective special linear group of dimension two over the Galois field $GF(p_i)$ of order p_i. If $G = \sum_{i\in\mathbb{N}} F_i$, we have $G^{xx} = G_T = bG$.*

Example 4.3 (Heyer [**35**]). *Let $\mathbb{Z}_3 \rtimes \mathbb{Z}_2 = S_3$ the permutation group. Define $G_i = \mathbb{Z}_3 \rtimes \mathbb{Z}_2$ for all $i \in \mathbb{N}$ and take $G = \sum_{i\in\mathbb{N}} G_i$. Then $G = G^{xx}$ and $G_T = \prod_{i\in\mathbb{N}} \mathbb{Z}_3 \rtimes \sum_{i\in\mathbb{N}} \mathbb{Z}_2$.*

More recently, we have the following results (cf. [**32**]).

Example 4.4. *Let p a prime number greater than 2, and let F_i be the projective special linear group of dimension two over the Galois field $GF(p)$ of order p. If $G = \sum_{i\in\mathbb{N}} F_i$, we have $G^{xx} = G$ and $G_T = bG$.*

Proposition 4.5. *Let G be a simple MAP discrete group (which implies $G' = G$). Then the following conditions are equivalent:*

(i) $G^{xx} = G_T$;

(ii) $\widehat{G}_n$ *is discrete for all* $n \in \mathbb{N}$;

(iii) $G^{xx} = bG$.

Proposition 4.6. *Let G be a discrete MAP group that is nilpotent of length two, and such that for each positive integer n there are only finitely many co-finite normal subgroups H of G' whose index is less or equal than n. Then $G^{xx} \cong G_T$.*

Proposition 4.6 is a variation of the following nice result due to Poguntke [**47**, **48**].

Corollary 4.7 (Poguntke, 1976). *The Heisenberg integral group H, satisfies that $H^{xx} \cong H_T$.*

Theorem 4.8. *Let G be a discrete MAP group that is an FC group and, for each positive integer n, there are only finitely many co-finite normal subgroups H of G' such that G'/H accepts faithful representations into $U(n)$. Then $G^{xx} \cong G_T$.*

Corollary 4.9. *Let $G = \sum_{n\in\mathbb{N}} F_n$, where each F_n' is simple and $\lim_{n\to\infty} \exp(F_n') = \infty$. Then $G^{xx} \cong G_T$.*

Furthermore, if the symbol $\mathrm{mdus}(G)$ denotes the minimal natural number n such that the unitary representations of dimension less or equal than n separate the points in G, then for FC groups we have:

Theorem 4.10. *Let G be an FC group and suppose there is $N \in \mathbb{N}$ such that $\exp(G') \leq N$ and $\mathrm{mdus}(G/H) \leq N$ for all normal subgroup H of G that is co-finite in G'. Then the group G is Chu reflexive.*

Examples 4.3 and 4.2 also follow from Theorem 4.10. Finally, next example shows that the Chu quasi-dual group G^{xx} need not be locally compact even for a countable discrete group G.

Example 4.11. *Let $\{p_n\}$ be an infinite sequence of distinct prime numbers ($p_n > 2$), and let $G_n = PSL(2, p_n)$ be the projective special group of dimension two over the finite field of order p_n. For each n, let $G_{n,m}$ be a copy of G_n, for $m = 1, 2, \ldots$. Let $G = \sum_{n=1}^{\infty}(\sum_{m=1}^{\infty} G_{n,m})$ with the discrete topology. The group G^{xx} is not locally compact.*

We have already mentioned the most recent results about the unitary or Chu duality (cf. [**18**, **32**, **52**]). Nevertheless, the subject is far from being settled. There are already too many open question that obstruct the progress in this area of research.

The following two questions concern the very basic structure of Chu duality; their solution would be of much importance for the study of Chu duality.

Question 3 (Poguntke, 1976 [**48**]). *Let G be a locally compact MAP group with evaluation map $\epsilon\colon G \to G^{xx}$. Is $\epsilon(G)$ dense in G^{xx}?* 966?

Chu [**5**] asserted that $bG = bG^{xx}$ for every locally compact group G. The proof in [**5**] of this fact is however incomplete, and this remains indeed as one of the main open questions.

Question 4 (Wu, 2000). *Let G be a locally compact MAP group. Is it true that $bG = bG^{xx}$?* 967?

A positive solution to Question 3 would imply a positive solution to Question 4.

Question 5 (Chu, 1966 [**5**]). *Does the free group with two generators, equipped with the discrete topology, satisfy Chu duality?* 968?

This question appears as one of the major difficulties for a full understanding of Chu duality. Avoiding it does not however answer all questions.

Question 6. *Characterize the MAP, discrete groups without free nonabelian subgroups that satisfy Chu duality.* 969?

Among groups with no free (nonabelian) subgroups, amenable groups are especially important. A topological group G is *amenable* when the Banach space $\ell^\infty(G)$ admits a left-invariant mean, that is, a continuous linear functional Λ on $\ell^\infty(G)$ with $\Lambda(1) = 1$ and $\Lambda(L_x f) = \Lambda(f)$, for every $x \in G$ and every $f \in \ell^\infty(G)$ (here L_x denotes as usual the left action of x on f, $(L_x f)(g) = f(x^{-1}g)$). For discrete G, this is equivalent, to the existence of a finitely-additive left-invariant probability measure on G. Amenability has a strong impact on the representation-theoretic properties of a locally compact group, see [**44**] for instance. Compact and abelian groups are amenable while any group having a discrete free nonabelian subgroup is not.

Question 7. *Characterize the MAP amenable, locally compact groups that satisfy Chu duality.* 970?

5. How is G placed in bG? Interpolation sets

Let G be a topological group and let X denote a point-separating, uniformly closed, self-adjoint subalgebra of $CB(G)$ (continuous, complex-valued, bounded functions on G). A subset S of G is said to be an *X-interpolation set* provided that every bounded function $f: S \to \mathbb{C}$ (continuous or not) can be extended to a function $\bar{f}: G \to \mathbb{C}$ with $\bar{f} \in X$.

Each closed subalgebra X as above is a commutative C^*-algebra and we can apply the full-strength of Gelfand's duality to it, see for instance [**55**]. Let $\sigma(X)$ denote the space of multiplicative linear functionals on X (i.e., linear functionals $T: X \to \mathbb{C}$ with $T(fg) = T(f)T(g)$, for all $f, g \in X$). The set $\sigma(X)$ with the topology of pointwise convergence on X is a compact topological space called the *spectrum* of X. Every element $f \in X$ can then be identified with a function $E_f \in C(\sigma(X), \mathbb{C})$ via evaluations ($E_f(T) = T(f)$ for every $T \in \sigma(X)$). The main consequence of Gelfand's duality is that this identification establishes an isomorphism of C^*-algebras.

The compact space $\sigma(X)$ also defines a compactification of G. Taking into account that the elements of X are continuous functions on G, we have an evaluation mapping $j: G \to \sigma(X)$ (given by $j(g)(T) = T(g)$) that defines a one-to-one continuous map with dense range. From this point of view X-interpolation sets are those subsets of G that are discrete and C^*-embedded in $\sigma(X)$.

The Bohr compactification can be obtained in the preceding way by considering $X = AP(G)$, the algebra of almost periodic functions on G. A bounded function $f: G \to \mathbb{C}$ is *almost periodic* if the set of translates $\{L_x f: x \in G\}$ is a compact subset of $CB(G)$ (for the topology of uniform convergence). A function $f: G \to \mathbb{C}$ turns to be almost periodic if and only if it is the uniform limit of matrix coefficients[1] of finite-dimensional unitary representations. Thus the almost periodic functions are precisely the functions that admit a continuous extension to bG. The spectrum $\sigma(AP(G))$ of $AP(G)$ can then be identified with the Bohr compactification of G.

The X-interpolation sets for $X = AP(G)$ are called I_0-sets. I_0-sets were first studied by Hartman and Ryll-Nardzewski in the 1960s in a series of papers starting with [**28**, **29**]. The fact that lacunary sequences of integers are I_0-sets was first proved in [**57**] (see [**38**] for a recent proof).

5.1. Existence and abundance of interpolation sets.

Existence problems on interpolation sets are amenable to topological techniques as the proof of van Douwen's theorem 1.1 [**11**] shows (see [**16**] for a simpler proof). We recast here Theorem 1.1 in terms of I_0-sets:

Theorem 5.1. *Every infinite subset A of a discrete abelian group G contains an I_0-set S with $|S| = |A|$.*

---

[1] If π is a unitary representation of a group G on a Hilbert space $\mathbb{H}$, a matrix coefficient of π is a complex-valued function $g \mapsto \langle \pi(g)\xi, \eta \rangle$, with $\xi, \eta \in \mathbb{H}$.

A sequence $S = (x_n : n \in \omega)$ of a Banach space E is said to be an (or equivalent to the) ℓ^1-basis if the map sending x_n to the canonical basis (e_n) of ℓ^1 extends to a linear homeomorphism on the closed linear span of S. Interpolation sets share many properties with ℓ^1-bases (see for instance [**3**], and what follows). This relation can be a very fruitful one, mainly because the existence of an ℓ^1-basis is neatly characterized by Rosenthal's well-known theorem.

Theorem 5.2 (Rosenthal 1971, [**54**]). *A bounded sequence in a Banach space either has a weakly Cauchy subsequence or has a subsequence which is an ℓ^1-basis.*

If E is a Banach space, a sequence (x_n) in E is a *weakly Cauchy* sequence if $f(x_n)$ is convergent for every continuous linear functional f on E.

Rosenthal's theorem relates the presence of an ℓ^1-basis to the absence of weakly convergent sequences. It can be adapted to provide a similar relation with interpolation sets, this is the Rosenthal-type theorem for locally compact groups that appears in [**18**].

Theorem 5.3. *Let G be a metrizable locally compact group. A sequence in G admitting no Bohr Cauchy subsequence (i.e., no subsequence converging to an element of bG), must contain an infinite I_0-subset.*

Observe that the combination of Glicksberg's Theorem 3.1 and Theorem 5.3 implies Theorem 1.1 for countable abelian groups. As indicated in Section 2, it is also true that Theorem 3.1 follows from an appropriate extension of Theorem 1.1 to nondiscrete groups, see [**17**], a fact that was used there to prove Glicksberg's-type theorems for some abelian nonlocally compact groups. These theorems are mainly based on the following analog of Theorem 1.1 that appears in [**17**]:

Theorem 5.4. *Let G be abelian, locally connected and Čech-complete. Every subset A of $\widehat{G}$ that is not equicontinuous as a set of $\mathbb{T}$-valued functions on G must contain an infinite I_0-set.*

Both approaches have failed so far to provide a general answer for the simplest questions about I_0 sets in the case of nonabelian locally compact (even discrete) groups. The relevant question here therefore is:

Question 8. *Which (countable) discrete groups contain no nontrivial Bohr convergent sequences? Or equivalently, which groups G have infinite I_0-sets inside every infinite subset $A \subset G$?* 971?

As far as we know the first noncommutative theorem related to Question 8 was given by Moran [**43**]. We need the concept of direct integral of a representation to understand his result. Roughly, the direct integral $\pi = \int_A^{\oplus} \pi_\alpha \, d\mu(\alpha)$ of a family of unitary representations π_α, where $\alpha \in A$ runs on a measure space (A, μ), is another representation such that its matrix coefficients are obtained as ordinary integrals, of the matrix coefficients of the representations π_α.

Theorem 5.5 (Moran, 1971 [**43**]). *Let G be a locally compact group and suppose its left regular representation can be decomposed as a direct integral of representations* almost all *of which are finite-dimensional. Then every Bohr convergent sequence of G is also convergent in the locally compact topology.*

Corollary 5.6. *Let G be a group that satisfies the hypothesis of the theorem above. Then every subset A of G either has compact closure or contains an infinite I_0-set.*

It happens that *every* unitary representation of a locally compact group may be obtained (often in several unrelated ways) as a direct integral of irreducible representations. With this fact in mind, Theorem 5.5 applies directly to those groups G whose irreducible representations are all finite dimensional, so-called *Moore groups*. In this line, Remus and Trigos-Arrieta have proved the following result that avoids direct integrals.

Theorem 5.7 (Remus and Trigos-Arrieta, 1999 [**51**]). *If the locally compact group G is Moore then G respects compactness.*

In the opposite direction we have.

Theorem 5.8 (Wu and Riggins, 1996 [**62**]). *Let G be a maximally almost periodic FC group that contains no nontrivial convergent sequences. Then G is abelian by finite (that is, has a normal Abelian subgroup of finite index).*

These results leave open the following main question.

972–973? **Question 9.** *Let G be a discrete group that contains no nontrivial convergent sequences.*

(a) *Is G abelian by finite?*
(b) *Can the left regular representation be decomposed as direct integral of finite dimensional representations?*

In the positive direction, we have the following result that appears in [**31**].

Theorem 5.9. *Let G be a finitely generated discrete group without nonabelian free subgroups. Then G has no non-trivial Bohr convergent sequences if and only if G is abelian by finite.*

Theorem 5.9 displays some examples of discrete groups that have Bohr convergent sequences despite having some good commutativity properties. For instance, the Heisenberg integral group, the *lamplighter group* $(\sum_{\mathbb{Z}} \mathbb{Z}/2\mathbb{Z}) \rtimes \mathbb{Z}$ or the direct sum $G = \sum_{n\in\mathbb{N}} F_n$, with F_n finite, simple and nonabelian. In order to solve Question 8, one has to overcome an important obstacle, namely that of dealing with the free group with two generators.

974? **Question 10.** *Does $F(a,b)$ contain non-trivial Bohr convergent sequences?*

The question may be extended to:

975? **Question 11.** *Characterize the MAP locally compact groups whose compact subsets are the same in the original and Bohr topologies.*

In connection with the Bohr topology of locally compact groups, several authors have considered the so-called *van der Waerden* (or *self-bohrifying*) groups. That is, compact groups G satisfying that $bG_d = G$, where G_d denotes the same algebraic group G equipped with the discrete topology. In this direction van der Waerden proved that every (algebraic) homomorphism from a compact connected semisimple Lie group into a compact group is continuous (cf. [**60**]). We mention here the following question along this direction, see [**26**].

Question 12. *What are the direct products $G = \prod_{i\in I} F_i$ of finite groups F_i such that the Bohr compactification of G_d is topologically isomorphic to G?* 976?

5.2. Sidon sets. The matrix coefficients of all (finite- or infinite-dimensional) unitary representations also constitute an algebra. This is the Fourier–Stieltjes algebra B(G) introduced by Eymard in [**14**]. When G is abelian B(G) reduces to the set of Fourier–Stieltjes transforms of measures of the dual group, see [**12**].

Let $\overline{\mathrm{B(G)}}$ denote the uniform closure of the Fourier–Stieltjes algebra of G. $\overline{\mathrm{B(G)}}$-interpolation sets in discrete (and locally compact) abelian groups G have been deeply studied under the name of *Sidon* sets, see for instance [**41**]. It should be remarked that $\overline{\mathrm{B(G)}}$-interpolation sets on noncommutative groups also appear in the literature as *weak Sidon* sets, see [**45**] for instance.

With these definitions in mind, one has that Sidon subsets of a group G are discrete and C^*-embedded in the spectrum of $\overline{\mathrm{B(G)}}$. Following [**42**] we will refer to this spectrum as the *Eberlein compactification* of G and denote it by eG. There are two main differences between bG and eG. Firstly, eG is no longer a topological group, only a semitopological semigroup, secondly eG is a proper compactification of G, the embedding of G in eG is a homeomorphism and thus a Glicksberg's-type theorem makes no sense for eG. The question on which sequences contained in G converge to some point in eG (we will refer to this property as being *eG-Cauchy*) does however make sense, and is the truly relevant one. After adapting Rosenthal's theorem to Sidon sets (as it was done in Theorem 5.3 to I_0-sets) the question that corresponds to Question 8 is:

Question 13. *Can discrete groups contain nontrivial eG-Cauchy sequences? Or equivalently, does every infinite subset of a discrete group G have infinite Sidon subsets?* 977?

Sidon sets are far more abundant than I_0-sets in noncommutative groups. We have for instance the following counterpart to Theorem 5.8.

Theorem 5.10 (de Michele and Soardi [**9**]). *Any infinite subset of a discrete FC group contains an infinite Sidon subset.*

It should be noticed, and this goes in the same direction of the preceding theorem, that, contrarily to the I_0-case, Sidon subsets of subgroups of a discrete group G are necessarily Sidon subsets of G. Using this fact, it is easy to see that infinite subsets of solvable groups always contain infinite Sidon sets. The following questions should by the same reason be far easier than Question 8 or 10:

978? **Question 14.** *Does every infinite subset of the free group on two generators contain an infinite Sidon subset?*

979? **Question 15.** *Does every infinite subset of an amenable discrete group contain an infinite Sidon subset?*

Finally, the main question here is

980? **Question 16** (López and Ross, 1975 [**41**]). *Does every discrete group contain some infinite Sidon set?*

It is important to remark that a discrete group with no infinite Sidon sets must necessarily be a torsion group. Even more, no subgroup of a torsion group may contain infinite Sidon sets at all since Sidon subsets of subgroups of *discrete* groups are Sidon.

The weakness of the algebraic structure of eG is also important for the very existence of Sidon sets. Note that the mere existence of an interpolation set implies that the cardinality of the space eG be at least $2^{\mathfrak{c}}$. Thence the interest on knowing which groups admit some interpolation set and which have none at all. Thanks to the Bourgain–Fremlin–Talagrand theorem, that question has a satisfactory answer for I_0-sets:

Theorem 5.11 ([**18**]). *Let G be a maximally almost periodic second countable topological group. The following assertions are equivalent.*

(1) *G has no I_0-sets.*
(2) *bG is Rosenthal compact (that is, bG is homeomorphic to a compact subset of $B_1(X)$, the space of all first class Baire functions defined on some Polish space X).*
(3) *The Bohr compactification bG of G is metrizable.*
(4) *$|bG| = \mathfrak{c}$.*
(5) *G has at most countably many inequivalent finite dimensional unitary representations.*

Countable groups always have a continuum of pairwise inequivalent irreducible representations (cf. [**1**]) and the same is true for every connected second countable locally compact group. Since $\overline{B(G)}$ is made from matrix coefficients of general unitary representations, just as $AP(G)$ is made from matrix coefficients of finite dimensional ones, it could be expected that the arguments leading to Theorem 5.11 also imply that every discrete or second countable connected locally compact group contains an infinite Sidon set. Notice that the absence of Sidon sets in G implies that eG is Rosenthal-compact and thus of cardinality $\mathfrak{c}$. But the failure of eG to be a topological group makes statements (3), (4) and (5) of Theorem 5.11 nonequivalent. Take for instance $G = SL(2,\mathbb{R})$, all nonconstant functions in $B(G)$ vanish at infinity, i.e., $B(G) = C_0(G) \oplus \mathbb{C}$, and eG can be identified with the one-point compactification of G. The Eberlein compactification of this group is therefore metrizable despite having uncountably many inequivalent irreducible representations. Concerning the equivalence between (2) and (3) the absence of better examples (in particular of discrete ones) leaves without answer the following question.

Question 17. *Can a non-metrizable Eberlein compactification eG be Rosenthal compact?* 981?

A negative answer would simplify and provide a higher lever of applicability of topological techniques for the Questions 14–16.

5.3. Other compactifications. The questions discussed in the preceding subsections regarding $\mathrm{AP}(G)$- and $\overline{\mathrm{B}(\mathrm{G})}$-interpolation sets (i.e., I_0- and Sidon sets, respectively) have easier answers in the case of X-interpolation sets with bigger X. The most immediate case is the algebra of weakly almost periodic functions $X = \mathrm{WAP}(G)$. A bounded function $\phi\colon G \to \mathbb{C}$ is *weakly almost periodic* if the set of translates $\{L_x f\colon x \in G\}$ is a weakly compact subset of $CB(G)$. Ruppert defines in [**56**] *translation-finite* sets as those sets $A \subset G$, G discrete, such that every bounded $f\colon G \to \mathbb{C}$ that vanishes off A is weakly almost periodic. These sets, called R_W-sets by Chou [**4**], are $\mathrm{WAP}(G)$-interpolation sets and by [**56**, Proposition 13], every infinite subset of a discrete group contains an infinite translation-finite subset. See [**15**] for an extension of this fact to more general (not necessarily locally compact) topological groups.

5.4. The structure of Sidon and I_0-sets. Perhaps the oldest question regarding interpolation sets is whether a Sidon set may be dense in the Bohr compactification. This is open even in the simplest groups:

Question 18 ([**36**, **41**]). *Can a Sidon subset of $\mathbb{Z}$ be dense in $b\mathbb{Z}$?*[2] 982?

A probabilistic argument due to Katznelson [**36**] seems to suggest a negative answer for Question 18. A theorem of Ramsey [**49**] shows that Question 18 is equivalent to the following one:

Question 19. *Can a Sidon subset of $\mathbb{Z}$ cluster (in the Bohr topology) at some point of $\mathbb{Z}$?*

Although the equivalence between Questions 18 and 19 could point towards a positive answer to the former, the converse conjecture gains strength if we compare with I_0-sets. By a theorem of Hartman and Ryll-Nardzewski [**28**] no point of G can be a Bohr-cluster point of an I_0-subset of G (the union of an I_0 set and a point is again I_0 and therefore discrete in bG). It is clear in this regard that a deeper knowledge of the relations between I_0 sets and Sidon sets (see [**50**] and the references therein) would help with Question 18. In particular an affirmative answer to the following question implies a negative to Question 18:

---

[2] A negative answer to the preceding question would leave some space to this one: is the closure of a Sidon subset of G a Helson subset of the Bohr compactification?, see [**55**] for the definition of a Helson set in a compact group K, needless to say that the whole group is not Helson.

983? **Question 20** (Grow, 1987 [**22**]). *Is every Sidon subset of $\mathbb{Z}$ a finite union of I_0-sets?*

Bourgain had already shown in [**2**] that the answer to Question 20 is positive for groups G of bounded order (groups with $mg = 0$ for some $m \in \mathbb{Z}$ and all $g \in G$).

Lefèvre and Rodríguez-Piazza have shown that interpolation sets with a lower degree of lacunarity, namely Rosenthal-sets, can be dense in the Bohr compactification (see [**40**]). This somehow shows how the case of Sidon sets consitutes a *limiting case.*

We finally mention two questions that appear in [**21**] and concern the structure of I_0-sets, just as the preceding questions concern the structure of Sidon sets. We need here the concepts of $I_0(U)$-set and ϵ-Kronecker set. If G is a compact group and $U \subseteq G$, we say $E \subset \widehat{G}$ is $I_0(U)$ if every bounded function on E is the restriction of the Fourier–Stieltjes transform of a discrete measure supported on U. A set $E \subset \widehat{G}$ is *ϵ-Kronecker* for some $\epsilon > 0$, if for every continuous function $\phi\colon E \to \mathbb{T}$ there exists $x \in G$ such that $|\gamma(x) - \phi(\gamma)| < \epsilon$ for all $\gamma \in E$.

984? **Question 21** (Graham, Hare and Körner, [**21**]). *Is every I_0-set a finite union of sets in a more limited class? Perhaps a finite union of ϵ-Kronecker sets?*

985–986? **Question 22** (Graham, Hare and Körner, [**21**]). *Is every I_0 set $I_0(U)$ for all $U \subset G$? (this assumes G to be connected) What about when $G = \mathbb{T}$?*

Lack of space has refrained us from referring to another whole lot of problems on interpolation sets that might respond to topological treatment. These concern interpolation sets in dual objects of compact groups, see [**23**] for a recent account and references to previous results.

Acknowledgement

We would like to thank Dikran Dikranjan and the referees for several suggestions and comments that have helped us to improve parts of this paper.

References

[1] L. Baggett, *A separable group having a discrete dual space is compact*, J. Funct. Anal. **10** (1972), 131–148.

[2] J. Bourgain, *Propriétés de décomposition pour les ensembles de Sidon*, Bull. Soc. Math. France **111** (1983), no. 4, 421–428.

[3] J. Bourgain, *ℓ^1 sequences generated by Sidon sets*, J. London Math. Soc. **29** (1984), no. 2, 283–288.

[4] C. Chou, *Weakly almost periodic functions and thin sets in discrete groups*, Trans. Amer. Math. Soc. **321** (1990), no. 1, 333–346.

[5] H. Chu, *Compactification and duality of topological groups*, Trans. Amer. Math. Soc. **123** (1966), 310–324.

[6] W. W. Comfort, S. Hernández, and F. J. Trigos-Arrieta, *Relating a locally compact abelian group to its Bohr compactification*, Adv. Math. **120** (1996), no. 2, 322–344.

[7] W. W. Comfort, S. Hernández, and F. J. Trigos-Arrieta, *Cross sections and homeomorphism classes of abelian groups equipped with the Bohr topology*, Topology Appl. **115** (2001), no. 2, 215–233.

[8] W. W. Comfort, D. Remus, and H. Szambien, *Extending ring topologies*, J. Algebra **232** (2000), no. 1, 21–47.

[9] L. De-Michele and P. M. Soardi, *Existence of Sidon sets in discrete FC-groups*, Proc. Amer. Math. Soc. **55** (1976), no. 2, 457–460.

[10] D. Dikranjan and S. Watson, *A solution to van Douwen's problem on Bohr topologies*, J. Pure Appl. Algebra **163** (2001), no. 2, 147–158.

[11] E. K. van Douwen, *The maximal totally bounded group topology on G and the biggest minimal G-space, for abelian groups G*, Topology Appl. **34** (1990), no. 1, 69–91.

[12] C. F. Dunkl and D. E. Ramirez, *Topics in harmonic analysis*, Appleton-Century-Croft, New York, 1971.

[13] M. Enock and J.-M. Schwartz, *Kac algebras and duality of locally compact groups*, Springer, Berlin, 1992.

[14] P. Eymard, *L'algèbre de Fourier d'un groupe localement compact*, Bull. Soc. Math. France **92** (1964), 181–236.

[15] S. Ferri and J. Galindo, *On the WAP-compactification of a topological group*, 2006, Preprint.

[16] J. Galindo and S. Hernández, *On a theorem of van Douwen*, Extracta Math. **13** (1998), no. 1, 115–123.

[17] J. Galindo and S. Hernández, *The concept of boundedness and the Bohr compactification of a MAP abelian group*, Fund. Math. **159** (1999), no. 3, 195–218.

[18] J. Galindo and S. Hernández, *Interpolation sets and the Bohr topology of locally compact groups*, Adv. Math. **188** (2004), no. 1, 51–68.

[19] I. Glicksberg, *Uniform boundedness for groups*, Canad. J. Math. **14** (1962), 269–276.

[20] M. Goto, *Note on a topology of a dual space*, Proc. Amer. Math. Soc. **12** (1961), 41–46.

[21] C. C. Graham and K. E. Hare, *ϵ-Kronecker and I_0 sets in abelian groups. III. Interpolation by measures on small sets*, Studia Math. **171** (2005), no. 1, 15–32.

[22] D. Grow, *Sidon sets and I_0-sets*, Colloq. Math. **53** (1987), no. 2, 269–270.

[23] K. E. Hare and L. T. Ramsey, *I_0 sets in non-abelian groups*, Math. Proc. Cambridge Philos. Soc. **135** (2003), no. 1, 81–98.

[24] P. de la Harpe and A. Valette, *La propriété (T) de Kazhdan pour les groupes localement compacts*, Astérisque (1989), no. 175, 158 pp.

[25] J. E. Hart and K. Kunen, *Bohr compactifications of discrete structures*, Fund. Math. **160** (1999), no. 2, 101–151.

[26] J. E. Hart and K. Kunen, *Bohr compactifications of non-abelian groups*, Topology Proc. **26** (2001/02), no. 2, 593–626.

[27] J. E. Hart and K. Kunen, *Bohr topologies and compact function spaces*, Topology Appl. **125** (2002), no. 2, 183–198.

[28] S. Hartman and C. Ryll-Nardzewski, *Almost periodic extensions of functions*, Colloq. Math. **12** (1964), 23–39.

[29] S. Hartman and C. Ryll-Nardzewski, *Almost periodic extensions of functions. II*, Colloq. Math. **15** (1966), 79–86.

[30] S. Hernández, *The dimension of an LCA group in its Bohr topology*, Topology Appl. **86** (1998), no. 1, 63–67.

[31] S. Hernández and T. S. Wu, *The Bohr topology of non abelian groups*, 2005, Preprint.

[32] S. Hernández and T. S. Wu, *Some new results on the Chu duality of discrete groups*, Monatsh. Math. (2006), Article in press.

[33] E. Hewitt and K. A. Ross, *Abstract harmonic analysis. Vol. II*, Springer, Berlin, 1970.

[34] H. Heyer, *Dualität lokalkompakter Gruppen*, Lecture Notes in Math., vol. 150, Springer, Berlin, 1970.

[35] H. Heyer, *Groups with Chu duality*, Probability and information theory, II, Lecture Notes in Math., vol. 296, Springer, Berlin, 1973, pp. 181–215.

[36] Y. Katznelson, *Sequences of integers dense in the Bohr group*, Proc. Roy. Inst. of Tech. (1973), no. June, 79–86.

[37] K. Kunen, *Bohr topologies and partition theorems for vector spaces*, Topology Appl. **90** (1998), 97–107.

[38] K. Kunen and W. Rudin, *Lacunarity and the Bohr topology*, Math. Proc. Cambridge Philos. Soc. **126** (1999), no. 1, 117–137.

[39] M. B. Landstad, *On the Bohr topology in amenable topological groups*, Math. Scand. **28** (1971), 207–214.

[40] P. Lefèvre and L. Rodríguez-Piazza, *The union of a Riesz set and a Lust–Piquard set is a Riesz set*, J. Funct. Anal. **233** (2006), no. 2, 545–560.

[41] J. M. López and K. A. Ross, *Sidon sets*, Marcel Dekker Inc., New York, 1975.

[42] M. Mayer, *Strongly mixing groups*, Semigroup Forum **54** (1997), no. 3, 303–316.

[43] W. Moran, *On almost periodic compactifications of locally compact groups*, J. London Math. Soc. **3** (1971), 507–512.

[44] A. L. T. Paterson, *Amenability*, American Mathematical Society, Providence, RI, 1988.

[45] M. A. Picardello, *Lacunary sets in discrete noncommutative groups*, Univ. Genova Pubbl. Ist. Mat. (2) (1973), no. 60, ii+24.

[46] D. Poguntke, *A universal property of the Takahashi quasi-dual*, Canad. J. Math. **24** (1972), 530–536.

[47] D. Poguntke, *Chu-Dualität und zwei Klassen maximal fastperiodischer Gruppen*, Monatsh. Math. **82** (1976), no. 1, 31–50.

[48] D. Poguntke, *Zwei Klassen lokalkompakter maximal fastperiodischer Gruppen*, Monatsh. Math. **81** (1976), no. 1, 15–40.

[49] L. T. Ramsey, *Bohr cluster points of Sidon sets*, Colloq. Math. **68** (1995), no. 2, 285–290.

[50] L. T. Ramsey, *Comparisons of Sidon and I_0 sets*, Colloq. Math. **70** (1996), no. 1, 103–132.

[51] D. Remus and F. J. Trigos-Arrieta, *The Bohr topology of Moore groups*, Topology Appl. **97** (1999), no. 1–2, 85–98.

[52] L. Riggins, *On infinite groups and unitary duality*, Ph.D. thesis, Case Western Reserve University, 1998.

[53] D. W. Roeder, *A characterization of unitary duality*, Trans. Amer. Math. Soc. **148** (1970), 129–135.

[54] H. P. Rosenthal, *A characterization of Banach spaces containing ℓ^1*, Proc. Natl. Acad. Sci. **71** (1974), 2411–2413.

[55] W. Rudin, *Fourier analysis on groups*, John Wiley & Sons Inc., New York, 1990.

[56] W. A. F. Ruppert, *On weakly almost periodic sets*, Semigroup Forum **32** (1985), no. 3, 267–281.

[57] E. Strzelecki, *On a problem of interpolation by periodic and almost periodic functions*, Colloq. Math. **11** (1963), 91–99.

[58] S. Takahashi, *A duality theorem for representable locally compact groups with compact commutator subgroup*, Tôhoku Math. J. (2) **4** (1952), 115–121.

[59] J. Tits, *Free subgroups in linear groups*, J. Algebra **20** (1972), 250–270.

[60] B. L. van der Waerden, *Stetigkeitssätze für halbeinfache Liesche Gruppen*, Math. Z. **36** (1933), no. 1, 780–786.

[61] P. S. Wang, *On isolated points in the dual spaces of locally compact groups*, Math. Ann. **218** (1975), no. 1, 19–34.

[62] T. S. Wu and L. Riggins, *Maximally almost periodic groups and a theorem of Glicksberg*, Papers on general topology and applications (Gorham, ME, 1995), Ann. New York Acad. Sci., vol. 806, New York Acad. Sci., 1996, pp. 454–464.

Open Problems in Topology II – Edited by E. Pearl

Topological transformation groups: selected topics

Michael Megrelishvili

1. Introduction

In this paper all topological spaces are Tychonoff. A topological transformation group, or a G-space, as usual, is a triple (G, X, π), where $\pi\colon G \times X \to X$, $\pi(g, x) := gx$ is a continuous action of a topological group G on a topological space X. Let G act on X_1 and on X_2. A continuous map $f\colon X_1 \to X_2$ is a *G-map* (or, an *equivariant map*) if $f(gx) = gf(x)$ for every $(g, x) \in G \times X_1$.

The Banach algebra of all continuous real valued bounded functions on a topological space X will be denoted by $C(X)$. Let (G, X, π) be a G-space. It induces the action $G \times C(X) \to C(X)$, with $(gf)(x) = f(g^{-1}x)$. A function $f \in C(X)$ is said to be *right uniformly continuous*, or also *π-uniform*, if the map $G \to C(X)$, $g \mapsto gf$ is norm continuous. The latter means that for every $\varepsilon > 0$ there exists a neighborhood V of the identity element $e \in G$ such that $\sup_{x \in X} |f(vx) - f(x)| < \varepsilon$ for every $v \in V$. The set $\mathrm{RUC}_G(X) := \mathrm{RUC}(X)$ of all right uniformly continuous functions on X is a uniformly closed G-invariant subalgebra of $C(X)$.

A *transitive* action in this paper means an action with a single orbit. Let H be a closed subgroup of G and G/H be the (left) coset space endowed with the quotient topology. In the sequel we will refer to G/H as a *homogeneous* G-space. In this particular case $f \in \mathrm{RUC}_G(X)$ iff f is a uniformly continuous bounded function with respect to the natural *right uniform structure* on G/H (this explains Fact 2.1.1 below). A G-space X will be called:

(1) *G-compactifiable*, or *G-Tychonoff*, if X is a G-subspace of a compact G-space.
(2) *G-homogenizable*, if there exists an equivariant embedding of (G, X) into a homogeneous space $(G', G'/H)$ (i.e., there exists a topological group embedding $h\colon G \hookrightarrow G'$ and a topological embedding $\alpha\colon X \hookrightarrow G'/H$ such that $h(g)\alpha(x) = gx$).
(3) *G-automorphic*, if X is a topological group and each $\tilde{g} = \pi(g, \cdot)\colon X \to X$ is a group automorphism. We say also that X is a *G-group*.
(4) *G-automorphizable*, if X is a G-subspace of an automorphic G-space. In particular, if Y is a locally convex G-space with a continuous linear action of G on Y then we say that X is *G-linearizable*.

2. Equivariant compactifications

A *G-compactification* of a G-space X is a G-map $\nu\colon X \to Y$ with a dense range into a compact G-space Y. A compactification is *proper* when ν is a topological embedding. The study of equivariant compactifications goes back to J. de Groot, R. Palais, R. Brook, J. de Vries, Yu. Smirnov and others.

The Gelfand–Raikov–Shilov classical functional description of compactifications admits a natural generalization for G-spaces in terms of G-subalgebras of $\mathrm{RUC}(X)$ (see for example [**4**, **12**, **76**]). The G-algebra $V := \mathrm{RUC}(X)$ defines the corresponding Gelfand (maximal ideal) space $\beta_G X \subset V^*$ and the, possibly improper, *maximal G-compactification* $i_{\beta_G}: X \to \beta_G X$. Consider the natural homomorphism $h: G \to \mathrm{Is}(\mathrm{RUC}(X))$, where $\mathrm{Is}(\mathrm{RUC}(X)$ is the group of all linear isometries of $\mathrm{RUC}(X)$ and $h(g)(f) := gf$. The pair (h, i_{β_G}) defines a *representation* (in the sense of Definition 7.1) of the G-space X on the Banach space $\mathrm{RUC}_G(X)$.

A G-space is G-Tychonoff iff it can be equivariantly embedded into a compact Hausdorff G-space iff i_{β_G} is proper iff $\mathrm{RUC}_G(X)$ separates points and closed subsets iff (G, X) is Banach representable (cf. Definition 7.1 and Fact 7.2).

Unless G is discrete, the usual maximal compactification $X \to \beta X$ (which always is a G_d-compactification for every G-space X, where G_d is the group G endowed with the discrete topology) fails to be a G-compactification, in general. However several standard compactifications are compatible with actions. For instance it is true for the one-point compactifications [**75**]. The Samuel compactification of an *equiuniform* G-spaces (X, μ) is a G-compactification (see [**18**, **36**, **75**]). Here 'μ is an equiuniformity on a G-space X' means that every translation $\tilde{g}: X \to X$ is μ-uniform and for every entourage $\varepsilon \in \mu$ there exists a neighborhood U of the identity e such that $(gx, x) \in \varepsilon$ for every $(g, x) \in U \times X$. Equiuniform precompact uniformities correspond to G-compactifications. For *G-proximities* see Smirnov [**12**]. It is easy to see that *Gromov's compactification*[1] of a bounded metric space (X, d) with a continuous G-invariant action is a proper G-compactification. The reason is that the function $f_z: X \to \mathbb{R}$ defined by $f_z(x) := d(z, x)$ is π-uniform for every $z \in X$.

By J. de Vries' well known result [**77**] if G is locally compact then every Tychonoff G-space is G-Tychonoff. See Palais [**55**] for the case of a compact Lie group G, and Antonyan [**12**] for compact G.

We call a group G a *V-group* if every Tychonoff G-space is G-Tychonoff. In [**75**], de Vries posed the 'compactification problem' which in our terms becomes: is every topological group G a V-group? Thus every locally compact group is a V-group. An example of [**38**] answers de Vries' question negatively: there exists a topological transformation group (G, X) such that both G and X are Polish and X is not G-Tychonoff.

Fact 2.1. *Recall some useful situations when G-spaces are G-Tychonoff:*

(1) *every coset G-space G/H* (de Vries [**75**]; see also Pestov [**61**])*;*
(2) *every automorphic G-space X (and, hence, every linear G-space X),* [**39**]*;*

[1] The corresponding algebra is generated by the set of functions $\{f_z: X \to \mathbb{R}\}_{z \in X}$, where $f_z(x) := d(z, x)$ (see for example [**4**, p. 112]).

(3) *every metric G-space (X,d), where G is second category and $\tilde{g}: X \to X$ is d-uniformly continuous for every $g \in G$,* [**39**];
(4) *every G-space X, where X is Baire, G is uniformly Lindelöf and acts transitively on X* (Uspenskij [**70**]).

For some results related to Fact 2.1(4) see Chatyrko and Kozlov [**20**].

A topological group G is *uniformly Lindelöf* (alternative names: $\aleph_0$-bounded, ω-bounded, ω-narrow, etc.) if for every nonempty open subset $O \subset G$ countably many translates $g_n O$ cover G. By a G-factorization theorem [**39**] every G-Tychonoff space X with uniformly Lindelöf G admits a proper G-compactification $X \hookrightarrow Y$ with the same weight and dimension $\dim Y \leq \dim \beta_G X$.

The following two results are proved in [**51**].

(1) If G is Polish then it is a V-group iff G is locally compact.
(2) If G is uniformly Lindelöf and *not* locally precompact, then G is not a V-group. Furthermore there exists a Tychonoff G-space X such that $i_{\beta_G}: X \to \beta_G X$ is not injective.

The following longstanding question remains open.

Question 2.2 (Yu.M. Smirnov, 1980). *Find a nontrivial Tychonoff G-space X such that every G-compactification of X is trivial.* 987?

The compactification problem is still open for many natural groups.

Question 2.3 ([**51**]). 988–989?

(1) *Is there a locally precompact group G which is not a V-group?*
(2) *What if G is the group $\mathbb{Q}$ of rational numbers? What if G is the precompact cyclic group $(\mathbb{Z}, \tau_p)$ endowed with the p-adic topology?*

Question 2.4 (Antonyan and Sanchis [**11**]). *Is every locally pseudocompact group a V-group?* 990?

Stoyanov gave (see [**21, 67**]) a geometric description of G-compactifications for the following natural action: $X := \mathbb{S}_H$ is the unit sphere of a Hilbert space H and $G := U(H)$ is the unitary group endowed with the strong operator topology. Then the maximal G-compactification is equivalent to the natural inclusion of X into the weak compact unit ball $\mathbb{B}_H$ of H.

Question 2.5. *Let V be a separable reflexive Banach space. Consider the natural action of the group* $\mathrm{Is}(V)$ *on the sphere $\mathbb{S}_V$. Is it true that the maximal G-compactification is equivalent to the natural inclusion of X into the weak compact unit ball $\mathbb{B}_V$ of V?* 991?

For more information about the question: 'whether simple geometric objects can be maximal equivariant compactifications?' we refer to Smirnov [**66**].

Question 2.6 (H. Furstenberg and T. Scarr). *Let X be a Tychonoff G-space with the transitive action. Is it true that X is G-Tychonoff?* 992?

Uspenskij's result (see Fact 2.1(4)) implies that the answer is 'yes' if X is Baire and G is uniformly Lindelöf.

Very little is known about the dimension of $\beta_G X$. Even in the case of the left regular action of G on $X := G$ the dimension of $\beta_G G$ (the so-called *greatest ambit* for G) may be greater than $\dim G$. Take a cyclic dense subgroup G of the circle group $\mathbb{T}$; then $\dim G = 0$ and $\dim \beta_G G = \dim \mathbb{T} = 1$. It is an old folklore result that $\dim \beta_G G = 0$ iff G is non-Archimedean[2] (see for example, [**52**, **58**]). It follows by [**35**, Thm 5.12] that in the case of the Euclidean group $G = \mathbb{R}^n$, we have $\dim \beta_G G = \dim G$.

993? **Question 2.7.** *Does the functor β_G preserve the covering dimension in case of compact Lie acting group G?*

If G is a compact Lie group then for every G-space X the inequality $\dim X/G \leq \dim X$ holds. For second countable X this is a classical result of Palais [**55**]. For general Tychonoff X this was done in [**39**] using a G-factorization theorem. This inequality does not remain true for compact (even 0-dimensional) groups. This led us [**38**] to an example of a locally compact Polish G-space X such that $\dim X = 1$ and $\dim \beta_G X \geq 2$, where G is a 0-dimensional compact metrizable group.

Fact 2.8 ([**6**, **12**]). *$(\beta_G X)/G = \beta(X/G)$ for every G-space X and compact G.*

994? **Question 2.9** (Zambakhidze). *Let G be a compact group, X a G-space, and $B(X/G)$ a proper compactification of the orbit space X/G. Does there exist a proper G-compactification $B_G(X)$ of X such that $B_G(X)/G = B(X/G)$?*

For some partial results see Antonyan [**7**] and Ageev [**1**].

995? **Question 2.10.** *Let G be a Polish group and X be a second countable G-Tychonoff G-space. Does there exist a metric G-completion of X with the same dimension?*

If G is not Polish then it is not true. The answer is affirmative if G is locally compact [**40**].

3. Equivariant normality

Definition 3.1 ([**37**, **51**, **53**]). Let (G, X, π) be a topological transformation group.

(1) Two subsets A and B in X are *π-disjoint* if $UA \cap UB = \emptyset$ for some neighborhood U of the identity $e \in G$.
(2) X is *G-normal* (or, *equinormal*) if for every pair of π-disjoint closed subsets A and B there exists a pair of π-disjoint neighborhoods $O_1(A)$ and $O_2(B)$. It is equivalent to say that every pair of π-disjoint closed subsets can be separated by a function from $\mathrm{RUC}_G(X)$ (*Urysohn lemma for G-spaces*).
(3) X is *weakly G-normal* if every pair of π-disjoint closed *G-invariant* subsets in X can be separated by a function from $\mathrm{RUC}_G(X)$.

[2] *Non-Archimedean* means having a local base at the identity consisting of open subgroups,

Another version of the Urysohn lemma for G-spaces appears in [**30**, Theorem 3.9].

Every G-normal space is G-Tychonoff. The action of $G := \mathbb{Q}$ on $X := \mathbb{R}$ is not G-normal. One can characterize locally compact groups in terms of G-normality.

Fact 3.2 ([**51**]). *For every topological group G the following are equivalent:*

(1) *Every normal G-space is G-normal.*
(2) *G is locally compact.*

It is unclear if 'G-normal' can be replaced by 'weakly G-normal'.

Question 3.3. *Is every second countable G-space weakly G-normal for the group $G := \mathbb{Q}$ of rational numbers?* 996?

If not, then by [**51**, Theorem 3.2] one can construct for $G := \mathbb{Q}$ a Tychonoff G-space X which is not G-Tychonoff. That is, it will follow that $\mathbb{Q}$ is not a V-group (see Question 2.3).

Fact 3.4. *Every coset G-space G/H is G-normal.*

Then the following *concrete* actions (being coset spaces) are equinormal:

(1) $(U(H), \mathbb{S}_H)$ for every Hilbert space H;
(2) $(\mathrm{Is}(\mathbb{U}), \mathbb{U})$ (where $\mathrm{Is}(\mathbb{U})$ is the isometry group of the Urysohn space $\mathbb{U}$ with the pointwise topology);
(3) $(\mathrm{GL}(V), V \setminus \{0\})$ for every normed space V (see [**44**]);
(4) $(\mathrm{GL}(V), \mathbb{P}_V)$ for every normed space V and its projective space $\mathbb{P}_V$.

It follows in particular by (4) that $\mathbb{P}_V$ is $\mathrm{GL}(V)$-Tychonoff. This was well known among experts and easy to prove (cf. e.g., Pestov [**60**]) using equiuniformities.

Question 3.5. *Is it true that the following (G-Tychonoff) actions are G-normal: $(U(\ell_2), \ell_2)$, $(\mathrm{Is}(\ell_p), \mathbb{S}_{\ell_p}))$, $p > 1$, ($p \neq 2$)?* 997?

4. Universal actions

Let $\mathcal{A}$ be some class of continuous actions (G, X). We say that a pair (G_u, X_u) from $\mathcal{A}$ is *(equivariantly) universal* for the class $\mathcal{A}$ if for every $(G, X) \in \mathcal{A}$ there exists an equivariant pair (h, f) such that $h\colon G \hookrightarrow G_u$ is a topological group embedding and $f\colon X \hookrightarrow X_u$ is a topological embedding. If, in addition we require that $G = G_u$ and $h = \mathrm{id}_G$ then we simply say that X_u is *G-universal.*

For a compact space X denote by $H(X)$ the group of all homeomorphisms of X endowed with the compact open topology.

Fact 4.1.

(1) (Antonyan and de Vries [**10**]; Tychonoff theorem for G-spaces) *For every locally compact σ-compact group G and a cardinal τ there exists a universal G-space of weight τ.*

(2) (Megrelishvili [**39**]; G-space version of Nagata's universal space theorem) *Let G be a locally compact sigma-compact group of weight $w(G) \leq \tau$. For every integer $n \geq 0$ there exists, in the class of metrizable G-spaces of dimension $\leq n$ and weight $\leq \tau$, a universal G-space.*

(3) (Hjorth [**33**]) *If G is a Polish group, then the class of Polish G-spaces has a G-universal object.*

(4) (Megrelishvili and Scarr [**52**]; Equivariant universality of the Cantor cube) *Let $K := \{0,1\}^{\aleph_0}$ be the Cantor cube. Then $(H(K), K)$ is equivariantly universal for the class of all 0-dimensional compact metrizable G-spaces, where G is second countable and non-Archimedean.*

See also results of Becker and Kechris [**14**, Section 2.2.6], Vlasov [**74**] and questions posed by Iliadis in [**34**, p. 502].

998? **Question 4.2.** *Let G be a Polish group. Is it true that there exists a universal G-space in the class of all Polish G-spaces with dimension $\leq n$?*

Fact 4.3 ([**42**]).

(1) *$(H(\mathbb{I}^{\aleph_0}), \mathbb{I}^{\aleph_0})$ is equivariantly universal for the class of all G-compactifiable actions (G, X) with second countable G and X.*

(2) *Let G be a uniformly Lindelöf group. Then every G-Tychonoff space X is equivariantly embedded into $(H(\mathbb{I}^{\tau}), \mathbb{I}^{\tau})$ where $\tau \leq w(X)w(G)$.*

A direct corollary of Fact 4.3(1) is a well known result of Uspenskij [**69**] about universality of the group $H(\mathbb{I}^{\aleph_0})$. Another proof of Uspenskij's result (see [**8**, Corollary 4]) follows by the following theorem of Antonyan.

Fact 4.4 ([**8**]). *Let G be a uniformly Lindelöf group. Then for every G-Tychonoff space X there exists a family of convex metrizable G-compacta $\{K_f\}_{f \in F}$ such that $|F| = w(X)$ and X possesses a G-embedding into the product $\prod_{f \in F} K_f$.*

The following natural question of Antonyan remains open (even for $\tau = \aleph_0$).

999? **Question 4.5** (Antonyan [**8, 9**]). *Let G be a uniformly Lindelöf group of weight $wG \leq \tau$. Does there exist a G-universal compact G-space of weight τ?*

1000–1001? **Question 4.6.** *Let τ be an uncountable cardinal.*

(1) *Is it true that there exists an equivariantly universal topological transformation group (G_u, X_u) in the class of all topological transformation groups (G, X) where X is G-Tychonoff and* $\max\{w(G), w(X)\} \leq \tau$*?*

(2) *What if G_u and G are abelian?*

A positive answer on (1) will imply the solution of the following question.

1002? **Question 4.7** (Uspenskij [**72**]). *Does there exist a universal topological group of every given infinite weight τ?*

Fact 4.8. *G-Compactifiable $\supset$ G-Homogenizable $\supset$ G-Automorphizable.*

Every G-group X is naturally identified with the coset P-space P/G, where $P := X \leftthreetimes G$ is the corresponding semidirect product. This explains the second inclusion. The first inclusion follows by Fact 3.4.

If G is locally compact then every G-space is *G-linearizable* (see for example, [**12**, **78**]) and all classes from Fact 4.8 coincide.

It is well known that the action of $H(\mathbb{I}^{\aleph_0})$ on $\mathbb{I}^{\aleph_0}$ is transitive. Using Effros' theorem one can show that Tychonoff cubes $\mathbb{I}^{\lambda}$ are coset $H(\mathbb{I}^{\lambda})$-spaces for every infinite power, [**42**]. Therefore Fact 4.3 leads to the equality G-Compactifiable = G-Homogenizable for every uniformly Lindelöf group G. It is unclear in general.

Question 4.9. *Is there a G-Tychonoff non-homogenizable G-space? Equivalently,* 1003?
is every compact G-space G-homogenizable?

5. Free topological G-groups

Let X be a Tychonoff G-space. By $F_G(X)$ we denote the corresponding *free topological G-group* in the sense of [**42**]. Recall a link with the *epimorphism problem*. Uspenskij has shown in [**71**] that in the category of Hausdorff topological groups epimorphisms need not have a dense range. This answers a longstanding problem by K. Hofmann. Pestov gave [**57**, **59**] a useful epimorphism criteria in terms of the free topological G-groups.

Fact 5.1 (Pestov [**57**]). *The natural inclusion $H \hookrightarrow G$ of a topological subgroup H into G is an epimorphism (in the category of Hausdorff groups) if and only if the free topological G-group $F_G(X)$ of the coset G-space $X := G/H$ is trivial (here the triviality means, 'as trivial as possible', isomorphic to the cyclic discrete group).*

For instance by results of [**42**], $F_G(X)$ is trivial in the following situation: the group $G := H(\mathbb{S})$ is the group of all homeomorphisms of the circle $\mathbb{S}$ which can be identified with the compact coset G-space $G/\operatorname{St}(z)$ (where z is a point of $\mathbb{S}$ and $\operatorname{St}(z)$ is the stabilizer of z). It follows that $\operatorname{St}(z) \hookrightarrow G$ is an epimorphism. This example shows also that not every compact G-space is G-automorphizable.

If G is locally compact then $F_G(X)$ canonically can be identified with the usual free topological group $F(X)$. This suggests the following questions.

Question 5.2. *Let X be G-automorphic (i.e., the canonical map $X \to F_G(X)$ is* 1004?
an embedding). Is it true that the natural map $F(X) \to F_G(X)$ is a homeomorphism?

Question 5.3. *Let X be a G-automorphic G-space. Is it true that $F_G(X)$ is* 1005?
algebraically free over X?

6. Banach representations of groups

A *representation* of a topological group G on a Banach space V is a homomorphism $h\colon G \to \operatorname{Is}(V)$, where $\operatorname{Is}(V)$ is the topological group of all linear surjective isometries $V \to V$ endowed with the *strong operator topology* inherited from V^V. Denote by V_w the space V in its weak topology. The corresponding topology on $\operatorname{Is}(V)$ inherited from V_w^V is the *weak operator topology*. By [**47**], for a wide class

PCP (*Point of Continuity Property*) of Banach spaces, including reflexive spaces, strong and weak operator topologies on $\mathrm{Is}(V)$ coincide.

Let $\mathcal{K}$ be a *well behaved* subclass of the class *Ban* of all Banach spaces. Important particular cases for such $\mathcal{K}$ are: *Hilb*, *Ref* or *Asp*, the classes of Hilbert, reflexive or Asplund spaces respectively. The investigation of *Asp* and the closely related *Radon–Nikodým property* are among the main themes in Banach space theory. Recall that a Banach space V is an *Asplund space* if the dual of every separable linear subspace is separable. It is equivalent to saying that every bounded subset A of the dual V^* is (weak*, norm)-*fragmented*, or that V^* has the Radon–Nikodým property. Reflexive spaces and spaces of the type $c_0(\Gamma)$ are Asplund. Namioka's Joint Continuity Theorem implies that every weakly compact set in a Banach space is norm fragmented. This explains why every reflexive space is Asplund. For more details cf. [**17**, **24**, **54**]. For some applications of the fragmentability concept for topological transformation groups, see [**27**, **43**, **47**, **48**].

We say that a topological group G is $\mathcal{K}$-*representable* if there exists a representation $h\colon G \to \mathrm{Is}(V)$ for some $V \in \mathcal{K}$ such that h is a topological embedding; notation: $G \in \mathcal{K}_r$. In the opposite direction, we say that G is $\mathcal{K}$-*trivial* if every continuous $\mathcal{K}$-representation of G is trivial. Of course, $\mathit{TopGr} = \mathit{Ban}_r \supset \mathit{Asp}_r \supset \mathit{Ref}_r \supset \mathit{Hilb}_r$. As to $\mathit{TopGr} = \mathit{Ban}_r$, it is an old observation due to Teleman [**68**] (see also [**59**]) that for every topological group G the natural representation $G \to \mathrm{Is}(V)$ on the Banach space $V := \mathrm{RUC}(G)$ is an embedding.

Every locally compact group is Hilbert representable (Gelfand–Raikov). (We say also, *unitarily representable.*) On the other hand, even for Polish groups very little is known about their representabilty in well behaved Banach spaces.

It is also well known that $\mathit{TopGr} \neq \mathit{Hilb}_r$. Moreover, there are examples of unitarily trivial, so-called *exotic* groups (Herer–Christensen [**32**] and Banasczyk [**13**]).

Classical results imply that a group is unitarily representable iff the positive definite functions separate the closed subsets and the neutral element. By results of Shoenberg the function $f(v) = e^{-\|x\|^p}$ is positive definite on $L_p(\mu)$ spaces for every $1 \leq p \leq 2$. An arbitrary Banach space V, as a topological group, cannot be exotic because the group V in the weak topology is unitarily representable. However $C[0,1], c_0 \notin \mathit{Hilb}_r$ (see Fact 6.6 below).

Fact 6.1 ([**46**]). *A topological group G is (strongly) reflexively representable (i.e., G is embedded into $\mathrm{Is}(V)$ endowed with the strong operator topology for some reflexive V) iff the algebra $\mathrm{WAP}(G)$ of all weakly almost periodic functions determines the topology of G.*

A weaker result replacing 'strong' by 'weak' appears earlier in Shtern [**65**]. The group $G := H_+[0,1]$ of orientation preserving homeomorphisms of the closed interval (with the compact open topology) is an important source for counterexamples.

Fact 6.2.

(1) ([**46**]) *$H_+[0,1]$ is reflexively (and hence also Hilbert) trivial.*
(2) ([**28**]) *Moreover, $H_+[0,1]$ is even Asplund trivial.*

The question if WAP(G) determines the topology of a topological group G was raised by Ruppert [**64**]. (1) means that every wap function on $H_+[0,1]$ is constant. The WAP triviality of $G := H_+[0,1]$ was conjectured by Pestov.

Question 6.3 (Glasner and Megrelishvili). *Is there an abelian group which is not reflexively representable?* 1006?

Equivalently: is it true that the algebra WAP(G) on an abelian group G separates the identity from closed subsets?

Question 6.4. *Is it true that every Banach space X, as a topological group, is reflexively representable?* 1007?

A separable Banach space U is *uniformly universal* if every separable Banach space, as a uniform space, can be embedded into U. Clearly, $C[0,1]$ is linearly universal and hence also uniformly universal. In [**2**] Aharoni proved that c_0 is uniformly universal. P. Enflo [**23**], in answer to a question by Yu. Smirnov, found in 1969 a countable metrizable uniform space which is not uniformly embedded into a Hilbert space. That is, ℓ_2 is not uniformly universal[3]. However, it is an open question if 'Hilbert' may be replaced by 'reflexive'.

Question 6.5. *Does there exist a uniformly universal reflexive Banach space?* 1008?

There is no linearly universal separable reflexive Banach space (Szlenk). Moreover, there is no Lipschitz embedding of c_0 into a reflexive Banach space (Mankiewicz). For more information on uniform classification of Banach spaces we refer to [**15**].

Fact 6.6 ([**41**, **47**]). *Let G be a (separable) metrizable group and let $\mathcal{U}_L$ denote its left uniform structure. If G is reflexively representable, than $(G, \mathcal{U}_L)$ as a uniform space is embedded into a (separable) reflexive space V. Moreover, if G is unitarily representable then G is uniformly embedded into a (separable) Hilbert space.*

As a corollary it follows that $C[0,1]$ and c_0 are not unitarily representable. A positive answer to the following question will imply a positive answer on Question 6.5.

Question 6.7. *Are the additive groups c_0 and $C[0,1]$ reflexively representable?* 1009?

A natural question arises about coincidence of Ref_r and Hilb_r. The positive answer was conjectured by A. Shtern [**65**]. By [**45**], $L_4[0,1] \in \mathit{Ref}_r$ and $L_4[0,1] \notin \mathit{Hilb}_r$. Chaatit [**19**] proved that every separable $L_p(\mu)$ space ($1 \le p < \infty$), is reflexively representable.

By [**3**], if a metrizable abelian[4] group, as a uniform space, is embedded into a Hilbert space then positive definite functions separate the identity and closed subsets. Combining this with Fact 6.6 we have the following[5].

---

[3]This result by Enflo has recently led to some exciting developments in geometric group theory, cf. Gromov [**31**].

[4]In fact, *metrizable amenable*, is enough.

[5]It was presented on Yaki Sternfeld Memorial International Conference (Israel, May 2002).

Fact 6.8. *A metric abelian group is unitarily representable if and only if it is uniformly embedded into a Hilbert space.*

The same observation (for second countable abelian groups) is mentioned by J. Galindo in a recent preprint [**25**]. Facts 6.6 and 6.8 suggest the following question.

1010? **Question 6.9** (See also [**45**]). *Let G be a metrizable group and it, as a uniform space $(G,\mathcal{U}_L)$, is uniformly embedded into a reflexive (Hilbert) Banach space. Is it true that G is reflexively (resp., unitarily) representable?*

Galindo announced [**25**] that for every compact space X the free abelian topological group $A(X)$ is unitarily representable. Uspenskij found [**73**] that in fact this is true for every Tychonoff space X. The case of $F(X)$ is open.

1011? **Question 6.10.** *Let X be a Tychonoff (or, even a compact) space.*

(1) *Is it true that the free topological group $F(X)$ is reflexively representable?*
(2) (see also Pestov [**56**]) *Is it true that $F(X)$ is unitarily representable?*

$U(\ell_2)$ clearly is universal for Polish unitarily representable groups.

1012? **Question 6.11.** *Does there exist a universal reflexively representable Polish group?*

1013? **Question 6.12.** *Is it true that if G is reflexively representable then the factor group G/H is also reflexively representable?*

It is impossible here to replace 'reflexively' by 'Hilbert' because every Abelian Polish group is a factor-group of a Hilbert representable Polish group (Gao and Pestov [**26**]). A positive answer to Question 6.12 will imply that every second countable Abelian group is reflexively representable. Also then we will get a negative answer to the following problem.

1014? **Question 6.13** (A.S. Kechris). *Is every Polish (nonabelian) topological group a topological factor-group of a subgroup of $U(\ell_2)$ with the strong operator topology?*

A natural test case by Fact 6.2 is the group $H_+[0,1]$. Fact 6.2 of course implies that every bigger group $G \supset H_+[0,1]$ is not reflexively representable. Moreover if G in addition is topologically simple then it is reflexively trivial. For instance the Polish group $\mathrm{Is}(\mathbb{U}_1)$[6] is reflexively trivial (as observed by Pestov [**62**], this fact follows immediately from results by Megrelishvili [**46**] and Uspenskij [**72**]). It follows that every Polish group is a subgroup of a reflexively trivial Polish group.

1015? **Question 6.14** (Glasner and Megrelishvili [**28**]). *Is it true that there exists a nontrivial Polish group which is reflexively (Asplund) trivial but does not contain a subgroup topologically isomorphic to $H_+[0,1]$?*

By a recent result of Rosendal and Solecki [**63**] every homomorphism of $H_+[0,1]$ into a separable group is continuous. Hence every representation (of a discrete group) $H_+[0,1]$ on a separable reflexive space is trivial.

[6] $\mathbb{U}_1$ is a sphere of radius $1/2$ in $\mathbb{U}$.

Question 6.15 (Glasner and Megrelishvili). *Find a Polish group G which is reflexively (Asplund) trivial but the discrete group G_d admits a nontrivial representation on a separable reflexive (Asplund) space.* 1016?

Question 6.16. *Is it true that the group $H(\mathbb{I}^{\aleph_0})$ is reflexively trivial?* 1017?

It is enough to show that the group $H(\mathbb{I}^{\aleph_0})$ is topologically simple.

Question 6.17 (Glasner and Megrelishvili). *Is it true that there exists a group G such that $G \in Asp_r$ and $G \notin Ref_r$.* 1018?

7. Dynamical versions of Eberlein and Radon–Nikodým compacta

Eberlein compacta in the sense of Amir and Lindenstrauss [**5**] are exactly the weakly compact subsets in Banach (equivalently, reflexive) spaces V. If X is a weak* compact subset in the dual V^* of an Asplund space V then, following Namioka [**54**], X is called *Radon–Nikodým compact* (in short: RN). Every reflexive Banach space is Asplund. Hence, every Eberlein compact is RN.

Definition 7.1 ([**48**]). A *(proper) representation* of (G, X) on a Banach space V is a pair (h, α) where $h\colon G \to \mathrm{Is}(V)$ is a continuous homomorphism of topological groups and $\alpha\colon X \to V^*$ is a weak star continuous bounded G-mapping (resp., *embedding*) with respect to the *dual action* $G \times V^* \to V^*$, $(g\varphi)(v) := \varphi(h(g^{-1})(v))$.

Note that the dual action is norm continuous whenever V is an Asplund space, [**43**]. It is well known that the latter does not remain true in general.

Fact 7.2. *A G-space X is properly representable on some Banach space V if and only if X is G-Tychonoff (consider the natural representation on $V := \mathrm{RUC}_G(X)$).*

The following dynamical versions of Eberlein and Radon–Nikodým compact spaces were introduced in [**48**]. A compact G-space X is *Radon–Nikodým*, RN for short, if there exists a proper representation of (G, X) on an Asplund Banach space V. If V is reflexive (Hilbert) then we get the definitions of reflexively (resp., Hilbert) representable G-spaces. In the first case we say that (G, X) is an *Eberlein* G-space.

Fact 7.3. *Let X be a metric compact G-space.*

(1) ([**48**]) *X, as a G-space, is Eberlein (i.e., reflexively G-representable) iff X is a weakly almost periodic G-space in the sense of Ellis [**22**].*
(2) ([**27**]) *X, as a G-space, is RN iff X is hereditarily nonsensitive.*

Compact spaces which are not Eberlein are necessarily nonmetrizable, while even for $G := \mathbb{Z}$, there are natural *metric* compact G-spaces which are not RN.

There exist compact a metric $\mathbb{Z}$-space which is reflexively but not Hilbert representable [**49**]. This answers a question of T. Downarowicz.

Question 7.4. *Is it true that Eberlein (that is, reflexively representable) compact G-spaces are closed under factors?* 1019?

For the trivial group G (i.e., in the purely topological setting) the answer is affirmative and this is just a well known result by Benyamini, Rudin and Wage [**16**]. The answer is 'yes' for compact *metric* G-spaces.

1020? **Question 7.5.** *Is it true that RN (that is, Asplund representable) compact G-spaces are closed under factors?*

For the trivial group one can recognize a longstanding open question by Namioka [**54**]. Again if X is metric then the answer is 'yes'. For Hilbert representable actions the situation is unclear even for the metric case.

1021? **Question 7.6.** *Is it true that Hilbert representable compact metric G-spaces are closed under factors?*

For a compact G-space X denote by $E := E(X)$ the corresponding (frequently, *huge*) compact right topological *(Ellis) enveloping semigroup*. It is the pointwise closure of the set of translations $\{\tilde{g}\colon X \to X\}_{g\in G}$ in the product space X^X.

The enveloping semigroup $E(X)$ of a metric compact RN G-space X is a separable Rosenthal compact (hence, $\mathrm{card}(E(X)) \leq 2^{\aleph_0}$), [**27**].

1022? **Question 7.7** (Glasner and Megrelishvili). *Is it true that for every compact metric RN G-space X the enveloping semigroup E(X) is metrizable?*[7]

A function $f \in \mathrm{RUC}(G)$ is *Asplund*, notation: $f \in \mathrm{Asp}(G)$, if f is a (generalized) matrix coefficient of an Asplund representation $h\colon G \to \mathrm{Is}(V)$. This means that V is Asplund and there exists a pair of vectors $(v, \psi) \in V \times V^*$ such that $f(g) = \psi(g^{-1}v)$. Similarly, $\mathrm{WAP}(G)$ is the set of all matrix coefficients of reflexive representations. Recall that if $\mathrm{RUC}(G) = \mathrm{WAP}(G)$ then G is precompact [**50**].

1023? **Question 7.8.** *Assume that* $\mathrm{RUC}(G) = \mathrm{Asp}(G)$. *Is it true that G is precompact?*

Acknowledgment

I am indebted to Yu. Smirnov who led me to the world of topological transformation groups. I thank S. Antonyan, D. Dikranjan, E. Glasner, V. Pestov and V. Uspenskij for helpful suggestions.

References

[1] S. M. Ageev, *On a problem of Zambakhidze and Smirnov*, Mat. Zametki **58** (1995), no. 1, 3–11, Translation: Math. Notes 58 (1995) no. 1, 679–684.

[2] I. Aharoni, *Every separable metric space is Lipschitz equivalent to a subset of* c_0^+, Israel J. Math. **19** (1974), 284–291.

[3] I. Aharoni, B. Maurey, and B. S. Mityagin, *Uniform embeddings of metric spaces and of Banach spaces into Hilbert spaces*, Israel J. Math. **52** (1985), no. 3, 251–265.

[4] E. Akin, *Recurrence in topological dynamics: Furstenberg families and Ellis actions*, Plenum Press, New York, 1997.

[5] D. Amir and J. Lindenstrauss, *The structure of weakly compact sets in Banach spaces*, Ann. of Math. (2) **88** (1968), 35–46.

[7] The answer is positive, see [**29**].

[6] N. Antonyan and S. A. Antonyan, *Free G-spaces and maximal equivariant compactifications*, Ann. Mat. Pura Appl. (4) **184** (2005), no. 3, 407–420.

[7] S. A. Antonyan, *A problem of L. G. Zambakhidze*, Uspekhi Mat. Nauk **41** (1986), no. 5(251), 159–160, Translation: Russian Math. Surveys 41 (1986) 125–126.

[8] S. A. Antonyan, *Equivariant embeddings and ω-bounded groups*, Vestnik Moskov. Univ. Ser. I Mat. Mekh. (1994), no. 1, 16–22, Translation: Moscow Univ. Math. Bull. 49 (1994) no. 1, 13–16.

[9] S. A. Antonyan, *Universal proper G-spaces*, Topology Appl. **117** (2002), no. 1, 23–43.

[10] S. A. Antonyan and J. de Vries, *Tychonov's theorem for G-spaces*, Acta Math. Hungar. **50** (1987), no. 3–4, 253–256.

[11] S. A. Antonyan and M. Sanchis, *Extension of locally pseudocompact group actions*, Ann. Mat. Pura Appl. (4) **181** (2002), no. 3, 239–246.

[12] S. A. Antonyan and Ju. M. Smirnov, *Universal objects and bicompact extensions for topological groups of transformations*, Dokl. Akad. Nauk SSSR **257** (1981), no. 3, 521–526, Translation: Soviet Math. Dokl. 23 (1981) 279–284.

[13] W. Banaszczyk, *Additive subgroups of topological vector spaces*, Lecture Notes in Math., vol. 1466, Springer, Berlin, 1991.

[14] H. Becker and A. S. Kechris, *The descriptive set theory of Polish group actions*, Cambridge University Press, Cambridge, 1996.

[15] Y. Benyamini and J. Lindenstrauss, *Geometric nonlinear functional analysis. Vol. 1*, American Mathematical Society, Providence, RI, 2000.

[16] Y. Benyamini, M. E. Rudin, and M. Wage, *Continuous images of weakly compact subsets of Banach spaces*, Pacific J. Math. **70** (1977), no. 2, 309–324.

[17] R. D. Bourgin, *Geometric aspects of convex sets with the Radon–Nikodým property*, Lecture Notes in Math., vol. 993, Springer, Berlin, 1983.

[18] R. B. Brook, *A construction of the greatest ambit*, Math. Systems Theory **4** (1970), 243–248.

[19] F. Chaatit, *A representation of stable Banach spaces*, Arch. Math. (Basel) **67** (1996), no. 1, 59–69.

[20] V. A. Chatyrko and K. L. Kozlov, *On G-compactifications*, Math. Notes **78** (2005), no. 5, 649–661.

[21] D. Dikranjan, I. R. Prodanov, and L. N. Stoyanov, *Topological groups: characters, dualities and minimal group topologies*, Marcel Dekker Inc., New York, 1990.

[22] R. Ellis, *Equicontinuity and almost periodic functions*, Proc. Amer. Math. Soc. **10** (1959), 637–643.

[23] P. Enflo, *On a problem of Smirnov*, Ark. Mat. **8** (1969), 107–109.

[24] M. J. Fabián, *Gâteaux differentiability of convex functions and topology*, John Wiley & Sons Inc., New York, 1997.

[25] J. Galindo, *On unitary representability of topological groups*, 2006, Preprint. `arXiv:math.GN/0607193`

[26] S. Gao and V. Pestov, *On a universality property of some abelian Polish groups*, Fund. Math. **179** (2003), no. 1, 1–15.

[27] E. Glasner and M. Megrelishvili, *Hereditarily non-sensitive dynamical systems and linear representations*, Colloq. Math. **104** (2006), no. 2, 223–283.

[28] E. Glasner and M. Megrelishvili, *Some new algebras of functions on topological groups arising from G-spaces*, 2006, Preprint. `arXiv:math.DS/0608575`

[29] E. Glasner, M. Megrelishvili, and V. V. Uspenskij, *On metrizable enveloping semigroups*, 2006, Preprint. `arXiv:math.DS/0606373`

[30] E. Glasner and B. Weiss, *Spatial and non-spatial actions of Polish groups*, Ergodic Theory Dynam. Systems **25** (2005), no. 5, 1521–1538.

[31] M. Gromov, *Random walk in random groups*, Geom. Funct. Anal. **13** (2003), no. 1, 73–146.

[32] W. Herer and J. P. R. Christensen, *On the existence of pathological submeasures and the construction of exotic topological groups*, Math. Ann. **213** (1975), 203–210.

[33] G. Hjorth, *A universal Polish G-space*, Topology Appl. **91** (1999), no. 2, 141–150.

[34] S. D. Iliadis, *Universal spaces and mappings*, Elsevier Science, Amsterdam, 2005.

[35] J. R. Isbell, *Uniform spaces*, American Mathematical Society, Providence, RI, 1964.

[36] M. Megrelishvili, *Equivariant completions and compact extensions*, Soobshch. Akad. Nauk Gruzin. SSR **115** (1984), no. 1, 21–24.

[37] M. Megrelishvili, *Uniformity and topological transformation groups*, Ph.D. dissertation, Tbilisi State University, 1985.

[38] M. Megrelishvili, *A Tikhonov G-space that does not have compact G-extension and G-linearization*, Uspekhi Mat. Nauk **43** (1988), no. 2(260), 145–146, Translation: Russian Math. Surveys 43 (1988) no. 2, 177–178.

[39] M. Megrelishvili, *Compactification and factorization in the category of G-spaces*, Categorical topology (J. Adamek and Mac Lane S., eds.), World Scientific Publishing, Singapore, 1989, pp. 220–237.

[40] M. Megrelishvili, *Equivariant completions*, Comment. Math. Univ. Carolin. **35** (1994), no. 3, 539–547.

[41] M. Megrelishvili, *Group representations and construction of minimal topological groups*, Topology Appl. **62** (1995), no. 1, 1–19.

[42] M. Megrelishvili, *Free topological G-groups*, New Zealand J. Math. **25** (1996), no. 1, 59–72.

[43] M. Megrelishvili, *Fragmentability and continuity of semigroup actions*, Semigroup Forum **57** (1998), no. 1, 101–126.

[44] M. Megrelishvili, *G-minimal topological groups*, Abelian groups, module theory, and topology (Padua, 1997), Lecture Notes in Pure and Appl. Math., vol. 201, Marcel Dekker Inc., New York, 1998, pp. 289–299.

[45] M. Megrelishvili, *Reflexively but not unitarily representable topological groups*, Topology Proc. **25** (2000), Summer, 615–625.

[46] M. Megrelishvili, *Every semitopological semigroup compactification of the group $H_+[0,1]$ is trivial*, Semigroup Forum **63** (2001), no. 3, 357–370.

[47] M. Megrelishvili, *Operator topologies and reflexive representability*, Nuclear groups and Lie groups (Madrid, 1999), Heldermann Verlag, Lemgo, 2001, pp. 197–208.

[48] M. Megrelishvili, *Fragmentability and representations of flows*, Topology Proc. **27** (2003), no. 2, 497–544.

[49] M. Megrelishvili, *Reflexively but not Hilbert representable compact flows and semitopological semigroups*, 2005, Preprint.

[50] M. Megrelishvili, V. G. Pestov, and V. V. Uspenskij, *A note on the precompactness of weakly almost periodic groups*, Nuclear groups and Lie groups (Madrid, 1999), Heldermann Verlag, Lemgo, 2001, pp. 209–216.

[51] M. Megrelishvili and T. Scarr, *Constructing Tychonoff G-spaces which are not G-Tychonoff*, Topology Appl. **86** (1998), no. 1, 69–81.

[52] M. Megrelishvili and T. Scarr, *The equivariant universality and couniversality of the Cantor cube*, Fund. Math. **167** (2001), no. 3, 269–275.

[53] M. Mergelishvili, *Equivariant normality*, Soobshch. Akad. Nauk Gruzin. SSR **111** (1983), no. 1, 17–19.

[54] I. Namioka, *Radon–Nikodým compact spaces and fragmentability*, Mathematika **34** (1987), no. 2, 258–281.

[55] R. S. Palais, *The classification of G-spaces*, Mem. Amer. Math. Soc. (1960), no. 36, 72 pp.

[56] V. Pestov, *Forty-plus annotated questions about large topological groups*, Open problems in topology II (E. Pearl, ed.), Elsevier Science.

[57] V. Pestov, *Epimorphisms of Hausdorff groups by way of topological dynamics*, New Zealand J. Math. **26** (1997), no. 2, 257–262.

[58] V. Pestov, *On free actions, minimal flows, and a problem by Ellis*, Trans. Amer. Math. Soc. **350** (1998), no. 10, 4149–4165.

[59] V. Pestov, *Topological groups: where to from here?*, Topology Proc. **24** (1999), Summer, 421–502.

[60] V. Pestov, *Remarks on actions on compacta by some infinite-dimensional groups*, Infinite dimensional Lie groups in geometry and representation theory (Washington, DC, 2000), World Scientific Publishing, Singapore, 2002, pp. 145–163.

[61] V. Pestov, *Dynamics of infinite-dimensional groups and Ramsey-type phenomena*, Publicações Matemáticas dos Colóquios de Matemática, IMPA, Rio de Janeiro, 2005.

[62] V. Pestov, *The isometry group of the Urysohn space as a Lévy group*, 2005, Preprint. To appear in Topology Appl. `arXiv:math.GN/0509402`

[63] C. Rosendal and S. Solecki, *Automatic continuity of homomorphisms and fixed points on metric compacta*, 2006, Preprint. To appear in Israel J. Math. `arXiv:math.LO/0604575`

[64] W. Ruppert, *Compact semitopological semigroups: an intrinsic theory*, Lecture Notes in Math., vol. 1079, Springer, Berlin, 1984.

[65] A. I. Shtern, *Compact semitopological semigroups and reflexive representability of topological groups*, Russian J. Math. Phys. **2** (1994), no. 1, 131–132.

[66] Yu. M. Smirnov, *Can simple geometric objects be maximal compact extensions for* $\mathbb{R}^n$*?*, Uspekhi Mat. Nauk **49** (1994), no. 6(300), 213–214, Translation: Russian Math. Surveys 49 (1994) no. 6, 214–215.

[67] L. N. Stoyanov, *On the infinite-dimensional unitary groups*, C. R. Acad. Bulgare Sci. **36** (1983), no. 10, 1261–1263.

[68] S. Teleman, *Sur la représentation linéaire des groupes topologiques*, Ann. Sci. Ecole Norm. Sup. (3) **74** (1957), 319–339.

[69] V. V. Uspenskij, *A universal topological group with a countable basis*, Funktsional. Anal. i Prilozhen. **20** (1986), no. 2, 86–87, Translation: Funct. Anal. Appl. 20 (1986) 160–161.

[70] V. V. Uspenskij, *Topological groups and Dugundji compacta*, Mat. Sb. **180** (1989), no. 8, 1092–1118, Translation: Sb. Math. 67 (1990) no. 2, 555–580.

[71] V. V. Uspenskij, *The epimorphism problem for Hausdorff topological groups*, Topology Appl. **57** (1994), no. 2–3, 287–294.

[72] V. V. Uspenskij, *On subgroups of minimal topological groups*, 1998, Preprint. `arXiv:math.GN/0004119`

[73] V. V. Uspenskij, *Unitary representability of free abelian topological groups*, 2006, Preprint. `arXiv:math.RT/0604253`

[74] S. V. Vlasov, *Universal bicompact G-spaces*, Uspekhi Mat. Nauk **49** (1994), no. 6(300), 199–200, Translation: Russian Math. Surveys 49 (1994) no. 6, 221–222.

[75] J. de Vries, *Can every Tychonoff G-space equivariantly be embedded in a compact Hausdorff G-space?*, Afd. Zuiver Wisk., vol. 36, Mathematisch Centrum, Amsterdam, 1975.

[76] J. de Vries, *Equivariant embeddings of G-spaces*, General topology and its relations to modern analysis and algebra, IV (Prague, 1976), Part B, Association of Czechoslovak Mathematicians and Physicists, Prague, 1977, pp. 485–493.

[77] J. de Vries, *On the existence of G-compactifications*, Bull. Acad. Polon. Sci. Sér. Sci. Math. Astronom. Phys. **26** (1978), no. 3, 275–280.

[78] J. de Vries, *Linearization of actions of locally compact groups*, Proc. Steklov Inst. Math. **154** (1984), 57–74.

Open Problems in Topology II – Edited by E. Pearl

Forty-plus annotated questions about large topological groups

Vladimir Pestov

This is a selection of open problems dealing with *large* (not locally compact) topological groups and concerning extreme amenability (fixed point on compacta property), oscillation stability, universal minimal flows and other aspects of universality, and unitary representations.

A topological group G is *extremely amenable*, or has the *fixed point on compacta property*, if every continuous action of G on a compact Hausdorff space has a G-fixed point. Here are some important examples of such groups.

Example 1. The unitary group $U(\ell^2)$ of the separable Hilbert space ℓ^2 with the strong operator topology (that is, the topology of pointwise convergence on ℓ^2) (Gromov and Milman [**31**]).

Example 2. The group $L^1((0,1),\mathbb{T})$ of all equivalence classes of Borel maps from the unit interval to the circle with the L^1-metric $d(f,g) = \int_0^1 |f(x) - g(x)|\,dx$ (Glasner [**24**], Furstenberg and Weiss, unpublished).

Example 3. The group $\mathrm{Aut}(\mathbb{Q},\leq)$ of all order-preserving bijections of the rationals, equipped with the natural Polish group topology of pointwise convergence on $\mathbb{Q}$ considered as a discrete space and, as an immediate corollary, the group $\mathrm{Homeo}_+[0,1]$ of all homeomorphisms of the closed unit interval, preserving the endpoints, equipped with the compact-open topology (the present author [**42**]).

The above property is not uncommon among concrete "large" topological groups coming from diverse parts of mathematics. In addition to the above quoted articles, we recommend [**23**, **37**] and the book [**45**].

The group in Example 2 is monothetic, that is, contains a dense subgroup isomorphic to the additive group of integers $\mathbb{Z}$. Notice that every abelian extremely amenable group G is *minimally almost periodic*, that is, admits no nontrivial continuous characters (the book [**14**] is a useful reference): indeed, if $\chi\colon G \to \mathbb{T}$ is such a character, then $(g,z) \mapsto \chi(g)z$ defines a continuous action of G on $\mathbb{T}$ without fixed points. The converse remains open.

Question 1 (Eli Glasner [**24**]). *Does there exist a monothetic topological group that is minimally almost periodic but not extremely amenable?* 1024?

An equivalent question is: does there exist a topology on the group $\mathbb{Z}$ of integers making it into a topological group that admits a free action on a compact space but has no nontrivial characters?

Suppose the answer to the above question is in the positive, and let τ be a minimally almost periodic Hausdorff group topology on $\mathbb{Z}$ admitting a free continuous action on a compact space X. Let $x_0 \in X$. Find an open neighbourhood V of x_0 with $1 \cdot V \cap V = \emptyset$. It is not difficult to verify that the set $S = \{n \in \mathbb{Z}\colon nx_0 \in V\}$

is *relatively dense* in $\mathbb{Z}$, that is, the size of gaps between two subsequent elements of S is uniformly bounded from above, and at the same time, the closure of $S - S = \{n - m : n, m \in S\}$ is a proper subset of $\mathbb{Z}$. The interior of $S - S$ in the Bohr topology on $\mathbb{Z}$ (the finest precompact group topology) is therefore not everywhere dense in $(\mathbb{Z}, \tau)$. Assuming this interior is nonempty, one can now verify that the τ-closures of elements of the Bohr topology on $\mathbb{Z}$ form a base for a precompact group topology that is nontrivial and coarser than τ, contradicting the assumed minimal almost periodicity of $(\mathbb{Z}, \tau)$. Thus, a positive answer to Glasner's question would answer in the negative the following very old question from combinatorial number theory/harmonic analysis, rooted in the classical work of Bogoliuboff, Følner [**18**], Cotlar and Ricabarra [**11**], Veech [**61**], and Ellis and Keynes [**16**]:

1025? **Question 2.** *Let S be a relatively dense subset of the integers. Is the set $S - S$ a Bohr neighbourhood of zero in $\mathbb{Z}$?*

We refer the reader to Glasner's original work [**24**] for more on the above. See also [**43, 45, 64**].

1026? **Question 3.** *Does there exists an abelian minimally almost periodic topological group acting freely on a compact space?*

This does not seem to be equivalent to Glasner's problem, because there are examples of minimally almost periodic abelian Polish groups whose every monothetic subgroup is discrete, such as $L^p(0,1)$ with $0 < p < 1$.

There are numerous known ways to construct monothetic minimally almost periodic groups [**1, 4, 14, 48**]. The problem is verifying their (non-) extreme amenability. The most general result presently known asserting non-extreme amenability of a topological group is:

Theorem (Veech [**62**]). *Every locally compact group admits a free action on a compact space.*

Since every locally compact abelian group admits sufficiently many characters, one cannot employ Veech's theorem to answer Glasner's question. Can the result be extended? Recall that a topological space X is called a k_ω*-space* (or: a *hemicompact* space) if it admits a countable cover K_n, $n \in \mathbb{N}$ by compact subsets in such a way that an $A \subseteq X$ is closed if and only if $A \cap K_n$ is closed for all n. For example, every countable CW-complex, every second-countable locally compact space, and the free topological group [**28**] on a compact space are such.

1027? **Question 4.** *Is it true that every topological group G that is a k_ω-space admits a free action on a compact space?*

1028? **Question 5.** *Same, for abelian topological groups that are k_ω-spaces.*

A positive answer would have answered in the affirmative Glasner's question because there are examples of minimally almost periodic k_ω group topologies on the group $\mathbb{Z}$ of integers [**48**].

Recall that the *Urysohn universal metric space* $\mathbb{U}$ is the (unique up to an isometry) complete separable metric space that is *ultrahomogeneous* (every isometry between two finite subsets extends to a global self-isometry of $\mathbb{U}$) and *universal* ($\mathbb{U}$ contains an isometric copy of every separable metric space) [**21**, **30**, **54**, **63**]. The group $\mathrm{Iso}(\mathbb{U})$ of all self-isometries of $\mathbb{U}$, equipped with the topology of pointwise convergence (which coincides with the compact-open topology), is a Polish topological group with a number of remarkable properties. In particular, $\mathrm{Iso}(\mathbb{U})$ is a universal second-countable topological group [**57**, **58**] and is extremely amenable [**44**].

Question 6. *Is the group* $\mathrm{Iso}(\mathbb{U})$ *divisible, that is, does every element possess roots of every positive natural order?*[1] 1029?

Returning to Glasner's Question 1, every element f of $\mathrm{Iso}(\mathbb{U})$ generates a monothetic Polish subgroup, so one can talk of *generic monothetic subgroups* of $\mathrm{Iso}(\mathbb{U})$ (in the sense of Baire category).

Question 7 (Glasner and Pestov, 2001, unpublished). *Is a generic monothetic subgroup of the isometry group* $\mathrm{Iso}(\mathbb{U})$ *of the Urysohn metric space minimally almost periodic?* 1030?

Question 8 (Glasner and Pestov). *Is a generic monothetic subgroup of* $\mathrm{Iso}(\mathbb{U})$ *of the Urysohn metric space extremely amenable?* 1031?

The concept of the universal Urysohn metric space admits numerous modifications. For instance, one can study the universal Urysohn metric space $\mathbb{U}_1$ *of diameter one* (it is isometric to every sphere of radius $1/2$ in $\mathbb{U}$). By analogy with the unitary group $U(\ell^2)$, it is natural to consider the *uniform topology* on the isometry group $\mathrm{Iso}(\mathbb{U}_1)$, given by the bi-invariant uniform metric $d(f,g) = \sup_{x\in\mathbb{U}_1} d_{\mathbb{U}_1}(f(x), g(x))$. It is strictly finer than the strong topology.

Question 9. *Is the uniform topology on* $\mathrm{Iso}(\mathbb{U}_1)$ *nondiscrete?*[2] 1032?

Question 10. *Does* $\mathrm{Iso}(\mathbb{U}_1)$ *possess a uniform neighbourhood of zero covered by one-parameter subgroups?* 1033?

Question 11. *Does* $\mathrm{Iso}(\mathbb{U}_1)$ *have a uniform neighbourhood of zero not containing nontrivial subgroups?* 1034?

Question 12. *Is* $\mathrm{Iso}(\mathbb{U}_1)$ *with the uniform topology a Banach–Lie group?* 1035?

The authors of [**50**] have established the following result as an application of a new automatic continuity-type theorem and Example 3 above.

Theorem (Rosendal and Solecki [**50**]). *The group* $\mathrm{Aut}(\mathbb{Q}, \leq)$, *considered as a discrete group, has the* fixed point on metric compacta property, *that is, every action of* $\mathrm{Aut}(\mathbb{Q}, \leq)$ *on a compact metric space by homeomorphisms has a common fixed point. The same is true of the group* $\mathrm{Homeo}_+[0,1]$.

---

[1] Recently Julien Melleray has announced a negative answer (private communication).

[2] According to Julien Melleray (a private communication), the answer is yes.

This is particularly surprising in view of the Veech theorem, or, rather, its earlier version established by Ellis [**15**]: every discrete group G acts freely on a suitable compact space by homeomorphisms (e.g., on βG). The two results seem to nearly contradict each other!

1036? **Question 13.** *Does the unitary group $U(\ell^2)$, viewed as a discrete group, have the fixed point on metric compacta property?*

1037? **Question 14.** *The same question for the isometry group of the Urysohn space $\mathbb{U}_1$ of diameter one.*

Extreme amenability is a strong form of *amenability*, an important classical property of topological groups. A topological group G is *amenable* if every compact G-space admits an invariant probability Borel measure. Another reformulation: the space $\mathrm{RUCB}(G)$ of all bounded right uniformly continuous real-valued functions on G admits a left-invariant mean, that is, a positive functional ϕ of norm 1 and the property $\phi({}^g f) = \phi(f)$ for all $g \in G$, $f \in \mathrm{RUCB}(G)$, where ${}^g f(x) = f(g^{-1}x)$. (Recall that the *right uniform structure* on G is generated by entourages of the diagonal of the form $V_R = \{(x,y) \in G \times G : xy^{-1} \in V\}$, where V is a neighbourhood of identity. For the *left* uniformity, the formula becomes $x^{-1}y \in V$.) For a general reference to amenability, see e.g., [**41**].

1038? **Question 15** (A. Carey and H. Grundling [**9**]). *Let X be a smooth compact manifold, and let G be a compact (simple) Lie group. Is the group $C^\infty(X,G)$ of all smooth maps from X to G, equipped with the pointwise operations and the C^∞ topology, amenable?*

This question is motivated by gauge theory models of mathematical physics [**9**].

1039? **Question 16.** *To begin with, is the group of all continuous maps $C([0,1], SO(3))$ with the topology of uniform convergence amenable?*

The following way to prove extreme amenability of topological groups was developed by Gromov and Milman [**31**]. A topological group G is called a *Lévy group* if there exists an increasing net (K_α) of compact subgroups whose union is everywhere dense in G, having the following property. Let μ_α denote the Haar measure on the group K_α, normalized to one ($\mu_\alpha(K_\alpha) = 1$). If $A \subseteq G$ is a Borel subset such that $\liminf_\alpha \mu_\alpha(A \cap K_\alpha) > 0$, then for every neighbourhood V of identity in G one has $\lim_\alpha \mu_\alpha(VA \cap K_\alpha) = 1$. (Such a family of compact subgroups is called a *Lévy family*.)

Theorem (Gromov and Milman [**31**]). *Every Lévy group is extremely amenable.*

Proof. We will give a proof in the case of a second-countable G, where one can assume the net (K_α) to be an increasing sequence. For every free ultrafilter ξ on $\mathbb{N}$ the formula $\mu(A) = \lim_{n\to\xi} \mu_n(A \cap K_n)$ defines a *finitely-additive* measure on G of total mass one, invariant under multiplication on the left by elements of the everywhere dense subgroup $G' = \bigcup_{n=1}^\infty K_n$. Besides, μ has the property that if $\mu(A) > 0$, then for every nonempty open V one has $\mu(VA) = 1$. Let now

G act continuously on a compact space X. Choose an arbitrary $x_0 \in X$. The push-forward, ν, of the measure μ to X along the corresponding orbit map, given by $\nu(B) = \mu\{g \in G\colon gx_0 \in B\}$, is again a finitely-additive Borel measure on X of total mass one, invariant under translations by G' and having the same blowing-up property: if $\nu(B) > 0$ and V is a nonempty open subset of G, then $\nu(VB) = 1$. Given a finite cover γ of X, an element W of the unique uniformity on X, and a finite subset F of G', there is at least one $A \in \gamma$ with $\nu(A) > 0$, consequently $\nu(W[A]) = 1$ and for all $g \in F$ the translates $g \cdot W[A]$, having full measure each, must overlap. This can be used to construct a Cauchy filter $\mathcal{F}$ of closed subsets of X with $A \in \mathcal{F}$, $g \in G'$ implying $gA \in \mathcal{F}$. The only point of $\bigcap \mathcal{F}$ is fixed under the action of G' and therefore of G as well. □

For instance, the groups in Examples 1 and 2 are Lévy groups, and so is the isometry group $\mathrm{Iso}(\mathbb{U})$ with the Polish topology [**46**].

The theorem of Gromov and Milman cannot be inverted, because the extremely amenable groups from Example 3 are not Lévy: they simply do not contain any nontrivial compact subgroups. What if such subgroups are present? The following is a reasonable general reading of an old question by Furstenberg discussed at the end of [**31**].

Question 17. *Suppose G is an extremely amenable topological group containing a net of compact subgroups (K_α) whose union is everywhere dense in G. Is G a Lévy group?*[3] 1040?

Question 18. *Provided the answer is yes, is the family (K_α) a Lévy family?*[4] 1041?

A candidate for a *natural* counterexample is the group $SU(\infty)$, the inductive limit of the family of special unitary groups of finite rank embedded one into the other via $SU(n) \ni V \mapsto \left(\begin{smallmatrix} I & 0 \\ 0 & V \end{smallmatrix}\right) \in SU(n+1)$. Equip $SU(\infty)$ with the inductive limit topology, that is, the finest topology inducing the given topology on each $SU(n)$.

Question 19. *Is the group $SU(\infty)$ with the inductive limit topology extremely amenable?* 1042?

If the answer is yes, then Questions 17 and 4 are both answered in the negative.

Historically the first example of an extremely amenable group was constructed by Herer and Christensen [**34**]. Theirs was an abelian topological group without strongly continuous unitary representations in Hilbert spaces (an *exotic group*).

Question 20. *Is the exotic group constructed in [**34**] a Lévy group?* 1043?

The following result shows that the properties of Lévy groups are diametrically opposed to those of locally compact groups in the setting of ergodic theory as well as topological dynamics.

[3] I. Farah and S. Solecki have announced a counterexample (May, 2006).

[4] See the previous footnote.

Theorem (Glasner–Tsirelson–Weiss [**25**]). *Let a Polish Lévy group act in a Borel measurable way on a Polish space X. Let μ be a Borel probability measure on X invariant under the action of G. Then μ is supported on the set of G-fixed points.*

1044? **Question 21** (Glasner–Tsirelson–Weiss, *ibid.*). *Is the same conclusion true if one only assumes that the measure μ is* quasi-invariant *under the action of G, that is, for all $g \in G$ and every null-set $A \subseteq X$, the set gA is null?*

Recall that a compact G-space X is called *minimal* if the orbit of every point is everywhere dense in X. To every topological group G there is associated the *universal minimal flow*, $\mathcal{M}(G)$, which is a minimal compact G-space uniquely determined by the property that every other minimal G-space is an image of $\mathcal{M}(G)$ under an equivariant continuous surjection. (See [**3**].) For example, G is extremely amenable if and only if $\mathcal{M}(G)$ is a singleton. If G is compact, then $\mathcal{M}(G) = G$, but for locally compact noncompact groups, starting with $\mathbb{Z}$, the flow $\mathcal{M}(G)$ is typically very complicated and highly nonconstructive, in particular it is never metrizable [**37**]. A discovery of the recent years has been that even nontrivial universal minimal flows of "large" topological groups are sometimes manageable.

Example 4. The flow $\mathcal{M}(\mathrm{Homeo}_+(\mathbb{S}^1))$ is the circle $\mathbb{S}^1$ itself, equipped with the canonical action of the group $\mathrm{Homeo}_+(\mathbb{S}^1)$ of orientation-preserving homeomorphisms, with the compact-open topology [**42**].

Example 5. Let S_∞ denote the infinite symmetric group, that is, the Polish group of all bijections of the countably infinite discrete space ω onto itself, equipped with the topology of pointwise convergence. The flow $\mathcal{M}(S_\infty)$ can be identified with the set of all linear orders on ω with the topology induced from $\{0,1\}^{\omega\times\omega}$ under the identification of each order with the characteristic function of the corresponding relation [**26**].

Example 6. Let $C = \{0,1\}^\omega$ stand for the Cantor set. The minimal flow $\mathcal{M}(\mathrm{Homeo}(C))$ can be identified with the space of all maximal chains of closed subsets of C, equipped with the Vietoris topology. This is the result of Glasner and Weiss [**27**], while the space of maximal chains was introduced into the dynamical context by Uspenskij [**59**].

1045? **Question 22** (Uspenskij). *Give an explicit description of the universal minimal flow of the homeomorphism group* $\mathrm{Homeo}(X)$ *of a closed compact manifold X in dimension* $\dim X > 1$ *(with the compact-open topology).*

1046? **Question 23** (Uspenskij). *The same question for the group of homeomorphisms of the Hilbert cube $Q = \mathbb{I}^\omega$.*

Note that both X and Q form minimal flows for the respective homeomorphism groups, but they are not universal [**59**]. Interesting recent advances on both Questions 22 and 23 belong to Yonatan Gutman [**32**].

1047? **Question 24** (Uspenskij). *Is the pseudoarc P the universal minimal flow for its own homeomorphism group?*

A recent investigation [**36**] might provide means to attack this problem.

Let G be a topological group. The completion of G with regard to the left uniform structure (the *left* completion), denoted by $\hat{G}^L$, is a topological semigroup with jointly continuous multiplication [**49**, Prop. 10.2(a)], but in general is not a topological group [**13**]. Note that every left uniformly continuous real-valued function f on G extends to a unique continuous function $\hat{f}$ on $\hat{G}^L$. Say that such an f is *oscillation stable* if for every $\varepsilon > 0$ there is a right ideal J in the topological semigroup $\hat{G}^L$ with the property that the values of $\hat{f}$ at any two points of J differ by $< \varepsilon$. If H is a closed subgroup of G, say that the homogeneous space G/H is *oscillation stable* if every bounded left uniformly continuous function f on G that factors through the quotient map $G \to G/H$ is oscillation stable. If G/H is not oscillation stable, we say that G/H has *distortion*.

Example 7. The unit sphere $\mathbb{S}^\infty$ in the separable Hilbert space ℓ^2, considered as the homogeneous factor-space of the unitary group $U(\ell^2)$ with the strong topology, has distortion. It means that there exists a uniformly continuous function $f\colon \mathbb{S}^\infty \to \mathbb{R}$ whose range of values on the intersection of $\mathbb{S}^\infty$ with every infinite-dimensional linear subspace contains the interval (say) $[0,1]$. This is a famous and very difficult result by Odell and Schlumprecht [**39**], answering a 30 year-old problem. The following question is well-known in geometric functional analysis.

Question 25. *Does there exist a direct proof of Odell and Schlumprecht's result, based on the intrinsic geometry of the unit sphere and/or the unitary group?* 1048?

Example 8. The set $[\mathbb{Q}]^n$ of all n-subsets of $\mathbb{Q}$, considered as a homogeneous factor-space of $\mathrm{Aut}(\mathbb{Q}, \leq)$, is oscillation stable if and only if $n = 1$. For $n = 1$, oscillation stability simply means that for every finite colouring of $\mathbb{Q}$, there is a monochromatic subset A order-isomorphic to $\mathbb{Q}$ (this is obvious). For $n \geq 2$, distortion of $[\mathbb{Q}]^n$ means the existence of a finite colouring of this set with $k \geq 2$ colours such that for every subset A order-isomorphic to the rationals the set $[A]^n$ contains points of all k colours. This follows easily from the classical Sierpiński partition argument [**53**], cf. [**45**, Example 5.1.27].

The above setting for analysing distortion/oscillation stability in the context of topological transformation groups was proposed in [**37**] and discussed in [**45**]. The most substantial general result within this framework is presently the following.

Theorem (Hjorth [**35**]). *Let G be a Polish topological group. Considered as a G-space with regard to the action on itself by left translations, G has distortion whenever $G \neq \{e\}$.*

Question 26 (Hjorth [**35**]). *Let E be a separable Banach space and let $\mathbb{S}_E$ denote the unit sphere of E viewed as an* $\mathrm{Iso}(E)$*-space, where the latter group is equipped with the strong operator topology. Is it true that the* $\mathrm{Iso}(E)$*-space $\mathbb{S}_E$ has distortion?* 1049?

Note of caution: this would not, in general, mean that E has distortion in the sense of theory of Banach spaces [**7**, Chapter 13], as the two concepts only coincide for Hilbert spaces.

For an ultrahomogeneous separable metric space X, oscillation stability of X equipped with the standard action of the Polish group of isometries $\mathrm{Iso}(X)$ is equivalent to the following property. For every finite cover γ of X, there is an $A \in \gamma$ such that for each $\varepsilon > 0$, the ε-neighbourhood of A contains an isometric copy of X. The following could provide a helpful insight into Question 25.

1050? **Question 27.** *Is the metric space $\mathbb{U}_1$ oscillation stable?*

The Urysohn metric space $\mathbb{U}$ itself has distortion, but for trivial reasons, just like any other unbounded connected ultrahomogeneous metric space.

The oscillation stability of a metric space X whose distance assumes a discrete collection of values is equivalent to the property that whenever X is partitioned into two subsets, at least one of them contains an isometric copy of X. The Urysohn metric space $\mathbb{U}_{\{0,1,2\}}$, universal for the class of metric spaces whose distances take values $0, 1, 2$, is oscillation stable, because it is isometric to the path metric space associated to the *infinite random graph* R, and oscillation stability is an immediate consequence of an easily proved property of R known as *indestructibility* (cf. [**8**]). Very recently, Delhomme, Laflamme, Pouzet, and Sauer [**12**] have established oscillation stability of the universal Urysohn metric space $\mathbb{U}_{\{0,1,2,3\}}$ with the distance taking values $0, 1, 2, 3$. The following remains unknown.

1051? **Question 28.** *Let $n \in \mathbb{N}$, $n \geq 4$. Is the universal Urysohn metric space $\mathbb{U}_{\{0,1,\ldots,n\}}$ oscillation stable?*

Resolving the following old question may help.

1052? **Question 29** (M. Fréchet [**19**], p. 100; P.S. Alexandroff [**55**]). *Find a model for the Urysohn space $\mathbb{U}$, that is, a concrete realization.*

Several such models are known for the random graph (thence, $\mathbb{U}_{\{0,1,2\}}$), cf. [**8**].

1053? **Question 30.** *Find a model for the metric space $\mathbb{U}_{\{0,1,2,3\}}$.*

In connection with Uspenskij's examples of universal second-countable topological groups [**56, 57**], including $\mathrm{Iso}(\mathbb{U})$, the following remains unresolved.

1054? **Question 31** (V.V. Uspenskij [**58**]). *Does there exist a universal topological group of every given infinite weight τ?*

1055? **Question 32** (V.V. Uspenskij). *The same, for* any *uncountable weight?*

1056? **Question 33** (A.S. Kechris). *Does there exist a* co-universal *Polish topological group G, that is, such that every other Polish group is a topological factor-group of G?*

In the abelian case, the answer is in the positive [**51**].

1057? **Question 34** (A.S. Kechris). *Is every Polish topological group a topological factor-group of a subgroup of $U(\ell^2)$ with the strong topology?*

Again, in the abelian case the answer is in the positive [**22**].

Question 35. *Is the free topological group $F(X)$ [**28**] on a metrizable compact space isomorphic to a topological subgroup of the unitary group $U(\mathcal{H})$ of a suitable Hilbert space $\mathcal{H}$, equipped with the strong topology?* 1058?

Galindo has announced [**20**] a positive answer for free abelian topological groups. Uspenskij [**60**] has given a very elegant proof of a more general result: the free abelian topological group $A(X)$ of a Tychonoff space embeds into $U(\mathcal{H})$ as a topological subgroup. This suggests a more general version of the same question:

Question 36. *The same question for an arbitrary Tychonoff space X.* 1059?

In connection with Questions 34, 35 and 36, let us remind the following old problem.

Question 37 (A.I. Shtern [**52**]). *What is the intrinsic characterization of topological subgroups of $U(\ell^2)$ (with the strong topology)?* 1060?

A unitary representation π of a topological group G in a Hilbert space $\mathcal{H}$ (that is, a strongly continuous homomorphism $G \to U(\mathcal{H})$) *almost has invariant vectors* if for every compact $F \subseteq G$ and every $\varepsilon > 0$ there is a $\xi \in \mathcal{H}$ with $\|\xi\| = 1$ and $\|\pi_g\xi - \xi\| < \varepsilon$ for every $g \in F$. A topological group G has *Kazhdan's property (T)* if, whenever a unitary representation of G almost has invariant vectors, it has an invariant vector of norm one. For an excellent account of this rich theory, see the book [**33**] and especially its many times extended and updated English version, currently in preparation and available on-line [**6**].

Most of the theory is concentrated in the locally compact case. Bekka has shown in [**5**] that the group $U(\ell^2)$ with the strong topology has property (T).

Question 38 (Bekka [**5**]). *Does the group $U(\ell^2)$ with the uniform topology have property (T)?* 1061?

Question 39 (Bekka [**5**]). *Does the unitary group $U(\ell^2(\Gamma))$ of a nonseparable Hilbert space ($|\Gamma| > \aleph_0$), equipped with the strong topology, have property (T)?* 1062?

Here is a remarkable "large" topological group that has been receiving much attention recently. Let $\|\cdot\|_2$ denote the Hilbert–Schmidt norm on the $n\times n$ matrices, $\|A\|_2 = (\sum_{i,j=1}^n |a_{ij}|^2)^{1/2}$, and let d_n be the *normalized* Hilbert–Schmidt metric on the unitary group $U(n)$, that is, $d_n(u,v) = \frac{1}{\sqrt{n}}\|u-v\|_2$. Choose a free ultrafilter ξ on the natural numbers and denote by $U(\xi)_2$ the factor-group of the direct product $\prod_{n\in\mathbb{N}} U(n)$ by the normal subgroup $N_\xi = \{(x_n) : \lim_{n\to\xi} d_n(e,x_n) = 0\}$.

The following question is a particular case of Connes' Embedding Conjecture [**10**], for a thorough discussion see [**40**] and references therein.

Question 40 (Connes' Embedding Conjecture for Groups). *Is every countable group isomorphic to a subgroup of $U(\xi)_2$ (as an abstract group)?* 1063?

Groups isomorphic to subgroups of $U(\xi)_2$ are called *hyperlinear*. Here are some of the most important particular cases of the above problem.

1064–1066? **Question 41.** *Are countable groups from the following classes hyperlinear:* (a) *one-relator groups;* (b) *hyperbolic groups [**29**];* (c) *groups amenable at infinity (a.k.a. topologically amenable groups, exact groups) [**2**]?*

Under the natural bi-invariant metric $d(x,y) = \lim_{n\to\xi} d_n(x_n, y_n)$, the group $U(\xi)_2$ is a complete nonseparable metric group whose left and right uniformities coincide, isomorphic to a topological subgroup of $U(\ell^2(\mathfrak{c}))$ with the strong topology. Understanding the topological group structure of $U(\xi)_2$ may prove important.

The Connes' Embedding Conjecture itself can be reformulated in the language of topological groups as follows. Say, following [**47**], that a topological group G has *Kirchberg's property* if, whenever A and B are finite subsets of G with the property that every elemant of A commutes with every element of B, there exist finite subsets A' and B' of G that are arbitrarily close to A and B, respectively, such that every element of A' commutes with every element of B', and the subgroups of G generated by A' and B' are relatively compact. As noted in [**47**], the deep results of [**38**], modulo a criterion from [**17**], immediately imply that the Connes Embedding Conjecture is equivalent to the statement that the unitary group $U(\ell^2)$ with the strong topology has Kirchberg's property.

1067–1068? **Question 42.** *Do the following topological groups have Kirchberg's property:* (a) *the infinite symmetric group S_∞,* (b) *the group* $\mathrm{Aut}(X,\mu)$ *of measure-preserving transformations of a standard Lebesgue measure space with the coarse topology?*

It was shown in [**47**] that $\mathrm{Iso}(\mathbb{U})$ has Kirchberg's property.

References

[1] M. Ajtai, I. Havas, and J. Komlós, *Every group admits a bad topology*, Studies in pure mathematics, Birkhäuser, Basel, 1983, pp. 21–34.

[2] C. Anantharaman-Delaroche and J. Renault, *Amenable groupoids*, L'Enseignement Mathématique, Geneva, 2000.

[3] J. Auslander, *Minimal flows and their extensions*, North-Holland, Amsterdam, 1988.

[4] W. Banaszczyk, *On the existence of exotic Banach–Lie groups*, Math. Ann. **264** (1983), no. 4, 485–493.

[5] B. Bekka, *Kazhdan's property (T) for the unitary group of a separable Hilbert space*, Geom. Funct. Anal. **13** (2003), no. 3, 509–520.

[6] B. Bekka, P. de la Harpe, and A. Valette, *Kazhdan's property (T)*, 2006, In preparation, `http://poncelet.sciences.univ-metz.fr/~bekka`.

[7] Y. Benyamini and J. Lindenstrauss, *Geometric nonlinear functional analysis. Vol. 1*, American Mathematical Society, Providence, RI, 2000.

[8] P. J. Cameron, *The random graph*, The mathematics of Paul Erdős, II, Springer, Berlin, 1997, pp. 333–351.

[9] A. Carey and H. Grundling, *On the problem of the amenability of the gauge group*, Lett. Math. Phys. **68** (2004), no. 2, 113–120.

[10] A. Connes, *Classification of injective factors. Cases II_1, II_∞, III_λ, $\lambda \neq 1$*, Ann. of Math. (2) **104** (1976), no. 1, 73–115.

[11] M. Cotlar and R. Ricabarra, *On the existence of characters in topological groups*, Amer. J. Math. **76** (1954), 375–388.

[12] C. Delhomme, C. Laflamme, M. Pouzet, and N. Sauer, *Divisibility of countable metric spaces*, 2005, Preprint. `arXiv:math.CO/0510254`

[13] J. Dieudonné, *Sur la complétion des groupes topologiques*, C. R. Acad. Sci. Paris **218** (1944), 774–776.

[14] D. Dikranjan, I. R. Prodanov, and L. N. Stoyanov, *Topological groups: characters, dualities and minimal group topologies*, Marcel Dekker Inc., New York, 1990.

[15] R. Ellis, *Universal minimal sets*, Proc. Amer. Math. Soc. **11** (1960), 540–543.

[16] R. Ellis and H. B. Keynes, *Bohr compactifications and a result of Følner*, Israel J. Math. **12** (1972), 314–330.

[17] R. Exel and T. A. Loring, *Finite-dimensional representations of free product C^*-algebras*, Internat. J. Math. **3** (1992), no. 4, 469–476.

[18] E. Følner, *Generalization of a theorem of Bogolioùboff to topological abelian groups. With an appendix on Banach mean values in non-abelian groups*, Math. Scand. **2** (1954), 5–18.

[19] M. Fréchet, *Les espaces abstraits*, Gauthier-Villars, Paris, 1928.

[20] J. Galindo, *On unitary representability of topological groups*, 2006, Preprint. `arXiv:math.GN/0607193`

[21] S. Gao and A. S. Kechris, *On the classification of Polish metric spaces up to isometry*, Mem. Amer. Math. Soc. **161** (2003), no. 766, 78 pp.

[22] S. Gao and V. Pestov, *On a universality property of some abelian Polish groups*, Fund. Math. **179** (2003), no. 1, 1–15.

[23] T. Giordano and V. Pestov, *Some extremely amenable groups*, C. R. Math. Acad. Sci. Paris **334** (2002), no. 4, 273–278.

[24] E. Glasner, *On minimal actions of Polish groups*, Topology Appl. **85** (1998), 119–125.

[25] E. Glasner, B. Tsirelson, and B. Weiss, *The automorphism group of the Gaussian measure cannot act pointwise*, Israel J. Math. **148** (2005), 305–329.

[26] E. Glasner and B. Weiss, *Minimal actions of the group $\mathbb{S}(\mathbb{Z})$ of permutations of the integers*, Geom. Funct. Anal. **12** (2002), no. 5, 964–988.

[27] E. Glasner and B. Weiss, *The universal minimal system for the group of homeomorphisms of the Cantor set*, Fund. Math. **176** (2003), no. 3, 277–289.

[28] M. I. Graev, *Theory of topological groups. I. Norms and metrics on groups. Complete groups. Free topological groups*, Uspekhi Mat. Nauk **5** (1950), no. 2(36), 3–56, Translation: Free topological groups, Amer. Math. Soc. Transl. 35 (1951).

[29] M. Gromov, *Hyperbolic groups*, Essays in group theory, Springer, New York, 1987, pp. 75–263.

[30] M. Gromov, *Metric structures for Riemannian and non-Riemannian spaces*, Birkhäuser, Boston, 1999.

[31] M. Gromov and V. D. Milman, *A topological application of the isoperimetric inequality*, Amer. J. Math. **105** (1983), no. 4, 843–854.

[32] Y. Gutman, *Minimal actions of homeomorphism groups*, 2005, Preprint.

[33] P. de la Harpe and A. Valette, *La propriété (T) de Kazhdan pour les groupes localement compacts*, Astérisque (1989), no. 175, 158 pp.

[34] W. Herer and J. P. R. Christensen, *On the existence of pathological submeasures and the construction of exotic topological groups*, Math. Ann. **213** (1975), 203–210.

[35] G. Hjorth, *An oscillation theorem for groups of isometries*, 2004, Preprint.

[36] T. Irwin and S. Solecki, *Projective Fraïssé limits and the pseudo-arc*, Trans. Amer. Math. Soc. **358** (2006), no. 7, 3077–3096.

[37] A. S. Kechris, V. G. Pestov, and S. Todorčević, *Fraïssé limits, Ramsey theory, and topological dynamics of automorphism groups*, Geom. Funct. Anal. **15** (2005), no. 1, 106–189.

[38] E. Kirchberg, *On nonsemisplit extensions, tensor products and exactness of group C^*-algebras*, Invent. Math. **112** (1993), no. 3, 449–489.

[39] E. Odell and T. Schlumprecht, *The distortion problem*, Acta Math. **173** (1994), no. 2, 259–281.

[40] N. Ozawa, *About the QWEP conjecture*, Internat. J. Math. **15** (2004), no. 5, 501–530.

[41] A. L. T. Paterson, *Amenability*, American Mathematical Society, Providence, RI, 1988.

[42] V. Pestov, *On free actions, minimal flows, and a problem by Ellis*, Trans. Amer. Math. Soc. **350** (1998), no. 10, 4149–4165.

[43] V. Pestov, *Some universal constructions in abstract topological dynamics*, Topological dynamics and applications (Minneapolis, MN, 1995), Contemp. Math., vol. 215, American Mathematical Society, Providence, Rl, 1998, pp. 83–99.

[44] V. Pestov, *Ramsey–Milman phenomenon, Urysohn metric spaces, and extremely amenable groups*, Israel J. Math. **127** (2002), 317–357.

[45] V. Pestov, *Dynamics of infinite-dimensional groups and Ramsey-type phenomena*, Publicações Matemáticas dos Colóquios de Matemática, IMPA, Rio de Janeiro, 2005.

[46] V. Pestov, *The isometry group of the Urysohn space as a Lévy group*, 2005, Preprint. To appear in Topology Appl. `arXiv:math.GN/0509402`

[47] V. Pestov and V. V. Uspenskij, *Representations of residually finite groups by isometries of the Urysohn space*, J. Ramanujan Math. Soc. **21** (2006), 189–203.

[48] I. V. Protasov and E. G. Zelenyuk, *Topologies on groups determined by sequences*, VNTL Publishers, Lviv, 1999.

[49] W. Roelcke and S. Dierolf, *Uniform structures on topological groups and their quotients*, McGraw-Hill, New York, 1981.

[50] C. Rosendal and S. Solecki, *Automatic continuity of homomorphisms and fixed points on metric compacta*, 2006, Preprint. To appear in Israel J. Math. `arXiv:math.LO/0604575`

[51] D. Shakhmatov, J. Pelant, and S. Watson, *A universal complete metric abelian group of a given weight*, Topology with applications (Szekszárd, 1993), Bolyai Soc. Math. Stud., vol. 4, János Bolyai Math. Soc., Budapest, 1995, pp. 431–439.

[52] A. Shtern, *Unitary representation*, Mathematical Encyclopaedia, vol. 5, Sov. Encycl., 1984, pp. 508–513 (Russian).

[53] W. Sierpiński, *Sur un problème de la théorie des relations*, Ann. Scuola Norm. Sup. Pisa **2** (1933), 285–287.

[54] P. S. Urysohn, *Sur un espace métrique universel*, C. R. Acad. Sci. Paris **180** (1925), 803–806.

[55] P. S. Urysohn, *On the universal metric space*, P. S. Urysohn. Selected Works, vol. 2 (P. S. Alexandroff, ed.), Nauka, Moscow, 1972, pp. 747–769 (Russian).

[56] V. V. Uspenskij, *A universal topological group with a countable basis*, Funktsional. Anal. i Prilozhen. **20** (1986), no. 2, 86–87, Translation: Funct. Anal. Appl. 20 (1986) 160–161.

[57] V. V. Uspenskij, *On the group of isometries of the Urysohn universal metric space*, Comment. Math. Univ. Carolin. **31** (1990), no. 1, 181–182.

[58] V. V. Uspenskij, *On subgroups of minimal topological groups*, 1998, Preprint. `arXiv:math.GN/0004119`

[59] V. V. Uspenskij, *On universal minimal compact G-spaces*, Topology Proc. **25** (2000), Spring, 301–308.

[60] V. V. Uspenskij, *Unitary representability of free abelian topological groups*, 2006, Preprint. `arXiv:math.RT/0604253`

[61] W. A. Veech, *The equicontinuous structure relation for minimal Abelian transformation groups*, Amer. J. Math. **90** (1968), 723–732.

[62] W. A. Veech, *Topological dynamics*, Bull. Amer. Math. Soc. **83** (1977), no. 5, 775–830.

[63] A. M. Vershik, *Random metric spaces and universality*, Uspekhi Mat. Nauk **59** (2004), no. 2(356), 65–104, Translation: Russian Math. Surveys 59 (2004) no. 2, 259–295.

[64] B. Weiss, *Single orbit dynamics*, American Mathematical Society, Providence, RI, 2000.

Part 5

Dynamical Systems

Open Problems in Topology II – Edited by E. Pearl

Minimal flows

William F. Basener, Kamlesh Parwani and Tamas Wiandt

1. Introduction

A *topological dynamical system* is a continuous group action $\varphi\colon X \times T \to X$ written $\varphi_t(x)$, where T is a group and X is a compact Hausdorff space. The most common cases are where $T = \mathbb{R}$, called a *continuous flow*, and $T = \mathbb{Z}$, called a *discrete flow*. In this paper, a flow means a continuous flow and discrete flows will be referred to as maps, homeomorphisms, and diffeomorphisms.

Flows arise most naturally as the set of solutions to a system of differential equations. A compact invariant subset $X \subset M$ is said to be a *minimal set* if it is a minimal compact invariant set under containment. Explicitly, X is a minimal set if the only compact invariant subsets of X are itself and the empty set. It is not difficult to show that a set is minimal if and only if the orbit of every point in X is dense in X. Another useful characterization of a minimal set is that a set is minimal if and only if every point in X is almost periodic, defined as follows. A point x is *almost periodic* for φ if for every neighborhood U of x there are times $t_0 < t_1 < \cdots < t_i < \cdots$ such that $\varphi_{t_i}(x) \in U$ for all i and the set of all $|t_{i+1} - t_i|$ is bounded. As Gottschalk put it, a periodic point returns every hour on the hour while an almost periodic point returns to its neighborhood every hour within the hour. It is also not hard to show that the orbit closure of any almost periodic point is a minimal set.

The notion of a minimal set was first introduced by G.D. Birkhoff in 1912 [**6**]. The motivation is that a minimal set is the smallest element of a dynamical system, and heuristically a dynamical system can be broken down into its minimal sets and the *transient* portion which moves between the minimal sets. A flow is said to be a minimal flow if the space M is itself a minimal set. This is the situation we focus on, putting aside questions about the behavior of flows around and near minimal sets.

The simplest examples of minimal flows are the trivial flow on a single point and a flow on a circle without fixed points. After these, there is the classical example of an irrational flow on a torus.

Example 1.1. Define $\mathbb{T}^n = \mathbb{R}^n/\mathbb{Z}^n$. Let $\alpha_1, \ldots, \alpha_n$ be real numbers. Define a flow on $\mathbb{T}^n$ by $\varphi_t(x_1, \ldots, x_n) = (x_1 + \alpha_1 t, \ldots, x_n + \alpha_n t) \mod 1$. The case where $n = 2$ corresponds to moving along lines of slope α_2/α_1. If α_2/α_1 is irrational then this flow is minimal. For the n–dimensional case, if each ratio α_i/α_j, $i \neq j$, is irrational then the flow is minimal. We call this flow an *irrational flow on the torus* if the ratios are all irrational. Note that the distance between points is preserved by the irrational flow.

It is important to determine when two flows are the *same*. Two flows $\varphi\colon X \times \mathbb{R} \to X$ and $\phi\colon Y \times \mathbb{R} \to Y$ are said to be *topologically conjugate* if there is a homeomorphism $h\colon X \to Y$ such that $h \circ \varphi = \phi \circ h$. This condition is often too strong. More generally, we say that φ and ϕ are *topologically equivalent* if there is a homeomorphism $h\colon X \to Y$ that takes orbits in X to orbits in Y. To illustrate the necessity for considering topological equivalence, observe that any two periodic orbits in a continuous flow are topologically equivalent, but are topologically conjugate only when they have the same period.

The primary questions regarding minimal flows are still the two questions Gottschalk put forth in [**24**].

(1) Construction Problem: Provide an explicit construction for all minimal flows

(2) Classification Problem: Classify all minimal flows either up to topologically equivalence or topological conjugacy.

Gottschalk considered the general case where T is an arbitrary group. He thus considered topological conjugacy but not topological equivelance.

A related question is to determine all manifolds that admit minimal flows. For $n = 1$, this is trivial. For $n = 2$, the question is also easy. The torus admits a minimal flow (Example 1.1), every flow on the Klein bottle has a periodic orbit [**30**], and every flow on any other 2-manifold is forced to have a fixed point because of nonzero Euler characteristic. The question for 3-manifolds is one of the most important open questions in dynamical systems. In [**24**], Gottschalk posed questions in the following way: "What compact metric spaces can be minimal sets under a discrete flow? Under a continuous flow? The universal curve of Sierpiński? The universal curve of Menger? A lens space? A polyhedra? ..." The question about the 3-sphere is usually called the Gottschalk conjecture:

1069? **Question 1.2** (Gottschalk Conjecture). *Does there exist a minimal flow on the 3-sphere?*

In [**35**], Steve Smale mentioned the problem in his list of the most important problems for the twenty-first century explicitly as "Is the three-sphere a minimal set? Can a C^∞ vector field be found on the three sphere so that every solution curve is dense?" The same question also appears in [**13**]. More generally, we may ask which 3-manifolds support minimal flows? In Section 2 we survey examples of minimal flows on 3-manifolds and present several related open questions. Minimal flows can also be studied by looking at the asymptotic behavior of orbits. Do they spread out? Are there orbits that get closer and closer? In Section 3, we take this point of view and discuss distal and proximal flows which are central in topological dynamics.

The choice of topics covered in this article is subjective, based on the authors' personal tastes. We ignore the analogous questions regarding minimal homeomorphisms and minimal actions of other non-compact Lie groups. However, we would be remiss in our duties if we failed to mention the following remarkable theorem:[1] if a compact manifold supports an (almost) free $\mathbb{T}^2$ action, then it supports a

[1] A. Katok proved the theorem for "free" hypothesis, and then later A. Fathi and M. Herman proved it under the "almost free" hypothesis."

minimal flow. The interested reader should consult [**29**] and [**15**] for more on this. Another important property, not discussed here, is the topology of individual orbits of a minimal flow. If we close an orbit segment with a short arc, what type of knot might we get? This is the approach of [**4, 26**].

Our main goal in this survey is to present enough material to whet the readers appetite and allow for further exploration through the references listed at the end.

The questions about the existence of minimal flows on manifolds are a special case of the following more general question.

Question 1.3. *Consider two integers n and m, with $n > m > 0$. Which n-dimensional manifolds support a minimal m-dimensional foliation? Here a minimal foliation is a foliation in which all leaves are dense.* 1070?

It is reasonable to assume that minimal flows on 3-manifolds are more tractable, and we now present many examples and related open questions.

2. Minimal flows on 3-manifolds

We provide examples of minimal flows on compact 3-manifolds. These examples fall into a few natural classes—suspensions and horocycle flows that are derived from Anosov flows and robustly transitive diffeomorphisms.

2.1. Suspensions. Let us first describe the suspension construction. Consider a compact manifold M and a homeomorphism $f\colon M \to M$. From this data we construct a manifold S_f and a flow f_t in the following manner. Define S_f as the manifold $M \times [0,1]/\{(x,1) \sim (f(x),0)\}$, that is, we glue the roof $(M \times 1)$ to the floor $(M \times 0)$ by the map f. Now define the flow f_t as the unit speed flow that moves points vertically up from $M \times 0$ to $M \times 1$. This construction is often referred to as the constant roof function suspension. The flow f_t and map f determine each other—f_t is constructed from f and the time-one map of f_t restricted to $M \times t$, for $t \in [0,1]$, is the homeomorphism f on a copy of M ($M \times t$ is a cross section to the flow and the return map is f).

Now it is easy to see that if the map f is minimal on M, then the flow f_t on S_f is a minimal flow.

Example 2.1. Let f be an irrational translation on the two-dimensional torus $\mathbb{T}^2$ defined by $f(x) = x + (\alpha, \beta) \mod 1$, where (α, β) is a vector with irrational slope. Then the suspension flow f_t on $\mathbb{T}^3$ is minimal. Note that f_t is equivalent to the irrational flow on $\mathbb{T}^3$ defined by $g_t = x + t(\alpha, \beta, 1) \mod 1$.

Question 2.2. *Does every minimal flow on $\mathbb{T}^3$ have a cross section?* 1071?

In order to obtain a minimal flow on a 3-manifold via the suspension construction, we need a minimal homeomorphism on a two-manifold. The only compact two-manifolds that support minimal maps are $\mathbb{T}^2$ and $\mathbb{K}^2$ (the Klein bottle). Minimal maps are easy to construct on the torus; consider irrational translations. On the Klein bottle, a theorem of Katok in [**29**] implies the existence of a minimal diffeomorphism isotopic to the identity; this follows from the existence of a free circle action on $\mathbb{K}^2$ and a nontrivial Baire category argument (also see [**15**]). Consequently, the only 3-manifolds that support minimal flows and arise from the suspension construction must be $\mathbb{T}^2$ or $\mathbb{K}^2$ bundles over the circle—like $\mathbb{T}^3$ and $\mathbb{K}^2 \times \mathbb{S}^1$. In any case, a finite cover of the manifold must be a torus bundle over the circle.

So we will focus on torus bundles over the circle. If a map on $\mathbb{T}^2$ is isotopic to an Anosov diffeomorphism, it must have infinitely many periodic orbits (since it is semi-conjugate to the Anosov diffeomorphism [**18**]), and hence, cannot be minimal. So we need concentrate only on non-hyperbolic maps (up to isotopy). If f is such a homeomorphism on $\mathbb{T}^2$, the induced action on the first homology group is a linear map that is not Anosov and has all eigenvalues of absolute value 1; the linear map is either of finite order or conjugate to $\left(\begin{smallmatrix} 1 & n \\ 0 & 1 \end{smallmatrix}\right)$, where n is a positive integer. The suspension of such maps produce 3-manifolds that are better known as circle bundles over the torus. These are also *Nil manifolds*—they are quotients of a nilpotent connected Lie group by a closed subgroup. The next example discusses minimal flows on circle bundles over the torus that arise from suspending maps of infinite order (up to isotopy). The reader may also consult [**2**, **3**] for examples of minimal flows on Nil manifolds.

Example 2.3. Define $f_n\colon \mathbb{T}^2 \to \mathbb{T}^2$ as $f_n(x,y) = (x, nx+y) + (\alpha, 0) \mod 1$, where α is irrational and n is an integer. Furstenberg proved that f_n is minimal on $\mathbb{T}^2$ for all n in [**20**]. Note that f_0 is just an irrational translation on $\mathbb{T}^2$.

For distinct positive n and m, since S_{f_n} is not homeomorphic to S_{f_m} (because the fundamenatal groups are not isomorphic), we obtain minimal flows on an infinite family of compact 3-manifolds via the suspension construction.

1072? **Question 2.4.** *Does every minimal flow on a circle bundle over $\mathbb{T}^2$ have a cross section?*

2.2. Derived from Anosov flows. A flow f_t on a compact manifold M associated to the vector field X is an *Anosov flow* if there is a splitting of the tangent bundle TM into the line field $\mathbb{R}X$ and two df_t-invariant subbundles E^s and E^u such that df_t uniformly contracts the vectors in E^s and uniformly expands the vectors in E^u as $t \to \infty$.

Generally, E^s and E^u are Hölder continuous subbundles of TM that are uniquely integrable and define foliations $\mathcal{F}^s$ and $\mathcal{F}^u$ (called the *stable foliation* and the *unstable foliation* respectively). Similarly, $E^s \oplus \mathbb{R}X$ and $E^u \oplus \mathbb{R}X$ also define foliations $\mathcal{F}^{cs}$ and $\mathcal{F}^{cu}$, called the *center-stable* and the *center-unstable* foliations respectively.

Example 2.5. Anosov flows are not minimal since they have periodic orbits. However, if an Anosov flow is transitive and not a suspension of an Anosov diffeomorphism, its stable and unstable foliations are minimal (see [**11**, **33**]), and these foliations readily provide minimal flows. If these foliations are orientable, we may obtain a vector field tangent to the leaves, which will then produce the required minimal flow when the all the leaves are dense. If the minimal foliation is not orientable, we can always obtain a minimal flow on the appropriate double cover.

The simplest examples of transitive Anosov flows are suspensions of Anosov diffeomorphisms on $\mathbb{T}^2$. In this case, the stable and unstable foliations are not minimal; however, after a reparametrization these foliations are minimal. Other standard examples of transitive Anosov flows are geodesic flows on unit tangent bundles over surfaces of constant negative curvature. It is a classical theorem of G.A. Hedlund that the horocycle foliations (the stable/unstable foliations) of these geodesic flows are minimal (see [**22**, **28**]), and the resulting horocycle flow along these leaves in a minimal flow. Handel and Thurston in [**27**] obtained more exotic examples of transitive Anosov flows on graph manifolds by doing surgery along closed orbits. The reader may also be interested in the example of Bonatti and Langevin in [**10**] and the examples of Fenley in [**17**] on non-orientable hyperbolic 3-manifolds.

We now list an example that is derived from an Anosov flow and deserves special attention.

Example 2.6. Goodman enhanced the surgery techniques of Handel and Thurston and in [**23**] constructed an Anosov flow on a 3-manifold by doing surgery around a closed orbit of an Anosov flow obtained by suspending the Cat map $\left(\begin{smallmatrix}2&1\\1&1\end{smallmatrix}\right)$. The resulting Anosov flow is on a hyperbolic 3-manifold that is not sufficiently large (see [**36**] for more details). This implies that it is a homology sphere, that is, it has the same homology groups as $\mathbb{S}^3$. Furthermore, the unstable/stable foliations are orientable and since the resulting manifold is hyperbolic, the Anosov flow on it must be transitive (see [**16**]). So we obtain a minimal flow on a homology 3-sphere!

The (horocycle) minimal flows that are derived from Anosov flows do not have cross sections. The existence of a cross section implies that the manifold (or a finite cover of the manifold) must be a circle bundle over $\mathbb{T}^2$, and therefore, the fundamental group has polynomial growth (the fundamental group is virtually nilpotent). If a 3-manifold supports an Anosov flow, its fundamental group has exponential growth and is not virtually nilpotent (see [**34**] or the appendix of [**1**]). So cross sections cannot exist for these minimal flows.

These examples illustrate that the problem of classifying 3-manifolds which support minimal flows is at least as intractable as the problem of classifying 3-manifolds that support transitive Anosov flows.

Question 2.7. *Which 3-manifolds support transitive Anosov flows?* 1073?

David Fried in [**19**] has shown that all transitive Anosov flows may be obtained by doing surgery along the singular orbits of flows obtained by suspending certain pseudo-Anosov maps. In general, given a 3-manifold (with exponential growth in its fundamental group), it is very difficult to determine if it supports an Anosov flow or not.

2.3. Derived from robustly transitive diffeomorphisms. A C^1 diffeomorphism f is robustly transitive if f is transitive and all maps in a C^1 neighborhood are also transitive. For instance, perturbations of transitive Anosov flows are known to be robustly transitive. Diaz, Pujals, and Ures in [**14**] have shown that in dimension 3, robustly transitive maps must necessarily be partially hyperbolic.

A C^1 diffeomorphism f on a compact manifold M is *partially hyperbolic* if there is a splitting of the tangent bundle TM $(E \oplus F)$ into two df-invariant bundles E and F such that either df uniformly contracts the vectors in E or df uniformly expands the vectors in F.

Bonatti, Diaz, Pujals, and Ures have proved that in any dimension, robustly transitive maps possess a dominated splitting; more precisely, every robustly transitive set of a C^1 diffeomorphism is volume hyperbolic ([**9**] is an excellent reference for this topic and for nonuniformly hyperbolic phenomena in general). However, very little is known about manifolds that support robustly transitive diffeomorphisms.

A C^1 diffeomorphism f on a compact manifold M is a *strong partially hyperbolic* diffeomorphism or *strongly partially hyperbolic* if there is a splitting of the tangent bundle TM $(E^s \oplus E^c \oplus E^u)$ into three df-invariant bundles E^s, E^c, and E^u such that df uniformly expands the vectors in E^u, uniformly contracts the vectors in E^s, and the vectors in E^c are expanded (respectively contracted) less than the vectors in E^u (respectively E^s).

In dimension 3, when f is strongly partially hyperbolic, Bonatti, Diaz and Ures in [**8**] have established the existence of minimal stable or unstable foliations. For example, perturbations of transitive Anosov flows and skew products over Anosov diffeomorphisms are examples of strong partially hyperbolic and robustly transitive diffeomorphisms (see [**7**]), and these examples either possess a minimal stable foliation or a minimal unstable foliation. So just like in the case of transitive Anosov flows, certain robustly transitive maps have minimal stable or unstable foliations, and we obtain minimal flows from these.

1074? **Question 2.8.** *Which 3-manifolds support robustly transitive diffeomorphisms and do not support transitive Anosov flows?*

1075? **Question 2.9.** *Which 3-manifolds support partially hyperbolic diffeomorphisms?*

Brin, Burago, and Ivanov have proved, under the assumption of dynamical coherence, that there are no strong partially hyperbolic maps on $\mathbb{S}^3$ (see [**12**]). Parwani has recently shown in [**32**] that if M supports a strong partially hyperbolic and dynamically coherent diffeomorphism, then the universal cover of M is homeomorphic to $\mathbb{R}^3$. Of course, since the time-one map of an Anosov flow is a strong partially hyperbolic diffeomorphism, the classification of manifolds that support partially hyperbolic maps is at least as difficult as the classification of manifolds that support Anosov flows.

1076? **Question 2.10.** *Is there a 3-manifold with exponential growth in its fundamental group that supports a partially hyperbolic diffeomorphism but does not support an Anosov flow?*

3. Asymptotic properties

We say that two points x, y are *positively asymptotic* in a flow $\varphi\colon M\times\mathbb{R} \to M$ if $d(\varphi_t(x), \varphi_t(y)) \to 0$ as $t \to \infty$. This defines a relation on $M \times M$ by $A = \{(x, y) \in M \times M : x$ and y are positively asymptotic$\}$. No two distinct points are positively asymptotic in a irrational flow since the flow preserves the distance between any two points. There are no positively asymptotic points in the classical horocycle flow. Choose $\epsilon > 0$ sufficiently small. If $d(x, y) < \epsilon$ then either x and y are on the same local leaf, in which case the distance between them will be unchanged by the flow, or they are on different local leaves in which case $d(\phi_t(x), \phi_t(y)) > \epsilon$ for some $t > 0$ (see Chapter 1 in [**31**]). Asymptotic points are in general characteristic of minimal sets for symbolic dynamics. We present an example below of a minimal homeomorphism on a Cantor set with asymptotic points. It is possible to suspend this map and obtain a minimal flow that enjoys the same property.

Example 3.1. Let Σ denote the set of all bi-infinite sequences of zeros and ones, with elements written as $\mathbf{x} = (\ldots, x_{-2}, x_{-1}, \underline{x_0}, x_1, x_2, \ldots)$, and with the metric

$$d(\mathbf{x}, \mathbf{y}) = \sum_{i\in\mathbb{Z}} \frac{|x_i - y_i|}{2^{|i|}}.$$

This metric makes Σ into a Cantor set. Define the shift map on Σ by $\sigma(\ldots, x_{-2}, x_{-1}, \underline{x_0}, x_1, x_2, \ldots) = (\ldots, x_{-1}, x_0, \underline{x_1}, x_2, x_3, \ldots)$. Define the element $\omega \in \Sigma$ as follows. Begin with 0. Make the substitution $0 \mapsto 01$ and $1 \mapsto 10$. Repeating this substitution several times we get

$$\begin{array}{l} \underline{0} \\ \underline{0}1 \\ \underline{0}110 \\ \underline{0}1101001 \\ \underline{0}110100110010110 \end{array}$$

Define ω to be the limit of this sequence concatenated to its mirror image,

$$\omega = (\ldots 0110100110010110\underline{0}110100110010110\ldots).$$

It is not hard to show that the orbit of ω is almost periodic but not periodic. Hence the closure of this orbit is a minimal set which we call X. Observe that any point $\omega' \in \Sigma$ that agrees with ω in all entries to the right is in X and is asymptotic to ω under the discrete flow,

$$d(\sigma^n(\omega), \sigma^n(\omega')) \to 0 \quad \text{as } n \to \infty. \tag{3.1}$$

This type of minimal set is called a substitution minimal set. Observe that the asymptotic behavior of Equation (3.1) is not present in either of the other two examples.

Question 3.2. *Does there exist a minimal flow on a smooth manifold with a* 1077?
nontrivial pair of positively asymptotic points?

Asymptotic behavior seems too much to ask in general. So we can consider a more general form of asymptotic–like behavior. We say that a pair of points $x, y \in M$ are *proximal* if there is a sequence $\{t_n\}$ in $\mathbb{R}$ such that $d(\varphi_{t_n}(x), \varphi_{t_n}(y) \to 0$ as $i \to \infty$. (In the case of a group action $X \times T \to X$ on a noncompact space X, two points x, y are proximal if there exists a point $z \in X$ and sequence $\{t_n\}$ such that $\varphi_{t_n}(x) \to z$ and $\varphi_{t_n}(y) \to z$ as $i \to \infty$.) We define the proximal relation on $M \times M$ by $P = \{(x, y) : x$ and y are proximal$\}$. In the irrational flow, the only proximal pairs are trivial, $P = \Delta$.

A pair of points that are not proximal is said to be *distal*. In other words, two points x, y are distal if there is a nonzero lower bound on the distance between the points under the flow; $\inf\{d(\varphi_t(x), \varphi_t(y) : t \in \mathbb{R}\} > 0$. A flow is said to be *distal* if nontrivial pairs of points are distal. That is, the flow is distal if and only if $P = \Delta$. The irrational flow on the torus is distal.

We can relax the proximal condition as follows. Let φ denote a flow on a compact manifold M. A pair of points x and y are *regionally proximal* if there are sequences $\{x_n\} \to x$ and $\{y_n\} \to y$ in M and $\{t_n\}$ in $\mathbb{R}$ such that $d(\varphi_{t_n}(x_n), \varphi_{t_n}(y_n)) \to 0$ as $n \to \infty$. (In the case of a group action $X \times T \to X$ on a noncompact space X, two points x, y are regionally proximal if there exists a point $z \in X$ and sequences $\{x_n\} \to x$ and $\{y_n\} \to y$ in M and $\{t_n\}$ in $\mathbb{R}$ such that $(\varphi_{t_n}(x_n), \varphi_{t_n}(y)) \to (z, z)$ as $n \to \infty$.) For flows on a compact manifold, this is equivalent to $d(\varphi_{t_n}(x), \varphi_{t_n}(y)) \to 0$ as $n \to \infty$. We define the regionally proximal relation on $M \times M$ by $Q = \{(x, y) : x$ and y are regionally proximal$\}$.

Also important is the property of equicontinuity. A flow is *equicontinuous* if the family of maps defining the flow is an equicontinuous family. That is, if for any $\epsilon > 0$, there exists a $\delta > 0$ such that $d(x, y) < \delta \Rightarrow d(\varphi_t(x), \varphi_t(y)) < \epsilon$ for all t. It is worth noting that the flow being not equicontinuous is equivalent to having sensitive dependence on initial conditions.

The proof of the following lemma is *easy*. (For details, see [**2**].)

Lemma. *A flow φ is equicontinuous if and only if $Q = \Delta$.*

The containment $A \subset P \subset Q$ implies the following relationship:

$$
\begin{aligned}
&Q = \Delta \Leftrightarrow \varphi \text{ is equicontinuous}\\
&\Downarrow\\
&P = \Delta \Leftrightarrow \varphi \text{ is distal}\\
&\Downarrow\\
&A = \Delta
\end{aligned}
$$

An obvious question is to determine which manifolds admit minimal flows with each type of asymptotic behavior. The only Riemannian manifolds that admit equicontinuous minimal flows are tori, which answers the question for $Q = \Delta$ (see [**5**]). L. Auslander, L. Green and F. Hahn showed that that typically minimal flows on Nil manifolds, such as the flows of Example 2.3, are distal but not equicontinuous in [**3**]. Their work led to Furstenberg's structure theorem for distal flows (see [**21**]).

Question 3.3. *Describe all 3-manifolds that admit distal minimal flows. (Perhaps* 1078?
only Nil manifolds?)

References

[1] D. V. Anosov and Ja. G. Sinaĭ, *Certain smooth ergodic systems*, Uspekhi Mat. Nauk **22** (1967), no. 5(137), 107–172.

[2] J. Auslander, *Minimal flows and their extensions*, North-Holland, Amsterdam, 1988.

[3] L. Auslander, L. Green, and F. Hahn, *Flows on homogeneous spaces*, Princeton University Press, Princeton, NJ, 1963.

[4] W. Basener, *Knots and topologically transitive flows on 3-manifolds*, Topology **43** (2004), no. 3, 697–703.

[5] W. Basener, *The geometry of minimal flows*, 2006, Preprint. To appear in Topology Appl.

[6] G. D. Birkhoff, *Quelques théorèmes sur le mouvement des systèmes dynamiques*, Bull. Soc. Math. France **40** (1912), 305–323.

[7] C. Bonatti and L. J. Díaz, *Persistent nonhyperbolic transitive diffeomorphisms*, Ann. of Math. (2) **143** (1996), no. 2, 357–396.

[8] C. Bonatti, L. J. Díaz, and R. Ures, *Minimality of strong stable and unstable foliations for partially hyperbolic diffeomorphisms*, J. Inst. Math. Jussieu **1** (2002), no. 4, 513–541.

[9] C. Bonatti, L. J. Díaz, and M. Viana, *Dynamics beyond uniform hyperbolicity*, Springer, Berlin, 2005.

[10] C. Bonatti and R Langevin, *Un exemple de flot d'Anosov transitif transverse à un tore et non conjugué à une suspension*, Ergodic Theory Dynam. Systems **14** (1994), no. 4, 633–643.

[11] R. Bowen and B. Marcus, *Unique ergodicity for horocycle foliations*, Israel J. Math. **26** (1977), no. 1, 43–67.

[12] M. Brin, D. Burago, and S. Ivanov, *On partially hyperbolic diffeomorphisms of 3-manifolds with commutative fundamental group*, Modern dynamical systems and applications, Cambridge University Press, Cambridge, 2004, pp. 307–312.

[13] F. Browder (ed.), *Mathematical developments arising from Hilbert problems*, American Mathematical Society, Providence, RI, 1976.

[14] L. J. Díaz, E. R. Pujals, and R. Ures, *Partial hyperbolicity and robust transitivity*, Acta Math. **183** (1999), no. 1, 1–43.

[15] A. Fathi and M. R. Herman, *Existence de difféomorphismes minimaux*, Astérisque (1977), no. 49, 37–59.

[16] S. R. Fenley, *Quasigeodesic Anosov flows and homotopic properties of flow lines*, J. Differential Geom. **41** (1995), no. 2, 479–514.

[17] S. R. Fenley, *The structure of branching in Anosov flows of* 3*-manifolds*, Comment. Math. Helv. **73** (1998), no. 2, 259–297.

[18] J. Franks, *Anosov diffeomorphisms*, Global analysis (Berkeley, 1968), American Mathematical Society, Providence, RI, 1970, pp. 61–93.

[19] D. Fried, *Transitive Anosov flows and pseudo-Anosov maps*, Topology **22** (1983), no. 3, 299–303.

[20] H. Furstenberg, *Strict ergodicity and transformation of the torus*, Amer. J. Math. **83** (1961), 573–601.

[21] H. Furstenberg, *The structure of distal flows*, Amer. J. Math. **85** (1963), 477–515.

[22] H. Furstenberg, *The unique ergodicity of the horocycle flow*, Recent advances in topological dynamics (New Haven, CT, 1972), Lecture Notes in Math., vol. 318, Springer, Berlin, 1973, pp. 95–115.

[23] S. Goodman, *Dehn surgery on Anosov flows*, Geometric dynamics (Rio de Janeiro, 1981), Lecture Notes in Math., vol. 1007, Springer, Berlin, 1983, pp. 300–307.

[24] W. H. Gottschalk, *Minimal sets: an introduction to topological dynamics*, Bull. Amer. Math. Soc. **64** (1958), 336–351.

[25] W. H. Gottschalk, *A survey of minimal sets*, Ann. Inst. Fourier (Grenoble) **14** (1964), no. 1, 53–60.

[26] C. Gutiérrez, *Knots and minimal flows on 3-manifolds*, Topology **34** (1995), no. 3, 679–698.

[27] M. Handel and W. P. Thurston, *Anosov flows on new three manifolds*, Invent. Math. **59** (1980), no. 2, 95–103.

[28] G. A. Hedlund, *Fuchsian groups and transitive horocycles*, Duke Math. J. **2** (1936), no. 3, 530–542.

[29] A. B. Katok, *Minimal diffeomorphisms on principal S^1-fiber spaces*, Proc. VI Soviet topological conference, Tbilisi, Metzniereba, Tbilisi, 1972 (Russian).

[30] N. G. Markley, *The Poincaré–Bendixson theorem for the Klein bottle*, Trans. Amer. Math. Soc. **135** (1969), 159–165.

[31] D. W. Morris, *Ratner's theorems on unipotent flows*, University of Chicago Press, Chicago, IL, 2005.

[32] K. Parwani, *On 3-manifolds that support partially hyperbolic diffeomorphisms*, 2006, Preprint.

[33] J. F. Plante, *Anosov flows*, Amer. J. Math. **94** (1972), 729–754.

[34] J. F. Plante and W. P. Thurston, *Anosov flows and the fundamental group*, Topology **11** (1972), 147–150.

[35] S. Smale, *Mathematical problems for the next century*, Mathematics: frontiers and perspectives, American Mathematical Society, Providence, RI, 2000, pp. 271–294.

[36] W. P. Thurston, *Lecture notes on the geometry and topology of 3-manifolds*, Princeton University, 1980, `http://msri.org/publications/books/gt3m/`.

Open Problems in Topology II – Edited by E. Pearl

The dynamics of tiling spaces

Alex Clark

In order to provide a topological structure for spaces of tilings, one uses a metric defined in analogy with a commonly used metric on symbolic spaces. Given a finite alphabet $\mathcal{A}$, one can define a metric on $\mathcal{A}^{\mathbb{Z}^d}$ by setting

$$d\left((x_{\mathbf{n}}),(y_{\mathbf{n}})\right) = \frac{1}{1+\min\{|\mathbf{n}| : x_{|\mathbf{n}|} \neq y_{|\mathbf{n}|}\}},$$

where $|\mathbf{n}|$ denotes the norm of $\mathbf{n} \in \mathbb{Z}^d$. With this metric $\mathcal{A}^{\mathbb{Z}^d}$ is a Cantor set that supports the continuous *shift* dynamical system: $\mathbb{Z}^d$ acts continuously by translation on index,

$$\mathbf{m} \,.\, (x_{\mathbf{n}}) = (x_{\mathbf{m}+\mathbf{n}}).$$

A *subshift* is the restriction of this action to a closed, shift-invariant subset.

A *tile* in $\mathbb{R}^d$ is subset of $\mathbb{R}^d$ homeomorphic to a closed d-dimensional ball, and a *tiling* of $\mathbb{R}^d$ is a covering by tiles that only intersect in their boundary. Given a finite set of polyhedral tiles $\mathcal{P}$, consider the collection $X_{\mathcal{P}}$ of all tilings of $\mathbb{R}^d$ by elements of $\mathcal{P}$ that meet only full edge to full edge (provided such exist). Then for two tiles T and T',

$$\begin{aligned} d\left(T,T'\right) = \inf \big\{\{1\} \cup \{\varepsilon > 0 : {}&\text{for some } u \in \mathbb{R}^d \text{ with } |u| < \varepsilon, \\ &T+u \text{ and } T' \text{ agree on } B\left(\mathbf{0}, \tfrac{1}{\varepsilon}\right)\}\big\}. \end{aligned}$$

This metric provides $X_{\mathcal{P}}$ a compact topology with respect to which the translation action $u \,.\, T = T - u$ is continuous. A *tiling space* is a closed subset of $X_{\mathcal{P}}$ that is invariant under this action. We shall focus on the dynamics of a particular type of tiling space: the tiling space $\mathcal{T}$ of a single tiling T, formed by taking the closure of the orbit of T. For a general survey, see, e.g., [**23**].

Topological rigidity

An especially well-behaved class of tilings are the self-similar tilings, see, e.g., [**23**, Section 4]. If T and T' are self-similar tilings with homeomorphic tiling spaces $\mathcal{T}$ and $\mathcal{T}'$, one should not expect the typical homeomorphism to be a conjugacy. But the structure of such tilings is so rigid, one might expect that this could almost be so. By considering a tiling T and the tiling T' obtained from T by inflating all tiles by a factor $\lambda > 1$, one obtains homeomorphic tiling spaces for which (in general) there can be no conjugacy of actions in the strictest sense.

$\mathbb{R}^d$ actions on X and Y are *linearly equivalent* if there is a homeomorphism $h\colon X \to Y$ and a linear map L of $\mathbb{R}^d$ satisfying $h\left(u \,.\, x\right) = L(u) \,.\, h(x)$. In general, not all homeomorphic tiling spaces are linearly equivalent, see [**10, 11**].

Question 1. *If T and T' are self-similar tilings with homeomorphic tiling spaces $\mathcal{T}$ and $\mathcal{T}'$, are $\mathcal{T}$ and $\mathcal{T}'$ linearly equivalent?* 1079?

Should the answer turn out to be negative, one might modify the question so as to apply to other classes of tilings; for example, to tilings with pure point discrete spectrum. Should the answer turn out to be positive, one may then ask whether any homeomorphism $\mathcal{T} \to \mathcal{T}'$ is homotopic to a homeomorphism that induces a linear equivalence.

The topological structure of tiling spaces

While any compact metric space is homeomorphic to the inverse limit of a sequence of compact polyhedra with PL-bonding maps (see, e.g., [**18**, Ch. I, §5.2]), the structure of tiling spaces leads to some especially natural inverse sequences. For substitution tiling spaces see [**1**], and for more general spaces see [**7, 8, 24**]. There has been extensive use of cohomology in the study tiling spaces, and in many cases well-chosen inverse sequences allow one to calculate the cohomology, see [**1**]. The occurrence of torsion in cohomology is still a bit mysterious, see [**15**].

However, much less is known about the role of homotopy and shape theory in tiling spaces. A well-studied class of sequences are the Sturmian sequences (see, e.g., [**12**]). Tiling spaces derived from Sturmian sequences (in other contexts known as Denjoy continua [**6**]) are homeomorphic to the inverse limit of a sequence $\{K_i, f_i\}$, where each K_i is a wedge of two circles and each f_i induces an isomorphism of fundamental groups. It follows that these tiling spaces have the shape of the wedge of two circles [**18**]. A natural generalization of this type of tiling space are the quasiperiodic tiling spaces formed by the cut and project technique, including the Penrose tiling space, see, e.g., [**23**, Section 8].

1080? **Question 2.** *Does the Penrose tiling space have the shape of a polyhedron?*

1081? **Question 3.** *Is there a natural class $\mathcal{Q}$ of quasiperiodic tiling spaces (metrically equivalent to toral Kronecker actions) so that each $\mathcal{T} \in \mathcal{Q}$ has the shape of a polyhedron?*

An answer to these questions could likely be revealed by understanding the homomorphisms on the homotopy groups induced by the bonding maps in the same inverse sequences used to calculate cohomology (when available).

Sadun and Williams [**25**] have shown that any tiling space of the type under consideration fibers over a torus with a totally disconnected fiber. Williams [**28**, Conjecture 2.4] conjectured that up to homotopy the fiber bundle of the Penrose tiling could be given in five different ways. Robinson has calculated the discrete spectrum of the Penrose tiling space P [**23**, Section 8] and found the group of eigenvalues to be isomorphic to $\mathbb{Z}^4$. To an element of this group there corresponds a map $g_i \colon P \to \mathbb{S}^1$ that factors the action on P onto a Kronecker action of $\mathbb{S}^1$ (one for which all maps $x \mapsto t \,.\, x$ are translations). Any choice of two distinct such maps leads to a bundle projection $g_i \times g_j \colon P \to \mathbb{T}^2$. It is not difficult to show that different choices of (i, j) lead to homotopically distinct bundle projections. In fact, there will be infinitely many homotopically distinct bundle projections, but the spirit of the conjecture can be conveyed by the following.

Question 4. *Is every bundle map $P \to \mathbb{T}^2$ homotopic to a map that factors the* 1082?
action on P to a Kronecker action of $\mathbb{T}^2$?

Question 5. *If $\mathcal{T}$ has pure point discrete spectrum and $p\colon \mathcal{T} \to \mathbb{T}^d$ is a bundle* 1083?
projection with totally disconnected fiber, is p homotopic to a map that factors the action on $\mathcal{T}$ to a Kronecker action of $\mathbb{T}^d$?

In their topological classification of one-dimensional tiling spaces, Barge and Diamond [**2**] made critical use of the asymptotic orbits of the tiling spaces. A homeomorphism carries a pair of topologically asymptotic orbits to a pair of topologically asymptotic orbits. Barge and Diamond have proved the coincidence conjecture for Pisot substitutions of two letters [**3**], and in the course of trying to construct a proof for the general case the weaker notion of proximality has proven key. The orbits of T and T' are *proximal* if there exists a sequence $u_n \in \mathbb{R}^d$ with $|u_n| \to \infty$ and $d\,(u_n \,.\, T, u_n \,.\, T') \to 0$.

Question 6 (Barge and Diamond). *If $h\colon \mathcal{T} \to \mathcal{T}'$ is a homeomorphism of one-* 1084?
dimensional tiling spaces, does h necessarily map a pair of proximal orbits to a pair of proximal orbits?

Deformations of tiling spaces

It the tiling T' is obtained from the tiling T by adjusting the size and shape of the tiles in T without changing the combinatorics of the tiling (which tiles border which others), the respective tiling spaces $\mathcal{T}$ and $\mathcal{T}'$ are homeomorphic [**25**]. However, the actions may not be linearly equivalent. We will refer to $\mathcal{T}'$ as a *deformation* of $\mathcal{T}$. In [**10, 11**] there are general results that allow one to determine when deformations change the dynamics. For large classes of substitution tiling spaces, these results suffice to completely determine how deformations effect the dynamics. However, the results are difficult to apply to tiling spaces that do not arise from substitutions.

Question 7. *If $\mathcal{T}'$ is a deformation of a Sturmian tiling space $\mathcal{T}$, are $\mathcal{T}'$ and $\mathcal{T}$* 1085?
linearly equivalent?

When $\mathcal{T}$ is Sturmian and a substitution tiling, then deformations are linearly equivalent [**22**]. But the general case is not as clear. For example, whether the irrational number α associated to the Sturmian has a bounded continued fraction expansion might be relevant. This leads naturally to the following.

Question 8. *If $\mathcal{T}'$ is a deformation of a quasiperiodic tiling space $\mathcal{T}$, when are* 1086?
$\mathcal{T}'$ and $\mathcal{T}$ linearly equivalent?

Again, the focus is on those tiling spaces that do not arise from substitutions. Recently, Harriss and Lamb [**16**] have found conditions that allow one to determine when a cut and project tiling is also a substitution tiling.

Mixing properties

A tiling space is (topologically) weakly mixing if it has no non-constant continuous eigenfunction, meaning it has no Kronecker action on a circle as a continuous factor. A tiling space is *topologically mixing* if for any pair of nonempty open sets U and V, there is a corresponding M so that if $|u| > M$, then $(u \,.\, U) \cap V \neq \emptyset$.

1087? **Question 9** ([**17**]). *If a primitive substitution has an associated matrix with no eigenvalues of modulus one, is topological mixing equivalent to weak mixing?*

This question applies to symbolic systems as well as to (one-dimensional) tiling spaces based on substitutions, and it is shown to have a positive answer in the case of substitutions on two letters in [**17**].

Much less is known about the mixing properties of tiling spaces that do not derive from substitutions.

1088? **Question 10.** *Can a tiling space based on a Sturmian sequence be weakly mixing? If so, is topological mixing equivalent to weak mixing?*

It is highly unlikely that a tiling space as we are currently considering could be (strong) mixing in the measure theoretic sense. However, it is still unknown whether more general tiling spaces with a larger group than the translation group acting on the tiling space, such as the pinwheel tiling investigated by Radin in [**21**], could be strongly mixing. As pointed out in [**17**], it is not even known whether the pinwheel tiling is topologically mixing.

1089? **Question 11.** *Is the pinwheel tiling topologically mixing?*

1090? **Question 12** ([**21**]). *Is the pinwheel tiling mixing?*

Tiling spaces that are not locally finite

To this point we have been considering tiling spaces arising from tilings by polyhedra meeting full edge to full edge. Given that there are well known tilings by fractals, this would seem to be a very restrictive class of tilings. However, Priebe [**20**] has shown with a Voronoi cell construction that any tiling space arising from a tiling with finite local complexity is conjugate to a tiling space with polyhedral tiles meeting full edge to full edge. A tiling has *finite local complexity* if up to translation there is a finite number of patches of two tiles. Solomyak [**27**] found arithmetic conditions for the weak mixing of self-similar tilings of $\mathbb{R}^2$ with finite local complexity. Little is known about tilings without finite local complexity. Some easy to digest examples of such tilings may be found in [**13**].

1091? **Question 13.** *Is there an arithmetic condition for the weak mixing of self-similar tilings of $\mathbb{R}^2$ without finite local complexity?*

In general, one may consider which of the known results can be generalized to tilings without finite local complexity.

Pisot conjecture

The Pisot conjecture is one of the most hotly pursued open problems in the theory of tiling spaces. It has connections to symbolic substitution systems, graph directed systems, β-shifts, and automorphisms of compact connected abelian groups. As a result, it has drawn the attention of a wide range of people. A survey of what is known and how the conjecture relates to tilings may be found in [**9**]. There are various formulations of the conjecture corresponding to the different perspectives.

Question 14. *If $\mathcal{T}$ is a tiling space associated to an irreducible, unimodular Pisot* 1092?
substitution, does $\mathcal{T}$ have pure point discrete spectrum?

There are various finiteness conditions on the associated Pisot number that ensure the conjecture holds. The first such condition seems to have been introduced in [**14**]. The most general conditions under which the conjecture is now known to hold are given in [**4**, **5**].

Some results of Siegel [**12**, **26**] indicate that it may not be necessary to assume that the substitution is unimodular.

Question 15. *If $\mathcal{T}$ is a tiling space associated to an irreducible, Pisot substitution,* 1093?
does $\mathcal{T}$ have pure point discrete spectrum?

New directions

In his thesis, Peach [**19**] gave a way of constructing an algebra associated to a tiling of the plane by rhombi. The questions he was most interested in were purely algebraic, and there is no apparent connection between the structure of this algebra and the dynamics of the tiling. However, by introducing quiver relations that reflect the nature of a substitution, it might possible to construct powerful invariants that reflect the dynamics.

Question 16. *Is it possible to construct a quiver algebra for a self-similar tiling* 1094?
of the plane that provides an important dynamical invariant?

Acknowledgments

I thank R. Fokkink and L. Sadun for their help.

References

[1] J. E. Anderson and I. F. Putnam, *Topological invariants for substitution tilings and their associated C^*-algebras*, Ergodic Theory Dynam. Systems **18** (1998), no. 3, 509–537.

[2] M. Barge and B. Diamond, *A complete invariant for the topology of one-dimensional substitution tiling spaces*, Ergodic Theory Dynam. Systems **21** (2001), no. 5, 1333–1358.

[3] M. Barge and B. Diamond, *Coincidence for substitutions of Pisot type*, Bull. Soc. Math. France **130** (2002), no. 4, 619–626.

[4] M. Barge and J. Kwapisz, *Geometric theory of unimodular Pisot substitutions*, 2004, Preprint.

[5] M. Barge and J. Kwapisz, *Elements of the theory of unimodular Pisot substitutions with an application to β-shifts*, Algebraic and topological dynamics, American Mathematical Society, Providence, RI, 2005, pp. 89–99.

[6] M. Barge and R. F. Williams, *Classification of Denjoy continua*, Topology Appl. **106** (2000), no. 1, 77–89.

[7] J. Bellissard, R. Benedetti, and J.-M. Gambaudo, *Spaces of tilings, finite telescopic approximations and gap-labeling*, Comm. Math. Phys. **261** (2006), no. 1, 1–41.

[8] R. Benedetti and J.-M. Gambaudom, *On the dynamics of G-solenoids: applications to Delone sets*, Ergodic Theory Dynam. Systems **23** (2003), no. 3, 673–691.

[9] V. Berthé and A. Siegel, *Tilings associated with beta-numeration and substitutions*, Integers **5** (2005), no. 3, 46 pp.

[10] A. Clark and L. Sadun, *When size matters: subshifts and their related tiling spaces*, Ergodic Theory Dynam. Systems **23** (2003), no. 4, 1043–1057.

[11] A. Clark and L. Sadun, *When shape matters: deformations of tiling spaces*, Ergodic Theory Dynam. Systems **26** (2006), no. 1, 69–86.

[12] N. P. Fogg, *Substitutions in dynamics, arithmetics and combinatorics*, Lecture Notes in Math., vol. 1794, Springer, Berlin, 2002.

[13] N. P. Frank and E. A. Robinson, Jr., *Generalized β-expansions, substitution tilings, and local finiteness*, Preprint.

[14] C. Frougny and B. Solomyak, *Finite beta-expansions*, Ergodic Theory Dynam. Systems **12** (1992), no. 4, 713–723.

[15] F. Gaehler, J. Hunton, and J. Kellendonk, *Torsion in tiling homology and cohomology*, 2005, Preprint. `arXiv:math-ph/0505048`

[16] E. O. Harriss, *On canonical substitution tilings*, Ph.D. thesis, Imperial College London, 2003.

[17] R. Kenyon, L. Sadun, and B. Solomyak, *Topological mixing for substitutions on two letters*, Ergodic Theory Dynam. Systems **25** (2005), no. 6, 1919–1934.

[18] S. Mardešić and J. Segal, *Shape theory*, North-Holland, Amsterdam, 1982.

[19] M. Peach, *Rhombal algebras*, Ph.D. thesis, University of Bristol, 2004.

[20] N. M. Priebe, *Towards a characterization of self-similar tilings in terms of derived Voronoï tessellations*, Geom. Dedicata **79** (2000), no. 3, 239–265.

[21] C. Radin, *The pinwheel tilings of the plane*, Ann. of Math. (2) **139** (1994), no. 3, 661–702.

[22] C. Radin and L. Sadun, *Isomorphism of hierarchical structures*, Ergodic Theory Dynam. Systems **21** (2001), no. 4, 1239–1248.

[23] E. A. Robinson, Jr., *Symbolic dynamics and tilings of* $\mathbb{R}^d$, Symbolic dynamics and its applications, American Mathematical Society, Providence, RI, 2004, pp. 81–119.

[24] L. Sadun, *Tiling spaces are inverse limits*, J. Math. Phys. **44** (2003), no. 11, 5410–5414.

[25] L. Sadun and R. F. Williams, *Tiling spaces are Cantor set fiber bundles*, Ergodic Theory Dynam. Systems **23** (2003), no. 1, 307–316.

[26] A. Siegel, *Représentation des systèmes dynamiques substitutifs non unimodulaires*, Ergodic Theory Dynam. Systems **23** (2003), no. 4, 1247–1273.

[27] B. Solomyak, *Corrections to: "Dynamics of self-similar tilings" [Ergodic Theory Dynam. Systems 17 (1997) no. 3, 695–738]*, Ergodic Theory Dynam. Systems **19** (1999), no. 6, 1685.

[28] R. F. Williams, *The Penrose, Ammann and DA tiling spaces are Cantor set fiber bundles*, Ergodic Theory Dynam. Systems **21** (2001), no. 6, 1883–1901.

Open Problems in Topology II – Edited by E. Pearl

Open problems in complex dynamics and "complex" topology

Robert L. Devaney

Complex dynamics is a field in which a large number of captivating structures from planar topology occur quite naturally. Of primary interest in complex dynamics is the Julia set of a complex analytic function. As we discuss below, these are the sets that often are quite interesting from a topological point of view. For example, we shall describe examples of functions whose Julia sets (or invariant subsets of the Julia sets) are Cantor bouquets, indecomposable continua, and Sierpiński curves. Because both the topology of and the dynamics on these Julia sets is so rich, it is little wonder that there are many open problems in this field. Our goal in this paper is to describe several of these problems. To keep the exposition accessible, we shall restrict attention to two very special families of functions, namely the complex exponential function and a particular family of rational maps. However, the problems and topological structures encountered in these families occur for many other types of complex analytic maps.

1. Cantor Bouquets and Indecomposable Continua

In this section we consider the dynamics of the complex exponential family $E_\lambda(z) = \lambda e^z$ where, for simplicity, λ is for the most part chosen to be real and positive. The Julia set for such an entire transcendental map has several equivalent definitions. For example, the Julia set may be defined as the closure of the set of points whose orbits escape to ∞ under iteration of E_λ. (Note that this is different from the definition of polynomial Julia sets, where it is the boundary and not the closure of the set of escaping points that forms the Julia set.) Equivalently, the Julia set is also the closure of the set of repelling periodic points. These two defintions show that the Julia set of E_λ is home to chaotic behavior: arbitrarily close to any point in the Julia set are points whose orbits tend off to ∞ as well as other points whose orbits are not only bounded, but in fact periodic. So the map depends quite sensitively on initial conditions near any point in the Julia set. In fact, much more can be said since the Julia set may also be defined as the set of points at which the family of iterates of E_λ fails to be a normal family. By Montel's Theorem, it then follows that, for any neighborhood U of a point in the Julia set, the union of the sets $E_\lambda^n(U)$ covers all of $\mathbb{C} \setminus \{0\}$. So arbitraily close to any point in the Julia set are points whose orbits visit any region whatsoever in $\mathbb{C}$. We denote the Julia set of a function F by $J(F)$.

The complement of the Julia set is called the Fatou set. Here the situation is quite different: the dynamics on the Fatou set is essentially completely understood. For example, all points in the basin of attraction af an attracting cycle clearly lie in the Fatou set: the orbits of all nearby points to a point in such a basin behave similarly. No nearby orbits tend to ∞ and none lie on repelling periodic cycles.

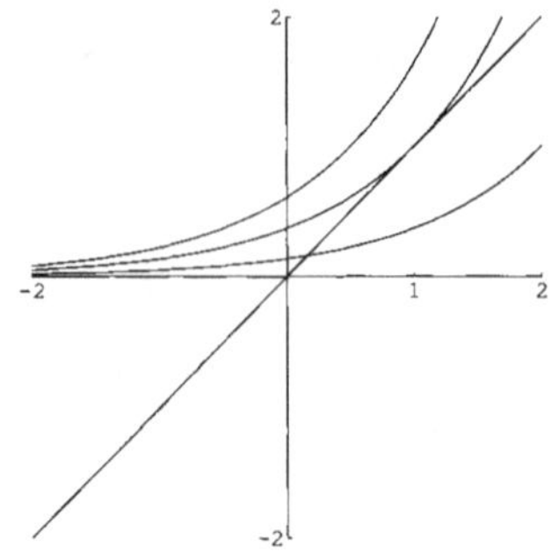

FIGURE 1. The graphs of E_λ for several λ-values

There are a few other possible types of behavior in the Fatou set, but none of these behaviors involve anything chaotic.

The graphs of E_λ on the real line (see Figure 1) show that there are two different types of dynamical behavior depending upon whether $\lambda < 1/e$ or $\lambda > 1/e$. When $\lambda < 1/e$, there are two fixed points on the real line, an attracting fixed point at $q = q_\lambda$ and a repelling fixed point at $p = p_\lambda$. All orbits to the right of p tend to ∞, so these points are in the Julia set, as is p. All points to the left of p are in the basin of attraction of q, so these points are not in the Julia set. In fact, let x be any point in $\mathbb{R}$ with $q < x < p$. Then one checks easily that the entire half plane $H_x = \operatorname{Re} z < x$ is wrapped infinitely often around a disk minus the origin, and this disk lies strictly inside the half plane H_x. By the Schwarz Lemma, all points in any of these half planes therefore have orbits that simply tend to q and hence lie in the Fatou set. So the Julia set must lie in the half-plane $\operatorname{Re} z \geq p$. This is essentially true when $\lambda = 1/e$, though now all orbits in the half-plane $\operatorname{Re} z < p$ now tend to the neutral fixed point at $p = q$.

To get a feeling for the structure of the Julia set when $\lambda \leq 1/e$, we paint the picture of its complement. Consider the preimage of H_x. This preimage must contain the lines $y = (2n+1)\pi$ for each $n \in \mathbb{Z}$, since these lines are mapped to the negative real axis. Hence there are open neighborhoods of each of these lines extending from H_x to ∞ in the right half plane and mapped onto H_x. This means that the Julia set is contained in infinitely many symmetrically located, simply connected, closed sets that extend to ∞ in the right half plane. Each of these sets is mapped one-to-one onto the entire half plane $\operatorname{Re} z \geq x$. As a consequence, there are points in each of these regions that map into each of the neighborhoods of the lines $y = (2n+1)\pi$ and hence these points are also in the Fatou set. So this breaks each of these complementary domains into infinitely many more sets, each of which extend off to ∞ to the right. And so the Julia set must lie in these regions. Continuing in this fashion, one can show that the Julia set is actually an uncountable collection of curves (called *hairs*) that extend to ∞ in the right half plane, and each of these hairs has a distinguished endpoint [**7**]. The set of all such hairs forms the Julia set and is an example of a *Cantor bouquet.* So each of these hairs consists of two subsets: the endpoint and the remainder of the hair that we call the stem. For example, one such hair is the half-line $[p, \infty) \subset \mathbb{R}$. The point p is the endpoint, which is fixed, and as we saw earlier, all points to the right of p simply tend to ∞. In general, it is known that, if a point lies on the stem, then, as in the case of (p, ∞), the orbit of this point necessarily tends to ∞ (though it usually jumps around between different hairs). Hence all of the bounded orbits must lie in the set of endpoints. But the repelling periodic points are bounded and hence they must lie in the set of endpoints. But this means that the set of endpoints is dense in this entire set, and so they accumulate on each point on any given stem.

Because of this, a Cantor bouquet has some very interesting topological properties. For example, Mayer [**13**] has shown that, in the Riemann sphere, the set of endpoints together with the point at ∞ forms a connected set, whereas the set consisting of just the endpoints (i.e., remove just one point from the previous set) is not just disconnected but totally disconnected. Moreover, Karpińska [**12**] has shown that the Hausdorff dimension of the set of stems is 1, whereas the Hausdorff dimension of the *much* smaller set of endpoints is actually 2.

When λ passes through $1/e$, E_λ undergoes a simple saddle node bifurcation in which the two fixed points q_λ and p_λ coalesce when $\lambda = 1/e$ and then reappear for $\lambda > 1/e$ above and below the real axis. Meanwhile, all points on the real axis now tend to ∞, so the entire real axis suddenly lies in the Julia set. But much more is happening in the complex plane.

The origin is what is known as an asymptotic value. It is the omitted value for E_λ. As such, it plays the same role as the critical values do in polynomial dynamics. In particular, via a result of Sullivan [**15**], as extended to the entire case by Goldberg and Keen [**11**], if the orbit of 0 tends to ∞, then the Julia set of E_λ must be the entire plane. Hence, when $\lambda \leq 1/e$, all of the repelling periodic points are constrained to lie in the half plane $\operatorname{Re} z \geq p$, whereas these points become dense in $\mathbb{C}$ for any $\lambda > 1/e$. Now no new repelling cycles are born as λ passes through $1/e$; all of these cycles simply move continuously, but the set of them migrates from occupying a small portion of the right half plane to suddenly filling all of $\mathbb{C}$.

However, even more is happening in this bifurcation. For example, consider what happens to the hair $[p, \infty)$ as soon as λ increases past $1/e$. Suddenly this hair is much longer: it becomes the entire real axis. But, in fact, it is longer still. Consider the set of points in the strip S defined by $0 \leq \operatorname{Im} z \leq \pi$ that eventually map onto $\mathbb{R}$. Clearly, the line $y = \pi$ maps into $\mathbb{R}$ after one iteration. So we can think of this hair through the origin as being extended by adjoining the point at $-\infty$ to the real axis and the line $y = \pi$. Now E_λ maps S one-to-one onto the upper half plane. So there is a unique curve in S that is mapped to $y = \pi$ and hence into $\mathbb{R}$ after two iterations. This curve actually tends to ∞ in the right half plane in both directions. So we can similarly adjoin a point at ∞ to the upper end of this preimage and the right end of $y = \pi$. Then the preimage of this curve in S is another curve that also extends to ∞ in the right half plane in both directions. In fact, all of the subsequent preimages of $y = \pi$ have this property. If we successively adjoin one endpoint of each curve with the corresponding endpoint of its preimage, we get a curve in S that can be shown to accumulate everywhere upon itself. If we compactify this picture by contracting S to the strip $-1 \leq \operatorname{Re} z \leq 1$ and again making these identifications, then this curve does not separate the plane. Using a result of Curry [**1**], the closure of this set can be shown to be an indecomposable continuum [**2**]. That is, as soon as the bifurcation occurs, the hair $[p, \infty)$ suddenly explodes into an indecomposable continuum.

Here is where a number of open problems arise. Let C_λ denote the indecomposable continuum in $J(E_\lambda)$ in S.

1095? **Problem 1.** *Suppose $\lambda, \mu > 1/e$. Are C_λ and C_μ homeomorphic?*

It is known that each of the maps E_λ and E_μ have the *same* symbolic dynamics on their Julia sets [**7**], but the maps themselves are not topologically conjugate [**10**]. This latter fact was proved by showing that certain collections of periodic points accumulate onto dynamically different points when $\lambda \neq \mu$. A more topological proof of this fact would ensue if Problem 1 were shown to be true.

The exact topology of these indecomposable continua is not known. There have been some piecewise linear models proposed [**9**], but so far a complete topological description of these sets has not been given.

1096? **Problem 2.** *Find a topological model for the sets C_λ.*

In contrast to the rich topology of these sets, the dynamical behavior on these sets is fairly well understood. There are only three types of orbits:

(1) The fixed point (which moves upward off the real axis after q and p merge);
(2) The points on any of the preimages of $\mathbb{R}$ whose orbits simply tend to ∞;
(3) The orbits of all other points which accumulate on the orbit of 0 together with the point at ∞.

In line with this, there are many other questions having to do with the relation between the dynamics and the topology of C_λ. For example:

1097? **Problem 3.** *What is the structure of the composant that contains the unique fixed point in C_λ?*

There are other indecomposable continua in the Julia set of E_λ. For example, one can associate an itinerary to any point in $J(E_\lambda)$ by watching how the orbit passes through the strips $S_n = \{z \mid (2n-1)\pi < \operatorname{Im} z < (2n+1)\pi\}$ at each iteration. Then we associate the infinite sequence of integers $s = (s_0 s_1 s_2 \ldots)$ to z if $E_\lambda^j(z) \in S_{s_j}$ for each j. Then, for $\lambda > 1/e$, consider the set of points whose itinerary is a given sequence s. For most sequences, this set of points remains a hair. However, if s terminates in all 0s, then this set is just a preimage of the indecomposable continuum (or its complex conjugate) constructed above and hence is homemorphic to this set. If the itinerary consists of blocks of 0s separated by non-zero entries and having the property that the lengths of the blocks of 0s goes to ∞ sufficiently quickly, then the corresponding set of points is also an indecomposable continuum which is presumably topologically diferent from the one constructed above. See [**5**]. A natural question is what other types of sets of points can correspond to a given itinerary.

Problem 4. *Identify which itineraries correspond to indecomposable continua when $\lambda > 1/e$ and which yield hairs. Are there any other possibilities for the types of sets corresponding to a given itinerary? And how does all of this depend on λ?* 1098?

Along this line, when λ is allowed to be complex and the orbit of 0 eventually lands on a repelling periodic orbit (as is the case when $\lambda = k\pi i$ with $k \neq 0$), then it is known that set of points corresponding to certain itineraries may be an indecomposable continuum together with a finite collection of curves that accumulates on the indecomposable continuum. But this is the only other type of set that is known to correspond to a given itinerary. See [**6**]. It seems strange that there is nothing *in-between*: either such a set is a simple curve or it is (or contains) an indecomposable continuum.

Problem 5. *Identify the types of sets of points that can correspond to a given itinerary under a complex exponential map.* 1099?

We have restricted to the complex exponential in this section for several reasons. First of all, this has been the most widely studied example of an entire transcendental dynamical system. Secondly, the corresponding results for other functions seem much more difficult. For example, consider the simple cosine family $i\mu \cos z$ where $\mu > 0$. It is known that, if $\mu \approx 0.67$, the cosine function undergoes a similar bifurcation as the exponential does when $\lambda = 1/e$. The Julia set is a pair of Cantor bouquets (one in the upper and one in the lower half plane) when $\mu < 0.67$, whereas the Julia set explodes to become $\mathbb{C}$ as soon as μ increases beyond 0.67. How this occurs is still a mystery. The hairs forming the Cantor bouquet do change after the bifurcation, but do they become indecomposable continua? The difficulty arises because the cosine function has critical points and not asymptotic values. This seems to cause a very different structure in the hairs when the critical points suddenly escape to ∞.

Problem 6. *Explain the bifurcation at $\mu = 0.67$ in the family $i\mu \cos z$. In particular, do hairs suddenly become indecomposable continua?* 1100?

Of course, there are many other instances of similar (and more complicated) bifurcations in transcendental dynamics. Perhaps other exotic topological structures arise in these bifurcations as well. Along these lines, there are examples of simple bifurcations in which the Julia set of an entire map migrates from a Cantor bouquet to a simple closed curve (passing through ∞) and also from a Cantor bouquet to a Cantor set. See [**4**].

2. Sierpiński Curve Julia Sets

In this section we turn to a very different type of topological structure that occurs often in complex dynamics, Sierpiński curves. A Sierpiński curve is any planar set that is homeomorphic to the well-known Sierpiński carpet fractal. This set is important in topology for several reasons. First, thanks to a result of Whyburn [**16**], there is a topological characterization of any set that is homeomorphic to the carpet. Any planar set that is compact, connected, locally connected, nowhere dense, and has the property that each complementary domain is bounded by a simple closed curve, any pair of which are disjoint, is homeomorphic to the Sierpiński carpet (and thus called a Sierpiński curve). More importantly, Sierpiński curves are universal plane continua since any planar, one-dimensional, compact, connected set may by embedded homeomorphically in a Sierpiński curve.

To see these sets in complex dynamics, we now turn to the family of rational maps given by $F_\lambda(z) = z^n + \lambda/z^n$ where $n \geq 2$ and $\lambda \in \mathbb{C} \setminus \{0\}$, although these types of sets occur in many other families of rational maps. For these maps, the definition of the Julia set is slightly different. The point at ∞ is no longer an essential singularity as in the case of the exponential map. Rather, since $n \geq 2$, the map F_λ is essentially given by z^n near ∞, so ∞ is an attracting fixed point for these maps and we have a basin of ∞ that we denote by B_λ. $J(F_\lambda)$ is still the closure of the set of repelling periodic points, but now it is the boundary of, not the closure of, the set of points whose orbits escape to ∞. Note that the origin is a pole and there is a neighborhood of 0 that is mapped n-to-1 onto a neighborhood of ∞ in B_λ. If this neighborhood of 0 does not intersect B_λ, then there is an open set containing 0 that is mapped n-to-1 onto the entire set B_λ. We then call this set the trap door and denote it by T_λ. T_λ is the trap door since any orbit that eventually reaches B_λ must in fact pass through T_λ exactly once.

These maps are special because, despite the high degree of the maps, there really is only one *free* critical orbit. Indeed the $2n$ critical points are given by $\lambda^{1/2n}$, but they are each mapped to one of the critical values $\pm 2\sqrt{\lambda}$ by F_λ. After that, the two critical values are mapped onto the same point (if n is even) or the orbits of these two points are arranged symmetrically under $z \mapsto -z$ (if n is odd). In either case, all the critical orbits behave in the same manner, so there is essentially only one critical orbit.

If one and hence all of the critical orbits end up in the basin of ∞, then the topology of the Julia set is completely determined. There are thre different ways that these orbits can reach B_λ. The following result is proved in [**8**]. Suppose the critical orbit tends to ∞.

(1) If the critical values lie in B_λ, the Julia set is a Cantor set;
(2) If the critical values lie in T_λ, the Julia set is a Cantor set of simple closed curves;
(3) If the crtical values do not lie in B_λ or T_λ but some subsequent iterate of these points does so, then the Julia set is a Sierpiński curve.

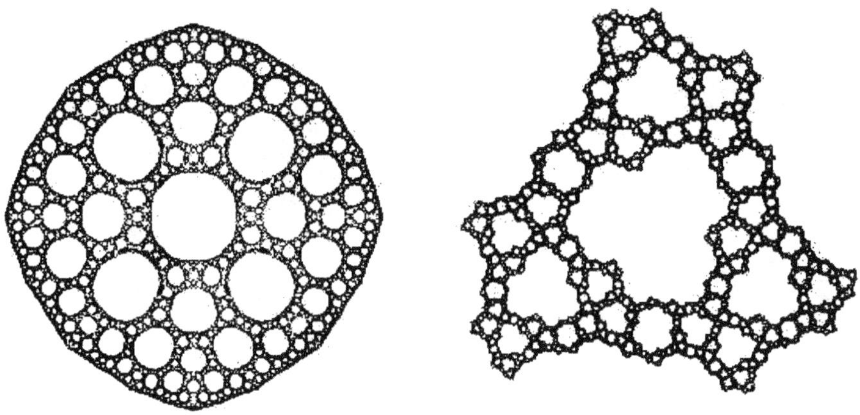

(a) $z^2 - 0.06/z^2$ (b) $z^2 + (-0.004 + 0.364i)/z$

FIGURE 2. Sierpiński curve Julia sets

As a remark, case (2) of this result was proved by McMullen [**14**]. This case cannot occur when $n = 2$. In Figure 2, we display several examples of Sierpiński curve Julia sets drawn from the family when $n = 2$.

The fact that there is essentially only one critical orbit for maps in these families says that the λ-plane is the natural parameter plane for these families. In Figure 3 we have displayed the parameter planes for the families when $n = 3$ and $n = 4$. The external white region consists of parameters for which the Julia set is a Cantor set; the central white region is the *McMullen domain* where the Julia set is a Cantor set of simple closed curves; and all of the other white regions contain parameters for which the Julia set is a Sierpiński curve. These regions are called Sierpiński holes. The region in parameter plane that excludes the Cantor set locus and the McMullen domain is called the connectedness locus; Julia sets whose parameters lie in this region are known to be connected sets.

For a parameter drawn from a Sierpiński hole, the complementary domains consist of B_λ and all of its preimages. It is known that if two parameters, λ and μ, lie in the same Sierpiński hole, then F_λ and F_μ are dynamically the same, i.e., F_λ is topologically conjugate to F_μ on their Julia sets. In particular, the critical orbits all land in B_λ under the same number of iterations under both of these maps. But if λ and μ are drawn from holes for which the number of iterations that it takes for the critical orbit to reach B_λ is different, then these maps are not conjugate. However, there are many different holes for which the critical values take the same number of iterations to reach B_λ. For example, when $n = 3$, it is

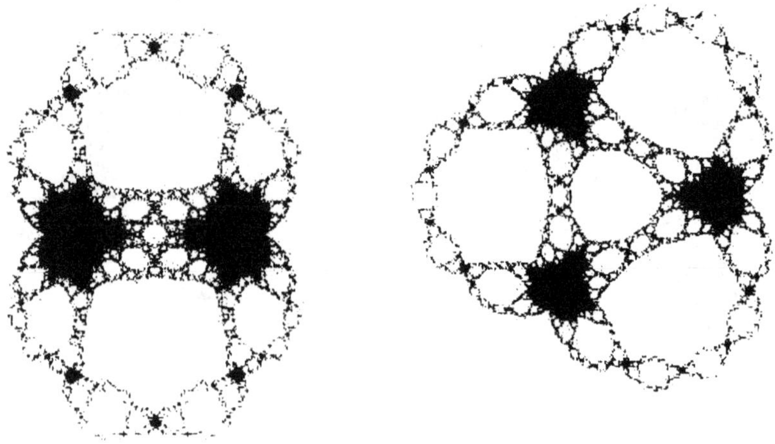

(a) $n = 3$ (b) $n = 4$

FIGURE 3. Parameter planes

known **[3]** that there are exactly $2 \cdot 6^j$ holes for which it takes the critical values $j + 2$ iterations to reach B_λ. This leads to a more dynamical type of problem:

1101? **Problem 7.** *Determine whether the dynamical behavior that occurs for parameters drawn from two different Sierpiński holes with the same escape time is the same or different.*

There are many types of parameters for which the corresponding Julia sets are Sierpiński curves. For example, a magnification of the parameter plane for $n = 2$ shown in Figure 4 shows that there are (in fact, infinitely many) *buried* small copies of Mandelbrot sets contained in the parameter plane. These are the Mandelbrot sets that do not touch the outer boundary of the connectedness locus. It is known that if λ lies in the main cardioid of such a Mandelbrot set, then again the Julia set is a Sierpiński curve. The dynamics on these types of sets are again different from the dynamics of maps drawn from Sierpiński holes, since there is an attracting cycle for such a map. So the complementary domains for these maps consist of all the preimages of this attracting basin as well as the preimages of B_λ. And there are other types of Sierpiński curve Julia sets: for example, it is known that there is a Cantor set of simple closed curves in the parameter plane that do not pass through any Sierpiński holes, yet all of the Julia sets corresponding to parameters on these curves are Sierpiński curves. As before, all but finitely many of these maps are dynamically distinct. So we have a huge number of Julia sets that are all the same from a topological point of view, but dynamically very different. This leads to a natural question:

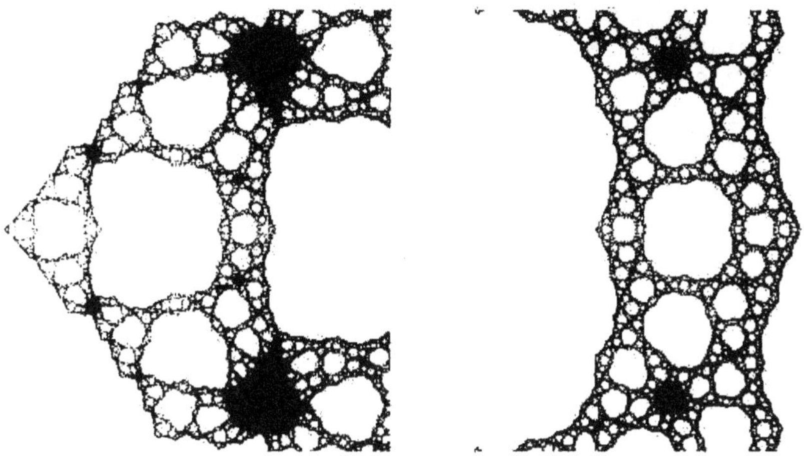

(a) $-0.4 \leq \operatorname{Re}\lambda \leq -0.06$ (b) $-0.2 \leq \operatorname{Re}\lambda \leq -0.15$

FIGURE 4. Two magnifications of the parameter plane for the family $z^2 + \lambda/z^2$ along the negative real axis

Problem 8. *Classify the dynamics of all the different types of Sierpiński curve* 1102?
Julia sets that arise in these families.

References

[1] S. B. Curry, *One-dimensional nonseparating plane continua with disjoint ϵ-dense subcontinua*, Topology Appl. **39** (1991), no. 2, 145–151.

[2] R. L. Devaney, *Knaster-like continua and complex dynamics*, Ergodic Theory Dynam. Systems **13** (1993), no. 4, 627–634.

[3] R. L. Devaney, *Se^x: dynamics, topology, and bifurcations of complex exponentials*, Topology Appl. **110** (2001), no. 2, 133–161.

[4] R. L. Devaney, *Topological bifurcations*, Topology Proc. **28** (2004), no. 1, 99–112.

[5] R. L. Devaney and X. Jarque, *Indecomposable continua in exponential dynamics*, Conform. Geom. Dyn. **6** (2002), 1–12.

[6] R. L. Devaney, X. Jarque, and M. Moreno Rocha, *Indecomposable continua and Misiurewicz points in exponential dynamics*, Internat. J. Bifurc. Chaos **15** (2005), 3281– 3293.

[7] R. L. Devaney and M. Krych, *Dynamics of* exp(z), Ergodic Theory Dynam. Systems **4** (1984), no. 1, 35–52.

[8] R. L. Devaney, D. M. Look, and D. Uminsky, *The escape trichotomy for singularly perturbed rational maps*, Indiana Univ. Math. J. **54** (2005), no. 6, 1621–1634.

[9] R. L. Devaney and M. Moreno Rocha, *A semilinear model for exponential dynamics and topology*, Topology Proc. **26** (2001/02), no. 1, 153–167.

[10] A. Douady and L. R. Goldberg, *The nonconjugacy of certain exponential functions*, Holomorphic functions and moduli (Berkeley, 1986), Springer, New York, 1988, pp. 1–7.

[11] L. R. Goldberg and L. Keen, *A finiteness theorem for a dynamical class of entire functions*, Ergodic Theory Dynam. Systems **6** (1986), no. 2, 183–192.

[12] B. Karpińska, *Hausdorff dimension of the hairs without endpoints for* $\lambda \exp z$, C. R. Acad. Sci. Paris Sér. I Math. **328** (1999), no. 11, 1039–1044.

[13] J. C. Mayer, *An explosion point for the set of endpoints of the Julia set of* $\lambda \exp(z)$, Ergodic Theory Dynam. Systems **10** (1990), no. 1, 177–183.

[14] C. McMullen, *Automorphisms of rational maps*, Holomorphic functions and moduli (Berkeley, 1986), Springer, New York, 1988, pp. 31–60.

[15] D. Sullivan, *Quasiconformal homeomorphisms and dynamics. I. Solution of the Fatou–Julia problem on wandering domains*, Ann. of Math. (2) **122** (1985), no. 3, 401–418.

[16] G. T. Whyburn, *Topological characterization of the Sierpiński curve*, Fund. Math. **45** (1958), 320–324.

Open Problems in Topology II – Edited by E. Pearl

The topology and dynamics of flows

Michael C. Sullivan

Flows

Let M be a compact connected Riemannian manifold without boundary. Let $\|\cdot\|$ be the norm on the tangent bundle TM and $d(\cdot,\cdot)$ the metric induced on M. By a *flow* on M we mean a smooth function $f\colon M \times \mathbb{R} \to M$ such that $f(f(x,s),t) = f(x,s+t)$ and $f(x,0) = x$. Much of what we describe in this sections for flows carries over with suitable modifications to diffeomorphisms.

The *chain recurrent set* of a flow f is

$$\begin{aligned}\mathcal{R} =\{&x \in M : \forall\epsilon > 0, \exists\{x_0 = x, x_1, x_2, \dots, x_k\} \subset M, \exists\{t_1, t_2, \dots, t_k\} \subset \mathbb{R}^+\\ &\text{such that } d(f(x_i,t_i), x_{i+1}) < \epsilon \text{ for } i = 1,\dots,k-1,\ d(f(x_k,t_k),x_0) < \epsilon\}.\end{aligned}$$

The chain recurrent set of a flow is said to have a *hyperbolic structure* if the tangent bundle of the manifold structure can be written as a Whitney sum $T_{\mathcal{R}} = E^u \oplus E^c \oplus E^s$ of sub-bundles invariant under Df where E^c_x is the subspace of TM_x corresponding to the orbit of x and such that there are constants $C > 0$ and $\lambda > 0$ for which $\|Df_t(v)\| \leq Ce^{-\lambda t}\|x\|$ for $v \in E^s$, $t \geq 0$ and $\|Df_t(v)\| \geq 1/Ce^{\lambda t}\|x\|$ for $v \in E^u$, $t \geq 0$.

Steve Smale showed that when $\mathcal{R}$ is hyperbolic it is the closure of the periodic orbits of the flow. Smale also showed that when $\mathcal{R}$ is hyperbolic it has a finite decomposition into compact invariant sets called *basic sets*:

Theorem (Spectral Decomposition Theorem). *Suppose that the chain recurrent set $\mathcal{R}$ of a flow has a hyperbolic structure. Then $\mathcal{R}$ is a finite disjoint union of compact invariant sets $\Lambda_1, \Lambda_2, \dots, \Lambda_k$ where each Λ_i contains an orbit that is dense in Λ_i.*

We define respectively the *stable and unstable manifolds* of an orbit $\mathcal{O}$ in a flow f.

$$\begin{aligned}W^s(\mathcal{O}) &= \{y \in M : d(f(y,t), f(x,t)) \to 0 \text{ as } t \to \infty \text{ for some } x \in \mathcal{O}\},\\ W^u(\mathcal{O}) &= \{y \in M : d(f(y,t), f(x,t)) \to 0 \text{ as } t \to -\infty \text{ for some } x \in \mathcal{O}\}.\end{aligned}$$

That these are manifolds is a classical result of Hirsch and Pugh [**29**] referred to as the Stable Manifold Theorem. A flow is *structurally stable* if it is *topologically equivalent*, i.e., there is a homeomorphism taking orbits to orbits preserving the flow direction, to flows obtained by small enough perturbations.

A flow with hyperbolic chain recurrent set $\mathcal{R}$ satisfies the *transversality condition* if the stable and unstable manifolds of $\mathcal{R}$ always meet transversally. A flow (or diffeomorphism) that has a hyperbolic chain recurrent set and satisfies the transversality condition is structurally stable; see [**19**, Theorem 1.10] for references. The converse—known as the C^1 Stability Conjecture—was proposed by

Palis and Smale in [**37**] has been proven by Hu [**30**] for dimension 3 and for arbitrary dimension by Hayashi [**28**]; see also [**52**].

For the three dimensional case the basic sets of C^1 structurally stable flows may be of the following types: isolated fixed points; isolated closed orbits; suspensions of nontrivial irreducible shifts of finite type (SFTs) (see [**31**] for definitions of terms for symbolic dynamics)—these have infinitely many periodic orbits but rational zeta functions; two-dimensional attractors or repellers, e.g., a suspension of Plykin's attractor—these are modeled by inverse limits of branched one-dimensional manifolds [**53**]; and lastly, if the invariant hyperbolic set is the whole of M, we have an *Anosov flow*.

If the chain recurrent set of a flow is hyperbolic, consists of a finite collection of periodic orbits and fixed points, and satisfies the transversality condition, we have a *Morse–Smale flow*. Daniel Asimov showed that for $n \neq 3$ all n-manifolds (possibly with boundary), subject to certain obvious Euler characteristic criteria, support nonsingular Morse–Smale flows [**1**]. (A *nonsingular flow* is just a flow without fixed points.) John Morgan has characterized which 3-manifolds (possibly with boundary) support nonsingular Morse–Smale flows [**36**] and Masaaki Wada has determined which links can be realized as the invariant set of a nonsingular Morse–Smale flow on $\mathbb{S}^3$ [**51**]; see also [**15**]. Wada's result shows, for example, that the figure-8 knots cannot be realized in a nonsingular Morse–Smale flow on $\mathbb{S}^3$. Thus, the existence of a figure-8 knot in a Morse–Smale flow on $\mathbb{S}^3$ forces a fixed point. Bifurcations of nonsingular Morse–Smale flows on $\mathbb{S}^3$ are studied in [**14**]. Given a link L in some orientable 3-manifold, Masahico Saito [**42**] shows how to modify the 3-manifold (by forming connected sums with $\mathbb{S}^2 \times \mathbb{S}^1$ pieces) so that the new 3-manifold has a nonsingular Morse–Smale flow with L as part of its chain recurrent set (actually he shows a bit more than this).

If the chain recurrent set of a flow is at most one-dimensional and satisfies the transversality condition the flow is known as a *Smale flow*. Basic sets which are not isolated fix points or closed orbits are suspensions of SFTs and must be saddle sets. There are no chaotic attractors or repellers.

Anosov flows arose from the study of geodesic flows on surfaces. Thus, "unit tangent bundles of all surfaces with genus greater than one" support Anosov flows. And so too do "all manifolds that can be obtained by suspending Anosov diffeomorphisms of T^2."[1] There are no Anosov flows on $\mathbb{S}^3$. It is known that a 3-manifold for which every co-dimension one foliation has a Reeb component does not support an Anosov flow. There are infinity many such manifolds [**41**]. In general, "[It is] not known at all which manifolds have Anosov flows."[2]

1103? **Question 1.** *Which 3-manifolds support Anosov flows?*[3]

[1]Keith Burns, private communication.

[2]Sergio Fenley, private communication.

[3]See the chapter by W.F. Basener, K. Parwani and T.Wiandt [**2**] for a more detailed discussion of Anosov flows.

There has been a great deal of interest in partially hyperbolic flows; see [**39**]. These can have singular hyperbolic invariant sets in which a saddle fixed point cannot be isolated from an invariant set with infinitely many periodic orbits. Together they form a compact invariant attractor or repeller. The Lorenz attractor is a standard example. The first return map of a cross section may be conjugate to a shift space with a countably infinite alphabet. They are not structurally stable. See [**35**]. Morales and Pacifico have shown that generically a flow on a 3-manifold either has infinitely many sinks or sources, or (exclusive) has a chain recurrent set that is hyperbolic or singular hyperbolic. Their result implies that a generic flow on a 3-manifold has an attractor or a repeller. This is done in [**34**], a paper that should be widely read. They raise the following question in Conjecture 1.3.

Question 2. *Can every C^1 vector field on a closed 3-manifold be approximated* 1104?
*by a vector field exhibiting a homoclinic tangency or by a singular Axiom A one without cycles? (See [**34**] for definitions and details.)*

The rest of this chapter is devoted to Smale flows except for the last section on Smale diffeomorphisms.

Templates for basic sets

Let B be a basic set of a Smale flow that is the suspension of a nontrivial irreducible SFT. We can pick a neighborhood of B that will be foliated by stable manifolds. If we form a quotient space by collapsing along the stable direction we derive a two dimensional branched manifold $\mathcal{T}_B$ known as a template. The original flow will induce a semi-flow on the template. A theorem of Joan Birman and Robert Williams asserts that there is a one-to-one correspondence between the periodic orbits of B and $\mathcal{T}_B$ that preserves the knot type of each periodic orbit and how they are linked [**7**]; see also [**25**, Theorem 2.2.4]. Templates allow us to see basic sets. The simplest example is the *Lorenz template* shown in Figure 1 on the left. One can recover the basic set by taking an inverse limit of the template's semi-flow. For the Lorenz template the basic set is a suspension of the full 2-shift.

Templates, slightly modified, are used to model singular hyperbolic attractors, see Figure 1 on the right [**8**], and Plykin-like attractors—here the templates have no boundary and are harder to draw [**25**, Figure 3.15].

It is natural to ask which knots and links exist on a given template. (For basic definitions of knot theory see [**13**]). Let $L(m,n)$ denote the *Lorenz-like template* constructed from the Lorenz template by adding m half twists in the left band and n in the right band; by symmetry $L(m,n) = L(n,m)$. It is known that for $n \geq 0$ that the knots in $L(0,n)$ are prime positive braids [**54**], while for $n < 0$ all knots and links are in $L(0,n)$ [**24**]. A template that contains all knots and links is called *universal*. A template is *positive* if it can be placed in a braid form with all crossings having the same orientation. For positive templates there is a bound on the number of prime factors of the supported knots [**48**]. For m and n positive the $L(m,n)$ knots have at most two prime factors, while for $L(-1,-1)$ the bound is three [**47**]. Even though $L(-1,-1)$ is not a positive template it can be

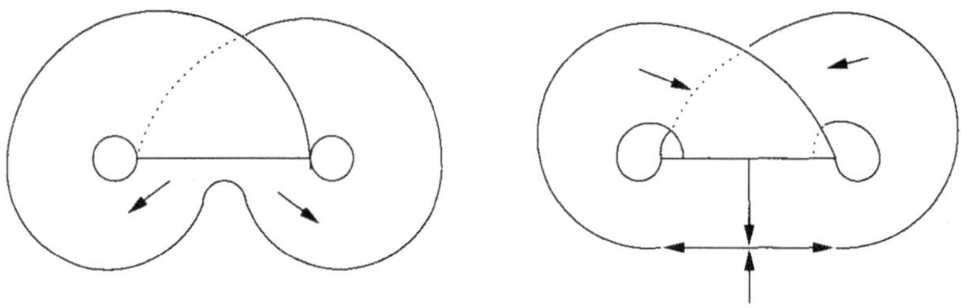

FIGURE 1. Left: Lorenz template for a saddle set. Right: Lorenz template for a singular hyperbolic Lorenz attractor.

presented so the all the crossing are positive, but not while it is in braid form [**45**]. When both m and n are negative it is known the $L(m,n)$ does not support all links [**25**, Proposition 3.2.21].

1105? **Question 3.** *Is there a general way to characterize which templates are universal? Is there a bound on the number of prime factors of knots in templates that can be presented with only one crossing type?*

When the "standard" suspension of the Plykin attractor is placed in a flow on $\mathbb{S}^3$ it contains a copy of $L(0,-1)$ [**25**, Proposition 3.2.18] and thus contains all knots and links [**25**, Proposition 3.2.18]. Rob Ghrist found that the same was true for every Plykin-like attractor he studied, but no general theorem is known here.

1106? **Question 4.** *Are there any attractors which are "standard" embeddings of Plykin-like attractors that do not have all knots and links?*

Twist-wise flow equivalence

Two flows are *topologically equivalent* if there is an orbit-wise homeomorphism between them that preserves the flow direction. Two SFTs are *flow equivalent* if their suspensions are topologically equivalent. Two non-negative square matrices are *flow equivalent* if they generate flow equivalent SFTs. In particular, incidence matrices of first return maps of any two cross sections to the same flow of this type are flow equivalent, although the two return maps need not be *topologically conjugate* (the usual equivalence relation for SFTs). Topologically conjugate SFTs are flow equivalent.

For nontrivial, irreducible non-negative square matrices, John Franks [**21**] has shown that flow equivalence is completely determined by two easy to compute invariants. They are the Parry–Sullivan number, denoted PS, and the Bowen-Franks group, denoted BF, derived in [**11**, **38**], respectively. If M is any non-negative integral $n \times n$ matrix then

$$PS = \det(I - M) \quad \text{and} \quad BF = \frac{\mathbb{Z}^n}{(I-M)\mathbb{Z}^n}.$$

We note that $|PS|$ is the order of BF if BF is finite, otherwise $PS = 0$. (The *trivial matrices* are the permutation matrices. These generate SFTs and suspension flows with a only a finite number of orbits, all closed.)

Flow equivalence only looks at basic sets, not at the ambient flows they may be embedded in. For example, the inverse limit flows of $L(0,0)$ and $L(0,1)$ are topologically equivalent since both are suspensions of the full 2-shift. Yet they look different: one has a (orientation reversing) twisted band, the other does not. To capture this *twist-wise flow equivalence* was introduced in [**43**]. We add additional information to the incidence matrices by using a t if the first return map is orientation reversing on a member of the Markov partition. This may require refining the Markov partition, which can always be done. Call these modified incidence matrices *twist matrices*. For $L(0,2n)$ and $L(0,2n+1)$ we get twist matrices $\left[\begin{smallmatrix}1&1\\1&1\end{smallmatrix}\right]$ and $\left[\begin{smallmatrix}1&1\\t&t\end{smallmatrix}\right]$, respectively. We take $t^2 = 1$ in all matrix calculations to mimic the fact the composition the first return map with itself is orientation preserving. Formally, twist matrices have entries of the form $at+b$, with a and b nonnegative integers, and are just matrices over the semi-group ring $\mathbb{Z}^+[\mathbb{Z}/2]$.

The topological interpretation is as follows. Take two basic sets of flows. Suppose they are flow equivalent. If we can extend the homeomorphism into the tangent bundles so that the stable and unstable sub-bundles are preserved, we say the embedded basic sets are *twist-wise flow equivalent* or sometimes *ribbon equivalent*; visually it is easier to extend the homeomorphism just a little, say $\epsilon > 0$, into the tangent bundle. Then the extended homeomorphism will take annuli to annuli, Mobius bands to Mobius bands, and infinite strips to infinite strips. Two twist matrices are *twist-wise flow equivalent* if they correspond to ribbon equivalent embedded basic sets.

There are several easy to compute invariants. If $T(t)$ is a twist matrix $T(\pm1)$ is defined by evaluating $T(t)$ at $t = \pm1$. Let

$$PS^{\pm} = \det(I - T(\pm1)) \quad \text{and} \quad BF^{\pm} = \frac{\mathbb{Z}^n}{(I - T(\pm1))\mathbb{Z}^n}.$$

Then PS^+ and BF^+ are clearly invariants since $T(1)$ is just the incidence matrix. It is shown in [**44, 45**] that PS^- and BF^- are also invariants that distinguish twist matrices not distinguished by PS^+ and BF^+.

Next, let $A = \left[\begin{smallmatrix}0&1\\1&0\end{smallmatrix}\right]$. Then regard $T(A)$ as the $2n \times 2n$ matrix obtained by replacing each t with A and each 1 with the 2×2 identity matrix: $a + bt \to \left[\begin{smallmatrix}a&b\\b&a\end{smallmatrix}\right]$. Define

$$PS^{\partial} = \det(I - T(A)) \quad \text{and} \quad BF^{\partial} = \frac{\mathbb{Z}^{2n}}{(I - T(A))\mathbb{Z}^{2n}}.$$

It is shown in [**44**] that PS^{∂} and BF^{∂} are invariants of twist-wise flow equivalence. While $PS^{\partial} = PS^+ \times PS^-$ there are examples of pairs of twist matrices which are not distinguished by $PS^{\pm}$ and $BF^{\pm}$ but are distinguished by BF^{∂}.

Example. *Let* $A = \left[\begin{smallmatrix}3&1+t\\1+t&3\end{smallmatrix}\right]$ *and* $B = \left[\begin{smallmatrix}3&1+t\\2&3\end{smallmatrix}\right]$. *We get* $PS^+ = 0$, $BF^+ = \mathbb{Z}\oplus\mathbb{Z}_2$, $PS^- = 4$, *and* $BF^- = \mathbb{Z}_2^2$ *for both matrices. But* $BF^{\partial}(A) = \mathbb{Z} \oplus \mathbb{Z}_4$ *while* $BF^{\partial}(B) = \mathbb{Z}_2^2$. *Thus, A and B are in distinct twist-wise flow equivalence classes.*

Example. *For* $\left[\begin{smallmatrix}1&1\\1&1\end{smallmatrix}\right]$ *and* $\left[\begin{smallmatrix}t&1\\1&1\end{smallmatrix}\right]$ *we get* $PS^{\pm} = -1$*, which implies all three BF groups are trivial. Yet, the suspension flow of the first matrix has no Mobius bands while the suspension flow of the second clearly does.*

Orientability, that is whether or not the ribbon set contains a Mobius band, is itself an invariant and easy to check for. If the twist matrix T is $n \times n$ then it is enough to check the diagonal entries of the first n powers of T. If no ts appear, then there are no Mobius bands in the ribbon set.

In [**12**] a complete algebraic invariant is produced. But, it is not known whether or not it is computable. An expository account is given in [**49**]. Given a matrix A over $\mathbb{Z}^+[\mathbb{Z}/2]$ let $(I-A)_\infty$ be the $\mathbb{N} \times \mathbb{N}$ matrix equal to $I-A$ in its upper left hand corner and the infinite identity everywhere else. The theorem below is Theorem 6.8 of [**49**] which is a special case of Theorem 6.3 of [**12**]; see either of these for the definition of *essentially irreducible*.

Theorem. *Let A and B be nontrivial essentially irreducible matrices over $\mathbb{Z}^+[\mathbb{Z}/2]$ and assume they are nonorientable. Then A and B are twist-wise flow equivalent if and only if there is an $SL(\mathbb{N}, \mathbb{Z}[\mathbb{Z}_2])$ equivalence from $(I-A)_\infty$ to $(I-B)_\infty$.*

Classifying matrices up to SL-equivalence over a PID is done by using an algorithm to convert them to a standard normal form (the *Smith normal form*). However, $\mathbb{Z}[\mathbb{Z}/2]$ is not a PID: $(1+t)(1-t) = 0$. It is unknown if an analogous algorithm exists for matrices over $\mathbb{Z}[\mathbb{Z}/2]$ or if SL-equivalence is decidable here.

1107? **Question 5.** *Is there an algorithm to classify square matrices over $\mathbb{Z}[\mathbb{Z}/2]$ up to SL-equivalence? This would then settle the problem of determining twist-wise flow equivalence of basic sets of Smale flows.*

Putting the pieces together and realization problems

Now we look at how the basic sets can be pieced together to form Smale flows, with an emphasis on nonsingular flows. This can be looked at from two prospectives. We will first review some results of John Franks that determine which basic sets can fit together to form a nonsingular flow on $\mathbb{S}^3$ and some generalizations. Next we ask just how the basic sets can fit together.

Suppose there is a single attracting closed orbit and a single repelling closed orbit in a nonsingular Smale flow on S^3. All other basic sets are saddle sets. Then we can compute the absolute value of the linking number of the attracting repelling pair as follows. Suppose there are n saddle sets and that for the ith one $\det T(t)$ is given by $a_i + tb_i$. Then the absolute value of the linking number is the product $|a_1 - b_1| \cdots |a_n - b_n|$ [**17**]. For example, the template $L(1,1)$ gives linking number 3.

The *structure matrix* of an embedded basic set is just its twist matrix evaluated at $t = -1$. In [**20**] the following are proved. If S is any structure matrix of a basic set, then there exists a nonsingular Smale flow ϕ_t on some 3-manifold with a basic set B whose structure matrix is A and every other basic set of ϕ_t consists of a single attracting or repelling closed orbit (Theorem 1). If there exists a nonsingular Smale follow on $\mathbb{S}^3$ with basic set B with structure matrix S then there exists a nonsingular Smale flow of $\mathbb{S}^3$ with a twist-wise flow equivalent basic set with all other basic sets being attracting or repelling closed orbits (Proposition 3.2). Furthermore, if $\det(I - S) \neq 0$ then the group $\mathbb{Z}^n/(I - S)\mathbb{Z}^n$ is cyclic (Theorem 3.3). Thus, $\left[\begin{smallmatrix}1&2\\2&1\end{smallmatrix}\right]$ cannot be realized as the structure matrix of a basic set in a nonsingular Smale flow on $\mathbb{S}^3$.

Question 6. *Are there any other obstructions (besides [**20**, Theorem 3.3]) to the* 1108?
realization of structure matrices in nonsingular Smale flows on $\mathbb{S}^3$?

Finally, in [**22**] we have an abstract classification of nonsingular Smale flows on $\mathbb{S}^3$. The major new tool is the *Lyapunov graph.* Given a Smale flow on a manifold there exist a smooth function from the manifold to the reals which is nonincreasing with respect to the flow parameter. Thus, each basic set goes to a point. This is called a *Lyapunov function.* The Lyapunov graph is defined by identifying connected components of the inverse images of points in the real line. Each vertex of the graph is a point whose connected component contained a basic set. A vertex is labeled by the basic set it is associated with. Edges are oriented by the flow direction.

Suppose Γ is an abstract Lyapunov graph whose sinks and sources are each labeled with a single attracting or repelling periodic orbit and suppose each remaining vertex is labeled with the suspension of a subshift of finite type. Then Γ is associated with a nonsingular Smale flow on $\mathbb{S}^3$, if and only if the following are satisfied. (1) The graph Γ is a tree with one edge attached to each source and each sink vertex. (2) If v is a saddle vertex whose basic set has incidence matrix M and with e_v^+ entering edges and e_v^- exiting edges then $e_v^+ \leq \mathbb{Z}_M + 1$, $e_v^- \leq \mathbb{Z}_M + 1$, and $\mathbb{Z}_M + 1 \leq e_v^+ + e_v^-$. Here $\mathbb{Z}_M$ is a the *Zeeman number* defined by $\dim \ker((I - M_2)\colon \mathbb{Z}_2^n \to \mathbb{Z}_2^n)$, where M_2 is the mod 2 reduction of M, $\mathbb{Z}_2$ is the integers mod 2, and n is the size of M. (Ketty de Rezende has generalized Lyapunov graphs to Smale flows with singularities [**40**].)

Thus, if there is a single attracting closed orbit and a single repelling closed orbit $\mathbb{Z}_M = 0$ or 1. The converse holds as well. Further, if $|a - b| = 1$ we know that the linking number is 1. But, we do not know whether or not they can or must form a Hopf link.

To see how the basic sets fit together involves mostly ad hoc cut-and-paste arguments. It is unlikely that a complete Wada like theorem will be found.

Smale flows on $\mathbb{S}^3$ where there is a single attracting and a single repelling closed orbit, and a single saddle set modeled by an embedding of the Lorenz template were studied in [**46**]. It was show that the attractor/repeller pair either formed a Hopf link or a trefoil and meridian, and that the template was $L(0, 2n)$ for some n.

Let ϕ_t be a Smale flow on M^3. We say a template T (we include the embedding in M^3 in the definition of the symbol T), is *realized* by ϕ is ϕ has a basic set modeled by a template isotopic to T in M^3. In his Ph.D. dissertation [**32**], Vadim Meleshuk studies realization of templates by Smale flows on $\mathbb{S}^3$. Without any other restriction, all templates are realizable with only fixed point basic sets [Theorem 3.3.1][4] A template can be realized in a flow whose only other basic sets are fixed point attractors and repeller if and only if certain easy to check topological criteria are meet [Theorem 3.3.2].

Meleshuk then switches his attention to nonsingular Smale flows. He shows that every template is realizable by a nonsingular Smale flow on some 3-manifold [Theorem 3.4.1]. On $\mathbb{S}^3$, he gives a complete criteria for when a template can be realized by a nonsingular Smale flow [Theorems 3.5.1, 3.5.6, 3.6.3]. In some cases a template T is realizable with only attractors and repellers, but other times T may force other saddle sets. For example, take a Lorenz template and tie a figure-8 knot in one band. By [**32**] it can be realized in a nonsingular Smale flow, but by [**46**] it cannot be realized with just a single repeller and attractor as the only other basic sets. What other basic sets could be forced?

Following [**23**] Meleshuk works with thickened templates. These are handle bodies whose boundaries are partitioned into 2-dimensional exit and entrance sets, separated by loops (the tangent set). They retract naturally to the branched manifold version of templates. He explores, using homological machinery, relations between the entrance and exit sets. For example, he shows that if T is realizable in a nonsingular Smale flow on $\mathbb{S}^3$ and the entrance and exit sets are connected, then they are diffeomorphic [Theorem 3.10.6], and conjectures that if a template can be realizable in a nonsingular Smale flow on $\mathbb{S}^3$ with only one attractor and one repeller, then the exit and entrance sets must be diffeomorphic [Conjecture 3.10.11].

There has been some work on Smale flows on manifolds beyond $\mathbb{S}^3$. Ketty de Rezende along with several collaborators has developed the theory of Lyapunov graphs of flows to other manifolds [**5**, **6**, **16**]. Sue Goodman has characterized when a flow on an arbitrary 3-manifold with a one-dimensional hyperbolic set has a transverse foliation noting the importance of transverse foliations in the study of Anosov flows; see [**27**] and also [**26**, **56**]. Indeed one of the motivations for the study of nonsignular Smale flows is their connection to Anosov flows; any Anosov flow can be turned into a nonsingular Smale flow via two surgery moves [**7**, **25**].

Bonatti's geometric type

A Smale diffeomorphism is a hyperbolic map with zero dimensional basic sets. A Smale flow always has Smale diffeomorphisms as cross sections. In a series of paper's Bonatti et al ([**3**, **4**, **10**, **50**]) have developed a new approach to the study of Smale flows on 3-manifolds and Smale diffeomorphisms on surfaces. The idea is to encode geometric information along with a Markov partition. This data includes twist data as in the twist matrices, but also includes order information; it is encoded as a *geometrized Markov partition* However, the geometrized Markov partition is not presented as a matrix but a mapping; whence it is not clear how to compute invariants from it.

[4]Meleshuk gives an independent proof, but the result can also be derived from a more general theorem of William Bloch [**9**].

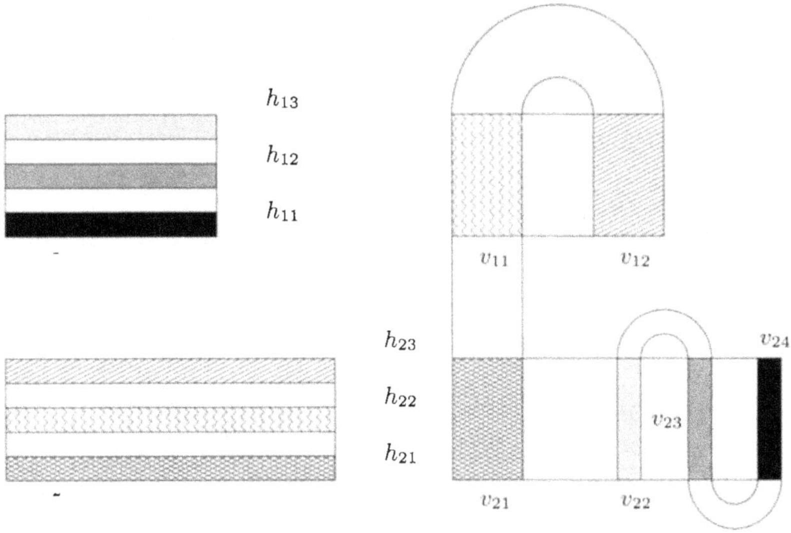

FIGURE 2. A Smale map

We shall give an example from a paper by Vago [**50**]. A Smale diffeomorphism f on a disk takes two large rectangles r_1 and r_2 to images shown in Figure 2. The horizontal strips, h_{11}, h_{12}, h_{13} in r_1 and h_{21}, h_{22}, h_{23} in r_2 are taken to the vertical strips $v_{24}, v_{23}, v_{22}, v_{21}, v_{11}, v_{12}$, respectively. From this one constructs the map ϕ, from $(1,2) \times (1,2,3)$ (more typically the subset of realized indices) into $(1,2) \times (1,2,3,4) \times (+,-)$ given by

$$\begin{array}{lll} \phi(1,1) = (2,4,-) & \phi(1,2) = (2,3,+) & \phi(1,3) = (2,2,-) \\ \phi(2,1) = (1,4,-) & \phi(2,2) = (1,3,+) & \phi(2,3) = (1,2,-), \end{array}$$

where the signs tells us whether the orientation has been reversed or not.

Geometrized Markov partitions have been used to prove theorems giving necessary and sufficient conditions for the existence of conjugacies. We shall state two, but shall not define all the terms, as we only intend to give the reader the flavor of this area.

Theorem ([**10**]; translated in [**50**]). *Let f and g be two Smale diffeomorphisms on compact surfaces, and let K and L be hyperbolic saturated sets of f and g respectively, without attractors or repellers. Then f and g are conjugate on* domains *of K and L if and only if (K, f) and (L, g) admit Markov partitions of the same geometrical type.*

Theorem ([**3**]). *Let X and Y be Smale vectors fields on compact orientable 3-manifolds. Let K and L be saturated saddle sets in X and Y respectively. Suppose that K and L admit good Markov partitions of the same geometrical type. Then there exist invariants neighborhoods of K and L where the restrictions of the field X and Y respectively, are equivalent.*

Order is nonabelian (obviously). What is needed is a nonabelian theory of symbolic dynamics. Bob Williams has developed a determinant for nonabelian matrices that contains some knot theoretic data for Lorenz attractors [**55**]. Could his matrices be modified to contain order data? They might capture part of the geometrized Markov partition in matrix form and thus facilitate the search for computable invariants. Another approach is to to use the skew-products systems in [**12**]. There the skew-products are of SFTs over finite groups. When the group is $\mathbb{Z}/2$ we get twist-wise flow equivalence. But the results in [**12**] hold for all finite groups including nonabelian groups. I have tried to find a way to use permutations groups to capture some of the order information in the geometrized Markov partition, but without success. When the map is iterated the order information does not seem to behave is a group-like manner. Could some non-group algebraic structure be used? But then would skew-products be meaningful??

1109? **Question 7.** *How can we get computable invariants that capture some of the order information in Bonatti's geometrized Markov partition?*

References

[1] D. Asimov, *Round handles and non-singular Morse–Smale flows*, Ann. of Math. (2) **102** (1975), no. 1, 41–54.

[2] W. Basener, K. Parwani, and T. Wiandt, *Minimal flows*, Open problems in topology II (E. Pearl, ed.), Elsevier Science.

[3] F. Béguin and C. Bonatti, *Flots de Smale en dimension 3: présentations finies de voisinages invariants d'ensembles selles*, Topology **41** (2002), no. 1, 119–162.

[4] F. Béguin, C. Bonatti, and J. L. Vieitez, *Construction de flots de Smale en dimension 3*, Ann. Fac. Sci. Toulouse Math. (6) **8** (1999), no. 3, 369–410.

[5] M. A. Bertolim, M. P. Mello, and K. A de Rezende, *Dynamical and topological aspects of Lyapunov graphs*, Qual. Theory Dyn. Syst. **4** (2003), no. 2, 181–203 (2004).

[6] M. A. Bertolim, M. P. Mello, and K. A. de Rezende, *Poincaré–Hopf and Morse inequalities for Lyapunov graphs*, Ergodic Theory Dynam. Systems **25** (2005), no. 1, 1–39.

[7] J. S. Birman and R. F. Williams, *Knotted periodic orbits in dynamical system. II. Knot holders for fibered knots*, Low-dimensional topology (San Francisco, CA, 1981), Contemp. Math., vol. 20, American Mathematical Society, Providence, RI, 1983, pp. 1–60.

[8] J. S. Birman and R. F. Williams, *Knotted periodic orbits in dynamical systems. I. Lorenz's equations*, Topology **22** (1983), no. 1, 47–82.

[9] W. L. Bloch, *Extending flows from isolated invariant sets*, Ergodic Theory Dynam. Systems **15** (1995), no. 6, 1031–1043.

[10] C. Bonatti and R. Langevin, *Difféomorphismes de Smale des surfaces*, Astérisque (1998), no. 250, viii+235, With the collaboration of E. Jeandenans.

[11] R. Bowen and J. Franks, *Homology for zero-dimensional nonwandering sets*, Ann. of Math. (2) **106** (1977), no. 1, 73–92.

[12] M Boyle and M. C. Sullivan, *Equivariant flow equivalence for shifts of finite type, by matrix equivalence over group rings*, Proc. London Math. Soc. **91** (2005), no. 1, 184–214.

[13] G. Burde and H. Zieschang, *Knots*, Walter de Gruyter & Co., Berlin, 2003.

[14] B. Campos, J. Martínez Alfaro, and P. Vindel, *Bifurcations of links of periodic orbits in non-singular Morse–Smale systems on S^3*, Nonlinearity **10** (1997), no. 5, 1339–1355.

[15] J. Casasayas, J. Martínez Alfaro, and A. Nunes, *Knots and links in integrable Hamiltonian systems*, J. Knot Theory Ramifications **7** (1998), no. 2, 123–153.

[16] R. N. Cruz and K. A. de Rezende, *Cycle rank of Lyapunov graphs and the genera of manifolds*, Proc. Amer. Math. Soc. **126** (1998), no. 12, 3715–3720.

[17] J. Franks, *Non-singular flows on S^3 with hyperbolic chain-recurrent set*, Rocky Mountain J. Math. **7** (1977), no. 3, 539–546.

[18] J. Franks, *Knots, links and symbolic dynamics*, Ann. of Math. (2) **113** (1981), no. 3, 529–552.

[19] J. Franks, *Homology and dynamical systems*, Conference Board of the Mathematical Sciences, Washington, DC, 1982.

[20] J. Franks, *Symbolic dynamics in flows on three-manifolds*, Trans. Amer. Math. Soc. **279** (1983), no. 1, 231–236.

[21] J. Franks, *Flow equivalence of subshifts of finite type*, Ergodic Theory Dynam. Systems **4** (1984), no. 1, 53–66.

[22] J. Franks, *Nonsingular Smale flows on S^3*, Topology **24** (1985), no. 3, 265–282.

[23] J. Franks and M. C. Sullivan, *Flows with knotted closed orbits*, Handbook of geometric topology, North-Holland, Amsterdam, 2002, pp. 471–497.

[24] R. W. Ghrist, *Branched two-manifolds supporting all links*, Topology **36** (1997), no. 2, 423–448.

[25] R. W. Ghrist, P. J. Holmes, and M. C. Sullivan, *Knots and links in three-dimensional flows*, Lecture Notes in Math., vol. 1654, Springer, Berlin, 1997.

[26] S. Goodman, *Vector fields with transverse foliations*, Topology **24** (1985), no. 3, 333–340.

[27] S. Goodman, *Vector fields with transverse foliations. II*, Ergodic Theory Dynam. Systems **6** (1986), no. 2, 193–203.

[28] S. Hayashi, *Connecting invariant manifolds and the solution of the C^1 stability and Ω-stability conjectures for flows*, Ann. of Math. (2) **145** (1997), no. 1, 81–137.

[29] M. W. Hirsch and C. C. Pugh, *Stable manifolds and hyperbolic sets*, Global analysis (Berkeley, CA, 1968), American Mathematical Society, Providence, RI, 1970, pp. 133–163.

[30] S. Hu, *A proof of C^1 stability conjecture for three-dimensional flows*, Trans. Amer. Math. Soc. **342** (1994), no. 2, 753–772.

[31] D. Lind and B. Marcus, *An introduction to symbolic dynamics and coding*, Cambridge University Press, Cambridge, 1995.

[32] V. Meleshuk, *Embedding templates in flows*, Ph.D. dissertation, Northwestern University, 2002.

[33] C. A. Morales, *3-manifolds supporting Anosov flows and Abelian normal subgroups*, Topology Appl. **143** (2004), 249–255.

[34] C. A. Morales and M. J. Pacifico, *A dichotomy for three-dimensional vector fields*, Ergodic Theory Dynam. Systems **23** (2003), no. 5, 1575–1600.

[35] C. A. Morales, M. J. Pacifico, and E. R. Pujals, *Singular hyperbolic systems*, Proc. Amer. Math. Soc. **127** (1999), no. 11, 3393–3401.

[36] J. W. Morgan, *Nonsingular Morse–Smale flows on 3-dimensional manifolds*, Topology **18** (1979), no. 1, 41–53.

[37] J. Palis and S. Smale, *Structural stability theorems*, Global analysis (Berkeley, 1968), American Mathematical Society, Providence, RI, 1970, pp. 223–231.

[38] B. Parry and D. Sullivan, *A topological invariant of flows on 1-dimensional spaces*, Topology **14** (1975), no. 4, 297–299.

[39] Y. B. Pesin, *Lectures on partial hyperbolicity and stable ergodicity*, European Mathematical Society, Zurich, 2004.

[40] K. de Rezende, *Smale flows on the three-sphere*, Trans. Amer. Math. Soc. **303** (1987), no. 1, 283–310.

[41] R. Roberts, J. Shareshian, and M. Stein, *Infinitely many hyperbolic 3-manifolds which contain no Reebless foliation*, J. Amer. Math. Soc. **16** (2003), no. 3, 639–679.

[42] M. Saito, *On closed orbits of Morse–Smale flows on 3-manifolds*, Bull. London Math. Soc. **23** (1991), no. 5, 482–486.

[43] M. C. Sullivan, *An invariant of basic sets of Smale flows*, Ergodic Theory Dynam. Systems **17** (1997), no. 6, 1437–1448.

[44] M. C. Sullivan, *Invariants of twist-wise flow equivalence*, Discrete Contin. Dyn. Syst. **4** (1998), no. 3, 475–484.

[45] M. C. Sullivan, *Positive knots and Robinson's attractor*, J. Knot Theory Ramifications **7** (1998), no. 1, 115–121.

[46] M. C. Sullivan, *Visually building Smale flows in S^3*, Topology Appl. **106** (2000), no. 1, 1–19.

[47] M. C. Sullivan, *Factoring families of positive knots on Lorenz-like templates*, 2005, Preprint.

[48] M. C. Sullivan, *Knots on a positive template have a bounded number of prime factors*, Algebr. Geom. Topol. **5** (2005), 563–576.

[49] M. C. Sullivan, *Twistwise flow equivalence and beyond ...*, Algebraic and topological dynamics, Contemp. Math., vol. 385, American Mathematical Society, Providence, RI, 2005, With an appendix by Mike Boyle and the author, pp. 171–186.

[50] G. M. Vago, *Conjugate unstable manifolds and their underlying geometrized Markov partitions*, Topology Appl. **104** (2000), 255–291.

[51] M. Wada, *Closed orbits of nonsingular Morse–Smale flows on S^3*, J. Math. Soc. Japan **41** (1989), no. 3, 405–413.

[52] L. Wen, *On the C^1 stability conjecture for flows*, J. Differential Equations **129** (1996), no. 2, 334–357.

[53] R. F. Williams, *Expanding attractors*, Inst. Hautes Études Sci. Publ. Math. (1974), no. 43, 169–203.

[54] R. F. Williams, *Lorenz knots are prime*, Ergodic Theory Dynam. Systems **4** (1984), no. 1, 147–163.

[55] R. F. Williams, *A new zeta function, natural for links*, From Topology to Computation: Proceedings of the Smalefest (Berkeley, 1990), Springer, New York, 1993, pp. 270–278.

[56] K. Yano, *Nonsingular Morse–Smale flows on 3-manifolds which admit transverse foliations*, Foliations (Tokyo, 1983), North-Holland, Amsterdam, 1985, pp. 341–358.

Part 6

Computer Science

Open Problems in Topology II – Edited by E. Pearl

Computational topology

Denis Blackmore and Thomas J. Peters

1. Introduction

The emphasis here will be upon how point-set topology can be applied to computing on geometric objects embedded in $\mathbb{R}^3$. The fundamental topological concept of a neighborhood generalizes limits over the reals, which inherently relies upon infinite precision arithmetic. Any specific computational representation of a real number is limited to being expressed in a finite number of bits. This cardinality disparity means that fundamental topological notions such as neighbhorhoods, dense sets and continuity are not well-expressed computationally, but can only be approximated. This presents novel opportunities for complementary research between topologists and numerical analysts.

The article *Computing over the reals: foundations for scientific computing* [**33**] begins,

> "The problems of scientific computing often arise from the study of continuous processes, and questions of computability and complexity over the reals are of central importance in laying the foundation for the subject."

The use of floating point numbers as an approximation of the reals entails a radically different perspective for classical point-set topologists, as the central topological notions regarding the interior, exterior and boundary of a set are based upon limits of infinite sequences of neighbhorhoods. These ideas are also crucial for geometric computations. Past practice can be somewhat tersely oversimplified as saying that the cardinality disparities have long been appreciated, but have been treated largely in an *ad hoc* fashion. Engineering practice and pragmatic programming, generally directed by heuristics, have been the dominant practice.

The definition adopted here for computational topology comes from the report *Emerging challenges in computational topology* [**21**]. (Also see Section 13.)

> "We intend the name *computational topology* to encompass both algorithmic questions in topology (for example, recognizing knots) and topological questions in algorithms (for example, whether a discrete construction preserves the topology of the underlying continuous domain)."

D. Blackmore gratefully acknowledges the partial support of this work from the NSF/DARPA CARGO grant CCR-0310619. T.J. Peters expresses appreciation for partial support from NSF grants DMS-0138098, CCR-022654, CCR-0429477, CCF-0533232 and by a 2005 IBM Faculty Award. All statements here are the responsibility of the authors, *not* of the National Science Foundation *nor* of the IBM Corporation.

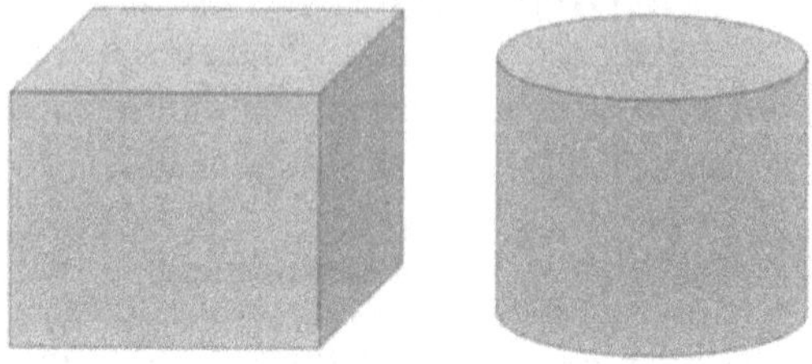

FIGURE 1. Box and cylinder

The broad definition is intended to prompt a "beneficial symbiosis" [**63**] between both subfields and "to extend computational geometry ... into contact with classical topology" [**21**] with expected benefits to both fields. The subdiscipline of computational topology is relatively young. This very immaturity provides an important opportunity to consider its foundations as well to explore pernicious specific problems that remain unresolved.

2. History

The first usage of the term 'computational topology' appears to have occurred in the dissertation of M. Mäntylä [**120**]. The focus there was upon the connective topology joining vertices, edges and faces in geometric models, frequently also informally described as the *symbolic* information of a solid model. These vertices, edges and faces are discussed as the operands for the classical Euler operations.

2.1. Elementary manifold examples. In Figure 1, the box depicted on the left would have 8 vertices, 12 edges and 6 faces. This should be obvious, while the cylinder shown on the right entails an additional minor subtlety. Namely, the cylinder can be considered to be composed of an open cylinder and a top disc and a bottom disc. To include explicitly vertices and edges, the open cylinder will often be considered to be formed from a flat rectangle which has been rolled into a cylinder, with two opposing edges identified as one. This one edge would be vertical in the image on the right and would have a vertex at each end. Each disc would then be seen as having a circular bounding edge that had its initial and terminal vertex at one of these points on the vertical edge. This representation would then have 2 vertices, 3 edges and 3 faces (though other variants are clearly possible).

2.2. Non-manifold topology. Manifolds have a rich history in topology. They provide extensions of the usual topology on Cartesian products of the reals. Moreover, manifolds provide a generalization whereby points, curves, surfaces and solids have a common abstraction, but vary in dimension (from 0 to 3, respectively). Within the Boolean algebra of regular-closed, compact 3-manifolds, curves and surfaces are nowhere dense sets – meaning that the interior of their closures is empty within $\mathbb{R}^3$. Hence, these sets are trivial operands within that algebra. So, strict adherence to a programming paradigm, based upon this Boolean algebra of

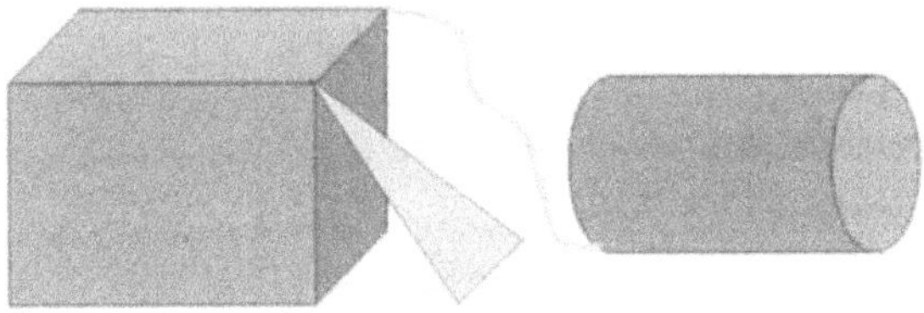

FIGURE 2. Model with non-manifold topology

3-manifolds would not admit directly the mixing of manifolds of differing dimension. Pioneering work by K.J. Weiler in his thesis [**176**] describing 'non-manifold topology' laid the intellectual framework for his initial prototype and extensive follow-up work by F. Printz in his 'Noodles' system [**88**].

Figure 2 shows how the manifolds of differing dimensions could be integrated in these systems to form one integrated geometric model. Each point need not have a neighborhood that is homeomorphic to a neighborhood in a 3-manifold (For 3-manifolds without boundary, these neighborhoods are just open neighborhoods of $\mathbb{R}^3$ and for 3-manifolds with boundary, the neighbhorhoods just have the usual relative topology associated with a boundary point). However, each point does have a neighborhood that is homeomorphic to an open neighborhood in an n-manifold, with n being equal to the lowest dimension of any of the manifolds joined at that point.

Question 2.1. *Is there a unifying topological abstraction covering manifolds, non-* 1110?
manifolds and other possible geometric models that might be useful to improve algorithmic design for geometric computations?

Some other relevant references in the development of computational topology are listed [**13, 50–52, 57, 59, 137–139, 172**].

3. Computation and the reals

Whenever computations are intended to be representative of operations on the reals, inherent concerns are the trade-offs required between algorithmic efficiency and sufficient numerical precision. This dilemma is discussed [**33**] relative to using a satisfactory number of terms from a Taylor's series approximation. The summarizing directive is "to take just enough terms *to satisfy our precision needs.*"

3.1. The role for tolerances. This same issue has been expressed within venues of the Society of Industrial and Applied Mathematics (SIAM) by the mathematician D.R. Ferguson and the engineer R. Farouki. Ferguson has observed that geometric models used in aeronautical and aerospace design require differing precisions depending upon the application software that is using such models for input [**75**]. His focus is upon the number of significant digits needed in order to have appropriate representations at topological boundaries formed from surface intersections. A broad overview of this concept is illustrated in the table in Figure 3. There the minimum precision needed for each application is given. Farouki

Application	Significant digits
visualization	2
computational fluid dynamics	3
multidisciplinary design optimization	4
computational electromagnetics	4
computational optics	7

FIGURE 3. Precision for differing applications

has espoused a similar point of view [**74**], based upon issues raised at a SIAM workshop that he and Ferguson organized with funding from the National Science Foundation (NSF).

This perspective raises several fundamental problems:

1111? **Question 3.1.** *What are the differing floating point precisions needed to capture accurately the topology along surface intersection boundaries in geometric models so that they can be reliably used in engineering simulations for visualization, computational fluid dynamics, stress analysis, computational optics, computational electromagnetics, etc.?*

1112? **Question 3.2.** *Are there crucially sensitive engineering applications that can be used to determine these precision needs? (For instance, are visualization and computational optics at extreme ends of the precision spectrum? Is understanding the needs of those two applications then sufficient for the conceptual framework for all modeling needs?)*

1113? **Question 3.3.** *Are some geometric intersection problems ill-conditioned?*

Question 3.4. *Will the process of finding the precision required for the models for these engineering simulations generalize to a mathematical methodology for being able to determine floating point precision needs for a wide variety of geometric models, inclusive of examples such as fractals and Julia sets?*

We note some abstractions that are appropriate for topologists. Individual computer science algorithms might be considered as specific functions, with distinctive domains and images. Even this level of abstraction is rarely articulated within computer science. Moreover, this view belies a cultural distinction between the computer science (CS) and mathematical communities. Topologists often focus attention of an entire family of functions, analyzing properties shared by an entire class of functions. For instance, homeomorphisms form an important such class within point-set topology, forming the basis for the traditional definition of topological equivalence. This broader approach, considering whole classes of algorithms, would be one way that topological perspectives can enrich CS. The process of going from one algorithm to another is then merely modeled by composition of functions. An example of how this view might also be useful in computer graphics is presented in Section 4.

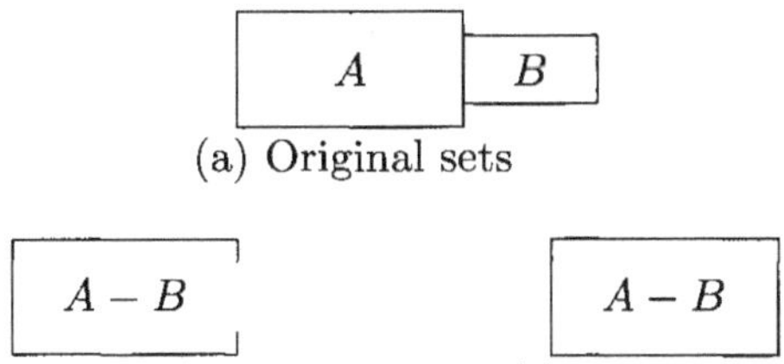

(a) Original sets

(b) Unexpected result (c) Expected result

FIGURE 4. Subtraction of two sets

3.2. Engineering examples for computational topology. The material of this subsection is largely extracted from a related article [**137**] in order to introduce topologists to prominent engineering examples for computational topology.

The Boolean algebra of regular closed sets is prominent in topology, particularly as a dual for the Stone–Čech compactification. This algebra is also central for the theory of geometric computation, as a representation for combinatorial operations on geometric sets. However, the issue of computational approximation introduces unresolved subtleties that do not occur within *pure* topology.

The standard algorithmic operators on regular closed set representations are those from its Boolean algebra. These Boolean operations have an elegant symbolic representation in a binary tree, but do not typically include error bounds on the leaf node operands, which appears to fall within Knuth's definition [**104**] of algorithms being "properly called *seminumerical* because they lie on the borderline between numeric and symbolic calculation." This disparity between the theory and practice on this Boolean algebra is a central aspect of the "geometric robustness" problem [**96**].

The regular closed sets discussed here will be assumed to be subsets of $\mathbb{R}^3$, with its usual topology. The Boolean algebra of regular closed sets in $\mathbb{R}^3$ will be denoted as $\mathcal{R}(\mathbb{R}^3)$. Furthermore, any regular closed set considered will be assumed to be compact. Any surfaces and curves considered will be assumed to be compact 2-manifolds and 1-manifolds, respectively. All neighborhoods will be assumed to be open subsets of $\mathbb{R}^3$.

The theoretical role for $\mathcal{R}(\mathbb{R}^3)$ was introduced into geometric computing to correct the unexpected output seen from combinatorial operations on geometric sets [**148**]. For instance, consider the two dimensional illustration shown in Figure 4. The original operands of A and B are indicated in Figure 4(a). The unexpected output is shown in Figure 4(b), where the expected result would have been what is shown in Figure 4(c).

The phenomenon shown in Figure 4(b) was informally described as "dangling edges" [**173**]. The formalism that was proposed to eliminate this behavior was that geometric combinatorial algorithms should accept only regular closed sets as input and then execute the Boolean operations of meet, join and complementation on these operands, thereby creating only regular closed sets as output [**170**].

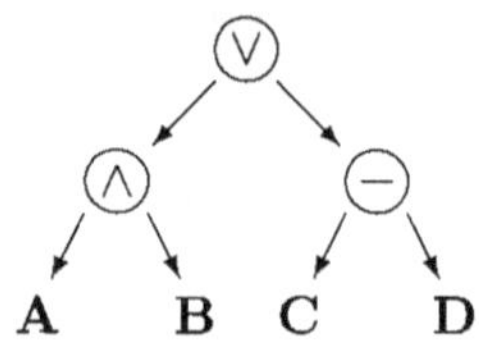

FIGURE 5. Tree for $(\mathbf{A} \wedge \mathbf{B}) \vee (\mathbf{C} - \mathbf{D})$.

The intent was to eliminate "dangling edges" and, in principle, this should have been sufficient.[1] However, each operand also has a geometric representation that depends upon the approximation methods used to compute the results. This additional subtlety raises issues in both theory and computation.

An earlier survey on topology in computer-aided geometric design [**139**] is recommended as introductory material for topologists. The texts [**96**, **136**] discuss the integration of computational geometry, shape modeling and topology.

3.2.1. *Theory versus computation.* One elegant computational representation for the combinatorial operators is to assign each object a symbol and then to indicate operations in a tree referencing those symbols. For instance, such a tree structure could be as depicted in Figure 5.

At this level of abstraction, the mathematical theory and the computational representation are completely consistent, and this representation became known as Constructive Solid Geometry (CSG). Difficulties arose in instantiating the basic geometric information that is represented by the operands at the leaf nodes and, sometimes, in computing geometric representations at the internal nodes of the tree. In CSG, the leaf nodes are restricted to a small set of specific geometric objects, known as primitives. A typical collection of primitives might consist of spheres, parallelepipeds, tori and right circular cylinders. The critical geometric algorithm underlying each Boolean operation is the pairwise intersection between the operands.

As the boundary of each of these primitives can be represented by linear or quadratic polynomials, the needed intersection between each pair of primitives was relatively simple and numerically stable, for most cases considered, although specific intersections could be problematic. For instance, suppose two cylinders of identical radius and height were created and then positioned so that the bottom of one cylinder was coincident with the top of the other cylinder. This special case was specifically considered in most intersection algorithms and could usually be processed without problem. However, if one then rotated the top cylinder a fraction of a degree about its axis (so that the planar coincidence remained intact) many software systems would fail to produce any output for this problem,

[1]The subtraction operation between two sets, shown as $A-B$ in Figure 4, is not *specifically* a Boolean operation. However, the use of $A-B$ should be understood to be conveniently shortened notation equivalent to the operations $A \wedge B'$, where B' represents the standard Boolean operation of complementation on the operand B.

sometimes even causing a catastrophic program failure. This particular problem became a celebrated test case and most systems developed *ad hoc* methods to solve this cylindrical intersection problem. Yet, this was just avoiding the more serious issue of the fragile theoretical foundations for many intersection algorithms. People using CSG systems became sensitive to their limitations and continued to use them effectively by avoiding these challenging circumstances, although the work-arounds were often tedious to execute.

The imperative, largely initiated by the aerospace and automotive industries, to model objects using polynomials of much higher degree than quadratic created a movement away from CSG systems. The alternative format was to represent compact elements of $\mathcal{R}(\mathbb{R}^3)$ by their boundaries, and this became known as the "boundary representation" approach, or "B-rep" for short. This has become the dominant mode today. Again, within this clean conceptual overview, the realities of computation pose some subtle problems. In most industrial practice, the modeling paradigm was further restricted so that the boundary of an object was a 2-manifold without boundary. However, it was difficult to create computer modeling tools that could globally define 2-manifolds without boundary, though there existed excellent tools for creating subsets of these 2-manifolds. For example, computational tools for creating splines were becoming prevalent [**141**]. Again, in principle, if each such spline subset was created with its boundaries, then the subsets could be joined along shared boundary elements to form a topological complex [**95**] for the bounding 2-manifold without boundary.

The inherent computational difficulty was to create separately two spline patches, each being a manifold with boundary, so that the corresponding boundary curves were identical and could be shared exactly between the patches. In some situations, algorithms for fitting spline patches were used successfully. In other cases, patches have been enlarged slightly and intersected so as to obtain improved fits. Indeed, such intersections are well-defined in pure mathematics, but, again, approximation in computation poses subtle variations from that theory, as described in the next section on pairwise surface intersection.

3.2.2. *Subtleties of pairwise spline surface intersection.* It is well known that unwanted gaps between spline surfaces or self-intersections within intended manifolds often appear as artifacts of various implemented intersection algorithms [**74**]. The mismatch between approximate geometry and exact topology has historically caused reliability problems in graphics, CAD, and engineering analyses, drawing the attention of both academia and industry. The severity of the problem increases with the complexity of the geometric data represented, both from high-degree nonlinearity and from the intricate interdependence of shape elements that should, but do not, fit together according to the specified topological adjacency information.

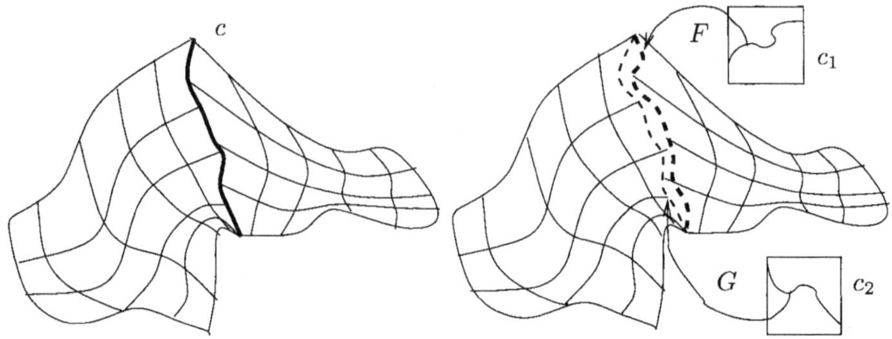

(a) Conceptual view (b) Computational reality

FIGURE 6. Spline surface interection

The conceptual view of these joining operations is illustrated in Figure 6(a), with an intersection curve[2] illustrated as a single curve. But this image only exemplifies the idealized, exact intersection curve. For practical computations, an approximation of the intersection set is often created [**85**, **86**] and, in many systems, an intersection curve will be approximated twice. These two approximations are created corresponding to each of the spline functions, denoted as $F\colon [0,1]^2 \to \mathbb{R}^3$, and $G\colon [0,1]^2 \to \mathbb{R}^3$, whose images are the surfaces being intersected. Specifically, a spline curve, denoted as c_1, is created so that $c_1 \subset [0,1]^2$ and the image of c_1 by F, denoted as $F(c_1)$ approximates the intersection curve, (with similar meaning given to $c_2 \subset [0,1]^2$ and $G(c_2)$). It is virtually certain that those approximations, $F(c_1)$ and $G(c_2)$, will not be exactly equal in $\mathbb{R}^3$, as shown in Figure 6(b).

The mismatch between concept and reality discussed above, can cause ambiguity, as the intersection representation is sometimes considered as a unique set, from the conceptual view, and at other times as two approximating sets, from an algorithmic view.

3.2.3. *Error bounds for topology from Taylor's theorem.* First, we present the Grandine–Klein (GK) intersection algorithm [**86**]. Referring to Figure 6, we note that the GK algorithm bases its error bounds on well-established numerical techniques in differential algebraic equations (DAE). While these DAE techniques provide rigorous error bounds, these bounds are expressed within the parameter space $[0,1]^2$, which serves as the domain of the spline functions (indicated as F and G, above). The code implementing the GK algorithm then has an interface that allows the user to specify an upper bound ϵ for the error within parameter space and the algorithm provides guarantees for meeting this error bound. However, the typical end user is often not fully aware of the details of the parametric definitions

[2]We focus on the generic case of an intersection curve, although isolated points and coincident areas can also arise, with similar complications.

of F and G, so selection of this parametric space error bound has often relied upon heuristics. It would be more convenient for the user to be able to specify an error bound within $\mathbb{R}^3$. One accomplishment within the I-TANGO [**137**] project has been to demonstrate a mathematical relation [**133**] between the error bounds in $\mathbb{R}^3$ and $[0,1]^2$, following from a straightforward application of Taylor's Theorem in two dimensions. The conversion between these error bounds has been implemented in a preprocessing interface to the GK algorithm and this new interface has been tested to be reliable, efficient and user-friendly.

Using the notation from Figure 12 for the spline function F, Taylor's Theorem provides a bound on the error of F evaluated at a particular point (u,v) versus F evaluated at a point (u_0,v_0), where (u,v) and (u_0,v_0) are within a sufficiently small neighborhood. This sufficiently small neighborhood will have radius given by the value in the parametric domain $[0,1]^2$ which was denoted as ϵ in the previous paragraph. Then it follows [**133**], with $\|\cdot\|$ being any convenient vector norm, that

$$\|F(u,v) - F(u_0,v_0)\| \le \epsilon M$$

for any M satisfying

$$\left\|\frac{\partial F}{\partial u}(u^*,v^*)\right\| + \left\|\frac{\partial F}{\partial v}(u^*,v^*)\right\| \le M,$$

for some point $[u^*,v^*]$ on the line segment joining $[u,v]$ and $[u_1,v_1]$.

For the single spline F, let $\gamma(F)$ be an upper bound for the acceptable error in $\mathbb{R}^3$ between the true intersection curve c and one of its approximants $F(c_1)$. In order to guarantee that this error is sufficiently small, it is sufficient that $\epsilon M \le \gamma(F)$, where an upper bound for M can be computed by using any standard technique for obtaining the maximums of the partials $\frac{\partial F}{\partial u}$ and $\frac{\partial F}{\partial v}$. For G, a similar relation between $\gamma(G)$ and ϵ exists.[3]

Then it is clear that a neighborhood can be defined that contains the true intersection curve c and both of its approximants. Let $N_{\gamma(F)}(F(c_1))$ be a tubular neighborhood of radius $\gamma(F)$ about $F(c_1)$, where c_1 has been generated from the GK intersector to satisfy the inequality presented in the previous paragraph. Similarly, define $N_{\gamma(G)}(G(c_2))$. Then, let $N(c) = N_{\gamma(F)}(F(c_1)) \cup N_{\gamma(G)}(G(c_2))$.

It is clear that $N(c)$ is a neighborhood of c, which contains both of its approximants, $F(c_1)$ and $G(c_2)$. However, there is both a theoretical and computational limitation to this approach.

- There is no theoretical guarantee that either approximant is topologically equivalent to c, and
- Any practical computation of $N(c)$ would depend upon an accurate computation of the set $N_{\gamma(F)}(F(c_1)) \cap N_{\gamma(G)}(G(c_2))$, which is likely to be as difficult as the original computation of $F \cap G$.

While the above bounds are often quite acceptable in practice to compute a reasonable approximant, further research has been completed into alternate methods

[3]This error bound assumed that the error due to algorithmic truncation within the numerical DAE methods dominated any other computational errors.

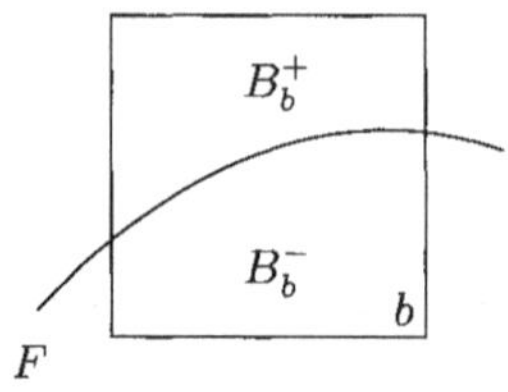

(a) Condition **C2**

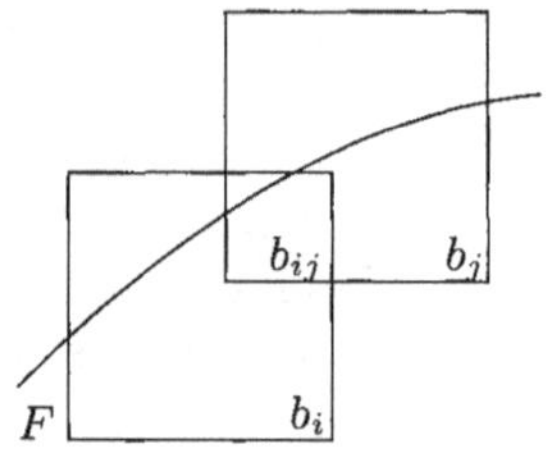

(b) Condition **C3**

FIGURE 7. 2D versions of properties of interval solids

to give guarantees of topological equivalence within a computationally acceptable neighborhood of the intersection set, as reported in the next subsection.

3.2.4. *Integrating error bounds and topology via interval solids.* Recent work by Sakkalis, Shen and Patrikalakis [**153**] emphasized that the numeric input to any intersection algorithm has an initial approximation in the coordinates used to represent points in $\mathbb{R}^3$, leading to their use of interval arithmetic [**136**]. The basic idea behind interval arithmetic is that any operation on a real value v is replaced by an operation of an interval of the form $[a, b]$, where $a, b \in \mathbb{R}$ and $a < v < b$. The result of any such interval operation is an interval, which is guaranteed to contain the true result of the operation on v. This led naturally to the concept of an *interval solid* and some of its fundamental topological and geometric properties were then proven, as summarized below.

Throughout this section, a *box* is a rectangular, closed parallelepiped in $\mathbb{R}^3$ with positive volume, whose edges are parallel to the coordinate axes.[4] Let F be a nonempty, compact, connected 2-manifold without boundary. Then the Jordan Surface Separation Theorem asserts that the complement of F in $\mathbb{R}^3$ has precisely two connected components, F_I, F_O; we may assume that F_I is bounded and F_O is unbounded. Let also $\mathcal{B} = \{b_j : j \in J\}$ be a finite collection of boxes that satisfies the following conditions:

C1: $\{\text{int}(b_j) : j \in J\}$ is a cover of F.

C2: Each member b of $\mathcal{B}$ intersects F generically; that is, $b \cap F$ is a nonempty closed disk that separates b into two (closed) balls, B_b^+ and B_b^-, with B_b^+, (B_b^-) lying in $F_I \cup F$ $(F_O \cup F)$, respectively.

C3: For any $b_i, b_j \in \mathcal{B}$, let $b_{ij} = b_i \cap b_j$. If $\text{int}(b_i) \cap \text{int}(b_j) \neq \emptyset$, then b_{ij} is also a box which satisfies **C2**.

Notice that condition **C2** indicates that every $b \in \mathcal{B}$ intersects F in a natural way (see Figure 7).

The following result summarizes several previously appearing results, where a solid is defined to be a nonempty compact, regular closed subset of $\mathbb{R}^3$.

[4]Enclosures other than boxes are quite possible and this is a subject of active research.

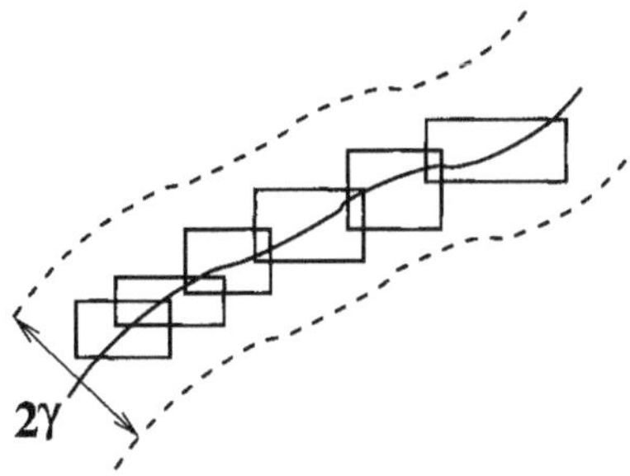

FIGURE 8. 2D version of proper subset condition

Theorem 3.1 ([**153**, Corollary 2.1, p. 165]). *If F is connected and $\mathcal{B}$ satisfies* **C1–C3**, *then $F \cup \bigcup_{j \in J} b_j$ is a solid.*

Bisceglio, Peters and Sakkalis [**151, 152**] have recently given sufficient conditions to show when the boundary of an interval solid is ambient isotopic to the well-formed solid that it is approximating, as described in the following theorem. To do so, they define a parameter, denoted here as, γ, which is based upon curvature and critical values of an energy function. This value of γ then permits the definition of non-self-intersecting tubular neighborhoods about the original object for all values of $r < \gamma$, when r is a positive number for a constant radius tubular neighbhorhood. For a positive number δ, define the open set $F(\delta) = \{x \in \mathbb{R}^3 : D(x, F) < \delta\}$, where $D(x, F) = \inf\{d(x, y) : y \in F\}$, with d being the Euclidean metric in $\mathbb{R}^3$.

Theorem 3.2. *Let F be a connected 2-manifold without boundary. For each $\epsilon > 0$, there exists δ, with $0 < \delta < \gamma$ so that whenever a family of boxes $\mathcal{B}$ satisfies conditions* **C1–C3**, *and for each b of $\mathcal{B}$, b is a subset of $F(\delta)$* (see Figure 8) *then, for $S = F \cup F_I$ and $S^{\mathcal{B}} = S \cup \bigcup_{j \in J} b_j$, the sets F and $\partial S^{\mathcal{B}}$ are ϵ-isotopic with compact support. Hence, they are also ambient isotopic.*

The quoted theorem depends upon results from Bing's book on PL topology [**22**, p. 214], and related literature [**103**], as is explained in full [**151, 152**]. The proof shows that normals to F do not intersect within the constructed tubular neighborhood, as is illustrated by the depiction of its planar cross-section in Figure 8.

If the boxes containing the true intersection curve can be made sufficiently small so that each such box fits inside $F(\rho)$, then the resulting intersection neighborhood will contain an object that is both close to the true solid and is ambient isotopic to it. Considerable success in meeting these constraints has already been achieved [**136, 152, 153**].

4. Correctly embedded approximations for graphics & applications

Papers on tolerances in engineering design [**31, 32, 166**] raised the issue of rigorous proofs for the preservation of topological form in geometric modeling,

FIGURE 9. Nonequivalent knots

but these papers did not specifically propose ambient isotopy as a criterion. The class of geometric objects considered was expanded appreciably by theorems for ambient isotopic perturbations of PL simplexes and splines [**14–16**].

As an elementary example, there is an exact computational representation of a unit circle centered at the origin, as $x^2 + y^2 = 1$. However, as soon as one goes to create a computer graphics image of this circle, some approximation is needed. The ultimate display on the screen is to 'turn on' a collection of pixels, each being some very small rectangle. If these pixels are sufficiently small and the approximation is sufficiently fine, then the user perceives a reasonable image of a circle. This has many parallels to a human rendition of a circle, such as a pen and ink image that approximates a circle. The criterion for success is largely subjective, though it has been successfully codified in standard algorithms for this simple case of the circle [**77**]. This technique does generalize to more difficult geometric shapes which also have nice differentiable properties [**37**], but there remain difficulties in the prevalent approximation paradigms, as will be discussed further, here.

However, there is a crucial distinction between the use of such images in classical mathematics and in computer science. The adage in pure mathematics is that 'A picture is not a proof.' Rather, the use of illustrations is meant to guide discovery and intuition in order to lead to formal proofs. The situation in computer science is quite different. Namely, the focus is upon the definition and properties in creating specific algorithms to work on particular abstract data types. Here, the data type is the equation of a circle, but this representation is then approximated for graphics rendering. So far, this offers little distinction to the classical case. However, the output of this approximation algorithm may often be used by another algorithm. The approximation becomes the object of interest. This could be translating the circle to another position or determining its circumference. Both these operations are quite successful for the circle.

Indeed, even at the graphics display level, the concern for a 'properly representative approximation' should not be dictated solely by subjective criterion, as

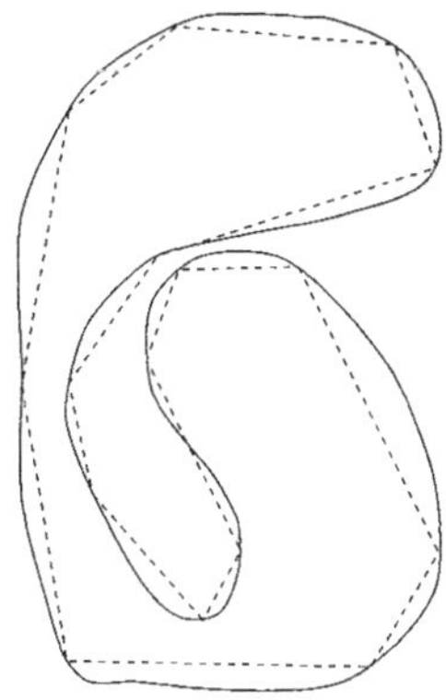

FIGURE 10. Ambient isotopic approximation

can be shown in the following knot approximation example, which summarizes a published example [**13**].

Many geometric approximation algorithms offer no guarantees about the topology of the output. Sometimes it is guaranteed that the output is homeomorphic to a desired manifold [**10**]. Indeed, in the simple circle example, essentially any reasonable PL approximation of the cirle will be homeomorphic to it. However, in graphics, any 3D image is projected onto a 2D display. One asks if this composition of functions will necessarily lead to a homeomorphic image. The answer can be easily shown to be 'no' and supports the argument for a stronger form of topological equivalence.

We argue here that a guarantee of homeomorphism is insufficient for many of the applications for which the algorithms are designed. Rather, examples are given for preferring a stronger equivalence relation based upon ambient isotopy.

Definition 4.1. If X and Y are subspaces of $\mathbb{R}^3$, then X and Y are *ambient isotopic* if there is a continuous mapping $F\colon \mathbb{R}^3 \times [0,1] \to \mathbb{R}^3$ such that for each $t \in [0,1]$, $F(\cdot,t)$ is a homeomorphism from $\mathbb{R}^3$ onto $\mathbb{R}^3$ such that $F(\cdot,0)$ is the identity and $F(X,1) = Y$.

For other fundamental terms, the reader is referred to the text [**94**].

Figure 9 shows an unknot, and its homeomorphic, *but non-isotopic* PL approximant, which is the knot with 4 crossings, known as 4_1^m. An improved approximation is shown in Figure 10.

All end points of the line segments in the approximation are also points on the original curve. In response to the example of Figure 9, a theorem was published for ambient isotopic PL approximations of 1-manifolds [**119**], with an illustrative outcome shown in Figure 10. The proof utilizes 'pipe surfaces' from classical differential geometry [**128**].

Although any two simple closed planar curves are ambient isotopic, this knotted curve as an approximant to the original unknot would be undesirable in many circumstances, such as graphics and engineering simulations [**15**]. For instance, projected images of this approximation could have self-intersections, whereas the original curve had none.

There is a related study of curves, comparing them to α-shapes [**69**] via ambient isotopies [**150**]. Recent work in support of molecular modeling appears in the doctoral thesis of E.L.F. Moore and related publications [**129–131**].

Other problems arise for surfaces (2-manifolds) in three dimensions. Some algorithms compute a triangulated surface C to approximate the boundary F of a closed, finite volume, with a guarantee that C is homeomorphic to F [**11**, **12**]. It is well known that this does *not* guarantee that the complement of C, denoted as $\mathbb{R}^3 \setminus C$, is homeomorphic to the complement of F, $\mathbb{R}^3 \setminus F$, meaning that there is no guarantee that F and C are equivalently embedded in $\mathbb{R}^3$. An ambient isotopy between C and F, on the other hand, provides such a guarantee.

The class of PL surfaces presents another domain in which topological guarantees are desirable. Even guaranteeing that the common *edge contraction* operator produces an object homeomorphic to its input requires some care for simplicial complexes [**52**]. Preservation of genus during approximation by a polygonal mesh [**171**] also requires considerable care.

Recent theorems [**4**, **5**, **13**] prove approximation techniques that preserve ambient isotopy over an important subclasses of 2-manifolds, covering cases both with and without boundary. The role for ambient isotopy has been recognized by the computer animation research community [**80**].

1114? **Question 4.1.** *Is ambient isotopy the appropriate topological equivalence relation for computational topology in computer graphics and animation?*

Question 4.2. *What geometric approximation algorithms can capture the topological equivalence needed in computer graphics and animation?*

1115? **Question 4.3.** *Are the known algorithms for ambient isotopic of parametric curves optimal with respect to performance and space requirements?*

1116? **Question 4.4.** *Are the known algorithms for ambient isotopic of parametric surfaces optimal with respect to performance and space requirements?*

Considerable work on isotopies in approximation has appeared, ranging over applications from computer graphics, geometric modeling and surface reconstruction [**38–40**].

5. The role for differentiability

Although computational topology is a relatively new discipline [**21**, **172**], it has grown and matured rapidly partially because of its increasing importance to many vital contemporary applications areas such as computer aided design and manufacturing, (CAD/CAM), the life sciences, image processing and virtual reality. It is leading to new techniques in algorithm and representation theory. These applications are evoking new connections between mathematical subdisciplines such as algebraic geometry, algebraic topology, differential geometry, differential topology, dynamical systems theory, general topology, and singularity and stratification theory [1–3]. The tender age of computational topology renders it fertile ground for a wide variety of challenging open problems—many of fundamental importance. While the primary focus of this book and, of course, this chapter is upon problems in general topology, the integrative nature of computational topology is expressed here with some attention to the role of differentiability.

5.1. Introduction. Computational geometry preceded computational topology as an indispensable theory and practice for solving difficult problems that have arisen in CAD/CAM and other contexts that rely on computationally powerful methods for analysis and accurate representation of various objects and configurations. On the other hand, computational topology has only considerably more recently risen to prominence in such applications [74]. The difference between these two disciplines is roughly analogous to the difference between geometry and topology, and can be rather effectively illustrated in the following terms: Whereas computational geometry is concerned essentially with algorithmic (and *a fortiori* computer implementable) methods for analyzing and producing representations of geometric objects that are close—usually in some Whitney-like (piecewise) C^2 sense—a primary focus of computational topology is to guarantee algorithmically that a computer generated representation of an object is equivalent to the actual object in an appropriate topological sense. In essence, computational geometry is concerned with insuring the (differential geometric) closeness of the representation of an object to the original, while computational topology takes care of the topological consistency of the rendering.

The importance of computational topology cannot be overestimated in certain contexts and applications—many of which have achieved significant prominence in the last few years. For example, suppose one wants to produce a computer generated representation of a water glass to be used in an automated manufacturing process. The glass can be viewed in ideal form as a smooth surface in space with a circular boundary, thus rendering it an object in a standard differential geometry or topology category. An algorithm can readily be found that produces a representation that is as close as desired (in some suitable Whitney-type topology) to the designed glass, but still have in it a very small hole. This may be considered satisfactory from the perspective of computational geometry, but certainly not from the computational topology viewpoint, and the result obviously would lead to shortcomings in the manufactured article.

In this section we shall identify several outstanding problems in computational (differential) topology—all of which are of a rather fundamental nature—and we also shall provide the necessary context and background for an appreciation of these problems, along with some insights that should prove helpful in their resolution. As computational topology is still an emerging discipline and is largely unknown to many in the computer aided geometric design, computer science, and mathematics communities, we shall present a brief outline of the elements of computational differential topology in Subsection 5.2, a description of the problem of identifying and classifying those objects in a category associated with computational differential topology in Subsection 5.3 and algebraic duals of previous problems now expressed as issues in isomorphism type in Subsection 5.4. In particular, we treat in Subsection 5.4 those that possess a complete set of effectively (algorithmically) computable topological invariants, i.e., those geometric objects that have sufficiently many algorithmically computable invariants to completely determine their isomorphism classes in an appropriate topological category.

5.2. Elements of computational differential topology. One unmistakable sign of a mature mathematical or scientific subdiscipline is the establishment and general acceptance of well defined mathematical categories that characterize and circumscribe the field. Such categories have yet to be universally embraced in the computational topology community, so we shall first describe the categories in which we shall work in order to frame the problems to be posed in this paper. This Subsection has its own Subsections 5.2.1 discussing the categorical structures needed; 5.2.2 raising the issue of shape equivalence within these categories and 5.2.3 emphasizing the interplay between topology and algorithms.

5.2.1. *Categories.* The sets of interest in computational topology are geometric objects in an Euclidean space, usually having certain differentiability properties, but they need not and should not be restricted to manifolds. Examples such as the locus of $x^2 + y^2 - z^2 = 0$ in $\mathbb{R}^3$ and geometric objects with self-intersections show that we need to include varieties. One approach to describing the objects in an appropriate category is to introduce special varieties (s-varieties) having the property that there are at most finitely many local regular (topological manifold) branches at each of the singular points [**26**, **27**]. However, a more efficient way to describe the objects in the computational topology categories is to employ *Whitney regular stratifications* [**19**, **27**, **28**, **41**, **84**, **123**, **168**, **178**]. First we fix an Euclidean space $\mathbb{R}^N$ to serve as the ambient space for the geometric objects and an order of differentiability k $(0 \leq k \leq \infty)$.

Definition 5.1. For a given Euclidean space $\mathbb{R}^N$ and order of differentiability $0 \leq k \leq \infty$, a computational differential topology object, denoted as cdt_N^k, is a subset V of $\mathbb{R}^N$ that can be represented in the form

$$V = M_1 \cup M_2 \cup \cdots \cup M_s, \tag{5.1}$$

where the collection $\mathfrak{S} := \{M_i : 1 \leq i \leq s\}$ is a Whitney regular stratification of V. This stratification is comprised of a finite disjoint set of strata M_i, which are open or closed C^k submanifolds of $\mathbb{R}^N$, called the strata of the stratification, and the strata have dimensions that can range from 0 (points) to N (open solid regions). The dimension of V in cdt_N^k is defined as $\dim V := \max\{\dim M_i : M_i \in \mathfrak{S}\}$.

Note that a cone is in cdt_3^∞, as is a closed cube. Since we shall be concentrating in this paper mainly on geometric objects that have some differential structure, most of our attention shall be directed to cases where $k \geq 1$.

We now have suitable objects for our categories, so it naturally remains to define the appropriate morphisms. It is clear that the more usual choice leading to homeomorphic or diffeomorphic equivalence will simply not do. For example, a circle S^1 and a smooth trefoil T knot embedded in $\mathbb{R}^3$ are obviously C^∞-diffeomorphic, 1-dimensional submanifolds, but can certainly not be viewed as equivalent in any reasonable computational topology sense since they are not equivalent as embeddings in the ambient space $\mathbb{R}^3$. In particular, the knot group for the circle is $\pi\left(\mathbb{R}^3 \setminus S^1\right) = \mathbb{Z}$, while the knot group for the trefoil knot $\pi\left(\mathbb{R}^3 \setminus T\right)$ is the group with two generators α and β and one relation, $\alpha\beta\alpha = \beta\alpha\beta$, where $\pi(X)$ denotes the fundamental group of a topological space X. Therefore, morphisms must be equivalent in some sense as embeddings in the ambient space, as well as having certain differentiability properties. The next definition attends to these requirements.

Definition 5.2. A morphism between two objects V and W in cdt_N^k is an embedding (in the topological sense) $\Phi\colon \mathbb{R}^N \to \mathbb{R}^N$ satisfying the following properties:

(i) $\Phi(V) \subseteq W$.

(ii) The restriction $\Phi_{|V}$ of Φ to V is of class C^k.

With this we have the last piece necessary for the definition of our computational topology categories for objects embedded in an Euclidean space $\mathbb{R}^N$:

Definition 5.3. For a given Euclidean space $\mathbb{R}^N$ and order of differentiability $0 \le k \le \infty$, the computational differential topology category, denoted as CDT_N^k, is comprised of all the objects in cdt_N^k as in Definition 5.1, and the morphisms as in Definition 5.2, with the usual composition of morphisms.

Observe that according to this definition, two objects V and W in CDT_N^k are isomorphic, denoted as $V \approx_N^k W$, iff there is a homeomorphism $\Phi\colon \mathbb{R}^N \to \mathbb{R}^N$ such that $\Phi(V) = W$, and the restrictions of Φ to V and its inverse Φ^{-1} to W are both of class C^k. We remark that in most cases when the ambient space and differentiability class are fixed, we simplify the above notation by omitting the subscript and superscript in the isomorphism notation, so that we shall simply write $V \approx W$. In the sequel we shall, for convenience, indulge in the harmless abuse of notation of referring to both objects and morphisms as being members of the category CDT_N^k rather than distinguishing between the set of objects and set of morphisms comprising this category.

Isomorphism in the categories CDT_N^k (which is sometimes referred to as embedding equivalence [**149**]) is obviously more restrictive than homeomorphic equivalence in the standard topological category TOP. More specifically, in addition to the usual homeomorphism type invariants such as homotopy, cohomotopy, homology, and cohomology that one needs to consider for equivalence in TOP, one needs also to verify the invariance of additional quantities such as linking numbers to verify equivalence in the computational differential topology categories. For future reference, we denote isomorphism in the TOP category as

$$V \overset{\mathrm{h}}{\approx} W. \tag{5.2}$$

Topological equivalence by isotopy [**13, 14, 16, 23, 139, 151**] is stronger than the isomorphic equivalence given in Equation 5.2, as has already been introduced in Definition 4.1. We remark here that for the case of smooth knotted and unknotted circles in $\mathbb{R}^3$, standard knot equivalence, ambient isotopy, and isomorphism in CDT_3^0 are all equivalent to one another [**91, 92, 122**].

One of the basic goals in computational topology is to create computer generated procedures for obtaining representations of objects having the same shape—at least in some acceptable approximate sense—as a given geometric object. This obviously begs the question of what is meant by shape, a question that we address in the next subsection.

5.2.2. *Shape of geometric objects.* What does it mean to say that two objects, V and W in CDT_N^k have the same shape? Naturally, to have the same shape, V and W ought to at least be isomorphic in the computational topology category, but intuition certainly requires more. A suitable definition is the following:

Definition 5.4. The objects V and W in CDT_N^k have the same shape if there exists an isomorphism $\varphi\colon V \to W$ that is a scaled C^k-isometry in the following sense: There exists a constant $c > 0$ such that $c^{-1}\varphi$ is an isometry. More particularly, recall that for φ to be an isomorphism in CDT_N^k it must be extendable to a homeomorphism $\Phi\colon \mathbb{R}^N \to \mathbb{R}^N$. Accordingly the definition requires that the restriction of Φ to V (which is φ) must be a C^k map such there exist a $c > 0$ and an isometric C^k-embedding $\psi\colon V \to \mathbb{R}^N$ (in the metric induced on V by the Euclidean metric on $\mathbb{R}^N$) with $\Phi(x) = c\psi(x)$ for all $x \in V$. We denote this property of having the same shape by $V \equiv_N^k W$, and omit the subscript and superscript for simplicity whenever the context is clear.

Computational representations of geometric objects—no matter what type of format is used to describe the rendered object—usually involve some approximation error, which necessitates the use of the following definition, or something of the same sort, for computational topology applications.

Definition 5.5. Given $\epsilon > 0$, we say that V and W in CDT_N^k have the same shape $(\mathrm{mod}\,\epsilon)$ if they are isomorphic in this category via $\varphi\colon V \to W$, and there are a positive number c and an isometric C^k-embedding $\psi\colon V \to \mathbb{R}^N$ such that φ is ϵ-close to $c\psi$ in the Whitney C^k-topology, which essentially means that derivatives of all orders less than or equal to k of φ and $c\psi$ differ by less than ϵ (in the appropriate operator norm) over all of V [**19, 84, 134**]. Having the same shape $(\mathrm{mod}\ \epsilon)$ is denoted as $V \equiv_N^k W\ (\mathrm{mod}\ \epsilon)$, where as usual we shall suppress the subscript and superscript when the context is clear.

We now are in possession of all the notation that we need to formulate the efficient approximation problem of computational differential topology, which we attend to in the succeeding subsection.

5.2.3. *The efficient approximation problem.* With the notation, it is simple to explain—at least in general terms—the nature of the essential problem confronting computational topologists. It begins with a given prototype object V_0 in CDT_N^k, which must be represented by computer generated means that are based upon an algorithm $\mathcal{A}$. The word 'given' here is somewhat of a misnomer that requires very broad interpretation: The prototype object may be defined exactly in terms of equations, or it may be a completely developed model of a geometric object, or represented by data sampled from an existing physical object such as a statue or building, or—in the worst case scenario—may be only partially and imprecisely known simply in terms of representative data, such as point-clouds, sampled according to some scheme [**161**].

An algorithm for representing and analyzing a geometric object with computational topology constraints should include an algorithmic subroutine for verifying that the computed object has the same isomorphism type as the given object—assuming that this much is known about the object to be represented. If we have only incomplete topological knowledge of the prototype object, an algorithm designed to produce computer generated representations, say at various levels of accuracy, should at least be capable of verifying that the isomorphism type remains constant as the accuracy is refined. When such an algorithm is available, such a constant 'limit' may serve as a good educated guess of the actual isomorphism type of the partially known prototype object. The following notion is useful in the investigation of such questions.

Definition 5.6. Let V_0 be a given object in CDT_N^k and let V be another such object. Then the isomorphism type of V is said to be V_0-*decidable* if there exist an algorithm $\mathcal{A}$ to determine if $V \approx V_0$. Such an algorithm is called a (V_0, V)-decider.

This brings us to the overarching focal point of any complete investigation of a problem in computational differential topology, which addresses both the mathematical and computer science aspects involved.

Efficient Approximation for Computational Differential Topology. *Given a prototype object V_0 in the category CDT_N^k, construct an algorithm $\mathcal{A}$ to be used for obtaining a computer generated representation V (in CDT_N^k) of V_0, which has the following properties:* (a) *For each sufficiently small positive ϵ, the algorithm generates a representation $V(\epsilon)$ of V_0 and includes a subalgorithm that is a $(V_0, V(\epsilon))$-decider;* (b) $V(\epsilon) \equiv V_0 \pmod{\epsilon}$ *for all such ϵ; and* (c) *the algorithm is optimally efficient to the degree that the computational complexity of $\mathcal{A}$, denoted as $CC(\mathcal{A})$, is minimal in some reasonable sense.*

It should be noted that, although not specifically included in the above definition of the efficient approximation problem, ease of implementation with regard to producing user-friendly software based on the algorithm is also an important consideration, especially when it comes to applications.

In general, a complete solution of the efficient approximation problem as stated may be extremely difficult—or even impossible—to achieve, so simplified versions of this problem, such as those we describe in the sequel, are highly desirable and often vigorously pursued. We note that if this efficient approximation problem is viewed from a computational geometry rather than a computational topology viewpoint, one should choose the differentiability class k to be greater or equal to two, so that the representations produced are acceptable in terms of differential geometry where second derivatives manifested in curvature tensors (or differential forms) are essential elements in determining good approximations.

5.3. The identification and classification problem. The reader is sure to have observed that the efficient approximation problem as presented in the preceding sections is somewhat lacking in rigor. Moreover, as Edelsbrunner pointed out when the version above was unveiled recently, it also is deficient in scope—especially as regards the wide range of possibilities in knowledge of the prototype object, means of obtaining data from the object for the algorithm, and methods available for rendering the computational representations. These observations constitute the core of the first few open problems that we pose here.

5.3.1. *Formulation of the identification and classification problem.* In order to pose this identification and classification problem with more precision, and to introduce sufficient rigor into supporting definitions and concepts so as to articulate which problems remain open, we shall first present a more detailed version than outlined in the preceding section. To begin, we develop more precise notation concerning the computational procedures embodied in the algorithm $\mathcal{A}$ devised to produce an approximate representation $V(\epsilon)$ of the prototype geometric object V_0 in CDT_N^k for a given error bound ϵ. We emphasize here that the error bound is on the geometry, not the topology, as invariance of the isomorphism type is an essential requirement for the algorithm. The input data from V_0, which we denote as $D(V_0)$, may assume any one of several possible forms such as the vertex points and connection relations for the elements of a triangulation of the prototype object, a global functional representation or a set of local functional expressions arising from exact mathematical models, or an approximate nonuniform rational B-spline (NURBS) decomposition of V_0, or points forming a point-cloud sampled in a manner designed to provide a good approximation of the given object, which is often the case when V_0 is not completely known or specified.

One can already see here that there is a problem in formulating an adequate characterization of the space $\mathfrak{D}$ in which the data obtained from the prototype object resides. A good definition of this data space is required so that we can consider D as a function from (the object set of) CDT_N^k to $\mathfrak{D}$, which can be expressed as $D\colon \mathsf{CDT}_N^k \to \mathfrak{D}$. Of course, the tolerance (geometric accuracy) ϵ must also be counted as an argument of the algorithm. With the notation developed, we may now view the algorithm as a recursive map of the form

$$\mathcal{A}\colon D(\mathsf{CDT}_N^k) \times \mathbb{R}_+ \to \mathsf{CDT}_N^k \qquad (D(V_0),\epsilon) \longmapsto V(\epsilon)$$

where $\mathbb{R}_+$ is the set of positive real numbers, and $V(\epsilon)$ is a graphical rendering of a (geometric) approximation of V_0, or more precisely, an algorithm for producing a computer generated approximate representation of the prototype object. We now have a more rigorous foundation for describing the identification and classification problem.

The Identification and Classification Problem in CDT_N^k. *Devise an algorithm $\mathcal{A} = \mathcal{A}(D(V_0),\epsilon)$ that*

(i) *is defined for all sufficiently small positive* ϵ,
(ii) *is defined for a suitably ample domain of prototype objects* V_0 *in* CDT_N^k,
(iii) *produces an output* $V(\epsilon) \equiv_N^k V_0 \pmod{\epsilon}$ *for all* ϵ *for which it is defined,*
(iv) *has minimal computational complexity* $CC(\mathcal{A})$ *in some sense.*

The above description of the identification and classification problem, although more precise than that which was presented in preceding section, is clearly still beset with deficiencies in several respects, two of which are embodied in the following posed problems.

Question 5.1. *Modify the description of the Identification and Classification Problem in* CDT_N^k *so that it more rigorously and completely encompasses the wide range of methods that can be used, and is better able to express the degree to which the prototype object is known.*

Question 5.2. *Find a way of better expressing the type of representation approach that is used to produce the output object* $V(\epsilon)$ *in the statement of the identification and classification problem in* CDT_N^k.

Note that in a case where the isomorphism class of the prototype object V_0 in CDT_N^k is mostly or partially unknown, it will be necessary to revise the requirement (iii) to something like

(iii)′ The outputs $V(\epsilon_1)$ and $V(\epsilon_2)$ with $0 < \epsilon_1, \epsilon_2 < \epsilon$ satisfy $V(\epsilon_1) \equiv_N^k V(\epsilon_2) \pmod{\epsilon}$ for all sufficiently small ϵ.

This suggests a possible notion of persistence of isomorphism type analogous to the basic ideas used to formulate persistent homology [**68**, **72**, **184**, **185**].

Question 5.3. *Reformulate and expand (iii) in the Identification and Classification Problem in* CDT_N^k *to include those cases where one only has incomplete knowledge of the isomorphism type of the prototype object—perhaps along the lines of persistence of isomorphism type for sufficiently small tolerances.*

Another inadequacy of our exposition of the identification and classification problem is manifested in the imprecision of the minimality statement for computational efficiency, which naturally leads to the following question.

Question 5.4. *Revise the definition of the Identification and Classification Problem in* CDT_N^k *so that it includes a more precise description of the computational cost that is consistent with the most important computational concerns arising in a broad spectrum of applications of computational topology.*

Resolving this minimality definition problem is bound to be quite challenging, partly owing to the extensive array of minimality criteria available for applications, but more likely to stem from the difficulty of actually proving minimality for an algorithm in most reasonable, nontrivial senses. As algorithms developed to render approximations of geometric objects possessing only a fair degree of complexity tend to be decidedly nontrivial, verifying minimality of computational complexity tends to be rather daunting.

In addition to the properties of the algorithm $\mathcal{A}$ delineated in the identification and classification problem, it is desirable for it to continue to generate representations satisfying property (iii) or (iii)′ when the data $D(V_0)$ and tolerance ϵ vary slightly in an appropriate sense. When the algorithm has this additional feature, it is natural to say that it is *stable*, and this leads to another problem.

Question 5.5. *Devise a rigorous definition of stability of computational topology algorithms, and develop methods for determining whether or not such an algorithm is stable.*

It should be clear to anyone with experience in solving problems in computational topology that it might help to ameliorate the inherent ambiguity of the identification and classification problem if some of the techniques for determining isomorphism type (at least approximately) were included in the above description. Most of the methods currently employed to analyze isomorphism type involve the algorithmic computation, where feasible, of isomorphism invariants such as characteristic classes, homology groups, and cohomology rings, along with approaches based upon tubular type neighborhoods, Morse theory, Morse–Floer theory, singularity/stratification theory, and obstruction theory [**1**, **3**, **100**, **120**, **126**, **133**, **137**, **140**, **151**, **158**, **184**, **185**]. However, there also is a fairly recent spate of articles employing innovative methods from general topology, such as [**73**, **81**, **83**, **93**, **105**, **107**, **112**, **121**], that appear to be applicable to the (complete or partial) computation of isomorphism type as well.

5.3.2. *Simplification of the identification and classification problem.* Owing to the impressive advances in the realm of computational geometry over the last several decades leading to the creation of several algorithms for generating very (geometrically) accurate representations of geometric objects, and the development of new tubular neighborhood based theorems, it now appears possible to recast the identification and classification problem in the following far more tractable simplified form.

Simplified Identification and Classification Problem. *Devise an algorithm* $\mathcal{A} = \mathcal{A}(D(V_0), \epsilon)$ *that*

(i) *is defined for all sufficiently small positive* ϵ,
(ii) *is defined for a suitably ample domain of prototype objects* V_0 *in* CDT_N^k,
(iii) *produces an output* $V(\epsilon)$ *that is* ϵ*-close to and has the same homeomorphism type as* V_0 *for all* ϵ *for which it is defined,*
(iv) *has minimal computational complexity* $CC(\mathcal{A})$ *in some sense.*

The basis for the above simplification is what has been called the *self-intersection precedes knotting principle* (SIPKP), which can be explained in the following way for compact objects V in CDT_N^k. Owing to the compactness, all of the strata in the regular stratification (Equation 5.1) of V_0 have compact closure. Each closed stratum has an arbitrarily thin, relatively compact tubular neighborhood, and the open strata also can be shown to have arbitrarily thin, relatively compact tubular-like neighborhoods. A tubular-like neighborhood for an open stratum has the form of a standard tubular neighborhood joined to open neighborhoods of the ends of the manifold, very much like the construction for manifolds-with-boundary employing boundary collars in [**4**, **5**]. Taking the union of these tubular and tubular-like neighborhoods for all the strata, we have an arbitrarily thin tubular-like neighborhood U. Then we can use an existing computational geometry algorithm to generate an approximation $V(\epsilon)$ contained in U. When the distance between images of a homeomorphism differ by no more than ϵ, some sufficient conditions are known to extend these homeomorphisms to ambient isotopies [**103**]. This known proof avoids the self-intersections mentioned, leading to the following open problem as to how far this technique can be extended.

Question 5.6. *Prove the SIPKP, or provide a counterexample.* 1117?

It goes almost without saying that there are obvious versions of Problems 5.1–5.6 for the simplified identification and classification problem, and these too are open problems of fundamental importance in computational topology.

5.4. Decidability of isomorphism type. The discussion of the identification and classification problem and a simplified version of it in the previous section raises the question of just what types of objects in CDT_N^k are amenable to algorithmic determination of their isomorphism types. We shall focus on this question in this section (assuming some familiarity with the basics of differential topology and such related fields as singularity and stratification theory as can be found in [**19**, **27**, **52**, **59**, **84**, **90**, **94**, **122**, **123**, **125**, **134**, **149**, **163**, **168**, **174**, **178**]), and will find it convenient to employ the following definition.

Definition 5.7. Let C be an arbitrary category, and suppose that X is an object in this category. If there is an algorithm for determining the isomorphism class of X, we say that X is C-*decidable*.

Bearing this in mind, we shall concentrate on identifying the properties that render a geometric object (embedded in an Euclidean space) decidable in the relevant categories for computational differential topology. To establish the overall theme of this section, we shall first summarize everything in one overarching problem, and then proceed to break this up into more manageable pieces. This unifying problem may be phrased in the following manner.

Unifying Topological Decidability Problem. *Determine all compact objects in* CDT_N^k *that are*

(a) *TOP-decidable,*

(b) *CDT_N^k-decidable,*

and determine the algorithm of minimum computational complexity capable of deciding the isomorphism type in each case.

We shall begin with compact submanifolds and submanifolds-with-boundary in CDT_N^k, with $1 \leq k$, as they are typically easier to classify in terms of the categories of interest here, namely CDT_N^k and TOP.

5.4.1. *Decidability of compact submanifolds.* In our discussion, we shall proceed in the order of increasing dimension N of the ambient Euclidean space. If $N = 1$, any compact submanifold, denoted as M, is *closed* (because by definition it has an empty boundary, i.e., $\partial M = \varnothing$). Hence, M is particularly simple, a finite set of points in the zero-dimensional case. There are no closed compact submanifolds of $\mathbb{R}^1$ of dimension one (or equivalently, of codimension zero, which is the dimension of the ambient space minus the dimension of submanifold). Even if we drop the compactness assumption, decidability is a simple matter owing to the fact that every connected, open, C^1 submanifold of $\mathbb{R}^1$ of codimension zero is an open interval. The compact submanifolds-with-boundary of $\mathbb{R}^1$ are also easy to classify algorithmically in CDT_N^k, for they must be one-dimensional and comprised of finitely many disjoint closed intervals. We note from these simple examples that we may assume that the submanifolds are connected, for if not, we can simply analyze the components one-by-one.

In $\mathbb{R}^2$, the situation is also essentially trivial, with the decidability of the homeomorphism type or isomorphism type in CDT_N^k being a simple matter indeed. For example, it follows from the Jordan curve theorem and other basic principles, that every connected, closed submanifold M of codimension-1 must be equivalent to the circle S^1 in either the category TOP or CDT_N^k. Moreover this can be determined by a single effectively computable invariant, which is the condition $H_1(M, \mathbb{Z}) = \mathbb{Z}$ for the first integral homology group, or equivalently described in terms of the Euler–Poincaré characteristic as

$$\chi(M) = \sigma_0 - \sigma_1 = \operatorname{rank} H_0(M, \mathbb{Z}) - \operatorname{rank} H_1(M, \mathbb{Z}) = 0,$$

where σ_j stands for the number of j-dimensional simplices in a triangulation, and the rank is defined in the usual way [**122**, **135**, **163**, **184**]. Note also that if we choose an algorithm $\mathcal{A}$ based on computation of χ, we readily find that $CC(\mathcal{A}) = O(n_s)$, where n_s is the number of top (=1)-dimensional simplices in a triangulation of M, and one cannot do much better than this with respect to computational efficiency. As a matter of fact, it follows readily that both the complete and simplified identification and classification problems are completely solved for compact submanifolds of $\mathbb{R}^2$, including the establishment of computational minimality for the algorithm assuming that the prototype submanifold is completely simplicially defined in terms of triangulations.

These simple results already provide an indication of the usefulness of algebraic topology in dealing with the decidability problem. In this vein, we include the following result for future reference. It can be proved using the stratification (Equation 5.1), the C^1 triangulation theorems of Munkres [**134**], and some basic results on the effective (algorithmic) computability of homology and cohomology for finite simplicial complexes (see [**100**, **135**, **184**]).

Theorem 5.1. *Let V be a compact object in CDT_N^k $(k \geq 1)$. Then V has a finite C^1 triangulation, and the homology $H_*(V, F)$, cohomology $H^*(V, F)$, and all of the applicable characteristic classes such as the Chern, Euler, Stiefel–Whitney, and*

Pontryagin classes (possibly just for the strata) for V are effectively computable, where the coefficient ring F is the integers $\mathbb{Z}$ or the integers mod 2.

It is in $\mathbb{R}^3$ that both the isomorphism classification and the decidability problem first assume nontrivial proportions.

Compact manifolds in Euclidean 3-space: Let M be a compact, connected submanifold (possibly with boundary) in CDT^k_N with $k \geq 1$. When $\dim M = 0$, both the classification and decidability problem are trivial in both TOP and CDT^k_N. For $\dim M = 1$, things begin to get very interesting and rather difficult. If M is closed, it must be diffeomorphic to a circle, but it can be embedded in $\mathbb{R}^3$ as a very complicated knot. Decidability in TOP is virtually the same as in the 2-space case described above, so the homeomorphism type can be algorithmically determined in linear time. In CDT^k_N, the isomorphism classes correspond to knot types. It follows from [**91**, **92**] that M is CDT^k_N-decidable, but may be NP-complete. This contrast is a very effective demonstration of how much more difficult it can be to solve the complete identification and classification problem than the simplified identification and classification problem.

An embedded closed surface M, must be orientable, and an easy solution of the decidability problem follows directly from the simple and elegant classical result [**122**, **163**] that the homeomorphism and diffeomorphism types of such a submanifold are completely determined by the Euler–Poincaré characteristic

$$\chi(M) = \sigma_0 - \sigma_1 + \sigma_2 = \operatorname{rank} H_0(M, \mathbb{Z}) - \operatorname{rank} H_1(M, \mathbb{Z}) + \operatorname{rank} H_2(M, \mathbb{Z}).$$

Accordingly the problem for TOP-decidability is solvable in linear time. Again, there is a huge difference in the degree of difficulty of the TOP- and CDT^k_N-decidability problems, as one can see by considering the thin toral surface of a smoothly thickened knotted curve. Once again, M is CDT^k_N-decidable—although there seems to be no proof of this in the literature—but the computational complexity of any associated algorithm appears to be very high, and may be NP-complete.

The homeomorphism or diffeomorphism types of a compact submanifold-with-boundary M of codimension-2 in $\mathbb{R}^3$—which may be nonorientable as in the case of a Möbius strip—is completely determined by $\chi(M)$, the orientability, and the number of boundary components [**122**]. Therefore, M is TOP-decidable in linear time. On the other hand, if M is CDT^k_3-decidable, then the computational complexity of the problem is bound to be of the order of knot decidability, but otherwise appears to be unknown.

Question 5.7. *Prove*[5] *that every compact, connected, C^1-submanifold of $\mathbb{R}^3$ of* 1118?
dimension less than or equal to 2 is CDT^k_3-decidable and obtain estimates for the computational complexity of any relevant algorithms that can be used to determine isomorphism type.

A compact, connected, 3-dimensional, C^1-submanifold M of $\mathbb{R}^3$ must have a nonempty boundary ∂M. It is easy to see that if ∂M is connected, it completely

[5]All problems of providing a proof include implicitly the option of finding a counterexample.

determines M; hence, M is decidable in both TOP and CDT_3^k. An analog of this ought to be true in the case when ∂M is not connected, but this still appears to be an open problem.

1119? **Question 5.8.** *Prove that every compact, connected, C^1-submanifold of $\mathbb{R}^3$ of dimension 3 is both TOP- and CDT_3^k-decidable (or provide a counterexample), and obtain estimates for the computational complexity of any relevant algorithms that can be used to determine isomorphism type in these categories.*

Compact manifolds in Euclidean 4-space: Of course there is a far more diverse and interesting range of compact submanifolds of $\mathbb{R}^4$ than $\mathbb{R}^3$, but we shall confine our attention to just a select few of the possible types of C^1-submanifolds of dimension two or higher. Moreover, in this and the higher dimensional cases in the sequel, we shall concentrate mainly on TOP-decidability, which is associated with the simplified identification and classification problem. We observe that all closed surfaces, or compact surfaces-with-boundary, including the nonorientable ones such the Klein bottle and the projective plane, can be embedded in $\mathbb{R}^4$.

We showed above how the decidability problem for oriented compact surfaces can be easily and very efficiently solved. This is also true for the nonorientable surfaces, all of which can be realized as two-dimensional, closed submanifolds and compact submanifolds-with-boundary of $\mathbb{R}^4$. For these cases the TOP and CDT_4^k isomorphism types also are completely determined by the orientability, or lack thereof, the Euler–Poincaré characteristic, and the number of boundary components. Moreover, the isomorphism type can be computed in linear time.

To summarize compact surfaces with regard to the decidability problem: they represent the lowest dimensional nontrivial submanifolds for which the problem becomes interesting, yet is easily solvable by simple classical means expressed, modulo orientability and possible boundary components, in terms of a single invariant that is computable in linear time. As such, they are excellent illustrative examples of some of the simplest solutions that provide direction for more general cases.

The 3-sphere S^3 is the simplest closed, connected, three-dimensional, submanifold of $\mathbb{R}^4$. It has been much in the mathematical news of late owing to the excitement created by the work of Perelman [**98**] on the famous and long-standing Poincaré Conjecture, which states that a connected, simply-connected (i.e., $\pi(M) = 0$) three-dimensional manifold M having the homology of a 3-sphere must, in fact, be homeomorphic with S^3 [**132**]. Perelman's work, which relies heavily upon Hamilton's Ricci flow methods, is still being studied by the experts, and at last look, the jury was still out. The recent paper by Cao and Zhu [**35**] sheds new light on this question. However, the opinions expressed so far are quite positive, and it looks very much like Perelman has finally affirmatively settled this amazingly difficult and influential conjecture. In the context of decidability questions, Perelman's work promises to have many important applications.

If Perelman is correct, this leads naturally to a very straightforward effective procedure for determining if a closed, three-dimensional, C^1-manifold M is a 3-sphere: First show that the fundamental group is trivial, which can be accomplished algorithmically by computing the edge-path group of a triangulation of M [**163**]. Using the same triangulation, it follows from Theorem 5.1 that the integral homology of M is effectively computable. Then if one computes that $H_0(M,\mathbb{Z}) = H_3(M,\mathbb{Z}) = \mathbb{Z}$, and $H_1(M,\mathbb{Z}) = H_2(M,\mathbb{Z}) = 0$, it follows that M is diffeomorphic, and *a fortiori* homeomorphic with S^3.

However, there already is an effective procedure, the Rubinstein–Thompson algorithm [**169**], for deciding within exponential time if a manifold is homeomorphic with S^3. This, of course, begs the question embodied in our next problem.

Question 5.9. *Develop an efficient algorithm based on the computation of the edge-path group and the integral homology as described above for deciding whether a closed manifold is homeomorphic with S^3. Then compare the computational complexity of this new algorithm with that of the Rubinstein–Thompson algorithm.* 1120?

Actually, Perelman's results claim to prove Thurston's Elliptization Conjecture for 3-manifolds (from which the Poincaré Conjecture follows immediately), which implies that all closed, connected, simply-connected, three-dimensional manifolds can be classified up to homeomorphism type. It appears that the elements of this classification theorem can be computed algorithmically, although this promises to be a daunting task owing to the techniques employed, not least of which are those generated by Hamilton's Ricci flow approach.

Question 5.10. *Within $\mathbb{R}^4$, prove that every closed, connected, simply-connected, three-dimensional C^1-submanifold is TOP-decidable and find estimates for the computational complexity of any relevant algorithms for deciding the homeomorphism types.* 1121?

Compact submanifolds of higher-dimensional Euclidean spaces: It follows from the Whitney Embedding Theorem [**149**] that every closed, four-dimensional C^1-manifold M can be embedded in $\mathbb{R}^N$ with $N \geq 9$. Four-manifolds provide some of the most intriguing and elegant TOP-decidable examples available, and they also yield important insights into the limitations of topological decidability. It follows from the work of Freedman, Donaldson, and others (as in [**56**, **79**]) that all closed, simply-connected, orientable, four-dimensional, C^1-manifolds M can be classified up to homeomorphism type. As a corollary, one obtains a proof of the Generalized Poincaré Conjecture for 4-spheres; namely, every simply-connected, homology 4-sphere is homeomorphic with the 4-sphere S^4.

One of the most beautiful aspects of this classification theory is the particularly simple criteria for determining the homeomorphism type, which comes out of the following observations. Elementary algebraic topology, Poincaré duality and the universal coefficient theorem for homology imply that $H_0(M,\mathbb{Z}) = H_4(M,\mathbb{Z}) = \mathbb{Z}$, $H_1(M,\mathbb{Z}) = H_3(M,\mathbb{Z}) = 0$, and $H_2(M,\mathbb{Z})$ is a free abelian group. This leads one to at least predict the important role in classification of 4-manifolds played by the bilinear, unimodular *intersection form* $\omega\colon H_2(M,\mathbb{Z}) \times H_2(M,\mathbb{Z}) \to \mathbb{Z}$. The classification theorem essentially states that the closed, oriented, simply-connected, differentiable four-dimensional manifolds are completely classified by their intersection forms. Consequently, we readily infer from Theorem 5.1 that these manifolds are also TOP-decidable. However, this result has, as far as we know, not appeared in the literature.

1122? **Question 5.11.** *Within Euclidean space $\mathbb{R}^N$, prove that all closed, orientable, simply-connected, four-dimensional C^1-submanifolds are TOP-decidable and estimate the computational complexity of the classifying algorithms.*

So 4-manifolds can lead to what may be considered to be among the best of times when it comes to topological decidability, but they also show us the worst of times—undecidability. It can be shown using simple manifold surgery techniques that every finitely presented group G (possibly very far removed from the trivial group) can be realized as the fundamental group of a closed, four-dimensional C^∞-manifold. Using this fact, and certain undecidability results for the isomorphism problem for groups, Markov proved that there exist certain 4-manifolds that are not TOP-decidable [**122**, **163**]. There are limits to the topological decidability of manifolds after all, and one need not look higher than four dimensions to find them. Naturally, this leads to several open problems that we leave to the reader to pose.

As higher dimensions provide more room for the techniques of differential topology to perform their mathematical magic, it is not surprising that the Generalized Poincaré Conjecture and the classification of closed, simply-connected, differentiable manifolds were proven by Smale [**162**], Stallings [**164**], Zeeman [**182**], and others more than a decade before Freedman's remarkable work. The earlier breakthroughs of Smale, Stallings and Zeeman employed a variety of differential topological techniques such as Morse theory, cobordism theory, and obstruction theory, all of which appear to be accessible to algorithmic formulations for manifolds in CDT_N^k and so we leave this subsection by posing the following (formidable) open problem.

1123? **Question 5.12.** *Prove that every closed, simply-connected, n-dimensional manifold in CDT_N^k, where $k \geq 1$ and $n \geq 5$, is TOP-decidable and estimate the computational complexity of any relevant classifying algorithms. In particular, consider the case of simply-connected, homology n-spheres.*

5.4.2. *Decidability of compact nonmanifolds.* Each of the decidability problems delineated for compact submanifolds in CDT_N^k have analogs—which are even more challenging—for compact varieties V that are not submanifolds. Taking our cue from the triviality of the decidability problems for manifolds embedded in Euclidean spaces of dimensions less than or equal to three, and expecting Thom–Mather theory (see [**19**, **26**, **27**, **84**, **123**, **168**, **178**] to reduce much of the work to submanifold strata in Equation 5.1 for which our previous observations provide much insight into decidability, we pose the following.

1124? **Question 5.13.** *Prove that every connected, compact subvariety V in CDT_N^k with $k \geq 1$ is TOP-decidable. Find tight upper bounds for the computational complexity of the resulting algorithms.*

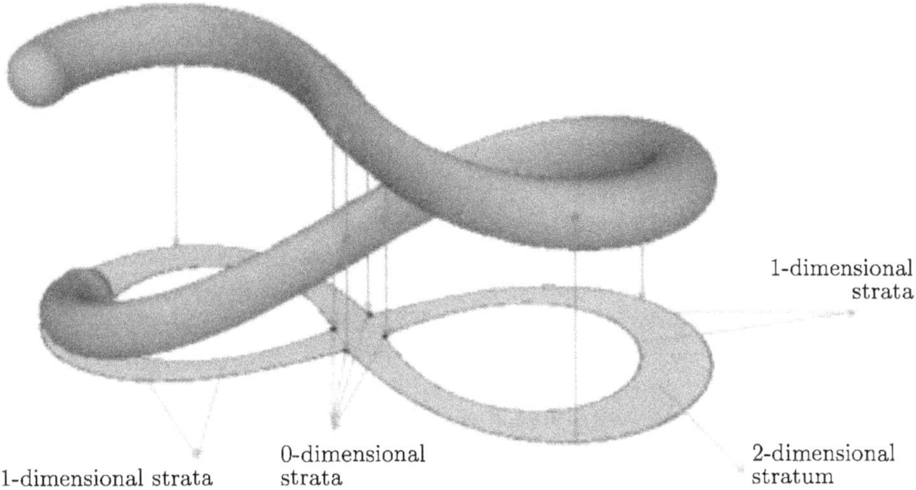

FIGURE 11. Manifold and projected sweep-like variety

It may be possible to show that the result in this theorem can be obtained in all higher dimensions as well, but clearly this would require some further restrictions on the homotopy type. Simple-connectedness might work, but this would severely restrict the types of nonmanifolds and many of the excluded ones would be apt to arise in a variety of applications. For, example, consider a thickened figure eight curve embedded in an Euclidean space of dimension four or higher. Another direction that one can pursue is to consider nonmanifolds obtained in a simple fashion from a compact manifold that is TOP-decidable. It is precisely this tack that we briefly follow in the remainder of this section, focusing upon compact sets that can be defined in terms of sweep-like operations.

Based upon extensive research on swept volumes [**1–3**, **24**, **25**, **27**, **28**, **175**], we are motivated make the following definition of a class of varieties that may yield to algorithmic classification.

Definition 5.8. A compact subvariety V of $\mathbb{R}^N$ is a sweep-like subvariety if there exists a compact submanifold M of $\mathbb{R}^{N+1} = \mathbb{R}^N \times \mathbb{R}$ such that $\Pi(M) = V$, where Π is the standard projection of $\mathbb{R}^N \times \mathbb{R}$ onto $\mathbb{R}^N = \mathbb{R}^N \times 0$, in which case V is said to be the projection of M.

A sweep-like variety is illustrated in Figure 11. Referring to this figure, we see that the self-intersection cell in the projection of the manifold has the appearance of an obstruction to lifting the variety to its regular preimage manifold of which it is the projection. This suggests that we can use a triangulation of the variety to identify this cell, in the manner of obstruction theory [**174**], in an algorithmic way. Thus, if the projecting manifold itself is topologically decidable, it appears that the same should be true of its image, which suggests that the following problem is solvable.

1125? **Question 5.14.** *Prove that every connected, compact, sweep-like subvariety V of $\mathbb{R}^N$ that is the projection of a compact, TOP-decidable, C^1-submanifold M of $\mathbb{R}^{N+1}$ is also TOP-decidable Find tight upper bounds for the computational complexity of any resulting algorithms.*

6. Computational topology resolution

A common practical concern is the development of algorithms that can produce appropriate topological representations upon models whose boundaries are formed by geometric intersections [**159**]. This is often known as 'topology resolution' and it affords many opportunities for additional research. The circumstances motivating this role for topological resolution have already been discussed in Section 3, particularly in Subsections 3.1 and 3.2.2.

One recent approach to managing the ill-formation of regular sets in computation [**146, 147**] utilizes tubular neighborhoods [**94**], but presents a very broad definition of a family of sets, each based upon an initial set. An overview is that each incomplete boundary is used to develop a new family of candidate sets by building offsets of each boundary element. New Boolean operations are then defined upon this family of sets. One of the authors conjectures that there is a relationship to the Čech topology.

This work provides a point-set topological characterization for a family of sets such that each member closely approximates the original set according to a precise criterion, where it is clear that the family has some similarities to sets defined via interval arithmetic. The methods presented are appealing and will work for simple cases. However, as the geometry becomes more complex it remains of interest to understand a general approach to formulate these tubular neighborhoods, along with guarantees upon the properties of the family of sets generated and operators used within that family. In order to obtain such a family from a specific instantiated boundary model, it becomes essential to understand which conditions must be satisfied by the approximants, where an argument is given for homotopy equivalence [**157**].

1126? **Question 6.1.** *Is there a characterization of those tubular neighborhoods which can be used to define useful families of regular closed sets as alternatives to ill-formed computational representations?*

1127? **Question 6.2.** *Does the construction provide some meaningful relation to the Čech topology?*

6.1. Integration with numerical analysis. Another approach relies upon more classical techniques from numerical analysis, specifically the Whitney Extension Theorem [**177**], as captured in a recent doctoral thesis of M. Zidini [**183**] and several related preprints [**17, 18, 167**].

The strategy presented is to take the ill-formed geometry and use the Whitney Extension Theorem to extrapolate the imperfectly fitting boundary elements until a satisfactory manifold boundary is created. The emphasis is to build a theoretical model, not necessarily one that would be instantiated in any specific computational

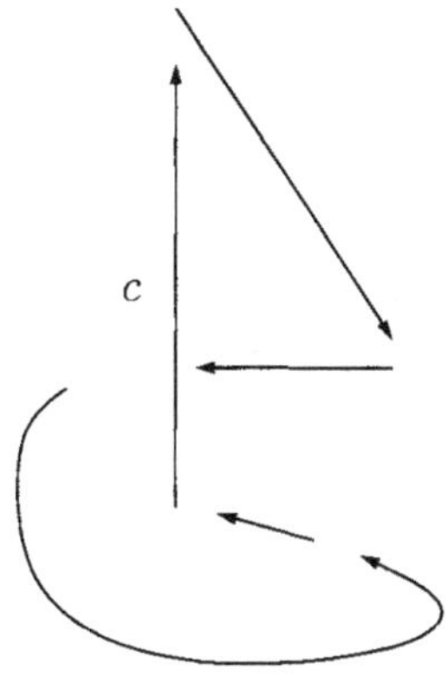

FIGURE 12. Gap analysis for Whitney extensions

representation. The intent would be to use this idealized model as a basis for developing rigorous error bounds as to how far any specific instantiation differed from this ideal. Some metrics are proposed for those measurements. Any two surface patches within the object boundary are said to be *adjacent* if they meet in a common boundary. Extensions of adjacent patches are proposed to ensure that they meet in a well-formed shared boundary, as a means to compensate for the numerical intersector errors previously discussed between surface patches.

The thesis makes the explicit assumption that "non-adjacent perturbed patches are disjoint." While this is standard, its use within computation raises the more subtle issue of the magnitude of the separation between these non-adjacent patches and the separation between adjacent patches.

Consider a finite set of surface patches whose union forms an object boundary, where the boundary curves for these surface patches have been created via the Grandine–Klein intersector [**86**] with an error preprocessor [**133**], so that the 'gaps' between adjacent surfaces were guaranteed to be no greater than λ in model space. One would hope that λ was chosen judiciously. Further, let δ denote the minimum distance between any two non-adjacent pathces. Generally, one would hope that $\lambda \ll \delta$. But, suppose, to the contrary, that $\lambda \gg \delta$.

To create a well-formed model from these surfaces, it would be appropriate to use the proposed Whitney extensions that would have perturbations on the order of λ, *but* perturbations of that magnitude could possibly introduce unwanted intersections between non-adjacent entities, as described below relative to the example of Figure 12. (Figure 12 uses curves for simplicity of exposition, but it should be clear that the example could easily be generalized to represent surfaces.[6] The curves should join to form a closed loop. This ideal connectivity is indicated by each arrowhead pointing to the end point of next segment.)

[6]For instance, each of the curves could serve as the spine of a swept surface having a generating curve of a circle of fixed radius.

For Figure 12, curve c lies in a plane perpendicular to the page, considered to be the plane $z = 0$. The image shown of c as a vertical line segment is merely its planar projection into the plane $z = 0$. The remaining curves shown all lie in the plane $z = 0$.

The other salient aspects of Figure 12 are summarized, as follows:

- The minimum distance δ is between c and the horizontal segment. This value of δ occurs at an interior point of c and at the left end point of the horizontal segment.
- The 'gap' between the left end point of the horizontal segment and its following free-form curve is the maximal value of λ.

Under these circumstances, the Whitney extensions to join properly the vertical segment and the free-form curve could create unintended intersections with c. If these gaps were the result of a construction process, such as a Boolean operation relying upon a numerical surface intersector, then a reasonable response might be to re-execute the procedures that generated the model, with tighter tolerances upon the numerical intersector so that one would have $\lambda \ll \delta$.

1128? **Question 6.3.** *Is it possible to provide practical criteria for the choice of λ, the separation distance between non-adjacent patches relative to δ, and the separation distance between adjacent patches?*

1129? **Question 6.4.** *If the errors resulting in these models being ill-formed as regular closed sets arose from some geometric construction process, such as a Boolean operation relying upon a numerical surface intersector, is a reasonable response to re-execute the procedures that generated the model, with tighter tolerances upon the numerical intersector so that one would have $\lambda \ll \delta$?*

1130? **Question 6.5.** *Are there implications that geometry should move from representation by specific instantiations into models that are more descriptive?*

1131? **Question 6.6.** *Might appropriate topological abstractions be more helpful than specific geometric coordinate based information?*

A very recent manuscript [**157**] argues that interpreting the inconsistencies between the geometric data and its connective information should rely upon a homotopy equivalence between the represented geometry and the intended exact set. The homotopy equivalent geometric sets are described as lying within the same tolerance zone. Additionally, graph theoretic and cell complex techniques are used to express and understand additional constraints that should be imposed upon these homotopies. In particular, it is proposed that these tolerance zones must be contractible for all cells that are homeomorphic to finite-dimensional Euclidean balls. Some further relationships are proposed to describe these homotopies in terms of the nerve of a collection of closed sets.

Question 6.7. *If homotopy equivalence is considered as a necessary condition for tolerant representations of geometry, what further conditions result in sufficiency?*

Question 6.8. *Can a practical algortihm be created to implement this theory regarding the homotopy equivalence between geometric representation and intent?*

6.2. The role of exact arithmetic. The study of 'exact arithmetic' arose from the computational geometry community [**181**] in recognition that many geometric predicates were critical to evaluation along boundaries. The question of being 'on' a boundary was equated to resolving whether numeric expressions were exactly equal to zero. In some cases, this can be done quite nicely. Assuming that all the input geometry coordinates are expressed exactly as rational numbers, then it is well known that roots for polynomials can be found within the field of algebraic numbers. So, one of the important aspects of exact arithmetic is to augment the typical floating point representation with additional data structures for radicals over the rationals.

Current language implementations for exact arithmetic have specific predicates for algebraic numbers [**142**]. So, solutions to $x^2 - 2$ can be represented exactly by these predicates. Then, these expression can be approximated to any number of bits specified by the user. Furthermore, algebraic operations are represented as directed acyclic graphics (DAG), with floating point values at the leaf nodes and algebraic operations at other nodes. In this sense, they are similar to CSG trees of Section 3.2.1. Since this DAG is the primary data structure, solutions can be adapted to user specified precision by just putting better approximations into the leaf nodes and being careful about error accumulation at the other nodes. There is a performance penalty for exact arithmetic. Efficient implementations are available for low-degree polynomial representations.

Question 6.9. *Can exact arithmetic be augmented to include non-algebraic nu-* 1132?
meric representations?

Question 6.10. *What happens when the assumption of exact rational input is not* 1133?
met?

The use of exact arithmetic can be contrasted with more classical techniques from numerical analysis. Specifically, the recent publication [**99**] presents a role for backward error analysis, with a reply included from proponents of exact arithmetic. This leads naturally to the next question.

Question 6.11. *What is the role of methods from numerical analysis, specifically* 1134?
backward error analysis, when there is uncertainty in the input data?

Question 6.12. *Can exact arithmetic have competitive performance with approximate floating point geometric algorithms over high-degree polynomial representations?*

7. Computational topology and surface reconstruction

A significant catalyst for computational topology has been the problem of constructing an approximating surface mesh given only a sample of points from the surface. This problem was formalized and brought to the attention of the computer graphics community in a seminal 1992 paper [**97**]. Amenta and Bern [**9**, **10**] described the *crust* algorithm for which they could show, under some conditions on the surface and the sample, that the output approximates, geometrically, the surface from which the samples were drawn. A later simplification [**11**] of this algorithm was shown to produce a PL (triangulated) manifold homeomorphic to the surface from which the samples were taken, using a somewhat complicated argument involving covering spaces. These results have been extended to prove isotopy equivalence, with the following being a representative theorem [**13**].

Theorem 7.1. *Let F be a compact, C^2 2-manifold without boundary. Let S be a set of sample points of F such that for each $x \in F$, there exists a point $s \in S$ such that $d(x,s) < k\,\mathrm{LFS}(x)$, where $k = 0.085$ and $\mathrm{LFS}(x)$ is the minimum distance between x and the medial axis of F. Then, there is an algorithm that will take S as input and produce a PL approximation of F that is ambient isotopic to F.*

1135? **Question 7.1.** *What are necessary and sufficient conditions on a low-dimensional manifold to permit an ambient isotopic approximant as the manifold reconstruction?*

1136? **Question 7.2.** *What criteria are necessary and sufficient on the density of the sampling set on a low-dimensional manifold to permit an ambient isotopic approximant as the manifold reconstruction?*

1137? **Question 7.3.** *What is the appropriate topological equivalence relation to consider for manifold reconstruction?*

1138? **Question 7.4.** *Specifically, for manifolds without boundary, what are necessary and sufficient conditions on the normal field on the boundary to permit an ambient isotopic approximant as the manifold reconstruction?*

Recent work that may be helpful references in considering these questions include [**4**, **42**, **43**, **50**, **51**, **54**, **78**, **127**, **152**], with recent theorems appearing for the cases with boundary [**4**].

8. Computational topology and low-dimensional manifolds

Many of the 1-manifolds and 2-manifolds for geometric computing are described as spline functions [**87**, **141**]. These splines are typically defined over very simple domains, such as $[0,1]$ and $[0,1]^2$. While low-dimensional manifolds have their own subdiscipline within topology, it is consistent here to consider these manifolds in relation to generalized spline functions.

8.1. Background. The basic approach we outline here uses two steps to construct a function. In the first step, we model the *domain* of the function as an abstract manifold (this manifold need not have geometry associated with it). In the second step we define an embedding or immersion of the domain, e.g., to produce a surface. This second step is done piecemeal by defining local embedding or immersion functions on subsets of the domain, then blending the results using a partition of unity.

More formally, given a manifold M, a method for defining charts $\alpha_c(M) \to c \subset \mathbb{R}^n$ on M, immersion $E_c\colon c \to \mathbb{R}^m$ and blend $B_c\colon c \to \mathbb{R}$ functions for each chart, we can define a function on the entire manifold as follows:

$$E(p) = \frac{\sum_c B_c(\alpha_c) E_c(\alpha_c(p))}{\sum_c B_c(\alpha_c)} \tag{8.1}$$

To ensure this equation is valid, we place some constraints on the chart α_c and blend B_c functions. First, the charts must cover the manifold, i.e., they are a finite atlas. Second, the blend functions are non-zero over c. This ensures that the denominator is not zero. (Note: There's nothing that prevents the support of B_c being smaller than c, but it makes it harder to prove that the denominator is non-zero.) The E_c functions can be any function of continuity C^k over the region c (the continuity outside of c does not matter).

The continuity of the above equation is the continuity of its constituent parts. Therefore, to have a C^k function the α_c, E_c, and B_c functions must be at least C^k. The blend functions must also have their value and first k derivatives go to zero near the boundary of c. This ensures continuity at the boundaries of each chart.

For surfaces, the manifolds that make sense are planes, spheres, and hyperbolic disks tiled with $4n$-sided polygons (with edge pairs identified). The latter is one possible domain for n-holed (genus n) surfaces. This domain simplifies to the tiled plane for a standard (1-holed) torus. The E_c functions are typically polynomials or spline functions.

For reinforcement learning, the manifold is a combination of all possible actions and sensor readings, and the E_c function is a number that says how good it is to take that action with those sensor readings (essentially, a height field).

In animation, the manifold depends on the movement. Suppose a character is throwing a ball. A manifold that describes this motion (in a simplistic way) consists of a periodic value (where in the throw they are) and a release point (x, y, z). The function on the manifold is a set of joint angles for every joint in the body.

8.2. Problem statement. The problem can be loosely stated as follows. There exist some number of samples d_i of what the surface or function should look like; those samples may contain noise. Additionally, the parameter values p_i for the samples (i.e., where they are on the manifold) may also be known. The goal is to minimize $\sum_i \|d_i - E(p_i)\|$ where the p_i are given or they give the closest point to d_i on E, $\min_{p_i} \|d_i - E(p_i)\|$.

If the goal is interpolation of the points d_i then the sum should be zero.

In addition to the above approximation constraints, there is usually some form of "smoothness" constraint to guide what happens between the sample points. This can take several forms. One option is to minimize some combination of the second derivatives, such as the bending energy. A second option is to bound how much the surface varies from, e.g., a linear approximation to the data points.

A related set of constraints concerns "features" in the data, such as sharp edges and corners. In this case, it may be desirable for the function E to correctly model the edge or corner, i.e., to exhibit a discontinuity in differentiability.

8.3. Solving the problem. There are two stages to solving the problem. The first is to decide the chart placement (the α_c), the second is to fit the individual functions E_c. Ideally, the E_c functions agree where they overlap, i.e., for all charts c_i overlapping a point p, $E_{c_i}(\alpha_{c_i}(p))$ evaluates to the same thing. In this case, the shape of the blend function doesn't matter. In practice, the shape of the blend function has little effect on the final shape, so we can simply define the same blend function shape for all charts.

There are two options for solving for the free parameters in the E_c functions. The first is to fit each E_c locally to an appropriate subset of the data. The second is to fit all of the E_c simultaneously. The latter is, in general, more computationally expensive, but has the potential to produce better results.

Some observations:

- The more the charts overlap, i.e., the more non-zero terms in Equation 8.1, the smoother the result tends to be, but this increases the computational expense. "Smoother" is not a well-defined term here; clearly, the surface has the same *continuity* regardless of the overlap. However, there is an averaging effect that reduces the effect of local variation in the individual E_c.
- The size of the chart and the corresponding required complexity of the E_c function are inversely related. As the chart size decreases, the variation in E that E_c is responsible for decreases. In the limit, with an infinite number of charts we could use piecewise constant functions for the E_c.
- Given a fixed number of degrees of freedom for E_c the desired local variation in E also determines the size of the chart. In large, flat areas, we can use a single chart, but in regions with more variation we need more charts.
- Features such as sharp edges, can be modeled using a function E_c that is capable of representing a discontinuity. In this case, all of the other E_c functions need to be "masked out" or they may unduly influence that area. However, it may be difficult to use a single chart for a feature that spans most of the manifold.

The following are important open questions:

1139? **Question 8.1.** *What are the optimum size, shape, and amount of overlap for the charts? The answer to this question depends both on the data and on the choice of E_c. Optimum is a measure both of the fit (including a definition of smoothness) and computational tractability.*

1140? **Question 8.2.** *Beyond questions of charts for a known manifold, there is also the question of figuring out what the underlying manifold is for a given set of data points. The assumption is that the data points d_i arise from samples of a low-dimensional manifold embedded in a high-dimensional space. This is the field of* manifold learning *in computer vision; most of the techniques (principal components analysis, isomap, simple linear embedding) currently work only for planar manifolds, or largely convex (geometrically) spherical or cylindrical data* *sets. What unifying theory is possible for determining the appropriate underlying manifold for a given set of data points?*

9. Skeletal structures

Many of the previously discussed approaches to surface reconstruction in Section 7 use the *medial axis*, which, under specific hypotheses, can be shown to be a deformation retract [**179**]. This is an important concept, but its reliable and efficient computation poses many theoretical [**46–48**] and practical [**45**] challenges.

Between any two points, $x, y \in \mathbb{R}^3$, let $d(x, y)$ denote the usual Euclidean distance and for any two sets $X, Y \subset \mathbb{R}^3$, let $d(X, Y) = \inf\{d(x, y) : x \in X, y \in Y\}$.

Definition 9.1. Let $x \in \mathbb{R}^n$ and $S \subset \mathbb{R}^n$. A point $s \in S$ is a *nearest point on* S to x if $d(x, s) = \inf\{d(x, t) : t \in S\}$. The *medial axis* of S, is the closure of the set of all points that have at least two distinct nearest points on S.

This concept was originally defined for object recognition in the life sciences [**29, 30**]. One investigation of the mathematical properties of the medial axis and its associated transform function [**44**] is restricted to geometry within the plane. More generally, there has been broad attention to the medial axis in $\mathbb{R}^n$ within the computer science literature, where the topological and differentiable investigations [**160, 179, 180**] are directly relevant to surface reconstruction work.

Both classical and contemporary research have emphasized the principle that many analytic attributes of surfaces can be determined using singularity theory and stratification theory [**1–3, 27, 28, 41, 46–48, 55, 144, 158, 168, 178**]. In particular, singularities can be shown to correspond to possible self-intersections or non-manifold points and can be organized in Thom–Boardman form [**19, 84, 123**]. However, computational solutions for the associated nonlinear equations can be prohibitively expensive using many variants of Newton's method. Furthermore, other relevant exponential algorithmic bounds [**41**] appear to pose daunting computational difficulties. Recent singularity publications do offer promising techniques that could lead to efficient algorithmic preservation of ambient isotopy type [**145, 149, 163, 174**], particularly in conjunction with recent findings by Blackmore [**27, 28**] of approximate methods. The "skin surfaces" introduced in the context of biological modeling [**58**] have been shown to have isotopic approximating meshes [**118**]. The authors of this last paper note that their algorithms presume that the geometric input set is fixed, but this raises a question about about whether a given output would be appropriate for other the input sets.

Question 9.1. *Once a mesh is created, does it remain valid for some deformations* 1141?
of the input set, if those deformations are suitably constrained?

The cut locus is similar to the medial axis and has been used in computational explorations of shape [**180**]. In particular, Wolter proves, for a rich class of surfaces, that C^2 continuity is not required to establish a positive distance between the surface and its cut locus, with a related corollary showing desirable smoothness properties of offsets of these surfaces.

Recognizing both the difficulties of approximating the medial axis and the sensitivity of the medial axis to small (though possibly inconsequential) changes in form, there has been recent mathematical work in proposing alternatives to the medial axis [**46–48**]. This work seeks the determination of relations between the skeletal structure proposed and the boundary of the original object, so that small changes in one will result in small changes in the other, where these alternative skeletal structures are often more topologically complex than the medial axis. The first of these publications [**46**] defines various geometric tools in support of these skeletal structures, and some of these tools hold promise for computational topology research, even as we note the distinction that the primary application of these skeletal generalizations has been to computer vision [**143**], as opposed to various simulation contexts.

The expected theory is likely to have some similarities to the use of the nerve simplicial complex technique previously invoked by Edelsbrunner and Shah [**70**]. There also appear to be similarities to the skeletal structures defined in the already cited papers by Damon and his coauthors [**46–48, 143**], for their consideration of robust variants of the medial axis and applications in computer vision. The envelope may also be regarded as one of the level sets generated by the normal flow, so there may be opportunities to leverage the extensive classical and contemporary literature on level sets. Similarly, the extensive existing literature on the Minkowski sum, deserves careful study for a variety of applications.

1142? **Question 9.2.** *What are the appropriate skeletal structures and algorithms to extract critical topological information while reducing the representation?*

10. Computational topology and Biology on simplicial complexes

Topology studies global properties of geometric objects, like the number of connected components, tunnels, or cavities. The work on computational topology led by Edelsbrunner has had many interesting applications to biology [**60**]. His more theoretically fundamental work on Delaunay triangulations [**57, 59**] is integral to these biological applications. This discussion presents those topics together. The Delaunay triangulations are typically classified as computational geometry, but the definition of their basic cells has a strong topological element. The triangulation is formed as a dual of a Voronoi diagram, which lies within a metric space, Z, having a metric $d\colon Z \to \mathbb{R}$. The Voronoi diagram presumes the existence of a finite set of points $Q = \{q_0, q_1, \ldots, q_n\}$ from Z. The Voronoi diagram is a collection of closed neighbhorhoods of the q_i, each containing one of the q_i. For each q_i its neighborhood is defined as the set of all $p \in Z$ such that $d(p, q_i) < d(p, q_j)$ for all $j \neq i$. Another related construct is that of α-shapes [**69, 150**].

An overview article has appeared [**60**]. The techniques are based largely on simplicial complexes, computing invariants such as Euler characteristics, Betti-numbers and writhing numbers [**8**]. Additionally, Morse theory is invoked [**61**] to develop novel data representations for visualization algorithms. These ideas were the subject of a *New Directions* short course at the Institute for Mathematics and Applications [**63**]. One outcome was to relate computational Morse theory to

Forman's discrete Morse theory. Some of the contributions to the literature along these themes appear in various venues [**7**, **20**, **34**, **52**, **53**, **58**, **61**, **62**, **64–69**, **71**, **72**]. However, even this list is only partially representative of the broad and deep impact this research has had within the computational topology community.

Some of the techniques evolve more from algebraic topology methods, which has become an independently rich area in computational topology under the leadership of Edelsbrunner, as well as that of Carlsson [**36**]. The latter endeavors have are also integrated with statistics, forming a very rich subject area, which can only be mentioned here for the benefit of the interested reader. Two conferences on *Algebraic Topological Methods in Computer Science* have been held.

A summarizing question becomes

Question 10.1. *What role can discrete Morse theory play for the theoretical basis for algorithms in computational topology?* 1143?

Additional work on simplicial complexes emphasizes recovering topological invariants of a space from a finite set of noisy samples, parameterized within a high-dimensional Euclidean space. In order to have robustness versus undersampling and noise, a multiscale view of the space is created that contains information at all granularities. A space is constructed incrementally using a geometric criterion, obtaining a family of spaces. The spaces are not independent, but are related by inclusion maps that induce maps between the topological attributes in the spaces. The theory of *persistent homology* captures these relationships as lifetimes for the evolving attributes [**68**]. These lifetimes translate into a measure of importance for topology. So, persistence is a robust mechanism for recovering topology as it separates topological noise from features.

The traditional approach is to approximate the space by placing small balls around the samples and characterizing the combinatorics of the ball set. The resulting complex is simple but very expensive to compute. Unfortunately, no effective techniques are known for computing small complexes for points in high-dimensional spaces.

Question 10.2. *Can local methods be used to take advantage of the geometry to yield small complexes that would be computationally tractable?* 1144?

Often, one can generate a multiple-parameter family of spaces that describes a point set. For example, one might wish to track the topology of isosurfaces of both pressure and temperature of a jet flow across time. Recent progress in persistent homology indicates that a simple description is not possible for multiple parameters [**185**]. There is need for an approximation theory that allows access to the topological information contained in such a family.

Question 10.3. *Can robust invariants be computed for these multiparameter spaces?* 1145?

This summary represents recent issues posed largely from the joint work of Zomorodian and Carlsson and earlier work of Zomordian with Edelsbrunner.

There is much emphasis upon homological invariants, which lies beyond the articulated scope of this article. Nonetheless these aspects are included here because of their nascent state, portending that there may remain unresolved issues about the underlying topological spaces as this work matures further.

11. Finite approximation and (non-Hausdorff) topology

It has been known for almost 80 years that every compact Hausdorff space is the subspace of closed points of an inverse limit of finite T_0-spaces, and that finite T_0-spaces are essentially finite posets. For many years this seemed an oddity; why would anyone approximate the best known and best understood topological spaces by spaces that were simultaneously trivial and nonintuitive?

11.1. Adapted inverse limit approximation by T_0-spaces. The development of computing and its need for information in bits, and more particularly the work on digital topology from a purely topological viewpoint led to much more intuition on these finite T_0-spaces. As a result, Kopperman and Wilson proved that these inverse systems can be assumed to have very special maps, which they called *calming maps*. If this is done, the following traditional knowledge can be recast, as stated, below.

11.2. Topological invariants. The association between an abstract simplicial complex, which can be seen as a finite T_0-space, and its polytope in a finite-dimensional Euclidean space can be used as follows: The topological spaces that most often occur in science and engineering are the metric continua. These are often viewed as inverse limits of polyhedra and simplicial maps. The work by Kopperman and Wilson [**113**] has shown that these inverse systems of polyhedra and simplicial maps can be replaced by inverse systems of abstract simplicial complexes and calming maps in such a way that the inverse limit of the former is exactly the subspace of closed points of the inverse limit of the latter. Rather than the Euclidean polytopes and simplicial maps, which are determined by vertices and subject to round-off error, one can use precisely given finite posets and special order-preserving maps, also precisely given. Here are some issues that arise before these methods can be applied: While it has been known for about three years that the above approximation can be done, no algorithm for finding these finite posets and calming maps has been described and this method has not been used to approximate spaces. But the digital topology needed to understand the finite spaces was learned over a dozen years ago, in part by Kopperman and co-workers [**101**, **102**, **106**, **108**, **111**, **117**].

11.3. Topological consistency. Much of the relationship between this approximation method and basic general topology has been resolved. For example, if the finite T_0-spaces are connected, then so is their limit [**114**, **116**], and so is this subspace of closed points. Also, the relationship between the separation axioms (particularly complete regularity, normality and hereditary normality) and properties of the finite spaces and maps has been determined [**115**, **116**]. The authors are now preparing for publication results on replacing maps between the original spaces with maps between inverse systems of finite approximants of these original spaces. These results yield characterizations of the Stone–Čech and Wallman compactification in terms of such finite approximations (some of this was noted earlier [**76**]). But much more knowledge is needed about such replacement and its use in computation and the preservation of invariants of algebraic topology.

A primary view from domain theory is that many important computational topology properties correspond to open sets and not to specific Euclidean points or scalar values [**6, 82, 83**]. More specifically, Kopperman and his collaborators have characterized those topological spaces that are computable in the sense of domain theory [**109**]. A special case involves those that are inverse limits of polyhedra, creating an opportunity to include domain theoretic results into computational topology investigations.

A summarizing research question becomes:

Question 11.1. *What are the essential topological relations for visualization* and *how can reliance upon domain theory and these approximating systems improve upon the state-of-the-art to preserve key embedding (homotopy and homology) invariants of the models and spaces as they become visualized, both statically and dynamically?* 1146?

12. Algorithmic topology and computational topology

The work of creating KnotPlot [**154, 155**] has been described as "topological drawing". By programs based upon Gaussian energy functionals, KnotPlot animates the process of unknotting and knot simplification on *specific examples* of knots. A key criterion is that the class of the knot is known *a priori*. This is an important aspect, as it is known [**91**] that the elementary problem of recognition of the piecewise linear unknot is in NP. Practical algorithms for knot recognition have proven elusive, but the problem remains an important stimulus for theoretical research [**89**].

This theoretical result provided valuable guidance to the work mentioned in Section 4 on isotopic approximations. Namely, it directed attention to just *preserving* the isotopy class of the original object *even when that classification was not known*. This is an example of the "beneficial symbiosis" anticipated [**63**] with algorithmic topology [**124**]. It leads to whether similar benefit can be gained by consideration of other algorithmic topology recognition problems, such as these summarized here.

The 3-sphere recognition problem starts with a given triangulation T and attempts to answer whether the underlying space $|T|$ is homeomorphic to the 3-sphere. It is shown that this problem lies in NP [**156**].

Question 12.1. *Is the 3-sphere recognition problem NP-hard?* 1147?

13. Computational topology workshop of 1999

To the best of our knowledge, the first broad workshop on Computational Topology was held in June, 1999 [**21**]. Its purpose was *direction finding* and its majority attendance was by scientists who are primarily recognized as computer scientists, though some pure mathematicians did attend and many of the participants are interdisciplinary in their work.

The report of this workshop is highly recommended for its broad coverage. Its impetus from computational geometers is reflected in the very applied nature of many of the topics and problems described. The report was not merely a description of technical problems, but also an attempt to identify areas, build community and develop an agenda for future research. As such, its purposes were somewhat different from the present article. Furthermore, because of the large number (22) of contributing coauthors, the report covers many subjects that are not intimately related to point-set topology. However, many of its findings are relevant for setting context. Some are quoted here. Furthermore, some specific problems do relate directly to this topology community, broadly considered. For instance, the definition of neighborhoods for differing topologies is a common problem of interest to many in the point-set topology community. Some problems, quoted below, mention the definition and representation of neighborhoods. Within the mathematics community, the specialty of low-dimensional topology is often viewed as being quite separate from that of point-set topology. However, problems from low-dimensional computational topology are presented here, because they depend upon such fundamental topological notions that it is hard to separate the fields. It is hoped that this blending of the subjects within computational topology might lead to more interaction among these communities within more established mathematical communities, hopefully to the benefit of mathematics at large. Those have been abstracted and updates provided, where relevant. Noticeably, several of these topics and problems are well-integrated with problems already posed within this article and that integration has been previously mentioned and is also noted, below.

The report begins with an emphasis upon the role of geometric computing to support the simulation of phsyical objects—"on scales that vary from the atomic to the astronomical."—emphasizing the role of topology in "Modeling the shapes of these objects and the space surrounding them." The role of information visualization is expressed as relying upon "shapes and motions" with obvious topological implications. The emphasis is upon support for geometric computing in that,

> "Some of the most difficult and least understood issues in geometric computing involve topology. Up until now, work on topological issues has been scattered among a number of fields, and its level of mathematical sophistication has been rather uneven. This report argues that a conscious focus on computational topology will accelerate progress in geometric computing."

While this specific focus on the benefits to geometric computing are understandable, this present article presents the point of view that topologists can make significant contributions to many aspects of computing. The "scattered" distribution throughout the literature is evident in the bibliography for this article, with cited publications appearing in mathematical and computer science venues, as well as within many different fields of engineering.

That the use of classical topology can "accelerate progress" has already been quite well documented in the literature. One notable success story has been the theorems generated by digital topologists. The wide-spread application, within the image processing community, of the Jordan curve and surface theorems to identify boundaries in images and partition viewed objects into parts lying inside and outside of those boundaries led to a contemporary study of these classical theorem, providing new proofs to apply to spaces that did not have T_2 separation properties [**110**]. These configurations of pixels on a computer screen were named *digital spaces.* While this seemed like an obvious use of a classical theorem, various unexpected subtleties occurred in algorithms in which this theory was applied. While it was unlikely that these difficulties would provide counterexamples that would invalidate such well-established theorems, it took the perspective of topologists to realize that

- the proofs of the classical Jordan separation theorems relied upon an assumption that the underlying topology was T_2,
- the digital spaces were discrete when modeled as T_2 topologies, and
- that weaker topologies were more descriptive of digital spaces.

These topologists then proved that the classical T_2 assumption was not needed and developed non-T_2 topologies for digital spaces. The rigorous consideration of these applied image problems led to extensions of classical theory and improved algorithms.

Some summarizing perspectives from this report state that "Topology separates global shape properties from local geometric attributes and provides a precise language for discussing these properties" and that "Mathematical abstraction can also unify similar concepts from different fields." These notions are, of course, well known to topologists, but it is of interest to understand that these aspects are now seen to be attractive in furthering the development of algorithms in robotics, molecular docking and geometric computing in general.

Some broad questions resulting from this report are summarized below, followed by more detailed sections with specific questions under each broader item. Again, the emphasis is upon topological issues that are most closely related to point-set topology, ignoring others that may have more of an algebraic topology or combinatorial topology emphasis.

13.1. Summary of broad questions.

Question 13.1. *How should shape be represented?*

Question 13.2. *How can topology preservation be ensured in converting from one shape representation to another?*

Question 13.3. *How can physical measurements, with sampling error and noise, be algorithmically converted into topologically valid shape representations, particularly for physical simulations that rely upon meshed geometry?*

Question 13.4. *How can "the development of algorithmic tools implementing topological concepts" [**63**] and "algorithmic questions in topology" be integrated for the benefit of both fields?'*

Because of preceding material, the key questions for each are tersely summarized.

13.2. Shape representation. This is consistent with the earlier remarks (Sections 2, 3 and 6) about the role of regular closed sets in solid modeling. Since this has already been discussed at some length, the relevant problems will be tersely stated, below.

Question 13.5. *Current shape representations include unstructured collections of polygons (with no specific connectivity information among geometric entities—often dubbed as 'polygon soup'), "polyhedral models, subdivision surfaces, spline surfaces, implicit surfaces, skin surfaces, alpha shapes", solid models, procedural models, digital and voxel models. What are the unifying topological constructs and how should they be expressed and implemented for efficient and robust algorithms?*

13.3. Topologically correct shape conversion. These issues have been discussed in some depth in Section 4 on approximation.

Question 13.6. *While there exist some methods for converting from one type of shape representation to another, these are mostly for polyhedral models and they are not totally rigorous or robust. How can topological principals be included in these shape conversions to both provide broad theory and improved algorithms? (We note that Section 4 has already discussed the inclusion of isotopy equivalence as a criterion for approximations (often PL ones) of smooth shapes in conjuction with traditional criteria of error bounds on the distance between one shape and its approximant.)*

1148? **Question 13.7.** *While classical topology has relied upon homeomorphisms for its primary equivalence relation, the geometric models in computing appear to need a stronger equivalence relation that includes correctness of the embedding within some low-dimensional topological (usually Euclidean) space. Is isotopy the preferred equivalence or is there need for even stronger equivalences such as diffeotopy?*

13.4. Shape acquisition algorithms and measurement error. Some of the dominant approaches here have avoided the issues of measurement error and noise. Recent abstractions [**152**] have proven theorems that leave open the opportunity to consider sample points with bounded measurement errors on a par with those that are exact samples. Several questions have already been articulated in previous sections.

Question 13.8. *How can these differing mathematical perspectives, across point-set and differentiable topology be best integrated for optimal shape-acquisition algorithms?*

13.5. Shape smoothness criteria. These issues range from unexpected appearance of non-smoothness in engineering design models to the need to represent non-smoothness in animation figures.

Question 13.9. *In some cases singularities arise because of numerical approxi-* 1149?
*mations made, which are inherent to a finite word length for numeric representations. In other cases, particularly for the motion picture industry, there are needs to model sharp changes in differentiability [**49**]. Some promising techniques have been presented that allow flexibility in moving gracefully between these needs [**165**]. Is there an appropriate topological abstraction that can be mapped easily to abstract data types that will permit appropriate representations of smoothness for differing applications?*

14. Conclusion

The bibliography is indicative of the breadth of interest in this subject, even though many references do not necessarily include the terminology "computational topology". As with any article presenting open problems, this one necessarily is reflective of the tastes and interests of the coauthors, where Sections 10, 12 and 13 are terse. This is not reflective of their scientific importance or impact, but rather an attempt to appeal to the expected point-set topology readership of this volume. In particular, the material presented here in those sections was directed towards emphasizing their general topology content, while showing their broader connections to other branches of topology for readers who might be interested in further consideration of these related subjects.

15. Acknowledgements

The authors wish to thank Herbert Edelsbrunner for some constructive criticism and insightful comments—most of which were made after lectures and during informal conversations at the recent Workshop on Computational Topology held on July 14, 2005 at Denison University, Granville. Thanks are also due to Ralph Kopperman and Yuriy Mileyko for numerous helpful conversations.

The following colleagues are acknowledged for their significant contributions to the indicated sections: C. Grimm (Section 8), R.D. Kopperman (Section 11) and V. Shapiro and N.F. Stewart (Section 6).

References

[1] K. Abdel-Malek, W. Seaman, and D. Blackmore, *On swept volume formulations: implicit surfaces*, Computer-Aided Design **33** (2001), 113–121.

[2] K. Abdel-Malek, W. Seaman, and H. J. Yeh, *NC verification of up to 5-axis machining processes using manifold stratification*, ASME J. Manufact. Sci. Eng. **122** (2000), 121–135.

[3] K. Abdel-Malek and H. J. Yeh, *Geometric representation of swept volume using Jacobian rank-deficiency conditions*, Computer-Aided Design **29** (1997), 457–468.

[4] K. Abe, J. Bisceglio, D. R. Ferguson, T. J. Peters, A. C. Russell, and T. Sakkalis, *Computational topology for isotopic surface reconstruction*, Theoret. Comput. Sci. (2006), Article in press.

[5] K. Abe, J. Bisceglio, T. J. Peters, A. C. Russell, D. R. Ferguson, and T. Sakkalis, *Computational topology for reconstruction of surfaces with boundary: integrating experiments and theory*, International Conference on Shape Modeling and Applications (M. Spagnuolo, A. Pasko, and A. Belyaev, eds.), IEEE Computer Society, 2005, pp. 290–299.

[6] S. Abramsky and A. Jung, *Domain theory*, Handbook of logic in computer science, Vol. 3, Oxford University Press, New York, 1994, pp. 1–168.

[7] P. K. Agarwal, H. Edelsbrunner, J. Harer, and Y. Wang, *Extreme elevation on a 2-manifold*, SCG'04: Proc. 20th Symposium on Computational Geometry, ACM Press, New York, 2004, pp. 357–365.

[8] P. K. Agarwal, H. Edelsbrunner, and Y. Wang, *Computing the writhing number of a polygonal knot*, Discrete Comput. Geom. **32** (2004), no. 1, 37–53.

[9] N. Amenta and M. Bern, *Surface reconstruction by Voronoi filtering*, Discrete Comput. Geom. **22** (1999), no. 4, 481–504.

[10] N. Amenta, M. W. Bern, and M. Kamvysselis, *A new Voronoi-based surface reconstruction algorithm*, ACM SIGGRAPH (1998), 415–421.

[11] N. Amenta, S. Choi, T. K. Dey, and N. Leekha, *A simple algorithm for homeomorphic surface reconstruction*, Proc. 16th Symposium on Computational Geometry, ACM Press, New York, 2000, pp. 213–222.

[12] N. Amenta, S. Choi, and R. K. Kolluri, *The power crust, unions of balls, and the medial axis transform*, Comput. Geom. **19** (2001), no. 2–3, 127–153.

[13] N. Amenta, T. J. Peters, and A. C. Russell, *Computational topology: ambient isotopic approximation of 2-manifolds*, Theoret. Comput. Sci. **305** (2003), 3–15.

[14] L.-E. Andersson, S. M. Dorney, T. J. Peters, and N. F. Stewart, *Polyhedral perturbations that preserve topological form*, Comput. Aided Geom. Design **12** (1995), no. 8, 785–799.

[15] L.-E. Andersson, T. J. Peters, and N. F. Stewart, *Selfintersection of composite curves and surfaces*, Comput. Aided Geom. Design **15** (1998), no. 5, 507–527.

[16] L.-E. Andersson, T. J. Peters, and N. F. Stewart, *Equivalence of topological form for curvilinear geometric objects*, Internat. J. Comput. Geom. Appl. **10** (2000), no. 6, 609–622.

[17] L.-E. Andersson, N. F. Stewart, and M. Zidani, *Error analysis for operations in solid modeling in the presence of uncertainty*, 2004, Preprint.

[18] L.-E. Andersson, N. F. Stewart, and M. Zidani, *Conditions for use of a non-selfintersection conjecture*, Comput. Aided Geom. Design (2006), To appear.

[19] V. I. Arnol'd, S. M. Guseĭn-Zade, and A. N. Varchenko, *Singularities of differentiable maps. Vol. I*, Birkhäuser, Boston, 1985.

[20] U. Axen and H. Edelsbrunner, *Auditory Morse analysis of triangulated manifolds*, Mathematical visualization (Berlin, 1997), Springer, Berlin, 1998, pp. 223–236.

[21] M. Bern et al, *Emerging challenges in computational topology*, 1999, Workshop Report. arXiv:cs.CG/9909001

[22] R H Bing, *The geometric topology of 3-manifolds*, American Mathematical Society, Providence, RI, 1983.

[23] J. Bisceglio, C. Mow, and T. J. Peters, *Boolean algebraic operands for engineering design*, 2006, Preprint.

[24] D. Blackmore, M. C. Leu, and Shih, *Analysis and modelling of deformed swept volumes*, Computer-Aided Design **26** (1994), 315–326.

[25] D. Blackmore, M. C. Leu, and L. P Wang, *The sweep-envelope differential equation algorithm and its application to NC machining verification*, Computer-Aided Design **29** (1997), 629–637.

[26] D. Blackmore and Y. Mileyko, *Computational differential topology*, Appl. Gen. Topol., To appear.

[27] D. Blackmore, Y. Mileyko, M. C. Leu, W. C. Regli, and W. Sun, *Computational topology and swept volumes*, Geometric and algorithmic aspects of computer-aided design and manufacturing, American Mathematical Society, Providence, RI, 2005, pp. 53–78.

[28] D. Blackmore, R. Samulyak, and M.C. Leu, *A singularity theory approach to swept volumes*, Int. J. Shape Modeling **6** (2000), 105–129.

[29] H. Blum, *A transformation for extracting new descriptors of shape*, Models for the Perception of Speech and Visual Form (W. Wathen-Dunn, ed.), MIT Press, Cambridge, 1967, pp. 362–380.

[30] H. Blum, *Biological shape and visual science (Part I)*, J. Theoret. Biol. **38** (1973), 205–287.

[31] M. Boyer and N. F. Stewart, *Modeling spaces for toleranced objects*, Internat. J. of Robotics Research **10** (1991), no. 5, 570–582.

[32] M. Boyer and N. F. Stewart, *Imperfect-form tolerancing on manifold objects: a metric approach*, Internat. J. of Robotics Research **11** (1992), no. 5, 482–490.

[33] M. Braverman and S. Cook, *Computing over the reals: foundations for scientific computing*, Notices Amer. Math. Soc. **53** (2006), no. 3, 318–329.

[34] P.-T. Bremer, V. Pascucci, H. Edelsbrunner, and B Hamann, *A topological hierarchy for functions on triangulated surfaces*, IEEE Trans. Vis. Comput. Graphics **10** (2004), 385–396.

[35] H.-D. Cao and X.-P. Zhu, *A complete proof of the Poincaré and geometrization conjectures—an application of the Hamilton–Perelman Theory of the Ricci flow*, Asian J. Math. **10** (2006), 165–492.

[36] G. Carlsson, *TMSCSCS: topological methods in scientific computing, statistics and computer science*, `http://math.stanford.edu/comptop/`.

[37] S.-L. Chang and M. S. R. Rocchetti, *Rendering cubic curves and surfaces with integer adaptive forward differencing*, ACM SIGGRAPH (1989), 157–166.

[38] F. Chazal and D. Cohen-Steiner, *A condition for isotopic approximation*, Graphical Models and Image Processing **67** (2005), 390–404.

[39] F. Chazal and A. Lieutier, *Topology guaranteeing surface reconstruction and medial axis approximation from noisy data*, Tech. Report 429, Institut de Mathématiques de Bourgogne, 2005, `http://math.u-bourgogne.fr/topolog/prepub/ApproxNoisyData3.pdf`.

[40] F. Chazal, A. Lieutier, and J. Rossignac, *Projection-homeomorphic surfaces*, SPM'05: Proc. ACM Symposium on Solid and Physical Modeling, ACM Press, New York, 2005, pp. 9–14.

[41] B. Chazelle, H. Edelsbrunner, L. J. Guibas, and M. Sharir, *A singly-exponential stratification scheme for real semi-algebraic varieties and its applications*, Automata, languages and programming (Stresa, 1989), Lecture Notes in Comput. Sci., vol. 372, Springer, Berlin, 1989, pp. 179–193.

[42] S.-W. Cheng, T. K. Dey, and E. A. Ramos, *Manifold reconstruction from point samples*, SODA'05: Proc. 16th ACM-SIAM Symposium on Discrete Algorithms, SIAM, Philadelphia, 2005, pp. 1018–1027.

[43] S.-W. Cheng, T. K. Dey, E. A. Ramos, and T. Ray, *Sampling and meshing a surface with guaranteed topology and geometry*, SCG'04: Proc. 20th Symposium on Computational Geometry, ACM Press, New York, 2004, pp. 280–289.

[44] H. I. Choi, S. W. Choi, and H. P. Moon, *Mathematical theory of medial axis transform*, Pacific J. Math. **181** (1997), no. 1, 57–88.

[45] T. Culver, J. Keyser, and D. Manocha, *Exact computation of the medial axis of a polyhedron*, Comput. Aided Geom. Design **21** (2004), no. 1, 65–98.

[46] J. Damon, *Smoothness and geometry of boundaries associated to skeletal structures. I. Sufficient conditions for smoothness*, Ann. Inst. Fourier (Grenoble) **53** (2003), no. 6, 1941–1985.

[47] J. Damon, *Smoothness and geometry of boundaries associated to skeletal structures. II. Geometry in the Blum case*, Compositio Math. **140** (2004), no. 6, 1657–1674.

[48] J. Damon, *Determining the geometry of boundaries of objects from medial data*, Int. J. Comp. Vision **63** (2005), no. 1, 45–64.

[49] T. DeRose, M. Kass, and T. Truong, *Subdivision surfaces in character animation*, ACM SIGGRAPH (1998), 85–94.

[50] T. K. Dey, *Curve and surface reconstruction*, Handbook of discrete and computational geometry (J. E. Goodman and J. O'Rourke, eds.), CRC Press, 2004, Second edition, pp. 677–692.

[51] T. K. Dey, *Curve and surface reconstruction: algorithms with mathematical analysis*, Cambridge University Press, Cambridge, 2006.

[52] T. K. Dey, H. Edelsbrunner, and S. Guha, *Computational topology*, Advances in discrete and computational geometry (South Hadley, MA, 1996), Contemp. Math., vol. 223, American Mathematical Society, Providence, RI, 1999, pp. 109–143.

[53] T. K. Dey, H. Edelsbrunner, S. Guha, and D. V. Nekhayev, *Topology preserving edge contraction*, Publ. Inst. Math. (Beograd) (N.S.) **66(80)** (1999), 23–45.

[54] T. K. Dey and P. Kumar, *A simple provable algorithm for curve reconstruction*, SODA'99: Proc. 10th ACM-SIAM Symposium on Discrete Algorithms, SIAM, Philadelphia, 1999, pp. 893–894.

[55] P. Dimitrov, J. Damon, and K. Siddiqi, *Flux invariants for shape*, Computer Vision and Pattern Recognition, IEEE Computer Society, 2003, pp. 835–841.

[56] S. K. Donaldson, *Yang–Mills invariants of four-manifolds*, Geometry of low-dimensional manifolds, 1 (Durham, 1989), Cambridge University Press, Cambridge, 1990, pp. 5–40.

[57] H. Edelsbrunner, *Algorithms in combinatorial geometry*, Springer, Berlin, 1987.

[58] H. Edelsbrunner, *Deformable smooth surface design*, Discrete Comput. Geom. **21** (1999), no. 1, 87–115.

[59] H. Edelsbrunner, *Geometry and topology for mesh generation*, Cambridge University Press, Cambridge, 2001.

[60] H. Edelsbrunner, *Biological applications of computational topology*, Handbook of discrete and computational geometry (J. E. Goodman and J. O'Rourke, eds.), CRC Press, Boca Raton, FL, 2004, pp. 1395–1412.

[61] H. Edelsbrunner, P.-T. Bremer, V. Pascucci, and B. Hamann, *A multi-resolution data structure for two-dimensional Morse–Smale functions*, VIS'03: Proceedings of the 14th IEEE Visualization 2003, IEEE Computer Society, Washington, DC, 2003, pp. 139–146.

[62] H. Edelsbrunner, M. A. Facello, P. Fu, J. Qian, and D. V. Nekhayev, *Wrapping 3D scanning data*, Three-dimensional image capture and applications, Proc. SPIE, vol. 3313, SPIE Press, Bellingham, WA, 1998, pp. 148–158.

[63] H. Edelsbrunner and J. Harer, *IMA workshop, July 6–16*, 2004, `http://www.ima.umn.edu/new-directions/2004NDshort-course/`.

[64] H. Edelsbrunner and J. Harer, *Jacobi sets of multiple Morse functions*, Foundations of computational mathematics: Minneapolis, 2002, Cambridge University Press, Cambridge, 2004, pp. 37–57.

[65] H. Edelsbrunner, J. Harer, A. Mascarenhas, and Pascucci V., *Time-varying Reeb graphs for continuous space-time data*, SCG'04: Proc. 20th Symposium on Computational Geometry, ACM Press, New York, 2004, pp. 366–372.

[66] H. Edelsbrunner, J. Harer, V. Natarajan, and V. Pascucci, *Morse–Smale complexes for piecewise linear 3-manifolds*, SCG'03: Proc. 19th Symposium on Computational Geometry, ACM Press, New York, 2003, pp. 361–370.

[67] H. Edelsbrunner, J. Harer, and A. Zomorodian, *Hierarchical Morse–Smale complexes for piecewise linear 2-manifolds*, Discrete Comput. Geom. **30** (2003), no. 1, 87–107.

[68] H. Edelsbrunner, D. Letscher, and A. Zomorodian, *Topological persistence and simplification*, Discrete Comput. Geom. **28** (2002), no. 4, 511–533.

[69] H. Edelsbrunner and E. P. Mücke, *Three-dimensional alpha shapes*, ACM Transactions on Graphics **13** (1994), no. 1, 43–72.

[70] H. Edelsbrunner and N. R. Shah, *Triangulating topological spaces*, Internat. J. Comput. Geom. Appl. **7** (1997), no. 4, 365–378.

[71] H. Edelsbrunner and R. Waupotitsch, *A combinatorial approach to cartograms*, Comput. Geom. **7** (1997), no. 5–6, 343–360.

[72] H. Edelsbrunner and A. Zomorodian, *Computing linking numbers of a filtration*, Algorithms in bioinformatics (Århus, 2001), Lecture Notes in Comput. Sci., vol. 2149, Springer, Berlin, 2001, pp. 112–127.

[73] A. V. Evako, R. Kopperman, and Y. V. Mukhin, *Dimensional properties of graphs and digital spaces*, J. Math. Imaging Vision **6** (1996), no. 2–3, 109–119.

[74] R. Farouki, *Closing the gap between CAD model and downstream application*, SIAM News **32** (1999).

[75] D. R. Ferguson, *An industrial look at CAGD within Boeing*, Invited plenary presentation, Sixth SIAM conference on geometric design, 1999.

[76] J. Flachsmeyer, *Zur Spektralentwicklung topologischer Räume*, Math. Ann. **144** (1961), 253–274.

[77] J. D. Foley, A. van Dam, S. K. Feiner, and J. F. Hughes, *Computer graphics: principles and practice*, Addison Wesley, 1997, Second Edition in C.

[78] D. Freedman, *Combinatorial curve reconstruction in Hilbert spaces: a new sampling theory and an old result revisited*, Comput. Geom. **23** (2002), no. 2, 227–241.

[79] M. H. Freedman, *The topology of four-dimensional manifolds*, J. Differential Geom. **17** (1982), no. 3, 357–453.

[80] A. Gain and A. Dodgson, *Preventing self-intersection under free-form deformation*, IEEE Trans. Vis. Comput. Graphics **7** (2001), no. 4, 289–298.

[81] E. Garduño and G. T. Herman, *Implicit surface visualization of reconstructed biological molecules*, Theoret. Comput. Sci. **346** (2005), no. 2–3, 281–299.

[82] G. Gierz, K. H. Hofmann, K. Keimel, J. D. Lawson, M. W. Mislove, and D. S. Scott, *A compendium of continuous lattices*, Springer, Berlin, 1980.

[83] G. Gierz, K. H. Hofmann, K. Keimel, J. D. Lawson, M. W. Mislove, and D. S. Scott, *Continuous lattices and domains*, Cambridge University Press, Cambridge, 2003.

[84] M. Golubitsky and V. Guillemin, *Stable mappings and their singularities*, Springer, New York, 1973.

[85] T. A. Grandine, *Applications of contouring*, SIAM Rev. **42** (2000), no. 2, 297–316.

[86] T. A. Grandine and F. W. Klein, IV, *A new approach to the surface intersection problem*, Comput. Aided Geom. Design **14** (1997), no. 2, 111–134.

[87] C. M. Grimm, *Simple manifolds for surface modeling and parameterization*, SMI'02: Proceedings of the Shape Modeling International 2002, IEEE Computer Society, Washington, DC, 2002, p. 237.

[88] E. L. Gursoz, Y. Choi, and F. B. Prinz, *Vertex-based representation of non-manifold boundaries*, Geometric modeling for product engineering, North-Holland, 1990, pp. 107–130.

[89] W. Haken, *Theorie der Normalflächen*, Acta Math. **105** (1961), 245–375.

[90] R. Hartshorne, *Algebraic geometry*, Springer, New York, 1977.

[91] J. Hass, J. C. Lagarias, and N. Pippenger, *The computational complexity of knot and link problems*, J. ACM **46** (1999), no. 2, 185–211.

[92] G. Hemion, *The classification of knots and 3-dimensional spaces*, The Clarendon Press Oxford University Press, New York, 1992.

[93] G. Herman, *3D imaging in Medicine*, CRC Press, 2000, 2nd Edition.

[94] M. W. Hirsch, *Differential topology*, Springer, New York, 1976.

[95] J. G. Hocking and G. S. Young, *Topology*, Addison Wesley, 1961.

[96] C. M. Hoffmann, *Geometric & solid modeling*, Morgan Kaufmann, San Mateo, CA, 1989.

[97] H. Hoppe, T. DeRose, T. Duchamp, J. McDonald, and W. Stuetzle, *Surface reconstruction from unorganized points*, ACM SIGGRAPH (1992), 71–78.

[98] A. Jackson, *Conjectures no more?: consensus forming on the proof of the Poincaré and geometrization conjectures*, Notices Amer. Math. Soc. **53** (2006), no. 8, 897–901.

[99] D. Jiang and N. F. Stewart, *Backward error analysis in computational geometry*, Sixth Annual International Workshop on Computational Geometry and Applications (CGA '06) (Glasgow, UK), Lecture Notes in Comput. Sci., vol. 3980, Springer, Berlin, 2006, pp. 50–60.

[100] T. Kaczynski, K. Mischaikow, and M. Mrozek, *Computational homology*, Springer, New York, 2004.

[101] E. Khalimsky, R. Kopperman, and P. R. Meyer, *Boundaries in digital planes*, J. Appl. Math. Stochastic Anal. **3** (1990), no. 1, 27–55.

[102] E. Khalimsky, R. Kopperman, and P. R. Meyer, *Computer graphics and connected topologies on finite ordered sets*, Topology Appl. **36** (1990), no. 1, 1–17.

[103] J. Kister, *Small isotopies in Euclidean spaces and 3-manifolds*, Bull. Amer. Math. Soc. **65** (1959), 371–373.

[104] D. E. Knuth, *The art of computer programming. Vol. 2*, Addison Wesley, 1981.

[105] T. Y. Kong and E. Khalimsky, *Polyhedral analogs of locally finite topological spaces*, General topology and applications (Middletown, CT, 1988), Lecture Notes in Pure and Appl. Math., vol. 123, Marcel Dekker Inc., New York, 1990, pp. 153–164.

[106] T. Y. Kong, R. Kopperman, and P. R. Meyer, *A topological approach to digital topology*, Amer. Math. Monthly **98** (1991), no. 10, 901–917.

[107] T. Y. Kong and R. D. Kopperman, *Digital topology*, Encyclopedia of general topology (K. P. Hart, J. Nagata, and J. E. Vaughan, eds.), Elsevier Science, 2004, pp. 428–432.

[108] R. Kopperman and Y. Kong, *Using general topology in image processing*, Geometrical problems of image processing (Georgenthal, 1991), Akademie-Verlag, Berlin, 1991, pp. 66–71.

[109] R. Kopperman, H.-P. A. Künzi, and P. Waszkiewicz, *Bounded complete models of topological spaces*, Topology Appl. **139** (2004), 285–297.

[110] R. Kopperman, P. R. Meyer, and R. G. Wilson, *A Jordan surface theorem for three-dimensional digital spaces*, Discrete Comput. Geom. **6** (1991), no. 2, 155–161.

[111] R. D. Kopperman, *The Khalimsky line as a foundation for digital topology*, Shape in Picture (O. Ying-Lie et al, eds.), Springer, New York, 1994, pp. 3–20.

[112] R. D. Kopperman, *Topological digital topology (invited contribution)*, Discrete geometry for computer imagery (A. Nyström, G. Sanniti di Baja, and S. Svensson, eds.), Lecture Notes in Comput. Sci., vol. 2886, Springer, New York, 2003, pp. 1–15.

[113] R. D. Kopperman, V. V. Tkachuk, and R. G. Wilson, *The approximation of compacta by finite T_0-spaces*, Quaest. Math. **26** (2003), no. 3, 355–370.

[114] R. D. Kopperman and R. G. Wilson, *Finite approximation of compact Hausdorff spaces*, Topology Proc. **22** (1997), Summer, 175–200.

[115] R. D. Kopperman and R. G. Wilson, *On the role of finite, hereditarily normal spaces and maps in the genesis of compact Hausdorff spaces*, Topology Appl. **135** (2004), 265–275.

[116] R. D. Kopperman and R. G. Wilson, *Separation and connectedness in spectral compactification*, Houston J. Math., To appear.

[117] V. Kovalevsky and R. Kopperman, *Some topology-based image processing algorithms*, Papers on general topology and applications (Flushing, NY, 1992), Ann. New York Acad. Sci., vol. 728, New York Acad. Sci., 1994, pp. 174–182.

[118] N. Kruithof and G. Vegter, *Approximation by skin surfaces*, SM'03: Proc. 8th ACM Symposium on Solid Modeling and Applications, ACM Press, New York, 2003, pp. 86–95.

[119] T. Maekawa, N. M. Patrikalakis, T. Sakkalis, and G. Yu, *Analysis and applications of pipe surfaces*, Comput. Aided Geom. Design **15** (1998), no. 5, 437–458.

[120] M. Mäntyalä, *Computational topology: a study on topological manipulations and interrogations in computer graphics and geometric modeling*, Finnish Academy of Technical Sciences, Helsinki, 1983.

[121] R. Marabini, C. O. S. Sorzano, S. Matej, J. J. Fernández, J. M. Carazo, and G. T. Herman, *3d reconstruction of 2d crystals in real space*, IEEE Trans. Image Processing **13** (2004), 546–561.

[122] W. S. Massey, *Algebraic topology: An introduction*, Harcourt, Brace & World, Inc., New York, 1967.

[123] J. N. Mather, *Stratifications and mappings*, Dynamical systems (Bahia, 1971), Academic Press, New York, 1973, pp. 195–232.

[124] S. Matveev, *Algorithmic topology and classification of 3-manifolds*, Springer, Berlin, 2003.

[125] J. Milnor, *Lectures on the h-cobordism theorem*, Princeton University Press, Princeton, NJ, 1965.

[126] J. Mitchell, *Topological obstructions to blending algorithms*, Comput. Aided Geom. Design **17** (2000), no. 7, 673–694.

[127] N. Mitra and A. Nguyen, *Estimating surface normals in noisy point cloud data*, SCG'03: Proc. 19th Symposium on Computational Geometry, ACM Press, New York, 2003, pp. 322–328.

[128] G. Monge, *Application de l'analys à la geometrie*, Bachelier, Paris, 1850.

[129] E. L. F. Moore, *Computational topology of spline curves for geometric and molecular approximations*, Ph.D. thesis, The University of Connecticut, 2006.

[130] E. L. F. Moore and T. J. Peters, *Computational topology for geometric design and molecular design*, Mathematics for Industry: Challenges and Frontiers: A Process View: Practice and Theory (D. R. Ferguson and T. J. Peters, eds.), SIAM, 2005, pp. 125–139.

[131] E. L. F. Moore, T. J. Peters, and J. A. Roulier, *Preserving computatational topology by subdivision of quadratic and cubic Bézier curves*, Computing, To appear.

[132] J. W. Morgan, *Recent progress on the Poincaré conjecture and the classification of 3-manifolds*, Bull. Amer. Math. Soc. (N.S.) **42** (2005), no. 1, 57–78.

[133] C. Mow, T. J. Peters, and N. F. Stewart, *Specifying useful error bounds for geometry tools: an intersector exemplar*, Comput. Aided Geom. Design **20** (2003), no. 5, 247–251.

[134] J. R. Munkres, *Elementary differential topology*, Princeton University Press, Princeton, NJ, 1966.

[135] J. R. Munkres, *Elements of algebraic topology*, Addison Wesley, 1984.

[136] N. M. Patrikalakis and T. Maekawa, *Shape interrogation for computer aided design and manufacturing*, Springer, Berlin, 2002.

[137] T. J. Peters, J. Bisceglio, D. R. Ferguson, C. M. Hoffmann, T. Maekawa, N. M. Patrikalakis, T. Sakkalis, and N. F. Stewart, *Computational topology for regular closed sets (within the I-TANGO project)*, Topology Atlas Inv. Contrib. **9** (2004), no. 1, 12 pp. `arXiv:math.GN/0412557`

[138] T. J. Peters and D. W. Rosen, *The role of topology in engineering design research*, Research in Engineering Design **8** (1996), no. 2, 81–98.

[139] T. J. Peters, D. W. Rosen, and S. M. Dorney, *The diversity of topological applications within computer-aided geometric design*, Papers on general topology and applications (Flushing, NY, 1992), Ann. New York Acad. Sci., vol. 728, New York Acad. Sci., 1994, pp. 198–209.

[140] T. J. Peters and N. F. Stewart, *Computational topology for CFD: theorems, criteria and issues*, Proc. First MIT Conference on Computational Fluid and Solid Mechanics, Elsevier Science, 2001, pp. 131–134.

[141] L. Piegl and W. Tiller, *The NURBS book*, Springer, New York, 1997.

[142] S. Pion and C. K. Yap, *Constructive root bound for k-ary rational input numbers*, SCG'03: Proc. 19th Symposium on Computational Geometry, ACM Press, New York, 2003, pp. 256–263.

[143] S. Pizer et al, *Deformable M-reps for 3D medical image segmentation*, Int. J. Comp. Vision **55** (2003), no. 2–3, 85–106.

[144] S. Pizer, K. Siddiqi, G. Székely, J. Damon, and S. Zucker, *Multiscale medial loci and their properties*, Int. J. Comp. Vision **55** (2003), 155–179.

[145] S. Plantinga and G. Vegter, *Contour generators of evolving implicit surfaces*, SM'03: Proc. 8th ACM Symposium on Solid Modeling and Applications, ACM Press, New York, 2003, pp. 23–32.

[146] J. Qi and V. Shapiro, *Epsilon-regular sets and intervals*, International Conference on Shape Modeling and Applications, IEEE Computer Society, 2005, pp. 310–319.

[147] J. Qi and V. Shapiro, *Epsilon-topological formulation of tolerant solid modeling*, Computer-Aided Design **38** (2006), 367–377.

[148] A. A. G. Requicha, *Representations for rigid solids: theory, methods and systems*, ACM Computing Surveys **46** (1999), 185–211.

[149] T. B. Rushing, *Topological embeddings*, Academic Press, New York, 1973.

[150] T. Sakkalis and C. Charitos, *Approximating curves via alpha shapes*, Graphical Models and Image Processing **61** (1999), 165–176.

[151] T. Sakkalis and T. J. Peters, *Ambient isotopic approximations for surface reconstruction and interval solids*, SM'03: Proc. 8th ACM Symposium on Solid Modeling and Applications, ACM Press, New York, 2003, pp. 176–184.

[152] T. Sakkalis, T. J. Peters, and J. Bisceglio, *Application of ambient isotopy to surface approximation and interval solids*, Computer-Aided Design **36** (2004), 1089–1100.

[153] T. Sakkalis, G. Shen, and N. M. Patrikalakis, *Topological and geometric properties of interval solid models*, Graphical Models and Image Processing **63** (2001), 163–175.

[154] R. Scharein, *The KnotPlot Site*, `http://www.knotplot.com/`.

[155] R. Scharein, *Interactive topological drawing*, Ph.D. thesis, The University of British Columbia, 1998.

[156] S. Schleimer, *Sphere recognition lies in NP*, 2004, Preprint. `arXiv:math.GT/0407047`

[157] V. Shapiro, *Homotopy conditions for tolerant geometric queries*, Preprint.

[158] V. Shapiro, *Maintenance of geometric representations through space decompositions*, Internat. J. Comput. Geom. Appl. **7** (1997), no. 1–2, 21–56.

[159] G. Shen, T. Sakkalis, and N. M. Patrikalakis, *Analysis of boundary representation model rectification*, SMA'01: Proc. 6th ACM Symposium on Solid Modeling and Applications, ACM Press, New York, 2001, pp. 149–158.

[160] E. C. Sherbrooke, N. M. Patrikalakis, and F-E. Wolter, *Differential and topological properties of medial axis transforms*, Graphical Models and Image Processing **58** (1996), 574–592.

[161] *Point clouds in imaging science*, SIAM News **37** (2004), no. 7, 3 pp.

[162] S. Smale, *Generalized Poincaré's conjecture in dimensions greater than four*, Ann. of Math. (2) **74** (1961), 391–406.

[163] E. H. Spanier, *Algebraic topology*, McGraw-Hill, New York, 1966.

[164] J. R. Stallings, *Polyhedral homotopy-spheres*, Bull. Amer. Math. Soc. **66** (1960), 485–488.

[165] I. F. Stewart and A. R. Foisy, *Arbitrary-degree subdivision with creases and attribute*, J. Graphical Tools **9** (2004), no. 4, 3–18.

[166] N. F. Stewart, *A sufficient condition for correct topological form in tolerance specification*, Computer-Aided Design **25** (1993), 39–48.

[167] N. F. Stewart and M. Zidani, *Transfinite interpolation for well-definition in error analysis in solid modelling*, 2006, Dagstuhl Seminar 06021, January 8–13.

[168] R. Thom, *Ensembles et morphismes stratifiés*, Bull. Amer. Math. Soc. **75** (1969), 240–284.

[169] A. Thompson, *Thin position and the recognition problem for S^3*, Math. Res. Lett. **1** (1994), no. 5, 613–630.

[170] R. B. Tilove and A. A. G. Requicha, *Closure of Boolean operations on geometric entities*, Computer-Aided Design **12** (1980), 219–220.

[171] A. Varshney, *Hierarchical geometric approximations*, Ph.D. thesis, University of North Carolina–Chapell Hill, Department of Computer Science, 1994.

[172] G. Vegter, *Computational topology*, Handbook of discrete and computational geometry, CRC Press, Boca Raton, FL, 1997, pp. 517–536.

[173] H. B. Voelcker and A. A. G. Requicha, *Geometric modelling of mechanical parts and processes*, IEEE Computer **10** (1977), no. 12, 48–57.

[174] C. T. C. Wall, *Surgery on compact manifolds*, American Mathematical Society, Providence, RI, 1999.

[175] G. W. Wang, J. Sun, and X. Hua, *The sweep-envelope differential equation algorithm for general deformed swept volumes*, Comput. Aided Geom. Design **17** (2000), no. 5, 399–418.

[176] K. J. Weiler, *Topological structures for geometric modeling*, Ph.D. thesis, Rensselaer Polytechnic Institute, 1986.

[177] H. Whitney, *Analytic extensions of differentiable functions defined in closed sets*, Trans. Amer. Math. Soc. **36** (1934), no. 1, 63–89.

[178] H. Whitney, *Tangents to an analytic variety*, Ann. of Math. (2) **81** (1965), 496–549.

[179] F.-E. Wolter, *Cut loci in bordered and unbordered Riemannian manifolds*, Doctoral dissertation, Technical University of Berlin, Department of Mathematics, 1985.

[180] F.-E. Wolter, *Cut locus and medial axis in global shape interrogation and representation*, Design Laboratory Memorandum 92-2, MIT, 1993.

[181] C. Yap and T. Dubé, *The exact computation paradigm*, Computing in Euclidean geometry, World Scientific Publishing, Singapore, 1995, pp. 452–492.

[182] E. Zeeman, *The Poincaré conjecture for $n \geq 5$*, Topology of 3-manifolds and related topics, Prentice Hall, Englewood Cliffs, NJ, 1962, pp. 198–204.

[183] M. Zidini, *Validité d'un modèle QuasiNURBS interpolant des données géométriques incertaines*, Doctoral dissertation, Université de Montréal, 2006.

[184] A. Zomorodian, *Topology for computing*, Cambridge University Press, Cambridge, 2005.

[185] A. Zomorodian and G. Carlsson, *Computing persistent homology*, Discrete Comput. Geom. **33** (2005), no. 2, 249–274.

Part 7

Functional Analysis

Open Problems in Topology II – Edited by E. Pearl

Non-smooth analysis, optimisation theory and Banach space theory

Jonathan M. Borwein and Warren B. Moors

1. Weak Asplund spaces

Let X be a Banach space. We say that a function $\varphi\colon X \to \mathbb{R}$ is *Gâteaux differentiable at $x \in X$* if there exists a continuous linear functional $x^* \in X^*$ such that

$$x^*(y) = \lim_{\lambda \to 0} \frac{\varphi(x + \lambda y) - \varphi(x)}{\lambda} \quad \text{for all } y \in X.$$

In this case, the linear functional x^* is called the *Gâteaux derivative of φ at $x \in X$*. If the limit above is approached uniformly with respect to all $y \in B_X$, the closed unit ball in X, then φ is said to be *Fréchet differentiable at $x \in X$* and x^* is called the *Fréchet derivative of φ at x*.

A Banach space X is called a *weak Asplund space* [*Gâteaux differentiability space*] if each continuous convex function defined on it is Gâteaux differentiable at the points of a *residual subset* (i.e., a subset that contains the intersection of countably many dense open subsets of X) [dense subset] of its domain.

Since 1933, when S. Mazur [**55**] showed that every separable Banach space is weak Asplund, there has been continued interest in the study of weak Asplund spaces. For an introduction to this area, see [**61**] and [**32**]. Also see the seminal paper [**1**] by E. Asplund.

The main problem in this area is given next.

Question 1.1. *Provide a geometrical characterisation for the class of weak Asplund spaces.* 1150?

Note that there is a geometrical dual characterisation for the class of Gâteaux differentiability spaces, see [**67**, §6]. However, it has recently been shown that there are Gâteaux differentiability spaces that are not weak Asplund [**58**]. Hence the dual characterisation for Gâteaux differentiability spaces cannot serve as a dual characterisation for the class of weak Asplund spaces.

The description of the next two related problems requires some additional definitions.

Let $A \subseteq (0,1)$ and let $K_A := [(0,1] \times \{0\}] \cup [(\{0\} \cup A) \times \{1\}]$. If we equip this set with the order topology generated by the lexicographical (dictionary) ordering (i.e., $(s_1, s_2) \leq (t_1, t_2)$ if, and only if, either $s_1 < t_1$ or $s_1 = t_1$ and $s_2 \leq t_2$) then with this topology K_A is a compact Hausdorff space [**46**]. In the special case of $A = (0,1)$, K_A reduces to the well-known double arrow space.

The first named author was supported by NSERC and by the Canada Research Chair Program. The second named author was supported by a Marsden Fund research grant, UOA0422, administered by the Royal Society of New Zealand.

1151? **Question 1.2.** *Is $(C(K_A), \|\cdot\|_\infty)$ weak Asplund whenever A is perfectly meagre?*

Recall that a subset $A \subseteq \mathbb{R}$ is called *perfectly meagre* if for every perfect subset P of $\mathbb{R}$ the intersection $A \cap P$ is meagre (i.e., first category) in P. An affirmative answer to this question will provide an example (in ZFC) of a weak Asplund space whose dual space is not weak* fragmentable, see [**58**] for more information on this problem. For example, it is shown in [**58**] that if A is perfectly meagre then $(C(K_A), \|\cdot\|_\infty)$ is *almost weak Asplund* i.e., every continuous convex function defined on $(C(K_A), \|\cdot\|_\infty)$ is Gâteaux differentiable at the points of an everywhere second category subset of $(C(K_A), \|\cdot\|_\infty)$. Moreover, it is also shown in [**58**] that if $(C(K_A), \|\cdot\|_\infty)$ is weak Asplund then A is obliged to be perfectly meagre.

Our last question on this topic is the following well-known problem.

1152? **Question 1.3.** *Is $(C(K_{(0,1)}), \|\cdot\|_\infty)$ a Gâteaux differentiability space?*

The significance of this problem emanates from the fact that $(C(K_{(0,1)}), \|\cdot\|_\infty)$ is not a weak Asplund space as the norm $\|\cdot\|_\infty$ is only Gâteaux differentiable at the points of a first category subset of $(C(K_{(0,1)}), \|\cdot\|_\infty)$, [**32**]. Hence a positive solution to this problem will provide another, perhaps more natural, example of a Gâteaux differentiability space that is not weak Asplund.

2. The Bishop–Phelps problem

For a Banach space $(X, \|\cdot\|)$, with closed unit ball B_X, the *Bishop–Phelps set* is the set of all linear functionals in the dual X^* that attain their maximum value over B_X; that is, the set $\{x^* \in X^* : x^*(x) = \|x^*\| \text{ for some } x \in B_X\}$. The Bishop–Phelps Theorem [**4**] says that the Bishop–Phelps set is always dense in X^*.

1153? **Question 2.1.** *Suppose that $(X, \|\cdot\|)$ is a Banach space. If the Bishop–Phelps set is a residual subset of X^* is the dual norm necessarily Fréchet differentiable on a dense subset of X^*?*

The answer to this problem is known to be positive in the following cases:

(i) if X^* is weak Asplund, [**36**, Corollary 1.6(i)];
(ii) if X admits an equivalent weakly mid-point locally uniformly rotund norm and the weak topology on X is σ-fragmented by the norm, [**59**, Theorem 3.3 and Theorem 4.4];
(iii) if the weak topology on X is Lindelöf, [**49**].

The assumptions in (ii) can be slightly weakened, see [**37**, Theorem 2]. It is also known that each equivalent dual norm on X^* is Fréchet differentiable on a dense subset on X^* whenever the Bishop–Phelps set of each equivalent norm on X is residual in X^*, [**57**, Theorem 4.4]. Note that in this case X has the Radon–Nikodým property.

For an historical introduction to this problem and its relationship to local uniformly rotund renorming theory, see [**48**].

Next, we give an important special case of the previous question.

Question 2.2. *If the Bishop–Phelps set of an equivalent norm $\|\cdot\|$ defined on* 1154?
$(\ell^\infty(\mathbb{N}), \|\cdot\|_\infty)$ is residual, is the corresponding closed unit ball dentable?

Recall that a nonempty bounded subset A of a normed linear space X is *dentable* if for every $\varepsilon > 0$ there exists a $x^* \in X^* \setminus \{0\}$ and a $\delta > 0$ such that

$$\|\cdot\| - \operatorname{diam}\{a \in A : x^*(a) > \sup_{x \in A} x^*(x) - \delta\} < \varepsilon.$$

It is well-known that if the dual norm has a point of Fréchet differentiability then B_X is dentable [**75**].

3. The complex Bishop–Phelps property

For S a subset of a (real or complex) Banach space X, we may recast the notion of *support functional* as follows: a nonzero functional $\varphi \in X^*$ is a support functional for S and a point $x \in S$ is a *support point* of S if $|\varphi(x)| = \sup_{y \in S} |\varphi(y)|$.

Let us say a set is *supportless* if there is no such φ.

As Phelps observed in [**66**] while the Bishop–Phelps construction resolved Klee's question [**51**, p. 98] of the existence of support points in real Banach space, it remained open in the complex case. Lomonosov, in [**52**], gives the first example of a closed convex bounded convex set in a complex Banach space with no support functionals.

Question 3.1. *Characterise (necessarily complex) Banach spaces which admit* 1155?
supportless sets.

It is known that they must fail to have the Radon–Nikodým property [**52, 53**].

A Banach space X has the *attainable approximation property (AAP)* if the set of support functionals for any closed bounded convex subset $W \subseteq X$ is norm dense in X^*. In [**53**] Lomonosov shows that if a uniform dual algebra $\mathcal{R}$ of operators on a Hilbert space has the AAP then $\mathcal{R}$ is self-adjoint.

Question 3.2. *Characterise complex Banach spaces with the AAP. In particular* 1156?
do they include $L^1[0,1]$?

4. Biorthogonal sequences and support points

Uncountable biorthogonal systems provide the easiest way to produce sets with prescribed support properties.

4.1. Constructible convex sets and biorthogonal families. A closed convex set is *constructible* [**10**] if it is expressible as the countable intersection of closed half-spaces. Clearly every closed convex subset of a separable space is constructible. More generally:

Theorem 4.1 ([**10**]). *Let X be a Banach space, then the following are equivalent.*

(i) *There is an uncountable family $\{x_\alpha\} \subseteq X$ such that $x_\alpha \notin \overline{\operatorname{conv}}(\{x_\beta : \beta \neq \alpha\})$ for all α.*
(ii) *There is a closed convex subset in X that is not constructible.*
(iii) *There is an equivalent norm on X whose unit ball is not constructible.*

(iv) *There is a bounded uncountable system* $\{x_\alpha, \phi_\alpha\} \subseteq X \times X^*$ *such that* $\phi_\alpha(x_\alpha) = 1$ *and* $|\phi_\alpha(x_\beta)| \leq a$ *for some* $a < 1$ *and all* $\alpha \neq \beta$.

Example ([**10**]). *The sequence space* c_0 *considered as a subspace of* ℓ_∞ *is not constructible. Consequently, no bounded set with nonempty interior relative to* c_0 *is constructible as a subset of* ℓ_∞. *In particular the unit ball of* c_0 *is not constructible when viewed as a subset of* ℓ_∞.

In particular, if X admits an uncountable biorthogonal system then it admits a nonconstructible convex set. Under additional set-theoretic axioms, there are nonseparable Banach spaces in which all closed convex sets are constructible. These are known to include: (i) the $C(K)$ space of Kunen constructed under the Continuum Hypothesis (CH) [**64**], and (ii) the space of Shelah constructed under the diamond principle [**73**]. In consequence, these nonseparable spaces of Kunen and Shelah have the property that for each equivalent norm, the dual unit ball is weak*-separable, [**10**].

1157? **Question 4.1.** *Can one construct an example of a nonseparable space whose dual ball is weak* separable for each equivalent norm using only* ZFC*?*

In contrast, it is shown in [**10**] that there are general conditions under which nonseparable spaces are known to have uncountable biorthogonal systems. Suppose X is a nonseparable Banach space such that

(i) X is a dual space, or
(ii) $X = C(K)$, for K compact Hausdorff, and one assumes Martin's Axiom along with the negation of the Continuum Hypothesis (MA + ¬CH).

Then X admits an uncountable biorthogonal system. Part (ii) is a deep recent result of S. Todorcevic, see for example [**41**, p. 5].

1158? **Question 4.2.** *When, axiomatically, does a continuous function space always admit an uncountable biorthogonal system?*

4.2. Support sets. In a related light, consider the question:

1159? **Question 4.3.** *Does every nonseparable* $C(K)$ *contain a closed convex set entirely composed of support points (the tangent cone is never linear)?*

In [**9**] it is shown that this is equivalent to $C(K)$ admitting an uncountable *semi-biorthogonal system*, i.e., a system $\{x_\alpha, f_\alpha\}_{1\leq\alpha<\omega_1} \subseteq X \times X^*$ such that $f_\alpha(x_\beta) = 0$ if $\beta < \alpha$, $f_\alpha(x_\alpha) = 1$ and $f_\alpha(x_\beta) \geq 0$ if $\beta > \alpha$. Moreover, [**9**] observes that Kunen's space is an example where this happens without there being an uncountable biorthogonal system assuming CH. Thus, the answer is 'yes' except perhaps when MA fails (along with CH).

4.3. Supportless sets. For a set C in a normed space X, $x \in C$ is a *weakly supported point* of C if there is a linear functional f such that the restriction of f to C is continuous and nonzero. Fonf [**35**], extending work of Klee [**50**] (see also Borwein–Tingley[**8**]) proves the following result which is in striking contrast to the Bishop–Phelps theorem in Banach spaces: *Every incomplete separable normed* *space X contains a closed bounded convex set C such that the closed linear span of C is all of X and C contains no weakly supported points.*

Let us call such a closed bounded convex set *supportless*. It is known that there are *Fréchet spaces* (complete metrizable locally convex spaces) which admit supportless sets. In [**65**] Peck shows that *for any sequence of nonreflexive Banach spaces $\{X_i\}$, in the product space $E = \prod_{i=1}^{\infty} X_i$, there is a closed bounded convex set that has no E^*-support points.* Peck also provides some positive results.

Question 4.4. *Characterise when a Fréchet space contains a closed convex supportless convex set?* 1160?

5. Best approximation

Even in Hilbert spaces and reflexive Banach spaces some surprising questions remain open.

Question 5.1. *Is there a nonconvex subset A of a Hilbert space H with the property that every point in $H \setminus A$ has a unique nearest point?* 1161?

Such a set is called a *Chebyshev set* and must be closed and bounded. For a good up-to-date general discussion of best approximation in Hilbert space we refer to [**27**]. Asplund [**2**] shows that if nonconvex Chebysev sets exist then among them are so called *Asplund caverns*—complements of open convex bodies. In finite dimensions, the Motzkin–Klee theorem establishes that all Chebyshev sets are convex. Four distinct proofs are given in [**7**, §9.2] which highlight the various obstacles in infinite dimensions.

Question 5.2. *Is there a closed nonempty subset A of a reflexive Banach space X with the property that no point outside A admits a best approximation in A? Is this possible in an equivalent renorm of a Hilbert space?* 1162?

The *Lau–Konjagin Theorem* (see [**5**]) states that in a reflexive space, *for every closed set A there is a dense (or generic) set in $X \setminus A$ which admits best approximations if and only if the norm has the Kadec–Klee property.* Thus, any counter example must have a non-Kadec–Klee norm and must be unbounded—via the Radon–Nikodým property. In [**5**], a class of reflexive non-Kadec–Klee norms is exhibited for which some nearest points always exist.

By contrast, in every nonreflexive space, James' Theorem [**34**] provides a closed hyperplane H with no best approximation: equivalently $H + B_X$ is open. More exactingly, two closed bounded convex sets with nonempty interior are called *companion bodies* and *antiproximinal* if their sum is open. Such research initiates with Edelstein and Thompson [**31**].

Question 5.3. *Characterise Banach spaces (over $\mathbb{R}$) that admit companion bodies.* 1163?

Such spaces include c_0 [**6**, **22**, **31**] and again do not include any space with the Radon–Nikodým property [**5**].

6. Metrizability of compact convex sets

One facet of the study of compact convex subsets of locally convex spaces is the determination of their metrizability in terms of topological properties of their extreme points. For example, a compact convex subset K of a Hausdorff locally convex space X is metrizable if, and only if, the extreme points of K (denoted $\mathrm{Ext}(K)$) are *Polish* (i.e., homeomorphic to a complete separable metric space), [**23**].

Since 1970 there have been many papers on this topic (e.g., [**23, 24, 45, 54, 69**] to name but a few).

1164? **Question 6.1.** *Let K be a nonempty compact convex subset of a Hausdorff locally convex space (over $\mathbb{R}$). Is K metrizable if, and only if, $A(K)$, the continuous real-valued affine mappings defined on K, is separable with respect to the topology of pointwise convergence on* $\mathrm{Ext}(K)$*?*

The answer to this problem is known to be positive in the following cases:

(i) if $\mathrm{Ext}(K)$ is Lindelöf, [**60**];
(ii) if $\overline{\mathrm{Ext}(K)} \setminus \mathrm{Ext}(K)$ is countable, [**60**].

Question 6.1 may be thought of as a generalisation of the fact that a compact Hausdorff space K is metrizable if, and only if, $C_p(K)$ is separable. Here $C_p(K)$ denotes the space of continuous real-valued functions defined on K endowed with the topology of pointwise convergence on K.

7. The boundary problem

Let $(X, \|\cdot\|)$ be a Banach space. A subset B of the dual unit ball B_{X^*} is called a *boundary* if for any $x \in X$, there is $x^* \in B$ such that $x^*(x) = \|x\|$. A simple example of boundary is provided by the set $\mathrm{Ext}(B_{X^*})$ of extreme points of B_{X^*}. This notion came into light after James' characterisation of weak compactness [**44**], and has been studied in several papers (e.g., [**16–19, 38–40, 70, 74, 76**]). In spite of significant efforts, the following question is still open (see [**38**, Question V.2] and [**30**, Problem I.2]):

1165? **Question 7.1.** *Let A be a norm bounded and $\tau_p(B)$ compact subset of X. Is A weakly compact?*

The answer to the boundary problem is known to be positive in the following cases:

(i) if A is convex, [**74**];
(ii) if $B = \mathrm{Ext}(B_{X^*})$, [**12**];
(iii) if X does not contain an isomorphic copy of $\ell_1(\Gamma)$ with $|\Gamma| = \mathfrak{c}$, [**17, 18**];
(iv) if $X = C(K)$ equipped with their natural norm $\|\cdot\|_\infty$, where K is an arbitrary compact space, [**16**].

Case (i) can be also obtained from James' characterisation of weak compactness, see [**39**]. The original proof for (ii) given in [**12**] uses, among other things, deep results established in [**11**]. Case (iii) is reduced to case (i): if $\ell_1(\Gamma) \not\subset X$,

$|\Gamma| = \mathfrak{c}$, and $C \subset B_{X^*}$ is 1-norming (i.e., $\|x\| = \sup\{|x^*(x)| : x^* \in C\}$), it is proved in [**17, 18**] that for any norm bounded and $\tau_p(C)$-compact subset A of X, the closed convex hull $\overline{\mathrm{co}}^{\tau_p(C)}(A)$ is again $\tau_p(C)$-compact; the class of Banach spaces fulfilling the requirements in (iii) is a wide class of Banach spaces that includes: weakly compactly generated Banach spaces or more generally weakly Lindelöf Banach spaces and spaces with dual unit ball without a copy of $\beta\mathbb{N}$. The techniques used in case (iv) are somewhat different, and naturally extend the classical ideas of Grothendieck, [**42**], that led to the fact that norm bounded $\tau_p(K)$-compact subsets of spaces $C(K)$ are weakly compact. It should be noted that it is easy to prove that for any set Γ, the boundary problem has also positive answer for the space $\ell^1(\Gamma)$ endowed with its canonical norm, see [**16, 18**].

We observe that the solution in full generality to the boundary problem without the concourse of James' theorem of weak compactness would imply an alternative proof of the following version of James' theorem itself: a Banach space X is reflexive if, and only if, each element $x^* \in X^*$ attains its maximum in B_X.

Finally, we point out that in the papers [**71, 79**], it has been claimed that the boundary problem was solved in full generality. Unfortunately, to the best of our knowledge both proofs appear to be flawed.

8. Separate and joint continuity

If X, Y and Z are topological spaces and $f\colon X \times Y \to Z$ is a function then we say that f is *jointly continuous* at $(x_0, y_0) \in X \times Y$ if for each neighbourhood W of $f(x_0, y_0)$ there exists a product of open sets $U \times V \subseteq X \times Y$ containing (x_0, y_0) such that $f(U \times V) \subseteq W$ and we say that f is *separately continuous* on $X \times Y$ if for each $x_0 \in X$ and $y_0 \in Y$ the functions $y \mapsto f(x_0, y)$ and $x \mapsto f(x, y_0)$ are both continuous on Y and X respectively.

Since the paper [**3**] of Baire first appeared there has been continued interest in the question of when a separately continuous function defined on a product of *nice* spaces admit a point (or many points) of joint continuity and over the years there have been many contributions to this area (e.g., [**15, 20, 21, 25, 26, 49, 56, 63, 68, 72, 77**] etc.). Most of these results can be classified into one of two types. (I) The existence problem, i.e., if $f\colon X \times Y \to \mathbb{R}$ is separately continuous find conditions on either X or Y (or both) such that f has at least one point of joint continuity. (II) The fibre problem, i.e., if $f\colon X \times Y \to \mathbb{R}$ is separately continuous find conditions on either X or Y (or both) such that there exists a nonempty subset R of X such that f is jointly continuous at the points of $R \times Y$.

The main existence problem is, [**78**]:

Question 8.1. *Let X be a Baire space and let Y be a compact Hausdorff space.* 1166?
If $f\colon X \times Y \to \mathbb{R}$ is separately continuous does f have at least one point of joint continuity?

We will say that a topological space X has the *Namioka Property* if for every compact Hausdorff space Y and every separately continuous function $f\colon X \times Y \to \mathbb{R}$ there exists a dense G_δ-subset G of X such that f is jointly continuous at each point of $G \times Y$. Similarly, we will say that a compact Hausdorff space Y has the *co-Namioka Property* if for every Baire space X and every separately continuous function $f: X \times Y \to \mathbb{R}$ there exists a dense G_δ-subset G of X such that f is jointly continuous at each point of $G \times Y$.

The main fibre problems are:

1167? **Question 8.2.** *Characterise the class of Namioka spaces.*

There are many partial results.

(i) Every Namioka space is Baire, [**72**];
(ii) Every separable Baire space and every Baire p-space is a Namioka space, [**72**];
(iii) Not every Baire space is a Namioka space, [**78**];
(iv) Every Lindelöf weakly α-favourable space is a Namioka space, [**49**]
(v) Every space expressible as a product of hereditarily Baire metric spaces is a Namioka space, [**20**].

1168? **Question 8.3.** *Characterise the class of co-Namioka spaces.*

There are many partial results.

(i) $\beta\mathbb{N}$ is not a co-Namioka space, [**28**];
(ii) Every Valdivia compact is a co-Namioka space, [**13**, **29**];
(iii) The co-Namioka spaces are stable under products, [**15**];
(iv) All scattered compacts K with $K^{(\omega_1)} = \emptyset$ are co-Namioka, where $K^{(\alpha)}$ denotes the αth derived set of K, [**28**];
(v) There exists a non co-Namioka compact space K such that $K^{(\omega_1)}$ is a singleton, [**43**].

A partial characterisation, in terms of a topological game on $C_p(K)$, is given in [**47**] for the class of compact spaces K such that: *for every weakly α-favourable space X and every separately continuous mapping $f: X \times K \to \mathbb{R}$ there exists a dense G_δ subset G of X such that f is jointly continuous at each point of $G \times K$.*

For an introduction to this topic, see [**56**, **68**]. Also see the seminal paper [**62**] by I. Namioka, as well as, the paper [**63**].

Acknowledgements

The material on the Boundary problem (§ 7) was provided by B. Cascales.

References

[1] E. Asplund, *Fréchet differentiability of convex functions*, Acta Math. **121** (1968), 31–47.
[2] E. Asplund, *Čebyšev sets in Hilbert space*, Trans. Amer. Math. Soc. **144** (1969), 235–240.
[3] R. Baire, *Sur les fonctions des variable réelles*, Ann. Math. Pura Appl. **3** (1899), 1–122.
[4] E. Bishop and R. R. Phelps, *A proof that every Banach space is subreflexive*, Bull. Amer. Math. Soc. **67** (1961), 97–98.
[5] J. M. Borwein and S. Fitzpatrick, *Existence of nearest points in Banach spaces*, Canad. J. Math. **41** (1989), no. 4, 702–720.
[6] J. M. Borwein, M. Jiménez-Sevilla, and J. P. Moreno, *Antiproximinal norms in Banach spaces*, J. Approx. Theory **114** (2002), no. 1, 57–69.

[7] J. M. Borwein and A. S. Lewis, *Convex analysis and nonlinear optimization*, CMS Books in Mathematics / Ouvrages de Mathématiques de la SMC, vol. 3, Springer, New York, 2000.

[8] J. M. Borwein and D. W. Tingley, *On supportless convex sets*, Proc. Amer. Math. Soc. **94** (1985), no. 3, 471–476.

[9] J. M. Borwein and J. D. Vanderwerff, *Banach spaces that admit support sets*, Proc. Amer. Math. Soc. **124** (1996), no. 3, 751–755.

[10] J. M. Borwein and J. D. Vanderwerff, *Constructible convex sets*, Set-Valued Anal. **12** (2004), no. 1–2, 61–77.

[11] J. Bourgain, D. H. Fremlin, and M. Talagrand, *Pointwise compact sets of Baire-measurable functions*, Amer. J. Math. **100** (1978), no. 4, 845–886.

[12] J. Bourgain and M. Talagrand, *Compacité extrémale*, Proc. Amer. Math. Soc. **80** (1980), no. 1, 68–70.

[13] A. Bouziad, *Jeux topologiques et points de continuité d'une application séparément continue*, C. R. Acad. Sci. Paris Sér. I Math. **310** (1990), no. 6, 359–361.

[14] A. Bouziad, *Une classe d'espaces co-Namioka*, C. R. Acad. Sci. Paris Sér. I Math. **310** (1990), no. 11, 779–782.

[15] A. Bouziad, *The class of co-Namioka compact spaces is stable under product*, Proc. Amer. Math. Soc. **124** (1996), no. 3, 983–986.

[16] B. Cascales and G. Godefroy, *Angelicity and the boundary problem*, Mathematika **45** (1998), no. 1, 105–112.

[17] B. Cascales, G. Manjabacas, and G. Vera, *A Krein–šmulian type result in Banach spaces*, Quart. J. Math. Oxford Ser. (2) **48** (1997), no. 190, 161–167.

[18] B. Cascales and R. Shvydkoy, *On the Krein–šmulian theorem for weaker topologies*, Illinois J. Math. **47** (2003), no. 4, 957–976.

[19] B. Cascales and G. Vera, *Topologies weaker than the weak topology of a Banach space*, J. Math. Anal. Appl. **182** (1994), no. 1, 41–68.

[20] J. Chaber and R. Pol, *On hereditarily Baire spaces, σ-fragmentability of mappings and Namioka property*, Topology Appl. **151** (2005), 132–143.

[21] J. P. R. Christensen, *Joint continuity of separately continuous functions*, Proc. Amer. Math. Soc. **82** (1981), no. 3, 455–461.

[22] Ş. Cobzaş, *Antiproximinal sets in Banach spaces of continuous vector-valued functions*, J. Math. Anal. Appl. **261** (2001), no. 2, 527–542.

[23] H. H. Corson, *Metrizability of compact convex sets*, Trans. Amer. Math. Soc. **151** (1970), 589–596.

[24] G. Debs, *Quelques propriétés des espaces α-favorables et applications aux convexes compacts*, Ann. Inst. Fourier (Grenoble) **30** (1980), no. 2, 29–43.

[25] G. Debs, *Points de continuité d'une fonction séparément continue*, Proc. Amer. Math. Soc. **97** (1986), no. 1, 167–176.

[26] G. Debs, *Pointwise and uniform convergence on a Corson compact space*, Topology Appl. **23** (1986), no. 3, 299–303.

[27] F. Deutsch, *Best approximation in inner product spaces*, CMS Books in Mathematics / Ouvrages de Mathématiques de la SMC, vol. 7, Springer, New York, 2001.

[28] R. Deville, *Convergence ponctuelle et uniforme sur un espace compact*, Bull. Polish Acad. Sci. Math. **37** (1989), no. 7–12, 507–515 (1990).

[29] R. Deville and G. Godefroy, *Some applications of projective resolutions of identity*, Proc. London Math. Soc. **67** (1993), no. 1, 183–199.

[30] R. Deville, G. Godefroy, and V. Zizler, *Smoothness and renormings in Banach spaces*, Longman Scientific and Technical, Harlow, 1993.

[31] M. Edelstein and A. C. Thompson, *Some results on nearest points and support properties of convex sets in c_0*, Pacific J. Math. **40** (1972), 553–560.

[32] M. Fabián, *Gâteaux differentiability of convex functions and topology*, John Wiley & Sons Inc., New York, 1997.

[33] M. Fabián and G. Godefroy, *The dual of every Asplund space admits a projectional resolution of the identity*, Studia Math. **91** (1988), no. 2, 141–151.

[34] M. Fabiań, P. Habala, P. Hájek, V. Montesinos Santalucía, J. Pelant, and V. Zizler, *Functional analysis and infinite-dimensional geometry*, Springer, New York, 2000.

[35] V. P. Fonf, *On supportless convex sets in incomplete normed spaces*, Proc. Amer. Math. Soc. **120** (1994), no. 4, 1173–1176.

[36] J. R. Giles, P. S. Kenderov, W. B. Moors, and S. D. Sciffer, *Generic differentiability of convex functions on the dual of a Banach space*, Pacific J. Math. **172** (1996), no. 2, 413–431.

[37] J. R. Giles and W. B. Moors, *A selection theorem for quasi-lower semicontinuous set-valued mappings*, J. Nonlinear Convex Anal. **2** (2001), no. 3, 345–350.

[38] G. Godefroy, *Boundaries of a convex set and interpolation sets*, Math. Ann. **277** (1987), no. 2, 173–184.

[39] G. Godefroy, *Five lectures in geometry of Banach spaces*, Seminar on Functional Analysis (Murcia, 1987), Universidad de Murcia, Murcia, 1988, pp. 9–67.

[40] G. Godefroy, *Some applications of Simons' inequality*, Serdica Math. J. **26** (2000), no. 1, 59–78.

[41] G. Godefroy, *Banach spaces of continuous functions on compact spaces*, Recent progress in general topology, II, North-Holland, Amsterdam, 2002, pp. 177–199.

[42] A. Grothendieck, *Critères de compacité dans les espaces fonctionnels généraux*, Amer. J. Math. **74** (1952), 168–186.

[43] R. Haydon, *A counterexample to several questions about scattered compact spaces*, Bull. London Math. Soc. **22** (1990), no. 3, 261–268.

[44] R. C. James, *Weakly compact sets*, Trans. Amer. Math. Soc. **113** (1964), 129–140.

[45] J. E. Jayne, *Metrization of compact convex sets*, Math. Ann. **234** (1978), no. 2, 109–115.

[46] O. Kalenda, *Stegall compact spaces which are not fragmentable*, Topology Appl. **96** (1999), no. 2, 121–132.

[47] P. S. Kenderov, I. S. Kortezov, and W. B. Moors, *Weakly continuous mappings into Banach spaces*, C. R. Acad. Bulgare Sci. **54** (2001), no. 3, 7–10.

[48] P. S. Kenderov and W. B. Moors, *A dual differentiation space without an equivalent locally uniformly rotund norm*, J. Aust. Math. Soc. **77** (2004), no. 3, 357–364.

[49] P. S. Kenderov and W. B. Moors, *Separate continuity, joint continuity and the Lindelöf property*, Proc. Amer. Math. Soc. **134** (2006), no. 5, 1503–1512.

[50] V. L. Klee, Jr., *Convex sets in linear spaces*, Duke Math. J. **18** (1951), 443–466.

[51] V. L. Klee, Jr., *Extremal structures of convex sets. II*, Math. Z. **69** (1958), 90–104.

[52] V. Lomonosov, *A counterexample to the Bishop–Phelps theorem in complex spaces*, Israel J. Math. **115** (2000), 25–28.

[53] V. Lomonosov, *The Bishop–Phelps theorem fails for uniform non-selfadjoint dual operator algebras*, J. Funct. Anal. **185** (2001), no. 1, 214–219.

[54] B. MacGibbon, *A criterion for the metrizability of a compact convex set in terms of the set of extreme points*, J. Funct. Anal. **11** (1972), 385–392.

[55] S. Mazur, *Über konvexe Mengen in linearen normierten Raümen*, Studia Math. **4** (1933), 70–84.

[56] S. Mercourakis and S. Negrepontis, *Banach spaces and topology. II*, Recent progress in general topology (Prague, 1991), North-Holland, Amsterdam, 1992, pp. 493–536.

[57] W. B. Moors, *The relationship between Goldstine's theorem and the convex point of continuity property*, J. Math. Anal. Appl. **188** (1994), no. 3, 819–832.

[58] W. B. Moors, *Some more recent results concerning weak Asplund spaces*, Abstr. Appl. Anal. (2005), no. 3, 307–318.

[59] W. B. Moors and J. R. Giles, *Generic continuity of minimal set-valued mappings*, J. Austral. Math. Soc. Ser. A **63** (1997), no. 2, 238–262.

[60] W. B. Moors and I. Namioka, *Separability in $a(k)$*, 1996.

[61] W. B. Moors and S. Somasundaram, *Some recent results concerning weak Asplund spaces*, Acta Univ. Carolin. Math. Phys. **43** (2002), no. 2, 67–86.

[62] I. Namioka, *Separate continuity and joint continuity*, Pacific J. Math. **51** (1974), 515–531.

[63] I. Namioka and R. Pol, *Mappings of Baire spaces into function spaces and Kadec renorming*, Israel J. Math. **78** (1992), no. 1, 1–20.

[64] S. Negrepontis, *Banach spaces and topology*, Handbook of set-theoretic topology, North-Holland, Amsterdam, 1984, pp. 1045–1142.

[65] N. T. Peck, *Support points in locally convex spaces*, Duke Math. J. **38** (1971), 271–278.

[66] R. R. Phelps, *The Bishop–Phelps theorem in complex spaces: an open problem*, Function spaces (Edwardsville, IL, 1990), Lecture Notes in Pure and Appl. Math., vol. 136, Marcel Dekker Inc., New York, 1992, pp. 337–340.

[67] R. R. Phelps, *Convex functions, monotone operators and differentiability*, Lecture Notes in Math., vol. 1364, Springer, Berlin, 1993.

[68] Z. Piotrowski, *Separate and joint continuity. II*, Real Anal. Exchange **15** (1980), no. 1, 248–258.

[69] E. A. Reznichenko, *Convex compact spaces and their maps*, Topology Appl. **36** (1990), no. 2, 117–141.

[70] G. Rodé, *Superkonvexität und schwache Kompaktheit*, Arch. Math. (Basel) **36** (1981), no. 1, 62–72.

[71] V. I. Rybakov, *On the convergence on the boundary of the unit ball in a dual space*, Mat. Zametki **59** (1996), no. 5, 753–758, Translation: Math. Notes 59 (1996) no. 5, 543–546.

[72] J. Saint-Raymond, *Jeux topologiques et espaces de Namioka*, Proc. Amer. Math. Soc. **87** (1983), no. 3, 499–504.

[73] S. Shelah, *Uncountable constructions for B.A., e.c. groups and Banach spaces*, Israel J. Math. **51** (1985), no. 4, 273–297.

[74] S. Simons, *A convergence theorem with boundary*, Pacific J. Math. **40** (1972), 703–708.

[75] V. Šmulian, *Sur la dérivabilité de la norme dans l'espace de Banach*, C. R. (Doklady) Acad. Sci. URSS (N.S.) **27** (1940), 643–648.

[76] C. Stegall, *Applications of descriptive topology in functional analysis*, 1985.

[77] M. Talagrand, *Deux généralisations d'un théorème de I. Namioka*, Pacific J. Math. **81** (1979), no. 1, 239–251.

[78] M. Talagrand, *Espaces de Baire et espaces de Namioka*, Math. Ann. **270** (1985), no. 2, 159–164.

[79] X. Wang and L. Cheng, *A note on weakly compact sets in Banach space*, Northeast. Math. J. **12** (1996), no. 4, 466–468.

Open Problems in Topology II – Edited by E. Pearl

Topological structures of ordinary differential equations

V. V. Filippov

Basic elements of the topological structures in questions look as follows.

Let U be a subset of the product $\mathbb{R} \times \mathbb{R}^n$. When considering differential equations $y' = f(t, y)$ or inclusions $y' \in f(t, y)$ having a right-hand side defined on the set U, we call a function z a *solution* to the equation if the graph of the function z lies in U, and (letting $\mathrm{dom}(z)$ denote the domain of the function z)

(a) $\mathrm{dom}(z)$ is a segment, z is generalized absolutely continuous and $z'(t) = f(t, z(t))$ (respectively, $z'(t) \in f(t, z(t))$) for almost all $t \in \mathrm{dom}(z)$, or
(b) $\mathrm{dom}(z)$ a singleton.

For equations with a continuous right-hand side our definition gives only continuously derivable solutions. For equations with a right-hand side satisfying the Caratheodory conditions this definition gives Caratheodory solutions.

In this definition we do not fix domains of functions. We consider functions with various domains together. We take as a distance between two solutions the Hausdorff distance between their graphs.

Let us emphasize some basic properties of the set Z of so defined solutions.

(1) If $z \in Z$ and a segment I lies in $\mathrm{dom}(z)$ then $z\big|_I \in Z$.
(2) If $z_1, z_2 \in Z$, $I = \mathrm{dom}(z_1) \cap \mathrm{dom}(z) \neq \emptyset$ and $z_1\big|_I = z_2\big|_I$ then the function

$$z(t) = \begin{cases} z_1(t) & \text{if } t \in \mathrm{dom}(z_1) \\ z_2(t) & \text{if } t \in \mathrm{dom}(z_2) \end{cases}$$

(defined on the segment $\mathrm{dom}(z_1) \cup \mathrm{dom}(z_2)$) belongs to the set Z.
(c) The set $Z_K = \{z \in Z : \text{the graph of } z \text{ is a subset of } K\}$ is compact for every compact K subset of U.
(e) For each point (t, y) of the set U there exists a function $z \in Z$ such that t is in the interior of $\mathrm{dom}(z)$ and $z(t) = y$.
(u) If $z_1, z_2 \in Z$, $\mathrm{dom}(z_1) = \mathrm{dom}(z_2)$ and $z_1(t) = z_2(t)$ for some $t \in \mathrm{dom}(z_1)$ then $z_1 = z_2$.

It is easy to see that the properties (1) and (2) follow directly from our definitions. Conditions (e) and (u) correspond to the existence theorem and the uniqueness theorem. Condition (c) corresponds to (upper semi-)continuity of the dependence of solutions to Cauchy problems on initial values.

The property of solution sets which corresponds to the continuity of the dependence of solutions on parameters looks as follows. $R(U)$ denotes the set of all sets of functions which are defined on segments and singletons and whose graphs lie in U, satisfying conditions (1) and (2). We say that a sequence $\{Z_i : i =$

$1, 2, \dots\} \subseteq R(U)$ *converges* (in U) to a space $Z \in R(U)$ if for any compact K subset of U and for any sequence $z_i \in (Z_{j_i})_K$ $(j_1 < j_2 < \cdots)$ there is a subsequence $\{z_{i_m} : m = 1, 2, \dots\}$ converging to a function $z \in Z$.

The notation $R_*(U)$ denotes the set of all elements of $R(U)$ satisfying all conditions listed in the subscript ($*$ may include c, e or u as above). The sets $R_*(U)$ are called *classes of solution spaces* or *spaces of solution spaces.*

On the class $R_c(U)$ the introduced convergence corresponds to a non-T_1 first countable topology.

It is surprising that these simple conditions suffice to account for a considerable part of the theory of ordinary differential equations, replacing the usual conditions of the continuity of the right-hand side. See [**2**] for details.

The first topological ideas arose in Analysis and Geometry. They passed to general topological notions and constructions when mathematicians felt that topological relations appeared more often in Mathematics that was previously understood. This implied many important consequences. It may be that the most brilliant consequence between those subjects was the creation of Functional Analysis.

But when (General) Topology was created it had many internal reasons for its development. The initial motivation, related to the necessity to serve other domains of Mathematics, was largely forgotten. The feeling of this omission encouraged me to try to compare the contents of General Topology itself and the contents of other domains of Mathematics.

Perhaps the most interesting observation in this direction was made when I saw that the analysis of the continuity of the dependence of solutions of ordinary differential equations on initial values and parameters leads to a topological structure of S. Nedev's type[**4**]. It was the *space of solution spaces* $R_{ce}(U)$. This first interest was related to the observation that Nedev's results give real information about the structure of this space. In particular, Nedev's theorems imply the metrizability of some subspaces of $R_{ce}(U)$. Later I understood that this topological structure gives a powerful tool for the theory of ordinary differential equations itself. Notions such as *first approximation, asymptotically autonomous spaces* receive here their natural importance without loss of possibility of their application. See below.

The new topological structure allows us to develop efficiently a theory which deals easily with equations having singularities and with equations with multivalued right-hand sides (differential inclusions). It extends the majority of assertions of the central part of the theory of ordinary differential equations in the existing framework to equations with complicated discontinuities of right-hand sides. The simplest example on this direction looks as follows.

Example. *Let the functions* $f, g\colon \mathbb{R} \to [1, 2]$ *be measurable. The equation*

$$y' = f(t) + g(y)$$

is far from the classical theory because it may have discontinuities both in time and in space variable but it is covered well by our approach.

Our approach reinforces the theory in the case of equations with continuous right-hand sides too.

Here are some other consequences. The notion of a *dynamical system* and that of an autonomous solution space, in which a solution for every Cauchy problem exists, is unique and is defined on the whole real line, give a different axiomatic description of the same object. So our results may be applied to studies of dynamical systems too.

So this research of a non-traditional topological description of relations in a domain of Mathematics was successful. A general problem arises.

Problem 1. *Find other as yet undiscovered topological relations in Mathematics and to try to use them to reinforce existing mathematical theories.*

This invitation contains nothing new. Such a rôle of topological structures (so, of General Topology) in Mathematics was highly praised in the famous N. Bourbaki's article *The Architecture of Mathematics* [**1**]. So the problem is to show that Bourbaki's appreciation is not exhausted by known cases of applications of topological structures.

In particular, some parts of the theory of partial differential equations and of equations in Banach spaces are close to the theory of ordinary differential equations. So, I ask:

Problem 2. *For which problems of the theory of partial differential equations and of equations in Banach spaces can this method be applied?*

One of first consequences of the usage of new structures for ordinary differential equations was a method of investigation of singularities. In the domain U under consideration we find open subsets V, $V \in \gamma$, which do not contains singularities. Then we try to find estimates of the remainder $R = U \setminus \bigcup \gamma$ which show that singularities lying in R do not influence the properties of solutions. The level of new topological structures is very suitable for this consideration. The simplest example on this direction looks as follows.

Example. *Let $f(t, y)$ be a polynomial. Then solutions to the equation*

$$y' = f(t, y) + \frac{\alpha}{t^2 + y^2 + \alpha^2}$$

depend continuously on the parameter α, although for $t = y = 0$ the second term tends to infinity when $\alpha \to 0$.

Problem 3. *For which other problems of the theory of ordinary differential equations (and outside it) can this method of study of singularities (using partial mappings) be applied?*

When we investigate a particular equation with singularities we need to prove the fulfillment of the above-listed basic conditions. We get this purpose in whole measure if we prove that the solution space Z in question belongs to the closure of the class $R_{ceu}(U)$: $Z \in \operatorname{cl}_{R_c(U)} R_{ceu}(U)$. Really, the condition $Z \in \operatorname{cl}_{R_c(U)} R_{ceu}(U)$ assures that the equation under consideration is covered by the theory completely. The following question remains unanswered:

1169? **Problem 4.** *Suppose that the domain U is covered by a family γ of open subsets and $Z_V \in \mathrm{cl}_{R_{ceu}(V)} R_c(V)$ for every $V \in \gamma$. Is necessarily $Z \in \mathrm{cl}_{R_c(U)} R_{ceu}(U)$?*

Now many chapters of the theory of ordinary differential equations are covered by the axiomatic approach. But the theory is very large and the problem to cover the entire theory will remain current for a long time. I do not think that all the topological effects of the theory of ordinary differential equations have been discovered yet.

In this investigation the main problems are not technical. The main problems are to understand topological contents of corresponding notions, constructions, and theorems.

Example (First approximation). *Usually they consider a vector equation*

$$y' = Ay + g(y),$$

where A is a matrix and $g(y)$ is small with respect to $\|y\|$: $\|g(y)\| = o(\|y\|)$. Instead, we consider the family of changes of variables, corresponding to homotheties $y \to \lambda y$, where $\lambda > 0$. This change of variables transforms our equation to the equation

$$y' = f(y) + \lambda g\left(\tfrac{1}{\lambda}y\right),$$

Denote its solution space by Z_λ. In our approach we replace the usual condition of first approximation by the convergence of Z_λ to the solution space of the equation

$$y' = Ay$$

as $\lambda \to \infty$ in the topological space $R_{ce}(U)$. We return to the classical version of this notion when we prove this convergence using the classical theorem on continuous dependence of solution on parameter λ.

Example (Asymptotically autonomous space). *Usually they consider a vector equation*

$$y' = f(y) + g(t, y),$$

where the term $g(t,y)$ has an estimate $\|g(t,y)\| \leq \phi(t)$ where the function ϕ is small at infinity. This means that $\phi(t) \to 0$ when $t \to \infty$ or that the function ϕ is integrable on (a, ∞). Instead, we consider the family of changes of variables, corresponding to translation along the time axis. Such translation transforms our equation to equation

$$y' = f(y) + g(t + \tau, y),$$

Denote its solution space by Z_τ. In our approach we require the convergence of Z_τ to the solution space of the equation

$$y' = f(y),$$

as $\tau \to \infty$ in the topological space $R_{ce}(U)$.

Because methods of proof of convergence in the space $R_c(U)$ are largely developed, there is a larger volume of new versions of notions of first approximation and of asymptotically autonomous spaces. Each theorem of the classical theory where those notions are used obtains a broader scope (and a possibility of application to equations and inclusions with discontinuous right-hand sides).

One of the consequences of this study was the creation of a new version of the translation method in the theory of boundary value problems in which the continuation principle works. Before, in the corresponding situation they used the Leray–Schauder theory. But the Leray–Schauder theory in general does not work with equations with a discontinuous right-hand side. Our new approach works; see [**3**] and my other articles. The new version of the translation method is based on studies of Čech homology of solution spaces. Perhaps something similar may be made by investigating shape properties of solution spaces. So, I ask:

Problem 5. *Is it possible to create an approach to the theory of boundary value* 1170?
problems based on shape properties of solution spaces and on essential mappings? (Of course, without assuming the uniqueness of solutions of the Cauchy problem.)

It may be simpler than the application of Čech homology.

References

[1] N. Bourbaki, *The architecture of mathematics*, Amer. Math. Monthly **57** (1950), 221–232.

[2] V. V. Filippov, *Basic topological structures of ordinary differential equations*, Kluwer Academic Publishers, Dordrecht, 1998.

[3] V. V. Filippov, *Which is better for the theory of boundary value problems for ordinary differential equations, the Leray–Schauder method or translation along the trajectories?*, Differ. Uravn. **37** (2001), no. 8, 1049–1061, Translation: Differ. Equ. 37 (2001) no. 8, 1097–1110.

[4] S. Ĭ. Nedev, *o-metrizable spaces*, Trudy Moskov. Mat. Obshch. **24** (1971), 201–236, Translation: Trans. Moscow Math. Soc. 24 (1971), 213–247.

Open Problems in Topology II – Edited by E. Pearl

The interplay between compact spaces and the Banach spaces of their continuous functions

Piotr Koszmider

Introduction

We will consider compact (always Hausdorff and infinite) spaces K and the set $C(K)$ of all continuous functions from K into the reals. If $C(K)$ is equipped with the supremum norm, it is a Banach space[1]. An isomorphism between Banach spaces is a linear isomorphism which is continuous (necessarily both ways, by the open mapping theorem). A $C(K)$ will mean a *Banach space of the form* $C(K)$ *for some compact* K. As general references that might be useful while reading this article we suggest [**9**, **16**, **49**] on functional analysis, [**22**, **33**] on set theory and [**15**] on topology.

If $T\colon C(K) \to C(L)$ is an *isometry* (i.e., a linear isomorphism which preserves the norm), T induces a homeomorphism between K and L (Banach–Stone, see [**49**, 7.8.4]), but many nonhomeomorphic Ks have isomorphic $C(K)$s.

The simpliest examples are of two disjoint convergent sequences K, i.e., $(y_{2m}) \to \infty_1$ and $(y_{2m-1}) \to \infty_2$ for $m > 0$ with their respective distinct limit points ∞_1 and ∞_2 and one convergent sequence L, i.e., $(x_n) \to \infty$ with its limit point ∞ (see e.g. [**2**] for more). One can explicitly define an isomorphism T:

$$\begin{aligned} T(f)(x_0) &= f(\infty_1) - f(\infty_2), \\ T(f)(x_{2m}) &= f(y_{2m}) - \tfrac{1}{2}(f(\infty_1) - f(\infty_2)), \\ T(f)(x_{2m-1}) &= f(y_{2m-1}) + \tfrac{1}{2}(f(\infty_1) - f(\infty_2)), \\ T(f)(\infty) &= \tfrac{1}{2}(f(\infty_1) + f(\infty_2)). \end{aligned}$$

for $m > 0$. This is clearly also an example of a $C(K)$ which is both a quotient (image under an onto linear and continuous mapping between Banach spaces) and a subspace (for us always meaning a closed linear subspace) of a $C(L)$ such that K is neither a subspace nor a continuous image of L.

Working with $C(K)$s then seems to be a weaker environment than working with compact spaces, because we identify compact spaces with the same $C(K)$ (in the isomorphic sense). It is just one point of view. From another perspective, we get more *continuous* mappings from K to K, i.e., not all operators on $C(K)$ come from a *usual* continuous mapping on K. The problems of this article could

[1]Other structures include: Banach algebra, C^*-algebra, lattice, ring, topological vector space with various topologies.

The author was partially supported by a research fellowship *Produtividade em Pesquisa* from National Research Council of Brazil (Conselho Nacional de Pesquisa, Processo Número 300369/01-8).

be summed up as: *What can we say about $C(K)$ if we know some topological properties of K?* (impose properties on $C(K)$ by manipulating Ks); or: *If $C(K)$ and $C(L)$ are related in the isomorphic sense, how are K and L related in the topological sense?* (get different $C(K)$s knowing that Ks are very different). This is certainly a very special and topological point of view since most of the theory of $C(K)$ spaces is developed having in mind only Banach space theory (for example, asking similar questions about the dual ball with the weak* topology instead of K). However as history shows, a K with sufficiently strong topological properties can produce a striking example of a Banach space. Another bias of the article is focusing on the isomorphic structure of subspaces, quotients and complemented subspaces on the $C(K)$s.

The ideal results here would be describing which Ks have isomorphic $C(K)$s. There are only a few such results. For example, $C(K)$ is isomorphic to $C([0,1])$ if and only if K is an uncountable metrizable compact space ([**38**]). Similarly one could characterize $C(K)$s for countable Ks ([**6**]) or $C(K)$s isomorphic to $C(L)$ where L is the one point compactification of a discrete space, i.e., is isomorphic to a $c_0(\kappa)$ for some cardinal κ ([**36**]).

Another natural type of an interesting result is to prove that if $C(K)$ and $C(L)$ are isomorphic and K has some topological property, then L has it as well. This holds for properties such as being dispersed, being Eberlein, being ccc, being metrizable and many others. Probably the most well-known (see [**23**] for definitions and related theory) open problem here is the following:

1171? **Question 1.** *If K is a Corson compact and $C(L)$ is isomorphic to $C(K)$, must L be a Corson compact?*

This is true if we assume $\mathsf{MA} + \neg\mathsf{CH}$, (see [**4**]).

As noted above, the Banach–Stone theorem gives a special place to *isometries* between $C(K)$s, hence even though we are interested in the isomorphic theory, the isometries will be mentioned. We have isometries between ℓ_∞ and $C(\beta\mathbb{N})$, c and $C([0,\omega])$, $C(\omega^*)$ and ℓ_∞/c_0 (see [**35**]). Also c_0 and c are isomorphic (but not isometric).

The main tools of functional analysis include the use of the dual and the bidual of the Banach space. In the case of $C(K)$s we are in a very privileged situation, (like in the case of ℓ_p or $L_p(\mu)$ spaces), i.e., we can see the functionals and even the elements of the bidual with the naked eye. The Riesz representation theorem (see [**49**, 18]) says that any continuous functional ϕ on a $C(K)$ can be isometrically associated with a unique Radon[2] measure μ on K by the formula $\phi(f) = \int f d\mu$ for all $f \in C(K)$ where the integration is like in the Lebesgue theory. Even the norm of ϕ is nicely describable by μ; it is the variation of μ, i.e., the supremum of the expressions of the form $|\mu(A_1)| + \cdots + |\mu(A_n)|$ for A_i pairwise disjoint and Borel. The addition of measures and multiplication by a scalar is setwise. Thus,

[2] A Radon measure here is a Borel, countably additive, signed, regular measure. Regularity for a signed measure means that for any Borel $A \subseteq K$ the value of $|\mu|(U - F)$ is arbitrarily small for some open U and compact F such that $F \subseteq A \subseteq U$.

we will use letters μ, ν, λ for elements of the dual of K (as a Banach space) which will be identified with the Banach space $M(K)$ of the Radon measures on K with the variation norm. As usual, one can decompose the Radon measure μ into its positive and negative parts μ_+, μ_- and define $\mu_+ - \mu_- = |\mu|$ which is a positive Radon measure (see [**49**, 17.2.5, 18.2.5]). $M(K)$ has a natural topology where K with its original topology is a subspace, i.e., the weak* topology defined by the subbasis of the sets of the form $\{\mu \in M(K) : \mu(f) \in I\}$ where $f \in C(K)$ and I is an open interval of the reals. Here $\{\delta_x : x \in K\}$ is a copy of K where $\delta_x(A) = 1$ if $x \in A$ and is 0 otherwise. This copy of K is quite big; its span is weak* dense in $M(K)$.

With the help of the dual one can see how the $C(K)$ partially loses information about K. If $T\colon C(K) \to C(L)$ is an operator (i.e., a linear continous function), define $T^*\colon M(L) \to M(K)$ by $T^*(\nu) = \nu \circ T$. Deciphering it in terms of the integration we get $\int f d(T^*(\nu)) = \int T(f) d\nu$. For example, if ν is the simplest Radon measure, the Dirac measure δ_x concentrated on a point $x \in K$, we have $\int T(f) d\delta_x = T(f)(x)$. That is, $T(f)(x)$ is the value of the functional $T^*(\delta_x)$ on f. If T is given by $T(f) = f \circ F$ where $F\colon K \to K$ is continuous, then $T(f)(x) = f(F(x))$, i.e., $T^*(\delta_x) = \delta_{F(x)}$ In other words, T^* essentially is F. However, in general $T^*(\delta_x)$ may be some more complicated measure, and in this way it loses information about K. In the example from the beginning of this section we have $T^*(\delta_{x_0}) = \delta_{\infty_1} - \delta_{\infty_2}$. No continuous function from L into K sends a point onto a linear combination of two points. Thus, one way of proving negative properties of $C(K)$s is to strengthen the topological properties of K, taking care not only of point-to-point continuous functions but also taking care of point-to-measure weak* continuous functions. Note that by knowing that the span of the pointwise measures δ_x for $x \in K$ is dense in the weak* topology in $M(K)$ and that T^* is always continuous in the weak* topologies, we may really restrict our attention to $T^*(\delta_x)$s to obtain complete information about T^*.

If points of K can be considered as functionals on $C(K)$, then functions of K should be functionals on functionals, i.e., elements of the bidual. Indeed $C(K)$, as any Banach space, canonically emdeds in its bidual, but also any bounded Borel function $g\colon K \to \mathbb{R}$ defines a functional Ψ_g on $M(K)$ by[3] $\Psi_g(\mu) = \int g d\mu$. As points span a weak* dense set in $M(K)$, the Borel sets (i.e., their characteristic functions) span a weak* dense set in the bidual of $C(K)$. If $X \subseteq M(K)$ is a separable subspace, then by the Radon–Nikodým theorem there is an isometry of X and a subspace of $L_1(\mu)$ for some $\mu \in M(K)$ (just take $\mu = \sum |\mu_n|/2^n \|\mu_n\|$ where $\{\mu_n : n \in \mathbb{N}\}$ is dense in X) and so $L_\infty(\mu)$ is the bidual of a superspace of this separable piece of $M(K)$ ([**49**, 27.1.3]). More on entire biduals of $C(K)$s can be found in [**25**, **26**] and [**49**, 27.2].

This world of dualities can be even more tangible and combinatorial if we are allowed to think about K as a dual of something more primitive, namely the

[3]This observation may serve to note that a $C(K)$ is reflexive as a Banach space iff the characteristic function of any Borel set is continuous, i.e., if and only if K is finite if and only if $C(K)$ is finite-dimensional. So, for example, ℓ_2 is not isomorphic to any $C(K)$.

Boolean algebra of clopen sets, such as when K is totally disconnected and Stone duality (see [**28**]) may enter the game. Then, by the Weierstrass–Stone theorem (see [**49**, 7.3]) the finite linear combinations of characteristic functions of clopen sets form a norm-dense subset of the $C(K)$, i.e., the span of the Boolean algebra is dense in $C(K)$. (Then the Hanh–Banach theorem and the Tarski ultrafilter theorem become one.) Radon measures on totally disconnected K are completely determined by their restrictions to clopen sets and the dual space may be interpreted as the space of *finitely* additive bounded signed measures of the Boolean algebra ([**49**, 18.7]).

If K is metrizable, then $C(K)$ is isomorphic to a $C(L)$ for L totally disconnected. Indeed, we have a classification of separable $C(K)$s, i.e., those whose Ks are metrizable (see [**49**, 7.6.5]): such $C(K)$s are isomorphic to $C(2^\omega)$, $C([0,\omega])$ or $C([0,\alpha])$ for $\alpha < \omega_1$ such that $\beta^n < \alpha$ for all $\beta < \alpha$ and $n \in \omega$ (due to Milutin [**38**], Bessaga, Pełczyński [**6**]). Here we need to admit another bias of this article: we are mainly interested in nonseparable $C(K)$s. The issues related to separable $C(K)$s are more analytic and are presented in detail in [**47**]. One wonders if any $C(K)$ is isomorphic to a $C(L)$ for L totally disconnected; the latter type of a $C(K)$ will be called *Boolean* in the sequel. Only recently it turned out that it is not the case ([**30**, **43**]), and the reason is quite ad hoc; the $C(K)$ of [**30**] is indecomposable. So the question remains when and why a $C(K)$ is Boolean.

1172–1174? **Question 2.** *If K is small compact space, that is,* (a) *first countable,* (b) *of weight smaller than 2^ω under* $\mathsf{MA} + \neg\mathsf{CH}$, *or* (c) *Eberlein, is $C(K)$ isomorphic to a Boolean $C(L)$?*

For more on these kinds of questions and the methods that are being used to answer them, see the section on complemented subspaces. Regardless of the results of [**30**] one may still hope for some theorem which would explain the special role of Boolean $C(K)$s not just in heuristic terms. For example, one could hope for some transfer principle which would imply general statements about $C(K)$s from those proved about Boolean $C(K)$s. Consider the following:

1175? **Question 3.** *Does every nonseparable $C(K)$ have a subspace which has a quotient isomorphic to a nonseparable Boolean $C(L)$?*

If the answer were positive, one could obtain some uncountable objects (those which are preserved when going to superspaces and preimages of operators) in any nonseparable $C(K)$ just by knowing that they exist in those with K nonmetrizable and totally disconnected[4]. Note that there is a dual result ([**11**]): every $C(K)$ is complemented (and hence both a quotient and a subspace, see the section on complemented subspaces for definitions) in a Boolean $C(K)$ of the same density. See [**5**] for more research on these issues.

[4]For example, we knew since [**52**] that it is consistent that every nonseparable Boolean $C(K)$ has an uncountable biorthogonal system. However, only recently ([**53**]) we have a separate proof that it is consistent for any $C(K)$.

The above question however shouldn't be about complemented subspaces of $C(K)$s since the indecomposable $C(K)$ of [**30**] would be a counterexample. There is a topological version of the previous question:

Question 4. *Is it consistent that every nonmetrizable compact K has a continuous image with a nonmetrizable compact subspace which is totally disconnected?* 1176?

We mention here only the consistency since we have a counterexample under CH.

Subspaces

If $F\colon K \to L$ is a continuous onto mapping, then $T_F(f) = f \circ F$ is an isometry of $C(L)$ and a subspace of $C(K)$. That is, continuous images of topological spaces produce isometric subspaces of $C(K)$s.

The Banach–Mazur theorem ([**49**, 8.7]) says that any separable Banach space is isometric to a subspace of a $C([0,1])$. In particular, it is easy for a subspace of a $C(K)$ not to be isomorphic to any space of the form $C(L)$. In general, by simple duality any Banach space is a subspace of a $C(K)$ where K is its dual ball with the weak* topology. Just send x to x^{**} in the bidual and restrict it to the dual ball. The weight of this K in the weak* topology is the density of X. Thus every Banach space embeds isometrically in a $C(K)$ of the same density. Under CH, every compact K of weight $\leq 2^\omega$ is a continuous image of ω^*. Thus, we have that under CH any Banach space of density not bigger than 2^ω is a subspace of $C(\omega^*) \equiv \ell_\infty/c_0$. It is well known that without CH spaces like $[0,\omega_2]$ are not continuous images of ω^*. What about the $C(K)$ analog?

Question 5. (a) *Is it consistent that $C(\omega^*) \equiv \ell_\infty/c_0$ does not contain an isomorphic copy of some Banach space of density not bigger than 2^ω?* (b) *Can this space be $C([0,\omega_2])$?* 1177–1178?

Of course one may state the above question for other spaces instead of ω^*, for example, spaces which consistently are 2^ω-Parovichenko (see [**12**]), or in general:

Question 6. *Is it provable in* ZFC *that there is a compact K of weight $\leq 2^\omega$ such that every Banach space of density $\leq 2^\omega$ is isomorphic to a subspace of $C(K)$?* 1179?

This is related to many very deep and influential on Banach space theory successful attempts of characterizing the existence of copies of some Banach space inside a $C(K)$ space in terms of topological properties of K. For example, for many infinite cardinals κ it is consistent that $C(K)$ has a subspace isomorphic to $\ell_1(\kappa)$ if and only if K maps onto $[0,1]^\kappa$. An excellent survey of this gigantic project developed over a few decades by Argyros, Fremlin, Haydon, Pełczyński, Plebanek and others is [**42**]. There one can find many related references and open questions. Another simple example of this kind of inquiry could be that an isomorphic copy of $c_0(\omega_1)$ is a subspace of $C(K)$ if and only if K is not ccc: $c_0(\omega_1)$ has a weakly compact subset which is not separable, thus by a result of [**46**], which says that K is ccc if and only if weakly compact subsets of $C(K)$ are separable, K is not ccc. However one can easily check that there is no compact L for which it is true that for every compact K the space L is a continuous image of K if and only if K is not ccc.

Quotients

If K is a subspace of L then $T_K: C(L) \to C(K)$ given by $T(f) = f \upharpoonright K$ is an onto operator, i.e., topological subspaces produce quotients in function spaces, $C(L)/\operatorname{Ker}(T)$ is isometric to $C(K)$. Our example of one and two convergent sequences shows that there may be more quotients of a $C(K)$ than subspaces of K. Also spaces like ℓ_2, i.e., not isomorphic to $C(K)$s, may be quotients of $C(K)$s (see [**34**]). One should remember that the image of a linear operator defined on a Banach space does not have to be a quotient, as the image of a linear continuous bijection does not have to be an isomorphism. Simply, the images may not be Banach spaces; they may not be complete.

Two basic examples of compact spaces are a convergent sequence $[0, \omega]$ and $\beta\mathbb{N}$, the Čech–Stone compactification of the integers. Since for a compact space having a copy of $\beta\mathbb{N}$ as a subspace is equivalent to having $[0,1]^{2^\omega}$ as a continuous image, and for a Banach space having ℓ_∞ as a quotient is equivalent to having $\ell_1(2^\omega)$ as a subspace, the question of ℓ_∞ as a quotient is more related to the fragment of the previous section where we referred the reader to [**42**]. Thus we will concentrate on c_0 as a quotient.

For a $C(K)$ not having c_0 as a quotient is equivalent to a well-known Banach space theory property of Grothendieck (see [**48**, 5.1. ii) and 5.3]). A Banach space X is said to have the Grothendieck property if weak* convergent sequences in the dual X^* are weakly convergent. This is what we get in the Banach space language about $X = C(K)$ if we want to guarantee that K has no convergent sequences. Indeed, this roughly says that if a bounded sequence of measures, like for example $(\delta_{x_n})_{n\in\mathbb{N}}$, is separated into two parts in the weak topology on $M(K)$, e.g., by a Borel subset, then it is separated by a continuous function on K. In particular if a $C(K)$ has the Grothendieck property then K has no nontrivial convergent sequences. The Grothendieck property is stronger. However no perfect solution exists, i.e., having a convergent sequence is not a Banach space theory property. Consider K, the Stone space of the subalgebra of $\mathcal{P}(\mathbb{N})$ of all subsets a of $\mathbb{N}$ such that $2n \in a$ if and only if $2n+1 \in a$ for all but finitely many $n \in \mathbb{N}$. It is easy to see that, like $\beta\mathbb{N}$, it has no convergent sequences, but $P(f) = (f(2n) - f(2n-1))_{n\in\mathbb{N}}$ defines an operator from C(K) onto c_0 ([**48**, 4.10]). Also $C(K) \sim \ell_\infty \oplus c_0 \sim C(L)$ where L is a disjoint union of $\beta\mathbb{N}$ and $[0, \omega]$. That is, $C(K)$s can be isomorphic even though one K has convergent sequences and the other does not. One should also realize that a $C(K)$ without the Grothendieck property may verify it for all atomic measures; there are even $C(K)$s without the Grothendieck property such that for every separable $L \subseteq K$, the space $C(L)$ has the Grothendieck property ([**44**]). Since [**48**] the following question attributed to Lindenstrauss is left open:

1180? **Question 7.** *Can we characterize topologically compact Ks such that $C(K)$ has the Grothendieck property?*

There is another Banach space theoretic nonequivalent way of guaranteeing that a totally disconnected K has no convergent sequences: require that $C(K)$ has the Nikodým property, i.e., whenever elements μ_n of $M(K)$ have their values $\mu_n(a)$ bounded for each clopen $a \subseteq K$, then μ_n are all norm-bounded ([**48**, 4.6]). If K is the Stone space of the Boolean algebra of Jordan measurable subsets of $[0, 1]$ then K has the Nikodým property but does not have the Grothendieck property ([**48**, 3.2., 3.3]). Assuming CH Talagrand constructed a Boolean algebra whose Stone space has the Grothendieck property but lacks the Nikodým property ([**51**]). Here the main remaining question is the following:

Question 8. *Is it consistent that every Boolean $C(K)$ which has the Grothendieck* 1181?
property has the Nikodým property?

In Talagrand's construction the algebra A of clopen subsets of K has the following property (∗): *there is a countable subalgebra $A_0 \subseteq A$ such that for any $a \in A$ there are $a_n \in A_0$ such that $\lambda(a_n \bigtriangleup a) \to 0$ (i.e., λ has countable Maharam type) and the measures witnessing the failure of the Nikodým property are absolutely continuous with respect to λ.* We do not know if this must be true for all algebras like in [**51**]. Thus an easier version of Question 7 would be:

Question 9. *Is it consistent that every $C(K)$ for K totally disconnected with* 1182?
property (∗) does not have the Grothendieck property?

Requiring that a totally disconnected K has both the Nikodým and Grothendieck properties is thus even a stronger way of imposing that K has no convergent sequences. It turns out that it is equivalent to the known Vitali–Hahn–Saks property. For a Boolean algebraic and measure-theoretic treatment of these properties see [**48**].

One can multiply questions about $C(K)$s with the Grothendieck, Nikodým, or Vitali–Hahn–Saks property like questions on convergent sequences of compact spaces. For example, what about the weights of compact spaces without convergent sequences? We have many interesting topological results on it (see [**13**]) and just one $C(K)$ result of Brech ([**7**]) showing that it is consistent that there are $C(K)$s with the Grothendieck property of density less than the continuum. One could define cardinal invariants $\mathfrak{gr}$, $\mathfrak{ni}$, $\mathfrak{vhs}$ as minimal infinite densities of $C(K)$s (K totally disconnected in the second and third case) which have Grothendieck, Nikodým, Vitali–Hahn–Saks properties respectively and $\mathfrak{ncs}$ could stand for the minimal infinite weight of a compact space without a converging sequence. We have $\mathfrak{p} \leq \mathfrak{ncs} \leq \mathfrak{gr}$, $\mathfrak{ni} \leq \mathfrak{vhs} \leq 2^\omega$ in ZFC and the result of Brech shows that $\mathfrak{gr} < 2^\omega$ is consistent.

Question 10. (a) *What is the value of $\mathfrak{gr}$, $\mathfrak{ni}$, $\mathfrak{vhs}$ among known cardinal invari-* 1183–1185?
ants of the continuum. In particular do we have in ZFC (b) $\mathfrak{ncs} = \mathfrak{gr}$*?*, (c) $\mathfrak{gr} = \mathfrak{p}$*?*

Another cardinal invariant of Boolean algebras which enters the game is the cofinality of a Boolean algebra. In [**48**, 4.6] it is shown that $\mathrm{cf}(A) = \omega$ implies that $C(K)$ does not have the Nikodým nor the Grothendieck property where K is the Stone space of A.

Not all quotients of $C(K)$ spaces are $C(L)$ for some L. For example, a nondispersed K always has ℓ_2 as a quotient ([**34**]). In some cases, like for Grothendieck $C(K)$s, the only separable quotients are reflexive, thus the result saying that if K has no convergent sequences then it has no infinite metric subspaces has its corresponding version that Grothendieck $C(K)$s do not have infinite-dimensional separable quotients of the form $C(K)$.

The still-open Efimov problem for compact spaces is whether any infinite compact space contains a copy of a convergent sequence or a copy of $\beta\mathbb{N}$. It has negative answer (first obtained by Fedorchuk [**17**]) only under a special set-theoretic assumptions, but we do not know if a counterexample can be obtained in ZFC. However a $C(K)$ version of the problem is independent and actually may be suggesting the way of solving the topological version: Talagrand shows in [**50**] that under CH there is a Grothendieck $C(K)$ which doesn't have ℓ_∞ as a quotient. On the other hand Haydon, Levy and Odell show in [**21**] that under $\mathfrak{p} = 2^\omega > \omega_1$ every $C(K)$ which is Grothendieck (i.e., doesn't have c_0 as a quotient) has ℓ_∞ as a quotient.

One can even get in ZFC a Grothendieck $C(K)$ which has no subspace isomorphic to ℓ_∞ (see [**20**]). In this paper a lemma due to Argyros is proved which says that if a Boolean algebra has the subsequential completeness property (this implies the Grothendieck property and means that for any antichain (a_n) in the algebra there is an infinite M such that $\bigvee_{n\in M} a_n$ exists in the algebra), then it has an uncountable independent family.

1186? **Question 11.** *Is it consistent that there is a Boolean algebra with the subsequential completeness property without an independent family of cardinality 2^ω?*

Complemented subspaces

Combining quotients and subspaces we obtain complemented subspaces. A subspace Y of a Banach space X is said to be complemented in X if and only if there is a bounded operator $P\colon X \to Y$, called a projection, onto Y such that $P \restriction Y = \mathrm{Id}_Y$. It is equivalent to the fact that $P^2 = P$ and to the existence of a decomposition of X as $Y \oplus Z$. If P is a projection, $Z = \mathrm{Ker}(P)$ ([**49**, 12]). A topological operation corresponding to a complemented subspace is a retraction. If $F\colon K \to L$ is a retraction onto $L \subseteq K$ (i.e., $F \restriction L = \mathrm{Id}_L$), then $P\colon C(K) \to C(K)$ is a norm-one projection where $P(f) = f \circ F$. The image of this projection is *isometric* to the $C(L)$, namely the restriction to L is the isometry from $P[C(K)]$ onto $C(L)$. Projections play a more important role in Banach space theory than retractions in topology because they define decompositions of Banach spaces.

There are other canonical topological ways of defining complemented subspaces of $C(K)$s through the theory of averaging and extension operators whose classical period is depicted in Pełczyński's monograph [**41**]. The work of Ditor, Haydon, Koppelberg, Ščepin and others contributed to resolving the main questions left after [**41**]. See the introductions of [**5**, **19**, **29**] for references and glimpses of this story.

We saw in the previous sections that trivially not all quotients nor subspaces of $C(K)$ spaces are again of this form. The analogous fact about complemented subspaces of the $C(K)$s is unknown (cf. [**47**]).

Question 12. *Is every complemented subspace of any $C(K)$ of the form $C(L)$?* 1187?

It is even not known in the case of $C(2^\omega)$ or $C([0,\alpha])$ for $\alpha \geq \omega^{\omega^\omega}$. For partial results in this metric case of K see [**47**]. The deepest result describing the complemented subspaces of the $C(K)$s in general remains Pełczynski's theorem saying that such subspaces have isomorphic copies of c_0 ([**39**, Cor. 2]).

Now let us talk about the structure of complemented subspaces of a $C(K)$. A suprising result of [**30**] is that there are $C(K)$s which are indecomposable, i.e., whose only complemented subspaces are finite-dimensional or co-finite-dimensional (such subspaces are always complemented in any Banach space). This result is obtained by constructing K which admits few functions like T^*, i.e., we are able to control all operators on the $C(K)$. Such T^* are of the form $gI + S$ where gI is a multiplication of measures by a Borel function g and S is a weakly compact operator. Assuming CH we can get K such that every operator T on $C(K)$ is of the form $gI + S$ where $g \in C(K)$ (i.e., gI is the multiplication of functions by a continuous function g) and S is a weakly compact operator. In [**43**], Plebanek removed the need of CH from the last statement at the price of losing the separability of K and his $C(K)$ is not a subspace of ℓ_∞.

Question 13. *Is it true in* ZFC *that there is a $C(K)$ which is a subspace of ℓ_∞ where every operator is of the form $gI + S$ where S is weakly compact and $g \in C(K)$?* 1188?

The results of [**30**] show that Banach spaces of the type obtained by Gowers and Maurey may be of the form $C(K)$. One construction however, the Schröder–Bernstein problem, is left:

Question 14. *Are there compact K and L such that $C(K)$ and $C(L)$ are nonisomorphic but each is isomorphic to a complemented subspace of the other?* 1189?

As the reader must have noted, in this section we entered the realm of operators on Banach spaces since being complemented is equivalent to the existence of some operator. Weakly compact operators which are not of finite-dimensional range are Banach spaces theory strangers in the land of Ks. However in any $C(K)$ there is room for perturbating operators by a weakly compact operator, i.e., an operator which sends bounded sets to relatively weakly compact sets. As by Gantmacher's theorem T is weakly compact if an only if T^* is, so the measures $\mu_n = T^*(\delta_{x_n})$ for such T must form a weakly compact set in $M(K)$ for any sequence of points $x_n \in K$. By the Dieudonne–Grothendieck characterization of such sets in $M(K)$ the $\sup_{n\in\mathbb{N}} \mu_n(U_k)$ always goes to 0 when $k \to \infty$ for a pairwise disjoint sequence (U_n) of open subsets of K. So, indeed if T is weakly compact, then one cannot recover even from T^* any part reasonable in terms of mappings of K. A researcher of topological origin should feel more secure facing the weakly compact perturbations in $C(K)$s after seeing the list in [**10**] of properties of operators in $C(K)$ equivalent to weak compactness from which we mention just one: T on $C(K)$ is weakly compact if and only if it is not an isomorphism while restricted to any infinite-dimensional subspace, i.e., T is strictly singular ([**40**]).

In [**32**] we pushed the construction of a Boolean $C(K)$ with few operators to densities above 2^ω. This strongly answered in the negative the question whether any Banach space can have complemented subspaces of density $\leq 2^\omega$ above any separable subspace. The issues of the densities of complemented subspaces are excellently surveyed in [**45**] where several open problems are stated. However our K of [**32**] exists only consistently. Immediate questions which appear are:

1190? **Question 15.** *Is it consistent that any Banach space has a complemented subspace of density* $\leq 2^\omega$*?*

S. Argyros suggested the following:

1191? **Question 16.** *Does there exist any bound for the density of indecomposable Banach spaces?*

This is quite natural if one remembers that hereditarily indecomposable Banach spaces may have density at most 2^ω ([**45**]). Related topological questions also seem unanswered:

1192? **Question 17.** *Is there in* ZFC *a compact space without infinite retracts of weights* $\leq 2^\omega$*?*

The methods of [**32**] suggest that if its results are not true in ZFC, then large cardinals (cf. [**24**]) may provide tools necessary for obtaining the other consistency. As far as now we only know how to get decomposable subspaces of large Banach spaces ([**27**]), but no methods on decomposing entire large spaces seem available.

In [**31**] assuming CH, we proved that there is a scattered space K with a minimal space of operators, i.e., where every operator is of the form $cI + S$ where c is a real and S has its range included in a copy of c_0. This has deep implications with respect to the complemented subspaces.

1193? **Question 18.** *Is there in* ZFC *a compact nonmetrizable scattered* K *such that all operators on* $C(K)$ *are of the form* $cI + S$ *where* c *is a real and* S *has its range included in a copy of* c_0*?*

Argyros ([**3**]) constructed a separable nonisomorphic to c_0 (thus not a $C(K)$) Banach space X whose only decompositions are $c_0 \oplus X$.

Besides asking about complemented subspaces we may inquire about being complemented in superspaces. A Banach space is said to be *injective* if it is complemented in any superspace. All finite-dimensional spaces are injective by the Hahn–Banach theorem. Similarly $\ell_\infty \equiv C(\beta\mathbb{N})$ is injective; just extend the coordinate functionals by the Hahn–Banach theorem. For equivalent definitions, see [**37**]. One of them is that X is injective if and only if whenever $Z \supseteq Y$ are Banach spaces, $T\colon Y \to X$ is a bounded operator, then there is an extension $T'\colon Z \to X$ of T.

Another way to prove (see [**37**]) that a $C(K)$ is injective is to construct a projection from $\ell_\infty(K)$ (the space of all bounded not necessarily continuous functions on K) onto $C(K)$. If K is extremally disconnected (i.e., the Stone space of a complete Boolean algebra), one constructs such a projection by the Boolean algebraic Sikorski extension criterion ([**28**]). The stakes are high in the following question, namely knowing the injective objects in the category of Banach spaces with their isomorphisms:

Question 19. *If a Banach space $C(K)$ is injective is it isomorphic to a $C(L)$ where L is extremally disconnected?* 1194?

Much effort was done to settle this question in the 1970s but the results are very partial. Grothendieck ([**18**]) proved that injective $C(K)$s are Grothendieck; in particular, Ks have no convergent sequences, Amir ([**1**]) proved that such a K contains a dense open extremally disconnected subset. Rosenthal's results together with Pełczyński's decomposition method imply that a ccc K such that $C(K)$ is injective and not isomorphic to ℓ_∞ cannot be separable and Wolfe ([**54, 55**]) proved that such Ks must be totally disconnected and a union of finitely many extremally disconnected (not necessarily compact) spaces.

One of the problems is how to prove that a $C(K)$ is not isomorphic to any $C(L)$ for L extremally disconnected. We do not have strong isomorphic properties of such $C(L)$s other than the Grothendieck property which can be shared by very different spaces ([**7**, **20**, **50**]). One shouldn't be also discouraged to try a positive result. If a Banach space is 1-complemented in any Banach space (i.e., the projection is of norm one), then Goodner, Kelley and Nachbin managed to prove that it is isomorphic to a $C(K)$ for K extremally disconnected (see [**37**]).

We have to mention at the end the following:

Question 20. *Is it consistent that $\ell_\infty/c_0 \equiv C(\omega^*)$ is isomorphic to $A \oplus B$ and none of the spaces A nor B is isomorphic to ℓ_∞/c_0?* 1195?

This is related to a result of Drewnowski and Roberts [**14**] which says that it is impossible under CH. Again, an information on complemented subspaces of $C(\omega^*)$ is obtained by conquering some partial knowledge about *all* operators on $C(\omega^*)$. S. Todorčević suggested that one could develop a theory of operators on $C(\omega^*)$ corresponding to the theory of autohomeomorphisms of ω^*.

If the answer to the above question were positive, it would mean that ℓ_∞/c_0 may fail to have the Schroder–Bernstein property as suggested in [**8**], as ℓ_∞/c_0 must be complemented in one of the spaces A or B by the results of [**14**].

For $p \in K$ define $C_0(K,p)$ to be the set of all functions in $C(K)$ which are zero in p. A natural decomposition for answering Question 20 would be of the form $(R \oplus C_0(\overline{X},p)) \oplus C_0(\overline{Y},p)$ where $p \in \omega^*$ and X, Y are open subsets of ω^* that would satisfy the conditions of the following:

Question 21. *Is it consistent that there are $p \in \omega^*$ and open $X, Y \subseteq \omega^*$ such that $X \cap Y = \emptyset$, $\{p\} = \overline{X} \cap \overline{Y}$ and none of the $\overline{X}$ nor $\overline{Y}$ is homeomorphic to ω^*?* 1196?

To answer Question 20 it would be sufficient to prove that $C(\overline{X})$ and $C(\overline{Y})$ are not isomorphic to l_∞/c_0 instead of $\overline{X}$ and $\overline{Y}$ not being homeomorphic to ω^*. This follows from the fact that

$$C(\overline{Z}) \sim R \oplus C_0(\overline{Z},p) \sim C_0(\overline{Z},p)$$

for any open $Z \subseteq \omega^*$, since Z must contain a copy of $\beta\mathbb{N}$, $C(\beta\mathbb{N}) \sim \ell_\infty \sim \ell_\infty \oplus R$ and ℓ_∞ is complemented in any superspace as an injective Banach space.

Most of the issues, even these set-theoretic topological, in the isomorphic theory of the $C(K)$ were not mentioned in this article. For example the questions of unconditional, transfinite and Markushevich's bases, biorthogonal and semibiorthogonal sequences, irredundant sets in Boolean algebras, or the topology of the dual ball involving such questions as countable tightness or hereditary separability or hereditary Lindelöf degree. Analogously the weak topology of the $C(K)$, its relation to the pointwise convergence topology with its vast literature and open problems has been untouched.

Acknowledgement. Many thanks to A. Aviles, O. Kalenda, W. Marciszewski, G. Plebanek for trying to explain to the author some facts on $C(K)$s related to this article.

References

[1] D. Amir, *Projections onto continuous function spaces*, Proc. Amer. Math. Soc. **15** (1964), 396–402.

[2] D. Amir, *On isomorphisms of continuous function spaces*, Israel J. Math. **3** (1965), 205–210.

[3] S. Argyros, *On a Koszmider's problem*, 2005, Note.

[4] S. Argyros, S. Mercourakis, and S. Negrepontis, *Functional-analytic properties of Corson-compact spaces*, Studia Math. **89** (1988), no. 3, 197–229.

[5] S. A. Argyros and A. D. Arvanitakis, *A characterization of regular averaging operators and its consequences*, Studia Math. **151** (2002), no. 3, 207–226.

[6] C. Bessaga and A. Pełczyński, *Spaces of continuous functions. IV. On isomorphical classification of spaces of continuous functions*, Studia Math. **19** (1960), 53–62.

[7] C. Brech, *On the density of Banach spaces $C(K)$ with the Grothendieck property*, Proc. Amer. Math. Soc. **134** (2006), 3653–3663.

[8] P. G. Casazza, *The Schroeder–Bernstein property for Banach spaces*, Banach space theory (Iowa City, IA, 1987), Contemp. Math., vol. 85, American Mathematical Society, Providence, RI, 1989, pp. 61–77.

[9] R. Deville, G. Godefroy, and V. Zizler, *Smoothness and renormings in Banach spaces*, Longman Scientific and Technical, Harlow, 1993.

[10] J. Diestel and J. J. Uhl, Jr., *Vector measures*, American Mathematical Society, Providence, RI, 1977.

[11] S. Z. Ditor, *On a lemma of Milutin concerning averaging operators in continuous function spaces*, Trans. Amer. Math. Soc. **149** (1970), 443–452.

[12] A. Dow, *Saturated Boolean algebras and their Stone spaces*, Topology Appl. **21** (1985), no. 2, 193–207.

[13] A. Dow and D. H. Fremlin, *Compact sets without converging sequences in the random real model*, 2004, Preprint. To appear in Acta Math. Univ. Comenian.

[14] L. Drewnowski and J. W. Roberts, *On the primariness of the Banach space ℓ_∞/C_0*, Proc. Amer. Math. Soc. **112** (1991), no. 4, 949–957.

[15] R. Engelking, *General topology*, Heldermann Verlag, Berlin, 1989.

[16] M. Fabiań, P. Habala, P. Hájek, V. Montesinos Santalucía, J. Pelant, and V. Zizler, *Functional analysis and infinite-dimensional geometry*, Springer, New York, 2000.

[17] V. V. Fedorchuk, *The cardinality of hereditarily separable bicompacta*, Dokl. Akad. Nauk SSSR **222** (1975), no. 2, 302–305, Translation: Soviet Math. Dokl. 16 (1975) 651–655.

[18] A. Grothendieck, *Sur les applications linéaires faiblement compactes d'espaces du type* $C(K)$, Canad. J. Math. **5** (1953), 129–173.

[19] R. Haydon, *On a problem of Pełczyński: Milutin spaces, Dugundji spaces and AE(0-dim)*, Studia Math. **52** (1974), 23–31.

[20] R. Haydon, *A nonreflexive Grothendieck space that does not contain* ℓ_∞, Israel J. Math. **40** (1981), no. 1, 65–73.

[21] R. Haydon, M. Levy, and E. Odell, *On sequences without weak* convergent convex block subsequences*, Proc. Amer. Math. Soc. **100** (1987), no. 1, 94–98.

[22] T. Jech, *Set theory*, Springer, Berlin, 1997.

[23] O. Kalenda, *Valdivia compact spaces in topology and Banach space theory*, Extracta Math. **15** (2000), no. 1, 1–85.

[24] A. Kanamori, *The higher infinite*, Springer, Berlin, 1994.

[25] S. Kaplan, *The bidual of* $C(X)$. *I*, North-Holland, Amsterdam, 1985.

[26] S. Kaplan, *Lebesgue theory in the bidual of* $C(X)$, Mem. Amer. Math. Soc. **121** (1996), no. 579, 127 pp.

[27] J. Ketonen, *Banach spaces and large cardinals*, Fund. Math. **81** (1974), 291–303.

[28] S. Koppelberg, *Handbook of Boolean algebras. Vol. 1*, North-Holland, Amsterdam, 1989.

[29] S. Koppelberg, *Projective Boolean algebras*, Handbook of Boolean algebras, Vol. 3, North-Holland, Amsterdam, 1989, pp. 741–773.

[30] P. Koszmider, *Banach spaces of continuous functions with few operators*, Math. Ann. **330** (2004), no. 1, 151–183.

[31] P. Koszmider, *On decompositions of Banach spaces of continuous functions on Mrówka's spaces*, Proc. Amer. Math. Soc. **133** (2005), no. 7, 2137–2146.

[32] P. Koszmider, *A space* $C(K)$ *where all nontrivial complemented subspaces have big densities*, Studia Math. **168** (2005), no. 2, 109–127.

[33] K. Kunen, *Set theory*, North-Holland, Amsterdam, 1980.

[34] H. E. Lacey and P. D. Morris, *On spaces of type* $A(K)$ *and their duals*, Proc. Amer. Math. Soc. **23** (1969), 151–157.

[35] I. E. Leonard and J. H. M. Whitfield, *A classical Banach space:* ℓ_∞/c_0, Rocky Mountain J. Math. **13** (1983), no. 3, 531–539.

[36] W. Marciszewski, *On Banach spaces* $C(K)$ *isomorphic to* $c_0(\Gamma)$, Studia Math. **156** (2003), no. 3, 295–302.

[37] V. Mascioni, *Topics in the theory of complemented subspaces in Banach spaces*, Exposition. Math. **7** (1989), no. 1, 3–47.

[38] A. A. Miljutin, *Isomorphism of the spaces of continuous functions over compact sets of the cardinality of the continuum*, Teor. Funkciĭ Funkcional. Anal. i Priložen. Vyp. **2** (1966), 150–156.

[39] A. Pełczyński, *Projections in certain Banach spaces*, Studia Math. **19** (1960), 209–228.

[40] A. Pełczyński, *On strictly singular and strictly cosingular operators. I. Strictly singular and strictly cosingular operators in* $C(S)$*-spaces*, Bull. Acad. Polon. Sci. Sér. Sci. Math. Astronom. Phys. **13** (1965), 31–36.

[41] A. Pełczyński, *Linear extensions, linear averagings, and their applications to linear topological classification of spaces of continuous functions*, Dissertationes Math. **58** (1968), 92 pp.

[42] G. Plebanek, *On compact spaces carrying random measures of large Maharam type*, Acta Univ. Carolin. Math. Phys. **43** (2002), no. 2, 87–99.

[43] G. Plebanek, *A construction of a Banach space* $C(K)$ *with few operators*, Topology Appl. **143** (2004), 217–239.

[44] G. Plebanek, October, 2005, Email note.

[45] A. M. Plichko and D. Yost, *Complemented and uncomplemented subspaces of Banach spaces*, Extracta Math. **15** (2000), no. 2, 335–371.

[46] H. P. Rosenthal, *On injective Banach spaces and the spaces $C(S)$*, Bull. Amer. Math. Soc. **75** (1969), 824–828.

[47] H. P. Rosenthal, *The Banach spaces $C(K)$*, Handbook of the geometry of Banach spaces, Vol. 2, North-Holland, Amsterdam, 2003, pp. 1547–1602.

[48] W. Schachermayer, *On some classical measure-theoretic theorems for non-sigma-complete Boolean algebras*, Dissertationes Math. **214** (1982), 33 pp.

[49] Z. Semadeni, *Banach spaces of continuous functions*, PWN, Warsaw, 1971.

[50] M. Talagrand, *Un nouveau $C(K)$ qui possède la propriété de Grothendieck*, Israel J. Math. **37** (1980), no. 1–2, 181–191.

[51] M. Talagrand, *Propriété de Nikodým et propriété de Grothendieck*, Studia Math. **78** (1984), no. 2, 165–171.

[52] S. Todorčević, *Irredundant sets in Boolean algebras*, Trans. Amer. Math. Soc. **339** (1993), no. 1, 35–44.

[53] S. Todorčević, *Biorthogonal systems and quotient spaces via Baire category methods*, 2004, Preprint.

[54] J. Wolfe, *Injective Banach spaces of type $C(T)$*, Israel J. Math. **18** (1974), 133–140.

[55] J. Wolfe, *Injective Banach spaces of continuous functions*, Trans. Amer. Math. Soc. **235** (1978), 115–139.

Open Problems in Topology II – Edited by E. Pearl

Tightness and t-equivalence

Oleg Okunev

All spaces below are assumed to be Tychonoff (that is, completely regular Hausdorff). We use terminology and notation as in [**5**], with the exception that the tightness of a space X is denoted as $t(X)$.

Two spaces X and Y are called *M-equivalent* if their free topological groups $F(X)$ and $F(Y)$ in the sense of Markov [**7**] are topologically isomorphic. The spaces X and Y are *l-equivalent* if the spaces $C_p(X)$ and $C_p(Y)$ of real-valued continuous functions equipped with the topology of pointwise convergence are linearly homeomorphic, and *t-equivalent* if $C_p(X)$ and $C_p(Y)$ are homeomorphic (see [**3**]); Arhangel′skiĭ showed in [**2**] that M-equivalence of two spaces implies their l-equivalence; clearly, l-equivalent spaces are t-equivalent. We say that a topological property is preserved by an equivalence relation if whenever two spaces are in the relation, one of them has the property if and only if the other one does. Similarly, we say that a cardinal invariant is preserved by a relation if its values on two spaces are the same whenever the spaces are in the relation.

The article [**9**] contains an example that shows that the sequentiality and tightness are not preserved by the relation of M-equivalence. Tkachuk proved in [**15**] that the tightness is preserved by l-equivalence *in the class of compact spaces*, that is, if X and Y are l-equivalent *compact* spaces; this was later extended in [**12**] by showing that the tightness is preserved by t-equivalence in the class of compact spaces, and the same holds for sequentiality if $2^{\mathfrak{t}} > \mathfrak{c}$ (in fact it easily follows from the main theorem in [**12**] that if X and Y are t-equivalent spaces and X is a countable union of its compact sequential subspaces, then so is Y).

As for the Fréchet property, it is not preserved by the relation of M-equivalence even in the class of compact spaces [**13**].

The example in [**9**] that shows the non-preservation of the tightness by M-equivalence depends heavily on the fact that one of the two spaces is not normal. Indeed, the construction of the example uses the fact that if K is a retract of a space X, then the spaces X^+ obtained by adding an isolated point to X and the direct sum of the spaces K and X/K are M-equivalent [**9**]; here X/K is the partition of X whose elements are K and singletons equipped with the R-quotient topology (that is, the strongest completely regular topology that makes the natural mapping $p\colon X \to X/K$ continuous). It is easy to see that if X is normal, then the natural mapping $p\colon X \to X/K$ is in fact quotient, and therefore, closed; by Theorem 4.5 in [**1**], in this case the tightness of the space X is equal to the supremum of the tightnesses of the image space X/K and of the fibers of p; the only nontrivial fiber of p is K, so $t(X^+) = t(X) = t(X/K \oplus K)$.

Hence, the following question:

Problem 1. *Let X and Y be normal M-equivalent spaces. Is it true that $t(X) =$* 1197?
$t(Y)$?

Since t-equivalent compact spaces have the same tightness, it is natural to ask whether this fact may be generalized to a wider class of spaces.

1198? **Problem 2.** *Let X and Y be M-equivalent σ-compact spaces. Is it true that $t(X) = t(Y)$?*

Similar questions remain open for the relations of l-equivalence and t-equivalence.

It is known that compactness is preserved by l-equivalence (and hence by M-equivalence) [**17**], but not by t-equivalence [**6**]. Thus, the following question is specific for the relation of t-equivalence:

1199? **Problem 3.** *Let X and Y be t-equivalent spaces such that X is compact. Is it true that $t(X) = t(Y)$?*

Note that every space t-equivalent to a compact space is σ-compact [**8**]. It can easily be deduced from the main theorem in [**12**] that if X is compact (in fact, the Lindelöf property is sufficient) and Y is t-equivalent to X, then every free sequence in Y is of length at most $t(X)$; unfortunately, for a σ-compact space Y this is not sufficient to conclude that the tightness of Y is at most $t(X)$ [**14**].

Of course, the following versions of Problems 1 and 2 quite naturally arise:

1200? **Problem 4.** *Let X and Y be M-equivalent Lindelöf spaces. Is it true that $t(X) = t(Y)$?*

1201? **Problem 5.** *Let X and Y be M-equivalent paracompact spaces. Is it true that $t(X) = t(Y)$?*

as well as their versions for the relations of l-equivalence and of t-equivalence.

Unlike the compact case, the tightness of σ-compact spaces is not productive (see, e.g., [**11, 16**]). There may be more hope for positive answers (or more challenge for finding examples) for the versions of the above problems where the equality $t(X) = t(Y)$ is replaced by $t(Y) \le t^*(X) = \sup\{t(X^n) : n \in \omega\}$ The example in Section 2 in [**10**] shows that there are M-equivalent spaces X and Y where X is metrizable and Y has one nonisolated point such that the tightness of Y^2 is uncountable, so we cannot expect $t^*(X) = t^*(Y)$ for paracompact spaces.

While [**12**] contains the proof that there is a topological property of $C_p(X)$ that, assuming that X is compact, is equivalent to the countability of the tightness of X, there is no *internal* description of this property. Hence, the following (somewhat fuzzy) request:

1202? **Problem 6.** *Find an internally defined cardinal function ϕ such that whenever X is a compact space, $t(X) = \phi(C_p(X))$.*

The inequality $t(Y) \le t(X)$ for compact X and Y is proved in [**12**] under the assumption that $C_p(Y)$ is an image under an open mapping of a subspace of $C_p(X)$, so we might expect that the function ϕ in Problem 6 should be hereditary and not raise in continuous open images. The following modification of Problem 6 for the relation of l-equivalence also appears interesting:

Problem 7. *Find a cardinal function ϕ internally defined for locally convex linear topological spaces such that whenever X is a compact space, $t(X) = \phi(C_p(X))$.* 1203?

A similar problem arises for the sequentiality:

Problem 8. *Find an internally defined topological property $\mathcal{P}$ such that whenever X is a compact space, X is a countable union of closed sequential subspaces iff $C_p(X)$ has $\mathcal{P}$.* 1204?

Tkachuk essentially proved in [**15**] that a compact space X is a countable union of closed sequential subspaces iff so is $L_p(X)$, and a compact X has countable tightness iff $L_p(X)$ is a countable union of its subspaces of countable tightness; here $L_p(X)$ is the weak dual space of $C_p(X)$.

An interesting hypothesis related to Problem 6 was communicated by E. Reznichenko:

Problem 9. *Let X be a compact space of countable tightness. Is it true that every compact subspace of $C_p(C_p(X))$ has countable tightness?* 1205?

The negative answer to the next question would give a consistently positive answer to Problem 9 (for example, in the model of ZFC described in [**4**]).

Problem 10. *Is there a compact space of countable tightness X such that $\omega_1 + 1$ (with the order topology) is homeomorphic to a subspace of $C_p(C_p(X))$?* 1206?

References

[1] A. V. Arhangel'skiĭ, *The spectrum of frequencies of a topological space and the product operation*, Trudy Moskov. Mat. Obshch. **40** (1979), 171–206, Translation: Trans. Moscow Math. Soc. (1981) 163–200.

[2] A. V. Arhangel'skiĭ, *On linear homeomorphisms of function spaces*, 1982, Translation: Soviet Math. Dokl. 25 (1982) 852–855, pp. 1289–1292.

[3] A. V. Arhangel'skiĭ, *Topological function spaces*, Kluwer Academic Publishers, Dordrecht, 1992.

[4] A. Dow, *On the consistency of the Moore–Mrówka solution*, Topology Appl. **44** (1992), 125–141.

[5] R. Engelking, *General topology*, Heldermann Verlag, Berlin, 1989.

[6] S. P. Gul'ko and T. E. Khmyleva, *Compactness is not preserved by the relation of t-equivalence*, Mat. Zametki **39** (1986), no. 6, 895–903, Translation: Math. Notes 39 (1986) no. 6, 484–488.

[7] A. A. Markov, *On free topological groups*, Izv. Akad. Nauk SSSR Ser. Mat. **9** (1945), no. 1, 3–64, Translation: Amer. Math. Soc. Transl. 30 (1950).

[8] O. Okunev, *Weak topology of a dual space and a t-equivalence relation*, Mat. Zametki **46** (1989), no. 1, 53–59, Translation: Math. Notes 46 (1989) no. 1, 534–538.

[9] O. Okunev, *A method for constructing examples of M-equivalent spaces*, Topology Appl. **36** (1990), no. 2, 157–171.

[10] O. Okunev, *M-equivalence of products*, Trudy Moskov. Mat. Obshch. **56** (1995), 192–205, Translation: Trans. Moscow Math. Soc. (1995) 149–158.

[11] O. Okunev, *A remark on the tightness of products*, Comment. Math. Univ. Carolin. **37** (1996), no. 2, 397–399.

[12] O. Okunev, *Tightness of compact spaces is preserved by the t-equivalence relation*, Comment. Math. Univ. Carolin. **43** (2002), no. 2, 335–342.

[13] O. Okunev, *Fréchet property in compact spaces is not preserved by M-equivalence*, Comment. Math. Univ. Carolin. **46** (2005), no. 4, 747–749.

[14] O. Okunev, *A σ-compact space without uncountable free sequences can have arbitrary tightness*, Questions Answers Gen. Topology **23** (2005), no. 2, 107–108.

[15] V. V. Tkachuk, *Duality with respect to the functor* C_p *and cardinal invariants of the type of the Suslin number*, Mat. Zametki **37** (1985), no. 3, 441–451, Translation: Math. Notes 37 (1985) no. 3, 247–252.

[16] S. Todorčević, *Some applications of S and L combinatorics*, The work of Mary Ellen Rudin (Madison, WI, 1991), Ann. New York Acad. Sci., vol. 705, New York Acad. Sci., 1993, pp. 130–167.

[17] V. V. Uspenskij, *A characterization of compactness in terms of the uniform structure in a space of functions*, Uspekhi Mat. Nauk **37** (1982), no. 4(226), 183–184, Translation: Russian Math. Surveys 37 (1982) no. 4, 143–144.

Open Problems in Topology II – Edited by E. Pearl

Topological problems in nonlinear and functional analysis

Biagio Ricceri

In this note, I intend to collect some problems, conjectures and perspectives, of topological nature, arising from the research work I made in the last years.

I start recalling the following definition.

Let $(E, \|\cdot\|)$ be a real normed space. A nonempty set $A \subset E$ is said to be *antiproximinal* with respect to $\|\cdot\|$ if, for every $x \in E \setminus A$ and every $y \in A$, one has $\|x - y\| > \inf_{z \in A} \|x - z\|$.

I then propose the following

Conjecture 1. *There exists a noncomplete real normed space E with the following* 1207?
property: for every nonempty convex set $A \subset E$ which is antiproximinal with respect to each norm on E, the interior of the closure of A is nonempty.

The main reason for the study of Conjecture 1 is to give a contribution to open mapping theory in the setting of noncomplete normed spaces. Recall that in any vector space there exists the strongest vector topology of the space [**30**, p. 42]. Actually, making use of Theorem 4 of [**41**], one can prove the following result.

Theorem 1. *Let X, E be two real vector spaces, C a nonempty convex subset of X, F a multifunction from C onto E, with nonempty values and convex graph. Then, for every nonempty convex set $A \subseteq C$ which is open with respect to the relativization to C of the strongest vector topology on X, the set $F(A)$ is antiproximinal with respect to each norm on E.*

Now, I am going to present a problem about an unusual way of finding global minima of functionals in Banach spaces. A closed hyperplane in a real normed space X is any set of the type $T^{-1}(r)$, where T is a nonzero continuous linear functional on X and $r \in \mathbb{R}$. First, I recall the following result from [**48**] (see also [**43, 46, 49, 50**]):

Theorem 2 ([**48**, Theorem 2.1]). *Let $(T, \mathcal{F}, \mu)$ be nonatomic measure space, with $\mu(T) < +\infty$, E a real Banach space, and $f\colon E \to \mathbb{R}$ a bounded below Borel functional such that, for some $\gamma \in]0, 1[$,*

$$\sup_{x \in E} \frac{f(x)}{\|x\|^{\gamma} + 1} < +\infty.$$

Then, for every $p \geq 1$ and every closed hyperplane V of $L^p(T, E)$, one has

$$\inf_{u \in V} \int_T f(u(t))d\mu = \inf_{u \in L^p(T,E)} \int_T f(u(t))d\mu.$$

I then propose the following

1208? **Problem 1.** *Let X be an infinite-dimensional real Banach space, and let $J\colon X \to \mathbb{R}$ be a bounded below functional satisfying, for some $\gamma \in]0,1[$,*

$$(\star) \qquad \sup_{u\in X} \frac{J(u)}{\|u\|^{\gamma}+1} < +\infty.$$

Find conditions under which there exists a closed hyperplane V of X such that the restriction of J to V has a local minimum.

The motivation for the study of Problem 1 is as follows. Assume that we wish to minimize a bounded below Borel functional f on a real Banach space E satisfying the condition, for some $\gamma \in]0,1[$,

$$\sup_{x\in E} \frac{f(x)}{\|x\|^{\gamma}+1} < +\infty.$$

For each $u \in L^1([0,1], E)$, set

$$J(u) = \int_0^1 f(u(t))dt.$$

So, J is bounded below and satisfies $(\star)$ with $X = L^1([0,1], E)$. Assume that there is some closed hyperplane V of $L^1([0,1], E)$ such that the restriction of J to V has a local minimum, say u_0. By a result of Giner ([**28**]), u_0 is actually a global minimum of the restriction of J to V. On the other hand, by Theorem 2, we have

$$\inf_{u\in V} J(u) = \inf_{u\in L^1([0,1],E)} J(u)$$

and so u_0 is a global minimum of J in $L^1([0,1], E)$. This easily implies that f has a global minimum in E.

Now, I list a series of possible new proofs of the Brouwer fixed point theorem recalling the results from which they originate. $\langle\cdot,\cdot\rangle$ will denote the usual inner product in $\mathbb{R}^n$.

1209? **Problem 2.** *Let $X \subset \mathbb{R}^n$ $(n \geq 2)$ be a compact convex set and $f\colon X \to X$ a continuous function. Without using any result based on the Brouwer fixed point theorem, is it possible to find a continuous function $\alpha\colon X \to \mathbb{R}$ in such a way that the set $\{(x,y) \in X \times \mathbb{R}^n : \langle f(x) - x, y\rangle = \alpha(x)\}$ is disconnected?*

A positive answer to Problem 2 would provide a new proof of the Brouwer theorem via the following results ([**45**]; see also [**37**]):

Theorem 3 ([**45**, Theorem 2]). *Let X be a topological space, let E be a real topological vector space (with topological dual E^*), and let $A\colon X \to E^*$ be such that the set $\{y \in E : x \mapsto \langle A(x), y\rangle$ is continuous$\}$ is dense in E. Then, the following assertions are equivalent:* (i) *The set $\{(x,y) \in X \times E : A(x)(y) = 1\}$ is disconnected.* (ii) *The set $X \setminus A^{-1}(0)$ is disconnected.*

Proposition 1 ([**45**, Proposition 1]). *Let X be a topological space, let E be a real topological vector space (with algebraic dual E') and let $A\colon X \to E'$. Assume that, for some continuous function $\alpha\colon X \to \mathbb{R}$, the set $\{(x,y) \in X\times E : A(x)(y) = \alpha(x)\}$*

is disconnected. Then, either $A^{-1}(0) \neq \emptyset$ *or the set* $\{(x,y) \in X \times E : A(x)(y) = 1\}$ *is disconnected.*

Assume that Problem 2 admits a positive answer. Apply Proposition 1 taking $E = \mathbb{R}^n$ and $A(x) = f(x) - x$ for all $x \in X$ (of course, E' is identified with $\mathbb{R}^n$). Since X is connected, Proposition 1 (on the basis of Theorem 3) ensures that A has a zero, that is f has a fixed point.

A possible positive answer to Problem 2 could be rather difficult due to the fact that the function α does not depend on y. I then propose a variant of Problem 2 without such a restriction.

Problem 3. *Let* $X \subset \mathbb{R}^n$ $(n \geq 2)$ *be a compact convex set and* $f\colon X \to X$ *a continuous function. Without using any result based on the Brouwer fixed point theorem, is it possible, for each* $\epsilon > 0$, *to find a continuous function* $\alpha_\epsilon\colon X \times \mathbb{R}^n \to \mathbb{R}$, *with* $\alpha_\epsilon(x,\cdot)$ *Lipschitzian in* $\mathbb{R}^n$ *with Lipschitz constant less than or equal to* ϵ, *in such a way that the set* $\{(x,y) \in X \times \mathbb{R}^n : \langle f(x) - x, y\rangle = \alpha_\epsilon(x,y)\}$ *is disconnected?* 1210?

Problem 3 originates from the following

Theorem 4 ([**47**, Theorem 19]). *Let* X *be a connected topological space,* E *a real Banach space,* A *an operator from* X *into* E^*, α *a real function on* $X \times E$ *such that, for each* $x \in X$, $\alpha(x,\cdot)$ *is Lipschitzian in* E, *with Lipschitz constant* $L(x) \geq 0$. *Further, assume that the set* $\{y \in E : A(\cdot)(y) - \alpha(\cdot,y)$ *is continuous*$\}$ *is dense in* E *and that the set* $\{(x,y) \in X \times E : A(x)(y) = \alpha(x,y)\}$ *is disconnected. Then, there exists some* $x_0 \in X$ *such that* $\|A(x_0)\|_{E^*} \leq L(x_0)$.

Arguing as before, a positive answer to Problem 3 would produce a new proof of the Brouwer theorem via Theorem 4 and an approximation argument.

I also wish to propose the following

Conjecture 2. *Let* $X \subset \mathbb{R}^n$ $(n \geq 2)$ *be a compact convex set and* $f\colon X \to X$ *a continuous function. Let* $\epsilon > 0$ *be small enough. Denote by* Λ_ϵ *the set of all continuous functions* $\alpha\colon X \times \mathbb{R}^n \to \mathbb{R}$ *such that, for each* $x \in X$, $\alpha(x,\cdot)$ *is Lipschitzian in* $\mathbb{R}^n$, *with Lipschitz constant less than or equal to* ϵ. *Consider* Λ_ϵ *equipped with the relativization of the strongest vector topology on the space* $\mathbb{R}^{X \times \mathbb{R}^n}$. *Then, the set* $\{(\varphi,x,y) \in \Lambda_\epsilon \times X \times \mathbb{R}^n : \langle f(x) - x, y\rangle = \alpha(x,y)\}$ *is disconnected.* 1211?

On the basis of Theorem 5 below, it would be of interest to prove Conjecture 2 without using the Brouwer theorem.

Theorem 5 ([**47**, Theorem 21]). *Let* X *be a connected and locally connected topological space,* E *a real Banach space,* $A\colon X \to E^*$ *a continuous operator with closed range. For each* $\epsilon > 0$, *denote by* Λ_ϵ *the set of all continuous functions* $\alpha\colon X \times E \to \mathbb{R}$ *such that, for each* $x \in X$, $\alpha(x,\cdot)$ *is Lipschitzian in* E, *with Lipschitz constant less than or equal to* ϵ. *Consider* Λ_ϵ *equipped with the relativization of the strongest vector topology on the space* $\mathbb{R}^{X \times E}$, *and assume that the set* $\{(\alpha,x,y) \in \Lambda_\epsilon \times X \times E : A(x)(y) = \alpha(x,y)\}$ *is disconnected. Then,* $A^{-1}(0) \neq \emptyset$.

To introduce the last problem related to the Brouwer theorem, let me recall a further result. The spaces $C^0(X, E)$ and $C^0(X)$ that will appear are considered with the sup-norm. Recall that a subset D of a topological space S is a retract of S if there exists a continuous function $h\colon S \to D$ such that $h(s) = s$ for all $s \in D$.

Theorem 6 ([**55**, Theorem 6]). *Let X be a compact Hausdorff topological space, E a real Banach space, with $\dim(E) \geq 2$, and $A\colon X \to E^*$ a continuous operator. Then, at least one of the following assertions holds:* (a) *$A^{-1}(0) \neq \emptyset$.* (b) *There exists $\epsilon > 0$ such that, for every Lipschitzian operator $J\colon C^0(X, E) \to C^0(X)$, with Lipschitz constant less than ϵ, the set $\{\psi \in C^0(X, E) : A(x)(\psi(x)) = J(\psi)(x)$ for all $x \in X\}$ is an unbounded retract of $C^0(X, E)$.*

Theorem 6 gives the motivation for the following

1212? **Problem 4.** *Let $X \subset \mathbb{R}^n$ $(n \geq 2)$ be a compact convex set and $f\colon X \to X$ a continuous function. Without using any result based on the Brouwer fixed point theorem, is it possible to prove that, for each $\epsilon > 0$, there exists a Lipschitzian operator $J\colon C^0(X, \mathbb{R}^n) \to C^0(X)$, with Lipschitz constant less than ϵ, in such a way that the set $\{\psi \in C^0(X, \mathbb{R}^n) : \langle f(x) - x, \psi(x)\rangle = J(\psi)(x)$ for all $x \in X\}$ is either bounded or disconnected?*

Before formulating the next conjecture, I recall two more results:

Theorem 7 ([**42**, Theorem 2.2]). *Let X be a topological space and let $S \subseteq X \times [0, 1]$ be a connected set whose projection on $[0, 1]$ is the whole of $[0, 1]$. Then, S intersects the graph of any continuous function from X into $[0, 1]$.*

Theorem 8 ([**50**, Proposition 2.1]). *Let X be a normed space, let $T \in X^*$ and let $J\colon X \to \mathbb{R}$ be a Lipschitzian functional with Lipschitz constant $L < \|T\|_{X^*}$. Then, the functional $T + J$ is onto $\mathbb{R}$.*

I now state

1213? **Conjecture 3.** *Let $(X, \langle\cdot,\cdot\rangle)$ be an infinite-dimensional real Hilbert space and let $A\colon [0, 1] \to X$ be a continuous function such that, for some $\lambda \in]0, 1[$, one has*

$$\sup_{x \in X} \inf_{t \in [0,1]} (\langle A(t), x\rangle - \lambda\|A(t)\|\|x\|) < +\infty.$$

Then, there exist $\mu \in]0, 1[$ and a continuous function $g\colon X \to [0, 1]$ such that

$$\sup_{x \in X} (\langle A(g(x)), x\rangle - \mu\|A(g(x))\|\|x\|) < +\infty.$$

The motivation for the study of Conjecture 3 is as follows. Assume that it is true. I then claim that $A^{-1}(0) \neq \emptyset$. Indeed, put

$$M := \sup_{x \in X} (\langle A(g(x)), x\rangle - \mu\|A(g(x))\|\|x\|).$$

Consider the set

$$S := \{(t, x) \in [0, 1] \times X : \langle A(t), x\rangle = \mu\|A(t)\|\|x\| + M + 1\}.$$

If $t \in [0,1]$ does not belong to the projection of S on $[0,1]$, then, in view of Theorem 8, we clearly have $A(t) = 0$, and we are done. Therefore, assume that such a projection is the whole of $[0,1]$. In this case, we observe that S does not meet the graph of the continuous function g, and so, by Theorem 7, S must be disconnected. At this point, we can apply Theorem 4 which ensures the existence of $t_0 \in [0,1]$ such that $\|A(t_0)\| \leq \mu\|A(t_0)\|$. Since $\mu < 1$, one then has $A(t_0) = 0$, and the claim is proved.

Let me now recall the notion of Gâteaux differentiability. Let X be a real normed space. A functional $J\colon X \to \mathbb{R}$ is said to be Gâteaux differentiable at a point x if there is $T \in X^*$ such that

$$\lim_{\lambda \to 0^+} \frac{J(x + \lambda y) - J(x)}{\lambda} = T(y)$$

for all $y \in X$. The functional T is the Gâteaux derivative of J at x and is denoted by $J'(x)$. The functional J is said to be of class C^1 if it is Gâteaux differentiable at any point of X and the operator $J'\colon X \to X^*$ is continuous. The critical points of J are the zeros of J'. Of course, if, for some vector topology on X, the point x is a local minimum of J and J is Gâteaux differentiable at x, then $J'(x) = 0$.

The following problem seems to be fascinating.

Problem 5. *Let X be a real Banach space and let $J\colon X \to \mathbb{R}$ be a functional of class C^1. Is there a topology τ on X such that the critical points of J are exactly the τ-local minima of J?* 1214?

Clearly, a possible positive answer to Problem 5 would be of great theoretical interest. At this point, I would like to point that in [**51**–**53**, **56**, **58**] one can find various results on local minima that have been widely applied to nonlinear differential equations (see, for instance, [**1**–**23**, **25**–**27**, **29**, **31**–**36**, **38**, **39**, **54**, **59**, **60**]).

In [**57**], I got the following general result (see also [**24**]):

Theorem 9 ([**57**, Theorem 2]). *Let X be a real Hilbert space and let $J\colon X \to \mathbb{R}$ be a nonconstant functional of class C^1, with compact derivative, such that $\limsup_{\|x\| \to +\infty} \frac{J(x)}{\|x\|^2} \leq 0$. Then, for each $r \in \left]\inf_X J, \sup_X J\right[$ for which the set $J^{-1}([r, +\infty[)$ is not convex and for each convex set $S \subseteq X$ dense in X, there exist $x_0 \in S \cap J^{-1}(]-\infty, r[)$ and $\lambda > 0$ such that the equation $x = \lambda J'(x) + x_0$ has at least three solutions.*

On the basis of Theorem 9, I now propose

Problem 6. *Let X, Y be two topological spaces and let $f\colon X \to Y$ be a continuous function. Assume that there is an open cover $\mathcal{F}$ of X such that* $\operatorname{card}(f^{-1}(y) \cap A) \leq 2$ *for all $y \in Y$, $A \in \mathcal{F}$. Find sufficient conditions in order that* $\operatorname{card}(f^{-1}(y)) \leq 2$ *for all $y \in Y$.* 1215?

Here is the meaning of Problem 6. Let (P) be such a sufficient condition (concerning f). Let J satisfy the assumptions of Theorem 9 and let $J^{-1}([r, +\infty[)$ be nonconvex for some $r \in \left]\inf_X J, \sup_X J\right[$. Moreover, assume that, for each

$\lambda > 0$, there is some open cover $\mathcal{F}$ of X such that, for every $y \in X$ and every $A \in \mathcal{F}$, the equation $x = \lambda J'(x) + y$ has at most two solutions in A. Then, for some $\lambda > 0$, the operator $x \mapsto x - \lambda J'(x)$ does not satisfy condition (P). The conclusion, of course, is a direct consequence Theorem 9. Clearly, the interest of results of this kind fully depends on the quality of the answers given to Problem 6.

In the final part of this note, I wish to propose some specific topological problems on the energy functional associated to the Dirichlet problem

$$(P_f) \qquad -\Delta u = f(x,u) \text{ in } \Omega, \qquad u_{|\partial\Omega} = 0.$$

So, let $\Omega \subset \mathbb{R}^n$ $(n \geq 3)$ be an open bounded set. Let $X = W_0^{1,2}(\Omega)$, with the usual norm $\|u\| = (\int_\Omega |\nabla u(x)|^2 dx)^{\frac{1}{2}}$. For $q > 0$, denote by $\mathcal{A}_q$ the class of all Carathéodory functions $f\colon \Omega \times \mathbb{R} \to \mathbb{R}$ such that

$$\sup_{(x,\xi)\in\Omega\times\mathbb{R}} \frac{|f(x,\xi)|}{1+|\xi|^q} < +\infty.$$

For $0 < q \leq \frac{n+2}{n-2}$ and $f \in \mathcal{A}_q$, put

$$J_f(u) = \frac{1}{2}\int_\Omega |\nabla u(x)|^2 dx - \int_\Omega \left(\int_0^{u(x)} f(x,\xi)d\xi\right) dx$$

for all $u \in X$.

So, the functional J_f is of Class C^1 on X and one has

$$J_f'(u)(v) = \int_\Omega \nabla u(x)\nabla v(x)dx - \int_\Omega f(x,u(x))v(x)dx$$

for all $u, v \in X$. Hence, the critical points of J_f in X are exactly the weak solutions of problem (P_f).

To formulate the next problem, denote by τ_s the topology on X whose members are the sequentially weakly open subsets of X. That is, a set $A \subseteq X$ belongs to τ_s if and only if for each $u \in A$ and each sequence $\{u_n\}$ in X weakly convergent to u, one has $u_n \in A$ for all n large enough.

1216? **Problem 7.** *Is there some $f \in \mathcal{A}_q$, with $q < \frac{n+2}{n-2}$, such that, for each $\lambda > 0$ and $r \in \mathbb{R}$, the functional $J_{\lambda f}$ is unbounded below and the set $J_{\lambda f}^{-1}(r)$ has no isolated points with respect to the topology τ_s?*

The interest for the study of Problem 7 comes essentially from the following result:

Theorem 10 ([**53**, Theorem 3]). *Let $f \in \mathcal{A}_q$ with $q < \frac{n+2}{n-2}$. Then, there exists some $\lambda^* > 0$ such that the functional $J_{\lambda^* f}$ has local minimum with respect to the topology τ_s.*

In the light of Theorem 10, the relevance of Problem 7 is clear. Actually, if f was answering Problem 7 in the affirmative, then, by Theorem 10, for some $\lambda^* > 0$, the functional $J_{\lambda^* f}$ would have infinitely many local minima in the topology τ_s. Consequently, problem $(P_{\lambda^* f})$ would have infinitely many weak solutions.

It is also worth noticing that if $f \in \mathcal{A}_q$ with $q < \frac{n+2}{n-2}$ and $\lim_{\|u\| \to +\infty} J_f(u) = +\infty$, then the local minima of J_f in the strong and in the weak topology of X do coincide ([**40**, Theorem 1]). On the other hand, if $f(x, \xi) = |\xi|^{q-1}\xi$ with $1 < q < \frac{n+2}{n-2}$, then, for some constant $\lambda > 0$, it turns out that 0 is a local minimum of $J_{\lambda f}$ in the strong topology but not in the weak one ([**40**, Example 2]). However, I do not know any example of f for which J_f has a local minimum in the strong topology but not in τ_s.

I conclude presenting what I consider the most important of the problems of this note.

Problem 8. *Denote by τ the strongest vector topology of X. Is there some $f \in$* 1217?
$\mathcal{A}_{\frac{n+2}{n-2}}$ such that the set $\{(u, v) \in X \times X : J'_f(u)(v) = 1\}$ is disconnected in $(X, \tau) \times (X, \tau)$?

Assume that $f \in \mathcal{A}_{\frac{n+2}{n-2}}$ have the property required in Problem 8. Since J_f is of class C^1, clearly the operator $J'_f : X \to X^*$ is τ-weakly-star continuous. Hence, by Theorem 3, the set $X \setminus (J'_f)^{-1}(0)$ is τ-disconnected. Then, this implies, in particular, that the set $(J'_f)^{-1}(0)$ is not τ-relatively compact ([**44**, Proposition 3]), and hence is infinite. So, for such an f, problem (P_f) would have infinitely many weak solutions. Hence, a possible positive answer to Problem 8 would open a completely new chapter in the theory of the multiplicity of solutions for problem (P_f). Finally, in personal communication T.-C. Tang has remarked that if $f \in \mathcal{A}_q$ for some $q < \frac{n+2}{n-2}$, then f cannot satisfy the property required in Problem 8. In other words, Problem 8 concerns nonlinearities with critical growth.

References

[1] G. Anello, *Existence of infinitely many weak solutions for a Neumann problem*, Nonlinear Anal. **57** (2004), no. 2, 199–209.

[2] G. Anello, *A note on a problem by Ricceri on the Ambrosetti–Rabinowitz condition*, Proc. Amer. Math. Soc. (2006), Article in press.

[3] G. Anello and G. Cordaro, *Existence of solutions of the Neumann problem for a class of equations involving the p-Laplacian via a variational principle of Ricceri*, Arch. Math. (Basel) **79** (2002), no. 4, 274–287.

[4] G. Anello and G. Cordaro, *Three solutions for a perturbed sublinear elliptic problem in $\mathbb{R}^N$*, Glasg. Math. J. **47** (2005), no. 1, 205–212.

[5] D. Averna and G. Bonanno, *Three solutions for a quasilinear two-point boundary-value problem involving the one-dimensional p-Laplacian*, Proc. Edinburgh Math. Soc. (2) **47** (2004), no. 2, 257–270.

[6] D. Averna and R. Salvati, *Three solutions for a mixed boundary value problem involving the one-dimensional p-Laplacian*, J. Math. Anal. Appl. **298** (2004), no. 1, 245–260.

[7] Z. Bai and W. Ge, *Existence of three positive solutions for some second-order boundary value problems*, Comput. Math. Appl. **48** (2004), no. 5–6, 699–707.

[8] G. Barletta and R. Livrea, *Existence of three periodic solutions for a non-autonomous second order system*, Matematiche (Catania) **57** (2002), no. 2, 205–215.

[9] G. Bonanno, *Existence of three solutions for a two point boundary value problem*, Appl. Math. Lett. **13** (2000), no. 5, 53–57.

[10] G. Bonanno, *A minimax inequality and its applications to ordinary differential equations*, J. Math. Anal. Appl. **270** (2002), no. 1, 210–229.

[11] G. Bonanno, *Some remarks on a three critical points theorem*, Nonlinear Anal. **54** (2003), no. 4, 651–665.

[12] G. Bonanno and R. Livrea, *Multiplicity theorems for the Dirichlet problem involving the p-Laplacian*, Nonlinear Anal. **54** (2003), no. 1, 1–7.

[13] F. Cammaroto, A. Chinnì, and B. Di Bella, *Infinitely many solutions for the Dirichlet problem involving the p-Laplacian*, Nonlinear Anal. **61** (2005), no. 1–2, 41–49.

[14] F. Cammaroto, A. Chinnì, and B. Di Bella, *Infinitely many solutions for the Dirichlet problem via a variational principle of Ricceri*, Variational analysis and applications, Springer, New York, 2005, pp. 215–229.

[15] F. Cammaroto, A. Chinnì, and B. Di Bella, *Some multiplicity results for quasilinear Neumann problems*, Arch. Math. (Basel) **86** (2006), no. 2, 154–162.

[16] F. Cammaroto, A. Chinnì, and B. Di Bella, *Multiple solutions for a quasilinear elliptic variational system on strip-like domains*, Proc. Edinburgh Math. Soc. (2), To appear.

[17] F. Cammaroto, A. Chinnì, and B. Di Bella, *Multiplicity results for nonlinear Schrödinger equation*, Glasg. Math. J., To appear.

[18] P. Candito, *Infinitely many solutions to the Neumann problem for elliptic equations involving the p-Laplacian and with discontinuous nonlinearities*, Proc. Edinburgh Math. Soc. (2) **45** (2002), no. 2, 397–409.

[19] G. Cordaro, *On a minimax problem of Ricceri*, J. Inequal. Appl. **6** (2001), no. 3, 261–285.

[20] G. Cordaro, *Existence and location of periodic solutions to convex and non coercive Hamiltonian systems*, Discrete Contin. Dyn. Syst. **12** (2005), no. 5, 983-996.

[21] G. Cordaro, *Further results related to a minimax problem of Ricceri*, J. Inequal. Appl. (2005), no. 5, 523–533.

[22] A. G. Di Falco, *Infinitely many solutions to the Neuman problem for quasilinear elliptic systems*, Matematiche (Catania) **58** (2003), no. 1, 117–130 (2005).

[23] F. Faraci, *Bifurcation theorems for Hammerstein nonlinear integral equations*, Glasg. Math. J. **44** (2002), no. 3, 471–481.

[24] F. Faraci and A. Iannizzotto, *An extension of a multiplicity theorem by Ricceri with an application to a class of quasilinear equations*, Studia Math. **172** (2006), no. 3, 275–287.

[25] F. Faraci and A. Iannizzotto, *A multiplicity theorem for a perturbed second-order non-autonomous system*, Proc. Edinburgh Math. Soc. (2) **49** (2006), no. 2, 267–275.

[26] F. Faraci and R. Livrea, *Infinitely many periodic solutions for a second-order nonautonomous system*, Nonlinear Anal. **54** (2003), no. 3, 417–429.

[27] F. Faraci and V. Moroz, *Solutions of Hammerstein integral equations via a variational principle*, J. Integral Equations Appl. **15** (2003), no. 4, 385–402.

[28] E. Giner, *Minima sous contrainte, de fonctionnelles intégrales*, C. R. Acad. Sci. Paris Sér. I Math. **321** (1995), no. 4, 429–431.

[29] A. Iannizzotto, *A sharp existence and localization theorem for a Neumann problem*, Arch. Math. (Basel) **82** (2004), no. 4, 352–360.

[30] J. L. Kelley and I. Namioka, *Linear topological spaces*, D. Van Nostrand Co., Princeton, NJ, 1963.

[31] A. Kristály, *Existence of two non-trivial solutions for a class of quasilinear elliptic variational systems on strip-like domains*, Proc. Edinburgh Math. Soc. (2) **48** (2005), no. 2, 465–477.

[32] A. Kristály, *Multiplicity results for an eigenvalue problem for hemivariational inequalities in strip-like domains*, Set-Valued Anal. **13** (2005), no. 1, 85–103.

[33] A. Kristály, *Infinitely many solutions for a differential inclusion problem in* $\mathbb{R}^N$, J. Differential Equations **220** (2006), no. 2, 511–530.

[34] A. Kristály and C. Varga, *Multiple solutions for elliptic problems with singular and sublinear potentials*, Proc. Amer. Math. Soc., Article in press.

[35] A. Kristály and C. Varga, *On a class of quasilinear eigenvalue problems in* $\mathbb{R}^N$, Math. Nachr. **278** (2005), no. 15, 1756–1765.

[36] R. Livrea, *Existence of three solutions for a quasilinear two point boundary value problem*, Arch. Math. (Basel) **79** (2002), no. 4, 288–298.
[37] A. J. B. Lopes-Pinto, *On a new result on the existence of zeros due to Ricceri*, J. Convex Anal. **5** (1998), no. 1, 57–62.
[38] S. A. Marano and D. Motreanu, *Infinitely many critical points of non-differentiable functions and applications to a Neumann-type problem involving the p-Laplacian*, J. Differential Equations **182** (2002), no. 1, 108–120.
[39] S. A. Marano and D. Motreanu, *On a three critical points theorem for non-differentiable functions and applications to nonlinear boundary value problems*, Nonlinear Anal. **48** (2002), no. 1, 37–52.
[40] O. Naselli, *A class of functionals on a Banach space for which strong and weak local minima do coincide*, Optimization **50** (2001), no. 5–6, 407–411.
[41] B. Ricceri, *Images of open sets under certain multifunctions*, Rend. Accad. Naz. Sci. XL Mem. Mat. (5) **10** (1986), no. 1, 33–37.
[42] B. Ricceri, *Some topological mini-max theorems via an alternative principle for multifunctions*, Arch. Math. (Basel) **60** (1993), no. 4, 367–377.
[43] B. Ricceri, *A variational property of integral functionals on L^p-spaces of vector-valued functions*, C. R. Acad. Sci. Paris Sér. I Math. **318** (1994), no. 4, 337–342.
[44] B. Ricceri, *Applications of a theorem concerning sets with connected sections*, Topol. Methods Nonlinear Anal. **5** (1995), no. 2, 237–248.
[45] B. Ricceri, *Existence of zeros via disconnectedness*, J. Convex Anal. **2** (1995), no. 1–2, 287–290.
[46] B. Ricceri, *On the integrable selections of certain multifunctions*, Set-Valued Anal. **4** (1996), no. 1, 91–99.
[47] B. Ricceri, *Recent uses of connectedness in functional analysis*, Sūrikaisekikenkyūsho Kōkyūroku (1996), no. 939, 11–22.
[48] B. Ricceri, *More on a variational property of integral functionals*, J. Optim. Theory Appl. **94** (1997), no. 3, 757–763.
[49] B. Ricceri, *On a topological minimax theorem and its applications*, Minimax theory and applications (Erice, 1996), Kluwer Academic Publishers, Dordrecht, 1998, pp. 191–216.
[50] B. Ricceri, *Further considerations on a variational property of integral functionals*, J. Optim. Theory Appl. **106** (2000), no. 3, 677–681.
[51] B. Ricceri, *A general variational principle and some of its applications*, J. Comput. Appl. Math. **113** (2000), no. 1–2, 401–410.
[52] B. Ricceri, *On a three critical points theorem*, Arch. Math. (Basel) **75** (2000), no. 3, 220–226.
[53] B. Ricceri, *A further improvement of a minimax theorem of Borenshtein and Shul'man*, J. Nonlinear Convex Anal. **2** (2001), no. 2, 279–283.
[54] B. Ricceri, *Infinitely many solutions of the Neumann problem for elliptic equations involving the p-Laplacian*, Bull. London Math. Soc. **33** (2001), no. 3, 331–340.
[55] B. Ricceri, *Existence of zeros for operators taking their values in the dual of a Banach space*, Fixed Point Theory Appl. (2004), no. 3, 187–194.
[56] B. Ricceri, *Sublevel sets and global minima of coercive functionals and local minima of their perturbations*, J. Nonlinear Convex Anal. **5** (2004), no. 2, 157–168.
[57] B. Ricceri, *A general multiplicity theorem for certain nonlinear equations in Hilbert spaces*, Proc. Amer. Math. Soc. **133** (2005), no. 11, 3255–3261.
[58] B. Ricceri, *Minimax theorems for limits of parametrized functions having at most one local minimum lying in a certain set*, Topology Appl. **153** (2006), no. 17, 3308–3312.
[59] B. Ricceri, *A multiplicity theorem for the Neumann problem*, Proc. Amer. Math. Soc. **134** (2006), no. 4, 1117–1124.
[60] L. K. Shilgba, *Multiplicity of periodic solutions for a boundary eigenvalue problem*, Dyn. Syst. **20** (2005), no. 2, 223–232.

Open Problems in Topology II – Edited by E. Pearl

Twenty questions on metacompactness in function spaces

V.V. Tkachuk

1. Introduction

A space Z is called *metacompact* (or *weakly paracompact*) if any open cover of Z has a point-finite refinement. This notion is well-known and thoroughly studied. It is preserved by closed maps and closed subspaces and coincides with paracompactness in the class of collectionwise normal spaces; besides, any pseudocompact metacompact space is compact (see the survey of Burke [**4**] for proofs and detailed treatment). Unfortunately, the list of important results on metacompactness is too long to be presented here.

However, the importance of metacompactness in general topology is not reflected in C_p-theory at all; the results are scarce and almost nothing can be said even if we ask the most naïve questions about metacompactness in $C_p(X)$. The purpose of this paper is to draw attention to a significant amount of interesting open problems as well as to numerous possibilities of a breakthrough in this area.

The material of this paper is presented in Section 3 and Section 4 which cover the general case and the compact case respectively. In fact, every problem, formulated in Section 3, is open for compact spaces as well. However, its clone is formulated in Section 4 (this occurs four times) only if it is of special importance for the compact case.

2. Notation and terminology

The symbol $\mathbb{R}$ stands for the real line with it natural topology. All spaces are assumed to be Tychonoff. Given spaces X and Y the set $C(X,Y)$ consists of continuous maps from X to Y; we write $C(X)$ instead of $C(X,\mathbb{R})$. The expression $C_p(X,Y)$ denotes the set $C(X,Y)$ endowed with the pointwise convergence topology, i.e., $C_p(X,Y)$ (or $C_p(X)$ respectively) is $C(X,Y)$ (or $C(X)$ respectively) with the topology inherited from Y^X ($\mathbb{R}^X$). We also let $C_{p,0}(X) = X$ and $C_{p,n+1}(X) = C_p(C_{p,n}(X))$ for any $n \in \omega$. A space Z is said to have *countable tightness* (this is denoted by $t(Z) = \omega$) if, for any $A \subset Z$ and $z \in \overline{A}$ we have $z \in \overline{B}$ for some countable $B \subset A$.

3. Metacompactness in $C_p(X)$ for general spaces X

The most ambitious purpose would be to characterize metacompactness of $C_p(X)$ in terms of the space X. We do not formulate this as a question because

Research supported by Consejo Nacional de Ciencia y Tecnología (CONACYT) de México, Grant 400200-5-38164-E even for the Lindelöf property, its characterization in $C_p(X)$ seems to be a hopeless thing. The Souslin property of the spaces $C_p(X)$ implies that $C_p(X)$ is paracompact if and only if it is Lindelöf (see Chapter 0, Section 1 of Arhangel'skiĭ [**2**]) so the following problem seems to be the most important one.

1218? **Problem 1.** *Does metacompactness of $C_p(X)$ imply that $C_p(X)$ is Lindelöf?*

This question has long been a part of the folklore. The general theory of covering properties shows that if any closed discrete subspace of a metacompact $C_p(X)$ is countable then it is Lindelöf. It is a brilliant theorem of Reznichenko (see Theorem I.5.12 of Arhangel'skiĭ [**2**]) that any normal $C_p(X)$ is collectionwise normal so metacompactness of $C_p(X)$ together with its normality implies that $C_p(X)$ is Lindelöf.

1219? **Problem 2.** *Suppose that $C_{p,n}(X)$ is metacompact for all $n \geq 1$. Must $C_p(X)$ be Lindelöf?*

Maybe, the Lindelöf property is too much to ask from a metacompact $C_p(X)$. The following two problems present more humble expectations.

1220? **Problem 3.** *Suppose that $C_p(X)$ is metacompact. Must it be realcompact?*

1221? **Problem 4.** *Let X be a space such that $C_p(X)$ is metacompact. Is it true that the tightness of X has to be countable?*

If Problem 4 is answered positively then the answer to Problem 3 is also positive (see Chapter II, Section 4 of Arhangel'skiĭ [**2**]). Besides, if $C_p(X)$ is Lindelöf then $t(X) = \omega$ (see Asanov [**3**]) so Problem 4 asks whether it is possible to strengthen Asanov's result.

1222? **Problem 5.** *Suppose that $C_p(X)$ is metacompact. Must $C_p(X) \times C_p(X)$ be metacompact?*

This question is obligatory apart from reminding us the famous Arhangel'skiĭ problem (unanswered for several decades and published in many places, see e.g., Group of Problems C in Section 1 of Chapter 0 of Arhangel'skiĭ [**2**]) on whether the square of any Lindelöf $C_p(X)$ is Lindelöf. Maybe some weaker property of $C_p(X) \times C_p(X)$ can be derived from the Lindelöf property of $C_p(X)$ so it is worth to check for metacompactness.

1223? **Problem 6.** *Suppose that $C_p(X)$ is Lindelöf. Must $C_p(X) \times C_p(X)$ be metacompact?*

The following two problems are related to the results of Tkachuk [**6**] on countable additivity of pseudocharacter, tightness, Čech-completeness and some other properties in $C_p(X)$. Outside of C_p-theory, it is easy to give examples of non-metacompact spaces which are countable unions of their closed metacompact subspaces.

1224? **Problem 7.** *Suppose that $C_p(X) = \bigcup_{n\in\omega} Y_n$ and every Y_n is metacompact. Must $C_p(X)$ be metacompact?*

Problem 8. *Suppose that $C_p(X) = \bigcup_{n\in\omega} Y_n$ and every Y_n is closed in $C_p(X)$ and metacompact. Must $C_p(X)$ be metacompact?* 1225?

Since it is difficult to obtain any consequences of metacompactness in $C_p(X)$, we can suspect that it has some kind of universal presence in every function space. The following two problems formalize these suspicions.

Problem 9. *Is it true that every $C_p(X)$ has a dense metacompact subspace?* 1226?

Problem 10. *Is it true that every space X can be embedded in a space Y such that $C_p(Y)$ is metacompact?* 1227?

4. Metacompactness in $C_p(X)$ when X is compact

If X is compact then, to prove that $C_p(X)$ is Lindelöf, it suffices to establish some weaker properties of $C_p(X)$. For example, if $C_p(X)$ is normal then it is Lindelöf (see Theorem III.6.3 of Arhangel'skiĭ [**2**]). Therefore it is mandatory to formulate the following clone of Problem 1.

Problem 11. *Suppose that X is compact and $C_p(X)$ is metacompact. Must $C_p(X)$ be Lindelöf?* 1228?

The Lindelöf property of $C_p(X)$ has very strong consequences when X is compact so we could expect that metacompactness of $C_p(X)$ implies some restrictions on X. The following question is the clone of Problem 4.

Problem 12. *Let X be a compact space such that $C_p(X)$ is metacompact. Is it true that $t(X) = \omega$?* 1229?

The next two problems have positive answer if we replace metacompactness with the Lindelöf property. No counterexample exists for general spaces as well.

Problem 13. *Let X be a compact space such that $C_p(X)$ is metacompact. Is it true that $C_p(Y)$ is metacompact for any closed subspace $Y \subset X$?* 1230?

Problem 14. *Suppose that X is compact and $C_p(X, [0,1])$ is metacompact. Must $C_p(X)$ be metacompact?* 1231?

Since preservation of topological properties in finite powers is often a key matter, we also have to present the clones of Problem 5 and Problem 6.

Problem 15. *Suppose that X is compact and the space $C_p(X)$ is metacompact. Must $C_p(X) \times C_p(X)$ be metacompact?* 1232?

Problem 16. *Suppose that X is compact and $C_p(X)$ is Lindelöf. Must the space $C_p(X) \times C_p(X)$ be metacompact?* 1233?

Every dyadic compact space of countable tightness is metrizable according to Theorem 3.1.1 of Arhangel'skiĭ [**1**] so the Lindelöf property of $C_p(X)$ implies metrizability of any dyadic compact space X. The following question is again about what is left in X if $C_p(X)$ is metacompact.

1234? **Problem 17.** *Suppose that X is a dyadic compact space such that $C_p(X)$ is metacompact. Must X be metrizable?*

The space $\omega_1 + 1$ is a model for many matters concerned with countable tightness in compact spaces. It is known that $C_p(\omega_1 + 1)$ is very far from being Lindelöf; it does not even have a Lindelöf dense subspace (see Proposition IV.11.7 in Arhangel'skiĭ [**2**]); this easily implies that $\omega_1 + 1$ cannot be embedded in any X such that $C_p(X)$ has a Lindelöf dense subspace.

On the other hand, Dow, Junnila and Pelant proved in [**5**] that for any compact X of weight at most ω_1, the space $C_p(X)$ is hereditarily metaLindelöf. Therefore we can expect some kind of metacompactness in $C_p(\omega_1 + 1)$. Our last three questions are intended to express these expectations.

1235? **Problem 18.** *Is the space $C_p(\omega_1 + 1)$ metacompact?*

1236? **Problem 19.** *Is it true that $C_p(\omega_1 + 1)$ has a dense metacompact subspace?*

1237? **Problem 20.** *Can the space $\omega_1 + 1$ be embedded in a space Y such that $C_p(Y)$ is metacompact?*

References

[1] A. V. Arhangel'skiĭ, *The structure and classification of topological spaces and cardinal invariants*, Uspekhi Mat. Nauk **33** (1978), no. 6(204), 29–84, Translation: Russian Math. Surveys 33 (1978) 33–95.

[2] A. V. Arhangel'skiĭ, *Topological function spaces*, Kluwer Academic Publishers, Dordrecht, 1992.

[3] M. O. Asanov, *On cardinal invariants of spaces of continuous functions*, Modern topology and set theory, 2, Izhevskij Universitet, Izhevsk, 1980, pp. 1–7 (Russian).

[4] D. K. Burke, *Covering properties*, Handbook of set-theoretic topology, North-Holland, Amsterdam, 1984, pp. 347–422.

[5] A. Dow, H. Junnila, and J. Pelant, *Weak covering properties of weak topologies*, Proc. London Math. Soc. **75** (1997), no. 2, 349–368.

[6] V. V. Tkachuk, *The decomposition of $C_p(X)$ into a countable union of subspaces with "good" properties implies "good" properties of $C_p(X)$*, Trudy Moskov. Mat. Obshch. **55** (1994), 310–322, Translation: Trans. Moscow Math. Soc. (1994) 239–248.

Part 8

Dimension Theory

Open Problems in Topology II – Edited by E. Pearl

Open problems in infinite-dimensional topology

Taras Banakh, Robert Cauty and Michael Zarichnyi

Introduction

The development of Infinite-Dimensional Topology was greatly stimulated by three famous open problem lists: that of Geoghegan [**58**], West [**74**] and Dobrowolski, Mogilski [**55**]. We hope that the present list of problems will play a similar role for further development of Infinite-Dimensional Topology.

We expect that future progress will happen on the intersection of Infinite-Dimensional Topology with neighbor areas of mathematics: Dimension Theory, Descriptive Set Theory, Analysis, Theory of Retracts. According to this philosophy we formed the current list of problems. We tried to select problems whose solution would require creating new methods.

We shall restrict ourselves by separable and metrizable spaces. A *pair* is a pair (X, Y) consisting of a space X and a subspace $Y \subset X$. By ω we denote the set of non-negative integers.

1. Higher-dimensional descriptive set theory

Many results and objects of infinite-dimensional topology have zero-dimensional counterparts usually considered in Descriptive Set Theory. As a rule, "zero-dimensional" results have simpler proofs compared to their higher-dimensional counterparts. Some zero-dimensional results are proved by essentially zero-dimensional methods (like those of infinite game theory) and it is an open question to which extent their higher-dimensional analogues are true. We start with two problems of this sort.

For a class $\mathcal{C}$ of spaces and a number $n \in \omega \cup \{\infty\}$ consider the subclasses $\mathcal{C}[n] = \{C \in \mathcal{C} : \dim C \leq n\}$ and $\mathcal{C}[\omega] = \bigcup_{n\in\omega} \mathcal{C}[n]$. Following the tradition of Logic and Descriptive Set Theory, by Σ^1_1 we denote the class of *analytic* spaces, i.e., metrizable spaces which are continuous images of $\mathbb{N}^\omega$. Also Π^0_α and Σ^0_α stand for the multiplicative and additive classes of absolute Borel spaces corresponding to a countable ordinal α. In particular, Π^0_1, Π^0_2, and Σ^0_2 are the classes of compact, Polish, and σ-compact spaces, respectively. In topology those classes usually are denoted by $\mathcal{M}_0, \mathcal{M}_1$, and $\mathcal{A}_1$, respectively.

Following the infinite-dimensional tradition, we define a space X to be $\mathcal{C}$-*universal* for a class $\mathcal{C}$ of spaces, if X contains a closed topological copy of each space $C \in \mathcal{C}$. According to a classical result of Descriptive Set Theory [**62**, 26.12], an analytic space X is $\Pi^0_\xi[0]$-universal for a countable ordinal $\xi \geq 3$ if and only if $X \notin \Sigma^0_\xi$. This observation implies that a space X is $\Pi^0_\xi[0]$-universal if and only if the product $X \times Y$ is $\Pi^0_\xi[0]$-universal for some/any space $Y \in \Sigma_\xi$. The philosophy of this result is that a space X is $\mathcal{C}$-universal if $X \times Y$ is $\mathcal{C}$-universal for a relatively simple space Y.

The following theorem proved in [**25**, 3.2.12] shows that in some cases this philosophy is realized also on the higher-dimensional level.

Theorem. *Let $\mathcal{C} = \Pi^0_\xi[n]$ where $n \in \omega \cup \{\infty\}$ and $\xi \geq 3$ is a countable ordinal. A space X is $\mathcal{C}$-universal if and only if $X \times Y$ is $\mathcal{C}$-universal for some space $Y \in \Sigma^0_2$.*

However we do not know if the condition $Y \in \Sigma^0_2$ can be replaced with a weaker condition $Y \in \Sigma^0_3$ (which means that Y is an absolute $G_{\delta\sigma}$-space).

1238? **Question 1.1.** *Let $\mathcal{C} = \Pi^0_\xi[n]$ where $n \in \omega \cup \{\infty\}$ and $\xi \geq 3$ be a countable ordinal. Is a space X is $\mathcal{C}$-universal if $X \times Y$ is $\mathcal{C}$-universal for some space $Y \in \Sigma^0_3$? $Y \in \Sigma^0_\xi$?*

As we already know the answer to this problem is affirmative for $n = 0$.

In fact, an affirmative answer to Question 1.1 would follow from the validity of the higher-dimensional version of the Separation Theorem of Louveau and Saint-Raymond [**62**, 28.18]. Its standard formulation says that two disjoint analytic sets A, B in a Polish space X cannot be separated by a Σ^0_ξ-set with $\xi \geq 3$ iff the pair $(A \cup B, A)$ is $(\Pi^0_1[0], \Pi_\xi)$-universal.

A pair (X, Y) of spaces is defined to be $\vec{\mathcal{C}}$-*universal* for a class of pairs $\vec{\mathcal{C}}$ if for every pair $(A, B) \in \vec{\mathcal{C}}$ there is a closed embedding $f\colon A \to X$ with $f^{-1}(Y) = B$. For classes $\mathcal{A}$, $\mathcal{B}$ of spaces by $(\mathcal{A}, \mathcal{B})$ we denote the class of pairs (A, B) with $A \in \mathcal{A}$ and $B \in \mathcal{B}$. We recall that Π^0_1 stands for the class of compacta.

The Separation Theorem of Louveau and Saint-Raymond implies that for every $\Pi^0_\xi[0]$-universal subspace X of a space $Y \in \Sigma^0_\xi$ the pair (Y, X) is $(\Pi^0_1[0], \Pi^0_\xi)$-universal. The philosophy of this result is clear: if a $\mathcal{C}$-universal space X for a complex class $\mathcal{C}$ embeds into a "relatively simple" space Y, then the pair (Y, X) is $(\Pi^0_1[0], \mathcal{C})$-universal. If the "relatively simple" means "σ-compact", then the above zero-dimensional result has a higher-dimensional counterpart proved in [**25**, 3.1.2] (see also [**16**] and [**50**]).

Theorem. *Let $n \in \omega \cup \{\infty\}$ and $\mathcal{C} \in \{\Pi^0_\xi, \Sigma^0_\xi : \xi \geq 3\}$. For every $\mathcal{C}[n]$-universal subspace X of a space $Y \in \Sigma^0_2$ the pair (Y, X) is $(\Pi^0_1[n], \mathcal{C})$-universal.*

We do not know if $Y \in \Sigma^0_2$ in this theorem can be replaced with $Y \in \Sigma^0_3$.

1239? **Question 1.2.** *Let $n \in \omega \cup \{\infty\}$ and $\mathcal{C} = \Pi^0_\xi$ for a countable ordinal $\xi \geq 3$. Is it true that for each $\mathcal{C}[n]$-universal subspace X of a space $Y \in \Sigma^0_3$ the pair (Y, X) is $(\Pi^0_1[n], \mathcal{C})$-universal?*

As we already know the answer to this question is affirmative for $n = 0$. Using Theorem 3.2.12 of [**25**] on preservation of $\mathcal{C}$-universality by perfect maps one can show that an affirmative answer to Question 1.2 implies that to Question 1.1.

Our third problem with a higher-dimensional descriptive flavor asks if higher-dimensional Borel complexity can be concentrated on sets of a smaller dimension. First let us make two simple observations: the Hilbert cube $[0,1]^\omega$ is Π^0_1-universal while its pseudointerior $(0,1)^\omega$ is Π^0_2-universal. In light of these observations one could suggest that for each $\xi \geq 1$ there is a one-dimensional space X whose countable power X^ω is Π^0_ξ-universal. But this is not true: no finite-dimensional space X has Σ^0_2-universal countable power X^ω (see [**17**, **24**, **42**]). On the other hand, for every meager locally path-connected space X the $(2n+1)$-st power X^{2n+1} is $\Sigma^0_2[n]$-universal (which means that X^{2n+1} contains a closed topological copy of each n-dimensional σ-compact space), see [**18**].

Question 1.3. *Let $\mathcal{B} \in \{\Pi^0_\xi, \Sigma^0_\xi : \xi \geq 1\}$ is a Borel class. Is there a one-dimensional space X (in $\mathcal{B}$) with $\mathcal{B}[\omega]$-universal power X^ω?* 1240?

The answer to this problem is affirmative for the initial Borel classes $\mathcal{B} \in \{\Pi^0_1, \Pi^0_2, \Sigma^0_2\}$, see [**18**, **21**, **36**]. Moreover, for such a class $\mathcal{B}$ a space X with $\mathcal{B}[n]$-universal power X^{n+1} can be chosen as a suitable subspace of a dendrite with a dense set of end-points.

A related question concerns the universality in classes of compact spaces. It is well-known that the n-dimensional cube $[0,1]^n$ is not $\Pi^0_1[n]$-universal. On the other hand, for any dendrite D with dense set of end-points the power D^{n+1} is $\Pi^0_1[n]$-universal [**36**], and the product $D^{n+1} \times [0,1]^{2n}$ is $\Pi^0_1[2n]$-universal, see [**28**].

Question 1.4. *Is $X \times [0,1]^{2n}$ $\Pi^0_1[2n]$-universal for any $\Pi^0_1[n]$-universal space X? Equivalently, is $\mu^n \times [0,1]^{2n}$ $\Pi^0_1[2n]$-universal (where μ^n denotes the n-dimensional Menger cube)?* 1241?

For $n \leq 1$ the answer to this problem is negative. We expect that this is so for all n.

2. Z_n-sets and related questions

In this section we consider some problems related to Z_n-sets, where $n \in \omega \cup \{\infty\}$. By definition, a subset A of a space X is a *Z_n-set* in X if A is closed and the complement $X \setminus A$ is *n-dense* in X in the sense that each map $f\colon [0,1]^n \to X$ can be uniformly approximated by maps into $X \setminus A$. In particular, 0-density is equivalent to the usual density and a subset $A \subset X$ is a Z_0-set in X if and only if it is closed and nowhere dense in X.

A set $A \subset X$ is a *σZ_n-set* if A is the countable union of Z_n-sets in X. A subset $A \subset X$ is called *n-meager* if $A \subset B$ for some σZ_n-set B in X. A space X is a *σZ_n-space* (or else *n-meager*) if X is a σZ_n-set (equivalently n-meager) in itself. In particular, a space is 0-meager if and only if it is of the first Baire category.

According to a classical result of S. Banach [**6**], an analytic topological group either is complete or else is 0-meager. It is natural to ask about the infinite version of this result. Namely, Question 4.4 in [**55**] asks if any incomplete Borel pre-Hilbert space is ∞-meager. The answer to this question turned out to be negative: the linear span(E) of the Erdős set $E \subset \{(x_i) \in \ell_2 : \forall i\ x_i \in \mathbb{Q}\}$ is meager but not ∞-meager, see [**25**, 5.5.19]. Moreover, span(E) cannot be written as the countable union $\bigcup_{n\in\omega} Z_n$ where each set Z_n is a Z_n-set in span(E). On the other hand, for every $n \in \omega$, span(E) can be written as the countable union of Z_n-sets, see [**10**].

Question 2.1. *Is an (analytic) linear metric space X a σZ_∞-space if X can be written as the countable union $X = \bigcup_{n\in\omega} X_n$ where each set X_n is a Z_n-set in X.* 1242?

By its spirit this problem is related to Selection Principles, a branch of Combinatorial Set Theory considered in the papers [**69**, **72**].

Another feature of span(E) leads to the following problem, first posed in [**7**].

1243? **Question 2.2.** *Is every analytic non-complete linear metric space X a σZ_n-space for every $n \in \omega$? Is this true if X is a linear subspace of ℓ^2 or $\mathbb{R}^\omega$?*

With the help of the Multiplication Formula for σZ_n-spaces [**27**] or [**20**], the (second part of the) above problem can be reduced to the following one.

1244? **Question 2.3.** *Let X be a non-closed analytic linear subspace of the space $L = \ell^2$ or $L = \mathbb{R}^\omega$. Can L be written as the direct sum $L = L_1 \oplus L_2$ of two closed subspaces $L_1, L_2 \subset L$ so that for every $i \in \{1,2\}$ the projection X_i of X onto L_i is a proper subspace of L_i?*

Let us note that the zero-dimensional counterpart of this question has an affirmative solution: for each meager subset $H \subset \{0,1\}^\omega$ there is a partition $\omega = A \cup B$ of ω into two disjoint sets A, B such that the projections of H onto $\{0,1\}^A$ and $\{0,1\}^B$ are not surjective. This partition can be easily constructed by induction.

The following weaker problem related to Question 2.3 also is open.

1245? **Question 2.4.** *Let X be a non-closed analytic linear subspace in ℓ^2. Is there a closed infinite-dimensional linear subspace $L \subset \ell^2$ such that $X + L \neq \ell^2$?*

We recall that a space X is *(strongly) countable-dimensional* if X can be written as the countable union $X = \bigcup_{n=1}^\infty X_n$ of (closed) finite-dimensional subspaces of X. The space span(E) is countable-dimensional but not strongly countable-dimensional, see [**10**].

1246? **Question 2.5.** *Is each (analytic) strongly countable-dimensional linear subspace of ℓ^2 ∞-meager? equivalently, 2-meager?*

In light of this question it should be mentioned that each closed finite-dimensional subspace of the Hilbert space ℓ^2 is a Z_1-set. On the other hand, ℓ^2 contains closed zero-dimensional subsets failing to be Z_2-sets in ℓ^2. Yet, each finite-dimensional Z_2-set in ℓ^2 is a Z_∞-set in ℓ^2, see [**64**].

Let $\mathcal{M}_n$ be the σ-ideal consisting of n-meager subsets of the Hilbert cube Q. In particular, $\mathcal{M}_0$ coincides with the ideal $\mathcal{M}$ of meager subsets of Q well studied in Set Theory. For each nontrivial ideal $\mathcal{I}$ of subsets of a set X we can study four cardinal characteristics:

- $\mathrm{add}(\mathcal{I}) = \min\{|\mathcal{J}| : J \subset \mathcal{I}, \bigcup \mathcal{J} \notin \mathcal{I}\}$;
- $\mathrm{cov}(\mathcal{I}) = \min\{|\mathcal{J}| : J \subset \mathcal{I}, \bigcup \mathcal{J} = X\}$;
- $\mathrm{non}(\mathcal{I}) = \min\{|A| : A \subset X, A \notin \mathcal{I}\}$;
- $\mathrm{cof}(\mathcal{I}) = \min\{|\mathcal{C}| : \mathcal{C} \subset \mathcal{I}, (\forall A \in \mathcal{I})(\exists C \in \mathcal{C}) A \subset C\}$.

The cardinal characteristics of the ideal $\mathcal{M}_0 = \mathcal{M}$ are calculated in various models of ZFC and can vary between $\aleph_1$ and the continuum $\mathfrak{c}$, see [**73**]. In [**30**] it is shown that $\mathrm{cov}(\mathcal{M}_n) = \mathrm{cov}(\mathcal{M})$ and $\mathrm{non}(\mathcal{M}_n) = \mathrm{non}(\mathcal{M})$ for every $n \in \omega \cup \{\infty\}$.

Question 2.6. *Is* $\mathrm{add}(\mathcal{M}_n) = \mathrm{add}(\mathcal{M})$ *and* $\mathrm{cof}(\mathcal{M}_n) = \mathrm{cof}(\mathcal{M})$ *for every* $n \in \omega \cup \{\infty\}$*?* 1247?

It is well-known that Z_∞-sets in ANR-spaces can be characterized as closed sets with homotopy dense complement. A subset D of a space X is called *homotopy dense* if there is a homotopy $h\colon X \times [0,1] \to X$ such that $h(x,0) = x$ and $h(x,t) \in D$ for all $(x,t) \in X \times (0,1]$.

One of the problems from [**55**, **74**] asked about finding an inner characterization of homotopy dense subspaces of s-manifolds. In [**25**, 1.3.2] (see also [**8**] and [**52**]) it was shown that such subspaces can be characterized with help of SDAP, the Toruńczyk's Strong Discrete Approximation Property. This characterization allowed to apply powerful tools of the theory of Hilbert manifolds to studying spaces with SDAP.

Question 2.7. *Find an inner characterization of homotopy dense subspaces of Q-manifolds.* 1248?

The problem of the characterization of locally compact spaces homeomorphic to homotopy dense subsets of compact ANRs (or compact Q-manifolds) was considered in [**46**, **51**].

It is known that each homotopy dense subspace X of a locally compact ANR-space has LCAP, the Locally Compact Approximation Property. The latter means that for every open cover $\mathcal{U}$ of X the identity map of X can be uniformly approximated by maps $f\colon X \to X$ whose range $f(X)$ has locally compact closure in X.

Question 2.8. *Is each space X with LCAP homeomorphic to a homotopy dense subspace of a locally compact ANR.* 1249?

Let us note that LCAP appears as an important ingredient in many results of infinite-dimensional topology, see [**9**, **25**].

Question 2.9. *Let X be a convex set in a linear metric space. Has X LCAP? Has X LCAP if X is an absolute retract?* 1250?

The answer to the latter question is affirmative if the completion of X is an absolute retract, see [**25**, 5.2.5].

3. The topology of convex sets and topological groups

One of the classical applications of infinite-dimensional topology is detecting the topological structure of convex sets in linear metric spaces. As a rule, convex sets are absolute retracts and have many other nice features facilitating applications of powerful methods of infinite-dimensional topology. Among such methods let us recall the theory of Q- and ℓ^2-manifolds and the theory of absorbing and coabsorbing spaces. The principal notion unifying these theories is the notion of a strongly universal space.

A topological space X is defined to be *strongly $\mathcal{C}$-universal* for a class $\mathcal{C}$ of spaces if for every cover $\mathcal{U}$, every space $C \in \mathcal{C}$ and a map $f\colon C \to X$ whose restriction $f|B\colon B \to X$ onto a closed subset $B \subset C$ is a Z-embedding there is a

Z-embedding $\tilde{f}: C \to X$ which is $\mathcal{U}$-near to f and coincides with f on B. A map $f: C \to X$ is called a *Z-embedding* if it is a topological embedding and $f(C)$ is a Z_∞-set in X. A topological space X is *strongly universal* if it is strongly $\mathcal{Z}(X)$-universal for the class $\mathcal{Z}(X)$ of spaces homeomorphic to Z_∞-sets of X. Many natural spaces are strongly universal.

In [**8, 13, 15, 25, 37, 38, 41, 45, 48, 54**] many results on the strong universality of convex sets in linear metric space were obtained. Nonetheless the following problem still is open.

1251? **Question 3.1.** *Let X be an infinite-dimensional closed convex set in a locally convex linear metric space L. Is X strongly universal?*

The answer is not known even for the case when X is a pre-Hilbert space. A bit weaker question also is open.

1252? **Question 3.2.** *Let X be an infinite-dimensional closed convex set in a locally convex linear metric space L. Is X strongly $\mathcal{Z}_{tb}(X)$-universal for the class of spaces homeomorphic to totally bounded Z_∞-subsets of X?*

The answer to this problem is affirmative if X has an almost internal point $x_0 \in X$ (the latter means that the set $\{x \in X : (\exists z \in X)(\exists t \in (0,1))\ x_0 = tx + (1-t)z\}$ is dense in X), see [**13**].

Strong universality enters as one of important ingredients into the definition of a (co)absorbing space. A topological space X is called *absorbing* (resp. *coabsorbing*) if X is an ∞-meager (resp. ∞-comeager) strongly universal ANR with SDAP. We recall that a space is n-meager where $n \in \omega \cup \{\infty\}$ if it is a σZ_n-set in itself. A space X is defined to be *n-comeager* if X contains an absolute G_δ-subset G with n-meager complement $X \setminus G$ in X. In particular an analytic space is 0-comeager if and only if it is Baire.

1253? **Question 3.3.** *Let X be a closed convex subset of a locally convex linear metric space. Assume that X is 0-comeager. Is it ∞-comeager? Is X n-comeager for all $n \in \omega$?*

1254? **Question 3.4.** *Assume that $X \in AR$ is an ∞-(co)meager closed convex set in a linear metric space. Is X a (co)absorbing space?*

The principal result of the theory of (co)absorbing spaces is the Uniqueness Theorem [**25**, §1.6] asserting that two (co)absorbing spaces X,Y are homeomorphic if and only if X, Y are homotopically equivalent and $\mathcal{Z}(X) = \mathcal{Z}(Y)$. This fact helps to establish the topological structure of many infinite-dimensional (co)absorbing spaces appearing in *nature*, see [**25, 41, 68**].

In particular, in [**25**] it was shown that a closed convex subset X of a locally convex linear metric space is Π^0_ξ-(co)absorbing for $\xi \geq 2$ if and only if X is a Π^0_ξ-universal ∞-(co)meager space and $X \in \Pi^0_\xi$. The same result is true for additive Borel classes Σ^0_ξ with $\xi \geq 3$. Surprisingly, but for the class Π^0_1 of compacta we still have an open question.

Question 3.5. *Let X be a Π_1^0-universal convex (closed σ-compact) subset of ℓ^2. Is X strongly Π_1^0-universal?* 1255?

This question has an affirmative answer if X contains an almost internal point.

A similar situation holds for metrizable topological groups. It is shown in [**25**, 4.2.3] that a topological group G is a Π_ξ^0-absorbing space for $\xi \geq 2$ if and only if $G \in \Pi_\xi^0$ is a Π_ξ^0-universal ANR.

Question 3.6. *Let $G \in$ ANR be a Π_1^0-universal σ-compact metrizable group. Is G an Π_1^0-absorbing space?*

A similar question for the class $\Pi_1^0[\omega]$ of finite-dimensional compacta is also open, see [**55**, 5.7].

Question 3.7. *Let G be an infinite-dimensional σ-compact strongly countable-dimensional locally contractible group (containing a topological copy of each finite-dimensional compactum). Is G a $\Pi_1^0[\omega]$-absorbing space? Equivalently, is G an ℓ_f^2-manifold?*

In fact, the method of absorbing sets works not only for spaces and pairs but also for order-preserving systems $(X_\gamma)_{\gamma\in\Gamma}$ of topological spaces, indexed by a partially ordered set Γ with largest element $\max\Gamma$. The order-preserving property of (X_γ) means that $X_\gamma \subset X_{\gamma'}$ for any elements $\gamma \leq \gamma'$ in Γ. So each X_γ is a subspace of $X_{\max\Gamma}$. Such systems (X_γ) are called Γ-systems. For a Γ-system $\mathcal{X} = (X_\gamma)_{\gamma\in\Gamma}$, a subset $F \subset X_{\max\Gamma}$, and a map $f\colon Y \to X$ we let $F \cap \mathcal{X} = (F \cap X_\gamma)_{\gamma\in\Gamma}$ and $f^{-1}(\mathcal{X}) = (f^{-1}(X_\gamma))_{\gamma\in\Gamma}$.

The notion of strong universality extends to Γ-systems as follows: A Γ-system $\mathcal{X} = (X_\gamma)_{\gamma\in\Gamma}$ is called *strongly $\vec{\mathcal{C}}$-universal* for a class $\vec{\mathcal{C}}$ of Γ-systems if given: an open cover $\mathcal{U}$ of $X_{\max\Gamma}$, a Γ-system $\mathcal{A} = (A_\gamma)_{\gamma\in\Gamma} \in \vec{\mathcal{C}}$, a closed subset $F \subset A_{\max\Gamma}$ and a map $f\colon A_{\max\Gamma} \to X_{\max\Gamma}$ such that $f|F$ is a Z-embedding with $F \cap f^{-1}(\mathcal{X}) = F \cap \mathcal{A}$, there is a Z-embedding $\tilde{f}\colon A_{\max\Gamma} \to X_{\max\Gamma}$ such that $\tilde{f}$ is $\mathcal{U}$-near to f, $\tilde{f}|F = f$ and $\tilde{f}^{-1}(\mathcal{X}) = \mathcal{A}$.

A system $\mathcal{X}$ is called *$\vec{\mathcal{C}}$-absorbing* in E if $\mathcal{X}$ is strongly $\vec{\mathcal{C}}$-universal and $X_{\max\Gamma} = \bigcup_{n\in\omega} Z_n$ where each Z_i is a Z_∞-set in $X_{\max\Gamma}$ and $Z_n \cap \mathcal{X} \in \vec{\mathcal{C}}$. For more information on absorbing systems, see [**5**].

Given a class $\vec{\mathcal{C}}$ of Γ-systems and a non-negative integer number n consider the subclass

$$\vec{\mathcal{C}}[n] = \{\mathcal{X} \in \vec{\mathcal{C}} : \dim(X_{\max\Gamma}) \leq n\}.$$

The following question is related to the results on existence of absorbing sets for n-dimensional Borel classes [**76**].

Question 3.8. *For which classes $\vec{\mathcal{C}}$ of Γ-systems does the existence of a $\vec{\mathcal{C}}$-absorbing Γ-system implies the existence of a $\vec{\mathcal{C}}[n]$-absorbing Γ-system for every $n \in \omega$?* 1256?

One can formulate this question also for other types of dimensions, in particular, for extension dimension introduced by Dranishnikov [**57**].

4. Topological characterization of particular infinite-dimensional spaces

The theory of (co)absorbing spaces is applicable for spaces which are either ∞-meager or ∞-comeager. However some natural strongly universal spaces do not fall into either of these two categories. One of such spaces is $\mathrm{span}(E)$, the linear hull of the Erdős set in ℓ^2 which is a meager strongly universal AR with SDAP that fails to be ∞-meager, see [**10**, **53**].

1257? **Question 4.1.** *Give a topological characterization of* $\mathrm{span}(E)$. *Is* $\mathrm{span}(E)$ *homeomorphic to the linear hull* $\mathrm{span}(\mathbb{Q}^\omega)$ *of* $\mathbb{Q}^\omega$ *in* $\mathbb{R}^\omega$*? to the linear hull* $\mathrm{span}(E_p)$ *of the Erdős set* $E_p = \{(x_i) \in \ell^p : (x_i) \in \mathbb{Q}^\omega$ *and* $\lim_{i\to\infty} x_i = 0\}$ *in the Banach space* ℓ^p, $1 \le p \le \infty$*?*

Another problem of this sort concerns the countable products X^ω of finite-dimensional meager absolute retracts X. Using [**25**, 4.1.2] one can show that the countable product of such a space X is a strongly universal AR with SDAP. The space X^ω is a n-meager for all $n \in \omega$ but unfortunately is not ∞-meager.

1258? **Question 4.2.** *Let* X, Y *be finite-dimensional* σ*-compact absolute retracts of the first Baire category. Are* X^ω *and* Y^ω *homeomorphic?* (Applying [**18**] one can show that each of the spaces X^ω, Y^ω admits a closed embedding into the other space.)

Our third pathological (though natural) example is the hyperspace $\exp_H(\mathbb{Q}_I)$ of closed subsets of the space of rationals $\mathbb{Q}_I = [0,1] \cap \mathbb{Q}$ on the interval, endowed with the Hausdorff metric. This space has many interesting features similar to those of $\mathrm{span}\, E$: $\exp_H(\mathbb{Q}_I)$ is n-meager for all $n \in \omega$ but fails to be ∞-meager; $\exp_H(\mathbb{Q}_I)$ is homeomorphic to its square and belongs to the Borel class Π^0_3 of absolute $F_{\sigma\delta}$-subsets; $\exp_H(\mathbb{Q}_I)$ is $\Pi^0_3[\omega]$-universal but fails to be Π^0_3-universal, see [**23**].

1259? **Question 4.3.** *Give a topological characterization of the space* $\exp_H(\mathbb{Q}_I)$. *In particular, are the spaces* $\exp_H(\mathbb{Q}_I)$ *and* $\exp_H(\mathbb{Q}_I \times \mathbb{Q}_I)$ *homeomorphic?*

The three preceding examples were meager but not ∞-meager. Because of that they cannot be treated by the theory of absorbing spaces. Our other two problems concern spaces that are 0-comeager but not ∞-comeager and hence cannot be treated by the theory of coabsorbing spaces. These spaces are defined with help of the operation of weak product

$$W(X,Y) = \{(x_i) \in X^\omega : (\exists n \in \omega)(\forall i \ge n)\ x_i \in Y\}$$

where Y is a subspace of X. The classical space of the form $W(X,Y)$ is the Nagata space $N = W(\mathbb{R}, \mathbb{P})$ well-known in Dimension Theory as a universal space in the class of countable-dimensional (absolute $G_{\delta\sigma}$-)spaces. Here $\mathbb{P} = \mathbb{R} \setminus \mathbb{Q}$ stands for the space of irrationals. The countable product $\mathbb{P}^\omega$ is a dense absolute G_δ-set in $W(\mathbb{R},\mathbb{P})$. Nonetheless, $W(\mathbb{R},\mathbb{P})$ contains no ∞-dense absolute G_δ-set (because $W(\mathbb{R},\mathbb{P})$ is countable-dimensional) and thus $W(\mathbb{R},\mathbb{P})$ is not a coabsorbing space (but is strongly universal and has SDAP).

Question 4.4. *Give a topological characterization of the Nagata space $N = W(\mathbb{R}, \mathbb{P})$.* 1260?

To pose a (possibly) more tractable question, let us note that N is homeomorphic to $W(N, \mathbb{P}^\omega)$ (by a coordinate-permuting homeomorphism).

Question 4.5. *Is $N = W(\mathbb{R}, \mathbb{P})$ homeomorphic to $W(N, (\mathbb{P} \setminus \{\sqrt{2}\})^\omega)$?* 1261?

Next, we shall ask about the characterization of the pair $(I^\omega, \mathbb{P}_I^\omega)$ where $\mathbb{P}_I = I \cap \mathbb{P}$ is the set of irrational numbers on the interval $I = [0, 1]$. Topological characterizations of the Hilbert cube I^ω and irrational numbers $\mathbb{P}_I^\omega$ are well-known.

Question 4.6. *Give a topological characterization of the pair $(I^\omega, \mathbb{P}_I^\omega)$. In particular, is $(I^\omega, \mathbb{P}_I^\omega)$ homeomorphic to (I^ω, G) for every dense zero-dimensional G_δ-subset $G \subset I^\omega$ with homotopy dense complement in I^ω?* 1262?

A similar question concerns also the pair $(I^\omega, \mathbb{Q}_I^\omega)$ where $\mathbb{Q}_I = I \cap \mathbb{Q}$. Since $\mathbb{Q}_I^\omega$ is not ∞-meager in I^ω, this pair can not be treated by the theory of absorbing pairs, see [**25**].

Question 4.7. *Give a topological characterization of the pair $(I^\omega, \mathbb{Q}_I^\omega)$.* 1263?

5. Problems on ANRs

One of the principal problems on ANRs from the preceding two lists [**58**, **74**], the classical Borsuk's Problem on the AR-property of linear metric spaces, has been resolved in negative by R. Cauty in [**40**] who constructed a σ-compact linear metric space that fails to be an absolute retract. However, the "compact' version of Borsuk's problem still is open.

Question 5.1. *Let C be a compact convex set in a linear metric space. Is C an absolute retract?* 1264?

There are also many other natural spaces whose ANR-property is not established. Some of them are known to be divisible by the Hilbert space ℓ^2 in the sense that they are homeomorphic to the product with ℓ^2 (and hence are ℓ^2-manifolds if and only if they are ANRs). A classical example of this sort is the homeomorphism group of an n-manifolds for $n \geq 2$, see [**74**, HS4].

Another example is the space H_B of Brouwer homeomorphisms of the plane, endowed with the compact-open topology. A homeomorphism $h\colon \mathbb{R}^2 \to \mathbb{R}^2$ is a *Brouwer homeomorphism* if h preserves the orientation and has no fixed point.

Question 5.2. *Is the space H_B an ANR?* 1265?

It is known that H_B is locally contractible [**34**], is homotopically equivalent to the circle S^1 and is divisible by ℓ^2 [**67**]. So, H_B is homeomorphic to $S^1 \times \ell^2$ if and only if H_B is an ANR.

In spite of the existence of a linear metric space failing to be an AR, R. Cauty proved that each convex subset of a linear metric space is an algebraic ANR (algebraic ANRs are defined with help of a homological counterpart of the Lefschetz condition, see [**44**]). This follows from an even more general fact asserting that each metrizable locally equiconnected space is an algebraic ANR, see [**44**].

We recall that a topological space X is *locally equiconnected* if there are an open neighborhood $U \subset X \times X$ of the diagonal and a continuous function $\lambda\colon U \times [0,1] \to X$ such that $\lambda(x,y,0) = x$, $\lambda(x,y,1) = y$ and $\lambda(x,x,t) = x$ for every $(x,y,t) \in U \times [0,1]$. If $U = X \times X$, then X is called *equiconnected*. It is easy to see that each (locally) contractible topological group G is (locally) equiconnected and so is any retract of G. We do not know if the converse is true.

1266? **Question 5.3.** *Let X be a (locally) equiconnected metrizable space. Is X a (neighborhood) retract of a contractible metrizable topological group?*

It should be mentioned that this question has an affirmative answer for compact X, see [**39**]. The proof of this particular case exploits the fact that each metrizable equiconnected space X admits a Mal′tsev operation (which is a continuous map $\mu\colon X^3 \to X$ such that $\mu(x,x,y) = \mu(y,x,x) = y$ for all $x,y \in X$). Due to Sipacheva [**71**] we know that each compact space X admitting a Mal′tsev operation is a retract of the free topological group $F(X)$ over X. Therefore, each equiconnected compact metrizable space X has a Mal′tsev operation and hence is a retract of the free topological group $F(X)$. Moreover, it can be shown that the connected component of $F(X)$ containing X is contractible. Now it is easy to select a metrizable group topology τ on $F(X)$ inducing the original topology on X and such that X still is a retract of $(F(X),\tau)$ and the component of $(F(X),\tau)$ containing X is contractible. This resolves the "compact" version of Question 5.3. The non-compact version of this problem is related to the following question (discussed also in [**63**]):

1267? **Question 5.4.** *Is a metrizable space admitting a Mal′tsev operation a retract of a metrizable topological group?*

By definition, an *n-mean* on a topological space X is a continuous map $m\colon X^n \to X$ such that $m(x,\dots,x) = x$ for all $x \in X$ and $m(x_{\sigma(1)},\dots,x_{\sigma(n)}) = m(x_1,\dots,x_n)$ for any vector $(x_1,\dots,x_n) \in X^n$ and any permutation σ of $\{1,\dots,n\}$. Let us note that each convex subset C of a linear topological space admits an n-means and so does any retract of C.

1268? **Question 5.5.** *Let $n \geq 2$. Is there a metrizable equiconnected compact space X admitting no n-mean?*

If such a compact space X exists then it is a retract of a contractible group but fails to be a retract of a convex subset of a linear topological space.

For a compact space X let $L(X)$ be the free topological linear space over X and $P(X)$ be the convex hull of X in $L(X)$ (it can be shown that $P(X)$ is a free convex set over X). Let $(U(X),\lambda_X)$ be the free equiconnected space over X (where $\lambda\colon U(X)\times U(X)\times[0,1] \to U(X)$ is the equiconnected map of $U(X)$), see [**43**]. Let $\mathcal{T}_v(X)$ be the family of metrizable linear topologies on $L(X)$ inducing the original topology on X. (The family $\mathcal{T}_v(X)$ was essentially used in [**40**] for constructing the example of a linear metric space failing to be an AR). Let $\mathcal{T}_c(X) = \{\tau|P(X) :$

$\tau \in \mathcal{T}_v(X)\}$ be the family consisting of the restrictions of the topologies $\tau \in \mathcal{T}_v(X)$ onto $P(X)$, and $\mathcal{T}_u(X)$ be the family of metrizable topologies on $U(X)$ which induce the initial topology on X and preserve the continuity of the equiconnected map $\lambda_X : U(X) \times U(X) \times [0,1] \to U(X)$.

It is interesting to study the classes $\mathcal{A}_v$ ($\mathcal{A}_c$, $\mathcal{A}_u$) of metric compacta X such that the spaces $(L(X), \tau)$ ($(P(X), \tau)$, $(U(X), \tau)$) are absolute retracts for all topologies τ in $\mathcal{T}_v(X)$ ($\mathcal{T}_c(X)$, $\mathcal{T}_u(X)$). It is known that $\mathcal{A}_u \subset \mathcal{A}_c \subset \mathcal{A}_v$, see [**43**].

Question 5.6. *Is it true that* $\mathcal{A}_u = \mathcal{A}_c = \mathcal{A}_v$*?* 1269?

The class $\mathcal{A}_u$ contains all metrizable compact ANRs and all metrizable compact C-spaces, see [**43**].

Question 5.7. *Is it true that each weakly infinite-dimensional compact metrizable space belongs to* $\mathcal{A}_v$*? to* $\mathcal{A}_u$*?* 1270?

In light of this question let us mention that there is a strongly infinite-dimensional compact space D of finite cohomological dimension with $D \notin \mathcal{A}_v$, see [**40**]. In fact, the free linear space $L(D)$ over D, endowed with a suitable metrizable topology, gives the mentioned example of a linear metric space which is not an AR.

According to [**43**] the class $\mathcal{A}_c$ is closed with respect to countable products. We do not know if the same is true for the class $\mathcal{A}_u$.

Question 5.8. *Is the class* $\mathcal{A}_u$ *closed with respect to countable products?* 1271?

6. Infinite-dimensional problems from Banach space theory

In this section we survey some open problems lying in the intersection of infinite-dimensional topology and the theory of Banach spaces. Our principal object is the unit ball $B_X = \{x \in X : \|x\| \leq 1\}$ of a Banach space X, endowed with the weak topology. It is well-known that the weak ball B_X is metrizable (and separable) if and only if the Banach space X has separable dual. So, till the end of this section by a "Banach space" we understand an infinite-dimensional Banach space with separable dual. In [**12**] the following general problem was addressed:

Question 6.1. *Investigate the interplay between geometric properties of a Banach space* X *and topological properties of its weak unit ball* B_X*. Find conditions under which two Banach spaces have homeomorphic weak unit balls.* 1272?

It turns out that answers to these questions depend on (1) the class $\mathcal{W}(X)$ of topological spaces homeomorphic to closed bounded subsets of a Banach space X endowed with the weak topology, and (2) [in case of complex $\mathcal{W}(X)$] on properties of the norm of X. Let us remark that $\mathcal{W}(X)$ coincides with the class $\mathcal{F}_0(B_X)$ of topological spaces homeomorphic to closed subsets of the weak unit ball B_X of X.

In [**12**] it was observed that the class $\mathcal{W}(X)$ is not too large: it lies in the class Π^0_3 of absolute $F_{\sigma\delta}$-sets. For reflexive infinite-dimensional Banach spaces X the class $\mathcal{W}(X)$ coincides with the class Π^0_1 of compacta. On the other hand, for the Banach space $X = c_0$ the class $\mathcal{W}(X)$ is the largest possible and coincides with the Borel class Π^0_3. An intermediate case $\mathcal{W}(X) = \Pi^0_2$ happens if and only if X is a non-reflexive Banach space with PCP, the Point Continuity Property (which means that for each bounded weakly closed subset $B \subset X$ the identity map $(B, \text{weak}) \to B$ has a continuity point).

For Banach spaces with PCP the weak unit ball B_X is homeomorphic either to the Hilbert cube Q (if X is reflexive) or to the pseudointerior $s = (0,1)^\omega$ of Q (if X is not reflexive). In two latter cases, the topology of B_X does not depend on the particular choice of an equivalent norm on X. In this case we say that the Banach space X has BIP, the Ball Invariance Property. More precisely, X has BIP if the weak unit ball B_X of X is homeomorphic to the weak unit ball B_Y of any Banach space Y, isomorphic to X. It is known [**12**] that PCP implies BIP and BIP implies CPCP, the Convex Point Continuity Property, which means that each closed convex bounded subset of the Banach space has a point at which the norm topology coincides with the weak topology. It is known that the properties PCP and CPCP are different: the Banach space B_∞ constructed in [**59**] has CPCP but not PCP.

1273? **Question 6.2.** *Is BIP equivalent to PCP? To CPCP? Has the Banach space B_∞ BIP?*

In fact, the geometric properties PCP, BIP, and CPCP of a Banach space X can be characterized via topological properties of the weak unit ball B_X: X has PCP (resp. CPCP, BIP) if and only if B_X is Polish (0-comeager, ∞-comeager).

The norm of a Banach space X will be called *n-(co)meager* if the respective weak unit ball B_X is n-(co)meager. Let us remark that each Kadec norm is ∞-comeager (since the unit sphere is an ∞-dense absolute G_δ-subset in the weak unit ball). It is well-known that each separable Banach space admits an equivalent Kadec (and hence ∞-comeager) norm.

1274? **Question 6.3.** *Give a geometric characterization of Banach spaces admitting an equivalent ∞-meager norm.*

For n-meager norms with $n \in \omega$ the answer is known: a Banach space X admits an equivalent n-meager norm if and only if X fails to have CPCP. On the other hand, a Banach space X has an equivalent ∞-meager norm if X fails to be strongly regular, see [**12**]. We recall that a Banach space X is called *strongly regular* if for every $\varepsilon > 0$ and every nonempty convex bounded subset $C \subset X$ there exist nonempty relatively weak-open subsets $U_1, \dots, U_n \subset C$ such that the norm diameter of $\frac{1}{n}\sum_{i=1}^n U_i$ is less than ε. An example of a strongly regular Banach space S_*T_∞ failing to have CPCP was constructed in [**60**]. This space has an equivalent norm which is n-meager for every $n \in \omega$, see [**12**].

1275? **Question 6.4.** *Is there a strongly regular Banach space admitting an equivalent ∞-meager norm? Has the space S_*T_∞ an equivalent ∞-meager norm?*

If a Banach space admits a 0-meager norm (equivalently, X fails to have CPCP), then the class $\mathcal{W}(X)$ contains all finite-dimensional absolute $F_{\sigma\delta}$-spaces.

If, moreover, the norm of a Banach space X is ∞-meager, then $\mathcal{W}(X) = \Pi^0_3$ and the weak unit ball B_X is homeomorphic to the weak unit ball of the Banach space c_0 endowed with the standard sup-norm. We do not know if the Banach space S_*T_∞ has an equivalent ∞-meager norm, but we know that $\mathcal{W}(S_*T_\infty) = \Pi^0_3$ and the weak unit ball of S_*T_∞ endowed with a Kadec norm is homeomorphic to the weak unit ball of c_0 endowed with a Kadec norm. The space S_*T_∞ is an example of a strongly regular space with $\mathcal{W}(S_*T_\infty) = \Pi^0_3$. However, S_*T_∞ fails to have CPCP.

Question 6.5. *Is there a Banach space X with $\mathcal{W}(X) = \Pi^0_3$ admitting no ∞-* 1276?
meager norm? having CPCP?

In light of this question it should be mentioned that each Banach space with PCP has $\mathcal{W}(X) = \Pi^0_i$ for $i \in \{1,2\}$. Also a Banach space with CPCP admits no 0-meager norm. It is known that the Banach space c_0 contains no conjugate subspaces and has $\mathcal{W}(c_0) = \Pi^0_3$.

Question 6.6. *Suppose X is a Banach space with separable dual, containing no* 1277?
subspace isomorphic to a dual space. Is $\mathcal{W}(X) = \Pi^0_3$?

For a Banach space X with an ∞-meager norm the weak unit ball is an absorbing space (in fact, a Π^0_3-absorbing space).

Similarly, for a Banach space with ∞-comeager norm the weak unit ball B_X is a coabsorbing space and its topology is completely determined by the class $\mathcal{W}(X)$. The same concerns the topology of the pair (B^{**}_X, B_X), where B^{**}_X is the unit ball in the second dual Banach space X^{**}, endowed with the $*$-weak topology. The topology of this pair is completely determined by the class $\mathcal{W}(X^{**}, X)$ of pairs (K, C) homeomorphic to pairs of the form $(B, B \cap X)$ where $B \subset X^{**}$ is w^*-closed bounded subset of the second dual space $(X^{**}, weak^*)$.

More precisely, we have the following classification theorem of Cantor–Bernstein type proved in [**12**].

Theorem (Classification Theorem). *Let X, Y be Banach spaces with separable dual and ∞-comeager norms.*

(1) *The weak unit balls B_X and B_Y are homeomorphic if and only if $\mathcal{W}(X) = \mathcal{W}(Y)$.*
(2) *The pairs (B^{**}_X, B_X) and (B^{**}_Y, B_Y) are homeomorphic if and only if $\mathcal{W}(X^{**}, X) = \mathcal{W}(Y^{**}, Y)$.*

It is clear that the topological equivalence of the pairs (B^{**}_X, B_X) and (B^{**}_Y, B_Y) implies the topological equivalence of the weak unit balls B_X and B_Y. We do not know if the converse is also true.

Question 6.7. *Assume that X, Y are Banach spaces with homeomorphic weak* 1278?
unit balls B_X and B_Y. Are the pairs (B^{**}_X, B_X) and (B^{**}_Y, B_Y) homeomorphic?

The answer to this question is affirmative provided $\mathcal{W}(X) = \mathcal{W}(Y) = \Pi^0_\xi$ for some $\xi \in \{1,2,3\}$.

The classification Theorem suggests introducing the partially ordered set

$$\mathfrak{W}^s_\infty = \{\mathcal{W}(X) : X \text{ is an infinite-dimensional Banach space with separable dual}\}$$

inducing the following preorder of the family of Banach spaces: $X \leq_{\mathcal{W}} Y$ if $\mathcal{W}(X) \subset \mathcal{W}(Y)$ (equivalently, if the weak unit ball of X admits a closed embedding into the weak unit ball of Y).

Since each separable Banach space is isomorphic to a subspace of $C[0,1]$, the set $\mathfrak{W}^s_\infty$ contains at most continuum elements. Note that the set $\mathfrak{W}^s_\infty$ is partially ordered by the natural inclusion relation.

It is easy to see that the poset $\mathfrak{W}^s_\infty$ has the smallest and largest elements: $\Pi^0_1 = \mathcal{W}(\ell_2)$ and $\Pi^0_3 = \mathcal{W}(c_0)$ corresponding to classes $\mathcal{W}(X)$ of the Hilbert space ℓ^2 and the Banach space c_0. Also it is known that the class $\Pi^0_2 = \mathcal{W}(J)$ where J is the James quasireflexive space is a unique immediate successor of Π^0_1. For some time there was a conjecture that $\mathfrak{W}^s_\infty$ consists just of these three elements: Π^0_1, Π^0_2, and Π^0_3. However it was discovered in [**12**] that for the Banach space B_∞ (distinguishing the properties PCP and CPCP) the class $\mathcal{W}(B_\infty)$ is intermediate between $\mathcal{W}(J)$ and $\mathcal{W}(c_0)$. So the poset $\mathfrak{W}^s_\infty$ appeared to be richer than expected.

1279? **Question 6.8.** *Investigate the ordered set $\mathfrak{W}^s_\infty$. In particular, is it infinite? Is it linearly ordered?*

The pathological class $\mathcal{W}(B_\infty)$ contains the class $\Pi^0_3[0]$ of all zero-dimensional absolute $F_{\sigma\delta}$-spaces but not the class $\Pi^0_3[1]$, see [**12**]. This suggests the following (probably difficult)

1280? **Question 6.9.** *Let $n \in \omega$. Is there a Banach space X such that $\Pi^0_3[n] \subset \mathcal{W}(X)$ but $\Pi^0_3[n+1] \not\subset \mathcal{W}(X)$?* (Such a space X if exists has CPCP but not PCP.)

1281? **Question 6.10.** *Is there a Banach space X such that $\Pi^0_3[\omega] \subset \mathcal{W}(X)$ but $\Pi^0_3 \not\subset \mathcal{W}(X)$?* (Such a space X if exists is strongly regular but fails to have PCP.)

In fact, the pathological space B_∞ is one of the spaces $J_*T_{\infty,n}$, $n \geq 0$, constructed in [**60**].

1282? **Question 6.11.** *Is $\mathcal{W}(J_*T_{\infty,n}) \neq \mathcal{W}(J_*T_{\infty,m})$ for $n \neq m$?*

Another two questions concern the influence of operations over Banach spaces on the classes $\mathcal{W}(X)$.

1283? **Question 6.12.** *Is $\mathcal{W}(X \oplus Y) = \max\{\mathcal{W}(X), \mathcal{W}(Y)\}$ for infinite-dimensional Banach spaces X and Y with separable duals?*

1284? **Question 6.13.** *Let X be an infinite-dimensional Banach space. Is $\mathcal{W}(X \oplus X) = \mathcal{W}(X)$? Is $\mathcal{W}(X \oplus \mathbb{R}) = \mathcal{W}(X)$?*

Note that an infinite-dimensional Banach space X with separable dual need not be isomorphic to $X \oplus X$ or $X \oplus \mathbb{R}$, see [**61**].

In Figure 1 we collect all known information on the relationship between geometric properties of a Banach space X with separable dual, topological properties of the weak unit ball B_X and properties of the class $\mathcal{W}(X)$. In the first line of the

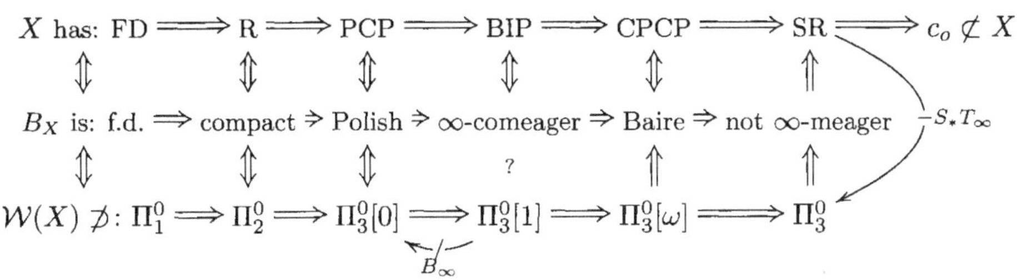

FIGURE 1. Relationship between geometric properties of a Banach space X with separable dual, topological properties of the weak unit ball B_X and properties of the class $\mathcal{W}(X)$

diagram FD means "finite-dimensional", R "reflexive", and SR "strongly regular". The second line of the diagram means that *every* equivalent weak unit ball B of X has the corresponding property; the third line means that the class $\mathcal{W}(X)$ *does not* contain the corresponding class of absolute $F_{\sigma\delta}$-spaces. The slashed and curved arrows indicate that the corresponding implication is false (with a counterexample written near the slashed arrow).

Finally we ask some questions on the topological structure of operator images. By an *operator image* we understand an infinite-dimensional normed space of the form TX for a suitable linear continuous operator $T\colon X \to Y$ between separable Banach spaces. In [**22**] it was shown that each operator image belongs to a Borel class $\Pi^0_{\alpha+2} \setminus \Sigma^0_{\alpha+2}$, $\Sigma^0_{\alpha+1} \setminus \Pi^0_{\alpha+1}$ or $D_2(\Pi^0_{\alpha+1}) \setminus (\Pi^0_\alpha \cup \Sigma^0_\alpha)$ for a suitable ordinal α (here $D_2(\Pi^0_{\alpha+1})$ is the class consisting of differences $X \setminus Y$ with $X, Y \in \Pi^0_{\alpha+1}$) and each such a Borel class contains an operator image. Moreover, up to a homeomorphism each class Π^0_2, Σ^0_2, $D_2(\Pi^0_2)$ contains exactly one operator image. On the other hand, the class $\Pi^0_3 \setminus \Sigma^0_3$ contains at least two topologically distinct operator images, see [**11, 22**].

Question 6.14. *Does the class $\Sigma^0_3 \setminus \Pi^0_3$ contain two topologically distinct operator* 1285?
images? The same question for other Borel classes.

The image $T\colon X \to Y$ of a Banach space under a compact operator T always is an absorbing space, see [**22**]. Moreover, for every countable ordinal $\alpha \geq 1$ the multiplicative Borel class $\Pi^0_{\alpha+2}$ contains an operator image which is a $\Pi^0_{\alpha+2}$-absorbing space. We do not know if the same is true for the additive Borel classes.

Question 6.15. *Is there an operator image which is a Σ^0_3-absorbing space? a* 1286?
$\Sigma^0_{\xi+1}$-absorbing space with $\xi \geq 1$?

For $\xi = 1$ the answer is affirmative: the image $T\colon X \to Y$ of any reflexive Banach space under a compact bijective operator $T\colon T \to Y$ is Σ^0_2-absorbing.

7. Some problems in dimension theory

In this section we address some problems related to distinguishing between certain classes of infinite-dimensional compacta intermediate between the class cd of countable-dimensional compacta and the class wid of weakly-infinite-dimensional compacta:

$$\mathsf{fd} \subset \mathsf{cd} \subset \sigma\mathsf{hd} \subset \mathsf{trt} \subset \mathsf{C} \subset \mathsf{wid}$$

In this diagram, by fd and C we denote the classes of finite-dimensional compacta and compact with the property C. The classes $\sigma\mathsf{hd}$ and trt are less known and consist of σ-hereditarily disconnected and trt-dimensional compacta, respectively. A topological space X is called *σ-hereditarily disconnected* if X can be written as the countable union of hereditarily disconnected subspaces.

The definition of trt-dimensional compacta is a bit longer and relies on the transfinite dimension trt introduced by Arenas, Chatyrko, and Puertas in [**4**]. For a space X they put

(1) $\mathsf{trt}(X) = -1$ iff $X = \emptyset$;
(2) $\mathsf{trt}(X) \le \alpha$ for an ordinal α iff each closed subset $A \subset X$ with $|A| \ge 2$ can be separated by a closed subset $B \subset A$ with $\mathsf{trt}(B) < \alpha$.
(3) $\mathsf{trt}(X) = \alpha$ if $\mathsf{trt}(A) \le \alpha$ and $\mathsf{trt}(A) \not\le \beta$ for any $\beta < \alpha$.

A space X is called trt-*dimensional* if $\mathsf{trt}(X) = \alpha$ for some ordinal α.

In [**4**] it was proved that each trt-dimensional compactum is a *C-space*, which gives the inclusion $\mathsf{trt} \subset \mathsf{C}$. The inclusion $\sigma\mathsf{hd} \subset \mathsf{trt}$ was proved in [**31**] with help of a game characterization of trt-dimensional spaces.

The classes cd and $\sigma\mathsf{hd}$ of countable-dimensional and σ-hereditarily disconnected compacta are distinguished by the famous Pol's compactum. We do not know if the other considered classes also are different.

1287? **Question 7.1.** *Is each* trt-*dimensional compactum σ-hereditarily disconnected? Is each C-compactum* trt-*dimensional?*

Recently, P. Borst [**35**] announced an example of a weakly infinite-dimensional compact metric space which fails to be a C-space, thus distinguishing the classes wid and C.

Some immediate questions still are open for the transfinite dimension trt.

1288? **Question 7.2.** *Is the ordinal* $\mathsf{trt}(X)$ *countable for each* trt-*dimensional metrizable compactum X?*

8. Homological methods in dimension theory

In this section we discuss some problems lying in the intersection of Infinite-Dimensional Topology, Dimension Theory, and Algebraic Topology. With help of (co)homologies we shall define two new dimension classes AZ_∞ and hsp of compacta including all trt-dimensional compacta.

The starting point is the homological characterization of Z_n-sets in ANRs due to Daverman and Walsh [**49**]: a closed subset A of an ANR-space X is a Z_n-set in X for $n \geq 2$ if and only if A is a Z_2-set in X and $H_k(U, U \setminus A) = 0$ for all $k \leq n$ and all open subsets $U \subset X$.

Having this characterization in mind we define a closed subset $A \subset X$ to be a *G-homological Z_n-set* in X for a coefficient group G if the singular relative homology groups $H_k(U, U \setminus A; G)$ are trivial for all $k \leq n$ and all open subsets $U \subset X$. If $G = \mathbb{Z}$, we shall omit the notation of the coefficient group and will speak about homological Z_n-sets. Thus a subset A of an ANR-space X is a Z_n-set for $n \geq 2$ if and only if it is a Z_2-set and a homological Z_n-set in X. Another characterization of Z_n-sets from [**20**] asserts that a closed subset A of an ANR-space is a homological Z_n-set in X if and only if $A \times \{0\}$ is a Z_{n+1}-set in $X \times [-1, 1]$.

It is more convenient to work with homological Z_n-sets than with usual Z_n-sets because of the absence of many wild counterexamples like wild Cantor sets in Q (these are topological copies of the Cantor set in Q that fail to be Z_2-sets, see [**75**]). According to an old result of Kroonenberg [**64**] any finite-dimensional closed subset $A \subset Q$ is a homological Z_∞-set in Q. A more general result was proved in [**20**]: *each closed* trt*-dimensional subset $A \subset Q$ is a homological Z_∞-set.* We do not know if the same is true for other classes of infinite-dimensional spaces like C or wid.

Question 8.1. *Is a closed subset $A \subset Q$ a homological Z_∞-set in Q if A is weakly infinite-dimensional? A is a C-space?* 1289?

This question is equivalent to the following one.

Question 8.2. *Let $W \subset Q$ be a closed weakly-infinite dimensional subset (with the property C). Is the complement $Q \setminus W$ homologically trivial?* 1290?

The preceding discussion suggests introducing new dimension classes AZ_n consisting of so-called absolute Z_n-compacta. Namely, we define a compact space K to be an *absolute Z_n-compactum* if for every embedding $e\colon K \to Q$ of K into the Hilbert cube Q the image $e(K)$ is a homological Z_n-set in Q. Among the classes AZ_n the most interesting are the extremal classes AZ_0 and AZ_∞. Both of them are hereditary with respect to taking closed subspaces.

In fact, the class AZ_0 coincides with the class of all compact spaces containing no copy of the Hilbert cube and thus AZ_0 is the largest possible nontrivial hereditary class of compact spaces. The class AZ_0 is strictly larger than the class AZ_1: the difference $\mathsf{AZ}_0 \setminus \mathsf{AZ}_1$ contains all hederitarily indecomposable continua $K \subset Q$ separating the Hilbert cube Q (such continua exist according to [**33**]). Observe also that $\mathsf{AZ}_\infty = \bigcap_{n \in \omega} \mathsf{AZ}_n$.

Question 8.3. *What can be said about the classes AZ_n for $n \in \mathbb{N}$. Are they hereditary with respect to taking closed subspaces? Are they pairwise distinct?* 1291?

The class AZ_∞ is quite rich and contains all trt-dimensional compacta. Besides being absolute Z_∞-compacta, trt-dimensional compacta have another interesting property: they contain many (co)homologically stable points. A point x of a space X will be called *homologically* (resp. *cohomologically*) *stable* if for some $k \geq 0$ the singular homology group $H_k(X, X \setminus \{x\})$ (resp. Čech cohomology group

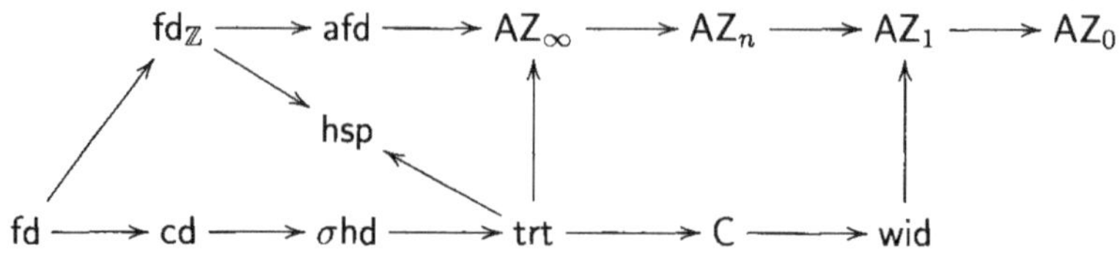

FIGURE 2. Inclusion relations among classes of compacta

$\check{H}^k(X, X \setminus \{x\}))$ is not trivial. For locally contractible spaces both notions are equivalent due to the duality between singular homologies and Čech cohomologies in such spaces. But it seems that Čech cohomologies work better beyond the class of locally contractible spaces.

According to [**31**], each trt-dimensional space contains a (co)homologically stable point and by [**19**] or [**20**] the same is true for every locally contractible C-compactum. Local contractibility is essential for the proof of the latter result and we do not known if it can be removed.

1292? **Question 8.4.** *Has each weakly infinite-dimensional (C-)compactum a cohomologically stable point?*

According to a classical result of Aleksandrov, each compact space X of finite cohomological dimension $\dim_\mathbb{Z}(X)$ contains a cohomologically stable point. This implies that the class $\mathsf{fd}_\mathbb{Z}$ of compacta with finite cohomological dimension lies in the class hsp of compacta whose any closed subspace has a cohomologically stable point. The class $\mathsf{fd}_\mathbb{Z}$ is also contained in the class afd of all almost finite-dimensional compacta, where a space X is called *almost finite-dimensional* if there is $n \in \omega$ such that each closed finite-dimensional subspace $F \subset X$ has dimension $\dim(F) \le n$. By [**14**], each almost finite-dimensional compactum is an absolute Z_∞-space. Figure 2 describes the (inclusion) relations between the considered classes of compacta (the arrow $\mathsf{x} \to \mathsf{y}$ means that $\mathsf{x} \subset \mathsf{y}$.

It follows from [**2**] (see also [**56**]) that the classes $\mathsf{fd}_\mathbb{Z}$ and C are *orthogonal* in the sense that $\mathsf{fd}_\mathbb{Z} \cap \mathsf{C} = \mathsf{fd}$. Is the same true for the intersection $\mathsf{fd}_\mathbb{Z} \cap \mathsf{wid}$?

1293? **Question 8.5** (Dranishnikov). *Is a weakly infinite-dimensional compact space finite-dimensional if it has finite cohomological dimension?*

A similar question concerns the class afd of almost finite-dimensional compacta. It is known [**14**] (and can be easily shown by transfinite induction) that $\mathsf{afd} \cap \mathsf{trt} = \mathsf{fd}$. Is the same true for the intersection $\mathsf{afd} \cap \mathsf{C}$? More precisely:

1294? **Question 8.6.** *Is a compact metrizable C-space finite-dimensional if it is almost finite dimensional?*

Another interesting class from the diagram is the class hsp of compacta all whose closed nonempty subspaces have cohomologically stable points.

Question 8.7. *What is the relation between the class* hsp *and other dimension classes from the diagram? In particular, has a (locally contractible) compact space X a cohomologically stable point if X is almost finite-dimensional? weakly infinite-dimensional? an absolute Z_∞-space?* 1295?

Question 8.8. *Is a compact space X an absolute Z_∞-space if* 1296?

- *all closed subspaces of X have a cohomologically stable point?*
- *all almost finite-dimensional closed subspaces of X are finite-dimensional?*

We have defined absolute Z_∞-compacta with help of their embedding into the Hilbert cube. What about embeddings into other spaces resembling the Hilbert cube?

Question 8.9. *Let A be a compact subset of an absolute retract X whose all points are homological Z_∞-points. Is A a homological Z_∞-set in X if A is an absolute Z_∞-space?* 1297?

Compact absolute retracts whose all points are homological Z_∞-points seem to be very close to being Hilbert cubes. By [**20**] all such spaces fail to be C-spaces and have infinite cohomological dimension with respect to any coefficient group.

Question 8.10. *Let X be a compact absolute retract whose all points are homological Z_∞-points. Is X strongly infinite-dimensional? Is $X \times [0,1]^2$ homeomorphic to the Hilbert cube? Is X homeomorphic to Q if X has DDP, the Disjoint Disks Property?* 1298?

In light of this question we should mention an example of a fake Hilbert cube constructed by Singh [**70**]. He constructed a compact absolute retract X such that (i) all points of X are homological Z_∞-points, (ii) $X \times [0,1]^2$ and $X \times X$ are homeomorphic to Q but (iii) X contains no proper closed ANR-subspace of dimension greater than one.

A bit weaker question of the same spirit asks if the Square Root Theorem holds for the Hilbert cube.

Question 8.11. *Is a space X homeomorphic to the Hilbert cube if X has DDP and X^2 is homeomorphic to Q.* 1299?

Let us note that for the Cantor and Tychonov cubes the Square Root Theorem is true, see [**26**].

The Singh's example shows that the class AZ_0 of absolute Z_0-compacta is not multiplicative. An analogous question for the class AZ_∞ is open.

Question 8.12. *Is the class AZ_∞ closed with respect to taking finite products?* 1300?

It should be noted that the product $X \times Y$ of a compact absolute Z_∞-space X and a trt-dimensional compact space Y is an absolute Z_∞-space, see [**14**].

9. Infinite-dimensional spaces in nature

The *Gromov–Hausdorff distance* between compact metric spaces (X_1, d_1) and (X_2, d_2) is the infimum of the Hausdorff distance between the images of isometric embeddings of these spaces into a metric space. Let GH denote the set of all compact metric spaces (up to isometry) endowed with the Gromov–Hausdorff metric. We call GH the Gromov–Hausdorff hyperspace. It is well-known that GH is a complete separable space.

1301? **Question 9.1.** *Is the Gromov–Hausdorff hyperspace homeomorphic to* ℓ^2*?*

1302? **Question 9.2.** *Is the subspace of the Gromov–Hausdorff hyperspace consisting of all finite metric spaces homeomorphic to* σ*?*

1303? **Question 9.3.** *What is the Borel type of the subspace of the Gromov–Hausdorff hyperspace consisting of all compact metric spaces of dimension* $\leq n$*?*

A *convex metric compactum* is a convex compact subspace of a normed space.

1304? **Question 9.4.** *Is the subspace of the Gromov–Hausdorff hyperspace consisting of all convex metric compacta homeomorphic to* ℓ^2*?*

1305? **Question 9.5.** *Is the subspace of the Gromov–Hausdorff hyperspace consisting of all convex finite polyhedra homeomorphic to* σ*?*

A *tree* is a connected acyclic graph endowed with the path metric.

1306? **Question 9.6.** *Is the subspace of the Gromov–Hausdorff hyperspace consisting of all finite trees homeomorphic to* σ*?*

For any metric space X, one can consider the Gromov–Hausdorff space $GH(X)$, the subspace of GH consisting of the (isometric copies of the) nonempty compact subsets of X. Note that the properties of $GH(X)$ can considerably differ from those of the Hausdorff hyperspace $\exp X$: as L. Bazylevych remarked, the space $GH(X)$ need not be zero-dimensional for zero-dimensional X.

1307? **Question 9.7.** *Is the Gromov–Hausdorff hyperspace* $GH([0,1])$ *homeomorphic to the Hilbert cube?*

Recall that the Banach–Mazur compactum $Q(n)$ is the space of isometry classes of n-dimensional Banach-spaces. The space $Q(n)$ is endowed with the distance $d(E,F) = \operatorname{Log}\inf\{\|T\| \cdot \|T^{-1}\| : T\colon E \to F \text{ is an isomorphism}\}$. Let $\{\mathrm{Eucl}\} \in Q(n)$ denote the Euclidean point to which corresponds the isometry class of standard n-dimensional Euclidean space. It is proved in [**1**] (see also [**3**]) that the space $Q_E(n) = Q(n) \setminus \{\mathrm{Eucl}\}$ is a Q-manifold.

1308? **Question 9.8.** *Are the Q-manifolds* $Q_E(n)$ *and* $Q_E(m)$ *homeomorphic for* $n \neq m$*?*

1309? **Question 9.9.** *Is the subspace of the Banach–Mazur compactum* $Q_{pol}(n)$ *consisting of classes of equivalence of polyhedral norms a* σ*-manifold? If so, is the pair* $(Q_E(n), Q_{pol}(n))$ *a* (Q,σ)*-manifold?*

1310? **Question 9.10.** *Is the subspace of* $Q_E(n) = Q(n) \setminus \{Eucl\}$ *consisting of classes of equivalence of (smooth) strictly convex norms an* ℓ^2*-manifold?*

Question 9.11. *Is there a topological field homeomorphic to the Hilbert space ℓ^2?* 1311?

Let (X,d) be a complete metric space. By $\mathrm{CL}_W(X)$ we denote the set of all nonempty closed subsets in X endowed with the Wijsman topology τ_W generated by the weak topology $\{d(x,\cdot) : x \in X\}$.

Question 9.12. *Let X be a Polish space. What is the Borel type of the subspace $\{A \in \mathrm{CL}_W(X) : \dim A \geq n\}$?* 1312?

Question 9.13. *Characterize metric spaces X whose hyperspace $\mathrm{CL}_W(X)$ is an ANR.* 1313?

Some partial results concerning the latter question can be found in [**65**].

For a metric space X by $\mathrm{Bdd}_H(X)$ we denote the hyperspace of closed bounded subsets of X endowed with the Hausdorff distance.

A metric space (X,d) is called *almost convex* if for any points $x, y \in X$ with $d(x,y) < s+t$ for some positive reals s,t there is a point $z \in X$ with $d(x,z) < s$ and $d(z,y) < t$. In particular, each subspace of the real line is almost convex.

By [**47**] or [**66**] for each almost convex metric space X the hyperspace $\mathrm{Bdd}_H(X)$ is an ANR. We do not know if the converse is true.

Question 9.14. *Let X be a metric space whose hyperspace $\mathrm{Bdd}_H(X)$ is an ANR. Is the topology (the uniformity) of X generated by an almost convex metric?* 1314?

Metric spaces X whose hyperspaces $\mathrm{Bdd}_H(X)$ are ANRs were characterized in [**29**]. This characterization implies that the hyperspace $\mathrm{Bdd}_H(X_\#)$ of the one-dimensional subspace

$$X_\# = \{(x_n) \in c_0 : (\exists n \in \omega)(\forall i \neq n)\ x_i \in \tfrac{1}{i}\mathbb{Z}\}$$

of the Banach space c_0 is an ANR.

Question 9.15. *Is the topology (the uniformity) of the space $X_\#$ generated by an almost convex metric?* 1315?

A metric space X is defined to be an *absolute neighborhood uniform retract* (briefly ANUR) if for any metric space $Y \supset X$ there is a uniformly continuous retraction $r\colon O_\varepsilon(X) \to X$ defined on an ε-neighborhood of X in Y. It is known that each uniformly convex Banach space X is an ANUR. In particular, the Hilbert space ℓ^2 is an ANUR.

Question 9.16. *Is $\mathrm{Bdd}_H(\ell^2)$ an absolute neighborhood uniform retract?* 1316?

It is known that $\mathrm{Bdd}_H(\ell^2)$ is an ANR [**47**] and the closed subspace of $\mathrm{Bdd}_H(\ell^2)$ consisting of closed bounded convex subsets of ℓ^2 is an ANUR, see [**32**].

References

[1] S. M. Ageev, S. A. Bogatyĭ, and D. Repovš, *The complement $Q_E(n)$ to the point Eucl of Euclidean space in the Banach–Mazur compactum $Q(n)$ is a Q-manifold*, Uspekhi Mat. Nauk **58** (2003), no. 3(351), 185–186, Translation: Russian Math. Surveys 58 (2003) no. 3, 607–609.

[2] F. D. Ancel, *The role of countable dimensionality in the theory of cell-like relations*, Trans. Amer. Math. Soc. **287** (1985), no. 1, 1–40.
[3] S. A. Antonyan, *The topology of the Banach–Mazur compactum*, Fund. Math. **166** (2000), no. 3, 209–232.
[4] F. G. Arenas, V. A. Chatyrko, and M. L. Puertas, *Transfinite extension of Steinke's dimension*, Acta Math. Hungar. **88** (2000), no. 1–2, 105–112.
[5] J. Baars, H. Gladdines, and J. van Mill, *Absorbing systems in infinite-dimensional manifolds*, Topology Appl. **50** (1993), no. 2, 147–182.
[6] S. Banach, *Über metrische Gruppen*, Studia Math. **3** (1931), 101–113.
[7] T. Banakh, *Some problems in infinite-dimensional topology*, Mat. Stud. **8** (1997), no. 1, 123–125.
[8] T. Banakh, *Characterization of spaces admitting a homotopy dense embedding into a Hilbert manifold*, Topology Appl. **86** (1998), no. 2, 123–131.
[9] T. Banakh, *The strongly universal property in convex sets*, Mat. Stud. **10** (1998), no. 2, 203–218.
[10] T. Banakh, *Some properties of the linear hull of the Erdős set in l^2*, Bull. Polish Acad. Sci. Math. **47** (1999), no. 4, 385–392.
[11] T. Banakh, *Operator images homeomorphic to Σ^ω*, Dissertationes Math. **387** (2000), 53–81.
[12] T. Banakh, *The topological classification of weak unit balls of Banach spaces*, Dissertationes Math. **387** (2000), 7–35.
[13] T. Banakh, *Toward a topological classification of convex sets in infinite-dimensional Fréchet spaces*, Topology Appl. **111** (2001), no. 3, 241–263.
[14] T. Banakh, *Absolute Z-spaces*, 2006, Preprint.
[15] T. Banakh and R. Cauty, *Universalité forte pour les sous-ensembles totalement bornés. Applications aux espaces $C_p(X)$*, Colloq. Math. **73** (1997), no. 1, 25–33.
[16] T. Banakh and R. Cauty, *Interplay between strongly universal spaces and pairs*, Dissertationes Math. **386** (2000), 38 pp.
[17] T. Banakh and R. Cauty, *On universality of countable and weak products of sigma hereditarily disconnected spaces*, Fund. Math. **167** (2001), no. 2, 97–109.
[18] T. Banakh and R. Cauty, *On universality of finite powers of locally path-connected meager spaces*, Colloq. Math. **102** (2005), no. 1, 87–95.
[19] T. Banakh and R. Cauty, *A homological selection theorem implying a division theorem for Q-manifolds*, 2006, Preprint.
[20] T. Banakh, R. Cauty, and A. Karasev, *On homotopical and homological Z_n-sets*, 2006, Preprint.
[21] T. Banakh, R. Cauty, K. Trushchak, and L. Zdomskyy, *On universality of finite products of Polish spaces*, Tsukuba J. Math. **28** (2004), no. 2, 455–471.
[22] T. Banakh, T. Dobrowolski, and A. Plichko, *The topological and Borel classification of operator images*, Dissertationes Math. **387** (2000), 37–52.
[23] T. Banakh, W. Kubiś, K. Sakai, and R. Voytsitsyy, *On the hyperspace of closed subsets of the space of rational numbers*, 2006, Preprint.
[24] T. Banakh and T. Radul, *On universality of countable powers of absolute retracts*, Ukr. Math J. **48** (1996), no. 4, 540–542.
[25] T. Banakh, T. Radul, and M. Zarichnyi, *Absorbing sets in infinite-dimensional manifolds*, VNTL Publishers, Lviv, 1996.
[26] T. Banakh and D. Repovš, *Division and k-th root theorems for Q-manifolds*, 2006, Preprint.
[27] T. Banakh and Kh. R. Trushchak, *Z_n-sets and the disjoint n-cells property in products of ANRs*, Mat. Stud. **13** (2000), no. 1, 74–78.
[28] T. Banakh and V. Valov, *Parametric general position properties and embedding of n-dimensional maps into trivial bundles*, 2006, Preprint.
[29] T. Banakh and R. Voytsitskyy, *Characterizing metric spaces whose hyperpsaces are absolute neighborhood retracts*, 2005, Preprint. To appear in Topology Appl. `arXiv:math.GT/0509395`

[30] T. Banakh and R. Voytsitskyy, *The cardinal characteristics of the ideal of n-meager sets in the Hilbert cube*, 2006, Preprint.
[31] T. Banakh and M. Zarichnyi, *On separation dimension of topological spaces*, 2006, Preprint.
[32] Y. Benyamini and J. Lindenstrauss, *Geometric nonlinear functional analysis. Vol. 1*, American Mathematical Society, Providence, RI, 2000.
[33] R H Bing, *Higher-dimensional hereditarily indecomposable continua*, Trans. Amer. Math. Soc. **71** (1951), 267–273.
[34] M. Bonino, *Sur l'espace des homéomorphismes de Brouwer*, Thèse, Grenoble, 1997.
[35] P. Borst, *A weakly infinite-dimensional compactum not having property C*, 2005, Preprint.
[36] P. L. Bowers, *General position properties satisfied by finite products of dendrites*, Trans. Amer. Math. Soc. **288** (1985), no. 2, 739–753.
[37] R. Cauty, *Une famille d'espaces préhilbertiens σ-compacts ayant la puissance du continu*, Bull. Polish Acad. Sci. Math. **40** (1992), no. 1, 41–43.
[38] R. Cauty, *Une famille d'espaces préhilbertiens σ-compacts de dimension dénombrable ayant la puissance du continu*, Bull. Polish Acad. Sci. Math. **40** (1992), no. 4, 309–313.
[39] R. Cauty, *Quelques problèmes sur les groupes contractiles et la théorie des rétractes*, Mat. Stud. **3** (1994), 111–116.
[40] R. Cauty, *Un espace métrique linéaire qui n'est pas un rétracte absolu*, Fund. Math. **146** (1994), no. 1, 85–99.
[41] R. Cauty, *The strong universality property and its applications*, Tr. Mat. Inst. Steklova **212** (1996), 95–122, Translation: Proc. Steklov Inst. Math. 212 (1996) 89–114.
[42] R. Cauty, *Sur l'universalité des produits de rétractes absolus*, Bull. Polish Acad. Sci. Math. **44** (1996), no. 4, 453–456.
[43] R. Cauty, *Foncteurs libres et rétractes absolus*, Mat. Stud. **24** (2005), no. 1, 89–98.
[44] R. Cauty, *Rétractes absolus de voisinage algébriques*, Serdica Math. J. **31** (2005), no. 4, 309–354.
[45] R. Cauty and T. Dobrowolski, *Applying coordinate products to the topological identification of normed spaces*, Trans. Amer. Math. Soc. **337** (1993), no. 2, 625–649.
[46] T. A. Chapman and L. C. Siebenmann, *Finding a boundary for a Hilbert cube manifold*, Acta Math. **137** (1976), no. 3–4, 171–208.
[47] C. Costantini and W. Kubiś, *Paths in hyperspaces*, Appl. Gen. Topol. **4** (2003), no. 2, 377–390.
[48] D. Curtis, T. Dobrowolski, and J. Mogilski, *Some applications of the topological characterizations of the sigma-compact spaces l^2_f and Σ*, Trans. Amer. Math. Soc. **284** (1984), no. 2, 837–846.
[49] R. J. Daverman and J. J. Walsh, *Čech homology characterizations of infinite-dimensional manifolds*, Amer. J. Math. **103** (1981), no. 3, 411–435.
[50] J. J. Dijkstra and J. van Mill, *On the dimension of Hilbert space remainders*, Proc. Amer. Math. Soc. **124** (1996), no. 10, 3261–3263.
[51] J. J. Dijkstra and J. van Mill, *Infinite-dimensional topology*, Recent progress in general topology, II, North-Holland, Amsterdam, 2002, pp. 115–130.
[52] T. Dobrowolski, *Enlarging ANRs with SDAP to ℓ_2-manifolds revisited*, Bull. Polish Acad. Sci. Math. **45** (1997), no. 4, 345–348.
[53] T. Dobrowolski and W. Marciszewski, *Failure of the factor theorem for Borel pre-Hilbert spaces*, Fund. Math. **175** (2002), no. 1, 53–68.
[54] T. Dobrowolski and J. Mogilski, *Sigma-compact locally convex metric linear spaces universal for compacta are homeomorphic*, Proc. Amer. Math. Soc. **85** (1982), no. 4, 653–658.
[55] T. Dobrowolski and J. Mogilski, *Problems on topological classification of incomplete metric spaces*, Open problems in topology, North-Holland, Amsterdam, 1990, pp. 409–429.
[56] A. Dranishnikov, *Stable cohomotopy dimension and weakly infinite-dimensional spaces*, Topology Appl. **47** (1992), no. 1, 79–81.

[57] A. Dranishnikov, *On the theory of extensions of mappings of compact spaces*, Uspekhi Mat. Nauk **53** (1998), no. 5(323), 65–72, Translation: Russian Math. Surveys 53 (1998) no. 5, 929–935.

[58] R. Geoghegan, *Open problems in infinite-dimensional topology*, Topology Proc. **4** (1979), no. 1, 287–338 (1980).

[59] N. Ghoussoub, B. Maurey, and W. Schachermayer, *A counterexample to a problem on points of continuity in Banach spaces*, Proc. Amer. Math. Soc. **99** (1987), no. 2, 278–282.

[60] N. Ghoussoub, B. Maurey, and W. Schachermayer, *Geometrical implications of certain infinite-dimensional decompositions*, Trans. Amer. Math. Soc. **317** (1990), no. 2, 541–584.

[61] W. T. Gowers, *Ramsey methods in Banach spaces*, Handbook of the geometry of Banach spaces, Vol. 2, North-Holland, Amsterdam, 2003, pp. 1071–1097.

[62] A. S. Kechris, *Classical descriptive set theory*, Springer, New York, 1995.

[63] K. Kozlov, E. Reznichenko, and O. Sipacheva, *Moscow questions on topological algebra*, Open problems in topology II (E. Pearl, ed.), Elsevier Science.

[64] N. Kroonenberg, *Characterization of finite-dimensional Z-sets*, Proc. Amer. Math. Soc. **43** (1974), 421–427.

[65] W. Kubiś, K. Sakai, and M. Yaguchi, *Hyperspaces of separable Banach spaces with the Wijsman topology*, Topology Appl. **148** (2005), 7–32.

[66] M. Kurihara, K. Sakai, and M. Yaguchi, *Hyperspaces with the Hausdorff metric and uniform ANRs*, J. Math. Soc. Japan **57** (2005), no. 2, 523–535.

[67] F. Le Roux, *Étude topologique de l'espace des homéomorphismes de Brouwer, I*, Topology **40** (2001), no. 5, 1052–1087.

[68] J. van Mill, *The infinite-dimensional topology of function spaces*, North-Holland, Amsterdam, 2001.

[69] M. Scheepers, *Topological games and Ramsey theory*, Open problems in topology II (E. Pearl, ed.), Elsevier Science.

[70] S. Singh, *Exotic ANRs via null decompositions of Hilbert cube manifolds*, Fund. Math. **125** (1985), no. 2, 175–183.

[71] O. V. Sipacheva, *Compacta with a continuous Mal'tsev operation and retracts of topological groups*, Vestnik Moskov. Univ. Ser. I Mat. Mekh. (1991), no. 1, 33–36, Translation: Moscow Univ. Math. Bull. 46 (1991) no. 1, 22–24.

[72] B. Tsaban, *Selection principles and special sets of reals*, Open problems in topology II (E. Pearl, ed.), Elsevier Science.

[73] J. E. Vaughan, *Small uncountable cardinals and topology*, Open problems in topology, North-Holland, Amsterdam, 1990, pp. 195–218.

[74] J. E. West, *Open problems in infinite-dimensional topology*, Open problems in topology, North-Holland, Amsterdam, 1990, pp. 523–597.

[75] R. Y. T. Wong, *A wild Cantor set in the Hilbert cube*, Pacific J. Math. **24** (1968), 189–193.

[76] M. M. Zarichnyi, *Absorbing sets for n-dimensional spaces in absolute Borel and projective classes*, Mat. Sb. **188** (1997), no. 3, 113–126, Translation: Sb. Math. 188 (1997) no. 3, 435–447.

Open Problems in Topology II – Edited by E. Pearl

Classical dimension theory

Vitalij A. Chatyrko

Introduction

Dimension theory has a long history which is very well described in different books and surveys (some of those published after 1990 are [**1**, **2**, **19**–**21**, **28**, **40**, **50**, **53**, **55**]). This appeared about 100 years ago as a theory of an integer-valued topological invariant whose values on each simple geometric figure coincided with the number assigned in geometry to this figure as a dimension. The invariant was called dimension and at first was considered on compact metrizable spaces, in brief CMS. At the beginning there were three approaches to define the dimension notated by dim, Ind and ind. They were based on very natural geometric observations and could be later easily extended outside the class of CMS. It was observed almost from the beginning that dim, Ind and ind could disagree for general spaces. Mathematicians started to talk about dimension functions or merely dimensions. But the main challenge for the dimension theory at this time was to study the dimension in the class of CMS. It was difficult to evaluate the dimension there. So topologists continued to look for more effective approaches which could involve the use of algebra. This led in particular to cohomological dimensions $\dim_G$. But would $\dim_G$ be the same dimension as defined by dim, Ind and ind for CMS? The answer was "no" and came from A. Dranishnikov [**12**]. After that one realized that there was not just one dimension in the class of CMS. In fact, there were many dimension functions which could be objects of study, and to restrict the dimension theory only to the class of CMS no longer made sense. Nowadays the dimension theory is huge and consists of different parts which have influence on each other. One can name as examples "classical dimension theory" (the references above), "algebraic dimension theory" ([**13**, **16**, **17**, **57**]), "asymptotic dimension theory" ([**14**]). This division is motivated by subjects of studies, different methods and applications outside the dimension theory. This survey is devoted to the classical dimension theory, strictly speaking, to some of its parts. We will recall results obtained after 1990 and consider problems (sometimes well known) which could probably be solved without the use of deep methods of algebra or geometry (one can find many other interesting problems in the references mentioned above).

Our terminology mostly follows [**18**, **19**] and most of our uncited remarks can be found in [**19**]. All spaces considered are assumed to be regular T_1. We define here the *covering dimension* of a completely regular space X, $\dim X$, as follows. $\dim X \leq n$ if each finite cover of X by functionally open sets has a finite refinement by functionally open sets such that each point belongs to at most $n+1$ of them. The *large inductive dimension* of a space X, $\operatorname{Ind} X$, is defined inductively by the following way. $\operatorname{Ind} X = -1$ if and only if $X = \emptyset$. $\operatorname{Ind} X \leq n$ if every closed subset A of X has arbitrarily tight open neighborhoods U with $\operatorname{Ind} \operatorname{Bd} U \leq n-1$, where $\operatorname{Bd} C$ denotes the boundary of a set C. (If X is normal then, equivalently, the set A

can be separated from the complement of U by a partition C with $\operatorname{Ind} C \leq n-1$.) One can get the definition of the *small inductive dimension* of a space X, $\operatorname{ind} X$, from the definition of $\operatorname{Ind} X$ by replacing the set A by a point. It is clear that $\operatorname{ind} X \leq \operatorname{Ind} X$; $\operatorname{Ind} X = 0$ implies the normality of X and $\dim X = 0$; $\dim X = 0$ implies $\operatorname{ind} X = 0$ and, if X is normal, even $\operatorname{Ind} X = 0$.

Coincidence of dimensions

Recall that for a space X we have the *Urysohn identity*

$$\text{(UID)} \qquad \dim X = \operatorname{Ind} X = \operatorname{ind} X$$

if X is separable metrizable (in brief, SM), and $\dim X \leq \operatorname{Ind} X = \operatorname{ind} X$ if X is Lindelöf and perfectly normal. A space X is called *cosmic* if X is a continuous image of a SM space. Any cosmic space X is evidently Lindelöf and perfectly normal. In [**5**] Charalambous, following the way developed by Delistathis and Watson [**10**], constructed in ZFC a cosmic space C that despite being the union of countably many of subspaces of the square I^2, has $\dim C = 1$ and $\operatorname{ind} C = 2$. (Independently, A. Dow and K.P. Hart ([**11**]) using the strategy from [**10**], presented under the assumption of Martin's Axiom for σ-centered partial orders a cosmic space with $\dim = 1$ and $\operatorname{ind} \geq 2$.)

1317? **Question 1.** *How large can the gap between* ind *and* dim *be for cosmic spaces?*

Recall (cf. [**10**] (resp. [**20**])) that by a theorem of Nagami (resp. Leibo) for a paracompact space X we have $\dim X = \operatorname{Ind} X$ if X is the union of countably many closed metrizable subspaces (resp. the closed image of a metrizable space). So UID is valid for any space which is either the union of countably many closed SM subspaces or the closed image of a SM space. Recall (cf. [**45**]) that every perfectly normal union of finitely many metrizable (resp. SM) subspaces is the union of countably many closed metrizable (resp. SM) subspaces.

1318? **Question 2.** *Does UID hold for quotient images of SM spaces?*

Due to Roy we know that there exist metrizable spaces with $\operatorname{ind} = 0$ and $\operatorname{Ind} = 1$. Later Mrówka, Ostaszewski and Kulesza simplified his construction. Thus Kulesza in [**33**] presented in ZFC a complete N-compact metric space K having weight $K = \omega_1$ such that $\operatorname{Ind} K^n = 1$ for any n. Recall (cf. [**42**]) that by a result of Katětov–Morita if every completion X^* of a metric space X has $\operatorname{ind} X^* \geq k$ for some integer k then $\operatorname{Ind} X \geq k$. In [**42**] Mrówka presented in ZFC a non-complete metric space $\nu\mu_0$ such that $\operatorname{ind} \nu\mu_0 = 0$, and showed that if we additionally assume his special set-theoretic axiom $S(\aleph_0)$ then for $n = 1, 2$ every completion of $(\nu\mu_0)^n$ contains an n-dimensional cube I^n, and we have $\operatorname{Ind}(\nu\mu_0)^n = n$ (the case $n \geq 3$ was proved by Kulesza [**34**]). On the contrary if we assume CH then $\operatorname{Ind}(\nu\mu_0)^n = 1$ for all n ([**43**]).

1319? **Question 3.** *Does there exist a complete (N-compact) metric space X (having weight $X = \omega_1$) such that* $\operatorname{Ind} X > 1$ *and* $\operatorname{ind} X = 0$?

Observe that it is still an open question if there exists in ZFC a (complete) (N-compact) metric space X (having weight $X = \omega_1$) such that $\dim X > 1$ and $\operatorname{ind} X = 0$. Many interesting problems on this subject one can find in [**43**].

Recall that for a compact space X we have $\dim X \leq \operatorname{ind} X$; $\dim X = \operatorname{ind} X = \operatorname{Ind} X$ if $\dim X = 0$; $\operatorname{ind} X = \operatorname{Ind} X$ if $\operatorname{ind} X = 1$, and (cf. [**20**, **21**]) there is a lot of examples of compact spaces with noncoinciding dimensions dim, ind and Ind, especially with dim $\neq$ ind. Recently Pasynkov in [49] presented for each $n \geq 2$ a Dugundji (resp. homogeneous) compact space D_n (resp. H_n) with dim $= 1$ and ind $= n$. A compact space X is *Dugundji* if for any compact space Y with $\dim Y = 0$ every continuous mapping from any its closed subset to X has a continuous extension to the whole Y. A space X is *homogeneous* if for each pair of points x, y in X there is a homeomorphism $h\colon X \to X$ such that $h(x) = y$. Recall (cf. [**49**]) that by a theorem of Fedorchuk for any Dugundji compact space we have ind $=$ Ind, but for a homogeneous compact space this is unknown. A compact space X is *algebraically homogeneous* if there exists a topological group G and its closed subgroup H such that $X = G/H$. By a theorem of Pasynkov (cf. [**50**]) if G is locally compact then UID holds in X.

Question 4. *Does UID hold for any algebraically homogeneous compact space?* 1320?

It is unknown if H_n is algebraically homogeneous.

Since late sixties of 20th century (cf. [**20**]) we know due to Filippov, Lifanov and Pasynkov that there are compact spaces with ind $\neq$ Ind. In particular, there exists a sequence of compact spaces $\{X_i\}_{i=2}^{\infty}$ ([**26**]) (resp. $\{Y_i\}_{i=2}^{\infty}$ ([**37**])) such that for each i, $\operatorname{ind} X_i = i$, $\operatorname{Ind} X_i = 2i - 1$ and $\dim X_i = 1$ (resp. $\operatorname{ind} Y_i = \dim Y_i = i$, $\operatorname{Ind} Y_i = i + 1$).

Question 5. *Does there exist for each $n \geq 5$ a compact space Z_n such that $\dim Z_n = 1$, $\operatorname{ind} Z_n = 2$ and $\operatorname{Ind} Z_n = n$?* 1321?

Such Z_4 was constructed by Kotkin ([**31**]). The positive answer to this question would show that for any integers n, m, p such that $1 \leq n \leq m \leq p$ there exists a compact space $X_{n,m,p}$ with $\dim X_{n,m,p} = n \leq \operatorname{ind} X_{n,m,p} = m \leq \operatorname{Ind} X_{n,m,p} = p$.

The definition of *Dimensionsgrad*, Dg, can be obtained from the definition of Ind for normal spaces via partitions by replacing the word "partition" with the word "cut". In [**25**] Fedorchuk, Levin, Scepin proved that for any metrizable compact space X we have $\operatorname{Dg} X = \dim X$, but surprisely for complete SM spaces the dimensions can differ as was showed by Fedorchuk and van Mill (cf. [**40**]). It is known (cf. [**21**]) that $\dim X \leq \operatorname{Dg} X \leq \operatorname{Ind} X$ for any compact space. A compact space X is *snake-like* if for any open cover α of X there exists an open refinement $\beta = \{U_i\}_{i=1}^{n}$ such that $U_i \cap U_j \neq \emptyset$ if and only if $|i - j| \leq 1$.

Question 6. *Does the equality $\operatorname{Dg} X = 1$ hold for any snake-like compact space X?* 1322?

In [**6**] Charalambous presented for each $n > 1$ a snake-like compact space C_n such that $1 \leq \operatorname{Dg} C_n < \operatorname{ind} C_n = \operatorname{Ind} C_n = n$ but the value $\operatorname{Dg} C_n$ remained unknown. There are few examples of compact spaces where dim, Dg and ind disagree. Thus Chatyrko and Fedorchuk ([**23**]) constructed a compact space X with $\dim X = 1 < \mathrm{Dg}\, X = 2 < \mathrm{ind}\, X = \mathrm{Ind}\, X = 3$. But we do not still know if there exists a compact space where ind < Dg or all dim, Dg, ind and Ind disagree.

Recall that if a space X is normal then $\dim X \leq \mathrm{Ind}\, X$. For a normal n-manifold M we have additionally $n = \mathrm{ind}\, M \leq \dim M$. It is known (cf. [**20**]) that if a manifold M is weakly paracompact then it is SM. The first example of a manifold with non-coinciding dimensions was presented by Fedorchuk and Filippov in [**24**]. Namely, they constructed in CH for each $n \geq 3$ a normal countably compact n-manifold M_n such that $n = \dim M_n < \mathrm{Ind}\, M_n = 2n - 2$. Later Fedorchuk (cf. [**20**]) constructed also in CH for any m, n such that $4 \leq n < m$, a perfectly normal separable n-manifold $M^{n,m}$ with $m - 1 \leq \dim M^{n,m} \leq m < m + n - 3 \leq \mathrm{Ind}\, M^{n,m} \leq m + n - 1$.

1323? **Question 7.** *Do there exist in* ZFC *(separable) (countably compact) n-manifolds M where UID does not hold* ($\dim M \neq n$)*?*

It is unknown if there exist a 2-manifold M with $\mathrm{Ind}\, M > 2$, a 3-manifold N with $\dim N > 3$ and an n-manifold M_n such that $\dim M_n = \mathrm{Ind}\, M_n > n$ for each n.

Addition theorems for dimensions dim, Ind, ind

Recall that if a normal space X is the union of two closed subsets A, B then $\mathrm{Ind}\, X \leq \mathrm{Ind}\, A + \mathrm{Ind}\, B$, and there exists a compact space $L = \bigcup_{i=1}^{2} L_i$ such that $\mathrm{Ind}\, L = 2$ and for each i the set L_i is closed in L and $\mathrm{Ind}\, L_i = 1$. In [**31**] Kotkin constructed a compact space $K = \bigcup_{i=1}^{3} K_i$ such that $\mathrm{ind}\, K = 3$ and for each i the set K_i is closed in K and $\mathrm{Ind}\, K_i = 1$.

1324? **Question 8.** *Does there exist for each* $n = 4, 5, \ldots$ *a compact space* $X_n = \bigcup_{i=1}^{n} Y_{i,n}$ *such that* $\mathrm{Ind}\, X_n = n$ *and for each* i *the set* $Y_{i,n}$ *is closed in* X_n *and* $\mathrm{Ind}\, Y_{i,n} = 1$*?*

In [**7**] Chatyrko proved that if a space X is the union of two closed subsets A, B then $\mathrm{ind}\, X \leq \max\{\mathrm{ind}\, A, \mathrm{ind}\, B\} + 1$. Recall that for any normal space X being the union of countably many closed subsets X_i, we have

$$(*) \qquad \dim X = \max\{\dim X_i\}.$$

So if $n \geq 4$, we would have $1 = \dim X_n < 2 \leq \mathrm{ind}\, X_n \leq p + 1 < \mathrm{Ind}\, X_n = n$ for the smallest p such that $n \leq 2^p$.

Let d will be either ind or Ind. We will say that the *finite sum theorem* for d holds in a space X (in dimension $k \geq 0$), in brief, FST(d) (respectively, FST(d, k)), if $\mathrm{d}(A \cup B) = \max\{\mathrm{d}\, A, \mathrm{d}\, B\}$ for any closed in X sets A and B (such that $\mathrm{d}\, A, \mathrm{d}\, B \leq k$). For any space X define $\mathrm{FST}(\mathrm{d}, X) = \infty$, if FST(d) holds in X; $\min\{k \geq 0$ such that FST(d, k) does not hold in $X\}$ otherwise.

1325? **Question 9.** *Does there exist for each integer* $m > 1$ *a space* X *such that* $\mathrm{FST}(\mathrm{d}, X) = m$*?*

Let now d be either dim, ind or Ind, and $X = A \cup B$. Recall that if X is hereditarily normal then we have the *Urysohn inequality*,

$$\text{(UIN)} \qquad \mathrm{d}\, X \le \mathrm{d}\, A + \mathrm{d}\, B + 1.$$

UIN could be useful when we want to evaluate d: if for a space $Z = \bigcup_{i=0}^{n} Z_i$, where for each i, $\mathrm{d}\, Z_i \le 0$, we have $\mathrm{d}\, Z = n \ge 1$, and for any subsets U, V of Z, $\mathrm{d}(U \cup V) \le \mathrm{d}\, U + \mathrm{d}\, V + 1$, then for any k such that $1 \le k \le n$, $\mathrm{d}(\bigcup_{i=0}^{k} Z_i) = k$. In [**41**] Mrówka showed that for any function $f\colon \{1,2,3\} \to \{0,1,2,\dots,\infty\}$ there exists a completely regular space X such that $\dim A = f(1)$, $\dim B = f(2)$, $\dim X = f(3)$, and the sets A, B are countable intersections of clopen sets (moreover, X is of type $N \cup R$, that is X is also the union of two discrete subspaces X_1, X_2, one of which is open, dense and countable, X is first countable, separable, locally compact, pseudocompact and all compact subspaces of X have $\dim = 0$). This statement witnesses some known facts by E. Pol, R. Pol and Terasawa about dim in completely regular spaces, concerning the failures of $(*)$, UIN and the monotonicity with respect to closed subsets (the last one was also demonstrated by the earlier mentioned example of Kulesza). However, if $F \subset Z \subset Y$, where Y is normal and F is closed in Y, then $\dim F \le \dim Z$. This implies by a standard method with help of $(*)$ that UIN holds for dim for any normal space X. In [**59**] Zambahidze proved that UIN is valid for Ind for any normal space X, where FST(Ind) holds. Let $\mathrm{Ind}\, A = n$, $\mathrm{Ind}\, B = m$ and $n, m \ge 0$. One can prove that $\mathrm{Ind}\, X \le mn + 2(m+n+1)$ if X is normal; $\mathrm{ind}\, X \le 2 \cdot (n+m+1)$; and $\mathrm{ind}\, X \le n+m+1$ if FST(ind) holds in X.

Question 10. *Does there exist a (normal) space $X = A \cup B$ such that* $\mathrm{ind}\, X = 2$ *(resp.,* $\mathrm{Ind}\, X = 2$*) and* $\mathrm{Ind}\, A = \mathrm{Ind}\, B = 0$*?* 1326?

In [**58**] Tsereteli constructed completely regular space $T = T_1 \cup T_2$ such that $\mathrm{Ind}\, T \ge 2$, T_1 is discrete, T_2 is dense and $\mathrm{Ind}\, T_i = 0$ for each i. Note that if $X = \bigcup_{i=0}^{n} X_i$, where for each i the subspace X_i is either discrete or dense and has $\mathrm{ind} = 0$ then $\mathrm{ind}\, X \le n$. We know due to Katětov (cf. [**21**], see also Oka [**45**] about different generalizations) that for a metrizable space X we have $\dim X \le n$ if and only if $X = \bigcup_{i=0}^{n} X_i$ where $\dim X_i = 0$ for each i.

Question 11. *Does there exist a metrizable space X such that* $0 < \mathrm{ind}\, X < \infty$ *which is not the union of* $\mathrm{ind}\, X + 1$ *many subspaces having* $\mathrm{ind} = 0$*?* 1327?

Observe (cf. [**21**]) that for each $n \ge 1$ (resp. ∞) there exists a first countable separable snake-like compact space S_n (where $S_1 = [0,1]$) such that $\mathrm{ind}\, S_n = n$ and each open subset of S_{n+1} contains a copy of S_n (resp. each closed non-trivial connected subset of S_∞ has $\mathrm{ind} = \infty$). So we can prove that for each $n \ge 2$ (resp. ∞), S_n is not the union of n (resp. countably many) subspaces having $\mathrm{ind} = 0$. In CH by a resulf of Odincov one can assume that all S_n are perfectly normal (cf. [**21**]).

Let d be either dim or Ind.

1328? **Question 12.** *Does there exist (for* dim *in* ZFC*) a hereditarily normal space* X *such that* $\mathrm{d}\,X < \infty$ *which is not the union of countably many (even* $\mathrm{d}\,X+1$*) many subspaces having* $\mathrm{ind} = 0$ *(even* $\mathrm{d} = 0$*)?*

In [**59**] Zambakhidze presented a homogeneous paracompact space Z with $\mathrm{Ind}\,Z = 2$ which is not the union of three subspaces of $\mathrm{Ind} = 0$.

Product theorems for dimensions dim, ind, Ind

One can show that for any regular spaces X, Y and any $k \geq 0$ such that $\mathrm{ind}\,X = n$, $\mathrm{ind}\,Y = m$ and $\mathrm{FST}(\mathrm{ind}, X)$, $\mathrm{FST}(\mathrm{ind}, Y) \geq k$ we have $\mathrm{ind}(X \times Y) \leq n+m$, if either $n = 0$, or $m = 0$, or $n, m \leq k$; $2(n+m) - k - 1$, otherwise. This is a combination of known facts by Pasynkov, Basmanov ($k = \infty$) and Chatyrko, Kozlov ($k = 0, 1$). Observe that for any compact space Z we have $\mathrm{FST}(\mathrm{ind}, Z) \geq 1$, and there exist due to Filippov two compact spaces X and Y such that $\mathrm{Ind}\,X = 2$, $\mathrm{Ind}\,Y = 1$ and $\mathrm{ind}(X \times Y) = 4$.

1329–1330? **Question 13.** *Does there exist two (compact) spaces* X, Y *such that* $\mathrm{ind}\,X = \mathrm{ind}\,Y = 1$ *(resp.,* 2*) and* $\mathrm{ind}(X \times Y) = 3$ *(resp.,* 6*)?*

Recall that for any completely regular space X we have $\mathrm{ind}(X \times I) \leq \mathrm{ind}\,X + 1$. In [**39**] D. Malykhin constructed a regular space M with $\mathrm{ind}\,M = 2$ such that $\mathrm{ind}(M \times I) = 4$.

Let $\Pi = X \times Y$ be the product of two completely regular spaces X, Y. Π is *piecewise rectangular* if for any finite functionally open cover of Π there exists a σ-locally finite refinement consisting of clopen subsets of functionally open rectangles (= the products of functionally open subsets of X and Y). Let d be either dim or Ind and $\mathrm{d}\,X = n$, $\mathrm{d}\,Y = m$. It is known (cf. [**50**]) that $\mathrm{d}\,\Pi \leq n + m$ if Π is piecewise rectangular and either $\mathrm{d} = \dim$ or $\mathrm{d} = \mathrm{Ind}$, Π is normal and FST(Ind) holds in X and Y. Recall (cf. [**50**]) that Π is piecewise rectangular if for example Π is normal and X is metrizable or the projection of Π onto X is closed or X is locally compact paracompact or Π is completely paracompact etc.

1331? **Question 14.** *Does the inequality* $\dim \Pi \leq \dim X + \dim Y$ *hold for a paracompact product* Π*?*

In [**48**] Pasynkov showed that $\mathrm{Ind}\,\Pi < \infty$ if Π is normal and X is either locally compact paracompact or metrizable.

1332? **Question 15.** *Is* $\mathrm{Ind}\,\Pi < \infty$ *if* Π *is normal and piecewise rectangular?*

Recall (cf. [**50**]) that $\dim \Pi = 0$ if and only if Π is piecewise rectangular and $\dim X = \dim Y = 0$, and there are due to Wage, Przymusiński, Tsuda, E. Pol, Engelking, Kozlov different examples of normal products Π with zero-dimensional in the sense of dim factors such that $\dim \Pi > 0$. Thus Kozlov in [**32**] applying Przymusiński's technique showed that for any positive integers k, m, n such that $k < m$ there exists a first countable space K satisfying: (i) K^s is Lindelöf if and only if $s < k$; (ii) K^s is collectionwise normal and countably paracompact if $s \leq m$; (iii) $\dim K^s = 0$ for $s < m$; (iv) $\dim K^m = \mathrm{Ind}\,K^m = n$; moreover for $k = 1$ we can assume that K is locally compact and locally countable.

Question 16. *Does there exist a normal product* Π *with* $\dim X = \dim Y = 0$ *where* dim *and* Ind *disagree?* 1333?

In [**27**] Hattori refined some earlier result of Kulesza: for each pair $n \leq d$ there is a non-complete subgroup $G_{n,d}$ of R^{n+1} satisfying: $\dim G_{n,d} = n$ and $\dim(G_{n,d})^{\omega} = d$. It is unknown if there are SM complete groups with such dimensional properties.

In [**15**] Dranishnikov improved UIN for compact metrizable spaces X such that $\dim X^2 = 2\dim X$. Namely, if such X is the union $X_1 \cup X_2$, then $\dim X \leq \dim(X_1 \times X_2) + 1$. He also showed that the weaker inequality with 1 replaced by 2 holds without the mentioned restriction.

Question 17. *Does there exist a (separable) metrizable space* $X = X_1 \cup X_2$ *such that* $\dim X > \dim(X_1 \times X_2) + 1$*?* 1334?

In [**15**] Dranishnikov presented a metrizable compact space $X = X_1 \cup X_2$ such that $\dim X > \dim(Y_1 \times Y_2) + 1$ for any compacta $Y_1 \subset X_1$ and $Y_2 \subset X_2$. Evidently, there exists a completely regular space X (the earlier mentioned space $N \cup R$ of Mrówka) such that $X = X_1 \cup X_2$ and $\dim X = \infty > \dim(X_1 \times X_2) + 1 = 0 + 1 = 1$.

Compactifications

It is known that $\dim \beta X = \dim X$ if X is completely regular, and $\operatorname{Ind} \beta X = \operatorname{Ind} X$ if X is normal. Moreover, there exists a preserving weight compactification bX (resp. cX) of X such that $\dim bX = \dim X$ (resp. $\operatorname{Ind} cX \leq \operatorname{Ind} X$) if X is completely regular (resp. normal). Recall that there exists a perfectly normal first countable space P with $\operatorname{ind} P = 1$ each Lindelöf extension of which has ind $= \infty$. Now it is natural to look for two classes A and B of spaces, where B is *better* than A, and properties such that for each element from A there exists its extension from B preserving the properties. Thus Kimura and Morishita in [**30**] showed that every metrizable space has a compactification that is Eberlein compact and preserves both dim and weight (in [**4**] Charalambous proved that this compactification preserves also Ind). A compact space E is said to be *Eberlein compact* if E is homeomorphic to a subset of a Banach space with its weak topology. A space U is *universal* for a class of spaces if each element of this class can be embedded in U. Recall (cf. [**50**]) that for a class M (resp. D or I) of all metrizable (resp. completely regular or normal) spaces with weight $\leq \tau$ and dim $\leq n$ (resp. dim $\leq n$ or Ind $\leq n$), where $n \geq 0$, there exists an element from this class that is universal for M (resp. D or I, and the element is compact). So the Kimura–Morishita result (and the result of Charalambous as well) implies the existence of an Eberlein compact space $E_{n,\tau}$ which is universal for all metrizable spaces with dim $\leq n$ and weight $\leq \tau$ and which has the same weight and dimensions dim and Ind.

Infinite-dimensional theory

All spaces considered here are SM. The inductive dimensions ind, Ind have natural transfinite extensions trind, trInd for which trind $\leq$ trInd. An infinite-dimensional space is *countable dimensional*, shortly c.d., if X is the union of countably many finite-dimensional subspaces. It is known that every space having trind $< \omega_1$ is c.d. and any c.d. compact space has trind $< \omega_1$. Recall ([**38**]) that there exist two functions $\phi, \psi\colon \{\alpha < \omega_1\} \to \{\alpha < \omega_1\}$ such that $\operatorname{trInd} X \le \phi(\operatorname{trind} X)$ and $\psi(\operatorname{trInd} X) \le \operatorname{trind} X$ for any compact space X. Moreover, for each $\alpha < \omega_1$ there exist compact spaces X_α and Y_α such that $\operatorname{trind} X_\alpha = \operatorname{trInd} Y_\alpha = \alpha$, $\operatorname{trInd} X_\alpha = \phi(\alpha)$ and $\operatorname{trind} Y_\alpha = \psi(\alpha)$. *Smirnov's compacta* S^α, $\alpha < \omega_1$, are defined as follows. For each integer $n \ge 0$ the space S^n is the Euclidean n-cube I^n, $S^{\alpha+1} = S^\alpha \times I$ and, for any limit α, S^α is the one-point compactification of the free union of S^β with $\beta < \alpha$. It is known that for each α, $\operatorname{trInd} S^\alpha = \alpha$. In [**7**] Chatyrko improved an earlier result of Luxemburg to the following effect: for each $m \ge 0$ and any limit $\lambda < \omega_1$, $\operatorname{trind} S^{\lambda+2^{m-1}} \le \lambda + m$. Recall ([**38**]) that for each λ such that $\phi(\lambda) = \lambda$ we have $\operatorname{trind} S^{\lambda+k} = \lambda + k$ for $k = 0, 1, 2$.

1335? **Question 18.** *What is* $\operatorname{trind} S^\alpha$ *for each* α*?*

One can decompose $S^{\lambda+n} = Y_0 \cup \cdots \cup Y_n$, where $n \ge 1$, into closed subsets Y_i such that for each i, $\operatorname{trInd} Y_i = \lambda$. Using the following sum theorems: if $X = X_1 \cup X_2$, where X_1, X_2 are closed in X, then $\operatorname{trind} X \le \max\{\operatorname{trind} X_i\} + 1$, $\operatorname{trInd} X \le \max\{\lambda_i\} + n_1 + n_2 + 1$, where $\operatorname{trInd} X_i = \lambda_i + n_i$ for each i, and the unions $Y_0 \cup \cdots \cup Y_k$, where $k \ge 3$, we get a variety of compact spaces with trind $\ne$ trInd.

1336? **Question 19.** *Does there exist for each* $\alpha < \omega_1$ *a compact space* X_α *such that* $\operatorname{trind} X_\alpha = \operatorname{trInd} X_\alpha = \alpha$*?*

Observe that if FST(trind) or FST(trInd) holds in a c.d. compact space X then $\operatorname{trind} X = \operatorname{trInd} X$.

Recall that by Luxemburg's results each space X having $\operatorname{trInd} X < \omega_1$ has a compactification preserving trInd, and there exists a complete space with trind $= \omega_0$ having no compactification preserving trind.

1337? **Question 20.** *Evaluate for each space* X, $\min\{\operatorname{trind} Y$*:* Y *is a compactification of* $X\}$.

We know due to R. Pol that for each $\alpha < \omega_1$, there exists a universal space in the class of spaces with trind $\le \alpha$. In [**47**] Olszewski proved that for any limit α there is no universal space neither in the class of spaces (resp. compacta) with trInd $\le \alpha$ nor in the class of compacta with trind $\le \alpha$. The existence of universal spaces for non-limit α for the mentioned cases is an open question. Let d be a transfinite dimension. A compact space C is an $(\alpha+1)$-dimensional *d-Cantor manifold* (resp. *infinite-dimensional Cantor manifold*) if $\operatorname{d} C = \alpha + 1$ (resp. $\dim C = \infty$) and no closed subspace F of C satisfying $\operatorname{d} F < \alpha$ (resp. $\dim F < \infty$) separates C. In [**46**] Olszewski presented for each $\alpha < \omega_1$ an $(\alpha+1)$-dimensional d-Cantor manifold for d $=$ trind or trInd (in [**56**] Renska constructed simpler examples of trInd-manifolds which are disjoint unions of countably many closed cells and irrationals). A continuum X is *hereditarily indecomposable*, shortly h.i., if for any subcontinua A, B in X with nonempty intersection, either $A \subset B$, or

$B \subset A$. We know due to Bing (cf. [**40**]) that for each $n = 1, 2, \ldots, \infty$ there exist h.i. continua with $\dim = n$, and in each such continuum X with $\dim X = n$ the set $B_n(X) = X \setminus \{$ the union of all non-trivial subcontinuum with $\dim < n\}$ is not empty if $n < \infty$. In [**54**] R. Pol and Renska showed that if X is a h.i. continuum with $2 \leq \dim X = n < \infty$ and $B_r(X)$ is the set of all points of X belonging to some continuum with $\dim = r$ but avoid any non-trivial continuum with $\dim < r$, where $1 \leq r \leq n$, then $\dim B_r(X) = n - (r - 1)$, moreover $B_n(X)$ is not of type $G_{\delta\sigma}$ (always $G_{\delta\sigma\delta}$-set). A space E is *weakly infinite-dimensional*, shortly w.i.d., if for each sequence of pairs (A_i, B_i), $i = 1, 2, \ldots$ of disjoint closed sets in E there are partitions L_i in E between A_i and B_i such that $\bigcap_{i=1}^{\infty} L_i = \emptyset$, otherwise E is *strongly infinite dimensional*, shortly s.i.d. In [**52**] E. Pol and Renska constructed for each infinite $\alpha < \omega_1$ h.i. continua with trind or trInd equal to α and demonstrated the diversity among types of the sets $B_\infty(X)$ for infinite-dimensional h.i. continua X ($B_\infty(X)$ can be any subset of the Cantor set, the set of irrational numbers, a 1-dimensional G_δ-subset of X), for s.i.d., h.i. continua X, $B_\infty(X)$ is always strongly infinte-dimensional that is a corollary of a theorem of Henderson (cf. [**40**]) or a more recent result of Levin [**35**].

Question 21. *Is there for each integer $n \geq 2$ an infinte-dimensional h.i. continuum X with $B_\infty(X) = n$?* 1338?

A space X is *hereditarily strongly infinite-dimensional*, shortly h.s.i.d., if every subspace of X is either 0-dimensional or s.i.d. We know due to Rubin (cf. [**40**]) that there are h.s.i.d. continua. Recall that such spaces have to contain infinite-dimensional h.s.i.d. Cantor manifolds. In [**51**] E. Pol constructed a family $\{Y_s : s \in S\}$, where $|S| = 2^{\aleph_0}$, of h.i., h.s.i.d. Cantor manifolds such that (a) no open subset of Y_s embeds in Y_p for every $s \neq p$, $s, p \in S$; (b) every embedding of Y_s into Y_s is the identity, for each $s \in S$. A space X is *C-space* if for every sequence $\{\alpha_i\}_{i=1}^{\infty}$ of open covers X there exist disjoint open collections $\beta_1, \beta_2, \ldots$ such that β_i refines α_i for each i and $\bigcup_{i=1}^{\infty} \beta_i = X$. Recall that every C-compact space is w.i.d. We know due to Hattori and Yamada that the product of two C-compact spaces is a C-space and the product of w.i.d. compact space and a C-compact space is w.i.d.

Question 22. *Is the product of two w.i.d. compact spaces w.i.d.?* 1339?

The dimension dim can be extended to transfinites by different ways. Usually one uses a characterization of dim which is possible to extend to transfinites and considers the extension as an extension of dim. In particular, we know due to Borst two extensions of dim, $\dim_w$ and $\dim_C$, such that for each compact space X, $\dim_w X < \omega_1$ (resp. $\dim_C X < \omega_1$) if and only if X is w.i.d. (resp. a C-space). Recently Borst [**3**] constructed for each $\alpha < \omega_1$ a compact space X_α such that $\dim_C X_\alpha = \alpha$ and $\dim_w X_\alpha = \omega_0$, then he embedded all X_α in a compact space Y which is w.i.d. This Y can not be a C-space.

Compactness degrees

All spaces considered here are SM. The *compactness deficiency* of a space X, $\operatorname{def} X$ is the least integer n for which X has a metrizable compactification Y with $\dim(Y \setminus X) \leq n$. It is known due to de Groot (cf. [**1**]) that $\operatorname{def} X = 0$ if and only if X is *rim-compact*, i.e., if every point of X has arbitrarily small neighborhoods with compact boundary. In spirit of definitions of ind and Ind one defined two extensions of the rim-compactness: a *small inductive compactness degree* of X, $\operatorname{cmp} X$, (replace the empty set in the definition of ind by a compact space) and a *large inductive compactness degree* of X, $\operatorname{Cmp} X$, (assume for $n = -1, 0$ that $\operatorname{Cmp} X = n$ if and only if $\operatorname{cmp} X = n$). Recall (cf. [**1**]) that $\operatorname{cmp} X \leq \operatorname{Cmp} X \leq \operatorname{def} X \leq \dim X$ for any space X. We knew due to R. Pol, Kimura, Hart (cf. [**1**]) that there were examples of SM spaces with cmp $\neq$ Cmp but these examples were rather complicated. Spaces Z_n, where $n \geq 1$, are geometric cubes I^{n+1} without one open n-dimensional face. It was known almost from the beginning (cf. [**1**]) that $\operatorname{Cmp} Z_n = \operatorname{def} Z_n = n$ for each $n \geq 1$ and for $n = 1, 2$, $\operatorname{cmp} Z_n = n$. Recently Chatyrko and Hattori ([**9**], $n \geq 5$), Nishiura ([**44**], $n = 4$), Fedorchuk ([**22**], $n = 3$) showed that for $n \geq 3$, $2 \leq \operatorname{cmp} Z_n \leq m$, where m is any integer satisfying $n + 1 \leq 2^m$.

1340? **Question 23.** *What is* $\operatorname{cmp} Z_n$ *equal to for* $n \geq 4$*?*

There are only two examples of spaces with Cmp $\neq$ def. Namely, Kimura in [**29**] constructed a subspace K of R^4 such that $\operatorname{Cmp} K = 1$ and $2 \leq \operatorname{def} K \leq 3$. In [**36**] Levin and Segal found a subspace E of R^3 such that $\operatorname{cmp} E = \operatorname{Cmp} E = 1$ and $\operatorname{def} E = 2$. We do not have any example of a space where all cmp, Cmp and def disagree.

1341? **Question 24.** *Does there exist for each* n *a SM space* X_n *such that* $\operatorname{cmp} X_n = \operatorname{Cmp} X_n = 1$ *and* $\operatorname{def} X_n = n$*?*

In [**8**] Chatyrko, following a way developed by Hart and Kimura (cf. [**1**]), showed the existence for each n, m such that $n \leq m$ a SM space $C_{n,m}$ with $\operatorname{cmp} C_{n,m} = n$ and $\operatorname{Cmp} C_{n,m} = \operatorname{def} C_{n,m} = m$. So the positive answer to this question would show that for any integers n, m, p such that $1 \leq n \leq m \leq p$ there exists a SM space $X_{n,m,p}$ with $\operatorname{cmp} X_{n,m,p} = n \leq \operatorname{Cmp} X_{n,m,p} = m \leq \operatorname{def} X_{n,m,p} = p$.

References

[1] J. M. Aarts and T. Nishiura, *Dimension and extensions*, North-Holland, Amsterdam, 1993.

[2] T. Banakh, T. Radul, and M. Zarichnyi, *Absorbing sets in infinite-dimensional manifolds*, VNTL Publishers, Lviv, 1996.

[3] P. Borst, *A weakly infinite-dimensional compactum not having property C*, 2005, Preprint.

[4] M. G. Charalambous, *The dimension of metrizable subspaces of Eberlein compacta and Eberlein compactifications of metrizable spaces*, Fund. Math. **182** (2004), no. 1, 41–52.

[5] M. G. Charalambous, *Resolving a question of Arhangel′skiĭ*, 2005, Preprint.

[6] M. G. Charalambous, *A note on the Brouwer dimension of chainable spaces*, Topology Appl. **153** (2006), no. 8, 1271–1278.

[7] V. A. Chatyrko, *On finite sum theorems for transfinite inductive dimensions*, Fund. Math. **162** (1999), no. 1, 91–98.

[8] V. A. Chatyrko, *On the relationship between cmp and def for separable metrizable spaces*, Topology Appl. **152** (2005), no. 3, 269–274.

[9] V. A. Chatyrko and Y. Hattori, *On a question of de Groot and Nishiura*, Fund. Math. **172** (2002), no. 2, 107–115.

[10] G. Delistathis and S. Watson, *A regular space with a countable network and different dimensions*, Trans. Amer. Math. Soc. **352** (2000), no. 9, 4095–4111.

[11] A. Dow and K. P. Hart, *Cosmic dimensions*, 2005, Preprint. `arXiv:math.GN/0509099`

[12] A. Dranishnikov, *Homological dimension with respect to a group of coefficients Z_p*, Uspekhi Mat. Nauk **41** (1986), no. 3(249), 179–180.

[13] A. Dranishnikov, *Homological dimension theory*, Uspekhi Mat. Nauk **43** (1988), no. 4(262), 11–55, Translation: Russian Math. Surveys 43 (1988) no. 4, 11–63.

[14] A. Dranishnikov, *Asymptotic topology*, Uspekhi Mat. Nauk **55** (2000), no. 6(336), 71–116, Translation: Russian Math. Surveys 55 (2000) no. 6, 1085–1129.

[15] A. Dranishnikov, *On the dimension of the product of two compacta and the dimension of their intersection in general position in Euclidean space*, Trans. Amer. Math. Soc. **352** (2000), no. 12, 5599–5618.

[16] A. Dranishnikov and J. Dydak, *Extension theory of separable metrizable spaces with applications to dimension theory*, Trans. Amer. Math. Soc. **353** (2001), no. 1, 133–156.

[17] J. Dydak, *Cohomological dimension theory*, Handbook of geometric topology (R. J. Daverman and R. B. Sher, eds.), North-Holland, Amsterdam, 2002, pp. 423–470.

[18] R. Engelking, *General topology*, Heldermann Verlag, Berlin, 1989.

[19] R. Engelking, *Theory of dimensions finite and infinite*, Heldermann Verlag, Lemgo, 1995.

[20] V. V. Fedorchuk, *The Urysohn identity and the dimension of manifolds*, Uspekhi Mat. Nauk **53** (1998), no. 5(323), 73–114, Translation: Russian Math. Surveys 53 (1998) no. 5, 937–974.

[21] V. V. Fedorchuk, *Fully closed mappings and their applications*, Fundam. Prikl. Mat. **9** (2003), no. 4, 105–235.

[22] V. V. Fedorchuk and V. A. Chatyrko, *On a question of de Groot*, Vestnik Moskov. Univ. Ser. I Mat. Mekh. (2004), no. 6, 50–52, Translation: Moscow Univ. Math. Bull. 59 (2004) no. 6, 34–36.

[23] V. V. Fedorchuk and V. A. Chatyrko, *On the Brouwer dimension of one-dimensional compact Hausdorff spaces*, Vestnik Moskov. Univ. Ser. I Mat. Mekh. (2005), no. 2, 22–27.

[24] V. V. Fedorchuk and V. V. Filippov, *Manifolds with noncoinciding inductive dimensions*, Mat. Sb. **183** (1992), no. 9, 29–44, Translation: Sb. Math. 77 (1994) no. 1, 25–36.

[25] V. V. Fedorchuk, M. Levin, and E. V. Ščepin, *On the Brouwer definition of dimension*, Uspekhi Mat. Nauk **54** (1999), no. 2(326), 193–194.

[26] V. V. Filippov, *Bicompacta with distinct inductive dimensions*, Dokl. Akad. Nauk SSSR **192** (1970), no. 2, 289–292, Translation: Soviet Math. Dokl. 11 (1970) 635–638.

[27] Y. Hattori, *Dimension and products of topological groups*, Yokohama Math. J. **42** (1994), no. 1, 31–40.

[28] M. Katětov and P. Simon, *Origins of dimension theory*, Handbook of the history of general topology, Vol. 1, Kluwer Academic Publishers, Dordrecht, 1997, pp. 113–134.

[29] T. Kimura, *A separable metrizable space X for which* $\mathrm{Cmp}X \neq \mathrm{def}X$, Bull. Polish Acad. Sci. Math. **37** (1989), no. 7–12, 487–495 (1990).

[30] T. Kimura and K. Morishita, *On Eberlein compactifications of metrizable spaces*, Fund. Math. **171** (2002), no. 3, 223–234.

[31] S. V. Kotkin, *On some questions in dimension theory*, Dissertation, Moscow State University, 1990.

[32] K. L. Kozlov and B. A. Pasynkov, *On the Lebesgue dimension of topological products*, 2005, To appear in J. Math. Sci.

[33] J. Kulesza, *Metrizable spaces where the inductive dimensions disagree*, Trans. Amer. Math. Soc. **318** (1990), no. 2, 763–781.

[34] J. Kulesza, *New properties of Mrowka's space $\nu\mu_0$*, Proc. Amer. Math. Soc. **133** (2005), no. 3, 899–904.

[35] M. Levin, *A short construction of hereditarily infinite-dimensional compacta*, Topology Appl. **65** (1995), no. 1, 97–99.

[36] M. Levin and J. Segal, *A subspace of $\mathbb{R}^3$ for which* Cmp $\neq$ def, Topology Appl. **95** (1999), no. 2, 165–168.

[37] I. K. Lifanov and V. V. Filippov, *Two examples for the theory of the dimension of bicompacta*, Dokl. Akad. Nauk SSSR **192** (1970), 26–29, Translation: Soviet Math. Dokl. 11 (1970) 573–576.

[38] L. A. Luxemburg, *On compact metric spaces with noncoinciding transfinite dimensions*, Pacific J. Math. **93** (1981), no. 2, 339–386.

[39] D. V. Malykhin, *Some properties of topological products*, Dissertation, Moscow State University, 1999.

[40] J. van Mill, *The infinite-dimensional topology of function spaces*, North-Holland, Amsterdam, 2001.

[41] S. Mrówka, *The total failure of the union theorem for covering dimension*, Bull. Polish Acad. Sci. Math. **43** (1995), no. 2, 87–100.

[42] S. Mrówka, *Small inductive dimension of completions of metric spaces*, Proc. Amer. Math. Soc. **125** (1997), no. 5, 1545–1554.

[43] S. Mrówka, *Small inductive dimension of completions of metric spaces. II*, Proc. Amer. Math. Soc. **128** (2000), no. 4, 1247–1256.

[44] T. Nishiura, *On the Chatyrko–Hattori solution of de Groot's question*, Topology Appl. **152** (2005), no. 3, 310–316.

[45] S. Oka, *Free patched spaces and fundamental theorems of dimension theory*, Bull. Acad. Polon. Sci. Sér. Sci. Math. **28** (1980), no. 11–12, 595–602 (1981).

[46] W. Olszewski, *Cantor manifolds in the theory of transfinite dimension*, Fund. Math. **145** (1994), no. 1, 39–64.

[47] W. Olszewski, *Universal spaces in the theory of transfinite dimension. I*, Fund. Math. **144** (1994), no. 3, 243–258.

[48] B. A. Pasynkov, *On the finite-dimensionality of topological products*, Topology Appl. **82** (1998), 377–386.

[49] B. A. Pasynkov, *On Dugundji and homogeneous compacta with noncoinciding dimensions dim and ind*, 2000, Manuscript.

[50] B. A. Pasynkov, *Dimension theory (general theory)*, Encyclopedia of general topology (K. P. Hart, J. Nagata, and J. E. Vaughan, eds.), Elsevier Science, 2004, pp. 308–313.

[51] E. Pol, *Collections of higher-dimensional hereditarily indecomposable continua, with incomparable Fréchet types*, Topology Appl. **146/147** (2005), 547–561.

[52] E. Pol and M. Reńska, *On Bing points in infinite-dimenisonal hereditarily indecomposable continua*, Topology Appl. **123** (2002), no. 3, 507–522.

[53] R. Pol, *Questions in dimension theory*, Open problems in topology, North-Holland, Amsterdam, 1990, pp. 279–291.

[54] R. Pol and M. Reńska, *On the dimensional structure of hereditarily indecomposable continua*, Trans. Amer. Math. Soc. **354** (2002), no. 7, 2921–2932.

[55] R. Pol and H. Toruńczyk, *Topics in dimension theory*, Recent progress in general topology, II, North-Holland, Amsterdam, 2002, pp. 395–421.

[56] M. Reńska, *On Cantor manifolds for the large transfinite dimension*, Topology Appl. **112** (2001), no. 1, 1–11.

[57] E. V. Ščepin, *Arithmetic of dimension theory*, Uspekhi Mat. Nauk **53** (1998), no. 5(323), 115–212, Translation: Russian Math. Surveys 53 (1998) no. 5, 975–1069.

[58] I. Tsereteli, *Counterexamples in dimension theory*, Questions Answers Gen. Topology **20** (2002), no. 2, 139–159.

[59] L. Zambakhidze, *On the properties of additivity and decomposability of dimension functions*, Questions Answers Gen. Topology **20** (2002), no. 2, 119–137.

Open Problems in Topology II – Edited by E. Pearl

Questions on weakly infinite-dimensional spaces

Vitalii V. Fedorchuk

Introduction

Weak infinite dimensionality was introduced by Alexandroff in 1948 [**2**]. The first results in this area were obtained by Sklyarenko [**20**] and Levshenko [**13**] in 1959. A great contribution to the theory of weakly infinite-dimensional spaces was made in 1981 by R. Pol [**15**], who constructed an example of a compact metrizable weakly infinite-dimensional space which is not countable-dimensional.

In 1974, Haver [**11**] introduced the C property for metric spaces and proved that every locally contractible metric space which is a union of countably many compact sets with property C is an ANR space. In 1978, Addis and Gresham [**1**] gave a topological definition of C-spaces.

The C-spaces proved to play an important role in topology. In particular, Ancel [**3**] showed that any cell-like map from a compact metrizable space onto a C-space is a hereditary shape equivalence. Consequently, every infinite-dimensional compact C-space has infinite cohomological dimension $c\text{-}\dim_{\mathbb{Z}}$.

One of the most important problems concerning infinite-dimensional spaces was whether any weakly infinite-dimensional compact space is a C-space. Recently, this problem was solved in the negative by Borst [**5**].

The questions considered here are related to new classes of spaces, which are intermediate between the classes of weakly infinite-dimensional spaces and C-spaces.

1. Definitions

For a topological space X, by $\operatorname{cov}(X)$ we denote the set of all open covers of X. A family $\mathcal{U} = \{u_\alpha : \alpha \in A\} \subset \operatorname{cov}(X)$ in said to be *essential* (in X) if, for any disjoint open families v_α, where $\alpha \in A$, such that v_α refines u_α for each α, the family $\bigcup\{v_\alpha : \alpha \in A\}$ does not cover X.

Let $\mathcal{P}$ be a class of open covers of topological spaces; for a space X, we set $\mathcal{P}(X) = \mathcal{P} \cap \operatorname{cov}(X)$. A normal space X is called a *$\mathcal{P}$-C-space* ($X \in \mathcal{P}\text{-}C$) if every countable family $\mathcal{U} \subset \mathcal{P}(X)$ is inessential.

This approach yields the following classes of spaces:

(1) *m-C-spaces*, where m is an integer ≥ 2 and $\mathcal{P}$ consists of all covers u with $|u| \leq m$;
(2) *∞-C-spaces*, where $\mathcal{P}$ consists of all finite covers;
(3) *lf-C-spaces*, where $\mathcal{P}$ consists of all locally finite covers;
(4) *C-spaces*, where $\mathcal{P}$ consists of all covers;

This work was financially supported by the Russian Foundation for Basic Research (project no. 06-01-00761).

and so on.

If $\mathcal{P}_1 \subset \mathcal{P}_2$, then $\mathcal{P}_2$-$C \subset \mathcal{P}_1$-C. Consequently,

$$C \subset \text{lf-}C \subset \infty\text{-}C \subset \cdots \subset m\text{-}C \subset \cdots \subset 2\text{-}C. \tag{1.1}$$

The largest member of this sequence coincides with the class wid of all weakly infinite-dimensional spaces. The space ω_1 of all countable ordinals is an lf-C-space but not a C-space.

1342? **Question 1.** *Does there exist an $X \in \infty$-$C \setminus$ lf-C?*

For compact spaces, the first three members of sequence (1.1) coincide.

1343? **Question 2.** *Does the equality $(m{+}1)$-C $=$ m-C hold in the class of compact metrizable spaces for all m?*

Let ω-$C = \bigcap\{m\text{-}C : m \in \mathbb{N}\}$.

1344? **Question 3.** *Does the equality $C = \omega$-C hold in the class of compact metrizable spaces?*

Because of Borst's example of a compact metrizable space $X \in 2$-$C \setminus C$, the answer to one of Questions 2 and 3 must be negative.

Yet another generalization of C-spaces is as follows. Let $\Phi = \{F_\alpha : \alpha \in A\}$ be a discrete family of closed subsets of a space X. A *neighborhood* $O\Phi$ of the family Φ is a disjoint collection $\{OF_\alpha : \alpha \in A\}$ of neighborhoods OF_α of the sets F_α. A set $\varphi = \{\Phi_\beta : \beta \in B\}$ of discrete families of closed subsets of X is said to be *essential* (in X) if, for any neighborhoods $O\Phi_\beta$, the family $\bigcup\{O\Phi_\beta : \beta \in B\}$ does not cover X. A collectionwise normal space X is called a *weak C-space* ($X \in$ w-C) if every countable family of discrete collections of closed subsets of X is inessential. Any collectionwise normal C-space is a w-C-space.

On the other hand, any finite-dimensional countably compact noncompact space is a w-C-space but not a C-space.

1345? **Question 4.** *Is it true that any paracompact w-C-space is a C-space?*

The answer to this question is unknown even for compact metrizable spaces.

If a disjoint family Φ of closed subsets of a space X consists of $\leq m$ members, where $m \in \mathbb{N}$, then we say that Φ is an *m-system* in X. An *∞-system* in X is any finite disjoint family of closed subsets of X. A normal space X is called a *w-m-C-space* ($X \in$ w-m-C), where $m \in \mathbb{N}$ or $m = \infty$, if any countable family of m-systems in X is inessential. Every m-C-space is a w-m-C-space. By definition, we have w-2-C = wid.

1346? **Question 5.** *Is it true that any compact metrizable w-m-C-space is an m-C-space?*

We have the following sequence of inclusions similar to (1.1):

$$\text{w-}C \subset \text{w-}\infty\text{-}C \subset \cdots \subset \text{w-}m\text{-}C \subset \cdots \subset \text{w-}2\text{-}C.$$

1347? **Question 6.** *Does the equality w-$(m{+}1)$-C = w-m-C hold in the class of compact metrizable spaces for all m?*

Let w-ω-$C = \bigcap\{$w-m-$C : m \in \mathbb{N}\}$.

Question 7. *Does the equality ω-C = w-ω-C hold in the class of compact metrizable spaces?* 1348?

Question 8. *Is it true that any closed subspace of an lf-C-space is an lf-C-space?* 1349?

In the class of all countably paracompact collectionwise normal spaces, the answer to Question 8 is positive.

If every G_δ-subset of a hereditarily normal space X is an m-C-space, then every subset of X is an m-C-space.

Question 9. *Is it true that every subset of a w-m-C-space X is a w-m-C-space provided that every G_δ-subset of X is a w-m-C-space?* 1350?

Any paracompact finite-dimensional space is a C-space [**1**].

Question 10. *Is it true that any weakly paracompact finite-dimensional space is a C-space?* 1351?

Recall that a normal space X is said to be 0-*countable-dimensional* if $X = \bigcup\{X_i : i \in \omega\}$, where $\dim X_i \leq 0$.

Proposition ([**8**]). *Any 0-countable-dimensional collectionwise normal hereditarily normal space is a w-C-space.*

Corollary ([**8**]). *Any subset of a linearly ordered continuum is a w-C-space.*

Question 11. *Is it true that any collectionwise normal finite-dimensional space is a w-C-space?* 1352?

Theorem 1 ([**8**]). *Any strongly paracompact space X for which* $\operatorname{ind} X$ *is defined is a C-space.*

Levshenko [**14**] proved that if X satisfies the assumptions of Theorem 1, then $X \in$ wid.

Question 12. *Is it true that if X is a paracompact space for which* $\operatorname{ind} X$ *is defined, then $X \in$ wid? $X \in C$?* 1353–1354?

This question can be strengthened as follows.

Question 13. *Is it true that if X is a metric space with* $\operatorname{ind} X = 0$, *then $X \in$ wid?* 1355?

Question 12 can also be strengthened in a different direction.

Question 14. *Is it true that if X is a completely paracompact space for which* $\operatorname{ind} X$ *is defined, then $X \in$ wid?* 1356?

It is known that if X is a completely paracompact metrizable space for which $\operatorname{ind} X$ is defined, then X is countable-dimensional and, consequently, $X \in C$ (this was proved by Smirnov in [**21**]).

2. Maps, products, and subsets

If $\mathcal{P}$ is class of a spaces and $f\colon X \to Y$ is a map, then $f \in \mathcal{P}$ means that $f^{-1}(y) \in \mathcal{P}$ for every $y \in Y$.

Theorem 2 ([**7**]). *Let $f\colon X \to Y$ be a closed map from a countably paracompact (or hereditarily normal) space X onto a C-space Y. Then the following assertions are valid:*

(1) *if $f \in m\text{-}C$, then $X \in m\text{-}C$;*
(2) *if $f \in w\text{-}m\text{-}C$, then $X \in w\text{-}m\text{-}C$.*

Assertion (1) was proved for $m = 2$ by Hattori and Yamada [**10**].

1357–1358? **Question 15.** *Let X be a countably paracompact or hereditarily normal space admitting a closed m-C-map (w-m-C-map) onto a w-C-space. Is it true that $X \in m\text{-}C$? Respectively, $X \in w\text{-}m\text{-}C$?*

1359? **Question 16.** *Given compact metrizable spaces X and Y, is it true that*

(1) *if $X \in m\text{-}C$ and $Y \in m\text{-}C$, then $X \times Y \in m\text{-}C$;*
(2) *if $X \in w\text{-}m\text{-}C$ and $Y \in w\text{-}m\text{-}C$, then $X \times Y \in w\text{-}m\text{-}C$;*
(3) *if $X \in w\text{-}C$ and $Y \in w\text{-}C$, then $X \times Y \in w\text{-}C$?*

1360? **Question 17.** *Let $f\colon X \to Y$ be a light map of compact metrizable spaces. Is it true that*

(1) *if $Y \in m\text{-}C$, then $X \in m\text{-}C$;*
(2) *if $Y \in w\text{-}m\text{-}C$, then $X \in w\text{-}m\text{-}C$;*
(3) *if $Y \in w\text{-}C$, then $X \in w\text{-}C$?*

A positive answer to Question 17 would imply a positive answer to Question 16 thanks to the following theorem.

Theorem 3 ([**8**]). *Suppose that $\mathcal{P}$ is one of the classes m-C, w-m-C, w-C, and C. Then any compact metrizable space $X \notin \mathcal{P}$ contains a compact space $Y \notin \mathcal{P}$ such that, for any $Z \subset Y$, either $\dim Z \leq 0$ or $Z \notin \mathcal{P}$.*

This theorem was proved by Rubin [**19**] for $\mathcal{P} = 2\text{-}C$, by R. Pol [**18**] for $\mathcal{P} = C$ and closed Z, and by Levin [**12**] for $\mathcal{P} = C$.

Definition. Let $\mathcal{P}$ be a topological property. A space X is said to be *hereditarily non-$\mathcal{P}$* if $X \notin \mathcal{P}$ and, for every closed set $Y \subset X$,

$$\text{either } \dim Y \leq 0 \text{ or } Y \notin \mathcal{P}.$$

If this alternative holds for all subsets $Y \subset X$, then we say that X is a *strongly hereditarily non-$\mathcal{P}$* space.

Let h-non-$\mathcal{P}$ (sh-non-$\mathcal{P}$) denote the class of all (strongly) hereditarily non-$\mathcal{P}$ spaces.

1361? **Question 18.** *Let $\mathcal{P}$ be one of the classes m-C, w-m-C, w-C, and C, and let X and Y be compact metrizable spaces. Is it true that*

(1) $X, Y \in$ h-non-$\mathcal{P} \Longrightarrow X \times Y \in$ h-non-$\mathcal{P}$;
(2) $X, Y \in$ sh-non-$\mathcal{P} \Longrightarrow X \times Y \in$ sh-non-$\mathcal{P}$;
(3) $X \in$ h-non-$\mathcal{P} \Longrightarrow X^2 \in$ h-non-$\mathcal{P}$;
(4) $X \in$ sh-non-$\mathcal{P} \Longrightarrow X^2 \in$ sh-non-$\mathcal{P}$?

Question 19. *Let $\mathcal{P}$ be one of the classes m-C, w-m-C, w-C, and C. Is it true that* Comp $\cap$(h-non-$\mathcal{P}$) $\subset$ sh-non-$\mathcal{P}$? 1362?

Definition. A space X is said to be *strongly hereditarily (hereditarily) non-*1 *dim-space*, $X \in$ sh-non-1 dim ($X \in$ h-non-1 dim), if $\dim X \geq 2$ and, for every (closed) set $Y \subset X$, either $\dim Y \leq 0$ or $\dim Y \geq 2$.

Question 20. *Let X and Y be compact metrizable spaces. Is it true that* 1363?
(1) $X, Y \in$ h-non-1 dim $\Longrightarrow X \times Y \in$ h-non-1 dim;
(2) $X, Y \in$ sh-non-1 dim $\Longrightarrow X \times Y \in$ sh-non-1 dim;
(3) $X \in$ h-non-1 dim $\Longrightarrow X^2 \in$ h-non-1 dim;
(4) $X \in$ sh-non-1 dim $\Longrightarrow X^2 \in$ sh-non-1 dim?

A positive answer to Question 20(4) would give a positive answer to van Mill's problem [**17**, Question 414], which can be formulated as follows.

Question 21. *Does there exist an infinite-dimensional compact space X such that $X^n \in$ sh-non-1 dim for all positive n?* 1364?

Note that there exists no example of infinite-dimensional compact metrizable spaces X and Y such that $X \times Y$ contains no one-dimensional compact sets. At the same time, under the continuum hypothesis, there exists an infinite compact space X such that, for any positive integer n, all infinite closed subspaces of X^n are strongly infinite-dimensional [**9**].

Question 22. *Is it true that any compact metrizable space containing one-dimensional subsets contains a compact one-dimensional subset?* 1365?

3. Transfinite dimensions

3.1. The ordinal number Ord. In this section, we recall Borst's definition from [**4**]. Let L be an arbitrary set. By Fin L we denote the collection of all finite nonempty subsets of L.

Let M be a subset of Fin L. For $\sigma \in \{\varnothing\} \cup \operatorname{Fin} L$, we set $M^\sigma = \{\tau \in \operatorname{Fin} L : \sigma \cup \tau \in M,\ \sigma \cap \tau = \varnothing\}$. For $a \in L$, we denote the set $M^{\{a\}}$ by M^a.

Definition. The *ordinal number* Ord M is defined by induction as follows.

- $\operatorname{Ord} M = 0$ if and only if $M = \varnothing$;
- $\operatorname{Ord} M \leq \alpha$ if and only if $\operatorname{Ord} M^a < \alpha$ for every $a \in L$;
- $\operatorname{Ord} M = \alpha$ if and only if $\operatorname{Ord} M \leq \alpha$ and it is not true that $\operatorname{Ord} M < \alpha$;
- $\operatorname{Ord} M = \infty$ if and only if $\operatorname{Ord} M > \alpha$ for every ordinal α.

For an integer $m \geq 2$, we set $\operatorname{cov}_m(X) = \{u \in \operatorname{cov}(X) : |u| \leq m\}$ and $\operatorname{cov}_\infty(X) = \bigcup_m \operatorname{cov}_m(X)$.

For an integer $m \geq 2$ and for $m = \infty$, we set $M_m(X) = \{\sigma \in \text{Fin cov}_m(X) : \sigma$ is essential$\}$.

For any normal space X, we have

$$\dim X \leq n \text{ if and only if } \operatorname{Ord} M_m(X) \leq n. \tag{3.1}$$

This is a generalization of Borst's theorems [**4**, **6**] for $m = 2$ and $m = \infty$.

3.2. Transfinite dimension $\dim_m$. For a normal space X, we set

$$\dim_m X = \operatorname{Ord} M_m(X). \tag{3.2}$$

If $\dim_m X = \infty$, then we say that the dimension $\dim_m X$ is not defined. Comparing (3.1) and (3.2), we see that each of the functions $\dim_m$ is a transfinite extension of Lebesgue covering dimension.

Theorem 4 ([**8**]). *For a compact space X, the dimension $\dim_m X$ is defined if and only if X is an m-C-space.*

For $m = 2$ and $m = \infty$, Theorem 4 was proved by Borst in [**4**, **6**]. Clearly, if $m_1 \leq m_2$ then $\dim_{m_1} X \leq \dim_{m_2} X$.

1366? **Question 23.** *Does the equality $\dim_m = \dim_{m+1}$ hold in the class of compact metrizable spaces for all m?*

In view of Theorem 4, a positive answer to Question 23 would give a positive answer to Question 2. For the compact space E_{ω_0} constructed by Borst in [**5**], we have $\dim_2 E_{\omega_0} = \omega_0 < \infty = \dim_\infty E_{\omega_0}$.

Using the ideas of R. Pol from [**16**], it is easy to prove the following theorem.

Theorem 5 ([**8**]). *Let $\mathcal{E}$ be a family of m-C-compact spaces. Then there exists an m-C-compact space into which all compact spaces from $\mathcal{E}$ can be embedded if and only if* $\sup\{\dim_m X : X \in \mathcal{E}\} < \omega_1$.

Thus, to give a negative answer to Question 2, it is sufficient to construct a family of $(m+1)$-C-compact spaces X_α, where $\alpha \in \omega_1$, such that $\sup\{\dim_m X_\alpha : \alpha \in \omega_1\} < \omega_1$ but $\sup\{\dim_{m+1} X_\alpha : \alpha \in \omega_1\} = \omega_1$.

1367? **Question 24.** *Does there exist a compact metrizable space X such that $\dim_m X < \dim_\infty X$ for all integer m?*

A negative answer to Question 3 would imply a positive answer to Question 24.

1368? **Question 25.** *Does there exist an infinite-dimensional metrizable C-compactum X containing no subcompacta Y of dimension $0 < \dim_\infty Y < \dim_\infty X$?*

3.3. Transfinite dimension $\dim_{wm}$. For a normal space X, we denote the set of all m-systems in X by $\varphi_m(X)$.

We set $L_m(X) = \{\sigma \in \operatorname{Fin} \varphi_m(X) : \sigma$ is essential$\}$. For a normal space X, we have $\dim X \leq n$ if and only if $\operatorname{Ord} L_m(X) \leq n$. Thus, it is natural to define $\dim_{wm} X = \operatorname{Ord} L_m(X)$. If $\dim_{wm} X = \infty$, then we say that the dimension $\dim_{wm} X$ is not defined.

The function $\dim_{wm}$, as well as $\dim_m$, is a transfinite extension of the covering dimension.

Theorem 6 ([**8**]). *For a compact space X, the dimension $\dim_{\mathrm{w}m} X$ is defined if and only if X is a w-m-C-space.*

Clearly, if $m_1 \leq m_2$ then $\dim_{\mathrm{w}m_1} X \leq \dim_{\mathrm{w}m_2} X$.

Question 26. *Does the equality $\dim_{\mathrm{w}m} = \dim_{\mathrm{w}(m+1)}$ hold in the class of compact* 1369?
metrizable spaces for all m?

By virtue of Theorem 6, a positive answer to Question 26 gives a positive answer to Question 6.

Question 27. *Does there exist a compact metrizable space X such that $\dim_{\mathrm{w}2} X <$* 1370?
$\dim_{\mathrm{w}\infty} X$?

A negative answer to Question 7 implies a positive answer to Question 27.

Proposition ([**8**]). *If X is a normal space and $m \geq 3$ is an integer or $m = \infty$, then $\dim_{\mathrm{w}m} X \leq \dim_m X$. Moreover, $\dim_{\mathrm{w}2} X = \dim_2 X$.*

Question 28. *Does there exist a compact metrizable space X such that $\dim_{\mathrm{w}m} X <$* 1371?
$\dim_m X$ for some $m \geq 2$ or for $m = \infty$?

The Borst compact space E_{ω_0} gives a positive answer to one of Questions 27 and 28. Borst's question of whether $\dim_2(X \times I) = \dim_2 X + 1$ for any compact metrizable space X can be generalized as follows.

Question 29. *Let X be a compact metrizable space. Is it true that* 1372?

(1) $\dim_m(X \times I) = \dim_m X + 1$;
(2) $\dim_{\mathrm{w}m}(X \times I) = \dim_{\mathrm{w}m} X + 1$?

3.4. Inductive dimensions. Borst's inequality $\dim_2 X \leq \operatorname{Ind} X$ [**4**, Theorem 3.2.4] can be strengthened as $\dim_{\mathrm{w}\infty} X \leq \operatorname{Ind} X$.

Question 30. *Does the inequality $\dim_\infty \leq \operatorname{Ind}$ hold in the class of all compact* 1373?
metrizable spaces?

This question has the following weak version.

Question 31. *Does the inequality $\dim_m \leq \operatorname{Ind}$ hold in the class of all compact* 1374?
metrizable spaces for some integer $m \geq 3$?

A pair (u, Φ), where $u = \{U_1, \dots, U_k\} \in \operatorname{cov}_m(X)$ and $\Phi = \{F_1, \dots, F_k\} \in \varphi_m(X)$, is called an *$m$-covering pair* if $F_i \subset U_i$ for each i. If $O\Phi = \{OF_1, \dots, OF_k\}$ is a neighborhood of Φ refining u, then the set $P = X \setminus \bigcup O\Phi$ is called a *partition* of the covering pair (u, Φ).

Definition. The *large transfinite inductive dimension* Ind_m (where m is an integer ≥ 2 or $m = \infty$) in the class of all normal spaces is defined as follows:

(a) $\operatorname{Ind}_m X = -1$ if and only if $X = \varnothing$;
(b) $\operatorname{Ind}_m X \leq \alpha$, where α is an ordinal, if, for every m-covering pair (u, Φ), there exists a partition P of (u, Φ) such that $\operatorname{Ind}_m P < \alpha$;
(c) $\operatorname{Ind}_m X = \alpha$ if $\operatorname{Ind}_m X \leq \alpha$ and $\operatorname{Ind}_m X \leq \beta$ for no $\beta < \alpha$;

(d) $\operatorname{Ind}_m X = \infty$ if $\operatorname{Ind}_m X \le \alpha$ for no ordinal α.

For every normal space X, we have

$$\operatorname{Ind} X = \operatorname{Ind}_2 X \le \cdots \le \operatorname{Ind}_m X \le \operatorname{Ind}_{m+1} X \le \cdots \le \operatorname{Ind}_\infty X.$$

Theorem 7 ([8]). *The dimension* Ind_m *is defined for any hereditarily normal compact space which can be represented as a countable union of subspaces for which the dimension* Ind_m *is defined.*

Corollary. *For any countable-dimensional compact metrizable space, the dimensions* $\operatorname{Ind}_\infty$ *and, therefore,* Ind_m *for all* m *are defined.*

Theorem 8 ([8]). *If the dimension* Ind_m *is defined for a compact space* X *with weight* $w(X) \le \omega_\alpha$, *then* $\operatorname{Ind}_m X \le \omega_{\alpha+1}$.

Theorem 9 ([8]). *For any normal space* X, $\dim_m X \le \operatorname{Ind}_m X$.

1375? **Question 32.** *Does the equality* $\operatorname{Ind}_m = \operatorname{Ind}_{m+1}$ *hold in the class of all compact metrizable spaces for all* m*?*

1376? **Question 33.** *Is it true that* $\operatorname{Ind} X = \operatorname{Ind}_\infty X$ *for an arbitrary compact metrizable space* X*?*

References

[1] D. F. Addis and J. H. Gresham, *A class of infinite-dimensional spaces. I. Dimension theory and Alexandroff's problem*, Fund. Math. **101** (1978), no. 3, 195–205.

[2] P. S. Alexandroff, *An introduction to the Russian translation*, W. Hurewicz and H. Wallman, Dimension Theory (Russian translation), Moscow, 1948, pp. 5–18.

[3] F. D. Ancel, *The role of countable dimensionality in the theory of cell-like relations*, Trans. Amer. Math. Soc. **287** (1985), no. 1, 1–40.

[4] P. Borst, *Classification of weakly infinite-dimensional spaces. I. A transfinite extension of the covering dimension*, Fund. Math. **130** (1988), no. 1, 1–25.

[5] P. Borst, *A weakly infinite-dimensional compactum not having property C*, 2005, Preprint.

[6] P. Borst, *Some remarks concerning C-spaces*, Topology Appl., To appear.

[7] V. V. Fedorchuk, *Certain classes of weakly infinite-dimensional spaces*, Submitted to *J. Math. Sci. (New York)*.

[8] V. V. Fedorchuk, *Weakly infinite-dimensional spaces*, Submitted to *Russian Math. Surveys*.

[9] V. V. Fedorchuk, A. V. Ivanov, and J. van Mill, *Intermediate dimensions of products*, Topology Appl. **153** (2006), no. 17, 3265–3276.

[10] Y. Hattori and K. Yamada, *Closed pre-images of C-spaces*, Math. Japon. **34** (1989), no. 4, 555–561.

[11] W. E. Haver, *A covering property for metric spaces*, Topology conference (Blacksburg, Va., 1973), Springer, Berlin, 1974, pp. 108–113.

[12] M. Levin, *Inessentiality with respect to subspaces*, Fund. Math. **147** (1995), no. 1, 93–98.

[13] B. T. Levšenko, *Strongly infinite-dimensional spaces*, Vestnik Moskov. Univ. Ser. Mat. Mekh. Astr. Fiz. Chim. **14** (1960), no. 5, 219–228.

[14] B. T. Levšenko, *On infinite-dimensional spaces.*, Dokl. Akad. Nauk SSSR **139** (1961), 286–289, Translation: Soviet Math. Dokl. 2 (1961) 915–918.

[15] R. Pol, *A weakly infinite-dimensional compactum which is not countable-dimensional*, Proc. Amer. Math. Soc. **82** (1981), no. 4, 634–636.

[16] R. Pol, *On classification of weakly infinite-dimensional compacta*, Fund. Math. **116** (1983), no. 3, 169–188.

[17] R. Pol, *Questions in dimension theory*, Open problems in topology, North-Holland, Amsterdam, 1990, pp. 279–291.
[18] R. Pol, *On light mappings without perfect fibers on compacta*, Tsukuba J. Math. **20** (1996), no. 1, 11–19.
[19] L. R. Rubin, *Hereditarily strongly infinite-dimensional spaces*, Michigan Math. J. **27** (1980), no. 1, 65–73.
[20] E. G. Skljarenko, *Dimensionality properties of infinite-dimensional spaces*, Izv. Akad. Nauk SSSR Ser. Mat. **23** (1959), 197–212.
[21] Yu. M. Smirnov, *On transfinite dimension*, Mat. Sb. **58(100)** (1962), 415–422.

Open Problems in Topology II – Edited by E. Pearl

Some problems in the dimension theory of compacta

Boris A. Pasynkov

Dedicated to the 70th anniversary of the publication of Alexandroff's problem on the dimensions of compacta.

All topological spaces considered in this paper are assumed to be Tychonoff and called simply spaces; by maps we mean continuous maps of topological spaces.

Almost all problems posed below concern compact spaces. Recall that the *dimension* Δ [**22**] of a paracompact space X is defined as follows: $\Delta X \le n$ if there exists a strongly zero-dimensional paracompact space X^0 and a surjective closed map $f\colon X^0 \to X$ such that $|f^{-1}x| \le n+1$ for any $x \in X$.

1. On the coincidence of dim, ind, Ind, and Δ for compact spaces

It is well known that the three basic dimension functions dim, ind, and Ind coincide for compact metrizable spaces, i.e.,

$$(*)\qquad \dim X = \operatorname{ind} X = \operatorname{Ind} X$$

for any compact metrizable space X. In 1936, Alexandroff [**1**] asked whether they coincide for arbitrary compact spaces. In 1941, he proved that $\dim X \le \operatorname{ind} X$ for any compact space X. Recall also that $\operatorname{ind} X \le \operatorname{Ind} X$ for any normal space X and (see [**22**]) $\operatorname{Ind} X \le \Delta X$ for any paracompact space X. Moreover, *for any metrizable space X, $\dim \beta X = \operatorname{ind} \beta X = \operatorname{Ind} \beta X = \Delta X$ and there exists a compactification cX of X such that $\dim cX = \operatorname{ind} cX = \operatorname{Ind} cX = \Delta cX$ and $w(cX) = w(X)$.*

In 1958, Pasynkov proved that $\dim G = \operatorname{ind} G = \operatorname{Ind} G$ for any compact group G; in 1962, he obtained the equalities $\dim G/H = \operatorname{ind} G/H = \operatorname{Ind} G/H$ for any locally compact group G and any closed subgroup H of G (in particular, they hold for compact coset spaces G/H). After that, the following definition and problem naturally arose.

A compact space X is called *algebraically homogeneous* if there exists a topological group G and its closed subgroup H such that X is homeomorphic to G/H.

Question 1. *Do all or some of the dimensions* dim, ind, *and* Ind *coincide for algebraically homogeneous compact spaces?* 1377?

The following problem is related to Question 1.

Question 2. *Describe the topological groups G, their closed subgroups H, and compact coset spaces G/H for which all or some of the dimensions $\dim G/H$, $\operatorname{ind} G/H$, and $\operatorname{Ind} G/H$ coincide. Do they coincide for first countable, perfectly normal, dyadic, hereditarily normal coset spaces?* 1378?

Note that a zero-dimensional perfectly normal (and, hence, first countable) compact coset space G/H may not be metrizable or dyadic (for example, Alexandroff's double arrow space is neither metrizable nor dyadic).

Partial answers to Questions 1 and 1 were given by Pasynkov in [**15**, **17**, **18**]. He called a topological group G *almost metrizable* if there exists a compact set $C \subset G$ and its neighborhoods O_i, where $i \in \mathbb{N}$, such that any neighborhood of C is contained in O_i for some i; a space X is *almost metrizable* if there exists a compact group G which acts continuously on X so that the orbit space X/G is metrizable (see [**15**, **17**]). All groups of pointwise countable type (in particular, all paracompact p, Čech complete, and locally compact groups) are almost metrizable. It was proved in [**17**] that $\dim X = \operatorname{Ind} X = \Delta X$ for any almost metrizable space X and that if $\dim X < \infty$, then X admits a perfect zero-dimensional (i.e., having zero-dimensional fibers) map f onto a metrizable space. Thus, *for any almost metrizable compact space X, relations $(*)$ hold,* $\dim X = \Delta X$, *and if* $\dim X < \infty$, *then X admits a zero-dimensional map onto a metrizable compact space.*

Since any coset space G/H of a closed subgroup H in an almost metrizable group G is almost metrizable (see [**18**]), it follows that *for any compact coset space G/H, where G is an arbitrary almost metrizable group (and H is its closed subgroup),* $\dim G/H = \operatorname{ind} G/H = \operatorname{Ind} G/H = \Delta G/H$ *and if* $\dim G/H < \infty$, *then G/H admits a zero-dimensional map onto a compact metrizable space.* On the other hand, *if a compact space X admits a zero-dimensional map onto a compact metrizable space, then relations $(*)$ hold and* $\dim X = \Delta X$ (the equality $\dim X = \operatorname{Ind} X$ was proved in [**14**] and $\dim X = \Delta X$ in [**6**, **17**]; as far as I know, the inequality $\dim X = \operatorname{ind} X$ is due to Katětov). Note that if a compact space X is a G_δ-subset of a topological group and $\dim X < \infty$, then X admits a zero-dimensional map onto a compact metrizable space (and hence relations $(*)$ hold).

It is not known whether there exist algebraically homogeneous compact spaces with noncoinciding dimensions. The situation with topologically homogeneous compacta is clearer.

In 1971, Fedorchuk [**8**] constructed a topologically homogeneous first countable compact space F with $\dim F = 1 < \operatorname{ind} F = 2$. Later, in 1990, Chatyrko [**2**] constructed first countable topologically homogeneous compact spaces C_n with $\dim C_n = 1$ and $\operatorname{ind} C_n = n$ for $n = 2, 3, \ldots$.

1379? **Question 3.** *Do the dimensions* ind *and* Ind *coincide for any (first countable, hereditarily normal, hereditarily paracompact, dyadic, (hereditarily) separable) topologically homogeneous compact space?*

1380? **Question 4** ([**20**]). *Does there exist a topologically homogeneous compact space T_α with* $\dim T_\alpha < \infty$ *and* $\operatorname{trind} X_\alpha = \alpha$ *for any ordinal $\alpha \geq \omega_0$?*

In this question, the topological homogeneity of T_α can be replaced by the weaker requirement that $\operatorname{trind}_x T_\alpha = \alpha$ for all $x \in T_\alpha$ (dimensional homogeneity).

Question 5. *Do there exist topologically homogeneous compact spaces that have infinite dimension* dim *and are weakly infinite-dimensional or weakly* (≡ *strongly (see* [**7**])) *countable-dimensional?* 1381?

2. Noncoincidence of dim and ind for compact spaces

In 1949, Lunc constructed a compact space L_1 with $\dim L_1 = 1 < \operatorname{ind} L_1 = 2$. Then (also in 1949), Lokutsievskii constructed a compact space L_2 with the same dimensions dim and ind being the union of two compact subspaces with dimensions ind and dim equal to 1. In 1958, for any positive integers n and $m > n$, Vopěnka constructed compact spaces X_{mn} and Y_{mn} such that $\dim X_{mn} = \operatorname{ind} Y_{mn} = m$, $\operatorname{ind} X_{mn} = n$, and $\dim Y_{mn} = n$. These results had completely clarified the relations between the dimensions dim and ind in the class of all compact spaces.

The first example of a first countable compactum with noncoinciding dimensions dim and ind was suggested by Fedorchuk in 1968. Then, in 1970, Filippov constructed first countable compacta F_{mn} with $\dim F_{mn} = m$ and $\operatorname{ind} F_{mn} = n$ for any positive integers m and $n > m$. General approaches to constructing compact spaces with noncoinciding dimensions dim and ind were suggested in [**3**, **4**].

Recall that a compact space X is *Dugundji* if X is the limit of an inverse system $\{X_\alpha, p_{\beta\alpha}; \alpha \in \mathcal{A}\}$ of compacta with the following properties: $\mathcal{A}$ is the set of all ordinals $< \tau$ for some τ; all of the bonding maps $p_{\beta\alpha}$ are open and surjective; X_1 is metrizable; there exists a metrizable space M and maps $q_\alpha\colon X_{\alpha+1} \to M$ for all $\alpha, \alpha+1 \in \mathcal{A}$ such that the diagonal $p_{\alpha+1\alpha} \Delta q_\alpha$ is a topological embedding; for any limit ordinal $\gamma \in \mathcal{A}$, X_γ is the limit of the system $\{X_\alpha, p_{\beta\alpha}; \alpha < \gamma\}$. Dugundji compacta are very close to topological products of compact metrizable spaces and have the following simple characterization: a compact space X is Dugundji if and only if, for any zero-dimensional compact space Z, any closed subset C of Z, and any map $f\colon C \to X$, there exists a map $F\colon X \to Z$ such that $F \restriction C = f$.

In 1977, Fedorchuk constructed a Dugundji compactum F of weight $\mathfrak{c}$ with $\dim F = 1$ and $\operatorname{ind} F = \operatorname{Ind} F = 2$. In 2002, Pasynkov and A.V. Odinokov constructed Dugundji compacta PO_n with $\dim PO_n = 1$ and $\operatorname{ind} PO_n = n$. In [**24**], Uspenskij considered *strongly homogeneous* (≡ *with rectifiable diagonal*) compact spaces. They are Dugundji and homogeneous.

Question 6. *Do the dimensions* dim *and* ind *coincide for homogeneous Dugundji compacta and for strongly homogeneous compacta?* 1382?

A few years ago, the study of the dimensional properties of Eberlein compacta was initiated. I can construct strong Eberlein compacta (that is, compact subsets of the σ-products $\{x = \{x_\alpha\}_{\alpha\in\mathcal{A}} \in \prod_{\alpha\in\mathcal{A}} I_\alpha : |\{\alpha \in \mathcal{A} : x_\alpha \neq 0\}| < \omega\}$ of unit intervals $I_\alpha = [0,1]$) P_n such that $\dim P_n = 1$ and $\operatorname{ind} P_n = n$ for $n = 2, 3, \ldots$.

Question 7. *Do the dimensions* dim *and* ind *coincide for homogeneous (or hereditarily normal, hereditarily paracompact, first countable, perfectly normal) (strong) Eberlein (Corson, Valdivia) compacta?* 1383?

3. On the noncoincidence of ind and Ind for compact spaces

In 1969, Filippov constructed a compactum F with $\dim F = \operatorname{ind} F = 2 < \operatorname{Ind} F = 3$ (the proofs were published in [**11**]). Then (in 1970 [**10**]), he explained (without detailed proofs) how to construct compacta F_i with $\dim F_i = 1$, $\operatorname{ind} F_i = i$, and $\operatorname{Ind} F_i = 2i - 1$ for $i = 2, 3, \ldots$. The following problem remains open.

1384? **Question 8.** *Do there exist a positive integer $m \geq 2$ and compact spaces A_{mn} for all integers $n > m$ such that* ($\dim A_{mn} = 1$ *and)* $\operatorname{ind} A_{mn} = m$ *and* $\operatorname{Ind} A_{mn} = n$*?*

Of course, the most interesting case is $m = 2$.

If the answer to Question 8 is "yes", then the one-point compactification $A_{m\omega}$ of the discrete union of all A_{mn} has the properties $\operatorname{ind} A_{m\omega} = m$ and $\operatorname{trInd} A_{m\omega} = \omega$. Thus, the following question makes sense.

1385? **Question 9.** *Do there exist a positive integer $m \geq 2$ and compact spaces $A_{m\alpha}$ for all transfinite numbers α such that* $\operatorname{ind} A_{m\alpha} = m$ *and* $\operatorname{trInd} A_{m\alpha} = \alpha$*?*

I can construct a strong Eberlein compactum Ψ with $\operatorname{ind}\Psi=2$ and $\operatorname{Ind}\Psi=3$. So, Question 8 makes sense for (strong) Eberlein (Corson, Valdivia) compacta.

1386? **Question 10.** *Do the dimensions* ind *and* Ind *coincide for a (strong) Eberlein (Corson, Valdivia) compactum provided that it is first countable (homogeneous, hereditarily normal, hereditarily paracompact)?*

Before Filippov's results, it was known that $\operatorname{ind} X = \operatorname{Ind} X$ for any perfectly normal compact space X (N. B. Vedenisov, 1939). Recall that a space X is said to be *perfectly $\varkappa$-normal* [**23**] (*quasi-perfectly normal* [**5**]) if the closure of every open (respectively, G_δ) subset of X is a zero-set. Obviously, any quasi-perfectly normal space is perfectly $\varkappa$-normal, and any perfectly normal space is quasi-perfectly normal. In 1977, Fedorchuk [**9**] asked the following question.

1387? **Question 11.** *Is it true that* $\operatorname{ind} X = \operatorname{Ind} X$ *holds for any perfectly $\varkappa$-normal compact space X?*

In 1982, Chigogidze [**5**] proved $\operatorname{ind} X = \operatorname{Ind} X$ for any quasi-perfectly normal compact space X. This gives a partial answer to Question 11, because any hereditarily normal quasi-perfectly normal space is perfectly $\varkappa$-normal (this was proved by Chigogidze). Earlier (in 1977), Fedorchuk [**9**] proved that $\operatorname{ind} X = \operatorname{Ind} X$ holds for any hereditarily perfectly $\varkappa$-normal space X (*hereditarily perfectly $\varkappa$-normal* means that every closed G_δ-subset of X is perfectly $\varkappa$-normal). In particular, $\operatorname{ind} X = \operatorname{Ind} X$ *holds for all Dugundji compacta and even for all $\varkappa$-metrizable compacta.* (Recall that a compact space X is *$\varkappa$-metrizable* if X is the limit of a countably directed inverse system $\{X_\alpha, p_{\beta\alpha}; \alpha \in \mathcal{A}\}$ of compact metrizable spaces with surjective open bonding maps $p_{\beta\alpha}$ such that, for any increasing sequence $\alpha(i) \in \mathcal{A}$, where $i \in \mathbb{N}$, the supremum $\beta = \sup\{\alpha(i) : i \in \mathbb{N}\}$ in $\mathcal{A}$ is defined and X_β is the limit of the inverse sequence $\{X_{\alpha(i)}, p_{\alpha(i+1)\alpha(i)} : i \in \mathbb{N}\}$.)

4. Dimensional properties of topological products

We start with the following old problem.

Question 12. *Is it true that* $\operatorname{Ind} X \times I \leq \operatorname{Ind} X + 1$ *for any (first countable) compact space* X? 1388?

In 1972 [**12**], Filippov constructed compact spaces X and Y such that $\operatorname{ind} X = \operatorname{Ind} X = 1$, $\operatorname{ind} Y = \operatorname{Ind} Y = 2$, and $\operatorname{ind} X + \operatorname{ind} Y = \operatorname{Ind} X + \operatorname{Ind} Y = 3 < \operatorname{ind} X \times Y \leq \operatorname{Ind} X \times Y$. In 1999 (in his Ph.D. thesis), D. V. Malykhin strengthened this result of Filippov. He constructed compact spaces X and Y such that they have the same dimensional properties as those constructed by Filippov and X has the additional property of being linearly ordered. Malykhin conjectured that, instead of his compact space X, the lexicographically ordered square can be taken.

Question 13. *How large can the gap between* $\operatorname{ind} X \times Y$ *and* $\operatorname{Ind} X \times Y$ *be for compact spaces* X *and* Y *with (given)* $\operatorname{ind} X = \operatorname{Ind} X$ *and* $\operatorname{ind} Y = \operatorname{Ind} Y$? *What if* X *and* Y *are first countable?* 1389–1390?

Question 13 is interesting for many special classes of compact spaces, including the classes of (strong) Eberlein, Corson, and Valdivia compacta.

In [**21**], an integer-valued function $f(k, l)$ for $k, l = 0, 1, 2, \ldots$ with the following property was defined: for any compact spaces X and Y with finite $\operatorname{Ind} X$ and $\operatorname{Ind} Y$, we have $\operatorname{Ind} X \times Y \leq f(\operatorname{Ind} X, \operatorname{Ind} Y)$ (this implies, in particular, that $X \times Y$ always has finite dimension Ind provided that both X and Y have finite dimension Ind).

Question 14. *Is it possible to improve the estimate* $\operatorname{Ind} X \times Y \leq f(\operatorname{Ind} X, \operatorname{Ind} Y)$? 1391?

Note that $\operatorname{Ind} X \times Y \leq \operatorname{Ind} X + \operatorname{Ind} Y$ for any compact spaces X and Y satisfying the conditions of the finite-sum theorem for Ind [**16**].

Question 15. *Do there exist perfectly normal (Dugundji, $\varkappa$-metrizable, perfectly $\varkappa$-normal) compact spaces* X *and* Y *such that* $\dim X = \operatorname{ind} X$, $\dim Y = \operatorname{ind} Y$, *and* $\dim X \times Y < \operatorname{ind} X \times Y$? *Can these relations hold if* Y *is metrizable or* $X \times Y$ *is perfectly normal?* 1392?

The next problem is about the dimensional properties of products of noncompact spaces; I believe, this is one of the most interesting problems concerning the dimensional properties of topological products.

Question 16 (see [**13, 19**]). *Is it true that* $\dim X \times Y \leq \dim X + \dim Y$ *if* $X \times Y$ *is paracompact (and* X *and* Y *are strongly paracompact)?* 1393?

5. A problem concerning the subset theorem

Question 17. *Is it true that* $\dim A \leq \dim X$ *for any metrizable subset* A *of a (strong) Eberlein (or Corson) compactum* X? 1394?

Obviously, it can be assumed without loss of generality that A is dense in X.

References

[1] P. S. Alexandroff, *Einige problemstellungen in der mengentheoretischen Topologie*, Rec. Math. [Mat. Sb.] **1** (1936), no. 5, 619–634.

[2] V. A. Chatyrko, *Compact spaces with noncoinciding dimensions*, Trudy Moskov. Mat. Obshch. **53** (1990), 192–228, Translation: Trans. Moscow Math. Soc. (1991) 199–236.

[3] V. A. Chatyrko, K. L. Kozlov, and B. A. Pasynkov, *On an approach to constructing compacta with different dimensions dim and ind*, Topology Appl. **107** (2000), no. 1–2, 39–55.

[4] V. A. Chatyrko, K. L. Kozlov, and B. A. Pasynkov, *On another approach for constructing compacta with different dimensions dim and ind*, Topology Proc. **25** (2000), Summer, 43–72.

[5] A. Chigogidze, *On a generalization of perfectly normal spaces*, Topology Appl. **13** (1982), no. 1, 15–20.

[6] M. M. Choban, *Topological structure of subsets of topological groups and their coset spaces*, Mat. Issled. **44** (1977), 117–163.

[7] R. Engelking, *Theory of dimensions finite and infinite*, Heldermann Verlag, Lemgo, 1995.

[8] V. V. Fedorchuk, *An example of a homogeneous compactum with non-coinciding dimensions.*, Dokl. Akad. Nauk SSSR **198** (1971), 1283–1286, Translation: Soviet Math. Dokl. 12 (1971) 989–993.

[9] V. V. Fedorchuk, *The dimension of κ-metrizable bicompacta, in particular of Dugundji spaces*, Dokl. Akad. Nauk SSSR **234** (1977), no. 1, 30–33, Translation: Soviet Math. Dokl. 18 (1977) 605–609.

[10] V. V. Filippov, *Bicompacta with distinct inductive dimensions*, Dokl. Akad. Nauk SSSR **192** (1970), no. 2, 289–292, Translation: Soviet Math. Dokl. 11 (1970) 635–638.

[11] V. V. Filippov, *A solution of a certain problem of P. S. Aleksandrov (a bicompactum with noncoinciding inductive dimensions)*, Mat. Sb. **83(125)** (1970), 42–60, Translation: Sb. Math. 12 (1970) 41–57.

[12] V. V. Filippov, *The inductive dimension of a product of bicompacta*, Dokl. Akad. Nauk SSSR **202** (1972), 1016–1019, Translation: Soviet Math. Dokl. 13 (1972) 250–254.

[13] K. Morita, *Dimension of general topological spaces*, Surveys in general topology (G. M. Reed, ed.), Academic Press, New York, 1980, pp. 297–336.

[14] B. A. Pasynkov, *On a class of mappings and on the dimension of normal spaces*, Sibirsk. Mat. Zh. **5** (1964), 356–376.

[15] B. A. Pasynkov, *Almost-metrizable topological groups*, Dokl. Akad. Nauk SSSR **161** (1965), 281–284, Translation: Soviet Math. Dokl. 7 (1966) 404–408.

[16] B. A. Pasynkov, *The dimension of rectangular products*, Dokl. Akad. Nauk SSSR **221** (1975), no. 2, 291–294, Translation: Soviet Math. Dokl. 16 (1975) 344–347.

[17] B. A. Pasynkov, *The dimension of spaces with a bicompact transformation group*, Uspekhi Mat. Nauk **31** (1976), no. 5(191), 112–120, Translation: Russian Math. Surveys 31 (1976) no. 5, 128–137.

[18] B. A. Pasynkov, *Spaces with a bicompact transformation group*, Dokl. Akad. Nauk SSSR **231** (1976), no. 1, 39–42, Translation: Soviet Math. Dokl. 17 (1976) 1522–1526.

[19] B. A. Pasynkov, *On the dimension of topological products and limits of inverse sequences*, Dokl. Akad. Nauk SSSR **254** (1980), no. 6, 1332–1336.

[20] B. A. Pasynkov, *On dimension theory*, Aspects of topology, Cambridge University Press, Cambridge, 1985, pp. 227–250.

[21] B. A. Pasynkov, *On the finite-dimensionality of topological products*, Topology Appl. **82** (1998), 377–386.

[22] V. I. Ponomarev, *Paracompacta: their projection spectra and continuous mappings*, Mat. Sb. **60(102)** (1963), 89–119, Translation: Amer. Math. Soc. Transl. 39 (1964) 133–164.

[23] E. V. Ščepin, *Topological products, groups, and a new class of spaces that are more general than metric spaces*, Dokl. Akad. Nauk SSSR **226** (1976), no. 3, 527–529.

[24] V. V. Uspenskij, *Topological groups and Dugundji compacta*, Mat. Sb. **180** (1989), no. 8, 1092–1118, Translation: Sb. Math. 67 (1990) no. 2, 555–580.

Part 9

Invited Problems

Open Problems in Topology II – Edited by E. Pearl

Problems from the Lviv topological seminar

Taras Banakh, Bohdan Bokalo, Igor Guran, Taras Radul, Michael Zarichnyi

Introduction

This collection of problems is formulated by participants and guests of the Lviv topological seminar held at the Ivan Franko Lviv National University (Ukraine).

1. Asymptotic dimension

We recall that a metric space X is *proper* if the distance $d(\cdot, x_0)$ to a fixed point is a proper map for any $x_0 \in X$. A map $f\colon X \to Y$ between metric spaces is called *coarse* if it satisfies the following two conditions [**34**]:

Coarse Uniformity: There is a monotone function $\lambda\colon [0,\infty) \to [0,\infty)$ such that $d_Y(f(x), f(x')) \leq \lambda(d_X(x,x'))$;

Metric Properness: The preimage $f^{-1}(B)$ is bounded for every bounded set $B \subset Y$.

Two maps f, g into a metric space Y are *close* if there exists a constant $C > 0$ such that $d_Y(f(x), g(x)) < C$, for every $C > 0$. Two metric spaces X, Y are said to be *coarse equivalent* if there exist coarse maps $f\colon X \to Y$ and $g\colon Y \to X$ such that the maps gf and 1_X are close and also fg and 1_Y are close.

For a proper metric space X the *Higson compactification* $\bar{X}$ is defined by means of the following proximity: $A\,\delta\,B$ if and only if $\lim_{r\to\infty} d(A \setminus B_r(x_0), B \setminus B_r(x_0)) < \infty$ if $\operatorname{diam} A = \operatorname{diam} B = \infty$ and $d(A,B) = 0$ otherwise. Here $x_0 \in X$ is a base point, $B_r(x_0)$ is the r-ball centered at x_0 and $d(A,B) = \inf\{d(a,b) : a \in A, b \in B\}$.

The remainder $\nu X = \bar{X} \setminus X$ of the Higson compactification is called the *Higson corona* [**34**].

The asymptotic dimension asdim of a metric space was defined by Gromov for studying asymptotic invariants of discrete groups [**25**]. This dimension can be considered as an asymptotic analogue of the Lebesgue covering dimension dim. Dranishnikov has introduced the dimension asInd which is analogous to the large inductive dimension Ind (see [**19**]). It is known that $\operatorname{asdim} X = \operatorname{asInd} X$ for each proper metric space with $\operatorname{asdim} X < \infty$. The problem of coincidence of asdim and asInd is still open in the general case [**19**].

The addition theorem for asdim is proved in [**13**]: suppose that a metric space X is presented as a union $A \cup B$ of its subspaces. Then $\operatorname{asdim} X \leq \max\{\operatorname{asdim} A, \operatorname{asdim} B\}$.

We have also a weaker result for the dimension asInd: let X be a proper metric space and $X = Y \cup Z$ where Y and Z are unbounded sets. Then $\operatorname{asInd} X \leq \operatorname{asInd} Y + \operatorname{asInd} Z$ (see [**32**]).

We do not know whether this estimate is the best possible.

1395? **Question 1.1.** *Let X be a proper metric space and $X = Y \cup Z$. Is it true that* $\operatorname{asInd} X \leq \max\{\operatorname{asInd} Y, \operatorname{asInd} Z\}$*?*

Let us note that the negative answer to this question gives us a negative answer to the problem of coincidence of asymptotic dimensions.

Extending codomain of Ind to ordinal numbers we obtain the transfinite extension trInd of the dimension Ind. It is known that there exists a space S_α such that $\operatorname{trInd} S_\alpha = \alpha$ for each countable ordinal number α [**22**]. This method does not work for asInd: the extension trasInd appears to be trivial: if $\operatorname{trasInd} X < \infty$, then $\operatorname{asInd} X < \infty$ (see [**32**]). However there exists a nontrivial transfinite extension trasdim of asdim (see [**33**]): there is a metric space X with $\operatorname{trasdim} X = \omega$.

1396? **Question 1.2.** *Find for each countable ordinal number ξ a metric space X_ξ with* $\operatorname{trasdim}, X_\xi = \xi$.

In the classical dimension theory of infinite dimensional spaces there is a special class of spaces that have property C. Properties of such spaces are close to those of finite-dimensional spaces. Dranishnikov defined an asymptotic analogue of property C [**18**].

1397? **Question 1.3.** *Let X and Y be two metric spaces with the asymptotic property C. Does $X \times Y$ have the asymptotic property C?*

It is known that the dimension trasdim classifies the class of metric spaces with the asymptotic property C. Hence a positive answer to the following question gives us the positive answer to the Question 1.3.

1398? **Question 1.4.** *Is there a function $\alpha\colon \omega_1 \to \omega_1$ such that* $\operatorname{trasdim} X \times Y \leq \alpha(\xi)$ *for each countable ordinal number ξ and two metric spaces X, Y with* $\operatorname{trasdim} X \leq \xi$ *and* $\operatorname{trasdim} Y \leq \xi$*?*

Arhangel′skiĭ introduced the dimension Dind (see [**21**]). This dimension has an asymptotic counterpart.

For a proper metric space (X, d) we let $\operatorname{asDind} X = -1$ if and only if X is bounded. Suppose that we have already defined the class of proper metric spaces for which $\operatorname{asDind} X \leq n - 1$. We say that $\operatorname{asDind} X \leq n$ if for every finite family $\mathcal{U}$ of open in the Higson compactification $\bar{X}$ sets there exists a finite family $\mathcal{V}$ of open subsets in $\bar{X}$ with the following property: the family $\{V \cap \nu X : V \in \mathcal{V}\}$ is a discrete in νX family which refines $\mathcal{U}$ and $\operatorname{asDind} X \setminus \bigcup \mathcal{V} \leq n - 1$.

1399? **Question 1.5.** *Find relations between the dimension* Dind *and the other asymptotic dimension functions.*

It is proved in [**20**] that every proper metric space of asymptotic dimension 0 is coarsely equivalent to an ultrametric space. Recall that a metric d on a set X is called an *ultrametric* if $d(x, y) \leq \max\{d(x, z), d(z, y)\}$ for every $x, y, z \in X$. The mentioned results from [**20**] is an asymptotic version of the classical de Groot's result characterizing zero-dimensional metric spaces as those admitting a compatible ultrametric.

Nagata [**30**] introduced a counterpart of the notion of ultrametric: a metric d on a set X is said to satisfy property $(*)_n$ if, for every $x, y_1, \dots, y_{n+2} \in X$, there exist i, j, $i \neq j$, such that $d(y_i, y_j) \leq d(x, y_i)$.

Question 1.6. *Is every proper metric space (X,d) with* $\operatorname{asdim} X \leq n$ *coarsely equivalent to a proper metric space whose metric satisfies $(as)_n$?* 1400?

There are another classes of metrics that characterize covering dimension (see, e.g., [**27**]).

Question 1.7. *Are there metrics that characterize as above the asymptotic dimension $n \geq 1$?* 1401?

2. Extension of metrics

The problem of existence of linear regular operators (i.e., operators of norm 1 that preserve linear combination with nonnegative coefficients), extending (pseudo)-metrics was formulated by C. Bessaga [**14**] and solved by T. Banakh [**2**].

Question 2.1. *Is there a linear operator that extends metrics from a compact metrizable space X to left invariant metrics on a free topological group of X?* 1402?

A similar question can be formulated for extension of metrics from a compact metrizable space X to norms on the free linear space over X.

Let (X,d) be a compact metric space. Given a subset A of X, we say that a pseudometric ϱ on A is *Lipschitz* if there is $C>0$ such that $d(x,y) \leq C\varrho(x,y)$, for any $x,y \in A$. Also, a function $f\colon A \to \mathbb{R}$ is *Lipschitz* if there is $C>0$ such that $|f(x)-f(y)| \leq Cd(x,y)$, for every $x,y \in A$. Denote by $\operatorname{lpm}(A)$ (resp. $\operatorname{lpf}(A)$) the set of all Lipschitz pseudometrics (resp. functions) on A. The set $\operatorname{lpm}(A)$ (resp. $\operatorname{lpf}(A)$) is a cone (resp. linear space) with respect to the operations of pointwise addition and multiplication by scalar. We endow $\operatorname{lpm}(A)$ with the norm $\|\cdot\|_{\operatorname{lpm}(A)}$,

$$\|\varrho\|_{\operatorname{lpm}(A)} = \sup\left\{\frac{\varrho(x,y)}{d(x,y)} : x \neq y\right\}$$

and $\operatorname{lpf}(A)$ with the seminorm $\|\cdot\|_{\operatorname{lpf}(A)}$,

$$\|f\|_{\operatorname{lpf}(A)} = \sup\left\{\frac{|f(x)-f(y)|}{d(x,y)} : x \neq y\right\}.$$

We say that a map $u\colon \operatorname{lpm}(A) \to \operatorname{lpm}(X)$ is an *extension operator* for Lipschitz pseudometrics if the following holds:

(1) u is linear (i.e., $u(\varrho_1+\varrho_2) = u(\varrho_1)+u(\varrho_2)$, $u(\lambda\varrho) = \lambda u(\varrho)$ for every $\varrho, \varrho_1, \varrho_2 \in \operatorname{lpm}(A)$, $\lambda \in \mathbb{R}_+$);
(2) $u(\varrho)|(A \times A) = \varrho$, for every $\varrho \in \operatorname{lpm}(A)$;
(3) u is continuous in the sense that $\|u\| = \sup\{\|u(\varrho)\|_{\operatorname{lpm}(X)} : \|\varrho\|_{\operatorname{lpm}(A)} \leq 1\}$ is finite.

This definition is a natural counterpart of those introduced in [**15**] for the extensions of Lipschitz functions. The following notation is introduced in [**15**]:

$$\lambda(A,X) = \inf\{\|u\| : u \text{ is a linear extension operator from } \operatorname{lpf}(A) \text{ to } \operatorname{lpf}(X)\}.$$

Similarly, we put

$$\Lambda(A,X) = \inf\{\|u\| : u \text{ is a linear extension operator from } \mathrm{lpm}(A) \text{ to } \mathrm{lpm}(X)\}.$$

In [**15**] the problem of existence of extension operators of Lipschitz functions is considered. It is natural to formulate the corresponding problem for pseudometrics.

1403? **Question 2.2.** *Let A be a closed subspace of a compact metric space X. Is there an extension operator for Lipschitz pseudometrics $u\colon \mathrm{lpm}(A) \to \mathrm{lpm}(X)$?*

1404? **Question 2.3.** *Compare $\Lambda(S,X)$ and $\lambda(S,X)$.*

3. Questions in general topology

All topological spaces in this section are assumed to be Hausdorff, see [**5, 6**] for undefined notions used below.

1405–1408? **Question 3.1.** *Is there an interplay between topological properties of a compact topological inverse semigroup S and those of the maximal Clifford semigroup $C \subset S$ and the maximal sublattice E? In particular:*

(a) *Is S countably cellular (or separable) if so is the space C?*
(b) *Is S countably cellular if the maximal semilattice E is second countable?*
(c) *Is S (hereditary) separable if all maximal groups of S are (hereditary) separable and the maximal semilattice is Lawson and (hereditary) separable?*
(d) *Is S fragmentable (resp. Corson, Eberlein, Gul'ko, Radon–Nikodým, or Rosenthal) compact if so is the Clifford semigroup C?*

By a *mean* on a space X we understand any commutative idempotent operation $m\colon X \times X \to X$. Associative means are also called *semilattice operations*. Each scattered metrizable compact space, being homeomorphic to an ordinal interval $[0,\alpha]$, admits a continuous associative mean (just take the operations min or max on $[0,\alpha]$).

1409? **Question 3.2.** *Does any scattered compact Hausdorff space X admit a (separately) continuous mean?*

It should be noted that there exist scattered compact Hausdorff spaces admitting no separately continuous *associative* mean, see [**7**].

Our other question is due to V. Maslyuchenko, V. Mykhaylyuk and O. Sobchuk and relates to the classical theorem of Baire on functions of the first Baire class. We recall that a function $f\colon X \to Y$ between topological spaces is called

- *of the first Baire class* if f is the pointwise limit of a sequence of continuous functions;
- *F_σ-measurable* if the preimage $f^{-1}(U)$ of any open set $U \subset Y$ is of type F_σ in X.

It is well-known that each function $f\colon X \to Y$ of the first Baire class with values in a perfectly normal space is F_σ-measurable. The converse is true if X is metrizable and the space Y is metrizable, separable, connected and locally path-connected, see [**24**, **37**].

Question 3.3. *Is each F_σ-measurable function $f\colon [0,1] \to C_p[0,1]$ a function of the first Baire class?* 1410?

This question is equivalent to the original question of V. Maslyuchenko, V. Mykhaylyuk, and O. Sobchuk [**29**]:

Question 3.4. *Let $f\colon [0,1] \times [0,1] \to \mathbb{R}$ be a function continuous with respect to the first variable and of the first Baire class with respect to the second variable. Is f the pointwise limit of separately continuous functions?* 1411?

Let $\mathcal{P}$ be a property of a subset in a topological space. A topological space is called an *$A\mathcal{P}$-space* (resp. *$WA\mathcal{P}$-space*) if for every subset $B \subset X$ and every (resp. some) point $x \in \bar{B} \setminus B$ there exists a subset $C \subset B$ with the property $\mathcal{P}$ in X such that $x \in \bar{C}$.

For example, a space X has countable tightness iff it is an $A\mathcal{P}$-space for the property $\mathcal{P}$ of being a countable subset. A space X is Fréchet–Urysohn (resp. sequential) if and only if X is an $A\mathcal{P}$-space (resp. $WA\mathcal{P}$-space) where $\mathcal{P}$ is the property of a subset $A \subset X$ to have compact metrizable closure. A space X is a *k'-space* in the sense of Arhangel'skiĭ [**1**] if and only if X is an $A\overline{\mathcal{C}}$-space where $\overline{\mathcal{C}}$ is the property of a subset $A \subset X$ to have compact closure in X.

Question 3.5. *Find an example of a countably compact $WA\overline{\mathcal{C}}$-space which is not an $A\overline{\mathcal{C}}$-space.* 1412?

Let $\mathcal{D}$ (resp. $\mathcal{M}$) denote the properties of a subspace to be discrete (resp. metrizable).

Question 3.6. *Is every topological group of countable tightness an $A\mathcal{M}$-space? $A\mathcal{D}$-space?* 1413?

Question 3.7. *Let X be an $A\mathcal{M}$-space. Is the free topological group of X an $A\mathcal{M}$-space?* 1414?

Question 3.8. *Characterize the class of monothetic $A\mathcal{M}$-groups ($A\mathcal{M}$-paratopological groups)?* 1415?

Question 3.9. *Is every countable regular space an $A\mathcal{M}$-space?* 1416?

4. Some problems in Ramsey theory

In this section we ask some problems on symmetric subsets in colorings of groups. By an r-coloring of a set X we understand any map $\chi\colon X \to \{1, \ldots, r\}$, which can be identified with a partition $X = \bigcup_r X_i$ of X into r disjoint pieces $X_i = \chi^{-1}(i)$. As a motivation for subsequent questions let us mention the following result of T. Banakh [**3**].

Theorem C. *For any n-coloring of the group $\mathbb{Z}^n$ there is an infinite monochromatic subset $S \subset \mathbb{Z}^n$ symmetric with respect to some point $c \in \{0,1\}^n$.*

A subset S of a group G is called *symmetric* with respect to a point $c \in G$ if $S = cS^{-1}c$.

This theorem suggests to introduce the cardinal function $\nu(G)$ assigning to each group G the smallest cardinal number r of colors for which there is an r-coloring of G without infinite monochromatic symmetric subsets.

In [**8**] the value $\nu(G)$ was calculated for any abelian group G:

$$\nu(G) = \begin{cases} r_0(G)+1 & \text{if } G \text{ is finitely generated} \\ r_0(G)+2 & \text{if } G \text{ is infinitely generated and } |G[2]| < \aleph_0 \\ \max\{|G_2|, \log|G|\} & \text{if } |G[2]| \geq \aleph_0 \end{cases}$$

where $r_0(G)$ is the free rank of G and $G[2] = \{x \in G : 2x = 0\}$ is the Boolean subgroup of G.

Much less is known for non-commutative groups.

1417? **Question 4.1.** *Investigate the cardinal $\nu(G)$ for non-commutative groups G. In particular, is $\nu(F_2)$ finite for the free group F_2 with two generators?*

The only information on $\nu(F_2)$ is that $\nu(F_2) > 2$, see [**26**].

1418? **Question 4.2.** *Has each finite coloring of an infinite group G a monochromatic symmetric subset $S \subset G$ of arbitrarily large finite size?* (The answer is affirmative if G is Abelian.)

For every uncountable abelian group G with $|G[2]| < |G|$ there is a 2-coloring of G without symmetric monochromatic subsets of size $|G|$, see [**31**].

1419? **Question 4.3.** *Is it true that for every 2-coloring of an uncountable abelian group G with $|G[2]| < |G|$ and for every cardinal $\kappa < |G|$ there is a monochromatic symmetric subset $S \subset G$ of size $|S| \geq \kappa$?* (The answer is affirmative under GCH, see [**26**].)

There is another interesting concept suggested by Theorem C on colorings of the group $\mathbb{Z}^n$. Let us define a subset $C \subset \mathbb{Z}^n$ to be *central* if for any n-coloring of $\mathbb{Z}^n$ there is an infinite monochromatic subset $S \subset \mathbb{Z}^n$ symmetric with respect to a point $c \in C$. A central set $C \subset \mathbb{Z}^n$ is called *minimal* if it does not lie in any smaller central set.

1420? **Question 4.4.** *Describe the geometric structure of (minimal) central subsets of $\mathbb{Z}^n$. Is each minimal central subset of $\mathbb{Z}^n$ finite? What is the smallest size $c(\mathbb{Z}^n)$ of a central set in $\mathbb{Z}^n$?*

It was proved in [**4**] that $\frac{n(n+1)}{2} \leq c(\mathbb{Z}^n) < 2^n$ and $c(\mathbb{Z}^n) = \frac{n(n+1)}{2}$ for $n \leq 3$.

1421? **Question 4.5.** *Calculate the number $c(\mathbb{Z}^4)$.* (It is known that $12 \leq c(\mathbb{Z}^4) \leq 14$, see [**4**].)

Concerning the first (geometric) part of Question 4.4 the following information is available for small n, see [**4**]:

(1) a subset $C \subset \mathbb{Z}$ is central if and only if C contains a point;
(2) a subset $C \subset \mathbb{Z}^2$ is central if and only if it contains a triangle $\{a, b, c\} \subset C$ (by which we understand a three-element affinely independent subset of $\mathbb{Z}^2$);
(3) each central subset $C \subset \mathbb{Z}^3$ of size $|C| = c(\mathbb{Z}^3) = 6$ is an octahedron $\{c \pm e_i : i \in \{1, 2, 3\}\}$ where $c \in \mathbb{Z}^n$ and $e_1, e_2, e_3 \in \mathbb{Z}^n$ are linearly independent vectors;
(4) there is a minimal central subset $C \subset \mathbb{Z}^3$ of size $|C| > 6$ containing no octahedron.

There is another numerical invariant $\mathrm{ms}(X, \mathcal{S}, r)$ related to colorings and defined for any space X endowed with a probability measure μ and a family $\mathcal{S}$ of measurable sets called symmetric subsets of X. By definition, $\mathrm{ms}(X, \mathcal{S}, r) = \inf\{\varepsilon > 0 :$ for every measurable r-coloring of X there is a monochromatic subset $S \in \mathcal{S}$ of measure $\mu(S) \geq \varepsilon\}$. The notation "ms" reads as the *maximal measure of a monochromatic symmetric subset* and was suggested by Ya. Vorobets. If the family $\mathcal{S}$ is clear from the context (as it is in case of groups), then we shall write $\mathrm{ms}(X, r)$ instead of $\mathrm{ms}(X, \mathcal{S}, r)$.

The numerical invariant $\mathrm{ms}(X, r)$ is defined for many natural algebraic and geometric objects: compact topological groups, spheres, balls etc. For such objects, typically, $\mathrm{ms}(X, r)$ is equal to $\frac{1}{r^2}$, see [**9, 11**]. For example, in the case of the ball B^n of the unit volume in the Euclidean space $\mathbb{R}^n$ of dimension $n \geq 2$ we get $\mathrm{ms}(B^n, \mathcal{S}, r) = \frac{1}{r^2}$ for any family $\mathcal{S}$ with $\mathcal{S}_0 \subset \mathcal{S} \subset \mathcal{S}_+$ where $\mathcal{S}_+$ is the family of measurable subsets of B^n that are symmetric with respect to some non-trivial isometry of $\mathbb{R}^n$ and $\mathcal{S}_0$ is the family of measurable subsets of B^n, symmetric with respect to some hyperplane passing through the center of the ball.

Moreover, for any measurable r-coloring of the ball B^n of dimension $n \geq 3$ there is a monochromatic subset $S \in \mathcal{S}_0$ of measure $> \frac{1}{r^2}$. This phenomenon does not hold in dimension 2: there is a 2-coloring of the two-dimensional disk B^2 such that all monochromatic symmetric subsets $S \in \mathcal{S}_0$ of B^2 have measure $\leq \frac{1}{4}$ (such an extremal coloring of the disk resembles the Chinese philosophical symbol "in-jan", see [**11**]). The situation with the 1-dimensional ball $[0, 1]$ is even worse: we known that $\frac{1}{r^2+r\sqrt{r^2-r}} \leq \mathrm{ms}([0, 1], \mathcal{S}_+, r) < \frac{1}{r^2}$ for $r > 1$ but the exact value of $\mathrm{ms}([0, 1], \mathcal{S}_+, r)$ is not known even for $r = 2$. However, we have some lower and upper bounds: $\frac{1}{4+\sqrt{6}} \leq \mathrm{ms}([0, 1], \mathcal{S}_+, 2) < \frac{5}{24}$, see [**11, 28**] for more information. Observe that for $n = 1$ the family $\mathcal{S}_+$ coincides with the family of subsets of $[0, 1]$, symmetric with respect to some point of $[0, 1]$. So, we shall write $\mathrm{ms}([0, 1], r)$ instead of $\mathrm{ms}([0, 1], \mathcal{S}_+, r)$.

Question 4.6. *Calculate the value* $ms([0, 1], r)$*, at least for* $r = 2$*. Can* $\mathrm{ms}([0, 1], 2)$ *be expressed via some known mathematic constants?* 1422?

It was proved in [**11**] that the limit $\lim_{r\to\infty} r^2 \cdot \mathrm{ms}([0, 1], r)$ exists and lies in the interval $[\frac{1}{2}, \frac{5}{6}]$.

1423? **Question 4.7.** *Calculate the constant* $c = \lim_{r\to\infty} r^2 \cdot \mathrm{ms}([0,1], r)$.

More detail information on these problems can be found in the surveys [**9**, **10**].

5. Questions on functors in the category of compact Hausdorff spaces

We denote by **Comp** the category of compact Hausdorff spaces and continuous maps. In the sequel, all the functors are assumed to be covariant endofunctors in **Comp**.

First, we mention few examples of functors. The *hyperspace functor* exp assigns to every compact Hausdorff space X the set $\exp X$ of nonempty closed subsets in X endowed with the Vietoris topology. A base of this topology consists of the sets of the form $\langle U_1, \dots, U_n\rangle = \{A \in \exp X : A \subset \bigcup_n U_i,\ (\forall i) A \cap U_i \neq \emptyset\}$, where $U_1, \dots, U_n$ are open subsets in X. Given a map $f\colon X \to Y$ in **Comp**, the map $\exp f\colon \exp X \to \exp Y$ is defined by $\exp f(A) = f(A)$.

The *probability measure functor* P assigns to every compact Hausdorff space X the set $P(X)$ of probability measures endowed with the weak* topology.

Let G be a subgroup of the permutation group S_n. The *G-symmetric power of* X, $SP^n_G(X)$, is the quotient space of X^n with respect to the natural action of G on X^n by permutation of coordinates. One can easily see that this construction determines a functor.

Some properties of the mentioned functors and other known functors were used by E.V. Shchepin [**35**] in order to introduce the notion of a normal functor. It turned out that normal functors, and some close to normal functors, found important applications in the topology of nonmetrizable compact Hausdorff spaces and other areas of topology (see, e.g., [**23**, **35**, **36**]).

If a functor F preserves embeddings, then, for a compact Hausdorff space X and a closed subspace A of X, we always identify the space $F(A)$ with a subspace in $F(X)$ along the embedding $F(i)$, where $i\colon A \to X$ is the inclusion map.

Let a functor F preserve embeddings. We say that F *preserves preimages* if $F\left(f^{-1}(A)\right) = F(f)^{-1}\left(F(A)\right)$ for every map $f\colon X \to Y$ and every closed subset A of Y. We say that F *preserves intersections* whenever $F\left(\bigcap_{\alpha\in\Gamma} A_\alpha\right) = \bigcap_{\alpha\in\Gamma} F(A_\alpha)$ for every family of closed subsets $\{A_\alpha : \alpha \in \Gamma\}$ in X.

An endofunctor F in **Comp** is called *normal* (Shchepin [**35**]) if F preserves embeddings, surjections, weight of infinite compacta, intersections, preimages, singletons, the empty set, and the limits of inverse systems $\mathcal{S} = \{X_\alpha, p_{\alpha\beta}; \mathcal{A}\}$ over directed sets $\mathcal{A}$. More precisely, the latter condition means that the map $h = (F(p_\alpha))_{\alpha\in\mathcal{A}}$ is a homeomorphism of $F(\varprojlim \mathcal{S})$ onto $\varprojlim F(\mathcal{S})$, where $p_\alpha\colon \varprojlim \mathcal{S} \to X_\alpha$ is the limit projection.

A functor F is said to be *weakly normal (almost normal)* if it satisfies all the properties from the previous definition except perhaps the property of being epimorphic (respectively, the preimage preserving property).

The hyperspace functor exp, the probability measure functor P, and the G-symmetric power functor SP^n_G are examples of normal functors.

For a functor F and a compact Hausdorff space X denote by $F_n(X)$ the subspace $\bigcup\{F(f)\left(F(n)\right) : f \in C(n, X)\}$ of $F(X)$ (here $C(n, X)$ denotes the set of all maps from the discrete space n to X). Clearly, such a construction determines a subfunctor F_n of F. A functor F is of *finite degree* if there exists $n \in \mathbb{N}$ such that $F = F_n$.

If $\varphi = (\varphi_X)\colon F \to F'$ is a natural transformation of functors then we say that F is a *subfunctor* of F' if all the components of φ are inclusion maps and we say that F' is a *quotient functor* of F if all the components of φ are onto maps.

The characteristic map of a commutative diagram

$$\begin{array}{ccc} X & \xrightarrow{f} & Y \\ {\scriptstyle g}\downarrow & & \downarrow{\scriptstyle h} \\ Z & \xrightarrow{u} & T \end{array}$$

in the category **Comp** is the map $\chi\colon X \to Y \times_T Z = \{(y, z) \in Y \times Z : h(y) = u(z)\}$ defined by the formula $\chi(x) = (f(x), g(x))$. A diagram is bicommutative if its characteristic map is onto. A diagram is open-bicommutative if its characteristic map is open and onto.

A functor $F\colon$ **Comp** $\to$ **Comp** is said to be *bicommutative* (resp. *open-bicommutative*) if F preserves the class of bicommutative (resp. open-bicommutative) diagrams.

A functor is *open* if it preserves the class of open surjective maps. E.V. Shchepin proved that every open functor is bicommutative.

Question 5.1 (Shchepin). *Is every normal bicommutative functor open?* 1424?

This problem was formulated more than 25 years ago. The notions of open and bicommutative functors were introduced by E.V. Shchepin [**35**]. The problem was solved in [**40**] for normal functors of finite degree.

Question 5.2. *Is every normal bicommutative (open) functor open-bicommutative?* 1425?

It is proved in [**36**] that natural transformations of (weakly, almost) normal functors form a set and therefore one can introduce the category of normal functors and their natural transformations.

A (weakly, almost) normal functor F is called *universal* if every (weakly, almost) normal functor is isomorphic to a subfunctor of F.

Question 5.3. *Is there a universal (weakly, almost) normal functor?* 1426?

A normal functor F is called *couniversal* if every normal functor F' is a quotient functor of F.

Question 5.4. *Is there a couniversal (weakly, almost) normal functor?* 1427?

A normal functor F is called *zero-dimensional* if $\dim F(X) = 0$ for every compact Hausdorff space X with $\dim X = 0$.

Question 5.5. *Is every normal functor a quotient functor of a zero-dimensional* 1428?
normal functor?

Let $\tau > \omega$ be a cardinal number. A functor F is called *τ-normal* if F satisfies all the properties from the definition of normality except the preserving of weight and, in addition, the weight of $F(X)$ is $\leq \tau$, for every compact metrizable X (the minimal τ for which a functor F is τ-normal is called the *weight* of F.

Actually, one can find the prototype of the notion of τ-normal functor in Shchepin's paper [**35**] as he considered the so-called normal functor-powers, i.e., the spaces of the form $F(X^\tau)$.

We say that a map $f\colon X \to Y$ satisfies the *homeomorphism-lifting property* if, for every homeomorphism $h\colon Y \to Y$ there exists a homeomorphism $h'\colon X \to X$ such that $fh' = hf$.

1429? **Question 5.6** (Shchepin). *Let X be a metric compact space and F a normal functor. Does the map $F\left((pr)^\tau\right) : F\left((X \times X)^\tau\right) \to F(X^\tau)$ satisfy the homeomorphism-lifting property?*

1430? **Question 5.7.** *Is every multiplicative τ-normal functor isomorphic to the power functor $(\cdot)^\tau$?*

For normal functors, this problem was posed by Shchepin and solved in [**38**].

Shchepin proved the so-called spectral theorem, which states that, under some reasonable conditions, if a nonmetrizable compact Hausdorff space is represented as the inverse limit of two systems consisting of spaces of smaller weight then these systems contain isomorphic cofinal subsystems (see [**35**] for details). One can consider representations of τ-normal functors as the limits of inverse systems consisting of functors of smaller weight and their natural transformations.

1431? **Question 5.8.** *Is there a counterpart of Shchepin's spectral theorem in the category of τ-normal functors?*

Of special interest are functors of finite degree that preserve the class of compact metric ANR spaces (i.e., absolute neighborhood retracts). Basmanov [**12**] established such a property for a wide enough class of functors. Such functors are known to preserve other classes of spaces too: Q-manifolds (i.e., manifolds modeled on the Hilbert cube $Q = [0,1]^\omega$) [**23**], n-movable spaces [**39**], compact metric absolute neighborhood extensors in dimension n [**16**].

We are going to formulate a few questions on the preservation of some geometric properties by functors of finite degree.

Let P be a CW-complex. For any compact metric space X the *Kuratowski notation* $X\,\tau\,P$ means the following: for every continuous map $f\colon A \to P$ defined on a closed subset A of X there is a continuous extension of f onto X.

Denote by $\mathcal{L}$ the class of all countable CW-complexes. Following [**17**], we define a preorder relation $\leq$ on $\mathcal{L}$. For $L_1, L_2 \in \mathcal{L}$, we have $L_1 \leq L_2$ if and only if $X\,\tau\,L_1$ implies $X\,\tau\,L_2$ for all compact metric spaces X. This preorder relation determines the following equivalence relation $\sim$ on $\mathcal{L}$: $L_1 \sim L_2$ if and only if $L_1 \leq L_2$ and $L_2 \leq L_1$. We denote by $[L]$ the equivalence class containing $L \in \mathcal{L}$.

For a compact metric space X, we say that its *extension dimension does not exceed* $[L]$ (briefly ext-dim $X \leq [L]$) whenever $X\,\tau\,L$.

A compact metric space X is said to be an absolute (neighborhood) extensor in extension dimension $[L]$ if for any compact metric pair (A, B) with ext-dim $A \leq [L]$ and any continuous map $f\colon B \to X$ there exists a continuous extension $\bar{f}\colon A \to X$ (respectively $\bar{f}\colon U \to X$, where U is a neighborhood of B in A) of f.

In the sequel, we suppose that F is a normal functor of finite degree that preserves the class of compact metrizable ANR-spaces.

Question 5.9. *Does F preserve the class of absolute (neighborhood) extensors in extension dimension $[L]$?* 1432?

Two maps $f_0, f_1\colon X \to Y$ are said to be $[L]$-*homotopic* if there exists a space Z with ext-dim $Z \leq [L]$, a map $\alpha\colon Z \to X \times [0,1]$ which is $[L]$-invertible (i.e., satisfies the property of lifting of maps from spaces of extension dimension $\leq [L]$), and a map $H\colon Z \to Y$ such that $f_i\alpha(z) = H(z)$, for every $z \in \alpha^{-1}(X \times \{i\})$, $i = \{0, 1\}$.

Question 5.10. *Does F preserve the relation of $[L]$-homotopy of maps?* 1433?

We finish with the following question.

Question 5.11. *Does F preserve the class of essential Q-M-factors, i.e., the class of spaces X such that $X \times A$ is a Q-manifold for some A with* $\dim A < \infty$? 1434?

References

[1] A. V. Arhangel′skiĭ, *Bicompact sets and the topology of spaces.*, Trudy Moskov. Mat. Obshch. **13** (1965), 3–55, Translation: Trans. Moscow Math. Soc. 13 (1965) 1–62.

[2] T. Banakh, *AE(0)-spaces and regular operators extending (averaging) pseudometrics*, Bull. Polish Acad. Sci. Math. **42** (1994), no. 3, 197–206.

[3] T. Banakh, *Solutions of certain problems of I. V. Protasov from the combinatorics of colorings*, Svītī Mat. **3** (1997), no. 1, 8–11.

[4] T. Banakh, *On a cardinal group invariant related to partitions of abelian groups*, Mat. Zametki **64** (1998), no. 3, 341–350, Translation: Math. Notes 64 (1998) no. 3, 295–302.

[5] T. Banakh, *On cardinal invariants and metrizability of topological inverse Clifford semigroups*, Topology Appl. **128** (2003), no. 1, 13–48.

[6] T. Banakh and B. Bokalo, *On cardinal invariants and metrizability of topological inverse semigroups*, Topology Appl. **128** (2003), no. 1, 3–12.

[7] T. Banakh, O. V. Gutik, and M. Rajagopalan, *Compatible algebraic structures on scattered compacta*, Topology Appl. **153** (2005), no. 5–6, 710–723.

[8] T. Banakh and I. V. Protasov, *Asymmetric partitions of abelian groups*, Mat. Zametki **66** (1999), no. 1, 10–19, Translation: Math. Notes 66 (1999) no. 1, 8–15.

[9] T. Banakh and I. V. Protasov, *Symmetry and colorings: some results and open problems*, Izv. Gomel Univ. Voprosy Algebry. **4** (2001), no. 17, 5–16.

[10] T. Banakh, O. Verbitsky, and Ya. Vorobets, *A Ramsey treatment of symmetry*, 2000, Research Paper 52, p. 25 pp.

[11] T. Banakh, Ya. B. Vorobets, and O. V. Verbitskiĭ, *Ramsey-type problems for spaces with symmetries*, Izv. Ross. Akad. Nauk Ser. Mat. **64** (2000), no. 6, 3–40.

[12] V. N. Basmanov, *Covariant functors, retracts, and dimension*, Dokl. Akad. Nauk SSSR **271** (1983), no. 5, 1033–1036, Translation: Soviet Math. Dokl. 28 (1983) 182–185.

[13] G. Bell and A. Dranishnikov, *On asymptotic dimension of groups*, Algebr. Geom. Topol. **1** (2001), 57–71.

[14] C. Bessaga, *Functional analytic aspects of geometry. Linear extending of metrics and related problems*, Progress in functional analysis (Peñíscola, 1990), North-Holland, Amsterdam, 1992, pp. 247–257.

[15] A. Brudnyi and Yu. Brudnyi, *Metric spaces with linear extensions preserving Lipschitz condition*, 2004, Preprint. `arXiv:math.MG/0404304`

[16] A. Dranishnikov, *Covariant functors and extensors in dimension n*, Uspekhi Mat. Nauk **40** (1985), no. 6(246), 133–134, Translation: Russian Math. Surveys 40 (1986) no. 6, 119–120.

[17] A. Dranishnikov, *Extension of mappings into CW-complexes*, Mat. Sb. **182** (1991), no. 9, 1300–1310, Translation: Sb. Math. 74 (1993) no. 1, 47–56.

[18] A. Dranishnikov, *Asymptotic topology*, Uspekhi Mat. Nauk **55** (2000), no. 6(336), 71–116, Translation: Russian Math. Surveys 55 (2000) no. 6, 1085–1129.

[19] A. Dranishnikov, *On asymptotic inductive dimension*, JP J. Geom. Topol. **1** (2001), no. 3, 239–247.

[20] A. Dranishnikov and M. Zarichnyi, *Universal spaces for asymptotic dimension*, Topology Appl. **140** (2004), no. 2–3, 203–225.

[21] V. Egorov and Ju. Podstavkin, *A definition of dimension*, Dokl. Akad. Nauk SSSR **178** (1968), 774–777, Translation: Soviet Math. Dokl. 9 (1968) 188–191.

[22] R. Engelking, *Theory of dimensions finite and infinite*, Heldermann Verlag, Lemgo, 1995.

[23] V. V. Fedorchuk, *Covariant functors in a category of compacta, absolute retracts and Q-manifolds*, Uspekhi Mat. Nauk **36** (1981), no. 3(219), 177–195, Translation: Russian Math. Surveys 36 (1981) no. 3, 211–233.

[24] M. Fosgerau, *When are Borel functions Baire functions?*, Fund. Math. **143** (1993), no. 2, 137–152.

[25] M. Gromov, *Asymptotic invariants of infinite groups*, Geometric group theory, Vol. 2 (Sussex, 1991), Cambridge University Press, Cambridge, 1993, pp. 1–295.

[26] Yu. Gryshko and A. Khelif, *Monochrome symmetric subsets of colored groups*, J. Combin. Theory Ser. A **112** (2005), no. 2, 212–221.

[27] Y. Hattori, *On special metrics characterizing topological properties*, Fund. Math. **126** (1986), no. 2, 133–145.

[28] G. Martin and K. O'Bryant, *Continuous Ramsey Theory and Sidon sets*, Preprint. `arXiv:math.NT/0210041`

[29] V. K. Maslyuchenko, V. V. Mikhaĭlyuk, and O. V. Sobchuk, *Construction of a separately continuous function of n variables with a given diagonal*, Mat. Stud. **12** (1999), no. 1, 101–107.

[30] J. Nagata, *On a special metric and dimension*, Fund. Math. **55** (1964), 181–194.

[31] I. V. Protasov, *Asymmetrically decomposable abelian groups*, Mat. Zametki **59** (1996), no. 3, 468–471, Translation: Math. Notes 59 (1996) no. 3, 98–99.

[32] T. Radul, *Addition and subspace theorems for asymptotic large inductive dimension*, Colloq. Math. **106** (2006), no. 1, 57–67.

[33] T. Radul, *On transfinite extension of asymptotic dimension*, 2006, Submitted. `arXiv:math.GN/0601235`

[34] J. Roe, *Coarse cohomology and index theory on complete Riemannian manifolds*, Mem. Amer. Math. Soc. **104** (1993), no. 497, 90 pp.

[35] E. V. Ščepin, *Functors and uncountable powers of compacta*, Uspekhi Mat. Nauk **36** (1981), no. 3(219), 3–62, Translation: Russian Math. Surveys 36 (1981) no. 3, 1–71.

[36] A. Teleiko and M. Zarichnyi, *Categorical topology of compact Hausdorff spaces*, VNTL Publishers, Lviv, 1999.

[37] L. Veselý, *Characterization of Baire-one functions between topological spaces*, Acta Univ. Carolin. Math. Phys. **33** (1992), no. 2, 143–156.

[38] M. M. Zarichnyi, *The multiplicative normal functor is a power functor*, Mat. Zametki **41** (1987), no. 1, 93–100, Translation: Math. Notes 41 (1987) no. 1, 58–61.

[39] M. M. Zarichnyi, *Preservation of some shape properties by functors in the category of compacta*, Bull. Polish Acad. Sci. Math. **39** (1991), no. 3–4, 235–239.

[40] M. M. Zarichnyi, *Characterization of functors of G-symmetric degree and extensions of functors to the Kleisli categories*, Mat. Zametki **52** (1992), no. 5, 42–48, Translation: Math. Notes 52 (1992) no. 5, 1107–1111.

Open Problems in Topology II – Edited by E. Pearl

Problems from the Bizerte–Sfax–Tunis Seminar

Othman Echi, Habib Marzougui and Ezeddine Salhi

Introduction

The Bizerte–Sfax–Tunis seminar [BST Seminar "Algebra, Dynamical Systems and Topology"] is organized by two Tunisian research groups "Algebra and Topology 03/UR/03-15" and "Dynamical Systems and Combinatorics 99/UR/15-15". Three meetings are held each academic year in one of the Faculties of Sciences of Bizerte, Sfax or Tunis. The Seminar has been founded, firstly, by Professor Ezzeddine Salhi since 1996 and has been called Bizerte–Sfax Meeting. In March 2001, Othman Echi has got the position of Professor at Faculty of Sciences of Tunis; and so the seminar is shared by Bizerte, Sfax and Tunis. The goal of this seminar is to shed light on the latest results obtained by the members of the two groups (in the areas of algebra, algebraic topology, combinatorics, complex analysis, dynamical systems, foliation theory, topology). It is worth noting that an interesting link between foliation theory and spectral topology has been discovered by three members of the above two research groups (see [**4**]).

This note deals with eight problems in Topology which are proposed by our seminar. These problems concern spectral spaces and some related topics; the space of leaves of a foliation; dynamics of groups of homeomorphisms and vector fields on surfaces.

Spectral spaces and related topics

This section is devoted to some problems related to the prime spectrum of a unitary commutative ring (equipped with the Zariski topology). However, no background of Algebra is needed: each concept used, here, has a translation into a *general* topological property. But, to motivate the reader, we will explain a little the origins of concepts.

Let $\mathrm{Spec}(R)$ denote the set of prime ideals of a commutative ring R with identity. Recall that, the *Zariski topology* or the *hull-kernel topology* for $\mathrm{Spec}(R)$ is defined by letting $C \subseteq \mathrm{Spec}(R)$ be closed if and only if there exists an ideal $\mathcal{A}$ of R such that $C = \{\mathcal{P} \in \mathrm{Spec}(R) : \mathcal{P} \supseteq \mathcal{A}\}$. The topological question of characterizing *spectral spaces* (that is, topological spaces homeomorphic to the prime spectrum of a ring equipped with the Zariski topology) was completely answered by Hochster in [**13**] (they are, precisely, compact sober spaces that have a basis of compact open sets closed under finite intersections).

We would like to thank the DGRST for its partial support of our research groups "Algebra and Topology: 03/UR/03-15" and "Dynamical Systems and Combinatorics: 99/UR/15-15".

Following Hochster [**13**], a continuous map of spectral spaces is said to be *spectral* if inverse images of compact open sets are compact. Hochster has extended the notion of spectral map to a larger class of spaces. Call a space *semispectral* if the intersection of any two compact open subsets is compact. Call an open subset U of a topological space X *intersection compact open*, or ICO, if for every compact open set Q of X, $U \cap Q$ is compact. Thus, X is semispectral if and only if the compact open sets are ICO. Then a continuous map f of semispectral spaces will be called *spectral* if f^{-1} carries ICO sets to ICO sets.

Recall that a topological space X is said to be a *Jacobson space* if the set $\mathcal{C}(X)$ of all closed points of X is strongly dense in X [**10**] (a subset of X is said to be *strongly dense* if it meets every nonempty locally closed subset of X). Obviously, when X is a topological space, $\mathrm{Jac}(X) = \{x \in X : \overline{\{x\}} = \overline{\overline{\{x\}} \cap \mathcal{C}(X)}\}$ is a Jacobson space; we call it the *Jacobson subspace* of X. It is easily seen that $\mathrm{Jac}(X)$ is the largest subset of X in which $\mathcal{C}(X)$ is strongly dense.

Let R be a ring, we denote by $\mathrm{Jac}(R)$ the Jacobson subspace of $\mathrm{Spec}(R)$. A prime ideal $\mathcal{P}$ of R is in $\mathrm{Jac}(R)$ if and only if it is the intersection of all maximal ideals $\mathcal{M}$ of R such that $\mathcal{P} \subseteq \mathcal{M}$. Jacobson spaces are linked with Hilbert rings; a ring in which every prime ideal is an intersection of some maximal ideals is called a *Hilbert ring* (or, also, a *Jacobson ring*). Clearly, a ring R is a Hilbert ring if and only if $\mathrm{Spec}(R)$ endowed with the hull-kernel topology is a Jacobson space.

According to [**2**], a *jacspectral space* is defined to be a topological space homeomorphic to the Jacobson space of $\mathrm{Spec}(R)$ for some ring R. The authors of [**2**] have given a topological characterization of jacspectral spaces(they are, precisely, compact Jacobson sober spaces).

Goldman ideals are important objects of investigation in algebra mostly because their role in the study of graded rings and some applications to algebraic geometry. Thus it is important to pay attention to the Goldman prime spectrum $\mathrm{Gold}(R)$ of a ring R.

Topologically speaking, G-ideals of a ring are the locally closed points for the hull-kernel topology.

It is worth noting that G-ideals have been used separately by Goldman [**9**] and Krull [**14**] for a short inductive proof of the Nullstellensatz.

Following [**7**], a topological space X is said to be *goldspectral* if there exists a ring R such that X is homeomorphic to $\mathrm{Gold}(R)$ (equipped with the topology inherited by that of Zariski on $\mathrm{Spec}(R)$). The main result of [**7**], provides an intrinsic topological characterization of goldspectral spaces:

> A topological space X is goldspectral if and only if it is a compact T_D−space and has a basis of compact open sets which is closed under finite intersections.

In [**13**], Hochster has introduced the notion of spectralifiable space: A *spectralification* of a semispectral space X is a spectral embedding g of X into a spectral space X' such that for every spectral space Y and spectral map f from X to Y there is a unique spectral map f' from X' to Y such that $f = f' \circ g$. The space X is said to be *spectralifiable* if it has a spectralification [**13**]. When a semispectral space has a spectralification (in the sense of Hochster), we will say that it is *H-spectralifiable*. A complete characterization of H-spectralifiable spaces has been given in [**13**]: A semispectral space is H-spectralifiable if and only if it can be spectrally embedded in some spectral space; or equivalently, it is a T_0-space and the ICO sets are an open basis.

In [**1**], Ayache and Echi have introduced the notion of $\mathcal{D}$-ifiable objects of a category: Let $\mathcal{C}$ be a category, $\mathcal{D}$ a subcategory of $\mathcal{C}$ and X an object of $\mathcal{C}$. A *$\mathcal{D}$-ification* of X, is a morphism $p\colon X \to X'$ where X' is an object of $\mathcal{D}$ such that for each morphism $f\colon X \to Z$ (with $Z \in \mathrm{ob}(\mathcal{D})$ there exists a unique morphism $\widetilde{f}\colon X' \to Z$ satisfying $\widetilde{f} \circ p = f$. The object X is said to be *$\mathcal{D}$-ifiable* if it has a $\mathcal{D}$-ification. Hence the full subcategory of $\mathcal{C}$ whose objects are the $\mathcal{D}$-ifiable objects of $\mathcal{C}$ is the largest subcategory in which $\mathcal{D}$ is reflective.

The previous new concept allows us to state the following problems.

Let $\mathcal{S}$ be the subcategory of **TOP** consisting of spectral spaces and spectral maps. The semispectral spaces and spectral maps form a full subcategory $\mathcal{U}$ of $\mathcal{S}$.

Problem BST 1. *It is clear that every H-spectralifiable space of $\mathcal{U}$ is $\mathcal{S}$-ifiable.* 1435?
When is a semispectral space $\mathcal{S}$-ifiable?

Let $\mathcal{JS}$ be the full subcategory of **TOP** whose objects are jacspectral spaces.

Problem BST 2. *When is a topological space $\mathcal{JS}$-ifiable?* 1436?

Let $\mathcal{GS}$ be the full subcategory of $\mathcal{U}$ whose objects are goldspectral spaces with spectral maps as morphisms.

Problem BST 3. *When is a semispectral space $\mathcal{GS}$-ifiable?* 1437?

The space of leaves and the space of leaves classes

Let M be a closed connected manifold, $\mathcal{F}$ a 1-codimensional transversally oriented foliation on M, of class C^r $(r \geq 0)$. Let F be a leaf; we define $\mathrm{Cl}(F)$, the class of F, as the union of all leaves G of F such that $\overline{F} = \overline{G}$. Let $\sim$ be the equivalence relation defined on M by $x \sim y$ if and only if, $\overline{F_x} = \overline{G_y}$, where F_x and F_y are the leaves of $\mathcal{F}$ containing respectively x and y. The quotient space, denoted by $X = M/\overline{\mathcal{F}} := \{\mathrm{Cl}(F) : F \text{ is a leaf}\}$, is called the *space of leaves classes*, it is a T_0-space, however the space of leaves $Z = M/\mathcal{F}$ is not in general a T_0-space. More precisely, the space X is the T_0-identification of Z [**4**].

In [**3**], the authors have proved that, if $\mathcal{F}$ has a height and X_0 denotes the union of open sets of X which are homeomorphic either to $\mathbb{R}$ or to the unit circle S^1, then $X \setminus X_0$ is a spectral space.

We set the following problem:

Problem BST 4. *Give an intrinsic topological characterization of the spaces* 1438–1439?
$X = M/\overline{\mathcal{F}}$ and $Z = M/F$ in purely topological terms.

This is, in fact, a very hard problem. However, in [**3, 4**], some topological properties of the quotient spaces $X = M/\overline{\mathcal{F}}$ and $Z = M/F$ are given. Note also that the authors of [**12**] have investigate weaker version of the above problem.

Dynamics of groups of homeomorphisms

Firstly, let us note that for more details concerning the material of this section, one may see [**18**].

A homeomorphism $f\colon \mathbb{R}^p \to \mathbb{R}^p$ defined on the Euclidean space $\mathbb{R}^p$ is said to be *regular* if the group $\{f^n : n \in \mathbb{Z}\}$ generated by $\{f\}$ is equicontinuous at any point of $\mathbb{R}^p$; it is the case if f is periodic or an isometry. The subset $\{f^n(x) : n \in \mathbb{Z}\}$ is the orbit of f at the point $x \in \mathbb{R}^p$.

If $p = 2$ and if all orbits of f are bounded, then f is topologically equivalent to an isometry [**5**].

1440? **Problem BST 5.** *When is a regular homeomorphism $f\colon \mathbb{R}^p \to \mathbb{R}^p$ topologically equivalent to an isometry on the Euclidean space $\mathbb{R}^p$?*

We define on $\mathbb{R}^p$ a metric by: $d^*(x,y) = \sup\{d(f^n(x), f^n(y)) : n \in \mathbb{Z}\}$. One may check that this metric d^* is equivalent to the Euclidean metric d and that f is an isometry when $\mathbb{R}^p$ is equipped with this new metric.

1441? **Problem BST 6.** *Let $f\colon \mathbb{R}^p \to \mathbb{R}^p$ be a regular homeomorphism such that f is equal to the identity map on a nonempty open subset. Is f equal to the identity map on the whole space?*

The answer is positive in the following cases: f is an isometry; f is periodic (by Newman's theorem [**6**]); $p = 2$ (in this case, from [**5**], f is topologically equivalent to an isometry).

A homeomorphism $f\colon E \to E$ on a metric space is *pointwise periodic* if it is periodic at any point of E.

We can construct a pointwise periodic, nonperiodic, regular homeomorphism $f\colon E \to E$ defined on a compact, arcwise connected and locally arcwise connected metric subspace E of $\mathbb{R}^3$ [**8**].

1442? **Problem BST 7.** *Can one construct a pointwise periodic nonregular homeomorphism $f\colon E \to E$ such that E is a compact, arcwise connected and locally arcwise connected metric subspace E of $\mathbb{R}^2$?*

Note that, from Montgomery–Zippin [**16**], any pointwise periodic homeomorphism on a connected manifold is periodic and so it is regular.

Vector fields on surfaces

The problem stated in this section concerns the qualitative behavior of orbits of vector fields on surfaces. Let X be a vector field without singularities on an open orientable surface M. We call a *quasi-minimal set* the closure of a nontrivial recurrent orbit of X. A nontrivial recurrent orbit which is nowhere dense is called *exceptional*.

If M is of *finite* genus, we proved in [**15**] that every nonclosed orbit which is contained in a quasi-minimal set is dense in it, moreover, we gave a dynamic characterization of the limit set of any orbit by the following theorem:

Theorem H ([**15**]). *Let X be a vector field without singularities on an open orientable surface M of* finite *genus. The ω-limit (resp. α-limit) set of an orbit of X (if it is nonempty), is a compact orbit or the union of closed and noncompact orbits or a quasi-minimal set.*

If M is of *infinite* genus, the behaviour of orbits is more complex; for instance, the above results are false in general:

Nikolaev–Zhuzhoma constructed in [**17**] an example of vector field X with a dense orbit on a surface of infinite genus such that X has a an exceptional orbit which is not dense in M. Also, Gutierrez, Hector and Lopez constructed in [**11**] a vector field X without singularities on a surface of infinite genus having each one of the following nontrivial dynamics:

(1) nontrivial recurrent orbits are exceptional and the union of them is a dense set;
(2) X has dense orbits and exceptional orbits;
(3) existence of dense sequence of exceptional orbits $(O_k)_{k\geq 1}$ such that $\overline{O_1} \subset \overline{O_2} \subset \cdots \overline{O_k} \subset \cdots$ (the inclusions are all strict).

Problem BST 8. *Find an analogue to Theorem H for surfaces of infinite genus.* 1443?

References

[1] A. Ayache and O. Echi, *The envelope of a subcategory in topology and group theory*, Int. J. Math. Math. Sci. (2005), no. 21, 3387–3404.

[2] E. Bouacida, O. Echi, G. Picavet, and E. Salhi, *An extension theorem for sober spaces and the Goldman topology*, Int. J. Math. Math. Sci. (2003), no. 51, 3217–3239.

[3] E. Bouacida, O. Echi, and E. Salhi, *Foliations, spectral topology, and special morphisms*, Advances in commutative ring theory (Fez, 1997), Lecture Notes in Pure and Appl. Math., vol. 205, Marcel Dekker Inc., New York, 1999, pp. 111–132.

[4] E. Bouacida, O. Echi, and E. Salhi, *Feuilletages et topologie spectrale*, J. Math. Soc. Japan **52** (2000), no. 2, 447–464.

[5] B. L. Brechner and R. D. Mauldin, *Homeomorphisms of the plane*, Pacific J. Math. **59** (1975), no. 2, 375–381.

[6] A. Dress, *Newman's theorems on transformation groups*, Topology **8** (1969), 203–207.

[7] O. Echi, *A topological characterization of the Goldman prime spectrum of a commutative ring*, Comm. Algebra **28** (2000), no. 5, 2329–2337.

[8] A. El Kacimi Alaoui, H. Hattab, and E. Salhi, *Remarques sur certains groupes d'homéomorphismes d'espaces métriques*, JP J. Geom. Topol. **4** (2004), no. 3, 225–242.

[9] O. Goldman, *Hilbert rings and the Hilbert Nullstellensatz*, Math. Z. **54** (1951), 136–140.

[10] A. Grothendieck and J. Dieudonné, *Eléments de géométrie algébrique*, Springer, New York, 1971.

[11] C. Gutierrez, G. Hector, and A. López, *Interval exchange transformations and foliations on infinite genus 2-manifolds*, Ergodic Theory Dynam. Systems **24** (2004), no. 4, 1097–1108.

[12] H. Hattab and E. Salhi, *Groups of homeomorphisms and spectral topology*, Topology Proc. **28** (2004), no. 2, 503–526.

[13] M. Hochster, *Prime ideal structure in commutative rings*, Trans. Amer. Math. Soc. **142** (1969), 43–60.

[14] W. Krull, *Jacobsonsche Ringe, Hilbertscher Nullstellensatz, Dimensionstheorie*, Math. Z. **54** (1951), 354–387.

[15] H. Marzougui, *Structure des feuilles sur les surfaces ouvertes*, C. R. Acad. Sci. Paris Sér. I Math. **323** (1996), no. 2, 185–188.

[16] D. Montgomery and L. Zippin, *Topological transformation groups*, Interscience Publishers, New York, 1955.

[17] I. Nikolaev and E. Zhuzhoma, *Flows on 2-dimensional manifolds*, Lecture Notes in Math., vol. 1705, Springer, Berlin, 1999.

[18] E. Salhi, *Dynamics of groups of homeomorphisms*, Differential geometry and topology, discrete and computational geometry (M. Boucetta and J.-M. Morvan, eds.), vol. 197, IOS Press, 2005, pp. 73–108.

[19] H. Wallman, *Lattices and topological spaces*, Ann. of Math. (2) **39** (1938), no. 1, 112–126.

Open Problems in Topology II – Edited by E. Pearl

Cantor set problems

Dennis J. Garity and Dušan Repovš

Introduction

A Cantor set is characterized as a topological space that is totally disconnected, perfect, compact and metric. Any two such spaces C_1 and C_2 are homeomorphic, but if C_1 and C_2 are subspaces of $\mathbb{R}^n$, $n \geq 3$, there may not be a homeomorphism of $\mathbb{R}^n$ to itself taking C_1 to C_2. In this case, C_1 and C_2 are said to be *inequivalent embeddings* of the Cantor set. There has been recent renewed attention to properties of embeddings of Cantor sets since these sets arise in the settings of dynamical systems, ergodic theory and group actions. The bibliography, while not complete, gives a sampling of the various mathematical areas where Cantor sets naturally arise.

A Cantor set C in $\mathbb{R}^n$ is *tame* if it is equivalent to the standard middle thirds Cantor set. If it is not tame, it is *wild*. A Cantor set C is *strongly homogeneously embedded* in $\mathbb{R}^n$ if every self homeomorphism of C extends to a self homeomorphism of $\mathbb{R}^n$. At the opposite extreme, a Cantor set C in $\mathbb{R}^n$ is *rigidly embedded* if the identity homeomorphism is the only self homeomorphism of C that extends to a homeomorphism of $\mathbb{R}^n$. A Cantor set C in $\mathbb{R}^n$ is *slippery* if for each Cantor set D in $\mathbb{R}^n$ and for each $\epsilon > 0$, there is a homeomorphism $h\colon \mathbb{R}^n \to \mathbb{R}^n$, within ϵ of the identity, with $h(C) \cap D = \emptyset$.

Željko [**28**] defines the genus of a Cantor set X in $\mathbb{R}^3$ and the local genus of points in X. A *defining sequence* for a Cantor set $X \subset \mathbb{R}^n$ is a sequence (M_i) of compact n-manifolds with boundary such that $M_{i+1} \subset \operatorname{int} M_i$ and $X = \bigcap_i M_i$. Let $\mathcal{D}(X)$ be the set of all defining sequences for X. For a disjoint union of handlebodies $M = \bigsqcup_{\lambda\in\Lambda} M_\lambda$, we define $g(M) = \sup\{\text{genus}(M_\lambda) : \lambda \in \Lambda\}$.

For any subset $A \subset X$, and for $(M_i) \in \mathcal{D}(X)$ we denote by M_i^A the union of those components of M_i which intersect A. The *genus of the Cantor set X with respect to the subset A*, $g_A(X) = \inf\{g_A(X;(M_i)) : (M_i) \in \mathcal{D}(X)\}$, where $g_A(X;(M_i)) = \sup\{g(M_i^A) : i \geq 0\}$. For $A = \{x\}$ we call the number $g_{\{x\}}(X)$ the *local genus of the Cantor set X at the point x* and denote it by $g_x(X)$. For $A = X$ we call the number $g_X(X)$ the *genus of the Cantor set X* and denote it by $g(X)$.

The problems

Antoine [**2**] produced the first example of a wild Cantor set in $\mathbb{R}^3$, the well-known Antoine's necklace. Blankinship [**6**] extended Antoine's construction to

---

The first author was supported in part by NSF grants DMS 0139678, DMS 0104325 and DMS 0453304. The second author was supported in part by Slovenian Research Agency program P1-0292-0101-04. Both authors were supported in part by Slovenian Research Agency grant SLO-US 2002/01 and BI-US/04-05/35.

higher dimensions, producing wild Cantor sets in Euclidean spaces of dimensions ≥ 4. Daverman [**8**] produced an example of a strongly homogenously embedded Cantor set if $\mathbb{R}^n$ for $n \geq 5$. His example relied on decomposition theory results that only applied in high dimensions and on the existence of non simply connected homology spheres in dimensions ≥ 3.

1444? **Question 1.** *Is there a strongly homogeneously embedded wild Cantor set in* $\mathbb{R}^3$ *or* $\mathbb{R}^4$*, or are such sets necessarily tame?*

The Antoine construction can be carefully done with sufficiently many tori at each stage so as to produce wild Cantor sets that are geometrically self similar and are Lipschitz homogenously embedded in $\mathbb{R}^3$. See [**12**, **15**, **29**] for definitions and details. It is not clear that the Blankinship construction in higher dimensions can be done so as to produce geometrically self similar Cantor sets.

1445? **Question 2.** *Is there a geometrically self similar wild Cantor set in* $\mathbb{R}^4$ *or in higher dimensions?*

1446? **Question 3.** *Are there Lipschitz homogenously embedded wild Cantor sets in* $\mathbb{R}^4$ *or in higher dimensions?*

Rushing [**18**] produced examples in $\mathbb{R}^3$ of wild Cantor sets of each possible Hausdorff dimension. At the end of his paper, he stated that a modification of the Blankinship construction would allow similar results in higher dimensions. Because of the difficulty in producing a self similar Blankinship construction, it is not clear how the generalization to higher dimensions would work.

1447? **Question 4.** *Are there wild Cantor sets in* $\mathbb{R}^n$, $n \geq 4$ *of arbitrary possible Hausdorff dimension?*

DeGryse and Osborne [**11**] produced an example of a wild Cantor set in $\mathbb{R}^3$ with simply connected complement. Later, Skora [**20**] produced such Cantor sets using a different construction. Rigid wild Cantor sets in $\mathbb{R}^3$ and in higher dimensions were produced by Wright [**24**] using variations on the Antoine and Blankinship constructions. Garity, Repovš, and Željko [**13**] recently produced examples of rigid wild Cantor sets in $\mathbb{R}^3$ that also had simply connected complement. However the latter examples necessarily used tori of arbitrarily high genus in the construction.

1448? **Question 5.** *Is there a rigid Cantor set in* $\mathbb{R}^3$ *with simply connected complement that has local genus* n *or less at every point, for some fixed* n*?*

Bing–Whitehead Cantor sets are a generalization of the Cantor sets produced by DeGryse and Osborne. Ancel and Starbird [**1**] and later Wright [**26**] characterized which Bing–Whitehead constructions actually yield Cantor sets.

1449? **Question 6.** *Is there a modification of the Bing–Whitehead Cantor set construction that yields rigid Cantor sets with simply connected complements?*

1450? **Question 7.** *Are Bing–Whitehead Cantor sets with infinite differences in the number of Whitehead constructions inequivalently embedded?*

Sher in [**19**] showed that two equivalent Antoine Cantor sets necessarily had the same number of components in each stage of their defining sequences. In [**12**], the authors and Željko show that Antoine Cantor sets with the same number of components at each stage can be inequivalent. Knot theory techniques are used in the proof. This leads to the following question.

Question 8. *Is it possible to completely classify Antoine Cantor sets using knot* 1451?
theory invariants?

The following questions deal with the possibility of classifying wild Cantor sets in $\mathbb{R}^3$ using various properties.

Question 9. *Is there a way of classifying wild Cantor sets in* $\mathbb{R}^3$ *using local genus* 1452?
and other geometric properties?

Question 10. *Can one use the volume of the hyperbolic 3-manifolds* $M^3 = S^3 \setminus X$ 1453?
where X *is a wild Cantor set to distinguish between classes of wild Cantor sets?*

The following questions are about the relationship of Hausdorff dimension to various types of Cantor sets.

Question 11. *Can two rigid Cantor sets have different Hausdorff dimensions?* 1454–1455?
How does Hausdorff dimension detect rigidity of Cantor sets?

Question 12. *Is there a rigid Cantor set of minimal Hausdorff dimension?* 1456?

Question 13. *Can two Cantor sets of different genus have the same Hausdorff* 1457–1458?
dimension? How are Hausdorff dimension and genus of Cantor sets related?

The final few questions deal with homotopy groups of the complement of wild Cantor sets.

Question 14. *Can two different (rigid) Cantor sets have the same fundamental* 1459?
groups of the complement?

Question 15. *Which groups can occur as the fundamental groups of (rigid) wild* 1460?
Cantor set complements?

References

[1] F. D. Ancel and M. P. Starbird, *The shrinkability of Bing–Whitehead decompositions*, Topology **28** (1989), no. 3, 291–304.

[2] M. L. Antoine, *Sur la possibilité d'étendre l'homómorphie de deux figures à leur voisinages*, C. R. Acad. Sci. Paris **171** (1920), 661–663.

[3] A. Babich, *Scrawny Cantor sets are not definable by tori*, Proc. Amer. Math. Soc. **115** (1992), no. 3, 829–836.

[4] M. Bestvina and D. Cooper, *A wild Cantor set as the limit set of a conformal group action on* S^3, Proc. Amer. Math. Soc. **99** (1987), no. 4, 623–626.

[5] R H Bing, *Tame Cantor sets in* E^3, Pacific J. Math. **11** (1961), 435–446.

[6] W. A. Blankinship, *Generalization of a construction of Antoine*, Ann. of Math. (2) **53** (1951), 276–297.

[7] J. W. Cannon and D. G. Wright, *Slippery Cantor sets in* E^n, Fund. Math. **106** (1980), no. 2, 89–98.

[8] R. J. Daverman, *Embedding phenomena based upon decomposition theory: wild Cantor sets satisfying strong homogeneity properties*, Proc. Amer. Math. Soc. **75** (1979), no. 1, 177–182.

[9] R. J. Daverman, *A strongly homogeneous but wildly embedded Cantor set in the Hilbert cube*, Bull. Acad. Polon. Sci. Sér. Sci. Math. **29** (1981), no. 5–6, 305–309.

[10] R. J. Daverman and R. D. Edwards, *Wild Cantor sets as approximations to codimension two manifolds*, Topology Appl. **26** (1987), no. 2, 207–218.

[11] D. G. DeGryse and R. P. Osborne, *A wild Cantor set in E^n with simply connected complement*, Fund. Math. **86** (1974), 9–27.

[12] D. Garity, D. Repovš, and M. Željko, *Uncountably many Inequivalent Lipschitz Homogeneous Cantor sets in R^3*, Pacific J. Math. **222** (2005), no. 2, 287–299.

[13] D. Garity, D. Repovš, and M. Željko, *Rigid Cantor sets in R^3 with simply connected complement*, Proc. Amer. Math. Soc. **134** (2006), no. 8, 2447–2456.

[14] J. W. Lamoreaux and D. G. Wright, *Rigid sets in the Hilbert cube*, Topology Appl. **22** (1986), no. 1, 85–96.

[15] J. Malešič and D. Repovš, *On characterization of Lipschitz manifolds*, New developments in differential geometry (Budapest 1996), Kluwer Academic Publishers, Dordrecht, 1999, pp. 265–277.

[16] R. P. Osborne, *Embedding Cantor sets in a manifold. I. Tame Cantor sets in E^n*, Michigan Math. J. **13** (1966), 57–63.

[17] R. P. Osborne, *Embedding Cantor sets in a manifold. II. An extension theorem for homeomorphisms on Cantor sets*, Fund. Math. **65** (1969), 147–151.

[18] T. B. Rushing, *Hausdorff dimension of wild fractals*, Trans. Amer. Math. Soc. **334** (1992), no. 2, 597–613.

[19] R. B. Sher, *Concerning wild Cantor sets in E^3*, Proc. Amer. Math. Soc. **19** (1968), 1195–1200.

[20] R. Skora, *Cantor sets in S^3 with simply connected complements*, Topology Appl. **24** (1986), 181–188.

[21] R. Y. T. Wong, *A wild Cantor set in the Hilbert cube*, Pacific J. Math. **24** (1968), 189–193.

[22] D. G. Wright, *Pushing a Cantor set off itself*, Houston J. Math. **2** (1976), no. 3, 439–447.

[23] D. G. Wright, *Cantor sets in 3-manifolds*, Rocky Mountain J. Math. **9** (1979), no. 2, 377–383.

[24] D. G. Wright, *Rigid sets in E^n*, Pacific J. Math. **121** (1986), no. 1, 245–256.

[25] D. G. Wright, *Rigid sets in manifolds*, Geometric and algebraic topology, PWN, Warsaw, 1986, pp. 161–164.

[26] D. G. Wright, *Bing–Whitehead Cantor sets*, Fund. Math. **132** (1989), no. 2, 105–116.

[27] M. Željko, *On defining sequences for Cantor sets*, Topology Appl. **113** (2001), 321–325, Geometric topology: Dubrovnik 1998.

[28] M. Željko, *Genus of a Cantor set*, Rocky Mountain J. Math. **35** (2005), no. 1, 349–366.

[29] M. Željko, *Minimal number of tori in geometric self-similar Antoine Cantor sets*, JP J. Geom. Topol. **5** (2005), no. 2, 109–113.

Open Problems in Topology II – Edited by E. Pearl

Problems from the Galway Topology Colloquium

Chris Good, Andrew Marsh, Aisling McCluskey and Brian McMaster

This article presents an overview of several groups of open problems that are currently of interest to researchers associated with the Galway Topology Colloquium. Topics include set and function universals, countable paracompactness, abstract dynamical systems, and the embedding ordering within families of topological spaces.

1. Universals: an introduction

Universals were introduced at the beginning of the last century in the study of classical descriptive set theory. They were used, for example, to show for an uncountable Polish space that the class of analytic sets is strictly greater than the class of Borel sets [**27**]. This work focused on universals for Polish spaces. In recent years a number of researchers, in particular Paul Gartside, have begun an investigation of universals in the more general setting of topological spaces. This research has a similar flavour to C_p-Theory, attempting to relate the topological properties of a space to those of some higher-order object.

The study of universals provides a suitable setting for other previous work. For example, the definition of a continuous function universal generalises the definition of an admissible topology on the ring of continuous functions on a space. A further example is continuous perfect normality, (see [**24**, **47**]) which can be defined in terms of zero-set universals.

A universal is a space that in some sense parametrises a collection of objects associated with a given topological space, such as the open subsets or the continuous real-valued functions. More precisely, we can define set or function universals as follows.

Given a space X we say that a space Y *parametrises a continuous function universal for X via the function F* if $F\colon X \times Y \to \mathbb{R}$ is continuous and for any continuous $f\colon X \to \mathbb{R}$ there exists some $y \in Y$ such that $F(x,y) = f(x)$ for all $x \in X$.

Let $\mathcal{T}$ be a function that assigns to each space X the set $\mathcal{T}(X) \subset \mathcal{P}(X)$. For example, $\mathcal{T}$ could take each space X to its topology.

Given a space X we say that a space Y *parametrises a $\mathcal{T}$-universal for X* if there exists $\mathcal{U} \in \mathcal{T}(X \times Y)$ such that for all $A \in \mathcal{T}(X)$ there exists $y \in Y$ such that $\mathcal{U}^y = \{x \in X : (x,y) \in \mathcal{U}\} = A$. Typically we are interested in open universals, Borel universals, zero-set universals or any other $\mathcal{T}$-universals when $\mathcal{T}$ has a natural definition as in these three examples.

The second author would like to acknowledge the support of the Irish Research Council for Science, Engineering and Technology.

For convenience we refer to X as the underlying space and we refer to Y as the parametrising space. If no separation axioms are specified we assume throughout, when dealing with continuous function universals or zero-set universals, that both the underlying space and the parametrising space are Tychonoff. For the other types of universals we assume that these spaces are regular and Hausdorff. Any property that is not defined in this section can be found in [**6**].

Most of the questions that we are interested in can usually be expressed as specific instances of the following metaquestion.

1461? **Question 1.1.** *For a fixed $\mathcal{T}$ and topological property $\mathcal{P}$ can we characterise those spaces that have a $\mathcal{T}$-universal parametrised by a space with property $\mathcal{P}$?*

Problems regarding the construction of universals. Before looking at those questions that are instances of Question 1.1 we will discuss the problem of constructing continuous function universals. Of course a discrete space of sufficient cardinality can always be used to parametrise a continuous function universal, however, in general we wish to find spaces with given global properties (e.g. spaces with a given weight or density) and so this construction is rarely of any use.

It is well known that if X is locally compact then $C_k(X)$ has an admissible topology and hence parametrises a continuous function universal for X via the evaluation map [**1**]. It is readily seen that, given any other continuous function universal for X, say Y, the obvious map from Y onto $C_k(X)$ is continuous. The result of this is that most of the problems regarding continuous function universals for locally compact spaces reduce down to the study of $C_k(X)$. In general, however, given a broader class of spaces there will not be a canonical continuous function universal.

Let τ, σ be two topologies on a set X with $\tau \subset \sigma$. We say that τ is a K-coarser topology if (X, σ) has a neighbourhood basis consisting of τ-compact neighbourhoods. In [**7**] it is shown that in this case we can refine the topology on $C_k(X, \sigma)$ to create a continuous function universal for (X, σ) without adding *too many* open sets.

There are many classes of spaces for which we might be able to find a similar type of construction. As an example, we pose the following (necessarily vague) question.

1462? **Question 1.2.** *Find a general method of constructing continuous function universals for k-spaces such that the parametrising space does not have* too many *more open sets than $C_k(X)$.*

Another way of expressing this is that the cardinal invariants of the parametrising space should be as close to the cardinal invariants of $C_k(X)$ as possible. For example, if $C_k(X)$ is separable and Lindelöf then the continuous function universal should also have these properties.

Problems regarding the cardinal invariants of universals. Via Question 1.1 we could construct a question for every known topological property. Here we mention those problems that have already been investigated with partial success. They relate to compactness type properties (compactness, Lindelöf property, Lindelöf-Σ spaces) and also hereditary cellularity, hereditarily Lindelöf spaces and hereditarily separable spaces. The interested reader should also see [**4**] an excellent survey of open problems in this area that focuses on those problems arising from [**9**, **10**]. There is some overlap between the questions mentioned here and those discussed in [**4**], specifically Question 1.10, Question 1.11 and Question 1.12.

It is worth noting that, in general, the following question remains unsolved.

Question 1.3. *Characterise those spaces that have a continuous function universal parametrised by a separable space.* 1463?

In [**7**] it is shown that if a space has a K-coarser separable metric topology then it has a continuous function universal parametrised by a separable space. This includes, for example, the Sorgenfrey line.

In [**11**] it is shown that one can characterise the metric spaces within the class of all Tychonoff spaces as those spaces with a zero-set universal parametrised by a compact (or even σ-compact) space. The same is true if we look at open regular F_σ universals. However, for open F_σ universals the results are inconclusive, leading to the following problem.

Question 1.4. *Characterise those spaces that have an open F_σ universal parametrised by a compact space.* 1464?

It is true that any Tychonoff space with an open F_σ universal parametrised by a compact space must be developable. Conversely, we know that every Tychonoff metacompact developable space has an open F_σ universal parametrised by a compact space. Is metacompactness necessary? Or can one find a developable, non-metacompact space X with an open F_σ universal parametrised by a compact space?

In [**11**] is also shown that every space with a zero-set universal parametrised by a second countable space must be second countable and hence metrisable. Recall that the class of Lindelöf-Σ spaces is the smallest class of spaces that contains all compacta, all second countable spaces and that is closed under countable products, continuous images and closed subspaces. Since every space with a zero-set universal parametrised by either a second countable space or a compact space must be metrisable, we might guess that this would hold true if the parametrising space were Lindelöf-Σ. This is not the case. However, the following question remains open.

Question 1.5. *If a space X has a zero-set universal parametrised by a space that is the the product of a compact space and a second countable space, then is X metrisable?* 1465?

We say a space X is *strongly quasidevelopable* if there exists a collection $\{\mathcal{G}_n : n \in \omega\}$ where each $\mathcal{G}_n$ is a collection of open subsets of X such that for all open U and $x \in U$ there exists open $V = V(x,U)$ with $x \in V \subset U$ and $n = n(x,V) \in \omega$ such that $x \in \bigcup \mathcal{G}_n$ and $x \in St(V, \mathcal{G}_n) \subset U$. It is known that if a space X

has a zero-set universal parametrised by a Lindelöf-Σ space, then X is strongly quasidevelopable. Yet this cannot be a sufficient condition. In [**11**] an example is given of a strongly quasidevelopable space with no zero-set universal parametrised by a Lindelöf-Σ space.

1466? **Question 1.6.** *Characterise the spaces with a zero-set universal parametrised by a Lindelöf-Σ space.*

There is a possibility that metrisable spaces are precisely those spaces with a continuous function universal parametrised by a Lindelöf-Σ space. A solution to the following question would go a long way towards proving this appealing conjecture.

1467? **Question 1.7.** *If a Tychonoff space has a continuous function universal parametrised by a Lindelöf-Σ space, then must it be metrisable?*

Restricting the class of spaces to the compacta gives us stronger results, as we would expect. For example, it is shown in [**10**] that if X is compact and has an open universal parametrised by a space whose square is hereditarily Lindelöf or hereditarily separable, then X must be metrisable. It is also shown that it is consistent that there is a zero-dimensional compact non-metrisable space with an open universal parametrised by a hereditarily separable space. But no example is known where the parametrising space is hereditarily Lindelöf.

1468? **Question 1.8.** *Is there a consistent example of a space X, such that X is compact and non-metrisable and has an open set universal parametrised by a space that is hereditarily separable?*

As regards hereditary ccc, in [**10**] it is shown that it is consistent that every compact zero-dimensional space with an open universal parametrised by a hereditarily ccc space is metrisable. It would be desirable to drop the restriction to zero-dimensional compacta and get a consistent result for all compacta, leading to the following question.

1469? **Question 1.9.** *Is it consistent that if X is compact and has an open set universal parametrised by a space that is hereditarily ccc, then X must be second countable?*

In the papers [**7**, **10**] it is shown that for open universals and zero-set universals, $hL(X) \leq hd(Y)$ and $hd(X) \leq hL(Y)$. In fact, for zero-set universals we get the stronger result that $hL(X^n) \leq hd(Y)$ for all $n \in \omega$. In both cases, however, we can construct consistent examples to show that $hL(Y)$ cannot bound $hL(X)$. Can we find a ZFC example?

1470? **Question 1.10.** *Is there a space X with either an open universal or a zero-set universal parametrised by Y such that $hL(X) > hL(Y)$?*

Our last two questions deal with Borel universals. Following the notation used in [**27**] we let $\Sigma^0_1(X)$ denote the open subsets of a space X. For every ordinal α, let $\Pi^0_\alpha(X)$ denote the complements of all sets in $\Sigma^0_\alpha(X)$. Finally we can define $\Sigma^0_\alpha(X) = \{\bigcup_{n\in\omega} A_n : A_n \in \Pi^0_\beta(X) \text{ for } \beta < \alpha\}$. In [**9**] it is shown that every compact space with a Σ_n^0 universal parametrised by a second countable space must be metrisable. This holds true for all finite n. But the situation for Σ_ω^0 universals has not been resolved.

Question 1.11. *Is it consistent that every compact space with a Σ_ω^0 universal* 1471?
parametrised by a second countable space is metrisable?

In [**9**] a consistent example is given of a compact, non-metrisable space with a Σ_ω^0 universal parametrised by the Cantor set. In addition it is shown that Question 1.11 has a positive answer if we assume that the underlying space is first countable and compact. If the space in question is compact and perfect with a Σ_ω^0 universal parametrised by a second countable space, then it is a ZFC theorem that it must be metrisable.

One approach to solving Question 1.11 is suggested in [**9**, Section 3.2] and this is also discussed in [**4**, Section 4]. A positive answer to the following question implies a positive answer to Question 1.11. The reasons for this are discussed in detail in both papers and so we will not repeat them here.

Question 1.12. *Is it consistent with $2^{\aleph_0} < 2^{\aleph_1}$ that every compact space X which* 1472?
is the disjoint union of two sets, A and B, where every point in A has countable character in X and B is hereditarily separable and hereditarily Lindelöf, must be hereditarily Lindelöf?

Regarding Question 1.12 it is worth mentioning a result of Eisworth, Nyikos and Shelah from [**5**]. They show that it is consistent with $2^{\aleph_0} < 2^{\aleph_1}$ that every compact, first countable, hereditarily ccc space must be hereditarily Lindelöf.

2. Embedding ordering among topological spaces: an introduction

The ordering by embeddability of topological spaces, although a fundamental notion in topology, has been remarkably little understood for some years. This ordering is that introduced into a family of topological spaces by writing $X \hookrightarrow Y$ whenever X is homeomorphic to a subspace of Y. Its subtlety and relative intractability are well illustrated by the problem of recognizing which order-types are those of collections of subspaces of the real line $\mathbb{R}$ (see [**31–35**]). A partially-ordered set (poset) $\mathbb{P}$ is *realized* (or *realizable*) within a family $\mathcal{F}$ of topological spaces whenever there is an injection $\theta\colon \mathbb{P} \to \mathcal{F}$ for which $p \leq q$ if and only if $\theta(p) \hookrightarrow \theta(q)$. Discussion of realizability in the powerset $\mathcal{P}(\mathbb{R})$ can be traced back to Banach, Kuratowski and Sierpiński ([**29**, **30**, **42**]), whose work on the extensibility of continuous maps over G_δ subsets (of Polish spaces) revealed *inter alia* that for a given Polish space X, it is possible to realize, within $\mathcal{P}(X)$, (i) the antichain of cardinality $2^{\mathfrak{c}}$ [**29**, p. 205] and (ii) the ordinal $\mathfrak{c}^+$ [**30**, p. 199]. Renewed interest in the problem was initiated in [**31**] for the special case of $\mathbb{R}$ in which it was shown that every poset of cardinality $\mathfrak{c}$ or less can be realized within $\mathcal{P}(\mathbb{R})$, and by the direct construction in [**34**] of a realization (by subspaces of *some* topological space) of an arbitrary quasiordered set. The question of precisely which posets of cardinalities exceeding $\mathfrak{c}$ can be realized within $\mathcal{P}(\mathbb{R})$ had been open until recently and exposed the question to be ultimately set-theoretic in nature. Article [**36**] establishes that it is consistent that all posets of cardinality $2^{\mathfrak{c}}$ can be realized within $\mathcal{P}(\mathbb{R})$ while [**28**] establishes—by exhibiting a consistent counterexample—that this statement is, in fact, independent of ZFC.

Problems involving the embedding ordering. In [**28**], forcing is used to construct a poset of cardinality $2^{\mathfrak{c}}$ which cannot be realized within $\mathcal{P}(\mathbb{R})$. In this model, the cardinal arithmetic is such that $\mathfrak{c} = \aleph_1$ and $2^{\aleph_1} = \aleph_3$, leaving open the following question:

1473? **Question 2.1.** *Is it true (in* ZFC*) that every poset of cardinality* $\mathfrak{c}^+$ *can be realized within* $\mathcal{P}(\mathbb{R})$*?*

Further, due to the nature of the construction in [**28**], it seems that for any space of cardinality $\mathfrak{c}$, such a consistent counterexample can be found. Of course, one does not need to take this trouble in the case of any discrete space as discrete spaces can only support linear orders. Another obvious question concerns $\mathbb{R}$ itself: just what aspect of its topological nature has influenced the order-theoretic structure of $\mathcal{P}(\mathbb{R})$? The following questions arise naturally:

1474? **Question 2.2.** *For which spaces X of cardinality $\mathfrak{c}$ is it consistent that every poset of cardinality $2^{\mathfrak{c}}$ can be realized by $\mathcal{P}(X)$?*

1475? **Question 2.3.** *For which spaces X of cardinality $\mathfrak{c}$ is it possible to find (in* ZFC*) a $2^{\mathfrak{c}}$-element poset which cannot be realized by $\mathcal{P}(X)$?*

1476? **Question 2.4.** *What can be said about the order-theoretic structure of $\mathcal{P}(X)$ where X is a Polish space of cardinality $\mathfrak{c}$?*

1477? **Question 2.5.** *What about spaces of higher cardinality? That is, given any cardinal κ where $\kappa > \mathfrak{c}$, if X is a (non-trivial) space of cardinality κ, which posets of cardinality 2^{κ} can be realized in $\mathcal{P}(X)$?*

Concerning representations within $\mathcal{P}(\mathbb{R})$, in the literature no particular demands have been made on the representative subsets of $\mathbb{R}$. In most cases they turn out to be Bernstein sets but, otherwise, *existence* of any representation has been key, rather than existence of a particularly 'nice' representation, such as by Borel sets or some such family. Thus, natural variations on the theme provide another question:

1478? **Question 2.6.** *For those posets which can be realized within $\mathcal{P}(\mathbb{R})$, is it possible to restrict the representative subspaces to some* nice *family of subsets of $\mathbb{R}$?*

Also in connection with the embedding ordering there is the *bottleneck* problem [**23**]. It is well known [**13**] that, in the family of all infinite topological spaces, every space contains a homeomorph of one or more of the five *minimal infinite* spaces created by imposing upon $\mathbb{N}$ the discrete, the trivial, the cofinite, the initial-segment and the final-segment topologies. Thus, these constitute a five-element *cross section* of the infinite spaces—a (very) small selection of spaces such that *every* space is comparable (*via* the embeddability ordering, that is, either as a subspace or as a superspace) with something in the selection. Is *five* the smallest possible cardinality of such a cross section?

Question 2.7. *Can there be four or fewer infinite spaces, with at least one of which every infinite topological space is embeddingwise comparable? Given an infinite cardinal κ, what can be determined about the least cardinality of a selection of spaces on κ-many points, with at least one of which every topological space on κ-many points is embeddingwise comparable?* 1479

3. Questions relating to countable paracompactness

A space X is *monotonically countably paracompact*, or *MCP* [**19**, **46**] if and only if there is an operator U assigning to each $n \in \omega$ and each closed set D an open set $U(n, D)$ containing D such that

(1) if $(D_i)_{i\in\omega}$ is a decreasing sequence of closed sets with $\bigcap_{n\in\omega} D_n = \emptyset$, then $\bigcap_{n\in\omega} \overline{U(n, D_n)} = \emptyset$;

(2) if $E \subseteq D$, then $U(n, E) \subseteq U(n, D)$.

Without condition (2), this is a characterization of countable paracompactness. Weakening the conclusion of condition (1) to $\bigcap_{n\in\omega} U(n, D_n) = \emptyset$ gives a characterization of β-spaces; strengthening (1) to $\bigcap_{n\in\omega} \overline{U(n, D_n)} = \bigcap D_n$, whenever $(D_i)_{i\in\omega}$ is a decreasing sequence of closed sets, characterizes stratifiability.

We have a reasonably complete picture of MCP as a generalized metric property closely related to stratifiability: for example, MCP Moore spaces are metrizable and there are monotonically normal spaces which fail to be MCP. In [**20**], however, we show that if an MCP space fails to be collectionwise Hausdorff, then there is a measurable cardinal and that, if there are two measurable cardinals, then there is an MCP space that fails to be collectionwise Hausdorff. We have been unable to decide:

Question 3.1. *Does the existence of a single measurable cardinal imply the existence of an MCP space that is not collectionwise Hausdorff?* 1480

In her thesis, Lylah Haynes [**26**] (see also [**17**, **18**]) makes a study of monotone versions of various characterizations of countable paracompactness. One possible monotone version of MCP, *nMCP*, arises from restricting condition (1) above to nowhere dense closed sets. Although it seems that most of the known results about MCP spaces hold for nMCP spaces as well, the following is not clear.

Question 3.2. *Is every nMCP space MCP?* 1481

Haynes did not consider monotonizations of countable paracompactness as a covering property. There are monotone versions of paracompactness about which one can say interesting things [**12**, **43**], so it is possible that the following is interesting.

Question 3.3. *Is there a sensible monotone version of the statement 'every countable open cover has a locally finite open refinement' or, indeed, any other characterization of countable paracompactness as a covering property?* 1482

A set D is a *regular* G_δ if and only if $D = \bigcap U_n = \bigcap \overline{U_n}$, where each U_n is open. A space is *δ-normal* if and only if every pair of disjoint closed sets, one of which is a regular G_δ, can be separated. Mack (see [**22**]) showed that a space is countably paracompact if and only if $X \times [0,1]$ is δ-normal.

Motivated by the Reed–Zenor theorem [**38**] that every locally connected, locally compact, normal Moore space is metrizable, and by Balogh and Bennett [**2**] who ask the same question for Moore manifolds, we ask:

1483? **Question 3.4.** *Is every locally connected, locally compact, countably paracompact Moore space metrizable?*

1484? **Question 3.5.** *Is every locally connected, locally compact, δ-normal Moore space metrizable?*

Haynes defines a space to be monotonically δ-normal, or mδn, if to each pair of disjoint closed sets C and D, one of which is a regular G_δ, one can assign an open set $H(C,D)$ such that

(1) $C \subseteq H(C,D) \subseteq \overline{H(C,D)} \subseteq X - D$ and
(2) $H(C,D) \subseteq H(C',D')$, whenever $C \subseteq C'$ and $D' \subseteq D$.

Neither MCP nor mδn imply one another but X is MCP whenever $X \times [0,1]$ is mδn. Every first countable, Tychonoff mδn space is monotonically normal.

1485? **Question 3.6.** *Is there an mδn space that is not monotonically normal?*

Of a similar flavour to the Reed–Zenor Theorem is Rudin's result that under MA + ¬CH every perfectly normal manifold is metrizable [**40**]. On the other hand, assuming ◇, Bešlagić [**3**] constructs a perfectly normal space with Dowker square and in [**14**] we construct a manifold with Dowker square, again using ◇. A number of related questions about countable paracompactness in product spaces seem natural here.

For a detailed survey of the Dowker space problem, see Paul Szeptycki's article in this volume.

1486? **Question 3.7.** *Is it consistent that there is a perfectly normal manifold M such that M^2 is a Dowker space?*

1487? **Question 3.8.** *Is there (in* ZFC*) a normal space with Dowker square?*

Rudin's ZFC Dowker space [**39**] is a subspace of a product and has been modified by Hart, Junnila and van Mill [**25**] to provide a Dowker group.

1488? **Question 3.9.** *Is there a topological group with Dowker square?*

Every monotonically normal space is countably paracompact (see [**41**]).

1489? **Question 3.10.** *Is there a monotonically normal space with Dowker square?*

1490? **Question 3.11.** *Is there a Dowker space with Dowker square?*

A base $\mathcal{B}$ for a space X is said to be *uniform* if, whenever $x \in X$ and $(B_n)_{n\in\omega}$ is a sequence of pairwise distinct elements of $\mathcal{B}$ each containing x, then $(B_n)_{n\in\omega}$ is a base at the point x. Then X has a uniform base if and only if it is metacompact and developable. Alleche, Arhangel′skiĭ and Calbrix introduced the notions of sharp base and weak development: $\mathcal{B}$ is said to be a *sharp* base if, whenever $x \in X$ and $(B_n)_{n\in\omega}$ is a sequence of pairwise distinct elements of $\mathcal{B}$ each containing x, the collection $\{\bigcap_{j\leq n} B_j : n \in \omega\}$ is a base at the point x. See [**21**] for more details. Since a space with a uniform base is both developable and has a sharp base, and since both of these notions imply that the space is weakly developable, it is natural to ask:

Question 3.12. *Is every collectionwise normal space with a sharp base metrizable?* 1491?

Question 3.13. *Does every Moore space with a sharp base have a uniform base?* 1492?

Presumably the answer to the next question is 'no.'

Question 3.14. *Is there a Dowker space with a sharp base?* 1493?

4. Abstract dynamical systems

Given a map $T\colon X \to X$ on a set X, there is a natural and obvious question one can ask.

Question 4.1. *When is there a* nice *topology on X with respect to which T is continuous?*

Substitute your own favourite definition of 'nice' in here.

With only the algebraic structure of T to work with, one has to consider the orbits of T. The equivalence relation $x \equiv y$ if and only if there are $n, m \in \mathbb{N}$ such that $T^n(x) = T^m(y)$ partitions X into the orbits of T. Let O be an orbit. Then O is an *n-cycle* if it contains points x_i, $0 \leq i < n$ such that $T(x_i) = x_{i+1}$, where $i + 1$ is taken modulo n. O is a *$\mathbb{Z}$-orbit* if it contains points x_i, $i \in \mathbb{Z}$ such that $T(x_i) = x_{i+1}$. An orbit that is neither an n-cycle nor a $\mathbb{Z}$-orbit is called an *$\mathbb{N}$-orbit*.

In [**16**], we prove that there is a compact, Hausdorff topology on X with respect to which T is continuous if and only if $T\big(\bigcap_{m\in\mathbb{N}} T^m(X)\big) = \bigcap_{m\in\mathbb{N}} T^m(X) \neq \emptyset$ and either:

(1) T has, in total, at least continuum many $\mathbb{Z}$-orbits or cycles; or
(2) T has both a $\mathbb{Z}$-orbit and a cycle; or
(3) T has an n_i-cycle, for each $i \leq k$, with the property that whenever T has an n-cycle, then n is divisible by n_i for some $i \leq k$; or
(4) the restriction of T to $\bigcap_{m\in\mathbb{N}} T^m(X)$ is not one-to-one.

We also prove that, if T is a bijection, then there is a compact metrizable topology on X with respect to which T is a homeomorphism if and only if one of the following holds.

(1) X is finite.

(2) X is countably infinite and either:
   (a) T has both a $\mathbb{Z}$-orbit and a cycle; or
   (b) T has an n_i-cycle, for each $i \leq k$, with the property that whenever T has an n-cycle, then n is divisible by n_i for some $i \leq k$.

(3) X has the cardinality of the continuum and the number of $\mathbb{Z}$-orbits and the number of n-cycles, for each $n \in \mathbb{N}$, is finite, countably infinite, or has the cardinality of the continuum.

One can obviously ask any number of questions here. For example, in [**15**] we show that there is a hereditarily Lindelöf, Tychonoff topology on X with respect to which T is continuous if and only if $|X| \leq \mathfrak{c}$.

.494–1496? **Question 4.2.** *Characterize continuity on compact metric spaces, on* $\mathbb{R}$, *or on* $\mathbb{R}^n$ *for some* n.

These are hard questions.

1497? **Question 4.3.** *Given a group* G *acting on a set* X, *under what circumstances is there a* nice *topology on* X *with respect to which each element of* G *is continuous?*

Aside from their intrinsic interest, such questions might provide useful examples in the study of permutation groups. For example, Mekler [**37**] characterizes the countable subgroups of the autohomeomorphism group of $\mathbb{Q}$ (see also [**45**]).

In the case of compact Hausdorff topologies on X, Rolf Suabedissen, in his impressive thesis [**44**], has made significant progress on the question of what happens with two or more commuting bijections on X. He also has a very neat characterization of continuous actions of compact Abelian Lie groups on compact Hausdorff spaces.

References

[1] R. Arens and J. Dugundji, *Topologies for function spaces*, Pacific J. Math. **1** (1951), 5–31.

[2] Z. T. Balogh and H. Bennett, *Conditions which imply metrizability in manifolds*, Houston J. Math. **15** (1989), no. 2, 153–162.

[3] A. Bešlagić, *Yet another Dowker product*, Topology Appl. **56** (1994), no. 1, 35–43.

[4] P. J. Collins, *Problems on universals*, Topology Appl. **140** (2004), no. 1, 33–36.

[5] T. Eisworth, P. J. Nyikos, and S. Shelah, *Gently killing S-spaces*, Israel J. Math. **136** (2003), 189–220.

[6] R. Engelking, *General topology*, Heldermann Verlag, Berlin, 1989.

[7] G. Fairey, P. M. Gartside, and A. Marsh, *Cardinal invariants of universals*, Comment. Math. Univ. Carolin. **46** (2005), no. 4, 685–703.

[8] P. Gartside, R. W. Knight, and J. T. H. Lo, *Parametrizing open universals*, Topology Appl. **119** (2002), no. 2, 131–145.

[9] P. Gartside and J. T. H. Lo, *The hierarchy of Borel universal sets*, Topology Appl. **119** (2002), no. 2, 117–129.

[10] P. Gartside and J. T. H. Lo, *Open universal sets*, Topology Appl. **129** (2003), no. 1, 89–101.

[11] P. Gartside and A. Marsh, *Compact universals*, Topology Appl. **143** (2004), 1–13.

[12] P. Gartside and P. J. Moody, *A note on proto-metrisable spaces*, Topology Appl. **52** (1993), no. 1, 1–9.

[13] J. Ginsburg and B. Sands, *Minimal infinite topological spaces*, Amer. Math. Monthly **86** (1979), no. 7, 574–576.

[14] C. Good, *Dowker spaces, anti-Dowker spaces, products and manifolds*, Topology Proc. **20** (1995), 123–143.

[15] C. Good and S. Greenwood, *Characterizing continuous functions on Lindelöf spaces*, In preparation.

[16] C. Good, S. Greenwood, R. W. Knight, D. W. McIntyre, and S. W. Watson, *Characterizing continuous functions on compact spaces*, Adv. Math. **206** (2006), no. 2, 695–728.

[17] C. Good and L. Haynes, *Monotone δ-normality*, 2006, Preprint.

[18] C. Good and L. Haynes, *Monotone versions of countable paracompactness*, 2006, Preprint.

[19] C. Good, R. Knight, and I. Stares, *Monotone countable paracompactness*, Topology Appl. **101** (2000), no. 3, 281–298.

[20] C. Good and R. W. Knight, *Monotonically countably paracompact, collectionwise Hausdorff spaces and measurable cardinals*, Proc. Amer. Math. Soc. **134** (2006), no. 2, 591–597.

[21] C. Good, R. W. Knight, and A. M. Mohamad, *On the metrizability of spaces with a sharp base*, Topology Appl. **125** (2002), no. 3, 543–552.

[22] C. Good and I. J. Tree, *On δ-normality*, Topology Appl. **56** (1994), no. 2, 117–127.

[23] M. G. Gormley and T. B. M. McMaster, *A problem concerning embeddability*, 2000, Questions and Answers, Topology Atlas, `http://at.yorku.ca/q/a/a/a/20.htm`.

[24] G. Gruenhage, *Continuously perfectly normal spaces and some generalizations*, Trans. Amer. Math. Soc. **224** (1976), no. 2, 323–338.

[25] K. P. Hart, H. Junnila, and J. van Mill, *A Dowker group*, Comment. Math. Univ. Carolin. **26** (1985), no. 4, 799–810.

[26] L. Haynes, *Monotonizations of countable paracompactness*, Ph.D. thesis, University of Birmingham, 2006.

[27] A. S. Kechris, *Classical descriptive set theory*, Springer, New York, 1995.

[28] R. W. Knight and A. E. McCluskey, *$\mathcal{P}(\mathbb{R})$, ordered by homeomorphic embeddability, does not represent all posets of cardinality $2^{\mathfrak{c}}$*, Preprint.

[29] C. Kuratowski, *Sur la puissance de l'ensemble des "nombres de dimension" au sens de M. Fréchet*, Fund. Math. **8** (1926), 201–208.

[30] C. Kuratowski and W. Sierpiński, *Sur un problème de M. Fréchet concernant les dimensions des ensembles linéaires*, Fund. Math. **8** (1926), 193–200.

[31] P. T. Matthews and T. B. M. McMaster, *Families of spaces having prescribed embeddability order-type*, Rend. Istit. Mat. Univ. Trieste **25** (1993), no. 1–2, 345–352 (1994).

[32] P. T. Matthews and T. B. M. McMaster, *On chains and posets within the power set of a continuum*, Internat. J. Math. Math. Sci. **18** (1995), no. 2, 305–309.

[33] A. E. McCluskey and T. B. M. McMaster, *Realizing quasiordered sets by subspaces of "continuum-like" spaces*, Order **15** (1998), no. 2, 143–149.

[34] A. E. McCluskey and T. B. M. McMaster, *Representing quasi-orders by embeddability ordering of families of topological spaces*, Proc. Amer. Math. Soc. **127** (1999), no. 5, 1275–1279.

[35] A. E. McCluskey, T. B. M. McMaster, and W. S. Watson, *Representing set-inclusion by embeddability (among the subspaces of the real line)*, Topology Appl. **96** (1999), no. 1, 89–92.

[36] A. E. McCluskey and D. Shakhmatov, *It is consistent that all posets of cardinality $2^{\mathfrak{c}}$ can be realized within $\mathcal{P}(\mathbb{R})$*, Preprint.

[37] A. H. Mekler, *Groups embeddable in the autohomeomorphisms of $\mathbb{Q}$*, J. London Math. Soc. **33** (1986), no. 1, 49–58.

[38] G. M. Reed and P. L. Zenor, *Metrization of Moore spaces and generalized manifolds*, Fund. Math. **91** (1976), no. 3, 203–210.

[39] M. E. Rudin, *A normal space X for which $X \times I$ is not normal*, Fund. Math. **73** (1972), no. 2, 179–186.

[40] M. E. Rudin, *The undecidability of the existence of a perfectly normal nonmetrizable manifold*, Houston J. Math. **5** (1979), no. 2, 249–252.

[41] M. E. Rudin, *Dowker spaces*, Handbook of set-theoretic topology, North-Holland, Amsterdam, 1984, pp. 761–780.

[42] W. Sierpiński, *Sur un problème concernant les types de dimensions*, Fund. Math. **19** (1932), 65–71.

[43] I. S. Stares, *Versions of monotone paracompactness*, Papers on general topology and applications (Gorham, ME, 1995), Ann. New York Acad. Sci., vol. 806, New York Acad. Sci., 1996, pp. 433–437.

[44] R. Suabedissen, *Topologizing groups actions*, D.Phil. thesis, Oxford University, 2006.

[45] J. K. Truss, *Embeddings of infinite permutation groups*, Proceedings of groups—St. Andrews 1985, Cambridge University Press, Cambridge, 1986, pp. 335–351.

[46] G. Ying and C. Good, *A note on monotone countable paracompactness*, Comment. Math. Univ. Carolin. **42** (2001), no. 4, 771–778.

[47] P. Zenor, *Some continuous separation axioms*, Fund. Math. **90** (1975), no. 2, 143–158.

Open Problems in Topology II – Edited by E. Pearl

The lattice of quasi-uniformities

Eliza P. de Jager and Hans-Peter A. Künzi

Introduction

Let X be a (nonempty) set. Recall that a filter $\mathcal{U}$ on $X \times X$ is called a *quasi-uniformity* on X if each member U of $\mathcal{U}$ contains the diagonal $\Delta = \{(x,x) : x \in X\}$ of X and for each $U \in \mathcal{U}$ there is $V \in \mathcal{U}$ such that $V \circ V \subseteq U$. Here $\circ$ denotes the usual composition of relations so that $V \circ V = \{(x,z) \in X \times X : \exists y \in X\ (x,y) \in V, (y,z) \in V\}$. If $\mathcal{U}$ is a quasi-uniformity on X, then $\mathcal{U}^{-1} = \{U^{-1} : U \in \mathcal{U}\}$ is the so-called *conjugate quasi-uniformity* of $\mathcal{U}$, where $U^{-1} = \{(x,y) \in X \times X : (y,x) \in U\}$ whenever $U \in \mathcal{U}$. A quasi-uniformity $\mathcal{U}$ will be called *symmetric* provided that $\mathcal{U} = \mathcal{U}^{-1}$, that is, if it is a *uniformity*. Otherwise it will be called *nonsymmetric*.

The topology $\tau(\mathcal{U})$ induced by a quasi-uniformity $\mathcal{U}$ on X is determined by the neighbourhood filters $\mathcal{U}(x) = \{U(x) : U \in \mathcal{U}\}$ of the points $x \in X$, where $U(x) = \{y \in X : (x,y) \in U\}$ whenever $x \in X$ and $U \in \mathcal{U}$.

As usual, a reflexive and transitive binary relation is called a *preorder*. For each preorder T on a set X we can consider the quasi-uniformity $\mathcal{U}(T)$ generated by the base $\{T\}$ on $X \times X$. (In the following $\mathcal{U}(T)$ is called the *quasi-uniformity generated by the preorder* T.) Note that in this sense, on a finite set X, any quasi-uniformity $\mathcal{U}$ is generated by the preorder $\bigcap \mathcal{U}$. If $\mathcal{U}_1$ and $\mathcal{U}_2$ are two quasi-uniformities on a set X and $\mathcal{U}_1 \subseteq \mathcal{U}_2$, then we say that $\mathcal{U}_2$ is *finer* than $\mathcal{U}_1$ respectively that $\mathcal{U}_1$ is *coarser* than $\mathcal{U}_2$.

We consider the set $q(X)$ of all quasi-uniformities on the set X, partially ordered under set-theoretic inclusion $\subseteq$. It is well known that $(q(X), \subseteq)$ is a complete lattice [**2**, p. 2]. We shall denote the smallest element of the lattice $(q(X), \subseteq)$, namely the *indiscrete uniformity* $\{X \times X\}$, by $\mathcal{I}$, or $\mathcal{I}_X$ for clarity. Similarly we shall denote the largest element of the lattice $(q(X), \subseteq)$, namely the *discrete uniformity* $\mathcal{U}(\Delta)$ by $\mathcal{D}$, or $\mathcal{D}_X$ for clarity. Of course, as usual, we put $\bigvee \emptyset = \mathcal{I}$ and similarly $\bigwedge \emptyset = \mathcal{D}$ in the lattice $(q(X), \subseteq)$.

Observe that for any set X the quasi-uniformities on X generated by preorders form a sublattice of $(q(X), \subseteq)$ (compare [**4**]).

A quasi-uniformity is called *transitive* if it has a base consisting of transitive relations. A quasi-uniformity $\mathcal{U}$ on a set X is *totally bounded* if its associated supremum uniformity $\mathcal{U}^s = \mathcal{U} \vee \mathcal{U}^{-1}$ is precompact, that is, for each $U \in \mathcal{U}$ the cover $\{(U \cap U^{-1})(x) : x \in X\}$ has a finite subcover. Each quasi-uniformity $\mathcal{U}$ on a set X contains a finest totally bounded quasi-uniformity $\mathcal{U}_\omega$ coarser than $\mathcal{U}$. For any quasi-uniformity $\mathcal{U}$ on a set X, the collection $\pi(\mathcal{U}) := \{\mathcal{V}$ is a quasi-uniformity

---

Both authors acknowledge partial support under the bilateral collaboration between South Africa and Flanders (period of 2005/6; project 2068799).

on X: $\mathcal{V}_\omega = \mathcal{U}_\omega\}$ is called the *quasi-proximity class of* $\mathcal{U}$.[1] In general $\pi(\mathcal{U})$ does not possess a finest member, but $\mathcal{U}_\omega$ is always the coarsest and only totally bounded member of $\pi(\mathcal{U})$. The quasi-uniformity $\mathcal{U}_\omega$ can be explicitly described as the filter on $X \times X$ generated by the subbase consisting of all relations of the form $[(X \setminus A) \times X] \cup [X \times (X \setminus B)]$, where $A \,\overline{\delta_\mathcal{U}}\, B$. A quasi-uniformity $\mathcal{U}$ on a set X such that $\mathcal{U}_\omega = \mathcal{D}_\omega$ is called *proximally discrete*; otherwise it is called *proximally nondiscrete*.

For any subset A of a set X we define the preorder $S_A = [(X \setminus A) \times X] \cup [A \times A]$. We recall that if (X, τ) is a topological space, then the compatible[2] *Pervin quasi-uniformity* $\mathcal{P}(\tau)$ of X is generated by the subbase $\{S_G : G$ is an open subset of $X\}$. It is always totally bounded and transitive. Note that for any set X, the uniformity $(\mathcal{D}_X)_\omega$ is equal to the Pervin quasi-uniformity of the discrete topology on X.

Interesting studies about the complete sublattice $u(X)$ of $q(X)$ consisting of all uniformities on X were conducted in [**6–9**]. Recently we completed related investigations about $q(X)$ [**3–5**], which led to the questions discussed below.

Solutions to our problems are likely to shed further light on two of the major topics studied in the theory of quasi-uniformities, namely the intriguing relationship between symmetry and asymmetry on the one hand, and the delicate connection between transitivity and non-transitivity on the other hand. Our questions, mainly dealing with anti-atoms and complements, are motivated by corresponding results that are known to hold in the lattice of preorders, or the lattice of uniformities (see in particular [**8–10**]). Often it remains unclear whether and how these results can be generalized to the (larger) lattice of quasi-uniformities. In our context the study of the concept of an anti-atom is closely related to deep set-theoretic investigations which try to clarify the fine structure (for instance selectivity properties) of ultrafilters on sets. This subject is discussed in some detail in [**8, 9**].

Adjacent quasi-uniformities

Two comparable distinct quasi-uniformities on a set X for which there does not exist another quasi-uniformity strictly in between will be called *adjacent* or *neighbours*. The concepts of "upper neighbour" and "lower neighbour" of a quasi-uniformity should now be self-explanatory. For an infinite set X, not much is known about the distribution of adjacent pairs in $(q(X), \subseteq)$.

Using the terminology from lattice theory, we call a quasi-uniformity on a set X an *atom* of $(q(X), \subseteq)$ if it is an upper neighbour of $\mathcal{I}$. We call it an *anti-atom* of $(q(X), \subseteq)$ if it is a lower neighbour of $\mathcal{D}$.

[1] In order to understand this terminology recall that any quasi-uniformity $\mathcal{U}$ on a set X induces a *quasi-proximity* $\delta_\mathcal{U}$ on X, which is defined as follows: For any subsets A, B of X, $A\delta_\mathcal{U}B$ provided that $(A \times B) \cap U \neq \emptyset$ whenever $U \in \mathcal{U}$. We shall use the symbol $\overline{\delta_\mathcal{U}}$ to denote the negation of the relation $\delta_\mathcal{U}$. The quasi-proximity class of a quasi-uniformity $\mathcal{U}$ on X contains as members exactly those quasi-uniformities on X that induce $\delta_\mathcal{U}$.

[2] A quasi-uniformity $\mathcal{U}$ on a topological space (X, τ) is said to be *compatible* if $\tau(\mathcal{U}) = \tau$.

Atoms of $(q(X), \subseteq)$ are readily characterized (compare [**4**]): A quasi-uniformity $\mathcal{U}$ on a set X is an atom of $(q(X), \subseteq)$ if and only if $\mathcal{U}$ is generated by some preorder S_A where A is a proper nonempty subset of X. In particular, each atom of $(q(X), \subseteq)$ is transitive and totally bounded. It follows that a quasi-uniformity on a set X is the supremum of a family of atoms in $(q(X), \subseteq)$ if and only if it is totally bounded and transitive [**5**]. An atom of $(q(X), \subseteq)$ cannot be a uniformity and there does not exist any atom of $(q(\mathbb{R}), \subseteq)$ that is coarser than the usual uniformity $\mathcal{R}$ on the set $\mathbb{R}$ of the reals.

The anti-atoms of $(q(X), \subseteq)$ are more difficult to describe (compare [**4**]). Let X be a set with at least two distinct elements x and y. Then the quasi-uniformity on X generated by the preorder $\Delta \cup \{(x, y)\}$ is a (proximally nondiscrete) anti-atom of $(q(X), \subseteq)$. For finite X, no other anti-atoms of $(q(X), \subseteq)$ exist.

On the other hand by Zorn's lemma each quasi-uniformity on a set X that does not contain a binary reflexive relation R on X is coarser than a maximal quasi-uniformity $\mathcal{M}_R$ on X not containing R. If $R = \Delta$, then $\mathcal{M}_R$ obviously is an anti-atom of $(q(X), \subseteq)$. Hence any quasi-uniformity on X that is distinct from $\mathcal{D}$ is contained in an anti-atom of $(q(X), \subseteq)$. One readily verifies that no anti-atom of $(q(X), \subseteq)$ is a uniformity (see [**4**]).

Let $\mathcal{F}$ and $\mathcal{G}$ be (ultra)filters on a set X and let $\mathcal{U}_{\mathcal{F},\mathcal{G}}$ be the quasi-uniformity on X having the base $\{\Delta \cup (F \times G) : F \in \mathcal{F},\ G \in \mathcal{G}\}$. It is known [**4**] that for each anti-atom $\mathcal{U}$ of $(q(X), \subseteq)$ there exist uniquely determined ultrafilters $\mathcal{F}$ and $\mathcal{G}$ on X such that $\bigcap \mathcal{F} \cap \bigcap \mathcal{G} = \emptyset$ and $\mathcal{U}_{\mathcal{F},\mathcal{G}} \subseteq \mathcal{U}$. An anti-atom of $(q(X), \subseteq)$ is proximally nondiscrete if and only if its associated ultrafilters $\mathcal{F}$ and $\mathcal{G}$ are distinct. Proximally nondiscrete anti-atoms of $(q(X), \subseteq)$ can be characterized as follows [**4**]: A quasi-uniformity $\mathcal{U}$ on a set X is a proximally nondiscrete anti-atom of $(q(X), \subseteq)$ if and only if there exists an ultrafilter $\mathcal{H}$ on $X \times X$ such that $\mathrm{pr}_1 \mathcal{H}$ and $\mathrm{pr}_2 \mathcal{H}$ are distinct and $\mathcal{U} = \{\Delta \cup H : H \in \mathcal{H}\}$. Here pr_i $(i = 1, 2)$ denote the projections from $X \times X$ to the first (resp. second) factor space X. It follows that each proximally nondiscrete anti-atom of $(q(X), \subseteq)$ is transitive. No comparable characterization of proximally discrete anti-atoms of $(q(X), \subseteq)$ is known.

It is natural to study anti-atoms of $(q(X), \subseteq)$ via properties of their associated pair of ultrafilters. For instance (see [**4**]) a proximally nondiscrete anti-atom $\mathcal{U}$ of $(q(X), \subseteq)$ is the finest quasi-uniformity of its quasi-proximity class if and only if for its associated ultrafilters $\mathcal{F}$ and $\mathcal{G}$ on X the filter generated by the base $\{F \times G : F \in \mathcal{F},\ G \in \mathcal{G}\}$ on $X \times X$ is an ultrafilter.

Little is known about proximally discrete anti-atoms of $(q(X), \subseteq)$. Some partial results related to the following two questions have been obtained for the sublattice $u(\omega)$ of $q(\omega)$ in [**8**, **9**] under CH, where ω denotes the first infinite ordinal. In particular it is known that under CH the lattice $u(\omega)$ possesses anti-atoms of $u(\omega)$ which are not transitive. (Since in [**8**, **9**] the authors use the dual order, they speak about atoms.)

Question 1. *Let $\mathcal{F}$ be a free ultrafilter (i.e., $\bigcap \mathcal{F} = \emptyset$) on a set X. Determine the number of anti-atoms $\mathcal{U}$ of $(q(X), \subseteq)$ such that $\mathcal{U}_{\mathcal{F},\mathcal{F}} \subseteq \mathcal{U}$.* 1498?

1499? **Question 2.** *Given a set X, is there a (necessarily proximally discrete) anti-atom of $(q(X), \subseteq)$ that is not transitive?*

Since each filter on a set can be represented as the intersection of a family of ultrafilters, the following question is very natural.

1500? **Question 3.** *Which quasi-uniformities on a set X are equal to the infimum of a family of anti-atoms of $(q(X), \subseteq)$?*

It is known that in general in the lattice $u(X)$ there are uniformities which are not the infimum of a family of anti-atoms of $u(X)$. Indeed according to [**8**, p. 7] for any separated uniformity $\mathcal{U}$ on an infinite set X with at least one nonisolated point x_0 the equivalence relation $\Delta \cup [(X \setminus \{x_0\}) \times (X \setminus \{x_0\})]$ is contained in any (uniform) anti-atom finer than $\mathcal{U}$ although it does not belong to $\mathcal{U}$. For the lattice $(q(X), \subseteq)$ no corresponding example is known. On the other hand a few positive partial answers have been found in [**5**], which we mention next:

Call a quasi-uniformity $\mathcal{U}$ on a set X *quasi-proximally maximal* if it a maximal element in its quasi-proximity class. Each quasi-proximally maximal quasi-uniformity on X is the infimum of a family of anti-atoms of $(q(X), \subseteq)$ (see [**5**, Corollary 7], where that result is proved, although a weaker result is stated). For instance the finest compatible quasi-uniformity of any topological space X is the infimum of a family of anti-atoms of $(q(X), \subseteq)$.

It has also been shown that each quasi-uniformity on a set X which has a linearly ordered base or is totally bounded is the infimum of a family of anti-atoms of $(q(X), \subseteq)$. Moreover each quasi-uniformity $\mathcal{U}$ on a set X such that $\tau(\mathcal{U}^s)$ is resolvable[3], is the infimum of a family of anti-atoms of $(q(X), \subseteq)$.

1501? **Question 4.** *Given any preorder T on a set X, is each quasi-uniformity $\mathcal{M}_T$ (as defined above) necessarily an anti-atom of $(q(X), \subseteq)$?*

Note that a counterexample $\mathcal{M}_T$ to the preceding question would yield a quasi-uniformity on X that is not the infimum of a family of anti-atoms of $(q(X), \subseteq)$.

1502? **Question 5.** *Given two distinct uniformities $\mathcal{U}$ and $\mathcal{V}$ on a set X such that $\mathcal{U} \subseteq \mathcal{V}$, does there exist a nonsymmetric quasi-uniformity $\mathcal{Q}$ on X such that $\mathcal{U} \subseteq \mathcal{Q} \subseteq \mathcal{V}$?*

It has been shown [**5**, Corollary 10] that if there exists a uniformity $\mathcal{U}$ on a set X that is not the infimum of a family of anti-atoms of $(q(X), \subseteq)$, then the preceding question has a negative answer.

According to [**4**], the answer to the preceding question is positive if $\mathcal{U}$ and $\mathcal{V}$ belong to different quasi-proximity classes. Furthermore it is also positive if there is a $\mathcal{V}$-discrete subset of X that is not $\mathcal{U}$-discrete, where a subset of a quasi-uniform space is called *discrete* if its subspace quasi-uniformity is the discrete uniformity. Finally, it is also positive if $\mathcal{V} = \mathcal{U} \vee \mathcal{U}(P)$ where P is an equivalence relation on X not belonging to $\mathcal{U}$.

---

[3] Recall that E. Hewitt called a topological space *resolvable* if it has two disjoint dense subsets.

Let us note that, in a certain sense, uniformities in $(q(X),\subseteq)$ are relatively rare. Indeed a quasi-proximity class does not contain any uniformities if its coarsest (unique totally bounded) member is not a uniformity.

There are quasi-uniformities that do not have any upper neighbour, as well as quasi-uniformities that do not have any lower neighbour [**4**]. For instance on any infinite set X, the quasi-uniformity $(\mathcal{D}_X)_\omega$ does not have an upper neighbour. On the other hand each *doubly point-symmetric quasi-uniformity*, that is, each quasi-uniformity $\mathcal{U}$ satisfying $\tau(\mathcal{U}) = \tau(\mathcal{U}^{-1})$, which is not indiscrete has a lower neighbour. Note that by Zorn's Lemma, for any preorder T on a set X, there is a maximal quasi-uniformity $\mathcal{N}_T$ on X not containing T and coarser than the quasi-uniformity $\mathcal{U}(T)$. Clearly $\mathcal{N}_T$ is a lower neighbour of $\mathcal{U}(T)$.

It is easy to see that if $\mathcal{U}$ and $\mathcal{V}$ are two quasi-uniformities on a set X and $\mathcal{U}$ is a lower neighbour of $\mathcal{V}$, then there exists a quasi-uniformity $\mathcal{W}$ with a countable base on X such that $\mathcal{V} = \mathcal{U} \vee \mathcal{W}$. In the light of the examples $\mathcal{N}_T$ just discussed it is natural to wonder whether in this statement we can even assume that $\mathcal{W}$ is generated by some preorder. However the answer to this question is negative: According to the result on double point-symmetry cited above the usual uniformity $\mathcal{R}$ on $\mathbb{R}$ possesses a lower neighbour $\mathcal{U}$ in $(q(\mathbb{R}),\subseteq)$. Observe that $\mathbb{R}^2$ is the only preorder that belongs to the uniformity $\mathcal{R}$, because the topology induced by $\mathcal{R}$ is connected. Hence $\mathcal{R}\setminus\mathcal{U}$ does not contain any preorder and the claim is verified.

Complements

In this section we make use of another concept from lattice theory. A quasi-uniformity $\mathcal{V}$ on a set X is called a *complement* of a quasi-uniformity $\mathcal{U}$ of $(q(X),\subseteq)$ provided that $\mathcal{V}\vee\mathcal{U} = \mathcal{D}$ and $\mathcal{V}\wedge\mathcal{U} = \mathcal{I}$. For instance, for each linear order $\leq$ on X the conjugate of $\mathcal{U}(\leq)$ is a complement of $\mathcal{U}(\leq)$. It is not always easy to decide whether a given quasi-uniformity on a set X has complements in $(q(X),\subseteq)$. However it is readily seen that each quasi-uniformity that has a complement, possesses a complement having a countable base [**4**]. Furthermore it is known that each quasi-uniformity $\mathcal{U}$ of $(q(X),\subseteq)$ that is generated by a preorder has a complement that is generated by a preorder (compare [**4**]).

Question 6. *Does each quasi-uniformity on a set X which possesses a complement in $(q(X),\subseteq)$ have a complement in $(q(X),\subseteq)$ that is generated by a preorder?* 1503?

Observe that the answer to the preceding question is positive for quasi-uniformities that have a transitive complement. Indeed the aforementioned question is equivalent to the following problem: Does each quasi-uniformity on a set X that possesses a complement in $(q(X),\subseteq)$ have a transitive complement in $(q(X),\subseteq)$?

It is known [**4**] that a nondiscrete quasi-uniformity on a set X which is proximally discrete does not have a complement in $(q(X),\subseteq)$. Thus for instance on an infinite set X, $(\mathcal{D}_X)_\omega$ does not have a complement in $(q(X),\subseteq)$. It follows that for a given set X, the lattice $(q(X),\subseteq)$ is complemented if and only if X is finite [**4**].

It can be shown [**3**] that the convergent sequence $S := \{\frac{1}{n+1} : n \in \omega\} \cup \{0\}$ equipped with its usual totally bounded complete metrizable uniformity does not have a complement in $(q(S), \subseteq)$. Also the unique compatible quasi-uniformity on an uncountable set X which carries the cofinite topology does not possess a complement in $(q(X), \subseteq)$ [**3**]. Recall that on the other hand the lattice of all topologies on any set X (with set-theoretic inclusion as partial order) is complemented (see for instance [**11**]).

The following positive results were established in [**3**, **4**]: The finest compatible quasi-uniformity of each countable T_1-space X has a complement in $(q(X), \subseteq)$. Every atom (resp. proximally nondiscrete anti-atom) of $(q(X), \subseteq)$ has a complement in $(q(X), \subseteq)$. Each compatible uniformity on a resolvable completely regular space X is complemented in $(q(X), \subseteq)$ (see [**4**, Proposition 8]).

Furthermore, in [**3**] the following results concerning the involved lattices of quasi-uniformities were obtained: If $\mathcal{U}$ and $\mathcal{V}$ are two complemented quasi-uniformities on sets X and Y respectively, then both their sum $\mathcal{U} \oplus \mathcal{V}$ and product $\mathcal{U} \times \mathcal{V}$ are complemented. If $(X, \mathcal{U})$ is a quasi-uniform space that has a complemented *doubly dense* subspace Y (that is, Y is dense in X with respect to the topologies $\tau(\mathcal{U})$ and $\tau(\mathcal{U}^{-1})$), then $\mathcal{U}$ has a complement.

Given a set X, in general complements in $(q(X), \subseteq)$ are highly non-unique. This fact motivates our last question (compare for instance with the studies [**1**, **10**, **11**]).

1504? **Question 7.** *On a given set X, determine the possible numbers of complements for an arbitrary element $\mathcal{U}$ of $(q(X), \subseteq)$.*

References

[1] J. I. Brown and S. Watson, *The number of complements of a topology on n points is at least 2^n (except for some special cases)*, Discrete Math. **154** (1996), 27–39.

[2] P. Fletcher and W. F. Lindgren, *Quasi-uniform spaces*, Marcel Dekker Inc., New York, 1982.

[3] E. P. de Jager and H.-P. A. Künzi, *Infima and complements in the lattice of quasi-uniformities*, 2005, Preprint.

[4] E. P. de Jager and H.-P. A. Künzi, *Atoms, anti-atoms and complements in the lattice of quasi-uniformities*, Topology Appl. **153** (2006), no. 16, 3140–3156.

[5] E. P. de Jager and H.-P. A. Künzi, *Infima of quasi-uniform anti-atoms*, Topology Appl. **153** (2006), no. 17, 3327–3337.

[6] N. Levine and L. J. Nachman, *Adjacent uniformities*, J. Math. Tokushima Univ. **5** (1971), 29–36.

[7] J. Pelant and J. Reiterman, *Complements in the lattice of uniformities*, Comment. Math. Univ. Carolin. **22** (1981), no. 2, 399–412.

[8] J. Pelant, J. Reiterman, V. Rödl, and P. Simon, *Ultrafilters on ω and atoms in the lattice of uniformities. I*, Topology Appl. **30** (1988), no. 1, 1–17.

[9] J. Pelant, J. Reiterman, V. Rödl, and P. Simon, *Ultrafilters on ω and atoms in the lattice of uniformities. II*, Topology Appl. **30** (1988), no. 2, 107–125.

[10] J. Steprāns and S. Watson, *Mutually complementary families of T_1 topologies, equivalence relations and partial orders*, Proc. Amer. Math. Soc. **123** (1995), no. 7, 2237–2249.

[11] S. Watson, *The number of complements in the lattice of topologies on a fixed set*, Topology Appl. **55** (1994), no. 2, 101–125.

Open Problems in Topology II – Edited by E. Pearl

Topology in North Bay: some problems in continuum theory, dimension theory and selections

Alexandre Karasev, Murat Tuncali and Vesko Valov

Dedicated to Ted Chase

1. Introduction

This article reports on some of the research activities in Topology at Nipissing University. Although our research areas encompass geometric topology, dimension theory, general topology, topological algebras, functional analysis, continuum theory and topological dynamics,in this article we only concentrate on some problems in dimension theory, selections and continuum theory. Section 2 is devoted to the problems on extension dimension. In the third section, problems concerning selections and C-spaces are discussed. The fourth section discusses questions concerning the parametric version of disjoint disks property. The last section is devoted to locally connected Hausdorff continua and rim-metrizablity.

Historically, since 1994, there have been regular Topology workshops in the month of May. Initially these workshops were organized by Tuncali. In 2000, Vesko Valov joined Nipissing University, and in 2003, Alexandre Karasev joined this group. The group has been organizing ongoing seminars and workshops at Nipissing. Since 1994, many topologists have visited Nipissing and participated in workshops and seminars. Among them, Nikolay Brodskiy, Dale Daniel, John C. Mayer, Jacek Nikiel and E.D. Tymchatyn have been visiting Nipissing regularly. During the last three years, Jacek Nikiel (2003–04), Taras Banakh (2004–2005) and Andriy Zagorodnyuk (2005–06) have visited Nipissing for entire academic years. Thus, this article is about some of the research interests of the topology group at Nipissing as well as some of the ongoing research programs of seminar/workshop participants.

2. Problems in dimension theory

All spaces in this section are assumed to be metrizable and separable, if not stated otherwise. Let G be an Abelian group. As usual, $K(G,n)$ is the *Eilenberg–Mac Lane complex*, i.e., a CW complex such that $\pi_n(K(G,n)) \approx G$ and $\pi_i(K(G,n)) \approx 0$ for all $i \neq n$. The *cohomological dimension* of a space X with respect to the coefficient group G is denoted by $\dim_G X$. Recall that a space Y is an *absolute (neighborhood) extensor* for X [notation: $Y \in A(N)E(X)$] if any map

The authors are partially supported by their NSERC Grants 257231-04, 141066-04 and 261914-03, respectively.

to Y, defined on an arbitrary closed subspace A of X, can be extended to a map of the whole X to Y (resp., to a map of some open neighborhood of A to Y).

Following Dranishnikov [**27**] we say that *extension dimension* of a space X does not exceed a complex L, notation $\text{e-dim}\, X \leq L$, if $L \in AE(X)$. Note that $\dim X \leq n \Leftrightarrow \text{e-dim}\, X \leq S^n$ and $\dim_G X \leq n \Leftrightarrow \text{e-dim}\, X \leq K(G,n)$. Further, we say [**27**] that $L \leq K$ if for each space X the condition $L \in AE(X)$ implies the condition $K \in AE(X)$. The relation $L \leq K$ leads to a preorder relation on complexes and generates an equivalence relation. Obviously, this equivalence relation can be described in terms of extension of maps: a complex L is equivalent to a complex K if $L \in AE(X) \Leftrightarrow K \in AE(X)$ for any space X. The equivalence class of complex L is called the extension type of L and is denoted by $[L]$. Due to the homotopy extension property of ANR-spaces, if L is homotopy equivalent to K, then $[L] = [K]$. The converse is not the case. For example, $[S^n] = [S^n \vee S^m]$, if $n \leq m$. It should be emphasized that there are many incomparable complexes. For instance, results of [**29**] imply that $\mathbb{R}P^2$ is not comparable with any sphere S^n, $n \geq 2$. It can be shown that for complexes L and K the minimum of their extension types is given by the extension type $[L \vee K]$. We refer the reader to the papers [**12**, **28**, **34**] for more information about extension dimension and extension types.

Extension and cohomological dimensions are related as follows.

Theorem 2.1 (Dranishnikov [**26**]). *Let X be a metrizable compactum and L be a complex. If $\text{e-dim}\, X \leq L$ then $\dim_{H_n(L)} \leq n$ for all positive integers n. If $\dim X < \infty$ and L is a simply connected complex then the following three conditions are equivalent: (1) $\text{e-dim}\, X \leq L$; (2) $\dim_{H_n(L)} X \leq n$ for all $n > 0$; (3) $\dim_{\pi_n(L)} X \leq n$ for all $n > 0$.*

Dydak generalized the above theorem on the case of metrizable noncompact spaces [**31**, **33**]. The condition $\dim X < \infty$ cannot be removed due to the existence of infinite-dimensional compacta of finite integral cohomological dimension [**24**, **38**]. Nevertheless, Dydak proved in [**31**] that condition (3) implies condition (1) if X is a metrizable space which is an absolute neighborhood extensor for metrizable spaces. There is a hope that Theorem 2.1 remains valid in the class of C-spaces (see next section for the definition of C-spaces).

1505? **Problem 2.1.** *Does Theorem 2.1 hold if the compactum X is a C-space?*

Another generalization of Theorem 2.1 belongs to Cencelj and Dranishnikov [**8**–**10**] and weakens the requirement of simply connectedness. Namely, Theorem 2.1 remains true for nilpotent complexes. The condition on the fundamental group of L cannot be dropped completely. Indeed, if X is a two-dimensional disk and L is a complex such that $\pi_1(L)$ is nontrivial and $\widetilde{H}_*(L) = 0$ then the implication (2) $\Rightarrow$ (1) of the above theorem does not hold [**35**]. An example of such L can be found, for instance, in [**47**, Example 2.38, p. 142].

Finally, consider the equivalence (1) $\Leftrightarrow$ (3) and take $\mathbb{R}P^2$ as a simplest example of a complex which is not nilpotent. Note that $H_1(\mathbb{R}P^2) \approx \pi_1(\mathbb{R}P^2) \approx \mathbb{Z}_2$ and hence, for any metrizable compactum X, $\text{e-dim}\, X \leq \mathbb{R}P^2$ implies $\dim_{\mathbb{Z}_2} X \leq 1$

due to Theorem 2.1. What can be said about the converse implication? The results of Levin [**51**] imply the existence of a metrizable compactum X such that $\dim_{\mathbb{Z}_2} X = 1$ and $\text{e-dim}\, X \not\leq \mathbb{R}P^2$. Nevertheless, this compactum is infinite-dimensional and the following well-known problem is still open.

Problem 2.2. *Does* $\dim_{\mathbb{Z}_2} X \leq 1$ *imply* $\text{e-dim}\, X \leq \mathbb{R}P^2$ *for a finite dimensional compactum* X*?* 1506?

Since the infinite projective space $\mathbb{R}P^\infty$ is the space of type $K(\mathbb{Z}_2, 1)$, the above problem can be restated as follows: does $\mathbb{R}P^\infty \in AE(X)$ imply $\mathbb{R}P^2 \in AE(X)$ for a finite dimensional compactum X? Dydak and Levin proved in [**35**] that the answer is positive if $\dim X \leq 3$.

The study of universal spaces of a given dimension occupies one of the central places in dimension theory. Recall that a space U is called *a universal space* for a class of topological spaces $\mathcal{C}$ if $U \in \mathcal{C}$ and any space X from $\mathcal{C}$ admits an embedding in U. Here are several classical papers devoted to the topic of universal spaces of a given dimension: [**5**, **55**, **56**, **61**, **63**, **66**, **73**]. In connection with the universal spaces theme one should mention the following unsolved problem.

Problem 2.3 (West [**72**]). *Does there exist a universal metric compactum of a given integral cohomological dimension?* 1507?

Employing the concept of extension dimension, the above problem can be generalized as follows.

Problem 2.4 (Chigogidze [**12**]). *Let* $\mathcal{C}_L$ *denote the class of all metrizable compacta* X *such that* $\text{e-dim}\, X \leq [L]$, *where* L *is a countable and locally finite complex. Characterize all such complexes* L *for which* $\mathcal{C}_L$ *contains a universal space.* 1508?

Problem 2.3 is obtained from Problem 2.4 by letting $L = K(\mathbb{Z}, n)$. Everywhere below, by $[L]$-universal compactum we mean a universal object for the class $\mathcal{C}_L$.

Problem 2.4 has partial solutions. Results of Chigogidze [**12**, Theorem 2.5] and Dydak [**33**] imply that a universal compactum exists in the case when L is finite or, more generally, finitely dominated. A standard way of obtaining such a universal compactum consists in construction of $[L]$-invertible map $f\colon X_L \to \mathbb{I}^\omega$, where X_L is a metric compactum with $\text{e-dim}\, X_L = [L]$.

Definition ([**12**]). A map $f\colon X \to Y$ is called *$[L]$-invertible* if for each space Z with $\text{e-dim}\, Z \leq [L]$ and for any map $g\colon Z \to Y$ there exists a map $h\colon Z \to X$ satisfying the conditions $f \circ h = g$.

Note that $[L]$-invertibility of L and universality of $\mathbb{I}^\omega$ for all metrizable compacta guarantees $[L]$-universality of X_L.

A universal object exists for any (countable and locally finite) complex L if we enlarge the class of spaces. Dydak and Mogilski [**36**] proved that for a given n there exists a Polish space X of integral cohomological dimension n which contains a topological copy of any separable metric space Y with $\dim_{\mathbb{Z}} Y \leq n$. This result was generalized by Olszewski [**62**], who proved the existence of a universal separable metric space of given extension dimension $[L]$ (where L is a countable and locally finite CW complex). As a corollary, this implies the existence of universal separable metrizable space of a given cohomological dimension with respect to any countable Abelian coefficient group.

There is an important connection between the existence of an $[L]$-universal compactum, $[L]$-invertible mappings, and the following question [**11**]: for a given complex L, does $\text{e-dim}\, X \leq L$ imply $\text{e-dim}\, \beta X \leq L$ for any space X? The study of this connection led to the introduction of a new class of CW complexes.

Definition ([**48**, **49**]). We say that a complex L is *quasi-finite* if for every finite subcomplex P of L there exists a finite subcomplex eP of L containing P such that the pair $P \subset eP$ is $[L]$-connected for Polish spaces. The latter means that any map $f\colon A \to P$, defined on a closed subset A of a Polish space X with $\text{e-dim}\, X \leq L$ can be extended to a map of X into eP.

The following theorem [**11**, **48**] reveals the relation between $[L]$-universality and quasi-finite complexes. Equivalences (ii)–(vi) of this theorem are due to Chigogidze [**11**].

Theorem 2.2. *Let L be a countable and locally finite CW complex. Then the following conditions are equivalent:*

(i) *L is quasi-finite;*
(ii) $\text{e-dim}\, \beta X \leq [L]$ *whenever X is a (Tychonoff) space with* $\text{e-dim}\, X \leq [L]$*;*
(iii) $\text{e-dim}\, \beta X \leq [L]$ *whenever X is normal and* $\text{e-dim}\, X \leq [L]$*;*
(iv) $\text{e-dim}\, \beta(\bigoplus\{X_t : t \in T\}) \leq [L]$ *whenever T is arbitrary and each X_t, $t \in T$, is a separable metrizable space with* $\text{e-dim}\, X_t \leq [L]$*;*
(v) $\text{e-dim}\, \beta(\bigoplus\{X_t : t \in T\}) \leq [L]$ *whenever T is arbitrary and each X_t, $t \in T$, is a Polish space with* $\text{e-dim}\, X_t \leq [L]$*;*
(vi) *There exists an $[L]$-invertible map $f\colon X \to \mathbb{I}^\omega$, where X is a metrizable compactum with* $\text{e-dim}\, X \leq [L]$.

From the results of Dranishnikov [**23**], Dydak [**32**], Dydak and Walsh [**37**], and Levin [**52**], we know that for each $n \geq 2$ there exists a (metrizable separable) space X with integral cohomological dimension $\dim_{\mathbb{Z}} X \leq n$ and $\dim_{\mathbb{Z}} \beta X > n$. Therefore the Eilenberg–Mac Lane complex $K(\mathbb{Z}, n)$ is not quasi-finite for all $n \geq 2$. In fact, the results of Levin [**52**] imply that $K(G, 2)$ is not quasi-finite for any nontrivial Abelian group G.

Quasi-finite complexes provide a negative answer to the following question by Chigogidze [**13**] and Dydak [**33**]: suppose that $[L]$-universal compactum exists; is it true that the extension type of L contains a finitely dominated complex? There exists a quasi-finite complex which is not equivalent to a finitely dominated complex [**48**].

In the light of the preceding discussion, a natural question to ask is the following.

1509? **Problem 2.5.** *Let L be a complex such that $[L]$-universal compactum exists. Is it true that L is quasi-finite?*

Further, the example of a quasi-finite complex in [**48**] is a bouquet of finite complexes. What are other possible examples?

Problem 2.6. *Is there a quasi-finite complex which is not equivalent to a bouquet* 1510?
of finite complexes?

Recalling main results about universal spaces in classical dimension n, one may vary Problem 2.4 as follows.

Problem 2.7. *Characterize all complexes L for which $\mathcal{C}_L$ contains a universal ob-* 1511?
ject which is an absolute extensor in dimension $[L]$ for Polish spaces (or metrizable compacta).

As usual, a space Y is called an absolute (neighborhood) extensor in dimension $[L]$, shortly $Y \in A(N)E([L])$, for a given class of spaces $\mathcal{C}$ if $Y \in A(N)E(X)$ for all X from $\mathcal{C}$ such that $\text{e-dim}\, X \leq [L]$.

If a complex L is finite, then $[L]$-universal compact absolute extensors in dimension L exist [**12**]. On the other hand, Zarichnyi [**74**] proved that there is no universal compactum of a given integral cohomological dimension which is an absolute extensor with respect to metrizable compacta of given cohomological dimension. Thus, in the Problem 2.7 the complex L cannot be the Eilenberg–Mac Lane complex $K(\mathbb{Z}, n)$, $n \geq 2$.

There is some hope that quasi-finite complexes may be candidates to provide a solution to Problem 2.7. Namely, it is shown in [**50**] that if there exists an $[L]$-universal compactum which is an absolute extensor in dimension $[L]$ for Polish spaces then L must be quasi-finite.

3. Selections and C-property

All spaces in this section are supposed to be paracompact and all maps continuous. By a perfect space we mean a space without isolated points.

Recall that a X is a *C-space* if for any sequence $\{\nu_n\}_{n=1}^{\infty}$ of open covers of X there exists a sequence $\{\gamma_n\}_{n=1}^{\infty}$ of disjoint open families in X such that each γ_n refines ν_n and $\bigcup_{n=1}^{\infty} \nu_n$ is a cover of X. Every countable-dimensional (a countable union of 0-dimensional subsets) metric space is a C-space [**40**]. R. Pol constructed a metrizable compact C-space which is not countable-dimensional [**65**].

Problem 3.1 (V. Gutev). *Let $f\colon X \to Y$ be an open surjective map between* 1512?
the metrizable spaces X and Y such that Y is a C-space and each fiber $f^{-1}(y)$, $y \in Y$ is (zero-dimensional) compact and perfect. Does there exist a surjective map $g\colon X \to Y \times \mathbb{I}$ such that $f = \pi_Y \circ g$, where $\pi_Y : Y \times \mathbb{I} \to Y$ is the projection?

According to a result of Bula [**7**, Theorem 1], Problem 3.1 has an affirmative answer in case X and Y are metrizable and Y is finite-dimensional. Gutev [**44**, Theorem 1.1] extended the Bula theorem for arbitrary metrizable X and countable-dimensional Y.

Observe that Problem 3.1 is equivalent to the following one:

1513? **Problem 3.2** (V. Gutev). *Is it true that under the hypotheses of Question 2.1 there exists a map $h\colon X \to \mathbb{I}$ with $h(f^{-1}(y)) = \mathbb{I}$ for every $y \in Y$, equivalently, $f(h^{-1}(t)) = Y$ for all $t \in \mathbb{I}$?*

Levin and Rogers [**53**, Theorem 1.3] proved that Problem 3.2 has a positive solution in the class of compact space. Therefore, the answer to Problem 3.1 is also "yes" for compact X and Y. In fact, one can try to solve a simplified version of Problem 3.1 and Problem 3.2. It is easily seen that a positive answer to one of this questions implies a positive answer to next question.

1514? **Problem 3.3** (V. Gutev). *Let X, Y and f be as in Problem 3.1. Are there closed sets $F, H \subset X$ such that $F \cap H = \emptyset$ and $f(F) = f(H) = Y$?*

Note that Dranishnikov [**25**] constructed an open surjection $f\colon X \to Y$ of metrizable compacta having all fibers homeomorphic to the Cantor set and such that there are no disjoint closed sets $F, H \subset X$ with $f(F) = f(H) = Y$. Hence, Problems 3.1 and 3.2 have a negative answer if there is no dimensional restrictions on Y.

There exists an equivalent version of Problem 3.3 in terms of semi-continuous selections. Recall that a set-valued map $\varphi\colon Y \to \mathcal{S}(X)$, where $\mathcal{S}(X)$ denotes the family of all nonempty subsets of X, is called lower (resp., upper) semi-continuous if for every open set $U \subset X$ the set $\{y \in Y : \varphi(y) \cap U \neq \emptyset\}$ (respectively, $\{y \in Y : \varphi(y) \subset U\}$) is open in Y. By $\mathcal{C}(X)$ we denote the family of compact nonempty subsets of X.

1515? **Problem 3.4** (V. Gutev). *Let Y be a metrizable C-space, X be metrizable and $\varphi\colon Y \to \mathcal{C}(X)$ be an l.s.c. map such that each $\varphi(y)$, $y \in Y$, is perfect. Does there exist a u.s.c. map $\theta\colon Y \to \mathcal{C}(X)$ with $\theta(y) \subset \varphi(y)$ and $\varphi(y) \setminus \theta(y) \neq \emptyset$ for every $y \in Y$?*

The existence of a u.s.c. map $\theta\colon Y \to \mathcal{C}(X)$ satisfying the conditions from Problem 3.4 is equivalent to the existence of two u.s.c. maps $\theta_i\colon Y \to \mathcal{C}(X)$, $i = 1, 2$, such that $\theta_1(y) \cap \theta_2(y) = \emptyset$ and $\theta_i(y) \subset \varphi(y)$ for all $y \in Y$ and $i = 1, 2$ (in such a case we say that φ admits disjoint u.s.c. selections). Actually, Dranishnikov's example mentioned above is based on this observation, he constructed an open surjection $f\colon X \to Y$ such that the map $\varphi(y) = f^{-1}(y)$ does not admit any disjoint u.s.c. selections.

On the other hand, there exist a few characterizations of paracompact C-spaces in terms of selections for set-valued maps. One of them was established by Uspenskij [**71**, Theorem 1.3] and another one by Gutev–Valov [**45**]. So, it is interesting whether the selection condition from Problem 3.4 also characterizes C-spaces.

1516? **Problem 3.5.** *Is it true that a metrizable space Y is a C-space if and only if any l.s.c. map $\varphi\colon Y \to \mathcal{C}(X)$ with perfect point-images $\varphi(y)$, $y \in Y$, and metrizable X admits disjoint u.s.c. selections?*

4. Parametrization of the disjoint n-disks property

In this section, unless stated otherwise, all spaces are Tychonoff.

The following property was introduced in [**4**] as a parametrization of the well know *disjoint n-disks property*. We say that a space X has the m-$\overline{\mathrm{DD}}^{\{n,k\}}$-*property*, where m, n, k are positive integers or infinity, if for if for any open cover $\mathcal{U}$ of X and two maps $f\colon \mathbb{I}^m \times \mathbb{I}^n \to X$, $g\colon \mathbb{I}^m \times \mathbb{I}^k \to X$ there exist maps $f'\colon \mathbb{I}^m \times \mathbb{I}^n \to X$, $g'\colon \mathbb{I}^m \times \mathbb{I}^k \to X$ such that $f' \sim_{\mathcal{U}} f$, $g' \sim_{\mathcal{U}} g$, and $f'(\{z\} \times \mathbb{I}^n) \times g'(\{z\} \times \mathbb{I}^k\} = \emptyset$ for all $z \in \mathbb{I}^m$. Here $f' \sim_{\mathcal{U}} f$ means that f' is $\mathcal{U}$-homotopic to f.

The importance of the m-$\overline{\mathrm{DD}}^{\{n,k\}}$-property is justified by the following results established in [**4**]:

(1) Let $X \in m$-$\overline{\mathrm{DD}}^{\{n,n\}}$ be a locally contractible and completely metrizable space and $p\colon K \to M$ be a perfect map between metrizable C-spaces with $\dim M \le m$ and $\dim f \le n$. Then the function space $C(K, X)$ equipped with the source limitation topology contains a dense G_δ-subset consisting of maps that are injective on each fiber of p.

(2) Let m, n, k, d, l be non-negative integers, L be a metrizable space with the 0-$\overline{\mathrm{DD}}^{\{0,0\}}$-property and D be a metrizable space with the 0-$\overline{\mathrm{DD}}^{\{0,d+l\}}$-property. If $m + n + k < 2d + l$, then the product $D^d \times L^l$ has the m-$\overline{\mathrm{DD}}^{\{n,k\}}$-property.

It follows from the above two results that $\mathbb{D}^d \times \mathbb{R}^l \in m$-$\overline{\mathrm{DD}}^{\{n,k\}}$ for any m, n, k, d, l with $m+n+k < 2d+l$, where $\mathbb{D}$ is a dendrite with a dense set of end-points and $\mathbb{R}$ is the real line. The last statement with $m = d = 0$ and $l = 2n+1$ is actually the Lefschetz–Menger–Pontrjagin embedding theorem; the case $m = l = 0$ and $d = n + 1$ is the embedding theorem of Bowers [**6**]; the case $m = 0$, $d = n$ and $l = 1$ is the embedding theorem of Sternfeld [**68**]; the case $m = 0$ and $d = 0$ is close to the embedding theorem from Banakh–Trushchak [**3**], while for $l = 0$ and $m = 0$ it is close to that one from Banakh–Cauty–Trushchak–Zdomskyy [**1**]; finally, letting $d = 0$ we obtain the Pasynkov theorem [**64**] asserting that for a map $p\colon X \to Y$ between compacta the function space $C(X, \mathbb{R}^{\dim Y + 2\dim(p)+1})$ contains a dense G_δ-set of maps that are injective on each fiber of the map p.

However, another generalization of Pasynkov's result due to H. Toruńczyk [**69**] is not covered by the statements (1) and (2):

> If $p\colon X \to Y$ is a map between compacta, then $C(X, \mathbb{R}^{\dim X + \dim(p)+1})$ contains a dense G_δ-set of maps that are injective on each fiber of the map p.

Since the Euclidean space $\mathbb{R}^d$ has the m-$\overline{\mathrm{DD}}^{\{n,k\}}$-properties for all m, n, k with $m + n + k < d$, we may ask whether the mentioned theorem of H. Toruńczyk [**69**] is true in the following more general form.

Problem 4.1 ([4]). *Does any map $p\colon K \to M$ between finite-dimensional compact* 1517?
metric spaces embed into the projection $\mathrm{pr}\colon M \times X \to M$ *along a Polish AR-space X possessing the m-$\overline{\mathrm{DD}}^{\{n,k\}}$-property for all m, n, k with $m + n + k \le \dim(K) + \dim(p)$?*

Let us also note that the above result of H. Toruńczyk would follow from [**4**, Theorem 1] if the following problem had an affirmative answer.

Problem 4.2 ([4]). *Let $f: X \to Y$ be a k-dimensional map between finite-dimensional metrizable compacta. Is it true that there is a map $g: Y \to Z$ to a compact space Z with* $\dim Z \leq \dim X - k$ *such that the map $g \circ f$ is still k-dimensional?*

Next two questions concern the minimal dimension of spaces with $m\text{-}\overline{\mathrm{DD}}^{\{n,n\}}$. It is known [**4**] that the smallest possible dimension of compact metrizable AR with $X \in m\text{-}\overline{\mathrm{DD}}^{\{n,n\}}$ is either $n+\lfloor \frac{m+1}{2} \rfloor$ or $n+\lceil \frac{m+1}{2} \rceil$, where $\lfloor r \rfloor = \max\{k \in \mathbb{Z} : k \leq r\}$ and $\lceil r \rceil = \min\{k \in \mathbb{Z} : k \geq r\}$.

1518? **Problem 4.3.** *What is the smallest possible dimension of Polish spaces with* $m\text{-}\overline{\mathrm{DD}}^{\{n,n\}}$? *Is it* $n + \lceil \frac{m+1}{2} \rceil$?

1519? **Problem 4.4.** *What is the smallest possible dimension of metrizable compacta X such that $X \times \mathbb{I}^n$ contains a copy of the n-dimensional Menger cube? Is it* $\lceil \frac{m}{2} \rceil$?

The last question in this section is a reformulation of the well known problem of finding a characterization of codimension one manifold factors, see [**22, 46**].

1520? **Problem 4.5.** *Let $X \times \mathbb{R}$ is an n-manifold with $n \geq 5$. Does X have the* $1\text{-}\overline{\mathrm{DD}}^{\{1,1\}}$-*property?*

5. Locally connected continua

By a *continuum*, we mean a compact connected Hausdorff space. A *compact ordered space* is a compact space with topology induced by a linear order. An *arc* is a compact ordered space which is connected. Equivalently, an arc is a continuum with exactly two non cut-points. Let P be a topological property. A topological space X is said to be *rim-P* if it has a basis of open sets with boundaries having the property P. Some of the spaces with natural rim-properties are *rim-finite* spaces, *rim-countable* spaces, *rim-metrizable* spaces, *rim-scattered* spaces and rim-compact spaces.

The Hahn–Mazurkiewicz Theorem (1914) characterizes the continuous images of the closed unit interval as locally connected metric continua. A theorem of Alexandroff characterizes the continuous images of the Cantor set as the class of compact metric spaces. In the non-metric case, continuous images of arcs and more generally, of compact ordered spaces are quite restricted and interesting. Mardešić (1960) gave an example of a locally connected continuum which is not a continuous image of an arc. Treybig (1964) showed that continuous images of compact ordered spaces do not contain a non-metric product of (infinite) compact spaces. In 1967, Mardešić [**54**] proved an important result: every continuous image of a compact ordered space is rim-metrizable. Heath, Lutzer and Zenor (1973) proved that continuous images of ordered compacta are monotonically normal. Nikiel (1988) characterized the continuous images of arcs in the non-metric case. He also proved that each hereditarily locally connected continuum is a continuous image of an arc and rim-countable, see [**57**]. Nikiel, Tymchatyn and Tuncali (1991) gave an example of a rim-countable locally connected continuum which is not a continuous image of an arc.

Following these results, the study of images of arcs/compact ordered spaces developed in several directions. One study focused on the behavior of images of arcs under inverse limits. Nikiel, Tymchatyn and Tuncali [**60**](1993) proved that the inverse limit of an inverse sequence of images of arcs with monotone bonding maps is a continuous image of an arc. They also proved that each one-dimensional continuous image of an arc can be obtained as an inverse limit of inverse sequence of rim-finite continua with monotone bonding maps. This result extends the similar theorem of Nikiel (1989) in the metric case, and indicates that in the 1-dimensional case, images of arcs behave like metric locally connected continua.

Tuncali [**70**] proved that continuous images of rim-metrizable continua do not contain a non-metric product of nondegenerate continua. These results suggest that some properties of images of compact ordered space/arcs depend on the boundary structure of basic open sets. Nikiel, Treybig and Tuncali [**58**] (1995) showed that continuous images of rim-metrizable locally connected continua are not necessarily rim-metrizable, hence Mardešić's 1967 result cannot be generalized. This result shows that the class of rim-metrizable continua is large. In 1989, Nikiel asked if every monotonically normal compactum is a continuous image of a compact ordered space. M.E. Rudin [**67**] (2001) answered that question affirmatively. Note that Ostaszewski (1978) proved that a separable, monotonically normal space is hereditarily Lindelöf. Therefore, each separable, monotonically normal, compact space is perfectly normal. On the other hand, under the continuum hypothesis, Filippov [**41**] (1969) constructed a perfectly normal and locally connected continuum which is nonmetrizable and has a basis of open sets with 0-dimensional metrizable boundaries. Gruenhage [**42**] (1990) also constructed an example of a perfectly normal locally connected continuum which is nonmetrizable, rim-metrizable, and not arcwise connected. Note that a product of $[0,1]$ with a Souslin line is a perfectly normal, localy connected continuum which is not rim-metrizable. Following these results, an interesting problem to consider is the following:

Problem 5.1. *Characterize locally connected, rim-metrizable, perfectly normal* 1521?
continua.

Recently, Daniel, Nikiel, Treybig, Tymchatyn and Tuncali in a sequence of papers have been investigating various properties of continuous images of arcs, Suslinian continua, rim-metrizable continua, and perfectly normal compact spaces, [**14–21**]. They proved that each Suslinian continuum is perfectly normal and rim-metrizable, [**16**]. A continuum is said to be *Suslinian* if it does not contain an uncountable collection of mutually disjoint continua. Lelek introduced Suslinian continua in 1971. Using inverse limit techniques, Daniel, Nikiel, Treybig, Tymchatyn and Tuncali showed that locally connected Suslinian continua must have weight ω_1 and under the Souslin Hypothesis such continua are metrizable. In [**2**], these results are improved. It is proved that all Suslinian continua must have weight ω_1, and under the Souslin Hypothesis, all Suslinian continua are metrizable. In [**18**], it is also proved that each homogenous Suslinian continuum X must be locally connected, and moreover, if X is seperable, then it must be metrizable. Another interesting result is that under Souslin Hypothesis each perfectly normal compact space of weight ω_1 contains an uncountable, upper semi-continuous, almost null family of nondegenerate, pairwise disjoint, closed subsets, [**17**]. Moreover, if X is locally connected continuum, the members of this family can be chosen to be continua.

Recently, Todd Eisworth [**39**] announced that each separable monotonically normal compact space admits two-to-one map onto a metric space. These results are related to the following well-known questions. First one is due to M.E. Rudin and the second one is due to D.H. Fremlin.

1522? **Problem 5.2.** *Is it consistent that each perfectly normal, locally connected continua is metrizable?*

1523? **Problem 5.3.** *Is it consistent that every pefectly normal compact space admits a two-to-one continuous map onto a metric space?*

It is not difficult to see that a locally connected, perfectly normal continuum X with small inductive dimension $\operatorname{ind}(X) = 1$ is rim-metrizable. Filippov's 1969 example is such a continuum. Moreover, the product $X \times [0,1]$ of a locally connected perfectly normal continuum X and $[0,1]$ is locally connected and perfectly normal again. However, $X \times [0,1]$ is rim-metrizable if and only if X is metrizable. This follows from the fact that rim-metrizable continua do not contain a nonmetric product of nondegenerate continua, [**70**]. These show why Problem 5.1 is natural to consider. Concerning Problem 5.2 and 5.3, readers are also referred to the article by Gary Gruenhage and Justin Moore titled "Perfect compacta and basis problems in topology" in this volume [**43**].

Daniel and Treybig, [**21**] showed that if there is an example of a locally connected Suslinian continuum which is not a continuous image of an arc, there is such a continuum X which is separable. Therefore, it will be interesting to know the answer to the following question.

1524? **Problem 5.4.** *Is a locally connected, rim-metrizable continuum X with no nondegenerate metric continuum rim-finite?*

In [**20**], it was shown that such a continuum X is rim-finite with the additional assumption that X contains no separable subcontinuum. It is known that each rim-finite continuum is a continous image of an arc. Also, Suslinian nonseparable continua are rim-finite on some open set, [**16**].

In addition, in [**16**, **21**] various interesting properties of Suslinian continua were investigated. There are some interesting problems concerning Suslinian continua remain to be answered.

1525? **Problem 5.5.** *Is a separable Suslinian continuum hereditarily separable?*

1526? **Problem 5.6.** *If X is a locally connected Suslinian continuum, is X connected by arcs (ordered continua)?*

Recently, in [**2**] a new cardinal invariant is introduced. Namely,

$$\mathrm{Sln}(X) = \sup\{|\mathcal{C}| : \mathcal{C} \text{ is a disjoint family of nondegenerate subcontinua of } X\}$$

defined for any continuum X and is called the *Suslinian number* of X. Thus a continuum X is Suslinian if and only $\mathrm{Sln}(X) \leq \aleph_0$. It is clear that $\mathrm{Sln}(X) \leq \mathrm{Sln}(Y)$ for any pair $X \subset Y$ of continua. It is convenient to extend the definition of $\mathrm{Sln}(X)$ to all Tychonov spaces by letting

$$\mathrm{Sln}(X) = \min\{\mathrm{Sln}(Y) : Y \text{ is a continuum containing } X\}$$

for a Tychonov space X. Like many other cardinal invariants the Suslinian number is monotone.

For any Tychonov space X, the hereditary Lindelöf number of any space X is bounded from above by the Suslinian number of X, *see also* ip-ktv-bft This generalizes the result of Daniel, Nikiel, Treybig, Tuncali and Tymchatyn that each Suslinian continua is perfectly normal, [**16**]. Since each Suslinian continua is rim-metrizable, it is natural to ask the following question.

Problem 5.7. *Is* $\text{rim-w}(X) \leq \mathrm{Sln}(X)$ *for any compact Hausdorff space?* 1527?

Note that for a given a topological space X $\text{rim-w}(X) = \min\{\sup_{U\in\mathcal{B}} w(\partial U) : \mathcal{B} \text{ is a base of the topology of } X\}$ is the *rim-weight* of X.

In addition to problems stated above, there are number of questions concerning rim-properties of locally connected continua. We list them below. For further reading, we refer readers to [**15**] and [**59**]. Note that some of these problems were listed before in various papers cited in this section.

Problem 5.8. *Is each rim-scattered locally connected continuum rim-metrizable?* 1528?

Note that Drozdovskiĭ and Filippov [**30**], gave an example of a rim-scattered, rim-metrizable locally connected continuum which is not rim-countable.

Problem 5.9. *Let X be a rim-metrizable locally connected continuum. Does X admit a basis of open F_σ-sets with metrizable boundaries?* 1529?

Problem 5.10. *Is a continous image of a rim-countable continuum rim-metrizable?* 1530?

Recall that, the continous images of rim-metrizable compact spaces are not necessarily rim-metrizable, [**58**].

In [**59**], it was proved that if X is a continous image of an arc, then the three classical dimension numbers $\mathrm{ind}(X)$, $\mathrm{Ind}(X)$ and $\dim(X)$ are equal. Moreover,

$$\mathrm{ind}(X) = \max\{1, \sup\{\mathrm{ind}(M) : M \subset X \text{ and } M \text{ is closed and metrizable}\}\}.$$

Problem 5.11. *If X is a locally connected rim-metrizable continuum, What is the relation among* $\mathrm{ind}(X)$, $\mathrm{Ind}(X)$, $\dim(X)$ *and* $\max\{1, \sup\{\mathrm{ind}(M) : M \subset X$ *and M is closed and metrizable*$\}\}$*?* 1531?

References

[1] T. Banakh, R. Cauty, K. Trushchak, and L. Zdomskyy, *On universality of finite products of Polish spaces*, Tsukuba J. Math. **28** (2004), no. 2, 455–471.

[2] T. Banakh, V. V. Fedorchuk, J. Nikiel, and M. Tuncali, *The Suslinian number and other cardinal invariants of continua*, 2006, Preprint.

[3] T. Banakh and Kh. R. Trushchak, *Z_n-sets and the disjoint n-cells property in products of ANRs*, Mat. Stud. **13** (2000), no. 1, 74–78.

[4] T. Banakh and V. Valov, *Parametric general position properties and embedding of n-dimensional maps into trivial bundles*, 2006, Preprint.

[5] M. Bestvina, *Characterizing k-dimensional universal Menger compacta*, Mem. Amer. Math. Soc. **71** (1988), no. 380, 110 pp.

[6] P. L. Bowers, *General position properties satisfied by finite products of dendrites*, Trans. Amer. Math. Soc. **288** (1985), no. 2, 739–753.

[7] W. Bula, *Open maps resemble projections*, Bull. Polish Acad. Sci. Math. **31** (1983), no. 3–4, 175–181.

[8] M. Cencelj and A. Dranishnikov, *Extension of maps to nilpotent spaces*, Canad. Math. Bull. **44** (2001), no. 3, 266–269.

[9] M. Cencelj and A. Dranishnikov, *Extension of maps to nilpotent spaces. II*, Topology Appl. **124** (2002), no. 1, 77–83.

[10] M. Cencelj and A. Dranishnikov, *Extension of maps to nilpotent spaces III*, Topology Appl. **153** (2005), no. 2-3, 208–212.

[11] A. Chigogidze, *Compactifications and universal spaces in extension theory*, Proc. Amer. Math. Soc. **128** (2000), no. 7, 2187–2190.

[12] A. Chigogidze, *Infinite dimensional topology and shape theory*, Handbook of geometric topology, North-Holland, Amsterdam, 2002, pp. 307–371.

[13] A. Chigogidze, *Notes on two conjectures in extension theory*, JP J. Geom. Topol. **2** (2002), no. 3, 259–264.

[14] D. Daniel, *On metrizability of images of ordered compacta*, Houston J. Math. **32** (2006), no. 4, 1047–1059.

[15] D. Daniel, J. Nikiel, L. B. Treybig, H. M. Tuncali, and E. D. Tymchatyn, *On rim-properties of continua*, Questions Answers Gen. Topology **19** (2001), no. 2, 187–193.

[16] D. Daniel, J. Nikiel, L. B. Treybig, H. M. Tuncali, and E. D. Tymchatyn, *On perfectly normal compacta*, Questions Answers Gen. Topology **23** (2005), no. 1, 1–14.

[17] D. Daniel, J. Nikiel, L. B. Treybig, H. M. Tuncali, and E. D. Tymchatyn, *On Suslinian continua*, Canad. Math. Bull. **48** (2005), no. 2, 195–202.

[18] D. Daniel, J. Nikiel, L. B. Treybig, M. Tuncali, and E. D. Tymchatyn, *Concerning local conectivity in Suslinian continua*, Preprint.

[19] D. Daniel, J. Nikiel, L. B. Treybig, M. Tuncali, and E. D. Tymchatyn, *Lifting maps of quotient spaces onto arcs*, Preprint.

[20] D. Daniel, J. Nikiel, L. B. Treybig, M. Tuncali, and E. D. Tymchatyn, *Concerning continua that contain no metric subcontinua*, Houston J. Math. **30** (2004), no. 3, 745–750.

[21] D. Daniel and B. Treybig, *A decomposition theorem for locally connected Suslinian continua*, Topology Proc. **23** (1998), Summer, 93–105.

[22] R. J. Daverman, *Problems about finite-dimensional manifolds*, Open problems in topology, North-Holland, Amsterdam, 1990, pp. 431–455.

[23] A. Dranishnikov, *The cohomological dimension is not preserved under the Stone–Čech compactification*, C. R. Acad. Bulgare Sci. **41** (1988), no. 12, 9–10.

[24] A. Dranishnikov, *On a problem of P. S. Aleksandrov*, Mat. Sb. **135(177)** (1988), no. 4, 551–557, Translation: Sb. Math. 63 (1989) no. 2, 539–546.

[25] A. Dranishnikov, *A fibration that does not accept two disjoint many-valued sections*, Topology Appl. **35** (1990), no. 1, 71–73.

[26] A. Dranishnikov, *Extension of mappings into CW-complexes*, Mat. Sb. **182** (1991), no. 9, 1300–1310, Translation: Sb. Math. 74 (1993) no. 1, 47–56.

[27] A. Dranishnikov, *The Eilenberg–Borsuk theorem for mappings in an arbitrary complex*, Mat. Sb. **185** (1994), no. 4, 81–90, Translation: Sb. Math. 81 (1995) no. 2, 467–475.

[28] A. Dranishnikov and J. Dydak, *Extension dimension and extension types*, Tr. Mat. Inst. Steklova **212** (1996), 61–94, Translation: Proc. Steklov Inst. Math. 212 (1996) 55–88.

[29] A. Dranishnikov and D. Repovš, *Cohomological dimension with respect to perfect groups*, Topology Appl. **74** (1996), 123–140.

[30] S. A. Drozdovskiĭ and V. V. Filippov, *An example of a rim-scattered locally connected continuum that is not rim-countable*, Mat. Sb. **185** (1994), no. 10, 27–38, Translation: Sb. Math. 83 (1995) no. 1, 177–187.

[31] J. Dydak, *Cohomological dimension and metrizable spaces*, Trans. Amer. Math. Soc. **337** (1993), no. 1, 219–234.

[32] J. Dydak, *Compactifications and cohomological dimension*, Topology Appl. **50** (1993), no. 1, 1–10.

[33] J. Dydak, *Cohomological dimension and metrizable spaces. II*, Trans. Amer. Math. Soc. **348** (1996), no. 4, 1647–1661.

[34] J. Dydak, *Cohomological dimension theory*, Handbook of geometric topology (R. J. Daverman and R. B. Sher, eds.), North-Holland, Amsterdam, 2002, pp. 423–470.

[35] J. Dydak and M. Levin, *Extensions of maps to the projective plane*, 2005, pp. 1711–1718.

[36] J. Dydak and J. Mogilski, *Universal cell-like maps*, Proc. Amer. Math. Soc. **122** (1994), no. 3, 943–948.

[37] J. Dydak and J. J. Walsh, *Spaces without cohomological dimension preserving compactifications*, Proc. Amer. Math. Soc. **113** (1991), no. 4, 1155–1162.

[38] J. Dydak and J. J. Walsh, *Infinite-dimensional compacta having cohomological dimension two: an application of the Sullivan conjecture*, Topology **32** (1993), no. 1, 93–104.

[39] T. Eisworth, *Elementary submodels and monotonically normal separable compacta*, 2006, Preprint. `arXiv:math.GN/0608376`

[40] R. Engelking, *Theory of dimensions finite and infinite*, Heldermann Verlag, Lemgo, 1995.

[41] V. V. Filippov, *Perfectly normal bicompacta*, Dokl. Akad. Nauk SSSR **189** (1969), 736–739, Translation: Soviet Math. Dokl. 10 (1969) 1503–1507.

[42] G. Gruenhage, *Two nonmetric locally connected continua*, Topology Appl. **35** (1990), no. 2–3, 209–216.

[43] G. Gruenhage and J. T. Moore, *Perfect compacta and basis problems in topology*, Open problems in topology II (E. Pearl, ed.), Elsevier Science.

[44] V. Gutev, *Open mappings looking like projections*, Set-Valued Anal. **1** (1993), no. 3, 247–260.

[45] V. Gutev and V. Valov, *Continuous selections and C-spaces*, Proc. Amer. Math. Soc. **130** (2002), no. 1, 233–242.

[46] D. M. Halverson, *Detecting codimension one manifold factors with the disjoint homotopies property*, Topology Appl. **117** (2002), no. 3, 231–258.

[47] A. Hatcher, *Algebraic topology*, Cambridge University Press, Cambridge, 2002.

[48] A. Karasev, *On two problems in extension theory*, Topology Appl. **153** (2006), no. 10, 1609–1613.

[49] A. Karasev and V. Valov, *Extension dimension and quasi-finite cw complexes*, Topology Appl. **153** (2006), no. 17, 3241–3254.

[50] A. Karasev and V. Valov, *Universal absolute extensors in extension theory*, Proc. Amer. Math. Soc. **134** (2006), 2473–2478.

[51] M. Levin, *Constructing compacta of different extensional dimensions*, Canad. Math. Bull. **44** (2001), no. 1, 80–86.

[52] M. Levin, *Some examples in cohomological dimension theory*, Pacific J. Math. **202** (2002), no. 2, 371–378.

[53] M. Levin and J. T. Rogers, Jr., *A generalization of Kelley's theorem for C-spaces*, Proc. Amer. Math. Soc. **128** (2000), no. 5, 1537–1541.

[54] S. Mardešić, *Images of ordered compacta are locally peripherally metric*, Pacific J. Math. **23** (1967), 557–568.

[55] K. Menger, *Über umfassendste n-dimensionalen Mengen*, Proc. Akad. Amsterdam **29** (1926), 1125–1128.

[56] J. Nagata, *On a universal n-dimensional set for metric spaces*, J. Reine Angew. Math. **204** (1960), 132–138.

[57] J. Nikiel, *The Hahn–Mazurkiewicz theorem for hereditarily locally connected continua*, Topology Appl. **32** (1989), no. 3, 307–323.

[58] J. Nikiel, L. B. Treybig, and H. M. Tuncali, *A rim-metrizable continuum*, Proc. Amer. Math. Soc. **123** (1995), no. 1, 281–286.

[59] J. Nikiel, H. M. Tuncali, and E. D. Tymchatyn, *On the rim-structure of continuous images of ordered compacta*, Pacific J. Math. **149** (1991), no. 1, 145–155.

[60] J. Nikiel, H. M. Tuncali, and E. D. Tymchatyn, *Continuous images of arcs and inverse limit methods*, Mem. Amer. Math. Soc. **104** (1993), no. 498, 80 pp.

[61] G. Nöbeling, *Über die Länge der Euklidischen Kontinuen. I*, Jber. Deutsch. Math. Verein. **52** (1942), 132–160.

[62] W. Olszewski, *Universal separable metrizable spaces of given cohomological dimension*, Topology Appl. **61** (1995), no. 3, 293–299.

[63] B. Pasynkov, *On universal bicompacta of given weight and dimension*, Dokl. Akad. Nauk SSSR **154** (1964), 1042–1043, Translation: Soviet Math. Dokl. 5 (1964) 245–246.

[64] B. A. Pasynkov, *On the geometry of continuous mappings of finite-dimensional metrizable compacta*, Tr. Mat. Inst. Steklova **212** (1996), 147–172, Translation: Proc. Steklov Inst. Math. 212 (1996) 138–162.

[65] R. Pol, *A weakly infinite-dimensional compactum which is not countable-dimensional*, Proc. Amer. Math. Soc. **82** (1981), no. 4, 634–636.

[66] L. Pontrjagin and G. Tolstowa, *Beweis des Mengerschen Einbettungssatzes*, Math. Ann. **105** (1931), no. 1, 734–745.

[67] M. E. Rudin, *Nikiel's conjecture*, Topology Appl. **116** (2001), no. 3, 305–331.

[68] Y. Sternfeld, *Mappings in dendrites and dimension*, Houston J. Math. **19** (1993), no. 3, 483–497.

[69] H. Torunczyk, *On a conjecture of T.R. Rushing and the structure of finite-dimensional mappings*, Geometric topology, discrete geometry and set-theory, International conference, Moscow, August 24–28, 2004.

[70] H. M. Tuncali, *Concerning continuous images of rim-metrizable continua*, Proc. Amer. Math. Soc. **113** (1991), no. 2, 461–470.

[71] V. V. Uspenskij, *A selection theorem for C-spaces*, Topology Appl. **85** (1998), 351–374.

[72] J. E. West, *Open problems in infinite-dimensional topology*, Open problems in topology, North-Holland, Amsterdam, 1990, pp. 523–597.

[73] A. Zarelua, *A universal bicompactum of given weight and dimension*, Dokl. Akad. Nauk SSSR **154** (1964), 1015–1018.

[74] M. Zarichnyi, *Universal spaces and absolute extensors for integral cohomological dimension*, Preprint.

Open Problems in Topology II – Edited by E. Pearl

Moscow questions on topological algebra

Konstantin L. Kozlov, Evgenii A. Reznichenko and Ol′ga V. Sipacheva

1. Topological groups

1.1. Unconditionally closed and algebraic sets. Markov [**23**] called a subset A of a group G *unconditionally closed* in G if it is closed in any Hausdorff group topology on G. Clearly, all solution sets of equations in G, as well as their finite unions and arbitrary intersections, are unconditionally closed.

Definition (Markov [**23**]). A subset A of a group G with identity element 1 is said to be *elementary algebraic* in G if there exists a word $w = w(x)$ in the alphabet $G \cup \{x^{\pm 1}\}$ (x is a variable) such that $A = \{x \in G : w(x) = 1\}$. Finite unions of elementary algebraic sets are *additively algebraic* sets. An arbitrary intersection of additively algebraic sets is called *algebraic*. Thus, the algebraic sets in G are the solution sets of arbitrary conjunctions of finite disjunctions of equations.

In his 1945 paper [**23**], A. A. Markov showed that any algebraic set is unconditionally closed and posed the problem of whether the converse is true. In [**24**] (see also [**22**]), he solved this problem for countable groups by proving that any unconditionally closed set in a countable group is algebraic. Recently, Sipacheva showed that the answer is also positive for subgroups of direct products of countable groups [**54**] and proved the following theorem [**52**].

Theorem 1.1. *Under* CH, *there exists a group containing a nonalgebraic unconditionally closed set.*

Such a group is the nontopologizable group M constructed by Shelah [**46**], which is an increasing union of topologizable (i.e., admitting nondiscrete Hausdorff group topologies) subgroups. The following general observation shows that this is sufficient for M to have a nonalgebraic unconditionally closed subset.

Lemma 1.2. *If G is a nontopologizable group and any finite subset of G is contained in a topologizable subgroup of G, then $G \setminus \{1\}$ is a nonalgebraic unconditionally closed subset of G.*

Indeed, since G admits no nondiscrete Hausdorff group topology, the set $A = G \setminus \{1\}$ is unconditionally closed in G. Suppose that it is algebraic. Then $A = \bigcap_{\gamma\in\Gamma} A_\gamma$, where Γ is an arbitrary index set and each A_γ is an additively algebraic set in G. Clearly, $A = A_\gamma$ for some γ. Thus, $A = \bigcup_{i\le k} A_i$, where $k \in \omega$ and each A_i is an elementary algebraic set. This means that there exist words $w_1(x), \dots, w_k(x)$ in the alphabet $G \cup \{x^{\pm 1}\}$ such that $A_i = \{x \in G : w_i(x) = 1\}$ for $i \le k$. Since the number of letters in each word is finite, we can find a topologizable subgroup

E.A. Reznichenko and O.V. Sipacheva acknowledge the support of the Russian Foundation for Basic Research (project no. 06-01-00761).

$H \subset G$ such that the $w_i(x)$ are words in the alphabet $H \cup \{x^{\pm 1}\}$. Thus, the $A_i \cap H$ are elementary algebraic sets in H, and $A \cap H = H \setminus \{1\}$ is an algebraic (and hence unconditionally closed) set in H, which contradicts the topologizability of H.

1532? **Question 1.3** (see also [**23**]). *Does there exist a group containing a nonalgebraic unconditionally closed set in* ZFC*?*

1533? **Question 1.4.** *Describe the groups in which all unconditionally closed sets are algebraic.*

1.2. Dimensions of metrizable groups. Many problems and results of the dimension theory of topological groups are given in [**43–45**]. In this section, we largely consider the dimensional properties of metrizable topological groups. The celebrated theorem of Katětov says that $\dim X = \operatorname{Ind} X$ for any metric space X; however, there exist examples of metrizable spaces with noncoinciding dimensions ind and dim. The first (very complicated) example of such a space was constructed by Roy in 1962 [**41**]. Since then, much simpler examples with various additional properties have been constructed (see, e.g., [**20, 29, 31**]), but the question about the coincidence of dimensions for metrizable topological groups had remained open; apparently, for the first time, it was asked by Mishchenko in 1964 [**26**].

The spaces embeddable in zero-dimensional topological groups occupy an intermediate position between zero-dimensional and strongly zero-dimensional metrizable spaces (any strongly zero-dimensional metrizable space X can be metrized by a non-Archimedean metric, and this metric can be assumed to take only rational values (see [**10**]). The Graev extension [**14**] of such a metric to the free group $F(X)$ takes only rational values as well; therefore, the group $F(X)$ with the Graev metric has dimension ind zero, and it contains X as a subspace). Recently, Sipacheva [**53**] constructed a space which can be embedded as a closed subspace in a zero-dimensional metrizable group but is not strongly zero-dimensional (this is a special case of Mrówka's space [**29**], namely, $\nu\mu_0(A)$, where A is the set $\sigma 2^\omega$ of binary sequences with only finitely many elements different from 0); thereby, she obtained an example of a metrizable group with noncoinciding dimensions ind and dim. She proved also that Kulesza's zero-dimensional metrizable space from [**19**] cannot be embedded in a metrizable zero-dimensional group [**53**].

The natural question arises: What properties of Kulesza's space obstruct its embedding in a zero-dimensional metrizable group? The most manifest difference between Mrówka's and Kulesza's spaces is that the latter is metrizable by a complete metric. This suggests the conjecture that a space metrizable by a complete metric can be embedded in a zero-dimensional metrizable group only if it is strongly zero-dimensional. This conjecture is based not only on purely formal grounds but also on some intuitive reasons; it seems rather likely to us. Even more likely is the following auxiliary conjecture: If (X, ρ) is a metric space with complete metric ρ, $A_\rho(X)$ is the free (Abelian) group of X metrized by the Graev extension of ρ, and $\operatorname{ind} A_\rho(X) = 0$, then $\dim X = 0$.

It is also unclear how the dimension of metrizable groups behaves under completion (this question is difficult even for general topological spaces). It is only clear that the free groups with Graev metrics (as well as the metrizable groups obtained by applying Sipacheva's construction) are never complete; we can always construct a fundamental sequence consisting of words with unboundedly increasing lengths which converges to no word of finite length.

Question 1.5. *Does there exist a complete metric group with noncoinciding dimensions* ind *and* dim*? In particular, can the completion of a free (Abelian) group with Graev metric be zero-dimensional?* 1534?

Question 1.6. *Is it true that any complete metric space which can be embedded into a zero-dimensional metrizable group is strongly zero-dimensional?* 1535?

These two questions are closely related to the following old problem of Mishchenko.

Question 1.7 ([**26**]). *Is it true that any complete metric space can be embedded into a complete metric group of the same dimension?* 1536?

Question 1.8. *How large can the gap between the dimensions* ind *and* dim *of a metrizable group be? What values can the dimension* dim *of a metrizable topological group* G *with* $\operatorname{ind} G = 0$ *take?* 1537?

Certainly, it makes sense to try to calculate the covering dimension of the zero-dimensional group G in which Mrówka's space $\nu\mu_0(A)$ embeds and the covering and inductive dimensions of the metric completions of G.

To conclude this section, we recall of the following old question of Arhangel′skiĭ, which is closely related to the problems considered above.

Question 1.9 ([**4**]). *Does there exist a topological group* G *with countable network weight for which* $\operatorname{ind} G \neq \dim G$*?* 1538?

Arhangel′skiĭ asked this question for general regular spaces with countable network in 1970 [**1**]. Delistathis and Watson suggested an approach to solving it, which enabled Charalambous to construct a counterexample [**7**, **8**]. Another counterexample was constructed by Dow and Hart [**9**] under MA for σ-centered partial orders. Apparently, the only progress towards answering Question 1.9 is due to Shakhmatov [**42**], who constructed a Lindelöf Σ-group G for which $\operatorname{ind} G \neq \dim G$. The following more general question may be easier to answer.

Question 1.10. *Does there exist an (algebraically) homogeneous space* X *with countable network for which* $\operatorname{ind} G \neq \dim G$*?* (A space X is *algebraically homogeneous* if there exists a topological group G and its closed subgroup H such that X is homeomorphic to G/H.) 1539?

1.3. The inequality $\dim X \times Y > \dim X + \dim Y$**.** The reverse (nonstrict) inequality holds in large classes of spaces, such as metrizable, compact, completely paracompact, paracompact p, and even paracompact Σ-spaces (see, e.g., [**18**]). However, spaces satisfying the inequality $\dim X \times Y > \dim X + \dim Y$ (they are known as Wage-type examples) are rather hard to construct; moreover, there is no Wage-type example of topological groups.

1540? **Question 1.11** ([**44**]). *Do there exist topological groups G and H such that* $\dim G \times H > \dim G + \dim H$*?* (The answer is not known even for Lindelöf topological groups.)

The next two questions refer to general topological spaces, but they can be formulated for groups. The first Wage-type example of separable locally compact spaces X and Y with perfectly normal $X \times Y$ was constructed under CH by Wage [**66**]. Przymusiński constructed the first ZFC Wage-type example of a separable first countable locally compact (or Lindelöf) strongly zero-dimensional space X with normal square [**35**]. Other Wage-type examples were constructed by Tsuda, E. Pol, Engelking, and Kozlov (a survey of related results is contained in [**18, 62**]).

1541? **Question 1.12.** *Is it true that, for any nondecreasing finite sequence $l_1 \leq l_2 \leq \cdots \leq l_m$ of nonnegative integers, there exists a space X such that* $\dim X^j = l_j$ *for $j = 1, \ldots, m$ and X^m is (perfectly) normal?*

Wage [**66**] constructed a Wage-type example of a zero-dimensional separable metric space M (a subset of the irrationals) and a zero-dimensional first countable Lindelöf space Y. Tsuda [**62**] showed that $\dim M \times Y = 1$ and asked whether there exists a Wage-type example with a complete metric factor (in particular, the space of irrationals). He posed also the following problem.

1542? **Question 1.13.** *How large can the gap between* $\dim M \times Y$ *and* $\dim M + \dim Y$ *be for metrizable M?*

For any space X and compact space K, we have $\dim X \times K \leq \dim X + \dim K$ (this is a result of Morita [**28**]). Tsuda [**63**] constructed a Wage-type example of a zero-dimensional Lindelöf separable space X and a strongly zero-dimensional pseudocompact space Y (Mrówka's space). The following problem is due to Pasynkov.

1543? **Question 1.14.** *Does the inequality* $\dim X \times Y \leq \dim X + \dim Y$ *hold for pseudocompact (countably compact) factors? What if the factors are equal?*

The classical theorem of Comfort and Ross (that any product of pseudocompact groups is pseudocompact) and Morita's theorem imply $\dim G \times H \leq \dim G + \dim H$ for pseudocompact groups G and H. Indeed, $\dim \beta G = \dim G$ and $\dim \beta H = \dim H$; on the other hand, by the Glicksberg theorem, $\beta(G \times H) = \beta G \times \beta H$.

1544? **Question 1.15.** *Does the inequality* $\dim G \times X \leq \dim G + \dim X$ *hold for a pseudocompact group G and a pseudocompact space X?*

Let $p\colon Y \to X$ be a locally trivial bundle with fiber F. In many cases (for example, if the base X is paracompact), $\dim Y = \dim X \times F$. However, as was shown in [**16**], Smirnov's n-dimensional modification [**55**] of Dowker's example of a zero-dimensional not strongly zero-dimensional space is a locally trivial bundle with base ω_1 and fiber $\mathbb{Q}$ under CH. No ZFC example of this kind is known.

Question 1.16. *Does the inequality $dim Y \leq \dim X + \dim F$ hold for a locally trivial bundle with compact fiber F?* 1545?

1.4. Values of metrics. A metric space X can very conveniently be embedded in a zero-dimensional metric group if the values of the metric belong to a zero-dimensional set of reals closed with respect to addition. Indeed, the set of values of the Graev extension of any metric is the semigroup (under addition) closure of the set of values of the metric; on the other hand, clearly, if the topology of a space is generated by a metric with zero-dimensional set of values, then this space is itself zero-dimensional. Thus, the free group $F(X)$ endowed with the Graev extension of a suitable metric on X is zero-dimensional.

Question 1.17. *Is it true that any zero-dimensional metrizable space admits a metric with zero-dimensional set of values? with rational values?* 1546?

It is easy to show that sets of values of complete metrics may be arbitrary (they must not be, say, closed). However, this is not so clear for metrics on groups generated by group norms; at least, it is likely that the sets of values of the completions of Graev extensions must have certain closedness properties. This might help in proving that the metric completions of free groups with Graev metrics are never zero-dimensional.

Question 1.18. *Describe the possible sets of values of the completions of Graev metrics on free groups of metric spaces.* 1547?

1.5. Free groups. The first question is closely related to the considerations of Sections 1.2 and 1.4.

Question 1.19. *Give a constructive description of the completion of the free (Abelian) group of a metric space endowed with the Graev extension metric.* 1548?

Markov invented free topological groups when trying to construct an example of a nonnormal group [**23**]. The converse problem of describing the spaces for which free topological groups are normal is still open. Clearly, the normality of a free topological group $F(X)$ implies the normality of all finite powers of X. There was the conjecture that this condition was also sufficient. However, Pavlov constructed a GCH example of a countably compact space X such that X^ω is normal (and even strictly collectionwise normal) but X^2 is not pseudocompact [**32**]. Okunev noticed that $F(X)$ cannot be normal for such X. Indeed, since X is countably compact, it must be functionally bounded in $F(X)$, and, by Tkachenko's theorem [**60**], the group product $X \cdot X$ in $F(X)$ must be bounded as well. On the other hand, $X \cdot X$ is homeomorphic to $X \times X$; hence there exists an unbounded continuous function $f\colon X \cdot X \to \mathbb{R}$. Since $X \cdot X$ is closed in $F(X)$, it follows that if $F(X)$ were normal, we could extend this function to a continuous function on $F(X)$, in contradiction to the boundedness of $X \cdot X$.

Question 1.20. *Describe the spaces X for which the free (Abelian) topological group $F(X)$ $(A(X))$ is normal. Does there exist a* ZFC *example of X such that X^n is strictly collectionwise normal for all n but $F(X)$ $(A(X))$ is not normal?* 1549?

For any set X and any positive integer n, the natural multiplication map $i_n\colon (X \oplus \{e\} \oplus X^{-1})^n \to F_n(X)$ taking $(x_1^{\varepsilon_1}, \dots, x_n^{\varepsilon_n})$ to $x_1^{\varepsilon_1} \dots x_n^{\varepsilon_n}$ is defined (here X^{-1} is a disjoint copy of X, e is the identity element of $F(X)$ (the empty word), and $F_n(X)$ is the set of words of length at most n). Let $i_\infty\colon \bigoplus_{n\in\omega}(X \oplus \{e\} \oplus X^{-1})^n \to F(X)$ be the map defined by the condition $i_\infty \restriction (X \oplus \{e\} \oplus X^{-1})^n = i_n$ for $n \in \omega$; it implements the natural factorization of the discrete union $\bigoplus_{n\in\omega}(X \oplus \{e\} \oplus X^{-1})^n$ which yields the group $F(X)$ (as a set). This map is continuous, because multiplication in the free topological group is continuous. It would be very convenient if it were also quotient, i.e., if the free topological group $F(X)$ were the topological quotient of the space $\bigoplus_{n\in\omega}(X \oplus \{e\} \oplus X^{-1})^n$. Unfortunately, this is not always so; i_∞ is quotient if and only if all i_n are quotient and $F(X)$ is the inductive limit of $\{F_n(X)\}$, which happens fairly rarely.

Very little is known about conditions under which the maps i_n are quotient. Mal′tsev noticed that i_∞ is quotient for any compact space [**21**]. Pestov characterized the spaces for which i_2 is quotient (these are precisely strictly collectionwise normal spaces [**33**]), and Fay, Ordman, and Thomas showed that i_3 is not quotient even for the space of rational numbers [**11**]. Tkachenko [**59**] proved that the map i_∞ (and, hence, all maps i_n) is quotient for all Lindelöf P-spaces and all C_ω-spaces (inductive limits of increasing sequences of subspaces X_n such that the X_n^k are countably compact and strictly collectionwise normal for any n and k). Arhangel′skiĭ proved that if X is Dieudonné complete and $F_n(X)$ is a k-space (for example, if X is paracompact and locally compact), then all maps i_n are quotient [**3**, §5]; in fact, if X is Dieudonné complete and $F(X)$ is a k-space, then i_∞ is quotient. Finally, in the joint paper [**38**] of Reznichenko with Sipacheva, the spaces for which i_3 or i_4 is quotient were described. Possibly, the question about i_4 being quotient plays a key role in solving the problem about all maps i_n being quotient. For the free Abelian groups of metric spaces, this is precisely the case: Yamada proved that, for a metric space X, all natural addition maps $i_n^+\colon (X \oplus \{e\} \oplus -X)^n \to A_n(X)$ are quotient if and only if the map i_4^+ is quotient [**67**]. A detailed survey of results related to the maps i_n and i_∞ being quotient is contained in [**51**].

1550? **Question 1.21.** *Characterize the topological spaces X for which the natural multiplication maps i_n (and i_∞) are quotient.*

This question is interesting, in particular, for countable spaces with only one nonisolated point; it is related to retral Mal′tsev spaces (see the next section).

Apparently, for the first time, Question 1.21 was asked by Pestov and Tkachenko in [**5**]. They posed also the following closely related problem.

1551? **Question 1.22** ([**5**]). *Characterize the topological spaces X for which the free topological group $F(X)$ is the inductive limit of its subspaces $F_n(X)$.*

(As mentioned, the $F(X)$ being the inductive limit of $\{F_n(X)\}$ is the second ingredient of the natural multiplication map i_∞ being quotient.) This question is studied better than Question 1.21. Apparently, the strongest result is due to Tkachenko [**59**], who proved that if X is a P-space or a C_ω-space, then $F(X)$ is the inductive limit of its subspaces $F_n(X)$. Moreover, Tkachenko characterized pseudocompact spaces X with the same property, namely, he proved that the free topological group $F(X)$ of a pseudocompact space X is the inductive limit of $F_n(X)$ if and only if all finite powers of X are normal and countably compact [**61**]. Pestov and Yamada [**34**] gave a complete description of metrizable spaces X for which $F(X)$ ($A(X)$) is the inductive limit of $F_n(X)$ ($A_n(X)$). Sipacheva characterized countable spaces X with only one nonisolated point for which $F(X)$ ($A(X)$) is the inductive limit of $F_n(X)$ ($A(X)$) [**48**]. A more detailed survey of related results is contained in [**51**]. Questions 1.21 and 1.22 are interesting for both free and free Abelian groups.

The following question is attributed to A. A. Markov (see [**5**]).

Question 1.23 (A. A. Markov). *What subgroups of a Markov free topological group are free topological groups?* 1552?

The subspaces Y of X for which the topological subgroup of $F(X)$ generated by Y is the free topological group of Y were described in [**49**].

1.6. An old problem. The following old problem seems very interesting to us.

Question 1.24 (V. I. Malykhin, [**5**]). *Does there exist a countable nonmetrizable Fréchet–Urysohn group in* ZFC*?* 1553?

Such a group exists, e.g., under Martin's axiom. Many interesting consistency results in this direction were obtained by Nyikos and Shibakov. Related ZFC results and problems can be found in [**15, 39, 50**].

2. Mal′tsev spaces and retracts of groups

Let X be a set. A map $m\colon X^3 \to X$ is called a *Mal′tsev operation* if $m(x,y,y) = m(y,y,x) = x$ for any $x, y \in X$. A topological space X is *Mal′tsev* if there exists a continuous Mal′tsev operation on X.

When we consider a Mal′tsev space X with Mal′tsev operation m as a universal topological algebra, we denote it by (X, m) and refer to it as a *Mal′tsev algebra*.

A subset Y of a Mal′tsev algebra (X, m) is said to be *M-closed* if Y is closed with respect to the operation m, i.e., $m(x,y,z) \in Y$ for all $x, y, z \in Y$. On such a set Y, we can consider the Mal′tsev operation $m_Y = m \restriction Y^3$, under which (Y, m_Y) is a Mal′tsev algebra. In what follows, we omit the subscript Y. A map $f\colon X \to Y$ of Mal′tsev algebras (X, m_X) and (Y, m_Y) is called a homomorphism if f respects the Mal′tsev operation, i.e., $f(m_X(x,y,z)) = m_Y(f(x), f(y), f(z))$. We refer to homomorphisms of Mal′tsev algebras as *M-homomorphisms*. Mal′tsev showed that *any quotient M-homomorphism is open* [**21**] (see also [**40, 64**]). Using this fact, Uspenskij proved that *any compact Mal′tsev space is Dugundji* [**64**].

On an arbitrary group, there is the natural *group Mal′tsev operation* defined by $m_g(x,y,z) = xy^{-1}z$. In what follows, we regard groups as Mal′tsev algebras under the group Mal′tsev operation.

A space X is said to be *retral* if X is a retract of a topological group. Any retract $r(X)$ of a Mal′tsev space (X, m) is a Mal′tsev space under the Mal′tsev operation $(x, y, z) \mapsto r(m(x, y, z))$. Thus, *any retral space is Mal′tsev.* For the class of compact spaces, Sipacheva succeeded in proving the converse: *any compact Mal′tsev space is retral* [**47**].

As above, we denote the free topological group (in the sense of Markov) of a Tikhonov space X by $F(X)$, the set of words of length at most n in $F(X)$ by $F_n(X)$, and the set of words of odd length by $F_{\rm odd}(X)$ (thus, $F_{\rm odd}(X) = \bigcup_n F_{2n+1}(X)$). We also put

$$M_{2n+1}(X) = \{x_1 x_2^{-1} x_3 \ldots x_{2n}^{-1} x_{2n+1} : x_i \in X\} \subset F_{2n+1}(X)$$

and $M(X) = \bigcup_n M_{2n+1}(X)$. By j_n we denote the map $X^n \to F_n(X)$ defined by $j_n(x_1, x_2, \cdots, x_n) = x_1 x_2^{-1} \cdots x_n^{-1^{n+1}}$. Clearly, $j_{2n+1}(X^{2n+1}) = M_{2n+1}(X)$, and it is easy to show that the map j_n is quotient if and only if so is the natural multiplication map i_n defined in the preceding section.

It is easy to prove that *a space X is retral if and only if X is a retract of $F(X)$* [**2, 13**].

For a topological group G, there is a natural retraction $r_g \colon F(G) \to G$; it is defined by $r_g(x_1^{\varepsilon_1} x_2^{\varepsilon_2} \cdots x_n^{\varepsilon_n}) = x_1^{\varepsilon_1} x_2^{\varepsilon_2} \cdots x_n^{\varepsilon_n}$, where $x_1^{\varepsilon_1} x_2^{\varepsilon_2} \cdots x_n^{\varepsilon_n}$ on the left-hand side denotes a word in $F(G)$ and the same expression on the right-hand side denotes a product in G.

We say that a space X is M_{2n+1}*-retral* (M*-retral*) if X is a retract of $M_{2n+1}(X)$ (of $M(X)$).

If $m \colon X^3 \to X$ is a Mal′tsev operation, then the map $r_3^m \colon M_3(X) \to X$ that takes $xy^{-1}z$ to $m(x, y, z)$ is well defined, and it is a retraction. The continuity of m does not generally imply that of r_3^m. Moreover, the only known example of a nonretral Mal′tsev space [**13**] is not a retract of $M_3(X)$. The map r_3^m is continuous if j_3 (i.e., i_3) is quotient. In this case, X is M_3-retral.

Sipacheva's construction of a retraction $F(X) \to X$ for a compact Mal′tsev space X with Mal′tsev operation m is as follows. First, retractions $r_{2n+1}^m \colon M_{2n+1} \to X$ such that $r_{2n+1}^m \restriction M_{2n-1}(X) = r_{2n-1}^m$ are defined recursively by

$$\begin{aligned} r_{2n+1}^m(x_1 x_2^{-1} \ldots x_{2n}^{-1} x_{2n+1}) = r_{2n-1}^m \Big(& m\big(r_1^m(x_1), x_2, r_{2n-1}^m(x_3 x_4^{-1} \ldots x_{2n+1})\big) \\ & m\big(r_3^m(x_1 x_2^{-1} x_3), x_3, r_{2n-1}^m(x_3 x_4^{-1} \ldots x_{2n+1})\big)^{-1} \\ \cdots\ & m\big(r_{2k-1}^m(x_1 x_2^{-1} \ldots x_{2k-1}), x_{2k}, r_{2n-2k+1}^m(x_{2k+1} x_{2k+2}^{-1} \cdots x_{2n+1})\big) \\ & m\big(r_{2k+1}^m(x_1 x_2^{-1} \ldots x_{2k+1}), x_{2k+1}, r_{2n-2k+1}^m(x_{2k+1} x_{2k+2}^{-1} \cdots x_{2n+1})\big)^{-1} \\ & \cdots\ m\big(r_{2n-1}^m(x_1 x_2^{-1} \ldots x_{2n-1}), x_{2n}, r_1^m(x_{2n+1})\big)\Big), \end{aligned}$$

where r_1^m is the identity self-map of X. Then, a retraction $r^m \colon M(X) \to X$ is defined by $r^m \restriction M_{2n+1}(X) = r_{2n+1}^m$. Finally, a retraction $s_\infty \colon F_{\rm odd}(X) \to M(X)$ is defined by the condition $s_\infty \restriction F_{2n+1}(X) = s_{2n+1}$, where $s_{2n+1} \colon F_{2n+1}(X) \to M_{2n+1}(X)$ is the map $x_1^{\varepsilon_1} x_2^{\varepsilon_1} \cdots x_{2n+1}^{\varepsilon_{2n+1}} \mapsto x_1 x_2^{-1} \cdots x_{2n}^{-1} x_{2n+1}$. Note that each map

s_{2n+1} is continuous; thus, s_∞ is continuous provided that $F(X)$ is the inductive limit of the sequence of $F_n(X)$. The map $R^m = r^m \circ s_\infty \colon F_{\text{odd}}(X) \to X$ is then a retraction; to obtain the required retraction $F(X) \to X$, it is sufficient to extend R_m by sending $F(X) \setminus F_{\text{odd}}(X)$ to any point of X.

The set $M(X)$ is M-closed in $F(X)$; hence it has the natural Mal′tsev structure. We denote the restriction of the group Mal′tsev operation to $M(X)$ by m_M.

The definitions of the maps r^m, R^m, and s_∞ do not depend on the topology of X and have nice categorical properties.

Theorem 2.1. *Let (X, m_X) and (Y, m_Y) be Mal′tsev algebras.*

(1) *If $f \colon X \to Y$ is an M-homomorphism, then the diagram*

$$\begin{array}{ccccc} F_{\text{odd}}(X) & \xrightarrow{s_\infty} & M(X) & \xrightarrow{r^{m_X}} & X \\ \downarrow{\scriptstyle F_{\text{odd}}(f)} & & \downarrow{\scriptstyle M(f)} & & \downarrow{\scriptstyle f} \\ F_{\text{odd}}(Y) & \xrightarrow{s_\infty} & M(Y) & \xrightarrow{r^{m_Y}} & Y \end{array}$$

is commutative; here $F_{\text{odd}}(f) = F(f) \restriction F_{\text{odd}}(X)$ and $M(f) = F(f) \restriction M(X)$, where $F(f) \colon F(X) \to F(Y)$ is the homomorphism extending f;

(2) *If $Y \subset X$ is M-closed and $m_Y = m_X \restriction Y^3$, then $r^{m_Y} = r^{m_X} \restriction M(Y)$ and $R^{m_Y} = R^{m_X} \restriction F_{\text{odd}}(Y)$;*

(3) *If $f \colon X \to Y$ and r^{m_X} are M-homomorphisms, then r^{m_Y} is an M-homomorphism;*

(4) *If X is M-closed in a topological group G and $m_X = m_g \restriction X^3$, then $r^{m_X} = r_g \restriction M(X)$ and r^{m_X} is an M-homomorphism;*

(5) *$r^{m_M} \colon M(M(X)) \to M(X)$ is an M-homomorphism.*

Question 2.2. *Let (X, m) be Mal′tsev algebra, and suppose that r^m is an M-homomorphism. Is it true that X can be embedded into a group so that m is the restriction of the group Mal′tsev operation to X^3?* 1554?

Definition. We say that X is *r_{2n+1}-retral* (*r-retral*) if there exists a Mal′tsev operation m on X for which the map r^m_{2n+1} (respectively, r^m) is continuous.

The maps r^m_{2n+1} may not be continuous, but the maps $r^m_{2n+1} \circ j_{2n+1} \colon X^{2n+1} \to X$ are always continuous. Thus, *if (X, m) is a Mal′tsev space, $k \in \{3, 5, 7, \ldots, \infty\}$, and j_k is quotient, then X is r_k-retral and, therefore, M_k-retral.*

Question 2.3. *What classes of spaces in the following diagram are different?* 1555?

$$\begin{array}{cccccccccc} \textit{Mal'tsev} & \supsetneq & M_3\text{-}\textit{retral} & \supset & M_5\text{-}\textit{retral} & \supset & \ldots \supset & M_\infty\text{-}\textit{retral} & \supset & \textit{retral} \\ & & \parallel & & \cup & & & \cup & & \\ & & r_3\text{-}\textit{retral} & \supset & r_5\text{-}\textit{retral} & \supset & \ldots \supset & r_\infty\text{-}\textit{retral} & & \end{array}$$

Question 2.4. *Suppose that X is a Mal′tsev space, n and m are positive integers, $n < m$, and the map j_{2n+1} is quotient. Is it true that X is M_{2m+1}-retral? r_{2m+1}-retral? r_∞-retral? M_∞-retral? retral?* 1556?

Theorem 2.5 ([**47**]). *If X is a Mal′tsev space with Mal′tsev operation m and the map i_∞ is quotient, then X is retral. Moreover, the retractions r^m and R^m and the map s_∞ are continuous.*

As mentioned in the preceding section, the map i_∞ is quotient, in particular, for k_ω spaces, Lindelöf P-spaces, and paracompact locally compact spaces. Therefore, the Mal′tsev spaces from these classes are retral.

A key role in the proof of Theorem 2.5 is played by the continuity of s_∞.

1557? **Question 2.6.** *Is it true that the map s_∞ is continuous for any space X?*

If the answer to this question is positive, then the answer to the following question is also positive.

1558? **Question 2.7.** *Suppose that a space X satisfies one of the following conditions:* (i) *X is M-retral;* (ii) *X is M-closed in some topological group;* (iii) *$X = M(Y)$ for some space Y. Is X retral?*

Recall that a base $\mathcal{B}$ for the topology of a space X is said to be *non-Archimedean* if, for any $U, V \in \mathcal{B}$, either $U \cap V = \varnothing$, $U \subset V$, or $V \subset U$. Spaces with non-Archimedean bases are called *non-Archimedean spaces.* All strongly zero-dimensional metrizable spaces and Lindelöf P-spaces of weight ω_1 are non-Archimedean; the non-Archimedean spaces are precisely subspaces of branch spaces of trees with the standard topology. It is known that *any non-Archimedean space X is a retract of its free Boolean group $B(X)$ and, therefore, retral* [**13**]. We denote the retraction $B(X) \to X$ by r_B^{nA}; it induces the retraction $r^{\mathrm{nA}} = \phi \circ r_B^{\mathrm{nA}}\colon F(X) \to X$, where $\phi\colon F(X) \to B(X)$ is the natural homomorphism. Let m^{nA} be the corresponding Mal′tsev operation (defined by $m^{\mathrm{nA}}(x, y, z) = r^{\mathrm{nA}}(xy^{-1}z)$).

1559? **Question 2.8.** *Is it true that any non-Archimedean space is r_∞-retral?*

1560? **Question 2.9.** *For which n does the relation $r_{2n+1}^{m^{\mathrm{nA}}} = r^{\mathrm{nA}} \restriction M_{2n+1}(X)$ hold? Is it true that $r^{m^{\mathrm{nA}}} = r^{\mathrm{nA}}$?* (It can be shown that $r_5^{m^{\mathrm{nA}}} = r^{\mathrm{nA}} \restriction M_5(X)$.)

Any space X with a topology τ admitting a coarser non-Archimedean topology σ is Mal′tsev, and if τ has a base $\mathcal{B}$ consisting of sets closed in σ, then X is a retract of $B(X)$ and the retraction r_B^{nA} is continuous (with respect to τ) [**13**]. In particular, any countable space is Mal′tsev, and any countable space with only one nonisolated point is retral.

1561? **Question 2.10.** *Suppose that X is a countable space (a separable metrizable Mal′tsev space, a metrizable Mal′tsev space, a Lindelöf Mal′tsev Σ-space) and $n \in \{3, 5, 7, \cdots, \infty\}$. Is it true that X is retral? M_n-retral? r_n-retral? Is X r_n-retral if it has only one nonisolated point? (See also Question 1.21.)*

Reznichenko and Uspenskij proved the following theorem.

Theorem 2.11 ([**40**]). *Let X be a pseudocompact space with a Mal′tsev operation m. Then m can be extended to a continuous Mal′tsev operation $\hat{m}\colon (\beta X)^3 \to \beta X$. Therefore, the maps r^m, R^m, and s_∞ are continuous, and X is a retral space.*

The proof of Theorem 2.11 consists of two parts. First, m is extended to a separately continuous Mal′tsev operation $\hat{m}\colon (\beta X)^3 \to \beta X$; then, it is proved that $\hat{m}$ is continuous. An important role is played by the following two assertions: (i) *If (X, m_X) and (Y, m_Y) are Mal′tsev algebras with separately continuous Mal′tsev operations, then any quotient M-homomorphism $f\colon X \to Y$ is open* [**40**]; (ii) *Any compact space with a separately continuous Mal′tsev operation is Dugundji* [**40**].

Question 2.12. *Is it true that any compact space X with a separately continuous Mal′tsev operation is Mal′tsev? What if X is a metrizable space? a manifold?* 1562?

If X is a countably compact space with a separately continuous Mal′tsev operation m, then m can be extended to a separately continuous Mal′tsev operation $\hat{m}\colon (\beta X)^3 \to \beta X$. Therefore, βX is a Dugundji compact space [**40**].

There exists a pseudocompact space X with a separately continuous Mal′tsev operation such that X is not Mal′tsev and βX is not Dugundji [**40**].

Interesting questions arise in considering retracts of groups satisfying certain topological and algebraic conditions. In [**13**], it was asked whether any Mal′tsev compact space is a retract of a compact group. This question was answered by Cauty, who proved that there exist finite CW-complexes that are Mal′tsev spaces but are not retracts of compact groups [**6**]. The following question remains open.

Question 2.13. *Let X be a Mal′tsev compact space. Is X a retract of a completely bounded group? Is the retraction R_∞ (r_∞) continuous with respect to (the topology induced by) the precompact free group topology on $F(X)$ ($M(X)$)?* 1563?

Question 2.14. *Characterize the (compact) retracts of Abelian (Boolean) topological groups.* 1564?

3. Convex compact spaces and affine functions

Let K be a convex compact subset of a locally convex space; by $\mathcal{E}(K)$ we denote the set of extreme points of K. In [**36**], Reznichenko studied the relation between the weights of K and $\mathcal{E}(K)$. He proved, in particular, that if K is a simplex, or $\mathcal{E}(K)$ is Lindelöf, or $hl(K) \le w(\mathcal{E}(K))$, then $w(K) = w(\mathcal{E}(K))$. However, the following main problem remains open.

Question 3.1 ([**36**])**.** *Is it true that, for any convex compact subset K of a locally convex space, $w(K) = w(\mathcal{E}(K))$? $w(K) = hl(K)$?* 1565?

Suppose that $\mathcal{E}(K) \subset X \subset K$. In [**27**], Moors and Reznichenko considered separable subspaces of the space $A_p^X(K)$ of continuous real-valued affine functions on K with the topology of pointwise convergence on X. The main question was: Is it true that such subspaces must have countable network weight? Moors and Reznichenko gave a negative answer to this question by constructing K and X such that $\mathcal{E}(K)$ and X is separable, the unit ball of $A_p^X(K)$ is separable but has uncountable network weight, $X \setminus \mathcal{E}(K)$ is second countable, and X^ω is Lindelöf. They also proved that (i) the one-point compactification of any Ψ-like (Mrówka–Isbell) space can be embedded in $A_p^{\mathcal{E}(K)}(K)$ for some K; (ii) if $\mathcal{E}(K)$ is Lindelöf, then $A_p^{\mathcal{E}(K)}(K)$ is $\aleph_0$-monolithic (i.e., any separable subspace of $A_p^{\mathcal{E}(K)}(K)$ has a countable network); and (iii) if X is a Lindelöf Σ-space, then $A_p^X(K)$ is $\aleph_0$-monolithic and any compact subset of $A_p^X(K)$ is an Eberlein compactum. The existence of separable nonmetrizable compact subspaces of $A_p^X(K)$ for Lindelöf X is independent of ZFC.

1566? **Question 3.2.** *Suppose that X^ω (or $X \times \omega^\omega$) is Lindelöf. Is it true in* ZFC *that any compact separable subspace of $A_p^X(K)$ is metrizable?*

1567? **Question 3.3.** *Suppose that Z is a compact $\aleph_0$-monolithic subspace of $A_p^{\mathcal{E}(K)}(K)$. Is it true that Z is an Eberlein compactum? What if $\mathcal{E}(K)$ is Lindelöf?*

It is known that any norm-bounded compact subspace of $A_p^X(K)$ is Eberlein and any compact subspace of $A_p^X(K)$ is σ-Eberlein and hence sequential (see [**27**]).

4. Stratifiable function spaces

In this section, all spaces are assumed to be separable and metrizable.

In [**12**], Gartside and Reznichencko proved that if X is a *Polish* (complete separable metric) space, then the space $C_k(X)$ of all real-valued continuous functions on X with the compact-open topology is stratifiable and asked whether the converse is true. We say that a space X is *C_k-stratifiable* if $C_k(X)$ is stratifiable. Recently, Nyikos [**30**] showed that $C_k(\mathbb{Q})$ is not stratifiable; hence any C_k-stratifiable space is hereditarily Baire. Reznichencko [**37**] proved that if a Baire space X is a continuous image of a C_k-stratifiable space, then X has a dense Polish subspace. Since C_k-stratifiability is closed- and open-hereditary and survives multiplication by compact spaces [**12**], it follows that every closed subset of a C_k-stratifiable space X contains a dense Polish subspace. The complements to *λ-spaces* (these are spaces in which all countable subsets are of type G_δ) in compact spaces have the same property. Note that if $|X| < \mathfrak{b}$, then X is a λ-space ($\mathfrak{b}$ is the minimum cardinality of an unbounded subset of ω^ω).

1568? **Question 4.1.** *Suppose that K is a metrizable compact space, $X \subset K$, and X is a λ-space (or $|X| < \mathfrak{b}$). Is it true that $C_k(K \setminus X)$ is stratifiable?*

1569? **Question 4.2.** *Is C_k-stratifiability finitely (countably) productive? Is the product of a C_k-stratifiable space and a Polish space C_k-stratifiable?*

More questions and results on stratifiable C_k-spaces (in the context of the $M_1 = M_3$ problem) are mentioned by Gruenhage in his chapter of this book.

5. Semilattices of compact G-extensions

A semilattice (lattice) is a partially ordered set in which any subset has least upper bound (and greatest lower bound); an introduction to G-spaces is contained in [**65**]. It is almost obvious that the semilattice $K_G(X)$ of compact G-extensions of a G-Tikhonov space X (with $\beta_G X$ being the maximal element) is a sublattice of the semilattice $K(X)$ of compactifications of X. Smirnov and Stoyanov [**57**] noticed that the semilattice $K_G(X)$ may have elements which are minimal but not least. It follows from their results that, for a compact group G, the following conditions are equivalent: $K_G(X)$ is a lattice; $K_G(X)$ has a least element; $K_G(X)$ has a minimal element; and X is locally compact. They conjectured that if G is a locally compact group, X is a Tikhonov G-space, and $K_G(X)$ is a lattice, then X is locally compact. Kozlov and Chatyrko [**17**] gave sufficient conditions for the semilattice $K_G(X)$ to be a lattice (or, equivalently, to have a least element); Sokolovskaya [**58**] constructed examples of G-Tikhonov spaces whose semilattices $K_G(X)$ have minimal elements but are not lattices. She showed that any semilattice $K_G(X)$ for a G-Tikhonov space can be realized as the semilattice $K_H(Y)$ for a pseudocompact H-Tikhonov space Y and a discrete group H.

Question 5.1. *When does the semilattice $K_G(X)$ for a G-Tikhonov space X have minimal elements? When is it a lattice?* 1570?

In [**17**], it was shown that any element of the semilattice $K_G(X)$ for a G-Tikhonov space X contains, in addition to X, the "completions" of all orbits (the action of any group on a compact space can be extended to an action of its Raikov completion [**25**]) and the orbits consisting of points at which the action is d-open (an action α is d-open at a point x if $x \in \operatorname{int}(\overline{\alpha(O,x)})$ for any neighborhood O of the identity element, or, equivalently, if the map $\alpha(\cdot,x)\colon G \to X$ is d-open in the sense of Uspenskij [**64**]).

Question 5.2. *Describe dense invariant subsets X and Y of a compact G-space for which the semilattices $K_G(X)$ and $K_G(Y)$ are isomorphic.* 1571?

Smirnov [**56**] studied compactifications of X that are not maximal compact G-extensions of X under any actions. In particular, he proved that $\mathbb{RP}^n \neq \beta_G\mathbb{R}^n$ for any action of any group G.

Question 5.3. *Let X be a Tikhonov space. What elements of $K(X)$ can be maximal (minimal, least) compact G-extensions of X under some actions?* 1572?

References

[1] A. V. Arhangel′skiĭ, *A survey of some recent advances in general topology, old and new problems*, 1971, pp. 19–26.

[2] A. V. Arhangel′skiĭ, *Classes of topological groups*, Uspekhi Mat. Nauk **36** (1981), no. 3(219), 127–146, Translation: Russian Math. Surveys 36 (1981) no. 3, 151–174.

[3] A. V. Arhangel′skiĭ, *Algebraic objects generated by a topological structure*, Algebra. Topology. Geometry, vol. 25, VINITI, Moscow, 1987, pp. 141–198 (Russian).

[4] A. V. Arhangel′skiĭ, *Topological homogeneity. Topological groups and their continuous images*, Uspekhi Mat. Nauk **42** (1987), no. 2(254), 69–105, Translation: Russian Math. Surveys 42 (1987) no. 2, 83–131.

[5] V. I Arnautov, A. V. Arhangel′skiĭ, P. I. Kirku, A. V. Mikhalev, Yu. N. Mukhin, I. V. Protasov, and M. M. Choban (eds.), *Unsolved problems of topological algebra*, Shtiintsa, Kishinev, 1985 (Russian), Preprint.

[6] R. Cauty, *Espaces de Maltsev compacts qui ne sont pas des rétractes de groupes compacts*, Bull. Polish Acad. Sci. Math. **46** (1998), no. 1, 67–70.

[7] M. G. Charalambous, *Resolving a question of Arhangel′skiĭ*, 2005, Preprint.

[8] M. G. Charalambous, *A note on the dimension of cosmic spaces*, 2006 International Conference on Topology and its Applications (Aegion, Greece). Abstracts, Municipal Library of Aegion, Aegion, 2006, p. 49.

[9] A. Dow and K. P. Hart, *Cosmic dimensions*, 2005, Preprint. `arXiv:math.GN/0509099`

[10] R. Engelking, *General topology*, Heldermann Verlag, Berlin, 1989.

[11] T. H. Fay, E. T. Ordman, and B. V. Smith Thomas, *The free topological group over the rationals*, General Topology and Appl. **10** (1979), no. 1, 33–47.

[12] P. Gartside and E. A. Reznichenko, *Near metric properties of function spaces*, Fund. Math. **164** (2000), no. 2, 97–114.

[13] P. Gartside, E. A. Reznichenko, and O. V. Sipacheva, *Mal'tsev and retral spaces*, Topology Appl. **80** (1997), no. 1–2, 115–129.

[14] M. I. Graev, *Theory of topological groups. I. Norms and metrics on groups. Complete groups. Free topological groups*, Uspekhi Mat. Nauk **5** (1950), no. 2(36), 3–56, Translation: Free topological groups, Amer. Math. Soc. Transl. 35 (1951).

[15] G. Gruenhage and P. J. Szeptycki, *Fréchet–Urysohn for finite sets*, Topology Appl. **151** (2005), 238–259.

[16] K. L. Kozlov, *On the dimension of spaces of locally trivial fiber bundles*, Vestnik Moskov. Univ. Ser. I Mat. Mekh. (1989), no. 3, 69–72, Translation: Moscow Univ. Math. Bull. 44 (1989) no. 3, 70–73.

[17] K. L. Kozlov and V. A. Chatyrko, *On compact G-extensions*, Mat. Zametki **78** (2005), no. 5, 695–709, Translation: Math. Notes 78 (2005) 649–661.

[18] K. L. Kozlov and B. A. Pasynkov, *On the Lebesgue dimension of topological products*, 2005, To appear in J. Math. Sci.

[19] J. Kulesza, *An example in the dimension theory of metrizable spaces*, Topology Appl. **35** (1990), no. 2–3, 109–120.

[20] J. Kulesza, *More examples of metric spaces where the inductive dimensions disagree*, Topology Appl. **122** (2002), no. 1–2, 297–307.

[21] A. I. Mal'tsev, *Free topological algebras*, Izv. Akad. Nauk SSSR Ser. Mat. **21** (1957), 171–198.

[22] A. A. Markov, *On unconditionally closed sets*, C. R. (Doklady) Acad. Sci. URSS (N.S.) **44** (1944), 180–181.

[23] A. A. Markov, *On free topological groups*, Izv. Akad. Nauk SSSR Ser. Mat. **9** (1945), no. 1, 3–64, Translation: Amer. Math. Soc. Transl. 30 (1950).

[24] A. A. Markov, *On unconditionally closed sets*, Rec. Math. [Mat. Sb.] **18(60)** (1946), 3–28, Translation: Amer. Math. Soc. Transl. 30 (1950).

[25] M. Megrelishvili, *Equivariant completions*, Comment. Math. Univ. Carolin. **35** (1994), no. 3, 539–547.

[26] A. Miščenko, *On the dimension of groups with left-invariant topologies*, Dokl. Akad. Nauk SSSR **159** (1964), 753–754, Translation: Soviet Math. Dokl. 5 (1965) 1603–1605.

[27] W. Moors and E. A. Reznichenko, *Separable subspaces of affine function spaces on convex compact spaces*, 2006, Preprint.

[28] K. Morita, *Dimension of general topological spaces*, Surveys in general topology (G. M. Reed, ed.), Academic Press, New York, 1980, pp. 297–336.

[29] S. Mrówka, *Small inductive dimension of completions of metric spaces*, Proc. Amer. Math. Soc. **125** (1997), no. 5, 1545–1554.

[30] P. J. Nyikos, *On some non-Archimedean spaces of Alexandroff and Urysohn*, Topology Appl. **91** (1999), no. 1, 1–23.

[31] A. Ostaszewski, *A note on the Prabir Roy space*, Topology Appl. **35** (1990), no. 2–3, 95–107.

[32] O. Pavlov, *A countably compact space X such that X^ω is strongly collectionwise normal and X^2 is not pseudocompact*, Special Session at AMS meeting #936 "Modern Methods in Set Theory and General Topology", 1998.

[33] V. Pestov, *Neighborhoods of identity in free topological groups*, Vestnik Moskov. Univ. Ser. I Mat. Mekh. (1985), no. 3, 8–10, Translation: Moscow Univ. Math. Bull. 40 (1985) no. 3, 8–12.

[34] V. Pestov and K. Yamada, *Free topological groups on metrizable spaces and inductive limits*, Topology Appl. **98** (1999), 291–301.

[35] T. C. Przymusiński, *Product spaces*, Surveys in general topology, Academic Press, New York, 1980, pp. 399–429.

[36] E. A. Reznichenko, *Convex compact spaces and their maps*, Topology Appl. **36** (1990), no. 2, 117–141.

[37] E. A. Reznichenko, *On stratifiable $C_k(X)$*, 2006, Preprint.

[38] E. A. Reznichenko and O. V. Sipacheva, *Quotient maps to words of bounded length in free topological groups*, General topology. Mappings, products, and dimension of spaces, Izd. Moskov. Univ., Moscow, 1995, pp. 98–119.

[39] E. A. Reznichenko and O. V. Sipacheva, *Properties of Fréchet–Uryson type in topological spaces, groups and locally convex spaces*, Vestnik Moskov. Univ. Ser. I Mat. Mekh. (1999), no. 3, 32–38, Translation: Moscow Univ. Math. Bull. 54 (1999) no. 3, 33–38.

[40] E. A. Reznichenko and V. V. Uspenskij, *Pseudocompact Mal′tsev spaces*, Topology Appl. **86** (1998), no. 1, 83–104.

[41] P. Roy, *Nonequality of dimensions for metric spaces*, Trans. Amer. Math. Soc. **134** (1968), 117–132.

[42] D. Shakhmatov, *A problem of coincidence of dimensions in topological groups*, Topology Appl. **33** (1989), no. 1, 105–113.

[43] D. Shakhmatov, *Imbeddings into topological groups preserving dimensions*, Topology Appl. **36** (1990), no. 2, 181–204.

[44] D. Shakhmatov, *A survey of current researches and open problems in the dimension theory of topological groups*, Questions Answers Gen. Topology **8** (1990), no. 1, 101–128.

[45] D. Shakhmatov, *A comparative survey of selected results and open problems concerning topological groups, fields and vector spaces*, Topology Appl. **91** (1999), no. 1, 51–63.

[46] S. Shelah, *On a problem of Kurosh, Jónsson groups, and applications*, Word problems, II (Oxford, 1976), North-Holland, Amsterdam, 1980, pp. 373–394.

[47] O. V. Sipacheva, *Compacta with a continuous Mal′tsev operation and retracts of topological groups*, Vestnik Moskov. Univ. Ser. I Mat. Mekh. (1991), no. 1, 33–36, Translation: Moscow Univ. Math. Bull. 46 (1991) no. 1, 22–24.

[48] O. V. Sipacheva, *On free topological groups with the inductive limit topologies*, Papers on general topology and applications (Amsterdam, 1994), Ann. New York Acad. Sci., vol. 788, New York Acad. Sci., 1996, pp. 188–196.

[49] O. V. Sipacheva, *Free topological groups of spaces and their subspaces*, Topology Appl. **101** (2000), no. 3, 181–212.

[50] O. V. Sipacheva, *Spaces Fréchet–Urysohn with respect to families of subsets*, Topology Appl. **121** (2002), no. 1–2, 305–317.

[51] O. V. Sipacheva, *The topology of a free topological group*, Fundam. Prikl. Mat. **9** (2003), no. 2, 99–204.

[52] O. V. Sipacheva, *Consistent solution of Markov's problem about algebraic sets*, 2006, Preprint. `arXiv:math.GR/0605558`

[53] O. V. Sipacheva, *Nonequality of dimensions for metric groups*, 2006, Preprint. `arXiv:math.GN/0606146`.

[54] O. V. Sipacheva, *Unconditionally closed and algebraic sets in groups*, 2006, Preprint.

[55] Yu. M. Smirnov, *An example of zero-dimensional normal space having infinite dimensions from the standpoint of covering*, Dokl. Akad. Nauk SSSR **123** (1958), 40–42.

[56] Yu. M. Smirnov, *Can simple geometric objects be maximal compact extensions for $\mathbb{R}^n$?*, Uspekhi Mat. Nauk **49** (1994), no. 6(300), 213–214, Translation: Russian Math. Surveys 49 (1994) no. 6, 214–215.

[57] Yu. M. Smirnov and L. N. Stoyanov, *On minimal equivariant compact extensions*, C. R. Acad. Bulgare Sci. **36** (1983), no. 6, 733–736.

[58] A. M. Sokolovskaya, *A method for constructing lattices of compact G-extensions*, 2006, Preprint.

[59] M. G. Tkachenko, *Strong collectionwise normality and countable compactness in free topological groups*, Sibirsk. Mat. Zh. **28** (1987), no. 5, 167–177, Translation: Sib. Math. J. 28 (1987) no. 5, 824–832.

[60] M. G. Tkachenko, *Compactness type properties in topological groups*, Czechoslovak Math. J. **38(113)** (1988), no. 2, 324–341.

[61] M. G. Tkachenko, *Free topological groups and inductive limits*, Topology Appl. **60** (1994), no. 1, 1–12.

[62] K. Tsuda, *Dimension theory of general spaces*, Dissertation, University of Tsukuba, 1985.

[63] K. Tsuda, *A Wage-type example with a pseudocompact factor*, Topology Appl. **20** (1985), no. 2, 191–200.

[64] V. V. Uspenskij, *Topological groups and Dugundji compacta*, Mat. Sb. **180** (1989), no. 8, 1092–1118, Translation: Sb. Math. 67 (1990) no. 2, 555–580.

[65] J. de Vries, *Topological transformation groups. 1*, Mathematisch Centrum, Amsterdam, 1975.

[66] M. L. Wage, *The dimension of product spaces*, Proc. Natl. Acad. Sci. **75** (1978), no. 10, 4671–4672.

[67] K. Yamada, *Characterizations of a metrizable space X such that every $A_n(X)$ is a k-space*, Topology Appl. **49** (1993), no. 1, 75–94.

Open Problems in Topology II – Edited by E. Pearl

Some problems from George Mason University

John Kulesza, Ronnie Levy and Mikhail Matveev

Introduction

This is a collection of our favorite problems. Unless otherwise stated, all spaces are assumed to be Tychonoff.

Kulesza's problems

These questions all come from the dimension theory of metric spaces. However, the study of dimension in non-separable spaces versus the study in separable ones is considerably different.

Problems in dimension theory of non-separable metric spaces. The behavior of dimension for non-separable metric spaces is not well understood, despite having been studied for well over half a century; there are no analogues for several important theorems regarding dimension in separable spaces and generally the results and examples are quite complicated.

Almost any new theorem or example relating to the covering dimension dim would be interesting; there are several problems in the papers mentioned below which are of interest. Here, we focus on two fundamental problems which remain largely unsolved. The relatively recent remarkable example $\nu\mu_0$ of Mrówka [**17**] gives a consistent solution to one of the great problems in dimension theory. Its finite powers give examples of metric spaces for which dim-ind, the discrepancy between covering and the small inductive dimension, can be any positive integer (see [**7, 17**]). However, a large cardinal assumption is necessary for this to happen; in fact, assuming CH (see [**17**]), all powers of $\nu\mu_0$ have all dimensions equal to zero. So, other than the examples with $\dim - \mathrm{ind} = 1$, as first demonstrated by Roy's example Δ [**18**], there are no other known gaps without the strong set theoretic assumption of Mrówka.

Demonstrating that Mrówka's example $\nu\mu_0$ has positive dim amounts to showing that, while $\mathrm{ind}\,\nu\mu_0 = 0$, every completion of it must contain an interval. Then, since there is a completion theorem for dim among all metric spaces (every metric space has a completion preserving dim), $\dim \nu\mu_0 > 0$ is immediate. Thus we have:

Question 1. *Are there real metric spaces for which the dimension gap* $\dim - \mathrm{ind}$ *is greater than 1? Even partial solutions, with less severe set theoretic assumptions would be interesting.* 1573?

This problem remains unsolved almost 50 years after Roy's ground breaking result was first announced.

Question 2. *Is there a real metric space which has no completion preserving its* ind*?* 1574?

Problems in dimension theory of separable metric spaces. The dimension of products of separable metric spaces has been extensively studied, particularly for compact spaces and especially for a product of two spaces. What can happen with noncompact spaces or in infinite products is less well understood. In this direction, Anderson and Keisler (in [**1**]) gave examples of n-dimensional spaces for which all powers are also n-dimensional and in [**7**], it is shown that spaces with these properties can also be made complete. An obvious restriction on the dimension in a product is given by the product theorem result that $\dim X \times Y \leq \dim X + \dim Y$, but it is not clear what other restrictions there might be. One very interesting problem, believed due originally due to Engelking, is this: If $dimX$ is finite, and $\dim X^\omega$ is infinite, can X^ω be countable dimensional? There are lots of other questions related to large products, whose solutions will likely require a deep understanding of the structure of these spaces. Here are two of them.

1575? **Question 3.** *Is there a finite dimensional space X which satisfies: for infinitely many positive integers n, $\dim X^n = \dim X^{n+1}$ while for infinitely many other integers m, $\dim X^m < \dim X^{m+1}$?*

1576? **Question 4.** *Given an $n \in \mathbb{N}$ is there an X_n with $\dim X = \dim X^n = 1$, but $\dim X^{n+1} = 2$?*

Levy's problems

A question about weak P-points. For a Tychonoff space X, let X^* denote the Stone–Čech remainder $\beta X \setminus X$ of X. If X is a space, an element x of X is a *weak P-point* if x is not a limit point of any countable subset of X, and X is a *weak P-space* if each of its elements is a weak P-point. If the topology of $\mathbb{R}$ is strengthened by declaring every countable subset to be closed, the resulting space will be a (non-regular) connected weak P-space. It seems difficult to determine the connectedness of some specific weak P-spaces. Kunen [**7**] proved that ω^* has weak P-points, and van Mill [**15**] observed that if N is a countably infinite closed discrete subset of $\mathbb{R}^n$, then for any positive integer n, $\mathrm{cl}_{\mathbb{R}^n} \mathbb{N} \setminus \mathbb{N}$ is a P-set in $(\mathbb{R}^n)^*$, that is, every G_δ subset of $(\mathbb{R}^n)^*$ which contains $\mathrm{cl}_{\mathbb{R}^n} \mathbb{N} \setminus \mathbb{N}$ is a neighborhood of $\mathrm{cl}_{\mathbb{R}^n} \mathbb{N} \setminus \mathbb{N}$. In particular, $(\mathbb{R}^n)^*$ has weak P-points.

1577? **Question 5.** *Is the set of weak P-points of $[0, \infty)^*$ connected? If n is an integer larger than 1, is the set of weak P-points of $(\mathbb{R}^n)^*$ connected?*

A question about closed subsets of products. Call a Tychonoff space X *image-realcompact* if every continuous image of X is realcompact. It is clear that every Lindelöf space is image-realcompact. An old question asked independently by Arhangel′skiĭ and Okunev [**2**] and Mrówka [**16**] asks whether or not a non-Lindelöf space can be image-realcompact. It is known that if certain additional conditions are put on the space X, then if X is image-realcompact, it is Lindelöf. In particular, it is shown in [**4**] that if X has weight at most $\mathfrak{c}$, then X is image-realcompact if and only if it is Lindelöf. Therefore, every image-realcompact subset of a product of $\mathfrak{c}$ copies of $\mathbb{R}$ is Lindelöf. Now suppose that X is an arbitrary image-realcompact space. Then X is realcompact, so it embeds as a closed subset of $\mathbb{R}^\kappa$ for some cardinal κ. Since every continuous image of an image-realcompact space is image-realcompact, the earlier result shows that every projection of X to a subproduct of at most $\mathfrak{c}$ factors of $\mathbb{R}^\kappa$ is Lindelöf. This observation leads to the following question.

Question 6. *Suppose that X is a closed subset of a product $\mathbb{R}^\kappa$ of copies of $\mathbb{R}$ and suppose that every projection of X onto a subproduct of at most $\mathfrak{c}$ factors is Lindelöf. Is X necessarily Lindelöf?* 1578?

We note that by the remarks before the question, an affirmative answer would give an affirmative answer to the Arhangel′skiĭ–Okunev–Mrówka problem. We also note that we cannot omit the assumption that X is closed in the product, because an ordinal of large cofinality can be embedded in a product of copies of $\mathbb{R}$ in such a way that the projection onto any subproduct of at most $\mathfrak{c}$ factors is compact.

Matveev's problems

Inverse compactness. A space is called *inversely compact* [**12**] if every independent family of closed sets has non empty intersection (recall that compactness is equivalent to the condition that every centered family of closed sets has non empty intersection, so inverse compactness is a generalization of compactness). The word "inversely" is motivated by the following: let $\mathcal{C}$ and $\mathcal{A}$ be families of subsets of a set X. Say that $\mathcal{A}$ is a *partial inversement* of $\mathcal{C}$ if there is an injection $f\colon \mathcal{A} \to \mathcal{C}$ such that for every $A \in \mathcal{A}$ either $f(A) = A$ or $f(A) = X \setminus A$ [**12**]. A space X is inversely compact iff every open cover has a finite partial inversement that covers X [**12**].

Question 7 ([**12**]). *Is every Hausdorff (regular, Tychonoff, normal) inversely compact space compact?* 1579?

A T_1 counterexample is given in [**12**].

The difficulty of constructing a Hausdorff example is shown in [**11, 12**]. A space is inversely countably compact (definition of inverse compactness reduced to countable families $\mathcal{C}$) iff it is countably compact. Many known examples of *good* countably compact spaces are shown not to be inversely compact. If X is not compact, then some power of X (very often X^2) is not inversely compact; and neither is the Alexandroff duplicate of X.

A Tychonoff inversely Lindelöf space (in the definition, one requires that the partial inversement is countable rather than finite) need not be Lindelöf [**10**]. The examples are all spaces of cardinality less than $\mathfrak{c}$, the ordinal space ω_1 (under CH), Ostaszewski space, $\omega^* \setminus \{p\}$ for some p.

Monotone compactness. A space is *monotonically Lindelöf* if there is an operator r that assigns to every cover $\mathcal{U}$ a countable open refinement $r(\mathcal{U})$ in such a way that $r(\mathcal{U})$ refines $r(\mathcal{V})$ whenever $\mathcal{U}$ refines $\mathcal{V}$. Replacing "countable" by "finite" we get the definition of *monotone compactness*. Various examples of non metrizable monotonically Lindelöf spaces are given in [**3**, **8**, **9**].

1580? **Question 8.** *Is every monotonically compact space metrizable?*

In particular: is the Alexandroff Double Arrow space monotonically compact? Every monotonically compact space is Fréchet [**19**], the Alexandroff duplicate of X is monotonically Lindelöf only if the space is countable [**19**].

Basic homogeneity. A space X is *basically homogeneous* if it has a base every element of which can be mapped onto any other by an autohomeomorphysm of X [**13**, **14**].

1581? **Question 9.** *Is every topological vector space basically homogeneous?*

For locally convex spaces the answer is affirmative [**14**].

References

[1] R. D. Anderson and J. E. Keisler, *An example in dimension theory*, Proc. Amer. Math. Soc. **18** (1967), 709–713.

[2] A. V. Arhangel'skiĭ and O. G. Okunev, *Characterization of properties of spaces by properties of their continuous images*, Vestnik Moskov. Univ. Ser. I Mat. Mekh. (1985), no. 5, 28–30, Translation: Moscow Univ. Math. Bull. 40 (1985) no. 5, 28-30.

[3] H. Bennett, D. Lutzer, and M. Matveev, *The monotone Lindelöf property and separability in ordered spaces*, Topology Appl. **151** (2005), 180–186.

[4] F. Eckertson, W. Fleissner, A. Korovin, and R. Levy, *Not realcompact images of not Lindelöf spaces*, Topology Appl. **58** (1994), no. 2, 115–125.

[5] J. Kulesza, *The dimension of products of complete separable metric spaces*, Fund. Math. **135** (1990), no. 1, 49–54.

[6] J. Kulesza, *New properties of Mrowka's space $\nu\mu_0$*, Proc. Amer. Math. Soc. **133** (2005), no. 3, 899–904.

[7] K. Kunen, *Weak P-points in $\mathbb{N}^*$*, Topology, Vol. II (Budapest, 1978), Colloq. Math. Soc. János Bolyai, vol. 23, North-Holland, Amsterdam, 1980, pp. 741–749.

[8] R. Levy and M. Matveev, *Monotone Lindelöfness for countable spaces*, Preprint.

[9] R. Levy and M. Matveev, *More examples of monotonically Lindelöf space*, Preprint.

[10] V. I. Malykhin, *Distinguishing Lindelöfness and inverse Lindelöfness*, Comment. Math. Univ. Carolin. **36** (1995), no. 1, 171–176.

[11] V. I. Malykhin and M. V. Matveev, *Inverse compactness versus compactness*, Papers on general topology and applications (Slippery Rock, PA, 1993), Ann. New York Acad. Sci., vol. 767, New York Acad. Sci., 1995, pp. 153–160.

[12] M. V. Matveev, *Inverse compactness*, Topology Appl. **62** (1995), no. 2, 181–191.

[13] M. V. Matveev, *Basic homogeneity in the class of zero-dimensional spaces*, Filomat **12** (1998), no. 2, 13–20.

[14] M. V. Matveev and S. A. Shkarin, *Some questions on topological homogeneity*, Questions Answers Gen. Topology **18** (2000), no. 1, 1–16.

[15] J. van Mill, *Weak P-points in Čech–Stone compactifications*, Trans. Amer. Math. Soc. **273** (1982), no. 2, 657–678.

[16] S. Mrówka, *Some properties of Q-spaces*, Bull. Acad. Polon. Sci. Cl. III **5** (1957), 947–950.

[17] S. Mrówka, *Small inductive dimension of completions of metric spaces*, Proc. Amer. Math. Soc. **125** (1997), no. 5, 1545–1554.

[18] P. Roy, *Nonequality of dimensions for metric spaces*, Trans. Amer. Math. Soc. **134** (1968), 117–132.

[19] J. Vaughan, Preprint.

Open Problems in Topology II – Edited by E. Pearl

Some problems on generalized metrizable spaces

Shou Lin

Generalized metrizable spaces are studied on the one hand to better understand the topology of metrizable spaces, and on the other to provide classes of non-metrizable spaces with some the desirable features of metrizable spaces. In the past years a number of excellent survey papers on generalized metrizable spaces have appeared. The paper [**12**] by Gruenhage was especially useful.

The theory of generalized metrizable spaces is closely related some questions about metrization theorems, mutual classifications of spaces and maps, countable product properties. Problems on generalized metrizable spaces are rich. In this chapter I shall pose only some problems about the images of metrizable spaces and connected spaces, and the spaces related hereditarily closure-preserving families. Some other problems about generalized metrizable spaces can be found, for examples, in G. Gruenhage's survey paper [**13**], and in G. Gruenhage's chapter, "Are stratifiable spaces M_1?" and C. Liu, Y. Tanaka's chapter, "Spaces and mappings, special networks" in this book.

All spaces are Hansdorff, and maps are continuous and onto. Readers may refer to [**9**] for unstated definitions and terminologies.

Sequence-covering maps

There are quite a few theorems about representing topological spaces as continuous images of spaces with additional properties. For examples, it is well-known that a space has a point-countable base if and only if it is an open and s-image of a metrizable space [**33**]. But, sometimes it is far from trivial to represent a space as an image of a metrizable space with some properties. Let $f\colon X \to Y$ be a map. f is called *sequence-covering* in the sense of Gruenhage, Michael and Tanaka [**14**] if in case S is a convergent sequence containing its limit point in Y then there is a compact subset K in X such that $f(K) = S$. Another definition about sequence-covering maps in the sense of Siwiec [**34**] is that $f\colon X \to Y$ is called *sequence-covering* if in case S is a convergent sequence in Y then there is a convergent sequence L in X such that $f(L) = S$, which is not used in this chapter. It was shown that every quotient and compact map of a metrizable space is sequence-covering [**24**], and every quotient and s-image of a metrizable space is a sequence-covering, quotient and s-image of a metrizable space [**14**]. Are those the best results? Let $f\colon (X,d) \to Y$ be a map with d a metric on X. f is a *π-map* with respect to d if for each $y \in Y$ and a neighborhood U of y in Y, $d(f^{-1}(y), X - f^{-1}(U)) > 0$ [**33**]. Every compact map of a metric space is a π-map. There is a quotient and π-map f from a metric space onto a compact metric space in which f is not sequence-covering [**24**].

This project was supported by NNSF of China (No. 10571151)

1582? **Question 1.** *Is every quotient and π-image of a metric space also a sequence-covering and π-image of a metric space?*

A map $f\colon X \to Y$ is *compact-covering* [**25**] if in case C is a compact subset in Y then there is a compact subset K in X such that $f(K) = C$. Every compact-covering map is sequence-covering. There is a sequence-covering, quotient and compact map $f\colon X \to Y$ from a separable metric space X onto a compact metric space Y in which f is not compact-covering [**26**]. The following classic problem posed by Michael and Nagami in [**27**] has been answered negatively: Is every quotient s-image of a metric space a compact-covering, quotient s-image of a metrizable space. Chen in [**6**] gave a (sequence-covering,) quotient and compact image of a locally separable metrizable space which is not any quotient, compact-covering s-image of a metric space. And in [**7**] Chen constructed a regular example of a (sequence-covering,) quotient s-image of a metric space which is not any quotient, compact-covering s-image of a metric space under the assumption that there exists a σ'-set.

583–1584? **Question 2.** *Let X be a regular space which is a (sequence-covering,) quotient and compact image of a metric space. Is X a compact-covering compact (resp. s-)image of a metrizable space?*

A map $f\colon X \to Y$ is called *bi-quotient* [**34**] if $f^{-1}(y)$ is covered by a family $\mathcal{U}$ consisting open subsets of X then there is a finite subset $\mathcal{U}'$ of $\mathcal{U}$ with $y \in \operatorname{int}(f(\bigcup\mathcal{U}'))$. Siwiec and Mancuso in [**35**] proved that a space Y is locally compact if and only if every compact-covering map onto Y is bi-quotient.

1585? **Question 3.** *Characterize the spaces Y such that every sequence-covering map onto Y is bi-quotient.*

A family $\mathcal{B}$ of subsets of a space X is called *point-regular* [**1**] if for every $x \in U$ with U open in X the set $\{B \in \mathcal{B} : x \in B \not\subset U\}$ is finite. It is a nice result that a space X is an open and compact image of a metrizable space if and only if X is a metacompact developable space [**15**], if and only if X has a point-regular base [**2**]. Let $\mathcal{P}$ be a family of subsets of a space X. $\mathcal{P}$ is called a *cs*-network* [**11**] for X if a sequence $\{x_n\}$ converges to a point $x \in U$ with U open in X, there exist $P \in \mathcal{P}$ and a subsequence $\{x_{n_i}\}$ of $\{x_n\}$ such that $\{x\} \cup \{x_{n_i} : i \in \mathbb{N}\} \subset P \subset U$. The following question was posed by Ikeda, Liu and Tanaka in [**16**]: For a sequential space X with a point-regular cs*-network, characterized X by means of a nice image of a metrizable space. It is easy to see that every sequence-covering and compact image of a metrizable space has a point-regular cs*-network. It was proved in [**41**] that a space X is a sequence-covering and compact image of a metrizable space if and only if X has a sequence $\{\mathcal{U}_n\}$ of point-finite covers satisfying that for each $n \in \mathbb{N}$ if S is a sequence converging to a point x in X then there are a $U_n \in \mathcal{U}_n$ for some $n \in \mathbb{N}$ and a subsequence L of S such that $\{x\} \cup L \subset U_n$.

1586? **Question 4.** *Is every space with a point-regular cs*-network a sequence-covering and compact image of a metrizable space?*

Connectedness is less closely related the properties of generalized metrizable spaces. A space is called *sequentially connected* if it cannot be expressed as the union of two non-empty disjoint sequentially open subsets [**10**]. Every connected and sequential space is sequentially connected, and every sequentially connected space is connected. Recently, it was shown in [**23**] that a space is sequentially connected if and only if it is a sequence-covering image of a connected metrizable space. Thus every connected and sequential (resp. Fréchet–Urysohn) space is a quotient (resp. pseudo-open) image of a connected metrizable space [**10, 23**]. It is known that a space is a k-space (resp. a first-countable space) if and only if it is a quotient (resp. an open) image of a paracompact locally compact (resp. a metrizable) space.

Question 5. *Are k-and connected spaces the quotient images of connected paracompact locally compact spaces?* 1587?

Question 6. *Are first-countable connected spaces the open images of connected metrizable spaces?* 1588?

σ-spaces and Σ-spaces

Let us recall some related generalized metrizable spaces. Let X be a topological space and $\mathcal{P}$ a cover of X. $\mathcal{P}$ is called a *quasi-(modk)-network* (resp. *(modk)-network*) [**32**] for X if there is a closed cover $\mathcal{K}$ by countably compact(resp. compact) subsets of X such that, whenever $K \in \mathcal{K}$ and $K \subset U$ with U open in X, then $K \subset P \subset U$ for some $P \in \mathcal{P}$. $\mathcal{P}$ is called a *network* for X if $\mathcal{P}$ is a (mod$k$)-network with $\mathcal{K} = \{\{x\} : x \in X\}$.

According to the Bing–Nagata–Smirnov metrization theorem, some generalized metrizable spaces were introduced. A space X is called a *σ-space* [**31**] if it is a regular space with a σ-locally finite network. A space X is called a *Σ-space* [**32**] (resp. a *strong Σ-space* [**30**]) if it has a σ-locally finite quasi-(mod$k$)-network (resp. (mod$k$)-network) by closed subsets. A space X is called *semi-stratifiable* [**8**] if, for each open set U of X, one can assign a sequence $\{F(n,U)\}_{n\in\mathbb{N}}$ of closed subsets of X such that

(1) $U = \bigcup_{n\in\mathbb{N}} F(n,U)$;
(2) $F(n,U) \subset F(n,V)$ whenever $U \subset V$.

Lašnev [**18**] proved that if X is metrizable and $f\colon X \to Y$ is a closed map, then $f^{-1}(y)$ is compact for all $y \in Y$ outside of some σ-closed discrete subset of Y. Some extensions of Lašnev's theorem to, e.g., σ-spaces [**5**], normal semi-stratifiable spaces [**36**], perfect pre-images of normal σ-spaces, are known to hold.

Question 7. *Is $f^{-1}(y)$ compact for all $y \in Y$ outside of some σ-closed discrete subset of Y if X is a perfect pre-image of a normal semi-stratifiable space and $f\colon X \to Y$ is a closed map?* 1589?

A family $\mathcal{P}$ of subsets of a space X is called *hereditarily closure-preserving* [**19**] if the family $\{H(P) : P \in \mathcal{P}\}$ is closure-preserving for each $H(P) \subset P \in \mathcal{P}$, i.e., $\overline{\bigcup\{H(P) : P \in \mathcal{P}'\}} = \bigcup\{\overline{H(P)} : P \in \mathcal{P}'\}$ for each $\mathcal{P}' \subset \mathcal{P}$. A space X is called a

Σ^**-space* (resp. a *strong* Σ^**-space*) [**32**] if it has a σ-hereditarily closure-preserving quasi-(mod$k$)-network (resp. (mod$k$)-network) by closed subsets. A regular space is a σ-space if and only if it is a semi-stratifiable and Σ^*-space [**17**]. Tanaka and Yajima in [**39**] proved the following a theorem for Σ-spaces. If X is a Σ-space and $f\colon X \to Y$ is a closed map, then $f^{-1}(y)$ is $\aleph_1$-compact for all $y \in Y$ outside of some σ-closed discrete subset of Y. In [**21**] the author tried to obtain a similar result as mentioned above for Σ^*-spaces, but its proof has a gap.

1590? **Question 8.** *Is $f^{-1}(y)$ $\aleph_1$-compact for all $y \in Y$ outside of some σ-closed discrete subset of Y if X is a Σ^*-space and $f\colon X \to Y$ is a closed map?*

It is a classic and important result that a regular space is a σ-space if and only if it has a σ-discrete network. Buhagiar and Lin in [**3**] showed that a space X is a strong Σ-space if and only if it has a σ-discrete (mod$k$)-network by closed subsets.

1591? **Question 9.** *Does every Σ-space have a σ-discrete quasi-(modk)-network by closed subsets?*

As for the product property of Σ-spaces, Okuyama in [**32**] proved that a space X is a Σ-space if and only if $X \times [0,1]$ is a Σ^*-space for a paracompact space X. It is known that a space X is a strong Σ-space if and only if it is a subparacompact Σ-space [**3**].

1592? **Question 10.** *Is X a strong Σ-space if $X \times [0,1]$ is a strong Σ^*-space?*

$\aleph_0$-spaces

A family $\mathcal{P}$ of subsets of a space X is called a *pseudo-base* if $\mathcal{P}$ is a (mod$k$)-network with $\mathcal{K} = \{K : K \text{ is compact in } X\}$. A space X is called an $\aleph_0$*-space* [**25**] if it is a regular space with a countable pseudo-base. It is easy to check that $\aleph_0$-spaces are preserved by closed maps. However, the regular image of an $\aleph_0$-space under an open map cannot be an $\aleph_0$-space [**25**].

1593? **Question 11.** *Is the regular image of an $\aleph_0$-space under an open and compact map an $\aleph_0$-space?*

Spaces related to pseudo-bases are special. For example, Lin in [**20**] obtained that a regular space is an $\aleph_0$-space if and only if it has a point-countable pseudo-base, and a regular space has a σ-hereditarily closure-preserving pseudo-base if and only if either it is an $\aleph_0$-spaces or it is a σ-closed discrete space in which all compact subsets are finite. On the other hand, some generalizations of the families about compact-finite families or hereditarily closure-preserving families were introduced by T. Mizokami in [**28**] as follows. A family $\mathcal{P}$ of subsets of a space X is called *CF* in X if $\mathcal{P}_{|K} = \{P \cap K : P \in \mathcal{P}\}$ is finite for each compact subset K of X, and called *CF** in X if additionally the family $\{P \in \mathcal{P} : P \cap K = P'\}$ is finite for each infinite subset $P' \in \mathcal{P}_{|K}$. It is easy to check that

Hereditarily closure-preserving family $\Rightarrow$ *CF** family $\Rightarrow$ CF family.

It was shown in [**29**] that a regular space X has a σ-CF^* pseudo-base consisting of perfect subsets of X if and only if X is either an $\aleph_0$-space or a space in which all compact subsets are finite.

Question 12. *Let X be a regular space with a σ-CF^* pseudo-base. Is X either an $\aleph_0$-space or a space in which all compact subsets are finite?* 1594?

A family $\mathcal{P}$ of subsets of a space X is called a *quasi-base* [**4**] if, whenever $x \in X$ and U is a neighborhood of x in X, then there exists a $P \in \mathcal{P}$ such that $x \in \text{int}(P) \subset P \subset U$. A regular space is metrizable if and only if it has a σ-compact finite base [**22**], if and only if it is a k-space with a σ-CF base [**28**], if and only if it is a k-space with a σ-CF^* quasi-base [**42**], if and only if it is a k'-space with a σ-CF quasi-base [**40**].

Question 13. *Let X be a regular space. Is X metrizable if it is a k-space with a σ-CF quasi-base?* 1595?

References

[1] P. S. Alexandroff, *On the metrisation of topological spaces*, Bull. Acad. Polon. Sci. Sér. Sci. Math. Astronom. Phys. **8** (1960), 135–140.

[2] A. V. Arhangel′skiĭ, *On mappings of metric spaces*, Dokl. Akad. Nauk SSSR **145** (1962), 245–247, Translation: Soviet Math. Dokl. 3 (1962) 953–956.

[3] D. Buhagiar and S. Lin, *A note on subparacompact spaces*, Mat. Vesnik **52** (2000), no. 3–4, 119–123.

[4] J. G. Ceder, *Some generalizations of metric spaces*, Pacific J. Math. **11** (1961), 105–125.

[5] J. Chaber, *Generalizations of Lašnev's theorem*, Fund. Math. **119** (1983), no. 2, 85–91.

[6] H. Chen, *Weak neighborhoods and Michael–Nagami's question*, Houston J. Math. **25** (1999), no. 2, 297–309.

[7] H. Chen, *Compact-covering maps and k-networks*, Proc. Amer. Math. Soc. **131** (2003), no. 8, 2623–2632.

[8] G. D. Creede, *Concerning semi-stratifiable spaces*, Pacific J. Math. **32** (1970), 47–54.

[9] R. Engelking, *General topology*, Heldermann Verlag, Berlin, 1989.

[10] A. Fedeli and A. Le Donne, *On good connected preimages*, Topology Appl. **125** (2002), no. 3, 489–496.

[11] Z. M. Gao, *ℵ-space is invariant under perfect mappings*, Questions Answers Gen. Topology **5** (1987), no. 2, 271–279.

[12] G. Gruenhage, *Generalized metric spaces*, Handbook of set-theoretic topology, North-Holland, Amsterdam, 1984, pp. 423–501.

[13] G. Gruenhage, *Metrizable spaces and generalizations*, Recent progress in general topology, II, North-Holland, Amsterdam, 2002, pp. 201–225.

[14] G. Gruenhage, E. Michael, and Y. Tanaka, *Spaces determined by point-countable covers*, Pacific J. Math. **113** (1984), no. 2, 303–332.

[15] R. W. Heath, *Screenability, pointwise paracompactness, and metrization of Moore spaces*, Canad. J. Math. **16** (1964), 763–770.

[16] Y. Ikeda, C. Liu, and Y. Tanaka, *Quotient compact images of metric spaces, and related matters*, Topology Appl. **122** (2002), no. 1–2, 237–252.

[17] H. J. K. Junnila, *On submetacompactness*, Topology Proc. **3** (1978), no. 2, 375–405 (1979).

[18] N. Lašnev, *On continuous decompositions and closed mappings of metric spaces*, Dokl. Akad. Nauk SSSR **165** (1965), 756–758, Translation: Soviet Math. Dokl. 6 (1965) 1504–1506.

[19] N. Lašnev, *Closed images of metric spaces*, Dokl. Akad. Nauk SSSR **170** (1966), 505–507, Translation: Soviet Math. Dokl. 7 (1966) 1219–1221.

[20] S. Lin, *A study of pseudobases*, Questions Answers Gen. Topology **6** (1988), no. 1, 81–97.

[21] S. Lin, *A decomposition theorem for Σ^*-spaces*, Topology Proc. **15** (1990), 117–120.

[22] S. Lin, *Generalized metric spaces and mappings*, Chinese Science Press, Beijing, 1995.

[23] S. Lin, *The images of connected metric spaces*, Chinese Ann. Math. Ser. A **26** (2005), no. 3, 345–350.

[24] S. Lin, *A note on sequence-covering mappings*, Acta Math. Hungar. **107** (2005), no. 3, 187–191.

[25] E. Michael, *$\aleph_0$-spaces*, J. Math. Mech. **15** (1966), 983–1002.

[26] E. Michael, *A problem*, Topological structures, II, Mathematisch Centrum, Amsterdam, 1979, pp. 165–166.

[27] E. Michael and K. Nagami, *Compact-covering images of metric spaces*, Proc. Amer. Math. Soc. **37** (1973), 260–266.

[28] T. Mizokami, *On CF families and hyperspaces of compact subsets*, Topology Appl. **35** (1990), no. 1, 75–92.

[29] T. Mizokami and S. Lin, *On spaces with a σ-CF* pseudo-base*, Math. Japon. **46** (1997), no. 3, 377–384.

[30] K. Nagami, *Σ-spaces*, Fund. Math. **65** (1969), 169–192.

[31] A. Okuyama, *Some generalizations of metric spaces, their metrization theorems and product spaces*, Sci. Rep. Tokyo Kyoiku Daigaku Sect. A **9** (1968), 236–254 (1968).

[32] A. Okuyama, *On a generalization of Σ-spaces*, Pacific J. Math. **42** (1972), 485–495.

[33] V. Ponomarev, *Axioms of countability and continuous mappings*, Bull. Acad. Polon. Sci. Sér. Sci. Math. Astronom. Phys. **8** (1960), 127–134.

[34] F. Siwiec, *Sequence-covering and countably bi-quotient mappings*, General Topology and Appl. **1** (1971), no. 2, 143–154.

[35] F. Siwiec and V. J. Mancuso, *Relations among certain mappings and conditions for their equivalence.*, General Topology and Appl. **1** (1971), no. 1, 33–41.

[36] R. A. Stoltenberg, *A note on stratifiable spaces*, Proc. Amer. Math. Soc. **23** (1969), 294–297.

[37] J. Suzuki, *On pre-σ-spaces*, Bull. Tokyo Gakugei Univ. (4) **28** (1976), 22–32.

[38] K. Tamano, *Definitions of Σ-spaces*, Topology Proc. **22** (1997), Summer, 529–532.

[39] Y. Tanaka and Y. Yajima, *Decompositions for closed maps*, Topology Proc. **10** (1985), no. 2, 399–411.

[40] L. Xie and Y. Liu, *On hyperspace of Lašnev space*, J. Math. Res. Exposition **17** (1997), no. 2, 197–202.

[41] P. F. Yan, *Compact images of metric spaces*, J. Math. Study **30** (1997), no. 2, 185–187.

[42] G. Ying, *Remarks on CF families*, Questions Answers Gen. Topology **18** (2000), no. 1, 103–110.

Open Problems in Topology II – Edited by E. Pearl

Problems from the Madrid Department of Geometry and Topology

José M. R. Sanjurjo

We present in this article several open problems which reflect the interests and the activity of some of the members of the Department of Geometry and Topology at the Universidad Complutense in Madrid. These interests and activities are shared, in many cases, with outside collaborators but we note that in each problem proposed at least one member of the department is present. The only exception is Problem 7, whose author does not belong to the department but which was nevertheless prompted by the work of one of its members. The author is grateful to all the colleagues who provided the necessary information for the writing of this paper.

We start with some problems belonging to the area of topological dynamics. Several authors have in recent years studied relations between the topological and the dynamical structure of attractors. An important property, establishing that every (asymptotically stable) attractor of a flow on a locally compact ANR has the shape of a finite polyhedron has been formulated in the papers [**2**, **8**, **10**, **19**] at various levels of generality using topological ideas connected with Borsuk's theory of shape. Moreover, in [**10**] it is proved that all finite-dimensional compacta with polyhedral shape can be represented in that way. There are, however, classes of isolated invariant compacta more general than attractors which have polyhedral shape, for example the class of non-saddle invariant compacta [**7**]. The following problem inquires about the existence of other classes of compacta with such a property.

Problem 1 (J.M.R. Sanjurjo). *Consider an isolated invariant compactum K for a flow on an ANR. Find dynamical properties (other than being an attractor) ensuring that K has polyhedral shape.* 1596?

The following problem asks about the role of movability (a shape invariant) in the context of topological dynamics.

Problem 2 (J.M.R. Sanjurjo). *Is there a dynamical condition C such that a finite-dimensional metric compactum K is movable if and only if K can be embedded as an invariant subset of a flow on a manifold in such a way that K satisfies condition C?* 1597?

The intrinsic topology of the unstable manifold of an isolated invariant set of a flow was introduced by Robbin and Salamon in [**16**]. They have proved that the shape of the Conley index of an isolated invariant set agrees with that of the one-point compactification of its unstable manifold endowed with the intrinsic topology. One of the drawbacks of the notion of Conley index is that a lot of information is lost when the exit set of an index pair is collapsed to a point (a disadvantageous feature which is transmitted to the one-point compactification just mentioned). In the following problem we conjecture that the use of the Freudenthal compactification could lead to more sophisticated topological and algebraic invariants.

1598? **Problem 3** (M.A. Morón, J.J. Sánchez-Gabites and J.M.R. Sanjurjo). *To what extent does the shape of the Freudenthal compactification of the unstable manifold of an isolated invariant set K endowed with its intrinsic topology carry more information than the Conley shape index of K? Is there a satisfactory theory relating the dynamical properties of K to the topological properties of such a compactification?*

The following problems are concerned with discrete dynamical systems instead of flows. The computation of the sequence of fixed point indices of a local homeomorphism in a neighborhood of an isolated fixed point is an important and non-trivial problem. In the plane, when the fixed point is an isolated invariant set, this problem was solved by Le Calvez and Yoccoz for orientation preserving planar homeomorphisms, and by Ruiz del Portal and Salazar in the orientation reversing case. The general problem was solved more recently in the orientation preserving case by Le Calvez and by Ruiz del Portal and Salazar for orientation reversing planar homeomorphisms.

We present here two related open problems. See [**17**] and [**18**] for more information about the notions involved.

1599? **Problem 4** (F.R. Ruiz del Portal and J.M. Salazar). *Let $f\colon U \subset \mathbb{R}^2 \to \mathbb{R}^2$ be a continuous map and p a fixed point that is an isolated invariant set. Is the sequence $i_{\mathbb{R}^2}(f^k, p)$ of fixed point indices periodic? Which sequences of integers satisfying Dold's congruences are reached?*

1600? **Problem 5** (F.R. Ruiz del Portal and J.M. Salazar). *Consider the sequence $i_{\mathbb{R}^2}(f^k, p)$ for homeomorphisms $f\colon \mathbb{R}^3 \to \mathbb{R}^3$ for which a fixed point p is an isolated invariant set. Is it periodic? What can be said for arbitrary $\mathbb{R}^3$-homeomorphisms? And for arbitrary $\mathbb{R}^n$-homeomorphisms?*

Another direction of research in the department focuses on low-dimensional topology. A closed set F in a *3*-manifold-with-boundary M is *tame* if there is a homeomorphism of M in itself sending F onto a subcomplex of some locally finite simplicial complex triangulating M. If there is no such homeomorphism, we say that F is *wild*. The set X is *locally tame* at a point x of X if there exist a neighbourhood U of x in M and a homeomorphism of U into M that takes $U \cap X$ onto a tame set. Otherwise we say that X is *locally wild* at x. A closed set is tame in S^3 if is locally tame at each of its points. The set of points of X at which it is locally wild is closed, and is called the *wild subset* of X.

A *knot* in $X = S^3$ (resp. *string* in $X = \mathbb{R}^3$) is a pair (X, N), where N is a subspace of X homeomorphic (resp. properly homeomorphic) to the 1-sphere S^1 (resp. real line $\mathbb{R}^1$). A *wild knot* (or *wild string*) has a non-empty wild subset. Otherwise it is a *tame knot* (or *tame string*). A knot (S^3, N) is the *unknot* if N bounds a tamely embedded disk in S^3. A string is the *unstring* if it bounds a tamely embedded half-plane in $\mathbb{R}^3$.

The following celebrated theorem was proved in [**15**]:

Smith Conjecture. *A tame knot K in the 3-sphere S^3 has a cyclic branched covering that is also S^3 only if K is the unknot.*

Problem 6 (J.M. Montesinos: Smith Conjecture for $\mathbb{R}^3$). *A tame string S in the real 3-space $\mathbb{R}^3$ has a cyclic branched covering that is also $\mathbb{R}^3$ only if S is the unstring.* 1601?

In [**14**] a *nontrivial, wild knot whose n-fold cyclic branched cover is S^3, for all* n, is constructed. Moreover, *there are uncountably many inequivalent knots with this property*, and the knots can be chosen to bound an embedded disk Δ whose wild subset is a tame Cantor subset of $\partial\Delta$. Allan Edmonds [**6**] in his review to [**14**] has conjectured that

Problem 7 (Conjecture (A. Edmonds)). *Any wild knot in S^3 whose nontrivial n-fold cyclic branched cover is S^3 bounds an embedded disk Δ that is tame in its interior.* 1602?

Problem 6 follows from a positive answer to this conjecture.

The following problem refers to topological aspects of real algebraic geometry. It is concerned with finiteness of topological operations. This is a basic problem for the understanding of the topology and the function theory of real analytic spaces, in particular real analytic manifolds (see [**1**] for information about these subjects). Let X be a real analytic space, and $S \subset X$ a subset that can be described by finitely many conjunctions and/or disjunctions of strict and/or relaxed equalities and/or inequalities of global analytic functions on X (we say in short, *finitely many global analytic inequalities*). Then:

Problem 8 (J.M. Ruiz). *Can the topological interior (resp. closure) of S in X be described using solely strict (resp. relaxed) global analytic inequalities? Can each union of connected components of S in X also be described by finitely many global analytic inequalities?* 1603?

Problem 8 is solved in the case in which the topological boundary of S in X is relatively compact. Without any compactness assumption, it is solved only for $\dim(X) \leq 2$. All known solutions are in the affirmative.

The following problems are motivated by the famous Borsuk's problem on intersection of ANRs ([**3**, p. 244 Problem 8.2]) and by Kolodziejczyk's work on related subjects. In a series of interesting papers (see [**11**] for instance), Kolodziejczyk solves some other Borsuk's problems on homotopy dominations of finite polyhedra. She uses algebraic techniques depending on deep results developed mainly by J.H.C. Whitehead, C.T.C. Wall and Hilton–Mislin–Roitberg. Concretely in [**11**] she proves:

> *There are finite polyhedra which homotopy dominate infinitely manypolyhedra with different homotopy types.*

1604? **Problem 9** (M.A. Morón). *Is there a finite polyhedron P dominating a sequence of finite polyhedra $\{P_n\}_{n\in\mathbb{N}}$, where P_n homotopy dominates P_{n+1} but P_n and P_{n+1} have different homotopy types for every $n \in \mathbb{N}$?*

1605? **Problem 10** (M.A. Morón). *Let P be a finite polyhedron which dominates infinitely many finite polyhedra $\{P_n\}_{n\in\mathbb{N}}$ of different homotopy types. Using a result of Borsuk–Oledzki, we can consider all of them as retracts of the Q-manifold $P \times Q$. Let us consider the hyperspace $2_H^{P\times Q}$ (with the Hausdorff metric). We can suppose, by compactness, that there exists $\lim_{n\to\infty} P_n$. What can be said about the movability of $K = \lim_{n\to\infty} P_n$?*

Related to this, note that K is not the limit of $\{P_n\}_{n\in\mathbb{N}}$ in the shape metric d_s as defined in [**12**]. In fact there is $\varepsilon > 0$ such that $d_s(P_n, P_k) \geq \varepsilon$ for $k \neq n$.

Another direction of research is connected with asymptotic topology. In [**9**], Gromov introduced the notions of coarse equivalence and asymptotic dimension for the study of group theory. The idea was to think of finitely generated groups as geometric objects in order to get algebraic properties.
Generalizing those ideas, Smith (see [**20**]) proved that a countable group is of asymptotic dimension zero if and only if it is locally finite. In [**4**] it was shown that every locally finite group is coarsely equivalent to one of the form $\bigoplus_{i=1}^{\infty} \mathbb{Z}_{p_i}$ and those groups are universal for proper metric spaces of bounded geometry and asymptotic dimension zero. The problem proposed asks about a classification of those universal groups.

1606? **Problem 11** (N. Brodskiy, J. Dydak, J. Higes and A. Mitra [**4**]). *Classify countable abelian torsion groups up to coarse equivalence.*

We give now a problem in the framework of topological abelian groups. In [**5**] an analog to Mackey–Arens theorem is intended for topological abelian groups. The lack of the notion of convexity in groups makes the question delicate. In fact, in the paper mentioned only some sufficient conditions are given in a topological abelian group G_τ in order that it admits a finest among all the locally quasi-convex topologies in G with the same set of continuous characters as the original topology τ. It is also an open problem whether such a topology always exists.

1607? **Problem 12** (E. Martín-Peinador and V. Tarieladze). *Let $G_\tau := (G, \tau)$ be an infinite abelian topological group, and let $\mathbb{T}$ be the multiplicative unit complex circle, with the euclidean topology. Assume that for any group topology ν on G such that $\tau < \nu$, the group of all continuous characters $\mathrm{CHom}(G_\tau, \mathbb{T})$ is properly contained in $\mathrm{CHom}(G_\nu, \mathbb{T})$, i.e., there is not a finer group topology ν in G with the condition $\mathrm{CHom}(G_\tau, \mathbb{T}) = \mathrm{CHom}(G_\nu, \mathbb{T})$. Is τ then the discrete topology?.*

Finally, another area of research is devoted to set theoretical topology. A topological space is said to be totally paracompact if every open basis contains a locally finite covering (R.M. Ford). It is known (D.W. Curtis) that every totally paracompact complete metric space is C-scattered and every $\sigma-$locally compact paracompact space is totally paracompact. Then every Banach space is totally paracompact if and only if it is finite dimensional. Thus, every infinite-dimensional

Banach space necessarily has an open basis which contains no locally finite covering (i.e., a "coarse open basis"). A theorem by Corson shows that for any covering $\mathcal{U}$ of a reflexive, infinite dimensional Banach space B, where $\mathcal{U}$ is formed by bounded, convex sets, there is a point x in B such that each neighborhood of x meets infinitely many members of $\mathcal{U}$. On the other hand, it is proved in [**13**] that c_0 does not satisfy such property. The following problem is connected with this situation.

Problem 13 (F.G. Lupiáñez). *Give an intrinsic description of those infinite-dimensional Banach spaces such that the open basis formed by all open balls is coarse.* 1608?

References

[1] C. Andradas, L. Bröcker, and J. M. Ruiz, *Constructible sets in real geometry*, Springer, Berlin, 1996.

[2] S. A. Bogatyĭ and V. I. Gutsu, *On the structure of attracting compacta*, Differ. Uravn. **25** (1989), no. 5, 907–909.

[3] K. Borsuk, *Theory of shape*, PWN, Warsaw, 1975.

[4] N. Brodskiy, J. Dydak, J. Higes, and A. Mitra, *Dimension zero at all scales*, 2006, Preprint. `arXiv:math.MG/0607241`

[5] M. J. Chasco, E. Martín-Peinador, and V. Tarieladze, *On Mackey topology for groups*, Studia Math. **132** (1999), no. 3, 257–284.

[6] A. Edmonds, *Review of J. M. Montesinos-Amilibia,* Uncountably many wild knots whose cyclic branched covering are S^3, *Rev. Mat. Complut. 16 (2003) 329–344*, Math. Reviews (2004), 2004m:57027.

[7] A. Giraldo, M. A. Morón, F. R. Ruiz del Portal, and J. M. R. Sanjurjo, *Some duality properties of non-saddle sets*, Topology Appl. **113** (2001), 51–59.

[8] A. Giraldo and J. M. R. Sanjurjo, *On the global structure of invariant regions of flows with asymptotically stable attractors*, Math. Z. **232** (1999), no. 4, 739–746.

[9] M. Gromov, *Asymptotic invariants of infinite groups*, Geometric group theory, Vol. 2 (Sussex, 1991), Cambridge University Press, Cambridge, 1993, pp. 1–295.

[10] B. Günther and J. Segal, *Every attractor of a flow on a manifold has the shape of a finite polyhedron*, Proc. Amer. Math. Soc. **119** (1993), no. 1, 321–329.

[11] D. Kołodziejczyk, *There exists a polyhedron dominating infinitely many different homotopy types*, Fund. Math. **151** (1996), no. 1, 39–46.

[12] V. F. Laguna, M. A. Morón, Nguyen To Nhu, and J. M. R. Sanjurjo, *Movability and limits of polyhedra*, Fund. Math. **143** (1993), no. 3, 191–201.

[13] F. G. Lupiañez, *Total paracompactness and Banach spaces*, Proc. Amer. Math. Soc. **103** (1988), no. 1, 210–214.

[14] J. M. Montesinos-Amilibia, *Uncountably many wild knots whose cyclic branched covering are S^3*, Rev. Mat. Complut. **16** (2003), no. 1, 329–344.

[15] W. Morgan and H. Bass (eds.), *The Smith conjecture*, Academic Press, Orlando, FL, 1984.

[16] J. W. Robbin and D. Salamon, *Dynamical systems, shape theory and the Conley index*, Ergodic Theory Dynam. Systems **8** (1988), 375–393, Charles Conley Memorial Issue.

[17] F. R. Ruiz del Portal, *Planar isolated and stable fixed points have index* $=1$, J. Differential Equations **199** (2004), no. 1, 179–188.

[18] F. R. Ruiz del Portal and J. M. Salazar, *Fixed point index of iterations of local homeomorphisms of the plane: a Conley index approach*, Topology **41** (2002), no. 6, 1199–1212.

[19] J. M. R. Sanjurjo, *Multihomotopy, Čech spaces of loops and shape groups*, Proc. London Math. Soc. **69** (1994), no. 2, 330–344.

[20] J. Smith, *On asymptotic dimension of countable abelian groups*, 2005, Preprint. `arXiv:math.GR/0504447`

Open Problems in Topology II – Edited by E. Pearl

Cardinal sequences and universal spaces

Lajos Soukup

In his contribution to this volume Joan Bagaria, [**1**], investigated the question whether certain sequences of cardinals can be obtained as cardinal sequences associated with locally compact scattered T_2 (in short: LCS) spaces in certain models of ZFC.

In this paper we try to characterize certain classes of cardinal sequences of scattered spaces and to formulate related problems. We also introduce the concept of "universal spaces" as a useful tool of such characterizations.

To start with we recall some definitions and introduce some notations.

Given a scattered T_2 space X the αth Cantor–Bendixson level will be denoted by $\mathrm{I}_\alpha(X)$. The *height* of X, $\mathrm{ht}(X)$, is the least ordinal α $\mathrm{ht}^-(X)$ is the smallest ordinal α such that $\mathrm{I}_\alpha(X)$ is finite. Clearly, one has $\mathrm{ht}^-(X) \leq \mathrm{ht}(X) \leq \mathrm{ht}^-(X)+1$. The *cardinal sequence* of X, denoted by $\mathrm{SEQ}(X)$, is the sequence of cardinalities of the infinite Cantor–Bendixson levels of X, i.e., $\mathrm{SEQ}(X) = \langle |I_\alpha(X)| : \alpha < \mathrm{ht}(X)^- \rangle$.

For an ordinal α we let $\mathcal{C}(\alpha)$ denote the class of all cardinal sequences of length α of LCS spaces. We also put, for any fixed infinite cardinal λ,

$$\mathcal{C}_\lambda(\alpha) = \{s \in \mathcal{C}(\alpha) : s(0) = \lambda,\ \forall \beta < \alpha[s(\beta) \geq \lambda]\}.$$

Because of the following theorem the characterization of $\mathcal{C}_\lambda(\alpha)$ for all λ yields a characterization of $\mathcal{C}(\alpha)$.

Theorem ([**4**, Theorem 2.1]). *For any ordinal α we have $f \in \mathcal{C}(\alpha)$ iff for some natural number n there is a decreasing sequence $\lambda_0 > \lambda_1 > \cdots > \lambda_{n-1}$ of infinite cardinals and there are ordinals $\alpha_0, \ldots \alpha_{n-1}$ such that $\alpha = \alpha_0 + \cdots + \alpha_{n-1}$ and $f = f_0{}^\frown f_1{}^\frown \cdots {}^\frown f_{n-1}$ with $f_i \in \mathcal{C}_{\lambda_i}(\alpha_i)$ for each $i < n$.*

In [**4**, Theorem 4.1] for any ordinal $\alpha < \omega_2$ and any infinite cardinal λ the authors gave the full description of $\mathcal{C}_\alpha(\lambda)$ under GCH. Using that theorem one can characterize even $\mathcal{C}_\lambda(\omega_2)$ under GCH provided $\lambda \neq \omega_1$.

To formulate that result we should introduce the following notions. If α is any ordinal, a subset $L \subset \alpha$ is called *κ-closed* in α, where κ is an infinite cardinal, iff $\sup \langle \alpha_i : i < \kappa \rangle \in L \cup \{\alpha\}$ for each increasing sequence $\langle \alpha_i : i < \kappa \rangle \in {}^\kappa L$. The set L is *$<\lambda$-closed* in α provided it is κ-closed in α for each cardinal $\kappa < \lambda$. We say that L is *successor-closed* in α if $\beta + 1 \in L \cup \{\alpha\}$ for all $\beta \in L$.

The constant λ-valued sequence of length α will be denoted by $\langle \lambda \rangle_\alpha$.

Theorem ([**4**, Theorem 4.1]). *Assume* GCH *and fix an ordinal $\alpha \leq \omega_2$.*

(1) $\mathcal{C}_\omega(\alpha) = \{s \in {}^\alpha\{\omega, \omega_1\} : s(0) = \omega\}$ *if $\alpha < \omega_2$ and $\mathcal{C}_\omega(\omega_2) = \emptyset$.*
(2) *If $\lambda > \mathrm{cf}(\lambda) = \omega$, then $\mathcal{C}_\lambda(\alpha) = \{s \in {}^\alpha\{\lambda, \lambda^+\}$: $s(0) = \lambda$ and $s^{-1}\{\lambda\}$ is ω_1-closed in $\alpha\}$.*
(3) *If $\mathrm{cf}(\lambda) = \omega_1$ and either $\lambda > \omega_1$ or $\alpha < \omega_2$ then $\mathcal{C}_\lambda(\alpha) = \{s \in {}^\alpha\{\lambda, \lambda^+\}$: $s(0) = \lambda$ and $s^{-1}\{\lambda\}$ is both ω-closed and successor-closed in $\alpha\}$.*

(4) *If* $\operatorname{cf}(\lambda) > \omega_1$, *then* $\mathcal{C}_\lambda(\alpha) = \{\langle\lambda\rangle_\alpha\}$.

The theorem above left open the characterization of $\mathcal{C}_{\omega_1}(\omega_2)$ under GCH.

For a cardinal λ and and ordinal δ put

$$\mathcal{D}_\kappa(\delta) = \{s \in {}^\delta\{\lambda, \lambda^+\} : s(0) = \lambda, s^{-1}\{\lambda\} \text{ is } {<}\lambda\text{-closed and successor-closed in } \delta\}.$$

In [4, Theorem 4.1] it was proved that if GCH holds then

$$\mathcal{C}_{\omega_1}(\delta) \subseteq \mathcal{D}_{\omega_1}(\delta), \tag{1}$$

for each ordinal $\delta < \omega_3$ and by [4, Theorem 4.1] above we have equality for $\delta < \omega_2$. By [8] not only for $\delta = \omega_2$ but even for each ordinal $\delta < \omega_3$ it is consistent with GCH that we have equality in (1).

To formulate our result we need to introduce some more notation.

An LCS space X is called $\mathcal{C}_\lambda(\alpha)$-*universal* if and only if $\mathrm{SEQ}(X) \in \mathcal{C}_\lambda(\alpha)$ and for each sequence $s \in \mathcal{C}_\lambda(\alpha)$ there is an open subspace Y of X with $\mathrm{SEQ}(Y) = s$.

The constructions of [4] imply that if GCH holds then for each ordinal $\delta < \omega_2$ and infinite cardinal $\lambda \geq \omega_1$ there is a $\mathcal{C}_\lambda(\delta)$-universal LCS space X.

For $\lambda = \omega$ and $\delta < \omega_2$ we do not know if GCH implies the existence of a $\mathcal{C}_\omega(\delta)$-universal space but [8, Theorem 1.3] below implies that the existence of such a space is consistent with GCH.

Theorem ([8, Theorem 1.3]). *If λ is a regular cardinal with $\lambda^{<\lambda} = \lambda$ then for each $\delta < \lambda^{++}$ there is a λ-complete λ^+-cc poset P of cardinality λ^+ such that in V^P there is a $\mathcal{C}_\lambda(\delta)$-universal LCS space and*

$$\mathcal{C}_\lambda(\delta) = \mathcal{D}_\lambda(\delta). \tag{2}$$

How do the universal spaces come into the picture? The first idea to prove the consistency of $\mathcal{C}_\lambda(\alpha) = \mathcal{D}_\lambda(\alpha)$ is to try to carry out an iterated forcing. For each $f \in \mathcal{D}_\lambda(\alpha)$ we can try to find a poset P_f such that

$$1_{P_f} \Vdash \text{There is an LCS space } X_f \text{ with cardinal sequence } f.$$

Since typically $|X_f| = \lambda^+$ if we want to preserve the cardinals and *CGH* the natural idea is to find a λ-closed, λ^+-cc poset P_f. In this case forcing with P_f introduces λ^+ new subsets of λ because P_f has cardinality λ^+. However $|\mathcal{D}_\lambda(\alpha)| = \lambda^{++}$ for $\lambda^+ \leq \alpha < \lambda^{++}$! So the length of the iteration should be at least λ^{++}, hence in the final model λ will have $\lambda^+ \cdot \lambda^{++} = \lambda^{++}$ many new subsets, i.e., GCH fails.

On the other hand, a $\mathcal{C}_\lambda(\delta)$-universal space has cardinality λ^+ so we may hope that there is a λ-closed, λ^+-cc poset P of cardinality λ^+ such that V^P contains a $\mathcal{C}_\lambda(\delta)$-universal space. In this case $(2^\lambda)^{V^P} \leq ((|P|^{<\lambda})^\lambda)^V = \lambda^+$. So in the the generic extension we might have GCH.

Unfortunately, for a fixed regular cardinal κ [8, Theorem 1.3] gives different posets for different ordinals $\delta < \kappa^{++}$. This raises the following questions:

1609? **Problem 1.** *Assume that κ is a regular cardinal. Is it consistent with* GCH *that we have equality in (2) for each for each ordinal $\delta < \kappa^{++}$?*

We can formulate another version of the problem above:

Problem 2. *Is it true under* GCH *that we have equality in (2) for each cardinal* κ *and ordinal* $\delta < \kappa^{++}$*?* 1610?

Juhász and Weiss proved in [**5**] that $\langle\omega\rangle_\delta \in \mathcal{C}(\delta)$ for each $\delta < \omega_2$. In [**6**] Martínez showed that for each $\delta < \omega_3$ it is consistent with GCH that $\langle\omega_1\rangle_\delta \in \mathcal{C}(\delta)$. However, the following question remained open.

Problem 3. *Is it consistent with* GCH *that we have* $\langle\omega_1\rangle_\delta \in \mathcal{C}(\delta)$ *for each* $\delta < \omega_3$*?* 1611?

If λ is singular and $\delta \geq \lambda^+$ then we do not know anything about $\mathcal{C}_\lambda(\delta)$.

Problem 4. *Characterize* $\mathcal{C}_\lambda(\alpha)$ *for singular cardinals* λ *and* $\alpha \geq \lambda^+$ *under* GCH*!* 1612?

So far we assumed GCH. What can we say if $2^\omega > \omega_1$?

By [**3**] it is consistent that $\langle\omega\rangle_{\omega_2} \in \mathcal{C}_\omega(\omega_2)$. By [**2**] it is consistent that $2^\omega = \omega_3$ and

$$\mathcal{C}_\omega(\omega_2) \supset \{s \in {}^{\omega_2}\{\omega,\omega_1\} : s(0) = \omega\}.$$

However, if $2^{\omega_0} = \omega_\alpha$ then a natural upper bound of $\mathcal{C}_\omega(\omega_2)$ is a much larger family of sequences:

$$\mathcal{C}_\omega(\omega_2) \subseteq \{s \in {}^{\omega_2}\{\omega_\nu : \nu \leq \alpha\} : s(0) = \omega\}. \tag{3}$$

In [**8**] the following result is proved.

Theorem. *It is consistent that* $2^\omega = \omega_2$ *and there is an* $\mathcal{C}_\omega(\omega_2)$*-universal LCS space witnessing that* $\mathcal{C}_\omega(\omega_2)$ *is large as possible, i.e.,* $\mathcal{C}_\omega(\omega_2) = \{s \in {}^{\omega_2}\{\omega,\omega_1,\omega_2\} : s(0) = \omega\}$.

Problem 5. *Assume that* $\langle\omega\rangle_{\omega_2} \in \mathcal{C}_\omega(\omega_2)$ *and* $2^\omega = \omega_2$. *Is it true that we have equality in (3)?* 1613?

So far we have seen that for different cardinals λ and different ordinals δ there may exist $\mathcal{C}_\lambda(\delta)$-universal LCS spaces in different models. However, we do not have a model, a cardinal λ and an ordinal δ such that $\mathcal{C}_\lambda(\delta)$ is nonempty but there is no $\mathcal{C}_\lambda(\delta)$-universal LCS space in that model. This fact yields the following question.

Problem 6. *Is it true for each cardinal* λ *and ordinal* δ *that either* $\mathcal{C}_\lambda(\delta) = \emptyset$ *or there is an* $\mathcal{C}_\lambda(\delta)$*-universal LCS space?* 1614?

Finally let us mention a stepping-up problem. In [**3**] Baumgartner and Shelah established the consistency of $\langle\omega\rangle_{\omega_2} \in \mathcal{C}(\omega_2)$. Martínez, [**7**], proved that a generalization of some ideas from [**3**] can give the consistency of $\langle\omega\rangle_\delta \in \mathcal{C}(\delta)$ for each $\delta < \omega_3$. Later it was proved in [**9**] that if there is a natural ccc poset P such that $\mathcal{C}_\omega(\omega_2) \in \mathcal{C}(\omega_2)$ in V^P then there is a natural ccc poset Q as well such that $\langle\omega\rangle_\delta \in \mathcal{C}(\delta)$ holds for each $\delta < \omega_3$ in V^Q. These theorems raise the the following question:

Problem 7. *Does* $\langle\omega\rangle_{\omega_2} \in \mathcal{C}_\omega(\omega_2)$ *imply that* $\langle\omega\rangle_\delta \in \mathcal{C}_\omega(\delta)$ *for each* $\delta < \omega_3$*?* 1615?

References

[1] J. Bagaria, *Thin-tall spaces and cardinal sequences*, Open problems in topology II (E. Pearl, ed.), Elsevier Science.

[2] J. Bagaria, *Locally-generic Boolean algebras and cardinal sequences*, Algebra Universalis **47** (2002), no. 3, 283–302.

[3] J. E. Baumgartner and S. Shelah, *Remarks on superatomic Boolean algebras*, Ann. Pure Appl. Logic **33** (1987), no. 2, 109–129.

[4] I. Juhász, L. Soukup, and W. Weiss, *Cardinal sequences of length* $< \omega_2$ *under GCH*, Fund. Math. **189** (2006), no. 1, 35–52.

[5] I. Juhász and W. Weiss, *On thin-tall scattered spaces*, Colloq. Math. **40** (1978), no. 1, 63–68.

[6] J. C. Martínez, *A forcing construction of thin-tall Boolean algebras*, Fund. Math. **159** (1999), no. 2, 99–113.

[7] J. C. Martínez, *A consistency result on thin-very tall Boolean algebras*, Israel J. Math. **123** (2001), 273–284.

[8] J. C. Martínez and L Soukup, *Universal locally compact scattered spaces*, In preparation.

[9] L. Soukup, *A lifting theorem on forcing LCS spaces,*, To appear in the Hajnal–T. Sós Conference volume.

List of contributors

Bagaria, Joan, *ICREA (Institució Catalana de Recerca i Estudis Avançats)* and *Universitat de Barcelona*, Barcelona, Spain .... § 11
Banakh, Taras, *Lviv Ivan Franko National University*, Lviv, Ukraine .... § 56, 60
Basener, William F., *Rochester Institute of Technology*, Rochester, NY, U.S.A. .... § 45
Bellamy, David P., *University of Delaware*, Newark, DE, U.S.A. .... § 30
Bennett, Harold, *Texas Tech University*, Lubbock, TX, U.S.A. .... § 1
Blackmore, Denis, *New Jersey Institute of Technology*, Newark, NJ, U.S.A. .... § 49
Bokalo, Bohdan, *Lviv Ivan Franko National University*, Lviv, Ukraine .... § 60
Bonanzinga, Maddalena, *Universita degli studi di Messina*, Messina, Italy .... § 2
Borwein, Jonathan M., *Dalhousie University*, Halifax, NS, Canada .... § 50
Bouziad, Ahmed, *Université de Rouen*, Saint-Étienne-du-Rouvray, France .... § 38
Cauty, Robert, *Université Paris 6 Pierre et Marie Curie*, Paris, France .... § 56
Chatyrko, Vitalij A., *Linköpings universitet*, Linköping, Sweden .... § 57
Clark, Alex, *University of North Texas*, Denton, TX, U.S.A. .... § 46
Clementino, Maria Manuel, *Universidade de Coimbra*, Coimbra, Portugal .... § 39
Comfort, W. W., *Wesleyan University*, Middletown, CT, U.S.A. .... § 40
Devaney, Robert L., *Boston University*, Boston, MA, U.S.A. .... § 47
Dikranjan, Dikran, *University of Udine*, Udine, Italy .... § 41
Dow, Alan, *University of North Carolina at Charlotte*, Charlotte, NC, U.S.A. ..... § 12
Echi, Othman, *University of Tunis El Manar*, Tunis, Tunisia .... § 61
Eisworth, Todd, *Ohio University*, Athens, OH, U.S.A. .... § 13
Farah, Ilijas, *York University*, Toronto, ON, Canada and *Matematicki Institut*, Belgrade, Serbia .... § 14
Fedorchuk, Vitalii V., *Moscow State University*, Moscow, Russia .... § 58
Filippov, V. V., *Moscow State University*, Moscow, Russia .... § 51
Galindo, Jorge, *Universidad Jaume I*, Castellón, Spain .... § 42
Garity, Dennis J., *Oregon State University*, Corvallis, OR, U.S.A. .... § 62
Georgiou, Dimitris N., *University of Patras*, Patras, Greece .... § 3
Good, Chris, *University of Birmingham*, Birmingham, United Kingdom .... § 63
Gruenhage, Gary, *Auburn University*, Auburn, AL, U.S.A. .... § 14, 15
Guran, Igor, *Lviv Ivan Franko National University*, Lviv, Ukraine .... § 60
Gutev, Valentin, *University of KwaZulu-Natal*, Durban, South Africa .... § 17
Hagopian, Charles L., *California State University*, Sacramento, CA, U.S.A. .... § 31
Hart, Klaas Pieter, *TU Delft*, Delft, Netherlands .... § 18
Hernández, Salvador, *Universidad Jaume I*, Castellón, Spain .... § 42
Hofmann, Dirk, *Universidade de Aveiro*, Aviero, Portugal .... § 39
Hrušák, Michael, *Universidad Nacional Autónoma de México, Morelia*, Morelia, Mexico .... § 10, 19
Iliadis, Stavros D., *University of Patras*, Patras, Greece .... § 3
Illanes, Alejandro, *Universidad Nacional Autónoma de México*, Mexico D.F., Mexico .... § 32
Ingram, W. T., *University of Missouri-Rolla*, Rolla, MO, U.S.A. .... § 33
de Jager, Eliza P., *University of Cape Town*, Rondebosch, South Africa .... § 64
Juhász, Istvan, *Alfréd Rényi Institute of Mathematics*, Budapest, Hungary .... § 20
Karasev, Alexandre, *Nipissing University*, North Bay, ON, Canada .... § 65
Koszmider, Piotr, *Universidade de São Paulo*, São Paulo, Brazil .... § 52

Kozlov, Konstantin L., *Moscow State University*, Moscow, Russia .... § 66
Kulesza, John, *George Mason University*, Fairfax, VA, U.S.A. .... § 67
Künzi, Hans-Peter A., *University of Cape Town*, Rondebosch, South Africa .... § 64
Levy, Ronnie, *George Mason University*, Fairfax, VA, U.S.A. .... § 67
Lewis, Wayne, *Texas Tech University*, Lubbock, TX, U.S.A. .... § 34
Lin, Shou, *Ningde Teachers' College*, Ninige, Fujian, P. R. China .... § 68
Liu, Chuan, *Ohio University-Zanesville*, Zanesville, OH, U.S.A. .... § 4
Lutzer, David, *College of William and Mary*, Williamsburg, VA, U.S.A. .... § 1
Marsh, Andrew, *National University of Ireland, Galway*, Galway, Ireland .... § 63
Martínez-de-la-Vega, Veronica, *Universidad Nacional Autónoma de México*, Mexico D.F., Mexico .... § 35
Martínez-Montejano, Jorge M., *Universidad Nacional Autónoma de México*, Mexico D.F., Mexico .... § 35
Marzougui, Habib, *Faculty of Sciences of Bizerte*, Jarzouna, Tunisia .... § 61
Matveev, Mikhail, *George Mason University*, Fairfax, VA, U.S.A. .... § 2, 67
McCluskey, Aisling, *National University of Ireland, Galway*, Galway, Ireland ...... § 63
McMaster, Brian, *Queen's University Belfast*, Belfast, Northern Ireland .... § 63
Megrelishvili, Michael, *Bar-Ilan University*, Ramat-Gan, Israel .... § 43
van Mill, Jan, *Vrije Universiteit*, Amsterdam, Netherlands .... § 21
Moore, Justin Tatch, *Boise State University*, Boise, ID, U.S.A. .... § 10, 16, 22, 23
Moors, Warren B., *University of Auckland*, Auckland, New Zealand .... § 50
Mynard, Frédéric, *Georgia Southern University*, Statesboro, GA, U.S.A. .... § 1
Nadler, Jr., Sam B., *West Virginia University*, Morgantown, WV, U.S.A. .... § 36
Nogura, Tsugunori, *Ehime University*, Matsuyama, Japan .... § 17
Nyikos, Peter J., *University of South Carolina*, Columbia, SC, U.S.A. .... § 24, 25
Ohta, Haruto, *Shizuoka University*, Shizuoka, Japan .... § 5
Okunev, Oleg, *Benemérita Universidad Autónoma de Puebla*, Puebla, Mexico .. § 6, 54
Parwani, Kamlesh, *The University of Houston*, Houston, TX, U.S.A. .... § 45
Pasynkov, Boris A., *Moscow State University*, Moscow, Russia .... § 59
Pavlov, Oleg, *University of North Carolina at Charlotte*, Charlotte, NC, U.S.A. .... § 7
Pearl, Elliott, Toronto, ON, Canada .... § 26
Pestov, Vladimir, *University of Ottawa*, Ottawa, ON, Canada .... § 44
Peters, Thomas J., *University of Connecticut*, Storrs, CT, U.S.A. .... § 49
Prajs, Janusz R., *California State University*, Sacramento, CA, U.S.A. .... § 37
Radul, Taras, *Lviv Ivan Franko National University*, Lviv, Ukraine .... § 60
Repovš, Dušan, *University of Ljubljana*, Ljubljana, Slovenia .... § 62
Reznichenko, Evgenii A., *Moscow State University*, Moscow, Russia .... § 66
Ricceri, Biagio, *University of Catania*, Catania, Italy .... § 54
Salhi, Ezeddine, *University of Sfax*, Sfax, Tunisia .... § 61
Sanjurjo, José M. R., *Universidad Complutense de Madrid*, Madrid, Spain .... § 69
Scheepers, Marion, *Boise State University*, Boise, ID, U.S.A. .... § 8
Shakhmatov, Dmitri, *Ehime University*, Matsuyama, Japan .... § 41
Simon, Petr, *Charles University*, Prague, Czech Republic .... § 19
Sipacheva, Ol′ga, *Moscow State University*, Moscow, Russia .... § 66
Soukup, Lajos, *Alfréd Rényi Institute of Mathematics*, Budapest, Hungary .... § 70
Sullivan, Michael C., *Illinois Southern University*, Carbondale, IL, U.S.A. .... § 48
Szeptycki, Paul J., *York University*, Toronto, ON, Canada .... § 27
Tall, Franklin D., *University of Toronto*, Toronto, ON, Canada .... § 28

Tanaka, Yoshio, *Tokyo Gakugei University*, Tokyo, Japan .... § 4
Tkachuk, V. V., *Universidad Autonoma Metropolitana-Iztapalapa*, Mexico D.F., Mexico .... § 55
Todorčević, Stevo, *University of Toronto*, Toronto, ON, Canada and *Université Paris 7*, Paris, France .... § 23
Troallic, Jean-Pierre, *Université du Havre*, Le Havre, France .... § 38
Tsaban, Boaz, *Weizmann Institute of Science*, Rehovot, Israel .... § 9
Tuncali, Murat, *Nipissing University*, North Bay, ON, Canada .... § 65
Valov, Vesko, *Nipissing University*, North Bay, ON, Canada .... § 65
Vaughan, Jerry E., *University of North Carolina*, Greensboro, NC, U.S.A. .... § 29
Weiss, William A. R., *University of Toronto*, Toronto, ON, Canada .... § 20
Wiandt, Tamas, *Rochester Institute of Technology*, Rochester, NY, U.S.A. .... § 45
Wu, Ta-Sun, *Case Western Reserve University*, Cleveland, OH, U.S.A. .... § 42
Yamazaki, Kaori, *University of Tsukuba*, Tsukuba, Japan .... § 5
Zarichnyi, Michael, *Lviv Ivan Franko National University*, Lviv, Ukraine .... § 56, 60

Index

CPSIA information can be obtained
at www.ICGtesting.com
Printed in the USA
LVHW082122010720
658557LV00009BA/49

9 780444 522085